Quantum Kinetic Theory
and Applications

Electrons, Photons, Phonons

Quantum Kinetic Theory and Applications

Electrons, Photons, Phonons

FEDIR T. VASKO
OLEG E. RAICHEV
Institute of Semiconductor Physics
NAS of Ukraine, Kiev

 Springer

Fedir T. Vasko
Institute of Semiconductor Physics, NAS
45 Prospekt Nauki
Kiev 03028 Ukraine

Oleg E. Raichev
Institute of Semiconductor Physics, NAS
45 Prospekt Nauki
Kiev 03028 Ukraine

Library of Congress Control Number: 2005926337

ISBN-10: 0-387-26028-5   e-ISBN: 0-387-28041-3
ISBN-13: 978-0387-26028-0

Printed on acid-free paper.

Printed in the United States of America. (HAM)

9 8 7 6 5 4 3 2 1

springeronline.com

Preface

Physical kinetics is the final section of the course of theoretical physics in its standard presentation. It stays at the boundary between general theories and their applications (solid state theory, theory of gases, plasma, and so on), because the treatment of kinetic phenomena always depends on specific structural features of materials. On the other hand, the physical kinetics as a part of the quantum theory of macroscopic systems is far from being complete. A number of its fundamental issues, such as the problem of irreversibility and mechanisms of chaotic responses, are now attracting considerable attention. Other important sections, for example, kinetic phenomena in disordered and/or strongly non-equilibrium systems and, in particular, phase transitions in these systems, are currently under investigation. The quantum theory of measurements and quantum information processing actively developing in the last decade are based on the quantum kinetic theory.

Because a deductive theoretical exposition of the subject is not convenient, the authors restrict themselves to a lecture-style presentation. Now the physical kinetics seems to be at the stage of development when, according to Newton, studying examples is more instructive than learning rules. In view of these circumstances, the methods of the kinetic theory are presented here not in a general form but as applications for description of specific systems and treatment of particular kinetic phenomena.

The quantum features of kinetic phenomena can arise for several reasons. One naturally meets them in strongly correlated systems, when it is impossible to introduce weakly interacting quasiparticles (for example, in a non-ideal plasma), or in more complicated conditions, such as in the vicinity of the phase transitions. Next, owing to complexity of the systems like superconductors, ferromagnets, and so on, the manifestations of kinetic phenomena change qualitatively. The theoretical consideration of these cases can be found in the literature. Another reason for studying quantum features of transport and optical phenomena has emerged in the past decades, in connection with extensive investigation of kinetic phenomena under strong external fields and in nanostructures. The quantum features of these phenomena follow from non-classical dynamics of quasiparticles, and these are the cases the present monograph takes care of, apart from consideration of standard problems of quantum transport theory. Owing to intensive development of the physics of nanostructures and wide application of strong external (both stationary and time-dependent) fields for studying various properties of solids, the theoretical methods presented herein are of current importance for analysis and interpretation of the experimental results of modern solid state physics.

This monograph is addressed to several categories of readers. First, it will be useful for graduate students studying theory. Second, the topics we cover should be interesting for postgraduate students of various specializations. Third, the researchers who want to understand the background of modern theoretical issues in more detail can find a number of useful results here. The phenomena we consider involve kinetics of electron, phonon, and photon systems in solids. The dynamical properties and interactions of electrons, phonons, and photons are briefly described in Chapter 1. Further, in Chapters 2−8, we present main theoretical methods: linear response theory, various kinetic equations for the quasiparticles under consideration, and diagram technique. The presentation of the key approaches is always accompanied by solutions of concrete problems, to illustrate applications of the theory. The remaining chapters are devoted to various manifestations of quantum transport in solids. The choice of particular topics (their list can be found in the Contents) is determined by their scientific importance and methodological value. The 268 supplementary problems presented at the end of the chapters are chosen to help the reader to study the material of the monograph. Focusing our attention on the methodical aspects and discussing a great diversity of kinetic phenomena in line with the guiding principle "a method is more important than a result," we had to minimize both detailed discussion of physical mechanisms of the phenomena considered and comparison of theoretical results to experimental data.

It should be emphasized that the kinetic properties are the important source of information about the structure of materials, and many peculiarities of the kinetic phenomena are used for device applications. These applied aspects of physical kinetics are not covered in detail either. However, the methods presented in this monograph provide the theoretical background both for analysis of experimental results and for device simulation. In the recent years, these theoretical methods were applied for the above-mentioned purposes so extensively that any comprehensive review of the literature seems to be impossible in this book. For this reason, we list below only a limited number of relevant monographs and reviews.

Fedir T. Vasko
Oleg E. Raichev
Kiev, December 2004

Monographs:

1. J. M. Ziman, *Electrons and Phonons, the Theory of Transport Phenomena in Solids*, Oxford University Press, 1960.

2. L. P. Kadanoff and G. Baym, *Quantum Statistical Mechanics*, W. A. Benjamin, Inc., New York, 1962.

3. A. A. Abrikosov, L. P. Gor'kov and I. E. Dzialoszynski, *Methods of Quantum Field Theory in Statistical Physics*, Prentice-Hall, 1963.

4. S. Fujita, *Introduction to Non-Equilibrium Quantum Statistical Mechanics*, Saunders, PA, USA, 1966.

5. D. N. Zubarev, *Nonequilibrium Statistical Thermodynamics*, Consultants Bureau, New York, 1974.

6. E. M. Lifshitz and L. P. Pitaevski, *Physical Kinetics*, Pergamon Press, Oxford, 1981.

7. H. Bottger and V. V. Bryksin, *Hopping Conduction in Solids*, VCH Publishers, Akademie-Verlag Berlin, 1985.

8. V. L. Gurevich, *Transport in Phonon Systems* (Modern Problems in Condensed Matter Sciences, Vol. 18), Elsevier Science Ltd., 1988.

9. V. F. Gantmakher and Y. B. Levinson, *Carrier Scattering in Metals and Semiconductors* (Modern Problems in Condensed Matter Sciences, Vol. 19), Elsevier Science Ltd., 1987.

10. A. A. Abrikosov, *Fundamentals of the Theory of Metals*, North-Holland, 1988.

11. H. Haug and S. W. Koch, *Quantum Theory of the Optical and Electronic Properties of Semiconductors*, World Scientific, Singapore, 1990.

12. N. N. Bogolubov, *Introduction to Quantum Statistical Mechanics*, Gordon and Breach, 1992.

13. G. D. Mahan, *Many Particle Physics*, Plenum, New York, 1993.

14. H. Haug and A.-P. Jauho, *Quantum Kinetics in Transport and Optics of Semiconductors*, Springer, Berlin, 1997.

15. Y. Imry, *Introduction to Mesoscopic Physics*, Oxford University Press, 1997.

16. D. K. Ferry and S. M. Goodnick, *Transport in Nanostructures*, Cambridge University Press, New York, 1997.

17. R. P. Feynmann, *Statistical Mechanics*, Addison-Wesley, 1998.

18. A. M. Zagoskin, *Quantum Theory of Many-Body Systems: Techniques and Applications*, Springer-Verlag, New York, 1998.

19. F. T. Vasko and A. V. Kuznetsov, *Electron States and Optical Transitions in Semiconductor Heterostructures*, Springer, New York, 1998.

20. J. Rammer, *Quantum Transport Theory* (Frontiers in Physics, Vol. 99), Westview Press, 1998.

21. T. Dittrich, P. Hänggi, G.-L. Ingold, B. Kramer, G. Schön, and W. Zverger, *Quantum Transport and Dissipation*, Wiley-VCH, Weinheim, 1998.

22. B. K. Ridley, *Quantum Processes in Semiconductors*, Oxford University Press, 1999.

23. D. Bouwmeester, A. Ekert, and A. Zeilinger, *The Physics of Quantum Information*, Springer, Berlin, Heidelberg, New York, 2000.

Reviews:

1. D. N. Zubarev, *Double-Time Green's Functions*, Sov. Phys. - Uspekhi **3**, 320 (1960).

2. R. N. Gurzhi and A. P. Kopeliovich, *Low-Temperature Electrical Conductivity of Pure Metals*, Sov. Phys. - Uspekhi **133**, 33 (1981).

3. T. Ando, A. B. Fowler, and F. Stern, *Electronic Properties of Two-Dimensional Systems*, Rev. Mod. Phys. **54**, 437 (1982).

4. J. Rammer and H. Smith, *Quantum Field-Theoretical Methods in Transport Theory of Metals*, Rev. Mod. Phys. **58**, 323 (1986); J. Rammer, *Quantum Transport Theory of Electrons in Solids: A Single-Particle Approach*, Rev. Mod. Phys. **63**, 781 (1991).

5. G. D. Mahan, *Quantum Transport Equation for Electric and Magnetic Fields*, Physics Reports **145**, 251 (1987).

6. W. R. Frensley, *Boundary Conditions for Open Quantum Systems Driven Far from Equilibrium*, Rev. Mod. Phys. **62**, 745 (1990).

7. B. Kramer and A. MacKinnon, *Localization: Theory and Experiment*, Rep. Prog. Phys. **56**, 1469 (1993).

8. C. H. Henry and R. F. Kazarinov, *Quantum Noise in Photonics*, Rev. Mod. Phys. **68**, 801 (1996).

9. C. W. J. Beenakker, *Random-Matrix Theory of Quantum Transport*, Rev. Mod. Phys. **69**, 731 (1997).

10. Ya. M. Blanter and M. Buttiker, *Shot Noise in Mesoscopic Conductors*, Physics Reports **336**, 1 (2000).

11. P. Lipavsky, K. Morawetz, and V. Spicka, *Kinetic Equation for Strongly Interacting Dense Fermi Systems*, Annales de Physique **26**, 1 (2001).

Contents

Chapter 1

ELEMENTS OF QUANTUM DYNAMICS

The dynamical equations for quantum systems, the Schroedinger equation for pure states and the density-matrix equation for mixed states, form the theoretical background for description of transport phenomena in systems with different kinds of elementary excitations (quasiparticles). Both single-particle formulation of these equations and many-particle formalism, which is required for the cases of interacting quasiparticles, are presented below. This chapter is not a systematic introduction to quantum theory. It contains only the description of some basic equations and definitions (probability of transitions, second quantization, and so on). The properties of concrete quasiparticles (electrons, phonons, and photons in different materials) and their interaction are also discussed in order to use the corresponding results in the next chapters.

1. Dynamical Equations

Let us start our consideration of the quantum dynamics with the simplest case of a single particle propagating along the x direction. The evolution of such a particle is described by the time-dependent Schroedinger equation:

$$ i\hbar \frac{\partial \Psi_{xt}^{(\delta)}}{\partial t} = \widehat{H} \Psi_{xt}^{(\delta)}, \qquad \Psi_{xt=t_0}^{(\delta)} = \Psi_x^{(\delta)}, \tag{1} $$

where the initial state at $t = t_0$ is determined by the wave function $\Psi_x^{(\delta)}$, which depends on the set of quantum numbers δ. The Hamiltonian $\widehat{H}$ can depend on time. A simple example of quantum evolution is a particle moving in a one-dimensional potential. The Hamiltonian $\widehat{H}_x$ for such a case is obtained from the classical expression for the energy after replacing the momentum by the operator proportional to the Planck constant $\hbar$:

$$\widehat{H}_x = \frac{\hat{p}^2}{2m} + U(x) , \qquad \hat{p} = -i\hbar\frac{\partial}{\partial x} , \tag{2}$$

where m is the mass of the particle. The character of the dynamics depends essentially on the potential energy $U(x)$. We mention, for example, formation of confined states in a potential well or tunneling penetration of the particle through a potential barrier. Different observable values (such as coordinate, velocity, and energy) of the system are determined by the quantum-mechanical average

$$\overline{Q}_t^{(\delta)} = \int dx\,\Psi_{xt}^{(\delta)*}\widehat{Q}\Psi_{xt}^{(\delta)}, \tag{3}$$

where the operator $\widehat{Q}$ corresponds to the classical expression for the observable value. Note that $\overline{Q}_t^{(\delta)}$ is expressed through a quadratic form of the Ψ-functions. Since $\overline{Q}_t^{(\delta)}$ is real, any operator $\widehat{Q}$ must be Hermitian. In particular, $\widehat{H} = \widehat{H}^+$, because the Hamiltonian corresponds to the energy of the system.

The operator nature of the characteristics of quantum systems makes it possible to rewrite Eqs. (1) and (3) in the integral representation. We introduce a kernel

$$H(x,x_1) = \left[\frac{\hat{p}_1^2}{2m}\delta(x-x_1)\right] + U(x_1)\delta(x-x_1) \tag{4}$$

containing Dirac's δ-function, and transform the Schroedinger equation (1) to the following integral form:

$$i\hbar\frac{\partial\Psi_{xt}^{(\delta)}}{\partial t} = \int dx_1 H(x,x_1)\Psi_{x_1t}^{(\delta)}, \qquad \Psi_{xt=t_0}^{(\delta)} = \Psi_x^{(\delta)}. \tag{5}$$

The kernel for the observable value, $Q(x,x_1) = [\widehat{Q}_1\delta(x-x_1)]$, is introduced in the same way (here $\widehat{Q}_1$ acts on the coordinate x_1 of the δ-function), and we obtain

$$\overline{Q}_t^{(\delta)} = \int dx \int dx_1 \Psi_{xt}^{(\delta)*} Q(x,x_1)\Psi_{x_1t}^{(\delta)}. \tag{6}$$

In these formulations, the state with quantum numbers δ is described by the wave function $\Psi_{xt}^{(\delta)}$ and by the operators of physical values appearing in Eqs. (1) and (3), or by the x-dependent kernels in Eqs. (5) and (6). Such a description is called the coordinate (or x-) representation.

In many cases, the description of quantum dynamics can be simplified by using the Fourier-transformed wave function introduced according to the relations

$$\Psi^{(\delta)}_{pt} = \int dx e^{-(i/\hbar)px} \Psi^{(\delta)}_{xt}, \qquad \Psi^{(\delta)}_{xt} = \frac{1}{L} \sum_p e^{(i/\hbar)px} \Psi^{(\delta)}_{pt}. \qquad (7)$$

In order to avoid the ambiguities due to δ-functions, the motion of the particle is considered here for an interval of length L, with the use of appropriate boundary conditions. In the limit $L \to \infty$, the momentum p in Eq. (7) is a quasi-discrete variable with values $(2\pi\hbar n/L)$, where n is an integer. The substitution $p \to (2\pi\hbar n/L)$ does not depend on the type of the boundary conditions used (hard-wall, periodic, etc.), provided that n is a large number. A Fourier transformation of Eq. (1) with the Hamiltonian (2) leads to the Schroedinger equation

$$i\hbar \frac{\partial \Psi^{(\delta)}_{pt}}{\partial t} = \sum_{p_1} H(p, p_1) \Psi^{(\delta)}_{p_1 t}, \qquad (8)$$

$$H(p, p_1) \equiv \frac{1}{L} \int dx e^{-(i/\hbar)px} \widehat{H}_x e^{(i/\hbar)p_1 x},$$

which is similar to Eq. (5). The kernel $H(p, p')$ depends on a pair of momenta. The initial condition to Eq. (8) is determined by the Fourier transformation of $\Psi^{(\delta)}_x$. In the above example of the particle in a one-dimensional potential, the Hamiltonian kernel is transformed to

$$H(p, p_1) = \frac{p^2}{2m} \delta_{p,p_1} + U(p, p_1), \qquad (9)$$

where δ_{p,p_1} is the Kronecker symbol (below we use two equivalent notations $\delta_{a,b}$ and δ_{ab} for such symbols). The kinetic energy acquires its classical form, while the action of the potential is described by the kernel $U(p, p_1)$. The expression for an observable through $\Psi^{(\delta)}_{pt}$ is written as

$$\overline{Q}^{(\delta)}_t = \frac{1}{L^2} \sum_{p_1 p_2} \Psi^{(\delta)}_{p_1 t}{}^* Q(p_1, p_2) \Psi^{(\delta)}_{p_2 t}, \qquad (10)$$

where the kernel $Q(p_1, p_2)$ can be written in terms of $\widehat{Q}$ in a similar way as the Hamiltonian kernel in Eq. (8). The structure of Eqs. (8) and (10) is analogous to that of Eqs. (5) and (6). This description is called the momentum (or p-) representation of the problem under consideration.

Obviously, the nature of quantum dynamics does not depend on the representation used. For this reason, it is convenient to consider the wave function as a projection of the ket-vector $|\delta, t\rangle$, which describes the state defined by the quantum numbers δ, onto the bra-vector, $\langle k|$, which determines the representation:

$$\Psi_{kt}^{(\delta)} = \langle k|\delta, t\rangle \qquad k \leftrightarrow x, p, \ldots \ . \tag{11}$$

It should be noted that the above-introduced bra- and ket-vectors are not usual functions. They are Hermitian conjugate elements of the Hilbert space satisfying the relations of orthogonality, normalization, and completeness:

$$\langle k| = |k\rangle^{+}, \qquad \langle k|k'\rangle = \delta_{k,k'}, \qquad \sum_{k} |k\rangle\langle k| = \hat{1}, \tag{12}$$

where $\hat{1}$ is the unit operator. Using these notations, one may formulate any dynamical problem in the operator form.

The Schroedinger equation for the state δ in this representation takes the following form:

$$i\hbar \frac{\partial |\delta, t\rangle}{\partial t} = \widehat{H} |\delta, t\rangle , \quad |\delta, t = t_0\rangle = |\delta\rangle , \tag{13}$$

with the initial condition determined by the ket-vector $|\delta\rangle$. A similar equation for the Hermitian conjugate vector $\langle \delta, t|$ contains $-\langle \delta, t|\widehat{H}$ on the right-hand side. Using Eq. (11) and rewriting the kernel $Q(k_1, k_2)$ as $\langle k_1|\widehat{Q}|k_2\rangle$, we define the observable $\overline{Q}_t^{(\delta)}$ as follows:

$$\overline{Q}_t^{(\delta)} = \sum_{k_1 k_2} \langle \delta, t|k_1\rangle\langle k_1|\widehat{Q}|k_2\rangle\langle k_2|\delta, t\rangle = \langle \delta, t|\widehat{Q}|\delta, t\rangle, \tag{14}$$

so that the classical observable is expressed through the diagonal matrix element. As a result, the dynamics of the system with a fixed initial state $|\delta, t = 0\rangle$ (such a system is said to be in the pure state) is described by Eqs. (13) and (14).

Transforming the double sum in Eq. (14) as $\sum_{k_1 k_2}\langle k_1|\widehat{Q}|k_2\rangle\langle k_2|\delta, t\rangle \times \langle \delta, t|k_1\rangle$, it is convenient to separate the operator $|\delta, t\rangle\langle \delta, t|$ there. This operator,

$$\hat{\eta}_t^{(\delta)} \equiv |\delta, t\rangle\langle \delta, t|, \tag{15}$$

known as the density matrix or as the statistical operator, describes the quantum dynamics of the system. The quantity $\langle k_2|\delta, t\rangle\langle \delta, t|k_1\rangle = \langle k_2|\hat{\eta}_t^{(\delta)}|k_1\rangle$ is also called the density matrix in the $|k\rangle$-representation. One may consider, for example, x- or p-representation, or a representation based upon discrete quantum numbers (problem 1.1). The description of the quantum dynamics based on the density matrix formalism is convenient for the cases when the initial state $|\delta, t = t_0\rangle$ of the quantum system is not defined (for example, because of the quantum-mechanical uncertainty).

Let us give a more general definition of the density matrix. Consider an ensemble of identical systems, which are distributed over the states δ with probabilities P_δ at the initial moment of time $t = t_0$ (such a system is called the mixed state, or the mixture of states). We introduce the observable quantity $\overline{Q}_t$ according to

$$\overline{Q}_t = \sum_\delta P_\delta \overline{Q}_t^{(\delta)}. \tag{16}$$

The probability for realization of δ-states is normalized as $\sum_\delta P_\delta = 1$. Since the operator $\widehat{Q}$ does not depend on the initial conditions, the density matrix for the mixed state is introduced as

$$\hat{\eta}_t \equiv \sum_\delta P_\delta |\delta, t\rangle\langle\delta, t|, \tag{17}$$

and the observable (16) is obtained from Eqs. (14) and (16) in the form

$$\overline{Q}_t = \sum_k \langle k|\widehat{Q}\hat{\eta}_t|k\rangle \equiv \mathrm{Sp}(\widehat{Q}\hat{\eta}_t). \tag{18}$$

Here and below $\mathrm{Sp}(\widehat{A})$ (or, equivalently, $\mathrm{Sp}\widehat{A}$), where $\widehat{A}$ is an arbitrary operator, denotes the sum of the diagonal matrix elements of this operator and is called the trace of the operator.

The equation of evolution for the density matrices (15) and (17) describing dynamics of pure and mixed states, respectively, is obtained in the following way. Let us take a derivative of the density matrix over time and use Eq. (13) together with the corresponding Hermitian conjugate equation. As a result,

$$i\hbar\frac{\partial}{\partial t}|\delta, t\rangle\langle\delta, t| = \widehat{H}|\delta, t\rangle\langle\delta, t| - |\delta, t\rangle\langle\delta, t|\widehat{H}. \tag{19}$$

Now, let us multiply this equation by P_δ and calculate the sums over δ of both its sides. Since $\widehat{H}$ does not depend on δ, we obtain, according to Eq. (17), the operator equation

$$i\hbar\frac{\partial\hat{\eta}_t}{\partial t} = [\widehat{H}, \hat{\eta}_t] \tag{20}$$

describing the evolution of the quantum system. The right-hand side of Eq. (20) is written using the commutator defined as $[\widehat{A}, \widehat{B}] = \widehat{A}\widehat{B} - \widehat{B}\widehat{A}$, where $\widehat{A}$ and $\widehat{B}$ are arbitrary operators. The initial condition for Eq. (20) in the case of a pure state may be expressed as $\hat{\eta}_{t=t_0} = |\delta\rangle\langle\delta|$, while for a mixed state one needs additional physical restrictions removing the uncertainty of the initial state.

2. *S*-Operator and Probability of Transitions

The evolution of the system with time-dependent Hamiltonian $\widehat{H}_t$ is described by the Schroedinger equation (1.13). The ket-vectors $|\delta t\rangle \equiv |\delta, t\rangle$ at the instants t and t' are connected through the evolution operator $\widehat{S}$ (also known as S-operator or scattering matrix):

$$|\delta t\rangle = \widehat{S}(t, t')|\delta t'\rangle. \tag{1}$$

Equation (1.13) leads to the operator equation for $\widehat{S}(t, t')$, with the initial condition at $t = t'$:

$$i\hbar\frac{\partial}{\partial t}\widehat{S}(t, t') = \widehat{H}_t\widehat{S}(t, t'), \quad \widehat{S}(t, t')_{t=t'} = 1. \tag{2}$$

For the case of a time-independent Hamiltonian, $\widehat{H}_t = \widehat{H}$, this equation is solved as

$$\widehat{S}(t, t') = \exp\left[-\frac{i}{\hbar}\widehat{H}(t - t')\right] \equiv \widehat{S}(t - t'), \tag{3}$$

and the temporal evolution is determined only by the difference $t - t'$. If the initial value of the ket-vector, $|\delta, t = 0\rangle = |\delta\rangle$, belongs to one of the vectors determined by the eigenstate problem $\widehat{H}|\delta\rangle = \varepsilon_\delta|\delta\rangle$, the evolution is harmonic:

$$|\delta t\rangle = \exp\left(-\frac{i}{\hbar}\varepsilon_\delta t\right)|\delta\rangle. \tag{4}$$

In the case of a mixed initial state, the evolution is described by a sum of oscillating factors with different energies ε_δ.

In the general case of the time-dependent Hamiltonian, it is convenient to transform Eq. (2) to the integral form:

$$\widehat{S}(t, t') = 1 - \frac{i}{\hbar}\int_{t'}^{t}d\tau\,\widehat{H}_\tau\widehat{S}(\tau, t'). \tag{5}$$

The solution of this equation is obtained by iterations and is written as

$$\widehat{S}(t, t') = 1 + \sum_{n=1}^{\infty}\left(-\frac{i}{\hbar}\right)^n\int_{t'}^{t}dt_1\ldots\int_{t'}^{t_{n-2}}dt_{n-1}\int_{t'}^{t_{n-1}}dt_n$$

$$\times\widehat{H}_{t_1}\ldots\widehat{H}_{t_{n-1}}\widehat{H}_{t_n}. \tag{6}$$

Introducing the operator of chronological ordering, $\hat{\mathcal{T}}$, we rewrite Eq. (6) as follows (problem 1.2):

$$\widehat{S}(t, t') = 1 + \sum_{n=1}^{\infty}\frac{(-i/\hbar)^n}{n!}\int_{t'}^{t}dt_1\ldots\int_{t'}^{t}dt_{n-1}\int_{t'}^{t}dt_n\hat{\mathcal{T}}\left\{\widehat{H}_{t_1}\widehat{H}_{t_2}\ldots\widehat{H}_{t_n}\right\},$$

$$\hat{\mathcal{T}}\left\{\widehat{H}_t\widehat{H}_{t'}\right\} = \left\{ \begin{array}{ll} \widehat{H}_t\widehat{H}_{t'}, & t > t' \\ \widehat{H}_{t'}\widehat{H}_t, & t < t' \end{array} \right. . \tag{7}$$

One can write $\widehat{S}(t,t')$ of Eqs. (6) and (7) as a chronologically ordered exponential operator

$$\widehat{S}(t,t') = \hat{\mathcal{T}}\left\{ \exp\left[-\frac{i}{\hbar}\int_{t'}^{t} d\tau \widehat{H}_\tau \right] \right\}. \tag{8}$$

This expression, together with Eq. (2), leads to the following properties of the evolution operator:

$$\widehat{S}(t,t') = \widehat{S}^+(t',t), \quad \widehat{S}^+(t,t')\widehat{S}(t,t') = 1, \quad \widehat{S}(t,t_1)\widehat{S}(t_1,t') = \widehat{S}(t,t'), \tag{9}$$

which can be checked by calculating the time derivatives (problem 1.3).

Below we consider a system with time-independent Hamiltonian $\widehat{H}$ in the presence of a weak harmonic perturbation. In other words, we discuss the evolution of the system with the Hamiltonian

$$\widehat{H} + (\hat{v}_\omega e^{-i\omega t} + H.c.) \equiv \widehat{H} + \widehat{V}_t, \tag{10}$$

where the operator $\hat{v}_\omega$ is small. The letters $H.c.$ in Eq. (10) indicate the Hermitian conjugate contribution to the perturbation. A solution of this problem not only describes a response of the system to the harmonic perturbation, but also allows one to consider a modification of stationary states under the time-independent perturbation $\hat{v}_\omega + \hat{v}_\omega^+$, where $\omega = 0$. It is convenient to use the interaction representation by introducing a new ket-vector $|\delta t\rangle$ according to $|\delta t\rangle = \widehat{S}(t)|\delta t\rangle$, where $\widehat{S}(t)$ is the S-operator (introduced by Eq. (3)) for the Hamiltonian $\widehat{H}$. Substituting $|\delta t\rangle = \widehat{S}(t)|\delta t\rangle$ into Eq. (1.13), we multiply the latter by $\widehat{S}^+(t)$ from the left and obtain the following Schroedinger equation in the interaction representation:

$$i\hbar\frac{\partial|\delta t\rangle}{\partial t} = \widehat{S}^+(t)\widehat{V}_t\widehat{S}(t)|\delta t\rangle, \quad |\delta, t = 0\rangle = |i\rangle. \tag{11}$$

To solve Eq. (11) with the accuracy of the first order in the perturbation $\widehat{V}_t$, we substitute the unperturbed ket-vector $|i\rangle$ to the right-hand side of this equation. Since we assume that the unperturbed system is in the initial state i, we have

$$|it\rangle \simeq |i\rangle + \frac{1}{i\hbar}\int_0^t dt'\widehat{V}(t')|i\rangle, \quad \widehat{V}(t) = \widehat{S}^+(t)\widehat{V}_t\widehat{S}(t). \tag{12}$$

The probability of finding the system at the instant t in the state f (described by the ket-vector $\widehat{S}(t)|f\rangle$ of zero-order approximation), calculated with the accuracy of the second order in the perturbation, is equal to $|\langle f|\widehat{S}^+(t)|it\rangle|^2 = |\langle f|it\rangle|^2$. The probability of transition between the states i and f is defined as a time derivative of this quantity:

$$W_{if} = \frac{d|\langle f|it\rangle|^2}{dt}. \tag{13}$$

We note that both $|i\rangle$ and $|f\rangle$ are the solutions of the eigenstate problem $\widehat{H}|i\rangle = \varepsilon_i|i\rangle$. An explicit expression for W_{if} is determined after a simple integration over time:

$$W_{if} = \frac{1}{\hbar^2}\frac{d}{dt}\left|v_{fi}\frac{e^{i(\omega_{fi}-\omega)t}-1}{i(\omega_{fi}-\omega)} + v_{if}^*\frac{e^{i(\omega_{fi}+\omega)t}-1}{i(\omega_{fi}+\omega)}\right|^2, \tag{14}$$

where $\omega_{fi} = (\varepsilon_f - \varepsilon_i)/\hbar$ is the frequency of transitions between the states i and f of the unperturbed system and $v_{fi} = \langle f|\hat{v}_\omega|i\rangle$. The probability of transitions has resonant behavior: at $t \to \infty$ it is not equal to zero only when ω coincides with one of the transition frequencies. We note that, on the small-time scale, a non-zero probability exists also for non-resonant conditions, owing to the energy-time uncertainty. In the case of time-independent perturbations ($\omega = 0$), the energy of the system is conserved, and the transitions occur between the degenerate states only.

Let us consider first the asymptotic behavior of W_{if} at large times under a time-independent perturbation $\widehat{V} \equiv 2\hat{v}$ (note that $\hat{v}$ is Hermitian). For this case, taking into account $|v_{if}|^2 = |v_{fi}|^2$, we obtain

$$W_{if} = \frac{4|v_{if}|^2}{\hbar^2}\frac{d}{dt}\frac{2-2\cos\omega_{fi}t}{\omega_{fi}^2} = \frac{8|v_{if}|^2}{\hbar^2}\frac{d}{dt}\left[\frac{2\sin^2(\omega_{fi}t/2)}{\omega_{fi}^2 t}t\right]. \tag{15}$$

If $\omega_{fi}t \gg 1$, the function $2\sin^2(\omega_{fi}t/2)/\omega_{fi}^2 t$ goes to $\pi\delta(\omega_{fi})$; see Fig. 1.1 and problem 1.4, where different presentations of Dirac's δ-function are discussed. As a result, the probability of transition becomes

$$W_{if} = \frac{2\pi}{\hbar}|\langle f|\widehat{V}|i\rangle|^2\delta(\varepsilon_f - \varepsilon_i). \tag{16}$$

This important result is known as Fermi's golden rule. We stress again that the energy of the system is conserved, and only the states with $\varepsilon_f = \varepsilon_i$ contribute into the probability (16) under a time-independent perturbation.

The probability of resonant transitions in the case of time-dependent perturbations is calculated in a similar way. If $\omega_{fi} \neq 0$, only the terms containing $\omega_{fi} - \omega$ in the factor $|\dots|^2$ of Eq. (14) are important at large t, and one obtains

$$W_{if}(\omega) = \frac{2\pi}{\hbar}|\langle f|\hat{v}_\omega|i\rangle|^2\delta(\varepsilon_f - \varepsilon_i - \hbar\omega). \tag{17}$$

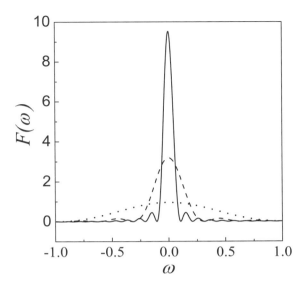

Figure 1.1. Function $F(\omega) = \sin^2(\omega t)/\pi\omega^2 t$ for $t = 3$, 10, and 30 (dotted, dashed, and solid curves, respectively).

The energy conservation law

$$\varepsilon_f = \varepsilon_i + \hbar\omega \tag{18}$$

is fulfilled for interlevel transitions excited by a harmonic perturbation with the energy of quantum $\hbar\omega$.

Equations (16) and (17) can be derived in an alternative way, under the assumption that the perturbation $\widehat{V}_t$ is adiabatically turned on at $t = -\infty$. The first-order solution of the time-dependent Schroedinger equation (1.13) with the Hamiltonian $\widehat{H} + \widehat{V}_t$ and boundary condition $|\delta, t = -\infty\rangle = |i\rangle$ is written as

$$|it\rangle \simeq |i\rangle + \frac{1}{i\hbar} \int_{-\infty}^{t} dt' e^{\lambda t'} \widehat{S}(t, t') \widehat{V}_{t'} \widehat{S}^+(t, t') |i\rangle \,, \tag{19}$$

where $\lambda \to +0$ describes the adiabatic turning-on. Consider, for example, a time-independent perturbation $\widehat{V}$. The integral in Eq. (19) is easily calculated by substituting $\tau = t' - t$. Since $\widehat{S}(t, t') = \widehat{S}(-\tau)$, we obtain

$$\langle f|it\rangle = e^{\lambda t} \frac{\langle f|\widehat{V}|i\rangle}{\varepsilon_i - \varepsilon_f + i\hbar\lambda}. \tag{20}$$

The transition probability $d|\langle f|it\rangle|^2/dt$ is reduced to Eq. (16) according to the first expression for the δ-function in problem 1.4. The case of time-dependent perturbation is considered in a similar way, leading to Eq. (17).

The probabilities of transitions, derived above in a pure quantum-mechanical approach, are the important characteristics determining kinetic properties of different systems. Indeed, let us introduce the occupation number n_{jt} for the state j, i.e., the average number of the particles in the state j at the instant t, according to (see Eq. (1.17))

$$n_{jt} = \langle j|\hat{\eta}_t|j\rangle = \sum_\delta P_\delta |\langle j|\delta t\rangle|^2. \tag{21}$$

One may expect that, under proper conditions (in the subsequent chapters this question will be considered in detail), the temporal evolution of the occupation numbers is determined by the balance equation

$$\frac{\partial n_{jt}}{\partial t} = \sum_{j'} W_{jj'} \left(n_{j't} - n_{jt}\right), \tag{22}$$

where $W_{jj'}$ is given by Eq. (16) for the case of time-independent perturbations and $W_{jj'} = W_{jj'}(\omega) + W_{jj'}(-\omega)$, see Eq. (17), for time-dependent harmonic perturbations. The first and second terms on the right-hand side of Eq. (22) describe incoming (arrival) and outgoing (departure) contributions to the balance of occupation, respectively. We note that the arrival rate from the state j' to the state j is equal to $W_{jj'}n_{j't}$, while the departure rate from the state j to all other states is equal to $n_{jt}\sum_{j'} W_{jj'}$. The balance equation (22) conserves the number of the particles and, for the case of time-independent perturbations, the energy of the system. In order to describe the temporal evolution of the other characteristics of the system (those which are sensitive to phase correlation), one has to consider quantum kinetic equations for $\hat{\eta}_t$; see the next chapters.

In the case of harmonic perturbations, one may express the power absorbed by the system through the transition probability (17). The absorbed power U_ω is defined as the energy of the quantum, $\hbar\omega$, multiplied by the difference between the rate of transition from the state j to the state j' (which corresponds to absorption of the quantum) and the rate of emission of the quantum associated with the transitions from j' to j:

$$U_\omega = \hbar\omega \sum_{jj'} \left[W_{jj'}(\omega)n_j(1 - n_{j'}) - W_{j'j}(-\omega)n_{j'}(1 - n_j)\right]$$

$$= \hbar\omega \sum_{jj'} W_{jj'}(\omega)(n_j - n_{j'}). \tag{23}$$

On the other hand, the absorbed power U_ω for the system excited by an electric field $\mathbf{E}_t$ is determined by the electrodynamical expression $\overline{\mathbf{I}_t \cdot \mathbf{E}_t}$, where $\mathbf{I}_t$ is the electric current density induced by the field and the line over the expression denotes the averaging over the period $2\pi/\omega$. Within the accuracy of $\mathbf{E}_t^2$, which corresponds to the perturbation theory applied above, one may consider $\mathbf{I}_t$ in the framework of the linear-response approximation and describe U_ω through the frequency-dependent conductivity of the system; see Chapter 3.

3. Photons in Medium

We begin our consideration of the quantum dynamics of concrete physical systems with the case of electromagnetic field in the spatially inhomogeneous medium described by the dielectric permittivity tensor $\hat{\epsilon}_{\mathbf{r}}$. Starting from an expression for the energy of the electromagnetic field in the absence of free electric charges, we derive the Hamiltonian of the field and, after a quantization procedure, describe the field as a set of oscillators corresponding to elementary quasiparticles known as photons. The photons are an example of bosons, the particles with a symmetric wave function corresponding to the Bose-Einstein statistics.

The energy of the field is determined by the expression

$$\mathcal{E}_f = \frac{1}{8\pi} \int_{(V)} d\mathbf{r} (\mathbf{E}_{\mathbf{r}t} \cdot \hat{\epsilon}_{\mathbf{r}} \mathbf{E}_{\mathbf{r}t} + \mathbf{H}_{\mathbf{r}t}^2)$$

$$= \frac{1}{8\pi} \int_{(V)} d\mathbf{r} \left\{ \frac{1}{c^2} \frac{\partial \mathbf{A}_{\mathbf{r}t}}{\partial t} \cdot \hat{\epsilon}_{\mathbf{r}} \frac{\partial \mathbf{A}_{\mathbf{r}t}}{\partial t} + ([\nabla \times \mathbf{A}_{\mathbf{r}t}])^2 \right\}, \tag{1}$$

where the integrals are taken over the normalization volume V. In Eq. (1) we assume a local relation between the electrostatic induction and the field: $\hat{\epsilon}_{\mathbf{r}} \mathbf{E}_{\mathbf{r}t}$. On the other hand, the magnetic induction is equal to $\mathbf{H}_{\mathbf{r}t}$ because the kinetic phenomena are considered in this book for non-magnetic materials only. The electric and magnetic field strengths, $\mathbf{E}_{\mathbf{r}t}$ and $\mathbf{H}_{\mathbf{r}t}$, which satisfy the Maxwell equations in medium, are expressed only through the vector potential $\mathbf{A}_{\mathbf{r}t}$, since we have chosen the Coulomb gauge $\nabla \cdot \hat{\epsilon}_{\mathbf{r}} \mathbf{A}_{\mathbf{r}t} = 0$ leading to zero scalar potential in the absence of free charges.

It is convenient to represent the electromagnetic field described by the vector potential $\mathbf{A}_{\mathbf{r}t}$ as

$$\mathbf{A}_{\mathbf{r}t} = \sum_\nu [q_\nu(t) \mathbf{A}_{\mathbf{r}}^\nu + q_\nu^*(t) \mathbf{A}_{\mathbf{r}}^{\nu*}], \tag{2}$$

where the modes $\mathbf{A_r^{\nu}}$ with frequencies ω_{ν} are determined by the wave equation following from the Maxwell equations:

$$[\nabla \times [\nabla \times \mathbf{A_r^{\nu}}]] - \left(\frac{\omega_{\nu}}{c}\right)^2 \hat{\epsilon}_{\mathbf{r}} \mathbf{A_r^{\nu}} = 0. \tag{3}$$

The modes satisfy the orthogonality and normalization conditions according to $\int_{(V)} d\mathbf{r} \mathbf{A_r^{\nu*}} \cdot \hat{\epsilon}_{\mathbf{r}} \mathbf{A_r^{\nu'}} = 2\pi c^2 \delta_{\nu\nu'}$. The coefficients $q_{\nu}(t)$ in the expression (2) can be considered as the generalized coordinates of ν-th mode. They satisfy the oscillator equation

$$\frac{d^2 q_{\nu}(t)}{dt^2} + \omega_{\nu}^2 q_{\nu}(t) = 0 \tag{4}$$

corresponding to the harmonic oscillations with eigenfrequencies ω_{ν}. In the presence of external sources described by the electric current density $\mathbf{I_{rt}}$, one must add the term $-c^{-1} \int_{(V)} d\mathbf{r} \, \mathbf{I_{rt}} \cdot \mathbf{A_{rt}}$ to the right-hand side of Eq. (1) and $(4\pi/c)\mathbf{I_r}$ to the right-hand side of Eq. (3). Such a contribution describes the interaction of the modes $\mathbf{A_r}$ with external charges.

Introducing the generalized momentum $p_{\nu}(t) \equiv dq_{\nu}(t)/dt$, we apply the orthogonality and normalization conditions for the modes of Eq. (3) to rewrite the energy of the field given by Eq. (1) as a sum of oscillator energies:

$$\mathcal{E}_f = \frac{1}{2} \sum_{\nu} \left\{ |p_{\nu}(t)|^2 + \omega_{\nu}^2 |q_{\nu}(t)|^2 \right\}. \tag{5}$$

Since the solutions of Eq. (4) are proportional to $\exp(-i\omega_{\nu}t)$, we have the relation $p_{\nu} = -i\omega_{\nu}q_{\nu}$. It is convenient to introduce the canonically conjugate variables

$$Q_{\nu}(t) = \frac{q_{\nu}(t) + q_{\nu}^*(t)}{2}, \qquad P_{\nu}(t) = -i\omega_{\nu} \frac{q_{\nu}(t) - q_{\nu}^*(t)}{2}, \tag{6}$$

which are used here in order to rewrite the energy of the field as $\mathcal{E}_f = \sum_{\nu} \left\{ |P_{\nu}(t)|^2 + \omega_{\nu}^2 |Q_{\nu}(t)|^2 \right\}/2$. The equations of motion acquire Hamiltonian form: $\dot{P}_{\nu} = \ddot{Q}_{\nu} = -\partial\mathcal{E}_f/\partial Q_{\nu}$, $\dot{Q}_{\nu} = \partial\mathcal{E}_f/\partial P_{\nu}$.

In order to quantize the electromagnetic field, we have to replace the canonically conjugate variables $Q_{\nu}(t)$ and $P_{\nu}(t)$ by the operators of generalized coordinate and momentum, $\widehat{Q}_{\nu}$ and $\widehat{P}_{\nu}$, which satisfy the commutation relation

$$[\widehat{Q}_{\nu}, \widehat{P}_{\nu'}] = i\hbar\delta_{\nu\nu'}. \tag{7}$$

Let us use the expression for the energy as a sum of the oscillatory contributions (5) and take into account the relation $\widehat{P}_{\nu} = -i\hbar\partial/\partial Q_{\nu}$. Then we write the Hamiltonian of quantized field in the Q-representation:

$$\widehat{H}_{ph} = \frac{1}{2} \sum_{\nu} \left\{ -\hbar^2 \frac{\partial^2}{\partial Q_{\nu}^2} + \omega_{\nu}^2 Q_{\nu}^2 \right\}. \qquad (8)$$

A solution of the eigenstate problem $\widehat{H}_{ph} \Psi_{\{n_{\nu}\}} = E_{\{n_{\nu}\}} \Psi_{\{n_{\nu}\}}$ determines a set of occupation numbers, $\{n_{\nu}\}$, for the given modes. The symmetrized wave function, corresponding to the Bose-Einstein statistics, is a product of the eigenfunctions of different modes, $\psi_{n_{\nu}}(Q_{\nu})$, while the total energy, $E_{\{n_{\nu}\}}$, is given by a sum of the oscillator energies:

$$\Psi_{\{n_{\nu}\}} = \prod_{\nu} \psi_{n_{\nu}}(Q_{\nu}), \qquad E_{\{n_{\nu}\}} = \sum_{\nu} \hbar \omega_{\nu} \left(n_{\nu} + \frac{1}{2} \right). \qquad (9)$$

The occupation numbers n_{ν} are integers ($n_{\nu} \geq 0$). As follows from Eq. (9), the wave function $\Psi_{\{n_{\nu}\}}$ is symmetric with respect to permutations of each oscillatory function $\psi_n(Q)$ (see Appendix A) with another oscillatory function. The matrix elements of the generalized coordinate for the transitions between the states with quantum numbers n_{ν} and n_{ν}' are equal to zero if $n_{\nu}' \neq n_{\nu} \pm 1$, while for the transition between adjacent levels these matrix elements are

$$\langle n_{\nu}' | \widehat{Q}_{\nu} | n_{\nu} \rangle = \sqrt{\frac{\hbar}{2\omega_{\nu}}} \left\{ \begin{array}{ll} \sqrt{n_{\nu} + 1}, & n_{\nu}' = n_{\nu} + 1 \\ \sqrt{n_{\nu}}, & n_{\nu}' = n_{\nu} - 1 \end{array} \right. . \qquad (10)$$

The matrix elements of the generalized momentum are $\langle n_{\nu}' | \widehat{P}_{\nu} | n_{\nu} \rangle = \pm i \omega_{\nu} \langle n_{\nu}' | \widehat{Q}_{\nu} | n_{\nu} \rangle$, where the signs $\pm$ correspond to the transitions between the states with occupation numbers $n_{\nu}' = n_{\nu} \pm 1$ and n_{ν}. This equation is consistent with the relation between the Fourier components of coordinate and momentum used in Eq. (6). Instead of a pair of canonically conjugate operators $\widehat{Q}_{\nu}$ and $\widehat{P}_{\nu}$, we introduce, by analogy to Eq. (A.11), two Hermitian conjugate creation and annihilation operators for the mode ν:

$$\hat{b}_{\nu} = \frac{\omega_{\nu} \widehat{Q}_{\nu} + i \widehat{P}_{\nu}}{\sqrt{2\hbar\omega_{\nu}}}, \qquad \hat{b}_{\nu}^{+} = \frac{\omega_{\nu} \widehat{Q}_{\nu}^{+} - i \widehat{P}_{\nu}^{+}}{\sqrt{2\hbar\omega_{\nu}}}. \qquad (11)$$

Representing the contribution of the state ν in the Hamiltonian (8) as $\{\ldots\} = (\omega_{\nu}\widehat{Q}_{\nu}^{+} - i\widehat{P}_{\nu}^{+})(\omega_{\nu}\widehat{Q}_{\nu} + i\widehat{P}_{\nu}) + \hbar\omega_{\nu}$, we rewrite $\widehat{H}_{ph}$ in the form

$$\widehat{H}_{ph} = \sum_{\nu} \hbar \omega_{\nu} \left(\hat{b}_{\nu}^{+} \hat{b}_{\nu} + \frac{1}{2} \right). \qquad (12)$$

The Hamiltonian of the field is given as a sum of the contributions $\hat{h}_{osc}$ determined by Eq. (A.12), with the oscillator frequencies ω_{ν}.

Therefore, the electromagnetic field in a medium is presented as a superposition of quantized normal vibrations with frequencies ω_ν and occupation numbers n_ν. It is convenient to use a representation described by the ket-vector $|\{n_\nu\}\rangle$ depending on the sets of occupation numbers $\{n_\nu\}$. Using these sets as independent variables of the problem (instead of the generalized coordinates Q_ν), one may define the creation and annihilation operators through their matrix elements

$$\langle n'_\nu|\hat{b}^+_\nu|n_\nu\rangle = \sqrt{n_\nu + 1}\delta_{n'_\nu,n_\nu+1},$$

$$\langle n'_\nu|\hat{b}_\nu|n_\nu\rangle = \sqrt{n_\nu}\delta_{n'_\nu,n_\nu-1}, \qquad (13)$$

instead of using Eq. (11). This means that the operators $\hat{b}^+_\nu$ and $\hat{b}_\nu$, while acting on the ket-vector $|\{n_\nu\}\rangle$, change the occupation number of the photons of the mode ν by ±1, respectively:

$$\hat{b}^+_\nu|n_1 n_2 \ldots n_\nu \ldots\rangle = \sqrt{n_\nu + 1}|n_1 n_2 \ldots n_\nu + 1 \ldots\rangle,$$

$$\hat{b}_\nu|n_1 n_2 \ldots n_\nu \ldots\rangle = \sqrt{n_\nu}|n_1 n_2 \ldots n_\nu - 1 \ldots\rangle. \qquad (14)$$

The commutation rules for these operators are obtained by using either the matrix elements (13) or the expressions of these operators through $\widehat{Q}_\nu$ and $\widehat{P}_\nu$, Eq. (11). For the Hermitian conjugate operators, one has

$$[\hat{b}_\nu, \hat{b}^+_{\nu'}] = \delta_{n_\nu,n_{\nu'}}, \qquad (15)$$

while the operators of the same kind (creation or annihilation) merely commute with each other. It is the commutation rule (15) that leads to the appearance of zero-field oscillation energy $\sum_\nu \hbar\omega_\nu/2$ in Eq. (12); see also Eqs. (A.11) and (A.12). By analogy to the case of a single oscillator, see Eq. (A.18), the set of ket-vectors $|\{n_\nu\}\rangle$ is presented as

$$|\{n_\nu\}\rangle = \prod_\nu \frac{(\hat{b}_\nu)^{n_\nu}}{\sqrt{n_\nu!}}|\{0\}\rangle, \qquad (16)$$

where $|\{0\}\rangle$ describes the vacuum state where only zero-field oscillations due to quantum-mechanical uncertainty are present. The set of ket-vectors also satisfies the completeness, orthogonality, and normalization conditions:

$$\sum_\nu |\{n_\nu\}\rangle\langle\{n_\nu\}| = \hat{1}, \quad \langle\{n_\nu\}|\{n'_\nu\}\rangle = \delta_{\{n_\nu\},\{n'_\nu\}}. \qquad (17)$$

The generalized Kronecker symbol $\delta_{\{n_\nu\},\{n'_\nu\}}$ is equal to unity only when all the occupation numbers from the sets $\{n_\nu\}$ and $\{n'_\nu\}$ coincide. The description of the electromagnetic field given by Eqs. (13)-(17) is called the occupation number representation or the second quantization. It is analogous to the description of a single oscillator given by Eqs. (A.16)-(A.19). In this representation, the sets of independent variables describing the system are the numbers of quanta of the field in each mode. These quanta are called the photons in medium, i.e., the system is described in terms of quasiparticles. The operator of the photon number for the mode ν is introduced as $\hat{n}_\nu = \hat{b}_\nu^+ \hat{b}_\nu$. The justification of this definition is the same as for a single oscillator, and the ket-vector $|\{n_\nu\}\rangle$ is the eigenvector of the operator $\hat{n}_\nu$ corresponding to the eigenvalue n_ν, according to $\hat{n}_\nu|\{n_\nu\}\rangle = n_\nu|\{n_\nu\}\rangle$; see Eq. (A.20).

Using the expansion (2) and expressing the amplitudes of vibrations according to Eqs. (6) and (11) through the creation and annihilation operators as $\hat{q}_\nu = \sqrt{\hbar/\omega_\nu}\,\hat{b}_\nu$ and $\hat{q}_\nu^+ = \sqrt{\hbar/\omega_\nu}\,\hat{b}_\nu^+$, we get the quantized operator of the vector potential

$$\hat{\mathbf{A}}_{\mathbf{r}} = \sum_\nu \sqrt{\frac{\hbar}{\omega_\nu}} \left(\mathbf{A}_{\mathbf{r}}^\nu \hat{b}_\nu + \mathbf{A}_{\mathbf{r}}^{\nu*} \hat{b}_\nu^+ \right), \tag{18}$$

where the modes $\mathbf{A}_{\mathbf{r}}^\nu$ are determined from Eq. (3). The operators of the second-quantized fields, $\hat{\mathbf{E}}_{\mathbf{r}}$ and $\hat{\mathbf{H}}_{\mathbf{r}}$, can be written by using Eq. (18) together with the relation $\dot{q}_\nu(t) = -i\omega_\nu q_\nu(t)$ and by expressing these fields through the vector potential according to $\mathbf{E}_{\mathbf{r}t} = -c^{-1}\dot{\mathbf{A}}_{\mathbf{r}t}$ and $\mathbf{H}_{\mathbf{r}t} = [\nabla \times \mathbf{A}_{\mathbf{r}t}]$. The classical vector of the radiation flux density (Poynting vector), $\mathbf{S}_{\mathbf{r}t} = (c/4\pi)[\mathbf{E}_{\mathbf{r}t} \times \mathbf{H}_{\mathbf{r}t}]$, is expanded in terms of the modes as follows:

$$\mathbf{S}_{\mathbf{r}t} = -\frac{1}{4\pi} \left[\frac{\partial \mathbf{A}_{\mathbf{r}t}}{\partial t} \times [\nabla \times \mathbf{A}_{\mathbf{r}t}] \right] = \frac{i}{4\pi} \sum_{\nu\nu'} \omega_{\nu'} \left[\left(q_{\nu'}(t) \mathbf{A}_{\mathbf{r}}^{\nu'} \right. \right.$$

$$\left. \left. - q_{\nu'}^*(t) \mathbf{A}_{\mathbf{r}}^{\nu'*} \right) \times \left(q_\nu(t) [\nabla \times \mathbf{A}_{\mathbf{r}}^\nu] + q_\nu^*(t) [\nabla \times \mathbf{A}_{\mathbf{r}}^{\nu*}] \right) \right]. \tag{19}$$

In the second quantization representation, the operator of the radiation flux density, $\hat{\mathbf{S}}_{\mathbf{r}}$, is written as

$$\hat{\mathbf{S}}_{\mathbf{r}} = \frac{i\hbar}{4\pi} \sum_{\nu\nu'} \sqrt{\frac{\omega_{\nu'}}{\omega_\nu}} \left[\left(\hat{b}_{\nu'} \mathbf{A}_{\mathbf{r}}^{\nu'} - \hat{b}_{\nu'}^+ \mathbf{A}_{\mathbf{r}}^{\nu'*} \right) \right.$$

$$\left. \times \left(\hat{b}_\nu [\nabla \times \mathbf{A}_{\mathbf{r}}^\nu] + \hat{b}_\nu^+ [\nabla \times \mathbf{A}_{\mathbf{r}}^{\nu*}] \right) \right] \tag{20}$$

after expressing the amplitudes $q_\nu(t)$ in Eq. (19) through the corresponding creation and annihilation operators.

In a homogeneous and isotropic medium with dielectric permittivity ϵ, the modes $\mathbf{A}_\mathbf{r}^\nu$ are the plane waves $\mathbf{A}_{\mathbf{q}\mu} \exp(i\mathbf{q} \cdot \mathbf{r})$ with wave vector $\mathbf{q}$ and polarization μ. The amplitudes $\mathbf{A}_{\mathbf{q}\mu}$ are determined from the vector equation

$$[\mathbf{q} \times [\mathbf{q} \times \mathbf{A}_{\mathbf{q}\mu}]] + \left(\frac{\omega_{\mathbf{q}\mu}}{c}\right)^2 \epsilon \mathbf{A}_{\mathbf{q}\mu} = 0 \tag{21}$$

following from Eq. (3) and from the gauge condition $(\mathbf{q} \cdot \mathbf{A}_{\mathbf{q}\mu}) = 0$. Equation (21) is equivalent to a system of three algebraic equations for the components of the vector $\mathbf{A}_{\mathbf{q}\mu}$. The requirement of orthogonality and normalization for the amplitudes is written as

$$\frac{V\epsilon}{2\pi c^2}(\mathbf{A}_{\mathbf{q}\mu}^* \cdot \mathbf{A}_{\mathbf{q}\mu'}) = \delta_{\mu\mu'}, \tag{22}$$

so that one can introduce the unit vectors of polarization, $\mathbf{e}_{\mathbf{q}\mu}$, according to $\mathbf{A}_{\mathbf{q}\mu} = \sqrt{2\pi c^2/\epsilon V}\,\mathbf{e}_{\mathbf{q}\mu}$. These vectors have the properties of transversity (following from the gauge conditions), orthogonality, and normalization (following from Eq. (22)), while Eq. (21) leads to a polarization-independent dispersion relation for the photon of frequency ω_q:

$$(\mathbf{q} \cdot \mathbf{e}_{\mathbf{q}\mu}) = 0, \quad (\mathbf{e}_{\mathbf{q}\mu}^* \cdot \mathbf{e}_{\mathbf{q}\mu'}) = \delta_{\mu\mu'}, \quad \omega_q = \tilde{c}q, \quad \tilde{c} = \frac{c}{\sqrt{\epsilon}}. \tag{23}$$

These relations describe propagation of the photons whose unit vectors of polarization, $\mathbf{e}_{\mathbf{q}\mu=1}$ and $\mathbf{e}_{\mathbf{q}\mu=2}$, are directed in the plane perpendicular to the wave vector $\mathbf{q}$. The dispersion of the photons is linear in q, and the proportionality coefficient $\tilde{c}$ is the velocity of light in the medium. The Hamiltonian and the radiation flux density operator for the homogeneous and isotropic medium are expressed, according to Eqs. (12) and (20), through the photonic creation and annihilation operators for the states $\nu = (\mathbf{q}, \mu)$. The polarization vectors and the frequency of these states are given by Eq. (23). The operator of electric field can be obtained from Eq. (18):

$$\hat{\mathbf{E}}_\mathbf{r} = i\sum_{\mathbf{q}\mu} \sqrt{\frac{2\pi\hbar\omega_q}{\epsilon V}}\,\mathbf{e}_{\mathbf{q}\mu} e^{i\mathbf{q}\cdot\mathbf{r}}\left(\hat{b}_{\mathbf{q}\mu} - \hat{b}_{-\mathbf{q}\mu}^+\right), \tag{24}$$

where we assumed that $\mathbf{e}_{\mathbf{q}\mu}^* = \mathbf{e}_{-\mathbf{q}\mu}$. The magnetic-field operator is given by a similar expression, which is obtained according to $\hat{\mathbf{H}}_\mathbf{r} = [\nabla \times \hat{\mathbf{A}}_\mathbf{r}]$ and contains the polarization factor $[\mathbf{q} \times \mathbf{e}_{\mathbf{q}\mu}]$ under the sum. The matrix element of $\hat{\mathbf{S}}_\mathbf{r}$ for the case of plane waves is determined according to Eqs. (20), (22), and (13) as $\langle n_\nu | \hat{\mathbf{S}}_\mathbf{r} | n_\nu \rangle = (\mathbf{q}/q)\tilde{c}V^{-1}\hbar\omega_q(n_\nu + 1/2)$. In this form, the Poynting vector has direct meaning of the flux of photon energy density with velocity $\tilde{c}$ in the direction of $\mathbf{q}$. In non-homogeneous media, the description of the modes based upon Eqs. (3) and (18) is more sophisticated, though relatively simple results exist for the case of one-dimensional inhomogeneities (problems 1.5 and 1.6).

Finally, let us calculate the averaged occupation number of the mode ν for the equilibrium distribution of photons with temperature T_{ph}. This distribution is described by the density matrix

$$\hat{\eta}_{eq} = Z^{-1} \exp(-\widehat{H}_{ph}/T_{ph}), \quad Z = \mathrm{Sp}\exp(-\widehat{H}_{ph}/T_{ph}). \qquad (25)$$

The partition function Z is expressed through the photon energy (9) written as $E_{\{n_\nu\}} = \sum_\nu \hbar\omega_\nu n_\nu + E_0$, where E_0 is the energy of zero vibrations, according to $Z = \exp(-E_0/T_{ph})\overline{Z}$ and

$$\overline{Z} = \sum_{\{n_\nu\}} \prod_\nu e^{-\hbar\omega_\nu n_\nu/T_{ph}} = \prod_\nu \sum_n e^{-\hbar\omega_\nu n/T_{ph}}$$

$$= \prod_\nu \left(1 - e^{-\hbar\omega_\nu/T_{ph}}\right)^{-1}. \qquad (26)$$

The mean value of the occupation number of the mode ν is defined as $\bar{n}_\nu = \mathrm{Sp}\hat{n}_\nu\hat{\eta}_{eq}$. It is expressed through $\overline{Z}$ as

$$\bar{n}_\nu = Z^{-1} \sum_{\{n_{\nu_1}\}} n_\nu e^{-E_{\{n_{\nu_1}\}}/T_{ph}} = -T_{ph}\frac{\partial \ln \overline{Z}}{\partial(\hbar\omega_\nu)}. \qquad (27)$$

Calculating the derivative in Eq. (27), we obtain the equilibrium Planck distribution

$$\bar{n}_\nu = \left[e^{\hbar\omega_\nu/T_{ph}} - 1\right]^{-1}. \qquad (28)$$

This distribution allows one to describe various equilibrium properties of the boson gas (problems 1.7 and 1.8). It is valid for all kinds of the bosons whose number is not fixed.

4. Many-Electron System

In contrast to the case of photons, the dynamics of a system of electrically charged particles depends on their interactions with external electric fields (created by different, with respect to the system under consideration, charges) and externally applied magnetic fields (note that we consider non-magnetic materials only), as well as on the interaction of these particles with each other. The existence of the spin variable leads to a further sophistication of such dynamics. Below we discuss the quantum dynamics for electrons, charged particles with two different spin states. The electrons are an example of fermions, the particles with an antisymmetric, with respect to particle permutation, wave function corresponding to the Fermi-Dirac statistics.

The Hamiltonian of the electron system in external fields is written as

$$\widehat{H}_e = \sum_j \hat{\mathrm{h}}_j + \widehat{H}_f. \tag{1}$$

Here $\hat{\mathrm{h}}_j$ is the one-electron (the index j numbers the electrons) operator of the kinetic energy. It is given by the equation

$$\hat{\mathrm{h}}_j = \frac{(\hat{\mathbf{p}}_j - e\mathbf{A}_{\mathbf{x}_jt}/c)^2}{2m}, \tag{2}$$

where $\hat{\mathbf{p}}_j - e\mathbf{A}_{\mathbf{x}_jt}/c$ is the operator of kinematic momentum expressed through the canonical momentum $\hat{\mathbf{p}}$ satisfying the ordinary commutation relations $[\hat{p}_\alpha, x_\beta] = -i\hbar\delta_{\alpha\beta}$ and through the vector potential $\mathbf{A}_{\mathbf{x}_jt}$. The second term of Eq. (1), $\widehat{H}_f$, is the operator of the field energy $\mathcal{E}_f$, the latter is given by the first part of Eq. (3.1). Using the expressions

$$\mathbf{E}_{\mathbf{r}t} = -\frac{1}{c}\frac{\partial \mathbf{A}_{\mathbf{r}t}}{\partial t} - \nabla\Phi_{\mathbf{r}t} , \quad \mathbf{H}_{\mathbf{r}t} = [\nabla \times \mathbf{A}_{\mathbf{r}t}] \tag{3}$$

relating the electric and magnetic fields to the vector potential $\mathbf{A}_{\mathbf{r}t}$ and scalar potential $\Phi_{\mathbf{r}t}$, we rewrite $\mathcal{E}_f$ as

$$\mathcal{E}_f = \int_{(V)} \frac{d\mathbf{r}}{8\pi} \left\{ \frac{1}{c^2}\frac{\partial \mathbf{A}_{\mathbf{r}t}}{\partial t} \cdot \hat{\epsilon}_{\mathbf{r}} \frac{\partial \mathbf{A}_{\mathbf{r}t}}{\partial t} + ([\nabla \times \mathbf{A}_{\mathbf{r}t}])^2 \right\}$$

$$+ \int_{(V)} \frac{d\mathbf{r}}{4\pi c} \nabla\Phi_{\mathbf{r}t}\hat{\epsilon}_{\mathbf{r}}\frac{\partial \mathbf{A}_{\mathbf{r}t}}{\partial t} + \int_{(V)} \frac{d\mathbf{r}}{8\pi}\nabla\Phi_{\mathbf{r}t}\hat{\epsilon}_{\mathbf{r}}\nabla\Phi_{\mathbf{r}t} . \tag{4}$$

This equation generalizes Eq. (3.1) to the case of non-zero gradient of the scalar potential. The tensor $\hat{\epsilon}_{\mathbf{r}}$ is assumed to be symmetric. Below we again employ the Coulomb gauge $\nabla \cdot (\hat{\epsilon}_{\mathbf{r}}\mathbf{A}_{\mathbf{r}t}) = 0$ and assume that the fields go to zero at the boundaries of the region V (one may also use the periodic boundary conditions). The first term of the expression (4) corresponds to the energy of transverse vibrations of the field and describes the photons in medium. After the quantization of the field done in the previous section, we can denote this term as $\widehat{H}_{ph}$. The second term on the right-hand side of Eq. (4) is equal to zero because

$$\nabla\Phi_{\mathbf{r}t}\hat{\epsilon}_{\mathbf{r}}\frac{\partial \mathbf{A}_{\mathbf{r}t}}{\partial t} = \nabla \cdot \left(\Phi_{\mathbf{r}t}\hat{\epsilon}_{\mathbf{r}}\frac{\partial \mathbf{A}_{\mathbf{r}t}}{\partial t} \right) \tag{5}$$

in the gauge used, and the integral over the volume V is reduced to a surface integral over an infinitely remote boundary where the fields are equal to zero. The third term of the expression (4) can be rewritten according to $\nabla\Phi\hat{\epsilon}\nabla\Phi = \nabla\cdot(\Phi\hat{\epsilon}\nabla\Phi) - \Phi\nabla\cdot(\hat{\epsilon}\nabla\Phi)$, and only the last term here remains finite after integrating over the volume. Next, by using the Poisson equation $\nabla\cdot(\hat{\epsilon}_{\mathbf{r}}\nabla\Phi_{\mathbf{r}t}) = -4\pi\rho_{\mathbf{r}t}$, where $\rho_{\mathbf{r}t}$ is the charge density, we obtain the following expression for this term:

$$-\int_{(V)}\frac{d\mathbf{r}}{8\pi}\Phi_{\mathbf{r}t}\nabla\cdot(\hat{\epsilon}_{\mathbf{r}}\nabla\Phi_{\mathbf{r}t}) = \frac{1}{2}\int_{(V)}d\mathbf{r}\Phi_{\mathbf{r}t}\rho_{\mathbf{r}t}. \tag{6}$$

Now we see that the third term on the right-hand side of Eq. (4) describes the interaction of electric charges with the longitudinal part of the electric field. We denote it below as $\mathcal{E}_{int}$. In a homogeneous and isotropic medium with constant dielectric permittivity ϵ, one can easily solve the Poisson equation as $\Phi_{\mathbf{r}t} = \epsilon^{-1}\int_{(V)}d\mathbf{r}'\rho_{\mathbf{r}'t}/|\mathbf{r} - \mathbf{r}'|$ so that $\mathcal{E}_{int}$ is expressed through the charge densities only:

$$\mathcal{E}_{int} = \frac{1}{2}\int_{(V)}\int_{(V)}d\mathbf{r}d\mathbf{r}'\frac{\rho_{\mathbf{r}t}\rho_{\mathbf{r}'t}}{\epsilon|\mathbf{r} - \mathbf{r}'|}. \tag{7}$$

One should remember that both $\Phi_{\mathbf{r}t}$ and $\rho_{\mathbf{r}t}$ include the contributions of the external fields and charges. To extract these contributions from $\mathcal{E}_{int}$, it is convenient to separate the contributions coming from the internal (i) and external (e) charges under the integrals of Eq. (7). Then,

$$\mathcal{E}_{int} = \frac{1}{2}\int\int_{(V)}d\mathbf{r}d\mathbf{r}'\frac{\rho_{\mathbf{r}t}^{(i)}\rho_{\mathbf{r}'t}^{(i)}}{\epsilon|\mathbf{r} - \mathbf{r}'|} + \frac{1}{2}\int\int_{(V)}d\mathbf{r}d\mathbf{r}'\frac{\rho_{\mathbf{r}t}^{(e)}\rho_{\mathbf{r}'t}^{(e)}}{\epsilon|\mathbf{r} - \mathbf{r}'|}$$

$$+ \int\int_{(V)}d\mathbf{r}d\mathbf{r}'\frac{\rho_{\mathbf{r}t}^{(e)}\rho_{\mathbf{r}'t}^{(i)}}{\epsilon|\mathbf{r} - \mathbf{r}'|}. \tag{8}$$

The first term on the right-hand side of Eq. (8) is the energy of Coulomb interaction between the electrons of the system (the electrostatic energy). The second term is the energy of interaction between the external charges. It should be omitted in the following, because such a contribution is not relevant to the dynamics of the system under consideration. Finally, the last term is the energy of interaction of electrons with the longitudinal part of the external field. It can be rewritten as $e^{-1}\int_{(V)}d\mathbf{r}U_{\mathbf{r}t}\rho_{\mathbf{r}t}^{(i)}$, where $U_{\mathbf{r}t}$ is the potential energy of an electron in the external field. One may introduce the potential of the external field as $U_{\mathbf{r}t}/e$. Below we omit the index i in $\rho_{\mathbf{r}t}^{(i)}$.

To transform $\mathcal{E}_{int}$ into the operator of the interaction, $\widehat{H}_{int}$, one should use the charge density operator $\hat{\rho}_{\mathbf{r}} = e\sum_j\delta(\mathbf{r} - \mathbf{x}_j)$ instead of $\rho_{\mathbf{r}t}$. As a result, we obtain

$$\widehat{H}_{int} = \frac{1}{2}\sum_{jj'}'\frac{e^2}{\epsilon|\mathbf{x}_j - \mathbf{x}_{j'}|} + \sum_j U_{\mathbf{x}_jt}. \tag{9}$$

Because the interaction of an electron with itself should not be considered, the prime sign at the sum denotes the exclusion of the terms whose indices coincide (in other words, $j \neq j'$ is assumed).

Combining $\widehat{H}_{int}$ with the kinetic-energy part $\sum_j \hat{h}_j$, we find that the total Hamiltonian of the system of interacting electrons in the presence of external fields is given as

$$\widehat{H}_e = \sum_j \hat{h}_j + \widehat{H}_{ee},$$

$$\hat{h}_j = \hat{\mathrm{h}}_j + U_{\mathbf{x}_j t}, \quad \widehat{H}_{ee} = \frac{1}{2} \sum_{jj'}{}' \frac{e^2}{\epsilon|\mathbf{x}_j - \mathbf{x}_{j'}|}, \tag{10}$$

where $\hat{h}_j$ is the single-particle Hamiltonian comprising both kinetic and potential energy operators, and $\widehat{H}_{ee}$ is the Hamiltonian of Coulomb interaction between the electrons in the medium with dielectric permittivity ϵ. It is represented as a binary sum over all particles. We stress that the vector potential $\mathbf{A}_{\mathbf{x}_j t}$ standing in $\hat{\mathrm{h}}_j$ includes a contribution of the external fields. This contribution, in particular, describes the interaction of electrons with a stationary magnetic field and with electromagnetic waves (photon field). In the above consideration, we have omitted the contribution corresponding to the Pauli interaction of the electron spin with the magnetic field. We have also neglected relativistic corrections, which are small if the energy of the particle is small in comparison to the energy $2mc^2$.

The evolution of a many-electron system in external fields is described by the Schroedinger equation analogous to Eq. (1.1):

$$i\hbar \frac{\partial \Psi_{\{\mathbf{x}_j\}t}}{\partial t} = \widehat{H}_e \Psi_{\{\mathbf{x}_j\}t}. \tag{11}$$

It determines the wave function of $\mathbf{x}$-representation, which depends on the set of coordinates $\{\mathbf{x}_j\}$. The charge density at the point $(\mathbf{r}, t)$ is expressed through $\hat{\rho}_{\mathbf{r}}$ according to the general rule (1.3) for observable values:

$$\rho_{\mathbf{r}t} = \int d\{\mathbf{x}_j\} \Psi^*_{\{\mathbf{x}_j\}t} \hat{\rho}_{\mathbf{r}} \Psi_{\{\mathbf{x}_j\}t}, \tag{12}$$

where the charge density operator $\hat{\rho}_{\mathbf{r}}$ is introduced above as a sum of the δ-functions multiplied by the electron charge. Calculating the time derivative of Eq. (12) with the use of Eq. (11), we take into account that $\widehat{H}_e$ is Hermitian and find

$$\frac{\partial \rho_{\mathbf{r}t}}{\partial t} = \frac{i}{\hbar} \int d\{\mathbf{x}_j\} \Psi^*_{\{\mathbf{x}_j\}t} \left[\widehat{H}_e, \hat{\rho}_{\mathbf{r}} \right] \Psi_{\{\mathbf{x}_j\}t}. \tag{13}$$

One can see that both the potential energy $U_{\mathbf{x}_j t}$ and Coulomb interaction energy $\widehat{H}_{ee}$ commute with $\hat{\rho}_{\mathbf{r}}$ and do not contribute to the right-hand side of Eq. (13). The calculation of the commutators $[\hat{h}_j, \delta(\mathbf{r} - \mathbf{x}_j)]$ (problem 1.9) gives us the continuity relation

$$\frac{\partial \rho_{\mathbf{r}t}}{\partial t} + \nabla \cdot \mathbf{I}_{\mathbf{r}t} = 0, \quad \mathbf{I}_{\mathbf{r}t} = \int d\{\mathbf{x}_j\} \Psi^*_{\{\mathbf{x}_j\}t} \hat{\mathbf{I}}_{\mathbf{r}} \Psi_{\{\mathbf{x}_j\}t}. \tag{14}$$

The current density $\mathbf{I}_{\mathbf{r}t}$ at the point $(\mathbf{r}, t)$ is introduced by analogy to Eq. (12), and the current density operator $\hat{\mathbf{I}}_{\mathbf{r}}$ is given by

$$\hat{\mathbf{I}}_{\mathbf{r}} = \frac{e}{2m} \sum_j \left[(\hat{\mathbf{p}}_j - \frac{e}{c} \mathbf{A}_{\mathbf{x}_j t}) \delta(\mathbf{r} - \mathbf{x}_j) + \delta(\mathbf{r} - \mathbf{x}_j)(\hat{\mathbf{p}}_j - \frac{e}{c} \mathbf{A}_{\mathbf{x}_j t}) \right]. \tag{15}$$

This equation allows us to represent the Hamiltonian $\widehat{H}_e$ introduced by Eq. (10) in the form

$$\widehat{H}_e = \hat{\mathcal{H}}_e - \frac{1}{c} \int d\mathbf{r} \, \hat{\mathbf{I}}_{\mathbf{r}} \cdot \mathbf{A}_{\mathbf{r}t} \tag{16}$$

so that $\hat{\mathcal{H}}_e$ accounts for the interaction with longitudinal fields only, while the second term gives us the interaction with transverse fields entering through the vector potential $\mathbf{A}_{\mathbf{r}t}$. These fields are described by the photon Hamiltonian $\widehat{H}_{ph}$ defined in Sec. 3 after the quantization of the field.

Because the many-electron wave function $\Psi_{\{\mathbf{x}_j\}t}$ is antisymmetric with respect to permutation of electrons (Fermi-Dirac statistics), one can introduce the occupation number representation with the aid of the following antisymmetric function of N particles:

$$\Phi_{\{\gamma_k\}}(\{\mathbf{x}_j\}) = \det \|\phi^{(\gamma_k)}_{\mathbf{x}_j}\| / \sqrt{N!} , \tag{17}$$

where $\phi^{(\gamma)}_{\mathbf{x}}$ is a complete set of one-electron wave functions numbered by the quantum numbers γ, and $\det\| \ldots \|$ denotes the determinant of the matrix $\| \ldots \|$. The set $\{\gamma_k\}$ in Eq. (17) determines the state of N electrons with coordinates $\{\mathbf{x}_j\}$ so that one can write N^2 functions of the kind of Eq. (17). A complete system of linearly independent functions is obtained from Eq. (17) with the aid of a set of quantum numbers, $\{\gamma_k\}$, ordered according to $\gamma_1 < \gamma_2 < \ldots$. The expansion of $\Psi_{\{\mathbf{x}_j\}t}$ in terms of the functions (17),

$$\Psi_{\{\mathbf{x}_j\}t} = \sum_{n_{\{\gamma_k\}}} \Psi_{\{n_{\gamma_k}\}t} \Phi_{\{\gamma_k\}}(\{\mathbf{x}_j\}), \tag{18}$$

contains the numbers n_γ which can be either 1 or 0. This property directly follows from the antisymmetry of the wave functions (17) and is known as the Pauli principle. The numbers are chosen in such a way that $n_\gamma = 1$ if the state γ belongs to $\Phi_{\{\gamma_k\}}(\{\mathbf{x}_j\})$ and $n_\gamma = 0$ for other γ, provided $\sum_\gamma n_\gamma = N$. Therefore, each state of the N-electron system corresponds to a set of occupation numbers $\{n_\gamma\}$. It is not necessary to specify the functions $\phi_\mathbf{x}^{(\gamma)}$ standing in Eq. (17) in order to introduce the ket-vector $|\{n_{\gamma_k}\}t\rangle$ instead of $\Psi_{\{n_{\gamma_k}\}t}$. Such ket-vectors form a complete, orthogonal, and normalized set: $\langle\{n'_\gamma\}|\{n_\gamma\}\rangle = \delta_{\{n'_\gamma\},\{n_\gamma\}}$ (here and below the argument t is omitted). The creation operator for the state $\bar\gamma$ is introduced by the relation

$$\hat{a}_{\bar\gamma}^+|\{n_\gamma\}\rangle = \sqrt{1-n_{\bar\gamma}}(-1)^{p(\bar\gamma)}|\ldots,n_{\bar\gamma}+1,\ldots\rangle, \qquad (19)$$

where $p(\bar\gamma) = \sum_{\gamma_k<\bar\gamma} n_{\gamma_k}$ is the number of occupied states preceding to the state $\bar\gamma$. If the state $\bar\gamma$ is not present in the N-particle ket-vector $|\{n_\gamma\}\rangle$ (i.e., $n_{\bar\gamma} = 0$), the operator $\hat{a}_{\bar\gamma}^+$ transforms it to the $(N+1)$-particle ket-vector standing on the right-hand side of Eq. (19), i.e., the creation operator adds an electron to the state $\bar\gamma$ of the system. If the state $\bar\gamma$ is already present in the set $\{n_\gamma\}$, one has $\hat{a}_{\bar\gamma}^+|\{n_\gamma\}\rangle = 0$. The annihilation operator is introduced in a similar way:

$$\hat{a}_{\bar\gamma}|\{n_\gamma\}\rangle = \sqrt{n_{\bar\gamma}}(-1)^{p(\bar\gamma)}|\ldots,n_{\bar\gamma}-1,\ldots\rangle. \qquad (20)$$

This operator connects N- and $(N-1)$-particle ket-vectors. The definitions of the creation and annihilation operators result in the anticommutation rules

$$[\hat{a}_\gamma^+,\hat{a}_{\gamma'}]_+ = \delta_{\gamma\gamma'}, \qquad [\hat{a}_\gamma^+,\hat{a}_{\gamma'}^+]_+ = [\hat{a}_\gamma,\hat{a}_{\gamma'}]_+ = 0, \qquad (21)$$

which can be checked by the action of the pairs of operators from Eq. (21) on an arbitrary ket-vector $|\{n_\gamma\}\rangle$. The anticommutator in Eq. (21) is introduced as $[\widehat{A},\widehat{B}]_+ = \widehat{A}\widehat{B} + \widehat{B}\widehat{A}$, where $\widehat{A}$ and $\widehat{B}$ are arbitrary operators.

The operator of the occupation number of the state γ is introduced as $\hat{n}_\gamma = \hat{a}_\gamma^+\hat{a}_\gamma$. One can easily check that

$$\hat{n}_\gamma|\ldots n_\gamma\ldots\rangle = n_\gamma|\ldots n_\gamma\ldots\rangle. \qquad (22)$$

In the occupation number representation, each operator of the form $\widehat{H} = \sum_j \hat{h}_j$ (the examples of such additive operators are the Hamiltonian of non-interacting electrons, the charge density operator, and the current density operator) is written as

$$\widehat{H} = \sum_{\gamma_1\gamma_2}\langle\gamma_1|\hat{h}|\gamma_2\rangle\hat{a}_{\gamma_1}^+\hat{a}_{\gamma_2}, \qquad \langle\gamma_1|\hat{h}|\gamma_2\rangle = \int d\mathbf{x}\,\phi_\mathbf{x}^{(\gamma_1)*}\hat{h}\phi_\mathbf{x}^{(\gamma_2)}. \qquad (23)$$

For the binary operators of the form $\widehat{H}_2 = (1/2) \sum_{jj'} \hat{v}_{jj'}$, where $\hat{v}_{jj'} = v(\mathbf{x}_j, \mathbf{x}_{j'})$, we have

$$\widehat{H}_2 = \frac{1}{2} \sum_{\gamma_1 \gamma_2 \gamma_3 \gamma_4} \langle \gamma_1 \gamma_2 | \hat{v} | \gamma_4 \gamma_3 \rangle \hat{a}_{\gamma_1}^+ \hat{a}_{\gamma_2}^+ \hat{a}_{\gamma_3} \hat{a}_{\gamma_4},$$

$$\langle \gamma_1 \gamma_2 | \hat{v} | \gamma_4 \gamma_3 \rangle = \int d\mathbf{x} \int d\mathbf{x}' \phi_{\mathbf{x}}^{(\gamma_1)*} \phi_{\mathbf{x}'}^{(\gamma_2)*} v(\mathbf{x}, \mathbf{x}') \phi_{\mathbf{x}}^{(\gamma_4)} \phi_{\mathbf{x}'}^{(\gamma_3)}. \qquad (24)$$

The operator $\widehat{H}_{ee}$ given by Eq. (10) is an example of such operators. Equations (23) and (24) are justified in view of the equivalence of their matrix elements calculated with arbitrary ket-vectors $|\{n_\gamma\}\rangle$ to the usual expressions for the matrix elements of $\widehat{H}$ and $\widehat{H}_2$ calculated in the anti-symmetric basis (17) (problem 1.10).

Instead of $\hat{a}_{\bar{\gamma}}^+$ and $\hat{a}_{\bar{\gamma}}$, one can use the field operators $\hat{\Psi}_{\mathbf{x}}^+$ and $\hat{\Psi}_{\mathbf{x}}$ defined by the relations

$$\hat{\Psi}_{\mathbf{x}}^+ = \sum_\gamma \phi_{\mathbf{x}}^{(\gamma)*} \hat{a}_\gamma^+, \qquad \hat{\Psi}_{\mathbf{x}} = \sum_\gamma \phi_{\mathbf{x}}^{(\gamma)} \hat{a}_\gamma. \qquad (25)$$

The operator $\hat{\Psi}_{\mathbf{x}}^+$ should be considered as the creation operator for the particle with coordinate $\mathbf{x}$, while $\hat{\Psi}_{\mathbf{x}}$ is the corresponding annihilation operator. The anticommutation rules for these operators are obtained from Eq. (21) according to the condition of completeness for the sets of one-electron wave functions $\phi_{\mathbf{x}}^{(\gamma)}$:

$$[\hat{\Psi}_{\mathbf{x}}^+, \hat{\Psi}_{\mathbf{x}'}^+]_+ = [\hat{\Psi}_{\mathbf{x}}, \hat{\Psi}_{\mathbf{x}'}]_+ = 0,$$

$$[\hat{\Psi}_{\mathbf{x}}^+, \hat{\Psi}_{\mathbf{x}'}]_+ = \sum_\gamma \phi_{\mathbf{x}}^{(\gamma)*} \phi_{\mathbf{x}'}^{(\gamma)} = \delta(\mathbf{x} - \mathbf{x}'). \qquad (26)$$

The additive and binary operators introduced by Eqs. (23) and (24) are written through the field operators in the following way:

$$\widehat{H} = \sum_{\gamma_1 \gamma_2} \int d\mathbf{x}\, \phi_{\mathbf{x}}^{(\gamma_1)*} \hat{h} \phi_{\mathbf{x}}^{(\gamma_2)}\, \hat{a}_{\gamma_1}^+ \hat{a}_{\gamma_2} = \int d\mathbf{x} \hat{\Psi}_{\mathbf{x}}^+ \hat{h} \hat{\Psi}_{\mathbf{x}},$$

$$\widehat{H}_2 = \frac{1}{2} \int d\mathbf{x} \int d\mathbf{x}' \hat{\Psi}_{\mathbf{x}}^+ \hat{\Psi}_{\mathbf{x}'}^+ v(\mathbf{x}, \mathbf{x}') \hat{\Psi}_{\mathbf{x}'} \hat{\Psi}_{\mathbf{x}}. \qquad (27)$$

These expressions are analogous to the matrix elements of Eqs. (23) and (24), where the one-electron functions are replaced by the field operators, as though the ψ-function is quantized again. This explains the origin of the term "second quantization." If the many-electron Hamiltonian $\widehat{H}_e$ depends on spin variable or/and band indices (see the next section and

Appendix B), one should consider **x** in Eqs. (25)−(27) as a combination of the coordinate and discrete indices so that the integral $\int d\mathbf{x}$ must include the sums over these indices.

To calculate the observable $\overline{Q}_t$, we substitute $\widehat{Q}$ in the form $\widehat{Q} = \sum_{\gamma\gamma'}\langle\gamma|\hat{q}|\gamma'\rangle\hat{a}_\gamma^+\hat{a}_{\gamma'}$ into Eq. (1.18):

$$\overline{Q}_t = \mathrm{Sp}\widehat{Q}\hat{\eta}_t = \sum_{\delta\nu}\langle\delta|\hat{q}|\nu\rangle\mathrm{Sp}\hat{a}_\delta^+\hat{a}_\nu\hat{\eta}_t \equiv \mathrm{sp}\hat{q}\hat{n}_t. \tag{28}$$

Here "Sp" and "sp" denote the traces over many-particle and single-particle states, respectively. We have introduced the average of the additive operator $\hat{q}$ through the one-particle density matrix $\hat{n}_t$ defined as

$$\langle\nu|\hat{n}_t|\delta\rangle = \mathrm{Sp}\hat{a}_\delta^+\hat{a}_\nu\hat{\eta}_t. \tag{29}$$

The dynamical equation for $\hat{n}_t$ follows from the general many-particle equation (1.20):

$$i\hbar\frac{\partial}{\partial t}\langle\nu|\hat{n}_t|\delta\rangle = \mathrm{Sp}\hat{a}_\delta^+\hat{a}_\nu[\widehat{H}, \hat{\eta}_t] = \mathrm{Sp}\hat{\eta}_t[\hat{a}_\delta^+\hat{a}_\nu, \widehat{H}]. \tag{30}$$

In the case of a system of non-interacting electrons described by the Hamiltonian $\widehat{H} = \sum_j \hat{h}_j$, one can write a closed equation for the single-particle operator $\hat{n}_t$. We use the commutator

$$[\hat{a}_\delta^+\hat{a}_\nu, \hat{a}_\gamma^+\hat{a}_\eta] = \hat{a}_\delta^+\hat{a}_\eta\delta_{\nu\gamma} - \hat{a}_\gamma^+\hat{a}_\nu\delta_{\delta\eta} \tag{31}$$

calculated according to the anticommutation relations (21) (problem 1.11). As a result, the dynamical equation for $\hat{n}_t$ has the same form as Eq. (1.20):

$$i\hbar\frac{\partial\hat{n}_t}{\partial t} = [\hat{h}, \hat{n}_t]. \tag{32}$$

Equation (32), however, contains one-electron variables only. According to Eq. (28), the observable values are found by averaging the operator $\hat{q}$ with the density matrix $\hat{n}_t$.

Let us find the averaged occupation number $\bar{n}_\gamma$ of the electron state γ for the equilibrium distribution of electrons with temperature T_e. This distribution is described by the density matrix

$$\hat{\eta}_{eq} = Z^{-1}\exp[-(\widehat{H}_e - \mu\hat{n})/T_e], \quad Z = \mathrm{Sp}\exp[-(\widehat{H}_e - \mu\hat{n})/T_e]. \tag{33}$$

This definition, in contrast to Eq. (3.25), accounts for the conservation of the total number N of electrons described by the particle number operator $\hat{n} = \sum_\gamma \hat{n}_\gamma$. The coefficient μ, called the chemical potential, is determined from the condition $\sum_\gamma \bar{n}_\gamma = N$. Under the assumption of ideal Fermi gas, i.e., neglecting the contribution of the electron-electron interaction to the equilibrium properties of the system, we use the Hamiltonian $\widehat{H}_e = \sum_\gamma \varepsilon_\gamma \hat{a}_\gamma^+ \hat{a}_\gamma$. Then, the total energy is $E_{\{n_\gamma\}} = \sum_\gamma \varepsilon_\gamma n_\gamma$, where n_γ is equal to 1 (0) for the occupied (empty) states. As a result, the partition function takes the following form:

$$Z = \sum_{\{n_\gamma\}} \prod_\gamma \exp[(\mu - \varepsilon_\gamma)n_\gamma/T_e]$$

$$= \prod_\gamma \sum_{n=0,1} \exp[(\mu - \varepsilon_\gamma)n/T_e] = \prod_\gamma \left[1 + e^{(\mu - \varepsilon_\gamma)/T_e} \right]. \qquad (34)$$

The mean value of the occupation number of the state γ is introduced as the average of the operator $\hat{n}_\gamma$ with the equilibrium density matrix (33), $\bar{n}_\gamma = \mathrm{Sp}\hat{n}_\gamma \hat{\eta}_{eq}$:

$$\bar{n}_\gamma = Z^{-1} \sum_{\{n_{\gamma_1}\}} n_\gamma \exp\left[\frac{1}{T_e}\left(\mu \sum_{\gamma_1} n_{\gamma_1} - E_{\{n_{\gamma_1}\}} \right) \right] = -T_e \frac{\partial \ln Z}{\partial \varepsilon_\gamma}. \qquad (35)$$

Substituting Z from Eq. (34) into Eq. (35), we find the equilibrium Fermi distribution

$$\bar{n}_\gamma = \left[e^{(\varepsilon_\gamma - \mu)/T_e} + 1 \right]^{-1}. \qquad (36)$$

If the temperature goes to zero, the electrons occupy only the states whose energies are below the chemical potential μ. This is the case of a degenerate electron gas. Another limiting case, the non-degenerate electron gas, takes place when the average occupation numbers are small, $\bar{n}_\gamma \ll 1$, and Eq. (36) is reduced to the Boltzmann distribution $\bar{n}_\gamma = e^{(\mu - \varepsilon_\gamma)/T_e}$.

5. Electrons under External Fields

After the general description of the system of interacting electrons given in Sec. 4, we are going to discuss the solutions of one-electron Schroedinger equations: $\hat{h}\Psi = E\Psi$ for stationary and $i\hbar\partial\Psi/\partial t = \hat{h}\Psi$ for time-dependent problems. The Hamiltonian $\hat{h}$ is given by Eqs. (4.10) and (4.2) with particle index j omitted. We will consider the wave functions and energy spectra of electrons for different kinds of external fields entering this Hamiltonian. These particular problems form a part of the quantum mechanics and are discussed in detail in the literature. For this reason, in this section we cover only the problems whose solutions will be used in the next chapters.

Free motion. Let us consider the electron states in the absence of any external fields. Formally, it is convenient to assume that the electrons are confined into a cubic volume L^3 with appropriate (zero or periodic) boundary conditions, and the length L is greater than any characteristic length of the problem. The dispersion law ε_p is obtained as a usual kinetic energy with quasi-discrete (for $L^3 \to \infty$) i-th component of the momentum, p_i, and the Ψ-function is the plane wave:

$$\psi_{\mathbf{r}}^{(\mathbf{p})} = L^{-3/2} \exp\left(\frac{i}{\hbar}\mathbf{p}\cdot\mathbf{r}\right),$$

$$p_i = \pm n_i \frac{2\pi\hbar}{L}, \quad n_i = 1, 2, \dots , \quad \varepsilon_p = \frac{p^2}{2m}. \tag{1}$$

The density of states (the number of electron states with energy E per unit volume) is defined as

$$\rho(E) = \frac{2}{L^3} \sum_\delta \delta(E - \varepsilon_\delta), \tag{2}$$

where the factor 2 takes into account double degeneracy of electron states with respect to spin (it is assumed that the quantum state indices δ do not include spin quantum numbers). Replacing the sum by the integral over the momentum (problem 1.12), one can obtain $\rho(E)$ in the form

$$\rho_{3D}(E) = \frac{m\sqrt{2mE}}{\pi^2\hbar^3}. \tag{3}$$

Therefore, the density of states in the bulk (three-dimensional) media is proportional to $\sqrt{E}$.

Electrons in crystals. To describe the electron states in crystals, one has to introduce a periodic potential energy $U_{cr}(\mathbf{r})$ in the Hamiltonian. Besides, it is necessary to take into account the relativistic corrections describing the spin-orbit interaction (the other relativistic terms are small in comparison to this one). The Hamiltonian has the following form:

$$\hat{h}_{cr} = \frac{\hat{\mathbf{p}}^2}{2m_e} + U_{cr}(\mathbf{r}) + \frac{\hbar}{(2m_ec)^2}\hat{\boldsymbol{\sigma}}\cdot[\nabla U_{cr}(\mathbf{r}) \times \hat{\mathbf{p}}], \tag{4}$$

where $\hat{\mathbf{p}} = -i\hbar\nabla$ is the momentum operator, m_e is the free electron mass, and $\hat{\boldsymbol{\sigma}}$ is the vector of Pauli matrices. Owing to periodicity of the potential in Eq. (4), $U_{cr}(\mathbf{r} + \mathbf{R}_i) = U_{cr}(\mathbf{r})$, where $\mathbf{R}_i$ is an arbitrary lattice vector, the solutions of the eigenstate problem are Bloch functions

$$\psi_{n\sigma\mathbf{p}}(\mathbf{r}) = L^{-3/2}e^{i\mathbf{p}\cdot\mathbf{r}/\hbar}u_{n\sigma\mathbf{p}}(\mathbf{r}), \quad u_{n\sigma\mathbf{p}}(\mathbf{r} + \mathbf{R}_i) = u_{n\sigma\mathbf{p}}(\mathbf{r}),$$

$$E_{n\sigma}(\mathbf{p} + \mathbf{G}) = E_{n\sigma}(\mathbf{p}), \tag{5}$$

where $\mathbf{p}$ is defined inside the first Brillouin zone and referred to as quasi-momentum, $\mathbf{G}$ is the reciprocal lattice vector, and the electron spectrum $E_{n\sigma}(\mathbf{p})$ depends on the band index n and spin quantum number σ.

The spectrum $E_{n\sigma}(\mathbf{p})$ and the Bloch amplitudes $u_{n\sigma\mathbf{p}}(\mathbf{r})$ are determined by the geometry of the lattice and by the type of interatomic bond of the crystal. Usually, one can separate the following types: the metallic bond, the hetero- and homeopolar bonds (also known as ionic and covalent bonds, respectively), and the molecular bond, when the crystal is formed due to van der Waals attraction between the molecules or atoms. Simple metals can be considered as ensembles of positive ions oscillating near their equilibrium positions in the crystal lattice and surrounded by a gas of nearly free electrons. The total energy of the bond is determined by a negative contribution of the electron-ion interaction ($\propto a^{-1}$, where a is the lattice period) and positive kinetic energy of strongly degenerate electrons ($\propto n^{2/3} \propto a^{-2}$, where n is the electron density). The value of a is determined by the condition of minimum energy. In the limiting cases considered above, the crystal is formed either due to long-range interaction between the molecules, or can be considered as a single macroscopic molecule with periodically placed ions (the metallic case). In the case of an ionic bond in biatomic crystals, the lattice is formed by periodically placed positive and negative ions appearing as a result of electron transfer between the neighboring atoms. Because each ion in such a lattice is surrounded by the ions of the opposite sign, the Coulomb interaction leads to the attraction that provides stability of the lattice. The covalent attractive bond appears between neutral atoms because of the formation of the pairs of collectivized electrons with antiparallel spins. This mechanism is completely analogous to the valence bond in a single molecule. We point out that a large group of insulators and semiconductors is characterized by the mixed ionic-covalent bond. Although the eigenstate problem for electrons in crystals is extremely complicated, a description of these particular materials can be considerably simplified because their energy bands are either almost empty or almost filled by electrons so that the electronic properties are determined by the states near the band extrema. The dynamics of these states is discussed below.

To describe the electron states near the band extrema (let us suppose that the latter are at $\mathbf{p} = 0$), we write the Hamiltonian (4) in the basis $\psi_{l\mathbf{p}}(\mathbf{r}) \equiv L^{-3/2} \exp(i\mathbf{p} \cdot \mathbf{r}/\hbar) u_{n\sigma}(\mathbf{r})$, where $l = (n, \sigma)$ and the Bloch amplitude in the center of the Brillouin zone, $u_{n\sigma}(\mathbf{r}) \equiv u_{n\sigma,\mathbf{p}=0}(\mathbf{r})$, is determined by the equation $\hat{h}_{cr} u_{n\sigma}(\mathbf{r}) = \varepsilon_n u_{n\sigma}(\mathbf{r})$. Here ε_n is the energy of n-th band extremum at $\mathbf{p} = 0$, and each band is doubly degenerate with respect to the spin σ. The matrix elements of the Hamiltonian (4) are

$$\langle l\mathbf{p}|\hat{h}_{cr}|l'\mathbf{p}'\rangle = \delta_{\mathbf{pp}'}\delta_{ll'}\left(\varepsilon_n + \frac{p^2}{2m_e}\right) + \delta_{\mathbf{pp}'}\mathbf{p}\cdot\mathbf{v}_{ll'} \equiv \delta_{\mathbf{pp}'}H_{ll'}(\mathbf{p}), \quad (6)$$

where the matrix elements of the velocity operator are non-diagonal in the band index:

$$\mathbf{v}_{ll'} = \left\langle l0\left|\frac{\hat{\mathbf{p}}}{m_e} + \frac{\hbar}{(2m_ec)^2}[\hat{\boldsymbol{\sigma}}\times\nabla U_{cr}(\mathbf{r})]\right|l'0\right\rangle. \quad (7)$$

The diagonal matrix elements of the velocity operator are equal to zero in the point of extrema. Expanding the wave functions $\Psi(\mathbf{r})$ over the complete set $\psi_{l\mathbf{p}}(\mathbf{r})$, we write

$$\Psi(\mathbf{r}) = \sum_{l\mathbf{p}}\varphi_{l\mathbf{p}}\psi_{l\mathbf{p}}(\mathbf{r}), \quad (8)$$

and the Schroedinger equation is transformed into a set of equations for the envelope functions $\varphi_{l\mathbf{p}}$:

$$\sum_{l'}H_{ll'}(\mathbf{p})\varphi_{l'\mathbf{p}} = E\varphi_{l\mathbf{p}}. \quad (9)$$

Equation (9) defines the many-band spectrum near the extremum. The expressions for the matrix elements of the Hamiltonian and velocity operator for the two-band model and the set of equations for envelope functions $\varphi_{l\mathbf{p}}$ in the presence of externally applied electric and magnetic fields are written in Appendix B.

To obtain the energy spectrum of electrons in the vicinity of the n-th band extremum, one can calculate the diagonal contribution to $E_{l\mathbf{p}}$ in the second order of the perturbation theory:

$$E_{l\mathbf{p}} = \varepsilon_n + \frac{p^2}{2m_e} + \sum_{l'(l'\neq l)}\frac{(\mathbf{p}\cdot\mathbf{v}_{ll'})(\mathbf{p}\cdot\mathbf{v}_{l'l})}{\varepsilon_n - \varepsilon_{n'}} \equiv \varepsilon_n + \frac{1}{2}\sum_{\alpha\beta}p_\alpha m_{\alpha\beta}^{-1}p_\beta. \quad (10)$$

The spectrum appears to be quadratic in $\mathbf{p}$. The right equation in Eq. (10) defines the inverse effective mass tensor of the band $l = (n, \sigma)$. This tensor is expressed through the velocity matrix elements (7) as

$$m_{\alpha\beta}^{-1} = \frac{\delta_{\alpha\beta}}{m_e} + \sum_{l'(l'\neq l)}\frac{v_{ll'}^\alpha v_{l'l}^\beta + v_{ll'}^\beta v_{l'l}^\alpha}{\varepsilon_n - \varepsilon_{n'}}. \quad (11)$$

In the general case, the surfaces of equal energy for $E_{l\mathbf{p}}$ of Eq. (10) are the ellipsoids characterized by three principal values of the effective masses along the main axes. In the uniaxial materials, there are two (longitudinal and transverse) effective masses, while in the cubic materials the tensor (11) becomes a scalar $\delta_{\alpha\beta}/m$. Depending on the sign of m, the dispersion relation (10) describes electron (at $m > 0$) or hole (at $m < 0$) states. This consideration is valid only in the vicinity of the extremum, where the kinetic energy $p^2/2m$ is small in comparison to the interband energies.

Therefore, the electrons moving in the periodic potential of a crystal can be considered as free electrons with scalar effective mass m (positive or negative near the band extrema) under certain conditions described above. The result (3) for the density of states of free electrons can be directly applied to this case by assuming that m denotes the effective mass. More complicated cases are realized when: *i)* the effective mass is a tensor; *ii)* several bands are close in energy and corresponding several branches of the spectrum with different effective masses have to be considered; *iii)* the kinetic energy is not small in comparison to the interband energies and the non-parabolicity effects (in particular, the deviation of the energy dispersion law from the quadratic form $p^2/2m$) have to be taken into account. Nevertheless, below we will concentrate on the simplest case of scalar effective mass.

Landau quantization. The electron states in the presence of a magnetic field **H** are described by the Schroedinger equation

$$\frac{(\hat{\mathbf{p}} - e\mathbf{A}/c)^2}{2m}\psi_{\mathbf{r}} = E\psi_{\mathbf{r}}, \quad \mathbf{A} = (0, Hx, 0), \tag{12}$$

where the vector potential **A** is written for $\mathbf{H}\|OZ$ and the magnetic field is supposed to be homogeneous and time-independent. Owing to the translational invariance of the problem (12) along OY and OZ, the motion along these directions is described by the plane waves with wave numbers $p_y/\hbar$ and $p_z/\hbar$. The Ψ-function for δ-state is written as a product

$$\psi_{\mathbf{r}}^{(\delta)} = (L_y L_z)^{-1/2} \exp\left[\frac{i}{\hbar}(p_y y + p_z z)\right] \varphi_x^{(Np_y)}, \tag{13}$$

and the quantum numbers δ include both the momenta (p_y and p_z) and the number of discrete level, N. The quantization of electron states is produced by the parabolic potential energy which appears from $\propto \mathbf{A}^2$ contribution in Eq. (12). The equation for $\varphi_x^{(Np_y)}$ standing in Eq. (13) is

$$\left[\frac{\hat{p}_x^2}{2m} + \frac{m\omega_c^2}{2}\left(x - X_{p_y}\right)^2\right] \varphi_x^{(Np_y)} = \varepsilon_N \varphi_x^{(Np_y)}. \tag{14}$$

The coordinate $X_{p_y} = -p_y c/|e|H$ determines the position of the center of harmonic-oscillator wave function. The electron energy E does not depend on p_y (i.e., the electron states under consideration are degenerate with respect to p_y). The momentum p_z along the direction of $\mathbf{H}$ determines the longitudinal kinetic energy. The dispersion law E_{Np_z} is written as a sum $\varepsilon_N + p_z^2/2m$, where ε_N is the energy of the N-th level of the harmonic oscillator (Appendix A) with frequency $\omega_c = |e|H/mc$ called the cyclotron frequency. The level numbers are the integers: $N = 0, 1, \ldots$. The eigenvalues of Eq. (14) and the corresponding wave functions are

$$\varepsilon_N = \hbar\omega_c(N + 1/2) ,$$

$$\varphi_x^{(Np_y)} = \frac{1}{\pi^{1/4} l_H^{1/2} \sqrt{2^N N!}} \exp\left[-\frac{1}{2}\left(\frac{x - X_{p_y}}{l_H} \right)^2 \right] H_N\left(\frac{x - X_{p_y}}{l_H} \right), \quad (15)$$

where $H_N(x)$ is the Hermite polynomial and $l_H = \sqrt{\hbar c/|e|H}$ is the magnetic length corresponding to the radius of cyclotron orbit in the classical dynamics. The electron levels with the energies ε_N given by Eq. (15) are called the Landau levels.

According to Eq. (2), the density of states is obtained by integrating $\delta(E - E_{Np_z})$ over p_y and p_z. Because of the degeneracy of the states with respect to p_y, we calculate $\sum_{p_y} \ldots$ under the condition $|X_{p_y}| < L_x/2$ corresponding to the requirement that the centers of the oscillator wave functions are inside the normalization volume. The result contains the sum over N:

$$\rho(E) = \frac{2}{L_x} \sum_{N=0}^{\infty} \int_{|X_{p_y}| < L_x/2} \frac{dp_y}{2\pi\hbar} \int \frac{dp_z}{2\pi\hbar} \delta(E - E_{Np_z})$$

$$= \frac{2|e|H}{c(2\pi\hbar)^2} \sum_{N=0}^{\infty} \sqrt{\frac{2m}{E - \varepsilon_N}} \theta(E - \varepsilon_N). \quad (16)$$

The density of states has inverse-square-root divergences at the bottoms of the Landau-level subbands, where E is close to ε_N.

Confinement. This is the simplest quantum-mechanical phenomenon, the quantization of low-energy electron states in a potential well. Consider first a one-dimensional confinement potential energy

$$U(z) = \begin{cases} 0, & |z| < d/2 \\ U_0, & |z| > d/2 \end{cases} \quad (17)$$

corresponding to the square well (U_0 is the depth of the potential well and d is the width). For the case $E < U_0$, the underbarrier part of the wave function is written as

$$\psi_z \propto \begin{cases} e^{-\kappa(z-d/2)}, & z > d/2 \\ e^{\kappa(d/2+z)}, & z < -d/2 \end{cases}, \tag{18}$$

where the underbarrier penetration length κ^{-1} is determined by $\hbar\kappa = \sqrt{2m(U_0 - E)}$. Eliminating the underbarrier part of the ψ-function with the aid of the boundary conditions $\hat{p}_z\psi_z|_{\pm d/2+0}^{\pm d/2-0} = 0$, we obtain the boundary condition of the third kind:

$$\left[\hat{p}_z \mp i\sqrt{2m(U_0 - E)}\right]\psi_z|_{z=\pm d/2} = 0, \tag{19}$$

and the eigenstate problem should be solved in the well region $|z| < d/2$ only. Owing to the symmetry of the system, the solution takes the form

$$\psi_z = \sqrt{\frac{2}{d}} \begin{cases} \cos(p_n z/\hbar), & n = 1, 3 \ldots \\ \sin(p_n z/\hbar), & n = 2, 4 \ldots \end{cases}, \quad \varepsilon_n = \frac{p_n^2}{2m}, \tag{20}$$

where p_n are determined by the dispersion relation following from the boundary conditions (19). For the case $U_0 \gg \varepsilon_n$, one can use zero boundary conditions for the ψ-function at $z = \pm d/2$ and obtain $p_n = n\pi\hbar/d$ (the difference between this p_n and p_i in Eq. (1) should not be confusing because p_n in Eq. (20) is positive, and one obtains the same result for the density of states if d approaches ∞). Thus, the dependence $\varepsilon_n \propto n^2$ is realized for the lowest energy levels. The approximation when one uses zero boundary conditions is called the hard-wall model of the quantum well.

Another important situation takes place for narrow quantum wells, when $(\pi\hbar/d)^2/2m \geq U_0$. In such a case, there is a single confined state with energy ε_0. The corresponding level is shallow, $U_0 - \varepsilon_0 \ll U_0$, and ψ_z is a weakly varying function inside the well, $\psi_{z=\pm d/2} \simeq \psi_{z=0}$. Because of this property, the Schroedinger equation is transformed to

$$\frac{\hat{p}_z^2}{2m}\psi_z = U_0\psi_{z=0}, \quad |z| < \frac{d}{2}. \tag{21}$$

By integrating this equation over the well region, we obtain the boundary condition

$$\frac{d\psi_z}{dz}\bigg|_{-d/2}^{d/2} = \frac{2md}{\hbar^2}U_0\psi_{z=0}. \tag{22}$$

Outside the well region, for $|z| > d/2$, we use the wave function (18) with $\hbar\kappa = \sqrt{2m(U_0 - \varepsilon_0)}$, and the energy of the level becomes

$$\varepsilon_0 = U_0 - \frac{mU_0^2}{2(\hbar/d)^2}. \tag{23}$$

This expression verifies the shallow level condition. The wave function is formed mostly by the tails outside the well; see Eq. (18).

Low-dimensional states. The dynamics of electrons is modified essentially due to the above-described confinement effect. Apart from the widely known case of attractive three-dimensional potentials describing the states in atoms or the localized states on the impurity centers in solids, there exist low-dimensional systems realized in solid-state nanostructures. In the two-dimensional (2D) systems realized in quantum wells, the electrons can move in a plane (say XOY), while the potential energy $U(\mathbf{r})$ provides their confinement along the direction perpendicular to this plane. In the one-dimensional (1D) systems realized in quantum wires, the electrons can move only in one dimension (say along OX) and are confined in two remaining dimensions. Accordingly, the Hamiltonian can be written as

$$\frac{\hat{\mathbf{p}}^2}{2m} + U(\mathbf{r}), \quad U(\mathbf{r}) = \left\{ \begin{array}{ll} U(z), & (2D) \\ U(y,z), & (1D) \end{array} \right. , \tag{24}$$

where m is the effective mass. The wave functions are written as products of the plane waves, which describe free motion of electrons with 2D momenta $\mathbf{p}$ or 1D momenta p, by the localized wave functions describing the confinement. The corresponding dispersion laws are written as sums of the kinetic energy $p^2/2m$ and the energies of the levels which depend on the discrete quantum numbers:

$$\psi_{\mathbf{r}}^{(n\mathbf{p})} = \psi_{x,y}^{(\mathbf{p})}\psi_z^{(n)}, \quad \varepsilon_{n\mathbf{p}} = \varepsilon_n + \frac{p^2}{2m},$$

$$\psi_{\mathbf{r}}^{(n_1 n_2 p)} = \psi_x^{(p)}\psi_{y,z}^{(n_1 n_2)}, \quad \varepsilon_{n_1 n_2 p} = \varepsilon_{n_1 n_2} + \frac{p^2}{2m}. \tag{25}$$

The corresponding densities of states per unit square (L^2 for the 2D case) or per unit length (L for the 1D case) are obtained in line with the general definition (2), after integrating over 2D or 1D momentum:

$$\rho_{2D}(E) = \frac{2}{L^2}\sum_{n\mathbf{p}}\delta(E - \varepsilon_{n\mathbf{p}}) = \rho_{2D}\sum_n \theta(E - \varepsilon_n),$$

$$\rho_{1D}(E) = \frac{2}{L}\sum_{n_1 n_2 p}\delta(E - \varepsilon_{n_1 n_2 p}) \tag{26}$$

$$= \frac{1}{\pi\hbar}\sum_{n_1 n_2}\theta(E - \varepsilon_{n_1 n_2})\sqrt{\frac{2m}{|E - \varepsilon_{n_1 n_2}|}},$$

where $\rho_{2D} = m/\pi\hbar^2$. The energy dependence of the density of states in the 2D case is step-like, while in the 1D case it has the inverse-square-root divergences, as in Eq. (16). Considering only the single-subband contributions for the 2D and 1D cases, and counting the energy E from the bottom of the corresponding subband, one may formally represent the ratios $\rho_{\mathcal{D}}(E)/\rho_{2D}$, where $\mathcal{D} = 1D, 2D, 3D$, as functions of a single parameter, $Em/\hbar^2$; see Fig. 1.2.

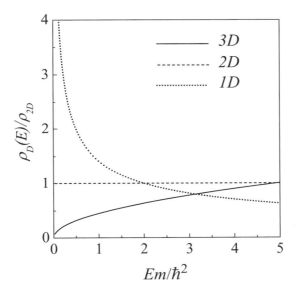

Figure 1.2. Density of states for 3D, 2D, and 1D electrons. Only the single-subband contributions are shown for the 2D and 1D cases, and the energy is counted from the bottom of the corresponding subband.

So far we have considered stationary problems. Below we analyze the Schroedinger equation $i\hbar\partial\Psi/\partial t = \widehat{H}_t\Psi$ with time-dependent Hamiltonian $\widehat{H}_t$.

Homogeneous electric field. To describe the temporal evolution of electron states in the electric field $\mathbf{E}_t$, one can express the field in the Hamiltonian $\widehat{H}_t$ either through the vector potential $\mathbf{A}_t = -c\int^t dt'\mathbf{E}_{t'}$ or through the scalar potential $-(\mathbf{E}_t \cdot \mathbf{x})$. For each of these variants, the Schroedinger equation in the momentum representation is written as

$$i\hbar\frac{\partial\psi_{\mathbf{p}t}}{\partial t} = \varepsilon(\boldsymbol{\pi}_t)\psi_{\mathbf{p}t}, \qquad \boldsymbol{\pi}_t = \mathbf{p} + e\int^t d\tau\mathbf{E}_\tau,$$

$$i\hbar\left(\frac{\partial}{\partial t} + e\mathbf{E}_t \cdot \frac{\partial}{\partial\boldsymbol{\pi}}\right)\psi_{\boldsymbol{\pi}t} = \varepsilon(\boldsymbol{\pi})\psi_{\boldsymbol{\pi}t}, \tag{27}$$

where $\mathbf{p}$ and $\boldsymbol{\pi}$ denote the canonical and kinematic momenta, respectively, and $\varepsilon(\mathbf{p}) = p^2/2m$. The initial conditions to Eq. (27) at t_0 are given by the Kronecker symbols $\delta_{\mathbf{pp}_0}$ and $\delta_{\boldsymbol{\pi}\boldsymbol{\pi}_0}$, where $\mathbf{p}_0$ and $\boldsymbol{\pi}_0$ are the quantum numbers of the canonical and kinematic momenta. The solution of the first equation of Eq. (27) is

$$\psi_{\mathbf{p}t}^{(\mathbf{p}_0)} = \exp\left[-\frac{i}{\hbar}\int_{t_0}^{t}dt'\varepsilon(\boldsymbol{\pi}_{t'})\right]\delta_{\mathbf{pp}_0}, \tag{28}$$

where $\boldsymbol{\pi}_t$ satisfies the classical equation of motion. The second equation of Eq. (27) is a differential equation of the first order with partial derivatives, and its solution is

$$\psi_{\boldsymbol{\pi}t}^{(\boldsymbol{\pi}_0)} = \exp\left[-\frac{i}{\hbar}\int_{t_0}^{t}dt'\frac{(\boldsymbol{\pi} + e\int_{t}^{t'}d\tau\mathbf{E}_\tau)^2}{2m}\right]\delta_{\boldsymbol{\pi}\boldsymbol{\pi}_0}. \tag{29}$$

The wave function given by Eq. (29) coincides with the one of Eq. (28) after a formal replacement of the kinematic momentum by the canonical one. Therefore, the use of either vector or scalar potentials corresponds to the formulation of the problem in terms of either $\mathbf{p}$ or $\boldsymbol{\pi}$, respectively. Under a time-independent electric field $\mathbf{E}$, the momentum is linear in time, $\mathbf{p} + e\mathbf{E}t$, while in the harmonic field $\mathbf{E}\cos\omega t$ it is convenient to expand $\psi_{\mathbf{p}t}^{(\mathbf{p}_0)}$ into Fourier series. Such an expansion (problem 1.14) demonstrates that a shift of time by $2\pi/\omega$ changes the wave function (28) by the phase factor $\exp[-2\pi i\overline{\varepsilon(\boldsymbol{\pi}_t)}/\hbar\omega]$, where the line over the expression denotes the averaging over the period $2\pi/\omega$. Some consequences of this transformation are discussed below for a more general case.

Quasienergy. If the potential energy is a periodic function of time, and the period is $2\pi/\omega$, one can write the wave function as

$$\psi(t) = \exp(-iEt/\hbar)u_E(t), \quad u_E(t + 2\pi/\omega) = u_E(t), \tag{30}$$

where E is referred to as quasienergy, since it is defined in the region $[0, \hbar\omega]$ (we note the analogy of Eq. (30) with the Bloch function in a spatially periodic potential considered above). The states with different E are orthogonal to each other. To find $u_E(t)$, it is convenient to represent the Hamiltonian containing a periodic potential as

$$\widehat{H}_t = \widehat{\overline{H}} + \sum_s \widehat{W}_s e^{is\omega t}, \tag{31}$$

where $\widehat{\overline{H}}$ is the part of the Hamiltonian averaged over the period $2\pi/\omega$, while $\widehat{W}_s$ describes the oscillating part and can be associated with the perturbation due to an external field. Introducing this field through the vector potential, it is easy to show that $s = \pm 1, \pm 2$, because the Hamiltonian is quadratic in the momentum. Further, with the Fourier expansion $u_E(t) = \sum_s \exp(-is\omega t)u_E(s)$, the Schroedinger equation is rewritten as a set of coupled equations

$$(E + s\hbar\omega - \hat{\overline{H}})u_E(s) = \sum_{s'} \widehat{W}_{s'}u_E(s - s'). \tag{32}$$

In many cases, the time-dependent part of the Hamiltonian (31) can be considered as a perturbation (of the first and of the second order in the field for $s = \pm 1$ and $s = \pm 2$, respectively). Within the accuracy of the second order, the perturbation theory gives us the following equation for $\overline{u_E(t)} = u_E(s = 0) \equiv u_E$:

$$(E - \hat{\overline{H}})u_E = \sum_{s=\pm 1} \widehat{W}_s(E - s\hbar\omega - \hat{\overline{H}})^{-1}\widehat{W}_s u_E. \tag{33}$$

This is the eigenstate problem determining the quasienergy spectrum of the system. For an electron in a homogeneous harmonic electric field, the quasienergy can be found from Eq. (28) by averaging $\varepsilon(\boldsymbol{\pi}_t)$ over the period. As a result, the parabolic spectrum is simply shifted in energy by $(eE/\omega)^2/4m$.

6. Long-Wavelength Phonons

The small-amplitude vibrations of the crystal lattice are described by a set of atomic displacement vectors $\mathbf{u}_s(\mathbf{R_n}t)$, where $\mathbf{R_n}$ is the radius-vector of the elementary crystal cell numbered by the integer vector $\mathbf{n}$ and the index s numbers the atoms in the cell. The expansion of the potential energy in the vicinity of the equilibrium positions of the atoms begins with the second-order terms, quadratic in the displacements. Accounting also for the anharmonic corrections described by the cubic terms, we write the total energy as

$$\mathcal{E} = \frac{1}{2}\sum_{\mathbf{n}sk} M_s \left[\dot{u}_s^k(\mathbf{R_n})\right]^2 + \frac{1}{2}\sum_{\mathbf{n}_1\mathbf{n}_2}\sum_{s_1 s_2 k_1 k_2} G_{s_1 s_2}^{k_1 k_2}(|\mathbf{R_{n_1}} - \mathbf{R_{n_2}}|)$$

$$\times u_{s_1}^{k_1}(\mathbf{R_{n_1}})u_{s_2}^{k_2}(\mathbf{R_{n_2}}) + \frac{1}{3!}\sum_{\mathbf{n}_{1-3}}\sum_{s_{1-3}k_{1-3}} A_{s_1 s_2 s_3}^{k_1 k_2 k_3}(|\mathbf{R_{n_1}} - \mathbf{R_{n_2}}|, |\mathbf{R_{n_1}} - \mathbf{R_{n_3}}|)$$

$$\times u_{s_1}^{k_1}(\mathbf{R_{n_1}})u_{s_2}^{k_2}(\mathbf{R_{n_2}})u_{s_3}^{k_3}(\mathbf{R_{n_3}}). \tag{1}$$

The index k numbers the Cartesian coordinates, $\dot{\mathbf{u}}_s(\mathbf{R_n})$ is the velocity of the atom s in the cell $\mathbf{n}$, and M_s is its mass. The matrices $G_{s_1 s_2}^{k_1 k_2}$

and $A^{k_1 k_2 k_3}_{s_1 s_2 s_3}$ are determined by the second and third derivatives of the potential energy in the equilibrium position. Owing to periodicity of the lattice, they depend only on the distances between the cells, $|\mathbf{R_n} - \mathbf{R_{n'}}|$.

Taking into account only the second-order contributions in Eq. (1), we obtain the classical equations of motion (problem 1.15)

$$M_s \ddot{u}^k_s(\mathbf{R_n}) + \sum_{\mathbf{n'}s'k'} G^{kk'}_{ss'}(|\mathbf{R_n} - \mathbf{R_{n'}}|) u^{k'}_{s'}(\mathbf{R_{n'}}) = 0 \qquad (2)$$

describing harmonic vibrations determined by the force constants $G^{kk'}_{ss'}$. The anharmonic contribution to the energy is described by the last term in the expression (1) and can be treated as a weak interaction between the vibrational modes. The translational invariance allows us to represent $\mathbf{u}_s(\mathbf{R_n}t)$ as a plane wave, $\mathbf{U}_s \exp(i\mathbf{q} \cdot \mathbf{R_n} - i\omega t)$. This substitution reduces the number of variables in Eq. (2) from $3N\bar{s}$ to $3\bar{s}$, where N is the number of elementary cells and $\bar{s}$ is the number of atoms in the cell. The amplitudes $\mathbf{U}_s$ obey the following set of linear algebraic equations:

$$M_s \omega^2 U^k_s - \sum_{k's'} \left[\sum_{\Delta \mathbf{n}} G^{kk'}_{ss'}(|\mathbf{R_{\Delta n}}|) e^{-i\mathbf{q} \cdot \mathbf{R_{\Delta n}}} \right] U^{k'}_{s'} = 0, \qquad (3)$$

where $\mathbf{R_{\Delta n}} \equiv \mathbf{R_n} - \mathbf{R_{n'}}$. There are $3\bar{s}$ solutions of this set, each corresponds to a branch (mode) of the vibrational spectrum.

It is convenient to express the displacements in terms of the polarization vectors $\mathbf{e}_s(\mathbf{q}l)/\sqrt{M_s}$ (the index $l = 1, \ldots, 3\bar{s}$ numbers the vibrational modes) found from the equations

$$\sum_{s'k'} \left[\omega^2 \delta_{ss'} \delta_{kk'} - \mathcal{G}^{kk'}_{ss'}(\mathbf{q}) \right] e^{k'}_{s'}(\mathbf{q}l) = 0,$$

$$\mathcal{G}^{kk'}_{ss'}(\mathbf{q}) = \sum_{\Delta \mathbf{n}} \frac{G^{kk'}_{ss'}(|\mathbf{R_{\Delta n}}|)}{\sqrt{M_s M_{s'}}} e^{-i\mathbf{q} \cdot \mathbf{R_{\Delta n}}}, \qquad (4)$$

which directly follow from Eq. (3). Because the matrix $\mathcal{G}^{kk'}_{ss'}(\mathbf{q})$ is periodic in $\mathbf{q}$, one should consider $\mathbf{q}$ inside the first Brillouin zone so that $\mathbf{q}$ is a quasi-wave vector. The set of solutions $\mathbf{e}_s(\mathbf{q}l)$ is normalized according to $\sum_s \mathbf{e}^*_s(\mathbf{q}l)\mathbf{e}_s(\mathbf{q}l') = \delta_{ll'}$, and $\mathbf{e}^*_s(\mathbf{q}l) = \mathbf{e}_s(-\mathbf{q}l)$. Employing the normal coordinates $Q_{\mathbf{q}l}(t)$, which include the exponential time-dependent factors, we write $\mathbf{u_{ns}}(t) \equiv \mathbf{u}_s(\mathbf{R_n}t)$ as

$$\mathbf{u_{ns}}(t) = (NM_s)^{-1/2} \sum_{\mathbf{q}l} Q_{\mathbf{q}l}(t)\mathbf{e}_s(\mathbf{q}l)e^{i\mathbf{q} \cdot \mathbf{R_n}}, \qquad (5)$$

where $Q_{\mathbf{q}l}^*(t) = Q_{-\mathbf{q}l}(t)$ because the displacements are real. In order to express the quadratic contributions of Eq. (1) through the normal coordinates, we take a sum over $\mathbf{n}$ with the aid of the relation $N^{-1} \sum_{\mathbf{n}} \exp[i(\mathbf{q} - \mathbf{q}') \cdot \mathbf{R_n}] = \Delta_{\mathbf{qq}'}$. The function $\Delta_{\mathbf{qq}'}$ is equal to 1 when $\mathbf{q}$ and $\mathbf{q}'$ either coincide or differ by a reciprocal lattice vector and is equal to zero otherwise (problem 1.16). The energy of small-amplitude vibrations takes the form

$$\frac{1}{2} \sum_{\mathbf{q}l} \left[\dot{Q}_{\mathbf{q}l}^*(t)\dot{Q}_{\mathbf{q}l}(t) + \omega_{\mathbf{q}l}^2 Q_{\mathbf{q}l}^*(t)Q_{\mathbf{q}l}(t) \right], \qquad (6)$$

which is quadratic in the normal coordinates and momenta $P_{\mathbf{q}l}(t) \equiv \dot{Q}_{\mathbf{q}l}(t)$. This is the energy of a set of harmonic oscillators with frequencies $\omega_{\mathbf{q}l}$.

The quantization of the lattice vibrations can be done by analogy with the case of photons described in Sec. 3. The normal coordinates and normal momenta are replaced by the operators $\widehat{Q}_{\mathbf{q}l}$ and $\widehat{P}_{\mathbf{q}l}$, which satisfy the commutation relations similar to Eq. (3.7):

$$[\widehat{Q}_{\mathbf{q}l}, \widehat{P}_{-\mathbf{q}'l'}] = i\hbar\delta_{\mathbf{qq}'}\delta_{ll'}. \qquad (7)$$

The energy (6), after such a substitution, becomes a Hamiltonian of the kind (3.8), where the index ν is replaced by the quantum numbers $\mathbf{q}$ and l. The quasiparticles with these quantum numbers are called the phonons. Their creation and annihilation operators, $\hat{b}_{\mathbf{q}l}^+$ and $\hat{b}_{\mathbf{q}l}$, satisfy the commutation relations

$$[\hat{b}_{\mathbf{q}l}, \hat{b}_{\mathbf{q}'l'}^+] = \delta_{\mathbf{qq}'}\delta_{ll'}. \qquad (8)$$

The Hamiltonian of phonons is written as a sum of the boson mode contributions according to Eq. (3.12), and the relation between the bosonic operators and normal coordinates and momenta is the same as in Eq. (3.11). It can be rewritten as

$$\widehat{Q}_{\mathbf{q}l} = \widehat{Q}_{-\mathbf{q}l}^+ = \sqrt{\frac{\hbar}{2\omega_{\mathbf{q}l}}}(\hat{b}_{\mathbf{q}l} + \hat{b}_{-\mathbf{q}l}^+),$$

$$\widehat{P}_{\mathbf{q}l} = \widehat{P}_{-\mathbf{q}l}^+ = -i\sqrt{\frac{\hbar\omega_{\mathbf{q}l}}{2}}(\hat{b}_{\mathbf{q}l} - \hat{b}_{-\mathbf{q}l}^+). \qquad (9)$$

Substituting the expression for $\widehat{Q}_{\mathbf{q}l}$ into Eq. (5), we obtain the second-quantized displacement operator

$$\hat{\mathbf{u}}_{\mathbf{n}s} = (NM_s)^{-1/2} \sum_{\mathbf{q}l} \sqrt{\frac{\hbar}{2\omega_{\mathbf{q}l}}}\mathbf{e}_s(\mathbf{q}l)e^{i\mathbf{q}\cdot\mathbf{R_n}}(\hat{b}_{\mathbf{q}l} + \hat{b}_{-\mathbf{q}l}^+) \qquad (10)$$

written through the linear combination of creation and annihilation operators. Therefore, the displacements $\mathbf{u}_s$ satisfy the oscillatory equations (2) analogous to the equations for the electromagnetic field in the absence of free charges. To quantize the lattice vibrations, one should express the displacements in terms of the generalized normal coordinates and momenta and, further, introduce the elementary excitations, phonons.

Below we consider the limit of long wavelengths (the region of small $\mathbf{q}$), when the lattice vibrations are described by a number of macroscopic parameters. This approach is convenient for description of electron-phonon and phonon-photon interactions. First we notice that under a shift of the lattice as a whole, when $\mathbf{u}_s(\mathbf{R_n}t)$ do not depend on $\mathbf{R_n}$, the force in Eq. (2) must be zero. This leads to the identity

$$\sum_{\Delta \mathbf{n} s'} G_{ss'}^{kk'}(|\mathbf{R}_{\Delta \mathbf{n}}|) = 0. \tag{11}$$

Using Eq. (11) together with the symmetry property of the force matrices, $G_{ss'}^{kk'} = G_{s's}^{k'k}$, we sum Eq. (3) over s. If $\mathbf{q} \to 0$, it gives us

$$\omega^2 \sum_s M_s \mathbf{u}_s = 0. \tag{12}$$

Equation (12) describes two kinds of vibrations pertinent to the long-wavelength limit: the acoustic phonons with $\omega \to 0$ at $\mathbf{q} \to 0$, and the optical phonons with $\omega \neq 0$. Under the optical vibrations, the center of mass of the cell remains at rest, and the atoms of the cell oscillate in antiphase, $\sum_s M_s \mathbf{u}_s = 0$. There are 3 acoustic modes and $3\bar{s} - 3$ optical modes.

Let us consider the long-wavelength optical vibrations in a biatomic ($\bar{s} = 2$) crystal with ionic bond, when the oppositely charged sublattices oscillate as a whole with respect to each other. In this approximation, the displacements $\mathbf{u_{ns}}$ do not depend on $\mathbf{n}$, and the vibrations are described by two variables $\mathbf{u}_\pm$ corresponding to two sublattices with effective charges $\pm e^*$ and atomic masses $M_\pm$. According to Eq. (12), $M_+\mathbf{u}_+ + M_-\mathbf{u}_- = 0$, and there is only one independent variable, the relative ionic displacement $\mathbf{u} = \mathbf{u}_+ - \mathbf{u}_-$. Under such vibrations, each elementary cell has a dipole moment $e^*\mathbf{u}$, and a significant contribution to the interatomic forces comes from the long-range dipole-dipole interaction. One has to subdivide the atomic force constant matrices $\mathcal{G}_{ss'}^{kk'}$ by a short-range part, proportional to the relative shift $\mathbf{u}$, and a long-range part, given through the electric field with local strength $\mathbf{E}_L$. In this way, the equations of motion become

$$M_+\ddot{\mathbf{u}}_+ = -k(\mathbf{u}_+ - \mathbf{u}_-) + e^*\mathbf{E}_L ,$$
$$M_-\ddot{\mathbf{u}}_- = k(\mathbf{u}_+ - \mathbf{u}_-) - e^*\mathbf{E}_L$$

(13)

where the coefficient k describes the short-range forces. The longitudinal field $\mathbf{E}_L$ in Eq. (13) induces the polarization $(N/V)\alpha\mathbf{E}_L$, where α is the polarizability of the cell and N/V is the number of the cells per unit volume. The total polarization $\mathbf{P}$ is the sum of this induced contribution and dipole moment $e^*\mathbf{u}$. Thus, for $\mathbf{u}$, $\mathbf{E}_L$, and $\mathbf{P}$, we have

$$\ddot{\mathbf{u}} = -\omega_{TO}^2\mathbf{u} + \frac{e^*}{\overline{M}}\mathbf{E}_L,$$

$$\mathbf{P} = \frac{N}{V}(e^*\mathbf{u} + \alpha\mathbf{E}_L).$$

(14)

The first equation, which follows from Eq. (13), contains the reduced mass $\overline{M} = M_+M_-/(M_+ + M_-)$. The contribution of the short-range forces to this equation is expressed through the transverse mode frequency ω_{TO}, the latter is introduced as $\omega_{TO} = (k/\overline{M})^{1/2}$. Since the long-range electric fields are not generated by the transverse displacements (see the picture of ionic crystal vibrations in Fig. 1.3), ω_{TO} is the frequency of transverse vibrations.

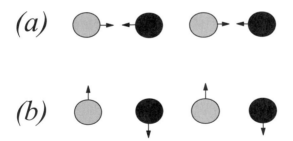

Figure 1.3. Pictures of the longitudinal (*a*) and transverse (*b*) vibrations of the ionic crystal sublattices.

To consider the longitudinal vibration, one has to use an additional, with respect to Eq. (14), relation between $\mathbf{E}_L$ and $\mathbf{P}$. Owing to electric neutrality of the lattice, the Poisson equation for the longitudinal fields gives us $\mathbf{E}_L + 4\pi\mathbf{P} = 0$. Before applying it for solution of Eq. (14), it is convenient to express the microscopic parameters e^* and α through the static and high-frequency dielectric constants, ϵ_0 and ϵ_∞. Let us introduce the temporal Fourier components of the polarization and field,

$\mathbf{P}_\omega$ and $\mathbf{E}_{L\omega}$. In the region of high frequencies, larger than the ionic vibration frequency but smaller than the frequencies of interband electron transitions, the sublattices do not shift, and at $\mathbf{u} \to 0$ these components are connected by the relation $\mathbf{P}_\infty = (N/V)\alpha\mathbf{E}_{L\infty}$. On the other hand, $\mathbf{P}_\infty = (\epsilon_\infty - 1)\mathbf{E}_{L\infty}/4\pi$. In this way we express $(N/V)\alpha$ through ϵ_∞. In the static case, one has $\dot{\mathbf{u}} = 0$, and Eq. (14) leads to $\mathbf{P}_0 = (N/V)[\alpha + e^{*\,2}/\omega_{TO}\overline{M}]$. On the other hand, $\mathbf{P}_0 = (\epsilon_0 - 1)\mathbf{E}_{L\omega=0}/4\pi$, and we find an expression for the charge e^*. As a result,

$$\alpha\frac{N}{V} = \frac{\epsilon_\infty - 1}{4\pi}, \qquad e^{*2}\frac{N}{\overline{M}V} = \omega_{TO}^2\frac{\epsilon_0 - \epsilon_\infty}{4\pi}. \tag{15}$$

Further, it is convenient to introduce a new variable $\mathbf{w} = (N\overline{M}/V)^{1/2}\mathbf{u}$ instead of $\mathbf{u}$ (this variable will be also called below as the relative ionic displacement, though its dimensionality is not a length). Using this definition and Eq. (15), we rewrite Eq. (14) as

$$\ddot{\mathbf{w}} = -\omega_{TO}^2\mathbf{w} + \omega_{TO}\sqrt{\frac{\epsilon_0 - \epsilon_\infty}{4\pi}}\mathbf{E}_L,$$

$$\mathbf{P} = \omega_{TO}\sqrt{\frac{\epsilon_0 - \epsilon_\infty}{4\pi}}\mathbf{w} + \frac{\epsilon_\infty - 1}{4\pi}\mathbf{E}_L. \tag{16}$$

Excluding $\mathbf{E}_L$ and $\mathbf{P}$ with the use of $\mathbf{E}_L + 4\pi\mathbf{P} = 0$, we arrive at the equation of motion $\ddot{\mathbf{w}} + \omega_{LO}^2\mathbf{w} = 0$ for the longitudinal vibrations. Their frequency, ω_{LO}, is expressed through ω_{TO} according to the Lyddane-Sachs-Teller relation

$$\omega_{LO} = \sqrt{\epsilon_0/\epsilon_\infty}\,\omega_{TO}. \tag{17}$$

The Hamiltonian of longitudinal optical (LO) and transverse optical (TO) phonons is given by Eq. (3.12) with $\mathbf{q}$-independent frequencies ω_{LO} and ω_{TO}. In the biatomic crystals there are two TO modes and one LO mode.

In the long-wavelength limit, it is convenient to write the contribution of the optical vibrations to the second-quantized displacement operator (10) after replacing the discrete vector $\mathbf{R_n}$ by the continuous variable $\mathbf{r}$. The unit vectors of polarization, $\mathbf{e}_\pm(\mathbf{q})$, are determined from the normalization conditions and from the requirement of zero displacement of the center of mass of the cell:

$$\begin{aligned}\mathbf{e}_+^2(\mathbf{q}) + \mathbf{e}_-^2(\mathbf{q}) &= 1 \\ \sqrt{M_+}\mathbf{e}_+(\mathbf{q}) + \sqrt{M_-}\mathbf{e}_-(\mathbf{q}) &= 0\end{aligned} \tag{18}$$

Introducing the total mass of the cell, $M_c = M_+ + M_-$, we write the solutions of Eq. (18) as $\mathbf{e}_\pm(\mathbf{q}) = \pm\sqrt{M_\mp/M_c}\mathbf{e_{q}}_l$, where $\mathbf{e_{q}}_l$ is the unit vector of polarization which depends on the mode index l ($l = LO$, TO_1, and TO_2). Therefore, the second-quantized operator $\widehat{\mathbf{w}}$ is given by

$$\widehat{\mathbf{w}}(\mathbf{r}) = \sum_{\mathbf{q}l} \sqrt{\frac{\hbar}{2\omega_l V}} \mathbf{e}_{\mathbf{q}l} e^{i\mathbf{q}\cdot\mathbf{r}} \left(\hat{b}_{\mathbf{q}l} + \hat{b}^+_{-\mathbf{q}l}\right). \tag{19}$$

The polarization induced by the longitudinal modes is expressed through $\mathbf{w}$ according to

$$\mathbf{P} = \sqrt{\frac{\omega^2_{LO}}{4\pi\epsilon^*}}\mathbf{w}, \qquad \frac{1}{\epsilon^*} \equiv \frac{1}{\epsilon_\infty} - \frac{1}{\epsilon_0}, \tag{20}$$

where the effective dielectric constant ϵ^* is introduced. The second-quantized polarization operator is given by

$$\hat{\mathbf{P}}(\mathbf{r}) = \sum_{\mathbf{q}} \sqrt{\frac{\hbar\omega_{LO}}{8\pi\epsilon^* V}} \mathbf{e}_{\mathbf{q}LO} e^{i\mathbf{q}\cdot\mathbf{r}} \left(\hat{b}_{\mathbf{q}LO} + \hat{b}^+_{-\mathbf{q}LO}\right). \tag{21}$$

The potential energy of interaction of electrons with the field induced by the longitudinal vibrations is obtained from the Poisson equation in the form $U(\mathbf{r}) = i4\pi e \sum_{\mathbf{q}} \exp(i\mathbf{q}\cdot\mathbf{r})(\mathbf{q}\cdot\mathbf{P}(\mathbf{q}))/q^2$. Choosing the unit vector of longitudinal displacement as $\mathbf{e}_{\mathbf{q}LO} = -i\mathbf{q}/q$ so that $\mathbf{e}^*_{\mathbf{q}LO} = \mathbf{e}_{-\mathbf{q}LO}$, we obtain the Hamiltonian of electron-phonon interaction:

$$\widehat{H}_{e,LO}(\mathbf{r}) = \sqrt{\frac{2\pi e^2 \hbar\omega_{LO}}{\epsilon^* V}} \sum_{\mathbf{q}} q^{-1} e^{i\mathbf{q}\cdot\mathbf{r}} \left(\hat{b}_{\mathbf{q}LO} + \hat{b}^+_{-\mathbf{q}LO}\right). \tag{22}$$

This expression is known as Froelich Hamiltonian. The electrons interact with the lattice vibrations through the long-range electric field generated by the longitudinal optical phonons.

Let us consider the interaction of the optical vibrations with the electric field of a transverse electromagnetic wave propagating in the isotropic ionic crystal. The energy of the interaction is expressed through the dipole moment $\mathbf{d_n}$ of $\mathbf{n}$-th cell according to $-\sum_{\mathbf{n}} \mathbf{d_n} \cdot \mathbf{E}(\mathbf{R_n}t)$, where $\mathbf{E}(\mathbf{R_n}t)$ is the field of the electromagnetic wave in the cell. Since $\mathbf{d} = e^*(\mathbf{u}_+ - \mathbf{u}_-)$, the second-quantized operator of the dipole moment is expressed, with the use of Eq. (10), as

$$\hat{\mathbf{d}}_\mathbf{n} = \sum_{\mathbf{q}l} \sqrt{\frac{\hbar e^{*\,2}}{2\omega_l \overline{M} N}} \mathbf{e}_{\mathbf{q}l} (\hat{b}_{\mathbf{q}l} + \hat{b}^+_{-\mathbf{q}l}) e^{i\mathbf{q}\cdot\mathbf{R_n}}. \tag{23}$$

The operator of interaction of phonons with the electric field described by the Fourier component $\mathbf{E_k}$ is given by

$$\widehat{H}_{\mathbf{k}} = -\sum_l \sqrt{\frac{\hbar\omega_{TO}^2(\epsilon_0 - \epsilon_\infty)V}{8\pi\omega_l}}(\mathbf{e}_{\mathbf{k}l} \cdot \mathbf{E_k})(\hat{b}_{-\mathbf{k}l} + \hat{b}_{\mathbf{k}l}^+). \qquad (24)$$

One can see that only TO phonons interact with the transverse field in the long-wavelength limit. To consider the phonon-photon interaction, one has to substitute the second-quantized electric field (3.24) into the operator of interaction. In this way we obtain the operator of phonon-photon interaction:

$$\widehat{H}_{ph,pht} = -i\hbar\sum_{\mathbf{q}l\mu} \sqrt{\omega_{TO}\omega_q \frac{\epsilon_0 - \epsilon_\infty}{4\epsilon_\infty}}(\mathbf{e}_{\mathbf{q}l} \cdot \mathbf{e}_{\mathbf{q}\mu})$$

$$\times (\hat{b}_{-\mathbf{q}l} + \hat{b}_{\mathbf{q}l}^+)(\hat{b}_{\mathbf{q}\mu} - \hat{b}_{-\mathbf{q}\mu}^+). \qquad (25)$$

We point out the appearance of a product of second-quantized operators of the photon and phonon modes denoted by the indices μ and l, respectively. Owing to the polarization factor $(\mathbf{e}_{\mathbf{q}l} \cdot \mathbf{e}_{\mathbf{q}\mu})$, the photons with polarization $\mu = 1$ interact with TO_1 phonons, while the photons with $\mu = 2$ interact with TO_2 phonons. The Hamiltonian (25) should be viewed as a perturbation coupling free photons to free phonons. For a correct description of the phonon-photon system, this perturbation has to be small in comparison to the unperturbed photon and phonon energies. This occurs when the parameter of coupling, $\epsilon_0/\epsilon_\infty - 1$, is much less than unity. A non-perturbative approach describing coupled phonon-photon modes at arbitrary coupling strength will be presented in Chapter 5.

When the long-wavelength acoustic phonons propagate in the crystal, the elementary cell oscillates as a whole, the atoms are moving in phase, and the displacement vector $\mathbf{u}_s$ determined by the equation of motion (3) does not depend on s. The equation for the long-wavelength Fourier components $\mathbf{u_q}$ of the displacements under acoustic vibration is obtained after a summation of Eq. (3) over s and subsequent expansion of the forces in series of q up to q^2 terms:

$$\omega^2 u_{\mathbf{q}}^k + \frac{1}{2M_c}\sum_{k'}\left[\sum_{\Delta nss'} G_{ss'}^{kk'}(|\mathbf{R}_{\Delta n}|)(\mathbf{q} \cdot \mathbf{R}_{\Delta n})^2\right]u_{\mathbf{q}}^{k'} = 0. \qquad (26)$$

The contribution linear in $\mathbf{q}$ is equal to zero since the function under the sum $\sum_{\Delta n}$ is odd with respect to Δn. In Eq. (26) we have introduced the total mass of the atoms in the cell, $M_c = \sum_s M_s$. It is convenient to introduce the elastic module tensor $\lambda_{kll'k'}$ and the crystal density $\rho = M_c N/V$ in order to rewrite Eq. (26) as an equation for the macroscopic displacement field:

$$\rho\omega^2 u_{\mathbf{q}}^k - \sum_{k'll'} \lambda_{kll'k'} q_l q_{l'} u_{\mathbf{q}}^{k'} = 0,$$

$$\lambda_{kll'k'} = \frac{\rho}{2M_c} \sum_{ss'\Delta\mathbf{n}} G_{ss'}^{kk'}(|\mathbf{R}_{\Delta\mathbf{n}}|) R_{\Delta\mathbf{n}}^l R_{\Delta\mathbf{n}}^{l'}. \tag{27}$$

In the isotropic media, the tensor $\lambda_{kll'k'}$ is expressed through two components, according to $\lambda_{kll'k'} = \mu\delta_{ll'}\delta_{kk'} + (\mu+\lambda)\delta_{kl}\delta_{k'l'}$. The parameters μ and λ are known as Lamé coefficients. The dispersion relation following from Eq. (27) gives us a longitudinal (LA) and a pair of transverse (TA) solutions:

$$\omega_{qLA,TA} = s_{l,t}q, \quad s_t = \sqrt{\mu/\rho}, \quad s_l = \sqrt{(\lambda+2\mu)/\rho}, \tag{28}$$

where we have introduced the longitudinal and transverse sound velocities, s_l and s_t. The dispersion laws for long-wavelength acoustic phonons are linear in q.

The second-quantized operator of the displacement field under acoustic vibrations is given by Eq. (10) with $\hat{\mathbf{u}}_{\mathbf{n}s}$ independent of s. One may replace $\mathbf{e}_s(\mathbf{q})/\sqrt{M_s}$ by $\mathbf{e}_{\mathbf{q}l}/\sqrt{M_c}$, where the index l numbers the acoustic modes and the mass of the cell stands here due to the normalization conditions $\sum_s |\mathbf{e}_s(\mathbf{q})|^2 = 1$ and $|\mathbf{e}_{\mathbf{q}l}|^2 = 1$. As a result, we obtain the operator

$$\hat{\mathbf{u}}_{ac}(\mathbf{r}) = \sum_{\mathbf{q}l} \sqrt{\frac{\hbar}{2\rho\omega_{\mathbf{q}l}V}} \mathbf{e}_{\mathbf{q}l} e^{i\mathbf{q}\cdot\mathbf{r}} \left(\hat{b}_{\mathbf{q}l} + \hat{b}_{-\mathbf{q}l}^+\right). \tag{29}$$

The relative change of the volume at the point $\mathbf{r}$, generated by the long-wavelength acoustic vibrations in the isotropic medium, is equal to $\nabla \cdot \mathbf{u}_{ac}(\mathbf{r})$. Only the longitudinal vibrations contribute to this quantity. This deformation changes the electron energy and, therefore, provides a mechanism for electron-phonon interaction. Near the electron energy band extrema in the cubic crystals, the interaction energy is equal to $\mathcal{D}(\nabla \cdot \mathbf{u}_{ac})$, where $\mathcal{D}$ is the deformation potential (see Appendix B). Again, applying $\mathbf{e}_{\mathbf{q}LA} = -i\mathbf{q}/q$, we obtain the Hamiltonian of electron-phonon interaction:

$$\widehat{H}_{e,LA}(\mathbf{r}) = \sum_{\mathbf{q}} \sqrt{\frac{\hbar\mathcal{D}^2 q}{2\rho s_l V}} e^{i\mathbf{q}\cdot\mathbf{r}} \left(\hat{b}_{\mathbf{q}LA} + \hat{b}_{-\mathbf{q}LA}^+\right). \tag{30}$$

This expression is similar to Eq. (22). However, the electron-phonon coupling energy is different: now it is expressed through the deformation potential, longitudinal sound velocity, and crystal density. Apart from the Froelich and deformation-potential interactions given by Eqs. (22) and (30), there exist other mechanisms of electron-phonon interactions, namely *i)* the interaction with the optical phonons due to deformation of the lattice, and *ii)* the interaction with the acoustic phonons due to piezoelectric fields generated by the lattice vibration. The former is important only in the non-ionic crystals where the Froelich interaction is absent, while the latter is often weaker than the deformation-potential interaction.

To complete this section, let us consider the phonon-phonon interaction appearing due to the anharmonic contributions in Eq. (1). Substituting the expression (10) for the displacement operators into the last (cubic) term of Eq. (1), we rewrite this term as

$$\widehat{H}_{ph,ph} = \frac{\rho V}{6} \left(\frac{\hbar}{2\rho V} \right)^{3/2} \sum_{\mathbf{q}_1 \mathbf{q}_2 \mathbf{q}_3} \sum_{l_1 l_2 l_3} \frac{\beta_{l_1 l_2 l_3}(\mathbf{q}_1, \mathbf{q}_2, \mathbf{q}_3)}{\sqrt{\omega_{\mathbf{q}_1 l_1} \omega_{\mathbf{q}_2 l_2} \omega_{\mathbf{q}_3 l_3}}}$$

$$\times (\hat{b}_{\mathbf{q}_1 l_1} + \hat{b}^+_{-\mathbf{q}_1 l_1})(\hat{b}_{\mathbf{q}_2 l_2} + \hat{b}^+_{-\mathbf{q}_2 l_2})(\hat{b}_{\mathbf{q}_3 l_3} + \hat{b}^+_{-\mathbf{q}_3 l_3}), \qquad (31)$$

where the anharmonic coefficients $\beta_{l_1 l_2 l_3}(\mathbf{q}_1, \mathbf{q}_2, \mathbf{q}_3)$ are introduced by the following expression:

$$\beta_{l_1 l_2 l_3}(\mathbf{q}_1, \mathbf{q}_2, \mathbf{q}_3) = \frac{1}{\rho V} \sum_{s_1 s_2 s_3} \frac{M_c^{3/2}}{\sqrt{M_{s_1} M_{s_2} M_{s_3}}}$$

$$\times \sum_{\mathbf{n}_1 \mathbf{n}_2 \mathbf{n}_3} \sum_{k_1 k_2 k_3} A^{k_1 k_2 k_3}_{s_1 s_2 s_3} (|\mathbf{R}_{\mathbf{n}_1} - \mathbf{R}_{\mathbf{n}_2}|, |\mathbf{R}_{\mathbf{n}_1} - \mathbf{R}_{\mathbf{n}_3}|) \qquad (32)$$

$$\times e^{k_1}_{s_1}(\mathbf{q}_1 l_1) e^{k_2}_{s_2}(\mathbf{q}_2 l_2) e^{k_3}_{s_3}(\mathbf{q}_3 l_3) \exp[i(\mathbf{q}_1 \cdot \mathbf{R}_{\mathbf{n}_1} + \mathbf{q}_2 \cdot \mathbf{R}_{\mathbf{n}_2} + \mathbf{q}_3 \cdot \mathbf{R}_{\mathbf{n}_3})].$$

The factor $1/\rho V$ is detached for the sake of convenience, to make the anharmonic coefficients defined by Eq. (32) independent of the volume V. In contrast to the operators (22) and (30), which are linear in $\hat{b}_{\mathbf{q}l}$, the operator (31) is cubic in $\hat{b}_{\mathbf{q}l}$. It describes the modification of the crystal energy due to vibrational anharmonicity. From the point of view of second quantization, the Hamiltonian (31) accounts for three-phonon processes corresponding to either a transformation (decay) of the phonon into two other phonons ($\hat{b}\hat{b}^+\hat{b}^+$ terms) or a fusion of two phonons into one phonon ($\hat{b}\hat{b}\hat{b}^+$ terms). The terms containing $\hat{b}\hat{b}\hat{b}$ and $\hat{b}^+\hat{b}^+\hat{b}^+$ do not correspond to any transitions since a creation or annihilation of three phonons in the absence of external perturbations is forbidden by the energy conservation requirement. A shift by an arbitrary lattice vector may not modify the coefficients $\beta_{l_1 l_2 l_3}(\mathbf{q}_1, \mathbf{q}_2, \mathbf{q}_3)$. Therefore, the sum

$\mathbf{q}_1 + \mathbf{q}_2 + \mathbf{q}_3$ is equal either to zero or to a reciprocal lattice vector. It means that $\beta_{l_1 l_2 l_3}(\mathbf{q}_1, \mathbf{q}_2, \mathbf{q}_3) \propto \Delta_{\mathbf{q}_1 + \mathbf{q}_2 + \mathbf{q}_3, 0}$, which expresses the conservation of quasimomentum in the phonon-phonon collisions. One can estimate the order of the absolute value of the anharmonic coefficients from the following consideration. The absolute value of $A^{k_1 k_2 k_3}_{s_1 s_2 s_3}$ is estimated as $\overline{M} \overline{v}^2 / \overline{a}^3$, where $\overline{M}$, $\overline{v}$, and $\overline{a}$ are the averaged atomic mass, sound velocity, and lattice constant. Therefore, if we assume that $\mathbf{q}_i \cdot \mathbf{R}_{\mathbf{n}_i} \sim 1$ $(i = 1, 2, 3)$ and account only for the nearest-neighbor interaction in the $\mathbf{n}$-sum in Eq. (32), we find $|\beta_{l_1 l_2 l_3}(\mathbf{q}_1, \mathbf{q}_2, \mathbf{q}_3)| \sim \overline{v}^2 / \overline{a}^3$ (here and below in this book the sign "$\sim$" defines an order-of-value estimate). If one, two, or all three phonons are the long-wavelength acoustic ones, the estimate should be written, respectively, as $q_1 \overline{v}^2 / \overline{a}^2$, $q_1 q_2 \overline{v}^2 / \overline{a}$, and $q_1 q_2 |q_1 + q_2| \overline{v}^2$. These estimates are not essentially modified for the ionic crystals with long-range interaction between the atoms. If only the long-wavelength acoustic phonons are important, one may write the anharmonic potential energy, i.e., the third term in the expression (1), in the elastic continuum approximation as

$$\frac{1}{3!} \int_{(V)} d\mathbf{r} \sum_{k_1 k_2 k_3} \sum_{\alpha \beta \gamma} \lambda_{k_1 \alpha, k_2 \beta, k_3 \gamma} \frac{\partial u^{k_1}_{ac}(\mathbf{r})}{\partial r_\alpha} \frac{\partial u^{k_2}_{ac}(\mathbf{r})}{\partial r_\beta} \frac{\partial u^{k_3}_{ac}(\mathbf{r})}{\partial r_\gamma}, \qquad (33)$$

where $\lambda_{k_1 \alpha, k_2 \beta, k_3 \gamma}$ are the third-order anharmonic elastic constants which have dimensionality of the energy density. To quantize this energy, one should simply replace the vectors of acoustic-phonon displacements by the operators of such displacements. Substituting these operators from Eq. (29) into Eq. (33), we find

$$\widehat{H}_{ph, ph} = \frac{1}{6} \sum_{\mathbf{q}_1 \mathbf{q}_2 \mathbf{q}_3} \sum_{l_1 l_2 l_3} B_{l_1 l_2 l_3}(\mathbf{q}_1, \mathbf{q}_2, \mathbf{q}_3)(\hat{b}_{\mathbf{q}_1 l_1} + \hat{b}^+_{-\mathbf{q}_1 l_1})$$

$$\times (\hat{b}_{\mathbf{q}_2 l_2} + \hat{b}^+_{-\mathbf{q}_2 l_2})(\hat{b}_{\mathbf{q}_3 l_3} + \hat{b}^+_{-\mathbf{q}_3 l_3}),$$

$$B_{l_1 l_2 l_3}(\mathbf{q}_1, \mathbf{q}_2, \mathbf{q}_3) = -\frac{i}{\sqrt{V}} \left(\frac{\hbar}{2\rho}\right)^{3/2} \sum_{k_1 k_2 k_3} \sum_{\alpha \beta \gamma} \lambda_{k_1 \alpha, k_2 \beta, k_3 \gamma}$$

$$\times \frac{e^{k_1}_{\mathbf{q}_1 l_1} e^{k_2}_{\mathbf{q}_2 l_2} e^{k_3}_{\mathbf{q}_3 l_3} q_{1\alpha} q_{2\beta} q_{3\gamma}}{\sqrt{\omega_{\mathbf{q}_1 l_1} \omega_{\mathbf{q}_2 l_2} \omega_{\mathbf{q}_3 l_3}}} \delta_{\mathbf{q}_1 + \mathbf{q}_2 + \mathbf{q}_3, 0} . \qquad (34)$$

Note that, instead of $\Delta_{\mathbf{q}_1 + \mathbf{q}_2 + \mathbf{q}_3, 0}$, we have obtained the conventional Kronecker symbol expressing the conservation of momentum. This is a consequence of the elastic continuum approximation. The coefficients

$\lambda_{k_1\alpha, k_2\beta, k_3\gamma}$ are related to $A_{s_1 s_2 s_3}^{k_1 k_2 k_3}$ in the following way:

$$\sum_{\mathbf{n_1 n_2 n_3}} \left[\sum_{s_1 s_2 s_3} A_{s_1 s_2 s_3}^{k_1 k_2 k_3} \left(|\mathbf{R_{n_1}} - \mathbf{R_{n_2}}|, |\mathbf{R_{n_1}} - \mathbf{R_{n_3}}| \right) \right]$$

$$\times \exp[i(\mathbf{q_1} \cdot \mathbf{R_{n_1}} + \mathbf{q_2} \cdot \mathbf{R_{n_2}} + \mathbf{q_3} \cdot \mathbf{R_{n_3}})] \tag{35}$$

$$= i^3 \lambda_{k_1\alpha, k_2\beta, k_3\gamma} q_{1\alpha} q_{2\beta} q_{3\gamma} \delta(\mathbf{q_1} + \mathbf{q_2} + \mathbf{q_3}).$$

This equation follows from a comparison of Eqs. (31) and (32) to Eq. (34) (note that for acoustic phonons $\mathbf{e_{ql}} = \mathbf{e}_s(\mathbf{q}l)\sqrt{M_c/M_s}$ is independent of the sort of the atom). Indeed, the right-hand side of Eq. (35) is the leading term in the expansion of the left-hand side of this equation in powers of small $\mathbf{q_1}$, $\mathbf{q_2}$, and $\mathbf{q_3}$. This term should be linear in each of these wave vectors, as follows from the invariance of the anharmonic energy with respect to the shift of the lattice as a whole. This invariance gives us a relation similar to Eq. (11):

$$\sum_{s_i \mathbf{n_i}} A_{s_1 s_2 s_3}^{k_1 k_2 k_3} \left(|\mathbf{R_{n_1}} - \mathbf{R_{n_2}}|, |\mathbf{R_{n_1}} - \mathbf{R_{n_3}}| \right) = 0, \tag{36}$$

where the sum is taken over either pair of variables s_i and $\mathbf{n}_i$, $i = 1, 2, 3$. This equation can be used to prove Eq. (35).

Problems

1.1. Write the dynamical equations (1.13) and (1.14) in the representation of the quantum numbers of harmonic oscillator, analogous to Eqs. (1.5) and (1.6) or to Eqs. (1.8) and (1.10).

Hint: Use the expansion of the wave function of x-representation in the oscillator wave functions (Appendix A). This leads to the Schroedinger equation in the N-representation,

$$i\hbar \frac{\partial \Psi_{Nt}^{(\delta)}}{\partial t} = \sum_{N'} H_{NN'} \Psi_{N't}^{(\delta)},$$

and the observable is written as $\overline{Q}_t^{(\delta)} = \sum_{NN'} \langle \delta, t|N\rangle \langle N|\widehat{Q}|N'\rangle \langle N'|\delta, t\rangle$.

1.2. Check the equivalence of Eqs. (2.6) and (2.7).

Solution: Consider first the double integral

$$\frac{1}{2} \int_{t'}^{t} dt_1 \int_{t'}^{t} dt_2 \widehat{\mathcal{T}} \left\{ \widehat{H}_{t_1} \widehat{H}_{t_2} \right\} = \frac{1}{2} \int_{t'}^{t} dt_1 \left[\int_{t'}^{t_1} dt_2 \widehat{H}_{t_1} \widehat{H}_{t_2} + \int_{t_1}^{t} dt_2 \widehat{H}_{t_2} \widehat{H}_{t_1} \right],$$

where the integrals are taken over the upper and lower triangles of the quadrant $t' < t_{1,2} < t$. If we change the order of the integrations in the second term and permute the variables according to $t_1 \leftrightarrow t_2$, we find that the contribution

$$\frac{1}{2} \int_{t'}^{t} dt_2 \int_{t'}^{t_2} dt_1 \widehat{H}_{t_2} \widehat{H}_{t_1} = \frac{1}{2} \int_{t'}^{t} dt_1 \int_{t'}^{t_1} dt_2 \widehat{H}_{t_1} \widehat{H}_{t_1}$$

coincides with the contribution of the first term. Thus, the transition from Eq. (2.6) to Eq. (2.7) is checked for a double integral. A generalization to the case of n integrals can be done by induction.

1.3. Prove the Hermiticity, unitarity, and multiplicativity of the S-operator, expressed by Eq. (2.9).

Hint: Use Eq. (2.2).

1.4. Check the representation of the δ-function $\delta(E)$ as a limit, at $\lambda \to +0$, of the following expressions:

$$\delta_\lambda(E) = \frac{1}{\pi} \frac{\lambda}{E^2 + \lambda^2}, \;\; \delta_\lambda(E) = \frac{1}{\sqrt{\pi}\lambda} e^{-(E/\lambda)^2}, \;\; \delta_\lambda(E) = \frac{\lambda \sin^2(E/\lambda)}{\pi E^2}.$$

Also, check the relation

$$(E \pm i\lambda)^{-1} = \frac{\mathcal{P}}{E} \mp i\pi\delta(E), \;\; \lambda \to +0,$$

where $\mathcal{P}$ is the symbol of the principal value.

Solution: The δ-function must satisfy the requirement $\lim_{\lambda \to +0} \delta_\lambda(E) = 0$ at $E \neq 0$ and go to infinity at $E = 0$, which is checked directly. Another requirement, $\int \delta(E)dE = 1$, is provided by appropriate normalization coefficients in the different representations of $\delta_\lambda(E)$; the integrals over E can be calculated exactly. The properties of the δ-function lead us to the equation $\int_\Delta \delta(E)F(E)dE = F(0)$, where the region of integration, Δ, includes the point $E = 0$, and $F(E)$ is assumed to be continuous in the vicinity of $E = 0$. To check the last relation, we use the first expression for the δ-function and obtain

$$\lim_{\lambda \to +0} \int \frac{F(E)}{E \pm i\lambda} dE = \int \frac{F(E)E}{E^2 + \lambda^2} dE \mp i\pi \int F(E)\delta(E)dE,$$

where $\int F(E)E dE/(E^2 + \lambda^2)$ can be replaced by the principal value of the integral, $\mathcal{P} \int [F(E)/E]dE$.

1.5. Write the equations describing the modes of electromagnetic field in the medium with one-dimensional inhomogeneity defined by the dielectric function ϵ_z.

Solution: Let us write $\mathbf{A_r^\nu}$ as a superposition of plane waves with two-dimensional wave vectors $\mathbf{q}$ ($\mathbf{q} \perp OZ$ and $\mathbf{r} = (\mathbf{x}, z)$):

$$\mathbf{A_r^\nu} = \sqrt{\frac{2\pi c^2}{\epsilon V}} e^{i\mathbf{q} \cdot \mathbf{x}} \left(e_{\mathbf{q}z}^s \frac{[\mathbf{n}_z \times \mathbf{q}]}{q} + \mathbf{e}_{\mathbf{q}z}^p \right),$$

where $\mathbf{n}_z$ is the unit vector along OZ, and z-dependent unit vectors $e_{\mathbf{q}z}^s$ and $\mathbf{e}_{\mathbf{q}z}^p$ describe two polarizations, s and p. The case $\mathbf{A_r^\nu} \parallel [\mathbf{n}_z \times \mathbf{q}]$ corresponds to s-polarization, and $e_{\mathbf{q}z}^s$ satisfies the wave equation

$$\frac{d^2 e_{\mathbf{q}z}^s}{dz^2} + \left[\left(\frac{\omega}{c}\right)^2 \epsilon_z - q^2 \right] e_{\mathbf{q}z}^s = 0$$

obtained directly from Eq. (3.3). For *p*-polarization, the vector $\mathbf{A}_\mathbf{r}^\nu$ lies in the plane defined by the vectors $\mathbf{q}$ and $\mathbf{n}_z$ so that one can define two *z*-dependent components of $\mathbf{e}_{\mathbf{q}z}^p$ directed along these two vectors. These components satisfy the equations similar to the one given above. These equations contain both ϵ_z and its logarithmic derivative.

1.6. Assuming that $\bar{e}_z$ is an arbitrary complex solution of the wave equation derived in the problem 1.5, one may define the flow Q:

$$Q = \frac{1}{i}\left(\bar{e}_z^* \frac{d\bar{e}_z}{dz} - \bar{e}_z \frac{d\bar{e}_z^*}{dz}\right).$$

Prove that Q does not depend on z. Relate the behavior of the modes $\bar{e}_z$ to the sign of the flow.

Solution: Let us calculate a derivative of Q over z, which gives us a sum of the terms containing the products of the first derivatives of $\bar{e}_z$ and the second derivatives. The products of the first derivatives vanish from dQ/dz. The second derivatives can be expressed from the wave equation and also lead to zero contribution so that $dQ/dz = 0$. Depending on the sign of Q, one may separate three kinds of solutions: left (l), right (r), and local (L) modes. If $Q > 0$ (l-mode), the waves propagate from the left to the right: in the region $z \to -\infty$ there are both incident and reflected waves, while in the region $z \to +\infty$ there are only transmitted waves (the dielectric permittivity is assumed to be constant at $z \to \pm\infty$). When $Q < 0$ (r-mode), the wave propagates from the right to the left. The case of $Q = 0$ (L-mode) corresponds to a localized solution characterized by $\bar{e}_{z\to\pm\infty} = 0$.

1.7. Determine the energy density of equilibrium photons with frequency ω.

Solution: Using Eq. (3.28), one can express the energy density of equilibrium photons in three-dimensional media. The energy dE (per unit volume of space) corresponding to the phase volume $\Omega_\mathbf{q} = 4\pi q^2 dq/(2\pi)^3$ of the state $\mathbf{q}$ is given as $dE = 2\hbar\omega_q \bar{n}_q \Omega_\mathbf{q}$, where $\bar{n}_q$ is given by Eq. (3.28) and the factor of 2 stands because of two polarizations of photons. Using Eq. (3.23) with $\omega_q = \omega$, we obtain

$$\frac{dE}{d\omega} = \frac{\hbar\omega^3}{\pi^2 \tilde{c}^3}\left[e^{\hbar\omega/T_{ph}} - 1\right]^{-1}.$$

This result is known as Planck's formula.

1.8. Determine *i)* the total energy of equilibrium photons and *ii)* the frequency corresponding to the maximum energy density, as functions of the photon temperature T_{ph}.

Hints: *i)* Calculate the integral of $dE/d\omega$ over ω. *ii)* Calculate the derivative of $dE/d\omega$ over ω and equate it to zero.

1.9. Calculate the commutator $[\widehat{\overline{H}}_e, \hat{\rho}_\mathbf{r}]$ in Eq. (4.13).

Hint: Use the operator equation $[\hat{a}, \hat{b}\hat{c}] = [\hat{a}, \hat{b}]\hat{c} + \hat{b}[\hat{a}, \hat{c}]$.

1.10. Check Eqs. (4.23) and (4.24).

Hint: Compare results of straightforward calculations based on the wave functions (4.17) and on the second quantization formalism.

1.11. Check the commutation relation (4.31).

Hint: Using Eqs. (4.19)–(4.21), show that the terms containing products of four operators disappear.

1.12. Prove the relation $\sum_{\mathbf{p}} \ldots = V \int d\mathbf{p} \ldots / (2\pi\hbar)^3$ at $V \to \infty$.

Hint: Use the definition of quasidiscrete momentum from Eq. (5.1) and consider the integral as a limit of the sum.

1.13. Find the spectrum and velocity of the electron in crossed (perpendicular to each other) electric and magnetic fields.

Solution: If we direct the electric field $\mathbf{E}$ along OX, the potential energy standing in the Schroedinger equation is $-eEx$. The solution of the Schroedinger equation is given by Eq. (5.13), where the oscillatory wave function is determined by the equation

$$\left[\frac{\hat{p}_x^2}{2m} + \frac{m\omega_c^2}{2} \left(x - X_{p_y}^E \right)^2 - eEX_{p_y}^E + \frac{m}{2} \left(\frac{eE}{m\omega_c} \right)^2 \right] \varphi_x^{(Np_y)} = E_{Np_y} \varphi_x^{(Np_y)},$$

and $X_{p_y}^E = X_{p_y} + eE/m\omega_c^2$ determines the shift of the oscillator center in the presence of the electric field (compare to Eq. (5.14)). The wave functions $\varphi_x^{(Np_y)}$ are the same as in Eq. (5.15), where $X_{p_y}^E$ stands instead of X_{p_y}, and the spectrum is given by

$$E_{Np_y} = \varepsilon_N + \frac{m}{2} \left(\frac{eE}{m\omega_c} \right)^2 - eEX_{p_y}^E.$$

It depends on p_y because of the presence of the electric field. The velocity is given by the diagonal matrix elements of the kinematic velocity operator $\hat{\mathbf{v}} = [-i\hbar\partial/\partial\mathbf{r} - e\mathbf{A_r}/c]/m$. Owing to the electric field effect, the velocity is no longer zero in the direction perpendicular to both $\mathbf{E}$ and $\mathbf{H}$: $v_y = (p_y + \hbar X_{p_y}^E/l_H^2)/m = eE/m\omega_c$. This is the classical drift velocity $\mathbf{v} = c[\mathbf{E} \times \mathbf{H}]/H^2$ for the particle in crossed electric and magnetic fields. The spectrum E_{Np_y} within a single Landau level can be presented as a sum of classical kinetic and potential energies of the particle with coordinate $X_{p_y}^E$ (corresponding to the center of the oscillatory wave function) and velocity v_y.

1.14. Consider an electron in the harmonic potential $U \cos \mathbf{Q} \cdot \mathbf{r}$ and harmonic electric field $\mathbf{E} \cos \Omega t$. Write the Schroedinger equation in the momentum representation for Fourier components of the wave function.

Solution: In the $\mathbf{p}$-representation, the potential term in the Schroedinger equation is transformed to $U(\psi_{\mathbf{p}+\hbar\mathbf{Q}t} + \psi_{\mathbf{p}-\hbar\mathbf{Q}t})/2$, while the kinematic momentum is equal to $\mathbf{p} + (e\mathbf{E}/\Omega) \sin \Omega t$. Carrying out the temporal Fourier transformation $\psi_{\mathbf{p}\omega} =$

$\int dt e^{i\omega t}\psi_{\mathbf{p}t}$, we obtain

$$\left[\hbar\omega - \frac{p^2}{2m} - \frac{1}{4m}\left(\frac{eE}{\Omega}\right)^2\right]\psi_{\mathbf{p}\omega} = \frac{U}{2}[\psi_{\mathbf{p}+\hbar\mathbf{Q}\omega} + \psi_{\mathbf{p}-\hbar\mathbf{Q}\omega}]$$

$$+\frac{e\mathbf{E}\cdot\mathbf{p}}{2m\Omega i}[\psi_{\mathbf{p}\omega+\Omega} - \psi_{\mathbf{p}\omega-\Omega}] - \frac{1}{8m}\left(\frac{eE}{\Omega}\right)^2[\psi_{\mathbf{p}\omega+2\Omega} + \psi_{\mathbf{p}\omega-2\Omega}].$$

This is a finite-difference equation.

1.15. Derive the equations of motion (6.2).

<u>Solution</u>: According to general principles, if the energy of the system depends on the variables q_i $(i = 1, 2, ...)$ and their temporal and spatial derivatives, $\dot{q}_i$ and $\nabla_\alpha q_i$, one may introduce the Lagrangian $\mathcal{L} = \sum_i \dot{q}_i p_i - \mathcal{E}$, where $p_i = \partial\mathcal{E}/\partial\dot{q}_i$, and write the Lagrange equations

$$\frac{\partial\mathcal{L}}{\partial q_i} - \frac{\partial}{\partial t}\frac{\partial\mathcal{L}}{\partial\dot{q}_i} - \sum_\alpha \nabla_\alpha\frac{\partial\mathcal{L}}{\partial\nabla_\alpha q_i} = 0,$$

which are identified with the equations of motion. Applying this procedure to the system of atoms in the crystal, when the energy is given by the expression (6.1), and assuming $q_i = u_s^k(\mathbf{R_n})$ so that the index i includes k, s, and $\mathbf{n}$, we obtain Eq. (6.2).

1.16. Prove that

$$N^{-1}\sum_{\mathbf{n}} e^{i\mathbf{q}\cdot\mathbf{R_n}} = \sum_{\mathbf{m}} \delta_{\mathbf{q},\mathbf{G_m}} \equiv \Delta_{\mathbf{q},0},$$

where $\mathbf{G_m}$ is a reciprocal lattice vector.

<u>Solution</u>: Only when $\mathbf{q}\cdot\mathbf{R_n}$ is equal to $2\pi l$ (l is integer) for all $\mathbf{n}$, the expression $N^{-1}\sum_{\mathbf{n}} e^{i\mathbf{q}\cdot\mathbf{R_n}}$ is equal to 1. Otherwise, it is equal to 0. Since the reciprocal lattice vector is defined as $\mathbf{G_m}\cdot\mathbf{R_n} = 2\pi l$, the expression under consideration satisfies the properties of the Kronecker symbol $\delta_{\mathbf{q},\mathbf{G_m}}$ for all integer vectors $\mathbf{m}$.

Chapter 2

ELECTRON-IMPURITY SYSTEM

Non-equilibrium states of the systems weakly coupled to a thermostat are well investigated because these systems are easily driven from equilibrium by relatively weak external fields. On the other hand, weakly coupled systems are usually described by means of kinetic equations, and this approach is developed below and in Chapters 4−7. In this chapter, the kinetic equation is derived and analyzed for the simplest case of electrons in a weak potential of static inhomogeneities of the crystal, for example, the impurity potential. A discussion of the approximations used below is not presented in detail because this subject is widely covered in the literature. The main features of the kinetic phenomena in solids are concerned with the complex nature of quasiparticle dynamics and interactions. For this reason, the standard results obtained for plasma have a limited applicability in solids. The quantum kinetic equation is obtained below in the operator form, which is more convenient for the cases under consideration. Such an approach allows one to formulate the kinetic equation in the most general way so that in each concrete case one has just to choose a proper representation for calculation of the matrix elements of the density matrix. Of course, one has to analyze the validity conditions for the approach of kinetic equation in each concrete case.

7. Kinetic Equation for Weak Scattering

We start from the equation for one-electron density matrix $\hat{n}_t$ under the assumption of weak interaction between the electrons and randomly distributed identical impurities. The one-electron Hamiltonian of the system is written as

$$\hat{h}_t + U_{im}(\mathbf{r}), \qquad U_{im}(\mathbf{r}) = \sum_{\alpha} v(\mathbf{r} - \mathbf{R}_{\alpha}). \tag{1}$$

The general form of the one-electron Hamiltonian $\hat{h}_t$ (the transition from many-electron to one-electron description is discussed in Sec. 4) allows one to consider the systems in arbitrary external fields. In Eq. (1) we have introduced the coordinate $\mathbf{R}_\alpha$ of α-th impurity ($\alpha = 1, 2, ..., N_{im}$, where N_{im} is the number of impurities in the volume V), and $v(\mathbf{r})$ is the potential of a single impurity placed at the origin of the coordinate system. Thus, we rewrite Eq. (4.32) in the form

$$i\hbar \frac{\partial \hat{n}_t}{\partial t} = \left[\hat{h}_t + \sum_\alpha v(\mathbf{r} - \mathbf{R}_\alpha), \hat{n}_t \right], \qquad (2)$$

where the one-electron density matrix $\hat{n}_t$ depends on coordinates of all impurities. The physical interest, however, is focused on the averaged, with respect to the impurity distribution, characteristics of the system. For this reason, the observable value $\overline{Q}_t$ introduced by Eq. (4.28) must be averaged. Since the one-particle operator $\hat{q}$ of the observable (charge density, current density, etc.) does not depend on $\mathbf{R}_\alpha$, the expression for this observable is written as

$$\overline{Q}_t = \langle\langle \mathrm{sp}\ \hat{q}\hat{n}_t \rangle\rangle = \mathrm{sp}\hat{q}\hat{\rho}_t, \qquad (3)$$

where $\langle\langle \ldots \rangle\rangle$ defines the averaging over the ensemble of randomly distributed impurities and $\hat{\rho}_t = \langle\langle \hat{n}_t \rangle\rangle$ is the averaged density matrix which does not depend on the set of coordinates $\mathbf{R}_\alpha$. Therefore, to derive a kinetic equation, one should average Eq. (2) over the variables $\mathbf{R}_\alpha$. Under the assumption of weak electron-impurity interaction, this procedure leads to a closed equation for $\hat{\rho}_t$.

If the concentration of the impurities is low enough, one can neglect the correlations between their positions. This means that the averaging over each $\mathbf{R}_\alpha$ should be done independently, as

$$\langle\langle \ldots \rangle\rangle = V^{-N_{im}} \int d\mathbf{R}_1 \ldots \int d\mathbf{R}_{N_{im}} \ldots \quad . \qquad (4)$$

The procedure of averaging introduced by Eq. (4) includes a possibility for two (or more) impurity centers to coincide. Thus, to improve the method for high impurity concentrations, one should insert a correlation function of impurity positions under the integral. This function depends on $\mathbf{R}_1, \mathbf{R}_2, \ldots, \mathbf{R}_{N_{im}}$ and is equal to zero if two or more impurity coordinates coincide. Averaging Eq. (2) with the use of Eq. (4), we encounter a new average taken over all impurities except the impurity α:

$$\langle\langle \ldots \rangle\rangle_\alpha = V^{-N_{im}+1} \int d\mathbf{R}_1 \ldots \int d\mathbf{R}_{\alpha-1} \int d\mathbf{R}_{\alpha+1} \ldots \int d\mathbf{R}_{N_{im}} \ldots \quad .$$

$$(5)$$

Again, applying this averaging to Eq. (2), we obtain the average over all impurities except those numbered by α and β. It is defined in a similar way and denoted as $\langle\langle\ldots\rangle\rangle_{\alpha\beta}$ ($\alpha \neq \beta$). One can continue in this way by introducing new averages of higher order. The quantities averaged over the impurity ensemble are not equal to their exact values, but the difference goes to zero at $N_{im} \to \infty$ (problem 2.1).

Equation (2), averaged with the use of Eq. (4), becomes

$$i\hbar\frac{\partial\hat{\rho}_t}{\partial t} = \left[\hat{h}_t, \hat{\rho}_t\right] + \sum_\alpha \int \frac{d\mathbf{R}_\alpha}{V} \left[v(\mathbf{r} - \mathbf{R}_\alpha), \langle\langle\hat{n}_t\rangle\rangle_\alpha\right], \tag{6}$$

where $\hat{\rho}_t$ depends on the coordinate of one electron. The operator $\langle\langle\hat{n}_t\rangle\rangle_\alpha$ appearing on the right-hand side of Eq. (6) describes the correlation of the electron with α-th impurity. The equation for $\langle\langle\hat{n}_t\rangle\rangle_\alpha$ is obtained after averaging Eq. (2) with the aid of Eq. (5):

$$i\hbar\frac{\partial\langle\langle\hat{n}_t\rangle\rangle_\alpha}{\partial t} = \left[\hat{h}_t + v(\mathbf{r} - \mathbf{R}_\alpha), \langle\langle\hat{n}_t\rangle\rangle_\alpha\right]$$
$$+ \sum_{\beta(\neq\alpha)} \int \frac{d\mathbf{R}_\beta}{V} \left[v(\mathbf{r} - \mathbf{R}_\beta), \langle\langle\hat{n}_t\rangle\rangle_{\alpha\beta}\right]. \tag{7}$$

The operator $\langle\langle\hat{n}_t\rangle\rangle_{\alpha\beta}$ describes the correlation of the electron with the impurities α and β. In turn, the equation for this operator contains the correlation function with three impurity variables. Proceeding in this way, one has an infinite chain of equations, which should be cut under the assumption of weak electron-impurity interaction (Born approximation) as described below.

Defining a correlation operator accounting for the α-th impurity contribution as

$$\hat{\kappa}_{\alpha t} = \langle\langle\hat{n}_t\rangle\rangle_\alpha - \hat{\rho}_t, \tag{8}$$

and noting that $\int d\mathbf{R}_\alpha v(\mathbf{r} - \mathbf{R}_\alpha)$ commutes with $\hat{\rho}_t$, one can transform the last term on the right-hand side of Eq. (6) in the following way:

$$\sum_\alpha \int \frac{d\mathbf{R}_\alpha}{V} \left[v(\mathbf{r} - \mathbf{R}_\alpha), \hat{\kappa}_{\alpha t}\right]. \tag{9}$$

Since $\hat{\kappa}_{\alpha t}$ is equal to zero when the impurity potential is absent, the contribution (9) is proportional to v^2. With this accuracy, one may neglect the last term in Eq. (7). Then, subtracting Eq. (6) from Eq. (7), we obtain an equation for $\hat{\kappa}_{\alpha t}$. In the first order in the electron-impurity interaction, this equation has the following form:

$$i\hbar\frac{\partial\hat{\kappa}_{\alpha t}}{\partial t} = \left[\hat{h}_t, \hat{\kappa}_{\alpha t}\right] + \left[v(\mathbf{r} - \mathbf{R}_\alpha), \hat{\rho}_t\right]. \tag{10}$$

The initial condition to Eq. (10) corresponds to the weakening of the electron-impurity correlations at $t \to -\infty$:

$$\hat{\kappa}_{\alpha t \to -\infty} = 0. \tag{11}$$

Using this initial condition, one can imagine that the impurity potential is adiabatically turned on at $t \to -\infty$.

Therefore, Eq. (6) with the last term given by Eq. (9) and Eq. (10) with the initial condition (11) form a closed system of equations. To solve it, we first exclude the correlation operator $\hat{\kappa}_{\alpha t}$. Expressing the solution of Eq. (10) with the initial condition (11) through the evolution operator defined by Eq. (2.2) with the Hamiltonian $\hat{h}_t$, we obtain (compare to Eq. (2.19))

$$\hat{\kappa}_{\alpha t} = \frac{1}{i\hbar} \int_{-\infty}^{t} dt' e^{\lambda t'} \widehat{S}(t,t') \left[v(\mathbf{r} - \mathbf{R}_\alpha), \hat{\rho}_{t'} \right] \widehat{S}^+(t,t'), \tag{12}$$

where $\lambda \to +0$. Substituting this result into Eq. (6) with the use of Eq. (9), we obtain the quantum kinetic equation

$$\frac{\partial \hat{\rho}_t}{\partial t} + \frac{i}{\hbar} \left[\hat{h}_t, \hat{\rho}_t \right] = \widehat{J}_{im}(\hat{\rho}|t) \tag{13}$$

for the one-electron density matrix averaged over the impurity ensemble. The left-hand side describes the evolution of the electron distribution in the absence of electron-impurity interactions, while the right-hand side gives us the electron-impurity collision integral in the operator form. The explicit expression for $\widehat{J}_{im}(\hat{\rho}|t)$ is written through the double commutator,

$$\widehat{J}_{im}(\hat{\rho}|t) = \frac{1}{\hbar^2} \int_{-\infty}^{t} dt' e^{\lambda t'} \sum_{\alpha} \int \frac{d\mathbf{R}_\alpha}{V}$$

$$\times \left[\widehat{S}(t,t') \left[v(\mathbf{r} - \mathbf{R}_\alpha), \hat{\rho}_{t'} \right] \widehat{S}^+(t,t'), v(\mathbf{r} - \mathbf{R}_\alpha) \right], \tag{14}$$

and the averaging over the impurity distribution is directly given by the integral over $\mathbf{R}_\alpha$. To calculate it, we transform the impurity potential into Fourier series according to standard relations

$$v(\mathbf{r}) = \frac{1}{V} \sum_{\mathbf{q}} e^{i\mathbf{q}\cdot\mathbf{r}} v(\mathbf{q}), \quad v(\mathbf{q}) = \int d\mathbf{r} e^{-i\mathbf{q}\cdot\mathbf{r}} v(\mathbf{r}). \tag{15}$$

Then, the impurity-dependent contribution in Eq. (14) is calculated as

$$v(\mathbf{q})v(\mathbf{q}') \sum_{\alpha} \int \frac{d\mathbf{R}_\alpha}{V} e^{i(\mathbf{q}\cdot\mathbf{R}_\alpha + \mathbf{q}'\cdot\mathbf{R}_\alpha)} = |v(\mathbf{q})|^2 N_{im} \delta_{\mathbf{q}+\mathbf{q}',0}, \tag{16}$$

where the identity $v(-\mathbf{q}) = v^*(\mathbf{q})$ is applied since $v(\mathbf{r})$ is real. Finally, the collision integral takes the following form:

$$\widehat{J}_{im}(\hat{\rho}|t) = \frac{n_{im}}{\hbar^2 V} \sum_{\mathbf{q}} |v(\mathbf{q})|^2 \int_{-\infty}^{t} dt' e^{\lambda t'}$$

$$\times \left[\widehat{S}(t,t') \left[e^{i\mathbf{q}\cdot\mathbf{r}}, \hat{\rho}_{t'} \right] \widehat{S}^+(t,t'), e^{-i\mathbf{q}\cdot\mathbf{r}} \right], \qquad (17)$$

where $n_{im} = N_{im}/V$ is the impurity concentration, and the sum is taken over the momentum $\hbar\mathbf{q}$ transmitted in the collisions.

The limiting transitions employed in the derivation of Eq. (17) are carried out in a standard, thermodynamic sequence

$$V \to \infty, \qquad N_{im}/V = const \qquad (I)$$

$$\lambda \to +0 \qquad (II). \qquad (18)$$

First (I) the volume goes to infinity at a fixed impurity concentration, and then (II) the interaction is adiabatically turned on. The physical reasons for this sequence are clear. First, to ensure that the electrons do not feel the boundaries, the volume is set at infinity, and then the moment of time when the interaction is turned on can be shifted to $-\infty$. The kinetic equation derived above is non-Markovian, i.e., the density matrix $\hat{\rho}_t$ is determined by the preceding evolution of the system. The main contribution to the integral over t in Eq. (17) comes from the time τ_c determined by the characteristic period of oscillations of the S-matrix. It can be estimated from Eqs. (2.2) and (2.3) as $\tau_c \sim \hbar/\bar{\varepsilon}$, where $\bar{\varepsilon}$ is the characteristic energy of electrons. If the evolution of electron distribution is slower than the quantum oscillations characterized by τ_c, one can neglect the non-Markovian contributions. This means that $\hat{\rho}_{t'}$ in the collision integral can be replaced by $\hat{\rho}_t$; see Sec. 8. The fast processes must be considered more accurately; see Sec. 10.

The kinetic equation should be accompanied by a normalization condition expressing the electron density conservation. This conservation is proved by calculating the trace of Eq. (13) whose left-hand side contains the commutator $[\hat{h}_t, \hat{\rho}_t]$ and the right-hand side (the collision integral) also contains a commutator; see Eq. (17). Since the trace of any commutator is equal to zero, one has

$$\frac{\partial}{\partial t}\text{sp}\,\hat{\rho}_t = 0, \qquad \frac{1}{V}\text{sp}\,\hat{\rho}_t = n. \qquad (19)$$

The second equation defines the electron density n. In some cases, for example, for the system described by a many-band Hamiltonian, where the valence band contains an infinite number of electrons, the normalization condition appears to be more complicated. The same procedure is applied to prove the energy conservation condition for the systems with time-independent Hamiltonian $\hat{h}$. After multiplying Eq. (13) by $\hat{h}$, we calculate the trace and find that both the collisionless contribution coming from $[\hat{h}, \hat{\rho}_t]$ and the contribution coming from the elastic-scattering collision integral do not change the energy (problem 2.2). Therefore,

$$\frac{\partial}{\partial t}\mathrm{sp}\ \hat{h}\hat{\rho}_t = 0, \qquad (20)$$

which means that sp $\hat{h}\hat{\rho}_t = const$ and the total energy of the electron system is conserved.

Since the observables introduced by Eq. (3) are real, the averaged density matrix must be Hermitian, $\hat{\rho}_t^+ = \hat{\rho}_t$. From Eqs. (13) and (17), it follows directly that $\hat{\rho}_t^+$ and $\hat{\rho}_t$ are governed by the same equation. In contrast to the classical Boltzmann equation, Eq. (13) exactly accounts for the quantum nature of the electron dynamics in external fields and even for the influence of the external fields on the scattering, since Eq. (17) contains S-operators defined by Eq. (2.2). However, like the Boltzmann equation, the quantum kinetic equation (13) is derived under the following assumptions: *i)* the approximation of weak interaction, which allows one to cut the chain of equations in the second order in v, and *ii)* the initial condition (11) for weakening of correlations, which enables one to solve Eq. (10) and obtain a closed equation for $\hat{\rho}_t$. These assumptions are valid when

$$\bar{\varepsilon} \gg \hbar/\bar{\tau}, \qquad (21)$$

i.e., when the characteristic relaxation time $\bar{\tau}$, which gives an estimate for the collision integral according to $\widehat{J}_{im}(\hat{\rho}|t) \sim -\hat{\rho}_t/\bar{\tau}$, is much greater than the time τ_c characterizing the period of oscillations of the S-matrix. Condition (21) corresponds to a weak collision-induced broadening of the electron states with energy $\bar{\varepsilon}$. It justifies the validity of the Born approximation for electron-impurity scattering. Generally, it is the main condition for applicability of the kinetic approach based on a reduction of various interactions of quasiparticles with their surrounding to the form of collision integrals. The kinetic equation (13) can be derived for the case of strong electron-impurity interaction when this interaction is a short-range one, so the condition (21) is still valid. The general case of strong interaction, when Eq. (21) is violated, is considered by using the diagram technique for linear response (Chapter 3) and non-equilibrium diagram technique (Chapter 8).

The derivation of the kinetic equation (13) can be generalized to the case of an arbitrary random static potential $U_{sc}(\mathbf{r})$. The averaging should be done over all realizations of the random potential, and this procedure is again denoted as $\langle\langle\ldots\rangle\rangle$. If the random potential obeys the Gaussian statistics, the results of the averaging are expressed through the binary correlation function $\langle\langle U_{sc}(\mathbf{r})U_{sc}(\mathbf{r}')\rangle\rangle = w(|\mathbf{r} - \mathbf{r}'|)$ (problem 2.3), which corresponds to a macroscopically homogeneous and isotropic case. The averaged potential energy $\langle\langle U_{sc}(\mathbf{r})\rangle\rangle$ can be set at zero, as a reference point of energy. The correlation functions of the Fourier-transformed potentials satisfy the following relation:

$$\langle\langle U_{sc}(\mathbf{q})U_{sc}(\mathbf{q}')\rangle\rangle = V\delta_{\mathbf{q},-\mathbf{q}'}w(q), \tag{22}$$

where $w(q)$ is the Fourier transform of the correlation function $w(|\mathbf{r}|)$. Although the potential $\sum_\alpha v(\mathbf{r} - \mathbf{R}_\alpha)$ of randomly distributed impurities does not satisfy the Gaussian statistics in the general case, only the pair correlation functions appear in the Born approximation, and we have ($\Delta\mathbf{r} = \mathbf{r} - \mathbf{r}'$)

$$w(q) = \int d\Delta\mathbf{r} e^{-i\mathbf{q}\cdot\Delta\mathbf{r}}\langle\langle\sum_{\alpha\beta} v(\mathbf{r} - \mathbf{R}_\alpha)v(\mathbf{r}' - \mathbf{R}_\beta)\rangle\rangle = n_{im}|v(\mathbf{q})|^2 . \tag{23}$$

Accordingly, the collision integral for the general case of an arbitrary random static potential $U_{sc}(\mathbf{r})$ is given by Eq. (17), where one should replace $n_{im}|v(\mathbf{q})|^2$ by $w(q)$. Similar as above, this collision integral is obtained under the assumption of weakness of the potential $U_{sc}(\mathbf{r})$.

8. Relaxation Rates and Conductivity

In this section we analyze the kinetic equation (7.13) for a relatively simple case, when the system is spatially homogeneous. The collision integral (7.17) can be applied to calculate the times describing the relaxation of initially anisotropic, with respect to momenta, electron distribution due to electron-impurity scattering. The electron energies are conserved in such scattering processes. In the absence of external fields, we use the Hamiltonian

$$\hat{h} = \frac{\hat{p}^2}{2m} , \qquad \hat{\mathbf{p}} = -i\hbar\frac{\partial}{\partial\mathbf{x}} , \tag{1}$$

which should be substituted to the operator kinetic equation (7.13). We write this equation in the momentum representation, by using the solutions of the eigenstate problem $\hat{h}|\mathbf{p}\rangle = \varepsilon_p|\mathbf{p}\rangle$ (see Sec. 5) and taking into account the translational invariance of the system. The Hamiltonian (1) commutes with the operator of translation, $\widehat{T}_{\mathbf{R}} = \exp(i\hat{\mathbf{p}}\cdot\mathbf{R}/\hbar)$, where $\mathbf{R}$ is an arbitrary vector. Therefore, both $\widehat{T}_{\mathbf{R}}\hat{\rho}_t\widehat{T}_{\mathbf{R}}^+$ and $\hat{\rho}_t$ satisfy the same equation. To check this statement, we act on Eq. (7.13) by $\widehat{T}_{\mathbf{R}}$

from the left and by $\widehat{T}_{\mathbf{R}}^{+}$ from the right (note that a shift of the phase by $\pm\mathbf{p}\cdot\mathbf{R}/\hbar$ in the factors $\exp(\pm i\mathbf{q}\cdot\mathbf{x})$ does not lead to a dependence of the collision integral on $\mathbf{R}$). If the density matrices $\hat{\rho}_t$ and $\widehat{T}_{\mathbf{R}}\hat{\rho}_t\widehat{T}_{\mathbf{R}}^{+}$ coincide at the initial moment of time $t = t_0$, one has to consider only the diagonal, with respect to $\mathbf{p}$, components of the density matrix, because the non-diagonal ones are equal to zero (problem 2.4).

As a result, the kinetic equation for the distribution function $f_{\mathbf{p}t} = \langle\mathbf{p}|\hat{\rho}_t|\mathbf{p}\rangle$ takes the form

$$\frac{\partial f_{\mathbf{p}t}}{\partial t} = J_{im}(f|\mathbf{p}t) \equiv \langle\mathbf{p}|\widehat{J}_{im}(\hat{\rho}|t)|\mathbf{p}\rangle. \qquad (2)$$

Since the Hamiltonian $\hat{h}$ is time-independent, the collision integral operator (7.17) is transformed to

$$\widehat{J}_{im}(\hat{\rho}|t) = \frac{n_{im}}{\hbar^2 V} \sum_{\mathbf{q}} |v(\mathbf{q})|^2 \int_{-\infty}^{0} d\tau e^{\lambda\tau}$$

$$\times \left[e^{i\hat{h}\tau/\hbar} \left[e^{i\mathbf{q}\cdot\mathbf{x}}, \hat{\rho}_{t+\tau} \right] e^{-i\hat{h}\tau/\hbar}, e^{-i\mathbf{q}\cdot\mathbf{x}} \right], \qquad (3)$$

where the S-operator is expressed according to Eq. (2.3). In the momentum representation, the collision integral (3) is rewritten as

$$J_{im}(f|\mathbf{p}t) = \frac{n_{im}}{\hbar^2 V} \sum_{\mathbf{p}'} |v[(\mathbf{p}-\mathbf{p}')/\hbar]|^2 \int_{-\infty}^{0} d\tau e^{\lambda\tau}$$

$$\times \left\{ \exp[i(\varepsilon_p - \varepsilon_{p'})\tau/\hbar] + c.c. \right\} (f_{\mathbf{p}'t+\tau} - f_{\mathbf{p}t+\tau}). \qquad (4)$$

To obtain this equation, one should consider the matrix element

$$\langle\mathbf{p}|e^{i\mathbf{q}\cdot\mathbf{x}}|\mathbf{p}'\rangle = \delta_{\mathbf{p},\mathbf{p}'+\hbar\mathbf{q}} \qquad (5)$$

and take the sum over the transferred momentum $\hbar\mathbf{q}$ expressed through $\mathbf{p}'$. As a result, the operator equation (7.13) is rewritten in the $\mathbf{p}$-representation as an integro-differential equation.

Below we use Eq. (7.21) and assume a locality of the collision integral with respect to time. Indeed, $f_{\mathbf{p}t}$ changes with time on the scale of the order of characteristic relaxation time $\bar{\tau}$ (the latter is calculated below), while the exponential factors in Eq. (3) oscillate with the characteristic time $\tau_c \sim \hbar/\bar{\varepsilon}$. Therefore, the non-Markovian nature of the collision integral is not essential under the conditions (7.21), and $f_{\mathbf{p}t+\tau} \simeq f_{\mathbf{p}t}$. In this approximation, the integral over τ in Eq. (4) gives us the energy conservation law (problem 2.5):

$$\frac{1}{\hbar} \int_{-\infty}^{0} d\tau e^{\lambda\tau} \left\{ \exp[i(\varepsilon_p - \varepsilon_{p'})\tau/\hbar] + c.c. \right\} = 2\pi\delta(\varepsilon_p - \varepsilon_{p'}). \qquad (6)$$

As a result of the transformations made, we obtain a simple equation describing the balance of occupation for the state with momentum $\mathbf{p}$:

$$\frac{\partial f_{\mathbf{p}t}}{\partial t} = \frac{2\pi}{\hbar} \frac{n_{im}}{V} \sum_{\mathbf{p}'} |v(|\mathbf{p} - \mathbf{p}'|/\hbar)|^2 \delta(\varepsilon_p - \varepsilon_{p'})(f_{\mathbf{p}'t} - f_{\mathbf{p}t})$$

$$\equiv \sum_{\mathbf{p}'} [W_{\mathbf{p}'\mathbf{p}} f_{\mathbf{p}'t} - W_{\mathbf{p}\mathbf{p}'} f_{\mathbf{p}t}] = \sum_{\mathbf{p}'} W_{\mathbf{p}\mathbf{p}'}(f_{\mathbf{p}'t} - f_{\mathbf{p}t}). \quad (7)$$

Thus, we have derived the explicit expression for the transition probability $W_{\mathbf{p}\mathbf{p}'}$ already introduced in Sec. 2:

$$W_{\mathbf{p}\mathbf{p}'} = \frac{2\pi}{\hbar} \frac{n_{im}}{V} |v(|\mathbf{p} - \mathbf{p}'|/\hbar)|^2 \delta(\varepsilon_p - \varepsilon_{p'}). \quad (8)$$

This expression is symmetric with respect to the permutation $\mathbf{p} \leftrightarrow \mathbf{p}'$. Equations (7) and (8) allow one to determine $\bar{\tau}$ and, therefore, the condition (7.21) takes an explicit form. Equation (7) does not contain any sources responsible for the anisotropy of the distribution in the $\mathbf{p}$-space. On the other hand, any stationary isotropic distribution $f(\varepsilon_p)$ satisfies Eq. (7) because of the elasticity property expressed by the δ-function of energies in the collision integral. If an anisotropic initial distribution $f_{\mathbf{p}t=0} = \bar{f}_{\mathbf{p}}$ is created by external perturbations, Eq. (7) can be employed to describe the relaxation of such distribution to the isotropic one.

Although the kinetic equation (7) is written above for 3D electrons, the same result is obtained for low-dimensional (2D or 1D) states if only one (the lowest) subband is considered; see the energy spectra (5.25). Since the condition (7.21) is assumed to be valid, the localization of low-dimensional electrons should be neglected; see the discussion after Eq. (21) below. Consider first the case of 1D-electrons, when the symmetric (s) and antisymmetric (a) parts of the distribution function can be separated as

$$f_{pt} = f_{pt}^s + f_{pt}^a, \quad f_{-pt}^s = f_{pt}^s, \quad f_{-pt}^a = -f_{pt}^a. \quad (9)$$

Note that f_{pt}^s does not depend on time (since the scattering is elastic) and already satisfies the kinetic equation (7), while f_{pt}^a should be found from this equation. Since the energy is conserved, only the backscattering processes with $p' = -p$ contribute to the collision integral. The factor $(f_{p't}^a - f_{pt}^a)$ is replaced by $-2f_{pt}^a$. As a result, the evolution of f_{pt}^a is described by the following equation:

$$\frac{\partial f_{pt}^a}{\partial t} = -\nu_p f_{pt}^a, \quad \nu_p = 2\sum_{p'} W_{pp'}|_{p \neq p'} = \frac{\pi n_{im}}{\hbar} |v(2p/\hbar)|^2 \rho_{1D}(\varepsilon_p), \quad (10)$$

where the density of states for one-dimensional electrons is given by Eq. (5.26). The relaxation rate of 1D momentum, denoted here as ν_p, is equal to the inverse time of departure from the state p.

In the 2D case, we introduce the polar angle ϕ and expand $f_{\mathbf{p}t}$ into Fourier series:

$$f_{\mathbf{p}t} = f_{\varepsilon t}^{(0)} + \sum_{l=1}^{\infty} \left[f_{\varepsilon t}^{(l+)} \cos l\phi + f_{\varepsilon t}^{(l-)} \sin l\phi \right]. \tag{11}$$

Once again, the isotropic contribution $f_{\varepsilon t}^{(0)}$ is not changed by the elastic scattering, while the coefficients $f_{\varepsilon t}^{(l\pm)}$ satisfy the following equations:

$$\frac{\partial}{\partial t} \left| \begin{array}{c} f_{\varepsilon t}^{(l+)} \\ f_{\varepsilon t}^{(l-)} \end{array} \right| = \frac{1}{\pi} \sum_{l'} \int_{-\pi}^{\pi} d\phi \left| \begin{array}{c} \cos l\phi \\ \sin l\phi \end{array} \right| \sum_{\mathbf{p}'} W_{\mathbf{pp}'}$$

$$\times \left[(\cos l'\phi' - \cos l'\phi) f_{\varepsilon t}^{(l'+)} + (\sin l'\phi' - \sin l'\phi) f_{\varepsilon t}^{(l'-)} \right]. \tag{12}$$

According to Eq. (8), the probability $W_{\mathbf{pp}'}$ depends on the energies ε and ε' and on the cosine of the angle between the momenta. This angle is given as $\phi'' = \phi - \phi' = \widehat{\mathbf{pp}'}$. Substituting $\phi' = \phi - \phi''$ into Eq. (12), we calculate the integrals over ϕ and rewrite the right-hand side of this equation as

$$\sum_{\mathbf{p}'} W_{\varepsilon\varepsilon'}(\cos \phi'') \left| \begin{array}{c} (\cos l\phi'' - 1) f_{\varepsilon t}^{(l+)} - \sin l\phi'' f_{\varepsilon t}^{(l-)} \\ (\cos l\phi'' - 1) f_{\varepsilon t}^{(l-)} + \sin l\phi'' f_{\varepsilon t}^{(l+)} \end{array} \right|. \tag{13}$$

The terms proportional to $\sin l\phi''$ give zero contribution, and Eq. (12) takes the form

$$\frac{\partial f_{\varepsilon t}^{(l\pm)}}{\partial t} = -\nu_\varepsilon^{(l)} f_{\varepsilon t}^{(l\pm)}, \quad \nu_\varepsilon^{(l)} = \sum_{\mathbf{p}'} W_{\mathbf{pp}'} \left[1 - \cos(l\widehat{\mathbf{pp}'}) \right]. \tag{14}$$

Therefore, the evolution of $f_{\varepsilon t}^{(l\pm)}$ is described by the l-dependent relaxation rate $\nu_\varepsilon^{(l)}$.

To describe the evolution of 3D electrons, it is convenient to expand $f_{\mathbf{p}t}$ with the aid of a complete set of spherical harmonics Y_{lk} according to

$$f_{\mathbf{p}t} = \sum_{lk} f_{\varepsilon t}^{(lk)} Y_{lk}(\theta, \varphi), \quad |k| \leq l, \quad l = 0, 1, \ldots, \tag{15}$$

where θ and φ are the angles of the spherical coordinate system. The equation for the coefficients $f_{\varepsilon t}^{(lk)}$, which describes the evolution of different components of the anisotropic distribution, is obtained by using the orthogonality and normalization condition for the spherical functions.

First we have

$$\frac{\partial f_{\varepsilon t}^{(lk)}}{\partial t} = \sum_{l'k'} \int d\Omega Y_{lk}^*(\theta, \varphi) \sum_{\mathbf{p}'} W_{\mathbf{pp}'}$$

$$\times \left[Y_{l'k'}(\theta', \varphi') - Y_{l'k'}(\theta, \varphi) \right] f_{\varepsilon t}^{(l'k')}, \tag{16}$$

where $d\Omega = \sin\theta d\theta d\varphi$ is the differential of the solid angle. The range of the angles is given by $\theta \in [0, \pi]$ and $\varphi \in [0, 2\pi]$. The angular integrals in the first term of the right-hand side of Eq. (16) can be calculated in the coordinate system where $\mathbf{p}$ is directed along OZ. The spherical function $Y_{l'k'}(\theta', \varphi')$ is written in this new coordinate system as a linear combination of $Y_{l'k_1}(\theta'', \varphi'')$ with the same l, because the Laplace operator generating the set of the spherical functions with a given l is not modified by the rotation under consideration:

$$Y_{l'k'}(\theta', \varphi') = \sum_{k_1=-l'}^{l'} \mathcal{D}_{k'k_1}^{l'}(\theta, \varphi) Y_{l'k_1}(\theta'', \varphi''). \tag{17}$$

The matrix $\mathcal{D}_{k'k_1}^{l'}$ describes the rotation and depends on the angles θ and φ defining the orientation of the vector $\mathbf{p}$ with respect to OZ axis. Substituting this expansion into the right-hand side of Eq. (16), we integrate this equation over φ'' and find that only the term with $k_1 = 0$ remains and the right-hand side is rewritten as

$$\sum_{l'k'} \int d\Omega Y_{lk}^*(\theta, \varphi) \sum_{\mathbf{p}'} W_{\varepsilon\varepsilon'}(\cos\theta'')$$

$$\times \left[\mathcal{D}_{k'0}^{l'}(\theta, \varphi) Y_{l'0}(\theta'', \varphi'') - Y_{l'k'}(\theta, \varphi) \right] f_{\varepsilon t}^{(l'k')}. \tag{18}$$

From the theory of spherical functions, it is known that $\mathcal{D}_{k'0}^{l'}$ is expressed as

$$\mathcal{D}_{k'0}^{l'}(\theta, \varphi) = \sqrt{\frac{4\pi}{2l'+1}} Y_{l'k'}(\theta, \varphi). \tag{19}$$

Finally, employing the equation $Y_{l'0}(\theta, \varphi) = \sqrt{(2l'+1)/(4\pi)} P_{l'}(\cos\theta)$, where P_l are the Legendre polynomials, and calculating the angular integrals, we obtain

$$\frac{\partial f_{\varepsilon t}^{(lk)}}{\partial t} = -\nu_{\varepsilon}^{(l)} f_{\varepsilon t}^{(lk)}, \quad \nu_{\varepsilon}^{(l)} = \sum_{\mathbf{p}'} W_{\mathbf{pp}'} \left[1 - P_l(\cos\widehat{\mathbf{pp}'}) \right], \tag{20}$$

where the momentum relaxation rates $\nu_{\varepsilon}^{(l)}$ do not depend on k. Equations (10), (14), and (20) describe exponential damping of the initially anisotropic distributions. The relaxation rates for the 2D and 3D cases depend on the numbers l.

A further simplification of the relaxation rates can be done for the case of scattering by the impurities with short-range potential (point defects), when $v(\mathbf{r} - \mathbf{R}_\alpha) \simeq v\delta(\mathbf{r} - \mathbf{R}_\alpha)$. This is the limit when $|v(q)|^2$ standing in the expression for the transition probability $W_{\mathbf{p}'\mathbf{p}}$ does not depend on q and can be replaced by $|v(0)|^2 = v^2$. Applying this substitution, we find that the arrival terms, containing $\cos \widehat{\mathbf{p}\mathbf{p}}'$ in the expressions of $\nu_\varepsilon^{(l)}$ for 2D and 3D problems, vanish after the integration over $\mathbf{p}'$. As a result, the relaxation rates do not depend on the number of harmonic and are given by the expression

$$\nu_\varepsilon = \sum_{\mathbf{p}'} W_{\mathbf{p}\mathbf{p}'} = \frac{\pi n_{im}}{\hbar} |v(0)|^2 \rho_{\mathcal{D}}(\varepsilon_p), \quad \mathcal{D} = 1D, 2D, 3D, \qquad (21)$$

unifying 1D, 2D, and 3D cases. Note that Eq. (21) for the 1D case directly follows from Eq. (10), since $v(2p/\hbar) = v(0)$ in the limit under consideration. The relaxation rates are proportional to the densities of states given by Eqs. (5.3) and (5.26) and essentially depend on the dimensionality. As the latter is reduced, the validity condition (7.21) becomes more rigid for low-energy electrons. Because of the inverse-square-root divergence of $\rho_{1D}(\varepsilon_p)$ at $\varepsilon_p \to 0$, the kinetic description of low-energy 1D electrons is not valid even if the impurity potential is weak. A more careful consideration demonstrates the role of quantum interference and the appearance of weak-localization effects in the 2D case (see Secs. 15 and 43), while non-interacting 1D electrons are found to be localized even in an infinitely weak random potential (Sec. 59).

Below we calculate the frequency dispersion of the conductivity by adding the potential energy $-e\mathbf{E}_t \cdot \mathbf{x}$, which describes the interaction of electrons with a homogeneous, time-dependent electric field $\mathbf{E}_t$, to the Hamiltonian (1). This contribution does not break the translational invariance of the problem, because an additional contribution $-e\mathbf{E}_t \cdot \mathbf{R}$ appearing in the Hamiltonian after a translation by the vector $\mathbf{R}$ is dropped out of the kinetic equation. However, in the presence of the field, the commutator $[\hat{h}_t, \hat{\rho}_t]$ is no longer equal to zero (problem 2.6) and, instead of Eq. (2), we have the following spatially-homogeneous kinetic equation:

$$\left(\frac{\partial}{\partial t} + e\mathbf{E}_t \cdot \frac{\partial}{\partial \mathbf{p}} \right) f_{\mathbf{p}t} = J_{im}(f|\mathbf{p}t). \qquad (22)$$

The field-induced current density is written as $\mathbf{I}_t = (2e/V) \sum_{\mathbf{p}} \mathbf{v}_{\mathbf{p}} f_{\mathbf{p}t}$, where $\mathbf{v}_{\mathbf{p}} = \mathbf{p}/m$ is the velocity and the factor of 2 accounts for the spin degeneracy. We consider weak electric fields and calculate a linear response to the Fourier component $\mathbf{E}\exp(-i\omega t)$ of the field. The frequency ω is assumed to be small in comparison to $\bar{\varepsilon}/\hbar$ so that the collision integral $J_{im}(f|\mathbf{p}t)$ is again considered in the Markovian approximation, with the use of Eq. (6). The distribution function is represented as a sum of the equilibrium distribution $f_\varepsilon^{(eq)}$ (here $\varepsilon = \varepsilon_p = p^2/2m$) and small non-equilibrium correction, according to

$$f_{\mathbf{p}t} \simeq f_\varepsilon^{(eq)} + [\Delta f_{\mathbf{p}}\exp(-i\omega t) + c.c.]. \qquad (23)$$

The correction is found from the non-homogeneous algebraic equation

$$(-i\omega + \nu_\varepsilon)\,\Delta f_{\mathbf{p}} = -e\mathbf{E}\cdot\frac{\partial f_\varepsilon^{(eq)}}{\partial \mathbf{p}} = \frac{e\mathbf{E}\cdot\mathbf{p}}{m}\left(-\frac{df_\varepsilon^{(eq)}}{d\varepsilon}\right), \qquad (24)$$

where $\nu_\varepsilon = \nu_\varepsilon^{(1)}$ and the right-hand side is determined by the field $\mathbf{E}$. Equation (24) is a simple example of a linearized kinetic equation. When transforming Eq. (22) to Eq. (24), one should take into account that $\Delta f_{-\mathbf{p}} = -\Delta f_{\mathbf{p}}$ and solve the linearized, with respect to $\Delta f_{\mathbf{p}}$, equation by assuming $\Delta f_{\mathbf{p}} \propto \cos\widehat{\mathbf{Ep}}$. Equivalently, one may search for $\Delta f_{\mathbf{p}}$ as $\Delta f_{\mathbf{p}} = (\mathbf{p}/m)\cdot\mathbf{g}_\varepsilon$, where $\mathbf{g}_\varepsilon$ is directed along $\mathbf{E}$. The collision integral is reduced to $-\nu_\varepsilon^{(1)}\Delta f_{\mathbf{p}}$, where the relaxation rate of the first $(l = 1)$ harmonic of the anisotropic distribution appears. This rate is defined by Eqs. (20) and (14) for the 3D and 2D cases, respectively. Since $P_1(x) = x$, the rate $\nu_\varepsilon^{(1)}$ for both these cases is described by formally equivalent expressions containing $(1 - \cos\widehat{\mathbf{pp}}')$ under the sum over momentum. The related relaxation time $\tau_{tr}(\varepsilon) = 1/\nu_\varepsilon$ is called the transport time or, equivalently, the momentum relaxation time. We note that in the case of several scattering mechanisms, for example, scattering by random static potential described by Eq. (7.23) or quasielastic scattering by acoustic phonons considered in Chapter 7, the relaxation rate standing in Eq. (24) is a sum of the relaxation rates for these mechanisms.

The induced Fourier-component of the current density $\mathbf{I}_t$ is $\mathbf{I}(\omega)e^{-i\omega t} +$ c.c., and the linear response is described by the conductivity tensor $\sigma_{\alpha\beta}(\omega)$ defined as

$$I_\alpha(\omega) = \sum_\beta \sigma_{\alpha\beta}(\omega)E_\beta. \qquad (25)$$

Solving Eq. (24), one finds an expression of $\sigma_{\alpha\beta}(\omega)$ for the system of dimensionality d in the form

$$\sigma_{\alpha\beta}(\omega) = 2\left(\frac{e}{m}\right)^2 \int \frac{d\mathbf{p}}{(2\pi\hbar)^d}\frac{p_\alpha p_\beta}{-i\omega + \nu_\varepsilon}\left(-\frac{df_\varepsilon^{(eq)}}{d\varepsilon}\right)$$

$$= \delta_{\alpha\beta}\frac{2e^2}{dm}\int d\varepsilon\rho_D(\varepsilon)\frac{\varepsilon}{-i\omega + \nu_\varepsilon}\left(-\frac{df_\varepsilon^{(eq)}}{d\varepsilon}\right), \qquad (26)$$

where $p_\alpha p_\beta \to \delta_{\alpha\beta} p^2/d$ after averaging over the angle of $\mathbf{p}$. The conductivity tensor appears to be diagonal, $\sigma_{\alpha\beta}(\omega) = \delta_{\alpha\beta}\sigma(\omega)$, because of the isotropy of the electron spectrum and macroscopic isotropy of the scattering potential. The frequency-dependent conductivity is a complex function of ω. The static conductivity (at $\omega = 0$) is written as

$$\sigma = \frac{e^2 n}{m}\tau_{tr}, \quad \tau_{tr} = \frac{2}{dn}\int d\varepsilon \rho_D(\varepsilon)\varepsilon\tau_{tr}(\varepsilon)\left(-\frac{df_\varepsilon^{(eq)}}{d\varepsilon}\right), \tag{27}$$

where $n = \int d\varepsilon \rho_D(\varepsilon) f_\varepsilon^{(eq)}$ is the electron density. The last equation of Eq. (27) should be considered as a definition of the averaged transport time. This time, in general, depends on the energy distribution function of electrons. However, if the time $\tau_{tr}(\varepsilon)$ does not depend on the energy ε, the result of the averaging is merely equal to this time, without regard to dimensionality. On the other hand, for degenerate electrons, when

$$-\frac{df_\varepsilon^{(eq)}}{d\varepsilon} = \delta(\varepsilon - \varepsilon_F) + O[(T/\varepsilon_F)^2] \tag{28}$$

and ε_F is the Fermi energy, we obtain $\tau_{tr} \simeq \tau_{tr}(\varepsilon_F)$. Therefore, the conductivity of degenerate electrons does not depend on the electron temperature T with the accuracy up to $(T/\varepsilon_F)^2$.

In the high-frequency limit $\omega\bar{\tau} \gg 1$ ($\bar{\tau}$ is the characteristic relaxation time), the denominator of the expression under the integral in Eq. (26) is expanded according to

$$(-i\omega + \nu_\varepsilon)^{-1} \simeq i/\omega + \nu_\varepsilon/\omega^2. \tag{29}$$

Using this equation, we take the integral in Eq. (26) by parts and obtain

$$\sigma(\omega) \simeq i\frac{e^2 n}{m\omega} + \frac{2e^2}{dm\omega^2}\int d\varepsilon \rho_D(\varepsilon)\varepsilon\nu_\varepsilon\left(-\frac{df_\varepsilon^{(eq)}}{d\varepsilon}\right). \tag{30}$$

The imaginary part of $\sigma(\omega)$ is expressed through the electron density n and proportional to ω^{-1}. The real part is proportional to the averaged, according to the definition given by Eq. (27), relaxation rate (in contrast, $\sigma(0)$ is proportional to the averaged transport time). Regardless of the scattering mechanism, this part is proportional to ω^{-2}. We stress that this behavior of the frequency dispersion is valid only in the classical frequency region $\omega \ll \bar{\varepsilon}/\hbar$ considered in this section. A generalization to the quantum region $\omega \sim \bar{\varepsilon}/\hbar$ will be given in Sec. 10.

A simple description of the frequency dispersion of the conductivity of degenerate electrons follows from Eqs. (26) and (28). Expressing the density of states $\rho_{3D}(\varepsilon)$ through the electron density n, we obtain

$$\sigma(\omega) = i\frac{e^2 n}{m(\omega + i\nu_{\varepsilon_F})}. \tag{31}$$

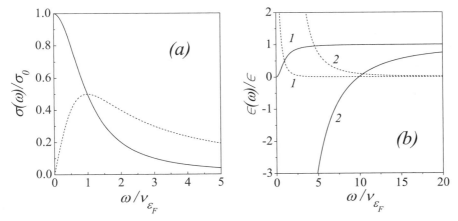

Figure 2.1. Spectral dependence of the real (solid) and imaginary (dashed) parts of the conductivity $\sigma(\omega)$ (a) and dielectric function $\epsilon(\omega)$ (b). In the panel (a), $\sigma_0 = e^2 n / m\nu_{\varepsilon_F}$. The curves 1 and 2 in the panel (b) correspond to $\omega_p/\nu_{\varepsilon_F} = 1$ and $\omega_p/\nu_{\varepsilon_F} = 10$, respectively.

This expression also describes the dielectric function (dielectric permittivity) of 3D electrons, since the latter is given by $\epsilon(\omega) = \epsilon + 4\pi i \sigma(\omega)/\omega$, where ϵ is the dielectric permittivity of the isotropic medium (it is not related to the electrons under consideration). In the collisionless approximation, $\epsilon(\omega) = \epsilon[1 - (\omega_p/\omega)^2]$, where $\omega_p = \sqrt{4\pi e^2 n/m\epsilon}$ is the plasma frequency. The spectral dependence of the real and imaginary parts of the conductivity and dielectric function is illustrated in Fig. 2.1.

9. Quasi-Classical Kinetic Equation

Let us consider the kinetic equation (7.13) for quasi-classical, i.e., slowly varying with time and smoothly varying in space, external fields. The interaction of an electron with the electric and magnetic fields $\mathbf{E}_{\mathbf{x}t}$ and $\mathbf{H}_{\mathbf{x}t}$ can be taken into account through the vector potential $\mathbf{A}_{\mathbf{x}t}$ standing in the one-electron Hamiltonian $\hat{h}_t$; see Eqs. (4.3) and (4.2). In the coordinate representation,

$$\langle \mathbf{x} | \hat{h}_t | \mathbf{x}' \rangle = \frac{[\hat{\mathbf{p}} - (e/c)\mathbf{A}_{\mathbf{x}t}]^2}{2m} \delta(\mathbf{x} - \mathbf{x}'). \tag{1}$$

We use the gauge where $\nabla_{\mathbf{x}} \cdot \mathbf{A}_{\mathbf{x}t} = 0$ and $U_{\mathbf{x}t} = 0$. A similar consideration is possible if the scalar potential is taken into account. The quantum kinetic equation (7.13) in the coordinate representation is written as

$$\frac{\partial \rho_t(\mathbf{x}_1, \mathbf{x}_2)}{\partial t} + \frac{i}{\hbar} \int d\mathbf{x}' \left\{ \langle \mathbf{x}_1 | \hat{h}_t | \mathbf{x}' \rangle \rho_t(\mathbf{x}', \mathbf{x}_2) \right.$$

$$\left. - \rho_t(\mathbf{x}_1, \mathbf{x}') \langle \mathbf{x}' | \hat{h}_t | \mathbf{x}_2 \rangle \right\} = \langle \mathbf{x}_1 | \widehat{J}_{im} | \mathbf{x}_2 \rangle, \tag{2}$$

where $\rho_t(\mathbf{x}_1, \mathbf{x}_2) \equiv \langle \mathbf{x}_1 | \hat{\rho}_t | \mathbf{x}_2 \rangle$. The response of the system at the point $(\mathbf{r}, t)$ is described by the macroscopic current density $\mathbf{I}_{\mathbf{r}t}$ defined by the general quantum-mechanical expressions (4.14) and (4.15) so that for the δ-state characterized by the one-electron wave function $\Psi_{\mathbf{r}t}^{(\delta)}$ one has

$$\mathbf{I}_{\mathbf{r}t}^{(\delta)} = i\frac{\hbar e}{2m} \left[\Psi_{\mathbf{r}t}^{(\delta)} \nabla \Psi_{\mathbf{r}t}^{(\delta)*} - \Psi_{\mathbf{r}t}^{(\delta)*} \nabla \Psi_{\mathbf{r}t}^{(\delta)} \right] - \frac{e^2}{mc} \mathbf{A}_{\mathbf{r}t} \Psi_{\mathbf{r}t}^{(\delta)} \Psi_{\mathbf{r}t}^{(\delta)*}. \qquad (3)$$

Using the density matrix (1.17) in the coordinate representation, we average Eq. (3) over the impurity ensemble and express $\mathbf{I}_{\mathbf{r}t}$ through $\rho_t(\mathbf{x}_1, \mathbf{x}_2)$ as

$$\mathbf{I}_{\mathbf{r}t} = \frac{e}{m} \lim_{\mathbf{x}_{1,2} \to \mathbf{r}} \left[\left(-i\hbar \frac{\partial}{\partial \mathbf{x}_2} - \frac{e}{c} \mathbf{A}_{\mathbf{x}_2 t} \right)^* + \left(-i\hbar \frac{\partial}{\partial \mathbf{x}_1} - \frac{e}{c} \mathbf{A}_{\mathbf{x}_1 t} \right) \right]$$
$$\times \rho_t(\mathbf{x}_1, \mathbf{x}_2), \qquad (4)$$

where the spin degeneracy leading to an extra factor of 2 is taken into account. This expression, together with the kinetic equation (2), completely describes the response of the system to the external fields.

To simplify the equations in the case of quasi-classical fields, we introduce new variables, a classical coordinate $\mathbf{r}$ and a differential coordinate $\Delta\mathbf{r}$, according to

$$\mathbf{r} = \frac{\mathbf{x}_1 + \mathbf{x}_2}{2}, \quad \Delta\mathbf{r} = \mathbf{x}_1 - \mathbf{x}_2, \quad \mathbf{x}_1 = \mathbf{r} + \frac{\Delta\mathbf{r}}{2}, \quad \mathbf{x}_2 = \mathbf{r} - \frac{\Delta\mathbf{r}}{2}. \qquad (5)$$

These coordinates are convenient for defining the Wigner distribution function $f_t(\mathbf{r}, \mathbf{p})$, which depends on the classical coordinate $\mathbf{r}$ and momentum $\mathbf{p}$ according to

$$f_t(\mathbf{r}, \mathbf{p}) = \int d\Delta\mathbf{r} \exp\left[-\frac{i}{\hbar} \mathbf{P}_{\mathbf{r}t} \cdot \Delta\mathbf{r} \right] \rho_t \left(\mathbf{r} + \frac{\Delta\mathbf{r}}{2}, \mathbf{r} - \frac{\Delta\mathbf{r}}{2} \right), \qquad (6)$$

where $\mathbf{P}_{\mathbf{r}t} \equiv \mathbf{p} + (e/c)\mathbf{A}_{\mathbf{r}t}$ differs from the kinematic momentum standing in Eq. (1) by the opposite sign at (e/c). The inverse Wigner transformation is

$$\rho_t \left(\mathbf{r} + \frac{\Delta\mathbf{r}}{2}, \mathbf{r} - \frac{\Delta\mathbf{r}}{2} \right) = \int \frac{d\mathbf{p}}{(2\pi\hbar)^3} \exp\left[\frac{i}{\hbar} \mathbf{P}_{\mathbf{r}t} \cdot \Delta\mathbf{r} \right] f_t(\mathbf{r}, \mathbf{p}). \qquad (7)$$

Although we have defined it for the 3D case, a similar expression containing $d\mathbf{p}/(2\pi\hbar)^2$ can be written for the 2D case. The normalization condition (7.19), $n = (2/V) \int d\mathbf{x} \rho_t(\mathbf{x}, \mathbf{x})$, is written for $f_t(\mathbf{r}, \mathbf{p})$ in the classical form (the factor of 2 accounts for the spin degeneracy):

$$n = \lim_{\Delta\mathbf{r} \to 0} \frac{2}{V} \int d\mathbf{r} \rho_t \left(\mathbf{r} + \frac{\Delta\mathbf{r}}{2}, \mathbf{r} - \frac{\Delta\mathbf{r}}{2} \right) = \frac{2}{V} \int d\mathbf{r} \int \frac{d\mathbf{p}}{(2\pi\hbar)^3} f_t(\mathbf{r}, \mathbf{p}). \qquad (8)$$

However, the exact Wigner distribution function is not necessarily positive. One can use the variables of Eq. (5) and the corresponding relations for the differential operators (problem 2.7),

$$\nabla_{\mathbf{x}_1} = \frac{1}{2}\nabla_{\mathbf{r}} + \nabla_{\Delta \mathbf{r}} , \quad \nabla_{\mathbf{x}_2} = \frac{1}{2}\nabla_{\mathbf{r}} - \nabla_{\Delta \mathbf{r}} , \tag{9}$$

to transform the current density (4) as

$$\mathbf{I}_{rt} = \frac{e}{m} \lim_{\Delta \mathbf{r} \to 0} \left[\left(-i\frac{\hbar}{2}\nabla_{\mathbf{r}} + i\hbar\nabla_{\Delta \mathbf{r}} - \frac{e}{c}\mathbf{A}_{\mathbf{r}-\Delta \mathbf{r}/2\ t} \right)^* \right.$$

$$\left. + \left(-i\frac{\hbar}{2}\nabla_{\mathbf{r}} - i\hbar\nabla_{\Delta \mathbf{r}} - \frac{e}{c}\mathbf{A}_{\mathbf{r}+\Delta \mathbf{r}/2\ t} \right) \right] \rho_t \left(\mathbf{r} + \frac{\Delta \mathbf{r}}{2}, \mathbf{r} - \frac{\Delta \mathbf{r}}{2} \right)$$

$$= \frac{2e}{m} \lim_{\Delta \mathbf{r} \to 0} \left(-i\hbar\nabla_{\Delta \mathbf{r}} - \frac{e}{c}\mathbf{A}_{\mathbf{r}t} \right) \int \frac{d\mathbf{p}}{(2\pi\hbar)^3} e^{(i/\hbar)\mathbf{P}_{\mathbf{r}t}\cdot\Delta \mathbf{r}} f_t(\mathbf{r}, \mathbf{p}). \tag{10}$$

After calculating the derivative over $\Delta \mathbf{r}$, we take the limit at $\Delta \mathbf{r} \to 0$ and obtain

$$\mathbf{I}_{rt} = 2e \int \frac{d\mathbf{p}}{(2\pi\hbar)^3} \mathbf{v}_{\mathbf{p}} f_t(\mathbf{r}, \mathbf{p}), \quad \mathbf{v}_{\mathbf{p}} = \frac{\mathbf{p}}{m}, \tag{11}$$

which is the standard classical expression for the current density.

Let us carry out the Wigner transformation of the kinetic equation (2). We multiply this equation by $\exp[-(i/\hbar)\mathbf{P}_{\mathbf{r}t}\cdot\Delta \mathbf{r}]$ from the left and integrate it over $\Delta \mathbf{r}$. The part containing the time derivative $\partial\rho_t(...)/\partial t$ is transformed to

$$\int d\Delta \mathbf{r} \left\{ \frac{\partial}{\partial t} + \frac{i}{\hbar}\frac{e}{c} \left(\frac{\partial \mathbf{A}_{\mathbf{r}t}}{\partial t} \cdot \Delta \mathbf{r} \right) \right\} \exp \left(-\frac{i}{\hbar}\mathbf{P}_{\mathbf{r}t}\Delta \mathbf{r} \right)$$

$$\times \rho_t \left(\mathbf{r} + \frac{\Delta \mathbf{r}}{2}, \mathbf{r} - \frac{\Delta \mathbf{r}}{2} \right), \tag{12}$$

where we have used the identity

$$\exp \left(-\frac{i}{\hbar}\mathbf{P}_{\mathbf{r}t}\cdot\Delta \mathbf{r} \right) \frac{\partial}{\partial t}$$

$$= \left[\frac{\partial}{\partial t} + \frac{i}{\hbar}\frac{e}{c} \left(\frac{\partial \mathbf{A}_{\mathbf{r}t}}{\partial t} \cdot \Delta \mathbf{r} \right) \right] \exp \left(-\frac{i}{\hbar}\mathbf{P}_{\mathbf{r}t}\cdot\Delta \mathbf{r} \right). \tag{13}$$

Since $(i/\hbar)\Delta \mathbf{r}$ in the second term can be written as $-\partial/\partial\mathbf{p}$, the contribution (12), which corresponds to the time derivative on the left-hand side of Eq. (2), is equal to (see Eq. (8.22))

$$\left(\frac{\partial}{\partial t} + e\mathbf{E}_{\mathbf{r}t} \cdot \frac{\partial}{\partial\mathbf{p}} \right) f_t(\mathbf{r}, \mathbf{p}). \tag{14}$$

This contribution describes the evolution of the electron distribution in the electric field $\mathbf{E}_{\mathbf{r}t}$.

The Wigner transformation of the contribution coming from the commutator $[\hat{h}_t, \hat{\rho}_t]$ is more complicated. We first rewrite this contribution as

$$\frac{i}{\hbar}\left\{\frac{[\hat{\mathbf{p}}_1 - (e/c)\mathbf{A}_{\mathbf{x}_1 t}]^2}{2m} - \frac{[\hat{\mathbf{p}}_2 - (e/c)\mathbf{A}_{\mathbf{x}_2 t}]^2}{2m}{}^*\right\}\rho_t(\mathbf{x}_1, \mathbf{x}_2)$$

$$= \frac{i}{2m\hbar}\left(\hat{\mathbf{p}}_1 - \frac{e}{c}\mathbf{A}_{\mathbf{x}_1 t} - \hat{\mathbf{p}}_2 - \frac{e}{c}\mathbf{A}_{\mathbf{x}_2 t}\right)\cdot\left(\hat{\mathbf{p}}_1 - \frac{e}{c}\mathbf{A}_{\mathbf{x}_1 t} + \hat{\mathbf{p}}_2 + \frac{e}{c}\mathbf{A}_{\mathbf{x}_2 t}\right)$$

$$\times \rho_t(\mathbf{x}_1, \mathbf{x}_2), \tag{15}$$

where the gauge relation $\nabla_{\mathbf{x}} \cdot \mathbf{A}_{\mathbf{x}t} = 0$ is taken into account. Using the condition of smooth fields, $\bar{\lambda} \gg \hbar/\bar{p}$, where $\bar{\lambda}$ is the characteristic spatial scale of the fields and $\bar{p}$ is the characteristic momentum, we have

$$\mathbf{A}_{\mathbf{r}+\Delta\mathbf{r}/2\ t} + \mathbf{A}_{\mathbf{r}-\Delta\mathbf{r}/2\ t} \simeq 2\mathbf{A}_{\mathbf{r}t},$$

$$\mathbf{A}_{\mathbf{r}+\Delta\mathbf{r}/2\ t} - \mathbf{A}_{\mathbf{r}-\Delta\mathbf{r}/2\ t} \simeq (\Delta\mathbf{r} \cdot \nabla_{\mathbf{r}})\mathbf{A}_{\mathbf{r}t}. \tag{16}$$

Therefore, the expression (15) is approximately equal to

$$\frac{i}{m\hbar}\left(-i\hbar\frac{\partial}{\partial\Delta\mathbf{r}} - \frac{e}{c}\mathbf{A}_{\mathbf{r}t}\right) \cdot \left[-i\hbar\nabla_{\mathbf{r}} - \frac{e}{c}(\Delta\mathbf{r}\cdot\underline{\nabla_{\mathbf{r}})\mathbf{A}_{\mathbf{r}t}}\right]$$

$$\times \rho_t(\mathbf{r}+\Delta\mathbf{r}/2, \mathbf{r}-\Delta\mathbf{r}/2), \tag{17}$$

where the limited region of action of the operator $\nabla_{\mathbf{r}}$ is underlined. Applying the Wigner transformation, we find that the expression in the first round brackets of Eq. (17) is transformed to the momentum $\mathbf{p} = m\mathbf{v}_{\mathbf{p}}$ and obtain the transformed expression (17) in the form

$$\int d\Delta\mathbf{r}\exp\left(-\frac{i}{\hbar}\mathbf{P}_{\mathbf{r}t}\cdot\Delta\mathbf{r}\right)\left(\mathbf{v}_{\mathbf{p}}\cdot\left[\nabla_{\mathbf{r}} - \frac{i}{\hbar}\frac{e}{c}(\Delta\mathbf{r}\cdot\underline{\nabla_{\mathbf{r}})\mathbf{A}_{\mathbf{r}t}}\right]\right)$$

$$\times \rho_t(\mathbf{r}+\Delta\mathbf{r}/2, \mathbf{r}-\Delta\mathbf{r}/2). \tag{18}$$

Using the commutation relation

$$\exp\left(-\frac{i}{\hbar}\mathbf{P}_{\mathbf{r}t}\cdot\Delta\mathbf{r}\right)\nabla_{\mathbf{r}}$$

$$= \left[\nabla_{\mathbf{r}} + \frac{i}{\hbar}\frac{e}{c}\underline{\nabla_{\mathbf{r}}(\mathbf{A}_{\mathbf{r}t}\cdot\Delta\mathbf{r})}\right]\exp\left(-\frac{i}{\hbar}\mathbf{P}_{\mathbf{r}t}\cdot\Delta\mathbf{r}\right), \tag{19}$$

we rewrite the expression (18) as

$$\int d\Delta\mathbf{r}\left(\mathbf{v}_{\mathbf{p}}\cdot\left[\nabla_{\mathbf{r}} + \frac{e}{c}\underline{\nabla_{\mathbf{r}}(\mathbf{A}_{\mathbf{r}t}\cdot}\frac{i}{\hbar}\Delta\mathbf{r}) - \frac{e}{c}(\frac{i}{\hbar}\Delta\mathbf{r}\cdot\underline{\nabla_{\mathbf{r}})\mathbf{A}_{\mathbf{r}t}}\right]\right)$$

$$\times \exp\left(-\frac{i}{\hbar}\mathbf{P_{rt}} \cdot \Delta\mathbf{r}\right) \rho_t(\mathbf{r} + \Delta\mathbf{r}/2, \mathbf{r} - \Delta\mathbf{r}/2). \qquad (20)$$

Under the integral, one can use the identity

$$\frac{i}{\hbar}\mathbf{p}\left[\nabla_\mathbf{r}(\mathbf{A_{rt}} \cdot \Delta\mathbf{r}) - (\Delta\mathbf{r} \cdot \nabla_\mathbf{r})\mathbf{A_{rt}}\right]$$

$$= [\mathbf{p} \times [\nabla_\mathbf{r} \times \mathbf{A_{rt}}]] \cdot \frac{\partial}{\partial\mathbf{p}} = [\mathbf{p} \times \mathbf{H_{rt}}] \cdot \frac{\partial}{\partial\mathbf{p}}, \qquad (21)$$

where the vector potential is expressed through the magnetic field $\mathbf{H_{rt}}$. Combining Eqs. (20) and (21), we find that the contribution of the commutator to the Wigner-transformed equation is given by

$$\left(\mathbf{v_p} \cdot \nabla_\mathbf{r} + \frac{e}{c}[\mathbf{v_p} \times \mathbf{H_{rt}}] \cdot \frac{\partial}{\partial\mathbf{p}}\right) f_t(\mathbf{r}, \mathbf{p}). \qquad (22)$$

From Eqs. (14) and (22) one can see that the left-hand side of the transformed kinetic equation contains the classical Lorentz force

$$\mathbf{F_{rpt}} = e\mathbf{E_{rt}} + \frac{e}{c}[\mathbf{v_p} \times \mathbf{H_{rt}}] \qquad (23)$$

expressed through the electric and magnetic field strengths, $\mathbf{E_{rt}}$ and $\mathbf{H_{rt}}$.

Now we turn to the Wigner transformation of the collision integral (7.17). This implies the transformation of the operator products $\hat{c}_t = \hat{a}_t \cdot \hat{b}_t$ and their commutators with the exponent $\exp(i\mathbf{q} \cdot \mathbf{x})$. As shown in Appendix C, the Wigner-transformed operator product $c_t(\mathbf{r}, \mathbf{p})$ is given by the expansion

$$c_t(\mathbf{r}, \mathbf{p}) = a_t(\mathbf{r}, \mathbf{p})b_t(\mathbf{r}, \mathbf{p}) + \frac{i\hbar}{2}\left\{\frac{\partial a}{\partial\mathbf{r}} \cdot \frac{\partial b}{\partial\mathbf{p}} - \frac{\partial a}{\partial\mathbf{p}} \cdot \frac{\partial b}{\partial\mathbf{r}}\right\} + \dots, \qquad (24)$$

where the dots ... correspond to $\propto \hbar^2$ and higher-order contributions. Below we neglect the quantum ($\propto \hbar$) corrections and write $c_t(\mathbf{r}, \mathbf{p})$ as a product of a_t and b_t. The collision integral (7.17) contains fast-oscillating expressions of the kind $\exp(i\mathbf{q} \cdot \mathbf{x})\widehat{F}\exp(-i\mathbf{q} \cdot \mathbf{x})$, where $\widehat{F}$ is an arbitrary operator. The contribution from such terms is calculated directly with the use of the definition (6) and leads to (problem 2.8)

$$\int d\Delta\mathbf{r} \exp\left(-\frac{i}{\hbar}\mathbf{P_{rt}} \cdot \Delta\mathbf{r}\right)$$

$$\times \left\langle \mathbf{r} + \frac{\Delta\mathbf{r}}{2}\left|\exp(i\mathbf{q} \cdot \mathbf{x})\widehat{F}\exp(-i\mathbf{q} \cdot \mathbf{x})\right|\mathbf{r} - \frac{\Delta\mathbf{r}}{2}\right\rangle = F_{\mathbf{r}, \mathbf{p} - \hbar\mathbf{q}}. \qquad (25)$$

Let us write the collision integral (7.17) in the coordinate representation through the variables (5) and use the relations (24) and (25). As a result,

$$J_{im}(f|\mathbf{r}\mathbf{p}t) = \frac{n_{im}}{\hbar^2 V} \sum_{\mathbf{p}'} |v[(\mathbf{p}-\mathbf{p}')/\hbar]|^2 \int_{-\infty}^{t} dt' e^{\lambda t'}$$

$$\times \left\{ S_{\mathbf{r}\mathbf{p}}(t,t') S^*_{\mathbf{r}\mathbf{p}'}(t,t') + c.c. \right\} [f_{t'}(\mathbf{r},\mathbf{p}') - f_{t'}(\mathbf{r},\mathbf{p})]. \tag{26}$$

The collision integral is expressed through the S-operators in the Wigner representation.

To calculate $S_{\mathbf{r}\mathbf{p}}(t,t')$, we write the operator equation (2.2) in the coordinate representation as

$$i\hbar \frac{\partial S_{\mathbf{x}_1,\mathbf{x}_2}(t,t')}{\partial t} = \frac{[\hat{\mathbf{p}}_1 - (e/c)\mathbf{A}_{\mathbf{x}_1 t}]^2}{2m} S_{\mathbf{x}_1,\mathbf{x}_2}(t,t'),$$

$$S_{\mathbf{x}_1,\mathbf{x}_2}(t,t')_{t=t'} = \delta(\mathbf{x}_1 - \mathbf{x}_2) \tag{27}$$

and apply the Wigner transformation to this equation. Using the relations (13) and (19), we obtain the equation of motion

$$\left(\frac{\partial}{\partial t} + e\mathbf{E}_{\mathbf{r}t} \cdot \frac{\partial}{\partial \mathbf{p}} \right) S_{\mathbf{r}\mathbf{p}}(t,t')$$

$$\simeq \frac{1}{2m} \left(\mathbf{p} - i\frac{\hbar}{2}\nabla_{\mathbf{r}} + i\frac{\hbar e}{2c} \left[\mathbf{H}_{\mathbf{r}t} \times \frac{\partial}{\partial \mathbf{p}} \right] \right)^2 S_{\mathbf{r}\mathbf{p}}(t,t') \tag{28}$$

with the initial condition $S_{\mathbf{r}\mathbf{p}}(t,t')_{t=t'} = 1$ (problem 2.9). We consider the case of smooth and slowly varying fields ($\bar{\lambda}$ and $\bar{t}$ are the characteristic spatial and temporal scales of the fields, respectively) satisfying the following conditions:

$$\bar{p} \gg \hbar/\bar{\lambda}, \quad \bar{\varepsilon} \gg \hbar/\bar{t}, \tag{29}$$

where $\bar{p}$ and $\bar{\varepsilon}$ are the characteristic momentum and energy. Therefore, one can neglect $\nabla_{\mathbf{r}}$ on the right-hand side of Eq. (28). To estimate the influence of the external fields $\mathbf{E}_{\mathbf{r}t}$ and $\mathbf{H}_{\mathbf{r}t}$ on $S_{\mathbf{r}\mathbf{p}}(t,t')$, we estimate the other derivatives in Eq. (28) as

$$\partial/\partial \mathbf{p} \sim 1/\bar{p}, \quad \partial/\partial t \sim \bar{\varepsilon}/\hbar. \tag{30}$$

We can neglect $\mathbf{H}_{\mathbf{r}t}$-contribution on the right-hand side of Eq. (28) under the condition $\bar{p}^2 \gg \hbar(|e|/c)H$ or $\bar{\varepsilon} \gg \hbar\omega_c/2$ (we note that $\omega_c \equiv |e|H/mc$ is the cyclotron frequency). On the other hand, the contribution of $\mathbf{E}_{\mathbf{r}t}$ in Eq. (28) can be neglected at $eE\hbar/\bar{\varepsilon} \ll \bar{p}$. Under these assumptions, the S-operator takes the form

$$S_{\mathbf{r}\mathbf{p}}(t,t') = S_{\mathbf{p}}(t-t') = \exp\left[-\frac{i}{\hbar}\varepsilon_p(t-t') \right] \tag{31}$$

and describes free motion of an electron. The same form of the S-operator is used in the collision integral (8.4).

Substituting the S-operator of Eq. (31) into Eq. (26), we obtain the collision integral

$$J_{im}(f|\mathbf{r}\mathbf{p}t) = \frac{n_{im}}{\hbar^2 V} \sum_{\mathbf{p}'} |v[(\mathbf{p}-\mathbf{p}')/\hbar]|^2 \int_{-\infty}^{0} d\tau e^{\lambda\tau}\{\exp[i(\varepsilon_p - \varepsilon_{p'})\tau/\hbar]$$

$$+ \exp[-i(\varepsilon_p - \varepsilon_{p'})\tau/\hbar]\}(f_{\mathbf{r}\mathbf{p}'t+\tau} - f_{\mathbf{r}\mathbf{p}t+\tau}), \qquad (32)$$

which differs from Eq. (8.4) only by a parametric dependence on $\mathbf{r}$ through the distribution function $f_{\mathbf{r}\mathbf{p}t} \equiv f_t(\mathbf{r}, \mathbf{p})$. Under the condition $\bar{\varepsilon} \gg \hbar/\bar{t}$, we replace $f_{\mathbf{r}\mathbf{p}t+\tau}$ by $f_{\mathbf{r}\mathbf{p}t}$ and calculate the integral over τ. Then, the coordinate-dependent collision integral becomes

$$J_{im}(f|\mathbf{r}\mathbf{p}t) = \sum_{\mathbf{p}'} \left[W_{\mathbf{p}'\mathbf{p}} f_{\mathbf{r}\mathbf{p}'t} - W_{\mathbf{p}\mathbf{p}'} f_{\mathbf{r}\mathbf{p}t} \right], \qquad (33)$$

where $W_{\mathbf{p}'\mathbf{p}} = W_{\mathbf{p}\mathbf{p}'}$ is given by Eq. (8.8). The quasi-classical kinetic equation, also known as Boltzmann equation, takes the following form:

$$\left(\frac{\partial}{\partial t} + \mathbf{v_p} \cdot \nabla_{\mathbf{r}} + \mathbf{F}_{\mathbf{r}\mathbf{p}t} \cdot \frac{\partial}{\partial \mathbf{p}} \right) f_{\mathbf{r}\mathbf{p}t} = J_{im}(f|\mathbf{r}\mathbf{p}t). \qquad (34)$$

Apart from the general condition (7.21) for the applicability of the kinetic approach, the validity of Eq. (34) is determined by the conditions (29) defining smooth and slow variations of the parameters of the system and by the additional conditions

$$eE\hbar/\bar{p} \ll \bar{\varepsilon}, \qquad \hbar\omega_c \ll \bar{\varepsilon} \qquad (35)$$

implying weak enough external fields. In terms of the characteristic time $\tau_c = \hbar/\bar{\varepsilon}$, one can rewrite the conditions (35) as $eE\tau_c/\bar{p} \ll 1$ and $\omega_c\tau_c \ll 1$. These conditions have clear physical meaning: the relative change of the momentum due to acceleration of an electron by the electric field during the time τ_c must be small, and the quantization of electron states by the magnetic field must be relatively weak. The conditions (29) and (35) form the main result of the above consideration, since they point out the limits of applicability of the quasi-classical kinetic equation. Equation (34) can be considered as a relation balancing the collision-induced arrival and departure of the electrons in the point $(\mathbf{r}, \mathbf{p}, t)$ (the right-hand side) with the collisionless change of the electron density in this point (the left-hand side). This interpretation of the left-hand side follows from the relation

$$\frac{\partial f_{\mathbf{r}\mathbf{p}t}}{\partial t} + \dot{\mathbf{r}}_t \cdot \nabla_{\mathbf{r}} f_{\mathbf{r}\mathbf{p}t} + \dot{\mathbf{p}}_t \cdot \frac{\partial f_{\mathbf{r}\mathbf{p}t}}{\partial \mathbf{p}} = \frac{df_{\mathbf{r}\mathbf{p}t}}{dt}, \qquad (36)$$

where the classical equations of motion (Newton's equations) $\dot{\mathbf{r}}_t = \mathbf{v_p}$ and $\dot{\mathbf{p}}_t = \mathbf{F}_{\mathbf{r}t}$ are used. The kinetic equations similar to Eq. (34) with the collision integral (33) can be written for the electrons with a more complicated energy spectrum, or when $f_{\mathbf{rp}t}$ is a matrix with respect to a discrete variable. The quasi-classical kinetic equation derived above for the 3D case can be applied, under the conditions (29) and (35), to the 2D case, provided that the Lorentz force is directed along the 2D plane. It means that the magnetic field, if present, is applied perpendicular to the 2D plane and the electric field is parallel to this plane.

10. Multi-Photon Processes

Once the exact S-operator (2.8) is used, the kinetic equation (7.13) with the collision integral (7.17) describes a response of the electron-impurity system to arbitrary external fields. As an example of the situation beyond the limits of applicability of the quasi-classical kinetic equation, we consider the response to a strong, high-frequency electric field (the first condition of Eq. (9.35) and the second condition of Eq. (9.29) are violated). If a spatially homogeneous electric field $\mathbf{E}\cos\omega t$ is applied to the system, the Hamiltonian is

$$\frac{[\hat{\mathbf{p}} + (e\mathbf{E}/\omega)\sin\omega t]^2}{2m} + U(\mathbf{x}), \tag{1}$$

where the electric field is introduced through the vector potential and $U(\mathbf{x})$ is a random static potential formed, for example, by randomly distributed impurities.

Below we apply a convenient formalism accounting for the influence of the electric field on the scattering and introduce the multi-photon processes. Instead of the one-electron density matrix $\hat{n}_t$ controlled by Eq. (7.2), we use a new operator $\hat{r}_t$ defined by the unitary transformation $\hat{r}_t = \widehat{V}_t^+ \hat{n}_t \widehat{V}_t$, where $\widehat{V}_t^+ \widehat{V}_t = 1$ and $\widehat{V}_t$ is introduced by the equation

$$i\hbar \frac{\partial \widehat{V}_t}{\partial t} = \left\{ \frac{[\hat{\mathbf{p}} + (e\mathbf{E}/\omega)\sin\omega t]^2}{2m} - \frac{\hat{\mathbf{p}}^2}{2m} \right\} \widehat{V}_t$$

$$= \left[\frac{e\mathbf{E}}{m\omega} \cdot \hat{\mathbf{p}}\sin\omega t + \frac{(e\mathbf{E}/\omega)^2}{2m}\sin^2\omega t \right] \widehat{V}_t \tag{2}$$

with $\widehat{V}_t = 1$ at $\mathbf{E} = 0$. This equation is solved as

$$\widehat{V}_t = \exp\left\{ -\frac{e\mathbf{E}}{m\omega} \int^t d\tau \sin\omega\tau \cdot \nabla - \frac{i}{\hbar}\frac{(e\mathbf{E}/\omega)^2}{2m} \int^t d\tau \sin^2\omega\tau \right\}. \tag{3}$$

Therefore, Eq. (7.2) is transformed to

$$i\hbar\frac{\partial\hat{r}_t}{\partial t} = \left[\frac{\hat{\mathbf{p}}^2}{2m} + \widehat{V}_t^+ U(\mathbf{x})\widehat{V}_t, \hat{r}_t\right],\tag{4}$$

where the kinematic momentum is changed to the canonical one. As a result, the kinetic energy in Eq. (4) becomes time-independent, while the transformed potential energy $\widehat{V}_t^+ U(\mathbf{x})\widehat{V}_t$ depends on time. This potential energy is rewritten as (problem 2.10)

$$\widehat{V}_t^+ U(\mathbf{x})\widehat{V}_t = U(\mathbf{x}_t), \quad \mathbf{x}_t = \mathbf{x} + \mathbf{v}_\omega \int^t d\tau \sin\omega\tau ,\tag{5}$$

where we have introduced a field-dependent velocity $\mathbf{v}_\omega = e\mathbf{E}/m\omega$.

Averaging Eq. (4) over the realizations of the random potential $U(\mathbf{x})$, we express $\hat{r}_t$ through the averaged density matrix $\hat{\rho}_t$ and correlation operator $\hat{\kappa}_t$ introduced by analogy with Eq. (7.8):

$$\hat{\rho}_t = \langle\langle\hat{r}_t\rangle\rangle, \quad \hat{r}_t = \hat{\rho}_t + \hat{\kappa}_t.\tag{6}$$

The correlation operator is determined by the equation

$$\hat{\kappa}_t = \frac{1}{i\hbar}\int_{-\infty}^t dt' e^{\lambda t'}\widehat{S}(t-t')\left[U(\mathbf{x}_{t'}), \hat{\rho}_{t'}\right]\widehat{S}^+(t-t')\tag{7}$$

analogous to Eq. (7.12), while $\hat{\rho}_t$ is expressed through $\hat{\kappa}_t$ by an equation similar to Eq. (7.6) with the last term (7.9). The field-independent S-operator $\widehat{S}(t-t')$ is determined by Eq. (2.3) with the Hamiltonian $\hat{p}^2/2m$. Excluding $\hat{\kappa}_t$, we obtain the quantum kinetic equation for the averaged density matrix $\hat{\rho}_t$:

$$\frac{\partial\hat{\rho}_t}{\partial t} + \frac{i}{\hbar}\left[\frac{\hat{p}^2}{2m}, \hat{\rho}_t\right] = \frac{1}{\hbar^2}\int_{-\infty}^t dt' e^{\lambda t'}$$

$$\times\left\langle\left\langle\left[\widehat{S}(t-t')\left[U(\mathbf{x}_{t'}), \hat{\rho}_{t'}\right]\widehat{S}^+(t-t'), U(\mathbf{x}_t)\right]\right\rangle\right\rangle.\tag{8}$$

The collision integral on the right-hand side of this equation differs from the one given by Eq. (7.17), since the external-field dependence is transferred from the S-operator to the scattering potential. Let us carry out a spatial Fourier transformation of the potential $U(\mathbf{x}_t)$ and use Eq. (7.22) for the correlation functions of random potentials. We transform the collision integral to its final form

$$\widehat{J}_{im}(\hat{\rho}|t) = \frac{1}{\hbar^2 V}\sum_{\mathbf{q}} w(q)\int_{-\infty}^t dt' e^{\lambda t'}\exp\left\{-i(\mathbf{v}_\omega\cdot\mathbf{q})\int_{t'}^t d\tau\sin\omega\tau\right\}$$

$$\times\left[\widehat{S}(t-t')\left[e^{i\mathbf{q}\cdot\mathbf{x}}, \hat{\rho}_{t'}\right]\widehat{S}^+(t-t'), e^{-i\mathbf{q}\cdot\mathbf{x}}\right].\tag{9}$$

This expression defines the right-hand side of the quantum kinetic equation (7.13) for the case under consideration. It differs from the collision integral of Eq. (8.3) by an additional exponential factor describing the oscillations of the electron in the field during the scattering process. Instead of the initial condition for this equation, we use the requirement of periodicity, $\hat{\rho}_{t+2\pi/\omega} = \hat{\rho}_t$, where $2\pi/\omega$ is the period of the field oscillations. The normalization condition (7.19) is also imposed on $\hat{\rho}_t$.

In the momentum representation, the distribution function $f_{\mathbf{p}t} = \langle \mathbf{p} | \hat{\rho}_t | \mathbf{p} \rangle$ satisfies the equation $\partial f_{\mathbf{p}t} / \partial t = J_{im}(f|\mathbf{p}t)$, since the commutator contribution on the left-hand side of Eq. (8) vanishes. In the high-frequency case $\omega \bar{\tau} \gg 1$ (here $\bar{\tau}^{-1}$ is the scattering rate estimating the collision integral $\widehat{J}_{im}$), we search for the distribution function in the following form:

$$f_{\mathbf{p}t} = \overline{f}_{\mathbf{p}} + \Delta f_{\mathbf{p}t}, \qquad \Delta f_{\mathbf{p}t} = \Delta f_{\mathbf{p},t+2\pi/\omega}, \tag{10}$$

where the time-independent part $\overline{f}_{\mathbf{p}}$ is separated from the oscillating part $\Delta f_{\mathbf{p}t}$. This leads to a linearization of the kinetic equation, because the oscillating contribution is small and can be neglected in the collision integral. As a result, $\overline{f}_{\mathbf{p}}$ is determined by the period-averaged collision integral

$$\frac{\omega}{2\pi} \int_{-\pi/\omega}^{\pi/\omega} dt\, J_{im}(\overline{f}|\mathbf{p}t) = 0, \tag{11}$$

while $\Delta f_{\mathbf{p}t}$ is determined by the equation

$$\frac{\partial \Delta f_{\mathbf{p}t}}{\partial t} = J_{im}(\overline{f}|\mathbf{p}t). \tag{12}$$

This last equation shows us that $\Delta f_{\mathbf{p}t}$ is small as $(\omega\bar{\tau})^{-1}$. Using the periodicity with respect to time, we expand both $\Delta f_{\mathbf{p}t}$ and $J_{im}(\overline{f}|\mathbf{p}t)$ into Fourier series

$$\Delta f_{\mathbf{p}t} = \sum_{k=-\infty}^{\infty} e^{-ik\omega t} \Delta f_{\mathbf{p}}^{(k)}, \qquad k \neq 0,$$

$$J_{im}(\overline{f}|\mathbf{p}t) = \sum_{k=-\infty}^{\infty} e^{-ik\omega t} J_{im}^{(k)}(\overline{f}|\mathbf{p}), \tag{13}$$

and rewrite Eq. (12) as

$$-ik\omega \Delta f_{\mathbf{p}}^{(k)} = J_{im}^{(k)}(\overline{f}|\mathbf{p}). \tag{14}$$

Equation (14) with $k = 0$ is equivalent to Eq. (11) determining the time-averaged distribution, while the Fourier harmonics of the time-dependent distribution are expressed through $J_{im}^{(k)}(\overline{f}|\mathbf{p})$ with $k \neq 0$.

To write the collision integral $J_{im}^{(k)}(\overline{f}|\mathbf{p})$, we carry out a serial expansion of the exponential factors describing the influence of the high-frequency field. We use the identity

$$\exp(iz\cos\omega t) = \sum_{k=-\infty}^{\infty} i^k e^{ik\omega t} \mathcal{J}_k(z), \tag{15}$$

where $\mathcal{J}_k(z)$ is the Bessel function of the first kind. Calculating the integrals over t and t', we find the k-th Fourier component of the collision integral in the following form:

$$J_{im}^{(k)}(\overline{f}|\mathbf{p}) = \frac{\omega}{2\pi} \int_{-\pi/\omega}^{\pi/\omega} dt\, e^{ik\omega t} J_{im}(\overline{f}|\mathbf{p}t)$$

$$= \frac{2\pi}{\hbar V} \sum_{\mathbf{q}} w(q)(\overline{f}_{\mathbf{p}+\hbar\mathbf{q}} - \overline{f}_{\mathbf{p}})\Delta_k(\varepsilon_{\mathbf{p}+\hbar\mathbf{q}} - \varepsilon_p). \tag{16}$$

This equation describes the balance of the departure and arrival terms for the state with canonical momentum $\mathbf{p}$. However, instead of the δ-function entering Eq. (8.7) for the case of free motion, there appears a field-dependent factor

$$\Delta_k(E) = \frac{\omega}{2\pi^2\hbar} \int_{-\pi/\omega}^{\pi/\omega} dt \int_{-\infty}^{0} d\tau\, e^{\lambda\tau + ik\omega t}$$

$$\times \cos\left\{\frac{\tau E}{\hbar} - \frac{\mathbf{q}\cdot\mathbf{v}_\omega}{\omega}[\cos\omega(t+\tau) - \cos\omega t]\right\} = \frac{1}{2\pi} \sum_{s=-\infty}^{\infty} \mathcal{J}_s\left(\frac{\mathbf{q}\cdot\mathbf{v}_\omega}{\omega}\right)$$

$$\times \left\{\frac{i^{-k-1}\mathcal{J}_{s-k}(\mathbf{q}\cdot\mathbf{v}_\omega/\omega)}{E - s\hbar\omega - i\lambda} + \frac{i^{-k+1}\mathcal{J}_{s+k}(\mathbf{q}\cdot\mathbf{v}_\omega/\omega)}{E - s\hbar\omega + i\lambda}\right\}, \tag{17}$$

where $\lambda \to +0$ and the Bessel functions come from the expansion (15).

Calculating the limit at $\lambda \to +0$ according to problem 1.4, we find that the averaged collision integral $J_{im}^{(0)}(\overline{f}|\mathbf{p})$ contains the factor

$$\Delta_0(E) = \sum_{k=-\infty}^{\infty} \left[\mathcal{J}_k\left(\frac{\mathbf{q}\cdot\mathbf{v}_\omega}{\omega}\right)\right]^2 \delta(E - k\hbar\omega) \tag{18}$$

so that the stationary kinetic equation is finally written as

$$J_{im}^{(0)}(\overline{f}|\mathbf{p}) = \frac{2\pi}{\hbar V} \sum_{\mathbf{q}} w(q)(\overline{f}_{\mathbf{p}+\hbar\mathbf{q}} - \overline{f}_{\mathbf{p}})\Delta_0(\varepsilon_{\mathbf{p}+\hbar\mathbf{q}} - \varepsilon_{\mathbf{p}}) = 0. \tag{19}$$

The scattering in the presence of a periodic electric field becomes inelastic because of the factors $\delta(\varepsilon_{\mathbf{p}+\hbar\mathbf{q}} - \varepsilon_{\mathbf{p}} - k\hbar\omega)$ accounting for the multiphoton processes. The contribution of k-photon process is described by the weight factor $[\mathcal{J}_k(\mathbf{q}\cdot\mathbf{v}_\omega/\omega)]^2$ in Eq. (18) (note that the Bessel functions are normalized according to $\sum_k[\mathcal{J}_k(x)]^2 = 1$). The electric field also makes the scattering anisotropic. The Bessel functions $\mathcal{J}_k$ rapidly decrease with increasing $|k|$ at $e\mathbf{E}\cdot\mathbf{q}/m\omega^2 \ll 1$. In these conditions one can use the terms with $k = 0, \pm 1$ only. However, if the field is strong enough, all terms in the sums of Eqs. (17) and (18) become significant. To find the non-equilibrium distribution function $\overline{f}_\mathbf{p}$ from Eqs. (18) and (19) (see Sec. 37), it is necessary to take into account inelastic relaxation of the electron distribution due to electron-phonon and electron-electron scattering.

The response with the frequency $k\omega$ is given by the complex function $\Delta f_\mathbf{p}^{(k)}$, which is written according to Eqs. (14), (16), and (17):

$$\Delta f_\mathbf{p}^{(k)} = \frac{i^{1-k}}{k\hbar\omega V} \sum_\mathbf{q} w(q)(\overline{f}_{\mathbf{p}+\hbar\mathbf{q}} - \overline{f}_\mathbf{p}) \sum_{s=-\infty}^{\infty} \mathcal{J}_s\left(\frac{\mathbf{q}\cdot\mathbf{v}_\omega}{\omega}\right)$$

$$\times \left\{ \pi\delta(\varepsilon_{\mathbf{p}+\hbar\mathbf{q}} - \varepsilon_\mathbf{p} - s\hbar\omega)\left[\mathcal{J}_{s+k}\left(\frac{\mathbf{q}\cdot\mathbf{v}_\omega}{\omega}\right) + \mathcal{J}_{s-k}\left(\frac{\mathbf{q}\cdot\mathbf{v}_\omega}{\omega}\right)\right] \right.$$

$$\left. + \frac{i\mathcal{P}}{\varepsilon_{\mathbf{p}+\hbar\mathbf{q}} - \varepsilon_\mathbf{p} - s\hbar\omega}\left[\mathcal{J}_{s+k}\left(\frac{\mathbf{q}\cdot\mathbf{v}_\omega}{\omega}\right) - \mathcal{J}_{s-k}\left(\frac{\mathbf{q}\cdot\mathbf{v}_\omega}{\omega}\right)\right] \right\}. \qquad (20)$$

It contains the contributions of both δ-functions and principal values appearing in the limiting transition $\lambda \to +0$ in Eq. (17). Since $k \neq 0$, the oscillating part $\Delta f_\mathbf{p}^{(k)}$ goes to zero at $\mathbf{E} \to 0$. In the case of weak electric field, one should retain only the terms linear in $\mathbf{E}$, with $k = \pm 1$ and $s = 0, \pm 1$. The current density is expressed through the distribution (10) over canonical momenta and electron velocity $[\mathbf{p}+(e/\omega)\mathbf{E}\sin\omega t]/m$ according to

$$\mathbf{I}_t = 2\frac{e}{m} \int \frac{d\mathbf{p}}{(2\pi\hbar)^3}\left(\mathbf{p} + \frac{e}{\omega}\mathbf{E}\sin\omega t\right) f_{\mathbf{p}t}. \qquad (21)$$

Using the Fourier expansion (see Eq. (13)) of $f_{\mathbf{p}t}$ together with the normalization condition $(2/V)\sum_\mathbf{p} f_{\mathbf{p}t} = n$, we rewrite Eq. (21) as

$$\mathbf{I}_t = \frac{e^2 n}{\omega m}\mathbf{E}\sin\omega t + \sum_{k=-\infty}^{\infty} e^{-ik\omega t}\Delta\mathbf{I}_{k\omega},$$

$$\Delta\mathbf{I}_{k\omega} = \frac{2e}{m} \int \frac{d\mathbf{p}}{(2\pi\hbar)^3}\mathbf{p}\Delta f_\mathbf{p}^{(k)}. \qquad (22)$$

The collisionless contribution with the frequency ω is written here as a separate term. The collision-dependent current is expressed through $\Delta f_\mathbf{p}^{(k)}$ given by Eq. (20). These contributions describe both the absorption of the electromagnetic radiation and nonlinear responses with the frequencies $k\omega$ determining nonlinear susceptibility of the electron system.

The absorbed power is obtained by averaging the product $\mathbf{I}_t \cdot \mathbf{E}_t$ over the period $2\pi/\omega$. The absorption coefficient α_ω is defined as a ratio of this power to the averaged absolute value of the Poynting vector describing the radiation flux in the medium with dielectric permittivity ϵ. In the limit of weak absorption, we obtain the absorption coefficient in the form (problem 2.11)

$$\alpha_\omega = \frac{4\pi}{c\sqrt{\epsilon}}\text{Re }\sigma(\omega). \tag{23}$$

The real part of the conductivity can be expressed through the first harmonics $\Delta\mathbf{I}_{\pm1\omega}$ defined by Eq. (22) as $\text{Re }\sigma(\omega) = \mathbf{e}\cdot(\Delta\mathbf{I}_{-1\omega}+\Delta\mathbf{I}_{1\omega})/|\mathbf{E}|$, where $\mathbf{e}$ is the unit vector in the direction of $\mathbf{E}$. Next, using Eq. (22) for $\Delta\mathbf{I}_{k\omega}$ and Eq. (20) for $\Delta f_\mathbf{p}^{(k)}$, we obtain

$$\alpha_\omega = \frac{(4\pi)^2 e}{c\sqrt{\epsilon}m|\mathbf{E}|\hbar\omega}\int\frac{d\mathbf{p}}{(2\pi\hbar)^3}(\mathbf{e}\cdot\mathbf{p})\int\frac{d\mathbf{q}}{(2\pi)^3}w(q)(\bar{f}_{\mathbf{p}+\hbar\mathbf{q}}-\bar{f}_\mathbf{p})$$

$$\times\sum_{k=-\infty}^{\infty}\delta(\varepsilon_{\mathbf{p}+\hbar\mathbf{q}}-\varepsilon_\mathbf{p}-k\hbar\omega)\mathcal{J}_k\left(\frac{\mathbf{q}\cdot\mathbf{v}_\omega}{\omega}\right) \tag{24}$$

$$\times\left[\mathcal{J}_{k+1}\left(\frac{\mathbf{q}\cdot\mathbf{v}_\omega}{\omega}\right)+\mathcal{J}_{k-1}\left(\frac{\mathbf{q}\cdot\mathbf{v}_\omega}{\omega}\right)\right].$$

The absorption coefficient α_ω determined by Eqs. (23) and (24) is expressed in units of cm^{-1}. Equation (24) is the general expression of nonlinear (field-dependent) absorption coefficient, which takes into account both the multi-photon processes described by the Bessel functions and non-equilibrium distribution of electrons caused by the interaction of the electrons with the field. We note that this distribution is determined by Eq. (19). In the linear regime, when $|\mathbf{E}|\to 0$, we can use the equilibrium Fermi distribution function $f_\varepsilon^{(eq)}$ (here ε stands for the kinetic energy $\varepsilon_p = p^2/2m$) instead of $\bar{f}_\mathbf{p}$ and expand the Bessel functions up to the lowest order in $|\mathbf{E}|$. In this approximation, only the terms with $k=\pm1$ have to be taken into account. After some transformations including the angular averaging $(\mathbf{e}\cdot\mathbf{q})(\mathbf{e}\cdot\mathbf{q})\to q^2/3$, we obtain

$$\alpha_\omega = \frac{8\pi^2 e^2}{3c\sqrt{\epsilon}m^2\omega^3}\int\frac{d\mathbf{p}}{(2\pi\hbar)^3}\int\frac{d\mathbf{q}}{(2\pi)^3}w(q)q^2$$

$$\times(f_\varepsilon^{(eq)}-f_{\varepsilon+\hbar\omega}^{(eq)})\delta(\varepsilon_{\mathbf{p}+\hbar\mathbf{q}}-\varepsilon_\mathbf{p}-\hbar\omega). \tag{25}$$

Equation (25) describes the linear absorption by free electrons. Further transformations of this equation can be done by introducing a new variable $\mathbf{p}' = \mathbf{p} + \hbar\mathbf{q}$ instead of $\mathbf{q}$. In the classical limit, we use $\varepsilon_{\mathbf{p}+\hbar\mathbf{q}} \simeq \varepsilon_{\mathbf{p}}$, $f_{\varepsilon}^{(eq)} - f_{\varepsilon+\hbar\omega}^{(eq)} \simeq \hbar\omega(-df_{\varepsilon}^{(eq)}/d\varepsilon)$, and $(\hbar q)^2 = p^2 + p'^2 - 2\mathbf{p}\cdot\mathbf{p}' \simeq 2p^2(1 - \cos\widehat{\mathbf{pp}'})$. Therefore, in this limit Eq. (25) is transformed into $\alpha_\omega = (4\pi/c\sqrt{\epsilon})\text{Re }\sigma(\omega)$, where $\sigma(\omega)$ is given by Eq. (8.30) (we note that Eqs. (24), (25), and (8.30) are valid at $\omega\bar{\tau} \gg 1$).

In the quantum region, Eq. (25) is considerably simplified under the assumption of short-range scattering potential, when $w(q) \simeq w$ does not depend on q. In this approximation, the term with $\mathbf{p} \cdot \mathbf{p}'$ gives zero after the integration over the angles of $\mathbf{p}$ and $\mathbf{p}'$. This means that the expression under the integrals over $\mathbf{p}$ and $\mathbf{p}'$ does not contain any angular dependence and, therefore, is proportional to the product of the densities of states at the energies ε and $\varepsilon + \hbar\omega$. We finally have

$$\alpha_\omega = \frac{4\pi^2 e^2 w}{3c\sqrt{\epsilon}m\hbar^2\omega^3} \int_0^\infty d\varepsilon(2\varepsilon+\hbar\omega)\rho_{3D}(\varepsilon)\rho_{3D}(\varepsilon+\hbar\omega)(f_{\varepsilon}^{(eq)} - f_{\varepsilon+\hbar\omega}^{(eq)}). \quad (26)$$

For strongly degenerate electrons, $T \ll \varepsilon_F$, the integral in Eq. (26) is taken easily. We obtain

$$\alpha_\omega = \frac{8\pi^2 e^2 \hbar w}{c\sqrt{\epsilon}m\varepsilon_F}[\rho_{3D}(\varepsilon_F)/3]^2\Phi(\hbar\omega/\varepsilon_F),$$

$$\Phi(x) = [(1+x)^{3/2} - \theta(1-x)(1-x)^{3/2}]/x^3, \quad (27)$$

where $\theta(x)$ is the step function. The spectral dependence of the absorption is determined by the function $\Phi(\hbar\omega/\varepsilon_F)$ (problem 2.12), which is shown in Fig. 2.2. We point out that the approximation $\Phi(x) \simeq 3/x^2$, which is valid for the quasi-classical region of frequencies ($x \ll 1$), appears to be good in the quantum region as well.

In the general case, beyond the short-range scattering model, the spectral dependence of the absorption is sensitive to the q-dependence of the correlation function $w(q)$. Therefore, the spectral dependence of the absorption in the quantum region of frequencies can be used for identification of the scattering mechanism determining $w(q)$. Direct information about $w(q)$ is obtained from the measurements of the frequency dispersion of $\alpha(\omega)$ at $\omega \gg \bar{\varepsilon}$. In this limit, Eq. (25) becomes

$$\alpha_\omega = \frac{8\pi^2 e^2 n}{3c\sqrt{\epsilon}m\hbar\omega^2}\rho_{3D}(\hbar\omega)w(\sqrt{2m\omega/\hbar}). \quad (28)$$

The examination of the frequency dispersion of conductivity allows one to determine both the scattering mechanism (from the spectral dependence of α_ω at $\hbar\omega \gg \bar{\varepsilon}$) and its contribution to the relaxation rate (from the transition point between the regimes $\omega\bar{\tau} \ll 1$ and $\omega\bar{\tau} \gg 1$; see Eq. (8.31)). Apart from this, the ratio of n/m is determined from the position of the plasma reflection peak. In conclusion, the measurements of

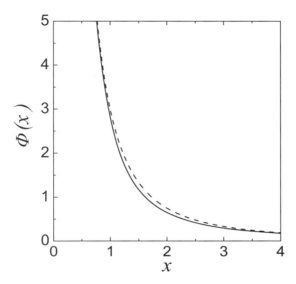

Figure 2.2. Function $\Phi(x)$ determining the spectral dependence of the absorption coefficient (the dashed line shows the dependence $3/x^2$).

the spectral dependence of both real and imaginary parts of the conductivity are useful for investigating the electron gas in solids.

11. Balance Equations

A direct method for determining the electron response to external fields assumes solution of a kinetic equation and subsequent calculation of the induced currents, charge densities, etc. When the spatial dependence of $f_{\mathbf{r}\mathbf{p}t}$ is essential, it is rather difficult to solve the kinetic equation even in the quasi-classical limit. In this section we discuss another method, which implies a reduction of the kinetic equation to a set of balance equations for macroscopic observables. Under certain approximations, the balance equations form a set which gives us an approximate description of electron response. These equations do not contain quantum numbers of electrons so that the response of the electron system is approximately described without a solution of the operator equation (7.13) or integro-differential equation (9.34). Below we derive the balance equations starting from the quasi-classical kinetic equation (9.34).

Summing both sides of Eq. (9.34) over $\mathbf{p}$, we have (the factor 2 is due to spin degeneracy)

$$\frac{\partial n_{\mathbf{r}t}}{\partial t} + \nabla_{\mathbf{r}} \cdot \mathbf{i}_{\mathbf{r}t} + \frac{2}{V} \sum_{\mathbf{p}} \mathbf{F}_{\mathbf{r}\mathbf{p}t} \cdot \frac{\partial f_{\mathbf{r}\mathbf{p}t}}{\partial \mathbf{p}} = \frac{2}{V} \sum_{k\mathbf{p}} J_k(f|\mathbf{r}\mathbf{p}t). \qquad (1)$$

The index k at the collision integral numbers scattering mechanisms. Equation (1) connects the local electron density and local electron flow density defined by

$$n_{\mathbf{r}t} = \frac{2}{V} \sum_{\mathbf{p}} f_{\mathbf{rp}t}, \quad \mathbf{i}_{\mathbf{r}t} = \frac{2}{V} \sum_{\mathbf{p}} \mathbf{v}_{\mathbf{p}} f_{\mathbf{rp}t}. \tag{2}$$

Using Eq. (9.33) for the collision integral, we find that the right-hand side of Eq. (1) vanishes:

$$\sum_{\mathbf{pp}'} \left[W_{\mathbf{p}'\mathbf{p}} f_{\mathbf{rp}'t} - W_{\mathbf{pp}'} f_{\mathbf{rp}t} \right] = 0. \tag{3}$$

This reflects the fact that the number of particles is conserved in the collisions (to prove Eq. (3), one has to permute $\mathbf{p}$ and $\mathbf{p}'$ in the second term). The action of the fields also conserves the number of particles:

$$\sum_{\mathbf{p}} \mathbf{F}_{\mathbf{rp}t} \cdot \frac{\partial f_{\mathbf{rp}t}}{\partial \mathbf{p}} = -\sum_{\mathbf{p}} f_{\mathbf{rp}t} \frac{\partial}{\partial \mathbf{p}} \cdot \mathbf{F}_{\mathbf{rp}t} = 0, \tag{4}$$

which is easy to check directly. As a result, we obtain the continuity equation

$$\frac{\partial n_{\mathbf{r}t}}{\partial t} + \operatorname{div} \mathbf{i}_{\mathbf{r}t} = 0. \tag{5}$$

We point out that the microscopic continuity equation connecting the charge density $\rho_{\mathbf{r}t} = e n_{\mathbf{r}t}$ and electric current density $\mathbf{I}_{\mathbf{r}t} = e \mathbf{i}_{\mathbf{r}t}$ has been derived in Sec. 4 for many-electron systems. The continuity equation is a general condition of compatibility of the Maxwell equations in media, and it must be valid for any approximate considerations.

One can write a balance equation for the flow density $\mathbf{i}_{\mathbf{r}t}$ by multiplying Eq. (9.34) by $\mathbf{v}_{\mathbf{p}}$ and summing it over $\mathbf{p}$:

$$\frac{\partial \mathbf{i}_{\mathbf{r}t}}{\partial t} + \frac{2}{V} \sum_{\mathbf{p}} \mathbf{v}_{\mathbf{p}} (\mathbf{v}_{\mathbf{p}} \cdot \nabla_{\mathbf{r}}) f_{\mathbf{rp}t} + \frac{2}{V} \sum_{\mathbf{p}} \mathbf{v}_{\mathbf{p}} \left(\mathbf{F}_{\mathbf{rp}t} \cdot \frac{\partial}{\partial \mathbf{p}} \right) f_{\mathbf{rp}t}$$

$$= \frac{2}{V} \sum_{k\mathbf{p}} \mathbf{v}_{\mathbf{p}} J_k(f|\mathbf{rp}t). \tag{6}$$

The second term, which contains $\nabla_{\mathbf{r}}$, is expressed through the tensor of the second rank,

$$Q_{\mathbf{r}t}^{\alpha\beta} = \frac{2}{V} \sum_{\mathbf{p}} v_{\mathbf{p}}^{\alpha} v_{\mathbf{p}}^{\beta} f_{\mathbf{rp}t} , \tag{7}$$

while the field contributions are expressed through the density and current (problem 2.13):

$$\frac{2}{V} \sum_{\mathbf{p}} \mathbf{v_p} \left(\mathbf{F_{rpt}} \cdot \frac{\partial}{\partial \mathbf{p}} \right) f_{\mathbf{rpt}} = -\frac{e}{m} \mathbf{E_{rt}} n_{\mathbf{rt}} - \frac{e}{mc} [\mathbf{i_{rt}} \times \mathbf{H_{rt}}]. \qquad (8)$$

The collision-integral contribution in Eq. (6) is transformed as

$$\frac{2}{V} \sum_{\mathbf{pp'}} \mathbf{v_p} \left[W_{\mathbf{p'p}} f_{\mathbf{rp't}} - W_{\mathbf{pp'}} f_{\mathbf{rpt}} \right] = \frac{2}{V} \sum_{\mathbf{pp'}} (\mathbf{v_{p'}} - \mathbf{v_p}) W_{\mathbf{pp'}} f_{\mathbf{rpt}} \equiv \mathbf{J_{rt}}.$$
$$\qquad (9)$$

Only the antisymmetric part $f^a_{\mathbf{rpt}}$ of the distribution function contributes to this expression, while the tensor $Q^{\alpha\beta}_{\mathbf{rt}}$ is expressed through the symmetric part $f^s_{\mathbf{rpt}}$. These parts are introduced as $f_{\mathbf{rpt}} = f^s_{\mathbf{rpt}} + f^a_{\mathbf{rpt}}$, where $f^s_{\mathbf{r},-\mathbf{pt}} = f^s_{\mathbf{rpt}}$ and $f^a_{\mathbf{r},-\mathbf{pt}} = -f^a_{\mathbf{rpt}}$. We find the current balance equation

$$\frac{\partial i^\alpha_{\mathbf{rt}}}{\partial t} + \sum_{\beta} \nabla^\beta_{\mathbf{r}} Q^{\alpha\beta}_{\mathbf{rt}} - \frac{e}{m} E^\alpha_{\mathbf{rt}} n_{\mathbf{rt}} + [\mathbf{i_{rt}} \times \boldsymbol{\omega}_c(\mathbf{rt})]_\alpha = J^\alpha_{\mathbf{rt}}, \qquad (10)$$

where $\boldsymbol{\omega}_c(\mathbf{rt})$ is a vector whose absolute value coincides with the cyclotron frequency in the magnetic field $\mathbf{H_{rt}}$ and the direction coincides with the direction of the field. Apart from $n_{\mathbf{rt}}$ and $\mathbf{i_{rt}}$, this equation contains new quantities introduced by Eqs. (7) and (9).

In a similar way, multiplying the kinetic equation (9.34) by $v^\alpha_{\mathbf{p}} v^\beta_{\mathbf{p}}$ and by $v^\alpha_{\mathbf{p}} v^\beta_{\mathbf{p}} v^\gamma_{\mathbf{p}}$, we obtain two more equations of the infinite chain of equations written below:

$$\frac{\partial n_{\mathbf{rt}}}{\partial t} + \sum_{\alpha} \nabla^\alpha_{\mathbf{r}} i^\alpha_{\mathbf{rt}} = 0,$$

$$\frac{\partial i^\alpha_{\mathbf{rt}}}{\partial t} + \sum_{\beta} \nabla^\beta_{\mathbf{r}} Q^{\alpha\beta}_{\mathbf{rt}} + \sum_{\beta\gamma} e_{\alpha\beta\gamma} i^\beta_{\mathbf{rt}} \omega^\gamma_c - (e/m) E^\alpha_{\mathbf{rt}} n_{\mathbf{rt}} = J^\alpha_{\mathbf{rt}},$$

$$\frac{\partial Q^{\alpha\beta}_{\mathbf{rt}}}{\partial t} + \sum_{\gamma} \nabla^\gamma_{\mathbf{r}} Q^{\alpha\beta\gamma}_{\mathbf{rt}} + \sum_{\gamma\delta} \left(e_{\alpha\gamma\delta} Q^{\beta\gamma}_{\mathbf{rt}} + e_{\beta\gamma\delta} Q^{\alpha\gamma}_{\mathbf{rt}} \right) \omega^\delta_c$$

$$- (e/m) \left(E^\alpha_{\mathbf{rt}} i^\beta_{\mathbf{rt}} + E^\beta_{\mathbf{rt}} i^\alpha_{\mathbf{rt}} \right) = J^{\alpha\beta}_{\mathbf{rt}}, \qquad (11)$$

$$\frac{\partial Q^{\alpha\beta\gamma}_{\mathbf{rt}}}{\partial t} + \sum_{\delta} \nabla^\delta_{\mathbf{r}} Q^{\alpha\beta\gamma\delta}_{\mathbf{rt}} + \sum_{\delta\nu} \left(e_{\alpha\delta\nu} Q^{\beta\gamma\delta}_{\mathbf{rt}} + e_{\beta\delta\nu} Q^{\alpha\gamma\delta}_{\mathbf{rt}} + e_{\gamma\delta\nu} Q^{\alpha\beta\delta}_{\mathbf{rt}} \right) \omega^\nu_c$$

$$- (e/m) \left(E^\alpha_{\mathbf{rt}} Q^{\beta\gamma}_{\mathbf{rt}} + E^\beta_{\mathbf{rt}} Q^{\alpha\gamma}_{\mathbf{rt}} + E^\gamma_{\mathbf{rt}} Q^{\alpha\beta}_{\mathbf{rt}} \right) = J^{\alpha\beta\gamma}_{\mathbf{rt}},$$

$$\cdots \quad \cdots \quad \cdots \quad \cdots \quad \cdots,$$

where $e_{\alpha\beta\gamma}$ is the antisymmetric unit tensor of the third rank (problem 2.14). The averages of the kind $Q^{\alpha\beta\cdots}_{\mathbf{rt}}$ appearing on the left-hand sides and the relaxation contributions standing on the right-hand sides are given by the expressions

$$Q_{\mathbf{r}t}^{\alpha_1 \ldots \alpha_l} = \frac{2}{V} \sum_{\mathbf{p}} v_{\mathbf{p}}^{\alpha_1} \ldots v_{\mathbf{p}}^{\alpha_l} f_{\mathbf{r}\mathbf{p}t},$$

$$J_{\mathbf{r}t}^{\alpha_1 \ldots \alpha_l} = \frac{2}{V} \sum_{\mathbf{p}} v_{\mathbf{p}}^{\alpha_1} \ldots v_{\mathbf{p}}^{\alpha_l} J_k(f|\mathbf{r}\mathbf{p}t), \qquad (12)$$

which define symmetric tensors of l-th rank. Some of their components have important physical meaning. For example, $\sum_\alpha Q_{\mathbf{r}t}^{\alpha\alpha} = (2/m)\mathcal{E}_{\mathbf{r}t}$, where $\mathcal{E}_{\mathbf{r}t}$ is the local energy density of the electron gas, and $\sum_\alpha J_{\mathbf{r}t}^{\alpha\alpha} = -(2/m)P_{\mathbf{r}t}$, where $P_{\mathbf{r}t}$ is the power loss term (the local energy density lost in the collisions in unit time). Equations (11) connect together the local electron density (zero moment), the current density (first moment), and the tensors $Q_{\mathbf{r}t}^{\alpha_1 \ldots \alpha_l}$ (higher moments). One can see that the l-th moment is connected with the $(l + 1)$-th moment through the spatial gradient of the latter. Therefore, under the approximation

$$\bar{l} \gg \bar{v} \max\{\bar{t}, \bar{\tau}\}, \qquad (13)$$

where $\bar{l}$, $\bar{v}$, $\bar{t}$, and $\bar{\tau}$ are the characteristic length of the spatial inhomogeneity, mean velocity of electrons, characteristic time, and mean scattering time, respectively, one may cut the infinite chain with a required accuracy with respect to the gradients. In this way one obtains a closed set of equations provided the collision-integral terms can be either calculated or neglected. Some important examples are given below.

Let us consider the collision-integral term $J_{\mathbf{r}t}^\alpha$, assuming the elastic scattering by impurities. As follows from Eq. (9),

$$J_{\mathbf{r}t}^\alpha = -\frac{2}{V} \sum_{\mathbf{p}} f_{\mathbf{r}\mathbf{p}t} \frac{2\pi}{\hbar} n_{im} \sum_{\mathbf{p}'} |v(|\mathbf{p} - \mathbf{p}'|/\hbar)|^2 \delta(\varepsilon_p - \varepsilon_{p'})(v_{\mathbf{p}}^\alpha - v_{\mathbf{p}'}^\alpha)$$

$$= -\frac{2}{V} \sum_{\mathbf{p}} \frac{v_{\mathbf{p}}^\alpha f_{\mathbf{r}\mathbf{p}t}}{\tau_{tr}(\varepsilon_p)}. \qquad (14)$$

If the energy dependence of the transport time $\tau_{tr}(\varepsilon)$ can be neglected (it is true, for example, for the 2D electrons scattering by the impurities with short-range potential), the term $\mathbf{J}_{\mathbf{r}t}$ is exactly equal to $-\mathbf{i}_{\mathbf{r}t}/\tau_{tr}$. In the general case of energy-dependent transport time, a similar reduction takes place for non-equilibrium distribution of degenerate electrons. Indeed, in these conditions the antisymmetric part of the distribution function contributing into Eq. (14) is essentially nonzero in the region

$|\mathbf{p}| \simeq p_F$ (only the electrons near the Fermi level contribute to the current). Therefore, one may neglect the energy dependence of $\tau_{tr}(\varepsilon)$ in this narrow region by approximating $\tau_{tr} = \tau_{tr}(\varepsilon_F)$ and taking it out of the integral. Then we again obtain $\mathbf{J_{rt}} = -\mathbf{i_{rt}}/\tau_{tr}$.

In the spatially homogeneous systems, the continuity equation ensures the conservation of the electron density, $n_{\mathbf{rt}} = n$. Since $\nabla_{\mathbf{r}}^{\beta} Q_{\mathbf{rt}}^{\alpha\beta} = 0$, the current balance equation gives us a closed description of the electron system if the collision-integral contribution $\mathbf{J}_t$ is known. Above we have demonstrated that for degenerate electrons one has $\mathbf{J}_t \simeq -\mathbf{i}_t/\tau_{tr}$, where the transport time $\tau_{tr} = \tau_{tr}(\varepsilon_F)$ is defined by Eq. (8.28). The current balance equation is reduced to the form

$$\frac{\partial \mathbf{i}_t}{\partial t} + [\mathbf{i}_t \times \boldsymbol{\omega}_c] - \frac{en}{m}\mathbf{E}_t = -\frac{\mathbf{i}_t}{\tau_{tr}}. \tag{15}$$

The electric current response $\mathbf{I}_t = e\mathbf{i}_t$ to the Fourier component of the field $\mathbf{E}_{\omega} \exp(-i\omega t)$ (the case of rapidly changing field is considered in problem 2.15) is written in the form $\mathbf{I}_{\omega} \exp(-i\omega t)$. The vector $\mathbf{I}_{\omega}$ is determined by the algebraic equation

$$(1 - i\omega\tau_{tr})\mathbf{I}_{\omega} + [\mathbf{I}_{\omega} \times \boldsymbol{\omega}_c]\tau_{tr} = \sigma_0 \mathbf{E}_{\omega}, \tag{16}$$

where $\sigma_0 = e^2 n \tau_{tr}/m$ is the static conductivity in the absence of magnetic fields; see Eq. (8.27). If $\boldsymbol{\omega}_c = 0$, Eq. (16) describes the frequency dispersion of the conductivity, which follows the law $\sigma(\omega) \propto (\tau_{tr}^{-1} - i\omega)^{-1}$; see Sec. 8. A general solution of Eq. (16) can be written after representing $\mathbf{I}_{\omega}$ through its component directed along the magnetic field (along $\mathbf{E}_{\omega}^{\|}\|\boldsymbol{\omega}_c$) and two components in the plane perpendicular to the magnetic field (along $\mathbf{E}_{\omega}^{\perp}$ and $[\boldsymbol{\omega}_c \times \mathbf{E}_{\omega}^{\perp}]$, where $\mathbf{E}_{\omega}^{\perp} \perp \boldsymbol{\omega}_c$):

$$\mathbf{I}_{\omega} = \sigma_d(\omega)\mathbf{E}_{\omega}^{\perp} + \sigma_{\perp}(\omega)[\boldsymbol{\omega}_c \times \mathbf{E}_{\omega}^{\perp}]/\omega_c + \sigma_{\|}(\omega)\mathbf{E}_{\omega}^{\|}. \tag{17}$$

The coefficients $\sigma_d(\omega)$, $\sigma_{\perp}(\omega)$, and $\sigma_{\|}(\omega)$ introduced in this equation define the components of the conductivity tensor in the general equation (8.25) connecting the current density and the electric field for the linear regime. If we assume that the magnetic field is directed along OZ, we obtain $\sigma_{xx} = \sigma_{yy} \equiv \sigma_d$, $-\sigma_{xy} = \sigma_{yx} \equiv \sigma_{\perp}$, and $\sigma_{zz} \equiv \sigma_{\|}$, while the other components are equal to zero. The coefficients $\sigma_d(\omega)$, $\sigma_{\perp}(\omega)$, and $\sigma_{\|}(\omega)$ can be easily determined after substituting Eq. (17) into Eq. (16). The longitudinal component $\sigma_{\|}(\omega)$ does not depend on the magnetic field and is given by Eq. (8.31), while the transverse components are given by the expressions

$$\sigma_d(\omega) = \frac{\sigma_0(1 - i\omega\tau_{tr})}{1 + (\omega_c^2 - \omega^2)\tau_{tr}^2 - 2i\omega\tau_{tr}}, \quad \sigma_{\perp}(\omega) = \frac{\sigma_d(\omega)\omega_c\tau_{tr}}{1 - i\omega\tau_{tr}}. \tag{18}$$

The power absorbed by the electron system is determined as Re $\mathbf{I}_\omega \cdot \mathbf{E}_\omega$. It is described by the diagonal (dissipative) components σ_d and $\sigma_\parallel$ of the conductivity tensor. On the other hand, $\sigma_\perp$ describes a non-dissipative contribution to the conductivity. According to Eq. (17), this contribution corresponds to the current perpendicular to both electric and magnetic fields.

The current density in the static limit ($\omega = 0$) is given by the expression

$$\mathbf{I} = \sigma_0 \left\{ \frac{\mathbf{E}^\perp + \tau_{tr}[\boldsymbol{\omega}_c \times \mathbf{E}]}{1 + (\omega_c \tau_{tr})^2} + \mathbf{E}^\parallel \right\}, \tag{19}$$

which directly demonstrates that a classically weak magnetic field modifies only the conductivity in the plane perpendicular to $\boldsymbol{\omega}_c$. The transverse dissipative conductivity σ_d decreases with increasing ω_c. If $\omega_c \tau_{tr} \gg 1$, the conductivity is suppressed by the field as $(\omega_c \tau_{tr})^{-2}$. The non-dissipative conductivity $\sigma_\perp$ first increases with increasing magnetic field, then starts to decrease. It is suppressed at $\omega_c \tau_{tr} \gg 1$ as $(\omega_c \tau_{tr})^{-1}$. Therefore, the component of the current perpendicular to $\boldsymbol{\omega}_c$ rotates from the direction of $\mathbf{E}^\perp$ at $\omega_c \to 0$ to the direction of $[\boldsymbol{\omega}_c \times \mathbf{E}]$ in the high-field regime (Hall effect). The non-dissipative conductivity $\sigma_\perp$ in this regime is equal to the Hall conductivity $|e|cn/H$ which does not depend on the scattering mechanisms.

In the high-frequency region, the magnetic-field-induced inequality of the diagonal components and the appearance of non-diagonal components of the conductivity tensor modify the polarization characteristics of the electromagnetic waves propagating in the system (Faraday and Voigt effects). These effects, as well as the absorption of power in the magnetic field, have resonant features when the electromagnetic wave frequency ω coincides with the cyclotron frequency, due to the factor $(\omega_c^2 - \omega^2)\tau_{tr}^2$ in the denominator of σ_d in Eq. (18). The solutions of the wave equations containing the high-frequency current (17), which completely describe these effects, are not discussed here. Below we present only the absorption coefficient for the wave polarized perpendicular to $\boldsymbol{\omega}_c$. According to Eqs. (10.23) and (18),

$$\alpha_\omega = \frac{4\pi\sigma_0}{c\sqrt{\epsilon}} \frac{1 + (\omega_c^2 + \omega^2)\tau_{tr}^2}{[1 + (\omega_c^2 - \omega^2)\tau_{tr}^2]^2 + 4\omega^2\tau_{tr}^2}. \tag{20}$$

The Lorentz peak of the absorbed power (the cyclotron resonance of absorption) is realized at $\omega_c \simeq \omega$ when $\omega_c \gg \tau_{tr}^{-1}$. The half-width of the peak at half-maximum is equal to τ_{tr}^{-1}. The spectral dependence of the cyclotron absorption coefficient, expressed in units of $\alpha_o = 4\pi\sigma_0/c\sqrt{\epsilon}$, is given in Fig. 2.3.

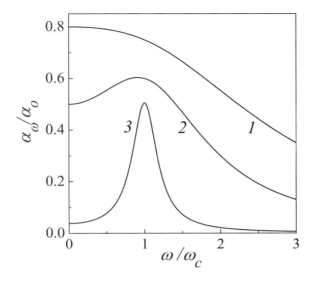

Figure 2.3. Spectral dependence of the cyclotron absorption coefficient α_ω at $\omega_c \tau_{tr} = 0.5$, 1, and 5 (curves 1, 2, and 3, respectively).

Let us apply the balance equations for calculating the linear response to a spatially-inhomogeneous electric field given by its Fourier component $\mathbf{E}_{\mathbf{q}\omega} e^{i\mathbf{q}\cdot\mathbf{r} - i\omega t}$. In the absence of the perturbation, the system is homogeneous and characterized by the equilibrium distribution function. In the linear approximation, we replace the electron density $n_{\mathbf{r}t}$ standing in the second equation of Eq. (11) by its equilibrium value n. The tensor $Q_{\mathbf{r}t}^{\alpha\beta}$ in the fourth equation of Eq. (11) is also replaced by its equilibrium value

$$Q^{\alpha\beta} = \frac{2}{V} \sum_{\mathbf{p}} v_{\mathbf{p}}^\alpha v_{\mathbf{p}}^\beta f_{\varepsilon_p}^{(eq)} = \delta_{\alpha\beta} \frac{2}{3m} \mathcal{E}_0, \tag{21}$$

where $\mathcal{E}_0$ is the equilibrium energy density of electron gas. On the other hand, the flow density $\mathbf{i}_{\mathbf{r}t}$ (as well as any tensor $Q_{\mathbf{r}t}^{\alpha\beta\cdots}$ of odd rank in Eq. (11)) is equal to zero in equilibrium. This means that the contribution proportional to $E^\alpha i^\beta$ in the third equation of the set (11) vanishes in the linear approximation. Below we consider the long-wavelength limit, when $\bar{v}q \ll \omega$, so that the condition (13) is valid. We neglect the gradients of $Q_{\mathbf{r}t}^{\alpha\beta\gamma\delta}$ in the fourth equation of the set (11), which corresponds to the accuracy up to $(\bar{v}q/\omega)^2$ in the calculation of the linear response; see Eqs. (24) and (28) below. On the other hand, we assume $\omega\bar{\tau} \gg 1$ and neglect the collision-integral contributions. In this way we obtain a closed set of linearized equations for the variables $\Delta n_{\mathbf{q}\omega} = n_{\mathbf{q}\omega} - n$,

$\mathbf{i_{q\omega}}$, $\Delta Q_{\mathbf{q}\omega}^{\alpha\beta} = Q_{\mathbf{q}\omega}^{\alpha\beta} - Q^{\alpha\beta}$, and $Q_{\mathbf{q}\omega}^{\alpha\beta\gamma}$. In the absence of magnetic fields (a more sophisticated set of equations characterizes the system in the homogeneous magnetic field, problem 2.16), we obtain

$$-i\omega\Delta n_{\mathbf{q}\omega} + i\sum_{\alpha} q_{\alpha} i_{\mathbf{q}\omega}^{\alpha} = 0,$$

$$-i\omega i_{\mathbf{q}\omega}^{\alpha} + i\sum_{\beta} q_{\beta}\Delta Q_{\mathbf{q}\omega}^{\alpha\beta} = (en/m)E_{\mathbf{q}\omega}^{\alpha}\ ,$$

$$-i\omega\Delta Q_{\mathbf{q}\omega}^{\alpha\beta} + i\sum_{\gamma} q_{\gamma}Q_{\mathbf{q}\omega}^{\alpha\beta\gamma} = 0, \tag{22}$$

$$-i\omega Q_{\mathbf{q}\omega}^{\alpha\beta\gamma} = (2e\mathcal{E}_0/3m^2)\left(E_{\mathbf{q}\omega}^{\alpha}\delta_{\beta\gamma} + E_{\mathbf{q}\omega}^{\beta}\delta_{\alpha\gamma} + E_{\mathbf{q}\omega}^{\gamma}\delta_{\alpha\beta}\right).$$

Excluding the tensors $\Delta Q_{\mathbf{q}\omega}^{\alpha\beta}$ and $Q_{\mathbf{q}\omega}^{\alpha\beta\gamma}$, we obtain a linear relation between the flow density $\mathbf{i_{q\omega}}$ and electric field $\mathbf{E_{q\omega}}$. Introducing a non-local conductivity tensor according to (see also Eq. (13.11) below)

$$ei_{\mathbf{q}\omega}^{\alpha} = \sum_{\beta} \sigma_{\alpha\beta}(\mathbf{q},\omega)E_{\mathbf{q}\omega}^{\beta}\ , \tag{23}$$

we find the following expression for it:

$$\sigma_{\alpha\beta}(\mathbf{q},\omega) = i\delta_{\alpha\beta}\frac{e^2 n}{m\omega} + i\frac{2e^2\mathcal{E}_0}{3m^2\omega^3}(\delta_{\alpha\beta}q^2 + 2q_{\alpha}q_{\beta}). \tag{24}$$

The first term is the same as in Eq. (8.31) for the collisionless limit, while the next term gives us the q^2-correction due to spatial dispersion.

The induced charge density $\Delta\rho_{\mathbf{q}\omega} = e\Delta n_{\mathbf{q}\omega}$ can be expressed through the electric field with the use of the continuity equation and Eq. (23):

$$\Delta\rho_{\mathbf{q}\omega} = \omega^{-1}\sum_{\alpha\beta} q_{\alpha}\sigma_{\alpha\beta}(\mathbf{q},\omega)E_{\mathbf{q}\omega}^{\beta}\ . \tag{25}$$

Substituting this expression into the Poisson equation, we write

$$i\mathbf{q}\cdot\mathbf{E_{q\omega}} = \frac{4\pi}{\epsilon\omega}\sum_{\alpha\beta} q_{\alpha}\sigma_{\alpha\beta}(\mathbf{q},\omega)E_{\mathbf{q}\omega}^{\beta}\ . \tag{26}$$

Since $\mathbf{q}$ is an arbitrary wave vector, we obtain a set of homogeneous linear equations for the components of the field. The solvability condition for these equations at nonzero $\mathbf{E_{q\omega}}$ is

$$\det\left|\left|i\delta_{\alpha\beta} - \frac{4\pi}{\epsilon\omega}\sigma_{\alpha\beta}(\mathbf{q},\omega)\right|\right| = 0, \tag{27}$$

where $\det||A_{\alpha\beta}||$ is the determinant of the matrix $A_{\alpha\beta}$. Equation (27) describes the oscillations of electron density (plasma oscillations, or plasmons). Substituting the expression (24) for the conductivity tensor into Eq. (27), we obtain the dispersion law of the plasma oscillations,

$$\omega^2 = \omega_p^2 + V_{pl}^2 q^2, \quad V_{pl}^2 = 2\mathcal{E}_0/mn, \tag{28}$$

where the first term is associated with the plasma frequency ω_p for infinitely long plasma waves, while the second term describes the corrections caused by the spatial dispersion. The characteristic velocity V_{pl} is equal to $\sqrt{3/5}v_F$ for degenerate electrons with Fermi velocity v_F.

12. Conductance of Microcontacts

The characteristic features of stationary response of a strongly inhomogeneous system are determined by the correlation between its spatial scale and the mean free path length $l_{tr} = \bar{v}\tau_{tr}$ introduced as a product of the averaged velocity $\bar{v}$ of electrons by the momentum relaxation time (transport time) τ_{tr}. Another important point is the inhomogeneity of the electric field and current in such systems. It is convenient to characterize the system by its conductance

$$G = I/V \tag{1}$$

introduced as a ratio of the total current I flowing through the system to the voltage V applied to the contacts to the system. In contrast to the local relations of the kind (8.25), the conductance characterizes the system as a whole and is expressed in units of Ohm^{-1}, while the units of conductivity are different for different dimensionalities. If the characteristic size of the inhomogeneities is large in comparison to l_{tr}, one may describe the linear response by introducing a local conductivity and by considering a steady-state electrodynamic problem in order to express the conductance through the local conductivity. On the other hand, if the characteristic size of the system is comparable to l_{tr}, a local conductivity cannot be introduced, and the conductance must be found from a solution of the quasi-classical kinetic equation, as far as we assume that the characteristic scale of the inhomogeneities is large in comparison to de Broglie wavelength. In this section we consider the conductance of a microcontact defined as a small-size conductor connecting the left $(z \to -\infty)$ and right $(z \to +\infty)$ macroscopic contact regions. The latter are attached to the voltage sources so that the voltage drop between them is equal to V. Since these contact regions are usually strongly doped or metallic, they are often called the lead banks or, merely, the leads. It is possible to create the microcontacts both in 3D and 2D electron gas environment. In Fig. 2.4 we show the microcontacts of different geometries discussed below in this section.

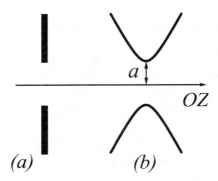

Figure 2.4. Different schemes of microcontacts: (*a*) a plane with a hole; (*b*) a hyperboloid of rotation defining a channel with circular constriction.

If the voltage applied between the contact regions is smaller than the Fermi energy of electrons in the microcontact, the distribution function $f_{\mathbf{rp}}$ is found from the kinetic equation (9.34) with the force $-\nabla \mathcal{V}_{\mathbf{r}}$:

$$\mathbf{v_p} \cdot \frac{\partial f_{\mathbf{rp}}}{\partial \mathbf{r}} - \nabla \mathcal{V}_{\mathbf{r}} \cdot \frac{\partial f_{\mathbf{rp}}}{\partial \mathbf{p}} = J(f|\mathbf{rp}), \qquad (2)$$

where $J(f|\mathbf{rp})$ is the collision integral, while the velocity is introduced as $\mathbf{v_p} = \partial \varepsilon_{\mathbf{p}}/\partial \mathbf{p}$ so that both non-parabolicity and anisotropy of the dispersion law (for example, in metals) can be taken into account. The potential energy distribution $\mathcal{V}_{\mathbf{r}}$ is determined by the Poisson equation

$$\Delta_{\mathbf{r}} \mathcal{V}_{\mathbf{r}} = -\frac{4\pi e^2}{\epsilon} \Delta n_{\mathbf{r}} , \quad \Delta n_{\mathbf{r}} = \frac{2}{V} \sum_{\mathbf{p}} \Delta f_{\mathbf{rp}} , \qquad (3)$$

which has to be solved with the boundary conditions $\mathcal{V}_{\mathbf{r}}|_{z \to \pm\infty} = \mp eV/2$. The current distribution $\mathbf{I_r}$ is determined by Eq. (9.11), and the current through the boundary Γ of the conducting region is absent: $\mathbf{n} \cdot \mathbf{I_r}|_{\Gamma} = 0$, where $\mathbf{n}$ is the unit vector normal to the boundary. If the reflection from the boundary is specular, this boundary condition can be replaced by a more detailed one: $[f_{\mathbf{rp}} - f_{\mathbf{rp}_R}]_{\Gamma} = 0$, where $\mathbf{p}_R$ is the momentum of the reflected electron which had the momentum $\mathbf{p}$ before the reflection at the point $\mathbf{r}$ of the boundary.

Below we consider the case of a metallic microcontact, when the screening length is much smaller than both the microcontact size and the mean free path length. Equation (3) in these conditions is replaced by the electric neutrality requirement $\sum_{\mathbf{p}} \Delta f_{\mathbf{rp}} = 0$. In semiconductor structures, one should compare the microcontact size and the screening length: if they are comparable, the response is described by the self-consistent set of equations (2) and (3).

We start from the description of a collisionless (ballistic) regime, when the microcontact size is much smaller than l_{tr}, and the right-hand side of Eq. (2) can be set to zero. The kinetic equation becomes a differential equation with partial derivatives of the first order, and the boundary conditions express the requirement of equilibrium far away from the contact, in the leads:

$$f_{\mathbf{rp}}|_{z \to \pm\infty} = f^{(eq)}(\varepsilon_{\mathbf{p}}). \tag{4}$$

Such a problem is solved by the method of characteristics. Let us introduce the paths $\mathcal{L}(\mathbf{r})$ of the electron motion under the electrostatic force $-\nabla_{\mathbf{r}} \mathcal{V}_{\mathbf{r}}$. These paths are determined by Newton's equations

$$\frac{d\mathbf{p}}{dt} = -\nabla_{\mathbf{r}} \mathcal{V}_{\mathbf{r}} , \quad \frac{d\mathbf{r}}{dt} = \mathbf{v}_{\mathbf{p}} . \tag{5}$$

The general solution of Eq. (2) with zero right-hand side is an arbitrary function of the total energy $\varepsilon_{\mathbf{p}} + \int_{\mathcal{L}} d\mathbf{l} \cdot \nabla\mathcal{V}$, where $d\mathbf{l}$ is the differential of coordinate along the path $\mathcal{L}$ in the direction of motion (problem 2.17). Accounting for the boundary conditions (4), we have the solution

$$f_{\mathbf{rp}} = f^{(eq)}\left(\varepsilon_{\mathbf{p}} + \int_{\mathcal{L}} d\mathbf{l} \cdot \nabla\mathcal{V}\right), \tag{6}$$

where the path $\mathcal{L}$ begins somewhere in the lead and ends at $\mathbf{r}$. Since the force is potential and the scattering from the surface Γ is elastic (specular), the integral in Eq. (6) depends only on the initial and final point positions. Using this fact, we rewrite Eq. (6) as

$$f_{\mathbf{rp}} = f^{(eq)}\left(\varepsilon_{\mathbf{p}} - \frac{eV}{2}\eta_{\mathbf{rp}} + \mathcal{V}_{\mathbf{r}}\right), \tag{7}$$

where $\eta_{\mathbf{rp}}$ is equal to 1 for the electrons coming from $z = -\infty$ and -1 for those from $z = +\infty$. To find the distribution function, one has to separate the electrons in the microcontact in two groups, depending on the direction of their momenta. This can be done easily for the simplest model of microcontact shown in Fig. 2.4 (*a*): the unpenetrable plane (or, in the 2D case, the unpenetrable line) at $z = 0$, with a hole in it. Since the microcontact is symmetric, one has $\mathcal{V}_{\mathbf{r}}|_{z=0} = 0$ and

$$\eta_{\mathbf{rp}}|_{z=0} = \operatorname{sgn} v_{\mathbf{p}}^{z} \tag{8}$$

in the region of the hole, $(x, y) \in S_\Gamma$. The function sgn here and below in this book denotes the sign of its argument (the sign of the velocity in Eq. (8)). We point out that Eq. (8) is not valid if there is a considerable number of the electrons which pass through the hole and return back due to the action of the electric field. However, this situation can be realized only in very strong electric fields (when eV exceeds the Fermi energy) and is not considered in the following. With the aid of Eq. (8), we find the distribution of electrons in the region of the hole from Eq. (7):

$$f_{\mathbf{rp}}|_{z=0} = f^{(eq)}\left(\varepsilon_{\mathbf{p}} - \frac{eV}{2}\operatorname{sgn} v_{\mathbf{p}}^z\right). \tag{9}$$

This equation describes a dynamical shift of the energy of an electron when the latter passes the potential $eV/2$, moving from one side, or $-eV/2$, moving from the other side. The current, according to Eq. (9.11), is expressed as $2eS \int v_{\mathbf{p}}^z f_{\mathbf{rp}} d\mathbf{p}/(2\pi\hbar)^d$, where S is the square of the hole S_Γ (in the 2D case, S is the length of the hole). With the use of Eq. (9), we obtain (problem 2.18)

$$I = 2eS \int_{v_{\mathbf{p}}^z > 0} \frac{ds_{\mathbf{p}}}{(2\pi\hbar)^d} \frac{v_{\mathbf{p}}^z}{v_{\mathbf{p}}^\perp} \int d\varepsilon_{\mathbf{p}} \left[f^{(eq)}(\varepsilon_{\mathbf{p}} - eV/2) - f^{(eq)}(\varepsilon_{\mathbf{p}} + eV/2)\right], \tag{10}$$

where $ds_{\mathbf{p}}$ is the differential of the surface of equal energy in the $\mathbf{p}$-space and $v_{\mathbf{p}}^\perp = |\partial\varepsilon_{\mathbf{p}}/\partial p_\perp|$ is the velocity normal to this surface. Equation (10) is valid at arbitrary voltage V and resembles the expression for the current due to tunneling through a barrier. This is not surprising, since in both cases the distribution functions for the right/left moving electrons are determined solely by the equilibrium distribution functions of the proper (left/right) lead.

In the linear regime ($eV \ll T$), the expression in the square brackets in Eq. (10) is equal to $eV\delta(\varepsilon_{\mathbf{p}} - \varepsilon_F)$, and Eq. (10) describes the ohmic current with the conductance

$$G = \frac{2e^2 SS_F}{(2\pi\hbar)^d} \langle\cos\theta\rangle, \tag{11}$$

where $S_F = \int ds_{\mathbf{p}}|_{\varepsilon_{\mathbf{p}}=\varepsilon_F}$ is the square of the Fermi surface in the momentum space, θ is the angle between $v_{\mathbf{p}_F}^\perp$ and OZ, and $\langle\ldots\rangle$ denote the averaging over the Fermi surface under the condition $v_{\mathbf{p}_F}^z > 0$. Using the isotropic quadratic dispersion law $\varepsilon_{\mathbf{p}} = \mathbf{p}^2/2m$, in the 3D case we have $ds_{\mathbf{p}} = 2\pi p_F^2 \sin\theta d\theta$ and $S_F = 4\pi p_F^2$. The averaging over the angle then gives us $\langle\cos\theta\rangle = \int_0^{\pi/2} d\theta \sin\theta\cos\theta / \int_0^\pi d\theta \sin\theta = 1/4$. As a result, the conductance is given by $G = e^2 Sp_F^2/4\pi^2\hbar^3$. In the 2D case, $ds_{\mathbf{p}} = p_F d\theta$,

$S_F = 2\pi p_F$, $\langle \cos\theta \rangle = (1/2\pi) \int_{-\pi/2}^{\pi/2} \cos\theta d\theta = 1/\pi$, and the conductance is $G = e^2 S p_F / \pi^2 \hbar^2$ (here S is the length). Equation (11) describes the ballistic limit, when the conductance is determined by the square of the Fermi surface and by the size of the hole but does not depend on scattering. For more complex geometries of microcontacts, one should multiply the conductance by a numerical coefficient determined by the geometry. We stress again the essential property used for deriving Eq. (11): the electron coming from one contact region to the other does not return back and goes to equilibrium with its new surrounding.

Let us study the microcontacts of the sizes comparable to the mean free path. The scattering in the system is assumed to be elastic, and we rewrite the collision integral in Eq. (2) as

$$J(f|\mathbf{rp}) = \int_{\varepsilon_\mathbf{p} = \varepsilon_{\mathbf{p}'}} \frac{ds_{\mathbf{p}'}}{(2\pi\hbar)^d} \frac{1}{v_{\mathbf{p}'}^\perp} W_{\mathbf{pp}'}[f_{\mathbf{rp}'} - f_{\mathbf{rp}}]. \tag{12}$$

It is convenient to solve the kinetic equation by representing the distribution function of electrons in the following way:

$$f_{\mathbf{rp}} = \alpha_{\mathbf{rp}} f^{(+)} + [1 - \alpha_{\mathbf{rp}}] f^{(-)}, \tag{13}$$

where

$$f^{(\pm)} = f^{(eq)}(\varepsilon_\mathbf{p} + \mathcal{V}_\mathbf{r} \mp eV/2) \tag{14}$$

and $\alpha_{\mathbf{rp}}$ is the probability for the electron to come into the point $(\mathbf{r}, \mathbf{p})$ of the phase space from $z = -\infty$. Accordingly, $1 - \alpha_{\mathbf{rp}}$ is the probability to come from $z = +\infty$. After a simple algebra, the kinetic equation for $f_{\mathbf{rp}}$ can be rewritten as an equation for $\alpha_{\mathbf{rp}}$:

$$\mathbf{v_p} \cdot \frac{\partial \alpha_{\mathbf{rp}}}{\partial \mathbf{r}} - \nabla \mathcal{V}_\mathbf{r} \cdot \frac{\partial \alpha_{\mathbf{rp}}}{\partial \mathbf{p}} = J(\alpha|\mathbf{rp}). \tag{15}$$

As we will see below, the function α changes with $\mathbf{p}$ on the scale $p \sim p_F$ and with $\mathbf{r}$ on the scale of the microcontact size. Therefore, the field term in Eq. (15) leads only to small corrections in the limit $eV \ll \varepsilon_F$. Neglecting it, we have the equation without the field:

$$\mathbf{v_p} \cdot \frac{\partial \alpha_{\mathbf{rp}}}{\partial \mathbf{r}} = J(\alpha|\mathbf{rp}). \tag{16}$$

The boundary condition (4) can be rewritten through the θ-function as

$$\alpha_{\mathbf{rp}}|_{|\mathbf{r}|\to\infty} = \theta(-z). \tag{17}$$

It must be accompanied by the requirement of elastic scattering from the boundary:

$$[\alpha_{\mathbf{rp}} - \alpha_{\mathbf{rp}_R}]_\Gamma = 0. \tag{18}$$

An analytical solution of Eq. (16) with the boundary conditions (17) and (18) is not available in the general case. Below we assume that the characteristic size of the contact is large in comparison to l_{tr}. This case is referred to as the diffusive limit. We also restrict ourselves by the isotropic dispersion law for electrons and isotropic scattering, $W_{\mathbf{pp'}} = W(|\mathbf{p} - \mathbf{p'}|)$. In these conditions, the elastic-scattering collision integral in Eq. (16) is reduced to $-\alpha^a_{\mathbf{rp}}/\tau_{tr}(\varepsilon_{\mathbf{p}})$, where $\alpha^a_{\mathbf{rp}}$ is the antisymmetric in $\mathbf{p}$ part of $\alpha_{\mathbf{rp}}$. Then we can solve Eq. (16) as

$$\alpha_{\mathbf{rp}} = \alpha_0(\mathbf{r}) - \tau_{tr}(\varepsilon_{\mathbf{p}})\mathbf{v_p} \cdot \frac{\partial \alpha_0(\mathbf{r})}{\partial \mathbf{r}} + \dots , \qquad (19)$$

where the dots denote the terms with higher-order derivatives. These terms can be neglected since they decrease as powers of the ratio of l_{tr} to the size of the contact. The term containing the first derivative of α_0 is antisymmetric in $\mathbf{p}$. For this reason, we retain this term in the expansion (19), though it is small in comparison to the symmetric part $\alpha_0(\mathbf{r})$. To find α_0, we substitute the expansion (19) into Eq. (16) and integrate the latter over $\mathbf{p}$. As a result, we obtain the Laplace equation $\Delta_{\mathbf{r}}\alpha_0(\mathbf{r}) = 0$, which should be solved with the boundary conditions following from Eqs. (17) and (18). In summary, we have a problem described by the following equations:

$$\Delta_{\mathbf{r}}\alpha_0(\mathbf{r}) = 0, \qquad \alpha_0(\mathbf{r})|_{|\mathbf{r}|\to\infty} = \theta(-z), \qquad \mathbf{n} \cdot \frac{\partial \alpha_0(\mathbf{r})}{\partial \mathbf{r}}\bigg|_{\mathbf{r}\in\Gamma} = 0, \qquad (20)$$

where the last two are the boundary conditions. The quantity $\alpha_0(\mathbf{r})$ determines not only the distribution function but also the distribution of the electrostatic potential, because of the electric neutrality requirement. The latter can be written as $\nabla_{\mathbf{r}} \sum_{\mathbf{p}} f_{\mathbf{rp}} = 0$. If we substitute $f_{\mathbf{rp}}$ given by Eqs. (13) and (14) into this equation and retain only the first term in the expansion (19), we obtain

$$\nabla_{\mathbf{r}}\mathcal{V}_{\mathbf{r}} = -eV\nabla_{\mathbf{r}}\alpha_0(\mathbf{r}). \qquad (21)$$

As seen from Eqs. (19), (21), and (13), the distribution of electrons in the momentum space is characterized by two concentric Fermi spheres:

$$\varepsilon_{\mathbf{p}} = \varepsilon_F - \mathcal{V} \pm eV/2. \qquad (22)$$

Inside the sphere of a smaller radius, the distribution function $f_{\mathbf{rp}}$ is equal to 1, and it is zero outside of the sphere of a larger radius. Between these spheres, the distribution is non-equilibrium and anisotropic.

The boundary problem described by Eq. (20) is often met in electrostatics, and its solutions for some simple boundaries are well known.

Below we consider the 3D case and use a solution for the boundary defined as a hyperboloid of rotation, $(x^2 + y^2)/a^2 - z^2/(b^2 - a^2) = 1$, where $b^2 > a^2$; see Fig. 2.4 (b). The microcontact is cylindrically symmetric and has a circular opening of a minimum radius a. The mentioned solution is (problem 2.19):

$$\alpha_0(\mathbf{r}) = \theta(-z) - \varphi_0(\mathbf{r}) \operatorname{sgn} z,$$

$$\varphi_0(\mathbf{r}) = \frac{1}{\pi} \arctan \left(\left\{ \frac{r^2}{2b^2} - \frac{1}{2} + \left[\left(\frac{r^2}{2b^2} - \frac{1}{2} \right)^2 + \frac{z^2}{b^2} \right]^{1/2} \right\}^{-1/2} \right), \quad (23)$$

where $r = |\mathbf{r}| = (x^2 + y^2 + z^2)^{1/2}$. Note that the solution depends only on the parameter b. Substituting the distribution function (13) with $\alpha_{\mathbf{rp}}$ from Eq. (19) into Eq. (9.11), we use the cylindrical symmetry of the problem and write the current as

$$I = 2e \int_0^{\rho_{\mathrm{r}}(z)} d\rho 2\pi\rho \int \frac{d\mathbf{p}}{(2\pi\hbar)^3} v_{\mathbf{p}}^z (f^{(+)} - f^{(-)})$$

$$\times \left[\alpha_0(\mathbf{r}) - \tau_{tr}(\varepsilon_{\mathbf{p}}) \mathbf{v_p} \cdot \frac{\partial \alpha_0(\mathbf{r})}{\partial \mathbf{r}} \right], \quad (24)$$

where $\rho = (x^2 + y^2)^{1/2}$ and $\rho_{\mathrm{r}}(z) = a[1 + z^2/(b^2 - a^2)]^{1/2}$ in our case. The current at $eV \ll \varepsilon_F$ is ohmic since $(f^{(+)} - f^{(-)}) = -eV[\partial f(\varepsilon_{\mathbf{p}})/\partial \varepsilon_{\mathbf{p}}]$. The first term in the square brackets in Eq. (24) gives zero contribution, and the current appears to be proportional to the bulk conductivity $\sigma_0 = e^2 n \tau_{tr}/m$. The conductance (1) is given by

$$G = -2\pi\sigma_0 \int_0^{\rho_{\mathrm{r}}(z)} d\rho\rho \frac{\partial \alpha_0(\mathbf{r})}{\partial z}. \quad (25)$$

Calculating the integral over ρ (the result, of course, does not depend on z because the continuity equation is satisfied), we find

$$G = 2\sigma_0 [b - (b^2 - a^2)^{1/2}]. \quad (26)$$

One can introduce the effective length L of the microcontact according to the definition $L = \sigma_0 S/\pi G$, where $S = \pi a^2$ is the minimum square of the cross-section of the contact. The parameter b is then expressed as

$$b = L + \frac{S}{4\pi L}. \quad (27)$$

The limit of a circular hole in a plane corresponds to the case $b = a$, when we have $G = 2\sigma_0 a$.

In conclusion, the conductance of a microcontact is always determined by the size and geometry of the latter. Besides, in the ballistic limit, when the size of the contact is much smaller than l_{tr}, the conductance depends on the Fermi momentum, while in the opposite limit it depends on the bulk conductivity. The intermediate situation can be studied by means of numerical solution of the kinetic equation. In this section we have considered only the microcontacts which are wide enough to ensure the applicability of the quasi-classical kinetic equation. The case of nanoscale contacts whose conductance shows quantum properties is considered in Sec. 58.

Problems

2.1. Let M be a quantity which depends on coordinates $\{\mathbf{R}_\alpha\}$ of N_{im} randomly placed impurities. Show that

$$\lim_{N_{im}\to\infty} \frac{\langle\langle[M - \langle\langle M\rangle\rangle]^2\rangle\rangle}{\langle\langle M\rangle\rangle^2} = \lim_{N_{im}\to\infty} \frac{\langle\langle M^2\rangle\rangle - \langle\langle M\rangle\rangle^2}{\langle\langle M\rangle\rangle^2} = 0.$$

Hint: Use the Fourier expansion

$$M = V^{-N_{im}} \sum_{\mathbf{q}_1\dots\mathbf{q}_{N_{im}}} \exp\{i\,(\mathbf{q}_1\cdot\mathbf{R}_1 + \dots + \mathbf{q}_{N_{im}}\cdot\mathbf{R}_{N_{im}})\}M(\mathbf{q}_1,\dots,\mathbf{q}_{N_{im}})$$

and average the exponential factors standing in the expressions $\langle\langle M^2\rangle\rangle$ and $\langle\langle M\rangle\rangle^2$.

2.2. Check that the quantum kinetic equation (7.13) with the collision integral (7.17) conserves the energy of electron system if the Hamiltonian $\hat{h}$ does not depend on time (relation (7.20)).

Solution: To prove Eq. (7.20), it is sufficient to show that $\mathrm{sp}\hat{h}\widehat{J}_{im}(\hat{\rho}|t) = 0$. Using Eq. (7.17) with $\widehat{S}(t,t') = \widehat{S}(t-t')$, we obtain

$$\mathrm{sp}\hat{h}\widehat{J}_{im}(\hat{\rho}|t) \propto \int_{-\infty}^{t} dt' e^{\lambda t'}\, \mathrm{sp}\left[\hat{h}\widehat{S}(t-t')[e^{i\mathbf{q}\cdot\mathbf{r}},\hat{\rho}_{t'}]\widehat{S}^+(t-t')e^{-i\mathbf{q}\cdot\mathbf{r}}\right.$$

$$\left. -\widehat{S}(t-t')[e^{i\mathbf{q}\cdot\mathbf{r}},\hat{\rho}_{t'}]\widehat{S}^+(t-t')\hat{h}e^{-i\mathbf{q}\cdot\mathbf{r}}\right]$$

Since $\hat{h}\widehat{S}(t-t') = -i\hbar\partial\widehat{S}(t-t')/\partial t'$ and $\widehat{S}^+(t-t')\hat{h} = i\hbar\partial\widehat{S}^+(t-t')/\partial t'$, we transform the expression above in the Markovian approximation ($\hat{\rho}_{t'} \simeq \hat{\rho}_t$) as

$$-i\hbar\int_{-\infty}^{t} dt' e^{\lambda t'}\frac{\partial}{\partial t'}\mathrm{sp}\widehat{S}(t-t')[e^{i\mathbf{q}\cdot\mathbf{r}},\hat{\rho}_t]\widehat{S}^+(t-t')e^{-i\mathbf{q}\cdot\mathbf{r}} = -i\hbar\mathrm{sp}[e^{i\mathbf{q}\cdot\mathbf{r}},\hat{\rho}_t]e^{-i\mathbf{q}\cdot\mathbf{r}} = 0,$$

where the identity $\widehat{S}(0) = 1$ is employed. The last expression is equal to zero because $e^{i\mathbf{q}\cdot\mathbf{r}}e^{-i\mathbf{q}\cdot\mathbf{r}} = 1$.

2.3. Show that if a random potential $U(\mathbf{r})$ obeys the Gaussian statistics, the correlation functions of this potential are expressed through the binary correlation function only.

Solution: Any random potential is characterized by a functional $\mathcal{P}[U]$, which describes the probability of realization of the function $U(\mathbf{r})$. The averaging of the quantity $M[U]$ over possible realizations of the random potential is defined as a functional integral $\int \delta U \mathcal{P}[U]M[U]$. By definition, the Gaussian-class potential U is characterized by the Gaussian distribution

$$\mathcal{P}[U] = N_B \exp\left\{-\frac{1}{2}\int d\mathbf{r}\int d\mathbf{r}' U(\mathbf{r})B(|\mathbf{r}-\mathbf{r}'|)U(\mathbf{r}')\right\},$$

where N_B is a coefficient determined from the normalization condition $\int \delta U \mathcal{P}[U] = 1$. Owing to macroscopic homogeneity and isotropy of the potential, the kernel B characterizing the distribution depends only on the difference between the coordinates $\mathbf{r}$ and $\mathbf{r}'$. Therefore, its Fourier component $B_\mathbf{q}$ has the symmetry property $B_\mathbf{q} = B_{-\mathbf{q}}$.

It is convenient to represent the product $U(\mathbf{r}_1)\ldots U(\mathbf{r}_k)$ in the form of a functional derivative

$$U(\mathbf{r}_1)\ldots U(\mathbf{r}_k) = \left.\frac{\delta^k \mathcal{A}[I]}{\delta I(\mathbf{r}_1)\ldots\delta I(\mathbf{r}_k)}\right|_{I=0}, \quad \mathcal{A}[I] = \exp\left(\int d\mathbf{r}I(\mathbf{r})U(\mathbf{r})\right).$$

The functional $\mathcal{A}$ is called the characteristic functional of the field U, and I is an arbitrary function. To solve the problem, we need to average $\mathcal{A}$. This is done easily if we carry out Fourier transformations of the functions U, B, and I standing in the definitions of $\mathcal{P}[U]$ and $\mathcal{A}[I]$. This leads to

$$\mathcal{P}[U] = N_B \exp\left(-\frac{1}{2}\sum_\mathbf{q} B_\mathbf{q} U_\mathbf{q} U_{-\mathbf{q}}\right), \quad \mathcal{A}[I] = \exp\left(\sum_\mathbf{q} U_\mathbf{q} I_{-\mathbf{q}}\right),$$

and the product $\mathcal{P}[U]\mathcal{A}[I]$ is represented as

$$N_B \exp\left(-\frac{1}{2}\sum_\mathbf{q} B_\mathbf{q}(U_\mathbf{q} - I_\mathbf{q}/B_\mathbf{q})(U_{-\mathbf{q}} - I_{-\mathbf{q}}/B_\mathbf{q}) - \frac{1}{2}\sum_\mathbf{q} B_\mathbf{q}^{-1} I_\mathbf{q} I_{-\mathbf{q}}\right).$$

Integrating this functional product, we take into account the normalization condition and finally obtain

$$\langle\langle\mathcal{A}[I]\rangle\rangle = \exp\left(-\frac{1}{2}\sum_\mathbf{q} B_\mathbf{q}^{-1} I_\mathbf{q} I_{-\mathbf{q}}\right) = \exp\left(-\frac{1}{2}\int d\mathbf{r}\int d\mathbf{r}' I(\mathbf{r})\Psi(|\mathbf{r}-\mathbf{r}'|)I(\mathbf{r}')\right),$$

where Ψ and B are connected through their Fourier components, $\Psi_\mathbf{q} = B_\mathbf{q}^{-1}$. Therefore, the correlation function $\langle\langle U(\mathbf{r}_1)\ldots U(\mathbf{r}_k)\rangle\rangle = \delta^k \langle\langle\mathcal{A}[I]\rangle\rangle/\delta I(\mathbf{r}_1)\ldots\delta I(\mathbf{r}_k)|_{I=0}$ of the order k is expressed through $\Psi(|\mathbf{r}-\mathbf{r}'|)$. Moreover, one may easily verify that $\Psi(|\mathbf{r}-\mathbf{r}'|) = \langle\langle U(\mathbf{r})U(\mathbf{r}')\rangle\rangle$ is the binary correlation function of the Gaussian random potential.

2.4. Show that any translation-invariant operator $\hat{A}$ is diagonal in the momentum representation.

Solution: Using $\widehat{T}_{\mathbf{R}}^{+}|\mathbf{p}\rangle = \exp{(-i\mathbf{p}\cdot\mathbf{R}/\hbar)}|\mathbf{p}\rangle$, we get $\langle\mathbf{p}|\,\hat{A}\,|\mathbf{p}'\rangle = \langle\mathbf{p}|\,\widehat{T}_{\mathbf{R}}\hat{A}\widehat{T}_{\mathbf{R}}^{+}\,|\mathbf{p}'\rangle$ $= \exp{[i(\mathbf{p}-\mathbf{p}')\cdot\mathbf{R}/\hbar]}\langle\mathbf{p}|\,\hat{A}\,|\mathbf{p}'\rangle$. Since $\mathbf{R}$ is an arbitrary vector of displacement, there must be $\langle\mathbf{p}|\,\hat{A}\,|\mathbf{p}'\rangle \propto \delta_{\mathbf{p}\mathbf{p}'}$.

2.5. Check Eq. (8.6).

Hints: Change the sign of τ in the complex conjugate contribution in the integral and use the representation $\delta(x) = (2\pi)^{-1}\int_{-\infty}^{\infty}d\kappa e^{i\kappa x}$. Another way: calculate the integral and employ the results of problem 1.4.

2.6. Calculate the field-induced contribution to the kinetic equation (8.22).

Hint: Use the operator of coordinate $\hat{\mathbf{x}} = i\hbar\partial/\partial\mathbf{p}$ to calculate the commutator $[\mathbf{E}\cdot\hat{\mathbf{x}},\hat{\rho}_t]$ in the $\mathbf{p}$-representation.

2.7. Prove the relations (9.9).

Hint: Consider a differential of an arbitrary function by using the coordinates introduced by Eq. (9.5).

2.8. Carry out the Wigner transformation in order to prove Eq. (9.25).

Hint: Use $\langle\mathbf{r}+\Delta\mathbf{r}/2|\ldots|\mathbf{r}-\Delta\mathbf{r}/2\rangle$ written in the coordinate representation.

2.9. Check the initial condition to Eq. (9.28).

Hint: Carry out the Wigner transformation of the δ-function in Eq. (9.27).

2.10. Carry out the unitary transformation to prove Eq. (10.5).

Hint: Prove the identity $e^{\mathbf{y}\cdot\nabla}F(\mathbf{x})e^{-\mathbf{y}\cdot\nabla} = F(\mathbf{x}+\mathbf{y})$ by expanding the exponents in series.

2.11. Express the absorption coefficient through the real part of the conductivity $\sigma(\omega)$.

Solution: The averaged absolute value of the Poynting vector is $c\sqrt{\epsilon}\,\overline{\mathbf{E}_t^2}/4\pi$. Since $\mathbf{I}_\omega = \sigma(\omega)\mathbf{E}_\omega$, we have $\overline{\mathbf{I}_t\mathbf{E}_t} = \mathrm{Re}\sigma(\omega)\overline{\mathbf{E}_t^2}$. Dividing the absorbed power $\overline{\mathbf{I}_t\mathbf{E}_t}$ by the averaged absolute value of the Poynting vector, we obtain $\alpha_\omega = (4\pi/c\sqrt{\epsilon})\overline{\mathbf{I}_t\mathbf{E}_t}/\overline{\mathbf{E}_t^2} = (4\pi/c\sqrt{\epsilon})\mathrm{Re}\sigma(\omega)$.

2.12. Generalize Eq. (10.27) for the 2D and 1D cases. Show that the absorption of electromagnetic radiation by degenerate 2D electrons interacting with point defects follows its classical expression up to the frequency $\varepsilon_F/\hbar$.

Solution: Assuming that in the 2D case $\mathbf{E}$ is directed in the 2D plane, while in the 1D case it is directed along the 1D line, we can repeat all steps of the derivation for the low-dimensional electrons. However, we have to use the phase space $(2\pi\hbar)^d$ ($d = 2,1$) and take into account that in Eq. (10.25) and in all subsequent equations the factor of 3 in the denominator is replaced by d because of appropriate angular averaging. Starting from Eq. (10.26), ρ_{3D} is replaced by ρ_{2D} or by ρ_{1D}. Finally, we obtain Eq. (10.27), where $[\rho_{3D}(\varepsilon_F)/3]^2\Phi(\hbar\omega/\varepsilon_F)$ is replaced by $[\rho_D(\varepsilon_F)/d]^2\Phi_d(\hbar\omega/\varepsilon_F)$, and $\Phi_d(x) = [(1+x)^{d/2} - \theta(1-x)(1-x)^{d/2}]/x^3$. If $\varepsilon_F > \hbar\omega$, this equation leads to a simple result for the 2D case, $\alpha_\omega = 4\pi e^2 n w/c\sqrt{\epsilon}\hbar^3\omega^2$, where $n = \rho_{2D}\varepsilon_F$ is the 2D electron density. This result coincides with the one given by the classical expression (8.31) for the conductivity in the approximation of point-defect scattering.

In the 2D electron systems, α_ω is dimensionless and defines the relative power loss of the electromagnetic wave normally incident on the 2D plane. In the 1D case, α_ω has dimensionality of cm and its physical meaning becomes clear if, for example, we consider a planar array of equivalent 1D electron systems (quantum wires) with $\mathcal{N}$ wires per unit length. The dimensionless quantity $\alpha_\omega\mathcal{N}$ defines the relative power loss for the wave polarized in the wire direction and transmitted through this array.

2.13. Prove Eq. (11.8) which expresses the field contribution to the balance equation for the flow density.

Solution: Let us integrate by parts in the left-hand side of Eq. (11.8). Then, taking into account that $\partial\mathbf{F}_{\mathbf{rpt}}/\partial\mathbf{p} = 0$, we rewrite this left-hand side as

$$-\frac{2}{V}\sum_{\mathbf{p}} f_{\mathbf{rpt}}\left(\mathbf{F}_{\mathbf{rpt}} \cdot \frac{\partial}{\partial\mathbf{p}}\right)\mathbf{v_p}.$$

This expression is directly transformed to the right-hand side of Eq. (11.8).

2.14. Write the vector product $[\mathbf{A} \times \mathbf{B}]$ in the Cartesian coordinates.

Solution: Since $[\mathbf{A} \times \mathbf{B}]$ is a vector perpendicular to both $\mathbf{A}$ and $\mathbf{B}$, and $[\mathbf{A} \times \mathbf{B}] = -[\mathbf{B} \times \mathbf{A}]$, the component α of the vector product can be written through the anti-symmetric unit tensor of the third rank, $e_{\alpha\beta\gamma}$, as $[\mathbf{A} \times \mathbf{B}]_\alpha = \sum_{\beta\gamma} e_{\alpha\beta\gamma}A_\beta B_\gamma$. Here $e_{xyz} = 1$ and $e_{\alpha\beta\gamma}$ is not changed under cyclic permutation of its indices. This tensor changes its sign under a permutation of two indices and is equal to 0 when at least two indices coincide.

2.15. Calculate the current response to an ultrashort (δ-shaped) pulse of electric field, $\mathbf{E}_t = \mathbf{E}\tau_E\delta(t)$.

Solution: Using either Fourier or Laplace transformation, one may reduce the differential equation (11.15) to a set of algebraic equations. After solving them and carrying out the inverse transformation of the current, the result is written in the following way:

$$\mathbf{I}_t = \frac{e^2 n\tau_E}{m}e^{-t/\tau_{tr}}\left(\mathbf{E}^\perp\cos\omega_c t + H^{-1}[\mathbf{H}\times\mathbf{E}^\perp]\sin\omega_c t + \mathbf{E}^\|\right),$$

where $\mathbf{E}^\perp$ and $\mathbf{E}^\|$ are the components of the field perpendicular and parallel to the magnetic field, respectively. The current along the magnetic field decreases with time as $e^{-t/\tau_{tr}}$, while the transverse current also rotates with the cyclotron frequency.

2.16. Write the set of equations (11.22) at $\omega_c \neq 0$. Determine the magnetoplasmon frequencies at $\mathbf{q} = 0$.

Solution: Equations (11.22) at $\omega_c \neq 0$ contain additional contributions to their left-hand sides: $\sum_{\beta\gamma} e_{\alpha\beta\gamma} i_{\mathbf{q}\omega}^{\beta} \omega_c^{\gamma}$ in the second, $\sum_{\gamma\delta} \left(e_{\alpha\gamma\delta} \Delta Q_{\mathbf{q}\omega}^{\beta\gamma} + e_{\beta\gamma\delta} \Delta Q_{\mathbf{q}\omega}^{\alpha\gamma} \right) \omega_c^{\delta}$ in the third, and $\sum_{\delta\nu} \left(e_{\alpha\delta\nu} Q_{\mathbf{q}\omega}^{\beta\gamma\delta} + e_{\beta\delta\nu} Q_{\mathbf{q}\omega}^{\alpha\gamma\delta} + e_{\gamma\delta\nu} Q_{\mathbf{q}\omega}^{\alpha\beta\delta} \right) \omega_c^{\nu}$ in the fourth equation. To describe the magnetoplasmon frequencies at $\mathbf{q} = 0$, one may neglect the terms proportional to $q\Delta Q_{\mathbf{q}\omega}^{\alpha\beta}$ and $q\Delta Q_{\mathbf{q}\omega}^{\alpha\beta\gamma}$. Then we have only two equations:

$$-i\omega \Delta n_{\mathbf{q}\omega} + i\sum_{\alpha} q_{\alpha} i_{\mathbf{q}\omega}^{\alpha} = 0 \quad \text{and} \quad -i\omega i_{\mathbf{q}\omega}^{\alpha} + \sum_{\beta\gamma} e_{\alpha\beta\gamma} i_{\mathbf{q}\omega}^{\beta} \omega_c^{\gamma} = (en/m) E_{\mathbf{q}\omega}^{\alpha}.$$

Solving them together with the Poisson equation, we obtain the equation

$$\left[(\omega^2 - \omega_p^2)\delta_{\alpha\beta} + i\omega \sum_{\gamma} e_{\alpha\beta\gamma} \omega_c^{\gamma} \right] q_{\alpha} i_{\mathbf{q}\omega}^{\beta} = 0,$$

which gives us three plasmon frequencies: $\omega = \omega_p$ for the wave polarized along the magnetic field and $\omega = \pm\omega_c/2 + \sqrt{\omega_p^2 + \omega_c^2/4}$ for the waves with perpendicular polarization. The splitting of the plasmon spectrum exists because of the magnetic-field-induced anisotropy.

2.17. Assuming that the function $F(\varepsilon_{\mathbf{p}} - g_{\mathbf{r}})$ satisfies Eq. (12.2) without the collision integral, find $g_{\mathbf{r}}$.

Hint: Substituting F in the kinetic equation, find a differential equation of the first order for $g_{\mathbf{r}}$ and integrate it.

2.18. Transform the integral $\int d\mathbf{p}...$ to the integral $\int ds_{\mathbf{p}} \int d\varepsilon_{\mathbf{p}}...$.

Hint: Express the differential of the phase volume $d\mathbf{p}$ through the differentials of the energy and of the surface of equal energy.

2.19. Check the solution given by Eq. (12.23) by a direct substitution of this solution into Eq. (12.20).

Hint: Use the cylindrical coordinate system.

Chapter 3

LINEAR RESPONSE THEORY

A general formalism for describing the response of a system to weak external perturbations can be developed in the linear approximation. In this chapter we consider the response of electrons to weak electromagnetic fields, while the case of non-mechanical perturbations (for example, temperature gradient) is discussed in Chapter 5. Using a solution of the linearized equation for the density matrix, one can write exact expressions for the kinetic coefficients connecting induced currents to the external fields. Such coefficients are expressed through the equilibrium characteristics of the system so that the linear-response problem is reduced to a statistical averaging (or to simple integrations in the case of non-interacting quasiparticles). The most effective and unified approach to such averaging is based upon diagrammatic expansion of the Green's function. The simplest variant of this method, developed for electron-impurity systems, is described in this chapter. Another approach assumes expression of the Green's function through the path integral, when the quantum-mechanical and statistical averaging can be done separately. Both these methods allow one to describe the case of strong scattering, when the quantum kinetic equation with the collision integral (7.17) is not valid.

13. Kubo Formula

To consider the linear response to the perturbation with frequency ω, we write the total Hamiltonian of the system as

$$\widehat{H} + \left(\widehat{\Delta H}_\omega e^{-i\omega t} + H.c. \right), \tag{1}$$

where $\widehat{H}$ is the Hamiltonian of the unperturbed system characterized by the density matrix $\hat{\eta}_{eq}$ which depends only on $\widehat{H}$. The non-equilibrium part of the density matrix, $\widehat{\Delta \eta}_\omega$, is introduced by the relations $\hat{\eta}_t = \hat{\eta}_{eq} +$

$\widehat{\Delta \eta}_t$ and $\widehat{\Delta \eta}_t = \widehat{\Delta \eta}_\omega \exp(-i\omega t) + H.c.$. It describes the linear response of the system to the perturbation introduced in Eq. (1). Since $\hat{\eta}_t$ satisfies Eq. (1.20), we linearize this equation and obtain an inhomogeneous operator equation for $\widehat{\Delta \eta}_t$:

$$\frac{\partial \widehat{\Delta \eta}_t}{\partial t} + \frac{i}{\hbar}[\widehat{H}, \widehat{\Delta \eta}_t] = \frac{1}{i\hbar}\left[(\widehat{\Delta H}_\omega e^{-i\omega t} + H.c.), \hat{\eta}_{eq}\right]. \tag{2}$$

The solution of Eq. (2) is expressed through the S-operator defined by Eq. (2.3):

$$\widehat{\Delta \eta}_t = \frac{1}{i\hbar}\int_{-\infty}^{t} dt' e^{\lambda t'} \widehat{S}(t-t')\left[(\widehat{\Delta H}_\omega e^{-i\omega t'} + H.c.), \hat{\eta}_{eq}\right]\widehat{S}^+(t-t') , \tag{3}$$

where $\lambda \to +0$ indicates that the perturbation is adiabatically turned on at $t \to -\infty$. Replacing $t' - t$ by τ, we express the Fourier component $\widehat{\Delta \eta}_\omega$ as

$$\widehat{\Delta \eta}_\omega = \frac{1}{i\hbar}\int_{-\infty}^{0} d\tau e^{\lambda\tau - i\omega\tau} e^{i\widehat{H}\tau/\hbar}\left[\widehat{\Delta H}_\omega, \hat{\eta}_{eq}\right] e^{-i\widehat{H}\tau/\hbar}. \tag{4}$$

The linear response is determined by a small deviation ΔQ of the macroscopic quantity $Q = Q_{eq} + \Delta Q$ from its equilibrium value Q_{eq}. The quantum-mechanical operator $\widehat{Q}$ corresponding to this quantity also can be represented as $\widehat{Q} = \widehat{Q}_0 + [\widehat{\Delta Q}_\omega \exp(-i\omega t) + H.c.]$, where $\widehat{\Delta Q}_\omega$ is caused by the small perturbation. Since Q is given as $Q = \mathrm{Sp}\widehat{Q}\hat{\eta}_t$, where $\mathrm{Sp}\ldots$ denotes the averaging over all variables of the system, the Fourier component of the deviation can be expressed as

$$\Delta Q(\omega) = \mathrm{Sp}\widehat{\Delta Q}_\omega \hat{\eta}_{eq} + \mathrm{Sp}\widehat{Q}_0 \widehat{\Delta \eta}_\omega. \tag{5}$$

Substituting $\widehat{\Delta \eta}_\omega$ from Eq. (4) to this equation, we find that the linear response for the quantity Q is expressed through the equilibrium characteristics of the system and through the quantum-mechanical perturbations $\widehat{\Delta H}_\omega$ and $\widehat{\Delta Q}_\omega$.

Below we consider the response of electron system to the electromagnetic field. Such a consideration can be applied to any system of interacting electrons described by the Hamiltonian $\widehat{H}$. According to Eq. (4.16), the first-order contribution for interaction of the electrons with the transverse electric field described by the Fourier component of the vector potential, $\mathbf{A}(\mathbf{r}, \omega) = (-ic/\omega)\mathbf{E}(\mathbf{r}, \omega)$, is

$$\widehat{\Delta H}_\omega = \frac{i}{\omega}\int d\mathbf{r}\, \hat{\mathbf{I}}(\mathbf{r}) \cdot \mathbf{E}(\mathbf{r}, \omega), \tag{6}$$

where $\mathbf{E}(\mathbf{r}, \omega)$ is the Fourier component of the electric field at the point $\mathbf{r}$ and $\hat{\mathbf{I}}(\mathbf{r})$ is the current density operator given by Eq. (4.15). Substituting $\hat{\mathbf{I}}(\mathbf{r})$ into Eq. (6) in the linear approximation, one should consider it in the absence of the perturbation, i.e., to neglect the vector potential $\mathbf{A} = (-ic/\omega)\mathbf{E}(\mathbf{r}, \omega)$ in Eq. (4.15). Owing to spatial dependence of both $\mathbf{E}$ and $\hat{\mathbf{I}}$, the approach developed here can be applied to inhomogeneous systems. Now, let Q in Eq. (5) be the current density. Using Eq. (4.15), one can write the field-induced correction to the current density operator as

$$\widehat{\Delta \mathbf{I}_\omega}(\mathbf{r}) = \frac{ie^2}{m\omega} \sum_j \delta(\mathbf{r} - \hat{\mathbf{x}}_j)\mathbf{E}(\mathbf{r}, \omega). \tag{7}$$

This operator is substituted as a quantum-mechanical perturbation $\widehat{\Delta Q}_\omega$ in Eq. (5). In place of $\widehat{Q}_0$, we substitute the unperturbed current density operator $\hat{\mathbf{I}}(\mathbf{r})$ given by Eq. (4.15) with $\mathbf{A} = 0$. Having done this, we write the Fourier component of non-equilibrium macroscopic current density as

$$\Delta \mathbf{I}(\mathbf{r}, \omega) = \frac{ie^2}{m\omega}n(\mathbf{r})\mathbf{E}(\mathbf{r}, \omega) + \frac{1}{i\hbar}\int_{-\infty}^0 d\tau e^{\lambda\tau - i\omega\tau}$$

$$\times \mathrm{Sp}\hat{\eta}_{eq}\left[e^{-i\widehat{H}\tau/\hbar}\hat{\mathbf{I}}(\mathbf{r})e^{i\widehat{H}\tau/\hbar}, \widehat{\Delta H}_\omega\right]. \tag{8}$$

The first term in this expression is obtained with the use of the expression for the local density, $n(\mathbf{r}) \equiv \mathrm{Sp}\sum_j \delta(\mathbf{r} - \hat{\mathbf{x}}_j)\hat{\eta}_{eq}$. This contribution corresponds to the collisionless current. The second term is $\mathrm{Sp}\hat{\mathbf{I}}(\mathbf{r})\widehat{\Delta\eta}_\omega$, transformed by permutations of the operators under the trace. According to Eq. (6), $\widehat{\Delta H}_\omega$ standing in this term is expressed through $\hat{\mathbf{I}}(\mathbf{r})$.

The non-local conductivity tensor is introduced by the definition

$$\Delta I_\alpha(\mathbf{r}, \omega) = \sum_\beta \int d\mathbf{r}' \sigma_{\alpha\beta}(\mathbf{r}, \mathbf{r}'|\omega)E_\beta(\mathbf{r}', \omega), \tag{9}$$

and the expression for $\sigma_{\alpha\beta}(\mathbf{r}, \mathbf{r}'|\omega)$ is obtained after substituting the perturbation (6) into Eq. (8):

$$\sigma_{\alpha\beta}(\mathbf{r}, \mathbf{r}'|\omega) = \frac{ie^2 n(\mathbf{r})}{m\omega}\delta_{\alpha\beta}\delta(\mathbf{r} - \mathbf{r}') + \frac{1}{\hbar\omega}\int_{-\infty}^0 d\tau e^{\lambda\tau - i\omega\tau}$$

$$\times \mathrm{Sp}\hat{\eta}_{eq}\left[e^{-i\widehat{H}\tau/\hbar}\hat{I}_\alpha(\mathbf{r})e^{i\widehat{H}\tau/\hbar}, \hat{I}_\beta(\mathbf{r}')\right]. \tag{10}$$

The second term on the right-hand side of this equation describes the correlation of the currents at the points $(\mathbf{r}, \tau)$ and $(\mathbf{r}', 0)$. The expressions relating the linear kinetic coefficients to the equilibrium correlation functions are called the Kubo formulas. After a spatial Fourier transformation of the current and field according to Eq. (7.15), one finds

$$\Delta I_\alpha(\mathbf{q}, \omega) = \frac{1}{V} \sum_{\mathbf{q}'\beta} \sigma_{\alpha\beta}(\mathbf{q}, \mathbf{q}'|\omega) E_\beta(\mathbf{q}', \omega), \qquad (11)$$

where the tensor $\sigma_{\alpha\beta}(\mathbf{q}, \mathbf{q}'|\omega)$ is obtained by the double Fourier transformation of Eq. (10) (problem 3.1). The expression for this tensor is

$$\sigma_{\alpha\beta}(\mathbf{q}, \mathbf{q}'|\omega) = \frac{ie^2 n(\mathbf{q} - \mathbf{q}')}{m\omega} \delta_{\alpha\beta} + \frac{1}{\hbar\omega} \int_{-\infty}^{0} d\tau e^{\lambda\tau - i\omega\tau}$$

$$\times \mathrm{Sp}\hat{\eta}_{eq} \left[e^{-i\hat{H}\tau/\hbar} \hat{I}_\alpha(\mathbf{q}) e^{i\hat{H}\tau/\hbar}, \hat{I}_\beta(-\mathbf{q}') \right], \qquad (12)$$

where $n(\mathbf{q} - \mathbf{q}')$ is the Fourier transform of electron density and $\hat{\mathbf{I}}(\mathbf{q})$ is the operator of $\mathbf{q}$-th Fourier component of the unperturbed current density. The expression for $\hat{\mathbf{I}}(\mathbf{q})$ follows from Eq. (4.15), where the δ-function is replaced by the plane wave with wave vector $\mathbf{q}$:

$$\hat{\mathbf{I}}(\mathbf{q}) = \frac{e}{2} \sum_j \left[\hat{\mathbf{v}}_j e^{-i\mathbf{q}\cdot\hat{\mathbf{x}}_j} + e^{-i\mathbf{q}\cdot\hat{\mathbf{x}}_j} \hat{\mathbf{v}}_j \right]. \qquad (13)$$

We remind that $\hat{\mathbf{v}}_j = \hat{\mathbf{p}}_j/m = -i(\hbar/m)\partial/\partial\mathbf{x}_j$ is the velocity operator for j-th electron.

In the case of a spatially-homogeneous system, the conductivity tensor (10) depends only on the difference of coordinates, $\sigma_{\alpha\beta}(\mathbf{r}, \mathbf{r}'|\omega) = \sigma_{\alpha\beta}(\mathbf{r} - \mathbf{r}'|\omega)$. Introducing the averaged and differential coordinates according to Eq. (9.5), we express the Fourier-transformed conductivity tensor as $\sigma_{\alpha\beta}(\mathbf{q}, \mathbf{q}'|\omega) = V\delta_{\mathbf{q}\mathbf{q}'}\sigma_{\alpha\beta}(\mathbf{q}, \omega)$. Now, instead of Eq. (11), we have an algebraic relation between the current and the field, $\Delta I_\alpha(\mathbf{q}, \omega) = \sum_\beta \sigma_{\alpha\beta}(\mathbf{q}, \omega) E_\beta(\mathbf{q}, \omega)$; see Eq. (11.23). The tensor $\sigma_{\alpha\beta}(\mathbf{q}, \omega)$ describes a reaction of the system on the inhomogeneous field (the effect of spatial dispersion). Since $n(\Delta\mathbf{q})|_{\Delta\mathbf{q}=0} = Vn$, where n is the electron density in the homogeneous system, the conductivity tensor is given by

$$\sigma_{\alpha\beta}(\mathbf{q}, \omega) = \frac{ie^2 n}{m\omega} \delta_{\alpha\beta} + \frac{1}{V\hbar\omega} \int_{-\infty}^{0} d\tau e^{\lambda\tau - i\omega\tau}$$

$$\times \mathrm{Sp}\hat{\eta}_{eq} \left[e^{-i\hat{H}\tau/\hbar} \hat{I}_\alpha(\mathbf{q}) e^{i\hat{H}\tau/\hbar}, \hat{I}_\beta(-\mathbf{q}) \right]. \qquad (14)$$

Although this expression contains ω in the denominators of both terms, the conductivity tensor $\sigma_{\alpha\beta}(\mathbf{q}, \omega)$ is not divergent at $\omega = 0$ if the collisions are taken into account. We already know this property from Sec. 8, where the conductivity of electrons interacting with impurities has been calculated; see Eq. (8.26). Next, Eq. (14) demonstrates that the conductivity tensor should go to zero at $\omega \to \infty$. Both these properties follow from the physical consideration that any response to a finite perturbation should be finite.

Considering analytical properties of $\sigma_{\alpha\beta}(\mathbf{q}, \omega)$ as a function of complex variable ω, one may derive integral relations between real and imaginary parts of the conductivity tensor (problem 3.2). In the absence of spatial dispersion, for $\sigma_{\alpha\beta}(\omega) = \sigma_{\alpha\beta}(\mathbf{q} = 0, \omega)$, these relations are written in the form

$$\mathrm{Re}\,\sigma_{\alpha\beta}(\omega) = \frac{2}{\pi}\mathcal{P}\int_0^\infty d\omega' \frac{\omega'\,\mathrm{Im}\,\sigma_{\alpha\beta}(\omega')}{\omega'^2 - \omega^2},$$

$$\mathrm{Im}\,\sigma_{\alpha\beta}(\omega) = -\frac{2}{\pi}\mathcal{P}\int_0^\infty d\omega' \frac{\omega\,\mathrm{Re}\,\sigma_{\alpha\beta}(\omega')}{\omega'^2 - \omega^2}. \tag{15}$$

Equations (15), known as Kramers-Kronig dispersion relations, are quite general because they follow just from the causality principle. Using them, one may find the imaginary part of the conductivity tensor at a given ω if the real part is known in the whole spectral range (for example, from the optical absorption measurements). Similar equations can be written for the dielectric permittivity tensor $\epsilon_{\alpha\beta}(\omega)$ which is directly related to the conductivity tensor; see Sec. 17.

Let us consider symmetry properties of the conductivity tensor. Taking into account that $\hat{\mathbf{I}}^+(\mathbf{r}) = \hat{\mathbf{I}}(\mathbf{r})$ (because the velocity and coordinate operators are Hermitian), one can also write $\hat{\mathbf{I}}^+(\mathbf{q}) = \hat{\mathbf{I}}(-\mathbf{q})$. Using these relations in Eqs. (10) and (14), we obtain the general symmetry properties

$$\sigma_{\alpha\beta}(\mathbf{r}, \mathbf{r}'|\omega) = \sigma_{\alpha\beta}^*(\mathbf{r}, \mathbf{r}'| - \omega), \qquad \sigma_{\alpha\beta}(\mathbf{q}, \omega) = \sigma_{\alpha\beta}^*(-\mathbf{q}, -\omega). \tag{16}$$

On the other hand, one can take into account that $\hat{H}^* = \hat{H}$, $\hat{\mathbf{I}}^*(\mathbf{r}) = -\hat{\mathbf{I}}(\mathbf{r})$, and $\hat{\mathbf{I}}^*(\mathbf{q}) = -\hat{\mathbf{I}}(-\mathbf{q})$, as follows from the explicit expressions of the time-independent Hamiltonian $\hat{H}$ and current density operator $\hat{\mathbf{I}}(\mathbf{r})$ in the absence of the vector potential. These relations are the consequences of the symmetry of classical equations of motion with respect to time reversal (in quantum mechanics, the wave functions are replaced by the complex conjugate ones under this reversal). Using them, we derive the symmetry properties $\sigma_{\alpha\beta}(\mathbf{r}, \mathbf{r}'|\omega) = \sigma_{\beta\alpha}^*(\mathbf{r}', \mathbf{r}| - \omega)$ and $\sigma_{\alpha\beta}(\mathbf{q}, \omega) = \sigma_{\beta\alpha}^*(\mathbf{q}, -\omega)$ from Eqs. (10) and (14). Combining these properties with those of Eq. (16), we find the important relations

$$\sigma_{\alpha\beta}(\mathbf{r}, \mathbf{r}'|\omega) = \sigma_{\beta\alpha}(\mathbf{r}', \mathbf{r}|\omega), \qquad \sigma_{\alpha\beta}(\mathbf{q}, \omega) = \sigma_{\beta\alpha}(-\mathbf{q}, \omega) \tag{17}$$

known as Onsager's symmetry. In the presence of magnetic fields created either by external magnets or by magnetization of the material itself, Eq. (17) should be modified. Indeed, the relations $\widehat{H}^* = \widehat{H}$ and $\hat{\mathbf{I}}^*(\mathbf{r}) = -\hat{\mathbf{I}}(\mathbf{r})$ are no longer valid if $\hat{\mathbf{p}}_j$ is replaced by the kinematic momentum operator $\hat{\mathbf{p}}_j - (e/c)\mathbf{A}_{\mathbf{x}_j}$, where $\mathbf{A}$ is the vector potential describing the magnetic field. If, however, the sign of the vector potential is changed simultaneously with complex conjugation of $\widehat{H}$ and $\hat{\mathbf{I}}(\mathbf{r})$, these relations again become true. Therefore, the relations (17) are valid if the signs of the magnetic field in the left- and right-hand sides of these relations are assumed to be opposite. The relation between the components $\sigma_{xy}(\omega)$ and $\sigma_{yx}(\omega)$ of the conductivity tensor calculated in Sec. 11 in the presence of a magnetic field is a manifestation of this symmetry principle. Onsager's symmetry takes place for any kind of equilibrium linear kinetic coefficients (generalized susceptibilities) introduced in a similar way as in Eq. (9); see below in this section.

In the case of a long-wavelength ($\mathbf{q} \to 0$) perturbation described by the field $\mathbf{E}(\omega)$, the induced current density is written as $\Delta\mathbf{I}(\mathbf{q},\omega) = V\delta_{\mathbf{q},0}\Delta\mathbf{I}(\omega)$, while the relation between the current and the field takes the form $\Delta I_\alpha(\omega) = \sum_\beta \sigma_{\alpha\beta}(\omega)E_\beta(\omega)$; see also Eq. (8.25). The conductivity tensor becomes

$$\sigma_{\alpha\beta}(\omega) = \frac{ie^2 n}{m\omega}\delta_{\alpha\beta} + \frac{e^2}{V\hbar\omega}\int_{-\infty}^{0} d\tau e^{\lambda\tau - i\omega\tau}$$

$$\times \mathrm{Sp}\hat{\eta}_{eq}\left[e^{-i\widehat{H}\tau/\hbar}\hat{v}_\alpha e^{i\widehat{H}\tau/\hbar}, \hat{v}_\beta\right], \tag{18}$$

where the current density operator is expressed through the velocity operator as $\hat{I}_\alpha(\mathbf{q} = 0) = e\hat{v}_\alpha$. The relations similar to Eqs. (10), (12), (14), and (18) can be also written for the polarizability α describing a linear response of electron charge distribution to the longitudinal field given by the scalar potential $\Phi(\mathbf{r},\omega)$ (problem 3.3). Owing to the gradient invariance, the expression of the current density $\Delta\mathbf{I}(\mathbf{r},\omega)$ through the electric field $\mathbf{E}(\mathbf{r},\omega) = -\nabla_\mathbf{r}\Phi(\mathbf{r},\omega)$ should be identical to that given by Eqs. (9) and (10). It is instructive to check this property directly (problem 3.4).

The expressions for the complex conductivity tensor $\sigma_{\alpha\beta}$ reduce the problem of linear response to a calculation of the correlation function of current densities (or electron velocities). Apart from Eqs. (10), (12), (14), and (18), there exist other representations of the Kubo formula for the conductivity (problem 3.5). Although the correlation functions are the equilibrium ones, the calculations according to Kubo formulas are often very complicated (for example, one has to take an average over the impurity distribution when the potential of electron-impurity interaction enters both S-operator and $\hat{\eta}_{eq}$). A convenient approach to such calculations assumes expression of the correlation functions through the Green's functions. This is done below for the conductivity of a macroscopically homogeneous and isotropic electron-impurity system, when the averaging over the impurity ensemble gives us $\sigma_{\alpha\beta} = \sigma\delta_{\alpha\beta}$. Employing a full set of the ket-vectors $|\delta\rangle$ determined from the eigenstate problem $\widehat{H}|\delta\rangle = \varepsilon_\delta|\delta\rangle$, we introduce the equilibrium distribution function $f(\varepsilon_\delta) = \langle\delta|\hat{\eta}_{eq}|\delta\rangle$. In this basis, $\text{Sp}\hat{\eta}_{eq}\ldots = 2\langle\langle\sum_\delta f(\varepsilon_\delta)\langle\delta|\ldots|\delta\rangle \rangle\rangle$, where $\langle\langle\ldots\rangle\rangle$ denotes the averaging over the impurity ensemble and the factor of 2 accounts for spin degeneracy (the spin index is not included in δ). Therefore, the real part of the conductivity given by Eq. (18) is rewritten as

$$\text{Re}\sigma(\omega) = \frac{2e^2}{\hbar\omega V}\text{Re}\int_{-\infty}^0 d\tau\ e^{\lambda\tau - i\omega\tau}\Big\langle\Big\langle\sum_{\delta\delta'} f(\varepsilon_\delta)|\langle\delta|\hat{v}_\alpha|\delta'\rangle|^2$$

$$\times\left\{e^{-i(\varepsilon_\delta - \varepsilon_{\delta'})\tau/\hbar} - e^{i(\varepsilon_\delta - \varepsilon_{\delta'})\tau/\hbar}\right\}\Big\rangle\Big\rangle. \tag{19}$$

Next, we permute δ and δ' in the second term inside $\{\ldots\}$ and calculate the integral $\int_{-\infty}^0 d\tau \exp[-i(\varepsilon_\delta - \varepsilon_{\delta'} + \hbar\omega + i\lambda)\tau/\hbar]$. The real part of this integral, according to the relations discussed in problem 1.4, leads to the δ-function $\delta(\varepsilon_\delta - \varepsilon_{\delta'} + \hbar\omega)$ so that the equation $\varepsilon_{\delta'} = \varepsilon_\delta + \hbar\omega$ expresses the energy conservation law. In this way we obtain the following expression for the frequency-dependent conductivity:

$$\text{Re}\sigma(\omega) = \frac{2\pi e^2}{\omega V}\Big\langle\Big\langle\sum_{\delta\delta'}[f(\varepsilon_\delta) - f(\varepsilon_\delta + \hbar\omega)]$$

$$\times|\langle\delta|\hat{v}_\alpha|\delta'\rangle|^2\delta(\varepsilon_\delta - \varepsilon_{\delta'} + \hbar\omega)\Big\rangle\Big\rangle. \tag{20}$$

The transition to the static limit is done with the use of the identity

$$\lim_{\omega\to 0}\frac{f(\varepsilon_\delta) - f(\varepsilon_\delta + \hbar\omega)}{\hbar\omega} = -\frac{df(\varepsilon_\delta)}{d\varepsilon_\delta}, \tag{21}$$

which means that at low temperatures only the electrons near the Fermi surface contribute to the static conductivity $\sigma = \lim_{\omega\to 0}\sigma(\omega)$. The imaginary part of the conductivity is equal to zero in the static limit. Finally, we obtain

$$\sigma = \frac{2\pi\hbar e^2}{V}\Big\langle\Big\langle\sum_{\delta\delta'}|\langle\delta|\hat{v}_\alpha|\delta'\rangle|^2\delta(\varepsilon_\delta - \varepsilon_{\delta'})\left(-\frac{df(\varepsilon_\delta)}{d\varepsilon_\delta}\right)\Big\rangle\Big\rangle. \tag{22}$$

This result is known as Greenwood-Peierls formula.

The equations obtained above describe an arbitrary many-electron system. In the case of non-interacting electrons scattering by randomly distributed impurities (see Chapter 2), it is convenient to employ the wave functions in the momentum representation, $\psi_{\mathbf{p}}^{(\delta)} \equiv \langle \mathbf{p}|\delta\rangle$, when the matrix elements of the velocity operator in Eq. (22) are

$$\langle \delta|\hat{v}_\alpha|\delta'\rangle = \sum_{\mathbf{p}} \psi_{\mathbf{p}}^{(\delta)*} \frac{p_\alpha}{m} \psi_{\mathbf{p}}^{(\delta')}. \tag{23}$$

We also use the identity (problem 3.6)

$$\delta(\varepsilon_\delta - \varepsilon_{\delta'}) = \int dE \delta(\varepsilon_\delta - E)\delta(E - \varepsilon_{\delta'}) \tag{24}$$

in order to avoid the averaging of the electron distribution function $f(\varepsilon_\delta)$ which depends on the characteristics of the state δ in a complicated way. As a result, Eq. (22) is transformed to

$$\sigma = \frac{2\pi\hbar e^2}{m^2 V} \int dE \left(-\frac{df(E)}{dE}\right) \sum_{\mathbf{pp'}} p_\alpha p'_\alpha$$

$$\times \left\langle\!\left\langle \left(\sum_{\delta\delta'} \psi_{\mathbf{p}}^{(\delta)*} \psi_{\mathbf{p'}}^{(\delta)} \delta(\varepsilon_\delta - E)\psi_{\mathbf{p}}^{(\delta')} \psi_{\mathbf{p'}}^{(\delta')*} \delta(E - \varepsilon_{\delta'})\right)\right\rangle\!\right\rangle, \tag{25}$$

where the impurity averaging is applied to the expression containing four ψ-functions.

Since the sums over δ and δ' in Eq. (25) are separated, one may rewrite the expression inside $\langle\langle\ldots\rangle\rangle$ through the spectral density function

$$A_E(\mathbf{p}, \mathbf{p'}) = \sum_\delta \psi_{\mathbf{p}}^{(\delta)} \psi_{\mathbf{p'}}^{(\delta)*} \delta(\varepsilon_\delta - E), \tag{26}$$

which depends on a pair of momenta and energy E. The expression of $A_E(\mathbf{p}, \mathbf{p'})$ through the Green's functions is discussed in the next section. Replacing $p_\alpha p'_\alpha$ by $(\mathbf{p} \cdot \mathbf{p'})/d$, according to the introduction of the isotropic conductivity as $\sigma = \sum_\alpha \sigma_{\alpha\alpha}/d$ (here $d = 2$ or 3, and the conductivity of 1D electrons will be discussed in Chapter 12), we finally obtain

$$\sigma = \frac{2\pi\hbar e^2}{m^2 V} \int dE \left(-\frac{df(E)}{dE}\right) \sum_{\mathbf{pp'}} \langle\langle A_E(\mathbf{p}, \mathbf{p'})A_E(\mathbf{p'}, \mathbf{p})\rangle\rangle \frac{(\mathbf{p} \cdot \mathbf{p'})}{d}, \tag{27}$$

where the normalization volume V is equal to L^3 or to L^2 in the cases of 3D or 2D electrons, respectively. Therefore, to calculate the conductivity, one should average the product $A_E(\mathbf{p}, \mathbf{p'})A_E(\mathbf{p'}, \mathbf{p})$ and integrate the result of such averaging over $\mathbf{p}$, $\mathbf{p'}$, and E.

Although only the electric current response has been considered above, it is not difficult to formulate the linear response theory in a general way. Let the perturbation be described by the linear relations

$$\widehat{\Delta H}_\omega = \sum_\alpha \widehat{Q}_\alpha \mathcal{F}_\alpha(\omega), \quad \widehat{\Delta Q}_{\alpha\omega} = \sum_\beta \widehat{B}_{\alpha\beta} \mathcal{F}_\beta(\omega), \qquad (28)$$

where $\mathcal{F}_\alpha(\omega)$ is the Fourier component of the generalized force causing the perturbation. The operators $\widehat{Q}_\alpha$ and $\widehat{B}_{\alpha\beta}$ are, in general, many-particle operators, and they can depend on ω. Introducing the generalized susceptibility $\chi_{\alpha\beta}(\omega)$ according to

$$\Delta Q_\alpha(\omega) = \sum_\beta \chi_{\alpha\beta}(\omega) \mathcal{F}_\beta(\omega), \qquad (29)$$

and using Eqs. (4) and (5), we find the susceptibility as

$$\chi_{\alpha\beta}(\omega) = \mathrm{Sp}\hat{\eta}_{eq}\widehat{B}_{\alpha\beta} + \frac{1}{i\hbar} \int_{-\infty}^{0} d\tau e^{\lambda\tau - i\omega\tau}$$

$$\times \mathrm{Sp}\hat{\eta}_{eq} \left[e^{-i\widehat{H}\tau/\hbar}\widehat{Q}_\alpha e^{i\widehat{H}\tau/\hbar}, \widehat{Q}_\beta \right]. \qquad (30)$$

Below we assume that only the second term in Eq. (30) contributes to the imaginary part of the susceptibility. Let us introduce the symmetric part of the generalized susceptibility according to $\chi_{\alpha\beta}^{(s)}(\omega) = [\chi_{\alpha\beta}(\omega) + \chi_{\beta\alpha}(\omega)]/2$. Employing the representation of exact eigenstates, we obtain

$$\mathrm{Im}\chi_{\alpha\beta}^{(s)}(\omega) = -2\pi \left\langle\!\!\left\langle \sum_{\delta\delta'} [f(\varepsilon_\delta) - f(\varepsilon_{\delta'})] \right.\right.$$

$$\left.\left. \times \mathrm{Re} \left[\langle\delta|\hat{q}_\alpha|\delta'\rangle\langle\delta'|\hat{q}_\beta|\delta\rangle \right] \ \delta(\varepsilon_\delta - \varepsilon_{\delta'} + \hbar\omega) \right\rangle\!\!\right\rangle. \qquad (31)$$

To derive Eq. (31), we have assumed that $\widehat{Q}_\alpha$ is an additive operator, $\widehat{Q}_\alpha = \sum_j \hat{q}_\alpha^j$. Equation (31) is similar to Eq. (20) describing the real part of the frequency-dependent conductivity. If the system is symmetric with respect to time reversal, there exists Onsager's symmetry $\chi_{\alpha\beta}(\omega) = \chi_{\beta\alpha}(\omega)$, and the symmetrization procedure is not necessary. The product $\langle\delta|\hat{q}_\alpha|\delta'\rangle\langle\delta'|\hat{q}_\beta|\delta\rangle$ in these conditions is real so that the sign Re in Eq. (31) can be omitted.

14. Diagram Technique

To describe the influence of the scattering on the spectral density function (13.26), it is convenient to introduce the retarded and advanced

Green's functions G_E^R and G_E^A by the relation

$$G_E^{R,A}(\mathbf{p}, \mathbf{p}') = \sum_\delta \frac{\psi_{\mathbf{p}}^{(\delta)} \psi_{\mathbf{p}'}^{(\delta)*}}{E - \varepsilon_\delta \pm i\lambda}, \qquad (1)$$

where the upper and lower signs correspond to the indices R and A, respectively, and $\lambda \to +0$. The wave functions $\psi_{\mathbf{p}}^{(\delta)}$ and energies ε_δ are determined from the Schroedinger equation in the momentum representation:

$$(\varepsilon_\delta - \varepsilon_p)\psi_{\mathbf{p}}^{(\delta)} - \frac{1}{V}\sum_{\mathbf{p}'} U_{im}(\mathbf{p} - \mathbf{p}')\psi_{\mathbf{p}'}^{(\delta)} = 0, \qquad (2)$$

where $\varepsilon_p = p^2/2m$ is the kinetic energy. The potential of the randomly distributed impurities, $U_{im}(\mathbf{r}) = \sum_\alpha v(\mathbf{r} - \mathbf{R}_\alpha)$, (as in Sec. 7, $\mathbf{R}_\alpha$ is the coordinate of α-th impurity) is written in this representation as

$$U_{im}(\mathbf{p} - \mathbf{p}') = \sum_\alpha \exp\left[-i(\mathbf{p} - \mathbf{p}') \cdot \mathbf{R}_\alpha/\hbar\right] v(|\mathbf{p} - \mathbf{p}'|/\hbar), \qquad (3)$$

where $v(q)$ is the Fourier transform of the potential of single impurity introduced by Eq. (7.15). According to Eq. (3), $U_{im}(\mathbf{p}-\mathbf{p}') = U_{im}^*(\mathbf{p}' - \mathbf{p})$ and Eq. (2) leads to $\psi_{\mathbf{p}}^{(\delta)} = \psi_{-\mathbf{p}}^{(\delta)*}$. This condition follows from the invariance of the electron-impurity system with respect to time reversal and causes the following symmetry property of the Green's functions:

$$G_E^s(\mathbf{p}, \mathbf{p}') = G_E^s(-\mathbf{p}', -\mathbf{p}), \qquad s = R, A. \qquad (4)$$

The definition (1) also implies a relation between the retarded and advanced functions, $G_E^R(\mathbf{p}, \mathbf{p}') = G_E^{A*}(\mathbf{p}', \mathbf{p})$. The spectral density function (13.26) is connected to G_E^R and G_E^A by the following relation:

$$A_E(\mathbf{p}, \mathbf{p}') = \frac{1}{2\pi i}\left[G_E^A(\mathbf{p}, \mathbf{p}') - G_E^R(\mathbf{p}, \mathbf{p}')\right]. \qquad (5)$$

Multiplying the Schroedinger equation (2) by $\psi_{\mathbf{p}}^{(\delta)*}$, we calculate the sum over δ and obtain equations for $G_E^{R,A}$. Next, using the orthogonality property $\sum_\delta \psi_{\mathbf{p}}^{(\delta)} \psi_{\mathbf{p}'}^{(\delta)*} = \delta_{\mathbf{p}\mathbf{p}'}$, we find

$$(E \pm i\lambda - \varepsilon_p)G_E^s(\mathbf{p}, \mathbf{p}') - \frac{1}{V}\sum_{\mathbf{p_1}} U_{im}(\mathbf{p} - \mathbf{p_1})G_E^s(\mathbf{p_1}, \mathbf{p}') = \delta_{\mathbf{p}\mathbf{p}'}, \qquad (6)$$

where the upper and the lower signs, as in Eq. (1), correspond to R and A. The Green's function $G_E^s(\mathbf{p}, \mathbf{p}')$ can be considered as a Fourier transform of the Green's function of time-dependent Schroedinger equation, $G_{tt'}^s(\mathbf{r}, \mathbf{r}')$ (problem 3.7). In the momentum representation, it is determined by a non-homogeneous integral equation. In the absence of scattering, this equation is reduced to the algebraic one:

$$(E \pm i\lambda - \varepsilon_p)g_E^s(\mathbf{p}, \mathbf{p}') = \delta_{\mathbf{pp}'}. \tag{7}$$

It determines the Green's function of free electron. Using Eq. (7), we rewrite Eq. (6) as

$$G_E^s(\mathbf{p}, \mathbf{p}') = g_E^s(\mathbf{p}, \mathbf{p}') + \frac{1}{V} \sum_{\mathbf{p}_1 \mathbf{p}_2} g_E^s(\mathbf{p}, \mathbf{p}_1)U_{im}(\mathbf{p}_1 - \mathbf{p}_2)G_E^s(\mathbf{p}_2, \mathbf{p}')$$

$$= \delta_{\mathbf{pp}'}g_E^s(\mathbf{p}) + \frac{1}{V} \sum_{\mathbf{p}_1} g_E^s(\mathbf{p})U_{im}(\mathbf{p} - \mathbf{p}_1)G_E^s(\mathbf{p}_1, \mathbf{p}'). \tag{8}$$

To obtain this equation, we have employed the expression of the free-electron Green's function:

$$g_E^s(\mathbf{p}, \mathbf{p}') = \delta_{\mathbf{pp}'}g_E^s(\mathbf{p}), \qquad g_E^s(\mathbf{p}) = (E \pm i\lambda - \varepsilon_p)^{-1}. \tag{9}$$

Equation (8) can be solved by iterations, leading to the following expansion of the Green's function into power series:

$$G_E^s(\mathbf{p}, \mathbf{p}') = \delta_{\mathbf{pp}'}g_E^s(\mathbf{p}) + V^{-1}g_E^s(\mathbf{p})U_{im}(\mathbf{p} - \mathbf{p}')g_E^s(\mathbf{p}')$$

$$+V^{-2} \sum_{\mathbf{p}_1} g_E^s(\mathbf{p})U_{im}(\mathbf{p} - \mathbf{p}_1)g_E^s(\mathbf{p}_1)U_{im}(\mathbf{p}_1 - \mathbf{p}')g_E^s(\mathbf{p}') + \dots . \tag{10}$$

Here and below we assume that the series of the kind (10) converge. Similar power series can be written for the Green's function $G_{tt'}^s(\mathbf{r}, \mathbf{r}')$ (problem 3.8).

To find the conductivity, one must average the product of the series (10). This requires a rather complicated procedure, which can be simplified by developing a diagram technique operating with graphic images of such series. Let the Green's functions (1) and (9) correspond to double and single solid lines, respectively, while the impurity potential corresponds to a vertical broken line attached to a vertex:

$$\underset{\mathbf{p} \quad \mathbf{p}'}{\bullet\!\!=\!\!=\!\!\bullet} = G_E^s(\mathbf{p}, \mathbf{p}') , \quad \underset{\mathbf{p}}{\rule{1cm}{0.4pt}} = g_E^s(\mathbf{p}) , \quad \underset{\mathbf{p} \quad \mathbf{p}_1}{\bullet} = U_{im}(\mathbf{p} - \mathbf{p}_1). \tag{11}$$

Using these images, one can represent Eq. (10) in the form of diagram series:

$$\underset{\mathbf{p} \quad \mathbf{p}'}{\bullet\!\!=\!\!=\!\!\bullet} = \delta_{\mathbf{pp}'} \underset{\mathbf{p}}{\rule{1cm}{0.4pt}} + \underset{\mathbf{p} \quad \mathbf{p}'}{\rule{1cm}{0.4pt}} + \underset{\mathbf{p} \quad \mathbf{p}_1 \quad \mathbf{p}'}{\rule{1.5cm}{0.4pt}} + \dots , \tag{12}$$

where summation over the inner momenta $\mathbf{p}_1, \dots$ is implied. In this section we apply the diagram technique in order to calculate the averaged Green's function

$$\langle\langle G_E^s(\mathbf{p}, \mathbf{p}') \rangle\rangle = \delta_{\mathbf{p}\mathbf{p}'} G_E^s(\mathbf{p}). \tag{13}$$

We note that the procedure of averaging is defined by Eq. (7.4). The averaged Green's function is diagonal in the momentum because of the translational invariance of the averaged problem (macroscopic translational invariance).

Since we consider power series of the impurity potential, one has to average the products of the exponential factors $\exp(-i\mathbf{q} \cdot \mathbf{R}_\alpha)$ from Eq. (3). The diagram with k impurity lines is averaged according to the equation

$$\langle\langle U_{im}(\hbar\mathbf{q}_1) \ldots U_{im}(\hbar\mathbf{q}_k) \rangle\rangle = v(\mathbf{q}_1) \ldots v(\mathbf{q}_k) V^{-k}$$

$$\times \sum_{\alpha_1 \ldots \alpha_k} \int d\mathbf{R}_{\alpha_1} \ldots \int d\mathbf{R}_{\alpha_k} \exp(-i\mathbf{q}_1 \cdot \mathbf{R}_{\alpha_1} \ldots - i\mathbf{q}_k \cdot \mathbf{R}_{\alpha_k}). \tag{14}$$

The linear contribution ($k = 1$) is proportional to the averaged impurity potential, since

$$\left\langle\left\langle \sum_\alpha e^{-i\mathbf{q}\cdot\mathbf{R}_\alpha} \right\rangle\right\rangle = 0\big|_{\mathbf{q}\neq 0} \tag{15}$$

and $U_{im}(\hbar\mathbf{q} = 0) = \int d\mathbf{r} U_{im}(\mathbf{r})$. This averaged impurity potential is chosen as a reference point (zero energy). The second-order averages ($k = 2$) lead to

$$\frac{1}{V}\left\langle\left\langle \sum_{\alpha_1\alpha_2} e^{-i\mathbf{q}_1\cdot\mathbf{R}_{\alpha_1}} e^{-i\mathbf{q}_2\cdot\mathbf{R}_{\alpha_2}} \right\rangle\right\rangle = n_{im}\delta_{\mathbf{q}_1,-\mathbf{q}_2} , \tag{16}$$

where non-zero terms in the sum correspond to $\alpha_1 = \alpha_2$ and the Kronecker symbol appears from $\int d\mathbf{R}_\alpha \exp(-i\mathbf{q} \cdot \mathbf{R}_\alpha) = V\delta_{\mathbf{q},0}$. As a result, the contribution of paired broken lines in the averaged diagram is given by the factor

$$\overset{\mathbf{q}}{\bullet \quad\quad\quad \bullet} \;=\; n_{im}|v(q)|^2 , \tag{17}$$

with transferred momentum $\hbar\mathbf{q}$. The third-order averages ($k = 3$) give rise to

$$\frac{1}{V}\left\langle\left\langle \sum_{\alpha_1\alpha_2\alpha_3} e^{-i(\mathbf{q}_1\cdot\mathbf{R}_{\alpha_1} + \mathbf{q}_2\cdot\mathbf{R}_{\alpha_2} + \mathbf{q}_3\cdot\mathbf{R}_{\alpha_3})} \right\rangle\right\rangle = n_{im}\delta_{\mathbf{q}_1+\mathbf{q}_2+\mathbf{q}_3,0} , \tag{18}$$

which is calculated analogous to Eq. (16), and non-zero terms correspond to $\alpha_1 = \alpha_2 = \alpha_3$. Next, the fourth-order contributions ($k = 4$)

contain both the terms proportional to n_{im}, coming from the integrals with $\alpha_1 = \alpha_2 = \alpha_3 = \alpha_4$, and the terms proportional to n_{im}^2, coming from the integrals with $\alpha_1 = \alpha_2$ and $\alpha_3 = \alpha_4$ ($\alpha_1 \neq \alpha_3$) and from similar integrals obtained by permutations of the α-indices (there are three such terms):

$$\frac{1}{V} \left\langle\left\langle \sum_{\alpha_1 - \alpha_4} e^{-i(\mathbf{q}_1 \cdot \mathbf{R}_{\alpha_1} + \mathbf{q}_2 \cdot \mathbf{R}_{\alpha_2} + \mathbf{q}_3 \cdot \mathbf{R}_{\alpha_3} + \mathbf{q}_4 \cdot \mathbf{R}_{\alpha_4})} \right\rangle\right\rangle$$

$$= n_{im} \delta_{\mathbf{q}_1 + \mathbf{q}_2 + \mathbf{q}_3 + \mathbf{q}_4, 0} + V n_{im}(n_{im} - 1/V) \tag{19}$$

$$\times \left(\delta_{\mathbf{q}_1, -\mathbf{q}_2} \delta_{\mathbf{q}_3, -\mathbf{q}_4} + \delta_{\mathbf{q}_1, -\mathbf{q}_3} \delta_{\mathbf{q}_2, -\mathbf{q}_4} + \delta_{\mathbf{q}_1, -\mathbf{q}_4} \delta_{\mathbf{q}_2, -\mathbf{q}_3} \right).$$

Since $N_{im} \gg 1$, one must replace $n_{im} - 1/V = (N_{im} - 1)/V$ in the second term by n_{im}. It is the $\propto n_{im}^2$ terms that give the main contribution in Eq. (19) for the case of weak impurity potential. Indeed, the first term of Eq. (19) and $\propto n_{im}$ term of Eq. (18) contain additional smallness in comparison to the contribution of Eq. (17), due to additional factors v^2 and v, respectively. One can neglect these contributions in the Born approximation for electron-impurity scattering (a consideration beyond the Born approximation is given in Sec. 49). Therefore, only the diagrams with even number ($k = 2n$) of impurity lines are essential in the expansion (12). To carry out the averaging, one should consider all possible pairings of the impurity lines, the total number of such pairings is equal to $(2n)!/2^n n!$ in the term of the order $2n$. Each pairing gives us the factor defined by Eq. (17) in the analytical expression of the averaged diagram. Owing to the δ-symbols in Eq. (16), the momentum conservation law is fulfilled in each connecting point of the averaged diagram.

One must point out the difference between the approximations used in the derivation of the kinetic equation in Sec. 7 and the approximations used here in the averaging of the products of impurity potentials. In both cases we used the Born approximation for the interaction of an electron with a single impurity. However, in Sec. 7 we restricted ourselves by the binary correlation functions only, which led us to the contributions of the order of $n_{im}|v(q)|^2$. The contributions of the order of $[n_{im}|v(q)|^2]^2$ coming from the fourth-order correlation functions (see Eq. (19)) and higher-order contributions were neglected. Now all such correlation functions are taken into account. The scattering by a single impurity is no longer considered separately from the scattering by the other impurities, thus making the validity of the consideration independent of the impurity concentration. This allows one to study kinetic properties for strong-scattering regimes, beyond the range of the conditions (7.21). Apart from the Born approximation, we implied that the concentration of the impurities was small enough to neglect the correlations between the positions of different impurities. Both these approximations allowed us to express the correlation functions of an arbitrary order k through the binary correlation functions only. The random potentials $U(\mathbf{r})$, whose statistical characteristics are completely expressed through the binary correlation functions $w(|\mathbf{r}|)$, form a class called the Gaussian potentials; see Sec. 7 and problem 2.3. The calculation of the averages and the diagram technique given in this chapter can be applied to any Gaussian potential, provided the product $n_{im}|v(q)|^2$ is replaced by the Fourier component $w(q)$ of the binary correlation function. Considering the impurity scattering, one can write $w(q)$ instead of $n_{im}|v(q)|^2$, keeping in mind that the correlation function $w(q)$ in this case is essentially determined by the nature of the impurity potential (problem 3.9).

Let us denote the averaged Green's function by a bold line so that Eq. (13) is rewritten as

$$\left\langle\!\!\left\langle \underset{\mathbf{p}\quad\mathbf{p'}}{\rule{0pt}{0pt}\bullet\!\!=\!\!=\!\!\bullet} \right\rangle\!\!\right\rangle = \delta_{\mathbf{pp'}} \;\underset{\mathbf{p}}{\rule{2em}{1pt}}\;. \tag{20}$$

The diagrammatic expansion for $G_E^s(\mathbf{p})$ is obtained by averaging the diagram equation (12) with the use of the definitions (17) and (20):

$$\tag{21}$$

Here we have two kinds of diagrams: the reducible diagrams, which can be divided in two parts by "cutting" only one electron line (for example, the third diagram on the right-hand side of Eq. (21)), and the irreducible ones, which cannot be divided in this way. The sum of all reducible diagrams gives us the averaged Green's function. Therefore, Eq. (21) can be presented as the Dyson equation for $G_E^s(\mathbf{p})$:

$$\tag{22}$$

The self-energy function $\Sigma_E^s(\mathbf{p})$, denoted by a semi-oval in Eq. (22), is given by the following diagram series involving the averaged Green's functions:

$$\Sigma_E^s(\mathbf{p}) = \overset{\frown}{\bullet\!\!-\!\!-\!\!\bullet} = \overset{\cdots}{-\!\!-\!\!-\!\!-} + \overset{\cdots}{-\!\!-\!\!-\!\!-} + \dots \quad . \quad (23)$$

This procedure corresponds to a partial summation of the series in Eq. (21) so that only bold lines appear on the right-hand side of Eq. (23). The Dyson equation (22) has the following analytical form:

$$G_E^s(\mathbf{p}) = g_E^s(\mathbf{p}) + g_E^s(\mathbf{p})\Sigma_E^s(\mathbf{p})G_E^s(\mathbf{p}), \qquad (24)$$

and $G_E^s(\mathbf{p})$ is expressed through $\Sigma_E^s(\mathbf{p})$ as

$$G_E^s(\mathbf{p}) = [E - \varepsilon_p - \Sigma_E^s(\mathbf{p})]^{-1}. \qquad (25)$$

The real part of $\Sigma_E^s(\mathbf{p})$ describes a "renormalization" of the kinetic energy ε_p, while the imaginary part introduces a finite imaginary contribution to the denominator. It describes the broadening of the electron states discussed at the end of this section. Equation (23), in the analytical form, becomes

$$\Sigma_E^s(\mathbf{p}) = \frac{n_{im}}{V} \sum_{\mathbf{q}} |v(q)|^2 G_E^s(\mathbf{p} - \hbar\mathbf{q}) + \frac{n_{im}^2}{V^2} \sum_{\mathbf{q}} \sum_{\mathbf{q}'} |v(q)|^2 |v(q')|^2$$

$$\times G_E^s(\mathbf{p} - \hbar\mathbf{q}) G_E^s(\mathbf{p} - \hbar(\mathbf{q} + \mathbf{q}')) G_E^s(\mathbf{p} - \hbar\mathbf{q}') + \ \dots \ . \qquad (26)$$

Therefore, the Dyson equation for $G_E^s(\mathbf{p})$ is written as a non-linear integral equation.

If the kinetic energy of electrons, ε_p, is large in comparison to $\Sigma_E^s(\mathbf{p})$, one can restrict the expansion (26) by the first term only, where $G_E^s(\mathbf{p} - \hbar\mathbf{q})$ is replaced by $g_E^s(\mathbf{p} - \hbar\mathbf{q})$. We obtain the self-energy function

$$\Sigma_E^{R,A}(p) \simeq \frac{n_{im}}{V} \sum_{\mathbf{p}'} \left| v(|\mathbf{p} - \mathbf{p}'|/\hbar) \right|^2 (E - \varepsilon_{p'} \pm i\lambda)^{-1}. \qquad (27)$$

The imaginary part of this expression, Im $\Sigma_E^{R,A}(\mathbf{p})$, is equal to

$$\mp\pi n_{im} \int \frac{d\mathbf{p}'}{(2\pi\hbar)^d} \left| v(|\mathbf{p} - \mathbf{p}'|/\hbar) \right|^2 \delta(\varepsilon_{p'} - E) = \mp\frac{\hbar}{2\tau_p(E)}. \qquad (28)$$

If ε_p is equal to E, the time $\tau_p(E)$ is equal to the departure time $\tau(E)$ of the electron with energy E (compare to the introduction of the relaxation rates in Sec. 8). The same statement is true for an arbitrary momentum $\mathbf{p}$ if the scattering potential can be treated as a short-range one (point-defect scattering), because in this case $\Sigma_E^s(\mathbf{p})$ is momentum-independent. In any case, the condition $E, \varepsilon_p \gg |\Sigma_E^s(\mathbf{p})|$ implies that $\Sigma_E^s(\mathbf{p})$ standing in Eq. (28) can be estimated at $E \simeq \varepsilon_p$ with the accuracy imposed by this condition. Therefore,

$$G_E^{R,A}(\mathbf{p}) \simeq [E - \varepsilon_p - \mathrm{Re}\Sigma_{E=\varepsilon_p}^s(\mathbf{p}) \pm i\hbar/2\tau(E)]^{-1}. \qquad (29)$$

A comparison of this expression to $g_E^{R,A}(\mathbf{p})$ of Eq. (9) makes it clear that a finite term $i\hbar/2\tau(E)$ replaces an infinitely small term $i\lambda$, and a renormalization of the kinetic energy, according to $\varepsilon_p \rightarrow \varepsilon_p + \mathrm{Re}\Sigma_{E=\varepsilon_p}^s(\mathbf{p})$, takes place. Therefore, Eq. (29) describes a quasiparticle formed as a result of electron-impurity interaction.

The formalism of Green's functions can be applied for calculating the equilibrium quantities of electron-impurity system characterized by the density of electron states. According to the general definition (5.2), the exact density of states, $\rho(E) = (2/V) \int d\mathbf{r} \, \langle\langle \sum_\delta |\psi_{\mathbf{r}}^{(\delta)}|^2 \delta(E - \varepsilon_\delta)\rangle\rangle$, is expressed as (problem 3.10)

$$\rho(E) = \mp \frac{2}{\pi V} \mathrm{Im} \int d\mathbf{r} \int d\Delta t \, e^{iE\Delta t} \langle\langle G_{\Delta t}^{R,A}(\mathbf{r}, \mathbf{r})\rangle\rangle$$

$$= \mp \frac{2}{\pi V} \mathrm{Im} \sum_{\mathbf{p}} G_E^{R,A}(\mathbf{p}), \qquad (30)$$

where we have used the normalization condition $\int d\mathbf{r}|\psi_{\mathbf{r}}^{(\delta)}|^2 = 1$. Note that the factor of 2 appears above due to spin degeneracy. Substituting the Green's function of Eq. (29) into Eq. (30), we rewrite the latter as

$$\rho(E) = \frac{2}{\pi} \int \frac{d\mathbf{p}}{(2\pi\hbar)^d} \frac{\hbar/2\tau(E)}{[E - \varepsilon_p - \mathrm{Re}\Sigma_E^s(\mathbf{p})]^2 + [\hbar/2\tau(E)]^2}. \qquad (31)$$

Therefore, the δ-function $\delta(E - \varepsilon_p)$ standing in the expression for the density of states of free electrons is replaced by the Lorentz factor with a characteristic broadening energy $\hbar/2\tau(E)$. The integral over $\mathbf{p}$ can be calculated easily under the assumption of short-range scattering potentials, because $\Sigma_E^s(\mathbf{p})$, estimated by the first term of the expansion in Eq. (26), becomes $\mathbf{p}$-independent in these conditions. In the 2D case, we obtain (note that $\mathrm{Im}\Sigma_E^A = \hbar/2\tau(E)$ is positive and $\mathrm{Re}\Sigma_E^A = \mathrm{Re}\Sigma_E^R$)

$$\rho(E) = \rho_{2D} \left[\frac{1}{\pi} \arctan\left(\frac{E - \mathrm{Re}\Sigma_E^A}{\mathrm{Im}\Sigma_E^A} \right) + \frac{1}{2} \right], \qquad (32)$$

which replaces the result of the collisionless approximation, $\rho(E) = \rho_{2D} \times \theta(E)$, given by Eq. (5.26). In other words, the scattering shifts and broadens the ideal step in the density of states near the threshold at $E = 0$. The shift $\mathrm{Re}\Sigma_E^A$ under the approximation of short-range scattering potential is given by $\mathrm{Re}\Sigma_E^A = V^{-1}n_{im}|v(0)|^2\mathrm{Re}\sum_{\mathbf{p}} G_E^A(\mathbf{p})$. Since this expression diverges due to the contribution of large p, one should restrict the summation by $p < \hbar/l_c$, where l_c is the characteristic spatial scale (correlation length) of the random potential. The scattering potential is of short range when the ratio $(\hbar/l_c)/\sqrt{2mE}$ is large. For the 2D case, $\mathrm{Re}\Sigma_E^A$ calculated in this way is proportional to a logarithm of this ratio.

Although Eq. (32) provides a reasonable qualitative estimate of the broadening, it cannot serve as a quantitative description of the density of states at $E \to 0$, because *i)* in this region we need to account all terms in the expansion (26) of the self-energy, since there is no small parameter allowing us to neglect them, and *ii)* the Born approximation is no longer applicable for low-energy electrons. To describe the Green's functions of low-energy electrons, another approach appears to be useful, when the Green's function is written in the path-integral form. The averaging over the random potential distribution in that case can be done in a more efficient way as compared to the diagram technique. The path-integral method will be considered in Sec. 16, while in the next section we directly apply the diagram technique in order to calculate the conductivity.

15. Bethe-Salpeter Equation

Let us consider the average of a pair of electron Green's functions, which stands in the expression (13.27) for the electrical conductivity. According to Eq. (14.5), one has to calculate the averages (correlation functions) of the products $G^R G^R$, $G^A G^A$, $G^R G^A$, and $G^A G^R$. Below we consider the most general form of such correlation functions,

$$K_{EE'}^{ss'}(\mathbf{p}_1, \mathbf{p}_3|\mathbf{p}_2, \mathbf{p}_4) \equiv \langle\langle G_E^s(\mathbf{p}_1, \mathbf{p}_2)G_{E'}^{s'}(\mathbf{p}_3, \mathbf{p}_4)\rangle\rangle , \qquad (1)$$

where s and s' may be R or A. Each Green's function is represented according to Eq. (14.12). In the Born approximation, when one should consider only pair correlation functions of the scattering potential, the function K is graphically reproduced as the following sum of diagrams:

We denote this function by a pair of extra bold lines. The bold lines in the right-hand part of Eq. (2) correspond to averaged one-particle Green's functions $G_E^s(\mathbf{p})$, each of them can be represented according to Eq. (14.25). Since the momentum transmitted between the lines is conserved, $K_{EE'}^{ss'}(\mathbf{p}_1\mathbf{p}_3|\mathbf{p}_2\mathbf{p}_4)$ is proportional to $\delta_{\mathbf{p}_1+\mathbf{p}_3,\mathbf{p}_2+\mathbf{p}_4}$.

In order to write a self-consistent equation for the correlation function (2), we again separate the diagrams of the series by reducible ones, which can be divided into two parts by a vertical cut of a pair of solid lines (for example, the third diagram in the second row of Eq. (2)), and irreducible ones, which cannot be divided in this way (for example, the last diagram in the second row of Eq. (2)). Summing the reducible diagrams, we obtain, instead of the infinite series of Eq. (2), the Bethe-Salpeter equation:

$$\text{(3)}$$

The vertex part, denoted by the rectangle in this equation, is introduced through the diagram series

$$\Gamma_{EE'}^{ss'}(\mathbf{p}, \mathbf{p}'|\mathbf{p} - \hbar\mathbf{q},\ \mathbf{p}' + \hbar\mathbf{q}) =$$

$$\text{(4)}$$

representing a sum of all irreducible diagrams without the outer electron lines. The total momentum $\hbar\mathbf{q}$ is transferred between the upper and lower electron lines. The analytical form of the Bethe-Salpeter equation (3) is

$$K_{EE'}^{ss'}(\mathbf{p}_1, \mathbf{p}_3|\mathbf{p}_2, \mathbf{p}_4) = \delta_{\mathbf{p}_1\mathbf{p}_2}\delta_{\mathbf{p}_3\mathbf{p}_4}G_E^s(\mathbf{p}_1)G_{E'}^{s'}(\mathbf{p}_3) + G_E^s(\mathbf{p}_1)G_{E'}^{s'}(\mathbf{p}_3)$$

$$\times \frac{1}{V}\sum_{\mathbf{q}}\Gamma_{EE'}^{ss'}(\mathbf{p}_1, \mathbf{p}_3|\mathbf{p}_1 - \hbar\mathbf{q},\ \mathbf{p}_3 + \hbar\mathbf{q})K_{EE'}^{ss'}(\mathbf{p}_1 - \hbar\mathbf{q},\ \mathbf{p}_3 + \hbar\mathbf{q}|\mathbf{p}_2, \mathbf{p}_4). \quad \text{(5)}$$

This is an integral equation with respect to the variables $\mathbf{p}_1$ and $\mathbf{p}_3$. The other variables ($\mathbf{p}_2$ and $\mathbf{p}_4$) are "idle" and, therefore, not essential here. The irreducible vertex part can be written in the analytical form corresponding to the expansion (4) (here and below we use $w(q)$ instead of $n_{im}|v(q)|^2$):

$$\Gamma_{EE'}^{ss'}(\mathbf{p}, \mathbf{p'}|\mathbf{p} - \hbar\mathbf{q}, \ \mathbf{p'} + \hbar\mathbf{q}) = w(q)$$

$$+\frac{1}{V}\sum_{\mathbf{q}_1} w(|\mathbf{q} - \mathbf{q}_1|)w(q_1)G_E^s(\mathbf{p} - \hbar\mathbf{q}+\hbar\mathbf{q}_1)G_{E'}^{s'}(\mathbf{p'} + \hbar\mathbf{q}_1)$$

$$+\frac{w(q)}{V}\sum_{\mathbf{q}_1} w(q_1)[G_{E'}^{s'}(\mathbf{p'} - \hbar\mathbf{q} + \hbar\mathbf{q}_1)G_{E'}^{s'}(\mathbf{p'} + \hbar\mathbf{q}_1) \qquad (6)$$

$$+G_E^s(\mathbf{p} - \hbar\mathbf{q} + \hbar\mathbf{q}_1)G_E^s(\mathbf{p} + \hbar\mathbf{q}_1)] + \ \cdots .$$

The static conductivity is directly expressed through the correlation function (1), according to Eq. (13.27) and (14.5):

$$\sigma = \frac{\hbar e^2}{2\pi m^2 V}\int dE \left(-\frac{df_E^{(eq)}}{dE}\right)\sum_{\mathbf{pp'}}\frac{(\mathbf{p}\cdot\mathbf{p'})}{d}\sum_{s,s'=R,A}(-1)^l K_E^{ss'}(\mathbf{p}, \mathbf{p'}),$$

$$K_E^{ss'}(\mathbf{p}, \mathbf{p'}) \equiv \langle\langle G_E^s(\mathbf{p}, \mathbf{p'})G_E^{s'}(-\mathbf{p}, -\mathbf{p'})\rangle\rangle = K_{EE}^{ss'}(\mathbf{p}, -\mathbf{p} \mid \mathbf{p'}, -\mathbf{p'}), \quad (7)$$

where $l = 1$ for $s = s'$ and $l = 0$ for $s \neq s'$. In Eq. (7) we have used the property (14.4) in order to shift the momenta $\mathbf{p}$ and $\mathbf{p'}$ to the left and right ends of the diagrams, respectively. For the coinciding energies and momenta, the vertex part (6) is reduced to $\Gamma_E^{ss'}(\mathbf{p}, \mathbf{p}-\hbar\mathbf{q}) \equiv \Gamma_{EE}^{ss'}(\mathbf{p}, -\mathbf{p}|\mathbf{p} - \hbar\mathbf{q}, -\mathbf{p}+\hbar\mathbf{q})$. Therefore, the Bethe-Salpeter equation for $K_E^{ss'}$ is rewritten as

$$K_E^{ss'}(\mathbf{p}, \mathbf{p'}) = G_E^s(\mathbf{p})G_E^{s'}(\mathbf{p})$$

$$\times \left[\delta_{\mathbf{pp'}} + \frac{1}{V}\sum_{\mathbf{q}}\Gamma_E^{ss'}(\mathbf{p}, \mathbf{p} - \hbar\mathbf{q})K_E^{ss'}(\mathbf{p} - \hbar\mathbf{q}, \mathbf{p'})\right]. \qquad (8)$$

Below we apply this equation in order to calculate the static conductivity of electron-impurity system in the limit when the characteristic energy of electrons, which is close to the Fermi energy ε_F (we consider the case of degenerate electrons), is large in comparison to $\hbar/\bar{\tau}$, i.e., under conditions when the kinetic equation considered in Sec. 7 is valid. In these conditions, the main contribution to $\Gamma_E^{ss'}(\mathbf{p}, \mathbf{p}-\hbar\mathbf{q})$ comes from the first term of the expansion in Eq. (6). To prove this, let us estimate the second term of this expansion. Owing to the presence of the product of Green's functions, $G_E^{s'}(-\mathbf{p}+\hbar\mathbf{q}_1)G_E^s(\mathbf{p}-\hbar\mathbf{q}+\hbar\mathbf{q}_1)$, the main contribution to the integral over $\mathbf{q}_1$ comes from the region $(\mathbf{p} - \hbar\mathbf{q}_1)^2/2m \simeq (\mathbf{p} - \hbar\mathbf{q} + \hbar\mathbf{q}_1)^2/2m \simeq E$. Since $\mathbf{p}$, $\mathbf{q}$, and E are fixed, this condition leaves very little space for $\mathbf{q}_1$ and, after the integral over $\mathbf{q}_1$ is calculated, it appears that this term contains an additional small factor of $\hbar/\bar{\tau}\varepsilon_F$ in comparison to the first term. A similar consideration can be done for the third and all other terms (see, however, the end of this section) so that $\Gamma_E^{ss'}(\mathbf{p}, \mathbf{p} - \mathbf{q}) \simeq w(q)$. This approach is known as the ladder approximation. Let us introduce the vector function

$$\mathbf{M}_E^{ss'}(\mathbf{p}) = \frac{1}{V} \sum_{\mathbf{p}_1} w(|\mathbf{p} - \mathbf{p}_1|/\hbar) \sum_{\mathbf{p}'} \mathbf{p}' K_E^{ss'}(\mathbf{p}_1, \mathbf{p}'). \qquad (9)$$

This function satisfies the following equation obtained from Eq. (8):

$$\mathbf{M}_E^{ss'}(\mathbf{p}) = \frac{1}{V} \sum_{\mathbf{p}'} w(|\mathbf{p} - \mathbf{p}'|/\hbar) G_E^s(\mathbf{p}') G_E^{s'}(\mathbf{p}') \left[\mathbf{p}' + \mathbf{M}_E^{ss'}(\mathbf{p}')\right]. \qquad (10)$$

Therefore, the vector $\mathbf{M}$ must be directed along $\mathbf{p}$, as follows from the isotropy of the Green's functions, $G_E^s(\mathbf{p}') = G_E^s(p')$. Accordingly, we search for the function (9) in the form $\mathbf{M}_E^{ss'}(\mathbf{p}) = \mathbf{p} M_E^{ss'}(p)$, where the scalar function $M_E^{ss'}(p)$ is isotropic. Substituting it into the integral equation (10), we approximately solve this equation, taking into account that the main contribution to the integral over $\mathbf{p}'$ comes from a narrow region around $p' \simeq p_E \equiv \sqrt{2mE}$, due to the presence of the Green's functions under the integral. Assuming that $M_E^{ss'}(p)$ weakly varies within this region (this assumption will be justified later), we multiply Eq. (10) by $\mathbf{p}$, put $|\mathbf{p}| = p_E$, and obtain

$$M_E^{ss'}(p_E) \simeq [1 + M_E^{ss'}(p_E)]\Lambda(E) \frac{1}{V} \sum_p G_E^s(p) G_E^{s'}(p), \qquad (11)$$

where $\Lambda(E)$ denotes the angular average of the correlation function $w(|\mathbf{p} - \mathbf{p}'|/\hbar)$ at $|\mathbf{p}| = |\mathbf{p}'| = p_E$, weighted with the factor $\cos\varphi$, where $\varphi = \widehat{\mathbf{p}\mathbf{p}'}$. It is expressed through the difference between the ordinary relaxation rate (inverse departure time) and transport relaxation rate:

$$\Lambda(E) \equiv \overline{w\left(2p_E|\sin(\varphi/2)|\right)\cos\varphi} = \frac{\hbar}{\pi\rho_{\mathcal{D}}(E)}\left(\frac{1}{\tau(E)} - \frac{1}{\tau_{tr}(E)}\right). \qquad (12)$$

The conductivity (7) is expressed through $M_E^{ss'}(p)$ as

$$\sigma = \frac{\hbar e^2}{2\pi dm^2 V} \int dE \left(-\frac{df_E^{(eq)}}{dE}\right) \sum_{s,s'=R,A} (-1)^l$$

$$\times \sum_{\mathbf{p}} G_E^s(p) G_E^{s'}(p) p^2 [1 + M_E^{ss'}(p)], \qquad (13)$$

where $l = 0$ for $s \neq s'$ and $l = 1$ for $s = s'$. Again, the factor $p^2[1+M_E^{ss'}(p)]$ in Eq. (13) can be approximated by $p_E^2[1 + M_E^{ss'}(p_E)]$. Therefore, both $M_E^{ss'}(p_E)$ and σ are expressed through $\sum_{\mathbf{p}} G_E^s(p)G_E^{s'}(p)$. Expressing the Green's functions according to Eq. (14.29), we calculate this sum as

$$\frac{1}{V} \sum_{\mathbf{p}} G_E^s(p)G_E^{s'}(p) \simeq \left\{ \begin{array}{cc} \pi\tau(E)\rho_{\mathcal{D}}(E)/\hbar & s \neq s' \\ 0 & s = s' \end{array} \right. . \tag{14}$$

The product $G_E^s(p)G_E^{s'}(p)$ at $s \neq s'$ is approximately equal to the Lorentz factor $\{(\varepsilon_p - E)^2 + [\hbar/2\tau(E)]^2\}^{-1}$, where $E \gg \hbar/2\tau(E)$. Indeed, the dependence of this product on ε_p has a sharp peak around $\varepsilon_p = E$. On the other hand, the dependence of $M_E^{ss'}(p)$ on ε_p and E is much weaker, since it is determined by the energy dependence of the density of states and relaxation times.

Expressing $M_E^{ss'}(p_E)$ from Eq. (11), we substitute it into Eq. (13) and obtain

$$\sigma \simeq \frac{2\hbar e^2}{\pi dm} \int dE \left(-\frac{df_E^{(eq)}}{dE} \right) E$$

$$\times \operatorname{Re}\left[\left(\frac{1}{V} \sum_{\mathbf{p}} G_E^A(p)G_E^R(p) \right)^{-1} - \Lambda(E) \right]^{-1} . \tag{15}$$

Finally, using Eqs. (12) and (14), we observe that the term with $\tau(E)$ vanishes and only the transport time contributes to Eq. (15). In this way we obtain Eq. (8.27) (problem 3.11). Although we have not found new results as compared to the results of Sec. 8, we have demonstrated a regular method, which can be applied for determination of higher-order corrections to the conductivity (with respect to the factor $\hbar/\overline{\tau}\varepsilon_F$) by taking into account the higher-order contributions to the irreducible vertex part.

Let us consider the contribution of the next terms in the diagrammatic expansion (4) and show that in the 2D case the conductivity is modified substantially due to backscattering processes described by the maximally crossed diagrams, the first of them is the second one on the right-hand side of Eq. (4). The irreducible vertex part corresponding to the maximally crossed diagrams is given by the following infinite series:

$$\Gamma_{EE'}^{ss'}(\mathbf{p}, -\mathbf{p}|\mathbf{p} - \hbar\mathbf{q}, -\mathbf{p} + \hbar\mathbf{q})$$

where the double broken line is introduced to denote this particular class of diagrams. The sum of the series (16) satisfies the diagram equation

$$\tag{17}$$

It can be verified easily that a solution of Eq. (17) by iterations leads to the expansion in Eq. (16). In the analytical form, Eq. (17) is written as

$$\Gamma_{EE'}^{ss'}(\mathbf{p}, -\mathbf{p}|\mathbf{p} - \hbar\mathbf{q}, -\mathbf{p} + \hbar\mathbf{q}) = w(q) + \frac{1}{L^2} \sum_{\mathbf{q}_1} w(|\mathbf{q} - \mathbf{q}_1|) \tag{18}$$

$$\times G_E^s(\mathbf{p} - \hbar\mathbf{q} + \hbar\mathbf{q}_1) G_{E'}^{s'}(-\mathbf{p} + \hbar\mathbf{q}_1) \Gamma_{EE'}^{ss'}(\mathbf{p} - \hbar\mathbf{q} + \hbar\mathbf{q}_1, -\mathbf{p}|\mathbf{p} - \hbar\mathbf{q}, -\mathbf{p} + \hbar\mathbf{q}_1).$$

Note that the second and the third arguments ($-\mathbf{p}$ and $\mathbf{p} - \hbar\mathbf{q}$) are idle and can be omitted in the vertex part. Below we assume equal energies for upper and lower electron lines and define $\widetilde{\Gamma}_E^{ss'}(\mathbf{p}, -\mathbf{p} + \hbar\mathbf{q}) \equiv \Gamma_{EE}^{ss'}(\mathbf{p}, -\mathbf{p}|\mathbf{p} - \hbar\mathbf{q}, -\mathbf{p} + \hbar\mathbf{q})$. The tilde indicates that the arguments of this function are defined in a different way as compared to $\Gamma_E^{ss'}$ in Eq. (8). Considering, for the sake of simplicity, the case of short-range correlated inhomogeneities, when $w(q)$ is approximated by the constant w, we rewrite Eq. (18) at $E = E'$ as

$$\widetilde{\Gamma}_E^{ss'}(\mathbf{p}, -\mathbf{p} + \hbar\mathbf{q}) = w + \frac{w}{L^2} \sum_{\mathbf{p}_1} G_E^s(2\mathbf{p} - \hbar\mathbf{q} + \mathbf{p}_1)$$

$$\times G_E^{s'}(p_1)\widetilde{\Gamma}_E^{ss'}(2\mathbf{p} - \hbar\mathbf{q} + \mathbf{p}_1, \mathbf{p}_1). \tag{19}$$

Since the right-hand side of this equation depends only on the difference between the arguments of $\widetilde{\Gamma}_E^{ss'}$, one can write $\widetilde{\Gamma}_E^{ss'}(\mathbf{p}_1, \mathbf{p}_2) = \widetilde{\Gamma}_E^{ss'}(\mathbf{p}_1 - \mathbf{p}_2)$. When this expression is substituted into the integral term of Eq. (19), the variable $\mathbf{p}_1$ drops out of $\widetilde{\Gamma}_E^{ss'}$ so that the irreducible vertex part is determined by a simple algebraic equation. Solving it, we obtain the result

$$\widetilde{\Gamma}_E^{ss'}(\Delta\mathbf{p}) = w\left[1 - \frac{w}{L^2} \sum_{\mathbf{p}_1} G_E^s(\mathbf{p}_1 + \Delta\mathbf{p})G_E^{s'}(p_1)\right]^{-1} \tag{20}$$

which is valid at $\hbar/\overline{\tau}\varepsilon_F \ll 1$.

Let us substitute the irreducible vertex part of Eq. (20) into the Bethe-Salpeter equation (8). Note that in the simple case of short-range correlated inhomogeneities one does not need to introduce the vector $\mathbf{M}_E^{ss'}(\mathbf{p})$, see Eq. (9), to solve Eq. (8). Instead, we introduce a scalar $K_E^{ss'}(\mathbf{p}) = \sum_{\mathbf{p}'} \mathbf{p} \cdot \mathbf{p}' K_E^{ss'}(\mathbf{p}, \mathbf{p}')$. Transforming Eq. (8) by the substitutions $\mathbf{p}_1 = \mathbf{p} - \hbar\mathbf{q}$ and $\Gamma_E^{ss'}(\mathbf{p}, \mathbf{p} - \hbar\mathbf{q}) = \widetilde{\Gamma}_E^{ss'}(\mathbf{p}, -\mathbf{p} + \hbar\mathbf{q}) = \widetilde{\Gamma}_E^{ss'}(\mathbf{p} + \mathbf{p}_1)$, we obtain the following equation:

$$K_E^{ss'}(\mathbf{p}) = G_E^s(p)G_E^{s'}(p) \tag{21}$$

$$\times \left[p^2 + \frac{1}{L^2} \sum_{\mathbf{p}_1} \widetilde{\Gamma}_E^{ss'}(\mathbf{p} + \mathbf{p}_1) \sum_{\mathbf{p}'} (\mathbf{p} \cdot \mathbf{p}') K_E^{ss'}(\mathbf{p}_1, \mathbf{p}') \right].$$

Since $K_E^{ss'}(\mathbf{p}, \mathbf{p}')$ depends on $\cos \widehat{\mathbf{p}\mathbf{p}'}$, we reduce Eq. (21) to a closed integral equation for $K_E^{ss'}(\mathbf{p})$:

$$K_E^{ss'}(\mathbf{p}) = G_E^s(p)G_E^{s'}(p) \tag{22}$$

$$\times \left[p^2 + \frac{1}{L^2} \sum_{\mathbf{p}_1} \widetilde{\Gamma}_E^{ss'}(\mathbf{p} + \mathbf{p}_1) \frac{p}{p_1} \cos \widehat{\mathbf{p}\mathbf{p}_1} K_E^{ss'}(\mathbf{p}_1) \right].$$

To find the conductivity, one needs to integrate $K_E^{ss'}(\mathbf{p})$ over $\mathbf{p}$. If $s = s'$, the second term in the denominator of the right-hand side of Eq. (20) is small and, with $\widetilde{\Gamma}_{\varepsilon_F}^{ss'}(\Delta\mathbf{p}) \simeq w$ into Eq. (22), the integral term there becomes small and can be neglected. Therefore, $K_E^{ss}(\mathbf{p}) \simeq p^2 G_E^s(p)G_E^s(p)$ and, according to Eq. (14), the terms with $s = s'$ do not modify the conductivity in the limit $\hbar/\overline{\tau}\varepsilon_F \ll 1$.

The terms with $s \neq s'$ have to be considered carefully, because the irreducible vertex part for them diverges at small momentum transfer. Indeed, using the Green's functions of Eq. (14.29), we rewrite Eq. (20) in the following form:

$$\widetilde{\Gamma}_{\varepsilon_F}^{ss'}(\Delta\mathbf{p}) = w[1 - I(\Delta p)]^{-1}, \tag{23}$$

$$I(\Delta p) = \frac{w}{L^2} \sum_{\mathbf{p}} (\varepsilon_F - \varepsilon_{\mathbf{p}+\Delta\mathbf{p}/2} + i\hbar/2\tau)^{-1} (\varepsilon_F - \varepsilon_{\mathbf{p}-\Delta\mathbf{p}/2} - i\hbar/2\tau)^{-1},$$

where E is replaced by ε_F since the electron gas is assumed to be strongly degenerate. In the region $\Delta p l_F/\hbar \ll 1$ one has (problem 3.12)

$$I(\Delta p) \simeq 1 - (\Delta p l_F/\hbar)^2/2, \tag{24}$$

where $l_F = v_F\tau$ is the mean free path length. If $\Delta p l_F/\hbar > 1$, the function $I(\Delta p)$ decreases with Δp. The solution of Eq. (22) by iterations is

$$K_{\varepsilon_F}^{RA}(p) = G_{\varepsilon_F}^R(p)G_{\varepsilon_F}^A(p) \tag{25}$$

$$\times \left[p^2 + \frac{1}{L^2} \sum_{\mathbf{p}_1} G_{\varepsilon_F}^R(p_1)G_{\varepsilon_F}^A(p_1)\widetilde{\Gamma}_{\varepsilon_F}^{RA}(\mathbf{p} + \mathbf{p}_1)(\mathbf{p} \cdot \mathbf{p}_1) \right],$$

and $K_{\varepsilon_F}^{AR}(p)$ is written in a similar way.

The first term on the right-hand side of Eq. (25) corresponds to the approximation $\widetilde{\Gamma}_E^{RA} = w$ and leads to the main part of the conductivity, $\sigma = e^2 n\tau/m$. The second term (and a similar term in the equation for $K_{\varepsilon_F}^{AR}(p)$) leads to a quantum correction to the conductivity, $\delta\sigma$. Calculating the conductivity according to Eq. (7) with the use of Eqs. (23) and (24), we write the quantum correction as

$$\delta\sigma = \frac{\hbar^3 e^2 w}{\pi m^2 l_F^2} \int \frac{d\mathbf{p}}{(2\pi\hbar)^2} \int \frac{d\mathbf{p}_1}{(2\pi\hbar)^2} \qquad (26)$$

$$\times G_{\varepsilon_F}^R(p) G_{\varepsilon_F}^A(p) G_{\varepsilon_F}^R(p_1) G_{\varepsilon_F}^A(p_1) \frac{(\mathbf{p} \cdot \mathbf{p}_1)}{(\mathbf{p} + \mathbf{p}_1)^2}.$$

The accuracy in determining $\widetilde{\Gamma}_{\varepsilon_F}^{ss'}(\Delta\mathbf{p})$ restricts the region of integration by the condition $|\mathbf{p} + \mathbf{p}_1| < \hbar/l_F$. However, this restriction is not essential, because the main contribution to the integral comes from the region of much smaller $|\mathbf{p} + \mathbf{p}_1|$. The quantum correction (26) appears to be negative, because at $\mathbf{p}_1 \simeq -\mathbf{p}$ one has $\mathbf{p} \cdot \mathbf{p}_1 \simeq -p^2$. It is caused by a strong renormalization of the irreducible vertex part in the region of momenta corresponding to backscattering processes. The divergence of $\widetilde{\Gamma}_{\varepsilon_F}^{RA}(\mathbf{p} + \mathbf{p}_1)$ at $\mathbf{p}_1 = -\mathbf{p}$ leads to a logarithmic divergence of the integral in Eq. (26). The cut-off for this divergence can be achieved if we take into account the factors which suppress the coherence of electron states, i.e., which lead to relaxation of the phase of electron wave function. This occurs, for example, in the inelastic scattering processes. Introducing a characteristic phase relaxation length l_φ, which restricts the integration by the region $|\mathbf{p}+\mathbf{p}_1| > \hbar/l_D$, where $l_D = \sqrt{l_F l_\varphi/2}$ is the diffusion length, we calculate the integrals in Eq. (26) (problem 3.13) and obtain

$$\delta\sigma \simeq -\frac{e^2}{2\pi^2\hbar} \ln\frac{l_\varphi}{l_F}. \qquad (27)$$

The reason why the cut-off of the momentum should be done at $\hbar/l_D$ (and not, say, at $\hbar/l_\varphi$) is based upon the strong inequality $l_F \ll l_\varphi$ implying that the mean distance passed by an electron between the inelastic collisions is l_D. Note that the factor at the logarithm in Eq. (27) depends only on the fundamental physical constants e and $\hbar$. The factor $e^2/2\pi\hbar$ is known as the fundamental conductance quantum (we note that in the 2D case the conductivity and conductance have the same dimensionality, Ohm^{-1}).

In summary, we have shown that $\delta\sigma$ is proportional to the conductance quantum $e^2/2\pi\hbar$ multiplied by a large logarithm. Since the correction $\delta\sigma$ is negative, its effect looks like a localization of electrons and is known as weak localization. A comparison of the classical conductivity given by Eq. (8.27) to the weak-localization correction $\delta\sigma$ shows us that the relative correction is small:

$$\frac{|\delta\sigma|}{\sigma} = \frac{\hbar}{2\pi\varepsilon_F\tau}\ln\frac{l_\varphi}{l_F} \ll 1. \tag{28}$$

Nevertheless, the presence of such small contributions can be easily verified experimentally, for example, from a logarithmic temperature dependence of the conductivity. If $l_\varphi \propto T^{-\alpha}$, where α is a positive constant determined by the mechanism of phase relaxation, one obtains $\delta\sigma \propto \alpha\ln T$. The logarithmic increase of the conductivity with increasing temperature, in contrast to its usual decrease with increasing T in the classical transport regime, is caused by the thermal suppression of the weak-localization correction. A magnetic field, since it changes the phases of electron wave functions, induces a similar suppression of $|\delta\sigma|$ leading to a negative magnetoresistance. The effect of external fields on the weak-localization correction is studied in Sec. 43. We stress that the consideration given above is restricted by the simplest case of weak scattering of electrons by short-range correlated inhomogeneities.

16. Green's Function as a Path Integral

In this section we consider a widely employed method based on a transformation of the exact Green's function in the time-coordinate representation into a path integral. For the electrons interacting with impurity potential, the Green's function satisfies the differential equation (see also problem 3.7)

$$\left[i\hbar\frac{\partial}{\partial t} + \frac{\hbar^2\nabla_\mathbf{r}^2}{2m} - U_{im}(\mathbf{r})\right]G_{t-t'}(\mathbf{r},\mathbf{r}') = \delta(t-t')\delta(\mathbf{r}-\mathbf{r}'), \tag{1}$$

where it is taken into account that the Green's function depends only on the difference between the times t and t', because the Hamiltonian is time-independent. Equation (1) can be viewed as a coordinate representation of the following general operator equation:

$$\left(i\hbar\frac{\partial}{\partial t} - \widehat{H}\right)\widehat{G}_t = \delta(t)\hat{1}. \tag{2}$$

Indeed, if we note that $\langle\mathbf{r}|\widehat{H}|\mathbf{r}\rangle = -\hbar^2\nabla_\mathbf{r}^2/2m + U_{im}(\mathbf{r})$ and $\langle\mathbf{r}|\widehat{G}_t|\mathbf{r}'\rangle = \widehat{G}_t(\mathbf{r},\mathbf{r}')$, we obtain Eq. (1). Equation (2) defines the Green's function in the operator form. The retarded and advanced Green's functions are obtained as formal solutions of this equation with the initial conditions

$\widehat{G}_{t=-\infty} = 0$ and $\widehat{G}_{t=\infty} = 0$, respectively, and are represented through the evolution operators as

$$\widehat{G}_t^R = -\frac{i}{\hbar}\theta(t)e^{-i\widehat{H}t/\hbar}, \quad \widehat{G}_t^A = \frac{i}{\hbar}\theta(-t)e^{-i\widehat{H}t/\hbar}, \qquad (3)$$

where $\theta(t)$ is the theta-function. Below we consider only the retarded Green's function, since $\widehat{G}_t^A = \widehat{G}_{-t}^{R+}$. In the coordinate representation,

$$G_t^R(\mathbf{r}, \mathbf{r}') = \langle \mathbf{r}|\widehat{G}_t^R|\mathbf{r}'\rangle = -\frac{i}{\hbar}\theta(t)\langle \mathbf{r}|e^{-i\widehat{H}t/\hbar}|\mathbf{r}'\rangle. \qquad (4)$$

In order to calculate the matrix element of the evolution operator in Eq. (4), we divide the interval $[0, t]$ in N small parts with the lengths $\tau_N = t/N$ and write the matrix element through a product of N factors integrated over the intermediate coordinates $\mathbf{r}_1, \ldots, \mathbf{r}_{N-1}$:

$$G_t^R(\mathbf{r}, \mathbf{r}') = -\lim_{\tau_N \to 0} \frac{i}{\hbar}\theta(t)\int d\mathbf{r}_{N-1}\ldots\int d\mathbf{r}_1$$

$$\times \langle \mathbf{r}|e^{-i\widehat{H}\tau_N/\hbar}|\mathbf{r}_1\rangle\ldots\langle \mathbf{r}_{N-1}|e^{-i\widehat{H}\tau_N/\hbar}|\mathbf{r}'\rangle. \qquad (5)$$

In the limit $\tau_N \to 0$ (i.e., $N \to \infty$), the matrix elements in Eq. (5) are calculated exactly, because now we can neglect the commutator of the kinetic and potential energies in $\exp(-i\widehat{H}\tau_N/\hbar)$. Within the accuracy of the order of τ_N^2, any matrix element in Eq. (5) takes the form

$$\langle \mathbf{r}_i|e^{-i\widehat{H}\tau_N/\hbar}|\mathbf{r}_{i+1}\rangle = e^{-i\tau_N U_{im}(\mathbf{r}_i)/\hbar}e^{i\hbar\tau_N \nabla_i^2/2m}\delta(\mathbf{r}_i - \mathbf{r}_{i+1}), \qquad (6)$$

where ∇_i acts on the coordinate $\mathbf{r}_i$. Accordingly, the contribution of the impurity potential in Eq. (5) is written as an infinite product of exponential factors. In order to calculate the contribution of the kinetic energy operator, we use the operator form of the integral

$$e^{\widehat{A}^2} = \frac{1}{\sqrt{\pi}}\int_{-\infty}^{\infty} d\xi e^{-\xi^2 + 2\xi\widehat{A}} \qquad (7)$$

for an arbitrary operator $\widehat{A}$. In our case, $\widehat{A} = \sqrt{i\hbar\tau_N/2m}\nabla_i$. The operator of shift, $e^{2\xi\widehat{A}}$, acts on the argument of the δ-function in Eq. (6). Since the Laplace operator ∇_i^2 is written as a sum of the second derivatives over each Cartesian coordinate, the contributions of each dimension can be calculated separately, and the total contribution is a product of these partial contributions. The contribution of one dimension (coordinate x) is obtained after calculating the integral of the δ-function with shifted argument according to Eq. (7):

$$\exp\left[\frac{i\hbar\tau_N}{2m}\frac{d^2}{dx_i^2}\right]\delta(x_i - x_{i+1}) = \sqrt{\frac{m}{2\pi i\hbar\tau_N}}\exp\frac{im(x_i - x_{i+1})^2}{2\hbar\tau_N}. \qquad (8)$$

In the d-dimensional case, the matrix element (6) is written as

$$\langle \mathbf{r}_i | e^{-i\hat{H}\tau_N/\hbar} | \mathbf{r}_{i+1} \rangle = \left(\frac{m}{2\pi i \hbar \tau_N} \right)^{d/2}$$

$$\times \exp \left[i \frac{m(\mathbf{r}_i - \mathbf{r}_{i+1})^2}{2\hbar \tau_N} - i\tau_N U_{im}(\mathbf{r}_i)/\hbar \right]. \tag{9}$$

Therefore, to calculate $G_t^R(\mathbf{r}, \mathbf{r}')$, one should evaluate a product of infinitely small contributions:

$$G_t^R(\mathbf{r}, \mathbf{r}') = - \lim_{N \to \infty} \frac{i}{\hbar} \theta(t) \int d\mathbf{r}_{N-1} \dots \int d\mathbf{r}_1 \left(\frac{m}{2\pi i \hbar \tau_N} \right)^{dN/2}$$

$$\times \exp \left\{ \frac{i}{\hbar} \tau_N \sum_{i=1}^{N} \left[\frac{m(\mathbf{r}_i - \mathbf{r}_{i-1})^2}{2\tau_N^2} - U_{im}(\mathbf{r}_i) \right] \right\}. \tag{10}$$

The infinite product standing in Eq. (5) is transformed to the sum in the exponent of this expression. In the limit $\tau_N \to 0$, this sum is replaced by the integral

$$\frac{i}{\hbar} \int_0^\tau d\tau \left[\frac{m}{2} \dot{\mathbf{r}}_\tau^2 - U_{im}(\mathbf{r}_\tau) \right], \qquad \mathbf{r}_0 = \mathbf{r}, \quad \mathbf{r}_t = \mathbf{r}', \tag{11}$$

where $\dot{\mathbf{r}}_\tau \equiv d\mathbf{r}_\tau/d\tau$. Introducing the functional differential as

$$\mathcal{D}\{\mathbf{r}_\tau\} \Rightarrow \left(\frac{m}{2\pi i \hbar \tau_N} \right)^{dN/2} d\mathbf{r}_{N-1} \dots d\mathbf{r}_1, \qquad N \to \infty, \tag{12}$$

we finally rewrite the Green's function in the following form:

$$G_t^R(\mathbf{r}, \mathbf{r}') = - \frac{i}{\hbar} \theta(t) \int_{\mathbf{r}_0=\mathbf{r}}^{\mathbf{r}_t=\mathbf{r}'} \mathcal{D}\{\mathbf{r}_\tau\} \exp \left\{ \frac{i}{\hbar} \int_0^t d\tau \left[\frac{m}{2} \dot{\mathbf{r}}_\tau^2 - U_{im}(\mathbf{r}_\tau) \right] \right\}$$

$$= - \frac{i}{\hbar} \theta(t) \int_{\mathbf{r}_0=\mathbf{r}}^{\mathbf{r}_t=\mathbf{r}'} \mathcal{D}\{\mathbf{r}_\tau\} \exp \left\{ \frac{i}{\hbar} \int_0^t d\tau \mathcal{L}(\mathbf{r}_\tau, \dot{\mathbf{r}}_\tau) \right\}. \tag{13}$$

The second equation employs the Lagrangian $\mathcal{L}(\mathbf{r}, \dot{\mathbf{r}}) = m\dot{r}^2/2 - U_{im}(\mathbf{r})$. Equation (13) is the exact expression of the Green's function through the path integral introduced with the aid of the above-described limiting procedure.

To find the averaged Green's function, one has to average the factor $\exp \left[-\frac{i}{\hbar} \int_0^t d\tau U_{im}(\mathbf{r}_\tau) \right]$. Thus, the approach developed above allows one to separate the statistical averaging over the random potential, described in Sec. 14, from the quantum-mechanical averaging. Let us average the characteristic functional (see problem 2.3) defined as $\exp[\int d\mathbf{r} f_{\mathbf{r}} U_{im}(\mathbf{r})]$, where $f_{\mathbf{r}}$ is an arbitrary function. First we expand the exponent in series. Then, in the Born approximation for the electron-impurity scattering [as well as for any Gaussian random potential $U_{im}(\mathbf{r})$], we consider only the binary correlation functions and obtain

$$\left\langle\!\!\left\langle e^{\int d\mathbf{r} f_{\mathbf{r}} U_{im}(\mathbf{r})}\right\rangle\!\!\right\rangle = 1 + \frac{1}{2!}\int d\mathbf{r}_1\int d\mathbf{r}_2 f_{\mathbf{r}_1} f_{\mathbf{r}_2}\left\langle\!\langle U_{im}(\mathbf{r}_1)U_{im}(\mathbf{r}_2)\rangle\!\right\rangle$$

$$+\frac{1}{4!}\int d\mathbf{r}_1\int d\mathbf{r}_2\int d\mathbf{r}_3\int d\mathbf{r}_4 f_{\mathbf{r}_1} f_{\mathbf{r}_2} f_{\mathbf{r}_3} f_{\mathbf{r}_4}$$

$$\times\left\langle\!\langle U_{im}(\mathbf{r}_1)U_{im}(\mathbf{r}_2)U_{im}(\mathbf{r}_3)U_{im}(\mathbf{r}_4)\rangle\!\right\rangle + \ldots$$

$$= 1 + \frac{1}{2}\int d\mathbf{r}\int d\mathbf{r}' f_{\mathbf{r}} w(|\mathbf{r}-\mathbf{r}'|) f_{\mathbf{r}'} + \ldots \tag{14}$$

$$+\frac{1}{n!}\left[\frac{1}{2}\int d\mathbf{r}\int d\mathbf{r}' f_{\mathbf{r}} w(|\mathbf{r}-\mathbf{r}'|) f_{\mathbf{r}'}\right]^n + \ldots\,,$$

where the correlation function $w(|\mathbf{r}-\mathbf{r}'|)$ introduced at the end of Sec. 7 is equal to $(n_{im}/V)\sum_{\mathbf{q}}|v(q)|^2 e^{i\mathbf{q}\cdot(\mathbf{r}-\mathbf{r}')}$. The transformation in Eq. (14) uses the fact that the number of possible binary correlation functions in the potential correlation function of the order $2n$ is equal to $(2n)!/2^n n!$; see Sec. 14. Since the infinite sum on the right-hand side of Eq. (14) is again transformed to an exponent, we have

$$\left\langle\!\!\left\langle e^{\int d\mathbf{r} f_{\mathbf{r}} U_{im}(\mathbf{r})}\right\rangle\!\!\right\rangle = \exp\left[\frac{1}{2}\int d\mathbf{r}\int d\mathbf{r}' f_{\mathbf{r}} w(|\mathbf{r}-\mathbf{r}'|) f_{\mathbf{r}'}\right]. \tag{15}$$

Substituting $f_{\mathbf{r}} = -(i/\hbar)\int_0^t d\tau\,\delta(\mathbf{r}_\tau - \mathbf{r})$ into Eq. (15), we average Eq. (13) with the aid of Eq. (15) and obtain

$$G_t^R(|\mathbf{r}-\mathbf{r}'|) = -\frac{i}{\hbar}\theta(t)\int_{\mathbf{r}_0=\mathbf{r}}^{\mathbf{r}_t=\mathbf{r}'}\mathcal{D}\{\mathbf{r}_\tau\} \tag{16}$$

$$\times\exp\left[\frac{im}{2\hbar}\int_0^t d\tau\,\dot{\mathbf{r}}_\tau^2 - \frac{1}{2\hbar^2}\int_0^t d\tau_1\int_0^t d\tau_2 w(|\mathbf{r}_{\tau_1} - \mathbf{r}_{\tau_2}|)\right].$$

One has to calculate the path integrals in this expression by taking into account the contributions to the effective action (which stands in the exponent) coming from the kinetic-energy term and from the non-local term describing the collision processes. In the general case, it is a complicated problem, which, however, can be formulated for numerical calculations in a straightforward way.

The density of states, according to Eqs. (14.30) and (16), is expressed through the contour path integral $\oint \mathcal{D}\{\mathbf{r}_\tau\}\ldots$ with $\mathbf{r}_0 = \mathbf{r}_t = \mathbf{r}$:

$$\rho(E) = \frac{2}{\pi\hbar}\text{Re}\int_0^\infty dt\, e^{iEt/\hbar}\oint\mathcal{D}\{\mathbf{r}_\tau\} \tag{17}$$

$$\times\exp\left[\frac{im}{2\hbar}\int_0^t d\tau\,\dot{\mathbf{r}}_\tau^2 - \frac{1}{2\hbar^2}\int_0^t d\tau_1\int_0^t d\tau_2 w(|\mathbf{r}_{\tau_1} - \mathbf{r}_{\tau_2}|)\right],$$

where we have carried out a Fourier transformation in the time domain. An analytical result for the density of states can be obtained in the case of classically smooth potentials, when $w(|\mathbf{r} - \mathbf{r}'|)$ is replaced by the constant $w(0)$. This means that the path integral is determined only by the kinetic energy of free electrons and reduced to a known expression, which can be represented through the retarded Green's function of free d-dimensional electrons in the time-coordinate representation (problem 3.14):

$$g_t^R(|\mathbf{r}|) = -\frac{i}{\hbar}\theta(t)\left(\frac{m}{2\pi i\hbar t}\right)^{d/2}\exp\left(\frac{im\mathbf{r}^2}{2\hbar t}\right). \tag{18}$$

In view of the presence of $\theta(t)$ in Eq. (18), one can extend the lower limit of integration over time in Eq. (17) to $-\infty$ and write Eq. (17) as

$$\rho(E) = -\frac{2}{\pi}\text{Im}\int_{-\infty}^\infty dt\, e^{iEt/\hbar}g_t^R(0)e^{-w(0)t^2/2\hbar^2} \tag{19}$$

$$= \frac{1}{\pi\hbar}\left(\frac{m}{2\pi\hbar}\right)^{d/2}\int_{-\infty}^\infty \frac{dt}{(it+0)^{d/2}}e^{-w(0)t^2/2\hbar^2+iEt/\hbar}.$$

Note that the path of integration over time passes under the peculiar point at $t = 0$. This is reflected by an infinitely small real positive term added to the factor it in the denominator. In the absence of the impurity potential, the integral in Eq. (19) is calculated easily, and we obtain the density of states for free d-dimensional electrons; see Sec. 5. It $w(0) \neq 0$, a simple result is obtained for 2D electrons, because the point $t = 0$ in this case is a simple pole. We have

$$\rho_{2D}(E) = \frac{m}{2\pi\hbar^2}\left[1 + \text{erf}\left(\frac{E}{\sqrt{2w(0)}}\right)\right], \tag{20}$$

where $\text{erf}(x)$ is the error function. Equation (20) demonstrates that the presence of inhomogeneities leads to a symmetric smearing of the edge of the density of states over the energy $\Delta E \sim \sqrt{2w(0)}$. A similar smearing takes place for 3D and 1D electrons; see Fig. 3.1. Analytical expressions for ρ_{3D} and ρ_{1D} can be written through the confluent hypergeometric functions (problem 3.15). In the tail of the density of states, when $E < 0$ and $|E| \gg \sqrt{w(0)}$, one has

$$\rho(E) \propto \exp\left(-\frac{E^2}{2w(0)}\right), \tag{21}$$

i.e., for arbitrary dimensionality, the density of states follows the exponential law with E^2 in the exponent.

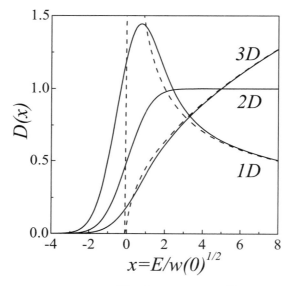

Figure 3.1. Function $D(x) = (2\pi)^{-d/2} \int_{-\infty}^{\infty} d\tau (i\tau + 0)^{-d/2} \exp[-\tau^2/2 + ix\tau]$, which defines the density of states of 1D, 2D, and 3D electron systems (d =1, 2, and 3, respectively) in a classically smooth random potential. The dashed lines show the function $(2\pi)^{-d/2} \int_{-\infty}^{\infty} d\tau (i\tau + 0)^{-d/2} \exp[ix\tau]$, which is proportional to the density of states in the absence of the random potential, for $d = 1$ and $d = 3$.

Below we study how the density of states of 2D electrons in the magnetic field $\mathbf{H}$ perpendicular to the 2D plane is modified by the random potential U in this plane. The density of states, again, is given by Eq. (14.30). Neglecting spin splitting, we write the Hamiltonian as $\widehat{H} = \hat{\pi}^2/2m + U(\mathbf{r})$, where $\mathbf{r} = (x, y)$ is the 2D coordinate and $\hat{\pi} = \hat{\mathbf{p}} - (e/c)\mathbf{A_r}$ is the kinematic momentum of 2D electrons. In contrast to Eq. (5.12), it is convenient to choose the vector potential as $\mathbf{A_r} = [\mathbf{H} \times \mathbf{r}]/2$ (symmetric gauge). The influence of the magnetic field in the Hamiltonian $\widehat{H}$ on the matrix elements standing in Eq. (5) is taken into account by the operator of shift, $\exp(\mathbf{a} \cdot \nabla)f(\mathbf{r}) = f(\mathbf{r} + \mathbf{a})$;

see problem 2.10. The result is written as (compare to Eq. (9))

$$\langle \mathbf{r}_i | e^{-i\hat{H}\tau_N/\hbar} | \mathbf{r}_{i+1} \rangle = \left(\frac{m}{2\pi i \hbar \tau_N} \right) \exp \left[i \frac{m(\mathbf{r}_i - \mathbf{r}_{i+1})^2}{2\hbar \tau_N} \right.$$

$$\left. -i \frac{e\mathbf{A}_{\mathbf{r}_i} \cdot (\mathbf{r}_i - \mathbf{r}_{i+1})}{\hbar c} - \frac{i}{\hbar} \tau_N U(\mathbf{r}_i) \right]. \tag{22}$$

In the limit $\tau_N \to 0$, instead of Eq. (13), we obtain

$$G_t^R(\mathbf{r}, \mathbf{r}') = -\frac{i}{\hbar} \theta(t) \int_{\mathbf{r}_0 = \mathbf{r}}^{\mathbf{r}_t = \mathbf{r}'} \mathcal{D}\{\mathbf{r}_\tau\}$$

$$\times \exp \left\{ \frac{i}{\hbar} \int_0^t d\tau \left[\frac{m}{2} \dot{\mathbf{r}}_\tau^2 - \frac{e}{c} \mathbf{A}_{\mathbf{r}_\tau} \cdot \dot{\mathbf{r}}_\tau - U(\mathbf{r}_\tau) \right] \right\}, \tag{23}$$

which corresponds to the Lagrangian $\mathcal{L}(\mathbf{r}, \dot{\mathbf{r}}) = m\dot{\mathbf{r}}^2/2 - (e/c)\mathbf{A}_{\mathbf{r}} \cdot \dot{\mathbf{r}} - U(\mathbf{r}) = m[\dot{x}^2 + \dot{y}^2 + \omega_c(x\dot{y} - y\dot{x})]/2 - U(x, y)$, where ω_c is the cyclotron frequency introduced in Sec. 5.

The procedure of averaging over the random potential is done according to Eqs. (14) and (15) and leads to a non-local term containing the correlation function $w(|\mathbf{r}-\mathbf{r}'|) = \langle\langle U(\mathbf{r})U(\mathbf{r}')\rangle\rangle$ in the exponential factor. We point out that the averaged Green's function $G_t^R(\mathbf{r}, \mathbf{r}')$ in a magnetic field, in contrast to the one given by Eq. (16), depends on both $\mathbf{r}$ and $\mathbf{r}'$, not only on $|\mathbf{r} - \mathbf{r}'|$. Substituting $\mathbf{r}_\tau \to \mathbf{r}_\tau + \mathbf{r}$ (so that in the new coordinates $\mathbf{r}_0 = 0$ and $\mathbf{r}_t = \mathbf{r}' - \mathbf{r}$), we obtain

$$G_t^R(\mathbf{r}, \mathbf{r}') = -\frac{i}{\hbar} \theta(t) \exp \left(-\frac{ie}{2\hbar c} \mathbf{H} \cdot [\mathbf{r} \times \mathbf{r}'] \right) \int_{\mathbf{r}_0 = 0}^{\mathbf{r}_t = \mathbf{r}' - \mathbf{r}} \mathcal{D}\{\mathbf{r}_\tau\} \tag{24}$$

$$\times \exp \left\{ \frac{i}{\hbar} \int_0^t d\tau \left[\frac{m}{2} \dot{\mathbf{r}}_\tau^2 - \frac{e}{c} \mathbf{A}_{\mathbf{r}_\tau} \cdot \dot{\mathbf{r}}_\tau - \frac{1}{2\hbar^2} \int_0^t d\tau_1 \int_0^t d\tau_2 w(|\mathbf{r}_{\tau_1} - \mathbf{r}_{\tau_2}|) \right] \right\}.$$

According to this equation, the Green's function in a magnetic field is represented as a product of a translation-invariant part, which depends only on $\mathbf{r} - \mathbf{r}'$, by a phase factor containing the vector product $[\mathbf{r} \times \mathbf{r}']$ (see also Chapter 10 and Appendix G). This factor, however, has no influence on the density of states:

$$\rho(E) = \frac{2}{\pi\hbar} \mathrm{Re} \int_0^\infty dt e^{iEt/\hbar} \oint \mathcal{D}\{\mathbf{r}_\tau\} \exp \left[\frac{i}{\hbar} \int_0^t d\tau \left(\frac{m\dot{\mathbf{r}}_\tau^2}{2} \right. \right. \tag{25}$$

$$\left. \left. -\frac{e\mathbf{H} \cdot [\mathbf{r}_\tau \times \dot{\mathbf{r}}_\tau]}{2c} \right) - \frac{1}{2\hbar^2} \int_0^t d\tau_1 \int_0^t d\tau_2 w(|\mathbf{r}_{\tau_1} - \mathbf{r}_{\tau_2}|) \right].$$

The expression under the path integral, as compared to Eq. (17), has an additional term in the exponent, due to the magnetic field. To calculate this integral, we again assume the case of smooth inhomogeneities, when $w(|\mathbf{r}|)$ is replaced by $w(0)$. The density of states is written according to the first equation of Eq. (19), where one should put the Green's function of free 2D electrons in the magnetic field:

$$g_t^R(0) = -\frac{i}{\hbar}\theta(t) \oint \mathcal{D}\{\mathbf{r}_\tau\} \tag{26}$$

$$\times \exp\left[\frac{i}{\hbar}\int_0^t d\tau \left(\frac{m\dot{\mathbf{r}}_\tau^2}{2} - \frac{e\mathbf{H}\cdot[\mathbf{r}_\tau \times \dot{\mathbf{r}}_\tau]}{2c}\right)\right].$$

Calculating the contour path integral in this equation (problem 3.16), we obtain

$$\rho(E) = \frac{m\omega_c}{4\pi^2 i\hbar^2} \int_C dt \frac{e^{-w(0)t^2/2\hbar^2 + iEt/\hbar}}{\sin(\omega_c t/2)}, \tag{27}$$

where the contour C goes along the real axis of complex variable t from $-\infty$ to ∞, passing under the poles $t_k = (2\pi k/\omega_c)$, where k is integer. Note that at $\mathbf{H} = 0$ we have only one such pole, $t = 0$. The integral over time in Eq. (27) is easily calculated at $w(0) = 0$ by shifting the contour to the upper half-plane (for positive energies E). The result is given as a sum of contributions from each pole, $\rho(E) = \pi^{-1}l_H^{-2}\sum_N \delta(E - \varepsilon_N)$, where l_H is the magnetic length and $\varepsilon_N = \hbar\omega_c(N + 1/2)$ is the Landau-level energy ($N = 0, 1, ...$); see also Eq. (5.15). In the case of finite $w(0)$, each of the peaks of the density of states acquires a finite broadening. The integral in Eq. (27) can be calculated if we assume that the characteristic broadening energy $\sqrt{w(0)}$ is small in comparison to the cyclotron energy $\hbar\omega_c$. Substituting $E = \varepsilon_N + \Delta E$, where $|\Delta E| \ll \hbar\omega_c$, we calculate the density of states in the vicinity of each Landau level (problem 3.17). The total density of states is reduced to a sum of the contributions from the Landau levels:

$$\rho(E) = \frac{2}{(2\pi)^{3/2}l_H^2 \sqrt{w(0)}} \sum_{N=0}^{\infty} \exp\left[-\frac{(E - \varepsilon_N)^2}{2w(0)}\right]. \tag{28}$$

Therefore, in the approximation of a classically smooth random potential the density of states of 2D electrons in a strong magnetic field is given as a sum of identical Gaussian peaks placed at the Landau level energies.

17. Dispersion of Dielectric Permittivity

The application of the Green's functions or path integrals to the analysis of the Kubo formula appears to be an efficient way for studying the influence of strong scattering on kinetic properties; see Secs. 15 and 18.

Moreover, even in the collisionless regime, the expressions for the complex conductivity tensor $\sigma_{\alpha\beta}(\mathbf{q}, \omega)$ obtained in Sec. 13 describe a number of non-trivial features of linear response related to both spatial and frequency dispersion of the conductivity tensor (i.e., to its dependence on $\mathbf{q}$ and ω, respectively). In this section we consider the contribution of interband electron transitions to the dielectric permittivity of crystals. The dielectric permittivity is introduced by the following relation

$$\epsilon_{\alpha\beta}(\mathbf{q}, \omega) = \kappa_{\alpha\beta}(\mathbf{q}, \omega) + i\frac{4\pi}{\omega}\sigma_{\alpha\beta}(\mathbf{q}, \omega), \tag{1}$$

where $\kappa_{\alpha\beta}(\mathbf{q}, \omega)$ describes the contribution of the crystal lattice due to ionic polarization, which can be considered in terms of interaction of the electromagnetic waves with TO phonons and will be studied in Sec. 27. The second term on the right-hand side of Eq. (1) is proportional to the complex conductivity, which takes into account the contribution of electron states near the conduction- and valence-band edges. The independent introduction of the lattice contribution and electron contribution to the crystal polarization is justified when the characteristic scales of their dispersion are essentially different. In particular, the consideration given below for constant $\kappa_{\alpha\beta}$ is not valid in the region of frequencies close to ω_{LO} and ω_{TO}, where the response is determined mostly by the phonon contribution.

Since the wave vectors $\mathbf{q}$ of the electromagnetic radiation are small in comparison to characteristic wave vectors of the crystal, up to the ultraviolet spectral region, we start our consideration with the case $\mathbf{q} = 0$, when one should substitute the conductivity tensor of Eq. (13.18) into Eq. (1). Since we consider the transitions of non-interacting electrons placed into the mean field of the crystal, it is possible to study the linear response by applying the linearized one-particle equation (4.32) for the density matrix, instead of using the general equation (13.2). This means that only one-particle operators remain under the trace in Eq. (13.18). It is convenient to separate the terms proportional to ω^{-1}, which describe the divergence of $\sigma_{\alpha\beta}$ at $\omega = 0$ in the collisionless approximation. Thus, we write the conductivity tensor as

$$\sigma_{\alpha\beta}(\omega) = \frac{e^2}{\omega V}\mathrm{Sp}\hat{\eta}_{eq}\left\{\frac{i}{m_e}\delta_{\alpha\beta} + \frac{1}{\hbar}\int_{-\infty}^{0}d\tau e^{\lambda\tau}\left[e^{-i\widehat{H}\tau/\hbar}\hat{v}_\alpha e^{i\widehat{H}\tau/\hbar}, \hat{v}_\beta\right]\right\}$$

$$+ \frac{e^2}{\hbar\omega V}\int_{-\infty}^{0}d\tau e^{\lambda\tau}\left(e^{-i\omega\tau} - 1\right)\mathrm{Sp}\hat{\eta}_{eq}\left[e^{-i\widehat{H}\tau/\hbar}\hat{v}_\alpha e^{i\widehat{H}\tau/\hbar}, \hat{v}_\beta\right], \tag{2}$$

where the second term is finite at $\omega = 0$. Calculating the trace in the first term by using the full basis of Bloch wave functions (5.5), we find that the integral over time in this term leads to

$$\frac{1}{\hbar} \int_{-\infty}^{0} d\tau e^{\lambda \tau} \mathrm{Sp} \hat{\eta}_{eq} \left[e^{-i\hat{H}\tau/\hbar} \hat{v}_\alpha e^{i\hat{H}\tau/\hbar}, \hat{v}_\beta \right] = i \sum_{\delta\delta'}{}' f_\delta \frac{v_{\delta\delta'}^\alpha v_{\delta'\delta}^\beta + v_{\delta\delta'}^\beta v_{\delta'\delta}^\alpha}{\varepsilon_\delta - \varepsilon_{\delta'}}.$$

(3)

Here $f_\delta = \langle \delta | \hat{\eta}_{eq} | \delta \rangle$ is the stationary distribution function of electrons over the Bloch states $|\delta\rangle \equiv |l\mathbf{p}\rangle$, $\mathbf{p}$ is the quasimomentum, and the index l numbers both band and spin states. By the prime sign at the sum, we indicate that the diagonal ($\delta = \delta'$) terms do not contribute to the expression under consideration. Employing the f-sum rule, which generalizes Eq. (5.11) to the case of electrons with non-parabolic spectrum (problem 3.18), we rewrite the first term on the right-hand side of Eq. (2) in the form

$$\frac{ie^2}{\omega} \frac{2}{V} \sum_{\mathbf{p}} f_{n\mathbf{p}} \frac{\partial^2 \varepsilon_{n\mathbf{p}}}{\partial p_\alpha \partial p_\beta},$$

(4)

where the contribution of filled bands with $f_{n\mathbf{p}} = 1$ becomes zero after calculating the sum over $\mathbf{p}$. Therefore, only free carriers (electrons or holes) contribute to $\propto \omega^{-1}$ response. In the case of parabolic electron spectrum with effective mass m, this response is equal to $i\delta_{\alpha\beta}e^2 n/m\omega$, which coincides with the collisionless contribution in Eq. (8.30).

Let us consider an insulating crystal, when the imaginary part of the dielectric function (1) is determined by the interband transitions described by the second term in Eq. (2). Again, we use the basis of Bloch states with the spin-degenerate dispersion laws $\varepsilon_{c\mathbf{p}}$ and $\varepsilon_{v\mathbf{p}}$ for the conduction (c-) and valence (v-) bands. The occupation numbers are $f_{v\mathbf{p}} = 1$ and $f_{c\mathbf{p}} = 0$. Integrating over time, as in Eq. (3), we find $\mathrm{Im}\epsilon_{\alpha\beta}(\omega)$ in the following form:

$$\mathrm{Im}\epsilon_{\alpha\beta}(\omega) = \frac{(2\pi e)^2}{\omega^2 V} \sum_{\mathbf{p}} M_{\alpha\beta}(\mathbf{p}) \delta(\varepsilon_{v\mathbf{p}} - \varepsilon_{c\mathbf{p}} + \hbar\omega),$$

$$M_{\alpha\beta}(\mathbf{p}) = \sum_{\sigma\sigma'} v_{v\sigma,c\sigma'}^\alpha v_{c\sigma',v\sigma}^\beta,$$

(5)

where the sum over the spin number σ is written explicitly. We also assume that the c- and v- bands are non-degenerate so that the sum over them does not appear. The expression (5) is evaluated below for the simplest two-band **kp**-model described by Eqs. (B.18)−(B.24). We first assume that the frequency ω is close to the threshold of interband transitions, $\varepsilon_g/\hbar$. The dispersion laws for the parabolic approximation give rise to $\varepsilon_{c\mathbf{p}} - \varepsilon_{v\mathbf{p}} = \varepsilon_g + p^2/2\mu^*$ in the argument of the δ-function. The reduced mass μ^* is expressed through the effective masses of electrons in conduction and valence bands, m_c and m_v, as $\mu^{*-1} = m_c^{-1} + m_v^{-1}$ (note that the sign of the valence-band electron mass is changed so that it is positive and has the meaning of hole mass). According to Eqs. (B.23) and (B.19), the interband velocity operator at small $\mathbf{p}$ is equal to $s\hat{\rho}_1\hat{\boldsymbol{\sigma}}$, and the factor $M_{\alpha\beta}$ in Eq. (5) is transformed to $M_{\alpha\beta} = s^2 \mathrm{tr}_\sigma \hat{\sigma}_\alpha \hat{\sigma}_\beta = \delta_{\alpha\beta} 2s^2$, where tr_σ denotes the trace over the spin variable and s is the interband velocity of the two-band model. Therefore, for the case of isotropic and homogeneous media we consider, the dielectric permittivity tensor becomes a scalar, ϵ_ω, and its spectral dependence is given by the joint density of states at the energy $\hbar\omega - \varepsilon_g > 0$. The joint density of states depends on the reduced mass μ^* rather than on m. The absorption coefficient α_ω is expressed through the imaginary part of the dielectric permittivity according to $\alpha_\omega = \omega \mathrm{Im}\epsilon_\omega / c\sqrt{\kappa}$ (see problem 3.19 as well as Eq. (10.23) and problem 2.11), where κ is the dielectric permittivity of the lattice at the frequency $\varepsilon_g/\hbar$. We obtain

$$\alpha_\omega = \frac{(2\pi e s)^2}{\omega c \sqrt{\kappa}} \rho_{3D}(\hbar\omega - \varepsilon_g), \tag{6}$$

where ρ_{3D} is given by Eq. (5.3) with m replaced by μ^*. The square-root spectral dependence of the absorption coefficient is valid only near the edge of absorption. In the region $\hbar\omega - \varepsilon_g \sim \varepsilon_g$, one should take into account both non-parabolicity of the electron spectrum and momentum dependence of the matrix elements of velocity. These factors lead to a more complicated absorption spectrum.

Considering the spectral dependence of $\mathrm{Re}\epsilon_\omega$ and $\mathrm{Im}\epsilon_\omega$ for strongly doped materials of n-type, we restrict ourselves to the case of low temperatures (smaller than both ε_g and Fermi energy ε_F in the c-band), when $f_{v\mathbf{p}} = 1$ and $f_{c\mathbf{p}} = \theta(\varepsilon_F - \varepsilon_{c\mathbf{p}})$. To find $M_{\alpha\beta}(\mathbf{p})$, we use Eq. (B.23) for the velocity operator and calculate the sum over the spin variables. Using the isotropy of the electron spectrum, we average $M_{\alpha\beta}(\mathbf{p})$ over the angle of $\mathbf{p}$ and obtain $\overline{M_{\alpha\beta}(\mathbf{p})} = \delta_{\alpha\beta} M_p$, where $M_p = 2s^2(1 + 2\eta_p^2)/3\eta_p^2$ and $\eta_p = \sqrt{1 + (p/ms)^2}$. To find $\mathrm{Im}\epsilon_\omega$ as a function of both ω and doping level determining the position of ε_F, one has to calculate the integral over p (or, equivalently, over η_p) in Eq. (5). The result is

$$\mathrm{Im}\epsilon_\omega = \frac{e^2}{\hbar s} \sqrt{\Omega^2 - 1} \frac{2\Omega^2 + 1}{6\Omega^3} \theta(\Omega - \eta_{p_F}), \quad \Omega = \frac{\hbar\omega}{\varepsilon_g}, \tag{7}$$

where the θ-function written through the Fermi momentum p_F describes the Pauli blocking effect, i.e., the shift of the interband transition edge due to occupation of the bottom of the conduction band by free electrons. If the conduction band is empty, the factor $\theta(\Omega - \eta_{p_F})$ in Eq. (7) is replaced by $\theta(\Omega - 1)$.

To calculate $\mathrm{Re}\epsilon_\omega$, we integrate over time in the general equation (2) and take a sum over spin variables. For insulators, when $f_{v\mathbf{p}} = 1$ and $f_{c\mathbf{p}} = 0$, we obtain

$$\mathrm{Re}\epsilon_\omega - \kappa = \frac{4\pi e^2}{\omega^2}\mathcal{P}\int \frac{d\mathbf{p}}{(2\pi\hbar)^3}M_p \tag{8}$$

$$\times \left[(\varepsilon_{cp} - \varepsilon_{vp} + \hbar\omega)^{-1} + (\varepsilon_{cp} - \varepsilon_{vp} - \hbar\omega)^{-1} - 2(\varepsilon_{cp} - \varepsilon_{vp})^{-1}\right]$$

$$= \frac{4\pi e^2}{\omega^2}\mathcal{P}\int \frac{d\mathbf{p}}{(2\pi\hbar)^3}\frac{2M_p(\hbar\omega)^2}{(\varepsilon_{cp} - \varepsilon_{vp})[(\varepsilon_{cp} - \varepsilon_{vp})^2 - (\hbar\omega)^2]},$$

where $\mathcal{P}$ is the symbol of principal value. The right-hand side of Eq. (8), which describes the contribution of virtual interband transitions, is logarithmic-divergent at $p \to \infty$. This divergent contribution is frequency-independent and can be made finite when finite widths of c- and v-bands are taken into account. Within the logarithmic accuracy, we use the condition $|\mathbf{p}| < p_m$, where $p_m \simeq \pi\hbar/a$ and a is the lattice constant, and include this contribution into the high-frequency dielectric constant ϵ_∞:

$$\epsilon_\infty = \kappa + 8\pi(e\hbar)^2 \int_{|\mathbf{p}|<p_m} \frac{d\mathbf{p}}{(2\pi\hbar)^3}\frac{M_p}{(\varepsilon_{cp} - \varepsilon_{vp})^3}$$

$$= \kappa + \frac{e^2}{3\pi\hbar s}\int_1^{\eta_m} d\eta\sqrt{\eta^2 - 1}\frac{1 + 2\eta^2}{\eta^4}. \tag{9}$$

As a result, the frequency dependence of $\mathrm{Re}\epsilon_\omega$ near the interband absorption edge is given by

$$\mathrm{Re}\epsilon_\omega = \epsilon_\infty - \frac{e^2\Omega^2}{3\pi\hbar s}\mathcal{P}\int_1^\infty d\eta\sqrt{\eta^2 - 1}\frac{1 + 2\eta^2}{\eta^4(\Omega^2 - \eta^2)}. \tag{10}$$

The variable of integration in Eq. (10) and in the second equation of Eq. (9) is $\eta = \eta_p$. The spectral dependences given by Eqs. (10) and (7) are presented in Fig. 3.2. One can see that the square-root spectral dependence of the absorption at $\Omega = 1$ corresponds to a non-analytic spectral dependence of $\mathrm{Re}\epsilon_\omega$. In the case of step-like absorption threshold, which is realized in strongly doped materials, $\mathrm{Re}\epsilon_\omega$ is logarithmic-divergent at the threshold. This property is easily checked with the use of the Kramers-Kronig dispersion relations (problems 3.20 and 3.21).

The general expressions (1) and (2) can be used to describe the contribution of non-equilibrium free carriers with an arbitrary steady-state distribution into the dielectric permittivity. Therefore, they can be applied for studying the optical properties of hot electrons whose distribution is

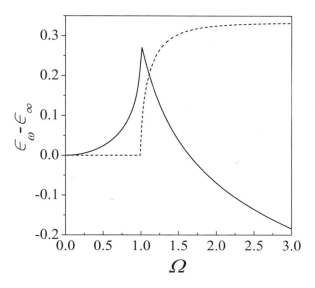

Figure 3.2. Spectral dependence of real and imaginary parts of the dielectric permittivity $\epsilon_\omega - \epsilon_\infty$ of non-doped material described by the two-band model, in units of $e^2/\hbar s$, according to Eqs. (17.7) and (17.10). Solid: $\mathrm{Re}(\epsilon_\omega - \epsilon_\infty)$, dashed: $\mathrm{Im}\epsilon_\omega$.

analyzed in Chapter 7. Indeed, the calculation of non-equilibrium part of the density matrix according to Eq. (13.4) essentially assumes the stationarity of the unperturbed statistical operator, but does not imply that this operator corresponds to thermodynamic equilibrium. Therefore, the expression for the complex conductivity is given by Eq. (13.18), where the statistical operator is assumed to be diagonal with respect to the band index and momentum. A stationary electric field applied to the crystal along OZ creates an anisotropic distribution of free carriers. As a result, the responses of the carriers to electromagnetic waves polarized parallel and perpendicular to the electric field become different. The diagonal contribution to the dielectric permittivity tensor due to free carriers in the conduction band is

$$\Delta\epsilon_{\alpha\alpha}(\omega) = -\frac{8\pi e^2}{\omega^2 V}\sum_{\mathbf{p}} f_{c\mathbf{p}}\frac{\partial^2\varepsilon_{c\mathbf{p}}}{\partial p_\alpha^2} - \frac{4\pi e^2}{\omega^2 V}\mathcal{P}\sum_{\sigma\sigma'\mathbf{p}} f_{c\mathbf{p}}|\langle v\sigma\mathbf{p}|\hat{v}_\alpha|c\sigma'\mathbf{p}\rangle|^2$$

$$\times \left[(\varepsilon_{c\mathbf{p}} - \varepsilon_{v\mathbf{p}} + \hbar\omega)^{-1} + (\varepsilon_{c\mathbf{p}} - \varepsilon_{v\mathbf{p}} - \hbar\omega)^{-1} - 2(\varepsilon_{c\mathbf{p}} - \varepsilon_{v\mathbf{p}})^{-1}\right]. \quad (11)$$

This equation is obtained in the way similar to that used in the derivation of Eqs. (4) and (8).

The induced optical anisotropy leading to the Kerr effect due to free carriers appears if one takes into account the non-parabolicity of the conduction-band spectrum and momentum dependence of the interband matrix elements of the velocity operator. To estimate the strength of the anisotropy, defined as $\delta\epsilon_\omega = \Delta\epsilon_{zz}(\omega) - \Delta\epsilon_\perp(\omega)$ with $\epsilon_\perp = \epsilon_{xx} = \epsilon_{yy}$, one may use the shifted Maxwell distribution, $f_\mathbf{p}$, obtained from the distribution function (31.25) in the limit of non-degenerate electron gas. Accounting for the drift corrections up to the second order, we have $f_{c\mathbf{p}} \simeq f_\varepsilon[1 + (\mathbf{p}\cdot\mathbf{u})/T_e + (\mathbf{p}\cdot\mathbf{u})^2/2T_e^2]$, where f_ε is the Boltzmann distribution of electrons with effective electron temperature T_e and $\mathbf{u}\|OZ$ is the drift velocity. Assuming that $T_e \ll \varepsilon_g$, we use the relations

$$\frac{\partial^2 \varepsilon_{cp}}{\partial p_\alpha^2} \simeq \frac{1}{m}\left(1 - \frac{p^2 + 2p_\alpha^2}{m\varepsilon_g}\right), \tag{12}$$

$$\sum_{\sigma\sigma'} |\langle v\sigma\mathbf{p}|\hat{v}_\alpha|co'\mathbf{p}\rangle|^2 \simeq 2s^2\left(1 - \frac{2p_\alpha^2}{m\varepsilon_g}\right),$$

which follow from the expansions of Eqs. (B.22) and (B.23) in powers of a small parameter $p^2/m\varepsilon_g$. As a result,

$$\delta\epsilon_\omega \simeq \frac{16\pi e^2}{m\omega^2}\int\frac{d\mathbf{p}}{(2\pi\hbar)^3}f_{c\mathbf{p}}\frac{p_z^2 - p_\perp^2}{m\varepsilon_g}$$

$$+\frac{8\pi(es)^2}{\omega^2}\int\frac{d\mathbf{p}}{(2\pi\hbar)^3}f_{c\mathbf{p}}\frac{p_z^2 - p_\perp^2}{m\varepsilon_g}\left[(\varepsilon_g + p^2/2\mu^* + \hbar\omega)^{-1}\right.$$

$$\left. +(\varepsilon_g + p^2/2\mu^* - \hbar\omega)^{-1} - 2(\varepsilon_g + p^2/2\mu^*)^{-1}\right], \tag{13}$$

where μ^* is the reduced mass used in Eqs. (5) and (6). A non-zero anisotropy results from the angular averaging of the contribution proportional to u^2 in $f_{c\mathbf{p}}$. The isotropic part of $f_{c\mathbf{p}}$ gives zero contribution into Eq. (13) because of the angular averaging. This averaging is done according to the following formula:

$$\int\frac{d\widetilde{\Omega}}{4\pi}(\mathbf{u}\cdot\mathbf{p})^2(\mathbf{e}_1\cdot\mathbf{p})(\mathbf{e}_2\cdot\mathbf{p}) = \frac{p^4}{15}[(\mathbf{e}_1\cdot\mathbf{e}_2)u^2 + 2(\mathbf{u}\cdot\mathbf{e}_1)(\mathbf{u}\cdot\mathbf{e}_2)], \tag{14}$$

where $d\widetilde{\Omega}$ is the differential of the solid angle of the vector $\mathbf{p}$ and $\mathbf{e}_{1,2}$ are the unit vectors of the Cartesian coordinate system. We use Eq. (14) with $\mathbf{e}_1 = \mathbf{e}_2$. The spectral dependence of $\delta\epsilon_\omega$ is given by the integral over a dimensionless momentum:

$$\delta\epsilon_\omega = \varepsilon_\infty\left(\frac{\omega_p}{\omega}\right)^2\frac{mu^2}{\varepsilon_g}\frac{32}{15\sqrt{\pi}}\int_0^\infty dx x^6 e^{-x^2}$$

$$\times\left\{1 + \frac{\varepsilon_g}{2}\left[(\varepsilon_g + x^2 T_e m/\mu^* + \hbar\omega)^{-1}\right.\right. \tag{15}$$

$$+(\varepsilon_g + x^2 T_e m/\mu^* - \hbar\omega)^{-1} - 2(\varepsilon_g + x^2 T_e m/\mu^*)^{-1}] \Big\}.$$

The plasma frequency is defined here as $\omega_p^2 = 4\pi e^2 n/\epsilon_\infty m$ (we have substituted $\epsilon = \epsilon_\infty$ into the definition of ω_p given in the end of Sec. 8 because ω considerably exceeds the optical phonon frequencies). The asymptotic behavior of the spectral dependence,

$$\delta\epsilon_\omega \simeq \epsilon_\infty \left(\frac{\omega_p}{\omega}\right)^2 \left\{ \begin{array}{ll} 2mu^2/[\varepsilon_g(1-\Omega^2)], & 1-\Omega \gg mT_e/\mu^*\varepsilon_g \\ 2\mu^* u^2/(5T_e), & 1-\Omega \ll mT_e/\mu^*\varepsilon_g \end{array} \right. , \quad (16)$$

demonstrates a considerable, determined by the factor $\varepsilon_g/T_e \gg 1$, enhancement of the anisotropy of dielectric permittivity near the fundamental absorption edge, when the small parameter mu^2/ε_g is replaced by $\mu^* u^2/T_e$.

If the spatial dispersion is taken into account, i.e., $\mathbf{q} \neq 0$, the dielectric permittivity becomes anisotropic due to the linear term in the drift-velocity expansion of $f_{c\mathbf{p}}$. We stress that in thermodynamic equilibrium the linear in $\mathbf{q}$ terms in the dielectric permittivity tensor are equal to zero due to Onsager's symmetry relation (problem 3.22). The presence of an electric current violates this relation and causes different responses for the electromagnetic waves propagating along the current and in the opposite direction (Fresnel drag of the radiation by the current). The characteristic feature of the case under consideration is the stationarity of the lattice. Only the drift of free electrons contributes to the drag. Therefore, a macroscopic consideration of the Fresnel drag by a moving medium is not applicable here, and one has to determine the dielectric permittivity from a microscopic calculation.

To find the linear in $\mathbf{q}$ contribution $\widetilde{\delta\epsilon}$ to the dielectric permittivity tensor, we use the general equation (13.14), where the linearized current density operator $e\hat{\mathbf{v}} - i(e/2)[\hat{\mathbf{v}}(\mathbf{q}\cdot\hat{\mathbf{x}}) + (\mathbf{q}\cdot\hat{\mathbf{x}})\hat{\mathbf{v}}]$ is derived from Eq. (13.13). As a result,

$$\widetilde{\delta\epsilon}_{\alpha\beta}(\mathbf{q},\omega) = \frac{2\pi e^2}{\hbar\omega^2 V} \int_{-\infty}^{0} d\tau e^{\lambda\tau - i\omega\tau}$$

$$\times \mathrm{Sp}\hat{\rho}_c \Big\{ \Big[e^{-i\widehat{H}\tau/\hbar}\{\hat{v}_\alpha(\mathbf{q}\cdot\hat{\mathbf{x}}) + (\mathbf{q}\cdot\hat{\mathbf{x}})\hat{v}_\alpha\} e^{i\widehat{H}\tau/\hbar}, \hat{v}_\beta \Big]$$

$$- \Big[e^{-i\widehat{H}\tau/\hbar}\hat{v}_\alpha e^{i\widehat{H}\tau/\hbar}, \hat{v}_\beta(\mathbf{q}\cdot\hat{\mathbf{x}}) + (\mathbf{q}\cdot\hat{\mathbf{x}})\hat{v}_\beta \Big] \Big\}, \quad (17)$$

where $\hat{\rho}_c$ is the stationary density matrix describing non-equilibrium electrons of c-band. Using the momentum representation of the velocity and coordinate operators, $\hat{\mathbf{v}}_{\mathbf{p}} = \hat{\mathcal{U}}_{\mathbf{p}}\hat{\mathbf{v}}\hat{\mathcal{U}}_{\mathbf{p}}^+$ and $\hat{\mathbf{x}}_{\mathbf{p}} = \hat{\mathcal{U}}_{\mathbf{p}}\hat{\mathbf{x}}\hat{\mathcal{U}}_{\mathbf{p}}^+$, see Eqs.

(B.19)−(B.23), we obtain the following expressions:

$$\hat{\mathbf{v}}_{\mathbf{p}} \simeq \frac{\mathbf{p}}{m}\hat{\rho}_3 + s\hat{\rho}_1\hat{\boldsymbol{\sigma}}, \qquad \hat{\mathbf{x}}_{\mathbf{p}} \simeq \hat{\mathbf{x}} + \frac{\hbar}{\sqrt{2m\varepsilon_g}}\hat{\rho}_2\hat{\boldsymbol{\sigma}}, \qquad (18)$$

where $\hat{\mathbf{x}} = i\hbar\nabla_{\mathbf{p}}$. The second terms in these expressions determine the non-diagonal parts of the operators and are responsible for interband transitions.

The contribution of the diagonal parts of the operators (18) to the commutators standing in Eq. (17) is given by the expression

$$i\frac{2\hbar}{m^2}(q_\alpha p_\beta + p_\alpha q_\beta). \qquad (19)$$

These terms give the following contribution to the tensor (17):

$$-\frac{4\pi e}{m\omega^3}(q_\alpha I_\beta + I_\alpha q_\beta), \qquad (20)$$

where the stationary current density caused by the drift is introduced in a standard way, as $\mathbf{I} = (e/mV)\mathrm{Sp}(\hat{\rho}_c\mathbf{p})$. The contribution of interband transitions into the expression in the braces $\{\ldots\}$ of Eq. (17) is written as

$$2s^2\hat{\sigma}_\alpha\hat{\sigma}_\beta\left(\mathbf{q}\cdot\frac{\partial\varepsilon_{vp}}{\partial\mathbf{p}}\right)\tau\left[e^{i(\varepsilon_{cp}-\varepsilon_{vp})\tau/\hbar} - e^{-i(\varepsilon_{cp}-\varepsilon_{vp})\tau/\hbar}\right]. \qquad (21)$$

Taking into account that the trace of $\hat{\sigma}_\alpha\hat{\sigma}_\beta$ over the spin variables is equal to $2\delta_{\alpha\beta}$, we calculate the integral over time and transform the interband contributions to $\widetilde{\delta\epsilon}_{\alpha\beta}(\mathbf{q},\omega)$ as

$$-\delta_{\alpha\beta}\frac{4\pi e^2}{m\omega}\int\frac{d\mathbf{p}}{(2\pi\hbar)^3}f_{c\mathbf{p}}\frac{(\mathbf{q}\cdot\mathbf{p})}{m}\frac{\varepsilon_g}{\hbar\omega}$$

$$\times\left[\left(\omega - \frac{\varepsilon_{cp}-\varepsilon_{vp}}{\hbar}\right)^{-2} - \left(\omega + \frac{\varepsilon_{cp}-\varepsilon_{vp}}{\hbar}\right)^{-2}\right]. \qquad (22)$$

In a similar way as in Eq. (16) at $|1 - \hbar\omega/\varepsilon_g| \gg (\mu^*/m)T_e/\varepsilon_g$, we obtain a simple expression for the anisotropic correction to the dielectric permittivity:

$$\widetilde{\delta\epsilon}_{\alpha\beta}(\mathbf{q},\omega) = -\frac{4\pi e}{m\omega^3}(q_\alpha I_\beta + I_\alpha q_\beta) - \delta_{\alpha\beta}\frac{8\pi e}{m\omega^3}(\mathbf{q}\cdot\mathbf{I})\frac{\Omega^2}{(1-\Omega^2)^2}. \qquad (23)$$

The contribution proportional to $\delta_{\alpha\beta}(\mathbf{q}\cdot\mathbf{I})$ in this equation describes the difference in the optical ways for the transverse electromagnetic waves propagating along the current and in the opposite direction. Although this difference contains a relativistic smallness u/c, it can be measured with the use of interference methods. Similar to the case of the quadratic electro-optical effect described above, the contribution of free carriers to the permittivity increases near the edge of fundamental absorption.

18. Interband Absorption under External Fields

In this section we consider the influence of external fields on the interband transitions near the fundamental absorption edge, when the energy of the photon, $\hbar\omega$, is close to the energy gap ε_g between the conduction and valence bands. The absorption coefficient is essentially modified if the energy $|\hbar\omega - \varepsilon_g|$ is comparable to the characteristic energies associated with the external fields applied to insulators or non-doped semiconductors. According to Eq. (10.23), the absorption coefficient is expressed through the real part of the conductivity tensor given by the general equation (17.2). Below we use the basis of eigenstates of the Hamiltonian $\widehat{H}$ describing the electrons in static external fields and integrate over time in Eq. (17.2) according to Eq. (8.6). The electron states near the edges of c- and v-bands are $|c\sigma\delta\rangle$, where $\sigma = \pm 1$ is the spin quantum number and δ describes the intraband motion under the external field. The field is assumed to be small enough to neglect the interband tunneling (see Sec. 60) so that we can use $f_{c\sigma\delta} = 0$ and $f_{v\sigma\delta} = 1$ and obtain

$$\mathrm{Re}\,\sigma_{\alpha\beta}(\omega) = \frac{\pi e^2}{\omega V} \sum_{\delta\delta'\sigma\sigma'} \langle v\sigma\delta|\hat{v}_\alpha|c\sigma'\delta'\rangle \langle c\sigma'\delta'|\hat{v}_\beta|v\sigma\delta\rangle \delta(\varepsilon_{v\delta} - \varepsilon_{c\delta'} + \hbar\omega), \quad (1)$$

where the δ-function describes the energy conservation.

If the fields acting on electrons are smooth on the scale of interband length, see the discussion of Eq. (B.24), the matrix element of the velocity operator can be written as $\langle v\sigma\delta|\hat{v}_\alpha|c\sigma'\delta'\rangle = v^\alpha_{v\sigma,c\sigma'}\langle v\delta|c\delta'\rangle$, where the factor $\langle v\delta|c\delta'\rangle$ describes the overlap of coordinate-dependent envelope wave functions of the c- and v-band states. Near the fundamental absorption edge in the material described by the two-band model, we use $\sum_{\sigma\sigma'} v^\alpha_{v\sigma,c\sigma'} v^\beta_{c\sigma',v\sigma} \simeq \delta_{\alpha\beta} 2s^2$ and find the absorption coefficient

$$\alpha_\omega = \frac{8(\pi e s)^2}{\omega c \sqrt{\kappa} V} \sum_{\delta\delta'} |\langle v\delta|c\delta'\rangle|^2 \delta(\varepsilon_{v\delta} - \varepsilon_{c\delta'} + \hbar\omega). \quad (2)$$

The overlap factor is written explicitly as $\langle v\delta|c\delta'\rangle = \int d\mathbf{r}\,\psi^{(v\delta)*}_{\mathbf{r}}\psi^{(c\delta')}_{\mathbf{r}}$, where the envelope functions satisfy the Schroedinger equations

$$\left[\mp \frac{\hbar^2 \nabla^2_{\mathbf{r}}}{2m_j} + U_j(\mathbf{r}) + \varepsilon_j - \varepsilon_{j\delta}\right] \psi^{(j\delta)}_{\mathbf{r}} = 0 \quad (3)$$

obtained from the many-band equations (B.6). Equation (3) contains the energy ε_j of j-band extremum and the effective mass m_j near this extremum. Note that we have changed the sign of the valence-band effective mass to make it positive. The upper (lower) sign in Eq. (3) and below corresponds to the conduction (valence) band. The potential energy $U_j(\mathbf{r})$, counted from the extremum of j-band, can depend on the band index (this occurs, for example, in non-homogeneous alloys or in non-ideal heterostructures).

In the absence of external fields, when the eigenstate indices δ and δ' are replaced by the momenta $\mathbf{p}$ and $\mathbf{p}'$, the overlap factor gives simply $\delta_{\mathbf{pp}'}$ and Eq. (2) is reduced to Eq. (17.6). In quantized magnetic fields and in quantum wells, the overlap factors are calculated with the wave functions (5.15) and (5.20), respectively, and are reduced to the Kronecker symbols of the corresponding quantum numbers. As a result, the absorption coefficient is again expressed according to Eq. (17.6), where the joint density of states is now determined by Eqs. (5.16) and (5.26).

The influence of a stationary electric field $\mathbf{E}$ on the interband absorption, known as Frantz-Keldysh effect, leads to a more complicated behavior of the absorption coefficient. In the case of a homogeneous field $\mathbf{E}$, each eigenstate is characterized by the transverse (perpendicular to the field) momentum $\mathbf{p}_\perp$ and continuous quantum number ξ describing the longitudinal motion. The overlap factor is $\delta_{\mathbf{p}_\perp \mathbf{p}'_\perp} \int_{-\infty}^{\infty} dp \psi_p^{(v\xi)*} \psi_p^{(c\xi')}$, where $\psi_p^{(j\xi)}$ are the wave functions of one-dimensional Schroedinger equation in the momentum representation:

$$\left(\pm \frac{p^2}{2m_j} + i\hbar|e|E\frac{d}{dp} - \xi \right) \psi_p^{(j\xi)} = 0. \tag{4}$$

The energies of the states in c- and v-bands are equal to $\varepsilon_c + p_\perp^2/2m_c + \xi'$ and $\varepsilon_v - p_\perp^2/2m_v + \xi$, respectively. The usage of independent equations for the envelope functions $\psi_p^{(c\xi)}$ and $\psi_p^{(v\xi)}$ necessarily implies that there is no interband tunneling. The solutions of these equations are

$$\psi_p^{(c\xi)} = N_E \exp\left\{ \frac{i}{\hbar|e|E} \int_0^p dp_1 (p_1^2/2m_c - \xi) \right\},$$

$$\psi_p^{(v\xi)} = N_E \exp\left\{ \frac{i}{\hbar|e|E} \int_0^p dp_1 (-p_1^2/2m_v - \xi) \right\}, \tag{5}$$

where the normalization factor $N_E = (2\pi\hbar|e|E)^{-1/2}$ is obtained from the condition $\int dp \psi_p^{(j\xi)*} \psi_p^{(j\xi')} = \delta(\xi - \xi')$. Substituting Eq. (5) into the overlap factor, we obtain

$$N_E^2 \int_{-\infty}^{\infty} dp \exp \left\{ \frac{i}{\hbar |e| E} \left[\frac{p^3}{6\mu^*} + (\xi - \xi')p \right] \right\} \tag{6}$$

$$= \frac{1}{2\pi \hbar \Omega_E} \int_{-\infty}^{\infty} du \exp \left\{ iu^3/3 + iu(\xi - \xi')/\hbar \Omega_E \right\} = \frac{1}{\hbar \Omega_E} \mathrm{Ai} \left(\frac{\xi - \xi'}{\hbar \Omega_E} \right).$$

The result is expressed through the frequency $\Omega_E = (|e|E)^{2/3} / (2\hbar\mu^*)^{1/3}$ and Airy function, $\mathrm{Ai}(x) \equiv (2\pi)^{-1} \int_{-\infty}^{\infty} du \exp(iu^3/3 + iux)$.

Substituting the overlap factor (6) and band spectra into Eq. (2), we find

$$\alpha_\omega = \frac{8(\pi e s)^2}{\omega c \sqrt{\kappa} V} \sum_{\mathbf{p}_\perp} \int d\xi \int d\xi' (\hbar \Omega_E)^{-2}$$

$$\times \left[\mathrm{Ai} \left(\frac{\xi - \xi'}{\hbar \Omega_E} \right) \right]^2 \delta(\hbar\omega - \varepsilon_g - p_\perp^2/2\mu^* + \xi - \xi')$$

$$= \frac{8(\pi e s)^2 |e| E}{\omega c \sqrt{\kappa} (\hbar \Omega_E)^2} \int \frac{d\mathbf{p}_\perp}{(2\pi\hbar)^2} \left[\mathrm{Ai} \left(\frac{\varepsilon_g - \hbar\omega + p_\perp^2/2\mu^*}{\hbar \Omega_E} \right) \right]^2, \tag{7}$$

where the have integrated over $\xi - \xi'$ with the use of the δ-function and taken into account that $\int d\xi = |e|EL$ (the energy in a homogeneous electric field is proportional to the normalization length). The cylindrical symmetry of the problem permits us to rewrite Eq. (7) as

$$\alpha_\omega = \frac{2\pi (e s)^2 \sqrt{\hbar \Omega_E} (2\mu^*)^{3/2}}{\omega c \sqrt{\kappa} \hbar^3} F \left(\frac{\hbar\omega - \varepsilon_g}{\hbar \Omega_E} \right), \quad F(x) = \int_{-x}^{\infty} dz \, \mathrm{Ai}^2(z). \tag{8}$$

The function $F(x)$ is shown in Fig. 3.3. The presence of the electric field leads to a finite absorption below the interband transition threshold. This absorption becomes exponentially small if $\varepsilon_g - \hbar\omega \gg \hbar\Omega_E$. In this limit, we use the asymptotic behavior of the Airy function, $\mathrm{Ai}(z) \simeq \exp(-2z^{3/2}/3)/(2\sqrt{\pi} z^{1/4})$ at $z \gg 1$, and obtain

$$\alpha_\omega \propto \exp \left[-\frac{4}{3} \left(\frac{\varepsilon_g - \hbar\omega}{\hbar \Omega_E} \right)^{3/2} \right]. \tag{9}$$

If $\hbar\omega > \varepsilon_g$, the function $F(x)$ in Eq. (8) oscillates and approaches the square-root dependence given by Eq. (17.6).

As an example of application of the path integral formalism, we consider the influence of random external fields on the fundamental absorption edge. We assume that the random potentials $U_j(\mathbf{r})$ are smooth on the scale of the crystal lattice period and use Eq. (3) for the envelope functions. The squared overlap factor $|\langle v\delta | c\delta' \rangle|^2$ is written as $\int d\mathbf{r} \int d\mathbf{r}' \psi_\mathbf{r}^{(v\delta)*} \psi_\mathbf{r}^{(c\delta')} \psi_{\mathbf{r}'}^{(v\delta)} \psi_{\mathbf{r}'}^{(c\delta')*}$. Our aim is to represent the absorption

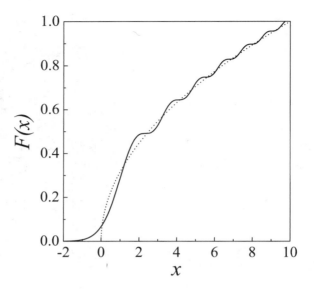

Figure 3.3. Function $F(x) = \int_{-x}^{\infty} dz \mathrm{Ai}^2(z)$ and $\sqrt{x}$ (dotted line).

coefficient through the products of two Green's functions averaged over the random potential distribution, as in Eq. (13.27). First, we rewrite the δ-function in Eq. (2) as $\int d\varepsilon \delta(\varepsilon - \varepsilon_{v\delta})\delta(\varepsilon - \varepsilon_{c\delta'} + \hbar\omega)$, in a similar way as in Eq. (13.24). Then we use the integral representation $2\pi\delta(E) = \int_{-\infty}^{\infty} dt e^{iEt}$ for each of these δ-functions and employ the expressions of the Green's functions given in problem 3.10. As a result, Eq. (2) is rewritten in the form

$$\alpha_\omega = \frac{8(\pi e s)^2}{\omega c \sqrt{\kappa} V} \sum_{s,s'=R,A} (-1)^l \int d\varepsilon \int_{-\infty}^{\infty} dt \int_{-\infty}^{\infty} dt' e^{i(\varepsilon + \hbar\omega)t/\hbar + i\varepsilon t'/\hbar}$$

$$\times \int d\mathbf{r} \int d\mathbf{r}' \left\langle \left\langle G_{t'}^{s(v)}(\mathbf{r}, \mathbf{r}') G_{t}^{s'(c)}(\mathbf{r}', \mathbf{r}) \right\rangle \right\rangle, \tag{10}$$

where $l = 0$ and 1 for $s \neq s'$ and $s = s'$, respectively. We stress that $G^{s(c)}$ and $G^{s(v)}$ standing in Eq. (10) are the Green's functions of the time-dependent Schroedinger equations with the Hamiltonians $\mp\hbar^2\nabla_{\mathbf{r}}^2/2m_j + U_j(\mathbf{r}) + \varepsilon_j$ $(j = c, v)$; see Eq. (3).

To calculate the absorption coefficient, we use the path-integral expression (16.13). The first step is to find the average of the characteristic functional $\exp\left[\int d\mathbf{r} f_{c\mathbf{r}} U_c(\mathbf{r}) + \int d\mathbf{r} f_{v\mathbf{r}} U_v(\mathbf{r})\right]$ appearing in the product of the Green's functions. For arbitrary Gaussian random potentials, it can be done by analogy with the calculations presented in Eqs. (16.14) and (16.15). We obtain

$$\left\langle\!\!\left\langle \exp\left\{-\frac{i}{\hbar}\int_0^t d\tau U_c(\mathbf{x}_\tau) - \frac{i}{\hbar}\int_0^{t'} d\tau U_v(\mathbf{y}_\tau)\right\}\right\rangle\!\!\right\rangle$$

$$= \exp\left\{-\frac{1}{2\hbar^2}\int_0^t d\tau \int_0^t d\tau' w_{cc}(|\mathbf{x}_\tau - \mathbf{x}_{\tau'}|)\right. \tag{11}$$

$$\left.-\frac{1}{2\hbar^2}\int_0^{t'} d\tau \int_0^{t'} d\tau' w_{vv}(|\mathbf{y}_\tau - \mathbf{y}_{\tau'}|) - \frac{1}{\hbar^2}\int_0^t d\tau \int_0^{t'} d\tau' w_{cv}(|\mathbf{x}_\tau - \mathbf{y}_{\tau'}|)\right\},$$

where $w_{jj'}(|\mathbf{r} - \mathbf{r}'|) = \langle\!\langle U_j(\mathbf{r})U_{j'}(\mathbf{r}')\rangle\!\rangle$ so that three kinds of pair correlation functions, the diagonal and non-diagonal ones, appear. The absorption can be expressed as

$$\alpha_\omega = \frac{(4\pi^2 es)^2}{\omega c\sqrt{\kappa}}\left[R_{AR} - R_{RR} + c.c.\right], \tag{12}$$

where

$$R_{AR} = \frac{1}{2\pi^2\hbar^2 V}\int d\varepsilon \int_{-\infty}^0 dt' \int_0^\infty dt e^{i(\varepsilon + \hbar\omega - \varepsilon_g)t/\hbar + i\varepsilon t'/\hbar}\int d\mathbf{r}\int d\mathbf{r}'$$

$$\times \int_{\mathbf{y}_0=\mathbf{r}}^{\mathbf{y}_{t'}=\mathbf{r}'}\mathcal{D}\{\mathbf{y}_\tau\}\int_{\mathbf{x}_0=\mathbf{r}'}^{\mathbf{x}_t=\mathbf{r}}\mathcal{D}\{\mathbf{x}_\tau\}\exp\left\{-\frac{im_v}{2\hbar}\int_0^{t'} d\tau\dot{\mathbf{y}}_\tau^2 + \frac{im_c}{2\hbar}\int_0^t d\tau\dot{\mathbf{x}}_\tau^2\right.$$

$$-\frac{1}{2\hbar^2}\int_0^{t'} d\tau_1 \int_0^{t'} d\tau_2 w_{vv}(|\mathbf{y}_{\tau_1} - \mathbf{y}_{\tau_2}|) - \frac{1}{2\hbar^2}\int_0^t d\tau_1 \int_0^t d\tau_2 w_{cc}(|\mathbf{x}_{\tau_1} - \mathbf{x}_{\tau_2}|)$$

$$\left.-\frac{1}{\hbar^2}\int_0^t d\tau_1 \int_0^{t'} d\tau_2 w_{cv}(|\mathbf{x}_{\tau_1} - \mathbf{y}_{\tau_2}|)\right\}. \tag{13}$$

The functional integrals are taken along the valence-band electron paths $\mathbf{y}_\tau$ and conduction-band electron paths $\mathbf{x}_\tau$. The expression for R_{RR} is similar, but the integrals over t and t' are taken in the same interval, from 0 to ∞. Integrating over ε in Eq. (13), we immediately obtain the delta-function $\delta(t + t')$. This means that $R_{RR} \simeq 0$, since it contains the contributions from positive t and t' only. Therefore, the absorption coefficient is proportional to $(R_{AR} + c.c.)$.

The approximation of classically smooth potentials $U_{c,v}(\mathbf{r})$ allows us to calculate R_{AR} analytically in the way similar to that described in Sec. 16. Replacing $w_{jj'}(|\mathbf{r} - \mathbf{r}'|)$ by $w_{jj'}(0)$, we calculate the path integral. It gives us a product of the free-electron Green's functions so that

$$R_{AR} = \frac{1}{\pi\hbar}\int_0^\infty dt e^{i(\omega - \varepsilon_g/\hbar)t}\int d\mathbf{r}\frac{(m_c m_v)^{3/2}}{(2\pi i\hbar t)^3}$$

$$\times \exp\left(\frac{i(m_c + m_v)\mathbf{r}^2}{2\hbar t}\right)\exp\left(-\frac{\widetilde{w}}{2\hbar^2}t^2\right), \tag{14}$$

where $\widetilde{w} = w_{cc}(0) + w_{vv}(0) - 2w_{cv}(0)$. The integral over $\mathbf{r}$ in Eq. (14) is taken easily. As a result, we find that $R_{AR} + c.c.$ is reduced to the expression for the density of states given by Eq. (16.19), where E, m, and $w(0)$ are replaced, respectively, by $\hbar\omega - \varepsilon_g$, $m_c m_v / (m_c + m_v) \equiv \mu^*$, and $\widetilde{w}$. Therefore, in the limit of classically smooth disorder, the absorption coefficient is determined by Eq. (17.6), where the joint density of states $\rho_{3D}(\hbar\omega - \varepsilon_g)$ differs from the density of states of Eq. (16.19) by substituting the reduced effective mass μ^* and correlation function $\widetilde{w}$ in place of m and $w(0)$, respectively. In the absence of the random potential, as well as in the region $\hbar\omega - \varepsilon_g \gg \widetilde{w}$, the frequency dependence of the absorption coefficient near the fundamental absorption edge ($|\hbar\omega - \varepsilon_g| \ll \varepsilon_g$) exactly follows the energy dependence of the density of states with the reduced effective mass. On the other hand, far below the fundamental edge, at $\varepsilon_g - \hbar\omega \gg \widetilde{w}$, one finds

$$\alpha_\omega \propto \exp\left[-(\hbar\omega - \varepsilon_g)^2 / 2\widetilde{w}\right]. \tag{15}$$

The broadening of the edge of the absorption spectrum cannot be described within the approximation of classically smooth potentials if $U_c(\mathbf{r}) = U_v(\mathbf{r}) = U(\mathbf{r})$, i.e., when the spatial variations of the potential energies of the valence- and conduction-band electrons do not result in the variations of the energy gap, and $\widetilde{w} = 0$. Quantum corrections are principally important in this case. To calculate them, we represent the paths as

$$\mathbf{x}_\tau = \mathbf{r}' + \frac{\tau}{t}(\mathbf{r} - \mathbf{r}') + \delta\mathbf{x}_\tau, \quad \mathbf{y}_\tau = \mathbf{r} + \frac{\tau}{t}(\mathbf{r} - \mathbf{r}') + \delta\mathbf{y}_\tau, \tag{16}$$

where $\delta\mathbf{x}_\tau$ and $\delta\mathbf{y}_\tau$ are the deviations from the straight paths. The integrals along $\delta\mathbf{x}_\tau$ and $\delta\mathbf{y}_\tau$ are the contour path integrals since $\delta\mathbf{x}_0 = \delta\mathbf{x}_t = 0$ and $\delta\mathbf{y}_0 = \delta\mathbf{y}_{-t} = 0$. Equation (13) can be rewritten as

$$R_{AR} = \frac{1}{\pi\hbar} \int_0^\infty dt\, e^{i(\hbar\omega - \varepsilon_g)t/\hbar} \int d\Delta\mathbf{r} \exp\left(\frac{i(m_c + m_v)\Delta\mathbf{r}^2}{2\hbar t}\right)$$

$$\times \oint \mathcal{D}\{\delta\mathbf{x}_\tau\} \oint \mathcal{D}\{\delta\mathbf{y}_\tau\} \exp\left\{\frac{i}{2\hbar}\int_0^t \left[m_c\delta\dot{\mathbf{x}}_\tau^2 + m_v\delta\dot{\mathbf{y}}_{\tau-t}^2\right]\right.$$

$$\left.-\frac{1}{2\hbar^2}\int_0^t d\tau \int_0^t d\tau' \left[w(|\mathbf{r}_{\tau-\tau'} + \delta\mathbf{y}_{\tau-t} - \delta\mathbf{y}_{\tau'-t}|)\right.\right. \tag{17}$$

$$\left.\left.+w(|\mathbf{r}_{\tau-\tau'} + \delta\mathbf{x}_\tau - \delta\mathbf{x}_{\tau'}|) - 2w(|\mathbf{r}_{\tau-\tau'} + \delta\mathbf{x}_\tau - \delta\mathbf{y}_{\tau'-t}|)\right]\right\},$$

where $\Delta\mathbf{r} = \mathbf{r} - \mathbf{r}'$ and $\mathbf{r}_{\tau-\tau'} \equiv \Delta\mathbf{r}(\tau - \tau')/t$. In the transformations done, we first calculated the integral over ε with the result $\delta(t' + t)$. Then we calculated the integrals over t' and $(\mathbf{r} + \mathbf{r}')$. Next, we took into account that $\int_0^t d\tau \delta\dot{\mathbf{x}}_\tau = \int_0^{-t} d\tau \delta\dot{\mathbf{y}}_\tau = 0$ and $w_{cc}(x) = w_{vv}(x) = w_{cv}(x) \equiv w(x)$ in order to transform, respectively, the kinetic and potential parts of the expression in the exponent. So far the transformations have been exact. Below we apply the approximation of small deviations and expand the correlation functions w in series of their arguments. The first non-vanishing contributions to the double integral over time in the exponent are quadratic in $\delta\mathbf{x}$ and $\delta\mathbf{y}$. The contributions of the higher order are neglected. In this approximation, the potential part is independent of $\Delta\mathbf{r}$ and characterized by the mean square of the potential gradient

$$\psi = \langle\langle [\nabla U(\mathbf{r})]^2 \rangle\rangle = \lim_{\Delta\mathbf{r}\to 0} \nabla_{\mathbf{r}} \cdot \nabla_{\mathbf{r}'} w(|\Delta\mathbf{r}|) = -3 \lim_{\mathbf{r}\to 0} \frac{d^2 w(|\mathbf{r}|)}{d|\mathbf{r}|^2}. \quad (18)$$

Calculating the integral over $\Delta\mathbf{r}$, we obtain

$$R_{AR} = \frac{1}{\pi\hbar} \int_0^\infty dt e^{i(\hbar\omega - \varepsilon_g)t/\hbar} \left(\frac{2\pi i \hbar t}{m_c + m_v} \right)^{3/2}$$

$$\times \oint \mathcal{D}\{\delta\mathbf{x}_\tau\} \oint \mathcal{D}\{\delta\mathbf{y}_\tau\} \exp\left\{ \frac{i}{2\hbar} \int_0^t d\tau \left[m_c \delta\dot{\mathbf{x}}_\tau^2 + m_v \delta\dot{\mathbf{y}}_{\tau-t}^2 \right] \right. \quad (19)$$

$$\left. - \frac{\psi}{6\hbar^2} \left[\int_0^t d\tau (\delta\mathbf{x}_\tau - \delta\mathbf{y}_{\tau-t}) \right]^2 \right\}.$$

In this expression we can replace $\delta\mathbf{y}_{\tau-t}$ by $\delta\mathbf{y}_\tau$ because the paths are closed. Next, since the potential part depends on the difference $\mathbf{z}_\tau^- = \delta\mathbf{x}_\tau - \delta\mathbf{y}_\tau$ only, it is convenient to use new coordinates, $\mathbf{z}_\tau^-$ and $\mathbf{z}_\tau^+ = (m_c\delta\mathbf{x}_\tau + m_v\delta\mathbf{y}_\tau)/(m_c + m_v)$, and carry out the transformation $\oint \mathcal{D}\{\delta\mathbf{x}_\tau\} \times \oint \mathcal{D}\{\delta\mathbf{y}_\tau\} \ldots \to \oint \mathcal{D}\{\mathbf{z}_\tau^+\} \oint \mathcal{D}\{\mathbf{z}_\tau^-\} \ldots$. The kinetic-energy part of the expression in the exponent of Eq. (19) is diagonal in these coordinates. It is written as

$$\frac{i}{2\hbar} \int_0^t \left[(m_c + m_v)(\dot{\mathbf{z}}_\tau^+)^2 + \mu^*(\dot{\mathbf{z}}_\tau^-)^2 \right], \quad (20)$$

which means that one can calculate the path integral over $\mathbf{z}_\tau^+$ separately from that over $\mathbf{z}_\tau^-$. The path integral over $\mathbf{z}_\tau^+$ has the same form as the path integral for a free electron. The remaining path integral,

$$\oint \mathcal{D}\{\mathbf{z}_\tau^-\} \exp\left\{ \frac{i\mu^*}{2\hbar} \int_0^t d\tau (\dot{\mathbf{z}}_\tau^-)^2 - \frac{\psi}{6\hbar^2} \left(\int_0^t d\tau \mathbf{z}_\tau^- \right)^2 \right\}, \quad (21)$$

depends on the reduced mass only. To calculate this integral, let us expand $\mathbf{z}_\tau^-$ into the sine Fourier series,

$$\mathbf{z}_\tau^- = \sum_{k=1}^{\infty} \mathbf{z}_k \frac{\pi k}{2t} \sin \frac{\pi k \tau}{t}. \tag{22}$$

The factor $\pi k/2t$ is included for convenience. The path integral becomes a multiple integral over the vectors $\mathbf{z}_k$. Calculating the integrals over time in the exponent of Eq. (21), we notice that only the terms with odd k ($k = 2l - 1$, $l = 1, 2, \ldots$) contribute to the potential-energy part. The contribution of even k ($k = 2l$, $l = 1, 2, \ldots$) enters the kinetic-energy part only, and is equal to $(i\mu^*\pi^4/4\hbar t^3) \sum_{l=1}^{\infty}(2l)^4 \mathbf{z}_{2l}^2$. Transforming the contour path integral $\oint \mathcal{D}\{\mathbf{z}_\tau^-\} \ldots$ to a multiple integral over $\mathbf{z}_k$, one may integrate out the variables $\mathbf{z}_{2l}$, including the result of such integration (together with the Jacobian of the transformation and with the result of integration over $\mathbf{z}_\tau^+$) into a time-dependent normalization factor $\mathcal{N}_t$. Obviously, this factor is independent of the potential energy. We do not need to search for this factor explicitly, since it can be found by comparison of our results in the limit $\psi = 0$ to the known result (17.6) for the absorption in the absence of the potential. Therefore, retaining the multiple integral over $\mathbf{z}_k$ with odd $k = 2l - 1$, we obtain

$$R_{AR} = \frac{1}{\pi\hbar} \int_0^{\infty} dt\, e^{i(\hbar\omega - \varepsilon_g)t/\hbar} \mathcal{N}_t \int d\mathbf{z}_1 \int d\mathbf{z}_3 \ldots \int d\mathbf{z}_{2l-1} \ldots$$

$$\times \exp\left\{ \frac{i\mu^*\pi^4}{\hbar t^3} \sum_{l=1}^{\infty} (l - 1/2)^4 \mathbf{z}_{2l-1}^2 - \frac{\psi}{6\hbar^2} \sum_{l,l'} \mathbf{z}_{2l-1}\mathbf{z}_{2l'-1} \right\}. \tag{23}$$

This integral contains a biquadratic form in the exponent and is calculated according to the relation

$$\prod_{l=1}^{\infty}\left\{ \left(\frac{A_l}{\pi}\right)^{3/2} \int d\tilde{\mathbf{z}}_l \exp\left[-A_l\tilde{\mathbf{z}}_l^2 - B\tilde{\mathbf{z}}_l \sum_{l'=1}^{\infty} \tilde{\mathbf{z}}_{l'} \right] \right\}$$

$$= \left(1 + \sum_{l=1}^{\infty} \frac{B}{A_l} \right)^{-3/2}, \tag{24}$$

where $A_l = \mu^*\pi^4(l - 1/2)^4/i\hbar t^3$, $B = \psi/6\hbar^2$, and $\tilde{\mathbf{z}}_l = \mathbf{z}_{2l-1}$. Equation (24) can be checked directly (problem 3.23). Finally, we make use of the identity $\sum_{l=1}^{\infty}(2l - 1)^{-4} = \pi^4/96$ and substitute the calculated R_{AR} into Eq. (12). The absorption coefficient is expressed as

$$\alpha_\omega = \frac{4\pi^2(es)^2}{\omega c\sqrt{\kappa}} \frac{1}{\pi\hbar} \left(\frac{\mu^*}{2\pi\hbar}\right)^{3/2} \int_{-\infty}^{\infty} dt \frac{e^{i(\hbar\omega - \varepsilon_g)t/\hbar}}{(it + 0)^{3/2}[1 - (iE_Bt/\hbar)^3]^{3/2}}, \tag{25}$$

where the characteristic broadening energy is $E_B = (\psi\hbar^2/36\mu^*)^{1/3}$. The quantum nature of the broadening becomes clear if one estimates the derivative $d^2 w(|\mathbf{r}|)/d|\mathbf{r}|^2$ as $w(0)/l_c^2$, where l_c is a characteristic length of random potential inhomogeneities. Then E_B is estimated as $[w(0)E_c]^{1/3}$, where $E_c = \hbar^2/12\mu^* l_c^2$ is equal, within the accuracy of a numerical coefficient, to the kinetic energy of a particle with effective mass μ^* and de Broglie wavelength l_c. This quantum broadening can be interpreted as Frantz-Keldysh effect in the electric field $-\nabla U(\mathbf{r})/e$.

The denominator in Eq. (25) goes to zero in four points: $t_1 = +i0$, $t_2 = -i\hbar/E_B$, and $t_{3,4} = \hbar(i \pm \sqrt{3})/2E_B$. Since these points are not simple poles, the integral over time in Eq. (25) cannot be calculated analytically. However, in the region far below the fundamental absorption edge, when $\varepsilon_g - \hbar\omega \gg E_B$, the main contribution to the integral comes from the vicinity of $t = -i\hbar/E_B$ and

$$\alpha_\omega \propto \exp(-|\hbar\omega - \varepsilon_g|/E_B) \qquad (26)$$

(compare this to Eq. (15)). Expression (26) describes the Urbach tail of the interband absorption observed in bulk semiconductors doped with impurities. Indeed, the impurity potential $U(\mathbf{r}) = U_{im}(\mathbf{r})$ is the same for c and v bands, and the assumption $U_c(\mathbf{r}) = U_v(\mathbf{r})$ leading to Eq. (26) is valid for this case.

Problems

3.1. Carry out the double Fourier transformation in Eqs. (13.9) and (13.10).

<u>Hint</u>: The double Fourier transformation is defined as

$$\sigma_{\alpha\beta}(\mathbf{q}, \mathbf{q}'|\omega) = \int d\mathbf{r} \int d\mathbf{r}' e^{-i\mathbf{q}\cdot\mathbf{r}} \sigma_{\alpha\beta}(\mathbf{r}, \mathbf{r}'|\omega) e^{i\mathbf{q}'\cdot\mathbf{r}'}.$$

3.2. Consider the analytical properties of the conductivity tensor $\sigma_{\alpha\beta}(\mathbf{q}, \omega)$ and derive the Kramers-Kronig dispersion relations (13.15).

<u>Solution</u>: From the formal point of view, one may consider ω in Eq. (13.14) as a complex variable, though only real and positive ω have physical meaning. The factor $e^{-i\omega\tau}$ is finite at $\tau < 0$ in the upper half-plane of the complex variable ω. According to the theory of the functions of complex variable, the conductivity tensor is analytical in the upper half-plane of ω and goes to zero in this half-plane at $|\omega| \to \infty$.

This analytical property, in fact, follows just from the causality principle. Introducing the generalized susceptibility $\chi_{\alpha\beta}(\mathbf{r}, \mathbf{r}'|t)$ characterizing a linear response of the observable quantity $\Delta Q_\alpha(\mathbf{r}, t)$ to the perturbation $\mathcal{F}_\beta(\mathbf{r}, t)$ according to (see also Eqs. (13.28)−(13.30)),

$$\Delta Q_\alpha(\mathbf{r}, t) = \sum_\beta \int d\mathbf{r}' \int dt' \chi_{\alpha\beta}(\mathbf{r}, \mathbf{r}'|t - t')\mathcal{F}_\beta(\mathbf{r}', t'),$$

one should demand that $\chi_{\alpha\beta}(\mathbf{r}, \mathbf{r}'|t - t')$ is non-zero only at $t > t'$. The temporal Fourier transformation defines the function $\chi_{\alpha\beta}(\mathbf{r}, \mathbf{r}'|\omega) = \int_{-\infty}^{\infty} dt e^{i\omega t}\chi_{\alpha\beta}(\mathbf{r}, \mathbf{r}'|t)$, which is analytical in the upper half-plane of ω and goes to zero in this half-plane at $|\omega| \to \infty$, i.e., has the analytical property mentioned above. One can also apply the inverse Fourier transformation to Eq. (13.10) (or to Eq. (13.14)) to see directly that $\sigma_{\alpha\beta}(\mathbf{r}, \mathbf{r}'|t)$ (or $\sigma_{\alpha\beta}(\mathbf{q}, t)$) is equal to zero for $t < 0$.

The analytical property of the conductivity tensor allows one to obtain integral relations between real and imaginary parts of $\sigma_{\alpha\beta}(\mathbf{q}, \omega)$. Indeed, using it, we can write

$$\oint d\omega' \frac{\sigma_{\alpha\beta}(\mathbf{q}, \omega')}{\omega' - \omega} = 0,$$

where the path of integration is an arbitrary closed contour in the upper half-plane of the complex variable ω'. Consider a contour which goes along the real axis from $-\infty$ to ∞, passing above the simple pole at $\omega' = \omega$, and then goes from ∞ to $-\infty$ along a semi-circle of infinite radius in the upper half-plane. Since the contribution of this upper part is zero, the equation can be rewritten as

$$\mathcal{P} \int_{-\infty}^{\infty} d\omega' \frac{\sigma_{\alpha\beta}(\mathbf{q}, \omega')}{\omega' - \omega} - i\pi\sigma_{\alpha\beta}(\mathbf{q}, \omega) = 0,$$

where the integral along the real axis is written as a sum of the principal value of this integral and the contribution from the infinitely small semi-circle over the pole $\omega' = \omega$. Separating the real and imaginary parts of the complex equation written above, we obtain the Kramers-Kronig dispersion relations:

$$\mathrm{Re}\sigma_{\alpha\beta}(\mathbf{q}, \omega) = \frac{1}{\pi}\mathcal{P} \int_{-\infty}^{\infty} d\omega' \frac{\mathrm{Im}\sigma_{\alpha\beta}(\mathbf{q}, \omega')}{\omega' - \omega},$$

$$\mathrm{Im}\sigma_{\alpha\beta}(\mathbf{q}, \omega) = -\frac{1}{\pi}\mathcal{P} \int_{-\infty}^{\infty} d\omega' \frac{\mathrm{Re}\sigma_{\alpha\beta}(\mathbf{q}, \omega')}{\omega' - \omega}.$$

If $\mathbf{q} = 0$, the symmetry property (13.16) in the form $\mathrm{Re}\sigma_{\alpha\beta}(\omega') = \mathrm{Re}\sigma_{\alpha\beta}(-\omega')$ and $\mathrm{Im}\sigma_{\alpha\beta}(\omega') = -\mathrm{Im}\sigma_{\alpha\beta}(-\omega')$ allows us to transform the integrals in these relations to the integrals over positive ω', and we obtain Eq. (13.15).

3.3. Consider the linear response of electrons to the longitudinal electric field described by a scalar potential.

Solution: The Hamiltonian of the perturbation is written as $\widehat{\Delta H}_\omega = \int d\mathbf{r}\hat{\rho}_{\mathbf{r}}\Phi(\mathbf{r}, \omega)$, where $\Phi(\mathbf{r}, \omega)$ is the scalar potential which determines the electric field $\mathbf{E}(\mathbf{r}, \omega) = -\nabla\Phi(\mathbf{r}, \omega)$, and $\hat{\rho}_{\mathbf{r}} = e\sum_j \delta(\mathbf{r} - \mathbf{x}_j)$ is the charge density operator. The induced charge density is $\Delta\rho(\mathbf{r}, \omega) = \mathrm{Sp}\hat{\rho}_{\mathbf{r}}\widehat{\Delta\eta}_\omega$. One may also introduce the vector of polarization, $\mathbf{\Delta P}(\mathbf{r}, \omega)$, according to $\nabla \cdot \mathbf{\Delta P}(\mathbf{r}, \omega) = -\Delta\rho(\mathbf{r}, \omega)$, and the corresponding operator $\widehat{\mathbf{\Delta P}_{\mathbf{r}}}$. Using Eq. (13.4) for non-equilibrium part of the density matrix, we obtain

$$\Delta\rho(\mathbf{r}, \omega) = \frac{1}{i\hbar} \int_{-\infty}^{0} d\tau e^{\lambda\tau - i\omega\tau} \mathrm{Sp}\hat{\eta}_{eq} \left[e^{-i\widehat{H}\tau/\hbar}\hat{\rho}_{\mathbf{r}}e^{i\widehat{H}\tau/\hbar}, \widehat{\Delta H}_\omega \right]$$

$$= \int d\mathbf{r}'\alpha(\mathbf{r}, \mathbf{r}'|\omega)\Phi(\mathbf{r}', \omega).$$

The last equation defines the non-local polarizability of electron system:

$$\alpha(\mathbf{r}, \mathbf{r}'|\omega) = \frac{1}{i\hbar} \int_{-\infty}^{0} d\tau e^{\lambda\tau - i\omega\tau} \mathrm{Sp}\hat{\eta}_{eq} \left[e^{-i\widehat{H}\tau/\hbar}\hat{\rho}_{\mathbf{r}}e^{i\widehat{H}\tau/\hbar}, \hat{\rho}_{\mathbf{r}'} \right].$$

It is expressed through the density correlation function. The spatial Fourier transform of the density perturbation is written as $\Delta\rho(\mathbf{q},\omega) = V^{-1}\sum_{\mathbf{q}}\alpha(\mathbf{q},\mathbf{q}'|\omega)\Phi(\mathbf{q}',\omega)$ (compare to Eq. (13.11)), where

$$\alpha(\mathbf{q},\mathbf{q}'|\omega) = \int d\mathbf{r}\int d\mathbf{r}' e^{-i\mathbf{q}\cdot\mathbf{r}+i\mathbf{q}'\cdot\mathbf{r}'}\alpha(\mathbf{r},\mathbf{r}'|\omega)$$

$$= \frac{1}{i\hbar}\int_{-\infty}^{0}d\tau e^{\lambda\tau-i\omega\tau}\mathrm{Sp}\hat{\eta}_{eq}\left[e^{-i\widehat{H}\tau/\hbar}\hat{\rho}_{\mathbf{q}}e^{i\widehat{H}\tau/\hbar},\hat{\rho}_{-\mathbf{q}'}\right].$$

For translation-invariant systems, $\alpha(\mathbf{q},\mathbf{q}'|\omega) = V\delta_{\mathbf{q}\mathbf{q}'}\alpha(\mathbf{q},\omega)$ and $\Delta\rho(\mathbf{q},\omega) = \alpha(\mathbf{q},\omega)$ $\times\Phi(\mathbf{q},\omega)$, where

$$\alpha(\mathbf{q},\omega) = \frac{1}{i\hbar V}\int_{-\infty}^{0}d\tau e^{\lambda\tau-i\omega\tau}\mathrm{Sp}\hat{\eta}_{eq}\left[e^{-i\widehat{H}\tau/\hbar}\hat{\rho}_{\mathbf{q}}e^{i\widehat{H}\tau/\hbar},\hat{\rho}_{-\mathbf{q}}\right].$$

3.4. Prove that the expression for the linear conductivity derived for the scalar potential perturbation $\widehat{\Delta H}_\omega = \int d\mathbf{r}\hat{\rho}_{\mathbf{r}}\Phi(\mathbf{r},\omega)$ coincides with Eq. (13.10) derived for $\widehat{\Delta H}_\omega$ of Eq. (13.6), i.e., the gradient invariance takes place.

<u>Solution:</u> For the scalar potential perturbation, we have

$$\Delta\mathbf{I}(\mathbf{r},\omega) = \frac{1}{i\hbar}\int_{-\infty}^{0}d\tau e^{\lambda\tau-i\omega\tau}\mathrm{Sp}\hat{\eta}_{eq}\left[e^{-i\widehat{H}\tau/\hbar}\hat{\mathbf{I}}(\mathbf{r})e^{i\widehat{H}\tau/\hbar},\widehat{\Delta H}_\omega\right]$$

instead of Eq. (13.8). Integrating over τ by parts, we transform this expression to

$$\Delta\mathbf{I}(\mathbf{r},\omega) = \frac{1}{\hbar\omega}\mathrm{Sp}\hat{\eta}_{eq}[\hat{\mathbf{I}}(\mathbf{r}),\widehat{\Delta H}_\omega] - \frac{i}{\hbar^2\omega}\int_{-\infty}^{0}d\tau e^{\lambda\tau-i\omega\tau}$$

$$\times\mathrm{Sp}\hat{\eta}_{eq}\left[e^{-i\widehat{H}\tau/\hbar}\hat{\mathbf{I}}(\mathbf{r})e^{i\widehat{H}\tau/\hbar},[\widehat{H},\widehat{\Delta H}_\omega]\right].$$

The commutators in the first and second terms are calculated directly, with the use of the expressions for $\widehat{H}$ (see Eqs. (4.10) and (4.2)), $\hat{\mathbf{I}}(\mathbf{r})$ (Eq. (4.15)), and $\widehat{\Delta H}_\omega$:

$$[\widehat{H},\widehat{\Delta H}_\omega] = i\hbar\int d\mathbf{r}\,\hat{\mathbf{I}}(\mathbf{r})\cdot\mathbf{E}(\mathbf{r},\omega), \qquad [\hat{\mathbf{I}}(\mathbf{r}),\widehat{\Delta H}_\omega] = \frac{ie^2\hbar}{m}\sum_{j}\delta(\mathbf{r}-\mathbf{x}_j)\mathbf{E}(\mathbf{r},\omega).$$

Substituting these results into the expression for $\Delta\mathbf{I}(\mathbf{r},\omega)$ presented above, we obtain Eqs. (13.9) and (13.10).

3.5. Write the Kubo formula for a spatially homogeneous perturbation in the following compact form:

$$\sigma_{\alpha\beta}(\omega) = \frac{ie^2 n}{m\omega}\delta_{\alpha\beta} + \int_{-\infty}^{0}d\tau\int_{0}^{1/T}d\lambda e^{-i\omega\tau}\mathrm{Sp}\left\{\hat{\eta}_{eq}\hat{I}_\alpha(-\tau)\hat{I}_\beta(-i\hbar\lambda)\right\},$$

where $\hat{I}_\alpha(t) = e^{i\widehat{H}t/\hbar}\hat{I}_\alpha e^{-i\widehat{H}t/\hbar}$ is the operator of the current density in the Heisenberg representation.

<u>Hints:</u> First use the expression $\hat{\eta}_{eq} = e^{-\widehat{H}/T}/\mathrm{Sp}(e^{-\widehat{H}/T})$ for the equilibrium density matrix and prove the operator identity $[\widehat{A},e^{-\widehat{H}/T}] = -e^{-\widehat{H}/T}\int_0^{1/T}d\lambda e^{\widehat{H}\lambda}[\widehat{A},\widehat{H}]e^{-\widehat{H}\lambda}$,

where $\widehat{A}$ is an arbitrary operator. Then put $\widehat{A} = \widehat{\Delta H}_\omega$ and apply this operator identity to Eq. (13.4). Expressing the perturbation through the scalar potential according to $\widehat{\Delta H}_\omega = -eE_\omega \cdot \hat{\mathbf{x}}$, use $[\hat{\mathbf{x}}, \widehat{H}] = i\hbar\hat{\mathbf{v}}$ to transform the commutator $[\widehat{\Delta H}_\omega, \widehat{H}]$.

3.6. Prove Eq. (13.24).

<u>Hint</u>: Use the properties of the δ-function, $\delta(E) = 0$ at $E \neq 0$, $\delta(0) = \infty$, and $\int dE\delta(E) = 1$, to prove that the integral in Eq. (13.24) also has all these properties.

3.7. Prove that the Green's function defined by Eq. (14.6) is given by

$$G_E(\mathbf{p}, \mathbf{p}') = \int d(t - t')\frac{1}{V} \int d\mathbf{r} \int d\mathbf{r}' e^{iE(t-t')/\hbar} e^{-i(\mathbf{p}\cdot\mathbf{r} - \mathbf{p}'\cdot\mathbf{r}')/\hbar} G_{tt'}(\mathbf{r}, \mathbf{r}'),$$

where $G_{tt'}(\mathbf{r}, \mathbf{r}')$ is the Green's function of the differential operator $i\hbar \times \partial/\partial t - \widehat{H}(\mathbf{r})$ with $\widehat{H}(\mathbf{r}) = -(\hbar^2/2m)\nabla_\mathbf{r}^2 + U_{im}(\mathbf{r})$.

<u>Solution</u>: In mathematics, the Green's function $G_{tt'}(\mathbf{r}, \mathbf{r}')$ of the differential equation $\widehat{M}(\mathbf{r}, t)y(\mathbf{r}, t) = 0$, where $\widehat{M}$ is a differential operator, is introduced according to $\widehat{M}(\mathbf{r}, t)G_{tt'}(\mathbf{r}, \mathbf{r}') = \delta(t - t')\delta(\mathbf{r} - \mathbf{r}')$. In our case, $\widehat{M}(\mathbf{r}, t) = i\hbar\partial/\partial t - \widehat{H}(\mathbf{r})$. Since $\widehat{H}(\mathbf{r})$ is time-independent, $G_{tt'}(\mathbf{r}, \mathbf{r}')$ depends only of $t - t'$. We point out that any solution of the equation for $G_{tt'}(\mathbf{r}, \mathbf{r}')$ is ambiguous in the sense that this function may contain the factors $\theta(t - t')$ and $\theta(t' - t)$. These cases correspond to retarded and advanced Green's functions, respectively.

Multiplying the equation for $G_{tt'}(\mathbf{r}, \mathbf{r}')$ by $\exp[-i(\mathbf{p} \cdot \mathbf{r} - Et)/\hbar]$ and $\exp[i(\mathbf{p}' \cdot \mathbf{r}' - Et')/\hbar]$, and calculating the integrals over times and coordinates with the aid of integration by parts, we transform this equation to Eq. (14.6).

3.8. Write an iterational solution for $G_{tt'}^s(\mathbf{r}, \mathbf{r}')$ by analogy to Eq. (14.10).

<u>Solution</u>: If $U_{im}(\mathbf{r}) = 0$, then $G_{tt'}^s(\mathbf{r}, \mathbf{r}') = g_{tt'}^s(\mathbf{r}, \mathbf{r}') = g_{t-t'}^s(|\mathbf{r} - \mathbf{r}'|)$. The double Fourier transformation of this function in space and time gives us $g_E^s(\mathbf{p}, \mathbf{p}')$ of Eq. (14.9). The fact that the free-electron Green's function in the coordinate representation depends only on the difference between the coordinates is caused by the translational invariance. This property is violated in inhomogeneous systems and in the systems of finite size, where the boundary conditions are imposed. Equations (14.8) and (14.10) in the time-coordinate representation have the following form:

$$G_{tt'}^s(\mathbf{r}, \mathbf{r}') = g_{t-t'}^s(\mathbf{r} - \mathbf{r}') + \int d\mathbf{r}_1 \int dt_1 g_{t-t_1}^s(\mathbf{r} - \mathbf{r}_1)U_{im}(\mathbf{r}_1)G_{t_1t'}^s(\mathbf{r}_1, \mathbf{r}'),$$

$$G_{tt'}^s(\mathbf{r}, \mathbf{r}') = g_{t-t'}^s(\mathbf{r} - \mathbf{r}') + \int d\mathbf{r}_1 \int dt_1 g_{t-t_1}^s(\mathbf{r} - \mathbf{r}_1)U_{im}(\mathbf{r}_1)g_{t_1-t'}^s(\mathbf{r}_1 - \mathbf{r}')$$

$$+ \int d\mathbf{r}_1 \int d\mathbf{r}_2 \int dt_1 \int dt_2 g_{t-t_1}^s(\mathbf{r} - \mathbf{r}_1)U_{im}(\mathbf{r}_1)g_{t_1-t_2}^s(\mathbf{r}_1 - \mathbf{r}_2)$$

$$\times U_{im}(\mathbf{r}_2)g_{t_2-t'}^s(\mathbf{r}_2 - \mathbf{r}') + \dots .$$

3.9. Find the pair correlation function $w(r)$ and its Fourier transform $w(q)$ for the impurities with screened Coulomb potential energy $v(r) = e^2 e^{-r/r_0}/\epsilon r$.

Solution: The Fourier component of the screened Coulomb potential is calculated directly: $v(q) = (4\pi e^2/\epsilon)(r_0^{-2} + q^2)^{-1}$. Then we have $w(q) = n_{im}(4\pi e^2/\epsilon)^2(r_0^{-2} + q^2)^{-2}$. Doing the inverse Fourier transformation of $w(q)$, we obtain

$$w(r) = n_{im}\frac{2\pi e^4}{\epsilon^2}r_0 e^{-r/r_0} .$$

In the absence of screening ($r_0 \to \infty$), the correlation function $w(q)$ diverges at small q and $w(r)$ goes to infinity.

3.10. Express the density of states by using different representations of the Green's function.

Hints: See problem 3.7, where the Green's function in the time-coordinate representation is defined. Prove that the retarded Green's function can be represented as

$$G^R_{\Delta t}(\mathbf{r}, \mathbf{r}') = -\frac{i}{\hbar}\theta(\Delta t)\sum_\delta e^{-i\varepsilon_\delta \Delta t/\hbar}\psi^{(\delta)}_{\mathbf{r}}\psi^{(\delta)*}_{\mathbf{r}'},$$

where δ is the index of exact eigenstates. Carrying out a temporal Fourier transformation, find the energy-coordinate representation

$$\int d\Delta t e^{iE\Delta t/\hbar}G^R_{\Delta t}(\mathbf{r}, \mathbf{r}') = G^R_E(\mathbf{r}, \mathbf{r}') = \sum_\delta \frac{\psi^{(\delta)}_{\mathbf{r}}\psi^{(\delta)*}_{\mathbf{r}'}}{E - \varepsilon_\delta + i\lambda}, \quad \lambda \to +0,$$

which is similar to Eq. (14.1). Using these equations together with the general definition of the density of states in terms of exact eigenstates, check Eq. (14.30).

3.11. Derive Eq. (8.27) from Eq. (15.15).

Hint: Using Eqs. (15.12) and (15.14), obtain $\left(V^{-1}\sum_{\mathbf{p}} G^A_E(p)G^R_E(p)\right)^{-1} - \Lambda(E) = \hbar[\pi\rho_{\mathcal{D}}(E)\tau_{tr}(E)]^{-1}$.

3.12. Calculate $I(\Delta p)$ given by Eq. (15.23) at small Δp.

Hints: Expand the Green's functions under the sum in series of Δp up to the terms $\propto (\Delta p)^2$. Take into account that $\tau = \hbar^3/mw$ in the case of short-range correlated inhomogeneities.

3.13. Calculate the integral in Eq. (15.26).

Solution: Introducing new variables $\mathbf{P} = (\mathbf{p} - \mathbf{p}_1)/2$ and $\Delta\mathbf{p} = \mathbf{p} + \mathbf{p}_1$, we rewrite this integral as

$$-\int_0^\infty \frac{d\Delta p}{\Delta p}\int_0^{2\pi}\frac{d\varphi}{2\pi}\int_0^\infty \frac{dP}{(2\pi\hbar^2)^2}$$
$$\times \frac{P(P^2 - \Delta p^2/4)}{[(\varepsilon_{\mathbf{P}+\Delta\mathbf{p}/2} - \varepsilon_F)^2 + (\hbar/2\tau)^2][(\varepsilon_{\mathbf{P}-\Delta\mathbf{p}/2} - \varepsilon_F)^2 + (\hbar/2\tau)^2]},$$

where $\varphi = \widehat{\mathbf{P}\,\Delta\mathbf{p}}$ and $\tau = \hbar^3/mw$. In the approximation $\varepsilon_p - \varepsilon_F \simeq v_F(p - p_F)$, which is valid at $\varepsilon_F \gg \hbar/2\tau$, one can calculate the integral over P by using the method of residue (two simple poles in the upper half-plane of the complex variable $P - p_F$ are $[\pm\Delta p \cos\varphi + i\hbar/\tau v_F]/2$). After this, the integral over φ is taken elementary, and the above expression becomes

$$-\frac{m^3 v_F \tau^2}{\pi\hbar^6} \int_0^\infty \frac{d\Delta p}{\Delta p \sqrt{(\Delta p)^2 + (\hbar/\tau v_F)^2}}.$$

This integral diverges at $\Delta p = 0$. Setting the lower limit of the integration at $\hbar/\sqrt{l_\varphi l_F/2}$, and substituting the result in Eq. (15.26), we obtain Eq. (15.27).

3.14. Obtain Eq. (16.18) in two different ways: by solving the differential equation (16.1) at $U_{im} = 0$, and by calculating the contour path integral in Eq. (16.17) at $w = 0$.

Hints: In the case of free electrons, it is easy to find solutions of Eq. (16.1) in the momentum representation, where the Hamiltonian $-\hbar^2 \nabla_\mathbf{r}^2/2m = \hat{\mathbf{p}}^2/2m$ is diagonal. Using Eq. (16.3), one has $g_t^R(\mathbf{p}) = -(i/\hbar)\theta(t)\exp(-ip^2t/2m\hbar)$. The Fourier transformation of this expression to the coordinate representation leads to Eq. (16.18). To calculate the path integral in Eq. (16.17), write the exponent $\exp\left[(im/2\hbar)\int_0^t d\tau \dot{\mathbf{r}}_\tau^2\right]$ in the form $\exp\left[(im/2\hbar\tau_N)\sum_{i=1}^N (\mathbf{r}_i - \mathbf{r}_{i-1})^2\right]$, as in Eq. (16.10), and use Eq. (16.12). Integrate out each intermediate coordinate with $i = 1, \ldots, N-1$ and take into account that $\mathbf{r}_0 = \mathbf{r}_N$.

3.15. Calculate the integrals over time in Eq. (16.19) in order to find ρ_{3D} and ρ_{1D}.

Results:

$$\rho_{3D}(E) = \frac{m^{3/2}[w(0)]^{1/4}}{2^{1/4}\pi^{5/2}\hbar^3}\left\{\frac{\Gamma(3/4)}{3}\left[4 - \Phi(3/4, 1/2; -E^2/2w(0))\right]\right.$$

$$\left. + \frac{E}{2\sqrt{2w(0)}}\Gamma(1/4)\Phi(1/4, 3/2; -E^2/2w(0))\right\},$$

$$\rho_{1D}(E) = \frac{m^{1/2}}{2^{3/4}\pi^{3/2}\hbar[w(0)]^{1/4}}\left\{\Gamma(1/4)\Phi(1/4, 1/2; -E^2/2w(0))\right.$$

$$\left. + \frac{2^{1/2}E}{\sqrt{w(0)}}\Gamma(3/4)\Phi(7/4, 3/2; -E^2/2w(0))\right\},$$

where $\Phi(\alpha, \gamma; z)$ is the confluent hypergeometric function and $\Gamma(x)$ is the Gamma function.

3.16. Calculate the contour path integral in Eq. (16.26).

Solution: Let us represent each time-dependent coordinate of the closed paths as Fourier series

$$\mathbf{r}_\tau = \sum_{k=0}^N \left[\mathbf{r}_k^c \cos(2\pi k\tau/t) + \mathbf{r}_k^s \sin(2\pi k\tau/t)\right],$$

where $\sum_{k=0}^N \mathbf{r}_k^c = 0$ and $N \to \infty$, and rewrite the path integral as a multiple integral over the coefficients $\mathbf{r}_k^c = (x_k^c, y_k^c)$ and $\mathbf{r}_k^s = (x_k^s, y_k^s)$. Substituting $\mathbf{r}_\tau$ defined in this way into the Lagrangian, we take the integral over time in the exponent and obtain the term in the exponent in the following form:

$$(im\pi^2/t) \sum_{k=1}^{N} k^2 \left[(x_k^c)^2 + (y_k^c)^2 + (x_k^s)^2 + (y_k^s)^2 + \omega_c t(x_k^c y_k^s - x_k^s y_k^c)/\pi k \right] .$$

Thus, the multiple integral over $4N$ variables is reduced to a product of N integrals over four variables, x_k^c, x_k^s, y_k^c, and y_k^s. These integrals are easily calculated if we write the term in the exponent as a full quadratic form. This procedure demonstrates that the result of the integration differs from that in the absence of the magnetic field by the factor

$$\prod_{k=1}^{N} \left[1 - (\omega_c t/\pi k)^2 \right]^{-1} .$$

Calculating this product at $N \to \infty$, we finally obtain the Green's function

$$g_t^R(0) = -\theta(t) \frac{m\omega_c}{4\pi\hbar^2 \sin(\omega_c t/2)}.$$

It is easy to observe that this function goes to $g_t^R(0)$ of Eq. (16.18) when $\omega_c \to 0$.

3.17. Calculate the integral over time in Eq. (16.27) by assuming that $\sqrt{w(0)} \ll \hbar\omega_c$.

Hints: For positive E, one can shift the contour in the upper half-plane by a finite time, $t = i\tau + t'$. The integral in Eq. (16.27) is written as an infinite sum of the pole contributions plus the integral over t'. The latter integral is calculated easily if we put $\tau\omega/2 \gg 1$ and appears to be small as $\exp[-(E + \hbar\omega/2)^2/2w(0)]$. The remaining infinite sum is transformed to the integral over the variable of summation if $\sqrt{w(0)} \ll \hbar\omega_c$ and $|E - \varepsilon_N| \ll \hbar\omega_c$. Calculating it, one obtains the density of states in the form of Eq. (16.28).

3.18. Prove the f-sum rule

$$\frac{\partial^2 \varepsilon_{n\mathbf{p}}}{\partial p_\alpha \partial p_\beta} = \frac{\delta_{\alpha\beta}}{m_e} + \sum_{\sigma\sigma'n'} \frac{v_{n\sigma,n'\sigma'}^\alpha v_{n'\sigma',n\sigma}^\beta + v_{n\sigma,n'\sigma'}^\beta v_{n'\sigma',n\sigma}^\alpha}{\varepsilon_{n\mathbf{p}} - \varepsilon_{n'\mathbf{p}}},$$

which takes into account non-parabolicity of the band structure.

Hints: Carry out the **kp**-expansion of the energy analogous to Eqs. (5.6)−(5.10) near the momentum **p**, with the accuracy up to the second order in the deviations. As a result, the inverse effective mass is given by the right-hand side of Eq. (5.11), taken in the point **p**. Since the free-electron mass m_e is large in comparison to the effective mass, the contribution proportional to m_e^{-1} can be neglected in Eq. (17.4).

3.19. Check the relation between α_ω and $\mathrm{Im}\,\epsilon_\omega$ used in Eq. (17.6).

Hints: One may use the results of problem 2.11 and Eq. (17.1). Another way is to write the wave equation in the medium with complex dielectric permittivity ϵ_ω and calculate the imaginary part of wave vector determining the exponential decrease of the intensity of radiation.

3.20. Using Eqs. (17.9) and (17.10), as well as Eq. (17.7) for non-doped materials ($\eta_{pF} = 1$), check the dispersion relation

$$\mathrm{Re}\,\epsilon_\omega = \kappa + \frac{2}{\pi}\mathcal{P}\int_0^\infty d\omega'\,\frac{\omega'\mathrm{Im}\epsilon_{\omega'}}{\omega'^2 - \omega^2},$$

which follows from Eqs. (13.15) and (17.1), by a direct calculation.

Hints: Take into account that $\mathrm{Im}\epsilon_\omega$ is non-zero at $\hbar\omega > \varepsilon_g$ and use new variables η and Ω instead of ω' and ω, respectively.

3.21. Check that $\mathrm{Re}\,\epsilon_\omega$ has a logarithmic divergence near the step-like threshold of absorption in doped materials.

Hint: Substitute the step-like $\mathrm{Im}\epsilon_\omega$ of Eq. (17.7) into the dispersion relation considered in the previous problem.

3.22. Prove that the terms linear in **q** in the dielectric permittivity tensor are equal to zero due to Onsager's symmetry relation.

Hint: This property is seen from Eqs. (13.17) and (17.1).

3.23. Prove Eq. (18.24) by a direct calculation of the multiple integral.

Solution: We need to take the multiple integral of the exponential function

$$\exp[-(A_1 + B)\widetilde{\mathbf{z}}_1^2 - (A_2 + B)\widetilde{\mathbf{z}}_2^2 - (A_3 + B)\widetilde{\mathbf{z}}_3^2 -$$

$$\ldots - 2B\widetilde{\mathbf{z}}_1\cdot(\widetilde{\mathbf{z}}_2 + \widetilde{\mathbf{z}}_3 + \ldots) - 2B\widetilde{\mathbf{z}}_2\cdot(\widetilde{\mathbf{z}}_3 + \widetilde{\mathbf{z}}_4 + \ldots) - \ldots].$$

Integrating this expression over $\widetilde{\mathbf{z}}_1$, we obtain

$$\left(\frac{\pi}{A_1 + B}\right)^{d/2}\exp\left[-\frac{A_1 A_2 + B(A_1 + A_2)}{A_1 + B}\widetilde{\mathbf{z}}_2^2 - \frac{2A_1 B}{A_1 + B}\widetilde{\mathbf{z}}_2\cdot(\widetilde{\mathbf{z}}_3 + \widetilde{\mathbf{z}}_4 + \ldots)\right.$$

$$\left. -\frac{B^2}{A_1 + B}(\widetilde{\mathbf{z}}_3 + \widetilde{\mathbf{z}}_4 + \ldots)^2 - (A_3 + B)\widetilde{\mathbf{z}}_3^2 - 2B\widetilde{\mathbf{z}}_3\cdot(\widetilde{\mathbf{z}}_4 + \widetilde{\mathbf{z}}_5 + \ldots) - \ldots\right],$$

where d is the dimensionality of the vectors $\widetilde{\mathbf{z}}_l$. Integrating over $\widetilde{\mathbf{z}}_2$, we obtain a new prefactor, $\left(\pi^2/[A_1 A_2 + B(A_1 + A_2)]\right)^{d/2}$, while $\widetilde{\mathbf{z}}_3^2$ will enter the exponent with the factor $-[A_1 A_2 A_3 + B(A_1 A_2 + A_2 A_3 + A_1 A_3)]/[A_1 A_2 + B(A_1 + A_2)]$. Acting by induction, we find that n integrations result in the following prefactor:

$$\pi^{nd/2}\left[A_1 A_2\ldots A_n + B\left(\frac{1}{A_1} + \frac{1}{A_2} + \ldots + \frac{1}{A_n}\right)A_1 A_2\ldots A_n\right]^{-d/2}.$$

Aiming n to infinity, substituting $d = 3$, and multiplying this expression by the factor $\prod_{l=1}^\infty (A_l/\pi)^{3/2}$ standing on the left-hand side of Eq. (18.24), we obtain the right-hand side of Eq. (18.24).

Chapter 4

BOSONS INTERACTING
WITH ELECTRONS

A consistent consideration of transport phenomena for different types of vibrational modes in solids (the branches of phonon spectrum and the modes of electromagnetic field introduced in Secs. 6 and 3, respectively) should be based upon the formalism of second quantization. Below we introduce the density matrix for bosons and derive kinetic equations for bosonic modes in the occupation number representation. The non-diagonal components of the boson density matrix describe phase correlations. In this chapter we study the case when the interaction of bosons with electron sub-system is described by the Hamiltonian linear in the creation and annihilation operators of bosons. This important case unifies the photon-electron and phonon-electron interactions in solids. Then we consider the processes of spontaneous and stimulated emission of bosons by non-equilibrium electrons, instabilities of phonon systems, and emission of three-dimensional boson modes by two-dimensional electrons. The kinetics of interacting bosons is considered in the next chapter for phonons.

19. Kinetic Equation for Boson Modes

A system of bosons (photons or phonons) interacting with electrons is described by the Hamiltonian comprising both types of quasiparticles and their interaction:

$$\widehat{H}_b + \widehat{H}_e + \widehat{H}_{e,b}. \qquad (1)$$

The Hamiltonian of free bosons, $\widehat{H}_b$, has the same form as $\widehat{H}_{ph}$ in Eq. (3.12). The Hamiltonian of electrons, $\widehat{H}_e$, in Eq. (1) does not include the electron-electron interaction term, in contrast to $\widehat{H}_e$ introduced by Eq. (4.10). Therefore, it is written as $\widehat{H}_e = \sum_j \hat{h}_j$, where $\hat{h}_j$ is the one-electron Hamiltonian containing, in the general case, a contribution of external fields. In the second quantization representation, the Hamiltonian $\widehat{H}_e$ has the same form as $\widehat{H}$ in Eq. (4.23). The electron-boson interaction is described by the Hamiltonian linear in the creation and annihilation operators of bosons, $\hat{b}_q^+$ and $\hat{b}_q$. It can be represented as

$$\widehat{H}_{e,b} = \sum_{jq} \hat{\chi}_q^{(j)} \hat{b}_q + H.c., \tag{2}$$

where the index q numbers a state of the boson system and the operator $\hat{\chi}_q^{(j)}$ describes the interaction of j-th electron with this state. In the case of electron-phonon interaction, an explicit form of this operator can be obtained from the expressions (6.30) and (6.22) describing interaction of a single electron at the point $\mathbf{r}$ with long-wavelength acoustic and optical phonon modes, respectively (problem 4.1). For electron-photon interaction, the Hamiltonian $\widehat{H}_{e,b}$ can be identified with the second term on the right-hand side of Eq. (4.16), where the current density operator is given by Eq. (4.15) and the vector potential $\mathbf{A}$ is quantized according to Eq. (3.18) (problem 4.2). The contributions quadratic in $\mathbf{A}$ should be neglected. In the second quantization representation, the electron part of the interaction operator (2) is written as $\sum_j \hat{\chi}_q^{(j)} = \sum_{\delta\eta} \langle \delta | \hat{\chi}_q | \eta \rangle \, \hat{a}_\delta^+ \hat{a}_\eta$, where $\hat{a}_\delta^+$ and $\hat{a}_\eta$ are the second-quantization operators for electron states $|\delta\rangle$ and $|\eta\rangle$ (problem 4.3). The average $\overline{Q}$ of any one-boson operator $\widehat{Q} = \sum_{qq_1} \langle q | \hat{q} | q_1 \rangle \hat{b}_q^+ \hat{b}_{q_1}$ is written according to the general formula (1.18):

$$\overline{Q} = \mathrm{Sp}\widehat{Q}\hat{\eta}_t = \sum_{qq_1} \langle q | \hat{q} | q_1 \rangle \, N_t(q_1, q). \tag{3}$$

The one-boson density matrix $N_t(q_1, q)$ introduced in this equation is defined as an average of the product of bosonic creation and annihilation operators with many-particle statistical operator $\hat{\eta}_t$:

$$N_t(q, q_1) \equiv \mathrm{Sp}\{\hat{b}_{q_1}^+ \hat{b}_q \hat{\eta}_t\}. \tag{4}$$

The trace Sp is taken over the quantum numbers of both bosons and electrons, and $N_t(q, q_1)$ depends on two bosonic variables, q and q_1.

To derive the kinetic equation for bosons, we use Eq. (1.20) for $\hat{\eta}_t$ with the Hamiltonian (1). After multiplying this equation by $\hat{b}_{q_1}^+ \hat{b}_q$, we take the traces of the both sides of the equation obtained and find

$$\left[i\hbar \frac{\partial}{\partial t} - \hbar(\omega_q - \omega_{q_1}) \right] N_t(q, q_1) = -\sum_{\delta\eta} \langle \delta | \hat{\chi}_{q_1 t} | \eta \rangle \, \langle\langle \hat{a}_\delta^+ \hat{a}_\eta \hat{b}_q \rangle\rangle_t$$

$$+ \sum_{\delta\eta} \langle \delta | \hat{\chi}_{qt} | \eta \rangle^* \, \langle\langle \hat{a}_\delta^+ \hat{a}_\eta \hat{b}_{q_1} \rangle\rangle_t^*, \tag{5}$$

where $\langle\langle \ldots \rangle\rangle_t \equiv \mathrm{Sp}\{\hat{\eta}_t \ldots\}$. The one-electron Hamiltonian $\widehat{H}_e$ does not contribute to this equation, while the free-boson Hamiltonian contributes to the left-hand side. Although the operators $\hat{\chi}_q$ considered in problems 4.1 and 4.2 are time-independent, in Eq. (5) they are written as functions of time. This corresponds to a more general case, when external, time-dependent electric or magnetic fields modify the electron-boson interaction. On the right-hand side of Eq. (5), we have the correlation functions of the kind $\langle\langle \hat{a}_\delta^+ \hat{a}_\eta \hat{b}_q \rangle\rangle_t$, which satisfy the equation

$$i\hbar \frac{\partial}{\partial t} \langle\langle \hat{a}_\delta^+ \hat{a}_\eta \hat{b}_q \rangle\rangle_t = \mathrm{Sp}\{[\hat{a}_\delta^+ \hat{a}_\eta \hat{b}_q, \widehat{H}_b + \widehat{H}_e + \widehat{H}_{e,b}]\hat{\eta}_t\} \qquad (6)$$

obtained from Eq. (1.20) in the way employed in the derivation of Eq. (5). It contains higher-order correlation functions on the right-hand side. They are evaluated according to the approximate equations

$$\langle\langle \hat{a}_\delta^+ \hat{a}_\eta \hat{b}_q^+ \hat{b}_{q_1} \rangle\rangle_t \simeq \langle\langle \hat{a}_\delta^+ \hat{a}_\eta \rangle\rangle_t \langle\langle \hat{b}_q^+ \hat{b}_{q_1} \rangle\rangle_t, \quad \langle\langle \hat{a}_\delta^+ \hat{a}_\eta \hat{b}_q \hat{b}_{q_1} \rangle\rangle_t \simeq 0,$$

$$\langle\langle \hat{a}_\delta^+ \hat{a}_\gamma^+ \hat{a}_\eta \hat{a}_\nu \rangle\rangle_t \simeq \langle\langle \hat{a}_\delta^+ \hat{a}_\nu \rangle\rangle_t \langle\langle \hat{a}_\gamma^+ \hat{a}_\eta \rangle\rangle_t - \langle\langle \hat{a}_\gamma^+ \hat{a}_\nu \rangle\rangle_t \langle\langle \hat{a}_\delta^+ \hat{a}_\eta \rangle\rangle_t, \qquad (7)$$

which would be exact in the absence of electron-boson interaction, because in that case the statistical operator $\hat{\eta}_t$ is written as a product of electron and boson statistical operators (we note that the second-quantized operators of electrons commute with those of bosons). The approximations (7) can be used to calculate the commutators on the right-hand side of Eq. (6). Thus, we rewrite Eq. (6) as

$$\left(i\hbar \frac{\partial}{\partial t} - \hbar\omega_q \right) \langle\langle \hat{a}_\delta^+ \hat{a}_\eta \hat{b}_q \rangle\rangle_t$$

$$- \sum_\gamma \left\{ \langle\eta|\hat{h}_t|\gamma\rangle \langle\langle \hat{a}_\delta^+ \hat{a}_\gamma \hat{b}_q \rangle\rangle_t - \langle\gamma|\hat{h}_t|\delta\rangle \langle\langle \hat{a}_\gamma^+ \hat{a}_\eta \hat{b}_q \rangle\rangle_t \right\}$$

$$= \sum_{\gamma\nu} \langle\gamma|\hat{\chi}_{qt}^+|\nu\rangle \left\{ \langle\langle \hat{a}_\gamma^+ \hat{a}_\nu \rangle\rangle_t \langle\langle \hat{a}_\delta^+ \hat{a}_\eta \rangle\rangle_t + \langle\langle \hat{a}_\delta^+ \hat{a}_\nu \rangle\rangle_t \left[\delta_{\gamma\eta} - \langle\langle \hat{a}_\gamma^+ \hat{a}_\eta \rangle\rangle_t \right] \right\}$$

$$+ \sum_{\gamma q'} \left\{ \langle\eta|\hat{\chi}_{q't}^+|\gamma\rangle \langle\langle \hat{a}_\delta^+ \hat{a}_\gamma \rangle\rangle_t - \langle\gamma|\hat{\chi}_{q't}^+|\delta\rangle \langle\langle \hat{a}_\gamma^+ \hat{a}_\eta \rangle\rangle_t \right\} \langle\langle \hat{b}_{q'}^+ \hat{b}_q \rangle\rangle_t. \qquad (8)$$

Equations (5) and (8) form a closed set of equations written with the accuracy up to the second order in electron-boson coupling.

It is convenient to introduce the correlation operator $\widehat{K}_{qt}$ and the one-electron density matrix $\hat{\rho}_t$ (cf. Eq. (4.29)) defined by their matrix elements:

$$\langle\eta|\widehat{K}_{qt}|\delta\rangle = \langle\langle \hat{a}_\delta^+ \hat{a}_\eta \hat{b}_q \rangle\rangle_t, \quad \langle\eta|\hat{\rho}_t|\delta\rangle = \langle\langle \hat{a}_\delta^+ \hat{a}_\eta \rangle\rangle_t. \qquad (9)$$

Equation (8) then acquires the following operator form:

$$\left(i\hbar\frac{\partial}{\partial t} - \hbar\omega_q\right)\widehat{K}_{qt} - \left[\hat{h}_t, \widehat{K}_{qt}\right] = \widehat{G}_{qt},$$

$$\widehat{G}_{qt} \equiv \langle\langle\hat{\chi}_{qt}^+\rangle\rangle_t\hat{\rho}_t + (1 - \hat{\rho}_t)\hat{\chi}_{qt}^+\hat{\rho}_t + \sum_{q'}[\hat{\chi}_{q't}^+, \hat{\rho}_t]N_t(q, q'), \qquad (10)$$

and Eq. (5) is written through $\widehat{K}_{qt}$ as

$$\left[i\hbar\frac{\partial}{\partial t} - \hbar(\omega_q - \omega_{q_1})\right]N_t(q, q_1) = \left(\mathrm{sp}\hat{\chi}_{qt}\widehat{K}_{q_1t}\right)^* - \mathrm{sp}\hat{\chi}_{q_1t}\widehat{K}_{qt}, \qquad (11)$$

where sp ... defines one-electron averaging.

Equations (10) and (11) form a closed system which can be used to determine $\widehat{K}_{qt}$ and $N_t(q, q_1)$. These equations must be, however, accompanied by the initial condition describing the requirement of weakening of correlations:

$$\widehat{K}_{qt}|_{t\to-\infty} = 0 . \qquad (12)$$

This condition is similar to Eq. (7.11) applied to electron-impurity system. Using the evolution operator defined in Sec. 2, one may exclude $\widehat{K}_{qt}$ by writing the solution of Eq. (10) as

$$\widehat{K}_{qt} = \frac{1}{i\hbar}\int_{-\infty}^{t} dt' e^{-i\omega_q(t-t')}\widehat{S}(t, t')\widehat{G}_{qt'}\widehat{S}^+(t, t'). \qquad (13)$$

As a result, the quantum kinetic equation for bosons becomes

$$\left[\frac{\partial}{\partial t} + i(\omega_q - \omega_{q_1})\right]N_t(q, q_1) = J_{b,e}(N|qq_1t). \qquad (14)$$

The boson-electron collision integral $J_{b,e}$ is expressed through the operator $\widehat{G}_{qt}$ defined in Eq. (10) as

$$J_{b,e}(N|qq_1t) = \frac{1}{\hbar^2}\int_{-\infty}^{t} dt' e^{\lambda t' - i\omega_q(t-t')}\mathrm{sp}\left[\hat{\chi}_{q_1t}\widehat{S}(t, t')\widehat{G}_{qt'}\widehat{S}^+(t, t')\right]$$

$$+(c.c., q \leftrightarrow q_1). \qquad (15)$$

It contains the contributions independent of N_t, which correspond to spontaneous emission of bosons by electrons, as well as the contributions proportional to N_t, which describe the evolution of bosons due to their interaction with the electrons.

Equation (14) can be simplified if the system has the translational symmetry. The set of quantum numbers q in this case defines the boson polarization (branch) μ and wave vector $\mathbf{q}$. The operator of translation, $\widehat{T}_{\mathbf{R}}$, transforms the operator $\hat{\chi}_{qt}$ according to

$$\widehat{T}_{\mathbf{R}}\hat{\chi}_{qt}\widehat{T}_{\mathbf{R}}^{+} = e^{i\mathbf{q}\cdot\mathbf{R}}\hat{\chi}_{qt}. \qquad (16)$$

Using this relation, one may prove, in a similar way as for the one-electron density matrix, see Sec. 8 and problem 2.4, that $N_t(q, q_1)$ is diagonal in the wave vector:

$$N_t(q, q_1) = \delta_{\mathbf{q}\mathbf{q}_1} N_{\mathbf{q}t}^{\mu\mu_1}. \qquad (17)$$

However, the bosonic distribution remains non-diagonal in the polarization indices μ and μ_1. Below we derive a kinetic equation for $N_{\mathbf{q}t}^{\mu\mu_1}$ in the case of degenerate and isotropic boson spectra, when the boson frequencies $\omega_{\mathbf{q}\mu} = \omega_q$ are independent of the polarization index μ as well as of the direction of $\mathbf{q}$. This consideration is directly applicable to photons. In the single-mode approximation, it can be also applied to the phonons in the isotropic crystals. Since $\langle\langle\hat{\chi}_{qt}^{+}\rangle\rangle_t = 0$ for a translation-invariant system, Eq. (10) for $\widehat{G}_{qt}$ contains only the second and the third terms on the right-hand side. As a result, we rewrite Eq. (14) as

$$\frac{\partial N_{\mathbf{q}t}^{\mu\mu_1}}{\partial t} = I_{\mu\mu_1}(\mathbf{q}t) + \sum_{\mu'} \int_{-\infty}^{t} dt' e^{\lambda t'} \qquad (18)$$

$$\times \left[N_{\mathbf{q}t'}^{\mu\mu'} \mathcal{R}_{\mu'\mu_1}(\mathbf{q}|t, t') + \mathcal{R}_{\mu'\mu}^{*}(\mathbf{q}|t, t') N_{\mathbf{q}t'}^{\mu'\mu_1} \right].$$

The first term on the right-hand side describes the spontaneous emission of bosons,

$$I_{\mu\mu_1}(\mathbf{q}t) = \frac{1}{\hbar^2} \int_{-\infty}^{t} dt' e^{\lambda t' - i\omega_q(t-t')} \qquad (19)$$

$$\times \mathrm{sp}\left[\widehat{S}^{+}(t, t')\hat{\chi}_{\mathbf{q}\mu_1}\widehat{S}(t, t')(1 - \hat{\rho}_{t'})\hat{\chi}_{\mathbf{q}\mu}^{+}\hat{\rho}_{t'} \right] + (c.c., \mu \leftrightarrow \mu_1),$$

while the non-Markovian evolution of bosons is described by the second term, where

$$\mathcal{R}_{\mu'\mu}(\mathbf{q}|t, t') = \hbar^{-2} e^{-i\omega_q(t-t')} \mathrm{sp}\hat{\rho}_{t'} \left[\widehat{S}^{+}(t, t')\hat{\chi}_{\mathbf{q}\mu}\widehat{S}(t, t'), \ \hat{\chi}_{\mathbf{q}\mu'}^{+} \right]. \qquad (20)$$

The second term on the right-hand side of Eq. (18) can be either negative or positive. The first case corresponds to the relaxation of bosons due to their scattering by electrons, while the second case corresponds to the regime of induced emission of bosons (also known as negative boson absorption). The second regime may be realized only if the electron distribution is non-equilibrium.

For a stationary electron system with the Hamiltonian $\hat{h}$, we use Eq. (2.3) for the evolution operator and substitute the time-independent electron density matrix $\hat{\rho}$ in Eqs. (19) and (20). Using the solutions of the eigenstate problem $\hat{h}|\delta\rangle = \varepsilon_\delta|\delta\rangle$, we integrate over time in Eq. (19), see Eq. (8.6) and problem 2.5, and obtain

$$I_{\mu\mu_1}(\mathbf{q}) = \frac{2\pi}{\hbar} \sum_{\delta\eta} \langle\eta|\hat{\chi}_{\mathbf{q}\mu}|\delta\rangle^* \langle\eta|\hat{\chi}_{\mathbf{q}\mu_1}|\delta\rangle \delta(\varepsilon_\delta - \varepsilon_\eta + \hbar\omega_q) f_\eta(1 - f_\delta), \quad (21)$$

where $f_\delta = \langle\delta|\hat{\rho}|\delta\rangle$ is the electron distribution function. Note that the principal-value contribution drops out of $I_{\mu\mu_1}$. Equation (21) contains the factor $f_\eta(1 - f_\delta)$, which accounts for the Pauli blocking effect for the transitions from the state η to the state δ. However, the conventional Fermi's golden rule for such transitions does not exist in the case of $\mu \neq \mu_1$, because the squared absolute value of the matrix element is obtained only for the diagonal component $I_{\mu\mu}(\mathbf{q})$. Similar operations applied to the second term on the right-hand side of Eq. (18) transform this term into

$$-\frac{1}{2} \sum_{\mu'} \left[\nu_{\mu\mu'}(\mathbf{q}) N_{\mathbf{q}t}^{\mu'\mu_1} + N_{\mathbf{q}t}^{\mu\mu'} \nu_{\mu'\mu_1}(\mathbf{q}) \right]$$

$$+i \sum_{\mu'} \left[N_{\mathbf{q}t}^{\mu\mu'} \Omega_{\mu'\mu_1}(\mathbf{q}) - \Omega_{\mu\mu'}(\mathbf{q}) N_{\mathbf{q}t}^{\mu'\mu_1} \right], \quad (22)$$

where the matrix $\nu_{\mu\mu_1}$ describes the relaxation of bosons,

$$\nu_{\mu\mu_1}(\mathbf{q}) = \frac{2\pi}{\hbar} \sum_{\delta\eta} \langle\eta|\hat{\chi}_{\mathbf{q}\mu}|\delta\rangle^* \langle\eta|\hat{\chi}_{\mathbf{q}\mu_1}|\delta\rangle \delta(\varepsilon_\delta - \varepsilon_\eta + \hbar\omega_q)(f_\delta - f_\eta), \quad (23)$$

and $\Omega_{\mu\mu_1}$ determines the renormalization of the boson modes due to interaction with electrons (we remind that $\mathcal{P}$ denotes principal value of the integral):

$$\Omega_{\mu\mu_1}(\mathbf{q}) = \frac{\mathcal{P}}{\hbar} \sum_{\delta\eta} \frac{\langle\eta|\hat{\chi}_{\mathbf{q}\mu}|\delta\rangle^* \langle\eta|\hat{\chi}_{\mathbf{q}\mu_1}|\delta\rangle}{\varepsilon_\delta - \varepsilon_\eta + \hbar\omega_q}(f_\delta - f_\eta). \quad (24)$$

To derive Eq. (22), we have also assumed slow variation of the boson density matrix with time so that the non-Markovian contributions have been neglected. Representing the boson distribution function $N_{\mathbf{q}t}^{\mu\mu_1}$ in the matrix form $\widehat{N}_{\mathbf{q}t}$ with respect to the indices μ, we find that it satisfies the matrix kinetic equation

$$\frac{\partial \widehat{N}_{\mathbf{q}t}}{\partial t} + i[\widehat{\Omega}_{\mathbf{q}}, \widehat{N}_{\mathbf{q}t}] = \hat{I}_{\mathbf{q}} - \frac{1}{2}[\hat{\nu}_{\mathbf{q}}, \widehat{N}_{\mathbf{q}t}]_+ \, . \tag{25}$$

The matrix elements of $\hat{I}_{\mathbf{q}}$, $\hat{\nu}_{\mathbf{q}}$, and $\widehat{\Omega}_{\mathbf{q}}$ are given by Eqs. (21), (23), and (24), respectively. If the electrons are in equilibrium, so that f_δ in Eqs. (21), (23), and (24) is the Fermi distribution (4.36), a stationary solution of Eq. (25) can be represented in the form $\delta_{\mu\mu_1} N_\omega$, where $N_\omega = [e^{\hbar\omega/T} - 1]^{-1}$ is the equilibrium Planck distribution function; see Eq. (3.28). Indeed, the time derivative, as well as the commutator on the left-hand side of Eq. (25) become zero, while in the right-hand side we can use the energy conservation law to obtain the factor $f(\varepsilon_\delta + \hbar\omega)[1 - f(\varepsilon_\delta)] - N_\omega[f(\varepsilon_\delta) - f(\varepsilon_\delta + \hbar\omega)]$ under the sum over δ and η. It is easy to see that this factor goes to zero if we use the Planck distribution N_ω as a solution.

The case of smooth (on the scale of boson wavelength) spatial inhomogeneity is described by the Wigner distribution function $N_{\mathbf{r}\mathbf{q}t}^{\mu\mu_1}$ defined according to the transformation

$$N_{\mathbf{r}\mathbf{q}t}^{\mu\mu_1} = \sum_{\mathbf{g}} e^{i\mathbf{g}\cdot\mathbf{r}} N_t \left(\mu\, \mathbf{q} + \frac{\mathbf{g}}{2},\ \mu_1\, \mathbf{q} - \frac{\mathbf{g}}{2} \right). \tag{26}$$

This transformation is similar to the one given by Eq. (9.6) for electrons, though Eq. (26) expresses the Wigner distribution function through the density matrix written in the momentum representation. The inverse transformation is done in a similar way as in Eq. (9.7):

$$N_t \left(\mu\, \mathbf{q} + \frac{\mathbf{g}}{2},\ \mu_1\, \mathbf{q} - \frac{\mathbf{g}}{2} \right) = \frac{1}{V} \int d\mathbf{r}\, e^{-i\mathbf{g}\cdot\mathbf{r}} N_{\mathbf{r}\mathbf{q}t}^{\mu\mu_1}. \tag{27}$$

Applying the transformation (26), we find that the term proportional to $\omega_q - \omega_{q_1}$ in the kinetic equation (14) is written as

$$i \sum_{\mathbf{g}} e^{i\mathbf{g}\cdot\mathbf{r}} \left(\omega_{|\mathbf{q}+\mathbf{g}/2|} - \omega_{|\mathbf{q}-\mathbf{g}/2|} \right) N_t \left(\mu\, \mathbf{q} + \frac{\mathbf{g}}{2},\ \mu_1\, \mathbf{q} - \frac{\mathbf{g}}{2} \right)$$

$$\simeq \frac{\partial \omega_q}{\partial \mathbf{q}} \cdot \frac{\partial N_{\mathbf{r}\mathbf{q}t}^{\mu\mu_1}}{\partial \mathbf{r}}. \tag{28}$$

To transform the right-hand side of Eq. (14), we make use of Eq. (9.24) for the operator products (see Appendix C, where Eq. (9.24) is derived in the coordinate representation). As a result, we obtain a quasi-classical kinetic equation for bosons in the matrix form:

$$\frac{\partial \widehat{N}_{\mathbf{r}\mathbf{q}t}}{\partial t} + \frac{\partial \omega_q}{\partial \mathbf{q}} \cdot \frac{\partial \widehat{N}_{\mathbf{r}\mathbf{q}t}}{\partial \mathbf{r}} + i[\widehat{\Omega}_{\mathbf{r}\mathbf{q}}, \widehat{N}_{\mathbf{r}\mathbf{q}t}] = \hat{I}_{\mathbf{r}\mathbf{q}} - \frac{1}{2}[\hat{\nu}_{\mathbf{r}\mathbf{q}}, \widehat{N}_{\mathbf{r}\mathbf{q}t}]_+. \tag{29}$$

The parametric coordinate dependence of the boson generation rate $\hat{I}_{\mathbf{rq}}$, as well as of the matrices $\hat{\nu}_{\mathbf{rq}}$ and $\widehat{\Omega}_{\mathbf{rq}}$, exists due to inhomogeneity of the electron distribution.

The local energy density of bosons is defined as

$$\mathcal{E}_{\mathbf{rt}} = \sum_{\mu} \int \frac{d\mathbf{q}}{(2\pi)^3} \hbar\omega_{\mathbf{q}\mu} N^{\mu}_{\mathbf{rqt}}. \tag{30}$$

This density is expressed through the diagonal elements $N^{\mu\mu}_{\mathbf{rqt}} \equiv N^{\mu}_{\mathbf{rqt}}$. Multiplying the diagonal part of Eq. (29) by $\hbar\omega_{\mathbf{q}\mu}$, integrating it over $\mathbf{q}$, and summing over μ, one can write the energy balance equation in the following form:

$$\frac{\partial \mathcal{E}_{\mathbf{rt}}}{\partial t} + \frac{\partial \mathbf{G}_{\mathbf{rt}}}{\partial \mathbf{r}} = P_{\mathbf{rt}} + \left(\frac{\partial \mathcal{E}_{\mathbf{rt}}}{\partial t} \right)_{sc}, \tag{31}$$

where the energy flow density is defined as

$$\mathbf{G}_{\mathbf{rt}} = \sum_{\mu} \int \frac{d\mathbf{q}}{(2\pi)^3} \frac{\partial \omega_{\mathbf{q}\mu}}{\partial \mathbf{q}} \hbar\omega_{\mathbf{q}\mu} N^{\mu}_{\mathbf{rqt}}, \tag{32}$$

and the power density due to boson generation,

$$P_{\mathbf{rt}} = \sum_{\mu} \int \frac{d\mathbf{q}}{(2\pi)^3} \hbar\omega_{\mathbf{q}\mu} I_{\mu\mu}(\mathbf{r}, \mathbf{q}), \tag{33}$$

is given through the boson generation rate (21). The term $(\partial \mathcal{E}_{\mathbf{rt}}/\partial t)_{sc}$ describes the dissipation of the energy density in unit time (power density loss) due to absorption of bosons by electrons. In the local equilibrium, the generation and absorption compensate each other, and the right-hand side of Eq. (31) vanishes. After multiplying Eq. (29) by the group velocity $\partial \omega_{\mathbf{q}\mu}/\partial \mathbf{q}$ and energy $\hbar\omega_{\mathbf{q}\mu}$, one can take the sums over μ and $\mathbf{q}$ to obtain a balance equation for $\mathbf{G}_{\mathbf{rt}}$, and so on. One may also define the momentum density of bosons and write for it similar equations. In this way one obtains a chain of equations similar to the balance equations for electrons discussed in Sec. 11.

20. Spontaneous and Stimulated Radiation

Below we apply Eq. (19.25) to describe the photons interacting with non-equilibrium electrons in a medium. In contrast to Chapters 2 and 3, where we considered the interaction of electrons with non-quantized external fields (it is a good approximation when photon occupation numbers are large), the kinetic equation used here also takes into account the spontaneous processes described by the generation rate (19.21). In the calculation of the generation rate (19.21) and other kinetic coefficients entering Eq. (19.25), we apply the dipole approximation. The operator $\hat{\chi}_{\mathbf{q}\mu}$ given in problem 4.2 is written as

$$\hat{\chi}_{\mathbf{q}\mu} = \frac{1}{2}\sqrt{\frac{2\pi\hbar e^2}{\omega_q\epsilon V}}\left[(\hat{\mathbf{v}}\cdot\mathbf{e}_{\mathbf{q}\mu})e^{i\mathbf{q}\cdot\mathbf{x}} + e^{i\mathbf{q}\cdot\mathbf{x}}(\hat{\mathbf{v}}\cdot\mathbf{e}_{\mathbf{q}\mu})\right]$$

$$\simeq \sqrt{\frac{2\pi\hbar e^2}{\omega_q\epsilon V}}(\hat{\mathbf{v}}\cdot\mathbf{e}_{\mathbf{q}\mu}), \tag{1}$$

where the photon polarization vector $\mathbf{e}_{\mathbf{q}\mu}$ is determined by the conditions $\mathbf{e}_{\mathbf{q}\mu}^*\cdot\mathbf{e}_{\mathbf{q}\mu_1} = \delta_{\mu\mu_1}$ and $\mathbf{q}\cdot\mathbf{e}_{\mathbf{q}\mu} = 0$; see Eq. (3.23). In the case of linear polarization of the electromagnetic waves, one may chose $\mathbf{e}_{\mathbf{q}\mu}$ real, and this is assumed below. We consider an isotropic medium with the dielectric permittivity ϵ. Therefore, the group velocity of photons in Eq. (19.29) is determined as $\partial\omega_q/\partial\mathbf{q} = \tilde{c}\mathbf{q}/q$, where $\tilde{c} = c/\sqrt{\epsilon}$ is the velocity of light in the medium under consideration.

The generation rate of photons is obtained from Eqs. (1) and (19.21) in the form

$$I_{\mu\mu_1}(\mathbf{q}) = \frac{(2\pi e)^2}{\omega_q\epsilon V}\sum_{\delta\eta}M_{\mu\mu_1}(\delta\eta|\mathbf{q})\delta(\varepsilon_\delta - \varepsilon_\eta + \hbar\omega_q)f_\eta(1 - f_\delta), \tag{2}$$

and the coefficients (19.23) and (19.24) become

$$\nu_{\mu\mu_1}(\mathbf{q}) = \frac{(2\pi e)^2}{\omega_q\epsilon V}\sum_{\delta\eta}M_{\mu\mu_1}(\delta\eta|\mathbf{q})\delta(\varepsilon_\delta - \varepsilon_\eta + \hbar\omega_q)(f_\delta - f_\eta) \tag{3}$$

and

$$\Omega_{\mu\mu_1}(\mathbf{q}) = \frac{2\pi e^2}{\omega_q\epsilon V}\sum_{\delta\eta}M_{\mu\mu_1}(\delta\eta|\mathbf{q})\frac{\mathcal{P}}{\varepsilon_\delta - \varepsilon_\eta + \hbar\omega_q}(f_\delta - f_\eta). \tag{4}$$

The matrix

$$M_{\mu\mu_1}(\delta\eta|\mathbf{q}) = \langle\eta|(\hat{\mathbf{v}}\cdot\mathbf{e}_{\mathbf{q}\mu})|\delta\rangle^*\langle\eta|(\hat{\mathbf{v}}\cdot\mathbf{e}_{\mathbf{q}\mu_1})|\delta\rangle \tag{5}$$

introduced in Eqs. (2)−(4) is Hermitian, $M_{\mu\mu_1}(\delta\eta|\mathbf{q}) = M_{\mu_1\mu}^*(\delta\eta|\mathbf{q})$. As a result, the matrices (2)−(4) are Hermitian as well, and the Hermiticity of the photon density matrix directly follows from Eq. (19.25) or Eq. (19.29).

In this section we restrict ourselves by the interband transitions of electrons and calculate the matrix elements in Eq. (5) by using the states of c- and v- bands. We denote these states as $|n\sigma\mathbf{p}\rangle$, $n = c, v$. Their energy spectra ε_{np} are assumed to be spin-degenerate and the corresponding distribution functions are $f_{n\mathbf{p}}$. Using the two-band model described in Appendix B, we substitute the interband velocity operators of Eq. (B.23) into Eq. (5). Calculating the sum over spin variables, we obtain $\sum_{\sigma\sigma'} M_{\mu\mu'}(v\sigma'\mathbf{p}', c\sigma\mathbf{p}|\mathbf{q}) = \delta_{\mathbf{pp'}} M_{\mu\mu'}(vc|\mathbf{p}, \mathbf{q})$, where

$$M_{\mu\mu'}(vc|\mathbf{p}, \mathbf{q}) = 2s^2 \left[\delta_{\mu\mu'} - \frac{(\mathbf{e}_{\mathbf{q}\mu} \cdot \mathbf{p})(\mathbf{e}_{\mathbf{q}\mu'} \cdot \mathbf{p})}{p^2} \frac{\eta_p^2 - 1}{\eta_p^2} \right]. \qquad (6)$$

If the electron distribution is isotropic in the $\mathbf{p}$-space, the expression (6) is reduced to $\delta_{\mu\mu'} 2s^2(2\eta_p^2 + 1)/3\eta_p^2$ after averaging over the angle of $\mathbf{p}$. For the transitions near the interband absorption edge, when $\eta_p \simeq 1$, this expression becomes $\delta_{\mu\mu'} 2s^2$. The commutator $[\widehat{\Omega}_{\mathbf{rq}}, \widehat{N}_{\mathbf{rqt}}]$ in Eq. (19.29) is equal to zero in this case, and the polarization of a photon mode is independent of time and coordinate. The generation rate (2) drops out of the kinetic equation if the electrons occupy the valence band only. Omitting the polarization indices, we rewrite Eq. (19.29) as

$$\mathbf{n_q} \cdot \nabla_{\mathbf{r}} N_{\mathbf{rq}} = -\alpha_\omega N_{\mathbf{rq}}, \quad \alpha_\omega = \frac{\nu_{\mathbf{q}}}{\tilde{c}}, \quad \mathbf{n_q} = \frac{\mathbf{q}}{q}, \qquad (7)$$

where $N_{\mathbf{rq}}$ is the photon distribution function. We also assume that the distribution of electrons is homogeneous so that the absorption coefficient α_ω is coordinate-independent and described by Eq. (17.6) near the edge of interband transitions. Equation (7) describes an exponential decrease of the intensity of radiation transmitted along OZ, according to $\exp(-\alpha_\omega z)$, provided the number of photons at $z = 0$ is fixed. This dependence is known as Buger's law (problem 4.4).

In the presence of free carriers, the interband transitions are modified. In doped materials of n- or p-type, the edge of interband transitions is shifted according to Eq. (17.7), while injection of additional electrons and holes leads to a spontaneous emission of photons (photoluminescence), since the generation rate (2) is no longer zero. The increase in the density of electron-hole pairs eventually leads to a stimulated radiation (laser effect), because the absorption coefficient α_ω, expressed according to Eqs. (7) and (3), becomes negative. A schematic representation of interband transitions for different occupations of the bands is given in Fig. 4.1.

Consider the photoluminescence in the p-type material with a small fraction of non-equilibrium conduction-band electrons described by the

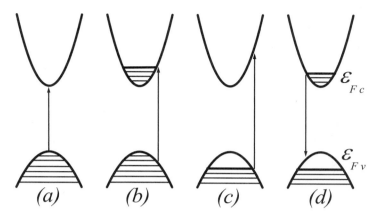

Figure 4.1. Different cases of conduction- and valence-band occupation: (*a*) intrinsic material (no doping), (*b*) *n*-doped material, (*c*) *p*-doped material, and (*d*) material with non-equilibrium electrons and holes. The arrows show interband optical transitions.

Maxwell distribution function $f_{cp} = (n_c/\mathcal{N}_c)\exp(-\varepsilon_{cp}/T_e)$ with effective temperature T_e. Here n_c is the electron density and $\mathcal{N}_c$ is a normalization coefficient. The holes are described by the Fermi distribution. Since this implies $1 - f_{vp} = \theta(\varepsilon_{vp} - \varepsilon_F)$, only the states with $p < p_F$ participate in the interband transitions. The electron density and Fermi energy are related to the density of acceptors (problem 4.5). Assuming that $\varepsilon_g \gg \varepsilon_F$, where $\varepsilon_g = (\varepsilon_{cp} - \varepsilon_{vp})_{p=0}$ is the gap energy, we transform the generation rate (2) to $\delta_{\mu\mu_1} I_\omega$ with

$$I_\omega = \frac{2(2\pi es)^2}{\omega \epsilon V} \sum_{\mathbf{p}} \delta(\varepsilon_g + p^2/2\mu^* - \hbar\omega) f_{cp}\theta(p_F - p), \qquad (8)$$

where μ^* is the reduced mass introduced in Secs. 17 and 18. Below we assume that the density of holes is high enough to have $|\varepsilon_F| \gg T_e$ and put $\theta(p_F - p) = 1$. Calculating the sum over $\mathbf{p}$ in Eq. (8), we obtain

$$I_\omega \simeq \widetilde{I} F\left(\frac{\mu^*}{m_c} \frac{\hbar\omega - \varepsilon_g}{T_e}\right), \qquad \widetilde{I} = \frac{8\pi^{3/2}\hbar(es)^2 n_c}{\epsilon \varepsilon_g T_e} \frac{\mu^*}{m_c}, \qquad (9)$$

where the spectral dependence is given by the function $F(x) = \sqrt{x}e^{-x}$. The intensity of the photoluminescence at the red (low-frequency) edge is determined by the density of non-equilibrium electrons, while its decrease in the spectral region above the edge is determined by the electron temperature. Therefore, measurements of the photoluminescence spectra allow one to characterize the distribution of non-equilibrium electrons. The intensity of the photoluminescence from a slab of width d is considered below by using the kinetic equation for the occupation numbers of the photons propagating along OZ axis:

$$\frac{dN_\omega(z)}{dz} = \frac{I_\omega}{\tilde{c}} - \alpha_\omega N_\omega(z), \quad 0 < z < d. \tag{10}$$

For the sake of convenience, we characterize the distribution function of photons by the frequency $\omega = \tilde{c}q$ rather than by the wave number q and put $N_{\mathbf{rq}} = N_\omega(z)$. Equation (10) takes into account both the spontaneous generation introduced by Eq. (8) and the absorption. The latter is described by the coefficient α_ω including interband transitions (this particular contribution is small due to the Pauli blocking effect because the energy of the photoluminescence peak is below the absorption edge) as well as other absorption mechanisms. If the width of the slab is small in comparison to the characteristic absorption length estimated as α_ω^{-1}, Eq. (10) gives us the number of the photons incident on the sample:

$$\overline{N}_\omega(z) = \frac{I_\omega d}{\tilde{c}}, \quad \alpha_\omega d \ll 1. \tag{11}$$

If the interference of photons in the sample is not essential, the number of outgoing photons is determined as a product of $\overline{N}_\omega(z)$ by the reflection coefficient r_ω. In the opposite case, the consideration should include the waveguide modes due to quantization of photons (instead of free photon modes in the medium).

As the density of electron-hole pairs increases, the equations for the generation rate and absorption coefficient are modified. Using the Fermi distributions with effective Fermi energies ε_{Fc} and ε_{Fv} at low temperatures, we obtain the absorption coefficient in the following form:

$$\tilde{\alpha}_\omega = \frac{2(2\pi es)^2}{\tilde{c}\omega\epsilon V} \sum_{\mathbf{p}} \delta(\varepsilon_g + p^2/2\mu^* - \hbar\omega)[\theta(p - p_{Fv}) - \theta(p_{Fc} - p)], \tag{12}$$

where p_{Fc} and p_{Fv} are the Fermi momenta in the bands; see Fig. 4.1 (d). Taking into account the redistribution of non-equilibrium carriers between c- and v-bands, we rewrite Eq. (12) as

$$\tilde{\alpha}_\omega = \alpha_\omega \begin{cases} -1 & 0 < \hbar\omega - \varepsilon_g < p_m^2/2\mu^* \\ 0 & p_m^2/2\mu^* < \hbar\omega - \varepsilon_g < p_M^2/2\mu^* \\ 1 & \hbar\omega - \varepsilon_g > p_M^2/2\mu^* \end{cases}, \tag{13}$$

where α_ω is given by Eq. (17.6). The momenta p_m and p_M are defined as minimal and maximal values of the Fermi momenta of electrons and holes, $p_m = \min\{p_{Fc}, p_{Fv}\}$ and $p_M = \max\{p_{Fc}, p_{Fv}\}$. If $p_{Fv} > p_{Fc}$, as in Fig. 4.1 (d), one has $p_m = p_{Fc}$ and $p_M = p_{Fv}$. In a similar way, the generation rate is given by

$$\widetilde{I}_\omega = \frac{(2\pi es)^2}{\omega\epsilon} \frac{2}{V} \sum_{\mathbf{p}} \delta(\varepsilon_g + p^2/2\mu^* - \hbar\omega)\theta(p_{Fc} - p)\theta(p_{Fv} - p)$$

$$= \frac{(2\pi es)^2}{\omega\epsilon}\rho_{3D}(\hbar\omega - \varepsilon_g), \quad 0 < \hbar\omega - \varepsilon_g < p_m^2/2\mu^*, \tag{14}$$

so that the spontaneous radiation is emitted in the spectral region where the absorption coefficient (13) is negative, and the generation rate is related to this negative absorption as $\widetilde{I}_\omega = -\tilde{c}\widetilde{\alpha}_\omega$. We note that ρ_{3D} in Eq. (14) is expressed through the reduced mass μ^*.

The occupation numbers of the photons propagating forward and backward along the OZ axis are denoted below as $N_\omega^{(+)}(z)$ and $N_\omega^{(-)}(z)$, respectively. They satisfy the following equation:

$$\frac{1}{\tilde{c}}\frac{\partial N_\omega^{(\pm)}(z,t)}{\partial t} \pm \frac{\partial N_\omega^{(\pm)}(z,t)}{\partial z} = \tilde{c}^{-1}\widetilde{I}_\omega - \widetilde{\alpha}_\omega N_\omega^{(\pm)}(z,t). \tag{15}$$

If the absorption coefficient $\widetilde{\alpha}_\omega$ is positive, this equation has a stationary and homogeneous solution $N_\omega = \widetilde{I}_\omega/\widetilde{\alpha}_\omega\tilde{c}$. In this case, the initial excitation decays on a characteristic time scale $(\widetilde{\alpha}_\omega\tilde{c})^{-1}$, and the contributions of the boundary conditions are not essential at the distances of the order of $\widetilde{\alpha}_\omega^{-1}$. If the absorption coefficient is negative, one has to consider time-dependent or spatially inhomogeneous solutions of Eq. (15). The homogeneous solution of this equation with the initial condition $N_\omega^{(\pm)}(t)|_{t=0} = 0$ describes an exponential growth of the number of photons:

$$N_\omega^{(\pm)}(t) = \frac{\widetilde{I}_\omega}{\nu_\omega}\left[e^{\nu_\omega t} - 1\right], \tag{16}$$

where $\nu_\omega = \tilde{c}|\widetilde{\alpha}_\omega|$ is the increment of amplification for the photons of frequency ω. We note that, under the conditions described by Eqs. (13) and (14), the factor $\widetilde{I}_\omega/\nu_\omega$ is equal to unity. The stationary solution is obtained with the aid of the boundary conditions

$$N_\omega^{(+)}(z)|_{z=0} = N_\omega^{(-)}(z)|_{z=0}, \quad r_\omega N_\omega^{(+)}(z)|_{z=d} = N_\omega^{(-)}(z)|_{z=d}, \tag{17}$$

which correspond to ideal reflection at $z = 0$ and partial reflection at $z = d$, where the reflection coefficient r_ω is smaller than unity. Solving Eq. (15) with the boundary conditions (17), we obtain

$$N_\omega^{(\pm)}(z) = \frac{\widetilde{I}_\omega}{\nu_\omega}\frac{r_\omega(e^{|\widetilde{\alpha}_\omega|d} - e^{\pm|\widetilde{\alpha}_\omega|z}) + e^{\pm|\widetilde{\alpha}_\omega|z} - e^{-|\widetilde{\alpha}_\omega|d}}{e^{-|\widetilde{\alpha}_\omega|d} - r_\omega e^{|\widetilde{\alpha}_\omega|d}}. \tag{18}$$

The nominator of this expression is always positive, while the denominator can be either negative or positive, depending on the values of r_ω and $|\widetilde{\alpha}_\omega|d$. If $r_\omega < e^{-2|\widetilde{\alpha}_\omega|d}$, the occupation numbers $N_\omega^{(\pm)}(z)$ of Eq. (18) are positive, which means that stationary solutions of Eq. (15) exist. In the opposite case, the occupation numbers continue to increase with time. To describe such an increase, one can write a more general solution of Eq. (15) by using both initial and boundary conditions (problem 4.6). A saturation of the amplification in lasers occurs due to nonlinear effects such as a decrease of the parameters r_ω and $|\widetilde{\alpha}_\omega|$ with increasing occupation numbers of photons, when the electromagnetic radiation modifies physical characteristics of the medium. As the stationary regime is reached, the parameters satisfy the following relation:

$$r_\omega \simeq e^{-2|\widetilde{\alpha}_\omega|d}, \tag{19}$$

and $N_\omega^{(\pm)}(z) \gg 1$ so that the spontaneous emission can be neglected. The number of outgoing photons, $N_{out} = (1-r_\omega)N_\omega^{(+)}(d)$, determines the intensity of laser radiation. Equation (19) can be obtained by equating N_{out} to the number of photons generated in the region of amplification whose length is equal to $2d$. This number is given by $N_\omega^{(-)}(d)(e^{2|\widetilde{\alpha}_\omega|d}-1)$. To write Eq. (19) for the case of two partially transparent boundaries, one should replace r_ω by the product of their reflection coefficients. Of course, the model considered here is oversimplified. To describe realistic laser devices, one should, in particular, quantize the waveguide modes and use more sophisticated boundary conditions.

Let us study the influence of anisotropy of non-equilibrium electrons on the characteristics of radiation. Consider the photoluminescence in p-type materials with shifted Maxwell distribution of non-equilibrium c-band electrons. The anisotropic contribution to the generation rate is given by

$$\Delta I_{\mu\mu'}(\mathbf{q}) = \frac{(2\pi e)^2}{\omega_q \epsilon V} \sum_{\mathbf{p}} \Delta M_{\mu\mu'}(\mathbf{p})\delta(\varepsilon_g + p^2/2\mu^* - \hbar\omega)\Delta f_{c\mathbf{p}}, \tag{20}$$

where $\Delta f_{c\mathbf{p}} = (\mathbf{p} \cdot \mathbf{u}/T_e)^2 f_{cp}/2$ is the quadratic correction to the distribution function (see Sec. 17), $\mathbf{u}$ is the drift velocity of electrons, and

$$\Delta M_{\mu\mu'}(\mathbf{p}) = -\frac{8s^4}{\varepsilon_g^2}(\mathbf{e}_{\mathbf{q}\mu} \cdot \mathbf{p})(\mathbf{e}_{\mathbf{q}\mu'} \cdot \mathbf{p}) \tag{21}$$

is the anisotropic correction to the matrix element (6) written in the approximation $T_e \ll \varepsilon_g$, when non-parabolic corrections are not essential. Averaging over the angle in Eq. (20) with the use of Eq. (17.14), we obtain

$$\Delta I_{\mu\mu'}(\mathbf{q}) = -(\mathbf{e_{q\mu}} \cdot \mathbf{u})(\mathbf{e_{q\mu'}} \cdot \mathbf{u}) \frac{m_c^2 s^2 \widetilde{I}}{\varepsilon_g^2} \mathcal{F}\left(\frac{\mu^*}{m_c} \frac{\hbar\omega - \varepsilon_g}{T_e}\right), \qquad (22)$$

where the spectral dependence is given by the function $\mathcal{F}(x) = (16/15)$ $\times x^{5/2} e^{-x}$. If the sample width d is small in comparison to the absorption length, the anisotropy of the photon distribution on the surface of the sample is determined similar to the case described by Eq. (11):

$$\overline{\Delta N} = \overline{N}_\perp - \overline{N}_\parallel = \frac{\widetilde{I}d}{\widetilde{c}} \frac{m_c^2 s^2 u^2}{\varepsilon_g^2} \mathcal{F}\left(\frac{\mu^*}{m_c} \frac{\hbar\omega - \varepsilon_g}{T_e}\right), \qquad (23)$$

where $\overline{N}_\parallel$ and $\overline{N}_\perp$ are the occupation numbers of the photons polarized parallel and perpendicular to the drift velocity $\mathbf{u}$. By measuring the polarization of photoluminescence, one can study the anisotropy of non-equilibrium electron distribution. In the case considered above, the polarization is weak due to a small parameter $m_c u^2/\varepsilon_g$.

21. Phonon Instabilities

Apart from the stimulated photon emission considered above, there exist several mechanisms of phonon-mode instabilities due to interaction of phonons with non-equilibrium electrons. In this section we study both the instabilities of the acoustic phonons interacting with the electrons drifting with a supersonic velocity and those of the optical phonons interacting with the electrons excited by the laser radiation (this mechanism of excitation is discussed in Sec. 10). Considering the deformation-potential interaction with long-wavelength longitudinal acoustic phonons (LA) and Froelich interaction with long-wavelength longitudinal optical phonons (LO), we use the factors

$$\hat{\chi}_\mathbf{q} = C_q e^{i\mathbf{q}\cdot\mathbf{x}}, \qquad C_q = \begin{cases} \mathcal{D}\sqrt{\hbar\omega_q/2s_l^2 \rho V}, & \omega_q = s_l q, \qquad LA \\ \sqrt{2\pi e^2 \hbar\omega_q/\epsilon^* q^2 V}, & \omega_q = \omega_{LO}, \quad LO \end{cases} \qquad (1)$$

in Eqs. (19.19) and (19.20); see Eqs. (6.22), (6.30), and problem 4.1. Since the electrons interact with the longitudinal modes only, the phonon density matrix is diagonal with respect to phonon branch indices, and Eq. (19.18) is rewritten as an equation for the phonon distribution function $N_{\mathbf{q}t}$:

$$\frac{\partial N_{\mathbf{q}t}}{\partial t} = I_{\mathbf{q}t} + \int_{-\infty}^t dt' e^{\lambda t'} R_\mathbf{q}(t, t') N_{\mathbf{q}t'}. \qquad (2)$$

The generation rate in Eq. (2) is obtained from Eq. (19.19) as

$$I_{\mathbf{q}t} = \frac{|C_q|^2}{\hbar^2} \int_{-\infty}^{t} dt' e^{\lambda t' - i\omega_q(t-t')} \tag{3}$$

$$\times \mathrm{sp}\{\widehat{S}^+(t,t')e^{i\mathbf{q}\cdot\mathbf{x}}\widehat{S}(t,t')(1-\hat{\rho}_{t'})e^{-i\mathbf{q}\cdot\mathbf{x}}\hat{\rho}_{t'}\} + c.c.$$

and the kernel in the non-Markovian relaxation contribution is obtained from Eq. (19.20) in the following form:

$$R_{\mathbf{q}}(t,t') = e^{-i\omega_q(t-t')}\frac{|C_q|^2}{\hbar^2}\mathrm{sp}[\widehat{S}^+(t,t')e^{i\mathbf{q}\cdot\mathbf{x}}\widehat{S}(t,t'), e^{-i\mathbf{q}\cdot\mathbf{x}}]\hat{\rho}_{t'} + c.c. \tag{4}$$

The non-equilibrium density matrix of electrons standing in Eqs. (3) and (4) is to be found from a quantum kinetic equation of the kind of Eq. (7.13).

Since we study a spatially homogeneous case, the traces in Eqs. (3) and (4) can be calculated by using the plane-wave eigenstates. This gives us

$$I_{\mathbf{q}t} = 2\frac{|C_q|^2}{\hbar^2} \int_{-\infty}^{t} dt' e^{\lambda t' - i\omega_q(t-t')} \sum_{\mathbf{p}} F_{\mathbf{p}}(\mathbf{q}|tt')f_{\mathbf{p}t'}(1-f_{\mathbf{p}-\hbar\mathbf{q}t'}) + c.c.,$$

$$R_{\mathbf{q}}(t,t') = 2e^{-i\omega_q(t-t')}\frac{|C_q|^2}{\hbar^2}\sum_{\mathbf{p}} F_{\mathbf{p}}(\mathbf{q}|tt')(f_{\mathbf{p}t'} - f_{\mathbf{p}-\hbar\mathbf{q}t'}) + c.c., \tag{5}$$

where $f_{\mathbf{p}t} = \langle \mathbf{p}|\hat{\rho}_t|\mathbf{p}\rangle$ is the distribution function in the momentum representation, the factor of 2 comes from the sum over electron spin, and

$$F_{\mathbf{p}}(\mathbf{q}|tt') = \langle \mathbf{p}|\widehat{S}^+(t,t')e^{i\mathbf{q}\cdot\mathbf{x}}\widehat{S}(t,t')e^{-i\mathbf{q}\cdot\mathbf{x}}|\mathbf{p}\rangle. \tag{6}$$

Employing Eq. (2.5) for the S-operator, one can write the following equation for F (problem 4.7):

$$i\hbar\frac{\partial F_{\mathbf{p}}(\mathbf{q}|tt')}{\partial t} = \langle \mathbf{p}|[e^{i\mathbf{q}\cdot\mathbf{x}}, \hat{h}_t]e^{-i\mathbf{q}\cdot\mathbf{x}}|\mathbf{p}\rangle F_{\mathbf{p}}(\mathbf{q}|tt'), \tag{7}$$

with the initial condition $F_{\mathbf{p}}(\mathbf{q}|tt) = 1$. To describe an electron in a homogeneous harmonic electric field, we employ the one-particle Hamiltonian $\hat{h}_t = \hat{\pi}_t^2/2m \equiv \varepsilon(\hat{\boldsymbol{\pi}}_t)$, where $\hat{\boldsymbol{\pi}}_t = \hat{\mathbf{p}} + (e/\omega)\mathbf{E}\sin\omega t$ is the kinematic momentum; see Eqs. (5.27) and (10.1). Then, on the right-hand side of Eq. (7) we have $\langle \mathbf{p}|[e^{i\mathbf{q}\cdot\mathbf{x}}, \hat{h}_t]e^{-i\mathbf{q}\cdot\mathbf{x}}|\mathbf{p}\rangle = \varepsilon(\boldsymbol{\pi}_t - \hbar\mathbf{q}) - \varepsilon(\boldsymbol{\pi}_t)$. The solution of Eq. (7) in this case is

$$F_{\mathbf{p}}(\mathbf{q}|tt') = \exp\left\{-\frac{i}{\hbar}\int_{t'}^{t} d\tau\,[\varepsilon(\boldsymbol{\pi}_\tau - \hbar\mathbf{q}) - \varepsilon(\boldsymbol{\pi}_\tau)]\right\}. \tag{8}$$

If the field is stationary, $\omega \to 0$, the kinematic momentum is given by $\boldsymbol{\pi}_\tau = \mathbf{p} + e\mathbf{E}\tau$. The electrons have a stationary distribution over kinematic momenta, $f_{\mathbf{p}t'} = f_{\boldsymbol{\pi}_{t'}}$. Substituting Eq. (8) into Eq. (5) and replacing $\boldsymbol{\pi}_{t'}$ by $\mathbf{p}$ to have $f_{\boldsymbol{\pi}_{t'}} \to f_{\mathbf{p}}$, we should simultaneously replace $\varepsilon(\boldsymbol{\pi}_\tau - \hbar\mathbf{q}) - \varepsilon(\boldsymbol{\pi}_\tau)$ by $\varepsilon(\boldsymbol{\pi}_{\tau-t'} - \hbar\mathbf{q}) - \varepsilon(\boldsymbol{\pi}_{\tau-t'})$ in the exponent. Next, we change the variable of integration in the exponent as $\tau \to \tau + t'$ so that this integral becomes $\int_0^{t-t'} d\tau [\varepsilon(\boldsymbol{\pi}_\tau - \hbar\mathbf{q}) - \varepsilon(\boldsymbol{\pi}_\tau)]$. We find the time-independent generation rate

$$I_{\mathbf{q}} = 2\frac{|C_q|^2}{\hbar^2} \sum_{\mathbf{p}} f_{\mathbf{p}}(1 - f_{\mathbf{p}-\hbar\mathbf{q}}) \int_{-\infty}^0 dt' e^{\lambda t' + i\omega_q t'}$$

$$\times \exp\left\{-\frac{i}{\hbar}\int_0^{-t'} d\tau \left[\varepsilon(\boldsymbol{\pi}_\tau - \hbar\mathbf{q}) - \varepsilon(\boldsymbol{\pi}_\tau)\right]\right\} + c.c. , \qquad (9)$$

while the function $R_{\mathbf{q}}$ depends only on $t - t'$:

$$R_{\mathbf{q}}(t - t') = 2e^{-i\omega_q(t-t')}\frac{|C_q|^2}{\hbar^2} \sum_{\mathbf{p}}(f_{\mathbf{p}} - f_{\mathbf{p}-\hbar\mathbf{q}})$$

$$\times \exp\left\{-\frac{i}{\hbar}\int_0^{t-t'} d\tau \left[\varepsilon(\boldsymbol{\pi}_\tau - \hbar\mathbf{q}) - \varepsilon(\boldsymbol{\pi}_\tau)\right]\right\} + c.c. \qquad (10)$$

Using this expression in the kinetic equation (2), we substitute $t' \to t' + t$ and rewrite Eq. (2) in the following form:

$$\frac{\partial N_{\mathbf{q}t}}{\partial t} = I_{\mathbf{q}} + 2\frac{|C_q|^2}{\hbar^2} \sum_{\mathbf{p}}(f_{\mathbf{p}} - f_{\mathbf{p}-\hbar\mathbf{q}})\left\{\int_{-\infty}^0 dt' e^{\lambda t' + i\omega_q t'}\right.$$

$$\left.\times \exp\left[-\frac{i}{\hbar}\int_0^{-t'} d\tau \left[\varepsilon(\boldsymbol{\pi}_\tau - \hbar\mathbf{q}) - \varepsilon(\boldsymbol{\pi}_\tau)\right]\right] N_{\mathbf{q}t+t'} + c.c.\right\}. \qquad (11)$$

The phonon distribution function changes with time on the scale $1/\overline{\nu}$ determined by the phonon relaxation rate, while the expression in the exponent in Eq. (11) oscillates with a characteristic period $t_o = (q\overline{v})^{-1}$, where the electron velocity $\overline{v}$ is much greater than the velocity of sound. Since $\overline{\nu}t_o \ll 1$, one may substitute $N_{\mathbf{q}t+t'} \simeq N_{\mathbf{q}t}$ so that Eq. (11) becomes Markovian. We rewrite it as

$$\frac{\partial N_{\mathbf{q}t}}{\partial t} = I_{\mathbf{q}} - \nu_{\mathbf{q}} N_{\mathbf{q}t}. \qquad (12)$$

If the characteristic electron momentum $\overline{p}$ is much greater than eEt_o, one can neglect the influence of the field on the electron-phonon interaction.

The relaxation rate of the mode $\mathbf{q}$ in Eq. (12) becomes

$$\nu_{\mathbf{q}} = 2\frac{|C_q|^2}{\hbar^2} \sum_{\mathbf{p}} f_{\mathbf{p}} \left\{ \int_{-\infty}^{0} dt e^{\lambda t + i\omega_q t} \left[e^{-i(\varepsilon_{\mathbf{p}+\hbar\mathbf{q}} - \varepsilon_p)t/\hbar} \right. \right.$$

$$\left. \left. - e^{i(\varepsilon_{\mathbf{p}-\hbar\mathbf{q}} - \varepsilon_p)t/\hbar} \right] + c.c. \right\}. \tag{13}$$

Integrating over time in Eq. (13) according to Eq. (8.6), we find

$$\nu_{\mathbf{q}} = \frac{4\pi}{\hbar} |C_q|^2 \sum_{\mathbf{p}} \delta(\varepsilon_p - \varepsilon_{\mathbf{p}+\hbar\mathbf{q}} + \hbar\omega_q)(f_{\mathbf{p}} - f_{\mathbf{p}+\hbar\mathbf{q}}). \tag{14}$$

A similar equation can be written for the generation rate (problem 4.8). However, our primary goal is to study the influence of non-equilibrium electron distribution on $\nu_{\mathbf{q}}$.

If the elastic scattering dominates, the electron distribution function is determined by Eqs. (8.23) and (8.24) with $\omega = 0$. Below, instead of using these equations, we employ the shifted Fermi distribution (see Eq. (31.25) below and its discussion)

$$f_{\mathbf{p}} = \left[\exp\left(\frac{\varepsilon_{\mathbf{p}-m\mathbf{u}} - \mu}{T_e} \right) + 1 \right]^{-1} \equiv f(\varepsilon_{\mathbf{p}-m\mathbf{u}}), \tag{15}$$

where $\mathbf{u}$ is the drift velocity of electron gas, which is proportional to the driving electric field and depends on the relaxation mechanisms, and T_e is the effective temperature of electron gas, which may differ from the equilibrium temperature T. The reliability of this approximation will be discussed in Secs. 31 and 36. Substituting the distribution function (15) into Eq. (14), we consider the interaction of electrons with LA phonons and obtain

$$\nu_{\mathbf{q}} = \frac{4\pi}{\hbar^2} V |C_q|^2 \int \frac{d\mathbf{p}_{\perp}}{(2\pi\hbar)^3} \int dp_{\parallel} \delta\left(s_l q - \frac{\mathbf{p}_{\perp} \cdot \mathbf{q}_{\perp} + p_{\parallel}q_{\parallel}}{m} - \frac{\hbar q^2}{2m} \right) \tag{16}$$

$$\times \left[f\left(\frac{\mathbf{p}_{\perp}^2}{2m} + \frac{(p_{\parallel} - mu)^2}{2m} \right) - f\left(\frac{(\mathbf{p}_{\perp} + \hbar\mathbf{q}_{\perp})^2}{2m} + \frac{(p_{\parallel} - mu + \hbar q_{\parallel})^2}{2m} \right) \right],$$

where the components of the vectors $\mathbf{p}$ and $\mathbf{q}$ parallel and perpendicular to the drift direction are separated. The integral over $p_{\parallel}$ is calculated with the use of the δ-function and gives us

$$\nu_{\mathbf{q}} = \frac{4\pi m \mathcal{D}^2 q}{\hbar s_l \rho q_{\parallel}} \int \frac{d\mathbf{p}_{\perp}}{(2\pi\hbar)^3} \left\{ f\left[(g_{\mathbf{q}} - \mathbf{p}_{\perp} \cdot \mathbf{q}_{\perp}/q_{\parallel} - \hbar q_{\parallel}/2)^2/2m + \mathbf{p}_{\perp}^2/2m \right] \right.$$

$$-f\left[(g_{\mathbf{q}} - \mathbf{p}_\perp \cdot \mathbf{q}_\perp/q_\parallel + \hbar q_\parallel/2)^2/2m + (\mathbf{p}_\perp + \hbar\mathbf{q}_\perp)^2/2m\right]\right\}, \qquad (17)$$

where $g_{\mathbf{q}} = m s_l \sqrt{q_\perp^2 + q_\parallel^2}/q_\parallel - \hbar q_\perp^2/2q_\parallel - mu$. The integral over $\mathbf{p}_\perp$ can be calculated analytically in two limits: for degenerate electrons, when $f(\varepsilon) = \theta(\mu - \varepsilon)$, and for non-degenerate electrons, when $f(\varepsilon) = (n/\mathcal{N}_c)\exp(-\varepsilon/T_e)$. Here n is the electron density and the normalization coefficient is $\mathcal{N}_c = (mT_e)^{3/2}/\sqrt{2}\pi^{3/2}\hbar^3$. In the first case, we have the following relaxation rate of acoustic phonons:

$$\nu_{\mathbf{q}} = \frac{\mathcal{D}^2 m^2}{2\pi\hbar^3 s_l \rho}\Omega_{\mathbf{q}}, \qquad (18)$$

while in the second case we obtain

$$\nu_{\mathbf{q}} = \frac{n\mathcal{D}^2}{\hbar s_l \rho}\sqrt{\frac{\pi m}{2T_e}}\exp\left[-\frac{(m\Omega_{\mathbf{q}}/q - \hbar q/2)^2}{2mT_e}\right]\left(1 - e^{-\hbar\Omega_{\mathbf{q}}/T_e}\right), \qquad (19)$$

where

$$\Omega_{\mathbf{q}} = s_l q - \mathbf{u} \cdot \mathbf{q}. \qquad (20)$$

Expression (18) remains valid as long as $(\hbar q/2 \pm m\Omega_{\mathbf{q}}/q)^2 < 2m\mu$. Beyond this region, $\nu_{\mathbf{q}}$ depends also on the chemical potential μ. Since $\Omega_{\mathbf{q}}/q$ is much smaller than the Fermi velocity, this validity condition can be violated only for the phonons whose momenta are close to $2p_F$.

Equations (18) and (19) contain a characteristic frequency $\Omega_{\mathbf{q}}$ which can be negative when the drift velocity is greater than the sound velocity s_l. The relaxation rate $\nu_{\mathbf{q}}$ is also negative in this case, and the number of the phonons moving at the angles smaller than $\arccos(s_l/u)$ with respect to the direction of $\mathbf{u}$ increases with time. A similar mechanism of photon emission takes place for fast electrons in media (Cherenkov emission). The emission appears when the velocity of electrons is higher than the phase velocity of bosons (photons or phonons). Since sound velocities in solids are small enough, the regime of acoustic-phonon instability is easily achievable both in metals and in semiconductors. The drift-induced instability of optical phonons, in principle, is also possible. However, it is very difficult to achieve this instability, since the phase velocities of optical phonons are large, and the drift velocities of electrons are limited by various relaxation processes.

Consider now the case of harmonic excitation. Calculating the integrals over τ in the exponential factor (8), we obtain

$$\int_{t'}^{t} d\tau[\varepsilon(\mathbf{p}_\tau - \hbar\mathbf{q}) - \varepsilon(\mathbf{p}_\tau)] = [-\hbar\mathbf{p} \cdot \mathbf{q}/m + \hbar^2 q^2/2m](t - t')$$

$$+(\hbar\mathbf{q} \cdot \mathbf{v}_\omega/\omega)[\cos\omega t - \cos\omega t'], \qquad (21)$$

where $\mathbf{v}_\omega$ is defined in Eq. (10.5). Using Eq. (10.15) to expand the exponent in terms of the Bessel functions, we express $R_\mathbf{q}(t,t')$ from Eqs. (5) and (8) as

$$R_\mathbf{q}(t,t') = 2e^{-i\omega_q(t-t')}\frac{|C_q|^2}{\hbar^2}\sum_\mathbf{p}(f_{\mathbf{p}t'} - f_{\mathbf{p}-\hbar\mathbf{q}t'})e^{-i(\varepsilon_{\mathbf{p}-\hbar\mathbf{q}}-\varepsilon_\mathbf{p})(t-t')/\hbar}$$

$$\times\sum_{kk'}i^{k-k'}e^{-ik\omega(t-t')+i(k-k')\omega t}\mathcal{J}_k(\mathbf{q}\cdot\mathbf{v}_\omega/\omega)\mathcal{J}_{k'}(\mathbf{q}\cdot\mathbf{v}_\omega/\omega) + c.c. \quad (22)$$

The distribution function of the electrons excited by the harmonic field is given by Eq. (10.10). The oscillating part $\Delta f_{\mathbf{p}t}$ of the distribution function is small due to parameter $1/\overline{\tau}\omega$, where $\overline{\tau}$ is the averaged time of electron relaxation. Below we consider the response of optical phonons whose energy $\hbar\omega_{LO}$ is much larger than $\hbar/\overline{\tau}$. This allows us to neglect the oscillating part in comparison to the time-independent part $\overline{f}_\mathbf{p}$. Since we are interested in the response averaged over a short period, $2\pi/\omega$, we put $k = k'$ in Eq. (22). Neglecting the non-Markovian contribution in the collision integral for the same reasons as in the stationary case described above, we replace $N_{\mathbf{q}t'}$ by $N_{\mathbf{q}t}$ and again obtain Eq. (12) for the phonon distribution function, where the generation and relaxation rates are given by

$$\left|\begin{array}{c}I_\mathbf{q}\\\nu_\mathbf{q}\end{array}\right| = \frac{4\pi}{\hbar}|C_q|^2\sum_{k=-\infty}^\infty\left[\mathcal{J}_k\left(\frac{\mathbf{q}\cdot\mathbf{v}_\omega}{\omega}\right)\right]^2\sum_\mathbf{p}\left|\begin{array}{c}\overline{f}_{\mathbf{p}+\hbar\mathbf{q}}(1-\overline{f}_\mathbf{p})\\(\overline{f}_\mathbf{p} - \overline{f}_{\mathbf{p}+\hbar\mathbf{q}})\end{array}\right|$$

$$\times\delta(\varepsilon_p - \varepsilon_{\mathbf{p}+\hbar\mathbf{q}} + \hbar\omega_{LO} + k\hbar\omega). \quad (23)$$

To calculate the integral over $\mathbf{p}$, one should know the distribution function $\overline{f}_\mathbf{p}$ governed by the kinetic equation (10.19). If the electric field is weak enough to have $\mathbf{q}\cdot\mathbf{v}_\omega/\omega \ll 1$, the distribution function is isotropic, $\overline{f}_\mathbf{p} = f(\varepsilon_p)$. We approximate it by the Fermi distribution (see Sec. 37), assuming that the temperature is small in comparison to both Fermi energy $\varepsilon_F \simeq \mu$ and phonon energy $\hbar\omega_{LO}$. The condition of weak electric field allows us to retain only the terms with $k = 0$ and $k = \pm 1$ in Eq. (23). We expand the corresponding Bessel functions according to $\mathcal{J}_0(x) \simeq 1 - (x/2)^2$ and $\mathcal{J}_{\pm 1}(x) \simeq \pm x/2$. Integrating over the angle of $\mathbf{p}$ in Eq. (23) by using the δ-function, we transform the relaxation rate to $\nu_\mathbf{q} = \nu_\mathbf{q}^{(0)} + \nu_\mathbf{q}^{(+)} + \nu_\mathbf{q}^{(-)}$, where

$$\nu_\mathbf{q}^{(0)} = |C_q|^2\frac{Vm^2}{\pi q\hbar^5}\left[1 - \frac{(\mathbf{q}\cdot\mathbf{v}_\omega)^2}{2\omega^2}\right]\int_{\varepsilon_0}^\infty d\varepsilon[f(\varepsilon) - f(\varepsilon + \hbar\omega_{LO})] \quad (24)$$

and

$$\nu_{\mathbf{q}}^{(\pm)} = |C_q|^2 \frac{Vm^2}{\pi q \hbar^5} \left(\frac{\mathbf{q} \cdot \mathbf{v}_\omega}{2\omega}\right)^2 \int_{\varepsilon_\pm}^\infty d\varepsilon[f(\varepsilon) - f(\varepsilon + \hbar\omega_{LO} \pm \hbar\omega)]. \quad (25)$$

The cut-off energies ε_0 and $\varepsilon_\pm$ appear after the angular integration, because of the restrictions imposed by the energy conservation law. They are given by the following expressions:

$$\varepsilon_0 = \frac{m\omega_{LO}^2}{2q^2} + \frac{\hbar^2 q^2}{8m} - \frac{\hbar\omega_{LO}}{2},$$

$$\varepsilon_\pm = \frac{m(\omega \pm \omega_{LO})^2}{2q^2} + \frac{\hbar^2 q^2}{8m} - \frac{\hbar(\omega_{LO} \pm \omega)}{2}. \quad (26)$$

Equations similar to Eqs. (24) and (25) can be written for the spontaneous generation rate $I_{\mathbf{q}}$.

Below we consider the case $\omega \gg \varepsilon_F/\hbar, \omega_{LO}$, when we have $\varepsilon_F < \varepsilon_-$ and can put $f(\varepsilon + \hbar\omega_{LO} + \hbar\omega) = 0$. Therefore, from Eqs. (25) and (26) we obtain

$$\nu_{\mathbf{q}}^{(\pm)} = \pm|C_q|^2 \frac{Vm^2}{\pi q \hbar^5} \left(\frac{\mathbf{q} \cdot \mathbf{v}_\omega}{2\omega}\right)^2$$

$$\times \left[\varepsilon_F + \frac{\hbar\omega \pm \hbar\omega_{LO}}{2} - \frac{\hbar^2 q^2}{8m} - \frac{(\omega \pm \omega_{LO})^2 m}{2q^2}\right], \quad (27)$$

where the expression in the square brackets must be positive, otherwise $\nu_{\mathbf{q}}^{(+)}$ and $\nu_{\mathbf{q}}^{(-)}$ are equal to zero. This restriction originating from the conservation laws shows us that the relaxation rates $\nu_{\mathbf{q}}^{(+)}$ and $\nu_{\mathbf{q}}^{(-)}$ are non-zero only for the phonons whose wave numbers are in a narrow interval around $q_\omega = \sqrt{2\omega m/\hbar}$. On the other hand, the relaxation rate $\nu_{\mathbf{q}}^{(0)}$ given by Eq. (24) corresponds to the wave numbers $q \sim \sqrt{2\omega_{LO}m/\hbar}$ or $q \sim \sqrt{2\varepsilon_F m/\hbar}$, which are much smaller than q_ω. Therefore, one can consider the kinetic equation (12) separately for two groups of phonons. The first group is the phonons with small wave numbers. For them we have the relaxation rate $\nu_{\mathbf{q}} = \nu_{\mathbf{q}}^{(0)}$, which is positive (since we assume the equilibrium Fermi distribution of electrons, there is no inversion of electron population). The second group includes the phonons with large wave numbers, for whom $\nu_{\mathbf{q}} \simeq \nu_{\mathbf{q}}^{(+)} + \nu_{\mathbf{q}}^{(-)}$. Since $\nu_{\mathbf{q}}^{(-)}$ is negative, $\nu_{\mathbf{q}}$ can be negative as well. Substituting $|C_q|^2$ for optical phonons from Eq. (1), and introducing a dimensionless constant of electron-phonon coupling

$$\alpha = \frac{e^2}{\epsilon^* \hbar} \sqrt{\frac{m}{2\hbar\omega_{LO}}}, \quad (28)$$

we obtain

$$\nu_{\mathbf{q}} \simeq \alpha \omega_{LO} \left(\sqrt{\frac{\omega_{LO}}{\omega}} \frac{q_\omega}{q} \right)^3 \left(\frac{\mathbf{q} \cdot \mathbf{v}_\omega}{2\omega} \right)^2 \left[1 - \left(\frac{q_\omega}{q} \right)^2 \right] \tag{29}$$

in the region of the wave numbers determined according to

$$\frac{\omega_{LO}}{\omega} - 2\sqrt{\frac{\varepsilon_F}{\hbar\omega}} < \frac{q^2}{q_\omega^2} - 1 < -\frac{\omega_{LO}}{\omega} + 2\sqrt{\frac{\varepsilon_F}{\hbar\omega}} . \tag{30}$$

To define this region, we have used the requirement of positiveness of the expression in the square brackets in Eq. (27) for both $\nu_{\mathbf{q}}^{(+)}$ and $\nu_{\mathbf{q}}^{(-)}$ and neglected the terms of the order of $\varepsilon_F/\hbar\omega$, $(\omega_{LO}/\omega)\sqrt{\varepsilon_F/\hbar\omega}$, and of the higher orders of smallness. According to Eq. (29), the optical-phonon instability occurs at $q < q_\omega$. Next, in the region

$$-\frac{\omega_{LO}}{\omega} - 2\sqrt{\frac{\varepsilon_F}{\hbar\omega}} < \frac{q^2}{q_\omega^2} - 1 < \min \left\{ -\frac{\omega_{LO}}{\omega} + 2\sqrt{\frac{\varepsilon_F}{\hbar\omega}}, \frac{\omega_{LO}}{\omega} - 2\sqrt{\frac{\varepsilon_F}{\hbar\omega}} \right\}, \tag{31}$$

we have $\nu_{\mathbf{q}}^{(+)} = 0$, and the relaxation rate

$$\nu_{\mathbf{q}} = \nu_{\mathbf{q}}^{(-)} = -\alpha \left(\sqrt{\frac{\omega_{LO}}{\omega}} \frac{q_\omega}{q} \right)^3 \left(\frac{\mathbf{q} \cdot \mathbf{v}_\omega}{2\omega} \right)^2$$

$$\times \left[\frac{\varepsilon_F}{\hbar} + \frac{\omega - \omega_{LO}}{2} - \frac{\hbar q^2}{8m} - \frac{(\omega - \omega_{LO})^2 m}{2\hbar q^2} \right] \tag{32}$$

is always negative. The region defined by Eq. (31) also corresponds to $q < q_\omega$. It is easy to show that the spontaneous generation rate $I_{\mathbf{q}}$ in this region is equal to $-\nu_{\mathbf{q}}$ (problem 4.9), and the solution of the kinetic equation (12) with the initial condition $N_{\mathbf{q}t=0} = 0$ is $N_{\mathbf{q}t} = \exp(|\nu_{\mathbf{q}}|t) - 1$. In other words, the occupation number for $\mathbf{q}$-th mode exponentially increases with time. If we add a phenomenological relaxation term $-N_{\mathbf{q}t}/\tau_{ph}$ due to phonon-phonon scattering (discussed in the next chapter) into the right-hand side of Eq. (12), we obtain a stationary solution $N_{\mathbf{q}} = \tau_{ph}|\nu_{\mathbf{q}}|/(1 - \tau_{ph}|\nu_{\mathbf{q}}|)$, which exists at $\tau_{ph}|\nu_{\mathbf{q}}| < 1$, when the relaxation overcomes the stimulated emission of phonons. In the opposite case, $N_{\mathbf{q}}$ continues to increase with time. Nevertheless, like in the case of photon instability in lasers considered in the previous section, the nonlinear effects eventually stabilize the distribution of phonons.

22. Boson Emission by 2D Electrons

A special consideration is necessary if we are going to describe the emission of bulk boson modes due to transitions of electrons between confined states (for example, between the states in quantum wells, wires, and dots). In the problems of this kind, the interaction between quasi-particles of different dimensionalities is essential in the kinetic phenomena. The most important examples of such phenomena are the emission of high-energy acoustic phonons by hot 2D electrons and the photoluminescence from quantum wells. Both these phenomena are described below on the basis of kinetic equation (19.14), where the factor $\hat{\chi}_q$ is given by the expressions (21.1) and (20.1) for acoustic phonons and photons, respectively (see also problems 4.1 and 4.2). The trace in Eq. (19.15) is calculated with the use of a complete set of electron eigenstates $|\delta\rangle$ with energies ε_δ:

$$J_{b,e}(N|qq_1) = \frac{1}{\hbar^2} \int_{-\infty}^{0} d\tau e^{\lambda\tau + i\omega_q \tau} \sum_{\delta\eta} \langle\eta|\hat{\chi}_{q_1}|\delta\rangle\langle\delta|\widehat{G}_q|\eta\rangle e^{-i(\varepsilon_\eta - \varepsilon_\delta)\tau/\hbar}$$

$$+(c.c., q \leftrightarrow q_1). \tag{1}$$

The matrix elements of the operator $\widehat{G}_q$ defined in Eq. (19.10) are given by the following expression:

$$\langle\delta|\widehat{G}_q|\eta\rangle = \delta_{\delta\eta} f_\delta \sum_{\gamma} \langle\gamma|\hat{\chi}_q^+|\gamma\rangle + (1 - f_\delta) f_\eta \langle\delta|\hat{\chi}_q^+|\eta\rangle$$

$$+\sum_{q'} \langle\delta|\hat{\chi}_{q'}^+|\eta\rangle (f_\eta - f_\delta) N(q, q'). \tag{2}$$

Let us represent the collision integral (1) as a sum of the generation rate $I(q, q_1)$ and the term $\overline{J}(q, q_1)$, the latter is proportional to the boson distribution function and describes the scattering of bosons by electrons. Then the stationary kinetic equation is written as

$$i(\omega_q - \omega_{q_1})N(q, q_1) = I(q, q_1) + \overline{J}(q, q_1). \tag{3}$$

Calculating the integral over time in Eq. (1), we write the generation rate on the right-hand side of Eq. (3) as

$$I(q, q_1) = \frac{1}{i\hbar^2 \omega_q} \sum_{\delta} \langle\delta|\hat{\chi}_{q_1}|\delta\rangle f_\delta \sum_{\gamma} \langle\gamma|\hat{\chi}_q^+|\gamma\rangle \tag{4}$$

$$+\sum_{\delta\eta} f_\eta(1 - f_\delta) \frac{\langle\eta|\hat{\chi}_q|\delta\rangle^* \langle\eta|\hat{\chi}_{q_1}|\delta\rangle}{i\hbar(\varepsilon_\delta - \varepsilon_\eta + \hbar\omega_q - i\lambda)} + (c.c., q \leftrightarrow q_1),$$

while the relaxation term $\overline{J}(q, q_1)$ becomes

$$\overline{J}(q,q_1) = \frac{1}{\hbar} \sum_{\delta\eta q'} (f_\eta - f_\delta)$$

$$\times \left\{ \frac{\mathcal{P}}{i} \left[\frac{\langle\eta|\hat{\chi}_{q'}|\delta\rangle^* \langle\eta|\hat{\chi}_{q_1}|\delta\rangle}{\varepsilon_\delta - \varepsilon_\eta + \hbar\omega_q} N(q,q') - \frac{\langle\eta|\hat{\chi}_q|\delta\rangle^* \langle\eta|\hat{\chi}_{q'}|\delta\rangle}{\varepsilon_\delta - \varepsilon_\eta + \hbar\omega_{q_1}} N(q',q_1) \right] \right.$$

$$+ \pi \left[\langle\eta|\hat{\chi}_{q'}|\delta\rangle^* \langle\eta|\hat{\chi}_{q_1}|\delta\rangle \delta(\varepsilon_\delta - \varepsilon_\eta + \hbar\omega_q) N(q,q') \right. \tag{5}$$

$$\left. \left. + \langle\eta|\hat{\chi}_q|\delta\rangle^* \langle\eta|\hat{\chi}_{q'}|\delta\rangle \delta(\varepsilon_\delta - \varepsilon_\eta + \hbar\omega_{q_1}) N(q',q_1) \right] \right\}.$$

The principal-value contribution in Eq. (5) describes the renormalization of boson frequency, while the contribution containing the energy conservation laws describes the relaxation of phonons or photons.

The general expressions (3)−(5) are used below to describe the photon emission by the 2D electrons localized in the plane $z = 0$. The electron eigenstates are written as $|l\mathbf{p}\rangle$, where $\mathbf{p}$ is the 2D momentum and l is the discrete quantum number including band index (c or v), spin number σ, and size-quantization subband number n. We consider the spin-degenerate case so that the corresponding energies $\varepsilon_{l\mathbf{p}}$ are $\varepsilon_{cn\mathbf{p}}$ and $\varepsilon_{vn\mathbf{p}}$, while the distribution functions $f_{l\mathbf{p}}$ are $f_{cn\mathbf{p}}$ and $f_{vn\mathbf{p}}$. If these distribution functions are isotropic in the 2D plane, the emission rate of the photons with wave vector $\mathbf{Q} = (\mathbf{q}, q_z)$ is written as $\delta_{\mathbf{qq'}} I_{\mathbf{q}}^{\mu\mu'}(q_z, q_z')$, where

$$I_{\mathbf{q}}^{\mu\mu'}(q_z, q_z') = \frac{2\pi e^2}{\epsilon V \sqrt{\omega_Q \omega_{Q'}}} \sum_{ll'\mathbf{p}} \langle l\mathbf{p}|\mathbf{e}_{\mathbf{Q}\mu} \cdot \hat{\mathbf{v}}|l'\mathbf{p}\rangle^* \langle l\mathbf{p}|\mathbf{e}_{\mathbf{Q}'\mu'} \cdot \hat{\mathbf{v}}|l'\mathbf{p}\rangle f_{l\mathbf{p}}(1 - f_{l'\mathbf{p}})$$

$$\times \left\{ \frac{\mathcal{P}}{i} \left[(\hbar\omega_Q + \varepsilon_{l'p} - \varepsilon_{lp})^{-1} - (\hbar\omega_{Q'} + \varepsilon_{l'p} - \varepsilon_{lp})^{-1} \right] \right. \tag{6}$$

$$\left. + \pi \left[\delta(\hbar\omega_Q + \varepsilon_{l'p} - \varepsilon_{lp}) + \delta(\hbar\omega_{Q'} + \varepsilon_{l'p} - \varepsilon_{lp}) \right] \right\}.$$

This equation employs the general expression (4) with $\hat{\chi}_{\mathbf{q}}$ written in the dipole approximation according to Eq. (20.1). The photon density matrix $N(q,q')$ satisfying Eq. (3) is diagonal in the 2D wave vectors and can be written as $\delta_{\mathbf{qq'}} N_{\mathbf{q}}^{\mu\mu'}(q_z, q_z')$. The partial Wigner transformation of $N_{\mathbf{q}}^{\mu\mu'}(q_z, q_z')$ is defined similar to Eqs. (9.6) and (19.26):

$$N_{\mathbf{q}}^{\mu\mu'}(z, q_z) = \sum_g e^{igz} N_{\mathbf{q}}^{\mu\mu'} \left(q_z + \frac{g}{2}, q_z - \frac{g}{2} \right). \tag{7}$$

The energy density and energy flow density can be expressed through the photon distribution function $N_{\mathbf{q}}^{\mu\mu'}(z, q_z)$ according to Eqs. (19.30) and (19.32).

Applying this Wigner transformation to Eq. (3), we should consider the equation

$$i \sum_g e^{igz} (\omega_{q,q_z+g/2} - \omega_{q,q_z-g/2}) N_{\mathbf{q}}^{\mu\mu'} \left(q_z + \frac{g}{2}, q_z - \frac{g}{2} \right) \qquad (8)$$

$$= \sum_g e^{igz} \left[I_{\mathbf{q}}^{\mu\mu'} \left(q_z + \frac{g}{2}, q_z - \frac{g}{2} \right) + \overline{J}_{\mathbf{q}}^{\mu\mu'} \left(q_z + \frac{g}{2}, q_z - \frac{g}{2} \right) \right],$$

taking into account that the quasi-classical approximation is not valid near the 2D plane $z = 0$. In the remote zone, $|z| \gg 2\pi/\overline{q}_z$, one may consider g small in comparison to the characteristic wave numbers $\overline{q}_z$, and the left-hand side of Eq. (8) is written as

$$v_\perp \frac{\partial N_{\mathbf{q}}^{\mu\mu'}(z, q_z)}{\partial z}, \qquad v_\perp = \frac{\partial \omega_Q}{\partial q_z}, \qquad (9)$$

while the right-hand side is transformed to

$$L_\perp \Delta(z) \left[I_{\mathbf{q}}^{\mu\mu'}(q_z, q_z) + J_{\mathbf{q}}^{\mu\mu'}(q_z, q_z) \right], \qquad (10)$$

where $L_\perp$ is the normalization length in z direction. The function $\Delta(z) = L_\perp^{-1} \sum_g \exp(igz)$ can be approximated by the δ-function of z (problem 4.10). Therefore, the contribution (10) goes to zero in the remote zone. Near the 2D plane, where $|z| < 2\pi/\overline{q}_z$, one has to analyze Eq. (8) more carefully, and the problem becomes complicated. However, to describe the emission of photons from the 2D layer, we can integrate Eq. (8) over the region $-z_0 < z < z_0$, where z_0 is placed somewhere in between the remote zone and $z = 0$. As a result, we obtain the boundary condition

$$v_\perp N_{\mathbf{q}}^{\mu\mu'}(z, q_z) \Big|_{z=-z_0}^{z=z_0} = \int_{-z_0}^{z_0} dz \sum_g e^{igz} \qquad (11)$$

$$\times \left[I_{\mathbf{q}}^{\mu\mu'} \left(q_z + \frac{g}{2}, q_z - \frac{g}{2} \right) + \overline{J}_{\mathbf{q}}^{\mu\mu'} \left(q_z + \frac{g}{2}, q_z - \frac{g}{2} \right) \right] \simeq \mathcal{I}_{\mathbf{Q}}^{\mu\mu'},$$

where on the right-hand side we have used $\int_{-z_0}^{z_0} dz \exp(igz) \simeq L_\perp \delta_{g0}$ and made the definition $\mathcal{I}_{\mathbf{Q}}^{\mu\mu'} = L_\perp I_{\mathbf{q}}^{\mu\mu'}(q_z, q_z)$ (we point out that $\mathcal{I}$ has the dimensionality of velocity). The term $\overline{J}$ is neglected in the second equation of Eq. (11), since we imply the condition $l_{ph} \gg 2\pi/\overline{q}_z$, where l_{ph} is the photon absorption length due to electron transitions in the 2D layer. The "glancing" photons, for which $q_z \to 0$, are excluded from the consideration.

Therefore, the emission of 3D photons by 2D electrons is described by a quasi-classical kinetic equation with a boundary condition at $z = 0$. This approach essentially implies that the photon wavelength is much larger than the localization length of electrons in z direction. The problem is written in the form

$$v_\perp \frac{\partial N_{\mathbf{q}}^{\mu\mu'}(z, q_z)}{\partial z} = \tilde{J}_{ph}, \quad v_\perp N_{\mathbf{q}}^{\mu\mu'}(z, q_z)\Big|_{z=-0}^{z=+0} = \mathcal{I}_{\mathbf{Q}}^{\mu\mu'}, \qquad (12)$$

where z_0 is aimed to zero. The phenomenologically introduced collision integral $\tilde{J}_{ph}$ describes the relaxation of photon distribution outside the 2D layer. If this relaxation is neglected, the photon distribution becomes spatially homogeneous, according to $N_{\mathbf{q}}^{\mu\mu'}(q_z) = \mathcal{I}_{\mathbf{Q}}^{\mu\mu'}/v_\perp$. The spectral, polarization, and angular dependence of the radiation is determined entirely by the generation rate (6). The intensity of photon generation is obtained after summing this rate over the indices of polarization, $I_0(\mathbf{Q}) = \sum_{\mu=1,2} \mathcal{I}_{\mathbf{Q}}^{\mu\mu}$. The polarization characteristics of the photoluminescence are determined by the matrix structure of $N^{\mu\mu'} \propto \mathcal{I}^{\mu\mu'}$. Let us express the 2×2 matrix $\mathcal{I}_{\mathbf{Q}}^{\mu\mu'}$ through the Pauli matrices $\hat{\sigma}_i$ ($i = x, y, z$) as

$$\hat{\mathcal{I}}_{\mathbf{Q}} = \frac{1}{2}\left[I_0(\mathbf{Q}) + \mathbf{s}(\mathbf{Q}) \cdot \hat{\boldsymbol{\sigma}}\right]. \qquad (13)$$

The coefficients s_x and s_z describe the degree and orientation of the linear polarization, while s_y describes the degree of the circular polarization. Since the intensity of spontaneous radiation is not directly related to the polarization characteristics, it is convenient to introduce the Stokes parameters according to $\boldsymbol{\xi}(\mathbf{Q}) = \mathbf{s}(\mathbf{Q})/I_0(\mathbf{Q})$. The explicit expressions for the components of this vector are

$$\xi_x(\mathbf{Q}) = \frac{\mathcal{I}_{\mathbf{Q}}^{12} + \mathcal{I}_{\mathbf{Q}}^{12*}}{I_0(\mathbf{Q})}, \quad \xi_y(\mathbf{Q}) = i\frac{\mathcal{I}_{\mathbf{Q}}^{12} - \mathcal{I}_{\mathbf{Q}}^{12*}}{I_0(\mathbf{Q})},$$

$$\xi_z(\mathbf{Q}) = \frac{\mathcal{I}_{\mathbf{Q}}^{11} - \mathcal{I}_{\mathbf{Q}}^{22}}{I_0(\mathbf{Q})}. \qquad (14)$$

We note that the unit vectors $\mathbf{e}_{1,2}$ describing the photon polarization are perpendicular to each other and are placed in the plane perpendicular to $\mathbf{Q}$. It is convenient to direct these vectors at the angles $\pm\pi/4$ with respect to the intersection line of this plane with the plane determined by the vectors $\mathbf{Q}$ and $\mathbf{n}$, the latter is the unit vector perpendicular to the 2D layer; see Fig. 4.2. Instead of ξ_x and ξ_z, we use the degree of polarization $l_{\mathbf{Q}}$ and the angle $\phi_{\mathbf{Q}}$ between the maximum of polarization and $\mathbf{e}_1$, according to the relations $\xi_x = l_{\mathbf{Q}} \sin 2\phi_{\mathbf{Q}}$ and $\xi_z = l_{\mathbf{Q}} \cos 2\phi_{\mathbf{Q}}$. The case $l_{\mathbf{Q}} = 1$ corresponds to completely polarized radiation. If $\phi_{\mathbf{Q}} =$

$-\pi/4$, the maximum of polarization is in the 2D plane. If $\phi_{\mathbf{Q}} = \pi/4$, the maximum is in the plane determined by the vectors $\mathbf{Q}$ and $\mathbf{n}$.

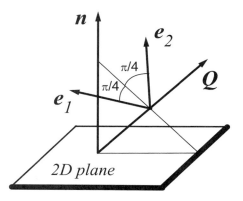

Figure 4.2. Orientation of the vectors characterizing the photoluminescence from the 2D layer.

Let us find the polarization characteristics defined above. Consider the intersubband transitions of the 2D electrons in the conduction band. We omit the band index so that $\varepsilon_{l\mathbf{p}} = \varepsilon_{n\mathbf{p}}$ and $f_{l\mathbf{p}} = f_{n\mathbf{p}}$. In the simple model given by Eqs. (5.24)–(5.25), only z-component of the velocity contributes to the matrix elements of the intersubband transitions, and from Eq. (6) we obtain

$$\mathcal{I}_{\mathbf{Q}}^{\mu\mu'} = 2\frac{(2\pi e)^2}{\omega_Q \epsilon} \sum_{nn'} \int \frac{d\mathbf{p}}{(2\pi\hbar)^2} (e_{\mathbf{Q}\mu}^z v_{nn'}^z)^* (e_{\mathbf{Q}\mu'}^z v_{nn'}^z)$$

$$\times \delta(\hbar\omega_Q + \varepsilon_{n'\mathbf{p}} - \varepsilon_{n\mathbf{p}}) f_{n\mathbf{p}}(1 - f_{n'\mathbf{p}}) \qquad (15)$$

$$= \frac{2(2\pi e)^2}{\omega_Q \epsilon} \int \frac{d\mathbf{p}}{(2\pi\hbar)^2} e_{\mathbf{Q}\mu}^{z\,*} e_{\mathbf{Q}\mu'}^z |v_{12}^z|^2 \delta_\gamma(\hbar\omega_Q - \varepsilon_{21}) f_{2\mathbf{p}}(1 - f_{1\mathbf{p}}),$$

where $v_{nn'}^z = m^{-1}\langle n|\hat{p}_z|n'\rangle$ and the factor of 2 comes from the sum over spin. The second equation in Eq. (15) is written in the resonance approximation, when $\hbar\omega_Q$ is close to the intersubband transition energy $\varepsilon_{21} = \varepsilon_{2\mathbf{p}} - \varepsilon_{1\mathbf{p}}$ so that the contribution of other subbands is neglected. The energy ε_{21} is independent of $\mathbf{p}$ since we assume the identical, parabolic dispersion for the electrons in the subbands. The function $\delta_\gamma(E)$ is defined as a "broadened" δ-function (see problem 1.4), where the broadening energy γ for intersubband transitions is introduced phenomenologically. This broadening is necessary to make the expression (15) finite. For non-degenerate electrons, the generation rate depends on the density of electrons in the second subband. For degenerate electrons, a non-zero generation takes place only in the case of negative interband absorption. Using Eqs. (14) and (15), we find $\xi_y(\mathbf{Q}) = 0$, while the other polarization characteristics are expressed as a solution of the following system of two equations:

$$l_\mathbf{Q} \sin 2\phi_\mathbf{Q} = \frac{2e_1^z e_2^z}{(e_1^z)^2 + (e_2^z)^2}, \quad l_\mathbf{Q} \cos 2\phi_\mathbf{Q} = \frac{(e_1^z)^2 - (e_2^z)^2}{(e_1^z)^2 + (e_2^z)^2}. \quad (16)$$

If the unit vectors $\mathbf{e}_{1,2}$ are chosen as shown in Fig. 4.2, we have $e_1^z = e_2^z$, and the solutions are $l_\mathbf{Q} = 1$ and $\phi_\mathbf{Q} = \pi/4$. This result is understandable: if the intersubband transitions are excited by the wave polarized perpendicularly to the 2D layer, the photoluminescence is linearly polarized in the plane determined by the vectors $\mathbf{Q}$ and $\mathbf{n}$.

A description of the photoluminescence due to interband transitions in quantum wells is more difficult. The spectral dependence of such photoluminescence in p-type quantum wells is given by an expression analogous to Eq. (20.9). However, it is determined by the two-dimensional joint density of states, and the square-root spectral dependence near the edge of interband transitions is replaced by a step-like dependence (compare Eqs. (5.3) and (5.26)). The polarization dependence of the photoluminescence is determined both by the interband matrix elements of the velocity operator (problem 4.11) and by the distribution of non-equilibrium electrons (problem 4.12). The photoluminescence is often used as a tool for optical characterization of heterostructures.

The consideration given above can be applied to describe the acoustic phonon emission by 2D electrons. We consider electron transitions within the conduction band states $|n\mathbf{p}\rangle$, and assume that the electrons are characterized by a non-equilibrium distribution and interact with the longitudinal acoustic (LA) phonons via deformation potential. Substituting the corresponding operator $\hat{\chi}_\mathbf{q}$ from Eq. (21.1) into Eq. (4), we obtain

$$I_\mathbf{q}(q_z, q_z') = 2C_Q^* C_{Q'} \frac{1}{\hbar} \sum_{nn'\mathbf{p}} f_{n\mathbf{p}}(1 - f_{n'\mathbf{p}-\hbar\mathbf{q}})\langle n|e^{iq_z z}|n'\rangle^* \langle n|e^{iq_z' z}|n'\rangle$$

$$\times \left\{ \frac{\mathcal{P}}{i} \left[(\hbar\omega_Q + \varepsilon_{n'\mathbf{p}-\hbar\mathbf{q}'} - \varepsilon_{np})^{-1} - (\hbar\omega_{Q'} + \varepsilon_{n'\mathbf{p}-\hbar\mathbf{q}} - \varepsilon_{np})^{-1} \right] \right.$$

$$\left. + \pi \left[\delta(\hbar\omega_Q + \varepsilon_{n'\mathbf{p}-\hbar\mathbf{q}'} - \varepsilon_{np}) + \delta(\hbar\omega_{Q'} + \varepsilon_{n'\mathbf{p}-\hbar\mathbf{q}} - \varepsilon_{np}) \right] \right\} \quad (17)$$

instead of Eq. (6). In contrast to the case of electron-photon interaction, the polarization index is omitted since we have only one polarization.

The Wigner distribution function of phonons, $N_{\mathbf{q}}(z, q_z)$, satisfies a kinetic equation with a boundary condition, according to Eq. (12). The boundary condition contains the generation velocity

$$
\mathcal{I}_{\mathbf{Q}} = \frac{4\pi}{\hbar} L_\perp |C_Q|^2 \sum_{nn'\mathbf{p}} |\langle n|e^{iq_z z}|n'\rangle|^2 \delta(\hbar\omega_Q + \varepsilon_{n'\mathbf{p}-\hbar\mathbf{q}} - \varepsilon_{n\mathbf{p}})
$$

$$
\times f_{n\mathbf{p}}(1 - f_{n'\mathbf{p}-\hbar\mathbf{q}}). \tag{18}
$$

The factor of 2 in Eqs. (17) and (18) comes from the sum over spin.

The homogeneous solution $N_{\mathbf{q}}(q_z) = \mathcal{I}_{\mathbf{Q}}/v_\perp$ is valid at the distances small in comparison to the lateral size L of the 2D layer. At the distances much larger than L, the 2D structure can be viewed as a local source of phonons placed at the origin of the coordinate system. This consideration allows one to neglect edge effects and leads to a simple stationary distribution

$$
N_{\mathbf{rQ}} = \mathcal{F}_{\mathbf{Q}}\left(x - \frac{q_x}{q_z}z, y - \frac{q_y}{q_z}z\right), \quad \mathcal{F}_{\mathbf{Q}}(x, y) = \begin{cases} \mathcal{I}_{\mathbf{Q}}/v_\perp, & (x, y) \in \Gamma \\ 0, & \text{outside} \end{cases},
$$

$$\tag{19}$$

where Γ is the area of the 2D layer covering the square L^2. Substituting this solution into the energy flow density (19.32), it is not difficult to calculate $\mathbf{G}$ at $R \gg L$, where R is the distance between the point of observation and the 2D structure. The tangential component of $\mathbf{G}$ vanishes, while the radial component G_r, which describes the energy flow from the 2D layer in the direction determined by the angles θ and φ of the spherical coordinate system, is given by

$$
G_r = \frac{L^2}{(2\pi s_l)^3 R^2} \int_0^\infty d\omega \, \hbar\omega^3 \mathcal{I}(\omega, \theta), \tag{20}
$$

where $\omega = \omega_Q = s_l Q = s_l |q_z/\cos\theta|$ and the function $\mathcal{I}_{\mathbf{Q}}$ is expressed in terms of the variables ω and θ (it is independent of the polar angle φ because of the averaging over the angle of $\mathbf{p}$ in Eq. (18)). The radial component G_r decreases as $1/R^2$ and is independent of the polar angle due to the axial symmetry of the problem at $R \gg L$. The result (20) can be expressed through the differential quantity δG defined as the energy flow in a unit frequency interval inside a unit solid angle in the direction determined by the angle θ:

$$
\delta G = \frac{\hbar\omega^3}{(2\pi s_l)^3} \mathcal{I}(\omega, \theta). \tag{21}
$$

The total intensity of the radiation emitted by a unit area of the 2D layer inside a unit solid angle in the direction determined by the angle θ is defined as $\int_0^\infty d\omega \, \delta G$.

The energy dependence of δG and angular distribution of emitted phonons are calculated below for the electrons occupying the lowest subband. The electrons are described by the quasi-equilibrium Fermi distribution with effective electron temperature T_e. This temperature can be controlled by an applied electric field which heats the electron system. The anisotropy of the electron distribution is neglected, since it is small as a ratio of the drift velocity to the Fermi velocity (we stress, however, that this anisotropy leads to an anisotropy of the emitted phonon distribution in the plane). From Eqs. (18) and (21) we find

$$\delta G = \frac{\hbar\omega^3}{(2\pi)^3 s_l^2 |\sin\theta|} |\langle 1|e^{iq_z z}|1\rangle|^2 \int_{\varepsilon_m}^{\infty} d\varepsilon \frac{f(\varepsilon)[1 - f(\varepsilon - \hbar\omega)]}{\sqrt{\varepsilon_0(\varepsilon - \varepsilon_m)}}. \qquad (22)$$

The energies entering this expression are

$$\varepsilon_0 = 2\left(\frac{\pi\hbar^3 \rho s_l^2}{\mathcal{D}^2 m^{3/2}}\right)^2 \qquad (23)$$

and

$$\varepsilon_m = \frac{m s_l^2}{2\sin^2\theta}\left(1 + \frac{\hbar\omega}{2m s_l^2}\sin^2\theta\right)^2. \qquad (24)$$

The latter is the cut-off energy appearing after the angular averaging, because of the energy conservation law. The squared overlap factor $|\langle 1|e^{iq_z z}|1\rangle|^2$ in Eq. (22) is determined by the shape of the wave function of the confined electron state and depends on the dimensionless parameter $q_z d$, where d is the width of the quantum well (problem 4.13). Under the conditions $\hbar\omega \ll \varepsilon_F$ and $T_e \ll \varepsilon_F$, the differential energy flow is independent of ε_m and given by

$$\delta G = \frac{\hbar\omega^3}{(2\pi)^3 s_l^2 |\sin\theta|} |\langle 1|e^{iq_z z}|1\rangle|^2 \frac{\hbar\omega}{\sqrt{\varepsilon_F \varepsilon_0}} \frac{1}{e^{\hbar\omega/T_e} - 1}. \qquad (25)$$

If $\hbar\omega \ll T_e$, this function increases linearly with increasing T_e. At low temperatures it is exponentially suppressed because of the Pauli blocking effect. Equation (25) is not valid for the phonons emitted perpendicular to the 2D layer, when $\theta \to 0$. The cut-off energy ε_m for such phonons becomes comparable to ε_F, in spite of smallness of the energy $m s_l^2$, and there appears an additional exponential suppression of δG.

Problems

4.1. Write the expressions for electron parts of the operators of electron-acoustic phonon and electron-optical phonon interactions standing in Eq. (19.2).

Solution: Let us substitute the electron coordinates $\mathbf{x}_j$ in place of $\mathbf{r}$ into the expressions for $\widehat{H}_{e,LO}(\mathbf{r})$ and $\widehat{H}_{e,LA}(\mathbf{r})$ given by Eqs. (6.22) and (6.30), respectively.

Summing these expressions over the particles, we obtain $\widehat{H}_{e,LO}$ and $\widehat{H}_{e,LA}$ in the form of Eq. (19.2), where the index q includes both the wave vector $\mathbf{q}$ and the mode index (LO or LA). The electron parts $\hat{\chi}_q^{(j)}$ are expressed as

$$\hat{\chi}_{\mathbf{q}LO}^{(j)} = \sqrt{\frac{2\pi e^2 \hbar \omega_{LO}}{\epsilon^* q^2 V}} e^{i\mathbf{q}\cdot\mathbf{x}_j}, \quad \hat{\chi}_{\mathbf{q}LA}^{(j)} = \sqrt{\frac{\hbar D^2 q}{2\rho s_l V}} e^{i\mathbf{q}\cdot\mathbf{x}_j}.$$

4.2. Do the same as in problem 4.1 for electron-photon interaction.
Result:

$$\hat{\chi}_{\mathbf{q}\mu}^{(j)} = \frac{1}{2}\sqrt{\frac{2\pi\hbar e^2}{\omega_{\mathbf{q}\mu}\epsilon V}}\left[(\hat{\mathbf{v}}_j \cdot \mathbf{e}_{\mathbf{q}\mu})e^{i\mathbf{q}\cdot\mathbf{x}_j} + e^{i\mathbf{q}\cdot\mathbf{x}_j}(\hat{\mathbf{v}}_j \cdot \mathbf{e}_{\mathbf{q}\mu})\right].$$

4.3. Rewrite the operator of electron-boson interaction in the second quantization representation.

Hint: Use the general formalism developed in Sec. 4 (see, for example, the discussion leading to Eq. (4.23)).

4.4. Consider absorption of the photons incident on a half-space $z > 0$ with absorption coefficient α_ω.

Solution: Solving Eq. (20.7) with the boundary condition $N_{z=0} = N_0$ at the surface, we obtain $N_z = N_0 e^{-\alpha_\omega z}$.

4.5. Consider the occupation of conduction and valence bands by electrons in the materials of n- and p-type at $T = 0$. Consider also a material with non-equilibrium electron-hole pairs.

Solution: The Fermi energies ε_{Fj} ($j = c, v$) at zero temperature are expressed through the densities n_j of electrons or holes according to $n_j = \int_0^{\varepsilon_{Fj}} d\varepsilon \rho_j(\varepsilon)$, where ρ_j is the density of states in the band j. The densities n_c and n_v in equilibrium are equal to the concentrations of ionized donors and acceptors, N_D and N_A, respectively. In the material with non-equilibrium electron-hole pairs, one should use the electric neutrality equation $n_v + N_D = n_c + N_A$. This equation allows one to express n_c through N_D, N_A, and ε_{Fv}.

4.6. Solve Eq. (20.15) with the initial condition $N_\omega^\pm(z,t)|_{t=0} = 0$ and boundary conditions (20.17).

Hints: Using the Laplace transformation $N_\omega^{(\pm)}(z,s) = \int_0^\infty dt e^{-st} N_\omega^{(\pm)}(z,t)$, reduce Eq. (20.15) to an ordinary differential equation of the first order. Solve it with the boundary conditions (20.17). Then apply the inverse Laplace transformation.

4.7. Derive Eq. (21.7).

Hint: Calculate the derivatives over time of both sides of Eq. (21.6) and use the definition of S-operators given in Sec. 2.

4.8. Write the generation term $I_{\mathbf{q}}$ in Eq. (21.12) under the assumption that the influence of electric field on electron-phonon interaction is neglected. Find a relation between $I_{\mathbf{q}}$ and $\nu_{\mathbf{q}}$ in the case of Fermi distribution of electrons.

Result:

$$I_{\mathbf{q}} = \frac{4\pi}{\hbar}|C_q|^2 \sum_{\mathbf{p}} \delta(\varepsilon_p - \varepsilon_{\mathbf{p}+\hbar\mathbf{q}} + \hbar\omega_q) f_{\mathbf{p}+\hbar\mathbf{q}}(1 - f_{\mathbf{p}}) = \frac{\nu_{\mathbf{q}}}{e^{\hbar\omega_{\mathbf{q}}/T_e} - 1}.$$

The last expression corresponds to the case of Fermi distribution.

4.9. Check that the generation rate (21.23) for short-wavelength phonons in the region (21.31) is equal to $-\nu_{\mathbf{q}}$ given by Eq. (21.32).

Hint: Take into account that the quasi-equilibrium function $\overline{f}_{\mathbf{p}} = f(\varepsilon_p)$ satisfies the following identity:

$$\overline{f}_{\mathbf{p}+\hbar\mathbf{q}}(1 - \overline{f}_{\mathbf{p}}) = \frac{\overline{f}_{\mathbf{p}+\hbar\mathbf{q}} - \overline{f}_{\mathbf{p}}}{1 - \exp[(\varepsilon_{\mathbf{p}+\hbar\mathbf{q}} - \varepsilon_p)/T_e]},$$

where T_e is the effective temperature of electrons. Prove that the exponent in this expression can be neglected in Eq. (21.23) at $k = -1$, if $\hbar\omega \gg T_e$.

4.10. Consider the function $\Delta(z)$ entering Eq. (22.10).

Solution: By definition, the region of summation in $\Delta(z) = L_{\perp}^{-1} \sum_{\eta} e^{i\eta z}$ is restricted by $|\eta| < \overline{q}_z$. The result of summation is written as $\Delta(z) = \sin(\overline{q}_z z)/\pi z$. This function has a maximum at $z = 0$, while at $z > \pi/\overline{q}_z$ it is small and oscillating. Next, one has $\int_{-z_0}^{z_0} dz \Delta(z) = 1$ at $z_0 \gg \pi/\overline{q}_z$. Therefore, $\Delta(z)$ has the properties of a broadened δ-function.

4.11. Consider the electron states in c- and v-bands of a symmetric quantum well described by the two-band model; see Eqs. (B.16)−(B.24). Calculate the matrix elements of the velocity operator for this system.

Solution: Let us use the Hamiltonian (B.18) with $M \to \infty$ and consider the case of symmetric c- and v-bands, assuming that the potentials of these bands change symmetrically at the boundaries $z = \pm d/2$ of the quantum well. Carrying out a unitary transformation of this Hamiltonian with the use of the operator $(1 + i\hat{\rho}_3\hat{\sigma}_z)/\sqrt{2}$, we obtain a Hamiltonian which can be easily diagonalized with respect to the spin variable. The eigenstates are the four-component wave functions written below through the two-component spinors:

$$\begin{pmatrix} \varphi|\sigma\rangle \\ \chi|\sigma\rangle \end{pmatrix} \frac{1}{L} \exp(i\mathbf{p} \cdot \mathbf{r}/\hbar),$$

where $\mathbf{r}$ and $\mathbf{p}$ are the two-dimensional coordinate and momentum. The spinors $|\sigma\rangle$ satisfy the eigenstate problem $\hat{\sigma}_z|\sigma\rangle = \sigma|\sigma\rangle$ with $\sigma = \pm 1$, while the coordinate-dependent functions φ and χ satisfy the following system of equations:

$$(\varepsilon_g(z)/2 - E)\varphi(z) + s(\sigma p + \hbar d/dz)\chi(z) = 0,$$

$$s(\sigma p - \hbar d/dz)\varphi(z) + (-\varepsilon_g(z)/2 - E)\chi(z) = 0,$$

where $\varepsilon_g(z) = \varepsilon_g^<$ inside the well ($|z| < d/2$), and $\varepsilon_g(z) = \varepsilon_g^>$ outside the well ($|z| > d/2$). The interband velocity s is assumed to be constant, it is not changed across the interfaces $z = \pm d/2$. The solutions for confined states are expressed through the functions exponentially decreasing with the increase of $|z|$ outside the wells. Using this property, we reduce the problem to the region inside the well, with the boundary conditions $\chi(-d/2) = -\alpha_l \varphi(-d/2)$ and $\chi(d/2) = \alpha_r \varphi(d/2)$, where α_l and α_r are the functions of $\varepsilon_g^>$, energy E, and momentum p. If the band offsets at the interfaces are large, $\varepsilon_g^> \gg E, sp$, one has simply $\alpha_l = \alpha_r = 1$. Assuming that this condition is fulfilled, we search for the solutions in the form $\varphi(z) = c_1 e^{ikz} + c_2 e^{-ikz}$ and $\chi(z) = c_3 e^{ikz} + c_4 e^{-ikz}$. The energy spectrum is written as $\varepsilon_{\pm n}(p) = \pm\sqrt{(\varepsilon_g^</2)^2 + (\hbar s k_n)^2 + (sp)^2}$, where we use the indices $+$ and $-$ as the band indices c and v, respectively. The discrete wave numbers $k_n > 0$ are determined from the dispersion relation $\tan(k_n d) = -2\hbar s k_n/\varepsilon_g^<$ (at $\varepsilon_g^< \gg \hbar s k_n$ it is reduced to the hard-wall quantization relation $k_n = \pi n/d$, $n = 1, 2, ...$). The electron states in the quantum well depend on the band index $\pm$, subband number n, 2D momentum $\mathbf{p}$, and spin number σ. The spectrum, however, is spin-degenerate because the quantum well is assumed to be symmetric. In the general case of an asymmetric quantum well, the spectrum is spin-split at $p \neq 0$; see Sec. 63.

Carrying out the unitary transformation of the velocity operator (B.16), and using the four-component wave functions defined above, we obtain the interband matrix elements of the velocity operator in the form

$$\mathbf{v}_{n\sigma,n'\sigma'}(\mathbf{p}) = s(\boldsymbol{\sigma}_{\sigma\sigma'} \cdot [\mathbf{n} \times \mathbf{p}]) \frac{[\mathbf{n} \times \mathbf{p}]}{p^2} \Psi_{n\sigma,n'\sigma'} + s\sigma \frac{\mathbf{p}}{p} \delta_{\sigma\sigma'} \Psi_{n\sigma,n'\sigma'} + is\mathbf{n}\delta_{\sigma\sigma'} \Phi_{n\sigma,n'\sigma'}.$$

In this equation, $\boldsymbol{\sigma}_{\sigma\sigma'} = \langle\sigma|\hat{\boldsymbol{\sigma}}|\sigma'\rangle$, $\mathbf{n}$ is the unit vector in the direction perpendicular to the 2D plane, and

$$\Psi_{\nu,\nu'} = \int_{-d/2}^{d/2} dz [\varphi_{+\nu p}(z)\chi_{-\nu' p}(z) + \chi_{+\nu p}(z)\varphi_{-\nu' p}(z)],$$

$$\Phi_{\nu,\nu'} = \int_{-d/2}^{d/2} dz [\varphi_{+\nu p}(z)\chi_{-\nu' p}(z) - \chi_{+\nu p}(z)\varphi_{-\nu' p}(z)],$$

where the multi-indices $\nu = n\sigma$ and $\nu' = n'\sigma'$ are used for the sake of brevity.

4.12. Calculate the Stokes parameters for photon emission by non-equilibrium electrons occupying the lowest ($n = 1$) conduction-band level in the quantum well of p-type. Use the model of quantum well described in problem 4.11.

Solution: Since the distribution of the 2D electrons and their energy spectra are isotropic and spin-degenerate, the averaging over the angle of $\mathbf{p}$ and spin summation in Eq. (22.6) is done in the following way:

$$\sum_{\sigma\sigma'} \int_0^{2\pi} \frac{d\varphi}{2\pi} \left(\mathbf{e}_{\mathbf{q}\mu} \cdot \mathbf{v}_{+1\sigma,-n'\sigma'}(\mathbf{p})\right)^* \left(\mathbf{e}_{\mathbf{q}'\mu'} \cdot \mathbf{v}_{+1\sigma,-n'\sigma'}(\mathbf{p})\right)$$

$$\simeq 2s^2 \left[\Psi_{1,n'}^2 \delta_{\mu\mu'} + (\Phi_{1,n'}^2 - \Psi_{1,n'}^2)(\mathbf{e}_{\mathbf{q}\mu} \cdot \mathbf{n})(\mathbf{e}_{\mathbf{q}'\mu'} \cdot \mathbf{n})\right].$$

The matrix elements $\Psi_{n\sigma,n'\sigma'}$ and $\Phi_{n\sigma,n'\sigma'}$ of the previous problem are written here in the approximation when their dependence on spin is neglected (accordingly, the spin indices σ and σ' are omitted). This approximation is valid if the momentum p is small enough, for example, when the non-equilibrium electrons are described by the Maxwell distribution with effective temperature $T_e \ll \varepsilon_g^<$. In the same approximation, we neglect the dependence of the factors $\Psi_{1,n'}$ and $\Phi_{1,n'}$ on p and express the Stokes parameters for the transitions $(+1) \to (-n)$ as

$$\xi_x(\mathbf{q}) = \frac{e_z^2(\Phi_{1,n}^2 - \Psi_{1,n}^2)}{\Psi_{1,n}^2 + e_z^2(\Phi_{1,n}^2 - \Psi_{1,n}^2)}, \quad \xi_y(\mathbf{q}) = \xi_z(\mathbf{q}) = 0,$$

where $e_z = (\mathbf{e}_{1,2} \cdot \mathbf{n}) = \sin \widehat{\mathbf{qn}}/\sqrt{2}$. If $\widehat{\mathbf{qn}} = 0$, the radiation is not polarized, while at non-zero $\widehat{\mathbf{qn}}$ the radiation is polarized parallel to the 2D layer. The degree of this linear polarization is determined by the parameters of the quantum well and by the number n of the valence subband involved in the optical transition.

4.13. Calculate the squared matrix element $\left|\langle 1|e^{iqz}|1\rangle\right|^2$ for the hard-wall quantum well.

<u>Result:</u> $\left|\langle 1|e^{iqz}|1\rangle\right|^2 = [\sin(qd/2)/(qd/2)]^2[(qd/2\pi)^2 - 1]^{-2}$, where d is the well width. This function is close to 1 at $qd < 2$ and rapidly decreases as q exceeds $2\pi/d$.

Chapter 5

INTERACTING PHONON SYSTEMS

In the absence of free charges, as in insulators and non-doped semiconductors, the transport phenomena in solids are determined by the kinetics of interacting phonon modes. Besides, the relaxation properties of the phonon system often control the energy and momentum transfer from non-equilibrium electrons to the lattice. Therefore, the kinetics of interacting phonons deserves a special consideration. Below we derive and analyze the kinetic equation for phonons involving phonon-phonon collisions and describe, on its basis, the thermal conduction of insulators, the thermal waves (second sound), and the features of non-equilibrium phonon relaxation determined by the dynamical properties of phonons. We consider perfect crystals, where the interaction of phonons with impurities can be neglected. The last section of this chapter contains a discussion of phonon-photon interaction and the calculation of the complex dielectric function of ionic crystals with the aid of the linear response theory and the formalism of double-time Green's functions.

23. Phonon-Phonon Collisions

A system of interacting phonons is described by the Hamiltonian

$$\widehat{H}_{ph} + \widehat{H}_{ph,ph}, \tag{1}$$

where $\widehat{H}_{ph}$ is the Hamiltonian of free phonons and $\widehat{H}_{ph,ph}$ is the interaction Hamiltonian originating from the anharmonicity of crystal lattice vibrations. The interaction Hamiltonian is given by Eqs. (6.31) and (6.32). Below, using the definition $B_{l_1 l_2 l_3}(\mathbf{q}_1, \mathbf{q}_2, \mathbf{q}_3) = (\hbar/2)^{3/2} \beta_{l_1 l_2 l_3}(\mathbf{q}_1, \mathbf{q}_2, \mathbf{q}_3) / \sqrt{\rho V \omega_{\mathbf{q}_1 l_1} \omega_{\mathbf{q}_2 l_2} \omega_{\mathbf{q}_3 l_3}}$, we rewrite this Hamiltonian as

$$\widehat{H}_{ph,ph} = \frac{1}{6} \sum_{\mathbf{q}_1 \mathbf{q}_2 \mathbf{q}_3} \sum_{l_1 l_2 l_3} B_{l_1 l_2 l_3}(\mathbf{q}_1, \mathbf{q}_2, \mathbf{q}_3) \tag{2}$$

$$\times (\hat{b}_{\mathbf{q}_1 l_1} + \hat{b}^+_{-\mathbf{q}_1 l_1})(\hat{b}_{\mathbf{q}_2 l_2} + \hat{b}^+_{-\mathbf{q}_2 l_2})(\hat{b}_{\mathbf{q}_3 l_3} + \hat{b}^+_{-\mathbf{q}_3 l_3}),$$

189

where l_i are the phonon branch numbers and $\mathbf{q}_i$ are the phonon wave vectors. The Hamiltonian (2) is cubic in the creation and annihilation operators of phonons, which means that we restrict ourselves by the cubic anharmonicity only. In some cases, however, one needs to account for a higher-order anharmonicity, when the interaction Hamiltonian contains products of four or more such operators.

Let us derive an equation for the one-phonon density matrix introduced by Eq. (19.4). To do this, we substitute the Hamiltonian (1) into Eq. (1.20), multiply this equation by $\hat{b}^+_{\mathbf{q}l}\hat{b}_{\mathbf{q}'l'}$, and calculate the trace over phonon variables. The left-hand side of the equation obtained is transformed to the same form as in Eq. (19.5). The right-hand side contains a trace of the product of the statistical operator $\hat{\eta}_t$ by the commutator $[\hat{b}^+_{\mathbf{q}l}\hat{b}_{\mathbf{q}'l'}, \widehat{H}_{ph,ph}]$. This commutator contains eight terms. Two of them are given below:

$$[\hat{b}^+_{\mathbf{q}l}\hat{b}_{\mathbf{q}'l'}, \hat{b}_{\mathbf{q}_1 l_1}\hat{b}_{\mathbf{q}_2 l_2}\hat{b}^+_{-\mathbf{q}_3 l_3}] = \hat{b}^+_{\mathbf{q}l}\hat{b}_{\mathbf{q}_1 l_1}\hat{b}_{\mathbf{q}_2 l_2}\delta_{l'l_3}\delta_{\mathbf{q}',-\mathbf{q}_3}$$

$$-\hat{b}_{\mathbf{q}_2 l_2}\hat{b}^+_{-\mathbf{q}_3 l_3}\hat{b}_{\mathbf{q}'l'}\delta_{ll_1}\delta_{\mathbf{q},\mathbf{q}_1} - \hat{b}_{\mathbf{q}_1 l_1}\hat{b}^+_{-\mathbf{q}_3 l_3}\hat{b}_{\mathbf{q}'l'}\delta_{ll_2}\delta_{\mathbf{q},\mathbf{q}_2} \ , \qquad (3)$$

$$[\hat{b}^+_{\mathbf{q}l}\hat{b}_{\mathbf{q}'l'}, \hat{b}_{\mathbf{q}_1 l_1}\hat{b}^+_{-\mathbf{q}_2 l_2}\hat{b}^+_{-\mathbf{q}_3 l_3}] = \hat{b}^+_{\mathbf{q}l}\hat{b}_{\mathbf{q}_1 l_1}\hat{b}^+_{-\mathbf{q}_3 l_3}\delta_{l'l_2}\delta_{\mathbf{q}',-\mathbf{q}_2}$$

$$+\hat{b}^+_{\mathbf{q}l}\hat{b}_{\mathbf{q}_1 l_1}\hat{b}^+_{-\mathbf{q}_2 l_2}\delta_{l'l_3}\delta_{\mathbf{q}',-\mathbf{q}_3} - \hat{b}^+_{-\mathbf{q}_2 l_2}\hat{b}^+_{-\mathbf{q}_3 l_3}\hat{b}_{\mathbf{q}'l'}\delta_{ll_1}\delta_{\mathbf{q},\mathbf{q}_1} \ , \qquad (4)$$

and the other are written in a similar way. After straightforward but cumbersome transformations, we find

$$\left[i\hbar\frac{\partial}{\partial t} - \hbar(\omega_{\mathbf{q}'l'} - \omega_{\mathbf{q}l})\right] N_t(l'\mathbf{q}', l\mathbf{q})$$

$$= -\frac{1}{6}\sum_{l_1\mathbf{q}_1 l_2\mathbf{q}_2}\left\{[B_{ll_1 l_2}(\mathbf{q},\mathbf{q}_1,\mathbf{q}_2) + \ldots]_3\, K_{(l'|l_2 l_1)}(\mathbf{q}'|-\mathbf{q}_2, -\mathbf{q}_1)\right.$$

$$+ [B_{ll_1 l_2}(\mathbf{q},\mathbf{q}_1,\mathbf{q}_2) + \ldots]_6\, K_{(l'l_2|l_1)}(\mathbf{q}',\mathbf{q}_2|-\mathbf{q}_1)$$

$$- [B_{l'l_1 l_2}(-\mathbf{q}',\mathbf{q}_1,\mathbf{q}_2) + \ldots]_6\, K_{(l_2|l_1 l)}(\mathbf{q}_2|-\mathbf{q}_1, \mathbf{q}) \qquad (5)$$

$$- [B_{l'l_1 l_2}(-\mathbf{q}',\mathbf{q}_1,\mathbf{q}_2) + \ldots]_3\, K_{(l_2 l_1|l)}(\mathbf{q}_2, \mathbf{q}_1|\mathbf{q})$$

$$+ [B_{ll_1 l_2}(\mathbf{q},\mathbf{q}_1,\mathbf{q}_2) + \ldots]_3\, \mathrm{Sp}\{\hat{b}_{\mathbf{q}_1 l_1}\hat{b}_{\mathbf{q}_2 l_2}\hat{b}_{\mathbf{q}'l'}\hat{\eta}_t\}$$

$$\left. - [B_{l'l_1 l_2}(-\mathbf{q}',\mathbf{q}_1,\mathbf{q}_2) + \ldots]_3\, \mathrm{Sp}\{\hat{b}^+_{-\mathbf{q}_1 l_1}\hat{b}^+_{-\mathbf{q}_2 l_2}\hat{b}^+_{\mathbf{q}l}\hat{\eta}_t\}\right\}.$$

The phonon density matrix $N_t(l'\mathbf{q}', l\mathbf{q})$ is defined by Eq. (19.4), and the functions $K\ldots$ are introduced as

$$K_{(l_3|l_2 l_1)}(\mathbf{q}_3|\mathbf{q}_2, \mathbf{q}_1) = \mathrm{Sp}\{\hat{b}^+_{\mathbf{q}_1 l_1}\hat{b}^+_{\mathbf{q}_2 l_2}\hat{b}_{\mathbf{q}_3 l_3}\hat{\eta}_t\},$$

$$K_{(l_3 l_2 | l_1)}(\mathbf{q}_3, \mathbf{q}_2 | \mathbf{q}_1) = \mathrm{Sp}\{\hat{b}^+_{\mathbf{q}_1 l_1} \hat{b}_{\mathbf{q}_2 l_2} \hat{b}_{\mathbf{q}_3 l_3} \hat{\eta}_t\}. \tag{6}$$

The dots in $[B_{l l_1 l_2}(\mathbf{q}, \mathbf{q}_1, \mathbf{q}_2) + \dots]_6$ denote the coefficients B obtained from the first one by five possible permutations of the indices $l\mathbf{q}$, $l_1 \mathbf{q}_1$, and $l_2 \mathbf{q}_2$, while the dots in $[B_{l l_1 l_2}(\mathbf{q}, \mathbf{q}_1, \mathbf{q}_2) + \dots]_3$ mean two possible permutations under the condition that the order of the indices $l_1 \mathbf{q}_1$ and $l_2 \mathbf{q}_2$ remains unchanged. Thus, the subindex p at the brackets $[\dots]_p$, where $p = 3$ or 6, denotes the number of the terms standing in the brackets. As seen from the definition (6.32), the anharmonic coefficients $\beta_{l_1 l_2 l_3}(\mathbf{q}_1, \mathbf{q}_2, \mathbf{q}_3)$ and, consequently, $B_{l_1 l_2 l_3}(\mathbf{q}_1, \mathbf{q}_2, \mathbf{q}_3)$ are invariant with respect to such permutations, since simultaneously one can permute the indices s_i, $\mathbf{n}_i$, and k_i under the sign of sum in Eq. (6.32). Therefore, $[B_{l_1 l_2 l_3}(\mathbf{q}_1, \mathbf{q}_2, \mathbf{q}_3) + \dots]_p = p B_{l_1 l_2 l_3}(\mathbf{q}_1, \mathbf{q}_2, \mathbf{q}_3)$.

To find the averages defined by Eq. (6), we multiply Eq. (1.20) by $\hat{b}^+_{\mathbf{q}_1 l_1} \hat{b}^+_{\mathbf{q}_2 l_2} \hat{b}_{\mathbf{q}_3 l_3}$ one time and by $\hat{b}^+_{\mathbf{q}_1 l_1} \hat{b}_{\mathbf{q}_2 l_2} \hat{b}_{\mathbf{q}_3 l_3}$ another time. Calculating the traces of the equations obtained in this way, we get

$$\left[i\hbar \frac{\partial}{\partial t} - \hbar(\omega_{\mathbf{q}_3 l_3} - \omega_{\mathbf{q}_2 l_2} - \omega_{\mathbf{q}_1 l_1}) \right] K_{(l_3 | l_2 l_1)}(\mathbf{q}_3 | \mathbf{q}_2, \mathbf{q}_1)$$

$$= \mathrm{Sp}\{[\hat{b}^+_{\mathbf{q}_1 l_1} \hat{b}^+_{\mathbf{q}_2 l_2} \hat{b}_{\mathbf{q}_3 l_3}, \widehat{H}_{ph,ph}] \hat{\eta}_t\} \tag{7}$$

and

$$\left[i\hbar \frac{\partial}{\partial t} - \hbar(\omega_{\mathbf{q}_3 l_3} + \omega_{\mathbf{q}_2 l_2} - \omega_{\mathbf{q}_1 l_1}) \right] K_{(l_3 l_2 | l_1)}(\mathbf{q}_3, \mathbf{q}_2 | \mathbf{q}_1)$$

$$= \mathrm{Sp}\{[\hat{b}^+_{\mathbf{q}_1 l_1} \hat{b}_{\mathbf{q}_2 l_2} \hat{b}_{\mathbf{q}_3 l_3}, \widehat{H}_{ph,ph}] \hat{\eta}_t\}. \tag{8}$$

Similar equations for the quantities $\mathrm{Sp}\{\hat{b}_{\mathbf{q}_1 l_1} \hat{b}_{\mathbf{q}_2 l_2} \hat{b}_{\mathbf{q}_3 l_3} \hat{\eta}_t\}$ and $\mathrm{Sp}\{\hat{b}^+_{\mathbf{q}_1 l_1} \times \hat{b}^+_{\mathbf{q}_2 l_2} \hat{b}^+_{\mathbf{q}_3 l_3} \hat{\eta}_t\}$ contain the factors $i\hbar\partial/\partial t \pm \hbar(\omega_{\mathbf{q}_3 l_3} + \omega_{\mathbf{q}_2 l_2} + \omega_{\mathbf{q}_1 l_1})$ on the left-hand sides.

The commutators on the right-hand sides of Eqs. (7) and (8) are reduced to sums of four-operator products. The next equation gives us an example of such calculations:

$$\left[\hat{b}^+_{\mathbf{q}l} \hat{b}_{\mathbf{q}'l'} \hat{b}_{\mathbf{q}''l''}, \hat{b}_{\mathbf{q}_1 l_1} \hat{b}^+_{-\mathbf{q}_2 l_2} \hat{b}^+_{-\mathbf{q}_3 l_3} \right] = \hat{b}^+_{\mathbf{q}l} \hat{b}_{\mathbf{q}'l'} \hat{b}_{\mathbf{q}_1 l_1} \left(\hat{b}^+_{-\mathbf{q}_3 l_3} \delta_{l'' l_2} \delta_{\mathbf{q}'', -\mathbf{q}_2} \right.$$

$$\left. + \hat{b}^+_{-\mathbf{q}_2 l_2} \delta_{l'' l_3} \delta_{\mathbf{q}'', -\mathbf{q}_3} \right) + \hat{b}^+_{\mathbf{q}l} \hat{b}_{\mathbf{q}_1 l_1} \left(\hat{b}^+_{-\mathbf{q}_3 l_3} \delta_{l' l_2} \delta_{\mathbf{q}', -\mathbf{q}_2} \right. \tag{9}$$

$$\left. + \hat{b}^+_{-\mathbf{q}_2 l_2} \delta_{l' l_3} \delta_{\mathbf{q}', -\mathbf{q}_3} \right) \hat{b}_{\mathbf{q}''l''} - \hat{b}^+_{-\mathbf{q}_2 l_2} \hat{b}^+_{-\mathbf{q}_3 l_3} \hat{b}_{\mathbf{q}'l'} \hat{b}_{\mathbf{q}''l''} \delta_{l l_1} \delta_{\mathbf{q}, \mathbf{q}_1}.$$

The four-operator products are averaged with the use of the following approximate equations:

$$\text{Sp}\{\hat{b}^+_{\mathbf{q}_1 l_1} \hat{b}^+_{\mathbf{q}_2 l_2} \hat{b}_{\mathbf{q}_3 l_3} \hat{b}_{\mathbf{q}_4 l_4} \hat{\eta}_t\} \simeq N_t(l_4 \mathbf{q}_4, l_1 \mathbf{q}_1) N_t(l_3 \mathbf{q}_3, l_2 \mathbf{q}_2)$$

$$+ N_t(l_4 \mathbf{q}_4, l_2 \mathbf{q}_2) N_t(l_3 \mathbf{q}_3, l_1 \mathbf{q}_1), \tag{10}$$

$$\text{Sp}\{\hat{b}_{\mathbf{q}_1 l_1} \hat{b}_{\mathbf{q}_2 l_2} \hat{b}_{\mathbf{q}_3 l_3} \hat{b}_{\mathbf{q}_4 l_4} \hat{\eta}_t\} \simeq \text{Sp}\{\hat{b}^+_{\mathbf{q}_1 l_1} \hat{b}^+_{\mathbf{q}_2 l_2} \hat{b}^+_{\mathbf{q}_3 l_3} \hat{b}^+_{\mathbf{q}_4 l_4} \hat{\eta}_t\} \simeq 0,$$

which would be exact in the absence of phonon-phonon interaction. Below we neglect the phase correlations of different modes, i.e., consider only the diagonal, with respect to the branch index l, phonon density matrices. This is justified in the case of non-degenerate phonon spectrum, when a characteristic difference in the phonon frequencies is much larger than the relaxation rate of the phonons. Let us consider first a spatially homogeneous (translation-invariant) case, when the density matrices are also diagonal with respect to the wave vector: $N_t(l'\mathbf{q}', l\mathbf{q}) = \delta_{l'l} \delta_{\mathbf{q}'\mathbf{q}} N^l_{\mathbf{q}t}$, where $N^l_{\mathbf{q}t}$ is the distribution function of phonons. The terms on the right-hand sides of Eqs. (7) and (8) become, respectively,

$$B_{l_3 l_2 l_1}(-\mathbf{q}_3, \mathbf{q}_2, \mathbf{q}_1) \left[N^{l_1}_{\mathbf{q}_1 t} N^{l_2}_{\mathbf{q}_2 t} - N^{l_1}_{\mathbf{q}_1 t} N^{l_3}_{\mathbf{q}_3 t} \right.$$

$$\left. - (1 + N^{l_2}_{\mathbf{q}_2 t}) N^{l_3}_{\mathbf{q}_3 t} \right] \equiv \hbar G^{(1)}_t \tag{11}$$

and

$$B_{l_1 l_2 l_3}(\mathbf{q}_1, -\mathbf{q}_2, -\mathbf{q}_3) \left[(1 + N^{l_2}_{\mathbf{q}_2 t}) N^{l_1}_{\mathbf{q}_3 t} + N^{l_1}_{\mathbf{q}_1 t} N^{l_3}_{\mathbf{q}_3 t} \right.$$

$$\left. - N^{l_2}_{\mathbf{q}_2 t} N^{l_3}_{\mathbf{q}_3 t} \right] \equiv \hbar G^{(2)}_t. \tag{12}$$

With the use of the definitions (11) and (12), Eqs. (7) and (8) can be rewritten as

$$\left[i \frac{\partial}{\partial t} - \Omega_i \right] K^{(i)}_t = G^{(i)}_t, \quad i = 1, 2, \tag{13}$$

where Ω_1 and Ω_2 are the shortcuts for frequency-difference terms in the round brackets in Eqs. (7) and (8), and $K^{(1)}_t$ and $K^{(2)}_t$ stand for $K_{(l_3|l_2 l_1)}(\mathbf{q}_3|\mathbf{q}_2, \mathbf{q}_1)$ and $K_{(l_3 l_2|l_1)}(\mathbf{q}_3, \mathbf{q}_2|\mathbf{q}_1)$, respectively. We already encountered and discussed such types of equations for the correlation functions in the theory of electron-impurity and electron-boson systems in Chapters 2 and 4. The solution of Eq. (13) satisfying the principle of the weakening of correlations, $K^{(i)}_{t \to -\infty} = 0$, is written according to

$$K^{(i)}_t = -i \int_{-\infty}^t dt' e^{\lambda t' - i\Omega_i(t-t')} G^{(i)}_{t'}. \tag{14}$$

The integral over t' can be calculated in the Markovian approximation, when it is assumed that the phonon distribution functions standing in $G^{(i)}_{t'}$ slowly vary on the time scale $2\pi/\overline{\omega}$, where $\overline{\omega}$ is a characteristic phonon frequency. Such integrals give both the terms proportional to $\delta(\Omega_i)$, expressing the energy conservation law, and the principal-value terms (the latter, however, disappear from the kinetic equation (5)). The solutions for $\mathrm{Sp}\{\hat{b}_{\mathbf{q}_1 l_1} \hat{b}_{\mathbf{q}_2 l_2} \hat{b}_{\mathbf{q}_3 l_3} \hat{\eta}_t\}$ and $\mathrm{Sp}\{\hat{b}^+_{\mathbf{q}_1 l_1} \hat{b}^+_{\mathbf{q}_2 l_2} \hat{b}^+_{\mathbf{q}_3 l_3} \hat{\eta}_t\}$ obtained in this way contain $\delta(\omega_{\mathbf{q}_3 l_3} + \omega_{\mathbf{q}_2 l_2} + \omega_{\mathbf{q}_1 l_1})$ and cannot satisfy the energy conservation law. Therefore, the two last terms on the right-hand side of Eq. (5) should be neglected and only the terms containing $K_t^{(i)}$ are considered there. Let us substitute the solutions of Eqs. (7) and (8) with the right-hand sides (11) and (12) into Eq. (5). Taking into account the property

$$B_{l_1 l_2 l_3}(-\mathbf{q}_1, -\mathbf{q}_2, -\mathbf{q}_3) = B^*_{l_1 l_2 l_3}(\mathbf{q}_1, \mathbf{q}_2, \mathbf{q}_3) \;, \qquad (15)$$

which is obvious from the definition of these coefficients and from Eq. (6.32), we finally obtain the kinetic equation which describes homogeneous phonon distribution:

$$\frac{\partial}{\partial t} N^l_{\mathbf{q}t} = J_{ph,ph}(N|\mathbf{q}lt). \qquad (16)$$

The collision integral in this equation comprises two terms:

$$J_{ph,ph}(N|\mathbf{q}lt) = \frac{1}{V} \sum_{l_1 l_2} \sum_{\mathbf{q}_1 \mathbf{q}_2} \frac{1}{2} \mathcal{W}_{(l|l_1 l_2)}(\mathbf{q}|\mathbf{q}_1, \mathbf{q}_2) \delta(\omega_{\mathbf{q}l} - \omega_{\mathbf{q}_1 l_1} - \omega_{\mathbf{q}_2 l_2})$$

$$\times [(N^l_{\mathbf{q}t} + 1) N^{l_1}_{\mathbf{q}_1 t} N^{l_2}_{\mathbf{q}_2 t} - N^l_{\mathbf{q}t}(N^{l_1}_{\mathbf{q}_1 t} + 1)(N^{l_2}_{\mathbf{q}_2 t} + 1)]$$

$$+ \frac{1}{V} \sum_{l_1 l_2} \sum_{\mathbf{q}_1 \mathbf{q}_2} \mathcal{W}_{(l_2|l_1 l)}(\mathbf{q}_2|\mathbf{q}_1, \mathbf{q}) \delta(\omega_{\mathbf{q}l} + \omega_{\mathbf{q}_1 l_1} - \omega_{\mathbf{q}_2 l_2}) \qquad (17)$$

$$\times [(N^l_{\mathbf{q}t} + 1)(N^{l_1}_{\mathbf{q}_1 t} + 1) N^{l_2}_{\mathbf{q}_2 t} - N^l_{\mathbf{q}t} N^{l_1}_{\mathbf{q}_1 t}(N^{l_2}_{\mathbf{q}_2 t} + 1)].$$

The scattering probabilities in Eq. (17) are expressed through the functions (problem 5.1)

$$\mathcal{W}_{(l_1|l_2 l_3)}(\mathbf{q}_1|\mathbf{q}_2, \mathbf{q}_3) = \frac{2\pi V}{\hbar^2} |B_{l_1 l_2 l_3}(\mathbf{q}_1, -\mathbf{q}_2, -\mathbf{q}_3)|^2 \propto \Delta_{\mathbf{q}_1, \mathbf{q}_2 + \mathbf{q}_3} \;, \quad (18)$$

which are proportional to the generalized Kronecker symbol introduced in Sec. 6. It is equal to 1 when $\mathbf{q}_1 = \mathbf{q}_2 + \mathbf{q}_3 + \mathbf{g}$, where $\mathbf{g}$ is either zero or one of the reciprocal lattice vectors, and equal to 0 otherwise. Thus, the kinetic equation accounts both for the normal processes describing transitions inside the Brillouin zone and for the umklapp processes with $\mathbf{g} \neq 0$. As follows from the energy conservation laws, the first term in Eq. (17) describes a decay of the phonon $\mathbf{q}l$ into two other phonons with quantum numbers $\mathbf{q}_1 l_1$ and $\mathbf{q}_2 l_2$, and the inverse process, when the phonons $\mathbf{q}_1 l_1$ and $\mathbf{q}_2 l_2$ fuse to form the phonon $\mathbf{q}l$. The second term corresponds to a fusion of the phonon $\mathbf{q}l$ with the phonon $\mathbf{q}_1 l_1$. As a result, the phonon $\mathbf{q}_2 l_2$ is created. The inverse process is a decay of the phonon $\mathbf{q}_2 l_2$ into the phonons $\mathbf{q}l$ and $\mathbf{q}_1 l_1$. The calculation has given the factor $1/2$ in the first term of the collision integral (17). This is explained by the fact that the sum there contains the contribution of two equivalent decay processes, formally differing in the permutation of $\mathbf{q}_2 l_2$ and $\mathbf{q}_1 l_1$, and they have to be considered as a single process.

If there is a smooth spatial inhomogeneity of the phonon distribution, one may introduce the Wigner distribution function $N^l_{\mathbf{r}\mathbf{q}t}$ according to Eq. (19.26). The factor $-(\omega_{\mathbf{q}'l'} - \omega_{\mathbf{q}l})N_t(l'\mathbf{q}', l\mathbf{q})$ on the left-hand side of Eq. (5) at $l = l'$ is exactly transformed to $i(\partial\omega_{\mathbf{q}l}/\partial\mathbf{q}) \cdot (\partial N^l_{\mathbf{r}\mathbf{q}t}/\partial\mathbf{r})$, while the collision integral contains operator products which are approximately replaced by corresponding products of the Wigner functions; see Appendix C. We obtain

$$\frac{\partial N^l_{\mathbf{r}\mathbf{q}t}}{\partial t} + \frac{\partial\omega_{\mathbf{q}l}}{\partial\mathbf{q}} \cdot \frac{\partial N^l_{\mathbf{r}\mathbf{q}t}}{\partial\mathbf{r}} = J_{ph,ph}(N|\mathbf{r}\mathbf{q}lt), \tag{19}$$

where the collision integral is given by Eq. (17). It depends on $\mathbf{r}$ parametrically, through the coordinate dependence of the distribution functions.

The dependence of the phonon-phonon scattering probabilities on the wave vectors cannot be expressed analytically in the general case, when the phonons are of short wavelength, i.e., when $|\mathbf{q}_i|$ are comparable to the size of the Brillouin zone. One can use the approximate expressions for the anharmonic coefficients A given in the end of Sec. 6 to find an order-of-value estimate

$$\mathcal{W} \sim \hbar\bar{s}/\overline{M}, \tag{20}$$

where $\bar{s}$ is the averaged velocity of sound. However, if the temperature of the crystal is small in comparison to the Debye temperature Θ_D defined as

$$\Theta_D \simeq \hbar\bar{s}\pi/\bar{a}, \tag{21}$$

the main contribution to kinetic phenomena often comes from the long-wavelength acoustic phonons only. If the three phonons participating in the transition are long-wavelength acoustic ones, the functions $\mathcal{W}_{(l_1|l_2l_3)}$ $(\mathbf{q}_1|\mathbf{q}_2,\mathbf{q}_3)$ can be expressed through the third-order anharmonic elastic constants, according to Eq. (6.34). Substituting the coefficients $B_{l_1 l_2 l_3}(\mathbf{q}_1,\mathbf{q}_2,\mathbf{q}_3)$ from Eq. (6.34) into Eq. (18), and taking into account a linear dispersion of long-wavelength acoustic phonon modes, one can see that the scattering probabilities in the collision integral (17) are proportional to a product of the wave vectors of three phonons participating in the transitions. These probabilities contain the usual Kronecker symbols, because the umklapp processes cannot be taken into account in the theory of elasticity for continuous media. On the other hand, owing to the conservation laws, at least two of the phonons participating in the umklapp processes should be of short wavelength. Since the number of these phonons exponentially decreases as the temperature goes down, the description of the long-wavelength acoustic phonons at $T \ll \Theta_D$ often can be done without a consideration of umklapp processes. Below we show, however, that the umklapp processes are necessary to establish thermal equilibrium in the interacting phonon system.

The phonon-phonon collisions lead to relaxation of non-equilibrium phonon distribution to quasi-equilibrium, when the solution of the stationary and homogeneous kinetic equation $J_{ph,ph}(N|\mathbf{q}l) = 0$ is $N_{\mathbf{q}}^l = [e^{\hbar\omega_{\mathbf{q}l}/T_{ph}} - 1]^{-1}$. The phonon temperature T_{ph} standing in this function is not, in general, equal to the equilibrium temperature T. Therefore, one cannot define the equilibrium temperature from a consideration of the kinetic equation with phonon-phonon collision integral. Indeed, the phonon-phonon collisions conserve the total energy of the phonon system, and the equilibrium state can be reached only due to energy exchange between the phonon system and a thermostat. We note that there exists another solution of the stationary and homogeneous kinetic equation (problem 5.2),

$$N_{\mathbf{q}}^l = c N_{\mathbf{q}}^{l(0)}(N_{\mathbf{q}}^{l(0)} + 1)\hbar\omega_{\mathbf{q}l}, \tag{22}$$

where $N_{\mathbf{q}}^{l(0)}$ is the Planck distribution with temperature T_{ph} and c is an arbitrary constant. This solution, which can be added to any other solution of the kinetic equation, is not an independent one: it is expressed through the solution mentioned above. To show this, let us consider the Planck function $N_{\mathbf{q}}^{l(0)\prime}$ with temperature T_{ph}', which is chosen under condition that $\Delta T = T_{ph}' - T_{ph}$ is small in comparison to T_{ph}. Expanding this function, we obtain $N_{\mathbf{q}}^{l(0)\prime} = N_{\mathbf{q}}^{l(0)} + (\partial N_{\mathbf{q}}^{l(0)}/\partial T_{ph})\Delta T = N_{\mathbf{q}}^{l(0)} + \Delta T(\hbar\omega_{\mathbf{q}l}/T_{ph}^2)N_{\mathbf{q}}^{l(0)}(N_{\mathbf{q}}^{l(0)} + 1)$. Now we see that the solution (22) is the difference of the solutions $N_{\mathbf{q}}^{l(0)\prime}$ and $N_{\mathbf{q}}^{l(0)}$ if one choses $c = \Delta T/T_{ph}^2$.

The ambiguity discussed above can be removed if we state the requirement that the energy density (19.30) of a stationary and homogeneous non-equilibrium phonon system should be equal to the energy density of the phonon system in thermodynamic equilibrium:

$$\mathcal{E} = \frac{1}{V}\sum_{\mathbf{q}l}\hbar\omega_{\mathbf{q}l}N_{\mathbf{q}}^l = \frac{1}{V}\sum_{\mathbf{q}l}\hbar\omega_{\mathbf{q}l}N_{l\mathbf{q}}^{(eq)}, \tag{23}$$

where $N_{l\mathbf{q}}^{(eq)} = [e^{\hbar\omega_{\mathbf{q}l}/T} - 1]^{-1}$ is the equilibrium distribution function. The normalization condition (23) allows us to avoid a consideration of the relaxation of phonon system due to interaction with a thermostat. The relation between $\mathcal{E}$ and T remains the same as in thermodynamic equilibrium. To demonstrate how the application of Eq. (23) removes the ambiguity with respect to the temperature, let us suppose that we have found a solution $N_{\mathbf{q}}^l$ of the kinetic equation. If this solution already satisfies Eq. (23), one should leave it as it is. If this solution does not satisfy Eq. (23), one should add it to the function (22) (at $T_{ph} = T$, for convenience) and determine the constant c by applying Eq. (23).

To describe non-stationary and/or non-homogeneous systems under the conditions

$$\bar{t} \gg \bar{\tau}, \quad \bar{l} \gg \bar{s}\bar{\tau} , \tag{24}$$

i.e., when the characteristic times and spatial scales of the inhomogeneities, $\bar{t}$ and $\bar{l}$, are large in comparison to the phonon-phonon scattering time $\bar{\tau}$ and phonon mean free path length $\bar{s}\bar{\tau}$, Eq. (23) is generalized as

$$\mathcal{E}_{\mathbf{r}t} = \frac{1}{V} \sum_{\mathbf{q}l} \hbar\omega_{\mathbf{q}l} N_{\mathbf{r}\mathbf{q}t}^l = \frac{1}{V} \sum_{\mathbf{q}l} \frac{\hbar\omega_{\mathbf{q}l}}{\exp(\hbar\omega_{\mathbf{q}l}/T_{\mathbf{r}t}) - 1}, \tag{25}$$

where $T_{\mathbf{r}t}$ is the local temperature. The introduction of this temperature can be justified under the conditions (24). The relation between the local quantities $\mathcal{E}_{\mathbf{r}t}$ and $T_{\mathbf{r}t}$ again remains the same as in thermodynamic equilibrium.

The normal processes alone cannot provide the relaxation of a phonon system to equilibrium, because they conserve the total quasimomentum of this system. This means that any stationary distribution function of the kind

$$N_{\mathbf{q}}^l = [e^{\hbar(\omega_{\mathbf{q}l} - \mathbf{q}\cdot\mathbf{u})/T_{ph}} - 1]^{-1}, \tag{26}$$

where $\mathbf{u}$ is an arbitrary velocity vector, satisfies the stationary and homogeneous kinetic equation $J_{ph,ph}(N|\mathbf{q}l) = 0$, where the umklapp processes in the collision integral are neglected. To show this, one may substitute the distribution function (26) into the collision integral (17) and check that each of its two terms vanishes because of the momentum conservation rule. The velocity $\mathbf{u}$ is the averaged group velocity of the phonon gas (problem 5.3), and it is also called the phonon drift velocity. On the other hand, let us define the momentum density of the phonon system according to

$$\mathbf{P}_{\mathbf{r}t} = \frac{1}{V} \sum_{\mathbf{q}l} \hbar\mathbf{q} N_{\mathbf{r}\mathbf{q}t}^l. \tag{27}$$

This momentum density is equal to zero in equilibrium. In the spatially homogeneous case, we multiply both sides of the kinetic equation (16) by $\hbar\mathbf{q}$ and sum the equation obtained over l and $\mathbf{q}$, which leads to

$$\frac{\partial \mathbf{P}_t}{\partial t} = \frac{\hbar}{V^2} \sum_{ll_1l_2} \sum_{\mathbf{q}\mathbf{q}_1\mathbf{q}_2} (\mathbf{q}/2 - \mathbf{q}_2) \mathcal{W}_{(l|l_1l_2)}(\mathbf{q}|\mathbf{q}_1,\mathbf{q}_2)\delta(\omega_{\mathbf{q}l} - \omega_{\mathbf{q}_1l_1} - \omega_{\mathbf{q}_2l_2})$$

$$\times [(N_{\mathbf{q}t}^l + 1)N_{\mathbf{q}_1t}^{l_1}N_{\mathbf{q}_2t}^{l_2} - N_{\mathbf{q}t}^l(N_{\mathbf{q}_1t}^{l_1} + 1)(N_{\mathbf{q}_2t}^{l_2} + 1)]. \tag{28}$$

To derive Eq. (28), we have permuted the indices $l\mathbf{q}$ and $l_2\mathbf{q}_2$ in the second term of the collision integral (17). As a result, this term has been united with the first one, and the factor $\mathbf{q}/2 - \mathbf{q}_2$ has appeared. If the umklapp processes are neglected, the momentum conservation term included in $\mathcal{W}_{(l|l_1l_2)}(\mathbf{q}|\mathbf{q}_1,\mathbf{q}_2)$ gives us $\mathbf{q} = \mathbf{q}_1 + \mathbf{q}_2$, and $\mathbf{q}/2 - \mathbf{q}_2 = (\mathbf{q}_1 - \mathbf{q}_2)/2$ becomes an odd function with respect to the permutation $l_1\mathbf{q}_1 \leftrightarrow l_2\mathbf{q}_2$. On the other hand, the remaining part of the expression under the sum in Eq. (28) is an even function with respect to this permutation. Therefore, the integral is equal to zero, and $\partial \mathbf{P}_t/\partial t = 0$. In conclusion, if only the normal processes are taken into account, the total momentum of the phonon system is conserved. This statement is true for an arbitrary phonon distribution function.

The scenario of phonon relaxation at low temperatures can be described as follows. First, the normal processes establish the distribution function (26). Then, the umklapp processes, which have a considerably lower probability, finally drive the system to the equilibrium state. In the next two sections we demonstrate the role of both these processes in thermal conduction and develop a hydrodynamical approach for description of inhomogeneous phonon systems.

24. Thermal Conductivity of Insulators

The linear response to a weak temperature gradient $\nabla T_{\mathbf{r}}$ created, for example, by means of contacting the crystal to two media having different temperatures, is described by a tensor $\kappa_{\alpha\beta}$ called the thermal conductivity. It is defined as

$$G_{\mathbf{r}}^\alpha = \frac{1}{V} \sum_{\mathbf{q}l} \frac{\partial \omega_{\mathbf{q}l}}{\partial q_\alpha} \hbar\omega_{\mathbf{q}l} N_{\mathbf{r}\mathbf{q}}^l = -\sum_\beta \kappa_{\alpha\beta} \frac{\partial T_{\mathbf{r}}}{\partial r_\beta}, \tag{1}$$

where $G_{\mathbf{r}}^\alpha$ is the energy flow density (19.32) in the stationary case considered in this section. The energy flows in the direction opposite to the thermal gradient, as reflected by the minus sign in Eq. (1). In this section we consider a homogeneous temperature gradient, which gives rise to a constant energy flow density $\mathbf{G_r} = \mathbf{G}$ and does not break the translational invariance of the system. To find the non-equilibrium distribution function $N_\mathbf{q}^l$ which determines $\mathbf{G}$, we linearize the kinetic equation (23.19) in the stationary case, assuming $N_\mathbf{q}^l = N_{l\mathbf{q}}^{(eq)} + \Delta N_\mathbf{q}^l$:

$$\frac{\partial N_{l\mathbf{q}}^{(eq)}}{\partial T} \frac{\partial \omega_{\mathbf{q}l}}{\partial \mathbf{q}} \cdot \frac{\partial T_\mathbf{r}}{\partial \mathbf{r}} = \Delta J_{ph,ph}(N|\mathbf{q}l). \tag{2}$$

The left-hand side of Eq. (2) is linear in the small temperature gradient, while the right-hand side contains the linearized collision integral, which is linear in the small non-equilibrium part $\Delta N_\mathbf{q}^l$ of the distribution function. It is convenient to search for $\Delta N_\mathbf{q}^l$ in the form

$$\Delta N_\mathbf{q}^l = N_{l\mathbf{q}}^{(eq)}(N_{l\mathbf{q}}^{(eq)} + 1)y_{l\mathbf{q}}. \tag{3}$$

The collision integral, expressed through the functions $y_{l\mathbf{q}} \equiv y$, is written as

$$\Delta J_{ph,ph}(N|\mathbf{q}l) = -\frac{1}{V}\sum_{l_1l_2}\sum_{\mathbf{q}_1\mathbf{q}_2}\frac{1}{2}(N_{l\mathbf{q}}^{(eq)} + 1)N_{l_1\mathbf{q}_1}^{(eq)}N_{l_2\mathbf{q}_2}^{(eq)}\mathcal{W}_{(l|l_1l_2)}(\mathbf{q}|\mathbf{q}_1,\mathbf{q}_2)$$

$$\times\delta(\omega_{\mathbf{q}l} - \omega_{\mathbf{q}_1l_1} - \omega_{\mathbf{q}_2l_2})(y - y_1 - y_2) - \frac{1}{V}\sum_{l_1l_2}\sum_{\mathbf{q}_1\mathbf{q}_2}(N_{l_2\mathbf{q}_2}^{(eq)} + 1)N_{l_1\mathbf{q}_1}^{(eq)}N_{l\mathbf{q}}^{(eq)}$$

$$\times\mathcal{W}_{(l_2|l_1l)}(\mathbf{q}_2|\mathbf{q}_1,\mathbf{q})\delta(\omega_{\mathbf{q}l} + \omega_{\mathbf{q}_1l_1} - \omega_{\mathbf{q}_2l_2})(y + y_1 - y_2), \tag{4}$$

where $y_1 = y_{l_1\mathbf{q}_1}$ and $y_2 = y_{l_2\mathbf{q}_2}$. Below we consider the collision integral as a sum of two parts,

$$\Delta J_{ph,ph}(N|\mathbf{q}l) = \Delta J_N(y) + \Delta J_U(y), \tag{5}$$

in order to take into account the normal (N) and umklapp (U) contributions separately.

The left-hand side of Eq. (2) can be rewritten as

$$-\frac{\omega_{\mathbf{q}l}}{T}\sum_\beta \frac{\partial N_{l\mathbf{q}}^{(eq)}}{\partial q_\beta}\frac{\partial T_\mathbf{r}}{\partial r_\beta}. \tag{6}$$

Therefore, if we multiply both sides of Eq. (2) by $\hbar q_\alpha$ and take a sum over $\mathbf{q}$ and l, the left-hand side is transformed to (problem 5.4)

$$-\frac{1}{V}\sum_{\beta\mathbf{q}l}\frac{\hbar\omega_{\mathbf{q}l}}{T}q_\alpha\frac{\partial N_{l\mathbf{q}}^{(eq)}}{\partial q_\beta}\frac{\partial T_\mathbf{r}}{\partial r_\beta} = -\frac{1}{V}\sum_{\beta\mathbf{q}l}q_\alpha\frac{\partial\sigma(N_{l\mathbf{q}}^{(eq)})}{\partial q_\beta}\frac{\partial T_\mathbf{r}}{\partial r_\beta}, \tag{7}$$

where

$$\sigma(N) = (N+1)\ln(N+1) - N\ln N. \tag{8}$$

Calculating the integral over $\mathbf{q}$ on the right-hand side of Eq. (7) by parts, we transform the expression (7) to the form

$$S\frac{\partial T_\mathbf{r}}{\partial r_\alpha}, \quad S = \frac{1}{V}\sum_{\mathbf{q}l}\sigma(N_{l\mathbf{q}}^{(eq)}). \tag{9}$$

The quantity S is the density of entropy of the phonon system. On the right-hand side of Eq. (2), the normal part of the collision integral vanishes after multiplication by $\hbar\mathbf{q}$ and summation over $\mathbf{q}$ and l, as described in the previous section. We obtain

$$S\frac{\partial T_\mathbf{r}}{\partial \mathbf{r}} = \frac{1}{V}\sum_{\mathbf{q}l}\hbar\mathbf{q}\Delta J_U(y). \tag{10}$$

To calculate the right-hand side of this expression, one needs to know the distribution function of phonons determined by Eq. (2). Since the left-hand side of this equation contains a small gradient, and the contribution of the umklapp processes into the linearized collision integral is also small, the distribution can be determined from the equation $\Delta J_N(y) = 0$. The solution of this equation is proportional to the drift velocity of phonons:

$$y = \hbar\mathbf{q}\cdot\mathbf{u}/T. \tag{11}$$

This solution follows from a linearization of the distribution function (23.26) at $T_{ph} = T$. The requirement $T_{ph} = T$ follows from Eq. (23.23) (problem 5.5). Substituting the solution (11) into $\Delta J_U(y)$ of Eq. (10), we find the equation

$$S\frac{\partial T_\mathbf{r}}{\partial r_\alpha} = -\sum_\beta \lambda_{\alpha\beta}u_\beta, \tag{12}$$

which expresses the drift velocity through the temperature gradient. As a result of this substitution, the symmetric tensor $\lambda_{\alpha\beta}$ is obtained in the form

$$\lambda_{\alpha\beta} = \frac{\hbar^2}{2V^2T}\sum_{ll_1l_2}\sum_{\mathbf{q}_1\mathbf{q}_2}(N_{l\mathbf{q}_1+\mathbf{q}_2+\mathbf{g}}^{(eq)}+1)N_{l_1\mathbf{q}_1}^{(eq)}N_{l_2\mathbf{q}_2}^{(eq)} \tag{13}$$

$$\times \mathcal{W}_{(l|l_1l_2)}(\mathbf{q}_1+\mathbf{q}_2+\mathbf{g}|\mathbf{q}_1,\mathbf{q}_2)\delta(\omega_{\mathbf{q}_1+\mathbf{q}_2+\mathbf{g},l}-\omega_{\mathbf{q}_1l_1}-\omega_{\mathbf{q}_2l_2})g_\alpha g_\beta,$$

where the reciprocal lattice vector $\mathbf{g}$ is chosen (depending on $\mathbf{q}_1$ and $\mathbf{q}_2$) to have $\mathbf{q}_1 + \mathbf{q}_2 + \mathbf{g}$ in the first Brillouin zone; see Eq. (23.18). One can see that the temperature gradient is proportional to the probability of umklapp processes, while the distribution function is controlled by the normal processes. This approach is valid at small temperatures. If T becomes comparable to the Debye temperature, the umklapp processes have influence on the phonon distribution function as well, and the kinetic equation cannot be solved analytically.

Let us calculate the energy flow density by using the solution for y obtained above. We have

$$G^\alpha = \frac{\hbar^2}{VT} \sum_{\beta \mathbf{q} l} \omega_{\mathbf{q} l} \frac{\partial \omega_{\mathbf{q} l}}{\partial q_\alpha} N_{l\mathbf{q}}^{(eq)}(N_{l\mathbf{q}}^{(eq)} + 1) q_\beta u_\beta. \tag{14}$$

Using the transformations similar to those employed in the transition from Eq. (7) to Eq. (9), we obtain

$$\mathbf{G} = TS\mathbf{u}. \tag{15}$$

Expressing $\mathbf{u}$ through $\nabla T_\mathbf{r}$, we substitute it to Eq. (15) and find the thermal conductivity tensor introduced by Eq. (1). It is proportional to the inverted tensor $\lambda_{\alpha\beta}$:

$$\kappa_{\alpha\beta} = TS^2 (\lambda^{-1})_{\alpha\beta}. \tag{16}$$

Let us find an estimate for the thermal conductivity, describing the phonon system by a model of three isotropic acoustic-phonon branches with velocities s_l (single LA branch) and s_t (two TA branches). Using the definition (9), one may calculate the entropy at low temperatures ($T \ll \Theta_D$):

$$S \simeq \frac{2\pi^2 T^3}{15\hbar^3} \left(\frac{1}{3s_l^3} + \frac{2}{3s_t^3} \right). \tag{17}$$

The estimates given below are done under a simplifying assumption $s_l \simeq s_t \simeq \bar{s}$ so that the factor in the brackets of Eq. (17) is written as $\bar{s}^{-3}$. It is convenient to introduce the umklapp scattering time τ_U according to the definition

$$\frac{1}{\tau_U} = \lambda_{\alpha\alpha} \left[\frac{\hbar^2}{VT} \sum_{\mathbf{q} l} q^2 N_{l\mathbf{q}}^{(eq)}(N_{l\mathbf{q}}^{(eq)} + 1) \right]^{-1}. \tag{18}$$

Calculating the sum in this equation, we obtain an order-of-value estimate $\tau_U^{-1} \sim \hbar^3 \bar{s}^5 \lambda_{\alpha\alpha}/T^4$ (problem 5.6) valid at low temperatures. As a result,

$$\kappa \sim \frac{T^3 \tau_U}{\hbar^3 \bar{s}}. \tag{19}$$

To estimate τ_U, one can calculate λ for a model of one-dimensional lattice (problem 5.7) and multiply it by $(\pi/\bar{a})^4$ to account for the change in the momentum-space dimensionality (note that the sum in Eq. (13) is taken over two wave vectors). In this way we obtain

$$\frac{1}{\tau_U} \sim \frac{\overline{\mathcal{W}}}{\hbar \bar{s} \bar{a}^2} \left(\frac{\Theta_D}{T}\right)^5 \exp\left(-\frac{\Theta_D}{T}\right). \tag{20}$$

The thermal conductivity exponentially increases with decreasing temperature.

Let us consider the case of high temperatures. The kinetic equation (2) cannot be solved analytically at $T \sim \Theta_D$. However, one can roughly approximate the linearized collision integral by the expression $-N_{l\mathbf{q}}^{(eq)}(N_{l\mathbf{q}}^{(eq)}+1)y/\tau$, where τ is the scattering time estimated according to

$$\frac{1}{\tau} = \frac{1}{V} \sum_{l_1 l_2} \sum_{\mathbf{q}_1 \mathbf{q}_2} \frac{N_{l_1\mathbf{q}_1}^{(eq)} N_{l_2\mathbf{q}_2}^{(eq)}}{2N_{l\mathbf{q}}^{(eq)}} \mathcal{W}_{(l|l_1 l_2)}(\mathbf{q}|\mathbf{q}_1, \mathbf{q}_2) \delta(\omega_{\mathbf{q}l} - \omega_{\mathbf{q}_1 l_1} - \omega_{\mathbf{q}_2 l_2}) \tag{21}$$

$$+\frac{1}{V} \sum_{l_1 l_2} \sum_{\mathbf{q}_1 \mathbf{q}_2} \frac{(N_{l_2\mathbf{q}_2}^{(eq)}+1)N_{l_1\mathbf{q}_1}^{(eq)}}{N_{l\mathbf{q}}^{(eq)}+1} \mathcal{W}_{(l_2|l_1 l)}(\mathbf{q}_2|\mathbf{q}_1, \mathbf{q}) \delta(\omega_{\mathbf{q}l} + \omega_{\mathbf{q}_1 l_1} - \omega_{\mathbf{q}_2 l_2}).$$

Then we have

$$y \sim \tau \frac{\hbar \omega_{\mathbf{q}l}}{T^2} \frac{\partial \omega_{\mathbf{q}l}}{\partial \mathbf{q}} \cdot \frac{\partial T}{\partial \mathbf{r}}. \tag{22}$$

Finally, estimating $\partial \omega_{\mathbf{q}l}/\partial \mathbf{q}$ as $\bar{s}$, we obtain

$$\kappa \sim \tau \frac{\bar{s}^2}{VT^2} \sum_{\mathbf{q}l} (\hbar \omega_{\mathbf{q}l})^2 N_{l\mathbf{q}}^{(eq)}(N_{l\mathbf{q}}^{(eq)}+1). \tag{23}$$

At high temperatures $T \gg \Theta_D$, one has $N_{l\mathbf{q}}^{(eq)} \simeq T/\hbar \omega_{\mathbf{q}l} \gg 1$. Therefore, according to Eq. (21), $\tau^{-1} \propto T$. Using these relations in Eq. (23), we find the estimate

$$\kappa \sim \frac{\tau \bar{s}^2}{\bar{a}^3} \propto T^{-1}, \tag{24}$$

which determines high-temperature behavior of the thermal conductivity.

25. Balance Equations for Phonons

The strong difference between the rates of normal and umklapp processes makes it possible to develop a hydrodynamical theory of phonons.

This theory operates with balance equations for macroscopic quantities and bears a similarity with the theory discussed in Sec. 11 for electrons. Consider a smoothly inhomogeneous phonon system described by the kinetic equation (23.19). After multiplying both sides of this equation by $\hbar\omega_{\mathbf{q}l}$, we sum it over the quantum numbers $\mathbf{q}$ and l and find that the collision-integral contribution is zero. This is not surprising, since the phonon-phonon interaction conserves the energy of the phonon system. We obtain an exact equation connecting the energy density $\mathcal{E}_{\mathbf{r}t}$ with the energy flow density $\mathbf{G}_{\mathbf{r}t}$ defined by Eqs. (23.25) and (24.1), respectively, as

$$\frac{\partial \mathcal{E}_{\mathbf{r}t}}{\partial t} + \frac{\partial \mathbf{G}_{\mathbf{r}t}}{\partial \mathbf{r}} = 0. \tag{1}$$

This is a continuity equation for the energy flow. In contrast to Eq. (19.31) for the bosons interacting with electrons, Eq. (1) does not contain any generation and relaxation terms. Let us multiply both sides of Eq. (23.19) by $\hbar\mathbf{q}$ and sum the equation obtained over $\mathbf{q}$ and l. We obtain the momentum balance equation

$$\frac{\partial P_{\mathbf{r}t}^{\alpha}}{\partial t} + \sum_{\beta} \frac{\partial G_{\mathbf{r}t}^{\alpha\beta}}{\partial r_{\beta}} = \frac{1}{V} \sum_{\mathbf{q}l} \hbar q_{\alpha} \Delta J_U(N|\mathbf{r}\mathbf{q}lt) \equiv J_{\mathbf{r}t}^{\alpha}. \tag{2}$$

The local momentum density $\mathbf{P}_{\mathbf{r}t}$ is introduced by Eq. (23.27), while $G_{\mathbf{r}t}^{\alpha\beta}$ is the tensor of the momentum flow density:

$$G_{\mathbf{r}t}^{\alpha\beta} = \frac{1}{V} \sum_{\mathbf{q}l} \hbar q_{\alpha} \frac{\partial \omega_{\mathbf{q}l}}{\partial q_{\beta}} \Delta N_{\mathbf{r}\mathbf{q}t}^{l}. \tag{3}$$

The right-hand side of Eq. (2) contains only the contribution of the umklapp part of the collision integral, since the momentum is conserved in the normal collision processes; see Eq. (23.28) and its discussion. Multiplying the kinetic equation (23.19) by $\hbar\omega_{\mathbf{q}l}(\partial\omega_{\mathbf{q}l}/\partial q_{\beta})$ and $\hbar q_{\alpha}(\partial\omega_{\mathbf{q}l}/\partial q_{\beta})$, one can obtain the balance equations expressing $\mathbf{G}_{\mathbf{r}t}$ and $G_{\mathbf{r}t}^{\alpha\beta}$ through the higher-order moments, and so on. Acting in this way, one gets an infinite chain of equations describing the phonon system (compare to the results of Sec. 11 for electrons).

Below we restrict ourselves by the balance equations (1) and (2). In order to solve the kinetic equation (23.19) by iterations, we assume that the inhomogeneities are smooth according to the conditions (23.24) and that the temperature is low enough to have a strong difference between the rates of normal and umklapp scattering processes. This allows us to express $\mathcal{E}_{\mathbf{r}t}$, $\mathbf{P}_{\mathbf{r}t}$, $\mathbf{G}_{\mathbf{r}t}$, and $G_{\mathbf{r}t}^{\alpha\beta}$ through the time- and coordinate-dependent local temperature $T_{\mathbf{r}t}$ and drift velocity $\mathbf{u}_{\mathbf{r}t}$ of the phonon system. First of all, we note that, according to Eq. (23.25), the energy density depends on time and coordinate only through the local temperature. Therefore, the first term in Eq. (1) in the linear approximation is written as $C(\partial T_{\mathbf{r}t}/\partial t)$, where

$$C = \frac{1}{V} \sum_{\mathbf{q}l} \frac{(\hbar\omega_{\mathbf{q}l})^2}{T^2} N_{l\mathbf{q}}^{(eq)}(N_{l\mathbf{q}}^{(eq)} + 1) = T\frac{\partial S}{\partial T} \tag{4}$$

is the specific heat. If we estimate the entropy density S according to Eq. (24.17), we have simply $C = 3S$. To determine the other terms, we use the procedure of iterations. In the first-order approximation, using the substitution (24.3) with $y_{l\mathbf{q}} = y^{(1)}$, we solve the linearized equation

$$\Delta J_N(N|\mathbf{r}\mathbf{q}lt) \equiv \Delta J_N(y^{(1)}) = 0. \tag{5}$$

The solution of this equation,

$$y^{(1)} = \hbar\omega_{\mathbf{q}l}\Delta T_{\mathbf{r}t}/T^2 + \hbar\mathbf{q}\cdot\mathbf{u}_{\mathbf{r}t}/T, \tag{6}$$

is a small, non-equilibrium part of the distribution function (23.26), obtained by its expansion in power series of small parameters $\Delta T_{\mathbf{r}t} = T_{ph} - T$ and $\mathbf{u} = \mathbf{u}_{\mathbf{r}t}$. Substituting $\Delta N_{\mathbf{r}\mathbf{q}t}^{l(1)} = N_{l\mathbf{q}}^{(eq)}(N_{l\mathbf{q}}^{(eq)} + 1)y^{(1)}$ into the definitions of the momentum density, energy flow density, and momentum flow density (Eqs. (23.27), (19.32), and (3), respectively), we obtain

$$P_{\mathbf{r}t}^{\alpha(1)} = \sum_\beta \chi_{\alpha\beta} u_{\mathbf{r}t}^\beta, \quad \chi_{\alpha\beta} = \frac{\hbar^2}{VT} \sum_{\mathbf{q}l} q_\alpha q_\beta N_{l\mathbf{q}}^{(eq)}(N_{l\mathbf{q}}^{(eq)} + 1) \tag{7}$$

and

$$G_{\mathbf{r}t}^{\alpha(1)} = STu_{\mathbf{r}t}^\alpha, \quad G_{\mathbf{r}t}^{\alpha\beta(1)} = \delta_{\alpha\beta}S\Delta T_{\mathbf{r}t}, \quad J_{\mathbf{r}t}^\alpha = -\sum_\beta \lambda_{\alpha\beta} u_{\mathbf{r}t}^\beta. \tag{8}$$

The collision-integral contribution $J_{\mathbf{r}t}^\alpha$ is written through the tensor $\lambda_{\alpha\beta}$ as in Eq. (24.12). One may express $\lambda_{\alpha\beta}$ through $(\kappa^{-1})_{\alpha\beta}$ according to Eq. (24.16). Substituting the results (7) and (8) into the balance equations (1) and (2), one finds the equations for $\Delta T_{\mathbf{r}t}$ and $\mathbf{u}_{\mathbf{r}t}$:

$$C\frac{\partial\Delta T}{\partial t} + TS\,\mathrm{div}\mathbf{u} = 0, \tag{9}$$

$$\sum_\beta \chi_{\alpha\beta}\frac{\partial u^\beta}{\partial t} + S\frac{\partial\Delta T}{\partial r_\alpha} = -TS^2 \sum_\beta (\kappa^{-1})_{\alpha\beta}u^\beta. \tag{10}$$

The appearance of the gradients of drift velocity and temperature gives rise to a modification of the phonon distribution function in the presence of normal collision processes. To find this modification, one has to consider the equation for the second-order correction to y:

$$\left(\frac{\partial}{\partial t} + \frac{\partial \omega_{\mathbf{q}l}}{\partial \mathbf{q}} \cdot \frac{\partial}{\partial \mathbf{r}}\right) y^{(1)} = \frac{1}{N_{l\mathbf{q}}^{(eq)}(N_{l\mathbf{q}}^{(eq)} + 1)} \Delta J_N(y^{(2)}). \qquad (11)$$

Substituting the solution (6) into the left-hand side of Eq. (11), we rewrite this side as

$$\frac{\hbar \omega_{\mathbf{q}l}}{T^2} \frac{\partial \Delta T_{\mathbf{r}t}}{\partial t} + \sum_\alpha \frac{\hbar q_\alpha}{T} \frac{\partial u_{\mathbf{r}t}^\alpha}{\partial t} + \sum_\alpha \frac{\partial \omega_{\mathbf{q}l}}{\partial q_\alpha} \left(\frac{\hbar \omega_{\mathbf{q}l}}{T^2} \frac{\partial \Delta T_{\mathbf{r}t}}{\partial r_\alpha} + \sum_\beta \frac{\hbar q_\beta}{T} \frac{\partial u_{\mathbf{r}t}^\beta}{\partial r_\alpha}\right)$$

$$= \sum_\alpha A_{\mathbf{q}l}^\alpha \frac{\partial \Delta T_{\mathbf{r}t}}{\partial r_\alpha} + \sum_{\alpha\beta} B_{\mathbf{q}l}^{\beta\alpha} \frac{\partial u_{\mathbf{r}t}^\beta}{\partial r_\alpha}, \qquad (12)$$

where

$$A_{\mathbf{q}l}^\alpha = \frac{\hbar \omega_{\mathbf{q}l}}{T^2} \frac{\partial \omega_{\mathbf{q}l}}{\partial q_\alpha} - \sum_\beta (\chi^{-1})_{\alpha\beta} \frac{\hbar q_\beta}{T} S,$$

$$B_{\mathbf{q}l}^{\beta\alpha} = \frac{\hbar q_\beta}{T} \frac{\partial \omega_{\mathbf{q}l}}{\partial q_\alpha} - \delta_{\alpha\beta} \frac{\hbar \omega_{\mathbf{q}l}}{CT} S. \qquad (13)$$

In the transformation of the temporal derivatives to the spatial ones in Eq. (12), we have used Eq. (9) and Eq. (10) without the right-hand side caused by the umklapp processes. The left-hand side of the kinetic equation (11) is split in two parts for the obvious reason: the vector $A_{\mathbf{q}l}^\alpha$ is an antisymmetric function of $\mathbf{q}$, while the tensor $B_{\mathbf{q}l}^{\alpha\beta}$ is symmetric in $\mathbf{q}$. Accordingly, let us search for a solution of the kinetic equation (11) in the form

$$y^{(2)} = \sum_\alpha a_{\mathbf{q}l}^\alpha \frac{\partial \Delta T_{\mathbf{r}t}}{\partial r_\alpha} + \sum_{\alpha\beta} b_{\mathbf{q}l}^{\beta\alpha} \frac{\partial u_{\mathbf{r}t}^\beta}{\partial r_\alpha}. \qquad (14)$$

Since the collision integral conserves the symmetry with respect to $\mathbf{q}$, we obtain two separate equations for the coefficients $a_{\mathbf{q}l}^\alpha$ and $b_{\mathbf{q}l}^{\beta\alpha}$. The first of them is written as

$$A_{\mathbf{q}l}^\alpha = -\frac{1}{V} \sum_{l_1 l_2} \sum_{\mathbf{q}_1 \mathbf{q}_2} \left\{ \frac{1}{2} \frac{N_{l_1 \mathbf{q}_1}^{(eq)} N_{l_2 \mathbf{q}_2}^{(eq)}}{N_{l\mathbf{q}}^{(eq)}} \mathcal{W}_{(l|l_1 l_2)}(\mathbf{q}|\mathbf{q}_1, \mathbf{q}_2) \right.$$

$$\times \delta(\omega_{\mathbf{q}l} - \omega_{\mathbf{q}_1 l_1} - \omega_{\mathbf{q}_2 l_2})(a_{\mathbf{q}l}^\alpha - a_{\mathbf{q}_1 l_1}^\alpha - a_{\mathbf{q}_2 l_2}^\alpha) + \frac{(N_{l_2 \mathbf{q}_2}^{(eq)} + 1) N_{l_1 \mathbf{q}_1}^{(eq)}}{N_{l\mathbf{q}}^{(eq)} + 1}$$

$$\times \mathcal{W}_{(l_2|l_1l)}(\mathbf{q}_2|\mathbf{q}_1,\mathbf{q})\delta(\omega_{\mathbf{q}l}+\omega_{\mathbf{q}_1l_1}-\omega_{\mathbf{q}_2l_2})(a^{\alpha}_{\mathbf{q}l}+a^{\alpha}_{\mathbf{q}_1l_1}-a^{\alpha}_{\mathbf{q}_2l_2})\Big\}, \quad (15)$$

and the second one is analogical: one should merely substitute $B^{\beta\alpha}$ in place of A^{α} and $b^{\beta\alpha}$ in place of a^{α}.

Having the general form of the correction $y^{(2)}$, we can express the corresponding corrections to the macroscopic quantities:

$$P^{\alpha(2)}_{\mathbf{r}t}=\sum_{\beta}\pi_{\alpha\beta}\frac{\partial\Delta T_{\mathbf{r}t}}{\partial r_{\beta}}, \quad \pi_{\alpha\beta}=\frac{\hbar}{V}\sum_{\mathbf{q}l}q_{\alpha}a^{\beta}_{\mathbf{q}l}N^{(eq)}_{l\mathbf{q}}(N^{(eq)}_{l\mathbf{q}}+1), \quad (16)$$

$$G^{\alpha(2)}_{\mathbf{r}t}=-\sum_{\beta}\mu_{\alpha\beta}\frac{\partial\Delta T_{\mathbf{r}t}}{\partial r_{\beta}}, \quad \mu_{\alpha\beta}=-\frac{\hbar}{V}\sum_{\mathbf{q}l}\omega_{\mathbf{q}l}\frac{\partial\omega_{\mathbf{q}l}}{\partial q_{\alpha}}a^{\beta}_{\mathbf{q}l}N^{(eq)}_{l\mathbf{q}}(N^{(eq)}_{l\mathbf{q}}+1),$$
$$(17)$$

$$G^{\alpha\beta(2)}_{\mathbf{r}t}=\sum_{\gamma\delta}\gamma_{\alpha\beta\gamma\delta}\frac{\partial^2V^{\gamma}_{\mathbf{r}t}}{\partial r_{\beta}\partial r_{\delta}}, \quad \gamma_{\alpha\beta\gamma\delta}=\frac{\hbar}{V}\sum_{\mathbf{q}l}q_{\alpha}\frac{\partial\omega_{\mathbf{q}l}}{\partial q_{\beta}}b^{\gamma\delta}_{\mathbf{q}l}N^{(eq)}_{l\mathbf{q}}(N^{(eq)}_{l\mathbf{q}}+1).$$
$$(18)$$

These gradient-containing terms should be added to $P^{\alpha(1)}$, $G^{\alpha(1)}$, and $G^{\alpha\beta(1)}$ given by Eqs. (7) and (8). Substituting the improved expressions for the momentum, energy flow, and momentum flow densities into the balance equations, we finally obtain

$$C\frac{\partial\Delta T}{\partial t}+TS\sum_{\alpha}\frac{\partial u^{\alpha}}{\partial r_{\alpha}}-\sum_{\alpha\beta}\mu_{\alpha\beta}\frac{\partial^2\Delta T}{\partial r_{\alpha}\partial r_{\beta}}=0, \quad (19)$$

$$\sum_{\beta}\chi_{\alpha\beta}\frac{\partial u^{\beta}}{\partial t}+S\frac{\partial\Delta T}{\partial r_{\alpha}}-\sum_{\beta\gamma\delta}\nu_{\alpha\beta\gamma\delta}\frac{\partial^2u^{\gamma}}{\partial r_{\beta}\partial r_{\delta}}+TS^2\sum_{\beta}(\kappa^{-1})_{\alpha\beta}u^{\beta}=0, \quad (20)$$

where $\nu_{\alpha\beta\gamma\delta}=-\gamma_{\alpha\beta\gamma\delta}+\pi_{\alpha\beta}\delta_{\gamma\delta}TS/C$. Equations (19) and (20) are called the equations of dissipative phonon hydrodynamics. They are valid when the characteristic spatial (temporal) scale of the inhomogeneity is much larger than the phonon mean free path length (time) with respect to normal scattering processes. The dissipation effects are determined by three tensors: the thermal conductivity tensor $\kappa_{\alpha\beta}$, the phonon-hydrodynamical viscosity tensor $\nu_{\alpha\beta\gamma\delta}$, and the phonon-hydrodynamical correction to the thermal conductivity, $\mu_{\alpha\beta}$ (the meaning of this correction is justified by its definition (17)). Another tensor entering the theory is $\chi_{\alpha\beta}$, which connects the averaged momentum to the drift velocity and has the dimensionality of a mass. All tensors of the second rank are positively defined and symmetric, while the fourth-rank tensor ν is symmetric with respect to the permutation of a pair of indices:

$\nu_{\alpha\beta\gamma\delta} = \nu_{\gamma\delta\alpha\beta}$. Let us estimate χ, μ, and ν (the tensor κ has been estimated in the previous section). A direct calculation according to Eq. (7) gives us

$$\chi_{\alpha\alpha} \simeq \frac{TS}{\bar{s}^2} \sim \frac{T^4}{\hbar^3 \bar{s}^5}. \tag{21}$$

To find the dissipative tensors, one may estimate a and b from the kinetic equations as $a_{\mathbf{q}l} \sim \tau_N A_{\mathbf{q}l}$ and $b_{\mathbf{q}l} \sim \tau_N B_{\mathbf{q}l}$, where τ_N is the scattering time with respect to normal processes. Knowing S, C, and χ, we can estimate the coefficients $A_{\mathbf{q}l}$ and $B_{\mathbf{q}l}$ according to Eq. (13). If $|\partial\omega_{\mathbf{q}l}/\partial\mathbf{q}| \sim \bar{s}$, we have $A_{\mathbf{q}l} \sim \hbar q\bar{s}^2/T^2$ and $B_{\mathbf{q}l} \sim \hbar q\bar{s}/T$. Combining these results, we finally get

$$\mu \sim \frac{\tau_N T^3}{\hbar^3 \bar{s}} \tag{22}$$

and

$$\nu \sim \frac{\tau_N T^4}{\hbar^3 \bar{s}^3}. \tag{23}$$

Note that the ratio μ/κ is estimated as τ_N/τ_U.

If the dissipation terms are neglected, Eq. (19) is reduced to Eq. (9) while Eq. (20) is reduced to Eq. (10) with zero right-hand side. Excluding the drift velocity from this system of equations, we obtain a single differential equation

$$\frac{\partial^2 \Delta T}{\partial t^2} - \frac{TS^2}{C} \sum_{\alpha\beta} (\chi^{-1})_{\alpha\beta} \frac{\partial^2 \Delta T}{\partial r_\alpha \partial r_\beta} = 0. \tag{24}$$

Equation (24) describes the waves of temperature, $\Delta T \propto e^{i(\mathbf{q}\cdot\mathbf{r} - \omega t)}$, propagating in the medium. The dispersion law for these waves,

$$\omega^2 = \frac{TS^2}{C}(\chi^{-1})_{\alpha\beta} q_\alpha q_\beta, \tag{25}$$

is similar to that for sound waves (acoustic phonons) in anisotropic media. Using Eqs. (21) and (24.17), one can find that the absolute values of the proportionality coefficients $(TS^2/C)(\chi^{-1})_{\alpha\beta}$ are close to the square of the averaged sound velocity, $\bar{s}^2$. For this reason, these temperature waves are also known as the second sound. The dissipation processes lead to attenuation of the second sound. To investigate it, let us substitute the wave solutions $\Delta T, \mathbf{u} \propto e^{i(\mathbf{q}\cdot\mathbf{r} - \omega t)}$ into Eqs. (19) and (20). The solvability condition of these equations is determined by the following dispersion relation:

$$\det ||(-i\omega C + \mu_{\gamma\delta} q_\gamma q_\delta)[-i\omega\chi_{\alpha\beta} + \nu_{\alpha\gamma\beta\delta} q_\gamma q_\delta$$

$$+TS^2(\kappa^{-1})_{\alpha\beta}] + TS^2 q_\alpha q_\beta|| = 0. \tag{26}$$

If the wave is propagating along one of the symmetry axes in cubic crystals, the tensors are reduced to scalars χ, κ, μ, and ν. In this case, Eq. (26) is transformed to

$$(v_{II}q)^2 = \omega^2 \left(1 + \frac{i\nu q^2}{\chi\omega} + \frac{iTS^2}{\kappa\chi\omega}\right)\left(1 + \frac{i\mu q^2}{C\omega}\right), \tag{27}$$

where the velocity of second sound is introduced according to $v_{II} = S\sqrt{T/\chi C}$. The attenuation of the intensity of second sound is determined by the quantity $\Gamma_{II} = 2\mathrm{Im}\,q$, which is found below under the approximation that this attenuation is weak, $\mathrm{Im}\,q \ll \mathrm{Re}\,q$:

$$\Gamma_{II} = \omega^2 \left(\frac{\mu}{Cv_{II}^3} + \frac{\nu}{\chi v_{II}^3}\right) + \frac{TS^2}{\kappa\chi v_{II}}. \tag{28}$$

The contribution of umklapp processes in Eq. (28) leads to a frequency-independent attenuation which is important at small ω. On the other hand, the contribution of normal processes increases with the increase of ω and suppresses the thermal waves at high frequencies. The favorable region, where the second sound is weakly damped, corresponds to intermediate frequencies. Using the above estimates for χ, μ, ν, and κ, we find an order-of-value estimate

$$\frac{\Gamma_{II}}{\mathrm{Re}\,q} \sim \omega\tau_N + \frac{1}{\omega\tau_U}, \tag{29}$$

which means that the second sound can exist at $1/\tau_U \ll \omega \ll 1/\tau_N$. This is a wide region, since at $T \ll \Theta_D$ one has $\tau_U \gg \tau_N$.

26. Relaxation of Long-Wavelength Phonons

The phonon-phonon interaction is responsible for relaxation of non-equilibrium acoustic and optical phonons generated by hot electrons in semiconductors. These phonons are always of long wavelength, since their characteristic wave numbers are of the order of $\sqrt{m\bar{\varepsilon}}/\hbar$, where $\bar{\varepsilon}$ is the mean energy of electrons and m is the effective mass of electrons. When electrons are excited by an intense laser radiation or accelerated by a strong electric field, see Chapter 7, the main channel of their energy exchange with the crystal lattice is known to be emission of longitudinal optical (LO) phonons. On the other hand, when electrons are heated by a moderate electric field at low temperatures, the electrons generate mostly the low-energy acoustic phonons. The consideration of the relaxation of long-wavelength phonons is instructive in the sense that the restrictions in the conservation laws caused by smallness of the momenta of relaxing phonons can simplify the kinetic equation even for the cases of strongly non-equilibrium phonons and allow one to solve it under physically reasonable assumptions.

The relaxation of non-equilibrium long-wavelength optical phonons occurs through their decay into a pair of longitudinal acoustic phonons: $LO \to LA_1 + LA_2$. The process when a LO phonon fuses with another phonon is forbidden by the conservation laws. Since the characteristic time of the decay is typically larger than the time of LO phonon emission, the decay is known as the bottleneck for relaxation of the electron-phonon system. The LA phonons produced are themselves nearly monoenergetic, with frequencies close to $\omega_{LO}/2$. Their wave vectors $\mathbf{q}_1$ and $\mathbf{q}_2$ are in the middle of the Brillouin zone and can be estimated by the absolute value as $\omega_{LO}/2\bar{s}$. The distribution of these LA phonons is controlled by their interaction with other acoustic phonons belonging to both longitudinal and transverse branches.

Consider a spontaneous emission of LO phonons by non-equilibrium electrons whose energy spectrum and distribution are isotropic, like in the case of photoexcited electrons. The generation rate I_q, see Sec. 21, is also isotropic in this case. Let us model this generation rate by a function which is equal to a constant I in the interval of wave numbers $q < q_0$ and zero elsewhere. The wave number q_0 can be roughly estimated as $q_0 \simeq \sqrt{m\omega_{LO}/\hbar}$. Thus, we have the kinetic equation for LO phonons,

$$\frac{\partial N_{\mathbf{q}}^{LO}}{\partial t} = I_q + J_{ph,ph}(N|\mathbf{q}\ LO\ t), \quad I_q = I\theta(q_0 - q), \tag{1}$$

where the collision integral is given by Eq. (23.17). Considering, as described above, only the first part of this collision integral, corresponding to $LO \leftrightarrow LA_1 + LA_2$, we obtain, in the stationary case,

$$I_q = \frac{1}{2V} \sum_{\mathbf{q}_1\mathbf{q}_2} \mathcal{W}_{(LO|LA,LA)}(\mathbf{q}|\mathbf{q}_1, \mathbf{q}_2)\delta(\omega_{\mathbf{q}LO} - \omega_{\mathbf{q}_1 LA} - \omega_{\mathbf{q}_2 LA})$$

$$\times [N_{\mathbf{q}}^{LO}(1 + N_{\mathbf{q}_1}^{LA} + N_{\mathbf{q}_2}^{LA}) - N_{\mathbf{q}_1}^{LA}N_{\mathbf{q}_2}^{LA}]. \tag{2}$$

Note that we have rewritten the factor containing the distribution functions in a more simple form, taking into account that triple products of the distribution functions drop out of this factor. The second term in the square brackets of Eq. (2) describes the fusion process $LA_1 + LA_2 \to LO$ and can be neglected if the phonon system is considerably out of equilibrium. Indeed, the volume in the reciprocal space occupied by the LO phonons is much less than that of LA phonons. The ratio of these volumes is of the order of $(q_0/q_1)^2 \sim m\bar{a}^2\omega_{LO}/\pi^2\hbar$, and this small parameter provides $N^{LA} \ll N^{LO}$. As a result, the kinetic equation (2) is reduced to

$$I_q = \frac{N_q^{LO}}{\tau_q}, \qquad \frac{1}{\tau_q} = \frac{1}{2V} \sum_{\mathbf{q}_1 \mathbf{q}_2} \mathcal{W}_{(LO|LA,LA)}(\mathbf{q}|\mathbf{q}_1, \mathbf{q}_2)$$

$$\times \delta(\omega_{\mathbf{q}LO} - \omega_{\mathbf{q}_1 LA} - \omega_{\mathbf{q}_2 LA})(1 + N_{\mathbf{q}_1}^{LA} + N_{\mathbf{q}_2}^{LA}), \tag{3}$$

where τ_q is the decay time of LO phonons. Below we use an isotropic approximation both for the acoustic phonon spectrum, $\omega_{\mathbf{q}LA} = sq$, and for the scattering probability. The latter approximation means that $\mathcal{W}_{(LO|LA,LA)}(\mathbf{q}|\mathbf{q}_1, \mathbf{q}_2) \simeq \mathcal{W}_{(LO|LA,LA)}(0|\mathbf{n}k_0, -\mathbf{n}k_0)$, where $\mathbf{n}$ is the unit vector in the direction of $\mathbf{q}_1$ and $k_0 = \omega_{LO}/2s$, is independent of the angle of $\mathbf{q}_1$ and constant. Therefore, the decay rate is expressed as

$$\frac{1}{\tau_q} = \Gamma \int_0^\infty q_1^2 dq_1 \int_0^\pi \sin\theta d\theta (1 + N_{q_1}^{LA} + N_{q_2}^{LA}) \delta[\omega_{LO} - s(q_1 + q_2)], \tag{4}$$

where $\Gamma = \mathcal{W}_{(LO|LA,LA)}(0|\mathbf{n}k_0, -\mathbf{n}k_0)/8\pi^2$ is constant and q_2 is fixed by the momentum conservation rule: $q_2 = \sqrt{q_1^2 + q^2 - 2q_1 q \cos\theta}$ with $\theta = \widehat{\mathbf{q}\mathbf{q}_1}$. Integrating over the angle in Eq. (4) under the condition $q \ll k_0$, we obtain

$$\frac{1}{\tau_q} = \frac{1}{\tau_{sp}}\left(1 + \frac{1}{q}\int_{k_0-q/2}^{k_0+q/2} dk[N_k^{LA} + N_{2k_0-k}^{LA}]\right), \tag{5}$$

where

$$\tau_{sp}^{-1} = \Gamma k_0^2/s \tag{6}$$

is the spontaneous LO phonon decay time.

Though the occupation number N_k^{LA} of acoustic phonons is much smaller than N_q^{LO}, it still can be large in comparison to unity, because the phase space for the generated LA phonons is limited by a narrow interval $[k_0 - q/2, k_0 + q/2]$. Therefore, a consideration of stimulated LO phonon decay processes described by the integral term in Eq. (5) is important. To determine the LA phonon distribution, one has to solve the kinetic equation which takes into account both the generation of LA phonons by LO phonon decay and the decay of LA phonons into acoustic phonons of larger wavelengths. The latter term will be written below through the acoustic phonon decay time τ_a, which is assumed to be independent of N_k^{LA} since the phase space for LA phonon decay is large. The stationary kinetic equation $J_{ph,ph}(N|\mathbf{k}\,LA) = 0$ is, therefore, written as $J_{ph,ph}^{(LO)}(N|\mathbf{k}\,LA) - N_k^{LA}/\tau_a = 0$, where $J_{ph,ph}^{(LO)}(N|\mathbf{k}\,LA)$ is the contribution of the processes involving LO phonons. As already discussed, the main contribution to such processes is the generation of LA phonons by LO phonon decay. Using Eq. (23.17), where now only the second part of the collision integral is retained and only the generation (arrival) term of this part is taken into account, we obtain

$$\frac{N_k^{LA}}{\tau_a} = \frac{1}{V} \sum_{\mathbf{q}\mathbf{k}'} \mathcal{W}_{(LO|LA,LA)}(\mathbf{q}|\mathbf{k}',\mathbf{k})\delta(\omega_{\mathbf{k}LA} + \omega_{\mathbf{k}'LA} - \omega_{\mathbf{q}LO})$$

$$\times N_q^{LO}[1 + N_k^{LA} + N_{k'}^{LA}]. \tag{7}$$

The integration over $\mathbf{k}'$ with the aid of the momentum conservation law $\mathbf{k}+\mathbf{k}' = \mathbf{q}$ and over the angle between $\mathbf{k}$ and $\mathbf{q}$ with the aid of the energy conservation law allows one to reduce this equation to

$$N_k^{LA} = \frac{2\tau_a[1 + N_k^{LA} + N_{2k_0-k}^{LA}]}{k_0^2 \tau_{sp}} \int_{2|k-k_0|}^{2k_0} dq q N_q^{LO}. \tag{8}$$

We again assume that the spectrum and transition probabilities are isotropic, which means that the phonon distribution functions depend only on the absolute values of the phonon wave vectors and the function $\mathcal{W}_{(LO|LA,LA)}(\mathbf{q}|\mathbf{k}',\mathbf{k})$ becomes a constant proportional to $1/\tau_{sp}$. Now we have two coupled equations: Eq. (3) (with τ_q given by Eq. (5)) and Eq. (8). Combining them, we obtain the integral equation for N_k^{LA}:

$$N_k^{LA} = \frac{2\tau_a[1 + N_k^{LA} + N_{2k_0-k}^{LA}]}{k_0^2}$$

$$\times \int_{2|k-k_0|}^{2k_0} dq \frac{q I_q}{1 + q^{-1} \int_{k_0-q/2}^{k_0+q/2} dk_1 [N_{k_1}^{LA} + N_{2k_0-k_1}^{LA}]}. \tag{9}$$

Its solution determines both the occupation numbers of LA phonons and the lifetime of LO phonons; see Eq. (5).

Using I_q from Eq. (1), we rewrite Eq. (9) in the form

$$\widetilde{N}(x) = \beta[1 + 2\widetilde{N}(x)] \int_{2|x|}^{1} \frac{y dy}{1 + (2/y)\int_{-y/2}^{y/2} dx_1 \widetilde{N}(x_1)}, \tag{10}$$

expressed through the dimensionless variables

$$x = (k - k_0)/q_0, \quad y = q/q_0, \quad \widetilde{N}(x) = N_{k_0+xq_0}^{LA}. \tag{11}$$

The dimensionless parameter

$$\beta = 2I\tau_a \frac{q_0^2}{k_0^2} \tag{12}$$

describes the intensity of LO phonon generation. In the transformation of Eq. (9) to Eq. (10), we have used the property $\widetilde{N}(x) = \widetilde{N}(-x)$ which follows from Eq. (9). For the case of $\beta \gg 1$, one can obtain a formal solution of Eq. (10) by taking into account that in this case the LA phonon distribution shrinks to the region of small x. This solution describes a Lorentz-shaped LA phonon distribution around the middle of the Brillouin zone:

$$\widetilde{N}(x) = \frac{N_0}{1 + 4\beta x^2}, \quad N_0 = \frac{2\beta^{3/2}}{3\pi}. \tag{13}$$

The inverse lifetime of LO phonons is given by

$$\frac{1}{\tau_q} = \frac{1}{\tau_{sp}} \begin{cases} 1 + 4\beta^{3/2}/3\pi, & y \ll \beta^{-1/2} \\ 1 + 2\beta q_0/3q, & \beta^{-1/2} \ll y < 1 \end{cases}, \tag{14}$$

and it decreases with increasing q. Both the increase of $1/\tau_q$ with increasing excitation β and the q-dependence of τ_q are caused by the stimulated LO phonon decay. They disappear for small excitation, when $\beta < 1$. In Fig. 5.1 we show numerically calculated $\widetilde{N}(x) = N_k^{LA}$ and τ_q as functions of the parameters x and y, respectively.

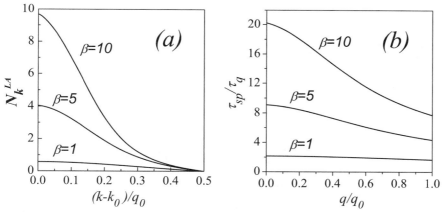

Figure 5.1. The distribution function of LA phonons and LO phonon decay time calculated from Eq. (26.10) for three different intensities of LO phonon generation.

Let us consider the relaxation of acoustic phonons. The energies of the acoustic phonons generated by non-equilibrium electrons are often considerably below the thermal energy. For example, if the phonons are generated by non-degenerate electrons ($\bar{\varepsilon} \sim T$), the mean energy of the phonons is estimated as $\sqrt{2m\bar{s}^2 T}$. This energy is considerably smaller than T at $T \gg 2m\bar{s}^2$. Owing to smallness of the sound velocities, this strong inequality can be satisfied even at $T \ll \Theta_D$. The relaxation of the low-energy phonons is determined by their collisions with thermal acoustic phonons, the latter occupy the region of much higher momenta, $\hbar q \sim T/\bar{s}$. The thermal phonons can be characterized by the equilibrium distribution function, since the collisions between them occur much faster than their collisions with low-energy phonons. This is because the phase space for the thermal phonons is larger, and the probabilities of their mutual scattering, which are proportional to the product of the wave numbers of three participating phonons, are higher. For the same reason, describing the relaxation of non-equilibrium phonons, one can neglect the processes when a low-energy phonon decays into two phonons of even lower energies. Such processes are of much smaller probability than the processes when the low-energy phonons fuse with thermal phonons or are emitted by these phonons. Therefore, the phonon-phonon collision integral standing in the stationary kinetic equation $J_{ph,e}(N|\mathbf{q}l) + J_{ph,ph}(N|\mathbf{q}l) = 0$ for low-energy phonons can be approximated by the second part of the collision integral (23.17). This approximation gives rise to the following symmetric form of the kinetic equation:

$$I_l^e(\mathbf{q}) + I_l^{ph}(\mathbf{q}) - [\nu_l^e(\mathbf{q}) + \nu_l^{ph}(\mathbf{q})]N_{\mathbf{q}}^l = 0. \tag{15}$$

In this equation, $I_l^e(\mathbf{q})$ and $\nu_l^e(\mathbf{q})$ are the generation and relaxation rates of the phonons due to their interaction with electrons (the same quantities as $I_{\mathbf{q}}$ and $\nu_{\mathbf{q}}$ of Sec. 21), while $I_l^{ph}(\mathbf{q})$ and $\nu_l^{ph}(\mathbf{q})$ are the phonon generation and relaxation rates due to phonon-phonon interaction:

$$\left| \begin{array}{c} I_l^{ph}(\mathbf{q}) \\ \nu_l^{ph}(\mathbf{q}) \end{array} \right| = \frac{1}{V} \sum_{l_1 l_2, \mathbf{q}_1 \mathbf{q}_2} \mathcal{W}_{(l_2|l_1 l)}(\mathbf{q}_2|\mathbf{q}_1, \mathbf{q})$$

$$\times \left| \begin{array}{c} N_{\mathbf{q}_2}^{l_2}(N_{\mathbf{q}_1}^{l_1} + 1) \\ (N_{\mathbf{q}_1}^{l_1} - N_{\mathbf{q}_2}^{l_2}) \end{array} \right| \delta(\omega_{\mathbf{q}l} + \omega_{\mathbf{q}_1 l_1} - \omega_{\mathbf{q}_2 l_2}). \tag{16}$$

If the temperature is small in comparison to the Debye temperature, the wave numbers $\mathbf{q}_1$ and $\mathbf{q}_2$ of the thermal acoustic phonons are much smaller than the size of the Brillouin zone. In this case, the phonon-pnonon scattering probability standing in Eq. (16) is described according to Eqs. (23.18) and (6.34):

$$\mathcal{W}_{(l_2|l_1 l)}(\mathbf{q}_2|\mathbf{q}_1, \mathbf{q}) = \delta_{\mathbf{q}+\mathbf{q}_1, \mathbf{q}_2} \frac{\pi \hbar q q_1 q_2}{4\rho^3 s_l s_{l_1} s_{l_2}}$$

$$\times \left| \sum_{\alpha\alpha'\beta\beta'\gamma\gamma'} \lambda_{\alpha\alpha',\beta\beta',\gamma\gamma'} e_{\mathbf{q}l}^{\alpha} e_{\mathbf{q}_1 l_1}^{\beta} e_{\mathbf{q}_2, l_2}^{\gamma} n_{\alpha'} n_{1\beta'} n_{2\gamma'} \right|^2, \tag{17}$$

where $\mathbf{n}$, $\mathbf{n}_1$, and $\mathbf{n}_2$ are the unit vectors directed along $\mathbf{q}$, $\mathbf{q}_1$, and $\mathbf{q}_2$, respectively. The dispersion relations for all three acoustic phonons are assumed to be linear and isotropic. Taking into account the momentum conservation rule $\mathbf{q}_2 = \mathbf{q} + \mathbf{q}_1$, we find that the delta-function in Eq. (16) is reduced to

$$\delta\left(s_l q + s_{l_1} q_1 - s_{l_2}[q^2 + q_1^2 + 2qq_1 \cos \widehat{\mathbf{q}\mathbf{q}_1}]^{1/2}\right). \tag{18}$$

Below we consider the transitions within one branch, $l = l_1 = l_2$, and omit the indices of phonon branches. In this case, the energy conservation law presented by the δ-function (18) tells us that $\widehat{\mathbf{q}\mathbf{q}_1}$ is zero, i.e., the wave vectors of three interacting phonons are aligned, $\mathbf{n} = \mathbf{n}_1 = \mathbf{n}_2$. Considering the longitudinal phonons ($l = LA$) which interact with conduction-band electrons via deformation potential, we also write the polarization vectors as $\mathbf{e}_{\mathbf{q}l} = \mathbf{e}_{\mathbf{q}_1 l_1} = \mathbf{e}_{\mathbf{q}_2 l_2} = -i\mathbf{n}$; see Eq. (6.30). The single-branch approximation is not unreasonable, since one can show (problem 5.8) that the longitudinal acoustic phonon relaxation involving transverse branch for one or two other participating phonons is impossible at $q \ll q_1$. Approximating $N_{\mathbf{q}_1}^{l_1}$ and $N_{\mathbf{q}_2}^{l_2}$ in Eq. (16) by the equilibrium distribution functions $N_{q_1}^{(eq)}$ and $N_{q_2}^{(eq)}$, we obtain isotropic relaxation and generation rates $\nu_l^{ph}(\mathbf{q}) \equiv \nu_q^{ph}$ and $I_l^{ph}(\mathbf{q}) \equiv I_q^{ph}$ in the form

$$\nu_q^{ph} = \frac{\hbar|\lambda_{LA}|^2}{16\pi\rho^3 s^4} \int_0^\infty dq_1 \, q_1^2 (q_1 + q)^2 (N_{q_1}^{(eq)} - N_{q_1+q}^{(eq)}), \tag{19}$$

$$I_q^{ph} = \frac{\nu_q^{ph}}{e^{\hbar sq/T} - 1}, $$

where s is the longitudinal sound velocity. The averaged anharmonic elastic constant introduced in this equation is given by

$$\lambda_{LA} = \sum_{\alpha\alpha'\beta\beta'\gamma\gamma'} \lambda_{\alpha\alpha',\beta\beta',\gamma\gamma'} n_\alpha n_\beta n_\gamma n_{\alpha'} n_{\beta'} n_{\gamma'}. \tag{20}$$

The main contribution to the integral in Eq. (19) comes from the thermal phonons, $\hbar sq_1 \sim T$. Calculating this integral at $\hbar sq \ll T$, we obtain

$$\nu_q^{ph} = \frac{\pi^3 |\lambda_{LA}|^2}{60\hbar^3 \rho^3 s^8} T^4 q. \tag{21}$$

The relaxation rate (21) is very sensitive to temperature.

Expressing I_q^{ph} through ν_q^{ph}, one may write the solution of Eq. (15) as

$$N_{\mathbf{q}} = \frac{I_{\mathbf{q}}^e + \nu_q^{ph} N_q^{(eq)}}{\nu_{\mathbf{q}}^e + \nu_q^{ph}}. \tag{22}$$

The anisotropy of the non-equilibrium phonon distribution is determined by the anisotropy of $I_{\mathbf{q}}^e$ and $\nu_{\mathbf{q}}^e$. Let us consider the case when the electrons interacting with the low-energy phonons can be described by the isotropic Fermi distribution with effective temperature T_e. According to the result of problem 4.8, $I_{\mathbf{q}}^e = I_q^e$ and $\nu_{\mathbf{q}}^e = \nu_q^e$ are related to each other as $I_q^e = \nu_q^e / (e^{\hbar s q / T_e} - 1)$, and Eq. (22) gives us

$$N_q = \frac{(1 + \nu_q^{ph}/\nu_q^e)^{-1}}{e^{\hbar s q/T_e} - 1} + \frac{(1 + \nu_q^e/\nu_q^{ph})^{-1}}{e^{\hbar s q/T} - 1}. \tag{23}$$

If $\nu_q^{ph} \gg \nu_q^e$, the low-energy phonons interacting with electrons are cooled to thermal equilibrium because of their interaction with thermal phonons. In the opposite limit, the low-energy phonons remain out of equilibrium, being heated to the electron temperature T_e. This is a bottleneck effect, when the energy transfer from electrons to the lattice is determined by phonon-phonon collisions. The acoustic-phonon bottleneck effect becomes important at low temperatures, when ν_q^{ph} is small (problem 5.9).

The processes of emission and absorption of non-equilibrium long-wavelength acoustic phonons by the thermal phonons also describe attenuation of sound waves in insulating crystals. The sound wave, whose amplitude is proportional to $e^{i\mathbf{k} \cdot \mathbf{r} - i\omega t}$, can be viewed as a flow of monochromatic acoustic phonons whose wave number k and energy $\hbar\omega$ are typically much smaller than those of the thermal phonons. Therefore, the results presented above are applicable to this case, and the relaxation rate given by Eq. (21) can be used for evaluation of the attenuation coefficient.

The above-used approximation of a single phonon branch with a linear dispersion cannot describe the angular relaxation of the momentum of phonon system, because the phonons do not change the direction of their motion in the collisions. Therefore, the anisotropic distributions of the low-energy phonons emitted, for example, by drifting electrons (see Sec. 21), cannot relax to equilibrium. The angular relaxation can occur via other processes, such as phonon-impurity collisions and four-phonon scattering, which are not considered here. However, a deviation of the acoustic phonon spectrum from linearity makes it possible to reach the isotropic distribution via three-phonon scattering within one branch. The momentum and energy conservation laws in these conditions permit such scattering only if the angles between the wave vectors of the participating phonons are small. If the spectrum is given by $\omega_{\mathbf{q}} = sq(1 + \xi(q))$, where $|\xi(q)| \ll 1$, the energy conservation law, $\delta(\omega_{\mathbf{q}} + \omega_{\mathbf{q_1}} - \omega_{|\mathbf{q}+\mathbf{q_1}|})$, becomes

$$s^{-1}\delta\left(\frac{qq_1}{q+q_1}\frac{\theta^2}{2} - (q+q_1)\xi(q+q_1) + q\xi(q) + q_1\xi(q_1)\right), \qquad (24)$$

where θ is the angle between $\mathbf{q}$ and $\mathbf{q}_1$. As seen from Eq. (24), this angle must be small. Another requirement necessary to satisfy the energy conservation law is $\xi(q) > 0$, i.e., the dispersion law must be superlinear. The process of angular relaxation of phonon distribution due to a weak dispersion of phonon velocities resembles a diffusion and is called the transverse relaxation.

27. Polaritons and Dielectric Function of Ionic Crystals

In Sec. 6 we considered the interaction of electromagnetic waves with transverse optical vibrations in ionic crystals. Below we show that this interaction leads to a reconstruction of the spectrum of bosonic elementary excitations of the crystal, when, instead of pure transverse phonons and photons, one has coupled excitations known as polaritons. We also calculate the dielectric function of the ionic crystal to describe the response of interacting phonons to the perturbation introduced by an electromagnetic wave.

The quantum theory of electromagnetic waves in classical ionic crystal is based upon the Maxwell equations in medium and Eq. (6.16) describing the lattice polarization. We use the Maxwell equation

$$\frac{1}{c}\frac{\partial \mathbf{D}}{\partial t} = [\nabla \times \mathbf{H}], \qquad (1)$$

where $\mathbf{D} = \mathbf{E} + 4\pi\mathbf{P} = \epsilon_\infty\mathbf{E} + 4\pi\Delta\mathbf{P}$ is the electrostatic induction and $\Delta\mathbf{P}$ is the polarization of the crystal due to ionic motion only:

$$\Delta\mathbf{P} = \mathbf{P} - \mathbf{P}_\infty, \quad \mathbf{P}_\infty = \frac{\epsilon_\infty - 1}{4\pi}\mathbf{E}. \qquad (2)$$

Taking into account that in the absence of longitudinal fields both $\mathbf{E}$ and $\mathbf{H}$ are expressed through the vector potential as $\mathbf{H} = [\nabla \times \mathbf{A}]$ and $\mathbf{E} = -c^{-1}\partial\mathbf{A}/\partial t$, we transform Eq. (1) to the wave equation

$$\frac{\epsilon_\infty}{c^2}\frac{\partial^2\mathbf{A}}{\partial t^2} + [\nabla \times [\nabla \times \mathbf{A}]] = \frac{4\pi}{c}\frac{\partial\Delta\mathbf{P}}{\partial t}, \qquad (3)$$

which replaces Eq. (3.3). On the other hand, Eq. (6.16) is reduced, after excluding $\mathbf{w}$ and expressing the electric field through the vector potential, to

$$\Delta\ddot{\mathbf{P}} + \omega_{TO}^2\Delta\mathbf{P} = -c^{-1}\omega_{TO}^2\beta\dot{\mathbf{A}}, \qquad (4)$$

where $\beta = (\epsilon_0 - \epsilon_\infty)/4\pi$ is the static polarizability. As usual, the dot and the double dot over the functions denote the first and the second derivative over time, respectively.

Now let us construct a Lagrange function expressed in terms of the variables (generalized coordinates) $\mathbf{A}$ and $\Delta\mathbf{P}$. This function should led to the Lagrange equations of motion coinciding with Eqs. (3) and (4). It can be checked directly (problem 5.10) that the Lagrangian density satisfying these properties is

$$\mathcal{L} = \frac{1}{8\pi}\left(\frac{\epsilon_\infty}{c^2}\dot{\mathbf{A}}^2 - [\nabla \times \mathbf{A}]^2\right) + \frac{1}{2\beta}(\omega_{TO}^{-2}\Delta\dot{\mathbf{P}}^2 - \Delta\mathbf{P}^2) - \frac{1}{c}\Delta\mathbf{P}\cdot\dot{\mathbf{A}}. \quad (5)$$

The first term in this expression is the Lagrangian density of electromagnetic field in the medium with dielectric permittivity ϵ_∞, while the second and the third terms describe the polarization and its interaction with the electromagnetic field. To quantize the field described by the Lagrangian density (5), we need to introduce canonically conjugate momenta $\mathbf{\Pi} = \partial\mathcal{L}/\partial\dot{\mathbf{A}}$ and $\mathbf{M} = \partial\mathcal{L}/\partial\Delta\dot{\mathbf{P}}$ and write the Hamiltonian density $\mathcal{H} = \mathbf{\Pi}\cdot\dot{\mathbf{A}} + \mathbf{M}\cdot\Delta\dot{\mathbf{P}} - \mathcal{L}$ (the same as the energy density; see problem 1.15) through the generalized coordinates and momenta. Having done this, we obtain

$$\mathcal{H} = \frac{1}{2}\left(\frac{4\pi c^2}{\epsilon_\infty}\mathbf{\Pi}^2 + \frac{1}{4\pi}[\nabla \times \mathbf{A}]^2\right)$$

$$+ \frac{1}{2}\left(\omega_{TO}^2\beta\mathbf{M}^2 + \frac{\epsilon_0}{\epsilon_\infty\beta}\Delta\mathbf{P}^2\right) + \frac{4\pi c}{\epsilon_\infty}\mathbf{\Pi}\cdot\Delta\mathbf{P}, \quad (6)$$

where Eq. (6.17) is also taken into account. Now we quantize the variables according to

$$\widehat{\Delta\mathbf{P}}(\mathbf{r}) = \sum_{\mathbf{q}l}\sqrt{\frac{\hbar\omega_{TO}^2\beta}{2\omega_{LO}V}}\mathbf{e}_{\mathbf{q}l}e^{i\mathbf{q}\cdot\mathbf{r}}\left(\hat{b}_{\mathbf{q}l} + \hat{b}_{-\mathbf{q}l}^+\right),$$

$$\widehat{\mathbf{M}}(\mathbf{r}) = -i\sum_{\mathbf{q}l}\sqrt{\frac{\hbar\omega_{LO}}{2\omega_{TO}^2\beta V}}\mathbf{e}_{\mathbf{q}l}e^{i\mathbf{q}\cdot\mathbf{r}}\left(\hat{b}_{\mathbf{q}l} - \hat{b}_{-\mathbf{q}l}^+\right), \quad (7)$$

and the photonic operators take the form

$$\hat{\mathbf{A}}(\mathbf{r}) = \sum_{\mathbf{q}\mu}\sqrt{\frac{2\pi\hbar c}{\epsilon_\infty^{1/2}qV}}\mathbf{e}_{\mathbf{q}\mu}e^{i\mathbf{q}\cdot\mathbf{r}}\left(\hat{a}_{\mathbf{q}\mu} + \hat{a}_{-\mathbf{q}\mu}^+\right),$$

$$\widehat{\mathbf{\Pi}}(\mathbf{r}) = -i\sum_{\mathbf{q}\mu}\sqrt{\frac{\hbar\epsilon_\infty^{1/2}q}{8\pi cV}}\mathbf{e}_{\mathbf{q}\mu}e^{i\mathbf{q}\cdot\mathbf{r}}\left(\hat{a}_{\mathbf{q}\mu} - \hat{a}_{-\mathbf{q}\mu}^+\right), \quad (8)$$

where $l = TO_1$ and TO_2 and $\mu = 1, 2$. For convenience, the creation and annihilation operators of photons are given by the letters $\hat{a}$ (not by $\hat{b}$ as in Chapters 1 and 4), while the letter $\hat{b}$ is reserved for the operators of phonons. The operators given by Eqs. (7) and (8) can be represented as $\widehat{\Delta\mathbf{P}}(\mathbf{r}) = \sum_l \widehat{\Delta\mathbf{P}_l}(\mathbf{r})$, $\widehat{\mathbf{M}}(\mathbf{r}) = \sum_l \widehat{\mathbf{M}_l}(\mathbf{r})$, $\hat{\mathbf{A}}(\mathbf{r}) = \sum_\mu \hat{\mathbf{A}}_\mu(\mathbf{r})$, and $\widehat{\mathbf{\Pi}}(\mathbf{r}) = \sum_\mu \widehat{\mathbf{\Pi}}_\mu(\mathbf{r})$, where the partial (single-mode) operators satisfy the commutation relations $[\widehat{\Delta\mathbf{P}_l}(\mathbf{r}), \widehat{\mathbf{M}_{l'}}(\mathbf{r}')] = i\hbar\delta_{ll'}\delta(\mathbf{r} - \mathbf{r}')$ and $[\hat{\mathbf{A}}_\mu(\mathbf{r}), \widehat{\mathbf{\Pi}}_{\mu'}(\mathbf{r}')] = i\hbar\delta_{\mu\mu'}\delta(\mathbf{r} - \mathbf{r}')$ (it is implied that the vectors standing in the commutators form the scalar products). These relations follow from the commutation relations for the bosonic operators $\hat{a}$ and $\hat{b}$ (problem 5.11). Because of orthogonality of the unit vectors of polarization, the phonon-photon interaction does not mix different modes of the bosons of the same kind (in other words, TO_1 phonon interacts only with $\mu = 1$ photon and TO_2 phonon only with $\mu = 2$ photon). For this reason, below we consider a single mode for each kind of bosons and omit the polarization indices.

Substituting the operators given by Eqs. (7) and (8) into Eq. (6), we obtain the Hamiltonian of phonon-photon system:

$$\widehat{H} = \sum_\mathbf{q} \hbar\omega_q \left(\hat{a}_\mathbf{q}^+ \hat{a}_\mathbf{q} + \frac{1}{2}\right) + \sum_\mathbf{q} \hbar\omega_{LO} \left(\hat{b}_\mathbf{q}^+ \hat{b}_\mathbf{q} + \frac{1}{2}\right)$$

$$+ i\hbar \sum_\mathbf{q} B_q(\hat{a}_{-\mathbf{q}}^+ \hat{b}_\mathbf{q}^+ - \hat{a}_{-\mathbf{q}} \hat{b}_\mathbf{q} + \hat{a}_\mathbf{q}^+ \hat{b}_\mathbf{q} - \hat{a}_\mathbf{q} \hat{b}_\mathbf{q}^+), \tag{9}$$

where $\omega_q = cq/\sqrt{\epsilon_\infty}$ and $B_q = \sqrt{\omega_{LO}\omega_q(\epsilon_0 - \epsilon_\infty)/4\epsilon_0}$. Formally, the expression (9) looks like a sum of free-photon, free-phonon, and interaction Hamiltonians. One may note that the "free-phonon" part contains the quasiparticle energy of longitudinal optical phonons, though only the transverse phonons interact with the electromagnetic radiation. Next, the "interaction" part does not coincide with the phonon-photon interaction Hamiltonian introduced by Eq. (6.25). These "inconsistencies" should not confuse the reader. They appear because the coupling of phonons with photons is considered here within a non-perturbative approach, and the effects of renormalization of the energy spectrum and interaction are already present in the Hamiltonian (9).

Since the Hamiltonian (9) contains binary products of creation and annihilation operators, it can be diagonalized to the form

$$\widehat{H} = \sum_{\mathbf{q}n} \hbar\Omega_{\mathbf{q}n} \left(\hat{c}_{\mathbf{q}n}^+ \hat{c}_{\mathbf{q}n} + \frac{1}{2}\right) \tag{10}$$

by a canonical transformation

$$\hat{c}_{\mathbf{q}n} = \lambda_n[\hat{a}_{\mathbf{q}} + x_n\hat{b}_{\mathbf{q}} + y_n\hat{a}^+_{-\mathbf{q}} + z_n\hat{b}^+_{-\mathbf{q}}]. \tag{11}$$

Here $n = 1, 2$ are the numbers of non-interacting polariton modes. The coefficients x_n, y_n, and z_n can be found from the equation of motion

$$\hbar\Omega_{\mathbf{q}n}\hat{c}_{\mathbf{q}n} = [\hat{c}_{\mathbf{q}n}, \widehat{H}], \tag{12}$$

which leads to a dispersion equation for the new modes (problem 5.12):

$$\Omega^4_{\mathbf{q}n} - \Omega^2_{\mathbf{q}n}(\omega^2_q + \omega^2_{LO}) + \omega^2_q\omega^2_{TO} = 0. \tag{13}$$

To determine the coefficients λ_n, x_n, y_n, and z_n, one should use, apart from Eq. (12), the commutation relation for new bosonic operators, $[\hat{c}_{\mathbf{q}n}, \hat{c}^+_{\mathbf{q}'n'}] = \delta_{nn'}\delta_{\mathbf{q}\mathbf{q}'}$. We obtain

$$x_n = i\left[\frac{\omega_q(\Omega_{\mathbf{q}n} - \omega_q)(\Omega_{\mathbf{q}n} + \omega_{LO})}{\omega_{LO}(\Omega_{\mathbf{q}n} + \omega_q)(\Omega_{\mathbf{q}n} - \omega_{LO})}\right]^{1/2},$$

$$y_n = -\frac{\Omega_{\mathbf{q}n} - \omega_q}{\Omega_{\mathbf{q}n} + \omega_q}, \quad z_n = \frac{\Omega_{\mathbf{q}n} - \omega_{LO}}{\Omega_{\mathbf{q}n} + \omega_{LO}}x_n, \tag{14}$$

and

$$|\lambda_n|^2 = \frac{(\Omega_{\mathbf{q}n} + \omega_q)^2(\Omega^2_{\mathbf{q}n} - \omega^2_{LO})}{4\Omega_{\mathbf{q}n}\omega_q(2\Omega^2_{\mathbf{q}n} - \omega^2_q - \omega^2_{LO})}. \tag{15}$$

The dispersion relation (13) has two solutions describing two polariton modes:

$$\Omega^2_{\mathbf{q}1,2} = \frac{c^2q^2 + \epsilon_0\omega^2_{TO} \pm [(c^2q^2 + \epsilon_0\omega^2_{TO})^2 - 4\epsilon_\infty\omega^2_{TO}c^2q^2]^{1/2}}{2\epsilon_\infty}, \tag{16}$$

where the signs $+$ and $-$ correspond to $n = 1$ and $n = 2$. At small q, the dispersion of the first polariton follows the *LO* phonon branch, $\Omega^2_{\mathbf{q}1} = \omega^2_{LO}$, while the second polariton behaves like a photon in the medium described by the static dielectric permittivity, $\Omega^2_{\mathbf{q}2} = c^2q^2/\epsilon_0$. At large q, the phonon-photon coupling becomes insignificant and the second polariton behaves like an ordinary *TO* phonon, while the first one behaves like a photon in the medium with high-frequency dielectric permittivity, $\Omega^2_{\mathbf{q}2} = \omega^2_{TO}$ and $\Omega^2_{\mathbf{q}1} = c^2q^2/\epsilon_\infty$. The spectra of polaritons for the cases of strong coupling ($\epsilon_0/\epsilon_\infty - 1 > 1$) and weak coupling ($\epsilon_0/\epsilon_\infty - 1 \ll 1$) are shown in Fig. 5.2. The correlation functions of the operators (7) and (8) have poles at polaritonic frequencies (problem 5.13).

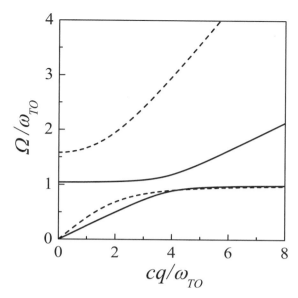

Figure 5.2. Spectrum of polaritons at $\epsilon_0 = 15.69$ and $\epsilon_\infty = 14.44$ (solid), and $\epsilon_0 = 5.62$ and $\epsilon_\infty = 2.25$ (dashed).

According to the wave equation, the dispersion law for the photons in dielectric crystals should follow the form $\omega^2 = c^2 q^2/\kappa(\mathbf{q}, \omega)$, where $\kappa(\mathbf{q}, \omega)$ is the dielectric permittivity whose dependence on the frequency and wave vector is caused by the lattice polarization. On the other hand, only the waves with frequencies $\omega^2 = \Omega^2_{\mathbf{q}1,2}$ can propagate in the crystal. Both these requirements are satisfied if we present the dielectric permittivity as

$$\kappa(\mathbf{q}, \omega) = \kappa(\omega) = \epsilon_\infty + \frac{\omega^2_{TO}(\epsilon_0 - \epsilon_\infty)}{\omega^2_{TO} - \omega^2}. \tag{17}$$

This equation can be formally obtained by substituting $c^2 q^2 = \kappa(\mathbf{q}, \omega)\omega^2$ and $\Omega^2_{\mathbf{q}1,2} = \omega^2$ into Eq. (16). The dielectric permittivity (17) does not depend on $\mathbf{q}$ and varies with ω in the infrared region of frequencies. In the interval $\omega_{TO} < \omega < \omega_{LO}$, the dielectric permittivity is negative, which means that electromagnetic waves cannot propagate in the crystal. This is clear, since the given interval of frequencies corresponds to the quasiparticle gap, where no excitations can exist. We stress that Eq. (17) can be obtained in a more simple way from Eq. (4), since the latter already gives us a relation between the polarization and electric field. Neither of these approaches, however, describes the region close to the resonance $\omega = \omega_{TO}$, where, according to Eq. (17), $\kappa(\omega)$ is infinitely large.

To overcome this difficulty, one should introduce a finite lifetime for *TO* phonons, which may result from phonon-phonon interaction. In the remaining part of this section we bring this interaction into consideration and calculate the dielectric function by using the linear response theory and double-time Green's functions for phonons; see Appendix D.

In the translation-invariant case, the dielectric function $\kappa(\mathbf{q}, \omega)$ is introduced as a proportionality coefficient in the linear relation between the Fourier components of electrostatic induction and electric field, $\mathbf{D} = \hat{\kappa}\mathbf{E}$. One may rewrite this linear relation, with the use of the lattice polarization vector $\Delta\mathbf{P}$, as

$$\Delta P_\alpha(\mathbf{q}, \omega) = \sum_\beta \frac{\kappa_{\alpha\beta}(\mathbf{q}, \omega) - \delta_{\alpha\beta}\epsilon_\infty}{4\pi} E_\beta(\mathbf{q}, \omega). \tag{18}$$

Now we apply the linear response theory of Sec. 13 to calculate the proportionality coefficient in Eq. (18). The perturbation Hamiltonian describing the interaction of ionic motion with electromagnetic field is

$$\widehat{\Delta H_\omega} = -\int d\mathbf{r}\widehat{\Delta\mathbf{P}}(\mathbf{r}) \cdot \mathbf{E}(\mathbf{r}, \omega) = -\frac{1}{V}\sum_\mathbf{q} \widehat{\Delta\mathbf{P}}(-\mathbf{q}) \cdot \mathbf{E}(\mathbf{q}, \omega). \tag{19}$$

To avoid misunderstandings, we have to point out that the lattice polarization operator $\widehat{\Delta\mathbf{P}}(\mathbf{r})$ used here and below is not the same operator as the one used above, in the theory of polaritons, though we apply the same letter for it. This is because the physical quantity $\Delta\mathbf{P}(\mathbf{r})$ corresponding to this operator is now governed by Eq. (4) without the right-hand side and describes the lattice polarization in the absence of perturbations. Therefore, this operator is no longer given by Eq. (7). It is given by Eq. (6.16) at $\mathbf{E}_L = 0$, where the relative displacement vector $\mathbf{w}$ is replaced by the corresponding operator (6.19). Now, let the quantity Q in Eq. (13.5) be $\Delta\mathbf{P}$. Then one can write an equation similar to Eq. (13.14):

$$\frac{\kappa_{\alpha\beta}(\mathbf{q}, \omega) - \delta_{\alpha\beta}\epsilon_\infty}{4\pi} = -\frac{1}{i\hbar V}\int_{-\infty}^0 d\tau e^{\lambda\tau - i\omega\tau}$$

$$\times \mathrm{Sp}\,\hat{\eta}_{eq}\left[e^{-i\widehat{H}\tau/\hbar}\widehat{\Delta P_\alpha}(\mathbf{q})e^{i\widehat{H}\tau/\hbar}, \widehat{\Delta P_\beta}(-\mathbf{q})\right]. \tag{20}$$

It is directly seen that $\kappa_{\alpha\beta}(\mathbf{q}, \omega)$ is expressed through the retarded double-time Green's function (see Appendix D) defined as a correlation function of polarization operators:

$$\kappa_{\alpha\beta}(\mathbf{q}, \omega) - \delta_{\alpha\beta}\epsilon_\infty = -\frac{4\pi}{V}\langle\langle\widehat{\Delta P_\alpha}(\mathbf{q})|\widehat{\Delta P_\beta}(-\mathbf{q})\rangle\rangle_{\hbar\omega}^R$$

$$= -\delta_{\alpha\beta}\frac{\hbar}{2}\omega_{TO}(\epsilon_0 - \epsilon_\infty)D^{TO,R}_\omega(\mathbf{q}). \tag{21}$$

In the second equation of Eq. (21), we have expressed $\widehat{\mathbf{\Delta P}}$ through $\hat{b}_\mathbf{q}$ and $\hat{b}_\mathbf{q}^+$ as described above and employed the definition of the Green's function of phonons given in the end of Appendix D. The usage of Eqs. (6.16) and (6.19) for describing the polarization implies consideration of small q, when the contribution comes from the long-wavelength phonons only. In these conditions, the tensor $\kappa_{\alpha\beta}$ is reduced to a scalar, $\kappa_{\alpha\beta}(\mathbf{q},\omega) = \delta_{\alpha\beta}\kappa(\mathbf{q},\omega)$. In the absence of phonon-phonon interaction, one should use the Green's function (D.25) of free phonons. Substituting it into Eq. (21), we obtain Eq. (17) for the real part of $\kappa(\mathbf{q},\omega)$, while the imaginary part is proportional to $\delta(\omega \pm \omega_{TO})$.

Now let us consider a system of interacting phonons described by the Hamiltonian (23.1). For a while, it is convenient to employ the operators of canonical variables defined by Eq. (6.9) instead of using the operators $\hat{b}_{\mathbf{q}l}$ and $\hat{b}_{-\mathbf{q}l}^+$. The canonical operators satisfy the commutation rule (6.7). Let us use the equations of motion (D.13) for Green's functions. Taking into account that $\widehat{Q}_{\mathbf{q}l}$ commutes with $\widehat{H}_{ph,ph}$ given by Eq. (23.2), we compose a pair of equations, $\partial\langle\langle\widehat{Q}_{\mathbf{q}l}|\widehat{Q}_{-\mathbf{q}l}\rangle\rangle_t^s/\partial t = \langle\langle\widehat{P}_{\mathbf{q}l}|\widehat{Q}_{-\mathbf{q}l}\rangle\rangle_t^s$ and $\partial\langle\langle\widehat{P}_{\mathbf{q}l}|\widehat{Q}_{-\mathbf{q}l}\rangle\rangle_t^s/\partial t = \delta(t) + \omega_{\mathbf{q}l}^2\langle\langle\widehat{Q}_{\mathbf{q}l}|\widehat{Q}_{-\mathbf{q}l}\rangle\rangle_t^s + (i/\hbar)\langle\langle[\widehat{P}_{\mathbf{q}l},\widehat{H}_{ph,ph}]|\widehat{Q}_{-\mathbf{q}l}\rangle\rangle_t^s$. The index s can be R, A, or c, as in Appendix D, though below we use only the retarded Green's functions, $s = R$. Combining these equations, we apply the energy representation of the Green's functions and obtain

$$(\omega^2 - \omega_{\mathbf{q}l}^2)\langle\langle\widehat{Q}_{\mathbf{q}l}|\widehat{Q}_{-\mathbf{q}l}\rangle\rangle_{\hbar\omega}^s - 1 = \sum_{\mathbf{q}_1\mathbf{q}_2,l_1l_2}\sqrt{2\omega_{\mathbf{q}_1l_1}\omega_{\mathbf{q}_2l_2}\omega_{\mathbf{q}l}/\hbar^3}$$

$$\times B_{l_1l_2l}(\mathbf{q}_1,\mathbf{q}_2,-\mathbf{q})\langle\langle\widehat{Q}_{\mathbf{q}_1l_1}\widehat{Q}_{\mathbf{q}_2l_2}|\widehat{Q}_{-\mathbf{q}l}\rangle\rangle_{\hbar\omega}^s . \tag{22}$$

The commutator $[\widehat{P}_{\mathbf{q}l},\widehat{H}_{ph,ph}]$ has been calculated with the aid of Eq. (23.2) for $\widehat{H}_{ph,ph}$. As a result, the phonon-phonon interaction couples $\langle\langle\widehat{Q}_{\mathbf{q}l}|\widehat{Q}_{-\mathbf{q}l}\rangle\rangle_{\hbar\omega}^s$ to the correlation function $C_{QQ} = \langle\langle\widehat{Q}_{\mathbf{q}_1l_1}\widehat{Q}_{\mathbf{q}_2l_2}|\widehat{Q}_{-\mathbf{q}l}\rangle\rangle_{\hbar\omega}^s$. We are going to take into account the effects of the interaction up to the second order in $\widehat{H}_{ph,ph}$, which means that we have to write an equation for this third-order correlation function. Let us denote the other third-order correlation functions as $C_{QP} = \langle\langle\widehat{Q}_{\mathbf{q}_1l_1}\widehat{P}_{\mathbf{q}_2l_2}|\widehat{Q}_{-\mathbf{q}l}\rangle\rangle_{\hbar\omega}^s$, $C_{PQ} = \langle\langle\widehat{P}_{\mathbf{q}_1l_1}\widehat{Q}_{\mathbf{q}_2l_2}|\widehat{Q}_{-\mathbf{q}l}\rangle\rangle_{\hbar\omega}^s$, and $C_{PP} = \langle\langle\widehat{P}_{\mathbf{q}_1l_1}\widehat{P}_{\mathbf{q}_2l_2}|\widehat{Q}_{-\mathbf{q}l}\rangle\rangle_{\hbar\omega}^s$. Applying Eq. (D.13) to them, we obtain four coupled equations

$$
\begin{aligned}
-i\omega C_{QQ} &= C_{QP} + C_{PQ} \\
-i\omega C_{QP} &= C_{PP} - \omega_{\mathbf{q}_2 l_2}^2 C_{QQ} - M_{QP} \\
-i\omega C_{PQ} &= C_{PP} - \omega_{\mathbf{q}_1 l_1}^2 C_{QQ} - M_{PQ} \\
-i\omega C_{PP} &= -\omega_{\mathbf{q}_1 l_1}^2 C_{QP} - \omega_{\mathbf{q}_2 l_2}^2 C_{PQ} - M_{PP}
\end{aligned}
\tag{23}
$$

where the interaction-dependent terms are expressed through the fourth-order correlation functions:

$$
M_{QP} = \sum_{\mathbf{q}_3 \mathbf{q}_4, l_3 l_4} \sqrt{2\omega_{\mathbf{q}_3 l_3} \omega_{\mathbf{q}_4 l_4} \omega_{\mathbf{q}_2 l_2}/\hbar^3} \, B_{l_3 l_4 l_2}(\mathbf{q}_3, \mathbf{q}_4, -\mathbf{q}_2)
$$

$$
\times \langle\langle \widehat{Q}_{\mathbf{q}_1 l_1} \widehat{Q}_{\mathbf{q}_3 l_3} \widehat{Q}_{\mathbf{q}_4 l_4} | \widehat{Q}_{-\mathbf{q}l} \rangle\rangle_{\hbar\omega}^s \,,
\tag{24}
$$

$$
M_{PP} = \sum_{\mathbf{q}_3 \mathbf{q}_4, l_3 l_4} \left\{ \sqrt{2\omega_{\mathbf{q}_3 l_3} \omega_{\mathbf{q}_4 l_4} \omega_{\mathbf{q}_2 l_2}/\hbar^3} \, B_{l_3 l_4 l_2}(\mathbf{q}_3, \mathbf{q}_4, -\mathbf{q}_2) \right.
$$

$$
\times \langle\langle \widehat{P}_{\mathbf{q}_1 l_1} \widehat{Q}_{\mathbf{q}_3 l_3} \widehat{Q}_{\mathbf{q}_4 l_4} | \widehat{Q}_{-\mathbf{q}l} \rangle\rangle_{\hbar\omega}^s + \sqrt{2\omega_{\mathbf{q}_3 l_3} \omega_{\mathbf{q}_4 l_4} \omega_{\mathbf{q}_1 l_1}/\hbar^3}
\tag{25}
$$

$$
\left. \times B_{l_3 l_4 l_1}(\mathbf{q}_3, \mathbf{q}_4, -\mathbf{q}_1) \langle\langle \widehat{Q}_{\mathbf{q}_3 l_3} \widehat{Q}_{\mathbf{q}_4 l_4} \widehat{P}_{\mathbf{q}_2 l_2} | \widehat{Q}_{-\mathbf{q}l} \rangle\rangle_{\hbar\omega}^s \right\},
$$

and M_{PQ} differs from M_{QP} by the permutation of the indices 1 and 2.

Solving the system (23), we express C_{QQ} through M_{QP}, M_{PQ}, and M_{PP} and substitute the result into Eq. (22). Thus, the right-hand side of Eq. (22) becomes of the second order in the interaction. Within the required accuracy, we can calculate the correlation functions entering Eqs. (24) and (25) in the free-phonon approximation, replacing $\widehat{H} = \widehat{H}_{ph} + \widehat{H}_{ph,ph}$ in $\exp(\pm i\widehat{H}t/\hbar)$ by $\widehat{H}_{ph}$. Below we consider the retarded Green's functions, $s = R$. After straightforward transformations, we obtain

$$
M_{QP}^R \simeq 2\omega_{\mathbf{q}l} \langle\langle \widehat{Q}_{\mathbf{q}l} | \widehat{Q}_{-\mathbf{q}l} \rangle\rangle_{\hbar\omega}^{R(0)} B_{l l_1 l_2}(\mathbf{q}, -\mathbf{q}_1, -\mathbf{q}_2)
$$

$$
\times (2N_{\mathbf{q}_1}^{l_1} + 1) \sqrt{\frac{\omega_{\mathbf{q}_2 l_2}}{2\hbar \omega_{\mathbf{q}_1 l_1} \omega_{\mathbf{q}l}}}
\tag{26}
$$

and $M_{PP}^R = 0$. The superscript (0) at $\langle\langle \widehat{Q}_{\mathbf{q}l} | \widehat{Q}_{-\mathbf{q}l} \rangle\rangle_{\hbar\omega}^{R(0)}$ standing in Eq. (26) indicates that this Green's function is calculated in the free-phonon approximation. Without a loss of accuracy, we replace it by the exact Green's function and omit the index (0). Let us substitute C_{QQ} found from Eq. (23) into the right-hand side of Eq. (22). This side becomes equal to

$$
2\omega_{\mathbf{q}l} \langle\langle \widehat{Q}_{\mathbf{q}l} | \widehat{Q}_{-\mathbf{q}l} \rangle\rangle_{\hbar\omega}^R \frac{1}{\hbar^2} \sum_{\mathbf{q}_1 \mathbf{q}_2, l_1 l_2} |B_{l_1 l_2 l}(\mathbf{q}_1, \mathbf{q}_2, \mathbf{q})|^2
$$

$$\times \frac{(2N_{\mathbf{q}_1}^{l_1} + 1)\omega_{\mathbf{q}_2 l_2}(\omega_{\mathbf{q}_1 l_1}^2 - \omega_{\mathbf{q}_2 l_2}^2 + \omega^2) + (\mathbf{q}_1 l_1 \leftrightarrow \mathbf{q}_2 l_2)}{d_{\mathbf{q}_1 l_1, \mathbf{q}_2 l_2}^{(++)} d_{\mathbf{q}_1 l_1, \mathbf{q}_2 l_2}^{(--)} d_{\mathbf{q}_1 l_1, \mathbf{q}_2 l_2}^{(-+)} d_{\mathbf{q}_1 l_1, \mathbf{q}_2 l_2}^{(+-)}} \qquad (27)$$

$$\equiv -2i\omega_{\mathbf{q}l}\gamma_{\mathbf{q}\omega}^l \langle\langle \widehat{Q}_{\mathbf{q}l} | \widehat{Q}_{-\mathbf{q}l}\rangle\rangle_{\hbar\omega}^R \, ,$$

where $d_{\mathbf{q}_1 l_1, \mathbf{q}_2 l_2}^{(\pm\pm)} = \omega + i0 \pm \omega_{\mathbf{q}_1 l_1} \pm \omega_{\mathbf{q}_2 l_2}$. We have added $+i0$ to ω in the denominator in order to satisfy the analytical properties of the retarded Green's functions. Now Eq. (22) can be solved with respect to $\langle\langle \widehat{Q}_{\mathbf{q}l} | \widehat{Q}_{-\mathbf{q}l}\rangle\rangle_{\hbar\omega}^R$. Since $\mathrm{D}_\omega^{l,R}(\mathbf{q}) = 2\omega_{\mathbf{q}l}\langle\langle \widehat{Q}_{\mathbf{q}l} | \widehat{Q}_{-\mathbf{q}l}\rangle\rangle_{\hbar\omega}^R/\hbar$, we obtain the Green's function of interacting phonons in the form

$$\mathrm{D}_\omega^{l,R}(\mathbf{q}) = \frac{1}{\hbar} \frac{2\omega_{\mathbf{q}l}}{\omega^2 - \omega_{\mathbf{q}l}^2 + 2i\omega_{\mathbf{q}l}\gamma_{\mathbf{q}\omega}^l}. \qquad (28)$$

Substituting $l = TO$ in this equation, we finally transform Eq. (21) to

$$\kappa(\mathbf{q}, \omega) = \epsilon_\infty + \frac{\omega_{TO}^2(\epsilon_0 - \epsilon_\infty)}{\omega_{TO}^2 - \omega^2 - 2i\omega_{TO}\gamma_{\mathbf{q}\omega}^{TO}}. \qquad (29)$$

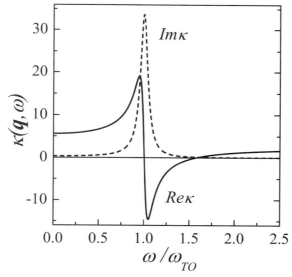

Figure 5.3. Real (solid) and imaginary (dashed) parts of $\kappa(\mathbf{q}, \omega)$ calculated at $\epsilon_0 = 5.62$ and $\epsilon_\infty = 2.25$ under the assumption that $\gamma_{\mathbf{q}\omega}^{TO} = 0.05 \, \omega_{TO}$.

The characteristic frequency $\gamma_{\mathbf{q}\omega}^l$ defined by Eq. (27) is complex. As follows from Eq. (28), its imaginary part leads to a small shift of the pole of the Green's function from $\omega = \pm\omega_{\mathbf{q}l}$. On the other hand, the real part of $\gamma_{\mathbf{q}\omega}^l$ defines a broadening energy so that the Green's function is finite everywhere. A simple transformation of the left-hand side of Eq. (27) gives us

$$2\mathrm{Re}\gamma_{\mathbf{q}\omega}^l = \frac{2\pi}{\hbar^2} \sum_{\mathbf{q}_1\mathbf{q}_2, l_1 l_2} |B_{l_1 l_2 l}(\mathbf{q}_1, \mathbf{q}_2, \mathbf{q})|^2 \left\{ \frac{1}{2}(1 + N_{\mathbf{q}_1}^{l_1} + N_{\mathbf{q}_2}^{l_2}) \right. \tag{30}$$

$$\left. \times\delta(\omega - \omega_{\mathbf{q}_1 l_1} - \omega_{\mathbf{q}_2 l_2}) + (N_{\mathbf{q}_2}^{l_2} - N_{\mathbf{q}_1}^{l_1})\delta(\omega - \omega_{\mathbf{q}_1 l_1} + \omega_{\mathbf{q}_2 l_2}) \right\} \equiv \frac{1}{\tau_{\mathbf{q}l}(\omega)}.$$

The time $\tau_{\mathbf{q}l}(\omega)$ introduced in Eq. (30) is the scattering time of the phonon in the state $\mathbf{q}l$ with energy $\hbar\omega$. This is seen directly, since the term proportional to $N_{\mathbf{q}}^l$ in the collision integral (23.17) is $-N_{\mathbf{q}}^l/\tau_{\mathbf{q}l}(\omega_{\mathbf{q}l})$. The two consecutive terms with different δ-functions in Eq. (30) correspond to a decay of the phonon $\mathbf{q}l$ and to its fusion with another phonon, respectively. In the case of long-wavelength TO phonons, only the decay processes remain, since they can simultaneously satisfy the momentum and energy conservation laws. This decay occurs in a similar way as the decay of the long-wavelength LO phonons considered in the previous section: a long-wavelength TO phonon decays into two short-wavelength acoustic phonons. The probability of this process is small enough to satisfy $\gamma_{\mathbf{q}\omega}^{TO} \ll \omega_{TO}$. Thus, one can substitute the resonance frequency ω_{TO} instead of ω in $\gamma_{\mathbf{q}\omega}^{TO}$. Next, since the momenta of photons are very small, one can put $\mathbf{q} = 0$ in $\gamma_{\mathbf{q}\omega}^{TO}$ with a high accuracy. Therefore, $\gamma_{\mathbf{q}\omega}^{TO}$ in Eq. (29) can be approximated by a constant. The imaginary part of $\kappa(\mathbf{q}, \omega)$, which describes the absorption of infrared radiation in ionic crystals, in this case is given by a Lorentzian function; see Fig. 5.3. The phonon-phonon interaction is not the only one mechanism responsible for broadening of the TO resonance in ionic crystals. A consideration of other broadening mechanisms is beyond the scope of this book.

Problems

5.1. Give an explicit expression of $\mathcal{W}_{(l_1|l_2l_3)}(\mathbf{q}_1|\mathbf{q}_2, \mathbf{q}_3)$ defined by Eq. (23.18) in terms of the anharmonic coefficients $A_{s_1 s_2 s_3}^{k_1 k_2 k_3}$ standing in Eq. (6.1).

Result:

$$\mathcal{W}_{(l_1|l_2l_3)}(\mathbf{q}_1|\mathbf{q}_2, \mathbf{q}_3) = \frac{\pi\hbar\Omega_0}{4\omega_{\mathbf{q}_1 l_1}\omega_{\mathbf{q}_2 l_2}\omega_{\mathbf{q}_3 l_3}} \left| \sum_{\mathbf{nn'}} \sum_{s_1 s_2 s_3} \frac{A_{s_1 s_2 s_3}^{k_1 k_2 k_3}(|\mathbf{R_n}|, |\mathbf{R_{n'}}|)}{\sqrt{M_{s_1} M_{s_2} M_{s_3}}} \right.$$

$$\left. \times e_{s_1}^{k_1}(\mathbf{q}_2 + \mathbf{q}_3 \, l_1)e_{s_2}^{k_2}(-\mathbf{q}_2 l_2)e_{s_3}^{k_3}(-\mathbf{q}_3 l_3)e^{-i\mathbf{q}_2\cdot\mathbf{R_n}-i\mathbf{q}_3\cdot\mathbf{R_{n'}}} \right|^2 \Delta_{\mathbf{q}_1, \mathbf{q}_2+\mathbf{q}_3},$$

where Ω_0 is the volume of the unit crystal cell and the other quantities are defined in Sec. 6.

5.2. Check, by a direct substitution, that the phonon distribution function (23.22) satisfies the kinetic equation (23.16).

Hint: Use the energy conservation laws expressed by the δ-functions standing in the collision integral (23.17).

5.3. Show that when the phonon distribution function is given by Eq. (23.26), the velocity **u** is the averaged group velocity of the phonons.

Hint: The averaged group velocity is defined as

$$\sum_{\mathbf{q}l} \frac{\partial \omega_{\mathbf{q}l}}{\partial \mathbf{q}} N_{\mathbf{q}}^l \bigg/ \sum_{\mathbf{q}l} N_{\mathbf{q}}^l \ .$$

Substitute $N_{\mathbf{q}}^l$ from Eq. (23.26) and calculate the integral over **q**, employing the integration by parts.

5.4. Check the relation (24.7).

Hint: Write the right-hand side through $d\sigma/dN = \ln[(N+1)/N]$ and substitute the equilibrium Planck distribution there.

5.5. Linearize the distribution function (23.26), where $T_{ph} = T + \Delta T$, with respect to small **u** and ΔT. In order to obtain a general solution, add the solution (23.22) at $T_{ph} = T$ to this linearized function. Find the constant c from the normalization condition (23.23).

Solution: The general linearized solution obtained in this way is

$$N_{\mathbf{q}}^l = N_{l\mathbf{q}}^{(eq)} + N_{l\mathbf{q}}^{(eq)}(N_{l\mathbf{q}}^{(eq)} + 1)\left[\hbar\mathbf{q} \cdot \mathbf{u}/T + \omega_{\mathbf{q}l}(\Delta T/T^2 + c)\right].$$

Substituting it into the normalization condition (23.23), we find that the first term in the square brackets does not contribute to the integral over **q**. Therefore, to satisfy Eq. (23.23), one should have $c = -\Delta T/T^2$, and we obtain the solution given by Eqs. (24.3) and (24.11).

5.6. Calculate the sum standing in Eq. (24.18) by assuming only one phonon branch with the linear dispersion law $\omega = sq$.

Hint: Substituting $q = \omega/s$, reduce the sum over **q** to the integral over $\hbar\omega/T$.

5.7. Calculate λ given by Eq. (24.13) in a one-dimensional lattice, assuming linear phonon dispersion ($\omega_q = sq$) up to the edge of the Brillouin zone.

Solution: The Brillouin zone is defined as $-\pi/a < q < \pi/a$ and the relevant reciprocal lattice vectors are $\pm 2\pi/a$. Two possible umklapp processes contributing to the integral are: *1)* $q_1, q_2 > 0$, $g = -2\pi/a$ and *2)* $q_1, q_2 < 0$, $g = 2\pi/a$. Their contributions are equal to each other. The δ-function of frequencies is reduced to $\delta[2s(|q_1| + |q_2| - \pi/a)]$. The factor $(N^{(eq)}_{q_1+q_2+g} + 1)N^{(eq)}_{q_1}N^{(eq)}_{q_2}$ is equal to $\exp(-\hbar\pi s/aT)$ everywhere except the regions of small q_1 or q_2. The integral over q_1 is taken by using the δ-function, and the remaining integral over q_2 defines the averaged factor

$$\overline{\mathcal{W}} = \frac{a}{\pi} \int_0^{\pi/a} dq_2 \mathcal{W}(-\pi/a \mid \pi/a - q_2, q_2).$$

The result is written as

$$\lambda = \frac{\pi\hbar\overline{\mathcal{W}}}{sa^3T} \exp\left(-\frac{\hbar\pi s}{aT}\right),$$

and one can see that λ exponentially increases with the increase of T.

5.8. Using the expression (26.18), find the conditions when the scattering of longitudinal acoustic phonons ($l = LA$) involving transverse branch for one or two other participating phonons is possible.

Result: For $l = LA$, the conservation law can be satisfied only if $l_1 = TA$ and $l_2 = LA$, at $q > q_1(1 - s_t/s_l)$.

5.9. Assuming the case of degenerate electrons, find the temperature when the ratio ν_q^{ph}/ν_q^e becomes equal to unity.

Solution: The relaxation rate ν_q^e is given by Eq. (21.18), where $\Omega_{\mathbf{q}} = s_l q$ (the drift is neglected). The ratio ν_q^{ph}/ν_q^e appears to be q-independent, and the required temperature is $T = (30^{1/4}s^2/\pi) \sqrt{\rho m \mathcal{D}/|\lambda_{LA}|}$.

5.10. Obtain the equations of motion (27.3) and (27.4) by using the Lagrangian density $\mathcal{L}$ of Eq. (27.5).

Hint: Apply the general procedure described in problem 1.15 to the Lagrangian (27.5) with $q_1 = \mathbf{A}$ and $q_2 = \Delta\mathbf{P}$.

5.11. Check the commutation relations $[\widehat{\Delta\mathbf{P}}_l(\mathbf{r}), \widehat{\mathbf{M}}_{l'}(\mathbf{r}')] = i\hbar\delta_{ll'}\delta(\mathbf{r} - \mathbf{r}')$ and $[\widehat{\mathbf{A}}_\mu(\mathbf{r}), \widehat{\Pi}_{\mu'}(\mathbf{r}')] = i\hbar\delta_{\mu\mu'}\delta(\mathbf{r} - \mathbf{r}')$.

Hints: Use the definitions (27.7) and (27.8) and the commutation rules for bosonic creation and annihilation operators. Then use the normalization requirement for the unit polarization vectors and take into account that the δ-function of coordinates is obtained as $V^{-1} \sum_{\mathbf{q}} \exp[i\mathbf{q} \cdot (\mathbf{r} - \mathbf{r}')] = \delta(\mathbf{r} - \mathbf{r}')$.

5.12. Derive the dispersion relation (27.13).

Hints: Substitute the solution (27.11) into Eq. (27.12). Calculating the commutators, obtain an equation linear in the operators $\hat{a}_{\mathbf{q}}$, $\hat{a}_{\mathbf{q}}^+$, $\hat{b}_{\mathbf{q}}$, and $\hat{b}_{\mathbf{q}}^+$. By demanding that the four coefficients at these four operators be zeros, find four linear equations connecting λ_n, $\lambda_n x_n$, $\lambda_n y_n$, and $\lambda_n z_n$. The solvability condition of this system leads to the dispersion relation (27.13).

5.13. Calculate the Green's function $\langle\langle\widehat{\Delta \mathbf{P}}(\mathbf{q})|\widehat{\Delta \mathbf{P}}(-\mathbf{q})\rangle\rangle^R_{\hbar\omega}$ for the polariton system.

Solution: To do this, one should express the polarization operator $\widehat{\Delta \mathbf{P}}(\mathbf{q})$ given by Eq. (27.7) through the creation and annihilation operators of polaritons. This can be done with the aid of the dispersion relation (27.13) and expressions (27.14) for the coefficients of the canonical transformation (27.11). The correlation functions of the polaritonic operators $c_{\mathbf{q}1}$ and $c_{\mathbf{q}2}$ are given by the same expressions as the correlation functions of free phonons (Appendix D), the difference is that, instead of the phonon frequency ω_{TO}, we put the polaritonic frequencies Ω_{qn} and consider two modes, $n=1$ and 2. After a simple algebra, we obtain

$$\langle\langle\widehat{\Delta \mathbf{P}}(\mathbf{q})|\widehat{\Delta \mathbf{P}}(-\mathbf{q})\rangle\rangle^R_{\hbar\omega} = \frac{\omega^2_{TO}(\omega^2 - \omega^2_q)(\epsilon_0 - \epsilon_\infty)}{[(\omega + i0)^2 - \Omega^2_{q1}][(\omega + i0)^2 - \Omega^2_{q2}]}.$$

This function has simple poles at the polaritonic frequencies and goes to zero at $\omega = \omega_q$.

Chapter 6

EFFECTS OF ELECTRON-ELECTRON INTERACTION

In this chapter we consider some features of transport phenomena caused by the interaction between electrons. As in the previous chapters, the kinetic equation is derived here under the assumption of weak electron-electron interaction. One should take into account that there are important physical situations when this assumption is not valid, and the electron-electron interaction rebuilds the ground state of the electron system and changes the nature of quasiparticles. The kinetic phenomena in such strongly correlated electron systems are not studied in this book. It is important that, in contrast to electron-impurity and electron-boson interactions, the terms linear in the electron-electron interaction (proportional to the square of the electron charge, e^2) contribute to the kinetic equation for electrons and describe both the self-consistent (mean) field and exchange effects. The kinetic equation written with this accuracy is used below to study the shift of intersubband resonance in quantum wells and the exciton absorption. The electron-electron scattering is described by the collision integral proportional to e^4. The matrix elements of electron-electron scattering have to be considered in more detail for the scattering with small momentum transfer, when the effects of dynamical screening are important. The electron-electron interaction can be directly probed by the Coulomb drag between the electrons in closely placed parallel 2D layers.

28. Hartree-Fock Approximation

To describe the electron systems with interaction between the particles, we use the Hamiltonian (4.10)

$$\sum_j \hat{h}_j + \widehat{H}_{ee}, \tag{1}$$

229

where $\hat{h}_j$ is the one-electron Hamiltonian and $\widehat{H}_{ee}$ is the Hamiltonian of the potential energy of electron-electron interaction. In the second quantization representation, the Hamiltonian $\widehat{H}_{ee}$ is written according to Eq. (4.24) in the form

$$\widehat{H}_{ee} = \frac{1}{2} \sum_{\gamma_1\gamma_2\gamma_3\gamma_4} \Phi_{\gamma_1\gamma_2\gamma_3\gamma_4} \hat{a}_{\gamma_1}^+ \hat{a}_{\gamma_2}^+ \hat{a}_{\gamma_3} \hat{a}_{\gamma_4}. \tag{2}$$

The matrix element $\Phi_{\gamma_1\gamma_2\gamma_3\gamma_4} = \int d\mathbf{x} \int d\mathbf{x}' \phi_{\mathbf{x}}^{(\gamma_1)*} \phi_{\mathbf{x}'}^{(\gamma_2)*} v(\mathbf{x},\mathbf{x}') \phi_{\mathbf{x}}^{(\gamma_4)} \phi_{\mathbf{x}'}^{(\gamma_3)}$ will be written below in the form valid for any dimensionality. Using the spatial Fourier transformation of the Coulomb energy $v(\mathbf{x},\mathbf{x}') = (e^2/\epsilon)|\mathbf{x}-\mathbf{x}'|^{-1}$, where ϵ is the dielectric permittivity of the medium, we have

$$\Phi_{\gamma_1\gamma_2\gamma_3\gamma_4} = \frac{1}{V} \sum_{\mathbf{q}} {}' v_q \langle\gamma_1|e^{-i\mathbf{q}\cdot\mathbf{x}}|\gamma_4\rangle \langle\gamma_2|e^{i\mathbf{q}\cdot\mathbf{x}}|\gamma_3\rangle, \quad v_q = \frac{4\pi e^2}{\epsilon q^2}. \tag{3}$$

We note that though the indices γ include the spin, the matrix element $\Phi_{\gamma_1\gamma_2\gamma_3\gamma_4}$ is spin-independent, i.e., contains the δ-symbols of spin states γ_1, γ_4 and γ_2, γ_3. In the case of low-dimensional electrons, the eigenstates $|\gamma\rangle$ describe confinement in one or two dimensions; see Eq. (5.25). Summing the matrix element (3) over the transverse components of the wave vector, one finds the contribution characterized by the momentum transferred in the 2D plane or along the 1D channel. As a result, v_q is replaced by the effective Fourier component of the Coulomb interaction potential in the 2D or 1D systems (problem 6.1).

Apart from the electron-electron interaction described by Eq. (2), the Hamiltonian of electrons must contain the interaction of electrons with positive static background charges (for example, the lattice ions in metals or doping impurities in semiconductors). If the electron distribution is spatially smooth (in comparison to the lattice constant or to the average distance between the impurities), the positive background can be treated as spatially homogeneous. In this case, the potential of this background exactly compensates the term with $\mathbf{q} = 0$ so that this term vanishes from the sum in Eq. (3). This fact is reflected by the prime sign at the sum in Eq. (3), which means that the sum is taken over non-zero $\mathbf{q}$. The contribution of small $\mathbf{q}$ in Eq. (3) appears to be ineffective in Eq. (2) because of screening effects. These effects, considered in the following on the basis of self-consistent description of many-electron system, lead to finite (proportional to e^2) terms in the denominator of v_q. The consideration given below is based upon iterations with respect to the interaction. It can account for the screening within a logarithmic accuracy (in the 3D case), when the sum over $\mathbf{q}$ is cut below $q \simeq r_{sc}^{-1}$ (here r_{sc} is the screening length).

Since the operator $\widehat{H}_{ee}$ is Hermitian, the matrix element (3) has the symmetry property $\Phi_{\gamma_1\gamma_2\gamma_3\gamma_4} = \Phi_{\gamma_3\gamma_4\gamma_1\gamma_2}^*$. Another symmetry property, $\Phi_{\gamma_1\gamma_2\gamma_3\gamma_4} = \Phi_{\gamma_2\gamma_1\gamma_4\gamma_3}$, directly follows from Eq. (3) and originates from the identity of electrons. Using the anticommutation property of the Fermi operators, we permute the indices as $\gamma_3 \leftrightarrow \gamma_4$ and find $\widehat{H}_{ee} = -(1/2)\sum_{\gamma_{1-4}} \Phi_{\gamma_1\gamma_2\gamma_4\gamma_3}\hat{a}_{\gamma_1}^+\hat{a}_{\gamma_2}^+\hat{a}_{\gamma_3}\hat{a}_{\gamma_4}$ so that the form (2) appears to be ambiguous. In some cases, it is convenient to use the form which is antisymmetric with respect to the indices γ_3 and γ_4:

$$\widetilde{\Phi}_{\gamma_1\gamma_2\gamma_3\gamma_4} = \frac{1}{2V}\sum_{\mathbf{q}}{}' v_q \left[\langle\gamma_1|e^{-i\mathbf{q}\cdot\mathbf{x}}|\gamma_4\rangle\langle\gamma_2|e^{i\mathbf{q}\cdot\mathbf{x}}|\gamma_3\rangle \right.$$

$$\left. -\langle\gamma_1|e^{-i\mathbf{q}\cdot\mathbf{x}}|\gamma_3\rangle\langle\gamma_2|e^{i\mathbf{q}\cdot\mathbf{x}}|\gamma_4\rangle \right]. \tag{4}$$

We stress that this ambiguity does not have any effect on the kinetic equation derived below (problem 6.2).

To obtain an equation for the one-electron density matrix $\hat{\rho}_t$, we average Eq. (1.20) according to the transformations given by Eqs. (4.29)−(4.31). We also use the definition (2) and the commutator

$$\left[\hat{a}_\alpha^+\hat{a}_\beta, \hat{a}_{\gamma_1}^+\hat{a}_{\gamma_2}^+\hat{a}_{\gamma_3}\hat{a}_{\gamma_4}\right] = \delta_{\gamma_1\beta}\hat{a}_\alpha^+\hat{a}_{\gamma_2}^+\hat{a}_{\gamma_3}\hat{a}_{\gamma_4}$$

$$+\delta_{\gamma_2\beta}\hat{a}_{\gamma_1}^+\hat{a}_\alpha^+\hat{a}_{\gamma_3}\hat{a}_{\gamma_4} - \delta_{\gamma_3\alpha}\hat{a}_{\gamma_1}^+\hat{a}_{\gamma_2}^+\hat{a}_\beta\hat{a}_{\gamma_4} - \delta_{\gamma_4\alpha}\hat{a}_{\gamma_1}^+\hat{a}_{\gamma_2}^+\hat{a}_{\gamma_3}\hat{a}_\beta. \tag{5}$$

As a result, we obtain the equation

$$\frac{\partial}{\partial t}\langle\beta|\hat{\rho}_t|\alpha\rangle + \frac{i}{\hbar}\langle\beta|[\hat{h}_t, \hat{\rho}_t]|\alpha\rangle$$

$$= -\frac{i}{2\hbar}\sum_{\gamma_{1-4}} \Phi_{\gamma_1\gamma_2\gamma_3\gamma_4} \left\{ \delta_{\gamma_1\beta}\langle\langle\hat{a}_\alpha^+\hat{a}_{\gamma_2}^+\hat{a}_{\gamma_3}\hat{a}_{\gamma_4}\rangle\rangle_t + \delta_{\gamma_2\beta}\langle\langle\hat{a}_{\gamma_1}^+\hat{a}_\alpha^+\hat{a}_{\gamma_3}\hat{a}_{\gamma_4}\rangle\rangle_t \right. \tag{6}$$

$$\left. -\delta_{\gamma_3\alpha}\langle\langle\hat{a}_{\gamma_1}^+\hat{a}_{\gamma_2}^+\hat{a}_\beta\hat{a}_{\gamma_4}\rangle\rangle_t - \delta_{\gamma_4\alpha}\langle\langle\hat{a}_{\gamma_1}^+\hat{a}_{\gamma_2}^+\hat{a}_{\gamma_3}\hat{a}_\beta\rangle\rangle_t \right\}$$

whose left-hand side describes one-electron evolution; see Eq. (4.32). The two-electron averages $\langle\langle\hat{a}_{\gamma_1}^+\hat{a}_{\gamma_2}^+\hat{a}_{\gamma_3}\hat{a}_{\gamma_4}\rangle\rangle_t$ on the right-hand side are defined as $\mathrm{Sp}\{\hat{a}_{\gamma_1}^+\hat{a}_{\gamma_2}^+\hat{a}_{\gamma_3}\hat{a}_{\gamma_4}\hat{\eta}_t\}$. It is convenient to separate the contribution of non-interacting electrons, $F_{\gamma_1\gamma_2\gamma_3\gamma_4}(t)$, from the electron-electron correlation function $g_{\gamma_1\gamma_2\gamma_3\gamma_4}(t)$ in these averages:

$$\langle\langle\hat{a}_{\gamma_1}^+\hat{a}_{\gamma_2}^+\hat{a}_{\gamma_3}\hat{a}_{\gamma_4}\rangle\rangle_t = F_{\gamma_1\gamma_2\gamma_3\gamma_4}(t) + g_{\gamma_1\gamma_2\gamma_3\gamma_4}(t). \tag{7}$$

The function $F_{\gamma_1\gamma_2\gamma_3\gamma_4}(t)$ is obtained according to Eq. (19.7):

$$F_{\gamma_1\gamma_2\gamma_3\gamma_4}(t) = \langle\langle \hat{a}^+_{\gamma_1}\hat{a}_{\gamma_4}\rangle\rangle_t \langle\langle \hat{a}^+_{\gamma_2}\hat{a}_{\gamma_3}\rangle\rangle_t - \langle\langle \hat{a}^+_{\gamma_1}\hat{a}_{\gamma_3}\rangle\rangle_t \langle\langle \hat{a}^+_{\gamma_2}\hat{a}_{\gamma_4}\rangle\rangle_t$$

$$\equiv \langle\gamma_4|\hat{\rho}_t|\gamma_1\rangle\langle\gamma_3|\hat{\rho}_t|\gamma_2\rangle - \langle\gamma_3|\hat{\rho}_t|\gamma_1\rangle\langle\gamma_4|\hat{\rho}_t|\gamma_2\rangle. \tag{8}$$

Substituting the expression (7) into the right-hand side of Eq. (6), we obtain

$$\frac{\partial}{\partial t}\langle\beta|\hat{\rho}_t|\alpha\rangle + \frac{i}{\hbar}\langle\beta|[\hat{h}_t,\hat{\rho}_t]|\alpha\rangle + \frac{i}{2\hbar}\sum_{\gamma_{1-4}}\Phi_{\gamma_1\gamma_2\gamma_3\gamma_4}\left\{\delta_{\gamma_1\beta}F_{\alpha\gamma_2\gamma_3\gamma_4}(t)\right.$$

$$\left. +\delta_{\gamma_2\beta}F_{\gamma_1\alpha\gamma_3\gamma_4}(t) - \delta_{\gamma_3\alpha}F_{\gamma_1\gamma_2\beta\gamma_4}(t) - \delta_{\gamma_4\alpha}F_{\gamma_1\gamma_2\gamma_3\beta}(t)\right\} = \mathcal{J}_{\beta\alpha}(t). \tag{9}$$

The collision integral $\mathcal{J}_{\beta\alpha}(t)$ standing on the right-hand side of Eq. (9) is written through the electron-electron correlation functions,

$$\mathcal{J}_{\beta\alpha}(t) = -\frac{i}{2\hbar}\sum_{\gamma_{1-4}}\Phi_{\gamma_1\gamma_2\gamma_3\gamma_4}\left\{\delta_{\gamma_1\beta}g_{\alpha\gamma_2\gamma_3\gamma_4}(t)\right.$$

$$\left. +\delta_{\gamma_2\beta}g_{\gamma_1\alpha\gamma_3\gamma_4}(t) - \delta_{\gamma_3\alpha}g_{\gamma_1\gamma_2\beta\gamma_4}(t) - \delta_{\gamma_4\alpha}g_{\gamma_1\gamma_2\gamma_3\beta}(t)\right\}, \tag{10}$$

and will be considered later in this chapter.

The last term on the left-hand side of Eq. (9) describes a renormalization of the one-electron Hamiltonian in the first order in the interaction. This contribution seems to be qualitatively different from the renormalization due to electron scattering by phonons and impurities, appearing in the second order in the interaction. The difference becomes less prominent if we take into account that the Coulomb interaction also can be viewed as a process of the second order caused by exchange of the longitudinal photons between the electrons. Employing Eqs. (3) and (8), we transform the first term of e^2-contribution from the left-hand side of Eq. (9) as

$$\sum_{\gamma_{2-4}}\Phi_{\beta\gamma_2\gamma_3\gamma_4}F_{\alpha\gamma_2\gamma_3\gamma_4}(t) = \frac{1}{V}\sum_{\mathbf{q}}{}' v_q\langle\beta|e^{i\mathbf{q}\cdot\mathbf{x}}\hat{\rho}_t|\alpha\rangle\mathrm{sp}\left(e^{-i\mathbf{q}\cdot\mathbf{x}}\hat{\rho}_t\right)$$

$$-\frac{1}{V}\sum_{\mathbf{q}}{}' v_q\langle\beta|e^{-i\mathbf{q}\cdot\mathbf{x}}\hat{\rho}_t e^{i\mathbf{q}\cdot\mathbf{x}}\hat{\rho}_t|\alpha\rangle, \tag{11}$$

and the other terms are transformed in a similar way. Note that we use the possibility to change the sign of $\mathbf{q}$ in the exponents under the sum because v_q is symmetric in $\mathbf{q}$. Introducing the Fourier component of the electron density,

$$n_{\mathbf{q}t} = \mathrm{sp}\left(e^{-i\mathbf{q}\cdot\mathbf{x}}\hat{\rho}_t\right), \tag{12}$$

we rewrite the e^2-contribution on the left-hand side of Eq. (9) in the following way:

$$\frac{i}{\hbar V} \sum_{\mathbf{q}}{}' v_q n_{\mathbf{q}t} \langle \beta | e^{i\mathbf{q}\cdot\mathbf{x}} \hat{\rho}_t - \hat{\rho}_t e^{i\mathbf{q}\cdot\mathbf{x}} | \alpha \rangle$$

$$-\frac{i}{\hbar V} \sum_{\mathbf{q}}{}' v_q \langle \beta | e^{-i\mathbf{q}\cdot\mathbf{x}} \hat{\rho}_t e^{i\mathbf{q}\cdot\mathbf{x}} \hat{\rho}_t - \hat{\rho}_t e^{-i\mathbf{q}\cdot\mathbf{x}} \hat{\rho}_t e^{i\mathbf{q}\cdot\mathbf{x}} | \alpha \rangle. \tag{13}$$

This contribution is expressed through the commutators of the density matrix and can be combined with the one-electron Hamiltonian $\hat{h}_t$. In this way we obtain the effective Hamiltonian

$$\widetilde{h}_t = \hat{h}_t + \frac{1}{V} \sum_{\mathbf{q}}{}' v_q n_{\mathbf{q}t} e^{i\mathbf{q}\cdot\mathbf{x}} - \frac{1}{V} \sum_{\mathbf{q}}{}' v_q e^{-i\mathbf{q}\cdot\mathbf{x}} \hat{\rho}_t e^{i\mathbf{q}\cdot\mathbf{x}} \tag{14}$$

describing the effects of Coulomb interaction in the Hartree-Fock approximation. With the use of Eq. (14), the quantum kinetic equation is written in its conventional form

$$\frac{\partial \hat{\rho}_t}{\partial t} + \frac{i}{\hbar} \left[\widetilde{h}_t, \hat{\rho}_t \right] = \widehat{J}_{ee}(\hat{\rho}|t), \tag{15}$$

where the operator of the collision integral is defined by its matrix elements $\mathcal{J}_{\beta\alpha}(t) = \langle \beta | \widehat{J}_{ee}(\hat{\rho}|t) | \alpha \rangle$ determined by Eq. (10). The last term of the Hamiltonian (14) describes the exchange effects, while the second one defines the mean-field potential energy $U_{\mathbf{r}t} = V^{-1} \sum_{\mathbf{q}}' v_q n_{\mathbf{q}t} \exp(i\mathbf{q}\cdot\mathbf{r})$ at the point $(\mathbf{r}, t)$. This energy as well can be determined from the Poisson equation

$$\Delta U_{\mathbf{r}t} = -\frac{4\pi e^2}{\epsilon} n_{\mathbf{r}t}, \tag{16}$$

where $n_{\mathbf{r}t} = \mathrm{sp}\{\delta(\mathbf{r}-\mathbf{x})\hat{\rho}_t\}$ and $n_{\mathbf{q}t}$ are related through a spatial Fourier transformation. Considering any electroneutral system, one should add the contribution of the positive background charge to the right-hand side of the Poisson equation.

Let us consider spatially smooth distributions of electrons and rewrite the left-hand side of Eq. (15) in the Wigner representation denoted below as $(...)_{\mathbf{rp}}$. According to Eq. (9.24), the commutators are written through the Poisson brackets

$$\left([\hat{a}, \hat{b}] \right)_{\mathbf{rp}} \simeq i\hbar \left(\frac{\partial a}{\partial \mathbf{r}} \cdot \frac{\partial b}{\partial \mathbf{p}} - \frac{\partial a}{\partial \mathbf{p}} \cdot \frac{\partial b}{\partial \mathbf{r}} \right), \tag{17}$$

and the mean field leads to the usual "force" term $(i/\hbar)([U_{\mathbf{x}t}, \hat{\rho}_t])_{\mathbf{rp}} = -\nabla U_{\mathbf{r}t} \cdot \partial f_{\mathbf{rp}t}/\partial \mathbf{p}$ in the kinetic equation. The exchange term can be transformed according to Eq. (9.25):

$$\left(e^{-i\mathbf{q}\cdot\mathbf{x}}\widehat{A}e^{i\mathbf{q}\cdot\mathbf{x}}\right)_{\mathbf{rp}} = A_{\mathbf{rp}+\hbar\mathbf{q}}, \tag{18}$$

and its contribution to the kinetic equation is written as

$$-\frac{i}{\hbar V}\sum_{\mathbf{q}} v_q \left(\left[e^{-i\mathbf{q}\cdot\mathbf{x}}\hat{\rho}_t e^{i\mathbf{q}\cdot\mathbf{x}}, \hat{\rho}_t\right]\right)_{\mathbf{rp}}$$

$$= \frac{1}{V}\sum_{\mathbf{q}} v_q \left(\frac{\partial f_{\mathbf{rp}+\hbar\mathbf{q}t}}{\partial \mathbf{r}} \cdot \frac{\partial f_{\mathbf{rp}t}}{\partial \mathbf{p}} - \frac{\partial f_{\mathbf{rp}+\hbar\mathbf{q}t}}{\partial \mathbf{p}} \cdot \frac{\partial f_{\mathbf{rp}t}}{\partial \mathbf{r}}\right). \tag{19}$$

Therefore, the left-hand side of the quasi-classical kinetic equation takes the form

$$\left\{\frac{\partial}{\partial t} + \left(\mathbf{v_p} - \frac{\partial \Delta\varepsilon_{\mathbf{rp}t}}{\partial \mathbf{p}}\right) \cdot \frac{\partial}{\partial \mathbf{r}} + \left(\mathbf{F}_{\mathbf{rp}t} - \nabla U_{\mathbf{r}t} + \frac{\Delta\varepsilon_{\mathbf{rp}t}}{\partial \mathbf{r}}\right) \cdot \frac{\partial}{\partial \mathbf{p}}\right\} f_{\mathbf{rp}t}, \tag{20}$$

where the exchange correction to the energy depends on coordinate through the Wigner distribution function:

$$\Delta\varepsilon_{\mathbf{rp}t} = \frac{1}{V}\sum_{\mathbf{q}} v_q f_{\mathbf{rp}+\hbar\mathbf{q}t} = \frac{4\pi e^2}{\epsilon}\int \frac{d\mathbf{q}}{(2\pi)^3 q^2} f_{\mathbf{rp}+\hbar\mathbf{q}t}. \tag{21}$$

In summary, the kinetic equation describes the quasiparticles with the spectrum $p^2/2m - \Delta\varepsilon_{\mathbf{rp}t}$. In the spatially homogeneous and equilibrium case, Eq. (21) is rewritten as

$$\Delta\varepsilon_p = \frac{e^2}{2\pi^2\hbar\epsilon}\int \frac{d\mathbf{p}_1\theta(\varepsilon_F - \varepsilon_{p_1})}{|\mathbf{p} - \mathbf{p}_1|^2}, \tag{22}$$

where we also assume that the electron gas is degenerate. The renormalization energy $\Delta\varepsilon_p$ is estimated as $e^2/\epsilon\lambda_F$, where $\lambda_F = 2\pi/k_F$ is the length of electron wave expressed here through the Fermi wave number k_F of degenerate electron gas. Since the system is isotropic, the function (22) depends only on p/p_F; see problem 6.3 and Fig. 6.1.

The consideration is more complicated for the low-dimensional case. The renormalization of the j-th subband energy is given by

$$\Delta\varepsilon_{jp} = \frac{1}{V}\sum_{\mathbf{q}}{}' v_q \langle j\mathbf{p}|e^{-i\mathbf{q}\cdot\mathbf{x}}\hat{\rho}e^{i\mathbf{q}\cdot\mathbf{x}}|j\mathbf{p}\rangle . \tag{23}$$

For example, in the case of electron confinement in one direction (along OZ) in a quantum well, we have

$$\Delta\varepsilon_{jp} = \frac{4\pi e^2}{\epsilon}\sum_{j'}\int \frac{dq_z d\mathbf{q}}{(2\pi)^3}\frac{\left|\langle j|e^{iq_z z}|j'\rangle\right|^2}{q_z^2 + q^2} f_{j'\mathbf{p}+\hbar\mathbf{q}}, \tag{24}$$

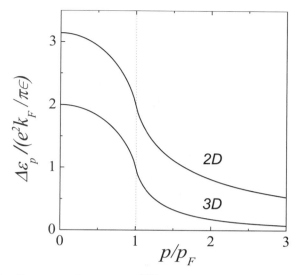

Figure 6.1. Functions $\Delta\varepsilon_p$ and $\Delta\varepsilon_p^{(2D)}$ given by Eqs. (28.22) and (28.25).

where $\mathbf{q} = (q_x, q_y)$ is the 2D wave vector and $f_{j'\mathbf{p}}$ is the distribution function for subband j'. Strictly speaking, Eq. (24) does not give us a solution of the problem of energy renormalization, because the electron-electron interaction deforms the confinement potential so that the wave functions $|j\rangle$ have to be found from a self-consistent calculation. Even when the exchange contribution is neglected, such self-consistent procedure is rather complicated since it requires solution of coupled Schroedinger and Poisson equations. However, in the 2D limit, when only one 2D subband is populated $(j' = j)$ and $pd/\hbar \ll 1$, where d is the quantum well width, the result appears to be independent of the explicit form of the wave functions:

$$\Delta\varepsilon_p^{(2D)} = \frac{e^2}{2\pi\hbar\epsilon} \int d\mathbf{p}_1 \frac{\theta(\varepsilon_F - \varepsilon_{p_1})}{|\mathbf{p} - \mathbf{p}_1|}. \tag{25}$$

The energy $\Delta\varepsilon_p^{(2D)}$, as a function of p/p_F, is also plotted in Fig. 6.1 (problem 6.3).

29. Shift of Intersubband Resonance

Consider the resonant response of electrons in quantum wells to electromagnetic radiation, when the transverse (perpendicular to the plane XOY of the quantum well) component of electric field, $E \exp(-i\omega t)$, excites intersubband transitions of the electrons. As a result of this excitation, the coordinate-dependent current density $I_{\omega z} \exp(-i\omega t)$ appears in the system. It is convenient to characterize the response by the Fourier component of the current density per quantum well, I_ω, defined as $I_\omega = \int dz I_{\omega z}$. Analogous to Eqs. (13.5)–(13.8), I_ω is given by

$$I_\omega = i\frac{e^2 n}{m\omega}E + \frac{e}{L^2}\mathrm{sp}\hat{v}_z\widehat{\delta\rho},\tag{1}$$

where $n = L^{-2}\mathrm{sp}\hat{\rho}$ is the density of electrons per unit square of the quantum well, L^2 is the normalization square, and $\hat{v}_z = \hat{p}_z/m$ is the transverse component of the velocity operator. The trace sp... in Eq. (1) is taken over electron variables including the spin. The response to the perturbation $(ie/\omega)E\hat{v}_z$ is described by the linearized kinetic equation for the non-equilibrium part $\widehat{\delta\rho}$ of the density matrix:

$$-i\omega\widehat{\delta\rho} + \frac{i}{\hbar}[\tilde{h}, \widehat{\delta\rho}] + \frac{i}{\hbar}[\widehat{\delta h}, \hat{\rho}] = \widehat{\Delta J},\tag{2}$$

where $\hat{\rho}$ is the equilibrium density matrix and $\widehat{\Delta J}$ is the linearized collision integral in the operator form. The perturbation $\widehat{\delta h}$ is obtained after linearizing the effective Hamiltonian $\tilde{h}$ given by Eq. (28.14):

$$\widehat{\delta h} = \frac{ie}{\omega}E\hat{v}_z + \frac{1}{V}\sum_{\mathbf{q}}{}' v_q\left[\delta n_{\mathbf{q}}e^{i\mathbf{q}\cdot\mathbf{x}} - e^{-i\mathbf{q}\cdot\mathbf{x}}\widehat{\delta\rho}e^{i\mathbf{q}\cdot\mathbf{x}}\right],\tag{3}$$

where $\delta n_{\mathbf{q}} = \mathrm{sp}(e^{-i\mathbf{q}\cdot\mathbf{x}}\widehat{\delta\rho})$ is the high-frequency Fourier component of the perturbation of electron density.

It is convenient to rewrite $\tilde{h}$ by separating the mean-field potential U_z, according to

$$\tilde{h} = \hat{h} + U_z - \frac{1}{V}\sum_{\mathbf{q}}{}' v_q e^{-i\mathbf{q}\cdot\mathbf{x}}\hat{\rho}e^{i\mathbf{q}\cdot\mathbf{x}}.\tag{4}$$

Below we use the basis $|j\sigma\mathbf{p}\rangle$ determined by the eigenstate problem $(\hat{h} + U_z)|j\sigma\mathbf{p}\rangle = \varepsilon_{j\mathbf{p}}|j\sigma\mathbf{p}\rangle$, where $\mathbf{p}$ is the 2D momentum, σ is the spin index, and the index j numbers the states of transverse motion (discrete subbands in the well and continuous spectrum above the well). Using the single-particle approach, one can obtain an explicit expression for the current density I_ω after solving Eq. (2) in the collisionless approximation. Such a consideration demonstrates that the second term in Eq. (1) decreases with increasing ω faster than ω^{-1} at high frequencies corresponding to electron transitions into the continuous spectrum (problem 6.4). Thus, the first term in Eq. (1) dominates in this spectral region, and the response remains the same as in the case of free electrons; see Eq. (8.30). On the other hand, if $\omega \to 0$, the response of localized electrons goes to zero. One can prove this statement by using a relation similar to the f-sum rule (see again problem 6.4).

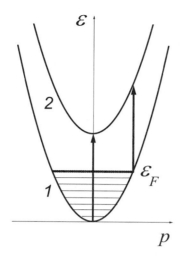

Figure 6.2. Energy spectrum of a two-level system in a quantum well. The arrows show the optical transitions.

The most interesting case is realized when the frequency ω is close to one of the frequencies of intersubband transitions. The effects of Coulomb interaction in these conditions become essential. We restrict ourselves by a pair of levels, $j = 1, 2$, between which the resonant transitions are excited, the lower level ($j = 1$) is occupied by electrons and the higher one ($j = 2$) is empty; see Fig. 6.2. The matrix elements of the velocity operator are diagonal in σ and $\mathbf{p}$: $\langle j\sigma\mathbf{p}|\hat{v}_z|j'\sigma'\mathbf{p}'\rangle = \delta_{\sigma\sigma'}\delta_{\mathbf{pp}'}\langle j|\hat{v}_z|j'\rangle$. Since we consider the localized states, the diagonal components $\langle j|\hat{v}_z|j\rangle$ are equal to zero, while the non-diagonal ones are imaginary: $\langle 2|\hat{v}_z|1\rangle \equiv v_{21}^z = -\langle 1|\hat{v}_z|2\rangle$. The induced current density is rewritten in the form

$$I_\omega = i\frac{e^2 n}{m\omega}E + \frac{2ev_{21}^z}{L^2}\sum_{\mathbf{p}}[\delta f_{12}(\mathbf{p}) - \delta f_{21}(\mathbf{p})], \qquad (5)$$

where $\delta f_{jj'}(\mathbf{p}) \equiv \langle j\mathbf{p}|\widehat{\delta\rho}|j'\mathbf{p}\rangle$ is the non-equilibrium part of the density matrix, and the factor of 2 comes from the sum over spin. From Eqs. (2)−(4) we find that $\delta f_{jj'}(\mathbf{p})$ is governed by the following system of equations:

$$\left(-i\omega + \frac{i\varepsilon_{21}(p)}{\hbar} + \nu\right)\delta f_{21}(\mathbf{p}) + \frac{i}{\hbar}\delta h_{21}f_{1p} = 0,$$

$$\left(-i\omega + \frac{i\varepsilon_{12}(p)}{\hbar} + \nu\right)\delta f_{12}(\mathbf{p}) - \frac{i}{\hbar}\delta h_{12}f_{1p} = 0, \qquad (6)$$

where f_{1p} is the equilibrium distribution function and $\delta h_{jj'}$ is the matrix element of the perturbation operator (3). The linearized collision integral $\langle j\mathbf{p}|\widehat{\Delta J}|j'\mathbf{p}\rangle$ is approximated in Eq. (6) as $-\nu\delta f_{jj'}(\mathbf{p})$, where ν is a phenomenological relaxation rate. The energy separation between the levels, $\varepsilon_{21}(p) = -\varepsilon_{12}(p)$, is renormalized by the exchange contribution according to Eq. (28.24):

$$\varepsilon_{21}(p) = \tilde{\varepsilon}_{21} - \frac{4\pi e^2}{\epsilon}\int \frac{dq_z d\mathbf{q}}{(2\pi)^3}\frac{|\langle 2|e^{iq_z z}|1\rangle|^2 - |\langle 1|e^{iq_z z}|1\rangle|^2}{q_z^2 + q^2}f_{1\mathbf{p}+\hbar\mathbf{q}}, \quad (7)$$

where $\tilde{\varepsilon}_{21}$ is the bare interlevel energy separation (determined by the quantum well confinement potential and by the self-consistent potential U_z) and $\mathbf{q} = (q_x, q_y)$ is the 2D wave vector.

Since the induced current (5) is expressed only through the difference of the non-diagonal components, it is convenient to rewrite the system of equations (6) by using $\delta f_{\mathbf{p}}^{(\pm)} = \delta f_{12}(\mathbf{p}) \pm \delta f_{21}(\mathbf{p})$, which leads to

$$(\omega + i\nu)\delta f_{\mathbf{p}}^{(+)} + \omega_{21}(p)\delta f_{\mathbf{p}}^{(-)} - f_{1p}[\delta h_{21} - \delta h_{12}]/\hbar = 0,$$

$$(\omega + i\nu)\delta f_{\mathbf{p}}^{(-)} + \omega_{21}(p)\delta f_{\mathbf{p}}^{(+)} + f_{1p}[\delta h_{21} + \delta h_{12}]/\hbar = 0, \quad (8)$$

where $\omega_{21}(p) = \varepsilon_{21}(p)/\hbar$. The matrix elements of the perturbation operator (3) are transformed to the following forms:

$$\delta h_{21} - \delta h_{12} = \frac{2ie}{\omega}Ev_{21}^z + \int\frac{d\mathbf{p}_1}{2\pi m}\sum_{ab}\{2\left[M_{12ba}(0) - M_{21ba}(0)\right]$$

$$+M_{1ab2}\left(|\mathbf{p} - \mathbf{p}_1|/\hbar\right) - M_{2ab1}\left(|\mathbf{p} - \mathbf{p}_1|/\hbar\right)\}\delta f_{ab}(\mathbf{p}_1),$$

$$\delta h_{21} + \delta h_{12} = \int\frac{d\mathbf{p}_1}{2\pi m}\sum_{ab}\{2[M_{12ba}(0) + M_{21ba}(0)]$$

$$-M_{1ab2}\left(|\mathbf{p} - \mathbf{p}_1|/\hbar\right) - M_{2ab1}\left(|\mathbf{p} - \mathbf{p}_1|/\hbar\right)\}\delta f_{ab}(\mathbf{p}_1), \quad (9)$$

where we have introduced a dimensionless kernel

$$M_{abcd}(q) = \frac{e^2 m}{\pi\epsilon\hbar^2}\int_{-\infty}^{\infty}dq_z\frac{\langle a|e^{-iq_z z}|b\rangle\langle c|e^{iq_z z}|d\rangle}{q_z^2 + q^2}. \quad (10)$$

Calculating the integral over q_z, we rewrite Eq. (10) as (see also problem 6.1)

$$M_{abcd}(q) = (a_B q)^{-1}\int dz\psi_z^{(a)}\psi_z^{(b)}\int dz'\psi_{z'}^{(c)}\psi_{z'}^{(d)}e^{-q|z-z'|}, \quad (11)$$

where $a_B = \epsilon\hbar^2/e^2 m$ is the Bohr radius and $\psi_z^{(a)}$ is the wave function describing confinement of the state a (the wave functions are chosen to be real). Using Eq. (7), we express the frequency of interlevel transitions standing in Eq. (8) through the kernel (11) according to

$$\omega_{21}(p) = \omega_{21} - \int \frac{d\mathbf{p}_1}{2\pi\hbar m} \left[M_{2112}\left(\frac{|\mathbf{p} - \mathbf{p}_1|}{\hbar}\right) - M_{1111}\left(\frac{|\mathbf{p} - \mathbf{p}_1|}{\hbar}\right) \right] f_{1p_1} \,,$$

(12)

where $\omega_{21} = \tilde{\varepsilon}_{21}/\hbar$.

Therefore, the linear response is described by a system of two non-homogeneous integral equations for $\delta f_{\mathbf{p}}^{(\pm)}$. If one uses $f_{1p} = \theta(\varepsilon_F - p^2/2m)$ corresponding to the Fermi distribution at zero temperature, $\delta f_{\mathbf{p}}^{(\pm)}$ are defined in the interval $0 < p < p_F$. Since the kernels of the kind $M_{abcd}(q)$ are invariant with respect to the permutations $a \leftrightarrow b$, $c \leftrightarrow d$, and $(ab) \leftrightarrow (cd)$, the contribution of $M_{...}(0)$ to $\delta h_{21} - \delta h_{12}$ vanishes and the system (8) is rewritten as

$$(\omega + i\nu)\delta f_{\mathbf{p}}^{(+)} + \omega_{21}\delta f_{\mathbf{p}}^{(-)} - \int \frac{d\mathbf{p}_1}{2\pi\hbar m} \left\{ \left[M_{1212}\left(\frac{|\mathbf{p} - \mathbf{p}_1|}{\hbar}\right) \right. \right.$$

$$\left. - M_{1111}\left(\frac{|\mathbf{p} - \mathbf{p}_1|}{\hbar}\right) \right] f_{1p_1}\delta f_{\mathbf{p}}^{(-)} - \left[M_{1212}\left(\frac{|\mathbf{p} - \mathbf{p}_1|}{\hbar}\right) \right.$$

$$\left. \left. - M_{1122}\left(\frac{|\mathbf{p} - \mathbf{p}_1|}{\hbar}\right) \right] f_{1p}\delta f_{\mathbf{p}_1}^{(-)} \right\} = \frac{2ie}{\hbar\omega} E v_{21}^z f_{1p} \,,$$

$$(\omega + i\nu)\delta f_{\mathbf{p}}^{(-)} + \omega_{21}\delta f_{\mathbf{p}}^{(+)} - \int \frac{d\mathbf{p}_1}{2\pi\hbar m} \left\{ \left[M_{1212}\left(\frac{|\mathbf{p} - \mathbf{p}_1|}{\hbar}\right) \right. \right.$$

$$\left. - M_{1111}\left(\frac{|\mathbf{p} - \mathbf{p}_1|}{\hbar}\right) \right] f_{1p_1}\delta f_{\mathbf{p}}^{(+)} + \left[M_{1212}\left(\frac{|\mathbf{p} - \mathbf{p}_1|}{\hbar}\right) \right.$$

$$\left. \left. + M_{1122}\left(\frac{|\mathbf{p} - \mathbf{p}_1|}{\hbar}\right) - 4M_{1212}(0) \right] f_{1p}\delta f_{\mathbf{p}_1}^{(+)} \right\} = 0.$$

(13)

The mean-field potential contribution has disappeared from the upper equation and remains only in the lower one, through the term proportional to $M_{1212}(0)$. We stress that, owing to isotropy of the electron energy spectrum, the functions $\delta f_{\mathbf{p}}^{(\pm)}$ do not depend on the direction of $\mathbf{p}$. Therefore, the integration of the kernels $M_{abcd}(|\mathbf{p} - \mathbf{p}_1|/\hbar)$ over the angle of $\mathbf{p}_1$ in Eq. (13) can be done independently, and Eq. (13) is reduced to a system of integral equations with a single variable of integration, $p_1 = |\mathbf{p}_1|$.

A simple solution of Eq. (13) can be obtained in the 2D limit, when the kernels $M_{...}(q)$ are expanded in power series of small parameter $p_F d/\hbar$ (d is the well width):

$$M_{abcd}(q) = \frac{\delta_{ab}\delta_{cd}}{a_B q} - \frac{L_{abcd}}{a_B} + O\left(\frac{d}{a_B}\frac{p_F d}{\hbar}\right),$$

$$L_{abcd} = \int dz\,\psi_z^{(a)}\psi_z^{(b)}\int dz'\,\psi_{z'}^{(c)}\psi_{z'}^{(d)}|z - z'|. \tag{14}$$

After this expansion, Eq. (13) is rewritten as

$$(\omega + i\nu)\delta f_{\mathbf{p}}^{(+)} + \omega_{21}\delta f_{\mathbf{p}}^{(-)} + \int \frac{d\mathbf{p}_1}{2\pi\hbar m a_B}\Bigg\{(L_{1212} - L_{1111})f_{1p_1}\delta f_{\mathbf{p}}^{(-)}$$

$$-(L_{1212} - L_{1122})f_{1p}\delta f_{\mathbf{p}_1}^{(-)} + \hbar\frac{f_{1p_1}\delta f_{\mathbf{p}}^{(-)} - f_{1p}\delta f_{\mathbf{p}_1}^{(-)}}{|\mathbf{p} - \mathbf{p}_1|}\Bigg\} = \frac{2ie}{\hbar\omega}Ev_{21}^z f_{1p},$$

$$(\omega + i\nu)\delta f_{\mathbf{p}}^{(-)} + \omega_{21}\delta f_{\mathbf{p}}^{(+)} + \int \frac{d\mathbf{p}_1}{2\pi\hbar m a_B}\Bigg\{(L_{1212} - L_{1111})f_{1p_1}\delta f_{\mathbf{p}}^{(+)}$$

$$-(3L_{1212} - L_{1122})f_{1p}\delta f_{\mathbf{p}_1}^{(+)} + \hbar\frac{f_{1p_1}\delta f_{\mathbf{p}}^{(+)} - f_{1p}\delta f_{\mathbf{p}_1}^{(+)}}{|\mathbf{p} - \mathbf{p}_1|}\Bigg\} = 0. \tag{15}$$

Defining the non-diagonal components of the electron density according to $\delta n_{\pm} = (2/L^2)\sum_{\mathbf{p}}\delta f_{\mathbf{p}}^{(\pm)}$, we find that the second term in the induced current (5) is written as $ev_{21}^z\delta n_-$. The equations for $\delta n_{\pm}$ are obtained from Eq. (15) after integrating the latter over the momentum:

$$(\omega + i\nu)\delta n_+ + \widetilde{\omega}_{21}\delta n_- = \frac{2ie}{\hbar\omega}Ev_{21}^z n,$$

$$(\omega + i\nu)\delta n_- + \overline{\omega}_{21}\delta n_+ = 0. \tag{16}$$

It is important that the integral contributions in Eq. (15) vanish after this procedure, and only the contributions containing L_{abcd} remain. They lead to a renormalization of the frequency of intersubband transitions according to the equations

$$\widetilde{\omega}_{21} = \omega_{21} + \frac{\varepsilon_F}{\hbar}\frac{L_{1122} - L_{1111}}{a_B},$$

$$\overline{\omega}_{21} = \omega_{21} + \frac{\varepsilon_F}{\hbar}\frac{L_{1122} - L_{1111} - 2L_{1212}}{a_B}. \tag{17}$$

Therefore, the system of two integral equations (15) is transformed to the system of algebraic equations (16). Solving it, we obtain the induced current in the form

$$I_\omega = i\frac{e^2 n}{m\omega}E + i\frac{e^2|v_{21}^z|^2}{\hbar\omega}\frac{2\overline{\omega}_{21}nE}{(\omega + i\nu)^2 - \Omega_{21}^2} \equiv \sigma_\omega E, \tag{18}$$

where $\Omega_{21} = \sqrt{\widetilde{\omega}_{21}\overline{\omega}_{21}}$ is the frequency of intersubband resonance shifted because of the Coulomb interaction. We have taken into account that $(v_{21}^z)^2 = -|v_{21}^z|^2$. Equation (18) also defines the complex conductivity σ_ω describing the excitation of electrons in a quantum well by the transverse field.

Let us consider a system of identical quantum wells placed with the period l in a medium with dielectric permittivity ϵ. The dielectric permittivity averaged over the layers is written by analogy to Eq. (17.1),

$$\epsilon_\omega = \epsilon - \frac{4\pi n_{3D}}{m\omega^2}\left[1 + \frac{m|v_{21}^z|^2}{\hbar}\frac{2\overline{\omega}_{21}}{(\omega + i\nu)^2 - \Omega_{21}^2}\right]$$

$$\simeq \epsilon - \frac{4\pi n_{3D}}{m\Omega_{21}^2}\left[1 + \frac{m|v_{21}^z|^2}{\hbar}\sqrt{\frac{\overline{\omega}_{21}}{\widetilde{\omega}_{21}}}\frac{\Delta\omega - i\nu}{(\Delta\omega)^2 + \nu^2}\right], \qquad (19)$$

where $n_{3D} = n/l$ is the effective 3D electron density and $\Delta\omega = \omega - \Omega_{21}$ is the frequency shift from the resonance (detuning frequency). The second equation is written in the resonance approximation, when $|\Delta\omega| \ll \Omega_{21}$. The imaginary part of ϵ_ω describes the absorption on the frequency Ω_{21}, while the contribution to Re $\epsilon_\omega - \epsilon$ caused by the intersubband transitions usually appears to be small. A typical multiple quantum well structure contains no more than one hundred layers so that its total thickness remains smaller than the length of electromagnetic wave. For this reason, the approximation of quasi-homogeneous medium is not adequate in this case, and one needs a more realistic approach to describe the intersubband transitions.

The expression (10.23) for the coefficient of optical absorption cannot be directly applied to the 2D systems, because it has been derived for the absorption on a finite length, while in the 2D systems the absorption coefficient should be attributed to the whole 2D layer. The influence of the 2D electron transitions on the propagation of electromagnetic waves can be described under the approximation that the wavelength $2\pi c/\omega\sqrt{\epsilon}$ is much larger than the width of the 2D layer. In these conditions, the field is constant across the layer, and its influence is described by the boundary conditions imposed on the spatial derivatives of the field. The transverse component of the field, $E_{\omega z}$, satisfies the one-dimensional wave equation (problem 6.5)

$$\left[\frac{d^2}{dz^2} + \left(\frac{\omega}{c}\right)^2 \epsilon\right]E_{\omega z} = -i\frac{4\pi\omega}{c^2}I_{\omega z}, \qquad (20)$$

where the contribution of longitudinal wave vector is neglected and the current density $I_{\omega z}$ is related to the response I_ω of Eq. (18) as $I_\omega = \int dz I_{\omega z}$. Let us assume that the layer is centered at $z = 0$. Taking into account that $E_{\omega z}$ is constant across the layer, we integrate Eq. (20) across the layer and obtain the following boundary condition:

$$\frac{dE_{\omega z}}{dz}\bigg|_{-0}^{+0} + iQ_\omega E_{\omega z=0} = 0, \qquad E_{\omega z}|_{-0}^{+0} = 0, \qquad (21)$$

where $Q_\omega = 4\pi\omega\sigma_\omega/c^2$ is a characteristic wave number.

The general solution of Eq. (20) for the wave incident from the left comprises incident (i) and reflected (r) waves at $z < 0$ and a transmitted (t) wave at $z > 0$:

$$E_{\omega z} = \begin{cases} E_i e^{iq_\omega z} + E_r e^{-iq_\omega z}, & z < 0 \\ E_t e^{iq_\omega z}, & z > 0 \end{cases}, \qquad (22)$$

where $q_\omega = (\omega/c)\sqrt{\epsilon}$. The boundary condition (21), together with the requirement of the field continuity, gives us two equations connecting the amplitudes E_i, E_r, and E_t:

$$\begin{aligned} q_\omega E_t - q_\omega (E_i - E_r) + Q_\omega E_t = 0 \\ E_t = E_i + E_r \end{aligned} \qquad (23)$$

The energy flows transmitted by each of the waves are given by the Poynting vectors $S_l = |E_l|^2 c\sqrt{\epsilon}/2\pi$, $l = i, r, t$. Multiplying Eq. (23) by E_t^*, we find the following relation between the Poynting vectors: $S_i = S_r + S_t(1 + \xi_\omega)$, where the contribution $S_t\xi_\omega$ determines the energy loss inside the quantum well. The dimensionless quantity ξ_ω is, therefore, the relative absorption coefficient of the 2D layer. It is expressed through the real part of the complex conductivity defined by Eq. (18) as

$$\xi_\omega = \frac{\text{Re}Q_\omega}{q_\omega} = \frac{4\pi}{c\sqrt{\epsilon}}\text{Re}\sigma_\omega . \qquad (24)$$

This equation has the same form as Eq. (10.23), and the dielectric constant ϵ in Eq. (24) corresponds to the medium surrounding the quantum well.

Using Eqs. (24) and (18), we obtain the relative absorption coefficient near the resonance frequency:

$$\xi_\omega = \frac{e^2}{\hbar c}\frac{4\pi|v_{21}^z|^2 n}{\sqrt{\epsilon}\Omega_{21}^2}\frac{\overline{\omega}_{21}\nu}{(\omega - \Omega_{21})^2 + \nu^2}, \qquad (25)$$

where we have taken into account that $\Omega_{21} \gg \nu$. Applying the hard-wall model of the quantum well of width d, described by the wave functions (5.20), we find that the characteristic lengths L_{abcd} are proportional to d and can be calculated analytically (problem 6.6). The renormalized frequencies $\widetilde{\omega}_{21}$ and $\overline{\omega}_{21}$ are estimated as $\widetilde{\omega}_{21} \simeq \omega_{21} + 0.063(\varepsilon_F d/\hbar a_B)$ and $\overline{\omega}_{21} \simeq \omega_{21} + 0.288(\varepsilon_F d/\hbar a_B)$, which means that the intersubband transition peak is shifted towards higher frequencies. If one neglects the exchange contributions and accounts only for the mean field in Eq. (13), $\widetilde{\omega}_{21}$ is equal to ω_{21}, while $\overline{\omega}_{21}$ is equal to $\omega_{21} - 4(\varepsilon_F/\hbar)(L_{1212}/a_B) \simeq \omega_{21} + 0.450(\varepsilon_F d/\hbar a_B)$. This leads to a stronger shift called the depolarization effect. Therefore, the exchange contribution weakens the shift of the intersubband resonance, though it does not change the sign of this shift.

30. Exciton Absorption

The square-root frequency dependence of interband absorption in dielectrics and non-doped semiconductors (see Sec. 17) is modified essentially near the fundamental absorption edge even in perfect crystals. Owing to Coulomb interaction between the electron and hole created by the absorbed photon, a new state is formed. This coupled electron-hole state can be viewed as a non-charged quasiparticle called the exciton. The contribution of such states into the absorption, which has not been taken into account in Sec. 17, leads to additional absorption peaks shifted from the fundamental absorption edge by the coupling energies. The absorption in these conditions is considered below both for bulk materials and for the structures with low-dimensional states, where the exciton coupling energies increase substantially.

The interband transitions excited by the homogeneous electric field $\mathbf{E} \exp(-i\omega t) + c.c.$ are described in the dipole approximation by a single-particle operator of perturbation, $\widehat{\delta h} \exp(-i\omega t) + H.c.$, where $\widehat{\delta h} = i(e/\omega)(\mathbf{E} \cdot \hat{\mathbf{v}})$ and $\hat{\mathbf{v}}$ is the interband velocity operator; see Appendix B. The induced current density is $\delta \mathbf{I}_\omega \exp(-i\omega t) + c.c.$, where the Fourier component $\delta \mathbf{I}_\omega$ is expressed through the matrix elements of the velocity operator and non-equilibrium density matrix:

$$\delta \mathbf{I}_\omega = \frac{e}{V} \sum_{\delta\eta} \mathbf{v}_{\delta\eta} \delta\rho_{\eta\delta}. \tag{1}$$

It is assumed that $\widehat{\delta\rho}$ is linear in the perturbation $\widehat{\delta h}$. In order to describe the exciton absorption, it is sufficient to determine the non-diagonal (with respect to the band indices) part of $\widehat{\delta\rho}$ from the quantum kinetic equation which takes into account the Coulomb interaction with the accuracy of e^2. Thus, one can use Eq. (28.15) linearized with respect to $\widehat{\delta h}$ (see also Eq. (29.2)) and neglect the collision integral on the right-hand side of this equation. Employing the basis determined by the eigenstate problem $\hat{h}|\delta\rangle = \varepsilon_\delta|\delta\rangle$, where $\hat{h}$ is the **kp**-Hamiltonian describing the conduction (c-) and valence (v-) band states, we obtain a linearized kinetic equation in the form

$$-i\omega\delta\rho_{\eta\delta} + \frac{i}{\hbar}(\varepsilon_\eta - \varepsilon_\delta)\delta\rho_{\eta\delta} + \frac{i}{\hbar}\delta h_{\eta\delta}(f_\eta - f_\delta) + \frac{i}{2\hbar}\sum_{\gamma_{1-4}}\Phi_{\gamma_1\gamma_2\gamma_3\gamma_4}$$

$$\times[\delta_{\gamma_1\eta}\delta F_{\delta\gamma_2\gamma_3\gamma_4} + \delta_{\gamma_2\eta}\delta F_{\gamma_1\alpha\gamma_3\gamma_4} - \delta_{\gamma_3\delta}\delta F_{\gamma_1\gamma_2\eta\gamma_4} - \delta_{\gamma_4\delta}\delta F_{\gamma_1\gamma_2\gamma_3\eta}] = 0. \quad (2)$$

The electron-electron correlations are described by the linearized function (28.8) as

$$\delta F_{\gamma_1\gamma_2\gamma_3\gamma_4} = \delta_{\gamma_2\gamma_3}\delta\rho_{\gamma_4\gamma_1}f_{\gamma_2} + \delta_{\gamma_4\gamma_1}\delta\rho_{\gamma_3\gamma_2}f_{\gamma_4}$$

$$-\delta_{\gamma_4\gamma_2}\delta\rho_{\gamma_3\gamma_1}f_{\gamma_2} - \delta_{\gamma_3\gamma_1}\delta\rho_{\gamma_4\gamma_2}f_{\gamma_3}, \quad (3)$$

and the Coulomb matrix element $\Phi_{\gamma_1\gamma_2\gamma_3\gamma_4}$ is given by the general formula (28.3); see also problem 6.1 for the case of low dimensions.

Below we describe the response near the absorption edge, when $|\hbar\omega - \bar{\varepsilon}_g| \ll \bar{\varepsilon}_g$. Here $\bar{\varepsilon}_g$ is the effective bandgap, renormalized by the confinement effects in low-dimensional structures and by the Coulomb interaction; see Eqs. (5)–(7) below. For such a case, the states η and δ correspond to c- and v-bands. Below we use the sets of indices $n\gamma$, where n is the band number (c or v) so that the Greek indices describe the intraband motion only. We consider non-doped structures, where $f_{v\gamma} = 1$ and $f_{c\gamma} = 0$. Retaining in Eq. (3) the contributions proportional to the resonant component $\delta\rho_{\eta\delta}^{cv}$, taking into account that $\Phi_{\gamma_1\gamma_2\gamma_3\gamma_4}^{n_1n_2n_3n_4}$ is non-zero if $n_1 = n_4$ and $n_2 = n_3$, and using the identity $\Phi_{\gamma_1\gamma_2\gamma_3\gamma_4}^{n_1n_2n_3n_4} = \Phi_{\gamma_2\gamma_1\gamma_4\gamma_3}^{n_2n_1n_4n_3}$, we transform the kinetic equation (2) to the following form:

$$(\varepsilon_{c\eta} - \varepsilon_{v\delta} - \hbar\omega)\delta\rho_{\eta\delta}^{cv} + \delta h_{\eta\delta}^{cv} =$$

$$= \sum_{\gamma\gamma'}\left[\Phi_{\eta\gamma\delta\gamma'}^{cvvc}\delta\rho_{\gamma'\gamma}^{cv} - \Phi_{\eta\gamma\gamma\gamma'}^{cvvc}\delta\rho_{\gamma'\delta}^{cv} + \Phi_{\gamma'\gamma\gamma\delta}^{vvvv}\delta\rho_{\eta\gamma'}^{cv} - \Phi_{\gamma'\gamma\gamma'\delta}^{vvvv}\delta\rho_{\eta\gamma}^{cv}\right]. \quad (4)$$

The second and the third Coulomb terms on the right-hand side give us the contributions proportional to $\sum_\gamma|\psi_{\gamma\mathbf{r}}^v|^2$ (here and below in this section we denote the coordinate-dependent wave function of single electron in the state $|n\gamma\rangle$ of the band n as $\psi_{\gamma\mathbf{r}}^n$). These contributions can be represented as

$$\int d\mathbf{r}\sum_{\gamma'}U_\mathbf{r}^{(v)}\left[-\psi_{\eta\mathbf{r}}^{c*}\psi_{\gamma'\mathbf{r}}^c\delta\rho_{\gamma'\delta}^{cv} + \psi_{\gamma'\mathbf{r}}^{v*}\psi_{\delta\mathbf{r}}^v\delta\rho_{\eta\gamma'}^{cv}\right], \quad (5)$$

where $U_\mathbf{r}^{(v)} = \int d\mathbf{r}' v_{|\mathbf{r}-\mathbf{r}'|}\sum_\gamma|\psi_{\gamma\mathbf{r}'}^v|^2$ (with $v_{|\mathbf{r}|} = e^2/\epsilon|\mathbf{r}|$) is the potential energy created by all valence-band electrons. Formally, $U_\mathbf{r}^{(v)}$ is infinite, because in the two-band model the valence band spreads in energy to $-\infty$ and contains an infinite number of states. However, if we take into account the positive-charge background, its potential compensates $U_{\mathbf{r}}^{(v)}$. In addition, since $U_{\mathbf{r}}^{(v)}$ is, in fact, coordinate-independent, the integral over $\mathbf{r}$ in Eq. (5) produces $\delta_{\eta\gamma'}$ in the first term of Eq. (5) and $\delta_{\delta\gamma'}$ in the second term, thereby making the sum of these terms equal to zero. Therefore, the second and the third Coulomb terms on the right-hand side of Eq. (4) are cancelled together. Using another identity,

$$\sum_{\gamma'} \Phi_{\gamma'\gamma\gamma'\delta}^{vvvv} = \delta_{\gamma\delta} v_{|\mathbf{r}|=0} \ , \tag{6}$$

we consider the last Coulomb term on the right-hand side of Eq. (4) as a contribution renormalizing the interband transition energy $\varepsilon_{c\eta} - \varepsilon_{v\delta}$ by a constant. This renormalized energy is denoted below as $E_{\eta\delta}^{cv}$. We see that only the first Coulomb term on the right-hand side of Eq. (4) remains essential, and finally obtain the following equation for the density matrix describing the interband polarization:

$$(E_{\eta\delta}^{cv} - \hbar\omega)\delta\rho_{\eta\delta}^{cv} + \delta h_{\eta\delta}^{cv} - \sum_{\gamma\gamma'} \Phi_{\eta\gamma\delta\gamma'}^{cvvc} \delta\rho_{\gamma'\gamma}^{cv} = 0. \tag{7}$$

The current density (1) is expressed through a solution of Eq. (7) according to $\delta\mathbf{I}_\omega = (e/V)\sum_{\delta\eta} \mathbf{v}_{\delta\eta}^{vc}\delta\rho_{\eta\delta}^{cv}$. Therefore, to describe the interband absorption spectra modified by Coulomb interaction, one needs to solve the non-homogeneous equation (7) and calculate the sum over the indices δ and η by using an appropriate model of the band structure.

For the case of transitions between the edges of spin-degenerate c- and v-bands described by the sets of quantum numbers $\eta = (\sigma, \bar{\eta})$ and $\delta = (\sigma', \bar{\delta})$, the interband matrix element of the velocity operator, $\mathbf{v}_{c\eta,v\delta} = \mathbf{v}_{c\sigma,v\sigma'} I_{\bar{\eta}\bar{\delta}}^{cv}$, is introduced in a similar way as in Secs. 17 and 18; see Eqs. (18.1) and (18.2) and their discussion. The overlap factor $I_{\bar{\eta}\bar{\delta}}^{cv} = \int d\mathbf{r}\psi_{\bar{\eta}\mathbf{r}}^{c*}\psi_{\bar{\delta}\mathbf{r}}^{v}$ is written through the conduction- and valence-band envelope wave functions $\psi_{\bar{\eta}\mathbf{r}}^{c}$ and $\psi_{\bar{\delta}\mathbf{r}}^{v}$ introduced below by Eq. (9). As a result, the matrix elements of the perturbation operator in Eq. (7) are given by

$$\delta h_{\eta\delta}^{cv} = i\frac{e}{\omega}\mathbf{E} \cdot \mathbf{v}_{c\sigma,v\sigma'} I_{\bar{\eta}\bar{\delta}}^{cv} \ . \tag{8}$$

For the sake of simplicity, we consider the transitions involving a single spin-degenerate valence-band state, in spite of the fact that the structure of the valence band in most materials is more complicated: it includes light-hole and heavy-hole states degenerate at zero quasimomentum. Nevertheless, the consideration presented below gives an appropriate description in view of a substantial difference in the effective masses for different branches of the valence-band spectrum. This difference gives rise to the different coupling energies for heavy- and light-hole excitons in the bulk materials and different confinement energies in low-dimensional structures. Using the effective-mass approximation near the band extrema, one can determine $\psi^c_{\bar\eta\mathbf{r}}$ and $\psi^v_{\bar\delta\mathbf{r}}$ from the single-particle Schroedinger equations (compare to Eq. (18.3)):

$$\left[\frac{\hat{\mathbf{p}}^2}{2m_c} + U_c(\mathbf{r}) - \varepsilon_{\bar\eta}\right]\psi^c_{\bar\eta\mathbf{r}} = 0, \quad \left[-\frac{\hat{\mathbf{p}}^2}{2m_v} + U_v(\mathbf{r}) - \varepsilon_{\bar\delta}\right]\psi^v_{\bar\delta\mathbf{r}} = 0. \quad (9)$$

The energies $\varepsilon_{\bar\eta}$ and $\varepsilon_{\bar\delta}$ are counted from the extrema of c- and v-bands, $\hat{\mathbf{p}} = -i\hbar\partial/\partial\mathbf{r}$ is the operator of momentum, and m_c and m_v are the effective masses of electrons and holes in the isotropic approximation. In low-dimensional systems, the potentials $U_c(\mathbf{r})$ and $U_v(\mathbf{r})$ also describe confinement of electron and hole states.

Since the Coulomb matrix element is diagonal with respect to spin variables, $\Phi^{cvvc}_{\sigma_1\bar\eta_1\sigma_2\bar\gamma_2\sigma_3\bar\gamma_3\sigma_4\bar\gamma_4} = \delta_{\sigma_1\sigma_4}\delta_{\sigma_2\sigma_3}\Phi^{cvvc}_{\bar\gamma_1\bar\gamma_2\bar\gamma_3\bar\gamma_4}$, it is convenient to introduce a spin-independent function $G^{cv}_{\bar\beta\bar\alpha}(\varepsilon_\omega)$ according to the relation

$$\delta\rho^{cv}_{\sigma\bar\eta\sigma'\bar\delta} = i\frac{e}{\omega}\mathbf{E}\cdot\mathbf{v}_{c\sigma,v\sigma'}G^{cv}_{\bar\eta\bar\delta}(\varepsilon_\omega), \quad (10)$$

where $\varepsilon_\omega = \hbar\omega - \bar\varepsilon_g$ is the excess energy of absorbed photons, which is counted from the renormalized gap $\bar\varepsilon_g$. Substituting the expression (10) into Eq. (7), we obtain an inhomogeneous equation

$$(\varepsilon_{\bar\eta} - \varepsilon_{\bar\delta} - \varepsilon_\omega - i\lambda)G^{cv}_{\bar\eta\bar\delta}(\varepsilon_\omega) - \sum_{\gamma\gamma'}\Phi^{cvvc}_{\bar\eta\bar\gamma\bar\delta\bar\gamma'}G^{cv}_{\bar\gamma'\bar\gamma}(\varepsilon_\omega) = -I^{cv}_{\bar\eta\bar\delta}. \quad (11)$$

The term $-i\lambda$ with $\lambda \to +0$ in this equation corresponds to the adiabatic turning-on of the excitation field at $t \to -\infty$. The conductivity tensor describing the response $\delta\mathbf{I}_\omega$ to the field $\mathbf{E}$ is expressed as follows:

$$\sigma_{\alpha\beta}(\omega) = \frac{ie^2}{\omega V}\sum_{\sigma\sigma'}v^\alpha_{v\sigma',c\sigma}v^\beta_{c\sigma,v\sigma'}\sum_{\bar\delta\bar\eta}I^{vc}_{\bar\delta\bar\eta}G^{cv}_{\bar\eta\bar\delta}(\varepsilon_\omega). \quad (12)$$

The polarization and spectral dependences of the interband absorption appear to be separated for the simple model under consideration. The polarization dependences are determined by the band structure, while the spectral ones essentially depend on the Coulomb contribution into Eq. (11). The factor depending on ε_ω in Eq. (12) is transformed as

$$\sum_{\bar\delta\bar\eta}I^{vc}_{\bar\delta\bar\eta}G^{cv}_{\bar\eta\bar\delta}(\varepsilon_\omega) = \int d\mathbf{r}\,G_{\varepsilon_\omega}(\mathbf{r},\mathbf{r}), \quad (13)$$

where the Green's function in the coordinate representation is introduced according to $G_{\varepsilon_\omega}(\mathbf{r}, \mathbf{r}') = \sum_{\bar{\delta}\bar{\eta}} \psi^c_{\bar{\eta}\mathbf{r}} \psi^{v*}_{\bar{\delta}\mathbf{r}'} G^{cv}_{\bar{\eta}\bar{\delta}}(\varepsilon_\omega)$. Using the one-particle Schroedinger equations (9) and the relation $\sum_{\bar{\delta}\bar{\eta}} \psi^{v*}_{\bar{\delta}\mathbf{r}'} \psi^c_{\bar{\eta}\mathbf{r}} I^{cv}_{\bar{\eta}\bar{\delta}} = \delta(\mathbf{r} - \mathbf{r}')$, we transform Eq. (11) to the following equation for the Green's function:

$$\left(\hat{h}_c - \hat{h}'_v - v_{|\mathbf{r}-\mathbf{r}'|} - \varepsilon_\omega - i\lambda \right) G_{\varepsilon_\omega}(\mathbf{r}, \mathbf{r}') = -\delta(\mathbf{r} - \mathbf{r}'), \qquad (14)$$

where $\hat{h}_c = -\hbar^2 \nabla^2_\mathbf{r}/2m_c + U_c(\mathbf{r})$ and $\hat{h}'_v = \hbar^2 \nabla^2_{\mathbf{r}'}/2m_v + U_v(\mathbf{r}')$ are the Hamiltonians of electrons and holes from Eq. (9).

The absorption coefficient is proportional to $\mathbf{E} \cdot \hat{\sigma}(\omega) \cdot \mathbf{E}$, and its spectral dependence is expressed through the function

$$\mathbf{E} \cdot \hat{\sigma}(\omega) \cdot \mathbf{E} \propto -\frac{1}{V} \mathrm{Im} \int d\mathbf{r} G_{\varepsilon_\omega}(\mathbf{r}, \mathbf{r}) \equiv \Psi(\varepsilon_\omega). \qquad (15)$$

Below we solve Eq. (14) and analyze $\Psi(\varepsilon_\omega)$ for bulk materials and 2D layers. In the first case, we use new coordinates

$$\mathbf{R} = \frac{m_c \mathbf{r} + m_v \mathbf{r}'}{m_c + m_v}, \qquad \Delta \mathbf{r} = \mathbf{r} - \mathbf{r}', \qquad (16)$$

so that $\mathbf{r} = \mathbf{R} + m_v \Delta \mathbf{r}/(m_c + m_v)$ and $\mathbf{r}' = \mathbf{R} - m_c \Delta \mathbf{r}/(m_c + m_v)$. Writing the kinetic energy in these new coordinates (problem 6.7), and assuming that there are no external fields, $U_{c,v}(\mathbf{r}) = 0$, we transform Eq. (14) to

$$\left(\frac{\hat{p}^2_\mathbf{R}}{2M} + \frac{\hat{p}^2_{\Delta \mathbf{r}}}{2\mu^*} - v_{|\Delta \mathbf{r}|} - \varepsilon_\omega - i\lambda \right) G_{\varepsilon_\omega}(\mathbf{R}, \Delta \mathbf{r}) = -\delta(\Delta \mathbf{r}), \qquad (17)$$

where the mass of electron-hole pair is introduced according to $M = m_c + m_v$, while the reduced mass $\mu^* = m_c m_v/(m_c + m_v)$ has been already introduced in Sec. 17. With these variables, the spectral dependence is written as $\Psi(\varepsilon_\omega) = -V^{-1} \mathrm{Im} \int d\mathbf{R} G_{\varepsilon_\omega}(\mathbf{R}, 0)$. To consider the absorption in a translation-invariant system, it is convenient to carry out a Fourier transformation with respect to the variable $\mathbf{R}$:

$$G_{\varepsilon_\omega}(\mathbf{P}, \Delta \mathbf{r}) = \frac{1}{V} \int d\mathbf{R} \exp\left(-\frac{i}{\hbar} \mathbf{P} \cdot \mathbf{R} \right) G_{\varepsilon_\omega}(\mathbf{R}, \Delta \mathbf{r}),$$

$$G_{\varepsilon_\omega}(\mathbf{R}, \Delta \mathbf{r}) = \sum_\mathbf{P} \exp\left(\frac{i}{\hbar} \mathbf{P} \cdot \mathbf{R} \right) G_{\varepsilon_\omega}(\mathbf{P}, \Delta \mathbf{r}). \qquad (18)$$

Therefore, $\Psi(E) = -\mathrm{Im} G_E(\mathbf{P} = 0, \Delta \mathbf{r} = 0)$. The Green's function $G_E(\Delta \mathbf{r}) \equiv G_E(\mathbf{P} = 0, \Delta \mathbf{r})$ satisfies the usual equation for a retarded

Green's function in the coordinate representation, similar to Eq. (16.1):

$$\left(\varepsilon_\omega + \frac{\hbar^2}{2\mu^*}\frac{\partial^2}{\partial\Delta\mathbf{r}^2} + v_{|\Delta\mathbf{r}|} + i\lambda\right)G_{\varepsilon_\omega}(\Delta\mathbf{r}) = \delta(\Delta\mathbf{r}), \qquad (19)$$

where $-v_{|\Delta\mathbf{r}|}$ plays the role of a potential energy. It is negative because the electron and hole attract each other. The function $G_{\varepsilon_\omega}(\mathbf{r})$ is expressed, according to problem 3.10, through the eigenfunctions of the Schroedinger equation

$$\left(-\frac{\hbar^2}{2\mu^*}\frac{\partial^2}{\partial\Delta\mathbf{r}^2} - v_{|\Delta\mathbf{r}|}\right)\psi_{\Delta\mathbf{r}}^{(\nu)} = \varepsilon_\nu\psi_{\Delta\mathbf{r}}^{(\nu)} \qquad (20)$$

containing the reduced effective mass and Coulomb potential energy. A similar equation describes the energy spectrum of the hydrogen atom, which is well known. The energies of confined states are given by $\varepsilon_\nu = -\varepsilon_B/n^2$, where $\varepsilon_B = \mu^* e^4/2\hbar^2\epsilon^2$ is the Bohr energy of exciton and n is an integer. The ground state ($n = 1$) is non-degenerate, its wave function is $\pi^{-1/2}a_B^{-3/2}e^{-|\Delta\mathbf{r}|/a_B}$, where $a_B = \hbar^2\epsilon/\mu^* e^2$ is the Bohr radius of exciton. The states with $n = 2, 3, \ldots$ are degenerate. Considering $\mathbf{r}$ as 2D or 1D coordinates, one may apply Eq. (20) for describing 2D or 1D excitons.

The spectral function is given by

$$\Psi(\varepsilon_\omega) = -\text{Im}\sum_\nu \frac{|\psi_{\Delta\mathbf{r}=0}^{(\nu)}|^2}{\varepsilon_\omega - \varepsilon_\nu + i\lambda} = \pi\sum_\nu |\psi_{\Delta\mathbf{r}=0}^{(\nu)}|^2\delta(\varepsilon_\nu - \varepsilon_\omega), \qquad (21)$$

where we have carried out a limiting transition leading to the δ-function of energy; see problem 1.4. Equation (21) differs from the usual expressions of interband absorption (containing the joint density of states) by an additional multiplier, $|\psi_{\Delta\mathbf{r}=0}^{(\nu)}|^2$, called the Sommerfeld factor. The quantum numbers ν include both the continuous-spectrum states modified by Coulomb interaction (problem 6.8) and the local, discrete states. For this reason, below the fundamental absorption edge one has a number of δ-peaks describing the exciton absorption. The first (lowest-energy) peak is shifted below the edge by the energy of the ground state, ε_B.

If the material is not homogeneous, the potential energies from Eq. (9) should be taken into account and included into Eq. (17). In the case of smooth inhomogeneities, whose characteristic length exceeds a_B, one can separate the relative motion of electron and hole from the motion of electron-hole pair in the random potential and describe this latter motion by using the path integral method (problem 6.9). As a result, the exciton peaks become broadened and appear on the background of the absorption tail described in Sec. 18 in the one-electron approximation.

Let us study now the exciton absorption in quantum wells, when the potentials for electrons and holes in Eq. (9) include the confinement potentials $U_h(z)$ and $U_c(z)$. It is convenient to express the Green's function by using the orbitals φ^c_{nz} and $\varphi^v_{n'z}$ describing the confinement in c- and v-band quantum wells:

$$G_{\varepsilon_\omega}(\mathbf{r}, \mathbf{r}') = \sum_{nn'} \varphi^c_{nz} \varphi^{v*}_{n'z'} G_{\varepsilon_\omega}(n\mathbf{x}, n'\mathbf{x}'). \qquad (22)$$

The sum is taken over the subband numbers. Let us substitute the expression (22) into Eq. (14), multiply the latter by $\varphi^{c*}_{n_1z} \varphi^v_{n'_1z'}$ from the left, and take the integrals over z and z'. We find that the Green's function $G_{\varepsilon_\omega}(n\mathbf{x}, n'\mathbf{x}')$ satisfies an equation similar to Eq. (14), where $\mathbf{r}$ and $\mathbf{r}'$ are replaced by the 2D coordinates $\mathbf{x}$ and $\mathbf{x}'$, the excess energy ε_ω is replaced by the energy $\varepsilon_\omega^{(nn')}$ depending on the quantization energies of electron and hole states, and the factor $\int dz \varphi^{c*}_{nz} \varphi^v_{n'z}$ appears on the right-hand side. The coordinates $\mathbf{X}$ and $\Delta\mathbf{x}$ are expressed through $\mathbf{x}$ and $\mathbf{x}'$ in the same way as in Eq. (16), and one should remember that the effective masses standing there are the effective masses for the in-plane (2D) motion of electron and holes. The Green's function is determined by the solutions $\psi^{(\nu)}_{\Delta\mathbf{x}}$ of the eigenstate problem (20) for the 2D case (problem 6.10), and the spectral function is given in the following way:

$$\Psi(\varepsilon_\omega) = \pi \sum_{nn'\nu} \Lambda_{nn'} |\psi^{(\nu)}_{\Delta\mathbf{x}=0}|^2 \delta(\varepsilon_\nu - \varepsilon_\omega^{(nn')}), \qquad (23)$$

$$\Lambda_{nn'} = \left| \int dz \varphi^{c*}_{nz} \varphi^v_{n'z} \right|^2.$$

The squared overlap factor $\Lambda_{nn'}$ is equal to $\delta_{nn'}$ for the hard-wall model of the quantum well. Therefore, the picture of the exciton absorption in low-dimensional systems is similar to that in bulk materials: the edge of interband absorption is modified due to Coulomb interaction, and the exciton absorption peaks appear below the fundamental edge. The ground energies of the excitons increase with lowering dimension. We point out that the two-dimensional approximation used here and in problem 6.10 is valid only if the characteristic radius of confinement is small in comparison to the exciton radius. If this requirement is not fulfilled, the Coulomb interaction between electron and hole essentially modifies the size quantization so that the mixing of in-plane motion and transverse motion occurs and the wave functions cannot be written in the forms $\varphi^c_{nz}\psi^c_{n\mathbf{x}}$ and $\varphi^v_{n'z}\psi^v_{n'\mathbf{x}}$. While the two-dimensional approximation for excitons in quantum wells gives reasonable results, the one-dimensional approximation for excitons in quantum wires fails completely, because it leads to an infinitely large energy of the ground excitonic state (problem 6.11), which corresponds to an exciton of infinitely small radius. Therefore, to find the exciton energies in quantum wires, one should consider the wires of finite widths and include the transverse motion into the Schroedinger equation with Coulomb potential.

31. Electron-Electron Collision Integral

To consider the contributions of the order of e^4 describing electron-electron collisions in the quantum kinetic equation (28.15), we introduce a two-particle correlation operator $\hat{\mathcal{G}}$ whose matrix elements are the correlation functions introduced in Eq. (28.7):

$$g_{\gamma_1\gamma_2\gamma_3\gamma_4} = \langle\gamma_4|\langle\gamma_3|\hat{\mathcal{G}}_t|\gamma_2\rangle|\gamma_1\rangle. \tag{1}$$

The matrix elements (28.10) of the collision integral standing on the right-hand side of Eq. (28.15) are given as (problem 6.12)

$$\langle\eta|\hat{J}_{ee}(\hat{\rho}|t)|\delta\rangle = \frac{1}{i\hbar V}\sum_{\mathbf{q}} v_q \sum_{\gamma_1\gamma_2\gamma_3} \left\{\langle\eta|e^{-i\mathbf{q}\cdot\mathbf{x}}|\gamma_1\rangle\langle\gamma_2|e^{i\mathbf{q}\cdot\mathbf{x}}|\gamma_3\rangle\right.$$

$$\times\langle\gamma_1|\langle\gamma_3|\hat{\mathcal{G}}_t|\gamma_2\rangle|\delta\rangle - \langle\eta|\langle\gamma_3|\hat{\mathcal{G}}_t|\gamma_2\rangle|\gamma_1\rangle\left.\langle\gamma_1|e^{-i\mathbf{q}\cdot\mathbf{x}}|\delta\rangle\langle\gamma_2|e^{i\mathbf{q}\cdot\mathbf{x}}|\gamma_3\rangle\right\}, \tag{2}$$

and the collision integral can be rewritten in the operator form

$$\hat{J}_{ee}(\hat{\rho}|t) = \frac{1}{i\hbar V}\sum_{\mathbf{q}} v_q \mathrm{sp}'\left[e^{-i\mathbf{q}\cdot(\mathbf{x}-\mathbf{x}')}, \hat{\mathcal{G}}_t\right]. \tag{3}$$

The definition of the "inner" trace $\mathrm{sp}'\ldots$ is clear from a comparison of Eqs. (2) and (3). To obtain an explicit expression for the collision integral with the accuracy e^4, we have to find $\hat{\mathcal{G}}$ from the equation of motion taking into account only the contributions of the order of e^2. Using the definition given by Eq. (28.8), we express $F_{\gamma_{1-4}}(t)$ through the one-particle density operators as $\hat{F} = (1-\hat{\mathcal{P}})\hat{\rho}_t\hat{\rho}'_t$, where the permutation operator $\hat{\mathcal{P}}$ is defined according to

$$\langle\gamma_4|\langle\gamma_3|\hat{\mathcal{P}}\hat{a}\hat{b}'|\gamma_2\rangle|\gamma_1\rangle = \langle\gamma_4|\hat{a}|\gamma_2\rangle\langle\gamma_3|\hat{b}|\gamma_1\rangle. \tag{4}$$

Without this operator, we have $\langle\gamma_4|\langle\gamma_3|\hat{a}\hat{b}'|\gamma_2\rangle|\gamma_1\rangle = \langle\gamma_4|\hat{a}|\gamma_1\rangle\langle\gamma_3|\hat{b}|\gamma_2\rangle$ (by convention, the operators with prime sign act on the "inner" states $|\gamma_2\rangle$ and $|\gamma_3\rangle$). Multiplying Eq. (1.20) by $\hat{a}^+_{\gamma_1}\hat{a}^+_{\gamma_2}\hat{a}_{\gamma_3}\hat{a}_{\gamma_4}$, taking the trace over the electron variables, and using Eq. (28.7), we obtain

$$\frac{\partial}{\partial t}\hat{\mathcal{G}}_t + \frac{i}{\hbar}\left[\hat{h}'_t + \hat{h}_t, \hat{\mathcal{G}}_t\right] = \frac{1}{i\hbar}\hat{\mathcal{K}}_t, \tag{5}$$

where the operator $\hat{\mathcal{K}}_t$ is defined by its matrix elements

$$\langle\gamma_4|\langle\gamma_3|\hat{\mathcal{K}}_t|\gamma_2\rangle|\gamma_1\rangle = \langle\gamma_4|\langle\gamma_3|(1-\hat{P})\left\{\hat{\rho}_t[\hat{h}'_t - \tilde{h}'_t, \hat{\rho}'_t] + [\hat{h}_t \right. \qquad (6)$$

$$\left. -\tilde{h}_t, \hat{\rho}_t]\hat{\rho}'_t\right\}|\gamma_2\rangle|\gamma_1\rangle + \frac{1}{2}\sum_{\delta_{1-4}}\Phi_{\delta_1\delta_2\delta_3\delta_4}\langle\langle[\hat{a}^+_{\gamma_1}\hat{a}^+_{\gamma_2}\hat{a}_{\gamma_3}\hat{a}_{\gamma_4}, \hat{a}^+_{\delta_1}\hat{a}^+_{\delta_2}\hat{a}_{\delta_3}\hat{a}_{\delta_4}]\rangle\rangle_t$$

and $\tilde{h}_t$ is introduced by Eq. (28.14). To obtain Eq. (6), the time derivatives of the one-particle density matrices have been calculated according to Eq. (28.15) without the collision integral. Within the same accuracy, one should neglect the interaction when calculating the four-particle averages in the last term of Eq. (6).

The commutator $[\hat{a}^+_{\gamma_1}\hat{a}^+_{\gamma_2}\hat{a}_{\gamma_3}\hat{a}_{\gamma_4}, \hat{a}^+_{\delta_1}\hat{a}^+_{\delta_2}\hat{a}_{\delta_3}\hat{a}_{\delta_4}]$ in Eq. (6) is transformed with the aid of the anticommutation relation (4.21):

$$\{(\delta_{\gamma_4\delta_1}\delta_{\gamma_3\delta_2} - \delta_{\gamma_3\delta_1}\delta_{\gamma_4\delta_2})\hat{a}^+_{\gamma_1}\hat{a}^+_{\gamma_2}\hat{a}_{\delta_3}\hat{a}_{\delta_4} - \delta_{\gamma_4\delta_1}\hat{a}^+_{\gamma_1}\hat{a}^+_{\gamma_2}\hat{a}^+_{\delta_2}\hat{a}_{\gamma_3}\hat{a}_{\delta_3}\hat{a}_{\delta_4}$$

$$+\delta_{\gamma_3\delta_1}\hat{a}^+_{\gamma_1}\hat{a}^+_{\gamma_2}\hat{a}^+_{\delta_2}\hat{a}_{\gamma_4}\hat{a}_{\delta_3}\hat{a}_{\delta_4} + \delta_{\gamma_4\delta_2}\hat{a}^+_{\gamma_1}\hat{a}^+_{\gamma_2}\hat{a}^+_{\delta_1}\hat{a}_{\gamma_3}\hat{a}_{\delta_3}\hat{a}_{\delta_4}$$

$$-\delta_{\gamma_3\delta_2}\hat{a}^+_{\gamma_1}\hat{a}^+_{\gamma_2}\hat{a}^+_{\delta_1}\hat{a}_{\gamma_4}\hat{a}_{\delta_3}\hat{a}_{\delta_4}\} - \{\gamma \leftrightarrow \delta\}. \qquad (7)$$

Therefore, only two- and three-particle correlation functions contribute to the right-hand side of Eq. (5). The two-particle correlation functions are averaged according to Eq. (19.7). The three-particle correlation functions are averaged in a similar way:

$$\langle\langle\hat{a}^+_{\gamma_1}\hat{a}^+_{\gamma_2}\hat{a}^+_{\gamma_3}\hat{a}_{\delta_1}\hat{a}_{\delta_2}\hat{a}_{\delta_3}\rangle\rangle_t \qquad (8)$$

$$\simeq \langle\langle\hat{a}^+_{\gamma_1}\hat{a}_{\delta_1}\rangle\rangle_t\langle\langle\hat{a}^+_{\gamma_2}\hat{a}_{\delta_3}\rangle\rangle_t\langle\langle\hat{a}^+_{\gamma_3}\hat{a}_{\delta_2}\rangle\rangle_t - \langle\langle\hat{a}^+_{\gamma_1}\hat{a}_{\delta_1}\rangle\rangle_t\langle\langle\hat{a}^+_{\gamma_2}\hat{a}_{\delta_2}\rangle\rangle_t\langle\langle\hat{a}^+_{\gamma_3}\hat{a}_{\delta_3}\rangle\rangle_t$$

$$+\langle\langle\hat{a}^+_{\gamma_1}\hat{a}_{\delta_2}\rangle\rangle_t\langle\langle\hat{a}^+_{\gamma_2}\hat{a}_{\delta_1}\rangle\rangle_t\langle\langle\hat{a}^+_{\gamma_3}\hat{a}_{\delta_3}\rangle\rangle_t - \langle\langle\hat{a}^+_{\gamma_1}\hat{a}_{\delta_2}\rangle\rangle_t\langle\langle\hat{a}^+_{\gamma_2}\hat{a}_{\delta_3}\rangle\rangle_t\langle\langle\hat{a}^+_{\gamma_3}\hat{a}_{\delta_1}\rangle\rangle_t$$

$$+\langle\langle\hat{a}^+_{\gamma_1}\hat{a}_{\delta_3}\rangle\rangle_t\langle\langle\hat{a}^+_{\gamma_2}\hat{a}_{\delta_2}\rangle\rangle_t\langle\langle\hat{a}^+_{\gamma_3}\hat{a}_{\delta_1}\rangle\rangle_t - \langle\langle\hat{a}^+_{\gamma_1}\hat{a}_{\delta_3}\rangle\rangle_t\langle\langle\hat{a}^+_{\gamma_2}\hat{a}_{\delta_1}\rangle\rangle_t\langle\langle\hat{a}^+_{\gamma_3}\hat{a}_{\delta_2}\rangle\rangle_t.$$

Since Eqs. (7) and (8) are rather cumbersome, let us first consider the case of non-degenerate electrons, when the terms proportional to n^3 (n is the electron density and $\hat{\rho}_t \propto n$) can be neglected in the expression for $\hat{\mathcal{K}}_t$. The three-particle correlation functions and the first term on the right-hand side of Eq. (6) are not essential in this situation. We obtain

$$\hat{\mathcal{K}}_t = \frac{1}{V}\sum_{\mathbf{q}}v_q(1-\hat{P})\left(e^{-i\mathbf{q}\cdot\mathbf{x}'}\hat{\rho}'_t e^{i\mathbf{q}\cdot\mathbf{x}}\hat{\rho}_t - \hat{\rho}'_t e^{-i\mathbf{q}\cdot\mathbf{x}'}\hat{\rho}_t e^{i\mathbf{q}\cdot\mathbf{x}}\right). \qquad (9)$$

Employing the initial condition of the correlation weakening,

$$\hat{\mathcal{G}}_{t\to-\infty} = 0, \qquad (10)$$

and the evolution operator introduced by Eq. (2.2), we solve Eq. (5) as

$$\hat{\mathcal{G}}_t = \frac{1}{i\hbar} \int_{-\infty}^{t} dt_1 e^{\lambda t_1} \widehat{S}(t,t_1) \widehat{S}'(t,t_1) \hat{\mathcal{K}}_{t_1} \widehat{S}'^{+}(t,t_1) \widehat{S}^{+}(t,t_1). \tag{11}$$

Substituting this solution into Eq. (3), we find the Coulomb collision integral for non-degenerate electrons:

$$\widehat{J}_{ee}(\hat{\rho}|t) = \frac{1}{\hbar^2 V^2} \sum_{\mathbf{q}\mathbf{q}_1} v_q v_{q_1} \int_{-\infty}^{t} dt_1 e^{\lambda t_1} \mathrm{sp}' \left[\widehat{S}(t,t_1)\widehat{S}'(t,t_1)(1-\hat{\mathcal{P}}) \right.$$

$$\times \left\{ e^{-i\mathbf{q}\cdot\mathbf{x}'} \hat{\rho}'_{t_1} e^{i\mathbf{q}\cdot\mathbf{x}} \hat{\rho}_{t_1} - \hat{\rho}'_{t_1} e^{-i\mathbf{q}\cdot\mathbf{x}'} \hat{\rho}_{t_1} e^{i\mathbf{q}\cdot\mathbf{x}} \right\} \tag{12}$$

$$\left. \times \widehat{S}'^{+}(t,t_1)\widehat{S}^{+}(t,t_1), e^{i\mathbf{q}_1\cdot(\mathbf{x}'-\mathbf{x})} \right].$$

The operator of permutations can be excluded from Eq. (12) with the use of the identities (problem 6.13)

$$\mathrm{sp}' \left[e^{-i\mathbf{q}\cdot(\mathbf{x}-\mathbf{x}')}, \widehat{S}(t,t_1)\widehat{S}'(t,t_1)\hat{\mathcal{P}}(\widehat{A}\widehat{B}')\widehat{S}'^{+}(t,t_1)\widehat{S}^{+}(t,t_1) \right]$$

$$= \left[e^{-i\mathbf{q}\cdot\mathbf{x}}, \widehat{S}(t,t_1)\widehat{A}\widehat{S}^{+}(t,t_1)e^{i\mathbf{q}\cdot\mathbf{x}}\widehat{S}(t,t_1)\widehat{B}\widehat{S}^{+}(t,t_1) \right] \tag{13}$$

and

$$\mathrm{sp}' \left[e^{-i\mathbf{q}\cdot(\mathbf{x}-\mathbf{x}')}, \widehat{S}(t,t_1)\widehat{S}'(t,t_1)\widehat{A}\widehat{B}'\widehat{S}'^{+}(t,t_1)\widehat{S}^{+}(t,t_1) \right]$$

$$= \left[e^{-i\mathbf{q}\cdot\mathbf{x}}, \widehat{S}(t,t_1)\widehat{A}\widehat{S}^{+}(t,t_1) \right] \mathrm{sp} \left(e^{i\mathbf{q}\cdot\mathbf{x}}\widehat{S}(t,t_1)\widehat{B}\widehat{S}^{+}(t,t_1) \right). \tag{14}$$

Thus, we rewrite the collision integral as

$$\widehat{J}_{ee}(\hat{\rho}|t) = \frac{1}{\hbar^2 V^2} \sum_{\mathbf{q}\mathbf{q}_1} v_q v_{q_1} \int_{-\infty}^{t} dt_1 e^{\lambda t_1} \left[\widehat{S}(t,t_1)\hat{\rho}_{t_1} e^{i\mathbf{q}_1\cdot\mathbf{x}}\widehat{S}^{+}(t,t_1) \right.$$

$$\left. \times \left\{ \left\{ e^{i\mathbf{q}\cdot\mathbf{x}}\widehat{S}(t,t_1)\hat{\rho}_{t_1} e^{-i\mathbf{q}_1\cdot\mathbf{x}}\widehat{S}^{+}(t,t_1) \right\} \right\}, e^{-i\mathbf{q}\cdot\mathbf{x}} \right] + H.c. , \tag{15}$$

where we have introduced $\{\{\widehat{A}\}\} \equiv \widehat{A} - \mathrm{sp}\widehat{A}$.

Let us calculate the matrix element $\langle\gamma_4|\langle\gamma_3|\hat{\mathcal{K}}_t|\gamma_2\rangle|\gamma_1\rangle$ defined by Eq. (6) for the general case of arbitrary degeneracy. After some technical efforts, we find that the sum of the first term on the right-hand side of Eq. (6) with the contribution of the second term containing the correlation functions $\langle\langle\hat{a}_{\gamma_j}^{+}\hat{a}_{\gamma_i}\rangle\rangle_t = \langle\gamma_i|\hat{\rho}_t|\gamma_j\rangle$ is equal to zero. Therefore, only the averages containing the "mixed" $\gamma\delta$ correlation functions from the second term on the right-hand side of Eq. (6) remain. Using the definition (28.4), we write the matrix element of $\hat{\mathcal{K}}_t$ in the form

$$\langle\gamma_4|\langle\gamma_3|\hat{\mathcal{K}}_t|\gamma_2\rangle|\gamma_1\rangle = 2\sum_{\delta_1\delta_2}[\widetilde{\Phi}_{\gamma_3\gamma_4\delta_1\delta_2}\langle\delta_1|\hat{\rho}|\gamma_1\rangle\langle\delta_2|\hat{\rho}|\gamma_2\rangle - \widetilde{\Phi}^*_{\gamma_1\gamma_2\delta_1\delta_2}$$

$$\times\langle\gamma_3|\hat{\rho}|\delta_1\rangle\langle\gamma_4|\hat{\rho}|\delta_2\rangle] + 2\sum_{\delta\delta_1\delta_2}\left\{(1-\mathcal{P}_{\gamma_3\gamma_4})\widetilde{\Phi}_{\gamma_3\delta\delta_1\delta_2}\langle\gamma_4|\hat{\rho}|\delta\rangle\langle\delta_2|\hat{\rho}|\gamma_1\rangle \quad (16)$$

$$\times \langle\delta_1|\hat{\rho}|\gamma_2\rangle + (1-\mathcal{P}_{\gamma_1\gamma_2})\widetilde{\Phi}^*_{\gamma_1\delta\delta_1\delta_2}\langle\delta|\hat{\rho}|\gamma_2\rangle\langle\gamma_3|\hat{\rho}|\delta_1\rangle\langle\gamma_4|\hat{\rho}|\delta_2\rangle\right\}.$$

Equation (16) can be rewritten in the operator form, which gives us a generalization of Eq. (9) to the case of degenerate electrons:

$$\hat{\mathcal{K}}_t = \frac{1}{V}\sum_{\mathbf{q}}v_q(1-\hat{\mathcal{P}})\left\{(1-\hat{\rho}_t-\hat{\rho}'_t)e^{-i\mathbf{q}\cdot\mathbf{x}'}\hat{\rho}'_t e^{i\mathbf{q}\cdot\mathbf{x}}\hat{\rho}_t\right.$$

$$\left.-\hat{\rho}'_t e^{-i\mathbf{q}\cdot\mathbf{x}'}\hat{\rho}_t e^{i\mathbf{q}\cdot\mathbf{x}}(1-\hat{\rho}_t-\hat{\rho}'_t)\right\}. \quad (17)$$

Let us substitute this expression into Eq. (11) and then substitute $\hat{\mathcal{G}}_t$ obtained in this way into Eq. (3). We obtain the collision integral in the form (12), where, however, the expression in the braces $\{\ldots\}$ is replaced by the expression in $\{\ldots\}$ of Eq. (17) taken at $t = t_1$. This expression contains the factors $(1-\hat{\rho}_{t_1}-\hat{\rho}'_{t_1})$ accounting for the Pauli principle. Further transformations are based upon the identities (13) and (14). We finally obtain the collision integral in the operator form

$$\hat{J}_{ee}(\hat{\rho}|t) = \frac{1}{\hbar^2 V^2}\sum_{\mathbf{q}\mathbf{q}_1}v_q v_{q_1}\int_{-\infty}^t dt_1 e^{\lambda t_1}\left[e^{-i\mathbf{q}\cdot\mathbf{x}},\right. \quad (18)$$

$$\widehat{S}(t,t_1)(1-\hat{\rho}_{t_1})e^{i\mathbf{q}_1\cdot\mathbf{x}}\hat{\rho}_{t_1}\widehat{S}^+(t,t_1)\left\{\left\{e^{i\mathbf{q}\cdot\mathbf{x}}\widehat{S}(t,t_1)e^{-i\mathbf{q}_1\cdot\mathbf{x}}\hat{\rho}_{t_1}\widehat{S}^+(t,t_1)\right\}\right\}$$

$$-\widehat{S}(t,t_1)\hat{\rho}_{t_1}e^{i\mathbf{q}_1\cdot\mathbf{x}}(1-\hat{\rho}_{t_1})\widehat{S}^+(t,t_1)\left\{\left\{e^{i\mathbf{q}\cdot\mathbf{x}}\widehat{S}(t,t_1)\hat{\rho}_{t_1}e^{-i\mathbf{q}_1\cdot\mathbf{x}}\widehat{S}^+(t,t_1)\right\}\right\}$$

$$+\widehat{S}(t,t_1)\left[\hat{\rho}_{t_1},e^{i\mathbf{q}_1\cdot\mathbf{x}}\right]\widehat{S}^+(t,t_1)\left\{\left\{e^{i\mathbf{q}\cdot\mathbf{x}}\widehat{S}(t,t_1)\hat{\rho}_{t_1}e^{-i\mathbf{q}_1\cdot\mathbf{x}}\hat{\rho}_{t_1}\widehat{S}^+(t,t_1)\right\}\right\}\right].$$

The Hermiticity of the operator (18) is checked directly (problem 6.14).

The expression for the collision integral given above is rather complicated. It is simplified essentially for the systems with time-independent Hamiltonian $\hat{h}$. We use below the basis of the eigenstate problem $\hat{h}|\gamma\rangle = \varepsilon_\gamma|\gamma\rangle$ and account for the diagonal part of the density matrix $\langle\gamma|\hat{\rho}_t|\gamma'\rangle$ only. The non-diagonal matrix elements are assumed to be either small due to $\hbar/\bar{\varepsilon}\bar{\tau} \ll 1$, see Eq. (7.21), or vanish due to symmetry properties (for example, when the Hamiltonian $\hat{h}$ is translation-invariant and the eigenstates $|\gamma\rangle$ are identified with eigenstates of momentum, $|\mathbf{p}\rangle$). The diagonal matrix element of the operator (18) are calculated by using the definition of S-operators given by Eq. (2.3). Then we take the integral over time and obtain

$$J_{ee}(f|\gamma t) = \frac{2\pi}{\hbar} \sum_{\nu\gamma'\nu'} V(\gamma\gamma'|\nu\nu')\delta(\varepsilon_\gamma + \varepsilon_{\gamma'} - \varepsilon_\nu - \varepsilon_{\nu'})$$

$$\times [f_{\nu t}f_{\nu' t}(1 - f_{\gamma t})(1 - f_{\gamma' t}) - f_{\gamma t}f_{\gamma' t}(1 - f_{\nu t})(1 - f_{\nu' t})], \qquad (19)$$

where $f_{\gamma t} = \langle\gamma|\hat{\rho}_t|\gamma\rangle$ is the distribution function. The principal-value contributions have disappeared, and only the δ-function expressing the energy conservation law remains. The probability of transition in Eq. (19) is expressed through the Coulomb matrix element defined as

$$V(\gamma\gamma'|\nu\nu') = \frac{1}{V^2} \sum_{\mathbf{q}\mathbf{q}_1} v_q v_{q_1} \left\{ \langle\gamma|e^{-i\mathbf{q}\cdot\mathbf{x}}|\nu\rangle\langle\nu|e^{i\mathbf{q}_1\cdot\mathbf{x}}|\gamma\rangle\langle\gamma'|e^{i\mathbf{q}\cdot\mathbf{x}}|\nu'\rangle \qquad (20)$$

$$\times \langle\nu'|e^{-i\mathbf{q}_1\cdot\mathbf{x}}|\gamma'\rangle - \mathrm{Re}\langle\gamma|e^{-i\mathbf{q}\cdot\mathbf{x}}|\nu\rangle\langle\nu|e^{i\mathbf{q}_1\cdot\mathbf{x}}|\gamma'\rangle\langle\gamma'|e^{i\mathbf{q}\cdot\mathbf{x}}|\nu'\rangle\langle\nu'|e^{-i\mathbf{q}_1\cdot\mathbf{x}}|\gamma\rangle \right\}.$$

The first part of this expression originates from the terms containing the trace in Eq. (18) (we remind that $\{\{\hat{A}\}\} \equiv \hat{A} - \mathrm{sp}\hat{A}$) and is called the direct Coulomb contribution to the probability of transition. The remaining terms on the right-hand side of Eq. (18) contribute to the second part of $V(\gamma\gamma'|\nu\nu')$, which is called the exchange contribution. The matrix element (20) has the symmetry properties $V(\gamma\gamma'|\nu\nu') = V(\gamma'\gamma|\nu'\nu) = V(\nu\nu'|\gamma\gamma') = V(\nu'\nu|\gamma'\gamma)$ following from the symmetry properties of $\Phi_{\gamma_1\gamma_2\gamma_3\gamma_4}$ discussed in Sec. 28.

The collision integral $J_{ee}(f|\mathbf{r}\mathbf{p}t)$ for the electron system in quasi-classical external fields is obtained from the operator expression (18) in the way described in Sec. 9:

$$J_{ee}(f|\mathbf{r}\mathbf{p}t) = \frac{2\pi}{\hbar} \sum_{\mathbf{p}'\mathbf{p}_1\mathbf{p}_1'} \delta_{\mathbf{p}+\mathbf{p}',\mathbf{p}_1+\mathbf{p}_1'} V(\mathbf{p}\mathbf{p}'|\mathbf{p}_1\mathbf{p}_1')\delta(\varepsilon_p + \varepsilon_{p'} - \varepsilon_{p_1} - \varepsilon_{p_1'})$$

$$\times \{ f_{\mathbf{r}\mathbf{p}_1 t}f_{\mathbf{r}\mathbf{p}_1' t}(1 - f_{\mathbf{r}\mathbf{p}t})(1 - f_{\mathbf{r}\mathbf{p}' t}) - f_{\mathbf{r}\mathbf{p}t}f_{\mathbf{r}\mathbf{p}' t}(1 - f_{\mathbf{r}\mathbf{p}_1 t})(1 - f_{\mathbf{r}\mathbf{p}_1' t}) \}. \quad (21)$$

The matrix elements of the exponential factors $\exp(-i\mathbf{q}\cdot\mathbf{x})$ are calculated in the basis $|\sigma\rangle|\mathbf{p}\rangle$, where the coordinate parts $|\mathbf{p}\rangle$ of the wave functions are plane waves, so that $\langle\mathbf{p}'|\langle\sigma'|\exp(-i\mathbf{q}\cdot\mathbf{x})|\sigma\rangle|\mathbf{p}\rangle = \delta_{\sigma\sigma'}\delta_{\mathbf{p},\mathbf{p}'+\hbar\mathbf{q}}$. The spin contributions to such matrix elements produces the orthogonality factor $\delta_{\sigma\sigma'}$, while the coordinate contributions lead to the momentum conservation law. The matrix element $V(\mathbf{pp}'|\mathbf{p}_1\mathbf{p}_1')$ is defined according to

$$\sum_{\sigma'\sigma_1\sigma_1'} V(\mathbf{p}\sigma\mathbf{p}'\sigma'|\mathbf{p}_1\sigma_1\mathbf{p}_1'\sigma_1') = \frac{1}{V^2}\delta_{\mathbf{p}+\mathbf{p}',\mathbf{p}_1+\mathbf{p}_1'}\{2|v_{|\mathbf{p}-\mathbf{p}_1|/\hbar}|^2$$

$$-v_{|\mathbf{p}-\mathbf{p}_1|/\hbar}v_{|\mathbf{p}-\mathbf{p}_1'|/\hbar}\} \equiv \delta_{\mathbf{p}+\mathbf{p}',\mathbf{p}_1+\mathbf{p}_1'}V(\mathbf{pp}'|\mathbf{p}_1\mathbf{p}_1'). \qquad (22)$$

The difference between the spin factors in the first (direct) and the second (exchange) terms appears to be essential. These factors are $\delta_{\sigma\sigma_1}\delta_{\sigma'\sigma_1'}$ and $\delta_{\sigma\sigma_1}\delta_{\sigma_1\sigma'}\delta_{\sigma'\sigma_1'}\delta_{\sigma_1'\sigma}$ for the first and for the second term, respectively. After the sums over spins in Eq. (22) are taken, a factor of 2 appears in the first term but does not appear in the second one. Because of the exchange terms, the scattering probabilities for the electrons with parallel and anti-parallel spins are different, which becomes essential for spin-polarized electrons.

The conservation of the number of particles in the electron-electron collisions directly follows from the equation $\mathrm{sp}\widehat{J}_{ee}(\hat{\rho}|t) = 0$. This means that Eq. (7.19) is not modified in the presence of these collisions. For the system with time-independent Hamiltonian, one can check that

$$\sum_{\gamma} \varepsilon_{\gamma} J_{ee}(f|\gamma) = 0 \qquad (23)$$

by using Eqs. (19) and (20) and the property $V(\gamma\gamma'|\nu\nu') = V(\gamma'\gamma|\nu'\nu)$. Therefore, the electron-electron collisions do not lead to energy relaxation. Finally, we use the explicit expression for the quasi-classical collision integral (21) and permute the variables according to $\mathbf{p} \leftrightarrow \mathbf{p}_1$ and $\mathbf{p}' \leftrightarrow \mathbf{p}_1'$. As a result, we obtain

$$\sum_{\mathbf{p}} \mathbf{p} J_{ee}(f|\mathbf{rp}) = 0, \qquad (24)$$

which means that the electron-electron collisions do not lead to momentum relaxation and the current is conserved. The energy and momentum conservation implies that any Fermi distribution function of the form

$$f_{\mathbf{p}} = [e^{(\varepsilon_{\mathbf{p}} - \mathbf{p}\cdot\mathbf{u} - \mu_e)/T_e} + 1]^{-1} = \{\exp[(\varepsilon_{\mathbf{p}-m\mathbf{u}} - \mu_e^*)/T_e] + 1\}^{-1}, \qquad (25)$$

where $\mathbf{u}$ is the electron drift velocity and T_e and μ_e are the effective electron temperature and chemical potential, satisfies the kinetic equation (compare this to a similar result for phonons, Eq. (23.26)). In the second expression of Eq. (25), we have used the parabolic electron spectrum $\varepsilon_{\mathbf{p}} = p^2/2m$ and made the definition $\mu_e^* = \mu_e + mu^2/2$. One may safely replace μ_e^* by μ_e, since the chemical potential is determined by the electron density. Besides, in the linear response problems, when the drift velocity is small, the quadratic term $mu^2/2$ should be neglected so that we have $\mu_e^* = \mu_e$ exactly. To provide the momentum and energy relaxation, when $\mathbf{u}$ goes to zero and T_e goes to the equilibrium temperature T, one must consider electron-impurity and electron-phonon scattering. The latter is responsible for the energy relaxation and also contributes to the relaxation of momentum. However, the momentum relaxation in metals and semiconductors at low temperatures occurs mostly due to electron-impurity scattering.

32. Coulomb Drag Between 2D Electrons

If two parallel 2D layers are placed close enough to each other, and a current flows through one of the layers (drive layer), the interlayer momentum transfer caused by the Coulomb interaction between the electrons of different layers leads to a net force acting on the electrons in the other layer (drag layer). This force can be calculated by using the electron-electron collision integral derived in Sec. 31. In the basis $|j\mathbf{p}\rangle$ described by the wave functions $\psi_z^{(j)} L^{-1} e^{i\mathbf{p}\cdot\mathbf{x}/\hbar}$, where $j = 1, 2$ is the layer number and $\mathbf{x}$ and $\mathbf{p}$ are the 2D coordinate and momentum, the stationary kinetic equation is written as

$$e\mathbf{E}_j \cdot \frac{\partial f_{j\mathbf{p}}}{\partial \mathbf{p}} = J_{ee}(f|j\mathbf{p}) + J_{im}(f|j\mathbf{p}), \tag{1}$$

where $\mathbf{E}_j$ is the electric field in the layer j. The right-hand side of this equation contains the electron-electron collision integral (31.19), written in the given basis as

$$J_{ee}(f|j\mathbf{p}) = \frac{2\pi}{\hbar L^4} \sum_{j'\mathbf{p}'\mathbf{p}_1\mathbf{p}_1'} [2|v_{|\mathbf{p}-\mathbf{p}_1|/\hbar}^{(jj')}|^2 - \delta_{jj'} v_{|\mathbf{p}-\mathbf{p}_1|/\hbar}^{(jj)} v_{|\mathbf{p}-\mathbf{p}_1'|/\hbar}^{(jj)}]$$

$$\times \delta_{\mathbf{p}+\mathbf{p}',\mathbf{p}_1+\mathbf{p}_1'} \delta(\varepsilon_{jp} + \varepsilon_{j'p'} - \varepsilon_{jp_1} - \varepsilon_{j'p_1'})$$

$$\times [f_{j\mathbf{p}_1} f_{j'\mathbf{p}_1'}(1 - f_{j\mathbf{p}})(1 - f_{j'\mathbf{p}'}) - f_{j\mathbf{p}} f_{j'\mathbf{p}'}(1 - f_{j\mathbf{p}_1})(1 - f_{j'\mathbf{p}_1'})], \tag{2}$$

and the electron-impurity collision integral (see Eq. (8.7))

$$J_{im}(f|j\mathbf{p}) = \frac{2\pi}{\hbar L^2} \sum_{\mathbf{p}'} w_j(|\mathbf{p} - \mathbf{p}'|/\hbar) \delta(\varepsilon_{jp} - \varepsilon_{jp'})(f_{j\mathbf{p}'} - f_{j\mathbf{p}}). \tag{3}$$

Assuming that the temperature is low enough, we consider the relaxation of momentum due to electron-impurity scattering and neglect electron-phonon scattering. The expression (2) is a generalization of the collision integral of Eqs. (31.21) and (31.22) to the case when several electron subbands are involved in the electron-electron collisions and there are no transitions between different subbands, since the layers are isolated from each other. The exchange contribution into the matrix element is diagonal in the layer number j. The matrix elements $v_q^{(jj')}$ standing in Eq. (2) are written according to problem 6.1 (see also Eq. (29.10)):

$$v_q^{(jj')} = \frac{4\pi e^2}{\epsilon} \int_{-\infty}^{\infty} \frac{dq_z}{2\pi} \frac{\langle j|e^{-iq_z z}|j\rangle\langle j'|e^{iq_z z}|j'\rangle}{q_z^2 + q^2}$$

$$= \frac{2\pi e^2}{\epsilon q} \int dz |\psi_z^{(j)}|^2 \int dz' |\psi_{z'}^{(j')}|^2 e^{-q|z-z'|}. \tag{4}$$

The symmetry $v_q^{(12)} = v_q^{(21)}$ is seen explicitly.

One may rewrite Eq. (1) as a set of two equations for the two layers. The interlayer ($j' \neq j$) and intralayer ($j' = j$) collisions are separated if we write the collision integral (2) in the form $J_{ee}(f|j\mathbf{p}) = J_{ee}^{intra}(f|j\mathbf{p}) + J_{ee}^{inter}(f|j\mathbf{p})$. The interlayer collisions typically have much smaller probability because of spatial separation of the layers. As seen from Eq. (4), $v_{|\mathbf{p}-\mathbf{p}_1|/\hbar}^{(12)} \propto e^{-|\mathbf{p}-\mathbf{p}_1|Z/\hbar}$, where Z is the distance between the layers, so that the electron transitions with large momentum transfer between the layers are exponentially suppressed. Therefore, we can consider the interlayer part $J_{ee}^{inter}(f|j\mathbf{p})$ of the electron-electron collision integral as the weakest contribution and solve the kinetic equation by iterations. In the zero-order approximation, we neglect $J_{ee}^{inter}(f|j\mathbf{p})$ and Eq. (1) splits into two uncoupled, single-layer equations. We have

$$e\mathbf{E}_j \cdot \frac{\partial f_{j\mathbf{p}}^{(0)}}{\partial \mathbf{p}} = J_{ee}^{intra}(f^{(0)}|j\mathbf{p}) + J_{im}(f^{(0)}|j\mathbf{p}). \tag{5}$$

Below we assume a linear transport regime and replace $f_{j\mathbf{p}}^{(0)}$ on the left-hand side of Eq. (5) by the equilibrium Fermi distribution function $f_{j\mathbf{p}}^{(eq)}$. In this way we obtain an inhomogeneous equation whose solution is determined by two scattering mechanisms. If the impurity scattering dominates, one can use the results of Chapter 2 to write this solution through the momentum-dependent, isotropic elastic-scattering transport time $\tau_{jp} = \tau_{tr}^{(j)}(\varepsilon_p)$:

$$f_{j\mathbf{p}}^{(0)} = f_{j\mathbf{p}}^{(eq)} - e\mathbf{E}_j \cdot \frac{\partial f_{j\mathbf{p}}^{(eq)}}{\partial \mathbf{p}} \tau_{jp}. \tag{6}$$

If the electron-electron collisions dominate, the function $f_{j\mathbf{p}}^{(0)}$ is given by the shifted Fermi distribution (31.25), where the electron drift velocity $\mathbf{u} = \mathbf{u}_j$ is determined by the electron-impurity interaction in the layer j

and $T_e = T$ since we assume no electron heating. If the momentum (or, equivalently, energy) dependence of τ_{jp} can be neglected, one has $\mathbf{u}_j = \tau_j e \mathbf{E}_j/m$, and the distribution function is again given by Eq. (6). We note that the transport time of 2D electrons is momentum-independent in the case of short-range impurity potential. Therefore, the expression (6) with $\tau_{jp} = \tau_j$ is the exact solution of the single-layer kinetic equation (5) in the case of energy-independent transport time. We will consider this regime in the following.

In the next step of the iteration procedure, we assume

$$f_{j\mathbf{p}} = f_{j\mathbf{p}}^{(0)} + \delta f_{j\mathbf{p}}, \tag{7}$$

where $\delta f_{j\mathbf{p}}$ is a small correction caused by the interlayer momentum transfer. Linearizing the collision integrals $J_{ee}^{intra}(f|j\mathbf{p})$ and $J_{im}(f|j\mathbf{p})$ with respect to this small correction, we obtain

$$\frac{4\pi}{\hbar L^4} \sum_{\mathbf{p}'\mathbf{p}_1\mathbf{p}_1'} |v_{|\mathbf{p}-\mathbf{p}_1|/\hbar}^{(12)}|^2 \delta_{\mathbf{p}+\mathbf{p}',\mathbf{p}_1+\mathbf{p}_1'} \delta(\varepsilon_{jp} + \varepsilon_{j'p'} - \varepsilon_{jp_1} - \varepsilon_{j'p_1'})$$

$$\times [f_{j\mathbf{p}_1}^{(0)} f_{j'\mathbf{p}_1'}^{(0)} (1 - f_{j\mathbf{p}}^{(0)})(1 - f_{j'\mathbf{p}'}^{(0)}) - f_{j\mathbf{p}}^{(0)} f_{j'\mathbf{p}'}^{(0)} (1 - f_{j\mathbf{p}_1}^{(0)})(1 - f_{j'\mathbf{p}_1'}^{(0)})]$$

$$= -\delta J_{ee}^{intra}(f|j\mathbf{p}) - \delta J_{im}(f|j\mathbf{p}), \tag{8}$$

where the interlayer part $(j' \neq j)$ of the electron-electron collision integral stands on the left-hand side. Substituting the expression (6) there, one should keep only the contributions linear in $\mathbf{E}_j$. After a set of transformations using the identity $\partial f_{j\mathbf{p}}^{(eq)}/\partial \mathbf{p} = -(\mathbf{p}/mT) f_{j\mathbf{p}}^{(eq)} (1 - f_{j\mathbf{p}}^{(eq)})$, the interlayer part of the collision integral is rewritten as

$$\frac{4\pi e}{L^4 mT} \sum_{\mathbf{p}'\mathbf{q}} |v_q^{(12)}|^2 \, \mathbf{q} \cdot [\mathbf{E}_{j'}\tau_{j'} - \mathbf{E}_j\tau_j] \delta(\varepsilon_{jp} + \varepsilon_{j'p'} - \varepsilon_{j\mathbf{p}-\hbar\mathbf{q}} - \varepsilon_{j'\mathbf{p}'+\hbar\mathbf{q}})$$

$$\times f_{j\mathbf{p}}^{(eq)} f_{j'\mathbf{p}'}^{(eq)} (1 - f_{j\mathbf{p}-\hbar\mathbf{q}}^{(eq)})(1 - f_{j'\mathbf{p}'+\hbar\mathbf{q}}^{(eq)}). \tag{9}$$

Let us multiply both sides of Eq. (8) by $2ep\tau_j/mL^2$ and sum the equation obtained over $\mathbf{p}$. Consider first the right-hand side of the equation found in this way. The contribution from the intralayer part of the electron-electron collision integral vanishes since the scattering within the layer conserves the momentum; see Eq. (31.24). The contribution from the electron-impurity term is

$$\frac{4\pi e}{\hbar L^4} \sum_{\mathbf{p}\mathbf{p}'} \frac{\mathbf{p}\tau_j}{m} w_j(|\mathbf{p} - \mathbf{p}'|/\hbar)[\delta f_{j\mathbf{p}} - \delta f_{j\mathbf{p}'}]\delta(\varepsilon_{jp} - \varepsilon_{jp'})$$

$$= \frac{4\pi e}{\hbar m L^4} \sum_{\mathbf{p}} \tau_j \delta f_{j\mathbf{p}} \sum_{\mathbf{p}'} (\mathbf{p} - \mathbf{p}') w_j(|\mathbf{p} - \mathbf{p}'|/\hbar) \delta(\varepsilon_{jp} - \varepsilon_{jp'}). \qquad (10)$$

According to Eq. (11.14), the sum over $\mathbf{p}'$ is proportional to $\mathbf{p}/\tau_j$ and the expression (10) is equal to $\delta \mathbf{I}_j$, where

$$\delta \mathbf{I}_j = 2e \int \frac{d\mathbf{p}}{(2\pi\hbar)^2} \frac{\mathbf{p}}{m} \delta f_{j\mathbf{p}} \qquad (11)$$

is the current density associated with the interlayer momentum transfer. Transforming the left-hand side of Eq. (8) (problem 6.15), we find

$$\delta I_{j\alpha} = \sum_{j'\beta} \delta\sigma_{\alpha\beta}^{jj'} E_{j'\beta}, \qquad (12)$$

where α and β are the coordinate indices. The conductivity tensor introduced by Eq. (12) contains both diagonal and non-diagonal, with respect to the layer index, contributions:

$$\delta\sigma_{\alpha\beta}^{jj'} = (-1)^l \tau_j \tau_{j'} \frac{4\pi e^2 \hbar}{m^2 T} \int \frac{d\mathbf{q}}{(2\pi)^2} q_\alpha q_\beta |v_q^{(12)}|^2 \int \frac{d\mathbf{p}}{(2\pi\hbar)^2} \int \frac{d\mathbf{p}'}{(2\pi\hbar)^2} \qquad (13)$$

$$\times \delta(\varepsilon_{jp} + \varepsilon_{j'p'} - \varepsilon_{j\mathbf{p}-\hbar\mathbf{q}} - \varepsilon_{j'\mathbf{p}'+\hbar\mathbf{q}}) f_{j\mathbf{p}}^{(eq)} f_{j'\mathbf{p}'}^{(eq)} (1 - f_{j\mathbf{p}-\hbar\mathbf{q}}^{(eq)})(1 - f_{j'\mathbf{p}'+\hbar\mathbf{q}}^{(eq)}),$$

where $l = 0$ for $j \neq j'$ and $l = 1$ for $j = j'$. The tensor $\delta\sigma_{\alpha\beta}^{jj'}$ is symmetric in the layer indices. Since the electron spectra and the matrix element $v_q^{(12)}$ are isotropic, this tensor is diagonal in the coordinate indices. Therefore, one may replace $q_\alpha q_\beta$ by $\delta_{\alpha\beta} q^2/2$ and omit the coordinate indices.

The total current density in the layer j is given as

$$\mathbf{I}_j = 2e \int \frac{d\mathbf{p}}{(2\pi\hbar)^2} \frac{\mathbf{p}}{m} f_{j\mathbf{p}}^{(0)} + \delta \mathbf{I}_j. \qquad (14)$$

The first term on the right-hand side of this equation is the usual expression for the current of electrons interacting with impurities; see Chapter 2. Therefore, we obtain the following solution of the linear response problem for the Coulomb-coupled layers:

$$\mathbf{I}_j = \sum_{j'} \sigma^{jj'} \mathbf{E}_{j'}. \qquad (15)$$

The conductivity tensor $\sigma^{jj'}$ is symmetric. Its non-diagonal part, which is entirely determined by the interlayer momentum transfer, is called the drag conductivity, $\sigma_D \equiv \delta\sigma^{jj'}$, $j' \neq j$. The diagonal components of the conductivity tensor have small negative corrections due to interlayer momentum transfer:

$$\sigma^{jj} = \frac{e^2 n_j \tau_j}{m} - \frac{\tau_j^2}{\tau_1 \tau_2} \sigma_D, \qquad (16)$$

where n_j is the electron density in the layer j. These corrections are neglected in the following, and we put $\sigma^{jj} \simeq e^2 n_j \tau_j / m$. The quantity usually measured in experiments is the drag resistivity ρ_D rather than σ_D. If the layer 1 is the drive layer and the layer 2 is the drag layer (see Fig. 6.3), this resistivity is defined as

$$\rho_D = -E_2 / I_1 \qquad (17)$$

under the condition that the current in the drag layer is equal to zero, $I_2 = 0$. One may introduce the drag voltage $V_2 = eE_2 L$, where L is the length of the layers. Solving Eq. (15) in these conditions, we neglect the corrections quadratic in σ_D and find

$$\rho_D = \frac{\sigma_D}{\sigma^{11} \sigma^{22}}. \qquad (18)$$

It is important that the sign of the drag voltage is opposite to the voltage applied to the drive layer, because the drag voltage develops to counteract the driving force. To make ρ_D positive by definition, the minus sign is introduced in Eq. (17).

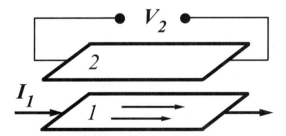

Figure 6.3. Schematic representation of the Coulomb drag effect.

Collecting the results given by Eqs. (13), (16), and (18), we write the expression for the drag resistivity:

$$\rho_D = \frac{2\pi\hbar}{e^2 T n_1 n_2} \int \frac{d\mathbf{q}}{(2\pi)^2} q^2 |v_q^{(12)}|^2 \int \frac{d\mathbf{p}}{(2\pi\hbar)^2} \int \frac{d\mathbf{p}'}{(2\pi\hbar)^2} \qquad (19)$$

$$\times \delta(\varepsilon_{1p} + \varepsilon_{2p'} - \varepsilon_{1\mathbf{p}-\hbar\mathbf{q}} - \varepsilon_{2\mathbf{p}'+\hbar\mathbf{q}}) f_{1\mathbf{p}}^{(eq)} f_{2\mathbf{p}'}^{(eq)} (1 - f_{1\mathbf{p}-\hbar\mathbf{q}}^{(eq)})(1 - f_{2\mathbf{p}'+\hbar\mathbf{q}}^{(eq)}).$$

As seen, τ_1 and τ_2 have disappeared from this equation, and ρ_D depends only on the interlayer electron-electron interaction. Below we rewrite Eq. (19) in a more convenient form by using the following identities:

$$\delta(\varepsilon_{1p} + \varepsilon_{2p'} - \varepsilon_{1\mathbf{p}-\hbar\mathbf{q}} - \varepsilon_{2\mathbf{p}'+\hbar\mathbf{q}})$$

$$= \hbar \int d\omega \delta(\varepsilon_{1p} - \varepsilon_{1\mathbf{p}-\hbar\mathbf{q}} - \hbar\omega)\delta(\varepsilon_{2p} - \varepsilon_{2\mathbf{p}'+\hbar\mathbf{q}} + \hbar\omega) \qquad (20)$$

and

$$f_\varepsilon^{(eq)}(1 - f_{\varepsilon\pm\hbar\omega}^{(eq)}) = \frac{f_\varepsilon^{(eq)} - f_{\varepsilon\pm\hbar\omega}^{(eq)}}{1 - \exp(\mp\hbar\omega/T)}. \qquad (21)$$

We also introduce the function

$$\Pi_j(\mathbf{q}, \omega) = 2 \int \frac{d\mathbf{p}}{(2\pi\hbar)^2} \frac{f_{j\mathbf{p}}^{(eq)} - f_{j\mathbf{p}-\hbar\mathbf{q}}^{(eq)}}{\varepsilon_{jp} - \varepsilon_{j\mathbf{p}-\hbar\mathbf{q}} - \hbar\omega - i\lambda}, \qquad \lambda \to +0, \qquad (22)$$

which has the symmetry property $\Pi_j(\mathbf{q}, \omega) = \Pi_j^*(-\mathbf{q}, -\omega)$ and depends, due to isotropy of the electron spectrum, only on the absolute value of $\mathbf{q}$. Using it, we find that Eq. (19) can be rewritten through the imaginary parts of Π_1 and Π_2:

$$\rho_D = \frac{\hbar^2}{8\pi^2 e^2 T n_1 n_2} \int_0^\infty dq q^3 |v_q^{(12)}|^2 \int_0^\infty d\omega \frac{\mathrm{Im}\Pi_1(q, \omega)\mathrm{Im}\Pi_2(q, \omega)}{\sinh^2(\hbar\omega/2T)}. \qquad (23)$$

The characteristic frequencies ω contributing to the integral in Eq. (23) are determined by the condition $\omega < 2T/\hbar$. Therefore, to describe the drag at low temperatures, we use the low-frequency asymptote of $\mathrm{Im}\Pi_j$ (problem 6.16):

$$\mathrm{Im}\Pi_j(\mathbf{q}, \omega) \simeq -\frac{m^2\omega}{\pi\hbar^2 q p_{Fj}}, \qquad q \ll p_{Fj}/\hbar, \qquad (24)$$

which allows us to calculate the integral over ω analytically. We obtain the expression

$$\rho_D = \frac{m^4 T^2}{6\pi^2\hbar^5 e^2 n_1 n_2 p_{F1} p_{F2}} \int_0^\infty dq q |v_q^{(12)}|^2 \qquad (25)$$

describing a quadratic temperature dependence of the drag resistance. The same behavior is seen experimentally. The drag resistance decreases with decreasing temperature because at low temperatures the probability of electron-electron scattering becomes smaller.

To complete the calculations, one should evaluate the integral over q in Eq. (25). However, a direct substitution of $v_q^{(12)}$ given by Eq. (4) into

Eq. (25) leads to a logarithmic-divergent expression for this integral, due to the contribution of small q (small-angle scattering). To overcome this difficulty, a self-consistent calculation of the interlayer Coulomb matrix element $v_q^{(12)}$ is necessary (problem 6.17). As a result, we obtain the "screened" matrix element

$$v_q^{(12)} \simeq \frac{2\pi e^2}{\epsilon} \frac{q}{(8/a_B^2)\sinh qZ + (4q/a_B + q^2)e^{qZ}}, \qquad (26)$$

which is derived under the approximation that the interlayer distance Z, defined as the distance between the weight centers $z_j = \int dz |\psi_z^{(j)}|^2 z$ of the wave functions in the 2D layers, considerably exceeds the layer widths d_j. If $qa_B \gg 1$, Eq. (26) gives us $v_q^{(12)} = (2\pi e^2/\epsilon q)e^{-qZ}$, which coincides, in the approximation $Z \gg d_j$, with the result of Eq. (4). On the other hand, $v_q^{(12)}$ of Eq. (26) is finite at small q. The main contribution to the integral in Eq. (25) comes from $q < 1/Z$. Assuming $Z \gg a_B/2$, we approximate the expression (26) at $q \ll 2/a_B$ as $v_q^{(12)} \simeq \pi e^2 a_B^2 q/4\epsilon \sinh qZ$. Substituting this expression into Eq. (25), we calculate the integral and finally obtain

$$\rho_D = \frac{\zeta(3)}{128\pi} \frac{T^2 m^2 a_B^2}{\hbar^3 e^2 (n_1 n_2)^{3/2} Z^4}, \qquad (27)$$

where $\zeta(p) = 1 + 1/2^p + 1/3^p + \ldots$ is the zeta-function and $\zeta(3) \simeq 1.2$. The drag resistivity decreases with increasing interlayer separation as Z^{-4}. It also decreases with increasing electron densities.

The Coulomb drag may occur between 1D electron systems, as well as between the systems of different dimensionalities (we do not consider such situations in this book). In the Coulomb drag effect, the electron-electron interaction manifests itself directly through the measured transport quantity, ρ_D. For this reason, the measurements of Coulomb drag are important for investigating the Coulomb interaction in low dimensions.

33. Dynamical Screening

Electron systems show collective behavior because of long-range nature of Coulomb interaction. One of the manifestations of such behavior is the plasma oscillations (plasmons) discussed in Sec. 11. Another important issue is the dynamical screening of the Coulomb potential. We have already seen the necessity of consideration of the static screening in the previous section, where the screening has removed the divergence of electron-electron collision integral caused by the divergence of the matrix element at small momentum transfer. A similar situation takes place for the electron-impurity collision integral when the impurity potential is the Coulomb one; see problem 3.9. Now it is time to take a closer look at the screening effect. In this section we discuss some useful approaches for considering the screening in the problems of kinetics.

The issue of dynamical screening appears when we consider a linear response of a system of interacting electrons to the charge density perturbation (often named as trial charge) with Fourier component $\delta\rho(\mathbf{q}, \omega)$. This perturbation causes the induced charge density $\Delta\rho(\mathbf{q}, \omega)$. The macroscopic electric field $\mathbf{E}(\mathbf{q}, \omega)$ arising in the electron system due to the perturbation is related to these densities according to the Poisson equation

$$i\mathbf{q} \cdot \mathbf{E}(\mathbf{q}, \omega) = \frac{4\pi}{\epsilon_\infty} \left[\delta\rho(\mathbf{q}, \omega) + \Delta\rho(\mathbf{q}, \omega)\right], \qquad (1)$$

where ϵ_∞ is the dielectric permittivity of the medium without the electrons whose response we study (the relativistic retardation effects are neglected). Note that in the previous sections of this chapter, where the dynamical response of electron system was neglected, we simply wrote $\epsilon_\infty = \epsilon$. If one considers an ionic crystal at ω comparable to or smaller than the optical phonon frequencies, ϵ_∞ should be replaced by $\kappa(\mathbf{q}, \omega)$ given by Eq. (27.29). Introducing the dielectric permittivity $\epsilon(\mathbf{q}, \omega)$ as

$$\mathbf{D}(\mathbf{q}, \omega) = \epsilon_\infty \mathbf{E}(\mathbf{q}, \omega) + 4\pi\Delta\mathbf{P}(\mathbf{q}, \omega) = \epsilon(\mathbf{q}, \omega)\mathbf{E}(\mathbf{q}, \omega), \qquad (2)$$

and taking into account that the polarization of electron system is related to the induced density as $\nabla \cdot \Delta\mathbf{P}(\mathbf{r}, \omega) = -\Delta\rho(\mathbf{r}, \omega)$ or, equivalently, $i\mathbf{q} \cdot \Delta\mathbf{P}(\mathbf{q}, \omega) = -\Delta\rho(\mathbf{q}, \omega)$, we use Eq. (1) and obtain the following relations:

$$i\mathbf{q} \cdot \mathbf{E}(\mathbf{q}, \omega) = \frac{4\pi\delta\rho(\mathbf{q}, \omega)}{\epsilon(\mathbf{q}, \omega)}, \qquad \frac{\epsilon_\infty}{\epsilon(\mathbf{q}, \omega)} = 1 + \frac{\Delta\rho(\mathbf{q}, \omega)}{\delta\rho(\mathbf{q}, \omega)}, \qquad (3)$$

which describe the response to the longitudinal field. A comparison of Eqs. (1) and (3) shows us that the effective longitudinal electric field in the system is determined by the Poisson equation containing "bare" charge perturbation only, while the effects of polarization are completely included into the dielectric permittivity $\epsilon(\mathbf{q}, \omega)$. To calculate the latter, we note that $\Delta\rho(\mathbf{q}, \omega) = \mathrm{Sp}\hat{\rho}_\mathbf{q}\widehat{\Delta\eta}_\omega$, see problem 3.3, and search for the perturbation of the density matrix of electron system following the external density perturbation $\delta\rho(\mathbf{q}, \omega)$. The Hamiltonian of this perturbation is written in the form of the last term in Eq. (4.8):

$$\widehat{\Delta H}_\omega = \int d\mathbf{r} \int d\mathbf{r}' \frac{\delta\rho(\mathbf{r}, \omega)\hat{\rho}_{\mathbf{r}'}}{\epsilon_\infty|\mathbf{r} - \mathbf{r}'|} = \sum_\mathbf{q} \frac{4\pi}{\epsilon_\infty q^2}\hat{\rho}_{-\mathbf{q}}\delta\rho(\mathbf{q}, \omega). \qquad (4)$$

Substituting this Hamiltonian into Eq. (13.4), we use the latter to calculate $\Delta\rho(\mathbf{q},\omega)$. Then we can write the second equation of Eq. (3) as

$$\frac{\epsilon_\infty}{\epsilon(\mathbf{q},\omega)} = 1 + \frac{4\pi\alpha(\mathbf{q},\omega)}{\epsilon_\infty q^2}, \qquad (5)$$

where $\alpha(\mathbf{q},\omega)$ is the polarizability of electron system.

The expression of $\alpha(\mathbf{q},\omega)$ in terms of the density-density correlation function is given in problem 3.3:

$$\alpha(\mathbf{q},\omega) = \frac{e^2}{i\hbar V} \int_{-\infty}^{0} d\tau\, e^{\lambda\tau - i\omega\tau} \mathrm{Sp}\hat{\eta}_{eq} \left[e^{-i\widehat{H}\tau/\hbar} \hat{n}_{\mathbf{q}} e^{i\widehat{H}\tau/\hbar}, \hat{n}_{-\mathbf{q}} \right], \qquad (6)$$

where $\hat{n}_{\mathbf{q}} = \hat{\rho}_{\mathbf{q}}/e$ is the Fourier transform of the operator of electron density. If $\widehat{H}$ standing in Eq. (6) is the Hamiltonian of many-electron system, this equation is exact. Let us calculate $\alpha(\mathbf{q},\omega)$ given by Eq. (6) in the Hartree-Fock approximation, when the Coulomb interaction in $\widehat{H}$ is neglected. This approach gives only $\propto e^2$ terms in $1/\epsilon(\mathbf{q},\omega)$, because $\alpha(\mathbf{q},\omega)$ is already proportional to e^2. Since we assume that the system is translation-invariant, the one-electron eigenstates are plane waves. In the second quantization representation, the operator $\hat{n}_{\mathbf{q}}$ can be written as $\hat{n}_{\mathbf{q}} = \sum_{\sigma\mathbf{p}} \hat{a}_{\sigma\mathbf{p}-\hbar\mathbf{q}}^{+} \hat{a}_{\sigma\mathbf{p}}$, where $\hat{a}_{\sigma\mathbf{p}}^{+}$ and $\hat{a}_{\sigma\mathbf{p}}$ are the creation and annihilation operators for the plane-wave eigenstates with momentum $\mathbf{p}$ and spin σ. In this approximation, the trace in Eq. (6) is equal to

$$\sum_{\sigma\mathbf{p}} e^{i(\varepsilon_{\mathbf{p}+\hbar\mathbf{q}} - \varepsilon_{\mathbf{p}})\tau/\hbar} \left[\langle\langle \hat{a}_{\sigma\mathbf{p}}^{+} \hat{a}_{\sigma\mathbf{p}+\hbar\mathbf{q}} \hat{a}_{\sigma\mathbf{p}+\hbar\mathbf{q}}^{+} \hat{a}_{\sigma\mathbf{p}} \rangle\rangle \right.$$

$$\left. - \langle\langle \hat{a}_{\sigma\mathbf{p}+\hbar\mathbf{q}}^{+} \hat{a}_{\sigma\mathbf{p}} \hat{a}_{\sigma\mathbf{p}}^{+} \hat{a}_{\sigma\mathbf{p}+\hbar\mathbf{q}} \rangle\rangle \right], \qquad (7)$$

where the averaging is carried out for the case of non-interacting electrons. As a result, $\langle\langle \hat{a}_{\sigma\mathbf{p}}^{+} \hat{a}_{\sigma\mathbf{p}+\hbar\mathbf{q}} \hat{a}_{\sigma\mathbf{p}+\hbar\mathbf{q}}^{+} \hat{a}_{\sigma\mathbf{p}} \rangle\rangle = f_{\mathbf{p}}(1 - f_{\mathbf{p}+\hbar\mathbf{q}})$, where $f_{\mathbf{p}}$ is the equilibrium distribution function of electrons. Calculating the integral over τ, we finally obtain

$$\alpha_{HF}(\mathbf{q},\omega) = e^2 \Pi_\omega^{(0)R}(\mathbf{q}),$$

$$\Pi_\omega^{(0)R}(\mathbf{q}) = \frac{2}{V} \sum_{\mathbf{p}} \frac{f_{\mathbf{p}+\hbar\mathbf{q}/2} - f_{\mathbf{p}-\hbar\mathbf{q}/2}}{\varepsilon_{\mathbf{p}+\hbar\mathbf{q}/2} - \varepsilon_{\mathbf{p}-\hbar\mathbf{q}/2} - \hbar\omega - i\lambda}, \qquad (8)$$

where the factor of 2 comes from the sum over the spin variable in Eq. (7). The function $\Pi_\omega^{(0)R}(\mathbf{q})$, which determines the Hartree-Fock polarizability α_{HF}, is called the retarded polarization function of free electrons or, simply, the polarization function. The full meaning of this definition will become clear later. At this point, we note that the word "retarded" reflects the fact that this function, just like any other generalized susceptibility, does not have any poles in the upper half-plane of the complex variable ω owing to the causality principle. This property is seen from Eq. (8) directly. The symmetry property $\Pi^{(0)R}_{-\omega}(-\mathbf{q}) = \Pi^{(0)R*}_{\omega}(\mathbf{q})$ is also obvious from Eq. (8). We point out that the polarization function of two-dimensional electrons has been introduced in the previous section by Eq. (32.22). The definition (32.22) is very similar to the one given by Eq. (8). The only difference is that in the 2D case $\mathbf{q}$ and $\mathbf{p}$ are the 2D vectors.

At zero temperature, one can reduce Eq. (6) to a simpler form, because the averaging $\mathrm{Sp}\hat{\eta}_{eq}\ldots$ is reduced to the averaging $\langle\ldots\rangle_o = \langle 0|\ldots|0\rangle$ over the ground state $|0\rangle$ of many-electron system. In these conditions, we apply the basis of the exact excited states $|\delta\rangle$ and, taking the integral over time, obtain

$$\alpha(\mathbf{q}, \omega) = e^2 \sum_\delta \left[\frac{|\langle\delta|\hat{n}^+_{\mathbf{q}}|0\rangle|^2}{\hbar\omega - E_{\delta 0} + i\lambda} - \frac{|\langle\delta|\hat{n}_{\mathbf{q}}|0\rangle|^2}{\hbar\omega + E_{\delta 0} + i\lambda} \right], \qquad (9)$$

where the energy $E_{\delta 0} = E_\delta - E_0$ is the difference in the energies of excited and ground states. To derive Eq. (9), we have used the transformations $\langle 0|\hat{n}_{\mathbf{q}}|\delta\rangle\langle\delta|\hat{n}_{-\mathbf{q}}|0\rangle = |\langle\delta|\hat{n}^+_{\mathbf{q}}|0\rangle|^2$, according to the symmetry property $\hat{n}^+_{\mathbf{q}} = \hat{n}_{-\mathbf{q}}$ following from the Hermiticity of the density operator in the coordinate representation, $\hat{n}^+_{\mathbf{r}} = \hat{n}_{\mathbf{r}}$. Evaluating the expression (9) in the Hartree-Fock approximation, we obtain the same result as follows from Eq. (8) in the case of zero temperature, when $f_{\mathbf{p}} = \theta(p_F - p)$ (problem 6.18).

Since $\Pi^{(0)R}_{\omega}(\mathbf{q})$ is finite at $q \to 0$, the function $1/\epsilon(\mathbf{q}, \omega)$ describing the response of electron system remains divergent at $q = 0$ in the Hartree-Fock approximation. To remove this divergence, let us discuss another way for evaluating the dielectric permittivity, based upon the diagram technique for interacting electron systems discussed in Appendix E. We take into account that $\alpha(\mathbf{q}, \omega) = \int d\mathbf{r} e^{-i\mathbf{q}\cdot\mathbf{r}}\alpha(\mathbf{r}, 0|\omega)$ and use the expression for $\alpha(\mathbf{r}, \mathbf{r}'|\omega)$ given in problem 3.3. The density operators are expressed through the field operators of electrons as $\hat{n}_{\mathbf{r}} = \hat{\Psi}^+_{\mathbf{r}}\hat{\Psi}_{\mathbf{r}}$, and we obtain

$$\mathrm{Sp}\hat{\eta}_{eq}\left[e^{-i\hat{H}\tau/\hbar}\hat{\rho}_{\mathbf{r}}e^{i\hat{H}\tau/\hbar}, \hat{\rho}_{\mathbf{r}'} \right] \qquad (10)$$

$$= e^2 \mathrm{Sp}\,\hat{\eta}_{eq}\hat{\Psi}^+_{\mathbf{r}}(-\tau)\hat{\Psi}_{\mathbf{r}}(-\tau)\hat{\Psi}^+_{\mathbf{r}'}(0)\hat{\Psi}_{\mathbf{r}'}(0) - c.c.\,,$$

where $\hat{\Psi}_{\mathbf{r}}(t)$ is the field operator in the Heisenberg representation; see Eq. (D.1). The commutator is transformed by using the Hermiticity of both $\hat{\eta}_{eq}$ and $\hat{n}_{\mathbf{r}}$ and the possibility for cyclic permutations of the

operators under the sign of the trace. Employing the definition (E.6) of
two-electron Green's functions and Eq. (10), we have

$$\alpha(\mathbf{q}, \omega) = i\hbar e^2 \int d\mathbf{r} e^{-i\mathbf{q}\cdot\mathbf{r}} \int_0^\infty dt e^{-\lambda t + i\omega t} \tag{11}$$

$$\times \left[G(\mathbf{r}t, \mathbf{r}'0; \mathbf{r}t + 0, \mathbf{r}' + 0)|_{\mathbf{r}'=0} - c.c. \right].$$

To obtain Eq. (11), one should substitute $\tau \to -t$ in the equation for
$\alpha(\mathbf{r}, \mathbf{r}'|\omega)$ in problem 3.3 and take into account that $\hat{\Psi}_{\mathbf{r}}^+(t + 0)\hat{\Psi}_{\mathbf{r}}(t)$
$\times\hat{\Psi}_{\mathbf{r}'}^+(+0)\hat{\Psi}_{\mathbf{r}'}(0) = \hat{\mathcal{T}}\hat{\Psi}_{\mathbf{r}}(t)\hat{\Psi}_{\mathbf{r}'}(0)\hat{\Psi}_{\mathbf{r}'}^+(+0)\hat{\Psi}_{\mathbf{r}}^+(t + 0)$ at $t > 0$, where $\hat{\mathcal{T}}$
is the operator of chronological ordering; see Eq. (2.7). Equation (11)
is written here and analyzed below for the case of zero temperature.
Nevertheless, it is formally valid at arbitrary temperatures if the many-
electron Green's functions are defined in such a way that the averaging
is carried out over the equilibrium electron distribution, i.e., $\langle\ldots\rangle_o$ is
replaced by $\langle\langle\ldots\rangle\rangle = \mathrm{Sp}\hat{\eta}_{eq}\ldots$. The diagram technique developed in
Appendix E, of course, is valid only at $T = 0$. In Chapter 8 we involve
a more sophisticated diagram technique which takes care of the case of
finite temperatures.

Equation (11) can be rewritten as

$$\alpha(\mathbf{q}, \omega) = e^2 \int_{-\infty}^\infty \frac{d\omega'}{2\pi i} \frac{K(\mathbf{q}, \omega') + K^*(-\mathbf{q}, -\omega')}{\omega' - \omega - i\lambda}, \tag{12}$$

$$K(\mathbf{q}, \omega) = i\hbar \int d\mathbf{r} \int_{-\infty}^\infty dt e^{i\omega t - i\mathbf{q}\cdot\mathbf{r}} G(\mathbf{r}t, \mathbf{r}'0; \mathbf{r}t + 0, \mathbf{r}' + 0)|_{\mathbf{r}'=0},$$

where $K(\mathbf{q}, \omega)$ is the Fourier transform of the two-particle Green's func-
tion multiplied by $i\hbar$. To calculate $G(\mathbf{r}t, \mathbf{r}'0; \mathbf{r}t + 0, \mathbf{r}' + 0)$, we use Eq.
(E.28):

$$G(\mathbf{r}t, \mathbf{r}'0; \mathbf{r}t + 0, \mathbf{r}' + 0) = G(\mathbf{r}t, \mathbf{r}t + 0)G(\mathbf{r}'0, \mathbf{r}' + 0) - G(\mathbf{r}t, \mathbf{r}'0)G(\mathbf{r}'0, \mathbf{r}t)$$

$$+ \int d\mathbf{r}_1 \int d\mathbf{r}_2 \int d\mathbf{r}_1' \int d\mathbf{r}_2' \int dt_1 \int dt_2 \int dt_1' \int dt_2' \tag{13}$$

$$\times G(\mathbf{r}t, \mathbf{r}_1 t_1)G(\mathbf{r}'0, \mathbf{r}_2 t_2)\mathcal{V}(\mathbf{r}_1 t_1, \mathbf{r}_2 t_2; \mathbf{r}_1' t_1', \mathbf{r}_2' t_2')G(\mathbf{r}_1' t_1', \mathbf{r}t)G(\mathbf{r}_2' t_2', \mathbf{r}'0),$$

where the integral term is expressed through the scattering amplitude.
The latter is expanded in series of the screened interaction potential
$V(\mathbf{r}_1 t_1, \mathbf{r}_2 t_2)$ defined by Eqs. (E.24) and (E.25):

$$\mathcal{V}(\mathbf{r}_1 t_1, \mathbf{r}_2 t_2; \mathbf{r}_1' t_1', \mathbf{r}_2' t_2') = i\hbar V(\mathbf{r}_1 t_1, \mathbf{r}_2 t_2)[\delta(\mathbf{r}_1 - \mathbf{r}_1')\delta(\mathbf{r}_2 - \mathbf{r}_2') \tag{14}$$

$$\times\delta(t_1 - t_1')\delta(t_2 - t_2') - \delta(\mathbf{r}_1 - \mathbf{r}_2')\delta(\mathbf{r}_2 - \mathbf{r}_1')\delta(t_1 - t_2')\delta(t_2 - t_1')] + \ldots,$$

the dots denote higher-order terms. Substituting this expansion into Eq. (13), we obtain the corresponding expansion of the two-electron Green's function. In the language of diagrams, the expansion of $i^2 G(\mathbf{r}t, \mathbf{r}'0; \mathbf{r}t + 0, \mathbf{r}' + 0)$ is represented as

$$\qquad (15)$$

The analytical expression of this equation can be written according to the diagram rules described in Appendix E. The first term on the right-hand side of Eq. (15), containing two separated, self-closed loops, is proportional to the product of electron densities. It is real and does not contribute into Eq. (11). The second and the fourth terms are the first and the second diagrams in the expansion (E.26) of the polarization function $i\hbar^{-1}\Pi(\mathbf{r}t, \mathbf{r}'0)$. The third term represents two "polarization loops" connected by a bold broken line, the latter corresponds to $-i\hbar V(\mathbf{r}_1 t_1, \mathbf{r}'_1 t'_1)$. The diagrams denoted by the dots "..." contain two or more bold broken lines. Some of these diagrams describe various corrections to $i\hbar^{-1}\Pi(\mathbf{r}t, \mathbf{r}'0)$. Such diagrams can be constructed if we add more broken lines to the loop in the fourth diagram or add more lines connecting two loops to the third diagram on the right-hand side of Eq. (15). The remaining diagrams describe various corrections to the third diagram on the right-hand side of Eq. (15) (this particular diagram is reducible and, by definition, does not enter the polarization function). Such diagrams, for example, contain the broken lines connecting the points within each of the two loops. Taking all the diagrams discussed above into account, one should replace the polarization loops in the second and third terms on the right-hand side of Eq. (15) by the exact polarization functions (the second and the fourth terms are unified in this way). Therefore, the analytical expression of Eq. (15) is written as

$$G(\mathbf{r}t, \mathbf{r}'0; \ \mathbf{r}t + 0, \mathbf{r}' + 0) = -\frac{n^2}{\hbar^2} - \frac{i}{\hbar}\Pi(\mathbf{r}t, \mathbf{r}'0) - \frac{i}{\hbar}\int d\mathbf{r}_1 \int d\mathbf{r}'_1$$

$$\times \int dt_1 \int dt'_1 \Pi(\mathbf{r}t, \mathbf{r}_1 t_1) V(\mathbf{r}_1 t_1, \mathbf{r}'_1 t'_1) \Pi(\mathbf{r}'_1 t'_1, \mathbf{r}'0). \qquad (16)$$

This expression is exact as far as the polarization function is assumed to be exact. Carrying out the Fourier transformations of $G(\mathbf{r}t, \mathbf{r}'0; \ \mathbf{r}t +$

$0, \mathbf{r}' + 0)$ in both time and space, we find the analytical expression of $K(\mathbf{q}, \omega)$ corresponding to Eq. (16):

$$K(\mathbf{q}, \omega) = -(i/\hbar)n^2\delta(\omega)\delta(\mathbf{q}) + \Pi_\omega(\mathbf{q})[1 + V_\omega(\mathbf{q})\Pi_\omega(\mathbf{q})]. \qquad (17)$$

To express $V_\omega(\mathbf{q})$ through the polarization function, we use Eq. (E.32) in the form $V_\omega(\mathbf{q}) = v_q[1 - v_q\Pi_\omega(\mathbf{q})]^{-1}$, where $v_q = 4\pi e^2/\epsilon_\infty q^2$ for the case of Coulomb interaction. Now we obtain

$$K(\mathbf{q}, \omega) + K^*(-\mathbf{q}, -\omega) = \frac{\Pi_\omega(\mathbf{q})}{1 - v_q\Pi_\omega(\mathbf{q})} + \frac{\Pi^*_{-\omega}(-\mathbf{q})}{1 - v_q\Pi^*_{-\omega}(-\mathbf{q})}. \qquad (18)$$

According to this equation, the polarizability given by Eq. (12) is expressed through the exact polarization function. Note that the term

$$\frac{\Pi_\omega(\mathbf{q})}{1 - v_q\Pi_\omega(\mathbf{q})} = \Pi_\omega(\mathbf{q}) + v_q[\Pi_\omega(\mathbf{q})]^2 + v_q^2[\Pi_\omega(\mathbf{q})]^3 + \dots \qquad (19)$$

is represented as a sum of infinite series of reducible diagrams containing chains of the exact polarization functions (denoted in Appendix E by circles) connected by the "bare" interaction lines. Owing to the symmetry property $\Pi_\omega(\mathbf{q}) = \Pi_{-\omega}(-\mathbf{q})$, the expression (18) is real. Therefore, using Eq. (18), we rewrite Eq. (12) as

$$\alpha(\mathbf{q}, \omega) = \int_{-\infty}^{\infty} \frac{d\omega'}{2\pi i} \frac{e^2}{\omega' - \omega - i\lambda} \left[\frac{\Pi^R_{\omega'}(\mathbf{q})}{1 - v_q\Pi^R_{\omega'}(\mathbf{q})} + \frac{\Pi^A_{\omega'}(\mathbf{q})}{1 - v_q\Pi^A_{\omega'}(\mathbf{q})} \right], \quad (20)$$

where $\Pi^A_\omega(\mathbf{q}) = \Pi^R_\omega{}^*(\mathbf{q})$. The retarded polarization function $\Pi^R_\omega(\mathbf{q})$ is defined in such a way that its real part and absolute value coincide with the real part and absolute value of $\Pi_\omega(\mathbf{q})$, and $\Pi^R_\omega(\mathbf{q})$ is analytical in the upper half-plane of the complex variable ω. Each term in the square brackets of Eq. (20) can be expanded according to Eq. (19). Thus, the function $\Pi^R_{\omega'}(\mathbf{q})/[1 - v_q\Pi^R_{\omega'}(\mathbf{q})]$ is analytical in the upper half-plane of the complex variable ω', while $\Pi^A_{\omega'}(\mathbf{q})/[1 - v_q\Pi^A_{\omega'}(\mathbf{q})]$ is analytical in the lower half-plane of ω'. The representation of the expression under the integral over ω' through the sum of retarded and advanced functions is very convenient. Indeed, when this integral is taken, only the retarded term, with $\omega' = \omega$, remains. Having calculated $\alpha(\mathbf{q}, \omega)$, we substitute it into Eq. (5) and find the main result of this section:

$$\alpha(\mathbf{q}, \omega) = \frac{e^2\Pi^R_\omega(\mathbf{q})}{1 - v_q\Pi^R_\omega(\mathbf{q})}, \qquad \epsilon(\mathbf{q}, \omega) = \epsilon_\infty - \frac{4\pi e^2}{q^2}\Pi^R_\omega(\mathbf{q}) . \qquad (21)$$

These expressions describe the dynamical response of electrons. One can see that $1/\epsilon(\mathbf{q}, \omega)$ is no longer divergent at small q, because the screening effects remove this divergence. The expressions (21) are exact if the polarization function $\Pi_\omega(\mathbf{q})$ is calculated by taking into account all relevant diagrams.

Let us consider an approximation when $\Pi_\omega(\mathbf{q})$ is evaluated in the lowest order in the interaction, i.e., $i\hbar^{-1}\Pi_\omega(\mathbf{q})$ is given by the second diagram of the expansion (15), where the exact one-particle Green's functions are replaced by the Green's functions of non-interacting electrons. This approach is called the polarization approximation, or, more often, the random phase approximation (RPA). The corresponding analytical expression is

$$\Pi_\omega^{(0)}(\mathbf{q}) = \frac{2}{V} \sum_{\mathbf{p}} \int \frac{d\varepsilon}{2\pi i} g_{\varepsilon+\hbar\omega/2}(\mathbf{p}+\hbar\mathbf{q}/2) g_{\varepsilon-\hbar\omega/2}(\mathbf{p}-\hbar\mathbf{q}/2). \quad (22)$$

The factor of 2 comes from the spin summation implicitly assumed above. The one-electron Green's functions $g_\varepsilon(\mathbf{p})$ are given by Eq. (E.33). Substituting them into Eq. (22), we calculate the integral over energy and, after some transformations, obtain

$$\Pi_\omega^{(0)}(\mathbf{q}) = \frac{2}{V} \sum_{\mathbf{p}} \frac{f_{\mathbf{p}+\hbar\mathbf{q}/2} - f_{\mathbf{p}-\hbar\mathbf{q}/2}}{\varepsilon_{\mathbf{p}+\hbar\mathbf{q}/2} - \varepsilon_{\mathbf{p}-\hbar\mathbf{q}/2} - \hbar\omega - i\lambda \, \mathrm{sgn}\omega}. \quad (23)$$

We note that we consider the case of $T = 0$ so that the distribution function is $f_{\mathbf{p}} = \theta(p_F - p)$. The fact that the analytical properties of $\Pi_\omega(\mathbf{q})$ at $T = 0$ are determined by the sign of the frequency ω is general and is not related to the approximation used in the derivation of Eq. (23) (problem 6.19). A direct comparison of $\Pi_\omega^{(0)}(\mathbf{q})$ of Eq. (23) to $\Pi_\omega^{(0)R}(\mathbf{q})$ of Eq. (8) shows us that the real parts and absolute values of these functions coincide, and the difference is only in the analytical properties. Therefore, $\Pi_\omega^{(0)R}(\mathbf{q})$ is in the same relationship to $\Pi_\omega^{(0)}(\mathbf{q})$ as $\Pi_\omega^R(\mathbf{q})$ from Eq. (21) to $\Pi_\omega(\mathbf{q})$. In summary, using the RPA, we replace $\Pi_\omega^R(\mathbf{q})$ in Eq. (21) by $\Pi_\omega^{(0)R}(\mathbf{q})$ given by Eq. (8). Comparing α_{RPA} obtained in this way to α_{HF} given by Eq. (8), we find that they differ by the factor $1 - v_q \Pi_\omega^{(0)R}(\mathbf{q})$ resulting from the screening effects. In the language of diagrams, the Hartree-Fock approximation corresponds just to the second diagram on the right-hand side of Eq. (15), where the exact one-particle Green's functions are replaced by the Green's functions of non-interacting electrons. Both Hartree-Fock and RPA polarizabilities are determined by the retarded free-electron polarization function $\Pi_\omega^{(0)R}(\mathbf{q})$ only. In view of its importance, let us present the expression of $\Pi_\omega^{(0)R}(\mathbf{q})$ for the electrons with isotropic parabolic dispersion $\varepsilon_p = p^2/2m$. This expression is known as the Lindhard formula:

$$\mathrm{Re}\Pi^{(0)R}_\omega(\mathbf{q}) = -\frac{mk_F}{2\pi^2\hbar^2}\left[1 + \frac{k_F^2 - (\frac{m\omega}{\hbar q} + \frac{q}{2})^2}{2k_F q}\ln\left(\frac{k_F + \frac{m\omega}{\hbar q} + \frac{q}{2}}{k_F - \frac{m\omega}{\hbar q} - \frac{q}{2}}\right)\right.$$

$$\left. + \frac{k_F^2 - (\frac{m\omega}{\hbar q} - \frac{q}{2})^2}{2k_F q}\ln\left(\frac{k_F - \frac{m\omega}{\hbar q} + \frac{q}{2}}{k_F + \frac{m\omega}{\hbar q} - \frac{q}{2}}\right)\right], \tag{24}$$

$$\mathrm{Im}\Pi^{(0)R}_\omega(\mathbf{q}) = \begin{cases} -m^2\omega/2\pi\hbar^3 q, & \omega < \Omega_q^{(-)} \\ -\left(\frac{mk_F^2}{4\pi\hbar^2 q}\right)\left[1 - \frac{(\omega - \hbar q^2/2m)^2}{q^2 v_F^2}\right], & \Omega_q^{(-)} < \omega < \Omega_q^{(+)} \\ 0, & \omega > \Omega_q^{(+)} \end{cases},$$

where $k_F = p_F/\hbar$ and $\Omega_q^{(\pm)} = v_F q \pm \hbar q^2/2m$. If $v_q\Pi^{(0)R}_\omega(\mathbf{q}) \ll 1$, the RPA and Hartree-Fock polarizabilities coincide. As follows from Eq. (24), this limit is realized with increasing q and ω. This corresponds to the situation when the dielectric permittivity is no longer determined by the electron system and goes to ϵ_∞.

In the static limit $\omega \to 0$, the imaginary part of $\Pi^{(0)R}$ goes to zero, while the real part is equal to $-mk_F/\pi^2\hbar^2$ at $q^2 \ll 12k_F^2$. Substituting this result into Eq. (21), we find the static dielectric permittivity $\epsilon(\mathbf{q})$ and the static Fourier component of the screened interaction potential energy, $V(\mathbf{q}) = \epsilon_\infty v_q/\epsilon(\mathbf{q})$, in the RPA:

$$\epsilon(\mathbf{q}) = \epsilon_\infty\left(1 + \frac{q_{TF}^2}{q^2}\right), \quad V(\mathbf{q}) = \frac{4\pi e^2}{\epsilon_\infty(q^2 + q_{TF}^2)}, \quad q_{TF} = \sqrt{\frac{4e^2 mk_F}{\pi\hbar^2\epsilon_\infty}}. \tag{25}$$

The length q_{TF}^{-1} is called the Thomas-Fermi screening length. If $\omega \gg v_F q$, the imaginary part of $\Pi^{(0)R}$ is again zero, while the real part gives us

$$\epsilon(\mathbf{q},\omega) = \epsilon_\infty - \frac{4\pi e^2 n}{m\omega^2} = \epsilon_\infty\left(1 - \frac{\omega_p^2}{\omega^2}\right),$$

$$V_\omega(\mathbf{q}) = \frac{4\pi e^2}{\epsilon_\infty q^2(1 - \omega_p^2/\omega^2)}. \tag{26}$$

One can see that $\epsilon(\mathbf{q},\omega)$ goes to zero and the Fourier component of the interaction potential energy diverges at the plasma frequency $\omega_p = \sqrt{4\pi e^2 n/\epsilon_\infty m}$. This is a manifestation of collective effects in the response of electron system. The plasmon spectrum $\omega(q) = \omega_p$ at $q \to 0$ is obtained as a solution of the dispersion relation $\epsilon(\mathbf{q},\omega) = 0$ which, according to Eq. (17.1), follows from Eq. (11.27) in the isotropic case. In the ionic crystals, where ϵ_∞ should be replaced by the frequency-dependent dielectric permittivity of the crystal lattice, $\kappa(\mathbf{q},\omega)$, the plasmons are transformed to coupled plasmon-phonon modes (problem 6.20).

The plasma waves become damped when the wave number q exceeds a critical value determined from the condition $\omega(q) = \Omega_q^{(+)}$, i.e., when the imaginary part of $\epsilon(\mathbf{q}, \omega)$ appears. This effect, known from the theory of plasma as Landau damping, is not related to the collision-induced damping of quasiparticles considered in the previous chapters. It can be viewed as a decay of plasmons into single-particle excitations, the electron above the Fermi surface and the "hole" below the Fermi surface. From the classical point of view (valid in the region $q \ll k_F$), the dissipation of the energy of the longitudinal wave $\mathbf{E}e^{i\mathbf{q}\cdot\mathbf{r}-i\omega t}$ occurs because of the electrons moving in phase with the wave. The velocity $\mathbf{v_p}$ of these electrons is determined by the relation $\mathbf{q} \cdot \mathbf{v_p} = \omega$. We stress that the existence of non-zero imaginary part of $\epsilon(\mathbf{q}, \omega)$ in the absence of any interaction, means, according to Eq. (17.1), the existence of the real part of frequency-dependent conductivity which is not related to any scattering. The conductivity describing the response to the longitudinal perturbation $(\mathbf{E} \| \mathbf{q})$ is written in the limit $q \ll k_F$ as

$$\text{Re } \sigma_l(\mathbf{q}, \omega) = \frac{(em\omega)^2}{2\pi\hbar^3 q^3}\theta(v_F q - \omega). \tag{27}$$

The presence of the theta-function reflects the fact that the resonant condition $\mathbf{q} \cdot \mathbf{v_p} = \omega$ can be fulfilled only at $v_F q > \omega$. Equation (27) can be as well derived from Eq. (13.14) or from the kinetic equation (9.34), where the collision integral is neglected. Besides, these equations allow us to find the response to the transverse perturbation, with $\mathbf{E} \perp \mathbf{q}$ (problem 6.21).

How good is the RPA? To obtain an answer on this question, one should consider the diagrams we have neglected and find the conditions when this neglect is justified. Let us consider, for example, the third and the fourth diagrams on the right-hand side of Eq. (15) and replace the exact Green's functions in these diagrams by the Green's functions of non-interacting electrons. Since both these diagrams are of the first order in the screened interaction V, they can be compared to each other. In the RPA, we take into account the third diagram, whose contribution to $K(\mathbf{q}, \omega)$ is $V_\omega(\mathbf{q})[\Pi_\omega^{(0)}(\mathbf{q})]^2$, and neglect the fourth one. The latter diagram results in the following contribution to $K(\mathbf{q}, \omega)$:

$$-\frac{2}{V^2}\sum_{\mathbf{pp'}}\int\frac{d\varepsilon}{2\pi i}\int\frac{d\varepsilon'}{2\pi i}V_{(\varepsilon-\varepsilon')/\hbar}\left[(\mathbf{p}-\mathbf{p'})/\hbar\right]g_{\varepsilon+\hbar\omega/2}(\mathbf{p}+\hbar\mathbf{q}/2)$$

$$\times g_{\varepsilon-\hbar\omega/2}(\mathbf{p}-\hbar\mathbf{q}/2)g_{\varepsilon'+\hbar\omega/2}(\mathbf{p'}+\hbar\mathbf{q}/2)g_{\varepsilon'-\hbar\omega/2}(\mathbf{p'}-\hbar\mathbf{q}/2). \tag{28}$$

We evaluate it below in the static limit, approximating also $V_{\omega'}(\mathbf{q'})$ standing in Eq. (28) by its static value $V(\mathbf{q'})$ at small q', given by Eq. (25). The integrals over ε and ε' in this approximation are calculated independently and the expression (28) becomes

$$-\frac{2}{V^2} \sum_{\mathbf{pp'}} \frac{4\pi e^2}{\epsilon_\infty [(\mathbf{p}-\mathbf{p'})^2/\hbar^2 + q_{TF}^2]}$$

$$\times \frac{f_{\mathbf{p}+\hbar\mathbf{q}/2} - f_{\mathbf{p}-\hbar\mathbf{q}/2}}{\varepsilon_{\mathbf{p}+\hbar\mathbf{q}/2} - \varepsilon_{\mathbf{p}-\hbar\mathbf{q}/2}} \frac{f_{\mathbf{p'}+\hbar\mathbf{q}/2} - f_{\mathbf{p'}-\hbar\mathbf{q}/2}}{\varepsilon_{\mathbf{p'}+\hbar\mathbf{q}/2} - \varepsilon_{\mathbf{p'}-\hbar\mathbf{q}/2}}. \tag{29}$$

The double sum is easily calculated at small q, when $[f_{\mathbf{p}+\hbar\mathbf{q}/2} - f_{\mathbf{p}-\hbar\mathbf{q}/2}]$ $/[\varepsilon_{\mathbf{p}+\hbar\mathbf{q}/2} - \varepsilon_{\mathbf{p}-\hbar\mathbf{q}/2}] \simeq -\delta(\varepsilon_p - \varepsilon_F)$. We obtain the contribution

$$-\left(\frac{mk_F}{\pi^2\hbar^2}\right)^2 \frac{\pi e^2}{2\epsilon_\infty k_F^2} \ln\left(1 + \frac{4k_F^2}{q_{TF}^2}\right). \tag{30}$$

We need to compare this contribution to $V(\mathbf{q})[\Pi_{\omega=0}^{(0)}(\mathbf{q})]^2$ estimated at small q. The absolute value of the ratio of these contributions is

$$\frac{q^2 + q_{TF}^2}{8k_F^2} \ln\left(1 + \frac{4k_F^2}{q_{TF}^2}\right). \tag{31}$$

If the wave numbers q are comparable to or smaller than q_{TF}, this ratio is small at $q_{TF}^2 \ll k_F^2$. The other diagrams we have neglected bring similar small factors. If, however, $q_{TF}^2 \sim k_F^2$, the ratio (31) is not small, without regard to q. Therefore, the validity of RPA at $q \ll k_F$ is determined by the relation $q_{TF}^2 \ll k_F^2$, which is equivalent to $a_B k_F \gg 1$. In other words, the mean distance between the electrons should be small in comparison to the screening length. It is possible to have $a_B k_F > 1$ both in metals and in doped semiconductors, but the strong inequality $a_B k_F \gg 1$ is difficult to achieve. Therefore, the RPA has a limited applicability for quantitative description of electron plasma in solids. In spite of this, it is widely used for describing the dynamical screening, because it reflects essential features of electron response.

To complete this section, we briefly discuss how the dynamical screening modifies the scattering matrix element $V(\mathbf{pp'}|\mathbf{p_1 p_1'})$ in the electron-electron collision integral $J_{ee}(f|\mathbf{p})$ given by Eq. (31.21). In contrast to the "bare" matrix element defined by Eq. (31.22), the "screened" matrix element is written as

$$V(\mathbf{pp'}|\mathbf{p_1 p_1'}) \simeq \frac{1}{V^2} \frac{2|v_{|\mathbf{p}-\mathbf{p_1}|/\hbar}|^2}{|\epsilon[(\mathbf{p}-\mathbf{p_1})/\hbar, (\varepsilon_p - \varepsilon_{p_1})/\hbar]/\epsilon_\infty|^2}. \tag{32}$$

Indeed, since $\hbar\omega = \varepsilon_p - \varepsilon_{p_1}$ and $\hbar\mathbf{q} = \mathbf{p} - \mathbf{p_1}$ are the energy and momentum transferred in the collision, one may expect that the effective interaction potential $v_q = v_{|\mathbf{p}-\mathbf{p_1}|/\hbar}$ should be replaced by $\epsilon_\infty v_q/\epsilon(\mathbf{q}, \omega)$. Of course, this simple physical consideration is not a justification of Eq. (32). Moreover, it tells us nothing about how to "screen" the exchange contribution which stands in Eq. (31.22) but omitted in Eq. (32). A rigorous derivation of the collision integral with the matrix element (32) is done in problem 8.10. It shows us that Eq. (32) is justified under conditions when the RPA is valid, which means that the dielectric permittivity standing in Eq. (32) should be taken in this particular approximation. In the RPA, the scattering matrix element (32) is largest at $q^2 < q_{TF}^2 \ll k_F^2$ and decreases as q exceeds q_{TF}. This means that the kinetics of interacting electrons in a single band is mostly determined by the small-angle electron-electron scattering, when the characteristic transferred momenta are small in comparison to the Fermi momentum. In these conditions, the exchange terms, indeed, can be neglected, and Eq. (32) is a good approximation. On the other hand, the electron-electron collisions involving interband transitions (Auger processes, see Sec. 66) may occur only with large momentum and energy transfer, due to conservation of the energy and momentum. In this case, the factor $v_q \Pi_\omega(\mathbf{q})$ is small, and one should use the collision integral (31.19) with the "bare" matrix element (31.20) including both direct and exchange terms. We also point out that the effect of dynamical screening on the electron-electron collision integral can be described by the theory of electron density fluctuations; see Sec. 69.

Problems

6.1. Find the Coulomb matrix elements $\Phi_{\gamma_1\gamma_2\gamma_3\gamma_4}$ for two-dimensional and one-dimensional electrons.

<u>Solution</u>: The wave functions of the 2D electrons in the subband j are written as $\psi_z^{(j)} L^{-1} \exp(i\mathbf{p} \cdot \mathbf{r}/\hbar)$, where $\mathbf{p}$ and $\mathbf{r}$ are 2D vectors. In a similar way, for 1D electrons we have $\psi_{y,z}^{(j)} L^{-1/2} \exp(ipx/\hbar)$. Substituting these functions into Eq. (28.3), we calculate the following integrals for the 2D and 1D cases, respectively:

$$\int \frac{dq_z}{2\pi} \frac{e^{iq_z(z-z')}}{q_z^2 + q_\perp^2} = \frac{e^{-q_\perp|z-z'|}}{2q_\perp},$$

where $q_\perp = \sqrt{q_x^2 + q_y^2}$, and

$$\int\int \frac{dq_z dq_y}{(2\pi)^2} \frac{e^{iq_z(z-z')+iq_y(y-y')}}{q_z^2 + q_y^2 + q_x^2} = (2\pi)^{-1} \mathcal{K}_0\left(q_x\sqrt{(z-z')^2 + (y-y')^2}\right).$$

Here $\mathcal{K}_0$ is the modified Bessel function of the second kind. The matrix elements are

$$\Phi_{j_1 j_2 j_3 j_4}^{(2D)}(\mathbf{p}_1, \mathbf{p}_2, \mathbf{p}_3, \mathbf{p}_4) = \frac{2\pi e^2 \hbar}{\epsilon|\mathbf{p}_1 - \mathbf{p}_4|} \int\int dz dz'$$

$$\times e^{-|\mathbf{p}_1 - \mathbf{p}_4||z-z'|/\hbar} \psi_z^{(j_4)} \psi_z^{(j_1)*} \psi_{z'}^{(j_3)} \psi_{z'}^{(j_2)*} \delta_{\mathbf{p}_1 + \mathbf{p}_2, \mathbf{p}_3 + \mathbf{p}_4},$$

$$\Phi_{j_1 j_2 j_3 j_4}^{(1D)}(p_1, p_2, p_3, p_4) = \frac{2e^2}{\epsilon} \int\int dz dz' \int\int dy dy'$$

$$\times \mathcal{K}_0\left(|p_1 - p_4|\sqrt{(y-y')^2 + (z-z')^2}/\hbar\right) \psi_{y,z}^{(j_4)} \psi_{y,z}^{(j_1)*} \psi_{y',z'}^{(j_3)} \psi_{y',z'}^{(j_2)*} \delta_{p_1 + p_2, p_3 + p_4}.$$

6.2. Prove that the ambiguity in the definition of Φ (Eq. (28.3) versus Eq. (28.4)) does not manifest itself in the quantum kinetic equation with the Hamiltonian $\widehat{H}_{ee}$ given by Eq. (28.2).

Hint: Prove, by a direct calculation, that the commutator of $\widehat{H}_{ee}$ with any combination of Fermi operators is the same, whether we use Eq. (28.3) or Eq. (28.4).

6.3. Calculate $\Delta\varepsilon_p$ given by Eqs. (28.22) and (28.25) for 3D and 2D electrons, respectively.

Solution: Calculating the integrals over the angles between $\mathbf{p}$ and $\mathbf{p}_1$, we obtain

$$\Delta\varepsilon_p = \frac{e^2}{\pi\hbar\epsilon p}\int_0^{p_F} dp_1 p_1 \ln\left|\frac{p_1+p}{p_1-p}\right|, \quad \Delta\varepsilon_p^{(2D)} = \frac{e^2}{\pi\hbar\epsilon}\int_0^{p_F} dp_1 \frac{2p_1}{p_1+p}K\left(\frac{2\sqrt{p_1 p}}{p_1+p}\right),$$

where $K(x)$ is the full elliptic integral of the first kind. Both $\Delta\varepsilon_p$ and $\Delta\varepsilon_p^{(2D)}$ are functions of p/p_F. The integral over p_1 in the expression for $\Delta\varepsilon_p$ can be taken analytically. The result is

$$\Delta\varepsilon_p = \frac{e^2 p_F}{\pi\hbar\epsilon}\left[1 + \frac{1-(p/p_F)^2}{2(p/p_F)}\ln\left|\frac{1+p/p_F}{1-p/p_F}\right|\right].$$

Note that $\partial\Delta\varepsilon_p/\partial p$ diverges logarithmically at $p \to p_F$.

6.4. Write the complex frequency-dependent conductivity of electrons in quantum wells and examine its behavior at $\omega \to \infty$ and $\omega \to 0$.

Solution: Substituting the solution of Eq. (29.2) in the exact eigenstate representation into Eq. (29.1), we obtain the conductivity $\sigma_\omega = I_\omega/E$ in the following form:

$$\sigma_\omega = i\frac{e^2 n}{m\omega} + i\frac{e^2}{\omega L^2}\sum_{\delta\delta'}\frac{|v_{\delta\delta'}^z|^2(f_\delta - f_{\delta'})}{\varepsilon_\delta - \varepsilon_{\delta'} - \hbar\omega - i\lambda},$$

where δ numbers the exact eigenstates of electrons (the spin number σ is included in δ) and $v_{\delta\delta'}^z$ is the matrix element of z-component of the velocity operator. The equation above can also be obtained from the Kubo formula. In the collisionless approximation, $|\delta\rangle = |j\sigma\mathbf{p}\rangle$, where j is the subband number and $\mathbf{p}$ is the 2D momentum. Therefore, the matrix element is expressed as $v_{\delta\delta'}^z = \delta_{\sigma\sigma'}\delta_{\mathbf{p}\mathbf{p}'}\langle j|\hat{p}_z/m|j'\rangle$. The conductivity takes the following form:

$$\sigma_\omega = i\frac{e^2 n}{m\omega} - i\frac{2e^2\hbar^2}{m^2\omega L^2}\sum_{jj'}\sum_{\mathbf{p}}\left|\int dz\psi_z^{(j')}\frac{\partial}{\partial z}\psi_z^{(j)}\right|^2\frac{f_{jp}-f_{j'p}}{\varepsilon_{j'}-\varepsilon_j - \hbar\omega - i\lambda}$$

$$\simeq i\frac{e^2 n}{m\omega}\left[1 - \frac{\hbar^2}{m}\sum_{j'}\left|\int dz\psi_z^{(j')}\frac{\partial}{\partial z}\psi_z^{(1)}\right|^2\frac{1}{\varepsilon_{j'}-\varepsilon_1 - \hbar\omega - i\lambda}\right],$$

where ε_j is the energy of size quantization and $\psi_z^{(j)}$ is the corresponding wave function satisfying the Schroedinger equation $(\hat{p}_z^2/2m + U_z)\psi_z^{(j)} = \varepsilon_j\psi_z^{(j)}$ with the confinement potential U_z. The second equation is written under a simplifying assumption that only the lowest ($j=1$) 2D subband is occupied by electrons and the terms containing $\varepsilon_{j'} - \varepsilon_1 + \hbar\omega$ in the denominator are neglected (resonance approximation). Now, let us assume that j' belongs to continuous spectrum above the well so that $\sum_{j'}$ is transformed to the integral over continuous wave number k and $\varepsilon_{j'} = \hbar^2 k^2/2m$. At large energies, $k \gg \pi/d$, where d is the well width, the second term in the square brackets is transformed to

$$-\frac{\hbar^2}{m} \int_0^\infty \frac{dk}{\pi} \left| \int dz e^{ikz} \frac{\partial}{\partial z} \psi_z^{(1)} \right|^2 \frac{1}{\hbar^2 k^2/2m - \hbar\omega - i\lambda}.$$

The squared matrix element in this expression is roughly estimated as $(2\pi^2/d)[1 - \cos(kd)]/(kd)^2$. Since the main contribution to the integral over k comes from the region $\hbar^2 k^2/2m \simeq \hbar\omega$, this expression decreases with increasing ω and can be neglected. Therefore, the conductivity at large ω is given by $\sigma_\omega \simeq i e^2 n/m\omega$.

To consider the low-frequency region, we use the following chain of identities:

$$\sum_{\delta\delta'} \frac{|v_{\delta\delta'}^z|^2 (f_\delta - f_{\delta'})}{\varepsilon_\delta - \varepsilon_{\delta'}} = \sum_{\delta\delta'} f_\delta \left(\frac{v_{\delta'\delta}^z v_{\delta\delta'}^z}{\varepsilon_\delta - \varepsilon_{\delta'}} - \frac{v_{\delta\delta'}^z v_{\delta'\delta}^z}{\varepsilon_{\delta'} - \varepsilon_\delta} \right)$$

$$= \frac{i}{\hbar} \sum_{\delta\delta'} f_\delta (z_{\delta\delta'} v_{\delta'\delta}^z - z_{\delta'\delta} v_{\delta\delta'}^z) = \frac{i}{\hbar m} \sum_\delta f_\delta \langle \delta | [z, \hat{v}_z] | \delta \rangle = -L^2 \frac{n}{m}.$$

Therefore,

$$\sigma_\omega = i \frac{e^2}{\omega L^2} \sum_{\delta\delta'} |v_{\delta\delta'}^z|^2 (f_\delta - f_{\delta'}) \left(\frac{1}{\varepsilon_\delta - \varepsilon_{\delta'} - \hbar\omega - i\lambda} - \frac{1}{\varepsilon_\delta - \varepsilon_{\delta'}} \right)$$

$$= i \frac{\hbar e^2}{L^2} \sum_{\delta\delta'} \frac{|v_{\delta\delta'}^z|^2 (f_\delta - f_{\delta'})}{(\varepsilon_\delta - \varepsilon_{\delta'} - \hbar\omega - i\lambda)(\varepsilon_\delta - \varepsilon_{\delta'})}.$$

Since $|v_{\delta\delta'}^z| = |v_{\delta'\delta}^z|$, the property $\lim_{\omega \to 0} \sigma_\omega = 0$ can be checked by permutation of the indices. In the intermediate region of frequencies, the frequency dispersion of the conductivity appears to be rather complicated.

6.5. Obtain Eq. (29.20) from the Maxwell equations.

<u>Solution:</u> The Maxwell equations lead to the wave equation $[\nabla \times [\nabla \times \mathbf{E}(\mathbf{r}, \omega)]] - (\omega/c)^2 \mathbf{D}(\mathbf{r}, \omega) = 0$, where the vector of electrostatic induction is written as $D_\alpha(\mathbf{r}, \omega) = \epsilon E_\alpha(\mathbf{r}, \omega) + i(4\pi/\omega) \sum_\beta \int d\mathbf{r}' \sigma_{\alpha\beta}(\mathbf{r}, \mathbf{r}'|\omega) E_\beta(\mathbf{r}', \omega) = \epsilon E_\alpha(\mathbf{r}, \omega) + i(4\pi/\omega) I_\alpha(\mathbf{r}, \omega)$. We substitute this expression into the wave equation and write the latter for z- components of the field and current density, $E_z(\mathbf{r}, \omega) \equiv E_{\omega z}$ and $I_z(\mathbf{r}, \omega) \equiv I_{\omega z}$, when it is reduced to Eq. (29.20).

6.6. Calculate L_{1111}, L_{1212}, L_{1122}, and L_{2222} by using Eq. (29.14) with the wave functions of hard-wall confinement.

<u>Results:</u> $L_{1111} = d(1/3 - 5/4\pi^2)$, $L_{1122} = d(1/3 - 5/8\pi^2)$, $L_{2222} = d(1/3 - 5/16\pi^2)$, $L_{1212} = -10d/9\pi^2$.

6.7. Write the kinetic energy of an electron-hole pair by using the coordinates defined by Eq. (30.16).

<u>Hint:</u> Take into account that the derivatives are expressed as

$$\frac{\partial}{\partial \mathbf{r}} = \frac{m_c}{m_c + m_v} \frac{\partial}{\partial \mathbf{R}} + \frac{\partial}{\partial \Delta \mathbf{r}}, \quad \frac{\partial}{\partial \mathbf{r}'} = \frac{m_v}{m_c + m_v} \frac{\partial}{\partial \mathbf{R}} - \frac{\partial}{\partial \Delta \mathbf{r}}.$$

6.8. Calculate the Sommerfeld factor for the states of continuous spectrum.

Solution: The solutions of the Schroedinger equation (30.20) for positive ε_ν can be represented as combinations of plane waves and spherical waves coming out from the origin of the coordinate system. The states are characterized by the wave vector $\mathbf{k}$ and have energies $\varepsilon_k = \hbar^2 k^2 / 2\mu^*$. The normalized wave functions are

$$\psi_{\mathbf{r}}^{(\mathbf{k})} = V^{-1/2} e^{\pi/2ka_B} \Gamma(1 - i/ka_B) e^{i\mathbf{k}\cdot\mathbf{r}} \Phi(i/ka_B, 1; i(kr - \mathbf{k}\cdot\mathbf{r})),$$

where a_B is the Bohr radius of the exciton, Φ is the confluent hypergeometric function, and Γ is the Gamma function. Taking into account that $\Gamma(1 - i/ka_B)\Gamma(1 + i/ka_B) = (\pi/ka_B)/\sinh(\pi/ka_B)$ and $|\Phi(i/ka_B, 1; 0)|^2 = 1$, we find the Sommerfeld factor

$$|\psi_{\mathbf{r}=0}^{(\mathbf{k})}|^2 = \frac{1}{V} \frac{(2\pi/ka_B)}{1 - \exp(-2\pi/ka_B)}.$$

This function is close to $1/V$ at large energies and increases as $1/\sqrt{\varepsilon_k}$ at small energies. This increase leads to enhanced absorption near the edge of interband transitions.

6.9. Consider the broadening of exciton absorption line in the potentials smooth on the scale of exciton radius.

Solution: If $U_{c,v}(\mathbf{r}) \neq 0$, we have an additional term $[U_c(\mathbf{R} + \Delta\mathbf{r}m_v/M) - U_v(\mathbf{R} - \Delta\mathbf{r}m_c/M)]G_{\varepsilon_\omega}(\mathbf{R}, \Delta\mathbf{r})$ on the left-hand side of Eq. (30.17). If the potentials are smooth on the scale of exciton radius, the expression in the square brackets is approximated as $\Delta U(\mathbf{R}) + \Delta\mathbf{r}[m_v \partial U_c(\mathbf{R})/\partial\mathbf{R} + m_c \partial U_v(\mathbf{R})/\partial\mathbf{R}]/M$, where $\Delta U(\mathbf{R}) = U_c(\mathbf{R}) - U_v(\mathbf{R})$. Let us assume that $U_c(\mathbf{R}) \neq U_v(\mathbf{R})$ and neglect the derivatives. It is convenient to introduce the Green's function $G_{\varepsilon_\omega}(\mathbf{R}, \mathbf{R}'|\Delta\mathbf{r})$ satisfying the following equation:

$$\left[\varepsilon_\omega - \widehat{H}_{\mathbf{R}} - \widehat{H}_{\Delta\mathbf{r}} + i\lambda\right] G_{\varepsilon_\omega}(\mathbf{R}, \mathbf{R}'|\Delta\mathbf{r}) = \delta(\mathbf{R} - \mathbf{R}')\delta(\Delta\mathbf{r}),$$

where $\lambda \to +0$ and

$$\widehat{H}_{\mathbf{R}} = -\frac{\hbar^2}{2M} \frac{\partial^2}{\partial\mathbf{R}^2} + \Delta U(\mathbf{R}), \qquad \widehat{H}_{\Delta\mathbf{r}} = -\frac{\hbar^2}{2\mu^*} \frac{\partial^2}{\partial\Delta\mathbf{r}^2} - v_{|\Delta\mathbf{r}|},$$

so that $G_{\varepsilon_\omega}(\mathbf{R}, \Delta\mathbf{r}) = \int d\mathbf{R}' G_{\varepsilon_\omega}(\mathbf{R}, \mathbf{R}'|\Delta\mathbf{r})$. The solutions of this equation are

$$G_{\varepsilon_\omega}(\mathbf{R}, \mathbf{R}'|\Delta\mathbf{r}) = \sum_{\delta\nu} \frac{\chi_{\mathbf{R}}^{(\delta)} \chi_{\mathbf{R}'}^{(\delta)*} \psi_{\Delta\mathbf{r}}^{(\nu)} \psi_0^{(\nu)*}}{\varepsilon_\omega - \varepsilon_\delta - \varepsilon_\nu + i0},$$

where $\chi_{\mathbf{R}}^{(\delta)}$ and $\psi_{\Delta\mathbf{r}}^{(\delta)}$ are the eigenfunctions of the Hamiltonians $\widehat{H}_{\mathbf{R}}$ and $\widehat{H}_{\Delta\mathbf{r}}$, respectively, and ε_δ and ε_ν are the corresponding eigenvalues. Introducing the retarded Green's function of the Schroedinger equation with the Hamiltonian $\widehat{H}_{\mathbf{R}}$ as

$$G_E^R(\mathbf{R}, \mathbf{R}') = \sum_\delta \frac{\chi_{\mathbf{R}}^{(\delta)} \chi_{\mathbf{R}'}^{(\delta)*}}{E - \varepsilon_\delta + i\lambda},$$

we express the spectral dependence of the absorption as

$$\Psi(\varepsilon_\omega) = -\frac{1}{V} \sum_\nu |\psi_{\Delta\mathbf{r}=0}^{(\nu)}|^2 \int d\mathbf{R} \int d\mathbf{R}' \mathrm{Im} G_{\varepsilon_\omega - \varepsilon_\nu}^R(\mathbf{R}, \mathbf{R}').$$

The function $G_E^R(\mathbf{R}, \mathbf{R}')$ is expressed through the path integral in the way described in Sec. 16. The double integral $V^{-1} \int d\mathbf{R} \int d\mathbf{R}' G_E^R(\mathbf{R}, \mathbf{R}')$ is equivalent to $\int d\Delta\mathbf{R} G_E^R(|\Delta\mathbf{R}|)$, where $G_E^R(|\Delta\mathbf{R}|)$ is the Green's function averaged over the random potential distribution. In the case of a Gaussian potential $\Delta U(\mathbf{R})$, the averaged Green's function depends on $w(|\mathbf{R} - \mathbf{R}'|) = \langle\langle \Delta U(\mathbf{R}) \Delta U(\mathbf{R}') \rangle\rangle$. If $\Delta U(\mathbf{R})$ is classically smooth, the correlation function $w(|\mathbf{R}-\mathbf{R}'|)$ is replaced by $w(0)$ and the Green's function is found exactly, see Sec. 16:

$$G_E^R(|\Delta\mathbf{R}|) = -\frac{i}{\hbar} \int_0^\infty dt e^{iEt/\hbar} \left(\frac{M}{2\pi i\hbar t} \right)^{3/2} \exp\left(\frac{iM|\Delta\mathbf{R}|^2}{2\hbar t} - \frac{w(0)t^2}{\hbar^2} \right).$$

Using this equation, we obtain

$$\Psi(\varepsilon_\omega) = \sqrt{\frac{\pi}{2w(0)}} \sum_\nu |\psi_{\Delta\mathbf{r}=0}^{(\nu)}|^2 \exp\left[-\frac{(\varepsilon_\nu - \varepsilon_\omega)^2}{2w(0)} \right],$$

instead of Eq. (30.21). Each exciton line is broadened, the sharp δ-peaks become Gaussian peaks.

6.10. Find the energy spectrum of 2D excitons and the Sommerfeld factor for the first exciton absorption peak.

Solution: Expressing the 2D coordinate vector $\mathbf{x} = (x, y)$ in the cylindrical coordinates ρ and φ, we search for the 2D exciton wave function $\psi_\mathbf{x}^{(\nu)}$ in the form $\psi_m(\rho)e^{im\varphi}$, where m is an integer. The Schroedinger equation (30.20) is then rewritten as an equation for $\psi_m(\rho)$:

$$\left[-\frac{\hbar^2}{2\mu^*} \left(\frac{d^2}{d\rho^2} + \frac{1}{\rho}\frac{d}{d\rho} - \frac{m^2}{\rho^2} \right) - \frac{e^2}{\epsilon\rho} - \varepsilon_\nu \right] \psi_m(\rho) = 0.$$

Considering the confined states ($\varepsilon_\nu < 0$), we substitute $\psi_m(\rho) = \rho^{|m|}e^{-\beta\rho}\chi(\rho)$, where $\beta = \sqrt{-2\mu^*\varepsilon_\nu}/\hbar$, and transform this equation to

$$\rho d^2\chi/d\rho^2 + (2|m| + 1 - 2\beta\rho)d\chi/d\rho - 2\beta\alpha\chi = 0, \quad \alpha = |m| + 1/2 - \mu^*e^2/\hbar^2\beta\epsilon,$$

whose solutions are confluent hypergeometric functions, $\chi(\rho) \propto \Phi(\alpha, 2m + 1; 2\beta\rho)$, and α must be either zero or negative integer, $\alpha = -l$. Therefore, the spectrum is given by $\varepsilon_\nu = -\varepsilon_B/(l + |m| + 1/2)^2$, where $\varepsilon_B = \mu^*e^4/2\hbar^2\epsilon^2$ is the Bohr energy of exciton. The energy of the ground state ($l = m = 0$) is $4\varepsilon_B$, which is four times greater than the ground state energy of 3D exciton. The corresponding normalized wave function is $\sqrt{8/\pi a_B^2}e^{-2\rho/a_B}$ so that the Sommerfeld factor is equal to $8/\pi a_B^2$, where $a_B = \hbar^2\epsilon/\mu^*e^2$.

6.11. Find the energy spectrum of 1D excitons. Prove that the energy of the ground state is infinite.

Solution: Searching for confined states, we introduce the dimensionless energy $\tilde{\varepsilon} = \varepsilon_\nu/\varepsilon_B$ and coordinate $\tilde{x} = 2x\sqrt{|\tilde{\varepsilon}|}/a_B$. With these variables, the Schroedinger equation (30.20) for 1D exciton is rewritten as

$$d^2\psi_x^{(\nu)}/d\tilde{x}^2 + |\tilde{\varepsilon}|^{-1/2}|\tilde{x}|^{-1}\psi_x^{(\nu)} - \psi_x^{(\nu)}/4 = 0.$$

Its solutions are either even (symmetric in x) or odd (antisymmetric), and we search for them in the forms $e^{-|\tilde{x}|/2}\chi_e(\tilde{x})$ and $\tilde{x}e^{-|\tilde{x}|/2}\chi_o(\tilde{x})$, respectively. It is easy to find (see the previous problem) that $\chi_e(\tilde{x}) \propto \Phi(-1/\sqrt{|\tilde{\varepsilon}_e|}, 0; \tilde{x})$ with $\tilde{\varepsilon}_e = -1/l^2$, $l = 0, 1, 2, \ldots$ and $\chi_o(\tilde{x}) \propto \Phi(1 - 1/\sqrt{|\tilde{\varepsilon}_o|}, 2; \tilde{x})$ with $\tilde{\varepsilon}_o = -1/(l+1)^2$, $l = 0, 1, 2, \ldots$. The energy of the ground state (the even state with $l = 0$) is infinite.

6.12. Check that the first and the third terms on the right-hand side of Eq. (28.10) are equal, respectively, to the second and the fourth ones.

Hint: Use the symmetry properties of $\Phi_{\gamma_1\gamma_2\gamma_3\gamma_4}$ discussed in Sec. 28.

6.13. Prove the operator identities (31.13) and (31.14).

Hint: It is convenient to write the expressions of the traces through the matrix elements of the operators.

6.14. Check the Hermiticity of the operator (31.18).

Hint: Use the identity $(\widehat{A}\widehat{B})^+ = \widehat{B}^+\widehat{A}^+$, the Hermiticity of $\hat{\rho}_{t_1}$, and the possibility to change the signs of $\mathbf{q}$ and $\mathbf{q}_1$ under the sum in Eq. (31.18).

6.15. Derive the expression for the drag conductivity tensor (32.13).

Hints: After multiplying Eq. (32.9) by $2e\mathbf{p}\tau_j/m$ and summing it over $\mathbf{p}$, make the substitutions $\mathbf{p} \to \mathbf{p} + \hbar\mathbf{q}/2$ and $\mathbf{p}' \to \mathbf{p}' - \hbar\mathbf{q}/2$. Show that the factor

$$\delta(\varepsilon_{j\mathbf{p}+\hbar\mathbf{q}/2} + \varepsilon_{j'\mathbf{p}'-\hbar\mathbf{q}/2} - \varepsilon_{j\mathbf{p}-\hbar\mathbf{q}/2} - \varepsilon_{j'\mathbf{p}'+\hbar\mathbf{q}/2})$$
$$\times f_{j\mathbf{p}+\hbar\mathbf{q}/2}^{(eq)} f_{j'\mathbf{p}'-\hbar\mathbf{q}/2}^{(eq)}(1 - f_{j\mathbf{p}-\hbar\mathbf{q}/2}^{(eq)})(1 - f_{j'\mathbf{p}'+\hbar\mathbf{q}/2}^{(eq)})$$

under the sum $\sum_{\mathbf{p}'\mathbf{q}}$ is symmetric in $\mathbf{q}$.

6.16. Calculate the low-frequency limit of $\mathrm{Im}\Pi_j(q, \omega)$ at $\hbar q \ll p_{Fj}$.

Solution: The imaginary part of the expression (32.22) contains the δ-function $\delta(\varepsilon_{j\mathbf{p}} - \varepsilon_{j\mathbf{p}-\hbar\mathbf{q}} - \hbar\omega)$ under the integral. Expanding the distribution function $f_{j\mathbf{p}-\hbar\mathbf{q}}^{(eq)}$ in series of small $\hbar\omega$, we obtain

$$\mathrm{Im}\Pi_j(q, \omega) = \frac{\omega}{\hbar} \int_0^\infty p\,dp \frac{\partial f_{j\mathbf{p}}^{(eq)}}{\partial \varepsilon_{j\mathbf{p}}} \int_0^{2\pi} \frac{d\varphi}{2\pi} \delta\left(\frac{\hbar pq}{m}\cos\varphi - \frac{\hbar^2 q^2}{2m}\right)$$

$$\simeq -\frac{m^2\omega}{\hbar^2 p_{Fj}q} \int_0^{2\pi} \frac{d\varphi}{2\pi} \delta(\cos\varphi - \hbar q/2p_{Fj}),$$

where $\varphi = \widehat{\mathbf{pq}}$. The second equation corresponds to the case of strongly degenerate electrons. Integrating over the angle φ at $\hbar q \ll p_{Fj}$, we obtain Eq. (32.24).

6.17. Derive the screened matrix element (32.26).

Solution: Let U_{qz} be the Fourier component of the potential energy created by a trial unit charge at z_0, and Δn_{jq} is the electron density perturbation created in the layer j by this charge. The Poisson equation for the double-layer system is

$$\left[\frac{d^2}{dz^2} - q^2\right] U_{qz} = -\frac{4\pi e^2}{\epsilon} \sum_{j=1,2} \Delta n_{jq}|\psi_z^{(j)}|^2 - \frac{4\pi e^2}{\epsilon}\delta(z - z_0).$$

Since the density of states $\rho_{2D} = m/\pi\hbar^2$ is constant, one has simply

$$\Delta n_{jq} = -\rho_{2D}U_{jq}, \quad U_{jq} = \int dz U_{qz}|\psi_z^{(j)}|^2.$$

For simplicity purpose, we assume that the layers are infinitely narrow so that $|\psi_z^{(j)}|^2$ is approximated as $\delta(z - z_j)$, where z_j are the centers of the layers and $|z_1 - z_2| = Z$. The solution of the Poisson equation with δ-like right-hand side is known, and we obtain a set of two coupled equations for U_{1q} and U_{2q}:

$$(1 + 2/a_B q)U_{1q} + (2/a_B q)e^{-qZ}U_{2q} = 2\pi e^2/q\epsilon\ ,$$

$$(1 + 2/a_B q)U_{2q} + (2/a_B q)e^{-qZ}U_{1q} = (2\pi e^2/q\epsilon)e^{-qZ},$$

where a_B is the Bohr radius. We have assumed that the trial charge is located in the layer 1, $z_0 = z_1$. This substitution identifies $v_q^{(12)}$ with U_{2q}. Solving the system of coupled equations, we obtain Eq. (32.26).

6.18. Calculate the polarizability (33.9) in the Hartree-Fock approximation.

Solution: If the interaction is neglected, the ground and excited states of many-electron system are antisymmetric products of the plane-wave states $|\mathbf{p}\rangle$. The operator $\hat{n}_\mathbf{q}^+$, when acting on the ground state, takes an electron in the state with momentum $\mathbf{p}$ inside the Fermi surface and places it in the state with momentum $\mathbf{p} + \hbar\mathbf{q}$ outside the Fermi surface. For such transitions, the absolute value of the matrix element $\langle\delta|\delta\hat{n}_\mathbf{q}^+|0\rangle$ is equal to unity. The energy of this transition is $E_{\delta 0} = \varepsilon_{\mathbf{p}+\hbar\mathbf{q}} - \varepsilon_\mathbf{p}$. If the transitions occur outside or inside the Fermi surface, the matrix element is zero, according to the Pauli principle. Therefore, the first term in the expression (33.9) is reduced to

$$\sum_\mathbf{p} \frac{\theta(p_F - p)\theta(|\mathbf{p} + \hbar\mathbf{q}| - p_F)}{\hbar\omega - (\varepsilon_{\mathbf{p}+\hbar\mathbf{q}} - \varepsilon_\mathbf{p}) + i\lambda},$$

where one can shift the momentum according to $\mathbf{p} \to \mathbf{p} - \hbar\mathbf{q}/2$. The second term is transformed in a similar way. Collecting both these terms, we obtain the polarizability in the form of Eq. (33.8), where $\theta(p_F - |\mathbf{p} \pm \hbar\mathbf{q}/2|)$ stand instead of $f_{\mathbf{p}\pm\hbar\mathbf{q}/2}$.

6.19. Prove that the analytical properties of $\Pi_\omega(\mathbf{q})$ at $T = 0$ are determined by the sign of the frequency ω.

Solution: Let us express the polarization function by using the one-electron Green's functions in the exact eigenstate representation described by the wave functions $\langle\mathbf{r}|\delta\rangle = \psi_\mathbf{r}^{(\delta)}$ and corresponding energies ε_δ. The exact polarization function in the energy-coordinate representation is written as

$$\Pi_\omega(\mathbf{r}, \mathbf{r}') = \int \frac{dE}{2\pi i} G_{E+\hbar\omega}(\mathbf{r}, \mathbf{r}')G_E(\mathbf{r}', \mathbf{r}),$$

where the Green's function is expressed through the coordinate-dependent wave functions as in problem 3.10:

$$G_E(\mathbf{r}, \mathbf{r}') = \sum_\delta \frac{\psi_\mathbf{r}^{(\delta)}\psi_{\mathbf{r}'}^{(\delta)*}}{E - \varepsilon_\delta + i\lambda\, \text{sgn}(\varepsilon_\delta - \varepsilon_F)}.$$

Calculating the integrals over E, we obtain

$$\Pi_\omega(\mathbf{r}, \mathbf{r}') = \sum_{\delta\delta'} \psi_\mathbf{r}^{(\delta)} \psi_{\mathbf{r}'}^{(\delta)*} \psi_{\mathbf{r}'}^{(\delta')} \psi_\mathbf{r}^{(\delta')*} \left[\frac{\theta(\varepsilon_F - \varepsilon_\delta)\theta(\varepsilon_{\delta'} - \varepsilon_F)}{\varepsilon_\delta - \varepsilon_{\delta'} - \hbar\omega + i0} - \frac{\theta(\varepsilon_F - \varepsilon_{\delta'})\theta(\varepsilon_\delta - \varepsilon_F)}{\varepsilon_\delta - \varepsilon_{\delta'} - \hbar\omega - i0} \right].$$

The nominators of the fractions can be rewritten as $[\theta(\varepsilon_F - \varepsilon_\delta) - \theta(\varepsilon_F - \varepsilon_{\delta'})]\theta(\varepsilon_{\delta'} - \varepsilon_\delta)$ and $-[\theta(\varepsilon_F - \varepsilon_\delta) - \theta(\varepsilon_F - \varepsilon_{\delta'})]\theta(\varepsilon_\delta - \varepsilon_{\delta'})$, respectively. Therefore, the real part of the expression in the square brackets is written in the form independent of the sign of $\varepsilon_\delta - \varepsilon_{\delta'}$:

$$\mathrm{Re}[\dots] = \frac{\theta(\varepsilon_F - \varepsilon_\delta) - \theta(\varepsilon_F - \varepsilon_{\delta'})}{\varepsilon_\delta - \varepsilon_{\delta'} - \hbar\omega}.$$

The imaginary part of the expression in the square brackets is proportional to

$$\delta(\varepsilon_\delta - \varepsilon_{\delta'} - \hbar\omega)[\theta(\varepsilon_\delta - \varepsilon_{\delta'}) - \theta(\varepsilon_{\delta'} - \varepsilon_\delta)] = \delta(\varepsilon_\delta - \varepsilon_{\delta'} - \hbar\omega)\mathrm{sgn}\omega .$$

Therefore, one can write

$$\Pi_\omega(\mathbf{r}, \mathbf{r}') = \sum_{\delta\delta'} \psi_\mathbf{r}^{(\delta)} \psi_{\mathbf{r}'}^{(\delta)*} \psi_{\mathbf{r}'}^{(\delta')} \psi_\mathbf{r}^{(\delta')*} \frac{\theta(\varepsilon_F - \varepsilon_\delta) - \theta(\varepsilon_F - \varepsilon_{\delta'})}{\varepsilon_\delta - \varepsilon_{\delta'} - \hbar\omega - i\lambda\,\mathrm{sgn}\omega}.$$

6.20. Find the spectrum of plasmon-phonon waves in ionic crystals.

Solution: Using Eq. (33.26), where ϵ_∞ is replaced by $\kappa(\mathbf{q}, \omega)$ given by (27.29), we obtain the following dispersion relation:

$$\epsilon(\mathbf{q}, \omega) = \epsilon_\infty + \frac{\omega_{TO}^2(\epsilon_0 - \epsilon_\infty)}{\omega_{TO}^2 - \omega^2} - \frac{4\pi e^2 n}{m\omega^2} = 0,$$

which is reduced to $\omega^4 - \omega^2(\omega_{LO}^2 + \omega_p^2) + \omega_{TO}^2\omega_p^2 = 0$ describing two branches with the frequencies

$$\omega_\pm^2 = \frac{\omega_{LO}^2 + \omega_p^2}{2} \pm \sqrt{\frac{(\omega_{LO}^2 + \omega_p^2)^2}{4} - \omega_{TO}^2\omega_p^2} .$$

6.21. Using the quasi-classical kinetic equation (9.34), calculate the real part of the conductivity $\sigma(\mathbf{q}, \omega)$ caused by the Landau damping.

Result: In the collisionless approximation, the solution of Eq. (9.34) for the Fourier transform of the distribution function is written in a straightforward way, and we obtain

$$\sigma_{\alpha\beta}(\mathbf{q}, \omega) = \frac{2e^2}{m^2} \int \frac{d\mathbf{p}}{(2\pi\hbar)^3} \left(-\frac{\partial f(\varepsilon_p)}{\partial \varepsilon_p} \right) \frac{p_\alpha p_\beta}{i(\mathbf{q} \cdot \mathbf{v_p} - \omega) + \lambda}, \quad \lambda \to +0.$$

The conductivity tensor is diagonal if one of the coordinate axes is directed along $\mathbf{q}$. The integrand determining the real part of the conductivity is proportional to $\delta(\mathbf{q} \cdot \mathbf{v_p} - \omega)$. The longitudinal ($O\alpha \parallel \mathbf{q}$) and transverse ($O\alpha \perp \mathbf{q}$) components of the conductivity tensor are different. The real part of the longitudinal component is given by Eq. (33.27), while the real parts of two transverse components are equal to $\mathrm{Re}\,\sigma_l(\mathbf{q}, \omega)[(v_F q/\omega)^2 - 1]/2$. The imaginary part of the conductivity at small q is given by Eq. (11.24). Note that the Landau damping effects are not captured by the formalism of balance equations given in Sec. 11.

Chapter 7

NON-EQUILIBRIUM ELECTRONS

The structure of the kinetic equation for the electrons interacting with bosons (phonons or photons) is similar to the one for the case of electron-impurity system considered in Chapter 2. However, a consideration of electron scattering with emission or absorption of bosons leads to a more sophisticated collision integral, because the Pauli principle is essential for inelastic scattering. This kinetic equation is derived by analogy to the case of non-equilibrium bosons considered in Chapter 4. The collision integral for inelastic scattering derived below describes the energy relaxation of electron system, the evolution of the occupation numbers of different electron states, etc. For this reason, it has numerous applications for strongly non-equilibrium electron systems. In this chapter we discuss the distribution of the electrons heated by a static electric field or high-frequency field of laser radiation in the cases when either electron-phonon or electron-electron collisions dominate. In the latter case, it is convenient to introduce the energy balance equations. Besides, we consider the interaction of non-equilibrium electrons with photons, in particular, in the process of radiative interband recombination. This process is described by the density balance equations. In this connection, we also study the phonon-assisted redistribution of electron population between the subbands in nanostructures of different dimensionalities.

34. Electron-Boson Collision Integral

In a similar way as in Chapter 4, the kinetic equation for electrons interacting with bosons is derived from Eq. (1.20) for the density matrix by using the Hamiltonian (19.1). It is convenient to subdivide the averaging Sp... over all variables of the system by the averaging $\mathrm{Sp_e}$... over electron variables only and another averaging, $\mathrm{Sp_b}$..., over boson variables so that $\mathrm{Sp}... = \mathrm{Sp_e Sp_b}...$. Applying the averaging $\mathrm{Sp_e}$...

to the definition (4.29), we introduce the one-electron density matrix $\hat{n}_t$ depending on the bosonic variables q. Then, multiplying Eq. (1.20) by $a_\delta^+ a_\gamma$ and taking the trace $\mathrm{Sp_e} \ldots$ of both its sides, we obtain the following equation for $\hat{n}_t$:

$$i\hbar\frac{\partial \hat{n}_t}{\partial t} = \left[\hat{h}_t + \widehat{H}_b + \sum_q (\hat{\chi}_{qt}\hat{b}_q + H.c.), \hat{n}_t\right] + \hat{\mathcal{L}}_t \;, \tag{1}$$

where $\hat{\mathcal{L}}_t$ appears because of the impossibility of permutation of $\widehat{H}_{e,b}$ with electron operators under the sign of $\mathrm{Sp_e}$. For this reason, Eq. (1) is not reduced to an equation for the one-electron density matrix. The operator $\hat{\mathcal{L}}_t$ is defined by its matrix elements

$$\langle\gamma|\hat{\mathcal{L}}_t|\delta\rangle \equiv \sum_{\nu\eta}\left\{\langle\nu|\sum_q(\hat{\chi}_{qt}\hat{b}_q + H.c.)|\eta\rangle\mathrm{Sp_e}(\hat{a}_\nu^+\hat{a}_\delta^+\hat{a}_\eta\hat{a}_\gamma\hat{n}_t)\right.$$

$$\left.-\mathrm{Sp_e}(\hat{a}_\delta^+\hat{a}_\nu^+\hat{a}_\gamma\hat{a}_\eta\hat{n}_t)\langle\nu|\sum_q(\hat{\chi}_{qt}\hat{b}_q + H.c.)|\eta\rangle\right\} \tag{2}$$

containing two-particle averages. Because of this contribution, the Pauli principle is satisfied for electron-boson scattering processes (for the case of elastic scattering considered in Chapter 2, this principle was satisfied automatically). Besides, the contribution $\hat{\mathcal{L}}_t$ describes electron-electron interaction through the boson field. This effect, however, should be neglected within the second-order accuracy with respect to electron-boson interaction, which is assumed below.

Averaging Eq. (1) over boson variables, we get the one-electron density matrix $\hat{\rho}_t = \mathrm{Sp_b}\hat{n}_t$ on the left-hand side. The contributions $[\widehat{H}_b, \hat{n}_t]$ and $\hat{\mathcal{L}}_t$ on the right-hand side disappear after this averaging, because (problem 7.1)

$$\mathrm{Sp_b}\left[\widehat{H}_b, \hat{n}_t\right] = 0, \qquad \mathrm{Sp_b}\hat{\mathcal{L}}_t = 0. \tag{3}$$

Therefore, we obtain the following equation for $\hat{\rho}_t$:

$$i\hbar\frac{\partial \hat{\rho}_t}{\partial t} = \left[\hat{h}_t, \hat{\rho}_t\right] + \mathrm{Sp_b}\left[\sum_q(\hat{\chi}_{qt}\hat{b}_q + H.c.), \hat{n}_t\right], \tag{4}$$

which is similar to Eq. (7.6). To evaluate the last term on the right-hand side of Eq. (4), we introduce the averages $\mathrm{Sp_b^{(q)}} \ldots$ over all boson modes except the q-th one, and $\mathrm{Sp_b^{(qq')}} \ldots$ over all modes except the q-th and q'-th. Further, we define the trace $\mathrm{tr}_q \ldots$ over the quantum numbers of q-th boson mode so that $\mathrm{Sp_b} \ldots = \mathrm{tr}_q \mathrm{Sp_b^{(q)}} \ldots$ and $\mathrm{Sp_b^{(q)}} \ldots = \mathrm{tr}_{q'} \mathrm{Sp_b^{(qq')}} \ldots$. Introducing the operator

$$\hat{n}_t^{(q)} = \mathrm{Sp_b^{(q)}} \hat{n}_t, \tag{5}$$

which depends on electron variables and q-th mode variables, we rewrite the last term on the right-hand side of Eq. (4) as $\sum_q \mathrm{tr}_q \left[(\hat{\chi}_{qt} \hat{b}_q + H.c.), \right.$ $\left. \hat{n}_t^{(q)} \right]$. The equation for $\hat{n}_t^{(q)}$ can be obtained after averaging Eq. (4) by application of $\mathrm{Sp_b^{(q)}} \ldots$. One should employ the following equations:

$$\mathrm{Sp_b^{(q)}} \left[\widehat{H}_b, \hat{n}_t \right] = \left[\hbar \omega_q \hat{b}_q^+ \hat{b}_q, \hat{n}_t^{(q)} \right],$$

$$\mathrm{Sp_b^{(q)}} \left[\sum_{q'} (\hat{\chi}_{q't} \hat{b}_{q'} + H.c.), \hat{n}_t \right] = \left[(\hat{\chi}_{qt} \hat{b}_q + H.c.), \hat{n}_t^{(q)} \right]$$

$$+ \sum_{q'(q' \neq q)} \mathrm{tr}_{q'} \left[(\hat{\chi}_{q't} \hat{b}_{q'} + H.c.), \mathrm{Sp_b^{(qq')}} \hat{n}_t \right], \tag{6}$$

and the term with $\mathrm{Sp_b^{(qq')}} \hat{n}_t$ should be neglected if we stay within the second-order accuracy. As a result, we arrive at the equation similar to Eq. (7.7):

$$i\hbar \frac{\partial \hat{n}_t^{(q)}}{\partial t} = \left[\hat{h}_t + \hbar \omega_q \hat{b}_q^+ \hat{b}_q, \hat{n}_t^{(q)} \right] + \left[\hat{\chi}_{qt} \hat{b}_q + \hat{\chi}_{qt}^+ \hat{b}_q^+, \hat{n}_t^{(q)} \right] + \mathrm{Sp_b^{(q)}} \hat{\mathcal{L}}_t , \tag{7}$$

where the last term describes electron pair correlations with the boson mode q. This term is determined by the matrix elements

$$\langle \gamma | \mathrm{Sp_b^{(q)}} \hat{\mathcal{L}}_t | \delta \rangle = \sum_{\nu \eta} \left\{ \langle \nu | \hat{\chi}_{qt} \hat{b}_q + \hat{\chi}_{qt}^+ \hat{b}_q^+ | \eta \rangle \mathrm{Sp_b^{(q)}} \mathrm{Sp_e} (\hat{a}_\delta^+ \hat{a}_\nu^+ \hat{a}_\eta \hat{a}_\gamma \hat{\eta}_t) \right.$$

$$\left. - \mathrm{Sp_b^{(q)}} \mathrm{Sp_e} (\hat{a}_\delta^+ \hat{a}_\nu^+ \hat{a}_\eta \hat{a}_\gamma \hat{\eta}_t) \langle \nu | \hat{\chi}_{qt} \hat{b}_q + \hat{\chi}_{qt}^+ \hat{b}_q^+ | \eta \rangle \right\} . \tag{8}$$

Considering Eq. (4) within the second-order accuracy with respect to electron-boson interaction, we introduce a correlation operator

$$\hat{\kappa}_{qt} = \hat{n}_t^{(q)} - \widetilde{n}_t^{(q)}, \tag{9}$$

where $\widetilde{n}_t^{(q)}$ is the density matrix $\hat{n}_t^{(q)}$ for non-interacting electron-boson system. It satisfies the following equation:

$$i\hbar \frac{\partial \widetilde{n}_t^{(q)}}{\partial t} = \left[\hat{h}_t + \hbar \omega_q \hat{b}_q^+ \hat{b}_q, \widetilde{n}_t^{(q)} \right]. \tag{10}$$

Within the zero-order accuracy in the interaction, the averages

$$g_{\eta\nu}^{\gamma\delta}(qt) = \mathrm{Sp}_{\mathrm{b}}^{(q)}\mathrm{Sp}_{\mathrm{e}}(\hat{a}_{\delta}^{+}\hat{a}_{\nu}^{+}\hat{a}_{\eta}\hat{a}_{\gamma}\hat{\eta}_{t}) \tag{11}$$

contributing into Eq. (8) are expressed through $\widetilde{n}_{t}^{(q)}$ according to

$$g_{\eta\nu}^{\gamma\delta}(qt) \simeq \langle\gamma|\widetilde{n}_{t}^{(q)}|\delta\rangle\langle\eta|\hat{\rho}_{t}|\nu\rangle - \langle\gamma|\widetilde{n}_{t}^{(q)}|\nu\rangle\langle\eta|\hat{\rho}_{t}|\delta\rangle. \tag{12}$$

To check this equation, one may write an equation of motion for $g_{\eta\nu}^{\gamma\delta}$ (i.e., consider $i\hbar\partial g_{\eta\nu}^{\gamma\delta}/\partial t$) and use Eqs. (4) and (10) (problem 7.2). It is convenient to apply again the operator of permutations $\hat{\mathcal{P}}$, introduced by Eq. (31.4), to rewrite the two-particle average in Eq. (8) in the operator form:

$$\mathrm{Sp}_{\mathrm{b}}^{(q)}\hat{\mathcal{L}}_{t} = \mathrm{sp}'\left[\hat{\chi}_{qt}'\hat{b}_{q} + \hat{\chi}_{qt}'^{+}\hat{b}_{q}^{+}, (1-\hat{\mathcal{P}})\widetilde{n}_{t}^{(q)}\hat{\rho}_{t}'\right]. \tag{13}$$

Using Eqs. (7) and (10), we obtain the following equation for the correlation operator:

$$i\hbar\frac{\partial\hat{\kappa}_{qt}}{\partial t} = \left[\hat{h}_{t} + \hbar\omega_{q}\hat{b}_{q}^{+}\hat{b}_{q}, \hat{\kappa}_{qt}\right] + \left[\hat{\chi}_{qt}\hat{b}_{q} + \hat{\chi}_{qt}^{+}\hat{b}_{q}^{+}, \widetilde{n}_{t}^{(q)}\right]$$

$$+\mathrm{sp}'\left[\hat{\chi}_{qt}'\hat{b}_{q} + \hat{\chi}_{qt}'^{+}\hat{b}_{q}^{+}, (1-\hat{\mathcal{P}})\widetilde{n}_{t}^{(q)}\hat{\rho}_{t}'\right]. \tag{14}$$

Equations (4) and (14) form a closed set, similar to Eqs. (7.6) and (7.10) for the electron-impurity system. One should use the initial condition of weakening of correlations,

$$\hat{\kappa}_{qt\to-\infty} = 0, \tag{15}$$

to solve Eq. (14). The solution is written through the evolution operator $\widehat{S}(t,t')$ defined by Eq. (2.2) with the one-electron Hamiltonian $\hat{h}_{t}$:

$$\hat{\kappa}_{q} = \frac{1}{i\hbar}\int_{-\infty}^{t}dt'e^{\lambda t'}\exp[-i\widehat{w}_{q}(t-t')]\widehat{S}(t,t')\left\{\left[\hat{\chi}_{qt'}\hat{b}_{q} + \hat{\chi}_{qt'}^{+}\hat{b}_{q}^{+}, \widetilde{n}_{t'}^{(q)}\right]\right.$$

$$\left.+\mathrm{sp}'\left[\hat{\chi}_{qt'}'\hat{b}_{q} + \hat{\chi}_{qt'}'^{+}\hat{b}_{q}^{+}, (1-\hat{\mathcal{P}})\widetilde{n}_{t'}^{(q)}\hat{\rho}_{t'}'\right]\right\}\widehat{S}^{+}(t,t')\exp[i\widehat{w}_{q}(t-t')], \tag{16}$$

where $\widehat{w}_{q} \equiv \omega_{q}\hat{b}_{q}^{+}\hat{b}_{q}$, and $\lambda\to+0$ means that the electron-boson interaction is turned on adiabatically. Since $\widetilde{n}_{t}^{(q)}$ depends on bosonic operators in the combination $\hat{b}_{q}^{+}\hat{b}_{q}$ only (this property follows from Eq. (10)), one has

$$\mathrm{tr}_{q}\left[\hat{\chi}_{qt}\hat{b}_{q} + \hat{\chi}_{qt}^{+}\hat{b}_{q}^{+}, \widetilde{n}_{t}^{(q)}\right] = 0. \tag{17}$$

Finally, Eq. (4) for $\hat{\rho}_t$ is reduced to the following kinetic equation:

$$\frac{\partial \hat{\rho}_t}{\partial t} + \frac{i}{\hbar} \left[\hat{h}_t, \hat{\rho}_t \right] = \frac{1}{i\hbar} \sum_q \mathrm{tr}_q \left[\hat{\chi}_{qt} \hat{b}_q + \hat{\chi}_{qt}^+ \hat{b}_q^+, \hat{\kappa}_q \right] \equiv \widehat{J}_{e,b}(\hat{\rho}|t), \qquad (18)$$

where the right-hand side is the collision integral in the operator form. To calculate the averages $\mathrm{tr}_q \ldots$ in this equation, we use the identities

$$\mathrm{tr}_q \hat{b}_q \hat{b}_q \widetilde{n}_t^{(q)} = \mathrm{tr}_q \hat{b}_q^+ \hat{b}_q^+ \widetilde{n}_t^{(q)} = 0,$$

$$\mathrm{tr}_q \hat{b}_q^+ \hat{b}_q \widetilde{n}_t^{(q)} = N_{qt} \hat{\rho}_t, \qquad \mathrm{tr}_q \hat{b}_q \hat{b}_q^+ \widetilde{n}_t^{(q)} = (N_{qt} + 1) \hat{\rho}_t, \qquad (19)$$

where $N_{qt} = N_t(q, q)$ is the diagonal part of the one-boson density matrix governed by Eq. (19.14). For equilibrium bosons, N_{qt} is time-independent and reduced to the Planck distribution function.

After substituting $\hat{\kappa}_{qt}$ into the right-hand side of Eq. (18), it is convenient to split the collision integral in two parts:

$$\widehat{J}_{e,b}(\hat{\rho}|t) = \widehat{J}_{e,b}^{(e)}(\hat{\rho}|t) + \widehat{J}_{e,b}^{(a)}(\hat{\rho}|t). \qquad (20)$$

The contribution $\widehat{J}_{e,b}^{(e)}$ describes the emission of bosons by electrons:

$$\widehat{J}_{e,b}^{(e)}(\hat{\rho}|t) = \frac{1}{\hbar^2} \sum_q \int_{-\infty}^t dt' e^{\lambda t'} (N_{qt'} + 1) \qquad (21)$$

$$\times \left\{ e^{-i\omega_q(t-t')} \left[\widehat{S}(t,t') \left(\hat{\chi}_{qt'}^+ \hat{\rho}_{t'} + \mathrm{sp}' \hat{\chi}_{qt'}'^+ (1 - \hat{\mathcal{P}}) \hat{\rho}_{t'} \hat{\rho}_{t'}' \right) \widehat{S}^+(t,t'), \hat{\chi}_{qt} \right] \right.$$

$$\left. - e^{i\omega_q(t-t')} \left[\widehat{S}(t,t') \left(\hat{\rho}_{t'} \hat{\chi}_{qt'} + \mathrm{sp}' \hat{\chi}_{qt'}' (1 - \hat{\mathcal{P}}) \hat{\rho}_{t'} \hat{\rho}_{t'}' \right) \widehat{S}^+(t,t'), \hat{\chi}_{qt}^+ \right] \right\},$$

while $\widehat{J}_{e,b}^{(a)}$ accounts for the absorption of bosons and is directly obtained from Eq. (21) after formal substitutions

$$(N_{qt'} + 1) \to N_{qt'}, \quad \omega_q \to -\omega_q, \quad \hat{\chi}_{qt}^+ \to \hat{\chi}_{qt}. \qquad (22)$$

The collision integral can be simplified if we exclude the permutation operator according to $\mathrm{sp}' \hat{b}' \hat{\mathcal{P}} \hat{a} \hat{c}' = \hat{a} \hat{b} \hat{c}$ and rewrite Eq. (21) as

$$\widehat{J}_{e,b}^{(e)}(\hat{\rho}|t) = \frac{1}{\hbar^2} \sum_q \int_{-\infty}^t dt' e^{\lambda t'} (N_{qt'} + 1) \qquad (23)$$

$$\times \left\{ e^{-i\omega_q(t-t')} \left[\widehat{S}(t,t') \left((1 - \hat{\rho}_{t'}) \hat{\chi}_{qt'}^+ \hat{\rho}_{t'} + \hat{\rho}_{t'} \mathrm{sp} \hat{\chi}_{qt'}^+ \hat{\rho}_{t'} \right) \widehat{S}^+(t,t'), \hat{\chi}_{qt} \right] \right.$$

$$\left. - e^{i\omega_q(t-t')} \left[\widehat{S}(t,t') \left(\hat{\rho}_{t'} \hat{\chi}_{qt'} (1 - \hat{\rho}_{t'}) + \hat{\rho}_{t'} \mathrm{sp} \hat{\chi}_{qt'} \hat{\rho}_{t'} \right) \widehat{S}^+(t,t'), \hat{\chi}_{qt}^+ \right] \right\}.$$

Thus, the quantum-kinetic description of the electrons interacting with boson modes, within the second-order accuracy in this interaction, is given by Eqs. (18), (20), (23), and (22).

In the translation-invariant case, we use the basis of plane waves $|\mathbf{p}\rangle$ with energies ε_p to describe the electron states and introduce the distribution function $f_{\mathbf{p}t} = \langle \mathbf{p}|\hat{\rho}_t|\mathbf{p}\rangle$; see a similar consideration in Sec. 8. We obtain the kinetic equation $\partial f_{\mathbf{p}t}/\partial t = J_{e,b}(f|\mathbf{p}t)$ with the collision integral $J_{e,b}(f|\mathbf{p}t) = \langle \mathbf{p}|\hat{J}_{e,b}(\hat{\rho}|t)|\mathbf{p}\rangle = J_{e,b}^{(e)}(f|\mathbf{p}t) + J_{e,b}^{(a)}(f|\mathbf{p}t)$, where

$$J_{e,b}^{(e)}(f|\mathbf{p}t) = \frac{1}{\hbar^2} \sum_{\mathbf{p}'q} \int_{-\infty}^{t} dt' e^{\lambda t'}(N_{qt'}+1)|\langle \mathbf{p}|\hat{\chi}_{qt'}|\mathbf{p}'\rangle|^2$$

$$\times \left\{ [e^{i(\varepsilon_{p'}-\varepsilon_p-\hbar\omega_q)(t-t')/\hbar} + c.c.]f_{\mathbf{p}'t'}(1-f_{\mathbf{p}t'}) \right. \tag{24}$$

$$\left. - [e^{i(\varepsilon_{p'}-\varepsilon_p+\hbar\omega_q)(t-t')/\hbar} + c.c.]f_{\mathbf{p}t'}(1-f_{\mathbf{p}'t'}) \right\},$$

and $J_{e,b}^{(a)}$ is obtained from this expression by means of the substitutions (22). We point out that the terms containing the averages $\mathrm{sp}\hat{\chi}_{qt}\hat{\rho}_t$ and $\mathrm{sp}\hat{\chi}_{qt}^{+}\hat{\rho}_t$ in Eq. (23) do not contribute to the collision integral since these terms are invariant with respect to translation. In the Markovian approximation, when the temporal dependence of $N_{qt'}$ and $f_{\mathbf{p}t'}$ is neglected on the time scale $\hbar/\bar{\varepsilon}$ determined by the characteristic energy of electrons, we calculate the integral over t' and obtain the collision integral in the form

$$J_{e,b}(f|\mathbf{p}t) = \frac{2\pi}{\hbar} \sum_{\mathbf{p}'\mathbf{q}} |\langle \mathbf{p}|\hat{\chi}_q|\mathbf{p}'\rangle|^2 \left\{ \delta(\varepsilon_{p'}-\varepsilon_p-\hbar\omega_q) \right.$$

$$\times \left[f_{\mathbf{p}'t}(1-f_{\mathbf{p}t})(N_{qt}+1) - f_{\mathbf{p}t}(1-f_{\mathbf{p}'t})N_{qt} \right] \tag{25}$$

$$\left. + \delta(\varepsilon_{p'}-\varepsilon_p+\hbar\omega_q) \left[f_{\mathbf{p}'t}(1-f_{\mathbf{p}t})N_{qt} - f_{\mathbf{p}t}(1-f_{\mathbf{p}'t})(N_{qt}+1) \right] \right\}.$$

It is not difficult to generalize this equation to the case of weakly inhomogeneous systems, when $f_{\mathbf{p}t}$ is replaced by the Wigner distribution function $f_{\mathbf{r}\mathbf{p}t}$. The corresponding transformations are completely analogous to those presented in Sec. 9. If several bands of electron spectrum are considered, the eigenstates $|l\mathbf{p}\rangle$ are described by both the momentum and the band index l. Nevertheless, one can often neglect the non-diagonal (interband) part of the density matrix for non-degenerate bands, because the interband energy considerably exceeds the energy broadening due to collisions. In this case, the collision integral again takes the form of Eq. (25), the only difference is that the band index stays together with the momentum. A similar collision integral can be written for the diagonal part of the density matrix of any quasi-stationary system described by the eigenstates $|\delta\rangle$ (problem 7.3).

Introducing the transition probabilities $W_{\mathbf{pp}'}$, one can rewrite the collision integral (25) in the form similar to Eq. (8.7), where, however, the Pauli principle is taken into account:

$$J_{e,b}(f|\mathbf{rp}t) = \sum_{\mathbf{p}'} \left[W_{\mathbf{p}'\mathbf{p}} f_{\mathbf{rp}'t}(1 - f_{\mathbf{rp}t}) - W_{\mathbf{pp}'} f_{\mathbf{rp}t}(1 - f_{\mathbf{rp}'t}) \right]. \quad (26)$$

The transition probabilities $W_{\mathbf{pp}'}$ and $W_{\mathbf{p}'\mathbf{p}}$ are given by

$$\left| \begin{matrix} W_{\mathbf{pp}'} \\ W_{\mathbf{p}'\mathbf{p}} \end{matrix} \right| = \frac{2\pi}{\hbar} \sum_q |\langle \mathbf{p} | \hat{\chi}_q | \mathbf{p}' \rangle|^2 \left[\left| \begin{matrix} N_q + 1 \\ N_q \end{matrix} \right| \delta\left(\varepsilon_{p'} - \varepsilon_p + \hbar\omega_q \right) \right.$$

$$\left. + \left| \begin{matrix} N_q \\ N_q + 1 \end{matrix} \right| \delta\left(\varepsilon_{p'} - \varepsilon_p - \hbar\omega_q \right) \right], \quad (27)$$

where the processes of emission and absorption of bosons correspond to the terms proportional to $N_q + 1$ and N_q, respectively. Note that $W_{\mathbf{pp}'} = W_{\mathbf{p}'\mathbf{p}} \exp[(\varepsilon_p - \varepsilon_{p'})/T]$ if N_q is the Planck distribution with temperature T. The stationary kinetic equation for the electrons interacting with the bosons described by the equilibrium Planck distribution is satisfied if we substitute the equilibrium Fermi distribution in place of $f_{\mathbf{rp}t}$. It is easy to check this property from Eqs. (26) and (27), since such a substitution makes the collision integral equal to zero (problem 7.4).

The electron-boson collision integral contains a sum over different types of bosons (acoustic and optical phonons or photons of different polarizations). Thus, we can rewrite Eq. (18) as a quantum kinetic equation containing the collision integrals caused by various scattering mechanisms:

$$\frac{\partial \hat{\rho}_t}{\partial t} + \frac{i}{\hbar} \left[\hat{h}_t, \hat{\rho}_t \right] = \sum_k \hat{J}_k(\hat{\rho}|t), \quad (28)$$

where $\hat{J}_k$ is the collision integral for the k-th scattering mechanism (in the operator form). To justify Eq. (28), let us show that, in the second order with respect to electron-phonon interaction, there is no interference between different scattering mechanisms induced by the same phonon branch. As already mentioned in Sec. 6, the long-wavelength phonons can interact with electrons through both macroscopic and microscopic (deformation) potentials. Thus, there are four kinds of electron-phonon interactions: deformation (Eq. (6.30)) and piezoelectric mechanisms for acoustic phonons and deformation and polarization (Eq. (6.22)) mechanisms for optical phonons. It is essential that the relation between the deformation potential and piezoelectric potential of acoustic phonons contains the phase factor $e^{i\pi/2} = i$. The same statement is true for the deformation and polarization potentials of optical phonons. Therefore, the squares of the absolute values of the matrix elements $|\langle \mathbf{p}|\hat{\chi}_{\mathbf{q}l}|\mathbf{p}'\rangle|^2$, where l numbers the phonon branches, are written as sums of macroscopic and microscopic field contributions for each branch. For example, the longitudinal optical branch generates both deformation optical (DO) and polar optical (PO) contributions, and one can write $|\langle \mathbf{p}|\hat{\chi}_{\mathbf{q}LO}|\mathbf{p}'\rangle|^2 = |\langle \mathbf{p}|\hat{\chi}_{\mathbf{q}DO}|\mathbf{p}'\rangle|^2 + |\langle \mathbf{p}|\hat{\chi}_{\mathbf{q}PO}|\mathbf{p}'\rangle|^2$. In other words, the microscopic and macroscopic fields do not interfere with each other. Each of these contributions gives a separate term in the collision integral and, for this reason, can be viewed as one of the scattering mechanisms for electrons.

The kinetic equation (28) satisfies the requirement of electron density conservation, $\partial(\mathrm{sp}\hat{\rho}_t)/\partial t = 0$. This requirement follows from $\mathrm{sp}\hat{J}_k(\hat{\rho}|t) = 0$, which is easy to prove by using Eq. (23) for the collision integral and taking into account that the traces of any commutators are equal to zero. On the other hand, the energy of electrons is not conserved because of energy exchange between electrons and bosons. From the formal point of view, this property is related to the inequality $\mathrm{sp}\hat{h}_t\hat{J}_k(\hat{\rho}|t) \neq 0$, in contrast to the case of electron-impurity interaction; see Eq. (7.20) and problem 2.2.

The transition probabilities (27) can be represented in the form (2.16) if the operator $\widehat{V}$ standing in Eq. (2.16) is assumed to be the Hamiltonian of electron-boson interaction, $\sum_q \hat{\chi}_q \hat{b}_q + H.c.$, and the initial and final states are treated as the states of electron-boson system in the absence of the interaction. Indeed, acting by the Hamiltonian $\widehat{V}$ on the state $|i\rangle = |\eta\rangle|N_1, N_2, \ldots, N_q, \ldots\rangle$ with the energy ε_i, we can obtain either $|\delta\rangle|N_1, N_2, \ldots, N_q - 1, \ldots\rangle$ with the energy $\varepsilon_f = \varepsilon_i + \varepsilon_\delta - \varepsilon_\eta - \hbar\omega_q$, or $|\gamma\rangle|N_1, N_2, \ldots, N_q + 1, \ldots\rangle$ with $\varepsilon_f = \varepsilon_i + \varepsilon_\gamma - \varepsilon_\eta + \hbar\omega_q$. According to Eqs. (3.13) and (3.14), the matrix elements for such transitions are $\sqrt{N_q}\langle\delta|\hat{\chi}_q|\eta\rangle$ and $\sqrt{N_q + 1}\langle\gamma|\hat{\chi}_q^+|\eta\rangle$, respectively. Therefore, applying Fermi's golden rule (2.16), we derive the transition probabilities standing in the kinetic equation of problem 7.3. If the electron states are the eigenstates of momentum, we obtain the transition probabilities in the form of Eq. (27). This approach can be applied to derive the kinetic equation for diagonal components of the density matrix in an arbitrary order in electron-boson interaction. This means that, instead of using the unperturbed initial and final states of electron-boson system in Eq. (2.16), one has to substitute there the states modified by electron-boson interaction, with a required accuracy. The first-order expansion of the wave function $|i\rangle$ in terms of the unperturbed states $|j\rangle_0$ is presented as $|i\rangle = |i\rangle_0 + \sum'_{j0}\langle j|it\rangle|j\rangle_0$, where the overlap factor $\langle j|it\rangle$ is given by Eq. (2.20) (it is time-independent at $\lambda \to 0$). Therefore, the next-order correction to the matrix element $\langle f|\widehat{V}|i\rangle$ is written as

$$\sum_j{}' \frac{\langle f|\widehat{V}|j\rangle\langle j|\widehat{V}|i\rangle}{\varepsilon_i - \varepsilon_j}, \tag{29}$$

where the prime sign at the sum indicates that the state j is different from i. The matrix elements in Eq. (29) are taken in the basis of non-perturbed states (the index 0 is omitted). The probability of two-boson transition, when an electron comes from the state δ to the state γ, is given by

$$W_{\delta\gamma}^{(2)} = \frac{2\pi}{\hbar} \sum_{qq_1} \sum_{s,s_1=\pm 1} \left| \sum_\nu \frac{\langle \gamma|\hat{\chi}_{q_1,s_1}|\nu\rangle\langle\nu|\hat{\chi}_{q,s}|\delta\rangle}{\varepsilon_\delta - \varepsilon_\nu - s\hbar\omega_q} + (q \leftrightarrow q_1,\ s \leftrightarrow s_1) \right|^2$$

$$\times \delta(\varepsilon_\delta - \varepsilon_\gamma - s\hbar\omega_q - s_1\hbar\omega_{q_1}) \left(N_{q_1} + \frac{1}{2} + \frac{s_1}{2}\right)\left(N_q + \frac{1}{2} + \frac{s}{2}\right), \tag{30}$$

where $\hat{\chi}_{q,+1} \equiv \hat{\chi}_q^+$ and $\hat{\chi}_{q,-1} \equiv \hat{\chi}_q$. The sum inside $|\ldots|$ is taken over the intermediate electron states ν and two different intermediate states of the boson system. Four different combinations of s and s_1 in Eq. (30) describe four possible kinds of transitions. For example, the process when both bosons are emitted corresponds to $s = s_1 = +1$. The two-boson contribution to the electron-boson collision integral is expressed through the two-boson transition probabilities in a similar way as in Eq. (26):

$$J_{e,b}^{(2)}(f|\delta t) = \sum_\gamma \left[W_{\gamma\delta}^{(2)} f_{\gamma t}(1 - f_{\delta t}) - W_{\delta\gamma}^{(2)} f_{\delta t}(1 - f_{\gamma t}) \right]. \tag{31}$$

This collision integral can be derived from Eq. (18) if the higher-order contributions to the correlation operator $\hat{\kappa}_q$ are taken into account. Such a derivation, however, requires a long and careful calculation, and we do not present it here. Though the probabilities (30) of two-boson transitions are of the fourth order in the electron-boson interaction, these transitions become important if, for some reasons, the one-boson transitions between the electron states δ and γ are suppressed or forbidden; see Sec. 38.

35. Quasi-Isotropic and Streaming Distributions

In this section we consider the electrons which are accelerating by a stationary electric field and scattering by equilibrium phonons (acoustic and optical). The current-voltage characteristics (i.e., the dependences of the current density on the strength of the applied electric field) for such hot-electron systems are determined by a competition between the increase in electron momentum and energy due to the acceleration and the transfer of momentum and energy from electrons to phonons via collisions. These processes are described by the quasi-classical kinetic equation (9.34). We consider this equation for a spatially-homogeneous case, by applying the collision integral (34.26) and assuming that the electron gas is non-degenerate. Even with these simplifying assumptions, it is impossible to obtain a general solution of such integro-differential equation, and the simulation procedures based upon the Monte-Carlo numerical method prove to be useful. Nevertheless, an analytical consideration can be carried out under some additional assumptions about the scattering. We study the cases of quasielastic and strongly inelastic scattering, when the distribution of electrons appears to be nearly isotropic and strongly anisotropic, respectively. We restrict ourselves by the transport of 2D electrons (the hot-electron problem for the 3D case is extensively studied in the literature), when the kinetic equations are simplified because of constant density of states. We also use the simplest model of bulk phonon modes to describe the scattering.

If the characteristic energy transmitted to the phonon system in the collision processes is small in comparison to the energy of electrons, one has a quasielastic regime of relaxation. This situation takes place for the scattering by acoustic phonons (if the temperature is high enough) and is justified because the sound velocities are small in comparison to the average electron velocity. In this case, it is convenient to write the general expression (34.26) through the half-sum $\overline{W}_{\mathbf{pp}'}$ and difference $\Delta W_{\mathbf{pp}'}$ of the transition probabilities defined by Eq. (34.27):

$$\left| \begin{array}{c} \overline{W}_{\mathbf{pp}'} \\ \Delta W_{\mathbf{pp}'} \end{array} \right| = \frac{2\pi}{\hbar} \sum_{q_z} |C_Q|^2 |\langle 1|e^{iq_z z}|1\rangle|^2 \tag{1}$$

$$\times \left| \begin{array}{c} (N_Q + 1/2)\left[\delta\left(\varepsilon_{p'} - \varepsilon_p + \hbar\omega_Q\right) + \delta\left(\varepsilon_{p'} - \varepsilon_p - \hbar\omega_Q\right)\right] \\ \left[\delta\left(\varepsilon_{p'} - \varepsilon_p + \hbar\omega_Q\right) - \delta\left(\varepsilon_{p'} - \varepsilon_p - \hbar\omega_Q\right)\right] \end{array} \right| .$$

Using the basis (5.25) for 2D electrons, we have written the matrix element $\langle 1\mathbf{p}'|\hat{\chi}_q|1\mathbf{p}\rangle$ in this equation as $C_Q\langle 1|e^{iq_z z}|1\rangle\delta_{\mathbf{p}',\mathbf{p}+\hbar\mathbf{q}}$, where C_Q is the matrix element of electron-phonon interaction given by Eq. (21.1) for LA phonons. Next, $\mathbf{Q} = (\mathbf{q}, q_z)$ is the 3D wave vector, and the overlap factor $\langle 1|e^{iq_z z}|1\rangle$ is calculated with the use of the wave functions describing the confinement in the ground-state subband, $\langle z|1\rangle = \psi_z^{(1)}$. The conservation law for the 2D momentum, following from the homogeneity of the system in the 2D plane, leads to the substitution $Q = \sqrt{(\mathbf{p} - \mathbf{p}')^2/\hbar^2 + q_z^2}$, which is assumed in Eq. (1). The phonon spectrum $\omega_Q = s_l Q$ is determined by the longitudinal sound velocity s_l. We also assume that the phonons are not heated by electrons so that N_Q is the Planck distribution function characterized by the equilibrium temperature T.

Using Eq. (1), we rewrite the collision integral (34.26) for non-degenerate electrons interacting with acoustic phonons in the form

$$J_{ac}(f|\mathbf{p}) = \sum_{\mathbf{p}'} \overline{W}_{\mathbf{p}\mathbf{p}'} (f_{\mathbf{p}'} - f_{\mathbf{p}}) - \frac{1}{2} \sum_{\mathbf{p}'} \Delta W_{\mathbf{p}\mathbf{p}'} (f_{\mathbf{p}'} + f_{\mathbf{p}}). \quad (2)$$

Following the quasielastic approximation, we expand the transition probabilities as

$$\overline{W}_{\mathbf{p}\mathbf{p}'} \simeq L^{-2} \mathcal{K}_{|\mathbf{p}-\mathbf{p}'|} \delta \left(\varepsilon_{p'} - \varepsilon_p \right) + L^{-2} \Delta \mathcal{K}_{|\mathbf{p}-\mathbf{p}'|} \delta'' \left(\varepsilon_{p'} - \varepsilon_p \right), \quad (3)$$

$$\Delta W_{\mathbf{p}\mathbf{p}'} \simeq L^{-2} \delta \mathcal{K}_{|\mathbf{p}-\mathbf{p}'|} \delta' \left(\varepsilon_{p'} - \varepsilon_p \right),$$

where L^2 is the normalization square in the 2D plane and the symbols $\delta'(\varepsilon_{p'} - \varepsilon_p)$ and $\delta''(\varepsilon_{p'} - \varepsilon_p)$ stand for the first and second derivatives of $\delta(\varepsilon_{p'} - \varepsilon_p)$ over $\varepsilon_{p'}$. The coefficients $\mathcal{K}_{|\mathbf{p}-\mathbf{p}'|}$, $\Delta \mathcal{K}_{|\mathbf{p}-\mathbf{p}'|}$, and $\delta \mathcal{K}_{|\mathbf{p}-\mathbf{p}'|}$ depend only on the transverse wave number q_z under the condition $\bar{p} \ll \hbar/d$, where d is the width of the 2D layer (note that the characteristic q_z contributing to the overlap factor are estimated as $1/d$). For the same reason, we can approximate ω_Q as ω_{q_z}. Using the equipartition condition for phonons, $\hbar \omega_{q_z} \ll T$, we approximate $N_Q + 1/2$ as $T/\hbar \omega_{q_z}$ and write the coefficients of the expansion (3) in the following form:

$$\mathcal{K} \simeq \frac{2\pi}{\hbar} L^2 \sum_{q_z} |C_{q_z}|^2 |\langle 1|e^{iq_z z}|1\rangle|^2 \frac{2T}{\hbar \omega_{q_z}},$$

$$\Delta \mathcal{K} \simeq \frac{2\pi}{\hbar} L^2 \sum_{q_z} |C_{q_z}|^2 |\langle 1|e^{iq_z z}|1\rangle|^2 \hbar \omega_{q_z} T \simeq \delta \mathcal{K} \frac{T}{2}. \quad (4)$$

Let us separate the symmetric (isotropic) part of the distribution function, $f_\varepsilon = \int_0^{2\pi} d\varphi f_{\mathbf{p}}/2\pi$, where $\varepsilon = \varepsilon_p$ and φ is the polar angle of the momentum $\mathbf{p}$ in the 2D plane, from the asymmetric part $\Delta f_{\mathbf{p}} = f_{\mathbf{p}} - f_\varepsilon$. This representation is convenient, since one can neglect a small non-elasticity in the relaxation of the asymmetric part and introduce the momentum relaxation rate ν_{ac} similar to the case of short-range static potential considered in Sec. 8:

$$J_{ac}(\Delta f|\mathbf{p}) \simeq -\nu_{ac} \Delta f_{\mathbf{p}}, \qquad \nu_{ac} = \mathcal{K} \rho_{2D}/2. \quad (5)$$

The symmetric part of the collision integral is expressed through the coefficients $\Delta \mathcal{K}$ and $\delta \mathcal{K}$. This part is reduced to a differential form of the second order:

$$J_{ac}(f|\varepsilon) \simeq \nu_{ac}\bar{\varepsilon}^2 \frac{d}{d\varepsilon}\left(\frac{df_\varepsilon}{d\varepsilon} + \frac{f_\varepsilon}{T}\right), \tag{6}$$

where we have introduced a characteristic energy $\bar{\varepsilon}$ through the relation $\nu_{ac}\bar{\varepsilon}^2 = \Delta\mathcal{K}\rho_{2D}/2$. According to Eqs. (4) and (5), the momentum relaxation rate is proportional to T, and $\bar{\varepsilon}$ is of the order of $\hbar s_l/d$. Explicit expressions for ν_{ac} and $\bar{\varepsilon}$ can be written if the eigenstates $|1\rangle$ are known (problem 7.5).

Using Eq. (5), we present the asymmetric part of the distribution function as $\Delta f_{\mathbf{p}} = -(e/\nu_{ac})\mathbf{E}\cdot\partial f_\varepsilon/\partial\mathbf{p}$ (see Eq. (8.24) and problem 7.6, where the elastic-scattering contribution is taken into account) and obtain a closed equation for the symmetric part of electron distribution:

$$\int_0^{2\pi}\frac{d\varphi}{2\pi}\left(e\mathbf{E}\cdot\frac{\partial}{\partial\mathbf{p}}\right)\frac{e}{\nu_{ac}}\mathbf{E}\cdot\frac{\partial f_\varepsilon}{\partial\mathbf{p}} + J_{ac}(f|\varepsilon) = 0. \tag{7}$$

The angular averaging transforms the first term on the left-hand side of Eq. (7) to

$$\frac{(eE)^2}{m\nu_{ac}}\left(\frac{df_\varepsilon}{d\varepsilon} + \varepsilon\frac{d^2f_\varepsilon}{d\varepsilon^2}\right) = \frac{(eE)^2}{m\nu_{ac}}\frac{d}{d\varepsilon}\left(\varepsilon\frac{df_\varepsilon}{d\varepsilon}\right). \tag{8}$$

Then, the symmetric part satisfies an ordinary differential equation of the second order. This equation is represented below as a continuity equation for the electron flow I_ε along the energy axis:

$$\frac{dI_\varepsilon}{d\varepsilon} = 0, \quad I_\varepsilon = \nu_{ac}\left[(\bar{\varepsilon}^2 + \varepsilon\varepsilon_E)\frac{df_\varepsilon}{d\varepsilon} + \frac{\bar{\varepsilon}^2}{T}f_\varepsilon\right]. \tag{9}$$

The characteristic energy $\varepsilon_E = (eE/\nu_{ac})^2/m$ introduced here determines the non-equilibrium part of the distribution function. For degenerate electrons, the last term in the expression for I_ε contains $f_\varepsilon(1-f_\varepsilon)$ instead of f_ε (problem 7.7). In zero electric field, $\varepsilon_E = 0$, the solution of Eq. (9) is merely the Boltzmann distribution with the temperature T. Equation (9) has the following analytical solution:

$$f_\varepsilon = N\left(1 + \varepsilon\varepsilon_E/\bar{\varepsilon}^2\right)^{-\bar{\varepsilon}^2/T\varepsilon_E} + C, \tag{10}$$

where the coefficients N and C are determined by two conditions. The first condition is the requirement of $f_\varepsilon = 0$ at $\varepsilon = \infty$. It gives us $C = 0$, which corresponds to zero flow along the energy axis. The second one is the normalization condition $\rho_{2D}\int_0^\infty d\varepsilon f_\varepsilon = n_{2D}$, where n_{2D} is the 2D electron density. In the case of $\varepsilon_E = 0$, the solution (10) is transformed to the Boltzmann distribution according to the limiting transition $(1+\xi)^{x/\xi}|_{\xi\to 0} = e^x$. The normalization condition is written as

$$\frac{n_{2D}}{\rho_{2D}} = \begin{cases} N(1/T - \varepsilon_E/\bar{\varepsilon}^2)^{-1}, & \bar{\varepsilon}^2/T\varepsilon_E > 1 \\ \infty, & \bar{\varepsilon}^2/T\varepsilon_E < 1 \end{cases} . \tag{11}$$

With the increase of the applied field, at $\varepsilon_E > \bar{\varepsilon}^2/T$, the distribution cannot be normalized because the integral $\int_0^\infty d\varepsilon\, f_\varepsilon$ starts to be divergent. Substituting $\Delta f_{\mathbf{p}}$ into the expression for the current, we have $\sigma = e^2 n_{2D}/m\nu_{ac}$ (compare to Eq. (8.27)). Therefore, both n_{2D} and σ appear to be divergent when the field E exceeds the critical value $(\bar{\varepsilon}\nu_{ac}/e)\sqrt{m/T}$. This non-physical situation is called the run-away effect. For an adequate description of the distribution function in high fields, one has to take into account the other scattering mechanisms preventing the electrons from running away. The spontaneous emission of optical phonons efficiently serves this purpose by keeping the electrons in the low-energy region. Its inclusion into the kinetic equation gives rise to $f_\varepsilon \simeq 0$ in the region $\varepsilon > \hbar\omega_{LO}$ (active region) if the temperature is low in comparison to the optical phonon energy $\hbar\omega_{LO}$. Before considering the optical-phonon scattering in more detail, we point out that the run-away effect is absent in the case of 3D electrons heated by a stationary electric field (problem 7.8), owing to the increase in the density of states with increasing energy, and, consequently, to the increase of the LA phonon scattering probability.

To describe the non-equilibrium distribution under the spontaneous emission of optical phonons by electrons, one has to consider the kinetic equation of the kind of Eq. (9.34), where the contribution

$$J_{LO}(f|\mathbf{p}) = \frac{2\pi}{\hbar} \sum_{\mathbf{Q}} |C_Q^{(LO)}|^2 |\langle 1|e^{iq_z z}|1\rangle|^2 \tag{12}$$

$$\times [\delta(\varepsilon_p - \varepsilon_{\mathbf{p}+\hbar\mathbf{q}} + \hbar\omega_{LO})f_{\mathbf{p}+\hbar\mathbf{q}} - \delta(\varepsilon_p - \varepsilon_{\mathbf{p}-\hbar\mathbf{q}} - \hbar\omega_{LO})f_{\mathbf{p}}]$$

is added to the collision integral (2); see Eq. (34.28). The matrix element $C_Q^{(LO)}$ is given by Eq. (21.1) for LO phonons. If the optical-phonon scattering dominates the active region, i.e., the characteristic rate of spontaneous emission, ν_{LO}, is much larger than ν_{ac} at $\varepsilon > \hbar\omega_{LO}$, and the field is not very strong, $|e|E/\nu_{LO} \ll p_{LO} \equiv \sqrt{2m\hbar\omega_{LO}}$, the distribution function of electrons is anisotropic in the passive region $\varepsilon < \hbar\omega_{LO}$ only. At low temperatures, only the spontaneous emission of optical phonons is possible, and $J_{LO}(\Delta f|\mathbf{p}) = 0$ in the passive region. Thus, the scattering by optical phonons does not modify the expression for $\Delta f_{\mathbf{p}}$ given above. Equation (7) for the symmetric part f_ε of the distribution function should be modified by adding there the collision integral (12) averaged over the angle:

$$\int_0^{2\pi} \frac{d\varphi}{2\pi} J_{LO}(f|\mathbf{p}) = \nu_\varepsilon^{(+)} f_{\varepsilon+\hbar\omega_{LO}} - \nu_\varepsilon^{(-)} f_\varepsilon, \tag{13}$$

$$\nu_\varepsilon^{(\pm)} = \frac{2\pi}{\hbar} \sum_{\mathbf{Q}} |C_Q^{(LO)}|^2 |\langle 1|e^{iq_z z}|1\rangle|^2 \delta(\varepsilon_p - \varepsilon_{\mathbf{p}\pm\hbar\mathbf{q}} \pm \hbar\omega_{LO}).$$

The relaxation rate $\nu_\varepsilon^{(-)}$ is the rate of spontaneous emission of optical phonons. Since $\nu_\varepsilon^{(-)} = 0$ in the passive region, it is convenient to write the kinetic equation for f_ε as

$$\frac{dI_\varepsilon}{d\varepsilon} + \nu_\varepsilon^{(+)} f_{\varepsilon+\hbar\omega_{LO}} = 0, \quad \varepsilon < \hbar\omega_{LO}$$

$$\frac{dI_\varepsilon^{(0)}}{d\varepsilon} + \nu_\varepsilon^{(+)} f_{\varepsilon+\hbar\omega_{LO}} - \nu_\varepsilon^{(-)} f_\varepsilon = 0, \quad \varepsilon > \hbar\omega_{LO}. \tag{14}$$

For the passive region, the flow I_ε introduced by Eq. (9) takes into account both the quasielastic acoustic-phonon scattering and the field contribution. To define the flow in the active region, $I_\varepsilon^{(0)}$, we neglect the field contribution, because the asymmetric part of the distribution function in this region is suppressed by strong optical-phonon scattering. In other words, $I_\varepsilon^{(0)}$ is given by Eq. (9) with $\varepsilon_E = 0$. Although this neglect requires an additional strong inequality, $(eE)^2 \hbar\omega_{LO}/(m\nu_{ac}\nu_{LO}) \ll \bar{\varepsilon}^2$, this assumption does not influence the final result given below by Eq. (18). The boundary conditions to the system (14) express the continuity of both f_ε and I_ε at the upper boundary of the passive region: $f_\varepsilon|_{\hbar\omega_{LO}-0}^{\hbar\omega_{LO}+0} = 0$ and $I_\varepsilon|_{\hbar\omega_{LO}-0}^{\hbar\omega_{LO}+0} = 0$.

In the active region, f_ε rapidly decreases because of $|e|E/\nu_{LO} \ll p_{LO}$. Therefore, we neglect the term $\nu_\varepsilon^{(+)} f_{\varepsilon+\hbar\omega_{LO}}$ and substitute the threshold energy $\varepsilon = \hbar\omega_{LO}$ into $\nu_\varepsilon^{(-)}$ in the second equation of the system (14). As a result, we transform this equation into a second-order differential equation with constant coefficients, whose solution, describing the active region, is written below as

$$f_\varepsilon = f_{\hbar\omega_{LO}} e^{-\lambda_{LO}(\varepsilon-\hbar\omega_{LO})}, \quad \lambda_{LO} = \frac{\sqrt{\nu_{\hbar\omega_{LO}}^{(-)}/\nu_{ac}}}{\bar{\varepsilon}}. \tag{15}$$

The expression for λ_{LO} is obtained with the use of the strong inequality $\nu_{\hbar\omega_{LO}}^{(-)}/\nu_{ac} \gg (\bar{\varepsilon}/T)^2$. This inequality remains valid even at $\nu_{\hbar\omega_{LO}}^{(-)} \sim \nu_{ac}$ because of the conditions $\bar{\varepsilon} \ll T$ ensuring the quasielasticity under acoustic-phonon scattering. Using the boundary conditions for f_ε and I_ε written above, we exclude the constant $f_{\hbar\omega_{LO}}$ and obtain the boundary condition of the third kind,

$$\left(\frac{df_\varepsilon}{d\varepsilon} + \frac{\lambda_{LO}}{1 + \varepsilon_E \hbar\omega_{LO}/\bar{\varepsilon}^2} f_\varepsilon\right)_{\varepsilon=\hbar\omega_{LO}} = 0, \tag{16}$$

which is to be used together with the first equation of the system (14) to find the distribution in the passive region. According to Eq. (15), the contribution of $f_{\varepsilon+\hbar\omega_{LO}}$ into the equations for the passive region exists in a narrow low-energy interval, $\varepsilon < \lambda_{LO}^{-1}$. Integrating the first equation of Eq. (14) over this interval, we use Eq. (15) and find the following effective boundary condition for low energies:

$$\left(I_\varepsilon + \frac{\nu_\varepsilon^{(+)}}{\lambda_{LO}} f_{\hbar\omega_{LO}}\right)_{\varepsilon=0} = 0. \tag{17}$$

Therefore, we again obtain Eq. (9) in the passive region, while the spontaneous emission of optical phonons provides the conditions (16) and (17) at the boundaries of this region. The solution of Eq. (9) in such conditions is given by Eq. (10), where, however, C is a non-zero constant proportional to N. The latter is to be found from the normalization condition $\rho_{2D} \int_0^{\hbar\omega_{LO}} d\varepsilon f_\varepsilon = n_{2D}$. In fact, to solve the problem, we need only one of the boundary conditions, either (16) or (17), since they are not independent of each other because of conservation of the flow along the energy axis (problem 7.9). Expressing C through N by using the boundary condition (16), we apply the normalizarion condition and obtain the normalized distribution function in the passive region:

$$f_\varepsilon = N\left[\left(1 + \varepsilon\varepsilon_E/\bar{\varepsilon}^2\right)^{-\bar{\varepsilon}^2/T\varepsilon_E} - \left(1 + \hbar\omega_{LO}\varepsilon_E/\bar{\varepsilon}^2\right)^{-\bar{\varepsilon}^2/T\varepsilon_E}\right], \tag{18}$$

$$N = \frac{n_{2D}}{\rho_{2D}}\left(\frac{1}{T} - \frac{\varepsilon_E}{\bar{\varepsilon}^2}\right)\left[1 - \left(1 + \frac{\hbar\omega_{LO}}{T}\right)\left(1 + \frac{\hbar\omega_{LO}\varepsilon_E}{\bar{\varepsilon}^2}\right)^{-\bar{\varepsilon}^2/T\varepsilon_E}\right]^{-1}.$$

In the transformation leading to this equation, we have used the expression for λ_{LO} given in Eq. (15) and the strong inequality $\nu_{\hbar\omega_{LO}}^{(-)}/\nu_{ac} \gg (\bar{\varepsilon}/T)^2$ (or, equivalently, $\lambda_{LO}T \gg 1$). We point out that f_ε appears to be independent of the optical-phonon scattering rate and goes to zero at $\varepsilon = \hbar\omega_{LO}$. The latter property corresponds to the case of zero boundary condition at the edge of the passive region. Indeed, since λ_{LO}^{-1} is small in comparison to the characteristic energy scale of the distribution function in the passive region, Eq. (16) effectively works as a zero boundary condition. The transition from the Boltzmann distribution $f_\varepsilon \propto e^{-\varepsilon/T}$ to the distributions spreading over the passive region is demonstrated in Fig. 7.1. Though the energy distribution remains monotonic as the electric field increases, the mean energy of electrons, determined as $\overline{E}_E = (\rho_{2D}/n_{2D}) \int_0^{\hbar\omega_{LO}} d\varepsilon \, \varepsilon f_\varepsilon$, increases. For the case of relatively strong fields, when $\varepsilon_E > \bar{\varepsilon}^2/T$, this energy is given by

$$\overline{E}_E \simeq \hbar\omega_{LO} \frac{\varepsilon_E T - \bar{\varepsilon}^2}{4\varepsilon_E T - 2\bar{\varepsilon}^2}. \tag{19}$$

This equation shows us that, with the increase of the electric field, $\overline{E}_E$ can rise up to $\hbar\omega_{LO}/4$.

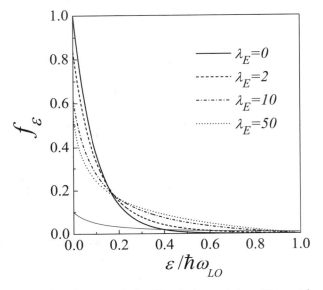

Figure 7.1. Energy distribution of the 2D electrons interacting with acoustic and optical phonons in a stationary electric field. The calculation is carried out according to Eq. (35.18) for several values of $\lambda_E = \varepsilon_E \hbar\omega_{LO}/\bar{\varepsilon}^2$ at $T = \hbar\omega_{LO}/10$. The distribution function f_ε is expressed in units of $f_{\varepsilon=0}$ at $\lambda_E = 0$. The narrow solid line at the bottom is the distribution function in the case when the optical-phonon scattering is neglected (Eq. (35.10) with $C = 0$) and the field is close to the threshold for the run-away effect, $\varepsilon_E = 0.9 \, \bar{\varepsilon}^2/T$.

The consideration given above is restricted by the case of quasi-isotropic distribution of electrons in the passive region, under the condition $|e|E/\nu_{ac} \ll p_{LO}$. As the field increases, the situation $\nu_{ac} \ll |e|E/p_{LO} \ll \nu_{LO}$ is possible (here we define the characteristic rate of spontaneous emission as $\nu_{LO} = \nu^{(-)}_{\hbar\omega_{LO}}$, problem 7.10). In these conditions, the electrons move through the passive region nearly ballistically and have a strongly anisotropic (streaming) distribution, while in the active region the scattering by optical phonons makes $f_{\mathbf{p}}$ rapidly decreasing with increasing energy. The stationary distribution is formed as follows. The electrons are accelerated by the field until their energy exceeds $\hbar\omega_{LO}$. Then they emit optical phonons, return to the low-energy region, and the cycle is repeated again. To describe the electron distribution quantitatively, we neglect the scattering by acoustic phonons and write the kinetic equation as

$$e\mathbf{E} \cdot \frac{\partial f_{\mathbf{p}}}{\partial \mathbf{p}} = J_{LO}(f|\mathbf{p}), \qquad (20)$$

where the collision integral is given by Eq. (12). Assuming that the field is applied along OX in the negative direction, we rewrite Eq. (20) in the passive region as

$$|e|E\frac{\partial f_{\mathbf{p}}}{\partial p_x} = \frac{2\pi}{\hbar} \sum_{\mathbf{p}'} w_{LO}(|\mathbf{p} - \mathbf{p}'|)\delta(\varepsilon_p - \varepsilon_{p'} + \hbar\omega_{LO})f_{\mathbf{p}'}, \quad p < p_{LO}, \quad (21)$$

where the effective squared matrix element of electron-phonon interaction is introduced as $w_{LO}(|\mathbf{p} - \mathbf{p}'|) = \sum_{q_z} |C_Q^{(LO)}|^2|\langle 1|e^{iq_z z}|1\rangle|^2$ with $Q = \sqrt{(\mathbf{p} - \mathbf{p}')^2/\hbar^2 + q_z^2}$. According to the energy conservation law, $f_{\mathbf{p}'}$ belongs to the active region. In the active region, Eq. (20) is written as

$$|e|E\frac{\partial f_{\mathbf{p}}}{\partial p_x} = -\nu_{LO}f_{\mathbf{p}}, \qquad p > p_{LO}. \qquad (22)$$

Since $f_{\mathbf{p}}$ rapidly decreases with increasing energy in this region, we have neglected the term corresponding to the arrival of electrons from higher-energy states in Eq. (22). For the same reason, we have approximated the energy-dependent relaxation rate $\nu_{\varepsilon}^{(-)}$ by its threshold value ν_{LO}. Equation (22) is solved as

$$f_{\mathbf{p}} = F(p_y)\exp[-(p_x - p_{x,LO})\nu_{LO}/|e|E], \qquad p > p_{LO}, \qquad (23)$$

where $p_{x,LO} = \sqrt{p_{LO}^2 - p_y^2}$ defines the boundary of the active region at a fixed p_y and $F(p_y)$ is the distribution function $f_{\mathbf{p}}$ at $p_x = p_{x,LO}$, which will be determined below. The rapid decrease of $f_{\mathbf{p}}$ into the active region along p_x is seen explicitly: only the interval where $p_x - p_{x,LO} \sim |e|E/\nu_{LO} \ll p_{LO}$ is essential. Below we also prove that the distribution $F(p_y)$ is very narrow. Therefore, after substituting the function given by Eq. (23) into the right-hand side of Eq. (21), we can put $|\mathbf{p}'| \simeq p_{LO}$ (and, consequently, $|\mathbf{p}| \simeq 0$) in $w_{LO}(|\mathbf{p} - \mathbf{p}'|)$, approximating it by the constant $w_{LO}(p_{LO})$. Equation (21) is rewritten as

$$|e|E\frac{\partial f_{\mathbf{p}}}{\partial p_x} = \frac{1}{\hbar}w_{LO}(p_{LO})\rho_{2D}\int_{-\infty}^{\infty} dp_y'F(p_y') \qquad (24)$$

$$\times \int_0^{\infty} d\Delta p_x'\exp(-\Delta p_x'\nu_{LO}/|e|E)\delta(2\Delta p_x'p_{x,LO}' - p^2),$$

where we have introduced a small deviation $\Delta p'_x = p'_x - p'_{x,LO}$ and neglected the term with $\Delta p'^2_x$ when transforming the argument of the δ-function. After calculating the integral over $\Delta p'_x$ on the right-hand side of Eq. (24), we integrate this equation over p_x and write its solution describing the passive region:

$$
f_{\mathbf{p}} = \frac{\nu_{LO}}{2\pi|e|E} \int_{-\infty}^{\infty} \frac{dp'_y F(p'_y)}{\sqrt{p^2_{LO} - p'^2_y}} \int_{-\infty}^{p_x} dp'_x \exp\left[-\frac{\nu_{LO}(p'^2_x + p^2_y)}{2|e|E\sqrt{p^2_{LO} - p'^2_y}} \right].
$$
(25)

The p_x-dependence of this function is essential only at small $|p_x| \sim \sqrt{2|e|Ep_{LO}/\nu_{LO}} \ll p_{LO}$, where $f_{\mathbf{p}}$ increases from zero to a function which depends on p_y only. This latter function is $F(p_y)$, because $f_{\mathbf{p}}$ is continuous at the boundary of the active region. Therefore, aiming p_x in Eq. (25) at $+\infty$, we obtain the following integral equation for $F(p_y)$:

$$
F(p_y) = \sqrt{\frac{\nu_{LO}}{2\pi|e|E}} \int_{-\infty}^{\infty} dp'_y \frac{F(p'_y)}{(p^2_{LO} - p'^2_y)^{1/4}} \exp\left(-\frac{\nu_{LO}p^2_y}{2|e|E\sqrt{p^2_{LO} - p'^2_y}} \right).
$$
(26)

It shows us that $F(p_y)$ is non-zero in the region of small $|p_y| \sim (2|e|Ep_{LO}/\nu_{LO})^{1/2} \ll p_{LO}$ so that one can approximate $p^2_{LO} - p'^2_y$ under the integral as p^2_{LO}. The solution is

$$
F(p_y) = C_1 \exp\left(-\frac{\nu_{LO}p^2_y}{2|e|Ep_{LO}} \right),
$$
(27)

and the distribution function in the passive region is given by

$$
f_{\mathbf{p}} = \frac{F(p_y)}{2} \left[1 + \mathrm{erf}\left(\sqrt{\frac{\nu_{LO}}{2|e|Ep_{LO}}} p_x \right) \right], \quad p < p_{LO},
$$
(28)

where $\mathrm{erf}(x)$ is the error function and $F(p_y)$ is given by Eq. (27). The factor $[1 + \mathrm{erf}\dots]/2$ describes the buildup of the distribution function from 0 to $F(p_y)$, which mostly takes place in the narrow region $-\sqrt{2|e|Ep_{LO}/\nu_{LO}} < p_x < \sqrt{2|e|Ep_{LO}/\nu_{LO}}$. The entire distribution given by Eqs. (28) and (23) can be approximately viewed as a narrow stripe beginning at $p_x = 0$ and ending at $p_x = p_{LO}$. The constant C_1 is determined from the normalization condition $n_{2D} = 2 \int d\mathbf{p} f_{\mathbf{p}}/(2\pi\hbar)^2 \simeq 2p_{LO} \int dp_y \times F(p_y)/(2\pi\hbar)^2$, and we find

$$
C_1 = \frac{\sqrt{2}\pi^{3/2}\hbar^2 n_{2D}}{\sqrt{|e|E/\nu_{LO}}p^{3/2}_{LO}}.
$$
(29)

One may often approximate the distribution function by a needle-like distribution

$$f_{\mathbf{p}} = \frac{(2\pi\hbar)^2}{2p_{LO}} n_{2D}\delta(p_y) \begin{cases} 1, & 0 < p_x < p_{LO} \\ \exp[-(p_x - p_{LO})\nu_{LO}/|e|E], & p_x > p_{LO} \end{cases} . \quad (30)$$

We stress again that the necessary condition to have the streaming distribution is $\nu_{ac} \ll \nu_{LO}$.

The streaming distribution of 3D electrons is not unlike the one discussed above. The main difference is that, owing to the square-root energy dependence of the 3D density of states, the rate of spontaneous emission of optical phonons goes to zero as p approaches p_{LO}. Near the threshold $\varepsilon_p = \hbar\omega_{LO}$, this rate is $\nu_{LO}(\varepsilon_p) \simeq 2\alpha\omega_{LO}\sqrt{\varepsilon_p/\hbar\omega_{LO} - 1}$, where α is the constant of electron-phonon coupling introduced by Eq. (21.28) (problem 7.11). Substituting this rate into Eq. (22) in place of ν_{LO}, we then integrate this equation and find that the function describing the decrease of the electron distribution in the active region contains $(p_x - p_{x,LO})^{3/2}$ in the exponent. Finally, approximating the distribution over the momenta $\mathbf{p}_\perp = (p_y, p_z)$ perpendicular to p_x by a δ-function, we write, in a similar way as in Eq. (30),

$$f_{\mathbf{p}} = \frac{(2\pi\hbar)^3}{2p_{LO}} n_{3D}\delta(\mathbf{p}_\perp) \begin{cases} 1, & 0 < p_x < p_{LO} \\ \exp\left[-\widetilde{\nu}(p_x - p_{LO})^{3/2}/|e|Ep_{LO}^{1/2}\right], & p_x > p_{LO} \end{cases} , \quad (31)$$

where $\widetilde{\nu} = 4\sqrt{2}\alpha\omega_{LO}/3$. Neglecting the exponential tail in the active region, one finds that the degree of anisotropy of the streaming distribution is not sensitive to the electric field under the approximations used in Eqs. (30) and (31). For this reason, the absolute value of the current density calculated with the aid of the distributions (30) or (31) is field-independent and equal to $en_D\hbar\omega_{LO}/p_{LO}$, while the direction of the current coincides with the direction of the electric field.

If the field is so strong that $|e|E/p_{LO} \gg \nu_{LO}$, the electrons penetrate deep into the active region and their average energy is much larger than $\hbar\omega_{LO}$. The scattering by optical phonons in this case becomes quasielastic and makes the electron distribution quasi-isotropic again. This distribution can be described in the way similar to the case of quasielastic scattering by acoustic phonons discussed in the beginning of this section.

36. Diffusion, Drift, and Energy Balance

The description of electron distribution in the external fields slowly varying with time and smoothly varying in space is simplified in the case when electron-electron scattering mostly controls the energy relaxation in electron system and the fields are not very strong so that the anisotropy of the electron distribution is weak. In this section we consider the quasi-classical kinetic equation (9.34), whose right-hand side contains a sum of the collision integrals for different scattering mechanisms, like in Eq. (34.28). We take into account electron-electron, electron-phonon, and electron-impurity scattering. If the mean free path lengths $\bar{v}\bar{\tau}$ and relaxation times $\bar{\tau}$ due to these scattering mechanisms are small in comparison to the characteristic spatial and temporal scales of the inhomogeneities, $\bar{l}$ and $\bar{t}$, i.e., when

$$\bar{l} \gg \bar{v}\bar{\tau}, \quad \bar{t} \gg \bar{\tau}, \tag{1}$$

we write the distribution function $f_{\mathbf{r}\mathbf{p}t}$ as a sum of the symmetric part $f_{\mathbf{r}t}^{(s)}(\varepsilon_p)$ and small antisymmetric part $\Delta f_{\mathbf{r}\mathbf{p}t}$. The symmetric part corresponds to a local equilibrium owing to electron-electron scattering and is represented as a Fermi distribution function whose parameters (temperature T and chemical potential μ) depend on both coordinate and time:

$$f_{\mathbf{r}t}^{(s)}(\varepsilon_p) = \left[\exp\left(\frac{\varepsilon_p - \mu_{\mathbf{r}t}}{T_{\mathbf{r}t}}\right) + 1\right]^{-1}. \tag{2}$$

This function is the symmetric part of the general expression (31.25) containing the local chemical potential and temperature. To find the antisymmetric part, we assume that the momentum relaxation is still controlled by the electron-impurity and electron-phonon scattering mechanisms. In these conditions, $\Delta f_{\mathbf{r}\mathbf{p}t}$ satisfies the following equation:

$$\frac{\partial \Delta f_{\mathbf{r}\mathbf{p}t}}{\partial t} + \frac{e}{c}[\mathbf{v_p} \times \mathbf{H}_{\mathbf{r}t}] \cdot \frac{\partial \Delta f_{\mathbf{r}\mathbf{p}t}}{\partial \mathbf{p}} + \mathbf{v_p} \cdot \frac{\partial f_{\mathbf{r}t}^{(s)}(\varepsilon_p)}{\partial \mathbf{r}} + e\mathbf{E}_{\mathbf{r}t} \cdot \frac{\partial f_{\mathbf{r}t}^{(s)}(\varepsilon_p)}{\partial \mathbf{p}}$$

$$= J_{im}(\Delta f|\mathbf{r}\mathbf{p}t) + J_{e,ph}(\Delta f|\mathbf{r}\mathbf{p}t), \tag{3}$$

where $\mathbf{E}_{\mathbf{r}t}$ and $\mathbf{H}_{\mathbf{r}t}$ are the strengths of the electric and magnetic fields, respectively, and the collision integral $J_{ee}(\Delta f|\mathbf{r}\mathbf{p}t)$ on the right-hand side is neglected. This neglect is justified because the electron-electron collisions themselves do not cause the relaxation of momentum, and an inclusion of $J_{ee}(\Delta f|\mathbf{r}\mathbf{p}t)$ would not essentially modify the transport time introduced below. Since we use the quasi-equilibrium distribution (2), the electron-electron contribution, $J_{ee}(f^{(s)}|\mathbf{r}\varepsilon_p t)$, on the right-hand side of the symmetric part of the kinetic equation is equal to zero, and we retain only the electron-phonon collision integral there (note that $J_{im}(f^{(s)}|\mathbf{r}\varepsilon_p t) = 0$ for any symmetric distribution):

$$\frac{\partial f_{\mathbf{r}t}^{(s)}(\varepsilon_p)}{\partial t} + \frac{e}{c}[\mathbf{v_p} \times \mathbf{H}_{\mathbf{r}t}] \cdot \frac{\partial f_{\mathbf{p}t}^{(s)}(\varepsilon_p)}{\partial \mathbf{p}} + \mathbf{v_p} \cdot \frac{\partial \Delta f_{\mathbf{r}\mathbf{p}t}}{\partial \mathbf{r}}$$

$$+ e\mathbf{E}_{\mathbf{r}t} \cdot \frac{\partial \Delta f_{\mathbf{r}\mathbf{p}t}}{\partial \mathbf{p}} = J_{e,ph}(f^{(s)}|\mathbf{r}\varepsilon_p t). \tag{4}$$

This equation is used below to find the slowly varying parameters $\mu_{\mathbf{r}t}$ and $T_{\mathbf{r}t}$.

If only the acoustic phonons contribute to electron-phonon scattering, we can estimate $J_{e,ph}(\Delta f|\mathbf{r}\mathbf{p}t)$ in Eq. (3) in the elastic approximation and rewrite the right-hand side as

$$J_{im}(\Delta f|\mathbf{r}\mathbf{p}t) + J_{e,ph}(\Delta f|\mathbf{r}\mathbf{p}t) = \sum_{\mathbf{p}'} W_{\mathbf{p}\mathbf{p}'}(\Delta f_{\mathbf{r}\mathbf{p}'t} - \Delta f_{\mathbf{r}\mathbf{p}t}),$$

$$W_{\mathbf{p}\mathbf{p}'} = \frac{2\pi}{\hbar} \left[n_{im}|v(|\mathbf{p} - \mathbf{p}'|/\hbar)|^2/V \right. \tag{5}$$

$$\left. + |C_{|\mathbf{p}-\mathbf{p}'|/\hbar}|^2 \coth(s_l|\mathbf{p} - \mathbf{p}'|/2T) \right] \delta(\varepsilon_p - \varepsilon_{p'}),$$

where the transition probability $W_{\mathbf{p}\mathbf{p}'}$ contains a sum of impurity and LA phonon contributions, which are obtained from Eqs. (8.8) and (34.27), respectively. Equation (5) is written here for the 3D case, under the assumption of equilibrium acoustic phonon distribution, when $2N_{\mathbf{q}} + 1 = \coth(\hbar\omega_q/2T)$. To consider the case of 2D electrons, one should replace $|C_{|\mathbf{p}-\mathbf{p}'|/\hbar}|^2 \coth(s_l|\mathbf{p} - \mathbf{p}'|/2T)$ by $\sum_{q_z} |\langle 1|e^{iq_z z}|1\rangle|^2 |C_q|^2 \coth(\hbar s_l q/2T)$, where $q = [(\mathbf{p} - \mathbf{p}')^2/\hbar^2 + q_z^2]^{1/2}$; see the previous section. In both cases, the elastic approximation is justified under the equipartition condition, when $\coth(\hbar s_l q/2T) \simeq 2T/\hbar s_l q$, and, since $|C_q|^2 \propto q$, the acoustic-phonon contribution to $W_{\mathbf{p}\mathbf{p}'}$ appears to be independent of electron momenta. For the scattering by the short-range potential of impurities, when the dependence of $v(|\mathbf{p} - \mathbf{p}'|/\hbar)$ on its argument can be neglected, the entire transition probability $W_{\mathbf{p}\mathbf{p}'}$ is independent of electron momenta. The elastic approximation for $J_{e,ph}(\Delta f|\mathbf{r}\mathbf{p}t)$ allows one to solve Eq. (3) exactly, by means of the substitution $\Delta f_{\mathbf{r}\mathbf{p}t} = \mathbf{v}_{\mathbf{p}} \cdot \mathbf{g}_{\mathbf{r}t}(\varepsilon_p)$; see also the consideration of Eq. (8.24). The vector-function $\mathbf{g}_{\mathbf{r}t}(\varepsilon)$ satisfies the algebraic equation

$$\frac{\partial \mathbf{g}_{\mathbf{r}t}}{\partial t} - [\boldsymbol{\omega}_c \times \mathbf{g}_{\mathbf{r}t}] + \frac{\mathbf{g}_{\mathbf{r}t}}{\tau_{tr}(\varepsilon)} = -\frac{\partial f_{\mathbf{r}t}^{(s)}(\varepsilon)}{\partial \mathbf{r}} - e\mathbf{E}_{\mathbf{r}t} \frac{\partial f_{\mathbf{r}t}^{(s)}(\varepsilon)}{\partial \varepsilon}. \tag{6}$$

The transport time $\tau_{tr}(\varepsilon)$ is introduced here according to the definitions given in Sec. 8. To apply them, one should merely replace the transition probability (8.8) by the one defined by Eq. (5) (see also problem 7.6). If the transition probability $W_{\mathbf{p}\mathbf{p}'}$ is independent of electron momenta, the energy dependence of the transport time is determined entirely by the energy dependence of the density of states, $\tau_{tr}(\varepsilon) \propto \rho_{\mathcal{D}}^{-1}(\varepsilon)$.

According to the condition $\bar{t} \gg \bar{\tau}$, one may neglect the time derivative in Eq. (6). This means that, in the absence of magnetic fields, $\mathbf{g}_{\mathbf{r}t}$ is equal to the product of $\tau_{tr}(\varepsilon)$ by the right-hand side of Eq. (6). Rewriting this right-hand side through the spatial gradients of the chemical potential and temperature, we obtain

$$\mathbf{g}_{\mathbf{r}t}(\varepsilon) = \tau_{tr}(\varepsilon)\frac{\partial f_{\mathbf{r}t}^{(s)}(\varepsilon)}{\partial\varepsilon}\left(e\nabla_{\mathbf{r}}w_{\mathbf{r}t} + \frac{\varepsilon - \mu_{\mathbf{r}t}}{T_{\mathbf{r}t}}\nabla_{\mathbf{r}}T_{\mathbf{r}t}\right), \qquad (7)$$

$$w_{\mathbf{r}t} = \mu_{\mathbf{r}t}/e + \Phi_{\mathbf{r}t},$$

where the electric field is expressed through the electrostatic potential $\Phi_{\mathbf{r}t}$ as $\mathbf{E}_{\mathbf{r}t} = -\nabla_{\mathbf{r}}\Phi_{\mathbf{r}t}$, and the local electrochemical potential $w_{\mathbf{r}t}$ is introduced. The local electric current density, $\mathbf{I}_{\mathbf{r}t} = e\mathbf{i}_{\mathbf{r}t}$, (see Eq. (11.2) which defines the local flow density $\mathbf{i}_{\mathbf{r}t}$) is determined by $\Delta f_{\mathbf{r}\mathbf{p}t}$ and equal to $(2e/V)\sum_{\mathbf{p}}\mathbf{v}_{\mathbf{p}}(\mathbf{v}_{\mathbf{p}} \cdot \mathbf{g}_{\mathbf{r}t})$. Therefore, the current density is expressed through the gradients of the electrochemical potential and temperature:

$$\mathbf{I}_{\mathbf{r}t} = \frac{2e^2}{dm}\int d\varepsilon\rho_{\mathcal{D}}(\varepsilon)\varepsilon\tau_{tr}(\varepsilon)\frac{\partial f_{\mathbf{r}t}^{(s)}(\varepsilon)}{\partial\varepsilon}\left(\nabla_{\mathbf{r}}w_{\mathbf{r}t} + \frac{\varepsilon - \mu_{\mathbf{r}t}}{eT_{\mathbf{r}t}}\nabla_{\mathbf{r}}T_{\mathbf{r}t}\right)$$
$$= -\sigma_{\mathbf{r}t}(\nabla_{\mathbf{r}}w_{\mathbf{r}t} + \alpha_{\mathbf{r}t}\nabla_{\mathbf{r}}T_{\mathbf{r}t}), \qquad (8)$$

where d is the dimensionality of the problem and $\rho_{\mathcal{D}}(\varepsilon)$ is the density of states; see Eqs. (5.3) and (5.26). The local conductivity $\sigma_{\mathbf{r}t}$ is introduced according to Eq. (8.27), where the quasi-equilibrium function $f_{\mathbf{r}t}^{(s)}(\varepsilon)$ stands instead of the equilibrium function $f_{\varepsilon}^{(eq)}$. The coefficient

$$\alpha_{\mathbf{r}t} = \frac{2e}{dm\sigma_{\mathbf{r}t}}\int_0^\infty d\varepsilon\rho_{\mathcal{D}}(\varepsilon)\tau_{tr}(\varepsilon)\varepsilon\frac{\varepsilon - \mu_{\mathbf{r}t}}{T_{\mathbf{r}t}}\left[-\frac{\partial f_{\mathbf{r}t}^{(s)}(\varepsilon)}{\partial\varepsilon}\right] \qquad (9)$$

introduced in Eq. (8) is called the thermo-electromotive force. It is expressed through the local temperature and chemical potential (problem 7.12). In the absence of temperature gradients, the current is determined entirely by the gradient of the electrochemical potential. In equilibrium, when the current and the temperature gradient are zero everywhere in the sample, the electrochemical potential is constant.

The expression for the current density can be written in another way, through the gradient of the electron density. Let us introduce the diffusion coefficient $D_{\mathbf{r}t}$ according to Einstein's relation

$$\sigma_{\mathbf{r}t} = e^2 D_{\mathbf{r}t}\frac{\partial n_{\mathbf{r}t}}{\partial\mu_{\mathbf{r}t}}, \qquad (10)$$

where the local electron density $n_{\mathbf{r}t}$ is defined by Eq. (11.2) and determined only by the symmetric part of the distribution function. For degenerate electrons, when $\partial f_{\mathbf{r}t}^{(s)}(\varepsilon)/\partial\varepsilon$ is non-zero only in a narrow energy interval near the local chemical potential $\mu_{\mathbf{r}t}$, the diffusion coefficient is given by a simple relation, $D_{\mathbf{r}t} = v_{\mathbf{r}t}^2 \tau_{tr}(\mu_{\mathbf{r}t})/d$, where $v_{\mathbf{r}t} = \sqrt{2\mu_{\mathbf{r}t}/m}$ is the local Fermi velocity. The expression of the flow in terms of the local density and temperature is obtained from Eq. (8):

$$\mathbf{i}_{\mathbf{r}t} = -D_{\mathbf{r}t}\nabla_{\mathbf{r}}n_{\mathbf{r}t} + \mathbf{u}_{\mathbf{r}t}n_{\mathbf{r}t} - \beta_{\mathbf{r}t}\nabla_{\mathbf{r}}T_{\mathbf{r}t} \ , \tag{11}$$

where the drift velocity

$$\mathbf{u}_{\mathbf{r}t} = \frac{\sigma_{\mathbf{r}t}\mathbf{E}_{\mathbf{r}t}}{en_{\mathbf{r}t}} = \frac{e\tau_{tr}(\mathbf{r},t)}{m}\mathbf{E}_{\mathbf{r}t} \tag{12}$$

is introduced to describe the response of the electron system to the electric field. The proportionality coefficient $|e|\tau_{tr}(\mathbf{r},t)/m$, where the local averaged transport time $\tau_{tr}(\mathbf{r},t)$ is introduced according to Eq. (8.27) with $f_{\mathbf{r}t}^{(s)}(\varepsilon)$ in place of $f_{\varepsilon}^{(eq)}$, is called the mobility. The drift velocity for electrons ($e < 0$) is directed against the electric field. The coefficient $\beta_{\mathbf{r}t}$ in Eq. (11) describes the response of electrons to the temperature gradient in the absence of drift and diffusion:

$$\beta_{\mathbf{r}t} = \frac{\sigma_{\mathbf{r}t}}{e^2}\left[e\alpha_{\mathbf{r}t} - \left(\frac{\partial n_{\mathbf{r}t}}{\partial T_{\mathbf{r}t}}\right)\bigg/\left(\frac{\partial n_{\mathbf{r}t}}{\partial \mu_{\mathbf{r}t}}\right)\right]. \tag{13}$$

According to Eq. (11), the flow of electrons is represented as a sum of the diffusion (due to the density gradient), drift, and thermoinduced currents. Substituting Eq. (11) into the continuity equation (11.5), we obtain

$$\frac{\partial n_{\mathbf{r}t}}{\partial t} = \nabla_{\mathbf{r}} \cdot (D_{\mathbf{r}t}\nabla_{\mathbf{r}} - \mathbf{u}_{\mathbf{r}t})n_{\mathbf{r}t} + \nabla_{\mathbf{r}} \cdot (\beta_{\mathbf{r}t}\nabla_{\mathbf{r}}T_{\mathbf{r}t}). \tag{14}$$

This equation describes the evolution of the electron density distribution in space and time.

Neglecting the drift term and temperature gradient, one can easily solve Eq. (14) in the linear approximation, assuming that $n_{\mathbf{r}t} = n + \Delta n_{\mathbf{r}t}$, where n is the equilibrium density and $\Delta n_{\mathbf{r}t}$ is the non-equilibrium correction to the density caused by small deviations of the chemical potential and temperature from their equilibrium values. This allows one to consider $D_{\mathbf{r}t}$ as a constant determined by the density n and temperature T. As a result, Eq. (14) is reduced to the linear diffusion equation $\partial\Delta n_{\mathbf{r}t}/\partial t = D\nabla_{\mathbf{r}}^2\Delta n_{\mathbf{r}t}$. With the initial condition $\Delta n_{\mathbf{r}t=0} = \Delta n_0\delta(\mathbf{r})$ describing a perturbation of the density at $\mathbf{r} = 0$ and $t = 0$, the solution of this equation is well-known:

$$\Delta n_{\mathbf{r}t} = \frac{\Delta n_0}{(4\pi Dt)^{d/2}}\exp\left(-\frac{r^2}{4Dt}\right). \tag{15}$$

It can be checked by a direct substitution into the diffusion equation. The solution (15) shows isotropic expansion of the narrow initial distribution due to diffusion. This solution describes electrically neutral systems, and cannot be applied to electron systems (or other systems of charged particles) because a redistribution of charged particles inevitably creates electric fields, and, consequently, drift currents, which are not taken into account in Eq. (15). To consider them, one has to solve Eq. (14) together with the Poisson equation. With the use of Eq. (12), the Poisson equation is rewritten for the case of 3D electrons as

$$\nabla_{\mathbf{r}} \cdot \frac{\epsilon_{\mathbf{r}} \mathbf{u}_{\mathbf{r}t}}{\tau_{tr}(\mathbf{r}, t)} = \frac{4\pi e^2}{m}(n_{\mathbf{r}t} - N_{\mathbf{r}}^{(+)}), \tag{16}$$

where we have introduced the coordinate-dependent density $N_{\mathbf{r}}^{(+)}$ of positive background charges and the dielectric permittivity $\epsilon_{\mathbf{r}}$. In the linear approximation (small deviations from equilibrium) and in the absence of the temperature gradient, Eqs. (14) and (16) form a closed system of two differential equations for non-equilibrium part of electron density, $\Delta n_{\mathbf{r}t}$, and drift velocity $\mathbf{u}_{\mathbf{r}t}$. The latter can be excluded, and we obtain a single equation,

$$\frac{\partial \Delta n_{\mathbf{r}t}}{\partial t} = \left(D\nabla_{\mathbf{r}}^2 - \tau_M^{-1}\right)\Delta n_{\mathbf{r}t}, \qquad \tau_M^{-1} = \omega_p^2 \tau_{tr}, \tag{17}$$

written under the assumption that $N_{\mathbf{r}}^{(+)}$ and $\epsilon_{\mathbf{r}}$ do not depend on $\mathbf{r}$. To exclude $N^{(+)}$, we have used the electric neutrality condition $n = N^{(+)}$. The characteristic time introduced in Eq. (17) is called the Maxwell relaxation time. It determines the temporal scale of the relaxation of charge density perturbations. This time is expressed through the plasma frequency introduced in the end of Sec. 8 and the averaged transport time defined by Eq. (8.27). For degenerate electrons, one may express the Maxwell relaxation time through the conductivity and dielectric permittivity, as $\tau_M = \epsilon/4\pi\sigma$. The diffusion length $L_D = \sqrt{\tau_M D}$ associated with this relaxation time determines the relaxation length of the charge density perturbation. For example, if one maintains a non-equilibrium density $n + \Delta n_0$ at the boundary $x = 0$, the perturbation Δn_x decreases inside the sample as $e^{-|x|/L_D}$. An example of non-stationary solutions of Eq. (17) is presented in problem 7.13.

In the general case, Eqs. (14) and (16) are non-linear. They do not form a closed system of equations even when $\nabla_{\mathbf{r}} T_{\mathbf{r}t} = 0$, because the diffusion coefficient and averaged transport time depend on the chemical potential and temperature (or, equivalently, on the electron density and temperature). Therefore, to complete the description, we should add one more equation describing the balance of temperature. It can be obtained from the symmetric part of the kinetic equation, i.e., from Eq. (4), where $\Delta f_{\mathbf{rp}t} = \mathbf{v_p} \cdot \mathbf{g}_{\mathbf{r}t}(\varepsilon_p)$ is expressed through $f_{\mathbf{r}t}^{(s)}$. However, since we have already written $f_{\mathbf{r}t}^{(s)}$ in terms of the local chemical potential and temperature, it is more convenient to use the third balance equation of Eq. (11.11) instead of using Eq. (4). The right-hand side of this balance equation is reduced to $\delta_{\alpha\beta}(2/3V) \sum_{\mathbf{p}} \mathbf{v_p^2} J_{e,ph}(f^{(s)}|\mathbf{r}\varepsilon_p t)$. After multiplying the third equation of Eq. (11.11) by $m/2$, we sum it over the coordinate index $\alpha = \beta$ and obtain the energy balance equation (problem 7.14)

$$\frac{\partial \mathcal{E}_{\mathbf{r}t}}{\partial t} + \nabla_{\mathbf{r}} \cdot \mathbf{G}_{\mathbf{r}t} - \mathbf{I}_{\mathbf{r}t} \cdot \mathbf{E}_{\mathbf{r}t} + P_{\mathbf{r}t} = 0. \tag{18}$$

The magnetic-field term has disappeared from this equation. In Eq. (18), $\mathcal{E}_{\mathbf{r}t} = (2/V) \sum_{\mathbf{p}} \varepsilon_p f_{\mathbf{rp}t}$ is the local energy density introduced in Sec. 11,

$$\mathbf{G}_{\mathbf{r}t} = \frac{2}{V} \sum_{\mathbf{p}} \mathbf{v_p} \varepsilon_p f_{\mathbf{rp}t} = \frac{2}{dm} \int_0^\infty d\varepsilon \rho_{\mathcal{D}}(\varepsilon) \varepsilon^2 \mathbf{g}_{\mathbf{r}t}(\varepsilon) \tag{19}$$

is the energy flow density for electrons (compare it to the energy flow density for bosons given by Eq. (19.32)), and

$$P_{\mathbf{r}t} = -\frac{2}{V} \sum_{\mathbf{p}} \varepsilon_p J_{e,ph}(f^{(s)}|\mathbf{r}\varepsilon_p t) = -\int_0^\infty d\varepsilon \rho_{\mathcal{D}}(\varepsilon) \varepsilon J_{e,ph}(f^{(s)}|\mathbf{r}\varepsilon t) \tag{20}$$

is the energy density absorbed by the phonon system in unit time (the power loss term).

Equation (18) shows us that the local energy of electrons changes with time due to the energy transfer in space described by the term $\nabla_{\mathbf{r}} \cdot \mathbf{G}_{\mathbf{r}t}$, the power gained by the electrons moving in the electric field, $(\mathbf{I}_{\mathbf{r}t} \cdot \mathbf{E}_{\mathbf{r}t})$, and the power transferred to phonons, $P_{\mathbf{r}t}$. For strongly degenerate 3D electrons, $\mathcal{E}_{\mathbf{r}t} = (3/5)n_{\mathbf{r}t}\mu_{\mathbf{r}t} = (3/5)n_{\mathbf{r}t}\hbar^2(3\pi^2 n_{\mathbf{r}t})^{2/3}/2m$, while for non-degenerate electrons $\mathcal{E}_{\mathbf{r}t} = (3/2)n_{\mathbf{r}t}T_{\mathbf{r}t}$. In zero magnetic field, when Eq. (7) is valid, the energy flow density can be written through the temperature gradient and current as (problem 7.15)

$$\mathbf{G}_{\mathbf{r}t} = (\alpha_{\mathbf{r}t}T_{\mathbf{r}t} + \mu_{\mathbf{r}t}/e)\mathbf{I}_{\mathbf{r}t} - \kappa_{\mathbf{r}t}\nabla_{\mathbf{r}}T_{\mathbf{r}t}, \tag{21}$$

where κ is the thermal conductivity of the electron system:

$$\kappa_{\mathbf{r}t} = \frac{2}{dm} \int_0^\infty d\varepsilon \rho_{\mathcal{D}}(\varepsilon) \tau_{tr}(\varepsilon) \varepsilon \frac{(\varepsilon - \mu_{\mathbf{r}t})^2}{T_{\mathbf{r}t}} \left[-\frac{\partial f_{\mathbf{r}t}^{(s)}(\varepsilon)}{\partial \varepsilon} \right] - \alpha_{\mathbf{r}t}^2 \sigma_{\mathbf{r}t} T_{\mathbf{r}t}. \tag{22}$$

The thermal conductivity of degenerate electron gas is determined by the first term on the right-hand side of Eq. (22), because the second one contains an extra smallness of $(T/\mu)^2$, and we obtain

$$\kappa_{\mathbf{rt}} = \frac{\pi^2}{3e^2}\sigma_{\mathbf{rt}}T_{\mathbf{rt}}. \tag{23}$$

According to this equation, the ratio of the thermal conductivity to the electric conductivity is equal to the temperature multiplied by a universal factor $\pi^2/3e^2$. For non-degenerate electrons, both terms in the expression (22) are essential (see problem 7.12, where α is calculated) and, assuming that the product $\rho_\mathcal{D}(\varepsilon)\tau_{tr}(\varepsilon)$ is energy-independent, we find that the thermal conductivity is equal to $2\sigma_{\mathbf{rt}}T_{\mathbf{rt}}/e^2$.

Let us consider the power loss term $P_{\mathbf{rt}}$. Substituting the collision integral defined by Eqs. (34.26) and (34.27) into Eq. (20), we find

$$P_{\mathbf{rt}} = \frac{4\pi}{V}\sum_{\mathbf{pq}}|C_q|^2\omega_q\delta(\varepsilon_{\mathbf{p}-\hbar\mathbf{q}} - \varepsilon_p + \hbar\omega_q) \tag{24}$$

$$\times\left\{(N_q+1)f_{\mathbf{rt}}^{(s)}(\varepsilon_p)[1 - f_{\mathbf{rt}}^{(s)}(\varepsilon_{\mathbf{p}-\hbar\mathbf{q}})] - N_q f_{\mathbf{rt}}^{(s)}(\varepsilon_{\mathbf{p}-\hbar\mathbf{q}})[1 - f_{\mathbf{rt}}^{(s)}(\varepsilon_p)]\right\}$$

for 3D electrons. Since the property of isotropy of $f_{\mathbf{rt}}^{(s)}(\varepsilon_p)$ has not been used in the derivation of Eq. (24), one may replace $f_{\mathbf{rt}}^{(s)}(\varepsilon_p)$ in this equation by the exact distribution function $f_{\mathbf{rp}t}$ (its anisotropic part $\Delta f_{\mathbf{rp}t}$ does not contribute to the power losses). Using Eq. (2), it is easy to show that the term $\{\dots\}$ in Eq. (24) is proportional to $[\exp(\hbar\omega_q/T) - \exp(\hbar\omega_q/T_{\mathbf{rt}})]$, where T is the equilibrium temperature. Therefore, the energy exchange between electrons and phonons occurs only if their effective temperatures are different. Considering the electron scattering by acoustic phonons in the quasielastic approximation (see problem 7.8 for the 3D case), we calculate the sum over $\mathbf{q}$ and represent Eq. (24) as

$$P_{\mathbf{rt}} = \nu_{\mathbf{rt}}^{(e)}n_{\mathbf{rt}}(T_{\mathbf{rt}} - T), \tag{25}$$

$$\nu_{\mathbf{rt}}^{(e)} = \frac{2^{3/2}\mathcal{D}^2 m^{5/2}}{\pi\hbar^4\rho n_{\mathbf{rt}}}\int_0^\infty d\varepsilon\,\rho_{3D}(\varepsilon)\varepsilon^{3/2}\left[-\frac{\partial f_{\mathbf{rt}}^{(s)}(\varepsilon)}{\partial\varepsilon}\right].$$

This equation defines the energy relaxation rate $\nu_{\mathbf{rt}}^{(e)}$. For degenerate (A) and non-degenerate (B) electrons, we have

$$\nu_{\mathbf{rt}}^{(e)} = \frac{3^{4/3}m^2\mathcal{D}^2}{\pi^{1/3}\hbar^3\rho}n_{\mathbf{rt}}^{1/3} = \frac{3ms_l^2}{\tau_{LA}(\mu_{\mathbf{rt}})T}, \qquad (A),$$

$$\nu_{\mathbf{rt}}^{(e)} = \frac{2^{7/2}m^{5/2}\mathcal{D}^2}{\pi^{3/2}\hbar^4\rho}T_{\mathbf{rt}}^{1/2} = \frac{8ms_l^2}{\pi^{1/2}\tau_{LA}(T_{\mathbf{rt}})T}, \quad (B). \tag{26}$$

The transport time $\tau_{LA}(\varepsilon)$ for quasielastic electron-phonon scattering is defined in problem 7.8.

In a stationary and homogeneous system, $T_{\mathbf{rt}}$ and $\mu_{\mathbf{rt}}$ should be replaced by the constant electron temperature T_e and chemical potential μ_e, respectively, and Eq. (18) is reduced to $(\mathbf{I} \cdot \mathbf{E}) = P$. Expressing the current through the conductivity, we use Eq. (25) for the power loss term and find the energy balance equation in the form

$$T_e = T + [eE\tau(\mu_e, T_e)]^2/m, \quad \tau(\mu_e, T_e) = \sqrt{\tau_{tr}/\nu^{(e)}}, \quad (27)$$

where both the averaged transport time τ_{tr} and the energy relaxation rate $\nu^{(e)}$ depend on T_e and μ_e because of the dependence of $f^{(s)}(\varepsilon_p)$ on these parameters. In the case of small deviations from equilibrium ("warm" electrons), one may use the equilibrium values of τ_{tr} and $\nu^{(e)}$ in Eq. (27). As a result, this equation demonstrates that the electric field E increases the effective electron temperature above the equilibrium temperature T according to E^2 law. If the deviations from equilibrium are not small, Eq. (27) becomes non-linear. It should be solved together with the electric neutrality condition $n(\mu_e, T_e) = N^{(+)}$, which follows from the Poisson equation (16) in the homogeneous case and allows one to express μ_e through T_e. If the effective temperature of electrons is not small in comparison to the optical phonon energy, or the electric field is strong enough to move a considerable amount of electrons to the active region $\varepsilon > \hbar\omega_{LO}$, the optical-phonon scattering also contributes to the energy losses (problem 7.16).

Equations (14), (16), and (18), where all terms are expressed through the quantities $n_{\mathbf{rt}}$, $\mathbf{u}_{\mathbf{rt}}$, and $T_{\mathbf{rt}}$, form a closed system which can be used for determining these quantities in non-homogeneous electron systems. In the general case, this system is non-linear and rather complicated. A considerable degree of simplifications is achieved for the stationary case in the linear approximation, when one considers small non-homogeneous deviations of the electron density and temperature ($\Delta n_{\mathbf{r}}$ and $\Delta T_{\mathbf{r}}$, respectively) from the equilibrium values n and T. Thus, the drift velocity $\mathbf{u}_{\mathbf{r}}$ can be excluded from Eq. (14), and the latter is reduced to

$$\left(D\nabla_{\mathbf{r}}^2 - \tau_M^{-1}\right)\Delta n_{\mathbf{r}} + \beta\nabla_{\mathbf{r}}^2\Delta T_{\mathbf{r}} = 0, \quad (28)$$

and Eq. (18), where the linearity of the problem leads to $P_{\mathbf{r}} = n\nu^{(e)}\Delta T_{\mathbf{r}}$, becomes a closed equation for $\Delta T_{\mathbf{r}}$:

$$\left(\kappa\nabla_{\mathbf{r}}^2 - n\nu^{(e)}\right)\Delta T_{\mathbf{r}} = 0. \quad (29)$$

According to Eq. (29), the temperature changes on a characteristic length $L_T = \sqrt{\kappa/n\nu^{(e)}}$, while the spatial distribution of the density deviation $\Delta n_\mathbf{r}$ is characterized by two lengths, $L_D = \sqrt{\tau_M D}$ and L_T. Since L_T, in contrast to L_D, is determined by slow energy relaxation, one has $L_T \gg L_D$ for typical parameters of electron-phonon systems. This means that large-scale inhomogeneities of the electron density can exist only in the presence of temperature gradients.

To complete this section, we discuss the situation when both energy and momentum relaxation are controlled by the electron-electron scattering mechanism. In this case, we search for the distribution function $f_{\mathbf{rp}t}$ in the form of shifted Fermi distribution (31.25), where we put $\mu_{\mathbf{r}t}$, $T_{\mathbf{r}t}$, and $\mathbf{u}_{\mathbf{r}t}$ instead of μ_e, T_e, and $\mathbf{u}$. To determine the temporal and spatial dependence of these parameters, one can use two first equations of the system (11.11) together with Eq. (18) (we remind that the latter follows from the third equation of this system). The quantities $\mathbf{i}_{\mathbf{r}t}$, $Q_{\mathbf{r}t}^{\alpha\beta}$, $\mathcal{E}_{\mathbf{r}t}$, and $\mathbf{G}_{\mathbf{r}t}$ standing in these three equations are determined with the aid of the shifted Fermi distribution as

$$\mathbf{i} = n\mathbf{u}, \qquad Q^{\alpha\beta} = nu_\alpha u_\beta + \delta_{\alpha\beta}\frac{2\mathcal{E}^*}{3m},$$

$$\mathcal{E} = n\frac{mu^2}{2} + \mathcal{E}^*, \qquad \mathbf{G} = \mathbf{u}\left(n\frac{mu^2}{2} + \frac{5}{3}\mathcal{E}^*\right), \tag{30}$$

where the time and coordinate indices are omitted for brevity. In this equation, $\mathcal{E}_{\mathbf{r}t}^*$ is the energy density calculated for a non-shifted Fermi distribution (like the one given by Eq. (2)), where the chemical potential is replaced by $\mu_{\mathbf{r}t}^* = \mu_{\mathbf{r}t} + mu_{\mathbf{r}t}^2/2$. The electron density is calculated in a similar fashion (see Eq. (31.25) and the discussion thereafter), and we have

$$\left|\begin{array}{c} n_{\mathbf{r}t} \\ \mathcal{E}_{\mathbf{r}t}^* \end{array}\right| = \frac{2}{V}\sum_\mathbf{p}\left(\exp\frac{\varepsilon_p - \mu_{\mathbf{r}t}^*}{T_{\mathbf{r}t}} + 1\right)^{-1}\left|\begin{array}{c} 1 \\ \varepsilon_p \end{array}\right|. \tag{31}$$

Employing Eq. (31), one can express the shifted chemical potential $\mu_{\mathbf{r}t}^*$ and energy density $\mathcal{E}_{\mathbf{r}t}^*$ through the electron density and temperature. As a result, $\mathbf{i}$, $Q^{\alpha\beta}$, $\mathcal{E}$, and $\mathbf{G}$ are determined by $n_{\mathbf{r}t}$, $\mathbf{u}_{\mathbf{r}t}$, and $T_{\mathbf{r}t}$. Note, however, that the expressions for $\mathbf{i}$ and $\mathbf{G}$ are different from Eqs. (11) and (21), since the antisymmetric part of the distribution function (31.25) differs from $\Delta f_{\mathbf{rp}t}$ introduced above. The power density $P_{\mathbf{r}t}$ transmitted to phonons and the velocity density $\mathbf{U}_{\mathbf{r}t} = -(2/V)\sum_\mathbf{p}\mathbf{v}_\mathbf{p}[J_{im}(f|\mathbf{rp}t) + J_{e,ph}(f|\mathbf{rp}t)]$ transmitted to impurities and phonons in unit time are also expressed through $n_{\mathbf{r}t}$, $\mathbf{u}_{\mathbf{r}t}$, and $T_{\mathbf{r}t}$. In the case of weak anisotropy (small $\mathbf{u}$), the expression for $P_{\mathbf{r}t}$ is the same as given above. To make the problem formally closed, we need to add the Maxwell equations to the three balance equation discussed above, in order to express the electric and magnetic fields through the charge and current densities. Under the conditions (1), one may neglect the inhomogeneous part of the magnetic field $\mathbf{H}_{\mathbf{r}t}$ and use only the Poisson equation written as $\nabla_{\mathbf{r}} \cdot \epsilon_{\mathbf{r}} \mathbf{E}_{\mathbf{r}t} = 4\pi e(n_{\mathbf{r}t} - N_{\mathbf{r}}^{(+)})$.

37. Heating under High-Frequency Field

The character of non-equilibrium distribution of electrons in the high-frequency field $\mathbf{E}\cos\omega t$ is different from that described in previous sections, because of equivalence of the directions along $\mathbf{E}$ and opposite to $\mathbf{E}$. Moreover, if $\hbar\omega$ exceeds the mean energy of electrons, $\bar{\varepsilon}$, one should take into account the discrete nature of the energy absorbed by electrons in the photoinduced transitions; see Sec. 10. We start our consideration from the classical case $\hbar\omega \ll \bar{\varepsilon}$, when the distribution is determined by the kinetic equation (9.34) for the homogeneous electron system, with the force term $\mathbf{F}_t = e\mathbf{E}\cos\omega t$. Introducing the canonical momentum $\mathbf{P}$ instead of $\mathbf{p} - (e/\omega)\mathbf{E}\sin\omega t$, we write this equation for the distribution function $f_{\mathbf{P}t}$ by using also the condition of periodicity over the time variable:

$$\frac{\partial f_{\mathbf{P}t}}{\partial t} = \sum_k J_k(f|\mathbf{P}t), \qquad f_{\mathbf{P}t+2\pi/\omega} = f_{\mathbf{P}t}. \tag{1}$$

The transformation of the collision integrals for electron-impurity and electron-phonon scattering (see Eqs. (8.8) and (34.27), respectively) to this representation shows us that the energy conservation terms in the transition probabilities contain the expressions $\varepsilon_P - \varepsilon_{P'} - (\mathbf{P}' - \mathbf{P}) \cdot \mathbf{v}_\omega \sin\omega t$, where the characteristic velocity $\mathbf{v}_\omega$ introduced in Sec. 10 describes the field contribution. As for the electron-electron collision integral, the field contribution does not enter the energy conservation terms because the total momentum is conserved in the collisions. Therefore, $J_{ee}(f|\mathbf{P}t)$ is given by the expression (31.21).

In the high-frequency limit, when ω exceeds the characteristic scattering rates, the distribution function $f_{\mathbf{P}t}$ is written as a sum of the contribution $\overline{f}_{\mathbf{P}}$, averaged over the period, and a small correction; see Eq. (10.10). The averaged contribution satisfies Eq. (10.11), whose left-hand side contains the sum of the collision integrals discussed above. The delta-functions $\delta(E)$ in the collision integrals for electron-impurity and electron-phonon scattering standing in this equation should be replaced by the averaged expressions

$$\Delta_0(E) = \frac{\omega}{2\pi}\int_{-\pi/\omega}^{\pi/\omega} dt\,\delta(E - A\sin\omega t) = \int_{-\infty}^{\infty} \frac{d\tau}{2\pi} e^{iE\tau}\mathcal{J}_0(A\tau)$$

$$= \begin{cases} \pi^{-1}(A^2 - E^2)^{-1/2}, & E^2 < A^2 \\ 0, & E^2 > A^2 \end{cases}, \tag{2}$$

where $A = (\mathbf{P}' - \mathbf{P}) \cdot \mathbf{v}_\omega$ is determined by the high-frequency field and transmitted momentum, and $\mathcal{J}_0(x)$ is the Bessel function; see Eq. (10.15). The function $\Delta_0(E)$ is independent of the signs of E and A. In weak fields ($A \to 0$), the expression (2) is reduced to $\delta(E)$, since any smooth function F_E satisfies the integral relation $\int dE F_E \Delta_0(E)|_{A\to 0} = F_0$. After averaging the collision integral standing in Eq. (8.7) over the period, we obtain

$$\frac{\omega}{2\pi} \int_{-\pi/\omega}^{\pi/\omega} dt J_{im}(\overline{f}|\mathbf{P}t) = \frac{2\pi}{\hbar V} \sum_{\mathbf{q}} w(q) \Delta_0(\varepsilon_P - \varepsilon_{\mathbf{P}+\hbar\mathbf{q}})(\overline{f}_{\mathbf{P}+\hbar\mathbf{q}} - \overline{f}_\mathbf{P}), \quad (3)$$

and a similar averaging of the collision integral (34.26) gives us

$$\frac{\omega}{2\pi} \int_{-\pi/\omega}^{\pi/\omega} dt J_{e,ph}(\overline{f}|\mathbf{P}t) = \frac{2\pi}{\hbar} \sum_{\mathbf{q}} |C_q|^2 \{\Delta_0(\varepsilon_P - \varepsilon_{\mathbf{P}+\hbar\mathbf{q}} + \hbar\omega_q)$$
$$\times [(N_q + 1)\overline{f}_{\mathbf{P}+\hbar\mathbf{q}} - N_q\overline{f}_\mathbf{P}] + \Delta_0(\varepsilon_P - \varepsilon_{\mathbf{P}-\hbar\mathbf{q}} - \hbar\omega_q) \quad (4)$$
$$\times [N_q\overline{f}_{\mathbf{P}-\hbar\mathbf{q}} - (N_q + 1)\overline{f}_\mathbf{P}]\}.$$

The collision integral of electron-phonon scattering is written here for the case of non-degenerate electron gas. Its generalization to the case of arbitrary degeneracy is obvious. It is easy to check that the electron density is conserved in the presence of collisions in the high-frequency field (problem 7.17).

Denoting the averaged collision integrals defined by Eqs. (3) and (4) as $\overline{J}_{im}(\overline{f}|\mathbf{P})$ and $\overline{J}_{e,ph}(\overline{f}|\mathbf{P})$, one can write the stationary kinetic equation as

$$\overline{J}_{im}(\overline{f}|\mathbf{P}) + \overline{J}_{e,ph}(\overline{f}|\mathbf{P}) + J_{ee}(\overline{f}|\mathbf{P}) = 0. \quad (5)$$

After multiplying Eq. (5) by the energy ε_P, we integrate it over the canonical momentum and obtain the energy balance equation

$$\frac{2}{V} \sum_{k\mathbf{P}} \varepsilon_P \frac{\omega}{2\pi} \int_{-\pi/\omega}^{\pi/\omega} dt J_k(\overline{f}|\mathbf{P}t) = 0, \quad (6)$$

where $k = im$ and e, ph, and the electron-electron collision integral is dropped out of Eq. (6) because of Eq. (31.23). The impurity contribution into the energy balance equation in transformed with the aid of the substitution $\varepsilon_P(\overline{f}_{\mathbf{P}'} - \overline{f}_\mathbf{P}) = (\varepsilon_{P'} - \varepsilon_P)\overline{f}_\mathbf{P}$. Next, $(\varepsilon_{P'} - \varepsilon_P)$ is replaced by the time-dependent contribution $(\mathbf{P} - \mathbf{P}') \cdot \mathbf{v}_\omega \sin\omega t$ from the argument of the δ-function of Eq. (2). As a result, the impurity contribution into the energy balance equation becomes

$$\frac{\omega}{2\pi} \int_{-\pi/\omega}^{\pi/\omega} dt \sin\omega t \frac{2}{V} \sum_{\mathbf{P}} (\mathbf{v}_\omega \cdot \mathbf{P}) \frac{2\pi}{\hbar V} \sum_{\mathbf{P}'} w\left(\frac{|\mathbf{P} - \mathbf{P}'|}{\hbar}\right) \Delta_0(\varepsilon_P - \varepsilon_{P'})$$

$$\times(\overline{f}_{\mathbf{P}} - \overline{f}_{\mathbf{P}'}) = -\frac{\omega}{2\pi} \int_{-\pi/\omega}^{\pi/\omega} dt \sin \omega t \frac{e}{\omega} \mathbf{E} \cdot \frac{2}{V} \sum_{\mathbf{P}} \mathbf{v}_{\mathbf{P}} J_{im}(\overline{f}|\mathbf{P}t). \quad (7)$$

Let us show that the contribution (7) is reduced to the averaged absorbed power $\mathcal{P}_{im} = \overline{\mathbf{I}_t \cdot \mathbf{E}_t} = (\omega/2\pi) \int_{-\pi/\omega}^{\pi/\omega} dt(\mathbf{E} \cdot \mathbf{I}_t) \cos \omega t$, where the current density $\mathbf{I}_t$ is expressed in the usual way, through the high-frequency correction $\Delta f_{\mathbf{P}t}$ to the distribution function in the presence of elastic scattering; see Eqs. (10.21) and (10.22). This correction is determined by Eq. (10.14). Taking into account that the power $\mathcal{P}_{im}$ is proportional to $(\omega/2\pi) \int_{-\pi/\omega}^{\pi/\omega} dt \Delta f_{\mathbf{P}t} \cos \omega t$, and using Eqs. (10.14) and (10.16), we see that the expression (7) is equal to $\mathcal{P}_{im}$. Next, considering the phonon contribution into the energy balance equation (6), we replace $\varepsilon_P - \varepsilon_{\mathbf{P}+\hbar\mathbf{q}}$ by $\hbar(\mathbf{q} \cdot \mathbf{v}_\omega \sin \omega t - \omega_q)$. The first term in this expression gives us the contribution to the averaged absorbed power due to electron-phonon scattering, and we denote this power as $\mathcal{P}_{ph}$. Now we may rewrite Eq. (6) as

$$\mathcal{P}_{im} + \mathcal{P}_{ph} = \frac{4\pi}{V} \sum_{\mathbf{Pq}} |C_q|^2 \omega_q \Delta_0(\varepsilon_{\mathbf{P}-\hbar\mathbf{q}} - \varepsilon_P + \hbar\omega_q)$$

$$\times \left[(N_q + 1)\overline{f}_{\mathbf{P}} - N_q \overline{f}_{\mathbf{P}-\hbar\mathbf{q}} \right] \equiv P^{(\omega)}. \quad (8)$$

The right-hand side of this equation, denoted as $P^{(\omega)}$, describes the power transmitted to phonons. In the static limit, when $\mathbf{P} = \mathbf{p}$, $\Delta_0(E) = \delta(E)$, and $\overline{f}_{\mathbf{P}} = f_{\mathbf{p}}$, this power is given by the expression (36.24), if we write the latter for non-degenerate electrons and substitute there $f_{\mathbf{p}}$ instead of $f_{\mathbf{rt}}^{(s)}(\varepsilon_p)$. Thus, $P^{(\omega)}$ differs from the power loss term P in the balance equation (36.18) because of an additional field dependence on the right-hand side of Eq. (8), where the energy conservation law is modified by the high-frequency field, according to Eq. (2).

The consideration we just presented can be easily repeated for the quantum region of frequencies. Under the condition $\hbar\omega \geq \overline{\varepsilon}$, Eq. (2) is replaced by Eq. (10.18), which takes into account the discrete energy $\hbar\omega$ (problem 7.18). Substituting Eqs. (3) and (4) with $\Delta_0(E)$ from Eq. (10.18) into Eq. (6), we obtain

$$\frac{4\pi}{V^2} \sum_{\mathbf{Pq}} w(q)\overline{f}_{\mathbf{P}} \sum_{k=-\infty}^{\infty} k\omega \left[J_k \left(\frac{\mathbf{q} \cdot \mathbf{v}_\omega}{\omega} \right) \right]^2 \delta(\varepsilon_{\mathbf{P}-\hbar\mathbf{q}} - \varepsilon_P - k\hbar\omega)$$

$$+ \frac{4\pi}{V} \sum_{\mathbf{Pq}} |C_q|^2 \left[(N_q + 1)\overline{f}_{\mathbf{P}} - N_q \overline{f}_{\mathbf{P}-\hbar\mathbf{q}} \right] \sum_{k=-\infty}^{\infty} (k\omega - \omega_q) \quad (9)$$

$$\times \left[J_k \left(\frac{\mathbf{q} \cdot \mathbf{v}_\omega}{\omega} \right) \right]^2 \delta(\varepsilon_{\mathbf{P}-\hbar\mathbf{q}} - \varepsilon_P + \hbar\omega_q - k\hbar\omega) = 0,$$

where the terms proportional to $k\omega$ in the impurity and phonon contributions are the absorbed powers $\mathcal{P}_{im}$ and $\mathcal{P}_{ph}$, respectively (problem 7.19), while the term proportional to ω_q is equal to $-P^{(\omega)}$ defined by Eq. (8), if we substitute the quantum expression of $\Delta_0(E)$ there. Once again, the population factor is written for the non-degenerate case. To rewrite Eq. (9) for the case of arbitrary degeneracy, one should replace the factor $(N_q+1)\overline{f}_{\mathbf{P}} - N_q\overline{f}_{\mathbf{P}-\hbar\mathbf{q}}$ by $(N_q+1)\overline{f}_{\mathbf{P}}(1-\overline{f}_{\mathbf{P}-\hbar\mathbf{q}}) - N_q\overline{f}_{\mathbf{P}-\hbar\mathbf{q}}(1-\overline{f}_{\mathbf{P}})$. The quantum expression for the absorbed power shows us that the energy of electromagnetic field is absorbed in quanta $k\hbar\omega$ ($k \neq 0$). Such multi-photon processes have been considered in Sec. 10 for the electrons interacting with a random static potential, which corresponds to the impurity contribution in Eq. (9). This contribution can be represented as $\mathrm{Re}\sigma(\omega)E^2/2$, where $\mathrm{Re}\sigma(\omega)$ is determined by Eqs. (10.23) and (10.24). The efficiency of multi-photon absorption with respect to single-photon absorption is characterized by a dimensionless parameter

$$\gamma \equiv \frac{m\mathrm{v}_\omega^2}{6\hbar\omega}. \tag{10}$$

If $\gamma \ll 1$, which is the case of large ω and small E, only single-photon absorption is essential.

If the electron-electron scattering gives the main contribution to the kinetic equation (5), $\overline{f}_{\mathbf{P}}$ is given by a quasi-equilibrium (Maxwell) distribution over canonical momenta, with the effective temperature T_e. To determine this temperature, one can use either the balance equation (8) or its quantum analog (9). After substituting the Maxwell distribution into Eq. (9), we calculate the integral over $\mathbf{P}$ and present both $\mathcal{P}_{ph}$ and $P^{(\omega)}$ as

$$\left| \begin{matrix} \mathcal{P}_{ph} \\ P^{(\omega)} \end{matrix} \right| = 2n\sqrt{\frac{2\pi m}{T_e}} \sum_{\mathbf{q}} (\hbar q)^{-1} |C_q|^2 (N_q + 1) \sum_{k=-\infty}^{\infty} \left| \begin{matrix} k\omega \\ \omega_q \end{matrix} \right|$$

$$\times \left[\mathcal{J}_k \left(\frac{\mathbf{q}\cdot\mathbf{v}_\omega}{\omega} \right) \right]^2 \exp\left\{ -\frac{(\hbar q/2 + m\omega_q/q - mk\omega/q)^2}{2mT_e} \right\} \tag{11}$$

$$\times \left[1 - \exp\left(-\hbar\omega_q/T + \hbar(\omega_q - k\omega)/T_e \right) \right],$$

where n is the electron density. Below we calculate the absorbed power in the case of quasielastic scattering of electrons by LA phonons, while the scattering by LO phonons is neglected. If the electric field is not very strong and only one-photon processes are essential, the Bessel function is approximated as $\mathcal{J}_{\pm 1}(x) \simeq \pm x/2$ and $\mathcal{P}_{ph}$ is proportional to E^2. Under the quasi-classical conditions $\hbar\omega \ll T_e$, Eq. (11) leads to ($\mathcal{P}_{ph} = \mathcal{P}_{LA}$)

$$\mathcal{P}_{LA} = \frac{4}{3\sqrt{\pi}} \frac{nm\mathrm{v}_\omega^2}{\tau_{LA}(T_e)}, \tag{12}$$

where $\tau_{LA}(\varepsilon)$ is defined in problem 7.8. Equation (12) can be rewritten in the form $\mathcal{P}_{LA} = \sigma_{LA}(\omega)E^2/2$, where the linear high-frequency conductivity $\sigma_{LA}(\omega)$ is calculated according to Eq. (8.26), where $\nu_\varepsilon = \tau_{LA}^{-1}(\varepsilon)$ and $\omega \gg \nu_{T_e}$. The term $\mathcal{P}_{im}$ is also represented in the form $\sigma_{im}(\omega)E^2/2$, where the high-frequency conductivity $\sigma_{im}(\omega)$ is expressed through the relaxation rate ν_ε for electron-impurity scattering calculated in Sec. 8. Under the conditions $\omega \gg \nu_{T_e}$, the total high-frequency conductivity $\sigma(\omega)$ is the sum of the partial contributions $\sigma_{LA}(\omega)$ and $\sigma_{im}(\omega)$. Therefore, we write $\mathcal{P}_{im} + \mathcal{P}_{LA} = \sigma(\omega)E^2/2$ under the quasi-classical conditions. The impurity contribution to $\sigma(\omega)$ is large if the scattering is caused by ionized impurities, because the momenta transferred in the collisions are small. However, with the increase of ω, the ionized impurity contribution decreases much faster than the LA phonon contribution. Calculating $\mathcal{P}_{LA}$ in the quantum limit $\hbar\omega \gg T_e$, we find

$$\mathcal{P}_{LA} = \frac{nm\mathrm{v}_\omega^2}{6\tau_{LA}(\hbar\omega)} \Phi\left(\sqrt{2ms_l^2\hbar\omega}/2T_e\right), \quad \Phi(x) = x\coth x, \tag{13}$$

where s_l is the longitudinal sound velocity. This expression is valid for electrons of arbitrary degeneracy. If $2ms_l^2\hbar\omega \ll T_e^2$, the function Φ in Eq. (13) is equal to unity.

Considering, under the same approximations, the power transmitted to LA phonons, we need to take into account both photonless ($k = 0$) and one-photon ($k = \pm 1$) contributions. The latter, however, appear to be small due to quasielasticity of the scattering. This means that the power $P_{LA}^{(\omega)}$ for this mechanism is equal to P of Eq. (36.25), where $\nu^{(e)}$ for non-degenerate electrons is presented by Eq. (36.26), case B. Equating this power to $\mathcal{P}_{LA}$ from Eq. (12), we find the relation $T_e - T = (T/6)(\mathrm{v}_\omega/s_l)^2$ demonstrating that the heating is sensitive to the frequency of electromagnetic field, $T_e - T \sim \omega^{-2}$. In the quantum limit, a similar procedure based upon Eq. (13) gives us $T_e \sim \omega^{-4/3}$ for strong heating, $T_e \gg T$. We point out that the high-frequency field not only determines the amount of power absorbed by electrons, but also modifies the rate of energy transmission to the crystal lattice.

If the scattering is essentially inelastic, the one-photon contributions to the power loss become important. Considering the scattering of electrons by optical phonons at $\hbar\omega_{LO} \gg T_e, T$, we may neglect the power loss due to photonless ($k = 0$) processes, because it is proportional to $e^{-\hbar\omega_{LO}/T_e}$ (see problem 7.16). This small factor does not appear for one-photon processes if the frequency ω is close to ω_{LO} or exceeds it.

Calculating $\mathcal{P}_{LO}$ and $P_{LO}^{(\omega)}$ from Eq. (11) in these conditions, we find that only the terms with $k = 1$ are essential:

$$\mathcal{P}_{LO} = \frac{\omega}{\omega_{LO}} P_{LO}^{(\omega)} = n \frac{e^2 m^{3/2} \omega_{LO} \mathrm{v}_\omega^2 \sqrt{|\Delta\omega|}}{3\sqrt{2}\hbar^{3/2}\epsilon^*\omega} \Psi(\hbar\Delta\omega/2T_e),$$

$$\Psi(x) = \sqrt{2|x|/\pi}\mathcal{K}_1(|x|)e^x, \qquad (14)$$

where $\Delta\omega = \omega - \omega_{LO}$ is the excess frequency and $\mathcal{K}_1$ is the modified Bessel function of the second kind. The expression (14) shows us that the absorbed power increases with the increase of $\Delta\omega$. The function Ψ in Eq. (14) is equal to unity if $\hbar\Delta\omega \gg T_e$. In this region, both $P_{LO}^{(\omega)}$ and $\mathcal{P}_{LO}$ are temperature-independent. The power absorption $\mathcal{P}_{LA}$ given by Eq. (13) is also temperature-independent at $2ms_l^2\hbar\omega \ll T_e^2$. Collecting the terms discussed above (we remind that $P_{LA}^{(\omega)}$ is frequency-independent and determined by Eqs. (36.25) and (36.26), case B), we write the energy balance equation

$$\mathcal{P}_{LA} + \mathcal{P}_{LO}(1 - \omega_{LO}/\omega) = n\frac{8ms_l^2(T_e - T)}{\sqrt{\pi}\tau_{LA}(T_e)T} \qquad (15)$$

valid in the high-frequency region $\hbar\Delta\omega \gg T_e$ at $\sqrt{2ms_l^2\hbar\omega} \ll T_e \ll \hbar\omega_{LO}$. The left-hand side of Eq. (15) is temperature-independent and proportional to n. However, its frequency dependence modifies the effective electron temperature which is to be determined from Eq. (15).

The distribution of electrons over quasienergies becomes more complicated when the electron density is low and electron-electron scattering is not essential. At low temperatures and in the frequency region $\omega > \omega_{LO}$, one has to take into account both quasielastic scattering and spontaneous emission of optical phonons, as in Sec. 35. Below we consider the case $\gamma \ll 1$, when only one-photon transitions are essential. The field-induced anisotropy of the distribution function $\overline{f}_{\mathbf{p}}$ in this case can be neglected, and we write this function as f_ε, where $\varepsilon = \varepsilon_P$. The kinetic equation (5) is written for this distribution function as

$$J_{ac}(f|\varepsilon) + J_{LO}(f|\varepsilon) + J_r(f|\varepsilon) = 0, \qquad (16)$$

where we have neglected the electron-electron and electron-impurity collision integrals as well as the part of the electron-LA phonon collision integral describing one-photon transitions. The remaining (zero-photon) part of this collision integral is the quasielastic collision integral $J_{ac}(f|\varepsilon)$ written in problem 7.8. The collision integrals $J_{LO}(f|\varepsilon)$ and $J_r(f|\varepsilon)$, which describe, respectively, zero-photon and one-photon transitions with spontaneous emission of optical phonons, are given below:

$$J_{LO}(f|\varepsilon) = \sqrt{\frac{\varepsilon + \hbar\omega_{LO}}{\varepsilon}} \nu_{LO}(\varepsilon + \hbar\omega_{LO})f_{\varepsilon + \hbar\omega_{LO}} - \nu_{LO}(\varepsilon)f_\varepsilon, \qquad (17)$$

$$J_r(f|\varepsilon) = \sqrt{\frac{\hbar\omega_{LO}}{\varepsilon}} \{\nu_r(\varepsilon, \varepsilon + \hbar\omega_{LO} + \hbar\omega)f_{\varepsilon + \hbar\omega_{LO} + \hbar\omega}$$

$$+ \nu_r(\varepsilon, \varepsilon + \hbar\omega_{LO} - \hbar\omega)f_{\varepsilon + \hbar\omega_{LO} - \hbar\omega} \qquad (18)$$

$$- [\nu_r(\varepsilon, \varepsilon - \hbar\omega_{LO} + \hbar\omega) + \nu_r(\varepsilon, \varepsilon - \hbar\omega_{LO} - \hbar\omega)]f_\varepsilon\} .$$

These contributions contain finite-difference terms. The rate of sponta-neous emission of LO phonons, $\nu_{LO}(\varepsilon)$, is given in problem 7.11. The one-photon contribution follows from the terms containing $[\mathcal{J}_{\pm 1}(\mathbf{q}\cdot\mathbf{v}_\omega/\omega)]^2 \simeq [(\mathbf{q}\cdot\mathbf{v}_\omega)/2\omega]^2$ in the spherically-symmetric part of Eq. (5). The optical absorption rate $\nu_r(\varepsilon, \varepsilon')$ is expressed through the matrix elements (21.1) of electron-LO phonon interaction according to (problem 7.20)

$$\nu_r(\varepsilon, \varepsilon') = \frac{\pi V}{\hbar} \frac{\rho(\varepsilon)\rho(\varepsilon')}{\rho(\hbar\omega_{LO})} \int \frac{d\widetilde{\Omega}}{4\pi} \int \frac{d\widetilde{\Omega}'}{4\pi} |C_q^{(LO)}|^2$$

$$\times \left[\frac{\mathbf{v}_\omega \cdot (\mathbf{p} - \mathbf{p}')}{2\hbar\omega}\right]^2 = 2\alpha\gamma\omega_{LO} \frac{\sqrt{\varepsilon\varepsilon'}}{\hbar\omega}, \qquad (19)$$

where $\rho(\varepsilon)$ is the density of states, α is the constant of electron-phonon coupling given by Eq. (21.28), and $d\widetilde{\Omega}$ is the differential of the solid angle of the vector $\mathbf{p}$. We have defined $\nu_r(\varepsilon, \varepsilon')$ in such a way that the factor $(\hbar\omega_{LO}/\varepsilon)^{1/2}$, inversely proportional to the density of states, is separated in Eq. (18). This definition is convenient because both $J_{ac}(f|\varepsilon)$ and $J_{LO}(f|\varepsilon)$ are also proportional to $\varepsilon^{-1/2}$ and this factor can be taken out of the kinetic equation. Besides, this definition makes $\nu_r(\varepsilon, \varepsilon')$ symmetric, $\nu_r(\varepsilon, \varepsilon') = \nu_r(\varepsilon', \varepsilon)$.

Since we consider one-photon transitions and $\nu_r \ll \nu_{LO}$, the photoex-cited electrons rapidly relax into the passive region $\varepsilon < \hbar\omega_{LO}$, where their distribution remains essentially non-equilibrium if $\nu_r > \nu_{qe}$, where ν_{qe} is introduced in problem 7.8. The character of this non-equilibrium distribution is determined by the energy transferred to the passive region in the process of relaxation. This energy is defined as $\delta\varepsilon_\omega = \hbar\omega - p\hbar\omega_{LO}$, where p is the integer part of the ratio ω/ω_{LO} determining the number of optical phonons emitted by the photoexited electrons. The sequential transitions leading to the shifts of electron energy by $\delta\varepsilon_\omega$ are shown in Fig. 7.2. The electrons excited into the region $\varepsilon > (p + 1)\hbar\omega_{LO}$ return to the passive region with the energies smaller than the energies of the electrons excited into the region $\hbar\omega < \varepsilon < (p+1)\hbar\omega_{LO}$. The kinetic equation for the passive region is reduced to

$$\sqrt{\frac{\varepsilon}{\hbar\omega_{LO}}} J_{ac}(f|\varepsilon) + \theta(\delta\varepsilon_\omega - \varepsilon)\nu_r(\varepsilon + p\hbar\omega_{LO}, \varepsilon + \hbar\omega_{LO} - \delta\varepsilon_\omega)f_{\varepsilon + \hbar\omega_{LO} - \delta\varepsilon_\omega}$$

$$+\theta(\varepsilon - \delta\varepsilon_\omega)\nu_r[\varepsilon + (p-1)\hbar\omega_{LO}, \varepsilon - \delta\varepsilon_\omega]f_{\varepsilon - \delta\varepsilon_\omega} \tag{20}$$

$$-\nu_r[\varepsilon, \varepsilon + (p-1)\hbar\omega_{LO} + \delta\varepsilon_\omega]f_\varepsilon = 0.$$

This equation can be formally derived in the following way. We represent f_ε as a sum of $f_\varepsilon^{(k)} = f_{\varepsilon + k\hbar\omega_{LO}}$, where ε belongs to the passive region: $f_\varepsilon = \sum_{k=0}^{\infty} f_\varepsilon^{(k)}$. Instead of Eqs. (16), (17), and (18), we write a chain of equations for $f_\varepsilon^{(k)}$, where the quasielastic contribution is taken into account in the passive region ($k=0$) only. This chain of equations is reduced to Eq. (20) if we assume that $\nu_{LO} \gg \nu_r$ in the active region, which means that the electrons absorb photons only in the passive region. Once coming to the active region, the electrons rapidly go back to the passive region by emitting a cascade of optical phonons (or just one optical phonon). That is why Eq. (20) does not contain the rate of spontaneous LO phonon emission, though essentially exploits the optical phonon energy.

To solve Eq. (20), we consider the case of small shift (detuning) from the photon-phonon resonance, when $0 < \delta\varepsilon_\omega \ll \hbar\omega_{LO}$ or $0 < \hbar\omega_{LO} - \delta\varepsilon_\omega \ll \hbar\omega_{LO}$. Since the energy shifts during each optical transition are small, one may replace the finite-difference terms in Eq. (20) by the differential ones. Before doing this, we introduce a small dimensionless variable $\delta\omega$ defined as $\delta\omega = \delta\varepsilon_\omega/\hbar\omega_{LO}$ if $\delta\varepsilon_\omega \ll \hbar\omega_{LO}$ and $\delta\omega = \delta\varepsilon_\omega/\hbar\omega_{LO} - 1$ if $\hbar\omega_{LO} - \delta\varepsilon_\omega \ll \hbar\omega_{LO}$. In contrast to $\delta\varepsilon_\omega$, which is positive by definition, $\delta\omega$ can be either positive or negative. It is convenient also to introduce the integer p' as $p' = p$ if $\delta\omega > 0$ and $p' = p+1$ if $\delta\omega < 0$ so that p' gives us the number of the integer photon-phonon resonance which we detune from by $\delta\omega$. Let us use $J_{ac}(f|\varepsilon)$ written in the problem 7.8 and express all energies in units of $\hbar\omega_{LO}$ according to $\xi = \varepsilon/\hbar\omega_{LO}$ and $t = T/\hbar\omega_{LO}$. We obtain the equation describing diffusion and drift along the energy axis:

$$\frac{d}{d\xi}\left(D_\xi \frac{df_\xi}{d\xi} - u_\xi f_\xi\right) = 0, \tag{21}$$

where the diffusion coefficient and drift velocity along the energy axis, D_ξ and u_ξ, contain both radiative and quasielastic scattering contributions:

$$D_\xi = \frac{1}{2}(\delta\omega)^2\nu_r(\xi, \xi + p' - 1) + t\xi^{3/2}\nu_{qe}(\xi),$$

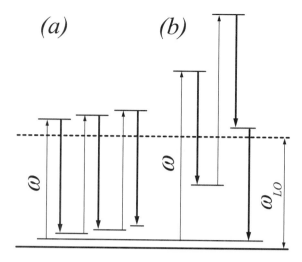

Figure 7.2. Transitions of electrons interacting with photons and optical phonons. (*a*) Upward drift of electrons for small positive detuning from the first integer photon-phonon resonance $\omega = \omega_{LO}$. (*b*) One cycle of electron evolution for the fractional photon-phonon resonance of the second order, $\omega = (3/2)\omega_{LO}$. The dashed line shows the boundary of the passive region.

$$u_\xi = \delta\omega\nu_r(\xi, \xi + p' - 1) - \xi^{3/2}\nu_{qe}(\xi). \tag{22}$$

Here and below in this section, the notations $\nu_r(\xi, \xi')$ and $\nu_{qe}(\xi)$ stand for the rates $\nu_r(\xi\hbar\omega_{LO}, \xi'\hbar\omega_{LO})$ and $\nu_{qe}(\xi\hbar\omega_{LO})$, respectively. In the transformation of the finite-difference terms to the differential ones, we have kept the factor $(\delta\omega)^2$ at the second derivatives only and neglected the terms proportional to higher powers of $\delta\omega$. However, in the case of $p' = 1$ all terms proportional to $(\delta\omega)^2$ are taken into account in Eq. (22), and the accuracy of Eq. (21) is improved in this way.

Equation (21) is considered in the interval $\xi \in [0, 1]$. The boundary condition at the edge of the passive region can be derived in the form similar to Eq. (35.16). However, since $\nu_{LO}(\varepsilon)$ rapidly overcomes $\nu_{qe}(\varepsilon)$ as ε exceeds $\hbar\omega_{LO}$, this boundary condition is reduced to $f_{\xi=1} = 0$ reflecting a rapid emptying of the active region due to spontaneous LO phonon emission. The flow $D_\xi(df_\xi/d\xi) - u_\xi f_\xi$ is conserved in the passive region. Denoting it as $-C$, where the constant C is to be determined from the normalization condition, we write the general solution of Eq. (21)

satisfying the boundary condition $f_{\xi=1} = 0$ as

$$f_\xi = C \int_\xi^1 \frac{d\xi'}{D_{\xi'}} \exp\left[\int_{\xi'}^\xi d\xi'' \frac{u_{\xi''}}{D_{\xi''}}\right]. \tag{23}$$

In the absence of photoexcitation, or at $\delta\omega = 0$, this solution is reduced to the Boltzmann distribution $f_\xi \propto e^{-\xi/t}$ (we remind that under the assumed strong inequality $t \ll 1$, the boundary condition $f_{\xi=1} = 0$ is not essential in this case). As $|\delta\omega|$ increases, different kinds of behavior are possible. Consider first the case of negative detuning, $\delta\omega < 0$, when the drift velocity u_ξ is negative in the entire passive region. Let us represent the integral in the exponent of Eq. (23) as

$$\int_{\xi'}^\xi d\xi'' \frac{u_{\xi''}}{D_{\xi''}} = F(\xi') - F(\xi), \qquad F(\xi) = -\int_0^\xi d\xi' \frac{u_{\xi'}}{D_{\xi'}}. \tag{24}$$

The derivative $dF(\xi)/d\xi = -u_\xi/D_\xi$ is always positive. Therefore, since the strong inequalities $|\delta\omega| \ll 1$ and $t \ll 1$ ensure $|u_\xi/D_\xi| \gg 1$, the function $e^{F(\xi')}$ exponentially increases to the end of the passive region and the main contribution to the integral over ξ' comes from $\xi' \simeq 1$, where $F(\xi')$ can be expanded as $F(1) + (u_1/D_1)(1-\xi')$. The integration over ξ' then gives us

$$f_\xi = \widetilde{C} e^{-F(\xi)} \left\{1 - \exp[(\xi - 1)|u_1|/D_1]\right\}, \qquad \widetilde{C} = \frac{C}{|u_1|} e^{F(1)}, \tag{25}$$

where the term inside the braces $\{\ldots\}$ can be replaced by unity if ξ is not too close to 1. To estimate the function $F(\xi)$, we use the explicit expressions for the quasielastic relaxation rate (see problem 7.8) and photon absorption rate (19), representing them as $\nu_{qe}(\xi) = \nu_{qe}(1)\sqrt{\xi}$ and $\nu_r(\xi, \xi') = \nu_r(1,1)\sqrt{\xi\xi'}$. We find

$$F(\xi) = \frac{\xi}{t}\left[1 - \left(1 + \frac{\delta\omega}{2t}\right)\frac{1}{\xi}\right.$$

$$\left. \times \int_0^\xi \frac{d\xi'}{(r/\delta\omega)\xi'^{3/2}/(\xi' + p' - 1)^{1/2} + \delta\omega/2t}\right], \tag{26}$$

where $r = \nu_{qe}(1)/\nu_r(1,1) \ll 1$. In the special case of $p' = 1$, the integral in Eq. (26) is taken easily (problem 7.21). Since $\delta\omega$ is negative, Eqs. (25) and (26) describe a cooling of the electron gas at $|\delta\omega| < 2t$ and its heating at $|\delta\omega| > 2t$. The distribution $e^{-F(\xi)}$ in this case is very close to the Boltzmann one, with the effective temperature $t \simeq |\delta\omega|/2$, because only the region $\xi \sim t$ or $\xi \sim |\delta\omega|$ is essential, where the integral over ξ' in Eq. (26) is approximately equal to $2t\xi/\delta\omega$.

Let us consider the case of positive detuning. The drift velocity u_ξ is determined by a competition of two terms with different signs. The quasielastic term, which gives a negative contribution to u_ξ, increases with ξ faster than the radiative term. Nevertheless, if $\delta\omega > r/\sqrt{p'}$, the latter term is more important and the drift velocity always remains positive in the passive region. The function $F(\xi')$ exponentially decreases with increasing ξ', and the main contribution to the integral over ξ' in Eq. (23) comes from the region $\xi' \simeq \xi$. In this region we may expand $F(\xi') - F(\xi)$ as $(u_\xi/D_\xi)(\xi - \xi')$. Integrating over ξ', we find

$$f_\xi = C/u_\xi. \qquad (27)$$

This simple solution is obtained under the assumption that ξ is not too close to the edge of the passive region. Otherwise, this solution should be multiplied by the factor $\{\ldots\}$ standing in Eq. (25). Equation (27) describes the distribution which spreads over the entire passive region. This distribution can be normalized because at small energies the velocity u_ξ is proportional to $\sqrt{\xi(\xi + p' - 1)}$, and the integral $\int_0^1 d\xi\rho(\xi)f_\xi$, where $\rho(\xi) \propto \sqrt{\xi}$ is the density of states, converges.

A completely different situation occurs at $\delta\omega < r/\sqrt{p'}$, where u_ξ changes its sign from positive to negative in a single point $\xi = \xi_s$ determined from the equation $u_{\xi_s} = 0$. This equation is transformed into

$$\xi_s^3 - (\delta\omega/r)^2\xi_s - (\delta\omega/r)^2(p' - 1) = 0. \qquad (28)$$

If $p' = 1$, we obtain $\xi_s = \delta\omega/r$. In the point ξ_s, the radiative drift flow and quasielastic drift flow compensate each other. As $\delta\omega$ increases from 0 to $r/\sqrt{p'}$, this point moves through the passive region, from 0 to 1. If ξ_s is not too close to the edge of the passive region, the integral over ξ' in Eq. (23) is, in a similar way as in the case of negative $\delta\omega$, determined by the region $\xi' \simeq 1$, where $e^{F(\xi')}$ is exponentially large. The result is again given by Eq. (25), i.e., the distribution function is proportional to $e^{-F(\xi)}$, where $F(\xi)$ is given by Eq. (26). However, the properties of this exponent are essentially different from those in the case of negative detuning. The function $F(\xi)$ has a minimum at $\xi = \xi_s$. Near this point, we have $F(\xi) = F(\xi_s) + \int_{\xi_s}^\xi d\xi' D_{\xi_s}^{-1}(du_\xi/d\xi)_{\xi=\xi_s}(\xi' - \xi_s)$ and the solution is written as

$$f_\xi = C_s \exp\left[-\frac{(\xi - \xi_s)^2}{2\Gamma_s^2}\right], \qquad \Gamma_s = \left(\frac{D_{\xi_s}}{|du_\xi/d\xi|_{\xi=\xi_s}}\right)^{1/2}, \qquad (29)$$

where C_s is to be found from a normalization condition. The distribution in the form of a Gaussian peak appears because the electrons, being driven towards higher energies by the radiation, are accumulated in the narrow region where their flow is stopped by the quasielastic relaxation. The width of the peak is given by

$$\Gamma_s = \left[\frac{\xi_s(t + \delta\omega/2)}{3/2 - (\delta\omega/r\xi_s)^2/2} \right]^{1/2}, \tag{30}$$

and can be estimated as $\Gamma_s \sim \sqrt{\xi_s(t + \delta\omega/2)}$. If $p' = 1$, this estimate becomes an exact relation. If $\xi_s \gg t + \delta\omega/2$, the peak is well-defined, because $\Gamma_s \ll \xi_s$. If $\delta\omega$ is so small that $\xi_s < t + \delta\omega/2$, we have a monotonic distribution of electrons, which is reduced to the equilibrium Boltzmann distribution as $\delta\omega$ goes to zero. In summary, the evolution of the distribution function with the increase of $\delta\omega$ can be represented as follows (see Fig. 7.3). At large enough negative detuning, we have a Boltzmann-like heated distribution, which becomes exact Boltzmann at $\delta\omega = -2t$. At $-2t < \delta\omega < 0$, we have a Boltzmann-like cooled distribution, which becomes again the exact Boltzmann distribution at $\delta\omega = 0$. In the region of small positive $\delta\omega$, the distribution is heated until a peak is formed at the bottom of the passive region. As $\delta\omega$ increases further, the peak moves towards higher energies and becomes broader. When $\delta\omega$ exceeds $r/\sqrt{p'}$, the peak is gone, and we have the distribution (27).

We have considered electron behavior in the vicinity of the integer photon-phonon resonance $\omega = p'\omega_{LO}$. Such resonance can be named as the first-order photon-phonon resonance. If ω is close to $(p + m/n)\omega_{LO}$, where n and m are integer numbers and $m < n$, one can consider fractional photon-phonon resonances of the n-th order. The simplest case of multiple-order resonances is the second-order resonance, when $\hbar\omega = (p + 1/2)\hbar\omega_{LO} + \delta\varepsilon_\omega$; see Fig. 7.2 (b). To study it quantitatively, let us use Eq. (20). We write this equation twice: first for the energy $\varepsilon < \hbar\omega_{LO}/2$ and then for the energy $\varepsilon + \hbar\omega_{LO}/2$. As a result, we obtain two coupled equations for f_ε and $g_\varepsilon = f_{\varepsilon+\hbar\omega_{LO}/2}$, where the energy ε is defined in the interval $0 < \varepsilon < \hbar\omega_{LO}/2$. Introducing a small deviation $\delta\varepsilon_\omega$ as explained above, we transform the finite-difference terms in these two equations to the terms containing derivatives of f_ε and g_ε. Finally, adding and subtracting these equations, we reduce them to the following system:

$$\frac{d}{d\xi}\left(\widetilde{D}_\xi \frac{df_\xi}{d\xi} - \widetilde{u}_\xi f_\xi + \widetilde{D}_{\xi+1/2}\frac{dg_\xi}{d\xi} - \widetilde{u}_{\xi+1/2}g_\xi \right) = 0, \tag{31}$$

$$\nu_r(\xi, \xi+p-1/2)\left(f_\xi - \frac{\delta\omega}{2}\frac{df_\xi}{d\xi} \right) = \nu_r(\xi+1/2, \xi+p)\left(g_\xi - \frac{\delta\omega}{2}\frac{dg_\xi}{d\xi} \right), \tag{32}$$

where we again use the dimensionless variables $\delta\omega$ and ξ. Equations (31) and (32) allow one to determine f_ξ and g_ξ defined in the interval

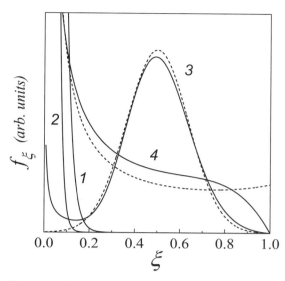

Figure 7.3. Modification of the distribution function of electrons interacting with photons, optical phonons, and acoustic phonons for small detuning $\delta\omega$ from the first-order photon-phonon resonance at $p' = 1$. The parameters used in the calculations are $T = 0.025 \ \hbar\omega_{LO}$ and $r = \nu_{qe}(\hbar\omega_{LO})/\nu_r(\hbar\omega_{LO}, \hbar\omega_{LO}) = 0.05$. The solid curves 1-4 show the Boltzmann distribution ($\delta\omega = 0$), cooled distribution ($\delta\omega = -0.025$), distribution with a peak ($\delta\omega = 0.025$), and broad distribution ($\delta\omega = 0.075$), respectively. The dashed curves 3 and 4 show the approximate solutions (37.29) and (37.27), respectively.

$\xi \in [0, 1/2]$. The diffusion coefficient and drift velocity standing in Eq. (31) are given by $\widetilde{D}_\xi = (\delta\omega)^2 \nu_r(\xi, \xi + p - 1/2)/2 + t\xi^{3/2}\nu_{qe}(\xi)$ and $\widetilde{u}_\xi = \delta\omega\nu_r(\xi, \xi + p - 1/2) - \xi^{3/2}\nu_{qe}(\xi)$. The boundary conditions to Eqs. (31) and (32) are $g_{\xi=1/2} = 0$ and $f_{\xi=1/2} = g_{\xi=0}$. The analysis of these coupled equations shows us that, for a positive detuning $\delta\omega$, one can get two peaks, one of them is placed in the region $[0, 1/2]$, while the other is shifted up by $\Delta\xi = 1/2$ and appears in the region $[1/2, 1]$ (problem 7.22). The origin of the single peak is the same as described above, while the doubling occurs because the points ξ and $\xi + 1/2$ are strongly coupled; see Fig. 7.2 (b). In a similar way, a system of n equidistant peaks, separated by an interval $1/n$, appears in the case of small detuning from the photon-phonon resonance of n-th order. The region of $\delta\omega$ where these peaks exist rapidly decreases with increasing n.

38. Relaxation of Population

Interaction of electrons with bosons determines not only the electron energy distribution, but also the population of different states by electrons, for example, the densities of electrons and holes in conduction and valence bands, or the occupation of different confined states (subbands or discrete levels) in nanostructures. If one excludes the case of zero-gap semiconductors, the phonon-assisted interband relaxation of non-equilibrium electrons and holes may occur due to multi-phonon processes only. These processes are not efficient at low temperatures in the materials with weak electron-phonon coupling. Therefore, the main contribution to the relaxation rate in the intrinsic semiconductors and insulators comes from the radiative recombination, when the electrons jump from the conduction band to the valence one by emitting photons, and from the Auger recombination due to interband electron-electron scattering. In quantum wells and quantum wires, where the electrons occupy 2D and 1D subbands, respectively, the phonon-assisted intersubband relaxation is important, because it is not forbidden by the energy conservation law even when the energy separation between the subbands is larger than the LO phonon energy; see Fig 7.4. In quantum dots, where the electrons occupy discrete levels, the phonon-assisted relaxation becomes important if the energy separation between the levels is small enough or close to the LO phonon energy. Below we consider the radiative interband recombination with one-photon emission in bulk materials, as well as phonon-assisted intersubband and interlevel relaxation of electrons in the nanostructures listed above.

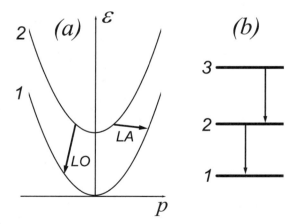

Figure 7.4. (*a*) Intersubband phonon-assisted transitions in quantum wells and wires. (*b*) Transitions between discrete levels in quantum dots.

The boson-assisted interband relaxation of the electrons occupying the conduction-band state $|c\delta\rangle$ is described by the interband collision integral obtained in a similar way as Eq. (34.25), see also problem 7.3:

$$J_{e,b}^{(inter)}(f|c\delta) \simeq -\frac{2\pi}{\hbar} \sum_{q\gamma} |\langle c\delta|\hat{\chi}_q|v\gamma\rangle|^2 \delta(\varepsilon_{c\delta} - \varepsilon_{v\gamma} - \hbar\omega_q)$$

$$\times f_{c\delta t}(1 - f_{v\gamma t}) = -\nu_R^{(\delta)} f_{c\delta t}. \tag{1}$$

Note that we have neglected the stimulated emission and absorption (since $N_q \ll 1$) and omitted the arrival term proportional to $f_{v\gamma t}(1 - f_{c\delta t})$, because $\varepsilon_{c\delta} - \varepsilon_{v\gamma} + \hbar\omega_q \neq 0$. In p-type materials, where the Fermi momentum of holes is greater than the momenta of non-equilibrium electrons (Fig. 4.1 (d)), one may neglect the Pauli blocking factor $1 - f_{v\gamma t}$ and write the recombination rate introduced in Eq. (1) as

$$\nu_R^{(\delta)} = \frac{2\pi}{\hbar} \sum_{q\gamma} |\langle c\delta|\hat{\chi}_q|v\gamma\rangle|^2 \delta(\varepsilon_{c\delta} - \varepsilon_{v\gamma} - \hbar\omega_q). \tag{2}$$

The operator $\hat{\chi}_q = \hat{\chi}_{\mathbf{q}\mu}$ for the electron-photon interaction is given by Eq. (20.1), and the dipole approximation is justified because the wave numbers of photons are much smaller than those of recombinating electron states.

To consider the transitions between spin-degenerate electron states near the extrema of c- and v-bands in bulk semiconductors, we use the matrix element of the two-band model (see Secs. 17 and 18 and Appendix B). Let us employ the plane-wave basis, when the quantum number δ is replaced by $\sigma\mathbf{p}$. The sum over the spin variable is reduced to

$$\sum_{\sigma'} |\langle c\sigma\mathbf{p}|\hat{\chi}_q|v\sigma'\mathbf{p}'\rangle|^2 = \frac{2\pi\hbar e^2}{V\omega_q\epsilon} \sum_{\sigma'} |\langle c\sigma\mathbf{p}|(\hat{\mathbf{v}} \cdot \mathbf{e}_{\mathbf{q}\mu})|v\sigma'\mathbf{p}'\rangle|^2$$

$$= \frac{2\pi\hbar(es)^2}{V\omega_q\epsilon} \delta_{\mathbf{p}\mathbf{p}'}, \tag{3}$$

where s is the interband velocity. The rate of interband recombination is written as

$$\nu_R^{(p)} \simeq 2 \sum_{\mathbf{q}} \frac{(2\pi es)^2}{V\omega_q\epsilon} \delta(\varepsilon_g + p^2/2\mu^* - \hbar\omega_q), \tag{4}$$

where the doubling occurs because of the sum over photon polarizations, and μ^* is the reduced mass defined in Sec. 17. Using the photon spectrum in the medium with dielectric permittivity ϵ, $\omega_q = cq/\sqrt{\epsilon}$, we calculate the sum over $\mathbf{q}$ and obtain the result $\nu_R = \nu_R^{(0)}$ for the electrons at the bottom of the conduction band ($p = 0$):

$$\nu_R \simeq \frac{(2es)^2\sqrt{\epsilon}}{\hbar^2 c^3} \int_0^\infty d\xi\, \xi\delta(\varepsilon_g - \xi) = 4\frac{e^2\sqrt{\epsilon}}{\hbar c}\left(\frac{s}{c}\right)^2 \frac{\varepsilon_g}{\hbar}. \tag{5}$$

If the kinetic energies of electrons and holes are small in comparison to ε_g, the rate of radiative interband recombination is independent of these energies and equal to ν_R given by Eq. (5). Therefore, this rate enters the density balance equation directly, as a constant. The interband recombination of the electrons occupying confined states in nanostructures is very similar to the one described above. One should merely renormalize the energy gap ε_g (due to confinement effects) and multiply ν_R of Eq. (5) by the squared overlap factor of electron and hole states defined in Eq. (30.23). If both electron and hole states belong to the ground-state subbands, the overlap factor is close to unity. In the model of hard-wall confinement, the overlap factor is equal to unity exactly so that Eq. (5) can be applied to low-dimensional structures without changes. This rule, however, is not valid in the heterostructures with spatially separated ground states of electrons and holes (so-called type II quantum well structures), where the overlap factor is small and the recombination is suppressed.

Let us consider the phonon-assisted transitions between two subbands of a quantum well. As in Sec. 35, we use the model of bulk (3D) phonon states and describe electrons in the basis $|n\mathbf{p}\rangle$ with coordinate-dependent wave functions $\psi_z^{(n)} L^{-1} e^{i\mathbf{p}\cdot\mathbf{r}/\hbar}$, where n is the subband number and $\mathbf{r} = (x, y)$ and $\mathbf{p}$ are the 2D coordinate and momentum. The rate of electron transitions from the state $|n\mathbf{p}\rangle$ to the subband n' is given by

$$\nu^{(nn')}(p) = \frac{2\pi}{\hbar} \sum_{\mathbf{Q}} |C_Q|^2 |\langle n'|e^{iq_z z}|n\rangle|^2 \left[(N_Q + 1)\delta(\varepsilon_{np} - \varepsilon_{n'\mathbf{p}-\hbar\mathbf{q}} - \hbar\omega_Q) \right.$$

$$\left. \times (1 - f_{n'\mathbf{p}-\hbar\mathbf{q}}) + N_Q \delta(\varepsilon_{np} - \varepsilon_{n'\mathbf{p}+\hbar\mathbf{q}} + \hbar\omega_Q)(1 - f_{n'\mathbf{p}+\hbar\mathbf{q}}) \right], \quad (6)$$

where $\mathbf{Q} = (\mathbf{q}, q_z)$. Equation (6) takes into account both emission and absorption of phonons. The collision integral for the distribution function of electrons in the subband n can be written according to problem 7.3. It contains the term $-\nu^{(nn')}(p)f_{n\mathbf{p}}$ describing the departure of electrons from the state $|n\mathbf{p}\rangle$ as well as the contribution describing the arrival of electrons to this state from all subbands.

Consider first the relaxation due to interaction of electrons with LO phonons. At low temperatures, one may neglect both absorption and stimulated emission of these phonons. Neglecting also the Pauli blocking, we replace $(1 - f_{n'\mathbf{p}-\hbar\mathbf{q}})$ by unity. Substituting $C_Q^{(LO)}$ from Eq. (21.1), we integrate over q_z in Eq. (6) and obtain

$$\nu_{LO}^{(nn')}(p) = \frac{2\pi^2 e^2 \omega_{LO}}{\epsilon^* L^2} \sum_{\mathbf{q}} \int dz \int dz' \frac{e^{-q|z-z'|}}{q} \psi_z^{(n)} \psi_z^{(n')} \psi_{z'}^{(n)} \psi_{z'}^{(n')}$$

$$\times \delta \left(\varepsilon_{nn'} - \hbar\omega_{LO} + \frac{\hbar \mathbf{p} \cdot \mathbf{q}}{m} - \frac{\hbar^2 q^2}{2m} \right), \tag{7}$$

where we have used the parabolic energy dispersion $\varepsilon_{np} = \varepsilon_n + p^2/2m$ and introduced the intersubband energy $\varepsilon_{nn'} = \varepsilon_n - \varepsilon_{n'}$ as in Secs. 22 and 29. Integrating over the angle of $\mathbf{q}$ in Eq. (7), we find

$$\nu_{LO}^{(nn')}(p) = 2\omega_{LO} \frac{\epsilon}{\epsilon^*} \int_{q_{\min}}^{q_{\max}} dq \frac{q M_{nn'nn'}(q)}{[(q^2 - q_{\min}^2)(q_{\max}^2 - q^2)]^{1/2}}, \tag{8}$$

$$\hbar q_{\min} = \left| p - \sqrt{p^2 + \widetilde{p}_{nn'}^2} \right|, \quad \hbar q_{\max} = p + \sqrt{p^2 + \widetilde{p}_{nn'}^2}, \quad \widetilde{p}_{nn'}^2 = p_{nn'}^2 - p_{LO}^2,$$

where $M_{abcd}(q)$ is defined by Eq. (29.11), $p_{nn'}^2 = 2m\varepsilon_{nn'}$, and $p_{LO} = \sqrt{2m\hbar\omega_{LO}}$ is introduced in Sec. 35. If $n' = n$, Eq. (8) describes the intrasubband transition rate (the intrasubband transitions exist at $p > p_{LO}$). The integral over q standing in Eq. (8) is determined by p, $\varepsilon_{nn'} - \hbar\omega_{LO}$, and quantum well parameters. If p is small in comparison to $\widetilde{p}_{nn'}$, the wave number q is fixed due to the energy conservation law. The matrix element $M_{nn'nn'}$ in this case is approximated by a constant, and the integral in Eq. (8) is transformed by using a new variable, $q^2 - (\widetilde{p}_{nn'}/\hbar)^2$, leading to the following momentum-independent relaxation rate:

$$\nu_{LO}^{(nn')} = \pi\omega_{LO}(\epsilon/\epsilon^*) M_{nn'nn'}(\widetilde{p}_{nn'}/\hbar). \tag{9}$$

The relaxation of electrons with small p is realized, for example, when these electrons are excited from the low-occupied ground-state subband to the subband n by light; see Sec. 29. If the intersubband energy is close to the optical phonon energy (in other words, $p_{nn'}$ is close to p_{LO}), the approximation $p^2 \ll \widetilde{p}_{nn'}^2$ is not valid. Nevertheless, Eq. (9) remains correct, because in this case one can expand $M_{nn'nn'}(q)$ in Eq. (8) in series of small q, which again leads to a constant $M_{nn'nn'}(q) \simeq -L_{nn'nn'}/a_B$ at $n \neq n'$; see Eq. (29.14).

Applying Eq. (6) to the acoustic phonon-assisted relaxation under the condition $T \gg \hbar s_l/d$, where d is the width of the quantum well, we use the elastic approximation and neglect the Pauli blocking. Substituting C_Q for acoustic phonons from Eq. (21.1), we integrate over q_z and find

$$\nu_{LA}^{(nn')}(p) = \frac{2\pi \mathcal{D}^2 T}{\hbar \rho s_l^2 L^2} \int dz |\psi_z^{(n)}|^2 |\psi_z^{(n')}|^2 \sum_{\mathbf{q}} \delta \left(\varepsilon_{nn'} + \frac{\hbar \mathbf{p} \cdot \mathbf{q}}{m} - \frac{\hbar^2 q^2}{2m} \right). \tag{10}$$

The sum over $\mathbf{q}$ in Eq. (10) can be written as $\sum_{\mathbf{p}'} \delta(\varepsilon_{nn'} + \varepsilon_p - \varepsilon_{p'})$, where $\mathbf{p}' = \mathbf{p} - \hbar\mathbf{q}$. Therefore, employing the 2D density of states introduced in Eq. (5.26), we obtain the momentum-independent relaxation rate

$$\nu_{LA}^{(nn')} = \frac{\mathcal{D}^2 T m}{\rho s_l^2 \hbar^3} \int dz |\psi_z^{(n)}|^2 |\psi_z^{(n')}|^2. \tag{11}$$

If the electron transitions occur between the ground-state subband $n' = 1$ and first excited subband $n = 2$, the overlap integral standing in this equation is equal to $1/d$ for the hard-wall model of the quantum well.

If the quantum well is made of ionic material, the rate $\nu_{LA}^{(nn')}$ is usually much smaller than $\nu_{LO}^{(nn')}$ and may become essential only at $\varepsilon_{n1} < \hbar\omega_{LO}$, when LO phonon-assisted relaxation of electrons with small momenta is forbidden. Typically, the intersubband energies in quantum wells exceed the optical phonon energies, and the non-equilibrium electrons occupying the excited subband n leave it with the total rate $\sum_{n'=1}^{n-1} \nu_{LO}^{(nn')}(p)$. If, for example, the non-equilibrium electrons occupy the second subband ($n = 2$) and their kinetic energy is smaller than $\hbar\omega_{LO}$, they relax to the first (ground-state) subband by emitting a LO phonon and cannot come back to the second subband either by spontaneous LO phonon emission or by quasielastic scattering. This condition justifies the neglect of the arrival terms in the intersubband collision integral, when the latter is written as $-\nu_{LO}^{(21)}(p)f_{2p}$. If, in addition, ε_{21} considerably exceeds $\hbar\omega_{LO}$, the rate $\nu_{LO}^{(21)}$ is momentum-independent, and the density balance equation is written as

$$\frac{dn_2}{dt} = -\nu_{LO}^{(21)} n_2 + G_\omega, \qquad G_\omega = \xi_\omega \frac{|E|^2 c\sqrt{\epsilon}}{2\pi\hbar\omega} \equiv \nu_\omega n_1. \tag{12}$$

On the right-hand side, we have added the photogeneration term G_ω describing the number of photons absorbed in unit time by unit square of the quantum well layer under the excitation by the electric field $Ee^{-i\omega t} +$ c.c. of monochromatic electromagnetic wave. This number is equal to the number of electrons photoexcited from the first subband to the second one. The relative absorption ξ_ω is given by Eq. (29.25). Since it is proportional to the electron density in the first subband, the generation term can be represented as $\nu_\omega n_1$, where ν_ω is the photogeneration rate. In the stationary case, Eq. (12) gives us a simple relation between the densities n_1 and n_2, namely $n_2 = (\nu_\omega/\nu_{LO}^{(21)})n_1$. One should remember that the consideration given above is valid if the electrons returning to the first subband have enough time to relax there before absorbing a photon. This assumes a weak photoexcitation, when $n_2 \ll n_1$.

The phonon-assisted electron relaxation between the 1D subbands of quantum wires is similar to the relaxation in quantum wells. Let us consider a wire along the OX direction and introduce the basis $|np\rangle$ with the wave functions $\psi_{\mathbf{r}_\perp}^{(n)} L^{-1/2} e^{ipx/\hbar}$, where p is the 1D momentum,

$\mathbf{r}_\perp = (y, z)$, and $n = (n_1, n_2)$; see Eq. (5.25). Instead of Eq. (6), we have

$$\nu^{(nn')}(p) = \frac{2\pi}{\hbar} \sum_{\mathbf{Q}} |C_Q|^2 |\langle n'|e^{i\mathbf{q}_\perp \cdot \mathbf{r}_\perp}|n\rangle|^2 \left[(N_Q + 1)\delta(\varepsilon_{np} - \varepsilon_{n'p-\hbar q}) \right.$$

$$\left. -\hbar\omega_Q)(1 - f_{n'p-\hbar q}) + N_Q \delta(\varepsilon_{np} - \varepsilon_{n'p+\hbar q} + \hbar\omega_Q)(1 - f_{n'p+\hbar q}) \right], \quad (13)$$

where $\mathbf{Q} = (q, \mathbf{q}_\perp)$. For *LO* phonon scattering, it is convenient to introduce the matrix elements

$$K_{abcd}(q) = \int d\mathbf{r}_\perp \int d\mathbf{r}'_\perp \mathcal{K}_0(q|\mathbf{r}_\perp - \mathbf{r}'_\perp|)\psi^{(a)}_{\mathbf{r}_\perp}\psi^{(b)}_{\mathbf{r}_\perp}\psi^{(c)}_{\mathbf{r}'_\perp}\psi^{(d)}_{\mathbf{r}'_\perp}. \quad (14)$$

This expression is written through the Bessel function $\mathcal{K}_0(x)$ (the same as in problem 6.1) after integrating over $\mathbf{q}_\perp$ in Eq. (13) with $|C_Q|^2 = |C_Q^{(LO)}|^2 \propto Q^{-2}$. The remaining integral over the scalar variable q is calculated by using the δ-functions. The result is

$$\nu^{(nn')}_{LO}(p) = \frac{\hbar\omega_{LO}}{a_B\sqrt{\tilde{p}^2_{nn'} + p^2}}\frac{\epsilon}{\epsilon^*}\left[K_{nn'nn'}(q_{\min}) + K_{nn'nn'}(q_{\max})\right], \quad (15)$$

where $q_{\min}$, $q_{\max}$, and $\tilde{p}^2_{nn'}$ are given by the same relations as in Eq. (8). Typically, the intersubband energies in artificial quantum wires are considerably smaller than the optical phonon energies, and the acoustic-phonon relaxation remains significant. In the quasielastic approximation, this mechanism leads to the following intersubband relaxation rate (compare to Eq. (11)):

$$\nu^{(nn')}_{LA}(p) = \frac{2\mathcal{D}^2 T m}{\rho s_l^2 \hbar^2 \sqrt{p^2_{nn'} + p^2}} \int d\mathbf{r}_\perp |\psi^{(n)}_{\mathbf{r}_\perp}|^2 |\psi^{(n')}_{\mathbf{r}_\perp}|^2, \quad (16)$$

which is momentum-independent at $p^2 \ll p^2_{nn'}$. The specific features of the transitions between 1D states can manifest themselves in the relaxation of the Landau level population (problem 7.23), because the dynamics of 3D electrons in strong magnetic fields resembles 1D motion, with a characteristic inverse-square-root divergence of the density of states at the bottom of each Landau level; see Sec. 5.

Let us consider the relaxation of electrons in quantum dots. A quantum dot is a small region of the crystal, where the electrons are localized due to a three-dimensional confinement potential. The electron transitions occur between the discrete levels of the localized states described by the wave functions $\langle \mathbf{r}|n\rangle = \psi^{(n)}_\mathbf{r}$ depending on the sets of discrete quantum numbers, n. If the dot contains two or more electrons, the state n should be considered as a state of many-electron system, like in atoms. The properties of the dots are essentially determined by Coulomb interaction between the electrons and depend on the number of electrons in the dot (we will discuss this issue later, in Sec. 61). Below we restrict ourselves by the single-electron approximation and write the probability of phonon-assisted transition between the states n and n' as

$$W_{nn'} = \frac{2\pi}{\hbar} \sum_{\mathbf{q}} |C_q|^2 |\langle n'|e^{i\mathbf{q}\cdot\mathbf{r}}|n\rangle|^2 \left[\delta(\varepsilon_n - \varepsilon_{n'} - \hbar\omega_q)(N_q + 1)\right.$$

$$\left. + \delta(\varepsilon_n - \varepsilon_{n'} + \hbar\omega_q)N_q\right]. \tag{17}$$

The phonon emission (first term) takes place when $\varepsilon_{nn'} = \varepsilon_n - \varepsilon_{n'}$ is positive. If it is negative, the phonon absorption (second term) occurs. Therefore, for a given pair of states n and n', one should consider either the first or the second term in Eq. (17). The sum over $\mathbf{q}$ in this equation is calculated easily, because the phonon wave number q is fixed by the energy conservation law, and only the matrix element $\langle n'|e^{i\mathbf{q}\cdot\mathbf{r}}|n\rangle$ depends on the angle of $\mathbf{q}$. The transition probability due to LA phonon emission is calculated as

$$W_{nn'}^{(LA)} = \frac{\mathcal{D}^2 q_{nn'}^3 \mathcal{L}_{nn'nn'}(q_{nn'})}{2\pi\hbar\rho s_l^2}(N_{q_{nn'}} + 1), \qquad q_{nn'} = \varepsilon_{nn'}/\hbar s_l > 0, \tag{18}$$

where the dimensionless factor

$$\mathcal{L}_{abcd}(q) = \int d\mathbf{r} \int d\mathbf{r}' \frac{\sin(q|\mathbf{r} - \mathbf{r}'|)}{q|\mathbf{r} - \mathbf{r}'|} \psi_{\mathbf{r}}^{(a)} \psi_{\mathbf{r}}^{(b)} \psi_{\mathbf{r}'}^{(c)} \psi_{\mathbf{r}'}^{(d)} \tag{19}$$

comes from the averaging of the squared matrix element $|\langle n'|e^{i\mathbf{q}\cdot\mathbf{r}}|n\rangle|^2$ over the angle. To introduce the relaxation rate and express it through the probability $W_{nn'}^{(LA)}$ calculated above, one should take into account possible degeneracy of the levels in the quantum dot. If we denote the energy level by the index $\mathcal{N}$, it can contain $P_{\mathcal{N}}$ different states n. The relaxation rate $\nu_{LA}^{(\mathcal{N}\mathcal{N}')}$ is defined as

$$\nu_{LA}^{(\mathcal{N}\mathcal{N}')} = P_{\mathcal{N}}^{-1} \sum_{n\in\mathcal{N}} \sum_{n'\in\mathcal{N}'} W_{nn'}^{(LA)}. \tag{20}$$

Considering an ensemble of equivalent isolated dots, one can introduce the occupation number $f_{\mathcal{N}t}$ as the number of the dots containing an electron in a state n belonging to the level $\mathcal{N}$ divided by the total number of the dots. The balance equation

$$\frac{\partial f_{\mathcal{N}t}}{\partial t} = \sum_{\mathcal{N}'} \left[\nu_{LA}^{(\mathcal{N}'\mathcal{N})} f_{\mathcal{N}'t} - \nu_{LA}^{(\mathcal{N}\mathcal{N}')} f_{\mathcal{N}t}\right] \tag{21}$$

reflects the evolution of the occupation numbers and provides the relation $f_{\mathcal{N}}/f_{\mathcal{N}'} = (P_{\mathcal{N}}/P_{\mathcal{N}'}) \exp[(\varepsilon_{\mathcal{N}'} - \varepsilon_{\mathcal{N}})/T]$ in the stationary case and when the phonons are in equilibrium.

The relaxation rate $\nu_{LA}^{(\mathcal{N}\mathcal{N}')}$ considerably depends on $\varepsilon_{nn'} = \varepsilon_{\mathcal{N}\mathcal{N}'}$, because the factor $\mathcal{L}_{nn'nn'}(q_{nn'})$ rapidly decreases as $q_{nn'}$ exceeds the inverse size of the dot, $1/\overline{d}$. Estimating $\varepsilon_{nn'}$ as $\hbar^2/m\overline{d}^2$, we find that $q_{nn'}\overline{d} \sim (\varepsilon_{nn'}/ms_l^2)^{1/2}$, which means that the relaxation becomes weak even at relatively small energies $\varepsilon_{nn'}$. For example, considering the transitions between the ground and first excited levels in a single-electron quantum dot with the parabolic confinement potential $m\Omega^2 r^2/2$ (problem 7.24), we obtain

$$\nu_{LA}^{(21)} = \frac{\mathcal{D}^2 \Omega^4 \exp(-\hbar\Omega/2ms_l^2)}{12\pi\rho ms_l^7 [1 - \exp(-\hbar\Omega/T)]}. \tag{22}$$

To derive this equation, we have used Eqs. (18) and (20) with $\varepsilon_{21} = \hbar\Omega$ for the model under consideration. The exponential decrease of $\nu_{LA}^{(21)}$ with increasing ε_{21} makes the interlevel relaxation inefficient. In this case, multi-phonon processes become important. Consider, for example, the relaxation with emission of two acoustic phonons. Substituting the two-phonon transition probability, calculated according to Eq. (34.30), in place of $W_{nn'}^{(LA)}$, we find

$$\nu_{2LA}^{(\mathcal{N}\mathcal{N}')} = \frac{2\pi}{\hbar} \sum_{\mathbf{q}\mathbf{q}_1} |C_q^{(LA)}|^2 |C_{q_1}^{(LA)}|^2 \frac{1}{P_{\mathcal{N}}} \sum_{n \in \mathcal{N}} \sum_{n' \in \mathcal{N}'}$$

$$\times \left| 2 \sum_m \frac{\langle n'|e^{-i\mathbf{q}_1 \cdot \mathbf{r}}|m\rangle \langle m|e^{-i\mathbf{q} \cdot \mathbf{r}}|n\rangle}{\varepsilon_{\mathcal{N}} - \varepsilon_m - \hbar\omega_q} \right|^2 \tag{23}$$

$$\times (N_{q_1} + 1)(N_q + 1)\delta[\varepsilon_{\mathcal{N}} - \varepsilon_{\mathcal{N}'} - \hbar s_l(q + q_1)].$$

Since we consider the emission of two equivalent bosons, the two terms forming the effective matrix element in Eq. (34.30) are equivalent, and we have a single term multiplied by 2 in Eq. (23). Formally, this reduction is justified by permutation of the variables $\mathbf{q}$ and $\mathbf{q}_1$ under the sign of sum.

Further analysis of Eq. (23) becomes convenient for the model of parabolic confinement potential. According to the result of problem 7.24, the effective matrix element in Eq. (23) is proportional to $\exp[-\ell^2(q^2 + q_1^2)/4]$, where $\ell = \sqrt{\hbar/m\Omega}$. Since both q and q_1 are smaller than $q_{nn'} = (\varepsilon_{\mathcal{N}} - \varepsilon_{\mathcal{N}'})/\hbar s_l$, this exponential factor is larger than the corresponding factor for single-phonon processes. Therefore, as the interlevel energy increases, the two-phonon transitions become more important than the single-phonon ones. This occurs at $q, q_1 \gg \ell^{-1}$, when the exponential factor is small. To estimate the relaxation rate (23), we restrict ourselves only by the states m belonging to the levels $\mathcal{N}$ and $\mathcal{N}'$, because they give the smallest denominators in the effective matrix element in Eq. (23). Assuming also that $q, q_1 \gg \ell^{-1}$, we obtain

$$\frac{1}{P_{\mathcal{N}}} \sum_{n \in \mathcal{N}} \sum_{n' \in \mathcal{N}'} \left| 2 \sum_{m \in \mathcal{N}, \mathcal{N}'} \frac{\langle n'|e^{-i\mathbf{q}_1 \cdot \mathbf{r}}|m\rangle \langle m|e^{-i\mathbf{q} \cdot \mathbf{r}}|n\rangle}{\varepsilon_{\mathcal{N}} - \varepsilon_m - \hbar s_l q} \right|^2$$

$$\simeq \frac{(\mathbf{q} \cdot \mathbf{q}_1)^2 \ell^6}{6\hbar^2 s_l^2} \exp[-\ell^2(q^2 + q_1^2)/2], \quad \mathcal{N} = 2, \ \mathcal{N}' = 1. \tag{24}$$

Substituting this approximate relation into Eq. (23), we integrate over $\mathbf{q}$ and $\mathbf{q}_1$. The rate of relaxation from the first excited state to the ground one is finally estimated as

$$\nu_{2LA}^{(21)} \propto \frac{\mathcal{D}^4 \Omega^8 \ell \exp(-\hbar\Omega/4m s_l^2)}{\rho^2 m^2 s_l^{15}[1 - \exp(-\hbar\Omega/2T)]^2}. \tag{25}$$

The main contribution to this relaxation rate comes from $q \simeq q_1$, when the exponential factor in Eq. (24) is maximal. The consideration given above can be generalized to take into account three-phonon processes and so on. In these conditions, the efficiency of multi-phonon processes increases with increasing temperature, similar to the two-phonon case described by Eq. (25)

The *LO* phonon-assisted relaxation in quantum dots can take place only if $\varepsilon_{\mathcal{N}\mathcal{N}'}$ comes in resonance with the *LO* phonon energy. This case is more complicated, because, due to discrete nature of electron spectrum, the relaxation rate diverges at $\varepsilon_{\mathcal{N}\mathcal{N}'} = \hbar\omega_{LO}$, and one has to introduce either a weak dispersion of *LO* phonons or a finite broadening of the levels, as in Sec. 29. If one considers an ensemble of dots, this broadening may be attributed to a random dispersion of the level energies. In contrast, for a single isolated quantum dot the broadening of the levels appears only due to interaction of electrons with bosons (problem 7.25). If $|\varepsilon_{\mathcal{N}\mathcal{N}'} - \hbar\omega_{LO}|$ exceeds the broadening energy but still remains small, an efficient relaxation is possible via two-phonon processes, when an optical phonon is emitted and an acoustic phonon is either emitted or absorbed. Using Eq. (34.30), we write the rate of such transitions as

$$\nu_{LO\pm LA}^{(\mathcal{N}\mathcal{N}')} = \frac{2\pi}{\hbar} \sum_{\mathbf{q}\mathbf{q}_1} |C_q^{(LA)}|^2 |C_{q_1}^{(LO)}|^2 \frac{1}{P_{\mathcal{N}}} \sum_{n \in \mathcal{N}} \sum_{n' \in \mathcal{N}'}$$

$$\times \left| \sum_m \frac{\langle n'|e^{-i\mathbf{q}_1\cdot\mathbf{r}}|m\rangle\langle m|e^{\mp i\mathbf{q}\cdot\mathbf{r}}|n\rangle}{\varepsilon_{\mathcal{N}} - \varepsilon_m \mp \hbar s_l q} + \sum_{m'} \frac{\langle n'|e^{\mp i\mathbf{q}\cdot\mathbf{r}}|m'\rangle\langle m'|e^{-i\mathbf{q}_1\cdot\mathbf{r}}|n\rangle}{\varepsilon_{\mathcal{N}} - \varepsilon_{m'} - \hbar\omega_{LO}} \right|^2$$

$$\times \left(N_q + \frac{1}{2} \pm \frac{1}{2} \right) \delta(\varepsilon_{\mathcal{N}} - \varepsilon_{\mathcal{N}'} - \hbar\omega_{LO} \mp \hbar s_l q). \tag{26}$$

We assume that the temperature is low enough and only the spontaneous emission of LO phonons is essential. The energy denominators in the effective matrix elements are small (equal to the acoustic phonon energy) only if m belongs to $\mathcal{N}$ and m' belongs to $\mathcal{N}'$. Taking this into account and using the parabolic potential model, we find ($\mathcal{N} = 2$, $\mathcal{N}' = 1$)

$$\frac{1}{P_{\mathcal{N}}} \sum_{n\in\mathcal{N}} \sum_{n'\in\mathcal{N}'} |\ldots|^2 = \frac{(\mathbf{q}\cdot\mathbf{q}_1)^2\ell^6}{24\hbar^2 s_l^2} \exp\left[-\frac{\ell^2(q^2 + q_1^2)}{2} \right]. \tag{27}$$

Substituting this expression into Eq. (26), one can calculate the integrals over q and q_1 separately. The first integral is calculated with the use of the δ-function, while the second one is taken elementary. We obtain

$$\nu_{LO\pm LA}^{(21)} = \frac{\mathcal{D}^2 e^2 \omega_{LO} \ell^3 |\Omega - \omega_{LO}|^5}{2^{9/2} 9\pi^{3/2} \hbar^2 \epsilon^* \rho s_l^9} \exp\left[-\frac{\hbar(\Omega - \omega_{LO})^2}{2m s_l^2 \Omega} \right]$$

$$\times \left| 1 - \exp\left[\frac{\hbar(\omega_{LO} - \Omega)}{T} \right] \right|^{-1}. \tag{28}$$

The relaxation rate $\nu_{LO+LA}^{(21)}$ exponentially decreases with increasing $|\Omega - \omega_{LO}|$, in a similar way as the single-phonon relaxation rate of Eq. (22) decreases with increasing Ω. Though the derivation of Eqs. (22), (25), and (28) has been based upon a model of the dots with parabolic confinement, the qualitative features of the interlevel energy dependence of the relaxation rate remain valid for any model, provided that the population of the dot ensemble is not high, less than one electron per dot in average, and the single-electron approximation is reliable.

Problems

7.1. Check the relations (34.3).

Hints: Take into account that $\widehat{H}_b$ is an operator with respect to bosonic variables only and can be permuted with $\hat{n}_t$ under the trace $\mathrm{Sp}_b\ldots$. To check the second relation, use Eq. (34.2) for $\hat{\mathcal{L}}_t$ and take into account that $\mathrm{Sp}_e\ldots$ is an operator with respect to bosonic variables only.

7.2. Prove the expression (34.12) for the average (34.11).

<u>Solution:</u> The equation of motion for $g_{\eta\nu}^{\gamma\delta}$ is written as

$$i\hbar\frac{\partial g_{\eta\nu}^{\gamma\delta}(qt)}{\partial t} = \sum_{\kappa}\{\langle\gamma|\hat{h}_t|\kappa\rangle g_{\eta\nu}^{\kappa\delta}(qt) - g_{\eta\nu}^{\gamma\kappa}(qt)\langle\kappa|\hat{h}_t|\delta\rangle$$

$$+\langle\eta|\hat{h}_t|\kappa\rangle g_{\gamma\delta}^{\kappa\nu}(qt) - g_{\gamma\delta}^{\eta\kappa}(qt)\langle\kappa|\hat{h}_t|\nu\rangle\} + [\hbar\omega_q\hat{b}_q^+\hat{b}_q, g_{\eta\nu}^{\gamma\delta}(qt)],$$

where we have used the identities $g_{\eta\nu}^{\gamma\delta}(qt) = g_{\gamma\delta}^{\eta\nu}(qt) = -g_{\eta\delta}^{\gamma\nu}(qt) = -g_{\eta\nu}^{\eta\delta}(qt)$ following from the anticommutation relations for Fermi operators. The same expression is obtained when we differentiate the right-hand side of Eq. (34.12) over time and multiply it by $i\hbar$.

7.3. Using the eigenstate problem $\hat{h}|\delta\rangle = \varepsilon_\delta|\delta\rangle$ with non-degenerate states $|\delta\rangle$, derive a kinetic equation for $f_{\delta t}$, where the collision integral is written in the form similar to Eq. (34.25).

<u>Solution:</u> The non-diagonal elements $\langle\delta|\hat{\rho}_t|\delta'\rangle$ of the density matrix are small if the interlevel energies considerably exceed the collision-induced broadening of the levels; see also problem 11.1. The terms proportional to $\mathrm{sp}\hat{\chi}_q^+\hat{\rho}_t$ and $\mathrm{sp}\hat{\chi}_q\hat{\rho}_t$ in Eq. (34.23) give zero contribution if only the diagonal matrix elements $\langle\delta|\hat{\rho}_t|\delta\rangle = f_{\delta t}$ are taken into account. Using an intermediate expression of the kind of Eq. (34.24), we calculate the integral over time and obtain

$$\frac{\partial f_{\delta t}}{\partial t} = \frac{2\pi}{\hbar}\sum_{\gamma q}\{|\langle\delta|\hat{\chi}_q^+|\gamma\rangle|^2\delta(\varepsilon_\delta - \varepsilon_\gamma + \hbar\omega_q)[f_{\gamma t}(1 - f_{\delta t})(N_{qt} + 1) - f_{\delta t}(1 - f_{\gamma t})N_{qt}]$$

$$+|\langle\delta|\hat{\chi}_q|\gamma\rangle|^2\delta(\varepsilon_\delta - \varepsilon_\gamma - \hbar\omega_q)[f_{\gamma t}(1 - f_{\delta t})N_{qt} - f_{\delta t}(1 - f_{\gamma t})(N_{qt} + 1)]\}.$$

7.4. Check that the equilibrium Fermi distribution is a solution of the stationary kinetic equation if the bosons are described by the Planck distribution.

<u>Solution:</u> The Planck distribution of the bosons with temperature T satisfies the identity $N_q + 1 = N_q e^{\hbar\omega_q/T}$. Employing it in Eq. (34.27), we get a relation between the transition probabilities, which is called the principle of detailed balance: $W_{\mathbf{pp'}} = W_{\mathbf{p'p}}\exp[(\varepsilon_p - \varepsilon_{p'})/T]$. With this relation, one may check that the collision integral (34.26) is equal to zero for equilibrium Fermi distribution of electrons. This result is valid for any stationary system, for example, the system considered in problem 7.3.

7.5. Calculate ν_{ac} and $\bar{\varepsilon}$: *a)* for a rectangular quantum well of width d in the hard-wall model, and *b)* for a triangular quantum well formed in a selectively-doped heterojunction (use the wave function $\psi_z = (z/b\sqrt{2b})e^{-z/2b}$ obtained by the variational method).

<u>Results:</u>

$$\nu_{ac} = \frac{\pi\mathcal{D}^2 T\rho_{2D}k_0}{\hbar s_l^2\rho}, \quad \bar{\varepsilon}^2 = \hbar^2 s_l^2 k_2^3/2k_0,$$

where the characteristic wave numbers k_0 and k_2 are defined as $k_0 = \int dz \psi_z^4$ and $k_2 = [\int dz (d\psi_z^2/dz)^2]^{1/3}$. For the models *(a)* and *(b)*, we have $k_0 = 3/2d$, $k_2 = \sqrt[3]{2\pi^2}/d$, and $k_0 = 3/16b$, $k_2 = 1/2\sqrt[3]{2}b$, respectively.

7.6. Consider the heating of 2D electrons by a stationary electric field, taking into account both quasielastic scattering by phonons and elastic scattering by the impurity potential.

<u>Solution:</u> The collision integral in the equation for the asymmetric part $\Delta f_{\mathbf{p}}$ contains two terms, the phonon-induced term (35.5) and the impurity-induced term $-\nu_\varepsilon \Delta f_{\mathbf{p}}$ standing in Eq. (8.24). In the stationary case, $\Delta f_{\mathbf{p}} = -\tau_{tr}(\varepsilon) e \mathbf{E} \cdot \mathbf{v_p} (df_\varepsilon/d\varepsilon)$, where the transport time $\tau_{tr}(\varepsilon) = (\nu_\varepsilon + \nu_{ac})^{-1}$ is determined by these two scattering mechanisms. As a consequence, the energy ε_E standing in Eq. (35.9) becomes proportional to $\tau_{tr}(\varepsilon)\nu_{ac}^{-1}$ instead of ν_{ac}^{-2}.

7.7. Check that, for an arbitrary degeneracy of the electron gas, the last term in the expression (35.9) for I_ε contains $f_\varepsilon(1 - f_\varepsilon)$ instead of f_ε.

<u>Hints:</u> Replace $[f_{\mathbf{p}} + f_{\mathbf{p}'}]$ by $[f_{\mathbf{p}} + f_{\mathbf{p}'} - 2f_{\mathbf{p}}f_{\mathbf{p}'}]$ in Eq. (35.2) and repeat the calculations leading to Eq. (35.9).

7.8. Consider the heating of 3D electrons by a stationary electric field, assuming that the electrons interact with acoustic phonons.

<u>Solution:</u> Using the quasielastic approximation in a similar way as in the 2D case, we find the collision integral ($\varepsilon = \varepsilon_p$, $\mathbf{p}' = \mathbf{p} + \hbar\mathbf{q}$):

$$J_{ac}(f|\varepsilon) = 2\pi \sum_{\mathbf{p}'} |C_q|^2 \omega_q \left[\frac{\partial A_\varepsilon}{\partial \varepsilon} \delta(\varepsilon_{p'} - \varepsilon_p) - 2A_\varepsilon \delta'(\varepsilon_{p'} - \varepsilon_p) \right],$$

where

$$A_\varepsilon = T \frac{\partial f_\varepsilon}{\partial \varepsilon} + f_\varepsilon(1 - f_\varepsilon).$$

This integral is transformed to

$$J_{ac}(f|\varepsilon) = \varepsilon^{-1/2} \frac{\partial}{\partial \varepsilon} \varepsilon^{3/2} \nu_{qe}(\varepsilon) A_\varepsilon, \quad \nu_{qe}(\varepsilon) = \frac{2^{3/2} m^{5/2} \mathcal{D}^2}{\pi \hbar^4 \rho} \varepsilon^{1/2},$$

where we have introduced the rate of quasielastic energy relaxation, $\nu_{qe}(\varepsilon)$. Calculating the field contribution on the left-hand side of the kinetic equation, we write the stationary kinetic equation as

$$-\frac{2e^2 E^2}{3m\varepsilon^{1/2}} \frac{\partial}{\partial \varepsilon} \left[\varepsilon^{3/2} \tau_{tr}(\varepsilon) \frac{\partial f_\varepsilon}{\partial \varepsilon} \right] = J_{ac}(f|\varepsilon).$$

The transport time, in general, includes different scattering mechanisms (as in problem 7.6). Considering only the acoustic-phonon scattering in the quasielastic approximation, we put $\tau_{tr}(\varepsilon) = \tau_{LA}(\varepsilon)$, where $\tau_{LA}(\varepsilon) = (2ms_l^2/T)\nu_{qe}^{-1}(\varepsilon)$. The kinetic equation is finally reduced to the form $dI_\varepsilon/d\varepsilon = 0$, as in Eq. (35.9), where

$$I_\varepsilon = \varepsilon^{3/2} \nu_{qe}(\varepsilon) \left[(T + \zeta_E^2/\varepsilon) \frac{df_\varepsilon}{d\varepsilon} + f_\varepsilon(1 - f_\varepsilon) \right], \quad \zeta_E^2 = \frac{4}{3} \left[\frac{eEs_l}{\nu_{qe}(T)} \right]^2.$$

The characteristic energy ζ_E is independent of ε. The structure of the flow I_ε is different from that of Eq. (35.9). Indeed, the term standing at $df_\varepsilon/d\varepsilon$ in the equation above decreases with increasing ε. By equating I_ε to zero (the flow is continuous), we find

$$f_\varepsilon = Ne^{-\varepsilon/T}(1 + \varepsilon T/\zeta_E^2)^{\zeta_E^2/T^2}$$

for the non-degenerate electron gas. The field effect is described by a multiplier to the Boltzmann distribution. This distribution function always can be normalized, which means that the electrons do not run away.

7.9. Check the compatibility of the boundary conditions (35.16) and (35.17).

Hint: Using the expressions for λ_{LO} and I_ε given by Eqs. (35.15) and (35.9), respectively, show that $I_{\varepsilon=0}$ determined by Eq. (35.17) is equal to $I_{\varepsilon=\hbar\omega_{LO}}$ within the accuracy of the approximations used in the derivation of the expression for λ_{LO}.

7.10. Calculate the rate of spontaneous emission of LO phonons by 2D electrons at the threshold ($\varepsilon = \hbar\omega_{LO}$).

Result: $\nu_{LO} = \pi\omega_{LO}(\epsilon/\epsilon^*)M_{1111}(p_{LO}/\hbar)$, where the matrix elements $M_{abcd}(q)$ are defined by Eq. (29.11).

7.11. Calculate the rate of spontaneous emission of LO phonons by 3D electrons.

Result:

$$\nu_{LO}(\varepsilon) = \alpha\omega_{LO}\sqrt{\frac{\hbar\omega_{LO}}{\varepsilon}}\ln\left|\frac{\sqrt{\varepsilon} + \sqrt{\varepsilon - \hbar\omega_{LO}}}{\sqrt{\varepsilon} - \sqrt{\varepsilon - \hbar\omega_{LO}}}\right|,$$

where ε is the electron energy and α is introduced by Eq. (21.28).

7.12. Calculate $\alpha_{\mathbf{r}t}$ introduced in Eq. (36.8) for degenerate and non-degenerate electron gases. Assume that the product $\rho_D(\varepsilon)\tau_{tr}(\varepsilon)$ is energy-independent.

Result: Calculating the integral (36.9), we obtain $\alpha = \pi^2 T/3e\mu$ for degenerate electrons and $\alpha = 2/e - \mu/eT$ for non-degenerate electrons.

7.13. Solve Eq. (36.17) with the initial condition $\Delta n_{\mathbf{r}t=0} = \Delta n_0\delta(\mathbf{r})$.
Result:

$$\Delta n_{\mathbf{r}t} = \frac{\Delta n_0}{(4\pi Dt)^{3/2}}\exp\left(-\frac{r^2}{4Dt} - \frac{t}{\tau_M}\right).$$

7.14. Obtain Eq. (36.18) directly from Eq. (36.4).

Hint: Multiply Eq. (36.4) by ε_p and integrate the equation obtained over $\mathbf{p}$.

7.15. Derive Eq. (36.21).

Hints: Applying Eq. (36.19) with $\mathbf{g}$ from Eq. (36.7), separate the term $(\mu/e)\mathbf{I}$ from $\mathbf{G}$ and write the remaining part of $\mathbf{G}$ through the gradients of the temperature and electrochemical potential, employing the kinetic coefficient α defined by Eq. (36.9). Use Eq. (36.8) to exclude the electrochemical potential.

7.16. Consider the energy balance of the electrons interacting with *LO* phonons. Approximate the electron distribution *a)* by the isotropic Maxwell distribution with the effective temperature T_e and *b)* by the streaming distribution.

<u>Solution:</u> Substituting $|C_q|^2 = 2\pi e^2 \hbar \omega_{LO}/V\epsilon^* q^2$ and $\omega_q = \omega_{LO}$ into Eq. (36.24), we calculate the sum over $\mathbf{q}$ and find the power loss term

$$P = \frac{2}{V}\sum_{\mathbf{p}} \hbar\omega_{LO}\nu_{LO}(\varepsilon_p)(N_{LO}+1)f(\varepsilon_p)\left[1-\exp(\hbar\omega_{LO}/T_e - \hbar\omega_{LO}/T)\right],$$

where $f(\varepsilon_p) = \exp[(\mu - \varepsilon_p)/T_e]$ and the rate of spontaneous *LO* phonon emission, $\nu_{LO}(\varepsilon)$, is given in problem 7.11. If $T_e \ll \hbar\omega_{LO}$, we expand the logarithm standing there near the threshold $\varepsilon = \hbar\omega_{LO}$ and find $P = 2n\alpha\hbar\omega_{LO}^2\left(e^{-\hbar\omega_{LO}/T_e} - e^{-\hbar\omega_{LO}/T}\right)$. Considering the streaming conditions, we can neglect the absorption of phonons and use the expression given above, where the factor in the square brackets is replaced by unity and the streaming distribution (35.31) stands instead of $f(\varepsilon_p)$. Again, after integrating over electron momenta, we obtain $P = n\hbar\omega_{LO}|e|E/p_{LO}$. This expression has clear physical meaning because $p_{LO}/|e|E$ is the time of ballistic motion of electrons to the active region, where they emit the phonons with energy $\hbar\omega_{LO}$.

In the case of Maxwell distribution, the energy balance equation $P = \mathbf{I}\cdot\mathbf{E}$, with the optical-phonon power loss term P calculated above, gives us a logarithmic increase of the electron temperature T_e with increasing electric field. In the case of streaming distribution, the balance equation coincides with the expression for the current density $I = en\hbar\omega_{LO}/p_{LO}$ obtained in the end of Sec. 35.

7.17. Check the particle conservation in the collisions described by the integrals $J_{im}(\overline{f}|\mathbf{P}t)$ and $J_{e,ph}(\overline{f}|\mathbf{P}t)$ in Sec. 37.

<u>Hint:</u> Integrate $J_{im}(\overline{f}|\mathbf{P}t)$ and $J_{e,ph}(\overline{f}|\mathbf{P}t)$ over $\mathbf{P}$.

7.18. Compare Eqs. (10.18) and (37.2) expressing $\Delta_0(E)$ in the quantum and quasi-classical cases, respectively.

<u>Solution:</u> Equation (10.18) follows from the general expression (10.17). Using this expression, we have

$$\Delta_0(E) = \frac{\omega}{2\pi^2\hbar}\int_{-\pi/\omega}^{\pi/\omega} dt \int_{-\infty}^{0} d\tau\, e^{\lambda\tau}\cos\left\{\frac{\tau E}{\hbar} - \frac{A}{\hbar\omega}[\cos\omega(t+\tau) - \cos\omega t]\right\},$$

where $A = \hbar\mathbf{q}\cdot\mathbf{v}_\omega$. Under the quasi-classical condition $E \gg \hbar\omega$, we carry out the expansion $\cos\omega(t+\tau) - \cos\omega t \simeq -\omega\tau\sin\omega t$. Calculating the integral over τ, we obtain the δ-function averaged over the period according to Eq. (37.2).

7.19. Using the results of Sec. 10, prove that the first term on the left-hand side of Eq. (37.9) is equal to the power of electromagnetic radiation absorbed by the electrons interacting with impurities.

Solution: According to Sec. 10, this power is equal to $\sigma(\omega)\overline{\mathbf{E}_t^2} = \sigma(\omega)E^2/2$, where $\sigma(\omega)$ is determined by Eqs. (10.23) and (10.24). Under the integrals over $\mathbf{p}$ and $\mathbf{q}$ in Eq. (10.24), one may replace $\mathbf{p}(\overline{f}_{\mathbf{p}+\hbar\mathbf{q}} - \overline{f}_{\mathbf{p}})$ by $\hbar\mathbf{q}\overline{f}_{\mathbf{p}}$. Finally, using the identity $[\mathcal{J}_{k-1}(x) + \mathcal{J}_{k+1}(x)]x = 2k\mathcal{J}_k(x)$, one proves the required statement.

7.20. Using Eq. (34.30), represent $J_r(f|\varepsilon)$ as a two-boson (one photon and one LO phonon) contribution to the electron-boson collision integral and reduce it to the form given by Eqs. (37.18) and (37.19).

Solution. Let q_1 in Eq. (34.30) numbers photon states and q numbers LO phonon states. Considering the monochromatic photons in the dipole approximation and at $N_{q_1} \gg 1$, we obtain, from Eqs. (34.30) and (34.31),

$$J_{e,b}^{(2)}(f|\delta) = \frac{2\pi}{\hbar}\left(\frac{eE}{2\omega}\right)^2 \sum_\gamma \sum_\mathbf{q} |C_q^{(LO)}|^2 \sum_{s=\pm1}\left\{\left|\sum_\nu \frac{\langle\gamma|\hat{\mathbf{v}}\cdot\mathbf{e}|\nu\rangle\langle\nu|e^{-i\mathbf{q}\cdot\mathbf{r}}|\delta\rangle}{\varepsilon_\delta - \varepsilon_\nu - \hbar\omega_{LO}}\right.\right.$$

$$+ \sum_\nu \frac{\langle\gamma|e^{-i\mathbf{q}\cdot\mathbf{r}}|\nu\rangle\langle\nu|\hat{\mathbf{v}}\cdot\mathbf{e}|\delta\rangle}{\varepsilon_\delta - \varepsilon_\nu - s\hbar\omega}\Bigg|^2 \delta(\varepsilon_\delta - \varepsilon_\gamma - s\hbar\omega - \hbar\omega_{LO})[f_\gamma(1-f_\delta)N_{LO}$$

$$-f_\delta(1-f_\gamma)(N_{LO}+1)] + \left|\sum_\nu \frac{\langle\gamma|\hat{\mathbf{v}}\cdot\mathbf{e}|\nu\rangle\langle\nu|e^{i\mathbf{q}\cdot\mathbf{r}}|\delta\rangle}{\varepsilon_\delta - \varepsilon_\nu + \hbar\omega_{LO}} + \sum_\nu \frac{\langle\gamma|e^{i\mathbf{q}\cdot\mathbf{r}}|\nu\rangle\langle\nu|\hat{\mathbf{v}}\cdot\mathbf{e}|\delta\rangle}{\varepsilon_\delta - \varepsilon_\nu - s\hbar\omega}\right|^2$$

$$\times\delta(\varepsilon_\delta - \varepsilon_\gamma - s\hbar\omega + \hbar\omega_{LO})[f_\gamma(1-f_\delta)(N_{LO}+1) - f_\delta(1-f_\gamma)N_{LO}]\Bigg\},$$

where N_{LO} is the occupation number of optical phonons and $\mathbf{e}$ is the unit vector along $\mathbf{E}$. Using the basis of plane waves, we have $\langle\mathbf{p}'|e^{i\mathbf{q}\cdot\mathbf{r}}|\mathbf{p}\rangle = \delta_{\mathbf{p}',\mathbf{p}+\hbar\mathbf{q}}$ and $\langle\mathbf{p}'|\hat{\mathbf{v}}\cdot\mathbf{e}|\mathbf{p}\rangle = (\mathbf{v}_\mathbf{p}\cdot\mathbf{e})\delta_{\mathbf{p}',\mathbf{p}}$. Assuming that the electron gas is non-degenerate, we find

$$J_{e,b}^{(2)}(f|\mathbf{p}) = \frac{v_\omega^2 e^2 \omega_{LO}}{2\omega^2\epsilon^*}\int_0^\infty dq \int \frac{d\widetilde{\Omega}_q}{4\pi}(\mathbf{q}\cdot\mathbf{e})^2$$

$$\times \sum_{s=\pm1}\{\delta(\varepsilon_p - \varepsilon_{\mathbf{p}-\hbar\mathbf{q}} - s\hbar\omega - \hbar\omega_{LO})[f_{\mathbf{p}-\hbar\mathbf{q}}N_{LO} - f_\mathbf{p}(N_{LO}+1)]$$

$$+\delta(\varepsilon_p - \varepsilon_{\mathbf{p}+\hbar\mathbf{q}} - s\hbar\omega + \hbar\omega_{LO})[f_{\mathbf{p}+\hbar\mathbf{q}}(N_{LO}+1) - f_\mathbf{p}N_{LO}]\},$$

where $d\widetilde{\Omega}_q$ is the differential of the solid angle of the vector $\mathbf{q}$. Assuming that $f_\mathbf{p} = f(\varepsilon_p)$, we average $J_{e,b}^{(2)}(f|\mathbf{p})$ over the angle of $\mathbf{p}$, $J_r(f|\varepsilon) = (4\pi)^{-1}\int d\widetilde{\Omega}_p J_{e,b}^{(2)}(f|\mathbf{p})$, and obtain Eqs. (37.18) and (37.19). Equation (37.18) is written at $N_{LO} \ll 1$, when the optical phonons are frozen out.

7.21. Using the expressions for $\nu_{qe}(\xi)$ and $\nu_r(\xi,\xi')$, calculate the integral standing in the exponent of Eq. (37.23) at $p' = 0$.

Result:

$$\int_{\xi'}^\xi d\xi'' \frac{u_{\xi''}}{D_{\xi''}} = \frac{\xi' - \xi}{t} + \frac{\delta\omega(1 + \delta\omega/2t)}{rt}\ln\left[\frac{\xi + (\delta\omega)^2/2rt}{\xi' + (\delta\omega)^2/2rt}\right].$$

7.22. Analyze Eqs. (37.31) and (37.32).

Solution: With the use of Eq. (37.32) and under the conditions $\nu_{qe} \ll \nu_r$, Eq. (37.31) can be rewritten as an equation for f_ξ:

$$\frac{d}{d\xi}\left[D(\xi)\frac{df_\xi}{d\xi} - u(\xi)f_\xi\right] = 0,$$

where

$$D(\xi) = (\delta\omega)^2 \nu_r(\xi, \xi + p - 1/2) + t\left[\xi^{3/2}\nu_{qe}(\xi) + (\xi + 1/2)^{3/2}\nu_{qe}(\xi + 1/2)\eta_\xi\right],$$

$$u(\xi) = 2\delta\omega\nu_r(\xi, \xi + p - 1/2) - \left[\xi^{3/2}\nu_{qe}(\xi) + (\xi + 1/2)^{3/2}\nu_{qe}(\xi + 1/2)\left(\eta_\xi + t\frac{\partial\eta_\xi}{\partial\xi}\right)\right],$$

and $\eta_\xi = \nu_r(\xi, \xi + p - 1/2)/\nu_r(\xi + 1/2, \xi + p)$. This drift-diffusion equation has the same form as Eq. (37.21) and can be analyzed in a similar way. If t is small and $0 < \delta\omega < r[1/4 + \sqrt{p/(2p + 1)}]/\sqrt{2p}$, the function $u(\xi)$ changes its sign in the interval $[0, 1/2]$ from positive to negative, and the distribution f_ξ has a peak. This peak is repeated in the upper half of the passive region because $g_\xi \simeq \eta_\xi f_\xi$ according to Eq. (37.32). Therefore, we have two peaks in the passive region. The interval of existence of these peaks is smaller than in the case of integer photon-phonon resonance. If $\delta\omega$ exceeds this interval, one gets a broad distribution.

7.23. Calculate the LA phonon-assisted relaxation rate between the Landau levels of 3D electrons.

Result: Consider the electron in the Landau level N with momentum p_z along the magnetic field (see the description of the Landau quantization in Sec. 5). The relaxation rate of this state due to quasielastic LA phonon-assisted transitions to the level N' is given by

$$\nu_{LA}^{(NN')}(p_z) = \frac{2\mathcal{D}^2 Tm}{\rho s_l^2 \hbar^2 \sqrt{p_{NN'}^2 + p_z^2}} \int_0^\infty \frac{du}{2\pi l_H^2}\Phi_{N'N}(u),$$

where $p_{NN'}^2 = 2m\hbar\omega_c(N - N')$, l_H is the magnetic length, and the function $\Phi_{N'N}(u)$ is defined by Eq. (48.14).

7.24. Calculate the matrix elements $\langle n|e^{i\mathbf{q}\cdot\mathbf{r}}|n'\rangle$ by using the eigenstates $|n\rangle$ of the Schroedinger equation with the spherically-symmetric potential energy $V(\mathbf{r}) = m\Omega^2 r^2/2$.

Solution: Considering the Schroedinger equation for the electron in the parabolic potential,

$$\left[-\frac{\hbar^2}{2m}\left(\frac{\partial^2}{\partial x^2} + \frac{\partial^2}{\partial y^2} + \frac{\partial^2}{\partial z^2}\right) + \frac{m\Omega^2}{2}(x^2 + y^2 + z^2) - \varepsilon_n\right]\psi_{\mathbf{r}}^{(n)} = 0,$$

we search for the wave function in the form $\psi_{\mathbf{r}}^{(n)} = \psi_x\psi_y\psi_z$. The three-dimensional problem is reduced to the one-dimensional harmonic oscillator problem considered in Appendix A. The normalized wave function is presented as a product of the eigenfunctions of the harmonic oscillator:

$$\psi_{\mathbf{r}}^{(n)} = \frac{H_{n_1}(x/\ell)H_{n_2}(y/\ell)H_{n_3}(z/\ell)}{\sqrt{n_1!n_2!n_3!2^{n_1+n_2+n_3}}\pi^{3/2}\ell^3}\exp(-r^2/2\ell^2),$$

where $\ell = (\hbar/m\Omega)^{1/2}$ estimates the size of the quantum dot. The quantum number n should be considered as a combination of the numbers n_1, n_2, and n_3, each of them is either zero or positive integer. The energy spectrum of the quantum dot depends on the sum of these numbers as $\varepsilon_n = \hbar\Omega(\mathcal{N} + 1/2)$, where $\mathcal{N} = n_1 + n_2 + n_3 + 1$ is the level number. The degeneracy of the level is given by $P_\mathcal{N} = \mathcal{N}(\mathcal{N}+1)/2$. The ground level ($\mathcal{N} = 1$) contains a single state $n = (0,0,0)$, while the first exited level ($\mathcal{N} = 2$) is triple degenerate, since it contains the states $(1,0,0)$, $(0,1,0)$, and $(0,0,1)$. The matrix elements are calculated according to Eq. (A.27):

$$\langle n|e^{i\mathbf{q}\cdot\mathbf{r}}|n'\rangle = \sqrt{\frac{n_1!n_2!n_3!}{n_1'!n_2'!n_3'!}}\left(\frac{iq_x\ell}{\sqrt{2}}\right)^{n_1'-n_1}\left(\frac{iq_y\ell}{\sqrt{2}}\right)^{n_2'-n_2}\left(\frac{iq_z\ell}{\sqrt{2}}\right)^{n_3'-n_3}e^{-q^2\ell^2/4}$$

$$\times L_{n_1}^{n_1'-n_1}(q_x^2\ell^2/2)L_{n_2}^{n_2'-n_2}(q_y^2\ell^2/2)L_{n_3}^{n_3'-n_3}(q_z^2\ell^2/2).$$

Taking into account Eq. (38.20), it is convenient to introduce the squared matrix element

$$|\langle\mathcal{N}'|e^{i\mathbf{q}\cdot\mathbf{r}}|\mathcal{N}\rangle|^2 \equiv P_\mathcal{N}^{-1}\sum_{n\in\mathcal{N}}\sum_{n'\in\mathcal{N}'}|\langle n'|e^{i\mathbf{q}\cdot\mathbf{r}}|n\rangle|^2.$$

For the transition between the ground and first excited states, we obtain

$$\langle 000|e^{i\mathbf{q}\cdot\mathbf{r}}|100\rangle = (iq_x\ell/\sqrt{2})e^{-q^2\ell^2/4},$$

while the matrix elements $\langle 000|e^{i\mathbf{q}\cdot\mathbf{r}}|010\rangle$ and $\langle 000|e^{i\mathbf{q}\cdot\mathbf{r}}|001\rangle$ differ from this expression by substitutions of q_y and q_z in place of q_x. Therefore, the average over all degenerate states gives us

$$|\langle 1|e^{i\mathbf{q}\cdot\mathbf{r}}|2\rangle|^2 = \frac{1}{3}\left[|\langle 000|e^{i\mathbf{q}\cdot\mathbf{r}}|100\rangle|^2 + |\langle 000|e^{i\mathbf{q}\cdot\mathbf{r}}|010\rangle|^2\right.$$

$$\left.+|\langle 000|e^{i\mathbf{q}\cdot\mathbf{r}}|001\rangle|^2\right] = (q^2\ell^2/6)e^{-q^2\ell^2/2}.$$

The same result is obtained if we average $|\langle 000|e^{i\mathbf{q}\cdot\mathbf{r}}|100\rangle|^2$ over the angle of $\mathbf{q}$. The probability of electron transitions between the levels is exponentially small at $(q\ell)^2 \gg 1$.

7.25. Consider the energy broadening of a non-degenerate discrete level due to interaction of electrons with equilibrium bosons.

Solution: Let us introduce the retarded double-time Green's function $G_{tt'}^R = \langle\langle\hat{a}|\hat{a}^+\rangle\rangle_{tt'}^R$, where $\hat{a}$ and $\hat{a}^+$ are the operators of annihilation and creation of electrons in the single state $|0\rangle$ of the non-degenerate level with the energy ε_0. The Hamiltonian of the problem is $\hat{H} = \hat{H}_e + \hat{H}_b + \hat{H}_{e,b}$, where $\hat{H}_e = \varepsilon_0\hat{a}^+\hat{a}$, $\hat{H}_b = \sum_q\hbar\omega_q\hat{b}_q^+\hat{b}_q$, and $\hat{H}_{e,b} = \hat{a}^+\hat{a}\sum_q[\langle 0|\hat{\chi}_q|0\rangle\hat{b}_q + \langle 0|\hat{\chi}_q^+|0\rangle\hat{b}_q^+]$. The problem is stationary and the function $G_{tt'}^R$ depends only on $t - t'$. Let us write Eq. (D.13) for this function and use the energy representation. We obtain the expression of G_ε^R through the higher-order correlation functions:

$$[\varepsilon - \varepsilon_0 + i\lambda]G_\varepsilon^R - 1 = \sum_q\left\langle\left\langle\hat{a}[\langle 0|\hat{\chi}_q|0\rangle\hat{b}_q + \langle 0|\hat{\chi}_q^+|0\rangle\hat{b}_q^+]\,\Big|\,\hat{a}^+\right\rangle\right\rangle_\varepsilon^R, \quad \lambda \to +0.$$

Applying Eq. (D.13) to the Green's functions $\langle\langle\hat{a}\hat{b}_q|\hat{a}^+\rangle\rangle_\varepsilon^R$ and $\langle\langle\hat{a}\hat{b}_q^+|\hat{a}^+\rangle\rangle_\varepsilon^R$, we retain the contributions linear in $\langle 0|\hat{\chi}_q|0\rangle$. This allows us to express these Green's functions through G_ε^R and rewrite the right-hand side of the equation above in the form $\Sigma_\varepsilon^R G_\varepsilon^R$, where the self-energy function is given by

$$\Sigma_\varepsilon^R \simeq \sum_q |\langle 0|\hat{\chi}_q|0\rangle|^2 \left[\frac{N_q}{\varepsilon - \varepsilon_0 + \hbar\omega_q + i\lambda} + \frac{N_q + 1}{\varepsilon - \varepsilon_0 - \hbar\omega_q + i\lambda} \right].$$

This corresponds to the second-order approximation with respect to electron-boson interaction. The occupation numbers of bosons are introduced in a standard way, $N_q = \langle\langle \hat{b}_q^+ \hat{b}_q \rangle\rangle$. The Green's function is expressed as in Sec. 14, $G_\varepsilon^R = [\varepsilon - \varepsilon_0 - \Sigma_\varepsilon^R]^{-1}$. The real part of Σ_ε^R determines a shift of the energy ε_0, while the imaginary part determines the broadening energy. Calculating the broadening energy caused by the interaction with LA phonons, we find

$$\text{Im}\, \Sigma_{\varepsilon_0+\delta\varepsilon}^R = -\frac{\mathcal{D}^2}{4\pi\rho s_l^2} \frac{|\delta\varepsilon/\hbar s_l|^3}{|1 - \exp(-\delta\varepsilon/T)|} \mathcal{L}_{0000}(\delta\varepsilon/\hbar s_l),$$

where $\mathcal{L}_{abcd}$ is introduced by Eq. (38.19) and $\delta\varepsilon = \varepsilon - \varepsilon_0$. Since the broadening energy goes to zero at $\delta\varepsilon \to 0$, the energy dependence of the density of states is rather complicated.

Chapter 8

NON-EQUILIBRIUM DIAGRAM TECHNIQUE

The description of transport phenomena under conditions far from equilibrium requires a more careful consideration for the systems with strong scattering, since both the reconstruction of energy spectrum and the distribution of quasiparticles have to be considered simultaneously. For these purposes, it is convenient to use the non-equilibrium diagram technique, which operates, apart from the retarded and advanced Green's functions, with an additional function determining non-equilibrium distribution of the quasiparticles. The one-electron formulation of this technique, given below for the electron-impurity system, is a generalization of the kinetic description developed in Chapter 2 for the case of weak scattering. The non-equilibrium diagram technique (NDT) allows one to derive kinetic equations for describing the response of a system to external fields (not necessarily in the linear regime) in a unified way, by considering the series of diagrams with a required accuracy with respect to the interaction. In this way one can describe, for example, the systems of interacting electrons and the interacting electron-phonon systems, which cannot be treated diagrammatically at non-zero temperature with the use of the technique given in Chapter 3. Since the applications of the NDT often require a complex and lengthy consideration, below we present only a limited number of examples of this kind.

39. Matrix Green's Function

Let us begin our consideration with the case of electron-impurity system. Equation (7.2) for the one-electron density matrix $\hat{n}_t$ is written below in the coordinate representation, analogous to Eq. (9.2):

$$i\hbar \frac{\partial}{\partial t} n_t(\mathbf{x}_1, \mathbf{x}_2) - \left(\widehat{H}_{\mathbf{x}_1 t} - \widehat{H}^*_{\mathbf{x}_2 t} \right) n_t(\mathbf{x}_1, \mathbf{x}_2) = 0, \tag{1}$$

where $n_t(\mathbf{x}_1, \mathbf{x}_2) \equiv \langle \mathbf{x}_1 | \hat{n}_t | \mathbf{x}_2 \rangle$. The Hamiltonian of the system is written as $\widehat{H}_{\mathbf{x}t} = \hat{h}_{\mathbf{x}t} + U_{im}(\mathbf{x})$. It contains the one-electron Hamiltonian $\hat{h}_{\mathbf{x}t}$ describing electron dynamics in external fields and the perturbation $U_{im}(\mathbf{x})$ describing the interaction of electrons with randomly distributed impurities. The procedure of averaging over the impurity distribution has been discussed in Secs. 7 and 14. To consider the evolution of $n_t(\mathbf{x}_1, \mathbf{x}_2)$ simultaneously with the scattering-induced modification of the electron spectrum, it is convenient to introduce a 2×2 matrix

$$\begin{bmatrix} \mathcal{G}^{--}(\mathbf{x}_1 t_1, \mathbf{x}_2 t_2) & \mathcal{G}^{-+}(\mathbf{x}_1 t_1, \mathbf{x}_2 t_2) \\ \mathcal{G}^{+-}(\mathbf{x}_1 t_1, \mathbf{x}_2 t_2) & \mathcal{G}^{++}(\mathbf{x}_1 t_1, \mathbf{x}_2 t_2) \end{bmatrix}. \tag{2}$$

The diagonal and non-diagonal components of this matrix satisfy, by definition, the following equations:

$$\left(i\hbar \frac{\partial}{\partial t_1} - \widehat{H}_1 \right) \mathcal{G}^{\mp\mp}(1, 2) = \pm \delta(1 - 2), \tag{3}$$

$$\left(i\hbar \frac{\partial}{\partial t_1} - \widehat{H}_1 \right) \mathcal{G}^{\mp\pm}(1, 2) = 0,$$

where either upper or lower pairs of signs are considered. The multi-indices 1 and 2 stand for $(\mathbf{x}_1, t_1)$ and $(\mathbf{x}_2, t_2)$. One can write similar equations with the operators acting on the variables $\mathbf{x}_2$ and t_2:

$$\left(i\hbar \frac{\partial}{\partial t_2} - \widehat{H}_2 \right)^* \mathcal{G}^{\mp\mp}(1, 2) = \pm \delta(1 - 2), \tag{4}$$

$$\left(i\hbar \frac{\partial}{\partial t_2} - \widehat{H}_2 \right)^* \mathcal{G}^{\mp\pm}(1, 2) = 0.$$

The multi-indices $1, 2, \ldots$ can also include discrete variables such as spin, band number, etc. For example, the function $\delta(1 - 2)$ denotes the δ-functions of spatial and temporal variables,

$$\delta(1 - 2) \Rightarrow \delta(\mathbf{x}_1 - \mathbf{x}_2) \delta(t_1 - t_2), \tag{5}$$

and can include, if necessary, the Kronecker symbols of discrete variables (spin, etc.). The diagonal ($--$ and $++$) components of the 2×2 matrix function (2) satisfy the equations with δ-source on the right-hand sides, in a similar way as in the equations for the ordinary Green's functions; see, for example, Eq. (14.6) and problem 3.7. On the other hand, the non-diagonal components satisfy the homogeneous equations. This difference can be taken into account by introducing a double time axis. We assume that the times t_1 and t_2 in the diagonal components of the matrix (2) are in the same axis so that the inhomogeneous contributions appear at $t_1 = t_2$. In the non-diagonal components, the times t_1 and t_2

are in the different axes and no δ-functions appear; see Fig. 1 in Sec. 41 below.

By comparing the Hermitian conjugate Eq. (3) to Eq. (4), we see that

$$\mathcal{G}^{--}(1,2) = -[\mathcal{G}^{++}(2,1)]^*, \qquad \mathcal{G}^{\pm\mp}(1,2) = -[\mathcal{G}^{\pm\mp}(2,1)]^*, \qquad (6)$$

which means that $\mathcal{G}^{--}(1,2)$ and $\mathcal{G}^{++}(1,2)$ are anti-Hermitian with respect to each other, while $\mathcal{G}^{-+}(1,2)$ and $\mathcal{G}^{+-}(1,2)$ are anti-Hermitian by themselves. It is also assumed that the relations like those in Eq. (6) can be applied to the initial conditions to Eqs. (3) and (4). Taking a sum of the equations for $\mathcal{G}^{++}$ and $\mathcal{G}^{--}$, and, separately, of the equations for $\mathcal{G}^{-+}$ and $\mathcal{G}^{+-}$, one finds that the sums $(\mathcal{G}^{++} + \mathcal{G}^{--})$ and $(\mathcal{G}^{+-} + \mathcal{G}^{-+})$ satisfy the same equation. As a result,

$$\mathcal{G}^{--}(1,2) + \mathcal{G}^{++}(1,2) = \mathcal{G}^{-+}(1,2) + \mathcal{G}^{+-}(1,2), \qquad (7)$$

and we should consider the expression (2) as a matrix Green's function with only three linearly-independent components, provided that Eq. (7) is true for the initial conditions to Eqs. (3) and (4) as well. To derive a relation between the exact density matrix (governed by Eq. (1)) and non-diagonal components of the matrix Green's function (2), we subtract the equation for non-diagonal component in Eq. (3) from the analogous equation in Eq. (4) and obtain

$$\left[i\hbar \left(\frac{\partial}{\partial t_1} + \frac{\partial}{\partial t_2} \right) - \widehat{H}_1 + \widehat{H}_2^* \right] \mathcal{G}^{\mp\pm}(1,2) = 0. \qquad (8)$$

At $t_{1,2} \to t$, this equation coincides with Eq. (1) for $n_t(\mathbf{x}_1, \mathbf{x}_2)$ (note that since t_1 and t_2 are in the different axes, the sum of the derivatives over these times is equal to $\partial/\partial t$). Since $\mathcal{G}^{-+}$ is anti-Hermitian and $\hat{n}_t$ is Hermitian, the relation between these functions should contain a factor of i. In the following, we specify this relation as

$$-i\hbar \lim_{t_{1,2} \to t} \mathcal{G}^{-+}(\mathbf{x}_1 t_1, \mathbf{x}_2 t_2) = n_t(\mathbf{x}_1, \mathbf{x}_2). \qquad (9)$$

Equation (9) completes the definition of the function $\mathcal{G}^{-+}$, which otherwise would contain an undefined numerical coefficient. The function $\mathcal{G}^{+-}$ is expressed through $\mathcal{G}^{-+}$ with the use of Eq. (7).

The one-electron density matrix $\rho_t(\mathbf{x}_1, \mathbf{x}_2)$, averaged over the impurity distribution, is introduced as

$$\rho_t(\mathbf{x}_1, \mathbf{x}_2) = \langle\langle n_t(\mathbf{x}_1, \mathbf{x}_2)\rangle\rangle. \qquad (10)$$

It is expressed through $\mathcal{G}^{-+}$ as

$$\rho_t(\mathbf{x}_1, \mathbf{x}_2) = -i\hbar\langle\langle \mathcal{G}^{-+}(\mathbf{x}_1 t, \mathbf{x}_2 t)\rangle\rangle \equiv -i\hbar G^{-+}(\mathbf{x}_1 t, \mathbf{x}_2 t), \qquad (11)$$

where the second equation introduces the averaged non-diagonal component of the matrix Green's function $\widehat{G}$. The latter is determined from the definition (2) as

$$\widehat{G}(1,2) = \left[\begin{array}{cc} \langle\langle \mathcal{G}^{--}(1,2)\rangle\rangle & \langle\langle \mathcal{G}^{-+}(1,2)\rangle\rangle \\ \langle\langle \mathcal{G}^{+-}(1,2)\rangle\rangle & \langle\langle \mathcal{G}^{++}(1,2)\rangle\rangle \end{array} \right]. \tag{12}$$

To avoid misunderstandings, we point out that $\widehat{G}$ is not a quantum-mechanical operator, since we have already employed the coordinate representation in the multi-indices. The hat over G merely reflects the matrix structure of the Green's function under consideration. On the other hand, the Hamiltonians below remain quantum-mechanical operators. Thus, the "hat" symbols have two different meanings. Below we derive a set of equations for the components of $\widehat{G}(1,2)$ for the case of electron interaction with randomly distributed impurities.

In order to carry out a diagrammatic expansion of $\widehat{G}(1,2)$, we rewrite Eqs. (3) and (4) for the matrix Green's function $\hat{\mathcal{G}}(1,2)$ in the form

$$\left(i\hbar\frac{\partial}{\partial t_1} - \widehat{H}_1 \right) \hat{\mathcal{G}}(1,2) = \hat{\sigma}_z \delta(1-2), \tag{13}$$

$$\left(i\hbar\frac{\partial}{\partial t_2} - \widehat{H}_2 \right)^* \hat{\mathcal{G}}(1,2) = \hat{\sigma}_z \delta(1-2).$$

The Pauli matrix $\hat{\sigma}_z$ on the right-hand sides reflects the difference in signs between the right-hand sides of the equations for $\mathcal{G}^{--}$ and $\mathcal{G}^{++}$. In the absence of scattering, the system is described by the Green's function of ideal electron gas, $\hat{g}(1,2)$, and

$$\left(i\hbar\frac{\partial}{\partial t_1} - \hat{h}_1 \right) \hat{g}(1,2) = \hat{\sigma}_z \delta(1-2), \tag{14}$$

$$\left(i\hbar\frac{\partial}{\partial t_2} - \hat{h}_2 \right)^* \hat{g}(1,2) = \hat{\sigma}_z \delta(1-2).$$

Using Eq. (14), one can reduce Eq. (13) to the following matrix integral equations:

$$\hat{\mathcal{G}}(1,2) = \hat{g}(1,2) + \int d1' \hat{g}(1,1') \hat{\sigma}_z U_{1'} \hat{\mathcal{G}}(1',2), \tag{15}$$

$$\hat{\mathcal{G}}(1,2) = \hat{g}(1,2) + \int d1' \hat{\mathcal{G}}(1,1') U_{1'} \hat{\sigma}_z \hat{g}(1',2),$$

which are valid for any time- and coordinate-dependent scattering potential $U_{1'} \equiv U_{\mathbf{x}'_1 t'_1}$, though in this section we consider time-independent (static) potentials. These equations are analogical to the scalar integral equation (14.8) employed in the diagram technique for retarded and advanced Green's functions in Chapter 3.

The diagram technique for the components of the matrix Green's function defined above is called the non-equilibrium diagram technique, and these components will be named below as the non-equilibrium Green's functions. The non-equilibrium diagram technique formulated here differs from the diagram technique of Chapter 3, in spite of the fact that in both cases we consider the electron-impurity interaction (or, more general, interaction of electrons with a random static potential). Now, the Green's function depends on two time variables and cannot be reduced to the Green's function of a stationary problem, which depends on the electron energy. The integral over the time variable enters the last terms in Eq. (15), and this makes the diagrammatic expansion more difficult. Each electron line connecting the points 1 and 2 corresponds to the matrix Green's function $\hat{\mathcal{G}}(1,2)$ multiplied by i (i.e., the present definition differs by the factor of i from that given by Eq. (14.11)). The vertices also have matrix structure. As seen from Eq. (15), one has to put

$$\vdots \atop \underset{1}{\bullet} = -i\hat{\sigma}_z U_1 \tag{16}$$

in contrast to Eq. (14.11). These modifications, however, do not change the basic structure of the diagrams, and the procedure of averaging over the random potential remains the same. Similar as in Sec. 14, we can introduce the self-energy matrix $\widehat{\Sigma}(1,2)$ and write the matrix Dyson equation:

$$\widehat{G}(1,2) = \hat{g}(1,2) + \int d1' \int d2' \hat{g}(1,1')\widehat{\Sigma}(1',2')\widehat{G}(2',2) \tag{17}$$

$$= \hat{g}(1,2) + \int d1' \int d2'\widehat{G}(1,1')\widehat{\Sigma}(1',2')\hat{g}(2',2).$$

This equation has the same form as the Dyson equation (14.22) written in the coordinate representation. For the purpose of graphical representation, we denote $i\widehat{G}$ and $i\hat{g}$ by bold and thin electron lines, respectively, and $-i\widehat{\Sigma}$ by a semi-oval, like in Eqs. (14.22) and (14.23). Then, the representation of Eq. (17) looks similar to Eq. (14.22). We remind that the definitions of the graphic elements used in the non-equilibrium diagram technique differ from those of Chapter 3 not only because of the matrix structure of the vertices and lines, but also due to the presence of the factors i or $-i$. The introduction of these factors is necessary because we are going to apply the diagram technique not only to the case of electron-impurity interaction, but also to the cases of electron-electron and electron-boson interaction. This will become clear in Sec. 41, where the general formulation of the non-equilibrium diagram technique justifies the convenience of such factors.

The self-energy matrix is given by a diagrammatic expansion analogous to Eq. (14.23):

$$-i\widehat{\Sigma}(1,2) = \underset{1 \qquad\qquad 2}{\overbrace{}} + \underset{1 \quad 2' \quad\; 1' \quad 2}{\overbrace{}} + \dots . \qquad (18)$$

The broken lines connecting vertices 1 and 2 imply the pair correlation functions $\langle\langle U_1 U_2 \rangle\rangle$ of the potentials at the vertices, in a similar way as in Eq. (14.17). In the analytical form, Eq. (18), multiplied by i, is written as

$$\widehat{\Sigma}(1,2) = \langle\langle U_1 U_2 \rangle\rangle \hat{\sigma}_z \widehat{G}(1,2)\hat{\sigma}_z \qquad (19)$$

$$+ \int d1' \int d2' \langle\langle U_1 U_{1'} \rangle\rangle \langle\langle U_{2'} U_2 \rangle\rangle \hat{\sigma}_z \widehat{G}(1,2')\hat{\sigma}_z \widehat{G}(2',1')\hat{\sigma}_z \widehat{G}(1',2)\hat{\sigma}_z + \dots .$$

It differs from Eq. (14.26) by the additional matrix factors $\hat{\sigma}_z$ coming from the vertices, as well as by the additional integrals over time. Note that, since the potentials U_i ($i = 1, 2, \dots$) are assumed to be time-independent, these integrals are applied to the Green's functions only.

Instead of writing the matrix diagram expressions, one may use the diagonality of the vertex matrix $\hat{\sigma}_z$ and attribute the indices $+$ or $-$ to the ends of electron lines, implying the sums over these indices in all vertices of the diagram. According to Eq. (16), the vertex 1 with the index s ($s = \pm$) gives the contribution siU_1 to the analytical expression of the diagram. Therefore, attributing the broken line connecting the vertices $1s$ and $2s'$ to the correlation function $\langle\langle U_1 U_2 \rangle\rangle$, we should simultaneously attach the factors si and $s'i$ to these vertices. Using the rules we just formulated, one may write the expression

$$-i\Sigma^{s_1 s_2}(1,2) = \langle\langle U_1 U_2 \rangle\rangle is_1 is_2 iG^{s_1 s_2}(1,2)$$

$$+ is_1 is_2 \int d1' \int d2' \langle\langle U_1 U_{1'} \rangle\rangle \langle\langle U_{2'} U_2 \rangle\rangle \qquad (20)$$

$$\times \sum_{s_1', s_2' = \pm} is_1' is_2' iG^{s_1 s_2'}(1,2') iG^{s_2' s_1'}(2',1') iG^{s_1' s_2}(1',2) + \dots$$

for each element of the matrix $-i\widehat{\Sigma}(1,2)$ given by the diagrammatic expansion (18). Here and below, the factors s_1, s_2, etc. in the expressions like Eq. (20) are equal to $+1$ or -1, corresponding to the indices $+$ or $-$

of the Green's functions. The equivalence of Eqs. (20) and (19) is easy to check by observing that the matrix element $[\hat{\sigma}_z \widehat{G}(1,2)\hat{\sigma}_z]_{s_1 s_2}$ is equal to $s_1 s_2 G^{s_1 s_2}(1,2)$. The general formulation of the diagram technique using the vertices with extra indices $+$ or $-$ is given in Sec. 41.

After acting by the operators $(i\hbar\partial/\partial t_1 - \hat{h}_1)$ and $(i\hbar\partial/\partial t_2 - \hat{h}_2)^*$ on the matrix integral equations (17), we use Eq. (14) and obtain the following integro-differential equations:

$$\left(i\hbar\frac{\partial}{\partial t_1} - \hat{h}_1\right)\widehat{G}(1,2) = \hat{\sigma}_z\delta(1-2) + \int d1'\hat{\sigma}_z\widehat{\Sigma}(1,1')\widehat{G}(1',2),$$

$$\left(i\hbar\frac{\partial}{\partial t_2} - \hat{h}_2\right)^*\widehat{G}(1,2) = \hat{\sigma}_z\delta(1-2) + \int d1'\widehat{G}(1,1')\widehat{\Sigma}(1',2)\hat{\sigma}_z. \quad (21)$$

Averaging Eq. (7) over the impurity distribution, we see that the relation between the components of $\widehat{G}(1,2)$ has the same form as Eq. (7):

$$G^{--}(1,2) + G^{++}(1,2) = G^{-+}(1,2) + G^{+-}(1,2). \quad (22)$$

A similar linear relation can be written for the components of Eq. (17). Doing this, one can see that the inhomogeneous contribution due to $\hat{g}(1,2)$ vanishes and only the integral term remains:

$$\int d1'[\Sigma^{--}(1,1') + \Sigma^{++}(1,1') + \Sigma^{-+}(1,1') + \Sigma^{+-}(1,1')]$$

$$\times[G^{--}(1',2) + G^{-+}(1',2)] = 0. \quad (23)$$

Generally speaking, $\widehat{\Sigma}$ and $\widehat{G}$ are not independent of each other. However, since they depend on the parameters of the system in different fashions, and Eq. (23) must be valid in the entire region of these parameters, there is only one possible way to satisfy Eq. (23), the following linear relation between the components of the self-energy matrix:

$$\Sigma^{--}(1,2) + \Sigma^{++}(1,2) + \Sigma^{-+}(1,2) + \Sigma^{+-}(1,2) = 0. \quad (24)$$

Therefore, there are only three independent components of the self-energy matrix, which have to be considered together with three independent components of the matrix (12).

40. Generalized Kinetic Equation

Using the linear relations (39.22) and (39.24), we can write Eqs. (39.17) and (39.21) through the three independent components of the matrices $\widehat{G}(1,2)$ and $\widehat{\Sigma}(1,2)$. The transition to the "triangular" representation of the Green's and self-energy matrices is realized by the unitary transformation

$$\widehat{R} = \frac{1 + i\hat{\sigma}_y}{\sqrt{2}}, \qquad \widehat{R}^{-1} = \frac{1 - i\hat{\sigma}_y}{\sqrt{2}}. \tag{1}$$

Using Eq. (39.22), we obtain the matrix

$$\widehat{R}^{-1}\widehat{G}(1,2)\widehat{R} = \begin{bmatrix} 0 & G^A(1,2) \\ G^R(1,2) & F(1,2) \end{bmatrix}, \tag{2}$$

whose components are

$$G^R(1,2) = G^{--}(1,2) - G^{-+}(1,2) = G^{+-}(1,2) - G^{++}(1,2),$$

$$G^A(1,2) = G^{--}(1,2) - G^{+-}(1,2) = G^{-+}(1,2) - G^{++}(1,2), \tag{3}$$

$$F(1,2) = G^{--}(1,2) + G^{++}(1,2) = G^{-+}(1,2) + G^{+-}(1,2).$$

The same transformation, when applied to $\widehat{\Sigma}(1,2)$, gives us the self-energy matrix

$$\widehat{R}^{-1}\widehat{\Sigma}(1,2)\widehat{R} = \begin{bmatrix} \Omega(1,2) & \Sigma^R(1,2) \\ \Sigma^A(1,2) & 0 \end{bmatrix} \tag{4}$$

with the components

$$\Sigma^R(1,2) = \Sigma^{--}(1,2) + \Sigma^{-+}(1,2),$$

$$\Sigma^A(1,2) = \Sigma^{--}(1,2) + \Sigma^{+-}(1,2), \tag{5}$$

$$\Omega(1,2) = \Sigma^{--}(1,2) + \Sigma^{++}(1,2).$$

To write the density matrix $\rho_t(\mathbf{x}_1, \mathbf{x}_2)$ through the components of the Green's function (3), one should carry out the inverse unitary transformation of the right-hand side of Eq. (2) according to $\widehat{R}[\ldots]\widehat{R}^{-1}$. Expressing G^{-+} through the components of the matrix (2), we obtain

$$\rho_t(\mathbf{x}_1, \mathbf{x}_2) = -\frac{i\hbar}{2} \lim_{t_1, t_2 \to t} [F(1,2) + G^A(1,2) - G^R(1,2)] \tag{6}$$

according to Eq. (39.11).

From the relations (39.6) we find

$$G^A(1,2) = G^{R*}(2,1), \qquad F(1,2) = -F^*(2,1), \tag{7}$$

which means that the component G^A is completely determined by G^R. Therefore, one may consider the equations for G^R and F only. Applying the unitary transformation (1) to Eq. (39.21), we transform the vertex with the use of the relation $\widehat{R}^{-1}\hat{\sigma}_z\widehat{R} = \hat{\sigma}_x$ (problem 8.1). This leads to the following equations for G^R:

$$\left(i\hbar\frac{\partial}{\partial t_1} - \hat{h}_1\right) G^R(1,2) = \delta(1-2) + \int d1'\Sigma^R(1,1')G^R(1',2), \quad (8)$$

$$\left(i\hbar\frac{\partial}{\partial t_2} - \hat{h}_2\right)^* G^R(1,2) = \delta(1-2) + \int d1' G^R(1,1')\Sigma^R(1',2).$$

These equations are similar to Eq. (14.6), which means that G^R and G^A are the conventional retarded and advanced Green's functions. These functions describe the dynamical properties of the system, including the case when external fields are present. The equations for F are obtained from Eq. (39.21) in the following form:

$$\left(i\hbar\frac{\partial}{\partial t_1} - \hat{h}_1\right) F(1,2) = \int d1'[\Omega(1,1')G^A(1',2) + \Sigma^R(1,1')F(1',2)], \quad (9)$$

$$\left(i\hbar\frac{\partial}{\partial t_2} - \hat{h}_2\right)^* F(1,2) = \int d1'[G^R(1,1')\Omega(1',2) + F(1,1')\Sigma^A(1',2)],$$

and the function $F(1,2)$ determines the kinetic properties of the system under consideration.

The explicit expressions for the impurity vertices has not been used in the derivation of Eqs. (8) and (9). They have to be considered only in the self-energy matrices, i.e., in the calculation of the matrix products like $\widehat{R}^{-1}\widehat{\Sigma}\widehat{R}$ appearing in the unitary transformation of Eq. (39.19). Any term of this kind contains products of the form

$$\hat{\sigma}_x\widehat{R}^{-1}\widehat{G}_I\widehat{R}\hat{\sigma}_x\widehat{R}^{-1}\widehat{G}_{II}\widehat{R}\hat{\sigma}_x \ldots \hat{\sigma}_x\widehat{R}^{-1}\widehat{G}_{...}\widehat{R}\hat{\sigma}_x, \quad (10)$$

where the indices $I, II, \ldots$ denote the sets of internal coordinates of the Green's functions. The integrals over these coordinates and the correlation functions of the impurity potentials (which are scalars) are not written in Eq. (10). By defining the matrix

$$\widetilde{G} \equiv \hat{\sigma}_x\widehat{R}^{-1}\widehat{G}\widehat{R}\hat{\sigma}_x = \begin{bmatrix} F & G^R \\ G^A & 0 \end{bmatrix}, \quad (11)$$

we rewrite the expression (10) as $\widetilde{G}_I\widehat{R}^{-1}\widehat{G}_{II}\widehat{R}\widetilde{G}_{III} \ldots$. The product of the matrices (2) and (11) gives us

$$\widetilde{G}_I\widehat{R}^{-1}\widehat{G}_{II}\widehat{R} = \begin{bmatrix} G_I^R G_{II}^R & F_I G_{II}^A + G_I^R F_{II} \\ 0 & G_I^A G_{II}^A \end{bmatrix}, \quad (12)$$

and the product of a pair of the matrices (12) becomes

$$
\left[
\begin{array}{cc}
G_I^R G_{II}^R G_{III}^R G_{IV}^R & G_I^R G_{II}^R \mathcal{F}_{III,IV} + \mathcal{F}_{I,II} G_{III}^A G_{IV}^A \\
0 & G_I^A G_{II}^A G_{III}^A G_{IV}^A
\end{array}
\right],
\qquad (13)
$$

where, by definition, $\mathcal{F}_{I,II} \equiv F_I G_{II}^A + G_I^R F_{II}$. Since the product (10) contains an odd number of the Green's functions, one should finally multiply the matrix of the kind of Eq. (13) by the matrix (11), which gives us

$$
\left[
\begin{array}{cc}
G_I^R G_{II}^R \cdots & \mathcal{F} \cdots \\
0 & G_I^A G_{II}^A \cdots
\end{array}
\right]
\left[
\begin{array}{cc}
F & G^R \\
G^A & 0
\end{array}
\right]
$$

$$
=
\left[
\begin{array}{cc}
G_I^R G_{II}^R \cdots F + \mathcal{F} \cdots G^A & G_I^R G_{II}^R \cdots G^R \\
G_I^A G_{II}^A \cdots G^A & 0
\end{array}
\right].
\qquad (14)
$$

The coordinates of the last matrix are not written here. One can see that Σ^s ($s = R, A$) contains the products $G^s G^s \cdots$, while Ω is linear in F. Note that the retarded Green's functions G^R in the expression for Ω stand left to F, while G^A stand right to F.

Introducing the diagram representation of the components of the "triangular" Green's function according to

$$
iG^{R(A)}(1,2) = \frac{\rule{3cm}{0.4pt}}{1 \ \ R(A) \ \ 2}
\qquad (15)
$$

$$
iF(1,2) = \frac{\rule{2.5cm}{0.4pt}}{1 \quad F \quad 2},
\qquad (16)
$$

we obtain the diagrammatic expansion of the self-energy component Ω:

$$
-i\Omega(1,2) = \frac{\rule{2.5cm}{0.4pt}}{F} \ + \ \Big[\ \frac{\rule{2.5cm}{0.4pt}}{R \ \ R \ \ F} \ \Big] + \big[\ RFA \ \big] + \big[\ FAA \ \big] + \ldots.
\qquad (17)
$$

The second-order term with line sequence $[RRF]$ is shown here explicitly, and the terms $[RFA]$ and $[FAA]$ differ from it only by the indices of the corresponding Green's functions. The whole expansion can be written by analogy. One should remember that the line F appears only one time per diagram, and the retarded (advanced) Green's functions always stand left (right) to F. Note that each vertex gives only a factor of i into analytical expressions, because the matrix multiplication is already done. The expansion of the components $-i\Sigma^{R(A)}(1,2)$ has the same form as Eq. (39.18), one has merely to replace the matrix vertices (39.16) by the conventional, scalar ones, and assume that the solid lines correspond to $iG^{R(A)}$ according to Eq. (15). The linear relation between the functions Ω and F exists because the Pauli principle in the case of electron-impurity scattering is automatically satisfied so that the factors of the kind $(1 - F)$ disappear.

In order to get a convenient set of equations for G^R and F, we employ Eqs. (8) and (9). Using Eq. (8), we add the second equation of the pair to the first one. Using Eq. (9), we subtract the second equation from the first one. As a result, we obtain

$$\left[i\hbar\left(\frac{\partial}{\partial t_1} - \frac{\partial}{\partial t_2}\right) - \hat{h}_1 - \hat{h}_2^*\right] G^R(1,2) = 2\delta(1-2)$$

$$+ \int d1' \left[\Sigma^R(1,1')G^R(1',2) + G^R(1,1')\Sigma^R(1',2)\right] \qquad (18)$$

(the equation for $G^A(1,2)$ differs from this one merely by the substitution of the index A in place of R), and

$$\left[i\hbar\left(\frac{\partial}{\partial t_1} + \frac{\partial}{\partial t_2}\right) - \hat{h}_1 + \hat{h}_2^*\right] F(1,2) = \int d1' \left[\Omega(1,1')G^A(1',2)\right.$$

$$\left. -G^R(1,1')\Omega(1',2) + \Sigma^R(1,1')F(1',2) - F(1,1')\Sigma^A(1',2)\right]. \qquad (19)$$

The equation for G^R is independent of F, i.e., the modification of electron dynamics due to scattering does not depend on the electron distribution. On the other hand, both the non-equilibrium distribution and the interaction-modified quasiparticle dynamics are important in Eq. (19) for F. This equation should be considered as a generalization of the quantum kinetic equation, and its right-hand side is the generalized collision integral describing both the relaxation of electron distribution and the renormalization of electron spectrum due to electron-impurity scattering. Therefore, Eqs. (18) and (19) describe the response of the quasiparticles formed in the electron-impurity system. In contrast to the ordinary kinetic equation, written for the density matrix with a single time variable, the generalized kinetic equation is written for the function F depending on two time variables.

However, if the external fields acting on the system are slowly varying with time, one may carry out a quasi-classical transformation of the time variables, analogous to the Wigner transformation of the coordinate variables used in Sec. 9. Introducing $t = (t_1 + t_2)/2$, $\tau = t_1 - t_2$, and carrying out the Fourier transformation over τ, we obtain

$$\widehat{G}^R_{\varepsilon t} = \int d\tau e^{i\varepsilon\tau/\hbar}\widehat{G}^R_{t+\tau/2,t-\tau/2}, \quad \widehat{G}^R_{t+\tau/2,t-\tau/2} = \int \frac{d\varepsilon}{2\pi\hbar} e^{-i\varepsilon\tau/\hbar}\widehat{G}^R_{\varepsilon t}. \quad (20)$$

The Green's function is written here in the operator form, and the coordinate representation is given as usual, $G^R_{\varepsilon t}(\mathbf{x}_1, \mathbf{x}_2) = \langle\mathbf{x}_1|\widehat{G}^R_{\varepsilon t}|\mathbf{x}_2\rangle$. If necessary, one may add discrete indices (spin, etc.) to the Green's function. The functions G^A, F, Σ^R, Σ^A, and Ω are transformed in the same way. One can see that $\widehat{G}^R_{\varepsilon t} = \widehat{G}^{A+}_{\varepsilon t}$ and $\widehat{F}_{\varepsilon t} = -\widehat{F}^+_{\varepsilon t}$. The expression for the density matrix follows from Eqs. (6) and (20):

$$\hat{\rho}_t = -i \int \frac{d\varepsilon}{4\pi} \left(\widehat{F}_{\varepsilon t} + \widehat{G}^A_{\varepsilon t} - \widehat{G}^R_{\varepsilon t} \right). \tag{21}$$

Let us write the equations for $\widehat{G}^R_{\varepsilon t}$ and $\widehat{F}_{\varepsilon t}$ in the case of slowly varying external fields, when the modifications of the Green's functions on the quantum scale of time $\hbar/\bar{\varepsilon}$ are small. This means that all functions standing under the integrals over time in Eqs. (18) and (19) are transformed into the functions with the same arguments ε and t after the Wigner transformation (20) (compare to the results of Appendix C, where the Wigner transformation of operator products is considered). Transforming Eq. (18), we also assume that $\hat{h}_{t\pm\tau/2} \simeq \hat{h}_t$ on the left-hand side, which is true for slowly varying external fields. The transformed Eq. (18) is written below in the operator form:

$$\varepsilon \widehat{G}^R_{\varepsilon t} - \frac{1}{2}[\widehat{G}^R_{\varepsilon t}, \hat{h}_t]_+ = 1 + \frac{1}{2}[\widehat{\Sigma}^R_{\varepsilon t}, \widehat{G}^R_{\varepsilon t}]_+. \tag{22}$$

In the transformation of Eq. (19), we use a more detailed expansion, $\hat{h}_{t\pm\tau/2} \simeq \hat{h}_t \pm (\partial \hat{h}_t/\partial t)\tau/2$, since $\widehat{F}_{\varepsilon t}$ may commute with $\hat{h}_t$. Therefore, Eq. (19) is transformed into

$$i\hbar \frac{\partial}{\partial t} \widehat{F}_{\varepsilon t} + [\widehat{F}_{\varepsilon t}, \hat{h}_t] + \frac{i\hbar}{2} \left[\frac{\partial \widehat{F}_{\varepsilon t}}{\partial \varepsilon}, \frac{\partial \hat{h}_t}{\partial t} \right]_+$$

$$= \widehat{\Omega}_{\varepsilon t} \widehat{G}^A_{\varepsilon t} - \widehat{G}^R_{\varepsilon t} \widehat{\Omega}_{\varepsilon t} + \widehat{\Sigma}^R_{\varepsilon t} \widehat{F}_{\varepsilon t} - \widehat{F}_{\varepsilon t} \widehat{\Sigma}^A_{\varepsilon t}. \tag{23}$$

The operator products of the kind $\widehat{\Omega}_{\varepsilon t} \widehat{G}^A_{\varepsilon t}$ assume the integrals over internal coordinates, according to $\langle \mathbf{x}_1 | \widehat{\Omega}_{\varepsilon t} \widehat{G}^A_{\varepsilon t} | \mathbf{x}_2 \rangle = \int d\mathbf{x}_3 \Omega_{\varepsilon t}(\mathbf{x}_1, \mathbf{x}_3) G^A_{\varepsilon t}(\mathbf{x}_3, \mathbf{x}_2)$. Equation (23), together with Eq. (22) and diagrammatic expansions of Σ^R, Σ^A, and Ω, generalizes the quantum kinetic equation (7.13) to the case when the electron-impurity interaction is strong and the condition (7.21) is not satisfied.

   In the stationary case described by the time-independent Hamiltonian $\widehat{H}_{\mathbf{x}}$, one has the exact relation (problem 8.2)

$$\widehat{F}_\varepsilon = \chi_\varepsilon \left(\widehat{G}^A_\varepsilon - \widehat{G}^R_\varepsilon \right), \tag{24}$$

where χ_ε is an arbitrary function of energy. One can demonstrate that a substitution of $\widehat{F}_\varepsilon$ from Eq. (24) to the right-hand side of Eq. (23) makes this side, i.e., the generalized collision integral, equal to zero (problem 8.3). In the case of thermodynamic equilibrium, when the electron system weakly interacts with a thermostat (for example, with acoustic phonons), $\chi_\varepsilon = 2f_\varepsilon^{(eq)} - 1$, where $f_\varepsilon^{(eq)}$ is the equilibrium distribution function of the electron system.

The quasi-classical kinetic equation can be derived from Eqs. (22) and (23) when the scattering potential is weak (under the condition (7.21)) and when the Green's functions are weakly varying in space on the scale of electron wavelength $\hbar/\bar{p}$. We specify the Hamiltonian $\hat{h}_{\mathbf{x}t}$ as $[\hat{\mathbf{p}} - (e/c)\mathbf{A}_{\mathbf{x}t}]^2/2m + U_{\mathbf{x}}$, where $\mathbf{A}_{\mathbf{x}t}$ is the vector potential of the external field and $U_{\mathbf{x}}$ is the potential energy, and use the spatial Wigner transformation defined by Eq. (9.6). This transformation, when applied to Eq. (22) or to a similar equation for the advanced Green's function ($s = R, A$ below), gives us

$$\left[\varepsilon - \frac{p^2}{2m} - U_{\mathbf{r}} + \frac{\hbar^2}{8m}\left(\frac{\partial}{\partial \mathbf{r}} + \frac{e}{c}\left[\mathbf{H}_{\mathbf{r}t} \times \frac{\partial}{\partial \mathbf{p}}\right]\right)^2\right] G_{\varepsilon t}^s(\mathbf{r}, \mathbf{p})$$

$$= 1 + \Sigma_{\varepsilon t}^s(\mathbf{r}, \mathbf{p})G_{\varepsilon t}^s(\mathbf{r}, \mathbf{p}), \qquad (25)$$

where $\mathbf{H}_{\mathbf{r}t}$ is the magnetic field expressed through the vector potential according to Eq. (4.3). To derive Eq. (25), we have neglected the quantum corrections when transforming the operator product on the right-hand side of Eq. (22). The quantum corrections in the calculation of the anticommutator give us the last (proportional to $\hbar^2$) term on the left-hand side of Eq. (25). Since we consider the case of smooth spatial inhomogeneity and weak magnetic field, this term also can be neglected under the conditions (9.29) and (9.35). As a result, Eq. (25) is rewritten as

$$G_\varepsilon^s(\mathbf{r}, \mathbf{p}) = [\varepsilon - \varepsilon_p - U_{\mathbf{r}} - \Sigma_\varepsilon^s(\mathbf{r}, \mathbf{p})]^{-1}, \quad s = R, A, \qquad (26)$$

where $\varepsilon_p = p^2/2m$. The dependence of G_ε and Σ_ε on $\mathbf{r}$ is parametric, it is caused by the presence of the smooth potential $U_{\mathbf{r}}$. In the absence of this potential, Eq. (26) coincides with Eq. (14.25).

The Wigner transformation of Eq. (23) is done in a similar way as in Sec. 9. The anticommutator $[(\partial \widehat{F}_{\varepsilon t}/\partial \varepsilon), (\partial \hat{h}_t/\partial t)]_+/2$ is transformed to $(e/m)(\mathbf{E}_{\mathbf{r}t} \cdot \mathbf{p})[\partial F_{\varepsilon t}(\mathbf{r}, \mathbf{p})/\partial \varepsilon]$. The quantum corrections in the transformation of the right-hand side of Eq. (23) are neglected. As a result, the transformed Eq. (23) is written as

$$i\left(\frac{\partial}{\partial t} + \mathbf{v} \cdot \frac{\partial}{\partial \mathbf{r}} + e\mathbf{E}_{\mathbf{r}t} \cdot \mathbf{v}\frac{\partial}{\partial \varepsilon} + e\mathbf{E}_{\mathbf{r}t} \cdot \frac{\partial}{\partial \mathbf{p}}\right.$$

$$\left. + \frac{e}{c}[\mathbf{v} \times \mathbf{H}_{\mathbf{r}t}] \cdot \frac{\partial}{\partial \mathbf{p}}\right) F_{\varepsilon t}(\mathbf{r}, \mathbf{p}) = \mathcal{J}_{\varepsilon t}(F|\mathbf{r}\mathbf{p}), \qquad (27)$$

$$\mathcal{J}_{\varepsilon t}(F|\mathbf{rp}) = \Omega_{\varepsilon t}(\mathbf{r}, \mathbf{p}) \left[G_\varepsilon^A(\mathbf{r}, \mathbf{p}) - G_\varepsilon^R(\mathbf{r}, \mathbf{p}) \right] / \hbar$$
$$+ F_{\varepsilon t}(\mathbf{r}, \mathbf{p}) \left[\Sigma_\varepsilon^R(\mathbf{r}, \mathbf{p}) - \Sigma_\varepsilon^A(\mathbf{r}, \mathbf{p}) \right] / \hbar,$$

where $\mathbf{v} = \mathbf{p}/m$. The collision integral $\mathcal{J}_{\varepsilon t}$ is evaluated below in the limit of weak scattering, when the functions Σ^R, Σ^A, and Ω are calculated in the lowest order with respect to the scattering potential. This implies that we take into account only the lowest terms in the diagrammatic expansions for these functions:

$$\Sigma_\varepsilon^s(\mathbf{r}, \mathbf{p}) = \frac{1}{V} \sum_{\mathbf{p}'} w(|\mathbf{p} - \mathbf{p}'|/\hbar) G_\varepsilon^s(\mathbf{r}, \mathbf{p}'),$$

$$\Omega_{\varepsilon t}(\mathbf{r}, \mathbf{p}) = \frac{1}{V} \sum_{\mathbf{p}'} w(|\mathbf{p} - \mathbf{p}'|/\hbar) F_{\varepsilon t}(\mathbf{r}, \mathbf{p}'), \tag{28}$$

and replace Σ^R and Σ^A in the expression (26) for the Green's functions by $-i0$ and $+i0$, respectively. As a result, $G_\varepsilon^A(\mathbf{r}, \mathbf{p}) - G_\varepsilon^R(\mathbf{r}, \mathbf{p}) = 2\pi i \delta(\varepsilon - \varepsilon_p - U_\mathbf{r})$ and

$$\mathcal{J}_{\varepsilon t}(F|\mathbf{rp}) = i \frac{2\pi}{\hbar V} \sum_{\mathbf{p}'} w(|\mathbf{p} - \mathbf{p}'|/\hbar) \tag{29}$$

$$\times \left[\delta(\varepsilon - \varepsilon_p - U_\mathbf{r}) F_{\varepsilon t}(\mathbf{r}, \mathbf{p}') - \delta(\varepsilon - \varepsilon_{p'} - U_\mathbf{r}) F_{\varepsilon t}(\mathbf{r}, \mathbf{p}) \right].$$

Equation (27) already resembles the quasi-classical kinetic equation. To complete the derivation of the latter, we integrate Eq. (27) over the energy ε. The third term on the left-hand side of Eq. (27) vanishes as a result of this integration. By using the identity $(2\pi i)^{-1} \int d\varepsilon [G_\varepsilon^A(\mathbf{r}, \mathbf{p}) - G_\varepsilon^R(\mathbf{r}, \mathbf{p})] = 1$, we transform Eq. (21) to

$$2 f_t(\mathbf{r}, \mathbf{p}) - 1 = \int \frac{d\varepsilon}{2\pi i} F_{\varepsilon t}(\mathbf{r}, \mathbf{p}), \tag{30}$$

where $f_t(\mathbf{r}, \mathbf{p})$ is the quasi-classical distribution function of electrons. With the use of Eq. (30), the integration of the left-hand side of Eq. (27) over energy becomes straightforward, and we obtain the quasi-classical kinetic equation

$$\left(\frac{\partial}{\partial t} + \mathbf{v} \cdot \frac{\partial}{\partial \mathbf{r}} + e\mathbf{E}_{\mathbf{r}t} \cdot \frac{\partial}{\partial \mathbf{p}} + \frac{e}{c} [\mathbf{v} \times \mathbf{H}_{\mathbf{r}t}] \cdot \frac{\partial}{\partial \mathbf{p}} \right) f_t(\mathbf{r}, \mathbf{p})$$

$$= -\frac{1}{4\pi} \int d\varepsilon \, \mathcal{J}_{\varepsilon t}(F|\mathbf{rp}). \tag{31}$$

To ensure that the right-hand side of this equation is reduced to the quasi-classical electron-impurity collision integral $J_{im}(f|\mathbf{rp}t)$ given by Eq. (9.33), one needs to express $F_{\varepsilon t}(\mathbf{r}, \mathbf{p})$ standing in $\mathcal{J}_{\varepsilon t}(F|\mathbf{rp})$ through

$f_t(\mathbf{r}, \mathbf{p})$. Under the approximation of weak scattering, the influence of the scattering potential on $F_{\varepsilon t}(\mathbf{r}, \mathbf{p})$ in the collision integral should be neglected. In these conditions, $F_{\varepsilon t}(\mathbf{r}, \mathbf{p})$ is to be found from Eq. (9) without the right-hand sides. Since the external fields are smooth in space and slowly varying with time, such a solution, satisfying Eq. (30), is

$$F_{\varepsilon t}(\mathbf{r}, \mathbf{p}) = 2\pi i[2f_t(\mathbf{r}, \mathbf{p}) - 1]\delta(\varepsilon - \varepsilon_p - U_{\mathbf{r}}). \tag{32}$$

Substituting this solution into Eq. (29), we integrate $\mathcal{J}_{\varepsilon t}(\mathbf{r}, \mathbf{p})$ over ε as specified in Eq. (31) and finally obtain the quasi-classical electron-impurity collision integral. In the absence of external potentials, Eq. (32) represents the exact relation between the non-equilibrium Green's function $F_\varepsilon(\mathbf{p})$ of the ideal electron gas and the distribution function $f_{\mathbf{p}}$. Using it, one can write the expressions for the elements of the matrix $\hat{g}_\varepsilon(\mathbf{p})$ in the following way:

$$g_\varepsilon^{-+}(\mathbf{p}) = 2\pi i f_{\mathbf{p}}\delta(\varepsilon - \varepsilon_{\mathbf{p}}),$$

$$g^{--} = g^R + g^{-+}, \quad g^{++} = -g^A + g^{-+}, \quad g^{+-} = g^R - g^A + g^{-+}, \tag{33}$$

where the retarded and advanced Green's functions of the ideal electron gas are $g_\varepsilon^R(\mathbf{p}) = g_\varepsilon^{A*}(\mathbf{p}) = (\varepsilon - \varepsilon_{\mathbf{p}} + i0)^{-1}$.

41. General Formulation of NDT

So far we considered the electrons interacting with impurities. In this section we show how the non-equilibrium diagram technique is introduced for all types of quasiparticles to take into account various interactions between them. The total Hamiltonian of the system is $\widehat{H} = \widehat{H}_0 + \widehat{\mathcal{H}}_{int}$, where $\widehat{H}_0$ describes unperturbed quasiparticles and $\widehat{\mathcal{H}}_{int}$ includes the interaction between these quasiparticles and their interaction with external fields. It is necessary to define the non-equilibrium Green's functions G^{--}, G^{++}, G^{-+}, and G^{+-} as correlation functions of a pair of Heisenberg field operators $\hat{\Psi}_{\mathbf{x}_1}(t_1)$ and $\hat{\Psi}_{\mathbf{x}_2}^+(t_2)$. The function G^{--} coincides with the causal Green's function introduced in Appendix D by Eq. (D.14):

$$G^{--}(1, 2) = -\frac{i}{\hbar}\theta(t_1 - t_2)\langle\langle\hat{\Psi}_{\mathbf{x}_1}(t_1)\hat{\Psi}_{\mathbf{x}_2}^+(t_2)\rangle\rangle \tag{1}$$

$$\pm\frac{i}{\hbar}\theta(t_2 - t_1)\langle\langle\hat{\Psi}_{\mathbf{x}_2}^+(t_2)\hat{\Psi}_{\mathbf{x}_1}(t_1)\rangle\rangle \equiv -\frac{i}{\hbar}\langle\langle\hat{\mathcal{T}}\hat{\Psi}(1)\hat{\Psi}^+(2)\rangle\rangle.$$

Here and below, we use the upper sign for fermions and the lower one for bosons. The operator of chronological ordering, $\hat{\mathcal{T}}$, is introduced by Eq. (2.7). The double angular brackets also include the averaging over random potentials of static inhomogeneities (if these potentials are present). It is easy to check that the function $G^{--}(1,2)$ of Eq. (1) satisfies the equations of motion (39.3) and (39.4), where, instead of the Hamiltonian of electron-impurity system, we should substitute the total Hamiltonian $\widehat{H}_0 + \widehat{\mathcal{H}}_{int}$. The corresponding equations for $G^{++}(1,2)$ have the opposite sign of the inhomogeneous terms (right-hand sides), which results in the inverted chronological ordering with respect to Eq. (1):

$$G^{++}(1,2) = -\frac{i}{\hbar}\theta(t_2 - t_1)\langle\langle\hat{\Psi}_{\mathbf{x}_1}(t_1)\hat{\Psi}^+_{\mathbf{x}_2}(t_2)\rangle\rangle \tag{2}$$

$$\pm\frac{i}{\hbar}\theta(t_1 - t_2)\langle\langle\hat{\Psi}^+_{\mathbf{x}_2}(t_2)\hat{\Psi}_{\mathbf{x}_1}(t_1)\rangle\rangle \equiv -\frac{i}{\hbar}\langle\langle\widetilde{\mathcal{T}}\hat{\Psi}(1)\hat{\Psi}^+(2)\rangle\rangle ,$$

where we have introduced the operator $\widetilde{\mathcal{T}}$ of inverse chronological ordering. The function G^{-+}, as we already know, is directly related to the density matrix of the quasiparticles; see Eq. (39.11). This relation is also evident if we represent this function in the form

$$G^{-+}(1,2) = \pm\frac{i}{\hbar}\langle\langle\hat{\Psi}^+_{\mathbf{x}_2}(t_2)\hat{\Psi}_{\mathbf{x}_1}(t_1)\rangle\rangle , \tag{3}$$

which satisfies the equations of motion (39.3) and (39.4) with the Hamiltonian $\widehat{H}_0 + \widehat{\mathcal{H}}_{int}$. Using Eq. (4.29) rewritten in the coordinate representation, we have $\rho_t(\mathbf{x}_1, \mathbf{x}_2) = \langle\langle\hat{\Psi}^+_{\mathbf{x}_2}(t)\hat{\Psi}_{\mathbf{x}_1}(t)\rangle\rangle \equiv \mathrm{Sp}\{\hat{\eta}\hat{\Psi}^+_{\mathbf{x}_2}(t)\hat{\Psi}_{\mathbf{x}_1}(t)\}$ (note that the statistical operator $\hat{\eta}$ is independent of time in the Heisenberg representation). Therefore, the function $G^{-+}(1,2)$ defined by Eq. (3) for fermions satisfies Eq. (39.11). For bosons, by convention, the sign of $G^{-+}(1,2)$ is altered. Therefore, the one-boson density matrix can be expressed as $\rho_t(\mathbf{x}_1, \mathbf{x}_2) = i\hbar G^{-+}(\mathbf{x}_1 t, \mathbf{x}_2 t)$, which is similar to Eq. (39.11) with the opposite sign of the right-hand side. Finally, we find the remaining function $G^{+-}(1,2)$ by using Eqs. (1)–(3) and the linear relation (39.22):

$$G^{+-}(1,2) = -\frac{i}{\hbar}\langle\langle\hat{\Psi}_{\mathbf{x}_1}(t_1)\hat{\Psi}^+_{\mathbf{x}_2}(t_2)\rangle\rangle. \tag{4}$$

Expressing the retarded and advanced Green's functions as $G^R = G^{--} - G^{-+}$ and $G^A = G^{--} - G^{+-}$, see Eq. (40.3), we find that they are the retarded and advanced double-time Green's functions introduced by Eq. (D.14). To write the non-equilibrium Green's function $G^{s_1 s_2}(1,2)$ ($s_1, s_2 = \pm$) in the explicit form, we use two equivalent notations, $G^{s_1 s_2}(\mathbf{x}_1 t_1, \mathbf{x}_2 t_2)$ and $G^{s_1 s_2}_{t_1 t_2}(\mathbf{x}_1, \mathbf{x}_2)$.

   The diagram technique is built as a perturbation theory with respect to the operator of interaction $\widehat{\mathcal{H}}_{int}$. Therefore, it is convenient to express the field operators in the interaction representation, using the basis $|\lambda t\rangle$ introduced in Sec. 2. We stress that the Green's functions (1)–(4) do not depend on the representation used for the field operators, provided that the statistical operator is written in the same representation. In the interaction representation, the statistical operator satisfies the equation of motion

$$ i\hbar\frac{\partial\hat{\eta}_t}{\partial t} = [\hat{\mathcal{H}}_{int}(t), \hat{\eta}_t], \quad \hat{\mathcal{H}}_{int}(t) = e^{i\hat{H}_0 t/\hbar}\hat{\mathcal{H}}_{int}e^{-i\hat{H}_0 t/\hbar}, \tag{5} $$

where we have introduced the operator $\hat{\mathcal{H}}_{int}(t)$. The equation expressing $\hat{\mathcal{H}}_{int}(t)$ through $\hat{\mathcal{H}}_{int}$ also demonstrates how to transform an arbitrary operator from the Schroedinger representation to the interaction representation. Equation (5) shows us that the statistical operator depends on time because of the interaction. It is desirable, however, to write Eqs. (1)–(4) by using the statistical operator $\hat{\eta}^{(0)}$ of the non-interacting system. It can be done with the aid of the following formal transformation. We assume that the interaction is adiabatically turned on at $t = -\infty$ and use Eq. (2.11), where $\hat{\mathcal{H}}_{int}$ stands in place of $\hat{V}_t$, with the initial condition $|\lambda, t = -\infty\rangle = |i\rangle$ to find $|\lambda t\rangle = \hat{S}_0(t, -\infty)|i\rangle$. Here $|i\rangle$ is the quantum state of the non-interacting (unperturbed) system of quasiparticles. The operator $\hat{S}_0(t, t')$ is introduced in a similar way as the evolution operator (2.8), the only difference is that instead of $\hat{H}_\tau$ we substitute the operator $\hat{\mathcal{H}}_{int}(\tau)$:

$$ \hat{S}_0(t, t') = \hat{\mathcal{T}}\exp\left[-\frac{i}{\hbar}\int_{t'}^{t}d\tau\hat{\mathcal{H}}_{int}(\tau)\right]. \tag{6} $$

With the use of this operator, the bra-vector $\langle\lambda t|$ is expressed as $\langle\lambda t| = \langle i|\hat{S}_0(-\infty, t) = \langle i|\hat{S}_0^+(t, -\infty)$. The average $\langle\langle\hat{\mathcal{T}}\hat{\Psi}(1)\hat{\Psi}^+(2)\rangle\rangle = \text{Sp}\{\hat{\eta}\hat{\mathcal{T}} \times\hat{\Psi}(1)\hat{\Psi}^+(2)\}$ can be rewritten as

$$ \sum_i P_i^{(0)}\left\{\theta(t_1 - t_2)\langle i|\hat{S}_0^+(t_1, -\infty)\hat{\Psi}_0(1)\hat{S}_0(t_1, t_2)\hat{\Psi}_0^+(2)\hat{S}_0(t_2, -\infty)|i\rangle \right. $$

$$ \left. \mp\theta(t_2 - t_1)\langle i|\hat{S}_0^+(t_2, -\infty)\hat{\Psi}_0^+(2)\hat{S}_0(t_2, t_1)\hat{\Psi}_0(1)\hat{S}_0(t_1, -\infty)|i\rangle\right\}, \tag{7} $$

where $P_i^{(0)} = \langle i|\hat{\eta}^{(0)}|i\rangle$ is the probability to find the unperturbed system in the state i; see Sec. 1. The subscripts 0 at the field operators indicate that these operators are written in the interaction representation. Next, applying the identity $\hat{S}_0^+(t, -\infty) = \hat{S}_0^+(\infty, -\infty)\hat{S}_0(\infty, t)$ to $\hat{S}_0^+(t_1, -\infty)$ and $\hat{S}_0^+(t_2, -\infty)$ in the first and in the second terms, respectively, we rewrite the expression (7) as

$$\sum_i P_i^{(0)} \left\{ \langle i| \widehat{S}_0^+(\infty, -\infty) \left[\theta(t_1 - t_2) \widehat{S}_0(\infty, t_1) \hat{\Psi}_0(1) \widehat{S}_0(t_1, t_2) \hat{\Psi}_0^+(2) \right. \right.$$

$$\left. \left. \times \widehat{S}_0(t_2, -\infty) \mp \theta(t_2 - t_1) \widehat{S}_0(\infty, t_2) \hat{\Psi}_0^+(2) \widehat{S}_0(t_2, t_1) \hat{\Psi}_0(1) \widehat{S}_0(t_1, -\infty) \right] |i\rangle \right\}$$

$$= \sum_i P_i^{(0)} \langle i| \widehat{S}_0^+(\infty, -\infty) [\hat{\mathcal{T}} \hat{\mathcal{S}} \hat{\Psi}_0(1) \hat{\Psi}_0^+(2)] |i\rangle, \tag{8}$$

where

$$\hat{\mathcal{S}} = \exp\left[-\frac{i}{\hbar} \int_{-\infty}^{\infty} dt \hat{\mathcal{H}}_{int}(t) \right] \tag{9}$$

denotes the operator $\widehat{S}_0(\infty, -\infty)$ without the chronological ordering. The term $\widehat{S}_0^+(\infty, -\infty) = \widehat{S}_0(-\infty, \infty)$ in Eq. (8) can be written as $\widetilde{\mathcal{T}} \hat{\mathcal{S}}^{-1}$. Therefore, we obtain

$$i\hbar G^{--}(1, 2) = \langle\langle [\widetilde{\mathcal{T}} \hat{\mathcal{S}}^{-1}][\hat{\mathcal{T}} \hat{\mathcal{S}} \hat{\Psi}_0(1) \hat{\Psi}_0^+(2)] \rangle\rangle_0 , \tag{10}$$

where $\langle\langle \ldots \rangle\rangle_0 = \mathrm{Sp}(\hat{\eta}^{(0)} \ldots)$ denotes the averaging over the states of the unperturbed system. The interaction terms are present only in the exponents $\hat{\mathcal{S}}$, which can be expanded in series to find the contributions of arbitrary order in the interaction. The other averages standing in Eqs. (2)–(4) are transformed in a similar way (problem 8.4):

$$i\hbar G^{++}(1, 2) = \langle\langle [\widetilde{\mathcal{T}} \hat{\Psi}_0(1) \hat{\Psi}_0^+(2) \hat{\mathcal{S}}^{-1}][\hat{\mathcal{T}} \hat{\mathcal{S}}] \rangle\rangle_0 , \tag{11}$$

$$i\hbar G^{-+}(1, 2) = \mp \langle\langle [\widetilde{\mathcal{T}} \hat{\Psi}_0^+(2) \hat{\mathcal{S}}^{-1}][\hat{\mathcal{T}} \hat{\mathcal{S}} \hat{\Psi}_0(1)] \rangle\rangle_0 , \tag{12}$$

$$i\hbar G^{+-}(1, 2) = \langle\langle [\widetilde{\mathcal{T}} \hat{\Psi}_0(1) \hat{\mathcal{S}}^{-1}][\hat{\mathcal{T}} \hat{\mathcal{S}} \hat{\Psi}_0^+(2)] \rangle\rangle_0 . \tag{13}$$

Instead of using two operators of direct and inverse chronological ordering, one may introduce the operator $\hat{\mathcal{T}}_C$ defining the ordering along the contour C, which starts at $-\infty$, goes to ∞, then turns back and ends at $-\infty$. This contour is shown in Fig. 8.1. The first its branch is indicated by the sign $-$, while the second (backward) branch is indicated by the sign $+$. Both field operators standing in G^{--} (G^{++}) belong to the first (second) branch. To emphasize this property, one may add the subscripts $-$ (or $+$) to these operators. The field operators standing in G^{-+} and G^{+-} belong to the different branches and bear the different subscripts. Introducing the operator

$$\hat{\mathcal{S}}_C = \exp\left[-\frac{i}{\hbar} \int_C dt \hat{\mathcal{H}}_{int}(t) \right], \tag{14}$$

one may write Eqs. (10)–(13) as a single equation

$$i\hbar G^{s_1 s_2}(1,2) = \langle\langle \hat{\mathcal{T}}_C \hat{\mathcal{S}}_C \hat{\Psi}_{0 s_1}(1) \hat{\Psi}^+_{0 s_2}(2) \rangle\rangle_0 \ , \quad s_1, s_2 = \pm. \tag{15}$$

This equation identifies the indices $\pm$ of the Green's functions with the contour branch indices of the field operators. Therefore, the four Green's functions of the non-equilibrium diagram technique can be considered as four different kinds of causal Green's functions defined on a special double-branch time contour C.

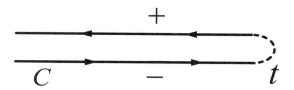

Figure 8.1. Double-time contour containing $+$ and $-$ branches. The arrows show the direction of chronological ordering.

Let us evaluate the Green's functions for different interactions. Consider first the case of external potential perturbation, when $\hat{\mathcal{H}}_{int}(t) = \int d\mathbf{x} \hat{\Psi}^+_{0\mathbf{x}}(t) U_{\mathbf{x}t} \hat{\Psi}_{0\mathbf{x}}(t)$. Substituting this form into $\hat{\mathcal{S}}$, we expand the exponents $\hat{\mathcal{S}}$ in Eqs. (10)–(13) in series. Consider, for example, $G^{--}(1,2)$. The expansion up to the first-order corrections gives us

$$iG^{--}(1,2) = ig^{--}(1,2) + \frac{i}{\hbar^2} \int d3 \langle\langle [\tilde{\mathcal{T}} \hat{\Psi}^+_0(3) U_3 \hat{\Psi}_0(3)][\hat{\mathcal{T}} \hat{\Psi}_0(1) \hat{\Psi}^+_0(2)] \rangle\rangle_0$$

$$- \frac{i}{\hbar^2} \int d3 \langle\langle [\hat{\mathcal{T}} \hat{\Psi}^+_0(3) U_3 \hat{\Psi}_0(3) \hat{\Psi}_0(1) \hat{\Psi}^+_0(2)] \rangle\rangle_0 + \dots , \tag{16}$$

where $g^{--}(1,2) = -(i/\hbar) \langle\langle \hat{\mathcal{T}} \hat{\Psi}_0(1) \hat{\Psi}^+_0(2) \rangle\rangle_0$ is the unperturbed Green's function and $U_3 = U_{\mathbf{x}_3 t_3}$. The averaging of the field-operator products standing in the perturbation terms is done in a similar way as the averaging of the products of creation and annihilation operators of electrons and phonons in Chapters 4–7. Namely, we make all possible pairings (contractions) of the operators, and write the correlation function of the product as a product of the pair correlation functions. Each pair includes one creation and one annihilation field operator (if both of them are creation or annihilation operators, their correlation function is zero). The order of the operators in the pairs is the same as in the original product (creation and annihilation operators are not permuted). For fermions, the final expression should contain a factor of $(-1)^n$, where n is the number of permutations done to bring the operators in the pairs together.

These rules can be derived from the knowledge of how the bosonic and fermionic operators act on the quantum states; see Eqs. (3.14), (4.19), and (4.20). Applying these rules to the chronologically-ordered products, one should keep the ordering intact. Thus, the two correlation functions standing in Eq. (16) are transformed as

$$\langle\langle[\tilde{\mathcal{T}}\hat{\Psi}_0^+(3)\hat{\Psi}_0(3)][\hat{\mathcal{T}}\hat{\Psi}_0(1)\hat{\Psi}_0^+(2)]\rangle\rangle_0 \tag{17}$$

$$= \mp\langle\langle\hat{\Psi}_0^+(3)\hat{\Psi}_0(1)\rangle\rangle_0\langle\langle\hat{\Psi}_0(3)\hat{\Psi}_0^+(2)\rangle\rangle_0 = (i\hbar)^2 G^{-+}(1,3)G^{+-}(3,2),$$

$$\langle\langle[\hat{\mathcal{T}}\hat{\Psi}_0^+(3)\hat{\Psi}_0(3)\hat{\Psi}_0(1)\hat{\Psi}_0^+(2)]\rangle\rangle_0 \tag{18}$$

$$= \mp\langle\langle\mathcal{T}\hat{\Psi}_0^+(3)\hat{\Psi}_0(1)\rangle\rangle_0\langle\langle\mathcal{T}\hat{\Psi}_0(3)\hat{\Psi}_0^+(2)\rangle\rangle_0 = (i\hbar)^2 G^{--}(1,3)G^{--}(3,2).$$

Note that we have omitted the correlation functions containing the pairing of $\hat{\Psi}_0(1)$ with $\hat{\Psi}_0^+(2)$, because they give zero contribution in Eq. (16). Substituting the transformations (17) and (18) into Eq. (16), we obtain

$$iG^{--}(1,2) = ig^{--}(1,2) + \int d3\,[(-iU_3)ig^{--}(1,3)ig^{--}(3,2)$$

$$+(iU_3)ig^{-+}(1,3)ig^{+-}(3,2)] + \dots\,, \tag{19}$$

where the integration is carried out over the internal variable 3. The calculations can be repeated for each Green's function. Let us carry out these calculations in a unified way, by using Eq. (15) with the ordering along the contour C:

$$iG^{s_1s_2}(1,2) = ig^{s_1s_2}(1,2) + \frac{1}{\hbar^2}\sum_{s_3=\pm}\int d3\,is_3U_3$$

$$\times\langle\langle\hat{\mathcal{T}}_C\hat{\Psi}_{0s_3}^+(3)\hat{\Psi}_{0s_3}(3)\hat{\Psi}_{0s_1}(1)\hat{\Psi}_{0s_2}^+(2)\rangle\rangle_0 + \dots\,. \tag{20}$$

As a result,

$$iG^{s_1s_2}(1,2) = ig^{s_1s_2}(1,2)$$

$$+\sum_{s_3=\pm}\int d3\,ig^{s_1s_3}(1,3)is_3U_3ig^{s_3s_2}(3,2) + \dots\,. \tag{21}$$

Expanding $\hat{S}_C$ up to the n-th order, one can see that the factor $1/n!$, which appears in the expansion of the exponent, vanishes after unification of the terms differing by permutations of internal variables. The correction of an arbitrary order looks like

$$ig^{s_1s_3}(1,3)is_3U_3\dots is_4U_4ig^{s_4s_2}(4,2) = \underset{1s_1\quad 3s_3}{\underbrace{\vdots\qquad\vdots}}\;\dots\;\underset{4s_4\quad 2s_2}{\underbrace{\vdots\qquad\vdots}}\,. \tag{22}$$

It is diagrammatically represented as a sequence of connected lines with vertices. Each $ig^{ss'}(X, X')$ is shown by a thin line with the indices Xs and $X's'$, and each vertex Xs brings the factor isU_X to the analytical expression; see also Eq. (39.16). The integrals over the internal variables (including the sums over the indices $s_3, \ldots, s_4$) are implied. These are the diagram rules for the interaction of quasiparticles with an external potential.

If the external potential perturbation $U_{\mathbf{x}t}$ contains a random static potential, the expression (22) should be averaged over realizations of this potential. The averaging procedure leads us to the diagram technique already discussed in Sec. 39. Therefore, the random and the regular potentials call for different treatment: the former enters the diagrammatic expansions in the form of binary correlation functions (broken lines), while the latter introduces separate vertices. By choosing a number of such vertices, one can calculate the response of the system to this regular potential perturbation in the corresponding order. The linear response theory can be derived as a particular case of the non-equilibrium diagram technique, when the applied potential contribution is treated in the first, linear order (problem 8.5). One should mention that the linear response of any kind of interacting quasiparticles also can be described within an approach called the temperature diagram technique. We do not consider this approach, since the non-equilibrium diagram technique is a more general method and can be formulated for the case of linear response in a straightforward way.

Consider now the interaction between the quasiparticles of the same kind. The Hamiltonian $\hat{\mathcal{H}}_{int}(t)$ has the same form as $\hat{H}_2$ given by Eq. (4.27), where the field operators should be written in the interaction representation. In the first order, we obtain

$$iG^{s_1 s_2}(1, 2) = ig^{s_1 s_2}(1, 2) + \frac{1}{2\hbar^3} \sum_{s=\pm} \int d3 \int d4 \; isU_{3-4}$$

$$\times \langle\langle \hat{\mathcal{T}}_C \hat{\Psi}_{0s}^+(3) \hat{\Psi}_{0s}^+(4) \hat{\Psi}_{0s}(4) \hat{\Psi}_{0s}(3) \hat{\Psi}_{0s_1}(1) \hat{\Psi}_{0s_2}^+(2) \rangle\rangle_0 + \ldots \quad (23)$$

By definition,

$$U_{3-4} = \hbar\delta(t_3 - t_4)v(\mathbf{x}_3 - \mathbf{x}_4). \quad (24)$$

The dimensionality of U_{3-4} is energy in power 2. The six-operator product in Eq. (23) brings us four possible ways to make the contractions (the pairing of $\hat{\Psi}_{0s_1}(1)$ with $\hat{\Psi}_{0s_2}^+(2)$ is not considered, since it gives zero contribution). The four terms obtained are equal in pairs, differing only by the permutation of the variables of integration (3 and 4). Therefore, only two terms remain, and the factor of 2 in the denominator vanishes (it is easy to see that this property takes place in an arbitrary order). Transforming the correlation function in Eq. (23), we find

$$iG^{s_1 s_2}(1,2) = ig^{s_1 s_2}(1,2) + \sum_{s=\pm} \int d3 \int d4 \, isU_{3-4} \qquad (25)$$

$$\times \left[-ig^{ss}(4,4)ig^{s_1 s}(1,3)ig^{ss_2}(3,2) + ig^{s_1 s}(1,3)ig^{ss}(3,4)ig^{ss_2}(4,2) \right] + \dots .$$

The two perturbation terms can be represented, respectively, as two diagrams

where the broken lines with vertex indices $3s$ and $4s$ denote the factor isU_{3-4}. The factor $-ig^{ss}(4,4)$ corresponds to the self-closed loop in the first diagram. If the spin variable is implicitly introduced into the multi-indices $1, 2, \dots$, and the interaction U_{3-4} is spin-independent, the sum over the spin variable of the multi-index 4 in the first term gives us the spin degeneracy factor (which is equal to 2 for electrons). On the other hand, there are no other spin sums in the expression (25), since we consider the contractions of the field operators with equal spins. Therefore, the spin of the multi-index 3 in the first term and the spins of the multi-indices 3 and 4 in the second term are equal to the spin of the multi-indices 1 and 2. The factor $-i\hbar g^{ss}(4,4)$, summed over the spin variable of the multi-index 4, is equal to the quasiparticle density $n_{\mathbf{x}_4 t_4}$, regardless of the vertex sign s. Thus, the first diagram of Eq. (26) corresponds to the mean-field correction, while the second one gives us the exchange correction to the Green's function $G^{s_1 s_2}(1,2)$ due to the interaction. The corrections of an arbitrary order are constructed diagrammatically from the solid lines ig and broken lines isU connecting the vertices with the same index s. Each closed loop of the solid lines brings us a factor of -1 (or -2, if the spin variables are not included in the multi-indices). The integrals over the internal indices are assumed. The order of the diagram is equal to the number of the broken lines.

We note that the Green's functions of the systems described by stationary Hamiltonians depend only on the difference of their time variables. For this reason, the energy representation becomes more convenient for them, and the exact eigenstate representation of $G^{s_1 s_2}(1,2)$ can be written in a rather simple way (problem 8.6).

The diagram technique allows us to operate with blocks of diagrams, introducing such graphic elements as the self-energy functions. We have already seen it on the example of electron-impurity interaction; see Eq. (39.17). A similar Dyson equation describes a system of interacting quasiparticles, one just needs to redefine the matrix $\widehat{\Sigma}$. The diagrammatic expansion of $-i\Sigma^{s_1 s_2}(1,2)$ is written in the same fashion as Eq. (E.23) for interacting electrons at zero temperature, where the signs $+$ or $-$ should be added to the internal vertices in all possible ways, and the signs s_1 and s_2 should be added to the external vertices 1 and 2. The thin broken line $-i\hbar v(1,2)$ introduced in Appendix E is equivalent to the thin broken line isU_{1-2} of the non-equilibrium diagram technique at $s=-$. Therefore, if one considers interacting electrons, U_{1-2} is equal to $\hbar v(1,2)$, though the notation U_{1-2} is more general in the sense that it can be applied to any kind of interacting quasiparticles. Since the thin broken lines denoting the interaction must have equal signs at their ends, the expansion of the non-diagonal elements $\Sigma^{-+}(1,2)$ and $\Sigma^{+-}(1,2)$ begins with the terms quadratic in the interaction, like the third and the fourth diagrams on the right-hand side of Eq. (E.23) (problem 8.7). The expansion of the diagonal elements $\Sigma^{--}(1,2)$ and $\Sigma^{++}(1,2)$ begins with the linear terms. One can also write matrix equations similar to Eq. (39.21), which lead to the generalized kinetic equation for interacting quasiparticles in the form

$$-\left[i\hbar\left(\frac{\partial}{\partial t_1}+\frac{\partial}{\partial t_2}\right)-\widehat{H}_0(1)+\widehat{H}_0^*(2)\right]G^{-+}(1,2)$$

$$=-\sum_{s=\pm}\int d1'\left[\Sigma^{-s}(1,1')G^{s+}(1',2)+G^{-s}(1,1')\Sigma^{s+}(1',2)\right].\qquad(27)$$

Using Eq. (27) with relevant self-energy functions, one can derive the expression for the electron-electron collision integral in the form given by Eqs. (31.21) and (31.22) (problem 8.8).

Other useful blocks are the polarization function $\Pi^{s_1 s_2}(1,2)$ and the screened interaction $V^{s_1 s_2}(1,2)$, which are also introduced and denoted in a similar way as in Appendix E. We denote $(i/\hbar)\Pi^{s_1 s_2}(1,2)$ and $-i\hbar V^{s_1 s_2}(1,2)$, respectively, by a circle and by a bold broken line, with the indices $1s_1$ and $2s_2$. The polarization function is a sum of all irreducible diagrams constructed from two or more Green's functions of electrons (solid lines) connecting the points $1s_1$ and $2s_2$. Its expansion in the diagram form is given by Eq. (E.26) with the indices $+$ or $-$ added to each vertex. In the analytical form,

$$(i/\hbar)\Pi^{s_1 s_2}(1,2)=-2iG^{s_1 s_2}(1,2)iG^{s_2 s_1}(2,1)+\dots\ .\qquad(28)$$

The factor of -2 appears because of the electron loop, where the spin degeneracy is taken into account. The components of the polarization function satisfy the linear relation

$$\Pi^{--}(1,2) + \Pi^{++}(1,2) = \Pi^{-+}(1,2) + \Pi^{+-}(1,2), \qquad (29)$$

which is similar to Eqs. (39.7) and (39.22). The screened interaction is expressed through the polarization function as described by the diagram equation (E.25), where we have to add the signs $+$ or $-$ to each vertex and take into account that the signs at the ends of the bare potential lines (thin broken lines) coincide. Therefore, instead of a single equation (E.24) for the zero-temperature case, we obtain four equations connecting $\Pi^{s_1 s_2}(1,2)$ and $V^{s_1 s_2}(1,2)$. We write this system of equations as

$$V^{s_1 s_1'}(1,1') = -s_1 \delta_{s_1 s_1'} v(1,1') - s_1 \int d2\, v(1,2) \int d2'$$

$$\times \sum_{s=\pm} \Pi^{s_1 s}(2,2') V^{s s_1'}(2',1'). \qquad (30)$$

The four equations are split in two pairs of equations, the first one connects V^{--} and V^{+-}, while the second one connects V^{++} and V^{-+}. The introduction of $\Pi^{s_1 s_2}(1,2)$ and $V^{s_1 s_2}(1,2)$ allows one to consider the problem of dynamical screening (see Sec. 33) for the case of finite temperatures (problem 8.9). It is also convenient for studying the influence of the dynamical screening on the electron-electron collision integral (problem 8.10).

42. NDT Formalism for Electron-Boson System

Let us apply the formalism of non-equilibrium diagram technique to study the interaction of electrons with phonons and photons. Since we have two kinds of quasiparticles, the Hamiltonian $\widehat{H}_0$ is presented as a sum of electron and boson Hamiltonians, $\widehat{H}_0 = \widehat{H}_e + \widehat{H}_b$, where $\widehat{H}_b$ is either the phonon Hamiltonian $\widehat{H}_{ph}$ or the photon Hamiltonian $\widehat{H}_{pht}$. The averaging in the correlation functions $\langle\langle \dots \rangle\rangle_0$, as in Eq. (41.15), is carried out over the unperturbed electron and boson states, which are the eigenstates of $\widehat{H}_e$ and $\widehat{H}_b$, respectively. In addition to the electron Green's function, we introduce the Green's functions of phonons and photons, and the operator of interaction is assumed to be linear in the bosonic coordinate.

Consider first the deformation-potential interaction of electrons with acoustic phonons; see Eq. (6.30). In the interaction representation discussed in the previous section, the operator of the interaction is written as

$$\hat{\mathcal{H}}_{int}(t) = \mathcal{D} \int d\mathbf{x} \hat{\Psi}_{0\mathbf{x}}^{+}(t) \nabla_{\mathbf{x}} \cdot \hat{\mathbf{u}}_0(\mathbf{x}, t) \hat{\Psi}_{0\mathbf{x}}(t), \tag{1}$$

where $\hat{\mathbf{u}}_0(\mathbf{x}, t) = \exp(i\widehat{H}_{ph}t/\hbar)\hat{\mathbf{u}}_{ac}(\mathbf{x})\exp(-i\widehat{H}_{ph}t/\hbar)$ is the operator of lattice displacement in the interaction representation. The displacement operator $\hat{\mathbf{u}}_{ac}(\mathbf{x})$ in the Schroedinger representation is defined by Eq. (6.29) as a linear combination of creation and annihilation operators. Therefore, when the operator $\hat{S}$ defined by Eq. (41.9) is expanded in series, only the terms containing products of even numbers of $\hat{\mathbf{u}}_0(\mathbf{x}, t)$ contribute to the correlation functions. According to Eqs. (41.14) and (41.15), the first perturbation term in the expansion of the electron Green's function $G^{s_1 s_2}(1, 2)$ is of the second order in electron-phonon interaction:

$$iG^{s_1 s_2}(1, 2) = ig^{s_1 s_2}(1, 2) + \frac{\mathcal{D}^2}{2\hbar^3} \sum_{s_3, s_4 = \pm} is_3 \, is_4 \int d3 \int d4$$

$$\times \langle\langle \hat{\mathcal{T}}_C \left(\nabla_{\mathbf{x}_3} \cdot \hat{\mathbf{u}}_{0s_3}(3)\right) \left(\nabla_{\mathbf{x}_4} \cdot \hat{\mathbf{u}}_{0s_4}(4)\right) \hat{\Psi}_{0s_3}^{+}(3)\hat{\Psi}_{0s_3}(3) \tag{2}$$

$$\times \hat{\Psi}_{0s_4}^{+}(4)\hat{\Psi}_{0s_4}(4)\hat{\Psi}_{0s_1}(1)\hat{\Psi}_{0s_2}^{+}(2)\rangle\rangle_0 + \dots,$$

where we use $\hat{\mathbf{u}}_0(i) \equiv \hat{\mathbf{u}}_0(\mathbf{x}_i, t_i)$ and attribute the contour branch indices s_3 and s_4 to the lattice displacement operators.

At this point, we need to introduce the non-equilibrium Green's functions $D^{\alpha\beta, s_1 s_2}(\mathbf{x}_1 t_1, \mathbf{x}_2 t_2) \equiv D_{t_1 t_2}^{\alpha\beta, s_1 s_2}(\mathbf{x}_1, \mathbf{x}_2) \equiv D^{\alpha\beta, s_1 s_2}(1, 2)$ according to

$$D^{\alpha\beta, --}(1, 2) = -\frac{i}{\hbar}\langle\langle \hat{\mathcal{T}}\hat{u}^{\alpha}(1)\hat{u}^{\beta}(2)\rangle\rangle,$$

$$D^{\alpha\beta, ++}(1, 2) = -\frac{i}{\hbar}\langle\langle \widetilde{\hat{\mathcal{T}}}\hat{u}^{\alpha}(1)\hat{u}^{\beta}(2)\rangle\rangle, \tag{3}$$

$$D^{\alpha\beta, +-}(1, 2) = -\frac{i}{\hbar}\langle\langle \hat{u}^{\alpha}(1)\hat{u}^{\beta}(2)\rangle\rangle,$$

$$D^{\alpha\beta, -+}(1, 2) = -\frac{i}{\hbar}\langle\langle \hat{u}^{\beta}(2)\hat{u}^{\alpha}(1)\rangle\rangle,$$

where α and β are the coordinate indices (so that $D^{\alpha\beta, s_1 s_2}$ are tensors) and $\hat{\mathbf{u}}(i)$ are the Heisenberg operators of the lattice displacement vectors. The averaging is carried out over the states of the interacting system. The set of equations (3) can be written as a single equation:

$$D^{\alpha\beta, s_1 s_2}(1, 2) = -\frac{i}{\hbar}\langle\langle \hat{\mathcal{T}}_C \hat{u}_{s_1}^{\alpha}(1)\hat{u}_{s_2}^{\beta}(2)\rangle\rangle, \tag{4}$$

with the chronological ordering along the contour C. The displacement operators are Hermitian, $\hat{\mathbf{u}}(i) = \hat{\mathbf{u}}^{+}(i)$. Therefore, there is no difference whether we put the cross superscript or not, and the definitions (3) formally coincide with Eqs. (41.1)–(41.4) for bosons if we replace $\hat{\Psi}(i)$ by the vectors $\hat{u}^\alpha(i)$. It is easy to find that $D^{\alpha\beta,s_1s_2}(1,2)$ satisfy the relation $D^{--}+D^{++} = D^{-+}+D^{+-}$ similar to Eqs. (39.7) and (39.22). The retarded and advanced Green's functions are expressed through the non-equilibrium Green's functions in a standard way, as $D^R = D^{--} - D^{-+}$ and $D^A = D^{--} - D^{+-}$. The Hermiticity of $\hat{u}(i)$ leads to the following symmetry property:

$$D^{\alpha\beta,s_1s_2}(1,2) = D^{\beta\alpha,s_2s_1}(2,1). \tag{5}$$

In the absence of interactions, Eq. (3) describes the Green's functions of free phonons, $d^{\alpha\beta,s_1s_2}(1,2) = -(i/\hbar)\langle\langle\hat{u}^\alpha_{s_1}(1)\hat{u}^\beta_{s_2}(2)\rangle\rangle_0$. The corresponding retarded and advanced Green's functions are the Green's functions of the elasticity theory; see further in Sec. 47.

Now we can express the contraction of the displacement operators $\hat{u}_{0s_3}(3)$ and $\hat{u}_{0s_4}(4)$ standing in Eq. (2) through the Green's function of free phonons. The contractions of six field operators of electrons can be done in six possible ways. Using Eq. (5), one can prove that the six corresponding terms are equal in pairs, differing only by the permutation of the indices $3s_3$ and $4s_4$. This cancels the factor of 2 in the denominator of Eq. (2), and only three terms remain. Two of them, containing the pairings of $\hat{\Psi}_{0s_1}(1)$ with $\hat{\Psi}^+_{0s_2}(2)$ and $\hat{\Psi}_{0s_3}(3)$ with $\hat{\Psi}^+_{0s_3}(3)$ give zero contribution in Eq. (2). Therefore, only one perturbation term of this order remains:

$$iG^{s_1s_2}(1,2) = ig^{s_1s_2}(1,2) + \hbar\mathcal{D}^2 \sum_{s_3,s_4=\pm} is_3\, is_4 \int d3 \int d4 \tag{6}$$

$$\times ig^{s_1s_3}(1,3)ig^{s_3s_4}(3,4)ig^{s_4s_2}(4,2) \sum_{\alpha\beta} \nabla^\alpha_{\mathbf{x}_3} \nabla^\beta_{\mathbf{x}_4} id^{\alpha\beta,s_3s_4}(3,4) + \cdots .$$

The diagram representation of the perturbation term in Eq. (6) can be written as

$$\frac{\overset{\alpha}{}\overset{\beta}{}}{1s_1 \quad 3s_3 \qquad 4s_4 \quad 2s_2} \quad, \tag{7}$$

where $id^{\alpha\beta,s_3s_4}(3,4)$ is denoted by a thin wavy line. The solid lines denote the electron Green's functions ig, which do not contain the Cartesian coordinate indices (for example, the line connecting the points $3s_3\alpha$ and $4s_4\beta$ corresponds to $ig^{s_3s_4}(3,4)$). With these definitions, one can formulate the following diagram rules. Each perturbation term contributing to the electron Green's function of the order of $2n$ contains $2n$

vertices, n phonon (wavy) lines, and $2n + 1$ electron (solid) lines. Two electron lines and one phonon line meet at each vertex, one electron line begins there, and the other ends. The vertex denoted as $1s\alpha$ brings the factor $is\hbar^{1/2}\mathcal{D}\nabla^{\alpha}_{\mathbf{x}_1}$, where the differential operator acts on the argument of the phonon Green's function associated with this vertex. One should take the sum over internal Cartesian indices. Each closed loop of electron lines brings the factor of -1 (or -2, if the electron spins are not included in the multi-indices). The diagrams containing the loops connected to the other parts by a single phonon line should be omitted, since their contribution is zero.

The same rules are used when we need to find the corrections to the phonon Green's function due to electron-phonon interaction. Using the formalism developed in the previous section, we can write (compare to Eq. (41.15))

$$i\hbar D^{\alpha\beta,s_1s_2}(1,2) = \langle\langle \hat{\mathcal{T}}_C \hat{\mathcal{S}}_C \hat{u}^{\alpha}_{0s_1}(1) \hat{u}^{\beta}_{0s_2}(2) \rangle\rangle_0 \ , \tag{8}$$

where the averaging is carried out over the states of the unperturbed phonon system. Expanding the exponent $\hat{\mathcal{S}}_C$ in series, we compose relevant contractions and obtain

$$iD^{\alpha\beta,s_1s_2}(1,2) = id^{\alpha\beta,s_1s_2}(1,2) - 2\hbar\mathcal{D}^2 \sum_{s_3,s_4=\pm} is_3 \ is_4 \int d3 \int d4 \tag{9}$$

$$\times ig^{s_3s_4}(3,4) ig^{s_4s_3}(4,3) \sum_{\gamma\delta} \nabla^{\gamma}_{\mathbf{x}_3} \nabla^{\delta}_{\mathbf{x}_4} id^{\alpha\gamma,s_1s_3}(1,3) id^{\delta\beta,s_4s_2}(4,2) + \dots \ .$$

This expansion is also written as

$$+ \dots \ , \tag{10}$$

where the bold wavy line corresponds to the dressed (i.e., modified by the electron-phonon interaction) Green's function iD. The factor of -2 in the first perturbation term of Eq. (9) appears because of the closed loop of electron lines (the spin variable is not included in the multi-indices). In such diagrams, each perturbation term of the order of $2n$ contains $n + 1$ phonon lines and $2n$ electron lines.

The consideration of the series (6) leads to the Dyson equation (39.17), where the self-energy function is given by the diagrammatic expansion similar to Eq. (39.18), with dressed phonon lines instead of impurity-potential correlation functions. In the analytical form,

$$-i\Sigma^{s_1 s_2}(1,2) = is_1 is_2 \hbar \mathcal{D}^2 iG^{s_1 s_2}(1,2) \sum_{\alpha\beta} \nabla_{\mathbf{x}_1}^{\alpha} \nabla_{\mathbf{x}_2}^{\beta} iD^{\alpha\beta,s_1 s_2}(1,2)$$

$$+is_1 is_2 \hbar^2 \mathcal{D}^4 \sum_{s_1' s_2'} is_1' is_2' \int d1' \int d2' iG^{s_1 s_2'}(1,2') iG^{s_2' s_1'}(2',1') iG^{s_1' s_2}(1',2)$$

$$\times \sum_{\alpha\beta\gamma\delta} \nabla_{\mathbf{x}_1}^{\alpha} \nabla_{\mathbf{x}_2}^{\beta} \nabla_{\mathbf{x}_1'}^{\gamma} \nabla_{\mathbf{x}_2'}^{\delta} iD^{\alpha\gamma,s_1 s_1'}(1,1') iD^{\delta\beta,s_2' s_2}(2',2) + \ldots . \quad (11)$$

This equation becomes formally equivalent to Eq. (39.20) if the vertex contributions $is\hbar^{1/2}\mathcal{D}\nabla_{\mathbf{x}_1}^{\alpha}$ are replaced by is and the phonon lines $iD^{\alpha\beta,s_1 s_2}(1,2)$ are replaced by the correlation functions $\langle\langle U_1 U_2 \rangle\rangle$.

To find the phonon Green's functions, we consider the expansion (9) leading to the matrix Dyson equation of the following form:

$$\widehat{D}^{\alpha\beta}(1,2) = \hat{d}^{\alpha\beta}(1,2) + \mathcal{D}^2 \sum_{\gamma\delta} \int d1' \int d2' \nabla_{\mathbf{x}_1'}^{\gamma} \hat{d}^{\alpha\gamma}(1,1')$$

$$\times \hat{\sigma}_z \widehat{\Pi}(1',2') \hat{\sigma}_z \nabla_{\mathbf{x}_2'}^{\delta} \widehat{D}^{\delta\beta}(2',2) = \hat{d}^{\alpha\beta}(1,2) \quad (12)$$

$$+ \sum_{\gamma\delta} \int d1' \int d2' \hat{d}^{\alpha\gamma}(1,1') \left[\mathcal{D}^2 \hat{\sigma}_z \nabla_{\mathbf{x}_1'}^{\gamma} \nabla_{\mathbf{x}_2'}^{\delta} \widehat{\Pi}(1',2') \hat{\sigma}_z \right] \widehat{D}^{\delta\beta}(2',2) .$$

In this equation, instead of writing the indices $+$ and $-$, we use the matrix form $\widehat{D}$ for the phonon Green's function. It is defined in a similar way as in Eq. (39.2). The phonon self-energy function, defined by the expression in the square brackets of Eq. (12), is directly related to the polarization function $\Pi^{s_1 s_2}$ introduced by Eq. (41.28). The expansion of this function in the presence of electron-phonon interaction is represented diagrammatically as

$$(i/\hbar)\Pi^{s_1 s_2}(1,2) \equiv \underset{1s_1}{\overset{}{\bigcirc}}\underset{2s_2}{} = \underset{1s_1}{\overset{}{\diamond}}\underset{2s_2}{} + \underset{1s_1}{\overset{3s_3\gamma}{\diamond}}\underset{4s_4\delta}{}\underset{2s_2}{} + \ldots . \quad (13)$$

The corresponding analytical expression can be written with the aid of the diagram rules given above (problem 8.11).

The function $D^{\alpha\beta,-+}(1,2)$ at $t_1 = t_2$ is not related directly to the phonon density matrix introduced in Chapter 4. Moreover, the presence of a differential operator at each vertex, such as in Eqs. (6), (9), and (11), is not convenient for calculations. Therefore, it is helpful to introduce another Green's function of phonons in the momentum representation. This function depends on the vibrational mode indices l_1 and l_2 and is defined as

$$D_{t_1 t_2}^{l_1 l_2, s_1 s_2}(\mathbf{q}_1, \mathbf{q}_2)$$

$$= -\frac{i}{\hbar} \langle\langle \hat{\mathcal{T}}_C [\hat{b}_{\mathbf{q}_1 l_1}(t_1) + \hat{b}^+_{-\mathbf{q}_1 l_1}(t_1)]_{s_1} [\hat{b}_{-\mathbf{q}_2 l_2}(t_2) + \hat{b}^+_{\mathbf{q}_2 l_2}(t_2)]_{s_2} \rangle\rangle . \tag{14}$$

One can see that that $D_{t_1 t_2}^{l_1 l_2, --}$ coincides with the causal Green's function of phonons, $D_{t_1 t_2}^{l_1 l_2, c}$, introduced by Eq. (D.21). In the limit $t_1 \to t_2$, we have

$$i\hbar \lim_{t_1, t_2 \to t} D_{t_1 t_2}^{l_1 l_2, -+}(\mathbf{q}_1, \mathbf{q}_2)$$

$$= N_t(l_1 \mathbf{q}_1, l_2 \mathbf{q}_2) + N_t(l_2 - \mathbf{q}_2, l_1 - \mathbf{q}_1) + \delta_{l_1 l_2} \delta_{\mathbf{q}_1 \mathbf{q}_2}, \tag{15}$$

where $N_t(l_1 \mathbf{q}_1, l_2 \mathbf{q}_2)$ is the density matrix of phonons; see Eq. (19.4) and Sec. 23. To express $D_{t_1 t_2}^{\alpha\beta, s_1 s_2}(\mathbf{x}_1, \mathbf{x}_2)$ through $D_{t_1 t_2}^{l_1 l_2, s_1 s_2}(\mathbf{q}_1, \mathbf{q}_2)$, we use Eq. (6.29) and obtain

$$D_{t_1 t_2}^{\alpha\beta, s_1 s_2}(\mathbf{x}_1, \mathbf{x}_2) = \frac{1}{V} \sum_{l_1 l_2} \sum_{\mathbf{q}_1 \mathbf{q}_2} \frac{\hbar e_{\mathbf{q}_1 l_1}^{\alpha} e_{\mathbf{q}_2 l_2}^{\beta*}}{2\rho \sqrt{\omega_{\mathbf{q}_1 l_1} \omega_{\mathbf{q}_2 l_2}}}$$

$$\times D_{t_1 t_2}^{l_1 l_2, s_1 s_2}(\mathbf{q}_1, \mathbf{q}_2) e^{i\mathbf{q}_1 \cdot \mathbf{x}_1 - i\mathbf{q}_2 \cdot \mathbf{x}_2}. \tag{16}$$

In the crystals with cubic symmetry there are one longitudinal and two transverse acoustic phonon modes. Since only the longitudinal modes interact with electrons via deformation-potential interaction, we omit the mode indices in the phonon Green's functions and take into account that $\mathbf{e}_{\mathbf{q}LA} = -i\mathbf{q}/q$ and $\omega_{\mathbf{q}LA} = s_l q$. The expressions with spatial derivatives standing in Eq. (11) can be represented as (see Eq. (21.1))

$$\mathcal{D}^2 \sum_{\alpha\beta} \nabla_{\mathbf{x}_1}^{\alpha} \nabla_{\mathbf{x}_2}^{\beta} \widehat{D}_{t_1 t_2}^{\alpha\beta}(\mathbf{x}_1, \mathbf{x}_2)$$

$$= \sum_{\mathbf{q}_1 \mathbf{q}_2} C_{q1}^{(LA)} C_{q2}^{(LA)} \widehat{D}_{t_1 t_2}(\mathbf{q}_1, \mathbf{q}_2) e^{i\mathbf{q}_1 \cdot \mathbf{x}_1 - i\mathbf{q}_2 \cdot \mathbf{x}_2}, \tag{17}$$

where $\widehat{D}$ is the matrix form of the Green's function (14), and it is assumed that $l_1 = l_2 = LA$.

In the absence of external fields and crystal inhomogeneities, when the system is homogeneous in space and time, the Green's function of electrons can be written as $G_{tt'}^{ss'}(\mathbf{x}_1, \mathbf{x}_2) = G_{t-t'}^{ss'}(\mathbf{x}_1 - \mathbf{x}_2)$. Doing both temporal and spatial Fourier transformations of this function, we obtain $G_\varepsilon^{ss'}(\mathbf{p})$. In the same conditions, $D_{tt'}^{ss'}(\mathbf{q}, \mathbf{q}') = \delta_{\mathbf{q}\mathbf{q}'} D_{t-t'}^{ss'}(\mathbf{q})$, and the temporal Fourier transformation of $D_{t-t'}^{ss'}(\mathbf{q})$ gives us $D_{\omega}^{ss'}(\mathbf{q})$. In this case, the Dyson equation for the electron Green's function is written in the most simple way, as a system of four algebraic equations, or as a single 2×2 matrix equation

$$\widehat{G}_{\varepsilon}(\mathbf{p}) = \hat{g}_{\varepsilon}(\mathbf{p}) + \hat{g}_{\varepsilon}(\mathbf{p})\widehat{\Sigma}_{\varepsilon}(\mathbf{p})\widehat{G}_{\varepsilon}(\mathbf{p}), \qquad (18)$$

where $\hat{g}_{\varepsilon}(\mathbf{p})$ is the Green's function of free electrons in the energy-momentum representation, and the self-energy function $\widehat{\Sigma}$ is given by the following expansion:

$$-i\Sigma_{\varepsilon}^{s_1 s_2}(\mathbf{p}) = is_1 is_2 \hbar \int \frac{d\omega}{2\pi} \int \frac{d\mathbf{q}}{(2\pi)^3} \mathcal{V}_q^2 i G_{\varepsilon-\hbar\omega}^{s_1 s_2}(\mathbf{p} - \hbar\mathbf{q}) i D_{\omega}^{s_1 s_2}(\mathbf{q})$$

$$+is_1 is_2 \hbar^2 \sum_{s_1' s_2' = \pm} is_1' is_2' \int \frac{d\omega}{2\pi} \int \frac{d\omega'}{2\pi} \int \frac{d\mathbf{q}}{(2\pi)^3} \int \frac{d\mathbf{q}'}{(2\pi)^3} \mathcal{V}_q^2 \mathcal{V}_{q'}^2 \qquad (19)$$

$$\times i G_{\varepsilon-\hbar\omega}^{s_1 s_2'}(\mathbf{p} - \hbar\mathbf{q}) i G_{\varepsilon-\hbar(\omega+\omega')}^{s_2' s_1'}[\mathbf{p} - \hbar(\mathbf{q} + \mathbf{q}')] i G_{\varepsilon-\hbar\omega'}^{s_1' s_2}(\mathbf{p} - \hbar\mathbf{q}')$$

$$\times i D_{\omega}^{s_1 s_1'}(\mathbf{q}) i D_{\omega'}^{s_2' s_2}(\mathbf{q}') + \dots .$$

For the sake of convenience, we have introduced the quantity $\mathcal{V}_q = \sqrt{V}C_q^{(LA)} = \sqrt{\hbar \mathcal{D}^2 q / 2\rho s_l}$, which is independent of the normalization volume V. Equation (19) is presented diagrammatically as

$$-i\Sigma_{\varepsilon}^{s_1 s_2}(\mathbf{p}) = \qquad\qquad + \qquad\qquad + \dots . \qquad (20)$$

The electron and phonon Green's functions are represented by the solid and wavy lines with energy and momentum indices. In each vertex, a phonon line is "emitted" or "absorbed" by an electron line in such a way that the momentum and energy conservation laws are fulfilled. The factor $is\hbar^{1/2}\mathcal{V}_q$ is attributed to the vertex s where a phonon line with wave vector $\mathbf{q}$ begins or ends. Then, the integrals over all transferred frequencies and wave vectors are taken. These are the diagram rules for electron-phonon interaction in the energy-momentum representation.

Equation (12) for the homogeneous systems can be rewritten as a matrix equation for $\widehat{D}_{\omega}(\mathbf{q})$ in the following way:

$$\widehat{D}_{\omega}(\mathbf{q}) = \hat{d}_{\omega}(\mathbf{q}) + \hat{d}_{\omega}(\mathbf{q}) \mathcal{V}_q^2 \hat{\sigma}_z \widehat{\Pi}_{\omega}(\mathbf{q}) \hat{\sigma}_z \widehat{D}_{\omega}(\mathbf{q}), \qquad (21)$$

where the polarization function in the energy-momentum representation is given by the expansion

$$\frac{i}{\hbar}\Pi_{\omega}^{s_1 s_2}(\mathbf{q}) = -2 \int \frac{d\varepsilon}{2\pi\hbar} \int \frac{d\mathbf{p}}{(2\pi\hbar)^3} i G_{\varepsilon+\hbar\omega}^{s_1 s_2}(\mathbf{p} + \hbar\mathbf{q}) i G_{\varepsilon}^{s_2 s_1}(\mathbf{p}) \qquad (22)$$

$$-2\hbar \sum_{s_3,s_4=\pm} is_3 is_4 \int \frac{d\varepsilon}{2\pi\hbar} \int \frac{d\mathbf{p}}{(2\pi\hbar)^3} \int \frac{d\omega'}{2\pi} \int \frac{d\mathbf{q}'}{(2\pi)^3} \mathcal{V}_{q'}^2 iG_{\varepsilon+\hbar\omega}^{s_1 s_3}(\mathbf{p}+\hbar\mathbf{q})$$

$$\times iG_{\varepsilon+\hbar(\omega-\omega')}^{s_3 s_2}[\mathbf{p}+\hbar(\mathbf{q}-\mathbf{q}')] iG_{\varepsilon-\hbar\omega'}^{s_2 s_4}(\mathbf{p}-\hbar\mathbf{q}') iG_{\varepsilon}^{s_4 s_1}(\mathbf{p}) iD_{\omega'}^{s_3 s_4}(\mathbf{q}') + \cdots .$$

It can be obtained either by Fourier transformations of the analytical form of Eq. (13) or by direct application of the diagram rules we just formulated. The first term of the expansion of $\Pi_\omega^{s_1 s_2}(\mathbf{q})$ has the same form as in problem 8.9, because it does not contain the phonon Green's function explicitly. Equations (18) and (21) with the self-energy (19) and polarization function (22) form a set of equations for the Green's functions of electrons and phonons interacting with each other. To complete this set, one should define $g_\varepsilon^{s_1 s_2}(\mathbf{p})$ and $\mathrm{d}_\omega^{s_1 s_2}(\mathbf{q})$. The former are given by Eq. (40.33), while the latter are

$$\mathrm{d}_\omega^{-+}(\mathbf{q}) = -2\pi i\hbar^{-1}\left[N_\mathbf{q}\delta(\omega-\omega_\mathbf{q}) + (N_{-\mathbf{q}}+1)\delta(\omega+\omega_\mathbf{q})\right],$$

$$\mathrm{d}^{--} = \mathrm{d}^R + \mathrm{d}^{-+}, \quad \mathrm{d}^{++} = -\mathrm{d}^A + \mathrm{d}^{-+}, \quad \mathrm{d}^{+-} = \mathrm{d}^R - \mathrm{d}^A + \mathrm{d}^{-+}, \quad (23)$$

where $\omega_\mathbf{q} = s_l q$ and $N_\mathbf{q}$ is the occupation number of acoustic phonons. The expressions (23) can be directly derived from Eq. (14). The retarded and advanced Green's functions, $\mathrm{d}_\omega^R(\mathbf{q})$ and $\mathrm{d}_\omega^A(\mathbf{q})$ in Eq. (23), are given by Eq. (D.25), where the mode index μ is omitted. In the equilibrium case, when $N_\mathbf{q} = [e^{\hbar\omega_q/T} - 1]^{-1}$, the function $\mathrm{d}_\omega^{-+}(\mathbf{q})$ can be expressed through $\mathrm{d}_\omega^R(\mathbf{q})$ and $\mathrm{d}_\omega^A(\mathbf{q})$ by using Eq. (D.11) with $\hat{A} = \hat{b}_\mathbf{q} + \hat{b}_{-\mathbf{q}}^+$ and $\hat{B} = \hat{b}_{-\mathbf{q}} + \hat{b}_\mathbf{q}^+$.

Although so far we discussed the deformation-potential interaction with acoustic phonons, it is not difficult to bring other mechanisms of electron-phonon interaction into consideration. All we have to do is to substitute the corresponding energy spectra and matrix elements of interaction. For example, in the case of polarization-potential interaction of electrons with LO phonons, when the interaction Hamiltonian is given by Eq. (6.22), one can apply Eqs. (18)–(22) with $\mathcal{V}_q = \sqrt{V}C_q^{(LO)} = \sqrt{2\pi e^2 \hbar\omega_{LO}/\epsilon^* q^2}$, see Eq. (21.1), and use $\omega_\mathbf{q} = \omega_{LO}$ in Eq. (23).

A similar diagram technique can be built to describe the interaction of electrons with photons. The operator of interaction of electrons with the electromagnetic field described by the vector potential $\mathbf{A}_{\mathbf{r}t}$ and scalar potential $\Phi_{\mathbf{r}t}$ is written as (see Eq. (4.16))

$$\hat{\mathcal{H}}_{int} = -\frac{1}{c}\int d\mathbf{r}\,\hat{\mathbf{I}}_\mathbf{r}\cdot\hat{\mathbf{A}}_{\mathbf{r}t} + \int d\mathbf{r}\hat{\rho}_\mathbf{r}\hat{\Phi}_{\mathbf{r}t} , \qquad (24)$$

where the vector and scalar potentials are quantized. If the gauge with zero scalar potential, only the first term of this expression remains. The current density operator is given by Eq. (4.15) and can be written through the field operators of electrons:

$$\hat{\mathbf{I}}_{\mathbf{x}} = -\frac{i\hbar e}{2m}\left[\hat{\Psi}_{\mathbf{x}}^{+}\nabla_{\mathbf{x}}\hat{\Psi}_{\mathbf{x}} - (\nabla_{\mathbf{x}}\hat{\Psi}_{\mathbf{x}}^{+})\hat{\Psi}_{\mathbf{x}}\right] - \frac{e^2}{mc}\mathbf{A}_{\mathbf{x}t}\hat{\Psi}_{\mathbf{x}}^{+}\hat{\Psi}_{\mathbf{x}}, \qquad (25)$$

while the charge density operator is written as $\hat{\rho}_{\mathbf{x}} = e\hat{\Psi}_{\mathbf{x}}^{+}\hat{\Psi}_{\mathbf{x}}$. The expansion of the exponential operator $\hat{\mathcal{S}}_C$ in the expressions like (41.15) contains products of the electron field operators as well as those of the vector and scalar potentials. The last term on the right-hand side of Eq. (25) is neglected because it brings non-linear terms in the interaction Hamiltonian. The non-linear interaction can be essential only in some special cases which are not considered here. We introduce the Green's functions of the vector potentials of electromagnetic field in a similar way as in Eq. (4), replacing the displacement operators by the vector-potential operators:

$$\mathcal{D}^{\alpha\beta,s_1s_2}(1,2) = -\frac{i}{\hbar}\langle\langle\hat{\mathcal{T}}_C\hat{A}_{s_1}^{\alpha}(1)\hat{A}_{s_2}^{\beta}(2)\rangle\rangle. \qquad (26)$$

In addition to this function, one may also introduce the Green's function of the scalar potentials, as well as the mixed Green's function describing the correlation of $\hat{\mathbf{A}}_{\mathbf{x}t}$ and $\hat{\Phi}_{\mathbf{x}t}$. The function $\mathcal{D}^{\alpha\beta,--}(1,2)$ coincides with the causal double-time Green's function $\langle\langle\hat{A}_{\mathbf{x}_1}^{\alpha}|\hat{A}_{\mathbf{x}_2}^{\beta}\rangle\rangle_{t_1t_2}^c$. The retarded and advanced functions, $\mathcal{D}^{\alpha\beta,R}(1,2)$ and $\mathcal{D}^{\alpha\beta,A}(1,2)$, are expressed through $\mathcal{D}^{\alpha\beta,s_1s_2}(1,2)$ in a standard way, as in Eq. (40.3) (problem 8.12).

Considering the interaction of electrons with transverse electromagnetic fields in the gauge $\nabla\cdot\mathbf{A}_{\mathbf{x}t} = 0$, we use Eq. (3.18) expressing the operator of the vector potential through the creation and annihilation operators of photons. Assuming $\mathbf{e}_{\mathbf{q}\mu}^{*} = \mathbf{e}_{-\mathbf{q}\mu}$, we obtain

$$\mathcal{D}_{t_1t_2}^{\alpha\beta,s_1s_2}(\mathbf{x}_1,\mathbf{x}_2) = \frac{1}{V}\sum_{\mu_1\mu_2}\sum_{\mathbf{q}_1\mathbf{q}_2}\frac{2\pi\hbar c^2}{\epsilon\sqrt{\omega_{q_1}\omega_{q_2}}}e_{\mathbf{q}_1\mu_1}^{\alpha}e_{\mathbf{q}_2\mu_2}^{\beta*}$$

$$\times \mathrm{D}_{t_1t_2}^{\mu_1\mu_2,s_1s_2}(\mathbf{q}_1,\mathbf{q}_2)e^{i\mathbf{q}_1\cdot\mathbf{x}_1 - i\mathbf{q}_2\cdot\mathbf{x}_2}, \qquad (27)$$

where $\mu = 1,2$ is the polarization index, $\mathbf{e}_{\mathbf{q}\mu}$ is the unit vector of polarization of the photon, $\mathrm{D}_{t_1t_2}^{\mu_1\mu_2,s_1s_2}(\mathbf{q}_1,\mathbf{q}_2)$ is defined by Eq. (14) with photon operators $\hat{b}_{\mathbf{q}\mu}$ standing in place of phonon operators, and ω_q is given by Eq. (3.23). Without repeating the calculations given above, we mention that the diagram technique in the homogeneous systems again leads to Eqs. (18)−(22) for $G_{\varepsilon}^{s_1s_2}(\mathbf{p})$ and $\mathrm{D}_{\omega}^{\mu,s_1s_2}(\mathbf{q})$, where, however, the polarization index enters the vertices, and $\mathcal{V}_q$ is replaced by

$$\mathcal{V}_\mu(\mathbf{p}, \mathbf{q}) = \sqrt{\frac{2\pi\hbar e^2}{\omega_q\epsilon}} (\mathbf{v_p} \cdot \mathbf{e_{q\mu}}), \qquad (28)$$

where $\mathbf{v_p} = \mathbf{p}/m$. The similarity of the description of electron-phonon and electron-photon interactions is explained by the fact that the operators of the interaction in both cases are written as (see Eq. (19.2) and below)

$$\sum_{\mathbf{pq}\mu} \langle \mathbf{p} + \mathbf{q} | \hat{\chi}_{\mathbf{q}\mu} | \mathbf{p} \rangle \hat{a}^+_{\mathbf{p}+\mathbf{q}} \hat{a}_{\mathbf{p}} \hat{b}_{\mathbf{q}\mu} + H.c. , \qquad (29)$$

where μ is either the mode index of phonons or the polarization index of photons, and the matrix element of the effective interaction potential is written as $\langle \mathbf{p} + \mathbf{q} | \hat{\chi}_{\mathbf{q}\mu} | \mathbf{p} \rangle$, where $\hat{\chi}_{\mathbf{q}\mu}$ is specified for each mechanism of electron-boson interaction in Chapter 4. The diagram technique can be generalized in a straightforward way for describing the electrons occupying several bands (or subbands in low-dimensional structures). In this case, one should add the band (or subband) index n to each electron Green's function and calculate the matrix elements of $\hat{\chi}_{\mathbf{q}\mu}$ by using the corresponding eigenstates $|n\mathbf{p}\rangle$. The kinetic equation for the electrons interacting with bosons, see Sec. 34, can be derived from the non-equilibrium diagram technique described above (problem 8.13).

At zero temperature, the diagram technique is simplified considerably, because all the diagrams for $--$ ($++$) Green's functions containing $+$ ($-$) vertices give zero contribution. This rule is general and valid for arbitrary interactions. Let us demonstrate it for the case of electron-phonon interaction, considering first the correction (7) to the electron Green's function in the energy-momentum representation, which is proportional to $g_\varepsilon^{s_1 s_3}(\mathbf{p}) g_{\varepsilon-\hbar\omega}^{s_3 s_4}(\mathbf{p} - \hbar\mathbf{q}) d_\omega^{s_3 s_4}(\mathbf{q}) g_\varepsilon^{s_4 s_2}(\mathbf{p})$. Assuming that $s_1 = s_2 = -$, we can write four diagrams differing by the signs s_3 and s_4. It is essential that Eq. (40.33) at $T = 0$ gives us $g_\varepsilon^{-+}(\mathbf{p}) = 0$ at $\varepsilon > \varepsilon_F$ and $g_\varepsilon^{+-}(\mathbf{p}) = 0$ at $\varepsilon < \varepsilon_F$ (we put $\varepsilon > 0$). This immediately makes the contribution with $s_3 = s_4 = +$ equal to zero. To analyze the contributions with $s_3 \neq s_4$, we note that $N_\mathbf{q} = 0$ at $T = 0$. Therefore, one has $d_\omega^{-+}(\mathbf{q}) = 0$ at $\omega > 0$ and $d_\omega^{+-}(\mathbf{q}) = 0$ at $\omega < 0$. At $s_3 = +$ ($s_4 = -$), the product of electron Green's function is non-zero only if $\varepsilon < \varepsilon_F$ and $\varepsilon - \omega > \varepsilon_F$, i.e., ω must be negative. Since in these conditions $d_\omega^{s_3 s_4}(\mathbf{q}) = 0$, the contribution is zero. The case $s_3 = -$ ($s_4 = +$) is considered in a similar way, again with zero contribution. Therefore, only the diagram with $s_3 = s_4 = -$ remains. Since the consideration can be repeated for the diagrams of arbitrary order, one should retain only the diagrams with "minus" vertices in the series for $G_\varepsilon^{--}(\mathbf{p})$ (and in the series for $D_\omega^{--}(\mathbf{q})$ as well). For the same reasons, the series for G^{++} and D^{++} would contain only the diagrams with "plus" vertices (a consideration of such diagrams, however, would not give us any additional information about the system at $T = 0$). We remind that G^{--} and D^{--} are the causal Green's functions of electrons and phonons. The diagram technique operating with these functions at $T = 0$ is called the zero-temperature diagram technique. An application of this technique to the case of electron-electron interaction is considered in Appendix E. Since the zero-temperature diagram technique appears to be a particular case of the non-equilibrium diagram technique, the diagram rules for this technique are, in fact, already formulated (problems 8.14 and 8.15).

43. Weak Localization under External Fields

To give an example of application of the formalism developed in Secs. 39 and 40, we calculate the quantum corrections to the linear response of electron-impurity systems in the presence of quasi-classical (slowly varying with time and smoothly varying in space) electric and magnetic fields, $\mathbf{E}_{\mathbf{x}t}$ and $\mathbf{H}_{\mathbf{x}t}$. The linear response to a stationary electric field in the absence of magnetic fields has been considered in Sec. 15 by using the diagram technique for electron-impurity systems. The non-equilibrium diagram technique provides a more general way of calculating the quantum corrections, based upon the generalized kinetic equation.

As we have seen in Sec. 40, the equations of the non-equilibrium diagram technique in quasi-classical fields are transformed to the quasi-classical (Boltzmann) kinetic equation when the self-energies standing in the collision integral of Eq. (40.27) are evaluated in the lowest order with respect to the impurity potential, see Eq. (40.28), which corresponds to the Born approximation. The non-equilibrium diagram technique allows one to take into account the higher-order diagrams describing backscattering processes. Below we again evaluate $\Sigma^{R,A}(\mathbf{x}_1 t_1, \mathbf{x}_2 t_2) \equiv \Sigma^{R,A}(1, 2)$ in the Born approximation, because, as in Sec. 15, the higher-order corrections appear to be small. However, in the diagrammatic expansion (40.17) of $\Omega(\mathbf{x}_1 t_1, \mathbf{x}_2 t_2) \equiv \Omega(1, 2)$, we take into account all maximally crossed diagrams corresponding to the backscattering processes with small total momentum transfer; see Sec. 15. The expression for $-i\Omega(1, 2)$ is represented as

$$
\underbrace{}_{1 \quad F \quad 2} + \underbrace{}_{1 \ R \ 2' \ F \ 1' A \ 2} + \underbrace{}_{1 \ R \ 3 \ R \ 2' \ F \ 1' A \ 3' A \ 2} + \ldots . \tag{1}
$$

The first diagram corresponds to the Born contribution $\Omega(1, 2) = w_{12}$

$\times F(1, 2)$, where $w_{12} \equiv \langle\langle U_{\mathbf{x}_1} U_{\mathbf{x}_2} \rangle\rangle = w(|\mathbf{x}_1 - \mathbf{x}_2|)$. The other diagrams correspond to a quantum correction $\delta\Omega(1, 2)$, which can be written as a product of $w_{11'} w_{22'} F(2', 1')$ by the infinite sum

$$G^R(1, 2')G^A(1', 2) + w_{33'} G^R(1, 3)G^R(3, 2')G^A(1', 3')G^A(3', 2) + \dots \quad (2)$$

integrated over the variables denoted by the primed indices. The sum (2) is easily identified with the correlation function $\langle\langle \mathcal{G}^R(1, 2')\mathcal{G}^A(1', 2) \rangle\rangle$ of retarded and advanced Green's functions written in the ladder approximation. Therefore,

$$\delta\Omega(1, 2) \simeq \int d1' \int d2' w_{11'} w_{2'2} \langle\langle \mathcal{G}^R(1, 2')\mathcal{G}^A(1', 2) \rangle\rangle F(2', 1'). \quad (3)$$

To find the Green's function F in quasi-classical fields, we consider the kinetic equation (40.27) in the local approximation, when the spatial gradient and time derivative on the left-hand side are neglected:

$$i\hbar \left(e\mathbf{E}_{\mathbf{r}t} \cdot \mathbf{v}_{\mathbf{p}} \frac{\partial}{\partial\varepsilon} + e\mathbf{E}_{\mathbf{r}t}\cdot\nabla_{\mathbf{p}} + \frac{e}{c}[\mathbf{v}_{\mathbf{p}} \times \mathbf{H}_{\mathbf{r}t}] \cdot \nabla_{\mathbf{p}} \right) F_{\varepsilon t}(\mathbf{r}, \mathbf{p}) \quad (4)$$

$$+[\Sigma_\varepsilon^A(\mathbf{r}, \mathbf{p}) - \Sigma_\varepsilon^R(\mathbf{r}, \mathbf{p})]F_{\varepsilon t}(\mathbf{r}, \mathbf{p}) = \Omega_{\varepsilon t}(\mathbf{r}, \mathbf{p})[G_\varepsilon^A(\mathbf{r}, \mathbf{p}) - G_\varepsilon^R(\mathbf{r}, \mathbf{p})].$$

Using the quasi-classical expression (9.11) for the current density and Eq. (40.21), one can write the induced current density as

$$\mathbf{I}_{\mathbf{r}t} = e \int \frac{d\mathbf{p}}{(2\pi\hbar)^3} \mathbf{v}_{\mathbf{p}} \int \frac{d\varepsilon}{2\pi i} \Delta F_{\varepsilon t}(\mathbf{r}, \mathbf{p}), \quad (5)$$

where ΔF is the non-equilibrium part of the Green's function F. The equilibrium part $F^{(eq)}$ of this function is given by Eq. (40.24), where $\chi_\varepsilon = 2f_\varepsilon^{(eq)} - 1$. Taking into account that both $F^{(eq)}$ and the right-hand side of Eq. (4) are proportional to $G^A - G^R$, we search for ΔF in the form

$$\Delta F_{\varepsilon t}(\mathbf{r}, \mathbf{p}) = \Delta s_{\varepsilon t}(\mathbf{r}, \mathbf{p}) \left[G_\varepsilon^A(\mathbf{r}, \mathbf{p}) - G_\varepsilon^R(\mathbf{r}, \mathbf{p}) \right]$$

$$\simeq 2\pi i \Delta s_{\varepsilon t}(\mathbf{r}, \mathbf{p})\delta_\tau(\varepsilon - \varepsilon_{p\mathbf{r}}). \quad (6)$$

To transform $G^A - G^R$ in this equation, we have used Eq. (40.26), where the self-energy is written through the coordinate-dependent relaxation time $\tau_{\mathbf{r}}$:

$$\Sigma_\varepsilon^A(\mathbf{r}, \mathbf{p}) \simeq -\Sigma_\varepsilon^R(\mathbf{r}, \mathbf{p}) \simeq i\hbar/2\tau_{\mathbf{r}}. \quad (7)$$

The broadened δ-function in Eq. (6) is defined as $\delta_\tau(E) = (\hbar/2\pi\tau)[E^2 + (\hbar/2\tau)^2]^{-1}$ (see problem 1.4), and $\varepsilon_{pr} = \varepsilon_p + U_{\mathbf{r}}$.

Neglecting the terms containing the derivatives of $1/\tau_{\mathbf{r}}$ over $\mathbf{p}$ and ε, one can find that the factor $G_\varepsilon^A(\mathbf{r}, \mathbf{p}) - G_\varepsilon^R(\mathbf{r}, \mathbf{p})$ commutes with the operator proportional to $\mathbf{E_{rt}}$ and $\mathbf{H_{rt}}$ on the left-hand side of Eq. (4). Therefore, Eq. (4) can be written as an equation for $\Delta s_{\varepsilon t}(\mathbf{r}, \mathbf{p})$:

$$\left(e\mathbf{E_{rt}} \cdot \mathbf{v_p}\frac{\partial}{\partial \varepsilon} + e\mathbf{E_{rt}} \cdot \nabla_{\mathbf{p}} + \frac{e}{c}[\mathbf{v_p} \times \mathbf{H_{rt}}] \cdot \nabla_{\mathbf{p}}\right)[2f_\varepsilon^{(eq)} - 1 + \Delta s_{\varepsilon t}(\mathbf{r}, \mathbf{p})]$$

$$= -\frac{1}{\tau_{\mathbf{r}}}\Delta s_{\varepsilon t}(\mathbf{r}, \mathbf{p}) - \frac{i}{\hbar}\Delta\Omega_{\varepsilon t}(\mathbf{r}, \mathbf{p}). \tag{8}$$

This equation is obtained by taking into account that the equilibrium part of F makes the collision integral equal to zero. The non-equilibrium part of Ω standing in Eq. (8) is a sum of the Born contribution and quantum correction $\delta\Omega$:

$$\Delta\Omega_{\varepsilon t}(\mathbf{r}, \mathbf{p}) = 2\pi i \int \frac{d\mathbf{p_1}}{(2\pi\hbar)^3}w(|\mathbf{p} - \mathbf{p_1}|/\hbar)\Delta s_{\varepsilon t}(\mathbf{r}, \mathbf{p_1})\delta_\tau(\varepsilon - \varepsilon_{p_1 \mathbf{r}})$$

$$+ \delta\Omega_{\varepsilon t}(\mathbf{r}, \mathbf{p}). \tag{9}$$

In this equation, as in Eq. (40.28), $w(q)$ is the Fourier transform of the impurity potential correlation function $w(r)$. To simplify the calculations, we assume the case of short-range impurity potential, when $w(q) \simeq w$ is independent of q.

To obtain a linearized quasi-classical kinetic equation, one should substitute $\Delta\Omega$ from Eq. (9) into Eq. (8), neglect the quantum correction, multiply Eq. (8) by $\delta(\varepsilon - \varepsilon_{pr})$, and integrate it over ε. After these transformations, the function $\Delta s_{\varepsilon_{pr}t}(\mathbf{r}, \mathbf{p})$ is identified with $2\Delta f_t(\mathbf{r}, \mathbf{p})$ and Eq. (5) is reduced to Eq. (9.11) for the current density. Below we solve Eq. (8) by iterations, assuming that $\Delta s_{\varepsilon t}(\mathbf{r}, \mathbf{p}) = \widetilde{\Delta s}_{\varepsilon t}(\mathbf{r}, \mathbf{p}) + \delta s_{\varepsilon t}(\mathbf{r}, \mathbf{p})$, where $\widetilde{\Delta s}_{\varepsilon t}(\mathbf{r}, \mathbf{p})$ satisfies Eq. (8) without the quantum correction. In the limit of classically weak magnetic fields, when the cyclotron frequency ω_c is small in comparison to $1/\tau_{\mathbf{r}}$, we have

$$\widetilde{\Delta s}_{\varepsilon t}(\mathbf{r}, \mathbf{p}) \simeq -2\tau_{\mathbf{r}}\left(e\mathbf{E_{rt}} \cdot \mathbf{v_p} + \tau_{\mathbf{r}}\frac{e^2}{mc}\mathbf{v_p} \cdot [\mathbf{E_{rt}} \times \mathbf{H_{rt}}]\right)\frac{df_\varepsilon^{(eq)}}{d\varepsilon}, \tag{10}$$

where we keep only the terms linear in $\mathbf{E_{rt}}$. The quantum correction $\delta s_{\varepsilon t}(\mathbf{r}, \mathbf{p})$ is determined from the following equation:

$$\left(e\mathbf{E_{rt}} \cdot \mathbf{v_p}\frac{\partial}{\partial \varepsilon} + e\mathbf{E_{rt}} \cdot \nabla_{\mathbf{p}} + \frac{e}{c}[\mathbf{v_p} \times \mathbf{H_{rt}}] \cdot \nabla_{\mathbf{p}}\right)\delta s_{\varepsilon t}(\mathbf{r}, \mathbf{p})$$

$$+ \frac{\delta s_{\varepsilon t}(\mathbf{r}, \mathbf{p})}{\tau_{\mathbf{r}}} = -\frac{i}{\hbar}\delta\Omega_{\varepsilon t}(\mathbf{r}, \mathbf{p}), \tag{11}$$

where the right-hand side is found by using $\Delta F_{\varepsilon t}(\mathbf{r}, \mathbf{p}) = 2\pi i \widetilde{\Delta s}_{\varepsilon t}(\mathbf{r}, \mathbf{p})$ $\times \delta_\tau(\varepsilon - \varepsilon_{p\mathbf{r}})$. Equation (11) should be solved by iterations. Since $\delta\Omega$ is already linear in $\mathbf{E}_{\mathbf{r}t}$, the solution linear in $\mathbf{E}_{\mathbf{r}t}$ and $\mathbf{H}_{\mathbf{r}t}$ is

$$\delta s_{\varepsilon t}(\mathbf{r}, \mathbf{p}) \simeq -\frac{i\tau_{\mathbf{r}}}{\hbar}\delta\Omega_{\varepsilon t}(\mathbf{r}, \mathbf{p})$$

$$-\tau_{\mathbf{r}}\frac{e}{c}[\mathbf{v_p} \times \mathbf{H}_{\mathbf{r}t}] \cdot \nabla_{\mathbf{p}}\left[-\frac{i\tau_{\mathbf{r}}}{\hbar}\delta\Omega_{\varepsilon t}(\mathbf{r}, \mathbf{p})\right]_{\mathbf{H}=0}. \tag{12}$$

Substituting the expressions (10) and (12) into Eq. (6), we calculate the current density according to Eq. (5) and obtain $\mathbf{I}_{\mathbf{r}t} = \mathbf{j}_{\mathbf{r}t} + \delta\mathbf{j}_{\mathbf{r}t}$, where the classical and quantum contributions, respectively, are given by

$$\mathbf{j}_{\mathbf{r}t} = \sigma_{\mathbf{r}}\left(\mathbf{E}_{\mathbf{r}t} + \tau_{\mathbf{r}}\frac{e}{mc}[\mathbf{E}_{\mathbf{r}t} \times \mathbf{H}_{\mathbf{r}t}]\right) \tag{13}$$

and

$$\delta\mathbf{j}_{\mathbf{r}t} = -\frac{i\tau_{\mathbf{r}}}{\hbar}e\int\frac{d\mathbf{p}}{(2\pi\hbar)^3}\int d\varepsilon\,\delta_\tau(\varepsilon - \varepsilon_{p\mathbf{r}})$$

$$\times\left[\mathbf{v_p}\delta\Omega_{\varepsilon t}(\mathbf{r}, \mathbf{p}) + \tau_{\mathbf{r}}\frac{e}{mc}[\mathbf{v_p} \times \mathbf{H}_{\mathbf{r}t}]\,\delta\Omega_{\varepsilon t}(\mathbf{r}, \mathbf{p})|_{\mathbf{H}=0}\right]. \tag{14}$$

The local conductivity is introduced according to $\sigma_{\mathbf{r}} = e^2 n_{\mathbf{r}}\tau_{\mathbf{r}}/m$, where $n_{\mathbf{r}}$ is the local electron density. The second term under the integral in Eq. (14) is transformed by using the integration by parts.

To calculate the quantum correction $\delta\mathbf{j}_{\mathbf{r}t}$, one has to write $\delta\Omega$ given by Eq. (3) in the Wigner representation. As a result of the Wigner transformations, we have

$$\delta\Omega_{\varepsilon t}(\mathbf{r}, \mathbf{p}) = w^2\int dt'\int\frac{d\varepsilon'}{2\pi\hbar}\int\frac{d\mathbf{p}'}{(2\pi\hbar)^3}\Delta F_{\varepsilon' t'}(\mathbf{r}, \mathbf{p}')$$

$$\times\int d\boldsymbol{\rho}\exp\left\{-\frac{i}{\hbar}[\mathbf{p} + \mathbf{p}' + \hbar\boldsymbol{\kappa}_{\mathbf{r}}(t, t')]\cdot\boldsymbol{\rho}\right\}C_{\varepsilon\varepsilon'}\left(\mathbf{r}+\frac{\boldsymbol{\rho}}{2}\,t, \mathbf{r}-\frac{\boldsymbol{\rho}}{2}\,t'\right), \tag{15}$$

where $\hbar\boldsymbol{\kappa}_{\mathbf{r}}(t, t') = e(\mathbf{A}_{\mathbf{r}t} + \mathbf{A}_{\mathbf{r}t'})/c$ and

$$C_{\varepsilon\varepsilon'}\left(\mathbf{r}t, \mathbf{r}'t'\right) = \int d\tau\int d\tau' e^{i(\varepsilon\tau+\varepsilon'\tau')/\hbar}$$

$$\times\langle\langle\mathcal{G}^R(\mathbf{r}\,t + \tau/2, \mathbf{r}'\,t' - \tau'/2)\mathcal{G}^A(\mathbf{r}\,t' + \tau'/2, \mathbf{r}'\,t - \tau/2)\rangle\rangle. \tag{16}$$

The correlation function of the Green's functions in Eq. (16) satisfies the Bethe-Salpeter equation in the ladder approximation, which has the following form [see also Eqs. (15.3) and (15.4)]:

$$\underset{\substack{rt_1' \qquad\qquad r't_2'}}{\overset{\substack{rt_1 \qquad\qquad r't_2}}{\rule{3cm}{2pt}}} = \underset{\substack{rt_1' \qquad r't_2'}}{\overset{\substack{rt_1 \qquad r't_2}}{\rule{2cm}{2pt}}} + \underset{\substack{rt_1' \qquad r_3t_3' \qquad r't_2'}}{\overset{\substack{rt_1 \qquad r_3t_3 \qquad r't_2}}{\rule{3cm}{1pt}}} \quad . \tag{17}$$

Owing to the assumed weak spatial and temporal variation of the fields, one should calculate the function (16) in the hydrodynamic region of parameters, $|\mathbf{r} - \mathbf{r}'| > l_F$ and $|t - t'| > \tau_F$, where l_F and τ_F are the mean free path length and scattering time at the Fermi surface (below we consider a strongly degenerate electron gas). Such an approximation considerably simplifies the calculation of the correlation function from Eq. (17). The details of this calculation are presented in Appendix F.

The function (16) becomes proportional to $\delta(\varepsilon - \varepsilon')$, according to

$$C_{\varepsilon\varepsilon'}\left(\mathbf{r}t, \mathbf{r}'t'\right) = 2\pi^2 \rho_{3D}\left(\varepsilon - U_\mathbf{r}\right)\delta(\varepsilon - \varepsilon')C\left(\mathbf{r}t, \mathbf{r}'t'\right). \tag{18}$$

The Cooperon $C\left(\mathbf{r}t, \mathbf{r}'t'\right)$ introduced by Eq. (18) satisfies the following equation:

$$\left[\frac{1}{2}\left(\frac{\partial}{\partial t} - \frac{\partial}{\partial t'}\right) + [-i\nabla_\mathbf{r} - \boldsymbol{\kappa}_\mathbf{r}(t, t')]\cdot D_\mathbf{r}[-i\nabla_\mathbf{r} - \boldsymbol{\kappa}_\mathbf{r}(t, t')] + \frac{1}{\tau_\varphi}\right]$$

$$\times C\left(\mathbf{r}t, \mathbf{r}'t'\right) = \delta(t - t')\delta\left(\mathbf{r} - \mathbf{r}'\right). \tag{19}$$

The term "Cooperon" is traditionally used because the pair correlation functions of the kind $\langle\langle \mathcal{G}(\mathbf{r}, \mathbf{r}')\mathcal{G}(\mathbf{r}, \mathbf{r}')\rangle\rangle$ have been encountered earlier in the theory of superconductivity. The local diffusion coefficient in Eq. (19) is introduced as $D_\mathbf{r} = v_F^2\tau_\mathbf{r}/3$. The phase relaxation time τ_φ is introduced phenomenologically, to provide a finite value of $C(\mathbf{r}t, \mathbf{r}'t')$ for the stationary and spatially homogeneous case at $\mathbf{A}_{\mathbf{r}t} = 0$. In the case of 2D electrons, the introduction of the phase relaxation length $l_\varphi = v_F\tau_\varphi$ leads to a cutoff of the logarithmic divergence of the static conductivity; see Sec. 15.

Substituting $C_{\varepsilon\varepsilon'}\left(\mathbf{r}t, \mathbf{r}'t'\right)$ given by Eq. (18) and $\Delta F_{\varepsilon t}(\mathbf{r}, \mathbf{p}) \simeq 2\pi i$ $\times \widetilde{\Delta s}_{\varepsilon t}(\mathbf{r}, \mathbf{p})\delta_\tau(\varepsilon - \varepsilon_{p\mathbf{r}})$ into Eq. (15), we find the quantum correction to the current in the form

$$\delta\mathbf{j}_{\mathbf{r}t} = \frac{2\pi}{\hbar}ew\int\frac{d\mathbf{p}}{(2\pi\hbar)^3}\int d\varepsilon[\delta_\tau(\varepsilon - \varepsilon_{p\mathbf{r}})]^2\int dt'C\left(\mathbf{r}t, \mathbf{r}t'\right)$$

$$\times\left[\mathbf{v}_\mathbf{p}\widetilde{\Delta s}_{\varepsilon t'}(\mathbf{r}, -\mathbf{p}) + \tau_\mathbf{r}\frac{e}{mc}[\mathbf{v}_\mathbf{p}\times\mathbf{H}_{\mathbf{r}t}]\widetilde{\Delta s}_{\varepsilon t'}(\mathbf{r}, -\mathbf{p})|_{\mathbf{H}=0}\right]. \tag{20}$$

If we substitute the function $\widetilde{\Delta s}_{\varepsilon t}(\mathbf{r}, \mathbf{p})$ given by Eq. (10) to this expression and take the integral over ε with the use of $-df_\varepsilon^{(eq)}/d\varepsilon = \delta(\varepsilon - \varepsilon_F)$, we also find

$$\delta\mathbf{j}_{\mathbf{r}t} = -\frac{4\pi}{\hbar}e^2 w\tau_{\mathbf{r}} \int \frac{d\mathbf{p}}{(2\pi\hbar)^3} \mathbf{v_p}[\delta_\tau(\varepsilon_F - \varepsilon_{p\mathbf{r}})]^2$$

$$\times \int dt' C\left(\mathbf{r}t, \mathbf{r}t'\right) \mathbf{v_p} \cdot \left(\mathbf{E}_{\mathbf{r}t'} + 2\tau_{\mathbf{r}}\frac{e}{mc}[\mathbf{E}_{\mathbf{r}t'}\times\mathbf{H}_{\mathbf{r}t'}]\right). \qquad (21)$$

The integral over the angle of $\mathbf{p}$ transforms $p_\alpha p_\beta$ to $p^2\delta_{\alpha\beta}/3$, and the integral over $|\mathbf{p}|$ is calculated easily. As a result, we obtain

$$\delta\mathbf{j}_{\mathbf{r}t} = -\frac{2}{\pi}\frac{\sigma_{\mathbf{r}}}{\hbar\rho_{\mathbf{r}}} \int dt' C\left(\mathbf{r}t, \mathbf{r}t'\right) \left(\mathbf{E}_{\mathbf{r}t'} + 2\tau_{\mathbf{r}}\frac{e}{mc}[\mathbf{E}_{\mathbf{r}t'}\times\mathbf{H}_{\mathbf{r}t'}]\right), \qquad (22)$$

where $\rho_{\mathbf{r}} \equiv \rho_{3D}\left(\varepsilon - U_{\mathbf{r}}\right)$ is the local density of states. Equations (13) and (22) are valid in the regions where $\varepsilon_F - U_{\mathbf{r}} > 0$ (the conductivity is of metallic type). The results presented above can be applied to the case of 2D electrons if $\sigma_{\mathbf{r}}$, $\rho_{\mathbf{r}}$, $D_{\mathbf{r}}$, and $\tau_{\mathbf{r}}$ are considered as 2D conductivity, density of states, diffusion coefficient, and relaxation time, respectively. The density of states $m/\pi\hbar^2$ in this case is coordinate-independent, and the diffusion coefficient is equal to $v_F^2\tau_{\mathbf{r}}/2$.

Consider the linear response to the homogeneous electric field $\mathbf{E}e^{-i\omega t} +$ c.c. at $\mathbf{H} = 0$. In the linear regime, one should neglect $\kappa_{\mathbf{r}}(t, t')$, and the Cooperon depends on $t-t'$. The quantum correction to the conductivity, $\delta\sigma$, is introduced as usual, according to $\delta\mathbf{j}_{\mathbf{r}t} = \delta\sigma_{\mathbf{r}}(\omega)\mathbf{E}e^{-i\omega t} +$ c.c. It is expressed through the temporal Fourier transform $C_\omega(\mathbf{r}, \mathbf{r}')$ of the Cooperon:

$$\delta\sigma_{\mathbf{r}}(\omega) = -\frac{2}{\pi}\frac{\sigma_{\mathbf{r}}}{\hbar\rho_{\mathbf{r}}}C_\omega\left(\mathbf{r}, \mathbf{r}\right),$$

$$\left(-i\omega - \nabla_{\mathbf{r}} \cdot D_{\mathbf{r}}\nabla_{\mathbf{r}} + \frac{1}{\tau_\varphi}\right) C_\omega\left(\mathbf{r}, \mathbf{r}'\right) = \delta\left(\mathbf{r} - \mathbf{r}'\right). \qquad (23)$$

In the spatially homogeneous case the Cooperon depends on $\mathbf{r} - \mathbf{r}'$. The equation for $C_\omega\left(\mathbf{r}, \mathbf{r}'\right)$ in Eq. (23) is transformed into an algebraic one by a spatial Fourier transformation, and the response is found in a straightforward way:

$$\delta\sigma(\omega) = -\frac{2\sigma}{\pi\hbar\rho_D} \int \frac{d\mathbf{q}}{(2\pi)^d} \left(Dq^2 + \frac{1}{\tau_\varphi} - i\omega\right)^{-1}, \qquad d = 2, 3. \qquad (24)$$

When calculating the integral in Eq. (24), one should introduce a cutoff at large q, of the order of $(v_F\tau_F)^{-1}$, because the Cooperon is large only in the hydrodynamic region of parameters. Equation (24) allows one to study the region of frequencies $\tau_F^{-1} > \omega > \tau_\varphi^{-1}$. In the 3D case, the relative correction $\delta\sigma/\sigma$ in this region is small and frequency-independent, while in the 2D case it contains a large logarithmic factor, $\ln(\omega\tau_F)$. In the static limit $\omega \to 0$, the integration over $\mathbf{q}$ in Eq. (24) leads to Eq. (15.27) (problem 8.16).

To study a transition between the 2D and 3D regimes, let us consider Eq. (23) at $\omega = 0$ for a film of width d. The film is placed at $-d/2 < z < d/2$, and the problem is translation-invariant in the plane XOY. Therefore, it is convenient to introduce a two-dimensional Fourier transform of the Cooperon $C(\mathbf{r}, \mathbf{r}') \equiv C_{\omega=0}(\mathbf{r}, \mathbf{r}')$ according to $C_q(z, z') = \int d\Delta\mathbf{x} e^{-i\mathbf{q}\cdot\Delta\mathbf{x}} C(\mathbf{x}z, \mathbf{x}'z')$, where $\mathbf{x} = (x, y)$ is the 2D coordinate and $\Delta\mathbf{x} = \mathbf{x} - \mathbf{x}'$. Taking into account that the diffusion coefficient goes to zero in the region $|z| > d/2$ (outside the film), we integrate Eq. (23) across the boundaries and find $[dC_q(z, z')/dz]_{z=\pm d/2} = 0$. To satisfy these boundary conditions, we write the following solution of Eq. (23):

$$C_q\left(z, z'\right) = \sum_{n=0}^{\infty} \frac{\chi_{nz}\chi_{nz'}^*}{Dq^2 + \nu_n + 1/\tau_\varphi}, \qquad (25)$$

where χ_{nz} are the eigenfunctions of the problem $-Dd^2\chi_{nz}/dz^2 = \nu_n\chi_{nz}$ with the boundary conditions $[d\chi_{nz}/dz]_{z=\pm d/2} = 0$ and $n = 0, 1, \ldots$ is the mode number. We find that $\chi_{nz} = \sqrt{2/d}\cos[\pi n(z/d - 1/2)]$ and $\nu_n = D(\pi n/d)^2$. The characteristic frequency interval between the modes is estimated as $D(\pi/d)^2 = (\pi l_F/d)^2/3\tau_F$. Introducing the sheet conductivity as $\delta\sigma^\square = \int_{-d/2}^{d/2} \delta\sigma dz$, we obtain

$$\delta\sigma^\square = -\frac{2\sigma}{\pi\hbar\rho_{3D}} \int \frac{d\mathbf{q}}{(2\pi)^2} \sum_{n=0}^{\infty} \left(Dq^2 + \nu_n + 1/\tau_\varphi\right)^{-1}. \qquad (26)$$

The two-dimensional regime is realized if the main contribution to the sum in Eq. (26) comes from $n = 0$. The condition for this regime is $\nu_n\tau_\varphi \gg 1$ for $n \neq 0$, which can be rewritten as $\sqrt{l_F l_\varphi} \gg d$. Since $l_\varphi \gg l_F$, this condition remains valid even at $l_F < d$. Therefore, the quantum corrections to the 2D conductivity also describe the classical thin films whose width d exceeds the mean free path length l_F but still remains smaller than $\sqrt{l_F l_\varphi}$.

Equations (19) and (22) describe various features of the linear response in the weak localization regime. One of them is the negative magnetoresistance of 2D electrons in weak stationary magnetic fields. To investigate it, let us substitute the vector potential $\mathbf{A} = (0, Hx, 0)$ in Eq. (19) and search for the Cooperon $C(\mathbf{x}, \mathbf{x}')$ in the form of a product of the phase factor $\exp[-i(x + x')(y - y')/l_H^2]$, where l_H is the magnetic length introduced in Sec. 5, by a translation-invariant part $\widetilde{C}(|\mathbf{x} - \mathbf{x}'|)$. The spatial Fourier transform $\widetilde{C}_q$ of this part satisfies the equation

$$\left(-\frac{D}{l_H^4}\frac{d^2}{dq^2} + Dq^2 + \frac{1}{\tau_\varphi}\right)\widetilde{C}_q = 1, \tag{27}$$

which is related to the equation for the Green's function of the harmonic oscillator (see Appendix A). Its solution is written as (problem 8.17)

$$\widetilde{C}_q = e^{-q^2 l_H^2/2}\sum_{n=0}^{\infty}\frac{(-1)^n L_n^0(q^2 l_H^2)}{D(2n+1)/l_H^2 + 1/2\tau_\varphi}, \tag{28}$$

where L_n^0 is the Laguerre polynomial. Integrating $\widetilde{C}_q$ over $\mathbf{q}$, we determine $\widetilde{C}(0) = C(\mathbf{x},\mathbf{x})$ and substitute it into Eq. (22) to find the current. The quantum correction to the diagonal component of the conductivity tensor is represented as (we use $D = v_F^2\tau_F/2$ for 2D electrons)

$$\delta\sigma_d = -\frac{e^2}{2\pi^2\hbar}\sum_{n=0}^{\infty}\frac{1}{(n+1/2) + l_H^2/(4D\tau_\varphi)}. \tag{29}$$

The sum over n is logarithmic-divergent at large n. We cut it off at $n = n_m$, where n_m is the integer part of $l_H^2/2l_F^2$, and obtain

$$\delta\sigma_d = \frac{e^2}{2\pi^2\hbar}\left[\psi\left(\frac{1}{2} + \frac{1}{D\tau_\varphi}\frac{\hbar c}{4|e|H}\right) + \ln\left(\frac{4|e|H}{\hbar c}D\tau_F\right)\right], \tag{30}$$

where $\psi(x) = d\ln\Gamma(x)/dx$ is the logarithmic derivative of the Gamma function. If H is so small that $(4|e|H/\hbar c)D\tau_\varphi \ll 1$, the function $\psi(1/2 + x)$ approaches $\ln x$ and we recover the result (15.27). Expanding ψ in series, we find that the quantum correction increases as H^2 in weak magnetic fields. When $(4|e|H/\hbar c)D\tau_\varphi > 1$, the second term in Eq. (30) dominates and $\delta\sigma_d$ increases as $\ln H$. The increase of $\delta\sigma_d$ with increasing H leads to a negative magnetoresistance which often serves as a signature of the weak localization effect. The quantum correction to the non-diagonal components of the conductivity tensor is expressed as $\delta\sigma_\perp = 2\omega_c\tau_F\delta\sigma_d$. It is much smaller than $\delta\sigma_d$ in the limit of small $\omega_c\tau_F$. Using this expression for $\delta\sigma_\perp$, one may check that there is no quantum correction to the Hall effect (problem 8.18). We also point out an unusual non-linear response due to the electric field entering Eq. (19). This effect, however, can take place only at $\omega \neq 0$, because the stationary electric field drops out of Eq. (19) (problem 8.19).

Problems

8.1. Prove the relation $\widehat{R}^{-1}\hat{\sigma}_z\widehat{R} = \hat{\sigma}_x$, where $\widehat{R}$ is given by Eq. (40.1).

<u>Hint</u>: Use $\hat{\sigma}_\alpha\hat{\sigma}_\beta = \delta_{\alpha\beta} + ie_{\alpha\beta\gamma}\hat{\sigma}_\gamma$, where the tensor $e_{\alpha\beta\gamma}$ is introduced in Sec. 11 and problem 2.14.

8.2. Prove Eq. (40.24) by using the definition of non-equilibrium Green's functions.

Solution: According to the definition of F, one may equivalently prove a similar property for $\widehat{G}_\varepsilon^{-+}$. Using the dynamical equations (39.3) and (39.4) for $\mathcal{G}^{-+}(1,2)$, we find the following general solution for the stationary case:

$$\mathcal{G}^{-+}(1,2) = \sum_\delta N_\delta \psi_{\mathbf{x}_1}^{(\delta)} \psi_{\mathbf{x}_2}^{(\delta)*} \exp[-i\varepsilon_\delta(t_1 - t_2)/\hbar],$$

where we have employed the exact eigenstates $\langle \mathbf{x}|\delta\rangle = \psi_{\mathbf{x}}^{(\delta)}$ and eigenvalues ε_δ of the Schroedinger equation $\widehat{H}_{\mathbf{x}}\psi_{\mathbf{x}}^{(\delta)} = \varepsilon_\delta\psi_{\mathbf{x}}^{(\delta)}$, and N_δ is a coefficient. The temporal Fourier transformation of $\mathcal{G}^{-+}(1,2)$ gives us, in the operator form,

$$\widehat{\mathcal{G}}_\varepsilon^{-+} = 2\pi\hbar \sum_\delta N_\delta |\delta\rangle\langle\delta| \ \delta(\varepsilon - \varepsilon_\delta).$$

The coefficient N_δ can be determined with the use of Eq. (39.9), which is rewritten as $\hat{n} = (2\pi i)^{-1} \int d\varepsilon \widehat{\mathcal{G}}_\varepsilon^{-+}$. After substituting $\widehat{\mathcal{G}}_\varepsilon^{-+}$ found above into this equation, we have $\langle\delta|\hat{n}|\delta\rangle \equiv f_\delta = (\hbar/i)N_\delta$, where f_δ is the distribution function of electrons over the quantum states δ. Therefore,

$$\mathcal{G}_\varepsilon^{-+}(\mathbf{x}_1, \mathbf{x}_2) = 2\pi i \sum_\delta f_\delta \psi_{\mathbf{x}_1}^{(\delta)} \psi_{\mathbf{x}_2}^{(\delta)*} \delta(\varepsilon - \varepsilon_\delta).$$

In the stationary case, f_δ can depend only on the energy ε_δ. Owing to the presence of the δ-function, we write $f_\delta = f(\varepsilon_\delta) = f(\varepsilon)$ under the sum. Employing Eqs. (13.26) and (14.5) to the equation above, averaging this equation over the impurity distribution, and using the operator representation for the Green's functions, we finally obtain

$$\widehat{G}_\varepsilon^{-+} = f(\varepsilon)[\widehat{G}_\varepsilon^A - \widehat{G}_\varepsilon^R],$$

which also proves that $\chi_\varepsilon = 2f(\varepsilon) - 1$ in Eq. (40.24). In thermodynamic equilibrium, $f(\varepsilon) = f_\varepsilon^{(eq)}$ is the equilibrium distribution function.

8.3. Prove that the Green's function of Eq. (40.24) makes the generalized collision integral (the right-hand side of Eq. (40.23)) equal to zero.

Hints: Use the diagrammatic expansion of Ω and Σ to prove this property in each order of the perturbation theory. Take into account that Ω is proportional to χ_ε because F appears only one time in each diagram for Ω.

8.4. Derive Eqs. (41.11)–(41.13), following the way we derived Eq. (41.10).

Hint: In the derivation of Eq. (41.12), replace $\widehat{S}_0(t_2, t_1)$ standing between the field operators by $\widehat{S}_0(t_2, \infty)\widehat{S}_0(\infty, t_1)$. Proceed in a similar way to derive Eq. (41.13).

8.5. Derive the Kubo formula by using the non-equilibrium diagram technique.

Solution: Let us consider the response of the electrons interacting with a random potential $U_{\mathbf{x}t}$ to the homogeneous external perturbation $\widehat{V}_t = i(e\mathbf{E} \cdot \hat{\mathbf{v}}/\omega)e^{-i\omega t}$ by taking into account the diagrams linear in $\widehat{V}_t$, i.e., the diagrams containing only one perturbation-potential vertex. According to the rules of the diagram technique explained in Secs. 39 and 41, the correction to the Green's function due to the perturbation is given by

$$\widehat{\delta G}_{t_1 t_2}^{-+} = \langle \langle \widehat{\delta \mathcal{G}}_{t_1 t_2}^{-+} \rangle \rangle = \int dt' \left\langle \left\langle i^2 \hat{\mathcal{G}}_{t_1 t'}^{-+} \widehat{V}_{t'} \hat{\mathcal{G}}_{t' t_2}^{++} - i^2 \hat{\mathcal{G}}_{t_1 t'}^{--} \widehat{V}_{t'} \hat{\mathcal{G}}_{t' t_2}^{-+} \right\rangle \right\rangle,$$

where the double angular brackets denote the averaging over the random potential. The non-averaged Green's functions $\hat{\mathcal{G}}_{t_1 t_2}^{s_1 s_2}$ describe the electron system in the absence of the perturbation potential. We use the operator form of these functions (as well as of the perturbation potential) so that one may choose a suitable representation later on. In the coordinate representation, $\langle \mathbf{x}_1 | \hat{\mathcal{G}}_{t_1 t_2}^{s_1 s_2} | \mathbf{x}_2 \rangle = \mathcal{G}^{s_1 s_2}(\mathbf{x}_1 t_1, \mathbf{x}_2 t_2) \equiv \mathcal{G}^{s_1 s_2}(1,2)$ satisfy Eq. (39.15). The current density is given by

$$\mathbf{I}_t = \frac{ie^2 n}{m\omega} \mathbf{E} e^{-i\omega t} + \frac{e}{V} \, \mathrm{Sp}\left(\hat{n}_t \hat{\mathbf{v}} \right),$$

where the trace includes the averaging over the random potential, $\hat{n}_t$ is the one-electron density matrix, and $n = V^{-1} \mathrm{Sp} \hat{n}_{eq}$ is the electron density expressed through the equilibrium density matrix $\hat{n}_{eq}$. The contribution to the current density comes from the correction $\widehat{\delta n}_t = \hat{n}_t - \hat{n}_{eq}$ linear in the perturbation. Therefore, expressing $\hat{n}_t$ through $\hat{\mathcal{G}}_{tt}^{-+}$ according to Eq. (39.9), we obtain

$$I_t^\alpha = \frac{ie^2 n}{m\omega} E_\alpha e^{-i\omega t} + \frac{e^2}{V\omega} E_\beta \int dt' e^{-i\omega t'} \mathrm{Sp}\hat{v}_\alpha \left[\hat{\mathcal{G}}_{tt'}^{--} \hat{v}_\beta \hat{\mathcal{G}}_{t't}^{-+} - \hat{\mathcal{G}}_{tt'}^{-+} \hat{v}_\beta \hat{\mathcal{G}}_{t't}^{++} \right].$$

If the random potential U is time-independent (static), we have $\hat{\mathcal{G}}_{tt'}^{ss'} = \hat{\mathcal{G}}_{t-t'}^{ss'}$. Carrying out the temporal Fourier transformation of the Green's functions, we can calculate the integral over t' and obtain the expression for the frequency-dependent conductivity tensor:

$$\sigma_{\alpha\beta}(\omega) = \frac{ie^2 n}{m\omega} \delta_{\alpha\beta} + \frac{e^2}{V\omega} \int \frac{d\varepsilon}{2\pi} \mathrm{Sp}\hat{v}_\alpha \left[\hat{\mathcal{G}}_\varepsilon^{--} \hat{v}_\beta \hat{\mathcal{G}}_{\varepsilon-\hbar\omega}^{-+} - \hat{\mathcal{G}}_\varepsilon^{-+} \hat{v}_\beta \hat{\mathcal{G}}_{\varepsilon-\hbar\omega}^{++} \right].$$

Let us express $\hat{\mathcal{G}}^{++}$ and $\hat{\mathcal{G}}^{--}$ through $\hat{\mathcal{G}}^{-+}$, $\hat{\mathcal{G}}^R$, and $\hat{\mathcal{G}}^A$ as $\hat{\mathcal{G}}^{++} = \hat{\mathcal{G}}^{-+} - \hat{\mathcal{G}}^A$ and $\hat{\mathcal{G}}^{--} = \hat{\mathcal{G}}^{-+} + \hat{\mathcal{G}}^R$. To find $\hat{\mathcal{G}}^{-+}$ in thermodynamic equilibrium, we can use the result of problem 8.2, $\hat{\mathcal{G}}_\varepsilon^{-+} = f_\varepsilon^{(eq)}(\hat{\mathcal{G}}_\varepsilon^A - \hat{\mathcal{G}}_\varepsilon^R)$. Since the averaging over the random static potential does not influence the energy distribution function $f_\varepsilon^{(eq)}$, this relation has the same form for the averaged Green's functions. The terms quadratic in $f_\varepsilon^{(eq)}$ drop out of the expression for $\sigma_{\alpha\beta}(\omega)$, and the final expression is

$$\sigma_{\alpha\beta}(\omega) = \frac{ie^2 n}{m\omega} \delta_{\alpha\beta}$$

$$+ \frac{e^2}{2\pi\omega V} \int d\varepsilon f_\varepsilon^{(eq)} \mathrm{Sp} \left[\hat{v}_\alpha \hat{\mathcal{G}}_{\varepsilon+\hbar\omega}^R \hat{v}_\beta \left(\hat{\mathcal{G}}_\varepsilon^A - \hat{\mathcal{G}}_\varepsilon^R \right) + \hat{v}_\alpha \left(\hat{\mathcal{G}}_\varepsilon^A - \hat{\mathcal{G}}_\varepsilon^R \right) \hat{v}_\beta \hat{\mathcal{G}}_{\varepsilon-\hbar\omega}^A \right].$$

We stress that the trace includes the averaging over the random potential distribution so that the expression for the conductivity tensor contains correlation functions of the retarded and advanced Green's functions. Representing the Green's functions in the operator form according to Eq. (16.3), one can find that the expression above is reduced to the Kubo formula (13.18); see also the beginning of Sec. 49 for more details. Using the momentum representation in this expression, it is easy to show that the diagonal conductivity tensor of a macroscopically homogeneous system at $\omega = 0$ is given by Eq. (13.27).

8.6. Write the Green's function G^{-+} as a function of energies and quantum numbers of exact one-particle eigenstates of the Hamiltonian $\widehat{H}$ which does not depend on time.

Solution For the case of time-independent Hamiltonian, the Green's function $G^{ss'}$ $(\mathbf{x}t, \mathbf{x}'t')$ depends on $t - t'$. Therefore, it is convenient to use the energy representation. Introducing the exact eigenstates described by the wave functions $\psi_{\mathbf{x}}^{(\delta)}$ and corresponding energies ε_δ, one may define the Green's functions $G_\varepsilon^{ss'}(\delta)$ in the exact eigenstate representation, according to

$$G^{ss'}(\mathbf{x}t, \mathbf{x}'t') = \sum_\delta \psi_{\mathbf{x}}^{(\delta)} \psi_{\mathbf{x}'}^{(\delta)*} \int \frac{d\varepsilon}{2\pi\hbar} e^{-i\varepsilon(t-t')/\hbar} G_\varepsilon^{ss'}(\delta).$$

If $\widehat{H}$ contains a random potential, the right-hand side of this equation should be averaged over this potential. It is easy to find that

$$G_\varepsilon^{-+}(\delta) = \pm 2\pi i n_\delta \delta(\varepsilon - \varepsilon_\delta),$$

where n_δ are the occupation numbers of Fermi or Bose quasiparticles (not necessarily the equilibrium ones). The other three Green's functions are expressed as $G^{--} = G^R + G^{-+}$, $G^{++} = -G^A + G^{-+}$, and $G^{+-} = G^R - G^A + G^{-+}$, where the retarded and advanced Green's functions are given by Eq. (D.20). If the system is translation-invariant, the exact eigenstates are the plane waves $\varphi_{\mathbf{x}}^{(\delta)} = V^{-1/2} e^{i\mathbf{p}\cdot\mathbf{x}/\hbar}$, and the quantum number δ is replaced by the momentum $\mathbf{p}$ (see also the case of ideal electron gas described by the Green's functions of Eq. (40.33)).

8.7. Write the analytical expressions of the leading-order correction to Σ^{-+} and Σ^{+-} with respect to the particle-particle interaction.
Result:

$$\Sigma^{-+}(1,2) = -G^{-+}(1,2) \int d3 \int d4 G^{-+}(3,4) G^{+-}(4,3) U_{1-3} U_{2-4}$$

$$+ \int d3 \int d4 G^{-+}(1,3) G^{+-}(3,4) G^{-+}(4,2) U_{1-4} U_{2-3}.$$

To obtain $\Sigma^{+-}(1,2)$, permute the indices + and − everywhere in this expression.

8.8. Using Eq. (41.27), derive the quasi-classical kinetic equation for interacting electrons.
Solution: Let us put $t_1 = t_2 = t$ in Eq. (41.27). If $\widehat{H}_0(1) = -(\hbar^2/2m)\partial^2/\partial\mathbf{x}_1^2$, the left-hand side of this equation is equal to

$$-i\hbar \left[\frac{\partial}{\partial t} - i\frac{\hbar}{m} \frac{\partial}{\partial \mathbf{r}} \cdot \frac{\partial}{\partial \Delta\mathbf{r}} \right] G^{-+}(\mathbf{r} + \Delta\mathbf{r}/2\, t, \mathbf{r} - \Delta\mathbf{r}/2\, t),$$

where $\mathbf{r} = (\mathbf{x}_1 + \mathbf{x}_2)/2$ and $\Delta\mathbf{r} = \mathbf{x}_1 - \mathbf{x}_2$. The Wigner transformation in space, according to Eq. (39.11), transforms this expression to the left-hand side of the quasi-classical kinetic equation, $(\partial/\partial t + \mathbf{v_p} \cdot \nabla_\mathbf{r})f_{\mathbf{rp}t}$. Carrying out the Wigner transformation of the products $\Sigma^{-s}(1,1')G^{s+}(1',2)$ and $G^{-s}(1,1')\Sigma^{s+}(1'2)$, we write the right-hand side of Eq. (41.27) as

$$-\int \frac{d\varepsilon}{2\pi\hbar} \sum_{s=\pm} \left[\Sigma_{\varepsilon t}^{-s}(\mathbf{r},\mathbf{p})G_{\varepsilon t}^{s+}(\mathbf{r},\mathbf{p}) + G_{\varepsilon t}^{-s}(\mathbf{r},\mathbf{p})\Sigma_{\varepsilon t}^{s+}(\mathbf{r},\mathbf{p})\right]$$

$$= \int \frac{d\varepsilon}{2\pi\hbar} \left[\Sigma_{\varepsilon t}^{+-}(\mathbf{r},\mathbf{p})G_{\varepsilon t}^{-+}(\mathbf{r},\mathbf{p}) - \Sigma_{\varepsilon t}^{-+}(\mathbf{r},\mathbf{p})G_{\varepsilon t}^{+-}(\mathbf{r},\mathbf{p})\right],$$

where the dependence of G and Σ on the coordinate $\mathbf{r}$ and time t is parametric. The right-hand side of this equation is obtained with the aid of Eqs. (39.22) ans (39.24). Using the expressions for $\Sigma^{-+}(1,2)$ and $\Sigma^{+-}(1,2)$ given in the previous problem, we find the corresponding $\Sigma_{\varepsilon t}^{-+}(\mathbf{r},\mathbf{p})$ and $\Sigma_{\varepsilon t}^{+-}(\mathbf{r},\mathbf{p})$ which have to be substituted in the expression above. Having done this, one can show that this expression is equal to the collision integral $J_{ee}(f|\mathbf{rp}t)$ of Eqs. (31.21) and (31.22). Indeed, since the self-energy functions are quadratic in the interaction, it is sufficient to employ the expressions $G_{\varepsilon t}^{-+}(\mathbf{r},\mathbf{p}) = 2\pi i f_{\mathbf{rp}t}\delta(\varepsilon - \varepsilon_p)$ and $G_{\varepsilon t}^{+-}(\mathbf{r},\mathbf{p}) = -2\pi i(1 - f_{\mathbf{rp}t})\delta(\varepsilon - \varepsilon_p)$ for the Green's functions of free electrons; see Eq. (40.33). This immediately gives us Eqs. (31.21) and (31.22), because the integral over ε is calculated with the use of the δ-function. The first terms in $\Sigma^{-+}(1,2)$ and $\Sigma^{+-}(1,2)$ (see problem 8.7), corresponding to the diagrams like the third one on the right-hand side of Eq. (E.23), give the direct Coulomb term in the matrix element (31.22). The second terms in $\Sigma^{-+}(1,2)$ and $\Sigma^{+-}(1,2)$, corresponding to the diagrams like the fourth one on the right-hand side of Eq. (E.23), give the exchange term in this matrix element.

8.9. Using the NDT formalism, express the dielectric permittivity of a macroscopically homogeneous electron system through the polarization functions $\Pi_\omega^{s_1 s_2}(\mathbf{q})$ and calculate it in the random phase approximation.

Solution: The dielectric permittivity of a macroscopically homogeneous electron system is expressed through the polarizability given by Eq. (33.12), where $K(\mathbf{q},\omega)$ is expressed through the Fourier transform of the causal two-electron Green's function. The diagram expansion of the latter in terms of single-electron Green's functions is given by Eq. (33.15). We may use this equation in the case of finite temperatures if we add the indices "$-$" to the initial and final vertices, $\mathbf{r}t$ and $\mathbf{r}'0$, and the indices "$+$" and "$-$" to all internal vertices in every possible way. Instead of Eq. (33.17), we obtain

$$K(\mathbf{q},\omega) = -(i/\hbar)n^2\delta(\omega)\delta(\mathbf{q}) + \Pi_\omega^{--}(\mathbf{q}) + \sum_{s_1 s_2} \Pi_\omega^{-s_1}(\mathbf{q})V_\omega^{s_1 s_2}(\mathbf{q})\Pi_\omega^{s_2 -}(\mathbf{q}),$$

where the functions $\Pi^{s_1 s_2}$ and $V^{s_1 s_2}$ are expressed in the energy-momentum representation, using the homogeneity of electron system. To find $V_\omega^{s_1 s_2}(\mathbf{q})$, we use Eq. (41.30) written in the energy-momentum representation as a system of algebraic equations. Its solutions are $V_\omega^{--}(\mathbf{q}) = v_q[1 + v_q\Pi_\omega^{++}(\mathbf{q})]/\mathcal{D}(\mathbf{q},\omega)$, $V_\omega^{++}(\mathbf{q}) = -v_q[1 - v_q\Pi_\omega^{--}(\mathbf{q})]/\mathcal{D}(\mathbf{q},\omega)$, and $V_\omega^{s_1 s_2}(\mathbf{q}) = -v_q^2\Pi_\omega^{s_1 s_2}(\mathbf{q})/\mathcal{D}(\mathbf{q},\omega)$ at $s_1 \neq s_2$. The determinant $\mathcal{D}(\mathbf{q}, \omega)$ is given by

$$\mathcal{D}(\mathbf{q}, \omega) = 1 - v_q[\Pi_\omega^{--}(\mathbf{q}) - \Pi_\omega^{++}(\mathbf{q})] + v_q^2[\Pi_\omega^{-+}(\mathbf{q})\Pi_\omega^{+-}(\mathbf{q}) - \Pi_\omega^{--}(\mathbf{q})\Pi_\omega^{++}(\mathbf{q})]$$

$$= [1 - v_q\Pi_\omega^R(\mathbf{q})][1 - v_q\Pi_\omega^A(\mathbf{q})],$$

$$\Pi_\omega^R(\mathbf{q}) = \Pi_\omega^{A*}(\mathbf{q}) = [\Pi_\omega^{--}(\mathbf{q}) - \Pi_\omega^{++}(\mathbf{q})]/2 - [\Pi_\omega^{-+}(\mathbf{q}) - \Pi_\omega^{+-}(\mathbf{q})]/2$$

$$= \mathrm{Re}\Pi_\omega^{--}(\mathbf{q}) - [\Pi_\omega^{-+}(\mathbf{q}) - \Pi_\omega^{+-}(\mathbf{q})]/2 ,$$

where we have employed the linear relation (41.29) and introduced the retarded and advanced polarization functions, Π^R and Π^A, by taking into account the properties $\Pi^{-+} = -(\Pi^{-+})^*$, $\Pi^{+-} = -(\Pi^{+-})^*$, and $\Pi^{++} = -(\Pi^{--})^*$. Substituting $V_\omega^{s_1 s_2}(\mathbf{q})$ into the expression for $K(\mathbf{q}, \omega)$, we find

$$K(\mathbf{q}, \omega) + K^*(-\mathbf{q}, -\omega) = \frac{\Pi_\omega^R(\mathbf{q})}{1 - v_q\Pi_\omega^R(\mathbf{q})} + \frac{\Pi_\omega^A(\mathbf{q})}{1 - v_q\Pi_\omega^A(\mathbf{q})}.$$

Using the exact eigenstate representation, one may check that $\Pi_\omega^R(\mathbf{q})$ is analytical in the upper half-plane of complex variable ω, while $\Pi_\omega^A(\mathbf{q})$ is analytical in the lower half-plane. As explained in Sec. 33, only the retarded term contributes to the integral over ω' in Eq. (33.12), and we again obtain the result (33.21):

$$\epsilon(\mathbf{q}, \omega)/\epsilon_\infty = 1 - v_q\Pi_\omega^R(\mathbf{q}),$$

where $\Pi_\omega^R(\mathbf{q})$ is now expressed through the components of the polarization function of NDT and can be found at finite temperatures with a required accuracy. After the Fourier transformations of Eq. (41.28) in space and time, we find

$$\Pi_\omega^{s_1 s_2}(\mathbf{q}) = \frac{2}{V} \sum_\mathbf{p} \int \frac{d\varepsilon}{2\pi i} G_{\varepsilon+\hbar\omega}^{s_1 s_2}(\mathbf{p} + \hbar\mathbf{q}) G_\varepsilon^{s_2 s_1}(\mathbf{p}) + \dots .$$

In particular, the RPA leads to (compare this to Eq. (33.22))

$$\Pi_\omega^{s_1 s_2}(\mathbf{q}) = \frac{2}{V} \sum_\mathbf{p} \int \frac{d\varepsilon}{2\pi i} g_{\varepsilon+\hbar\omega}^{s_1 s_2}(\mathbf{p} + \hbar\mathbf{q}) g_\varepsilon^{s_2 s_1}(\mathbf{p}),$$

where the Green's functions of the ideal electron gas are given by Eq. (40.33). Substituting them into the equation above, one can find the corresponding RPA expressions for each component of the polarization function:

$$\Pi_\omega^{--}(\mathbf{q}) = -[\Pi_\omega^{++}(\mathbf{q})]^*$$

$$= \frac{2}{V} \sum_\mathbf{p} \left[\frac{f_{\mathbf{p}+\hbar\mathbf{q}/2}(1 - f_{\mathbf{p}-\hbar\mathbf{q}/2})}{\varepsilon_{\mathbf{p}+\hbar\mathbf{q}/2} - \varepsilon_{\mathbf{p}-\hbar\mathbf{q}/2} - \hbar\omega + i\lambda} - \frac{f_{\mathbf{p}-\hbar\mathbf{q}/2}(1 - f_{\mathbf{p}+\hbar\mathbf{q}/2})}{\varepsilon_{\mathbf{p}+\hbar\mathbf{q}/2} - \varepsilon_{\mathbf{p}-\hbar\mathbf{q}/2} - \hbar\omega - i\lambda} \right],$$

$$\left| \begin{matrix} \Pi_\omega^{-+}(\mathbf{q}) \\ \Pi_\omega^{+-}(\mathbf{q}) \end{matrix} \right| = -2\pi i \frac{2}{V} \sum_\mathbf{p} \delta(\varepsilon_{\mathbf{p}+\hbar\mathbf{q}/2} - \varepsilon_{\mathbf{p}-\hbar\mathbf{q}/2} - \hbar\omega) \left| \begin{matrix} f_{\mathbf{p}+\hbar\mathbf{q}/2}(1 - f_{\mathbf{p}-\hbar\mathbf{q}/2}) \\ f_{\mathbf{p}-\hbar\mathbf{q}/2}(1 - f_{\mathbf{p}+\hbar\mathbf{q}/2}) \end{matrix} \right| ,$$

where $f_\mathbf{p}$ is the distribution function of electrons and $\lambda \to +0$. Using these expressions, we find that $\Pi_\omega^R(\mathbf{q})$ calculated in the random phase approximation is equal to $\Pi_\omega^{(0)R}(\mathbf{q})$ of Eq. (33.8). In conclusion, the generalization of the expressions for the RPA polarizability and dielectric permittivity to the case of finite temperatures is straightforward: one should write the distribution function $f_\mathbf{p}$ at finite temperature instead of the distribution function at zero temperature. The function $f_{\mathbf{p}}$ is assumed to be quasi-stationary but not necessarily the equilibrium one.

8.10. Derive the expression for the electron-electron collision integral accounting for the dynamical screening.

Solution: One has to act in the same way as in problem 8.8. However, instead of using the expressions for $\Sigma^{-+}(1,2)$ and $\Sigma^{+-}(1,2)$ given in problem 8.7 and corresponding to the second order with respect to the "bare" interaction potential, one should use their expressions in the leading order with respect to the screened interaction potential. In the diagram form,

$$-i\Sigma^{s_1 s_2}(1,2) \simeq \underset{1s_1 \qquad 2s_2}{\overline{}} \qquad s_1 \neq s_2 \ .$$

The corresponding analytical expression is $\Sigma^{s_1 s_2}(1,2) = i\hbar V^{s_1 s_2}(1,2)G^{s_1 s_2}(1,2)$. Carrying out the Wigner transformation of the self-energy functions, and employing the expressions for $V_\omega^{-+}(\mathbf{q})$ and $V_\omega^{+-}(\mathbf{q})$ from the previous problem, we find

$$\Sigma_\varepsilon^{-+}(\mathbf{p}) = -\frac{i\hbar}{V}\sum_{\mathbf{q}} v_q^2 \int \frac{d\omega}{2\pi} G_{\varepsilon-\hbar\omega}^{-+}(\mathbf{p}-\hbar\mathbf{q})\Pi_\omega^{-+}(\mathbf{q})/\mathcal{D}(\mathbf{q},\omega),$$

where $\mathcal{D}(\mathbf{q},\omega)$ is introduced in problem 8.9. The expression of the polarization function in terms of the Green's functions $G_\varepsilon^{s_1 s_2}(\mathbf{p})$ is also given in problem 8.9. Substituting it into the equation for $\Sigma_\varepsilon^{-+}(\mathbf{p})$ above, we can write the following expression for the collision integral:

$$\frac{2}{V^2}\int\frac{d\varepsilon}{2\pi}\int\frac{d\varepsilon'}{2\pi}\int\frac{d\omega}{2\pi}\sum_{\mathbf{p}'\mathbf{q}}\frac{v_q^2}{\mathcal{D}(\mathbf{q},\omega)}\left[G_{\varepsilon-\hbar\omega}^{-+}(\mathbf{p}-\hbar\mathbf{q})G_\varepsilon^{+-}(\mathbf{p})\right.$$

$$\times G_{\varepsilon'+\hbar\omega}^{-+}(\mathbf{p}'+\hbar\mathbf{q})G_{\varepsilon'}^{+-}(\mathbf{p}') - G_{\varepsilon-\hbar\omega}^{+-}(\mathbf{p}-\hbar\mathbf{q})G_\varepsilon^{-+}(\mathbf{p})G_{\varepsilon'+\hbar\omega}^{+-}(\mathbf{p}'+\hbar\mathbf{q})G_{\varepsilon'}^{-+}(\mathbf{p}')\right].$$

Finally, approximating G^{-+} and G^{+-} by g^{-+} and g^{+-}, we find that our expression for the collision integral differs from the direct Coulomb contribution in the collision integral (31.21) only by the denominator $\mathcal{D}[(\mathbf{p}-\mathbf{p}_1)/\hbar, (\varepsilon_p - \varepsilon_{p_1})/\hbar]$. As we know from the previous problem, $\mathcal{D}(\mathbf{q},\omega) = \epsilon(\mathbf{q},\omega)\epsilon^*(\mathbf{q},\omega)/\epsilon_\infty^2$. Therefore, we obtain the collision integral (31.21) with the scattering matrix element (33.32). The higher-order contributions to Σ^{-+} and Σ^{+-} can be neglected under the approximation $q < q_{TF}^2 \ll k_F^2$, where $\hbar\mathbf{q}$ is the momentum transferred in the electron-electron collisions. This corresponds to the range of validity of the RPA; see Sec. 33.

8.11. Write the analytical expression of the second diagram on the right-hand side of Eq. (42.13).

Hint: Take into account that the vertices $3s_3\gamma$ and $4s_4\delta$ bring the operator $is_3 is_4 \hbar\mathcal{D}^2\nabla_{\mathbf{x}_3}^\gamma\nabla_{\mathbf{x}_4}^\delta$ acting on the coordinates of the phonon Green's function.

8.12. Calculate the retarded Green's function $\mathcal{D}^{\alpha\beta,R}(1,2)$ of non-interacting photons in a homogeneous medium in the energy-momentum representation.

<u>Solution</u>: Consider the gauge $\nabla \cdot \mathbf{A} = 0$. One should express the vector potentials according to Eq. (3.18) and take into account that $\langle\langle \hat{b}_{\mathbf{q}\mu}(t_1) \hat{b}^+_{\mathbf{q}'\mu'}(t_2) \rangle\rangle = e^{-i\omega_{\mathbf{q}\mu}(t_1-t_2)} \delta_{\mathbf{qq}'} \delta_{\mu\mu'} (1 + N^\mu_{\mathbf{q}})$ and $\langle\langle \hat{b}^+_{\mathbf{q}\mu}(t_1) \hat{b}_{\mathbf{q}'\mu'}(t_2) \rangle\rangle = e^{i\omega_{\mathbf{q}\mu}(t_1-t_2)} \delta_{\mathbf{qq}'} \delta_{\mu\mu'} N^\mu_{\mathbf{q}}$. The Fourier transformation of the retarded Green's function

$$\mathcal{D}^{\alpha\beta, R}_{t_1 t_2}(\mathbf{x}_1, \mathbf{x}_2) = -(i/\hbar)\theta(t_1 - t_2)\langle\langle \hat{A}^\alpha_{\mathbf{x}_1}(t_1) \hat{A}^\beta_{\mathbf{x}_2}(t_2) - \hat{A}^\beta_{\mathbf{x}_2}(t_2) \hat{A}^\alpha_{\mathbf{x}_1}(t_1) \rangle\rangle$$

over the spatial and temporal variables, $(\mathbf{x}_1 - \mathbf{x}_2)$ and $(t_1 - t_2)$, leads to the following result $(\lambda \to +0)$:

$$\mathcal{D}^{\alpha\beta, R}_\omega(\mathbf{q}) = \sum_\mu \frac{2\pi c^2}{\epsilon \omega_{\mathbf{q}\mu}} \left[\frac{1}{\omega - \omega_{\mathbf{q}\mu} + i\lambda} - \frac{1}{\omega + \omega_{\mathbf{q}\mu} + i\lambda} \right] e^\alpha_{\mathbf{q}\mu} e^{\beta*}_{\mathbf{q}\mu}.$$

Since $\omega_{\mathbf{q}\mu} = cq/\sqrt{\epsilon}$ and $\mathbf{q} \cdot \mathbf{e}_{\mathbf{q}\mu} = 0$ for both polarizations, one has $\sum_\mu e^\alpha_{\mathbf{q}\mu} e^{\beta*}_{\mathbf{q}\mu} = \delta_{\alpha\beta} - q_\alpha q_\beta/q^2$. Finally, we obtain

$$\mathcal{D}^{\alpha\beta, R}_\omega(\mathbf{q}) = \mathcal{D}^{\alpha\beta, A *}_\omega(\mathbf{q}) = \frac{4\pi(\delta_{\alpha\beta} - q_\alpha q_\beta/q^2)}{(\omega + i\lambda)^2 \epsilon/c^2 - q^2}.$$

8.13. Derive the electron-boson collision integral standing in the quasi-classical kinetic equation (see Eq. (34.25) and problem 7.3) from the NDT formalism.

<u>Solution</u>: The expression of the collision integral in terms of $G^{s_1 s_2}_\varepsilon(\mathbf{p})$ and $\Sigma^{s_1 s_2}_\varepsilon(\mathbf{p})$ is given in problem 8.8. One should substitute there the corresponding self-energy function for electron-boson interaction, which, for the case of acoustic phonons, is given by Eq. (42.19). Below we rewrite it in the general form, valid for any electron-boson interaction described by the Hamiltonian (19.2). Taking into account only the diagonal contribution $D^{s_1 s_2}_{t_1 t_2}(q) \equiv D^{s_1 s_2}_{t_1 t_2}(q, q)$ to the boson Green's function $D^{s_1 s_2}_{t_1 t_2}(q, q') = -(i/\hbar)\langle\langle \hat{\mathcal{T}}_C [\hat{b}_q(t_1) + \hat{b}^+_q(t_1)]_{s_1} [\hat{b}_{q'}(t_2) + \hat{b}^+_{q'}(t_2)]_{s_2} \rangle\rangle$, we obtain

$$-i\Sigma^{s_1 s_2}_\varepsilon(\mathbf{p}) = is_1 is_2 \hbar \int \frac{d\omega}{2\pi} \sum_q \sum_{\mathbf{p}_1} |\langle \mathbf{p}|\hat{\chi}_q|\mathbf{p}_1\rangle|^2 iG^{s_1 s_2}_{\varepsilon-\hbar\omega}(\mathbf{p}_1) iD^{s_1 s_2}_\omega(q)$$

$$+ is_1 is_2 \hbar^2 \sum_{s'_1 s'_2} is'_1 is'_2 \int \frac{d\omega}{2\pi} \int \frac{d\omega'}{2\pi} \sum_{qq'} \sum_{\mathbf{p}_1 \mathbf{p}_2 \mathbf{p}_3} \langle \mathbf{p}|\hat{\chi}_q|\mathbf{p}_1\rangle$$

$$\times \langle \mathbf{p}_1|\hat{\chi}_{q'}|\mathbf{p}_2\rangle \langle \mathbf{p}_2|\hat{\chi}^+_q|\mathbf{p}_3\rangle \langle \mathbf{p}_3|\hat{\chi}^+_{q'}|\mathbf{p}\rangle$$

$$\times iG^{s_1 s'_2}_{\varepsilon-\hbar\omega}(\mathbf{p}_1) iG^{s'_2 s'_1}_{\varepsilon-\hbar(\omega+\omega')}(\mathbf{p}_2) iG^{s'_1 s_2}_{\varepsilon-\hbar\omega'}(\mathbf{p}_3) iD^{s_1 s'_1}_\omega(q) iD^{s'_2 s_2}_{\omega'}(q') + \dots .$$

To apply this equation to the case of electron-phonon interaction in the single-mode approximation, one should treat q as the wave vector $\mathbf{q}$ of the phonon belonging to the corresponding mode and take $\hat{\chi}_q$ from Eq. (21.1). This allows one to express $\mathbf{p}_1$, $\mathbf{p}_2$, and $\mathbf{p}_3$ through $\mathbf{p}$, $\mathbf{q}$, and $\mathbf{q}'$ according to the momentum conservation rule, which leads to Eq. (42.19). In the lowest order in the interaction, we take into consideration only the first term of the expression for $\Sigma^{s_1 s_2}_\varepsilon(\mathbf{p})$. To obtain Eq. (34.25), we need to substitute $\Sigma^{-+}_\varepsilon(\mathbf{p})$ and $\Sigma^{+-}_\varepsilon(\mathbf{p})$ into the collision integral presented in problem 8.8 and calculate the integrals over ε and ω by using the free-electron and free-boson

Green's functions given by Eqs. (40.33) and (42.23).

8.14. Using the Dyson equation (42.18) at zero temperature, find the energy shift and renormalization of the effective mass of low-energy electrons due to their interaction with LO phonons (the polaronic effect).

Solution: Let us calculate $\Sigma_\varepsilon^{--}(\mathbf{p})$ from the expression given in the previous problem by using the free-electron and free-phonon Green's functions, the parabolic dispersion law $\varepsilon_p = p^2/2m$, and $\hat{\chi}_q$ corresponding to the interaction of electrons with LO phonons (Eq. (21.1)). Substituting the result into the Dyson equation, we express the Green's function as

$$G_\varepsilon^{--}(\mathbf{p}) = [\varepsilon - \varepsilon_p + \alpha(\hbar\omega_{LO} + p^2/12m) + i\lambda\,\mathrm{sgn}(p - p_F)]^{-1},$$

where $\lambda \to +0$ and α is the constant of electron-phonon coupling defined by Eq. (21.28). Thus, the renormalized energy is $\varepsilon_p^* = -\alpha\hbar\omega_{LO} + p^2/2m^*$, where the renormalized mass is $m^* = m(1 + \alpha/6)$. These results are valid for $\alpha \ll 1$.

8.15. Using the Dyson equation (42.21) at zero temperature, find: *a)* the renormalization of the longitudinal sound velocity due to interaction of LA phonons with electrons, *b)* the lifetime of LA phonons due to this interaction, and *c)* the renormalization of the LO phonon frequency due to electron-phonon interaction.

Solution: According to the Dyson equation,

$$\mathrm{D}_\omega^{--}(\mathbf{q}) = \frac{1}{\hbar}\frac{2\omega_\mathbf{q}}{\omega^2 - \omega_\mathbf{q}^2 - 2\omega_\mathbf{q}\mathcal{V}_q^2\Pi_\omega^{--}(\mathbf{q})/\hbar}.$$

If $T = 0$, the function $\Pi_\omega^{--}(\mathbf{q})$, calculated within the RPA (see problem 8.9), coincides with the $\Pi_\omega^{(0)R}(\mathbf{q})$ given by Eq. (33.24) at positive ω (the case of $\omega > 0$ is assumed in the following). For acoustic phonons, $\omega_\mathbf{q} = s_l q$ and $\mathcal{V}_q^2 = \hbar\mathcal{D}^2 q/2\rho s_l$. In the limits $\omega \ll v_F q$ and $q \ll k_F$, we have $\Pi_\omega^{--}(\mathbf{q}) \simeq -mk_F/\pi^2\hbar^2 - im^2\omega/2\pi\hbar^3 q$. The real part of $2\omega_\mathbf{q}\mathcal{V}_q^2\Pi_\omega^{--}(\mathbf{q})$ is proportional to q^2 and can be appended to $\omega_\mathbf{q}^2$, leading to the renormalized spectrum $\omega_\mathbf{q}^* = s_l^* q$ with the reduced LA phonon velocity $s_l^* = \sqrt{s^2 - mk_F\mathcal{D}^2/\pi^2\hbar^2\rho}$. The phonon lifetime due to the imaginary part of $\Pi_\omega^{--}(\mathbf{q})$ (compare to Eq. (27.28)) is given by $\tau^{-1} = m^2\mathcal{D}^2 q/2\pi\hbar^3\rho$.

In the case of optical phonons, we have $\omega_\mathbf{q} = \omega_{LO}$ and $\mathcal{V}_q^2 = 2\pi\hbar e^2\omega_{LO}/\epsilon^* q^2$. We take the limit $\omega \gg v_F q$, which leads to $\Pi_\omega^{--}(\mathbf{q}) \simeq k_F^3 q^2/3\pi^2\omega^2 m$ (the imaginary part can be neglected). Using the identity $n = k_F^3/3\pi^2$ and the definition of the plasma frequency, $\omega_p = \sqrt{4\pi e^2 n/\epsilon_\infty m}$, we obtain $\mathrm{D}_\omega^{--}(\mathbf{q}) = 2\hbar^{-1}\omega_{LO}/[\omega^2 - \omega_{LO}^2 - \omega_p^2\omega_{LO}^2(1 - \epsilon_\infty/\epsilon_0)/\omega^2]$. Therefore, the LO phonon frequency becomes $\omega_{LO}^* \simeq \sqrt{\omega_{LO}^2 + \omega_p^2(1 - \epsilon_\infty/\epsilon_0)}$. The frequency ω_+ of the phonon-like plasmon-phonon mode (see problem 6.20) is given by the same expression in the limit $\omega_p^2 \ll \omega_{LO}^2$.

8.16. Obtain Eq. (15.27) by calculating the integral in Eq. (43.24) for the 2D case at $\omega = 0$.

Hint: The integral is proportional to $\ln(q^2 + (D\tau_\varphi)^{-1})|_{q=0}^{q=q_{\max}}$, where the cutoff at large q is introduced as explained in the discussion following Eq. (43.24).

8.17. Prove Eq. (43.28).

Solution: One may use Eq. (43.27) to obtain Eq. (43.28). It is more instructive, however, to start from Eq. (43.19), which is written in the steady-state limit as

$$\left[-D\frac{\partial^2}{\partial x^2} + D\left(-i\frac{\partial}{\partial y} + \frac{2|e|H}{\hbar c}x \right)^2 + \frac{1}{\tau_\varphi} \right] C(\mathbf{x}, \mathbf{x}') = \delta(\mathbf{x} - \mathbf{x}'),$$

where $\mathbf{x} = (x, y)$ and we use the gauge $\mathbf{A} = (0, Hx, 0)$. This equation formally coincides with the equation for the Green's function of electron in the magnetic field if we substitute $D \to -\hbar^2/2m$, $1/\tau_\varphi \to \varepsilon$, and $2|e| \to |e|$. Therefore, its solution is written according to the result of problem 3.10 with the aid of the eigenfunctions $\psi_{\mathbf{x}}^{(\delta)} = L_y^{-1/2} e^{ipy/\hbar} \varphi_x^{(np)}$, where the oscillatory functions $\varphi_x^{(np)}$ are given by Eq. (5.15):

$$C(\mathbf{x}, \mathbf{x}') = \sum_{n=0}^{\infty} \int_{-\infty}^{\infty} \frac{dp}{2\pi\hbar} \frac{\exp[ip(y - y')/\hbar]}{2^n n! \sqrt{\pi}\tilde{l}_H} \exp\left[-\frac{(x - X_p)^2}{2\tilde{l}_H^2} - \frac{(x' - X_p)^2}{2\tilde{l}_H^2} \right]$$

$$\times \frac{H_n[(x - X_p)/\tilde{l}_H]H_n[(x' - X_p)/\tilde{l}_H]}{2D(n + 1/2)/\tilde{l}_H^2 + 1/\tau_\varphi},$$

where $p = p_y$, $X_p = -pc/2|e|H$, and $\tilde{l}_H^2 = l_H^2/2 = \hbar c/2|e|H$. To calculate the integral over p, we use the identity $\int_{-\infty}^{\infty} dx e^{-x^2} H_n(x + a)H_n(x - b) = 2^n n! \sqrt{\pi} L_n^0(2ab)$ and obtain $C(\mathbf{x}, \mathbf{x}') = \exp[-(x + x')(y - y')/l_H^2]\tilde{C}(|\mathbf{x} - \mathbf{x}'|)$, where

$$\tilde{C}(r) = \pi^{-1} \exp(-r^2/2l_H^2) \sum_{n=0}^{\infty} \frac{L_n^0(r^2/l_H^2)}{4D(n + 1/2) + l_H^2/\tau_\varphi}.$$

To obtain Eq. (43.28), one should carry out a Fourier transformation of $\tilde{C}(r)$.

8.18. Prove that the quantum correction to the Hall resistance can be neglected in comparison to the quantum correction to the diagonal resistance.

Solution: The conductivity tensor is written in the form

$$\begin{bmatrix} \sigma_d + \delta\sigma_d & -\omega_c\tau_F(\sigma_d + 2\delta\sigma_d) \\ \omega_c\tau_F(\sigma_d + 2\delta\sigma_d) & \sigma_d + \delta\sigma_d \end{bmatrix}.$$

Assuming $\delta\sigma_d \ll \sigma_d$, we find the Hall resistivity:

$$\rho_\perp = (H/|e|nc)/\{1 + 2(\omega_c\tau_F)^2(\delta\sigma_d/\sigma_0) + [1 + (\omega_c\tau_F)^2](\delta\sigma_d/\sigma_0)^2\}.$$

At small fields, $\omega_c\tau_F \ll 1$, the correction to the classical Hall resistivity $H/|e|nc$ is quadratic in $\delta\sigma_d/\sigma_0$ and can be neglected.

8.19. Prove that a time-independent electric field $\mathbf{E}$ does not affect $C(\mathbf{r}t, \mathbf{r}'t')$.

Hint: Use the vector potential $\mathbf{A} = -c\mathbf{E}t$ in Eq. (43.19).

Chapter 9

KINETICS OF BOUNDED SYSTEMS

To derive the general kinetic equations in the previous chapters, we did not employ the condition of homogeneity, except for Chapter 5, where the basis of phonon wave vectors was used in the initial equations. Nevertheless, in each concrete case, a detailed analysis has been carried out either for homogeneous systems, when the momentum (or quasimomentum) conservation rule is valid, or for the systems with slowly varying macroscopic parameters, when one can effectively employ the Wigner representation of the density matrix. On the other hand, if a system has the abrupt inhomogeneities which can be defined in a regular way, for example, the interfaces between different materials, or just the surfaces, the quasi-classical kinetic equations should be considered together with the boundary conditions at such interfaces. In this chapter we derive the boundary conditions for the Wigner distribution functions of electrons and phonons in the presence of non-ideal (rough) surfaces and interfaces and study the influence of the surfaces and interfaces on transport phenomena.

44. Boundary Conditions at Non-Ideal Surface

The description of electron states in the **kp**-approximation discussed in Sec. 5 and Appendix B cannot be directly applied near the surfaces or interfaces where the potential energy changes abruptly, on the scale of the crystal lattice constant. Even in the simplest case of conduction-band electrons with isotropic effective mass m described by the bulk Hamiltonian $\hat{\mathbf{p}}^2/2m$, the envelope wave function $\psi_{\mathbf{r}}$ satisfies a boundary condition of the third kind; see Eq. (5.19). In the many-band approach, different components of the columnar function $\varphi_{l\mathbf{r}}$ (see Appendix B) are mixed due to the interface potential, and spin-flip transitions at the interface become possible (problem 9.1). Below we consider the simplest case of zero boundary conditions $\Psi|_\Gamma = 0$ at the surface Γ in the vicinity of the plane $z = 0$. Such non-ideal (rough) surface is shown in Fig. 9.1. It can be described by the equation $z = \xi(\mathbf{x})$, where $\xi(\mathbf{x})$ is the deviation of the surface from its average position $\langle\langle \xi(\mathbf{x}) \rangle\rangle = 0$ at the point $\mathbf{x}$ of the plane $z = 0$. The brackets $\langle\langle \ldots \rangle\rangle$ denote the averaging over the random surface. The scale of the roughness in the plane (the correlation length) is assumed to be large in comparison to the electron wavelength. We also assume that the mean square deviation $\bar{\xi} = \sqrt{\langle\langle \xi(\mathbf{x})\xi(\mathbf{x}) \rangle\rangle}$ is small in comparison to the transverse component $\lambda_\perp = 2\pi\hbar/\sqrt{2m(\varepsilon - \varepsilon_\parallel)}$ of de Broglie wavelength of the electron. Here ε is the total energy of the electron and $\varepsilon_\parallel$ is the kinetic energy associated with electron motion along the plane. The condition $\bar{\xi} \ll \lambda_\perp$ is easy to satisfy in semiconductors. In metals, where the wavelength of the electrons on the Fermi surface is comparable to the lattice constant, this condition can be satisfied only for the glancing electrons (for whom $\varepsilon - \varepsilon_\parallel \ll \varepsilon$) if $\bar{\xi}$ is of the order of the lattice constant. However, it is the glancing electrons that give the main contribution to the transport along the surface $z = \xi(\mathbf{x})$. Within the accuracy up to the first order in $\bar{\xi}$, one can transform the boundary condition $\Psi|_{z=\xi(\mathbf{x})} = 0$ to the following form:

$$\Psi|_{z=0} + \xi(\mathbf{x})\frac{\partial\Psi}{\partial z}\bigg|_{z=0} = 0. \tag{1}$$

This approximation allows us to calculate the averages within the accuracy up to $\bar{\xi}^2$.

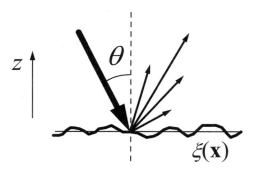

Figure 9.1. Diffuse scattering of particles on a rough surface. The angle of incidence θ does not coincide with the angle of reflection.

Before considering the density matrix in the near-surface region, let us discuss the problem of electron wave reflection from the non-ideal surface. This problem is described by the Schroedinger equation for a half-space $z > 0$, which is to be solved with the boundary condition (1).

It is convenient to reformulate the problem by using a mixed coordinate-momentum representation, when the wave function $\Psi_{\mathbf{p}z}$, which depends on the two-dimensional momentum $\mathbf{p}$, satisfies the Schroedinger equation

$$\left(\frac{\hat{p}_z^2}{2m} + \frac{p^2}{2m} \right) \Psi_{\mathbf{p}z} = \varepsilon \Psi_{\mathbf{p}z}. \tag{2}$$

The boundary condition (1) is transformed to the following integro-differential equation:

$$\Psi_{\mathbf{p}z=0} + \int \frac{d\mathbf{p}'}{(2\pi\hbar)^2} \xi(\mathbf{p} - \mathbf{p}') \frac{\partial \Psi_{\mathbf{p}'z}}{\partial z}\Big|_{z=0} = 0, \tag{3}$$

where $\xi(\mathbf{p}) = \int d\mathbf{x} \exp(-i\mathbf{p} \cdot \mathbf{x}/\hbar)\xi(\mathbf{x})$ is the 2D Fourier transform of the deviation. The averaging over the surface is done below with the use of the correlation function $\langle\langle \xi(\mathbf{p})\xi(\mathbf{p}') \rangle\rangle = (2\pi\hbar)^2 \delta(\mathbf{p} + \mathbf{p}')W(|\mathbf{p}|)$, introduced by analogy with Eq. (7.22). For concrete calculations, the function $W(|\mathbf{p}|)$ can be chosen Gaussian, $W(p) = \pi l_c^2 \bar{\xi}^2 \exp[-(pl_c/2\hbar)^2]$, where the correlation length l_c is introduced according to $\langle\langle \xi(\mathbf{x})\xi(\mathbf{x}') \rangle\rangle = \bar{\xi}^2 \exp[-(\mathbf{x} - \mathbf{x}')^2/l_c^2]$.

The solution of the problem defined by Eqs. (2) and (3) is written as a sum of the incident (−) and reflected (+) plane waves:

$$\Psi_{\mathbf{p}z} = \frac{1}{\sqrt{L}} \left[\psi_{\mathbf{p}p_\varepsilon}^{(-)} e^{-ip_\varepsilon z/\hbar} + \psi_{\mathbf{p}p_\varepsilon}^{(+)} e^{ip_\varepsilon z/\hbar} \right], \tag{4}$$

where $p_\varepsilon = \sqrt{2m\varepsilon - p^2}$ is the transverse momentum and L is the normalization length. The total energy ε is assumed to be greater than the kinetic energy of in-plane motion so that $p_\varepsilon^2 > 0$. The coefficients $\psi_{\mathbf{p}p_\varepsilon}^{(-)}$ and $\psi_{\mathbf{p}p_\varepsilon}^{(+)}$ are connected through the boundary condition (3), which is rewritten below as

$$\psi_{\mathbf{p}p_\varepsilon}^{(-)} + \psi_{\mathbf{p}p_\varepsilon}^{(+)} - \frac{i}{\hbar} \int \frac{d\mathbf{p}'}{(2\pi\hbar)^2} \xi(\mathbf{p} - \mathbf{p}')p_\varepsilon' \left(\psi_{\mathbf{p}'p_\varepsilon'}^{(-)} - \psi_{\mathbf{p}'p_\varepsilon'}^{(+)} \right) = 0. \tag{5}$$

To solve Eq. (5) by iterations, we use the approximation of weak scattering, when the amplitude $\psi_{\mathbf{p}p_\varepsilon}^{(+)}$ of the reflected wave is expressed through the amplitude $\psi_{\mathbf{p}p_\varepsilon}^{(-)}$ of the incident wave within the accuracy of the second order in the deviations:

$$\psi_{\mathbf{p}p_\varepsilon}^{(+)} \simeq -\psi_{\mathbf{p}p_\varepsilon}^{(-)} + \frac{i}{\hbar} \int \frac{d\mathbf{p}'}{(2\pi\hbar)^2} \xi(\mathbf{p} - \mathbf{p}')p_\varepsilon' 2\psi_{\mathbf{p}'p_\varepsilon'}^{(-)} \tag{6}$$

$$+ \frac{1}{\hbar^2} \int \frac{d\mathbf{p}'}{(2\pi\hbar)^2} \int \frac{d\mathbf{p}''}{(2\pi\hbar)^2} \xi(\mathbf{p} - \mathbf{p}')\xi(\mathbf{p}' - \mathbf{p}'')p_\varepsilon' p_\varepsilon'' 2\psi_{\mathbf{p}''p_\varepsilon''}^{(-)},$$

where $p'_\varepsilon = \sqrt{2m\varepsilon - p'^2}$ and $p''_\varepsilon = \sqrt{2m\varepsilon - p''^2}$. Equations (4) and (6) describe a solution of the second-order differential equation with a boundary condition at $z = 0$, and the amplitude $\psi^{(-)}_{\mathbf{p}p_\varepsilon}$ is to be determined from a boundary condition at $z \to \infty$.

According to the standard interpretation of quantum mechanics, the squared absolute values of the wave functions, $|\psi^{(-)}_{\mathbf{p}p_\varepsilon}|^2$ and $|\psi^{(+)}_{\mathbf{p}p_\varepsilon}|^2$, determine the densities of the incident and reflected electrons at the surface. To find a relation between the averaged densities of the incident and reflected electrons (the incident electron density is determined at $z \to \infty$ and does not depend on the properties of the rough surface), we write

$$\langle\langle|\psi^{(+)}_{\mathbf{p}p_\varepsilon}|^2\rangle\rangle \simeq |\psi^{(-)}_{\mathbf{p}p_\varepsilon}|^2 + \left(\frac{2}{\hbar}\right)^2 \int \frac{d\mathbf{p'}}{(2\pi\hbar)^2} \int \frac{d\mathbf{p''}}{(2\pi\hbar)^2} p'_\varepsilon p''_\varepsilon$$

$$\times\langle\langle\xi(\mathbf{p}-\mathbf{p'})\xi^*(\mathbf{p}-\mathbf{p''})\rangle\rangle\psi^{(-)}_{\mathbf{p'}p'_\varepsilon}\psi^{(-)*}_{\mathbf{p''}p''_\varepsilon} - \frac{2}{\hbar^2}\int \frac{d\mathbf{p'}}{(2\pi\hbar)^2}\int \frac{d\mathbf{p''}}{(2\pi\hbar)^2} p'_\varepsilon p''_\varepsilon$$

$$\times \left[\langle\langle\xi(\mathbf{p}-\mathbf{p'})\xi(\mathbf{p'}-\mathbf{p''})\rangle\rangle\psi^{(-)*}_{\mathbf{p}p_\varepsilon}\psi^{(-)}_{\mathbf{p''}p''_\varepsilon} + c.c.\right]. \qquad (7)$$

Note that the contributions linear in ξ have disappeared after the averaging. Taking into account that $\langle\langle\xi(\mathbf{p})\xi(\mathbf{p'})\rangle\rangle = (2\pi\hbar)^2\delta(\mathbf{p}+\mathbf{p'})W(|\mathbf{p}|)$, we obtain

$$\langle\langle|\psi^{(+)}_{\mathbf{p}p_\varepsilon}|^2\rangle\rangle \simeq \left[1 - \int \frac{d\mathbf{p'}}{(\pi\hbar)^2}\frac{p_\varepsilon p'_\varepsilon}{\hbar^2}W(|\mathbf{p}-\mathbf{p'}|)\right]|\psi^{(-)}_{\mathbf{p}p_\varepsilon}|^2$$

$$+ \int \frac{d\mathbf{p'}}{(\pi\hbar)^2}\frac{p'^2_\varepsilon}{\hbar^2}W(|\mathbf{p}-\mathbf{p'}|)|\psi^{(-)}_{\mathbf{p'}p'_\varepsilon}|^2. \qquad (8)$$

The factor $[1 - \ldots]$ on the right-hand side describes the departure of electrons from the state $(\mathbf{p}, \varepsilon)$, while the integral term describes the arrival of electrons to this state as a result of the scattering. The specular reflection is realized in the limiting case of ideal surface, when $\xi = 0$ and the amplitudes $|\psi^{(-)}_{\mathbf{p}p_\varepsilon}|^2$ and $|\psi^{(+)}_{\mathbf{p}p_\varepsilon}|^2$ are equal to each other since the 2D momentum is conserved.

The densities of the reflected (incident) electrons with momenta $\mathbf{p}, p_z$ $(\mathbf{p}, -p_z)$ are directly proportional to the occupation numbers $f(\mathbf{p}, p_z)$ $[f(\mathbf{p}, -p_z)]$ of these states at $z = 0$ and inversely proportional to the transverse component of the velocity. Thus, Eq. (8) can be rewritten as a boundary condition for the distribution function at the surface. Explicitly, we use the substitution

$$\langle\langle|\psi^{(\pm)}_{\mathbf{p}p_\varepsilon}|^2\rangle\rangle \propto \frac{f^{(\pm)}(\mathbf{p}, p_z)}{p_z} \qquad (9)$$

and assume that $p_z > 0$. It is convenient to use the variables $\mathbf{p}$ and p_z instead of $\mathbf{p}$ and ε, since the reflected waves (with transverse momentum

p_z) and the incident waves (with $-p_z$) are described by the distribution functions $f^{(+)}(\mathbf{p}, p_z)$ and $f^{(-)}(\mathbf{p}, p_z)$, respectively. Multiplying the equation connecting $f^{(+)}(\mathbf{p}, p_z)$ with $f^{(-)}(\mathbf{p}, p_z)$ by p_z, we obtain the following boundary condition:

$$f^{(+)}(\mathbf{p}, p_z) \simeq \left[1 - \int \frac{d\mathbf{p}'}{(\pi\hbar)^2} \frac{p_z p_z'}{\hbar^2} W(|\mathbf{p} - \mathbf{p}'|) \right] f^{(-)}(\mathbf{p}, p_z) \qquad (10)$$

$$+ \int \frac{d\mathbf{p}'}{(\pi\hbar)^2} \frac{p_z p_z'}{\hbar^2} W(|\mathbf{p} - \mathbf{p}'|) f^{(-)}(\mathbf{p}', p_z'),$$

where $p_z' = \sqrt{p_z^2 + p^2 - p'^2}$. One should mention that at $p_z \to 0$ the influence of the interface roughness can be neglected, i.e., the glancing electrons, for whom the reflection angles θ are close to $\pm\pi/2$, scatter specularly: $f^{(+)}(\mathbf{p}, p_z) = f^{(-)}(\mathbf{p}, p_z)$. The consideration given above is not, of course, a rigorous derivation of the boundary condition, and it does not give us validity conditions of Eq. (10). To formulate the problem rigorously, we introduce a Wigner density matrix in a half-space and derive the boundary conditions for it.

The one-electron density matrix in the $(\mathbf{p}, z)$-representation is

$$\rho_{\mathbf{p}\mathbf{p}'}(z, z') = \text{Sp}\hat{\eta}_t \hat{\Psi}^+_{\mathbf{p}'z'} \hat{\Psi}_{\mathbf{p}z} = \sum_{\delta\delta'} \Psi^\delta_{\mathbf{p}z} \Psi^{\delta'*}_{\mathbf{p}'z'} \rho_{\delta\delta'} , \qquad (11)$$

where $\text{Sp}\ldots$ denotes the averaging over all states (electron, phonon, etc.) of the system. The field operators $\hat{\Psi}_{\mathbf{p}z}$ and $\hat{\Psi}^+_{\mathbf{p}z}$ in Eq. (11) are expressed as $\sum_\delta \Psi^\delta_{\mathbf{p}z}\hat{a}_\delta$ and $\sum_\delta \Psi^{\delta*}_{\mathbf{p}z}\hat{a}^+_\delta$, respectively, and the definition $\rho_{\delta\delta'} = \text{Sp}\hat{\eta}_t\hat{a}^+_{\delta'}\hat{a}_\delta$ is applied; see Eq. (4.29). Taking into account the interface roughness scattering discussed above, we consider the wave function $\Psi^\delta_{\mathbf{p}z}$ as the eigenstate of the Schroedinger equation (2) corresponding to the eigenvalue $\varepsilon = \varepsilon_\delta$. The one-particle density matrix $\rho_{\delta\delta'}$ satisfies the quantum kinetic equation in the δ-state representation:

$$\frac{\partial \rho_{\delta\delta'}}{\partial t} + \frac{i}{\hbar}(\varepsilon_\delta - \varepsilon_{\delta'})\rho_{\delta\delta'} = J_{\delta\delta'} , \qquad (12)$$

where $J_{\delta\delta'} = \langle\delta|\hat{J}|\delta'\rangle$ is the collision integral describing the scattering of electrons by impurities, phonons, etc. Note that the properties of this scattering can be modified in the near-surface region.

Expressing $\Psi^\delta_{\mathbf{p}z}$ according to Eq. (4) with $p_\varepsilon = \sqrt{2m\varepsilon_\delta - p^2} \equiv p_z(\delta, p)$, we obtain

$$\rho_{\mathbf{p}\mathbf{p}'}(z, z') = \frac{1}{L} \sum_{\delta\delta'} \rho_{\delta\delta'} \sum_{rr'} \psi^{\delta(r)}_{\mathbf{p}p_z(\delta,p)} \psi^{\delta'(r')*}_{\mathbf{p}'p_z(\delta',p')}$$

$$\times e^{i[rp_z(\delta,p)z - r'p_z(\delta',p')z']/\hbar}, \qquad (13)$$

where $r, r' = \pm$ so that the sum over r and r' includes four terms. Equation (13) defines the density matrix in the mixed representation. It is convenient, however, to work with the density matrix $\rho_{\mathbf{pp'}}(p_z, p'_z)$ in the momentum representation. We write this matrix as $\rho_{\mathbf{pp'}}^{(rr')}(p_z, p'_z)$, where the arguments p_z and p'_z are assumed to be positive, while the directions of the momenta p_z and p'_z are determined by the indices r and r', respectively. The different representations are connected as

$$\rho_{\mathbf{pp'}}(z, z') = L \int_0^\infty \frac{dp_z}{2\pi\hbar} \int_0^\infty \frac{dp'_z}{2\pi\hbar} \sum_{rr'} \rho_{\mathbf{pp'}}^{(rr')}(p_z, p'_z) e^{i(rp_z z - ir' p'_z z')/\hbar} . \quad (14)$$

Comparing Eqs. (13) and (14), we get the expression

$$\rho_{\mathbf{pp'}}^{(rr')}(p_z, p'_z) = \left(\frac{2\pi\hbar}{L}\right)^2 \frac{p_z p'_z}{m^2} \sum_{\delta\delta'} \rho_{\delta\delta'} \delta[\varepsilon_\delta - (p^2 + p_z^2)/2m]$$

$$\times \delta[\varepsilon_{\delta'} - (p'^2 + p'^2_z)/2m] \psi_{\mathbf{p}p_z}^{\delta(r)} \psi_{\mathbf{p'}p'_z}^{\delta'(r')*} \quad (15)$$

with positive p_z and p'_z. To check this equation, one may substitute Eq. (15) into Eq. (14) and reduce the latter to Eq. (13) by integrating over p_z and p'_z. Let us use the boundary condition (3). The expansion of the wave function $\psi_{\mathbf{p}p_z}^{\delta(+)}$ in terms of $\psi_{\mathbf{p}p_z}^{\delta(-)}$ is obtained from Eq. (6) if we use the variable p_z instead of p_ε:

$$\psi_{\mathbf{p}p_z}^{\delta(+)} \simeq -\psi_{\mathbf{p}p_z}^{\delta(-)} + \frac{4i}{\hbar} \int \frac{d\mathbf{p}_1}{(2\pi\hbar)^2} \int_0^\infty dp_{1z} p_{1z}^2 \delta(2m\varepsilon_\delta - p_1^2 - p_{1z}^2) \quad (16)$$

$$\times \xi(\mathbf{p} - \mathbf{p}_1) \psi_{\mathbf{p}_1 p_{1z}}^{\delta(-)} + \frac{8}{\hbar^2} \int \frac{d\mathbf{p}_1}{(2\pi\hbar)^2} \int \frac{d\mathbf{p}_2}{(2\pi\hbar)^2} \int_0^\infty dp_{1z} \int_0^\infty dp_{2z} p_{1z}^2 p_{2z}^2$$

$$\times \delta(2m\varepsilon_\delta - p_1^2 - p_{1z}^2) \delta(2m\varepsilon_\delta - p_2^2 - p_{2z}^2) \xi(\mathbf{p} - \mathbf{p}_1) \xi(\mathbf{p}_1 - \mathbf{p}_2) \psi_{\mathbf{p}_2 p_{2z}}^{\delta(-)}.$$

We substitute this expansion into $\rho_{\mathbf{pp'}}^{(++)}(p_z, p'_z)$ of Eq. (15) and average the latter over the rough surface, taking into account the terms up to the second order in ξ. As a result, we obtain the integral equation connecting $f_{\mathbf{pp'}}^{(+)}(p_z, p'_z) = \langle\langle \rho_{\mathbf{pp'}}^{(++)}(p_z, p'_z)\rangle\rangle$ and $f_{\mathbf{pp'}}^{(-)}(p_z, p'_z) = \langle\langle \rho_{\mathbf{pp'}}^{(--)}(p_z, p'_z)\rangle\rangle$:

$$f_{\mathbf{pp'}}^{(+)}(p_z, p'_z) \simeq f_{\mathbf{pp'}}^{(-)}(p_z, p'_z) \left\{ 1 - \int \frac{d\mathbf{p}_1}{(\pi\hbar^2)^2} \int_0^\infty dp_{1z} p_{1z}^2 [p_z W(|\mathbf{p} - \mathbf{p}_1|) \right.$$

$$\times \delta(p^2 + p_z^2 - p_1^2 - p_{1z}^2) + p'_z W(\mathbf{p}' - \mathbf{p}_1) \delta(p'^2 + p'^2_z - p_1^2 - p_{1z}^2)] \Big\}$$

$$+ 4 \int \frac{d\mathbf{P}_1}{(\pi\hbar^2)^2} W(|\mathbf{P} - \mathbf{P}_1|) \int_0^\infty dp_{1z} \int_0^\infty dp'_{1z} p_z p'_z p_{1z} p'_{1z} \quad (17)$$

$$\times \delta(P_1^2 - P^2 + (\mathbf{P}_1 - \mathbf{P}) \cdot \Delta\mathbf{p} + p_{1z}^2 - p_z^2) \delta(P_1^2 - P^2 - (\mathbf{P}_1 - \mathbf{P}) \cdot \Delta\mathbf{p} + p_{1z}'^2 - p_z'^2)$$
$$\times f_{\mathbf{P}_1 + \Delta\mathbf{p}/2, \mathbf{P}_1 - \Delta\mathbf{p}/2}^{(-)}(p_{1z}, p_{1z}'),$$

where we have introduced the variables $\mathbf{P} = (\mathbf{p} + \mathbf{p}')/2$ and $\Delta\mathbf{p} = \mathbf{p} - \mathbf{p}'$ in the integral term. Let us transform Eq. (17) by using the Wigner distribution function

$$f_{\mathbf{px}}^{(\pm)}(p_z, z) = L \sum_{\Delta\mathbf{p}} \int_{-2p_z}^{2p_z} \frac{d\Delta p_z}{2\pi\hbar} \exp\left(\frac{i}{\hbar}\Delta\mathbf{p} \cdot \mathbf{x} + \frac{i}{\hbar}\Delta p_z z\right)$$
$$\times f_{\mathbf{p} + \Delta\mathbf{p}/2, \mathbf{p} - \Delta\mathbf{p}/2}^{(\pm)}(p_z + \Delta p_z/2, p_z - \Delta p_z/2), \tag{18}$$

where the limits of integration over Δp_z appear because of the requirement $p_z \pm \Delta p_z/2 > 0$. In the quasi-classical case, when $f_{\mathbf{px}}^{(\pm)}(p_z, z)$ weakly changes with z on the scale of $\hbar/p_z$, this integral converges at small Δp_z and the limits become insignificant. Applying the transformation (18) to Eq. (17), we assume that the characteristic spatial scales $\overline{|\mathbf{x}|}$ and $\overline{z}$ of the Wigner distribution functions $f_{\mathbf{px}}^{(\pm)}(p_z, z)$ are large in comparison to the longitudinal and transverse components of the electron wavelength:

$$\overline{|\mathbf{x}|} \gg \frac{\hbar}{|\mathbf{p}|}, \quad \overline{z} \gg \frac{\hbar}{p_z}. \tag{19}$$

This corresponds to the quasi-classical limit. The terms of the order of $\hbar^2 \partial^2 f_{\mathbf{px}}^{(-)}(p_z, z)/\partial\mathbf{x}^2$, $\hbar^2 \partial^2 f_{\mathbf{px}}^{(-)}(p_z, z)/\partial z^2$, and $\hbar^2 \partial^2 f_{\mathbf{px}}^{(-)}(p_z, z)/\partial\mathbf{x}\partial z$ should be neglected in comparison to the terms of the order of $(p^2 + p_z^2) f_{\mathbf{px}}^{(-)}(p_z, z)$. Therefore, we obtain

$$f_{\mathbf{px}}^{(+)}(p_z, z) \simeq \left[1 - \int \frac{d\mathbf{p}'}{(\pi\hbar^2)^2} p_z \sqrt{p_z^2 + p^2 - p'^2} W(|\mathbf{p} - \mathbf{p}'|)\right] f_{\mathbf{px}}^{(-)}(p_z, z)$$
$$+ \int \frac{d\mathbf{p}'}{(\pi\hbar^2)^2} p_z^2 W(|\mathbf{p} - \mathbf{p}'|) L \int_{-2p_z}^{2p_z} \frac{d\Delta p_z}{2\pi\hbar} e^{i\Delta p_z z/\hbar} \tag{20}$$
$$\times f_{\mathbf{p'x}}^{(-)}\left(\sqrt{p^2 - p'^2 + \left(p_z + \frac{\Delta p_z}{2}\right)^2}, \sqrt{p^2 - p'^2 + \left(p_z - \frac{\Delta p_z}{2}\right)^2}\right).$$

The distribution function in the integral term appears as a result of the Wigner transformation in the plane (x, y):

$$f_{\mathbf{px}}^{(\pm)}(p_{1z}, p_{2z}) = \sum_{\Delta\mathbf{p}} e^{i\Delta\mathbf{p} \cdot \mathbf{x}/\hbar} f_{\mathbf{p} + \Delta\mathbf{p}/2, \mathbf{p} - \Delta\mathbf{p}/2}^{(\pm)}(p_{1z}, p_{2z}). \tag{21}$$

One can substitute $p_z' = \sqrt{p_z^2 + p^2 - p'^2}$ in Eq. (20). Since the functions $f^{(\pm)}(\mathbf{p}, p_z)$ standing in Eq. (10) are identified with $f_{\mathbf{px}}^{(\pm)}(p_z, z)$ at $z = 0$, we see that Eq. (20) already resembles Eq. (10). Finally, we need to check the equivalence of the integral terms in Eqs. (10) and (20). Taking into account the second strong inequality of Eq. (19), we find that the integral over Δp_z in Eq. (20) converges at $\Delta p_z \ll p_z, p_z'$. Therefore, this integral can be approximated as

$$L \int_{-\infty}^{\infty} \frac{d\Delta p_z}{2\pi\hbar} e^{i\Delta p_z z/\hbar} f_{\mathbf{p'x}}^{(-)} \left(p_z' + \frac{p_z \Delta p_z}{2p_z'}, p_z' - \frac{p_z \Delta p_z}{2p_z'} \right) \qquad (22)$$

and estimated as $(p_z'/p_z) f_{\mathbf{p'x}}^{(-)} [p_z', (p_z'/p_z)z]$ after introducing a new variable of integration, $(p_z/p_z')\Delta p_z$, instead of Δp_z. The discussion presented above shows that Eq. (20) at $z = 0$ is reduced to Eq. (10). The latter, therefore, corresponds to the quasi-classical limit of the general boundary condition derived from the quantum kinetic consideration.

45. Size-Dependent Conductivity

According to the boundary condition (44.10), the collisions of electrons with non-ideal surfaces change the longitudinal momenta of the electrons. Such processes contribute to the resistance of thin films, whose widths are smaller than or comparable to the mean free path length, the latter is determined by various scattering mechanisms in the bulk; see Secs. 8, 11, and 35. The non-linear dependence of the conductivity on the width of the film is called the size effect. The frequency and magnetic-field dependences of the conductivity of thin films are also modified in comparison to those in bulk media. Below we consider such effects by using the quasi-classical kinetic equation derived in Sec. 9, together with the boundary conditions derived in the previous section.

We consider a linear response to the perturbation $\mathbf{E}e^{-i\omega t}$, assuming that the electric field $\mathbf{E}$ is applied in the plane of the film (XOY) and the magnetic field $\mathbf{H}$ is directed perpendicular to this plane. The film of width d is placed within $0 < z < d$ and assumed to be macroscopically homogeneous in the plane. The linearized kinetic equation is obtained from the general quasi-classical equation (9.34) in the form

$$\left(-i\omega + v_z \frac{\partial}{\partial z} + [\boldsymbol{\omega}_c \times \mathbf{p}] \cdot \nabla_{\mathbf{p}} - \widehat{J}_c \right) \Delta f_{z\mathbf{p}} = e\mathbf{E} \cdot \mathbf{v}\delta(\varepsilon_F - \varepsilon_p), \qquad (1)$$

where $\Delta f_{z\mathbf{p}}$ is the non-equilibrium part of the distribution function depending on the transverse coordinate z and 3D momentum $\mathbf{p}$. Next, v_z is the transverse component of the velocity $\mathbf{v} = \mathbf{p}/m$ and $\boldsymbol{\omega}_c = |e|\mathbf{H}/mc$ is the cyclotron frequency vector introduced in Sec. 11. The action of the operator $\widehat{J}_c$ on $\Delta f_{z\mathbf{p}}$ defines the linearized collision integral. The right-hand side of Eq. (1) is written in the zero-temperature approximation, since we consider a degenerate electron gas with Fermi energy

ε_F. The presence of two boundaries at $z = 0$ and $z = d$ implies two boundary conditions for $f_{z\mathbf{p}}$. These boundary conditions directly follow from Eq. (44.10) and can be rewritten in the same form for $\Delta f_{z\mathbf{p}}$, because of their linearity. For the sake of simplicity, we assume that the surface at $z = d$ is ideal, while the roughness of the surface at $z = 0$ is described by a small correlation length. For this case, $W(|\mathbf{p} - \mathbf{p}'|)$ can be approximated by a constant, and the integral term on the right-hand side of Eq. (44.10) vanishes when the distribution function is antisymmetric in $\mathbf{p}$. Thus, the boundary conditions are rewritten as algebraic equations

$$\Delta f|_{p_z>0} = (1 - P_\theta)\Delta f|_{p_z<0}, \quad z = 0,$$

$$\Delta f|_{p_z>0} = \Delta f|_{p_z<0}, \quad z = d, \qquad (2)$$

where P_θ is the diffusivity of scattering, which depends on the angle θ as $P_\theta = P_0|\cos\theta|$, according to Eq. (44.10).

One can represent the non-equilibrium part of the distribution function as

$$\Delta f_{z\mathbf{p}} = eEv_F\delta(\varepsilon_F - \varepsilon_p)\chi_{z\varphi p_z}, \qquad (3)$$

where v_F is the Fermi velocity determined by the electron density n. The function $\chi_{z\varphi p_z}$ depends on the coordinate z and on the angles φ and θ, the dependence on θ enters through p_z. Using the identity $[\boldsymbol{\omega}_c \times \mathbf{p}]\cdot\nabla_\mathbf{p} = \omega_c\partial/\partial\varphi$ (problem 9.2), we rewrite Eq. (1) as an equation for χ:

$$\left(-i\omega + v_z\frac{\partial}{\partial z} + \omega_c\frac{\partial}{\partial\varphi} - \widehat{J}_c\right)\chi_{z\varphi p_z} = (e_x\cos\varphi + e_y\sin\varphi)\sin\theta, \quad (4)$$

where e_x and e_y are the components of the unit vector in the direction of the electric field. Since the problem is periodic in φ, i.e., $\chi_{z,\varphi+2\pi,p_z} = \chi_{z\varphi p_z}$, one can search for $\chi_{z\varphi p_z}$ in the form

$$\chi_{z\varphi p_z} = \chi_{zp_z}^{(+)}e^{i\varphi} + \chi_{zp_z}^{(-)}e^{-i\varphi}. \qquad (5)$$

The functions $\chi_{zp_z}^{(\pm)}$ satisfy the ordinary differential equations of the first order:

$$\left[-i(\omega \mp \omega_c) + v_z\frac{\partial}{\partial z} + \nu_F\right]\chi_{zp_z}^{(\pm)} = \frac{e_x \mp ie_y}{2}\sin\theta, \qquad (6)$$

where the integral operator $-\widehat{J}_c$ describing the scattering in the bulk is replaced by the momentum relaxation rate ν_F on the Fermi surface. This substitution is rigorously justified in Sec. 8 for the case of elastic scattering. We note, however, that in the case of a strongly inhomogeneous system, when the symmetric part of the distribution function appears to be significant, this substitution becomes invalid and can be used for

estimates only. The boundary conditions to Eq. (6) are obtained from
Eq. (2):

$$\chi^{(\pm)}_{z=0,p_z} = (1 - P_\theta)\chi^{(\pm)}_{z=0,-p_z}, \quad \chi^{(\pm)}_{z=d,p_z} = \chi^{(\pm)}_{z=d,-p_z}. \tag{7}$$

The general solution of the inhomogeneous differential equation (6) is

$$\chi^{(\pm)}_{zp_z} = \frac{(e_x \mp i e_y)\sin\theta}{2\left[\nu_F - i(\omega \mp \omega_c)\right]}\left[1 + \mathcal{F}^{(\pm)}\exp\left\{-\frac{z}{v_z}\left[\nu_F - i(\omega \mp \omega_c)\right]\right\}\right]. \tag{8}$$

The second term in this expression satisfies Eq. (6) with zero right-
hand side. It depends on z exponentially and oscillates at non-zero ω
or $\mathbf{H}$. The first term, which appears due to the inhomogeneous part of
Eq. (6), is chosen to be z-independent. The coefficients $\mathcal{F}^{(\pm)}$ are to be
determined from the boundary conditions (7). The solution given by
Eq. (8) determines the distribution of the current density in the film,
$\mathbf{I_r} = \mathbf{I}_z$, in the usual way; see Eq. (9.11). Integrating the current density
over the transverse coordinate according to $\mathbf{I}_\square = \int_0^d dz \mathbf{I}_z$, we obtain

$$\mathbf{I}_\square = \frac{2e}{m}\int_0^d dz \int \frac{d\mathbf{p}}{(2\pi\hbar)^3}\mathbf{p}\Delta f_{z\mathbf{p}} = \hat{\sigma}^\square \mathbf{E}, \tag{9}$$

where the second equation defines the tensor of sheet conductivity. Note
that this conductivity is equal to the conductance of a square-shaped
film (the conductance does not depend on the size of such a film). The
explicit expressions for the components of the conductivity tensor are
obtained after substituting $\Delta f_{z\mathbf{p}}$ defined by Eqs. (3), (5), and (8) into
Eq. (9). The diagonal components $\sigma^\square_{xx} = \sigma^\square_{yy} \equiv \sigma^\square_d$ are given by

$$\sigma^\square_d = \frac{3e^2 n}{4m}\int_0^d dz \int_{-1}^1 d\zeta(1 - \zeta^2)\left\{\frac{1 + \mathcal{F}^{(+)}_\zeta\exp\left[-z\frac{\nu_F - i(\omega - \omega_c)}{v_F\zeta}\right]}{2\left[\nu_F - i(\omega - \omega_c)\right]}\right.$$

$$\left. + \frac{1 + \mathcal{F}^{(-)}_\zeta\exp\left[-z\frac{\nu_F - i(\omega + \omega_c)}{v_F\zeta}\right]}{2\left[\nu_F - i(\omega + \omega_c)\right]}\right\}, \tag{10}$$

while for the non-diagonal components $\sigma^\square_{yx} = -\sigma^\square_{xy} \equiv \sigma^\square_\perp$ we find

$$\sigma^\square_\perp = i\frac{3e^2 n}{4m}\int_0^d dz \int_{-1}^1 d\zeta(1 - \zeta^2)\left\{\frac{1 + \mathcal{F}^{(+)}_\zeta\exp\left[-z\frac{\nu_F - i(\omega - \omega_c)}{v_F\zeta}\right]}{2[\nu_F - i(\omega - \omega_c)]}\right.$$

$$\left. - \frac{1 + \mathcal{F}^{(-)}_\zeta\exp\left[-z\frac{\nu_F - i(\omega + \omega_c)}{v_F\zeta}\right]}{2[\nu_F - i(\omega + \omega_c)]}\right\}. \tag{11}$$

To obtain these expressions, we have calculated the integrals over the angle φ determining the symmetry of the conductivity tensor and over the absolute value of $\mathbf{p}$ by using the δ-function of Eq. (3). Instead of θ, we have introduced the variable $\zeta = \cos\theta$.

The coefficients $\mathcal{F}_\zeta^{(\pm)}$ are discontinuous functions of ζ. For this reason, the integrals in Eqs. (10) and (11) should be calculated separately in the regions $\zeta < 0$ and $\zeta > 0$. Using Eq. (8), we rewrite the boundary conditions (7) as linear equations for these coefficients:

$$\mathcal{F}_{\zeta>0}^{(\pm)} \exp\left\{-\frac{d\left[\nu_F - i(\omega \mp \omega_c)\right]}{v_F|\zeta|}\right\} = \mathcal{F}_{\zeta<0}^{(\pm)} \exp\left\{\frac{d\left[\nu_F - i(\omega \mp \omega_c)\right]}{v_F|\zeta|}\right\},$$

$$1 + \mathcal{F}_{\zeta>0}^{(\pm)} = (1 - P_\zeta)(1 + \mathcal{F}_{\zeta<0}^{(\pm)}). \tag{12}$$

The solutions of Eq. (12) are

$$\mathcal{F}_{\zeta>0}^{(\pm)} = \frac{-P_\zeta}{1 - (1 - P_\zeta)\exp\left\{-2d\left[\nu_F - i(\omega \mp \omega_c)\right]/v_F|\zeta|\right\}},$$

$$\mathcal{F}_{\zeta<0}^{(\pm)} = \frac{-P_\zeta \exp\left(-2d\left[\nu_F - i(\omega \mp \omega_c)\right]/v_F|\zeta|\right)}{1 - (1 - P_\zeta)\exp\left\{-2d\left[\nu_F - i(\omega \mp \omega_c)\right]/v_F|\zeta|\right\}}, \tag{13}$$

and their substitution into Eqs. (10) and (11) solves the problem after calculating the integrals over z and ζ. The integrals over z are evaluated easily, and we obtain

$$\sigma_d^\square = \sigma_0 d(A_+ + A_-)/2, \quad \sigma_\perp^\square = -i\sigma_0 d(A_+ - A_-)/2, \tag{14}$$

where $\sigma_0 = e^2 n/m\nu_F$,

$$A_\pm = r_\pm - r_\pm^2 \frac{3l_F}{2d}\int_0^1 d\zeta\ \zeta(1 - \zeta^2)\left(\frac{2}{P_\zeta} + \coth\frac{d}{r_\pm l_F \zeta} - 1\right)^{-1}, \tag{15}$$

$l_F = v_F/\nu_F$ is the mean free path length, and

$$r_\pm = [1 - i(\omega \pm \omega_c)/\nu_F]^{-1}. \tag{16}$$

One can see that the first terms in A_+ and A_- describe the bulk conductivity, see Eq. (11.18), while the second (integral) terms describe the surface-induced contribution. In order to calculate the integral in Eq. (15), one should specify P_ζ. Below we use $P_\zeta = P_0\zeta$. This form is valid in the limit of small correlation lengths, when the algebraic boundary conditions (2) are justified, and P_0 estimated according to Eq. (44.10) appears to be much smaller than unity (problem 9.3). We stress, however, that the conditions (2) are often considered as phenomenological boundary conditions, where P_ζ is a physically reasonable function. Therefore, we first present some results valid for arbitrary P_ζ.

In the limit of thick films, when $d \gg l_F$, the factor $\coth x - 1$ in the denominator of Eq. (15) should be neglected, and the integral over ζ is calculated with the following result:

$$A_{\pm} = r_{\pm}\left(1 - \frac{\beta l_F}{d}r_{\pm}\right), \quad \beta = \frac{3}{4}\int_0^1 d\zeta P_\zeta \zeta(1 - \zeta^2). \tag{17}$$

Substituting these A_+ and A_- into Eq. (14), one can see that the surface gives a small correction to the bulk conductivity. In the case of $P_\zeta = P_0\zeta$, the constant β is equal to $P_0/10$.

In the limit of thin films, $d \ll l_F$, we consider two cases. The first one corresponds to moderate frequencies and/or magnetic fields, $\omega, \omega_c \ll \tau_d^{-1}$, where $\tau_d = d/v_F$ is the time of passage of the electron with Fermi velocity across the film. In this case one still can have $\omega, \omega_c \gg \nu_F$. Since $d \ll |r_{\pm}|l_F$, we replace $\coth x$ by $1/x$ in Eq. (15). Next, using the identity $(3/2)\int_0^1 d\zeta(1 - \zeta^2) = 1$, we replace the first term in $A_{\pm}$ by $(3r_{\pm}/2)\int_0^1 d\zeta(1 - \zeta^2)$, unify it with the second term, and obtain

$$A_{\pm} = \frac{3d}{2l_F}\int_0^1 d\zeta \frac{(1 - \zeta^2)}{\zeta P_\zeta/(2 - P_\zeta) + d/l_F r_{\pm}}. \tag{18}$$

The result depends on P_ζ and $d/l_F r_{\pm}$. In the case of $P_\zeta = P_0\zeta$,

$$A_{\pm} = \frac{3d}{P_0 l_F}\left[(B_{\pm} + B_{\pm}^{-1})\arctan B_{\pm} - 1\right], \quad B_{\pm} = \sqrt{\frac{P_0 l_F}{2d}r_{\pm}}. \tag{19}$$

At non-zero ω and/or ω_c, the dimensionless functions $B_{\pm}$ are complex. Since $P_0 \ll 1$, the absolute values of these functions are not necessarily large in comparison to unity. If they are large, $A_{\pm}$ takes a simpler form:

$$A_{\pm} = \frac{3\pi}{4}\sqrt{\frac{2dr_{\pm}}{l_F P_0}}. \tag{20}$$

If $\omega = \omega_c = 0$, Eq. (19) is valid for an arbitrary ratio d/l_F. This property follows from the fact that the term $\coth(d/l_F\zeta)$ in Eq. (15) contributes to the denominator only when this term is large in comparison to 1 (because $2/P_\zeta \gg 1$), i.e., when the expansion $\coth(d/l_F\zeta) \simeq l_F\zeta/d$ is valid.

The second case corresponds to intermediate and large frequencies and/or magnetic fields, when $\omega, \omega_c \sim \tau_d^{-1}$ and $\omega, \omega_c \gg \tau_d^{-1}$, respectively. Since we consider thin films, this necessarily implies $\omega, \omega_c \gg \nu_F$. Calculating the integrals over ζ in Eq. (15), we take into account only the oscillating part of the exponents in $\coth x$ and obtain

$$A_{\pm} = r_{\pm} - r_{\pm}^2\frac{3l_F}{2d}\int_0^1 d\zeta \frac{\zeta(1 - \zeta^2)}{2/P_\zeta - 1 + i\cot[(\omega \pm \omega_c)\tau_d/\zeta]}. \tag{21}$$

The integral in Eq. (21) is taken analytically in the limit of large frequencies or magnetic fields, when one can average the fast-oscillating term under the integral over the period of the oscillations. This averaging gives us $[2/P_\zeta - 1 + i \cot y]^{-1} \to P_\zeta/2$. Therefore, $A_\pm$ is again given by Eq. (17), where $r_\pm$ is written at $\omega, \omega_c \gg \nu_F$:

$$A_\pm \simeq \frac{i\nu_F}{\omega \pm \omega_c} + \frac{\nu_F^2}{(\omega \pm \omega_c)^2}\left(1 + \frac{\beta l_F}{d}\right). \tag{22}$$

Let us apply the above analysis for describing the conductivity in some limiting cases, always assuming that $P_\zeta = P_0\zeta$ and $P_0 \ll 1$. The static conductivity at zero magnetic field depends on a single dimensionless parameter, the ratio of the film width to the mean free path length. Using Eqs. (14), (17), and (20) at $\omega = \omega_c = 0$, we obtain

$$\sigma_d^\square = \sigma_0 d \begin{cases} (1 - l_F P_0/10d), & d \gg l_F P_0 \\ (3\pi/4)\sqrt{2d/l_F P_0}, & d \ll l_F P_0 \end{cases}. \tag{23}$$

Both these limits can also be derived by substituting the result (19), which is generally applicable at $P_0 \ll 1$ and $\omega = \omega_c = 0$, into Eq. (14). In the limit of thin films, the conductivity decreases considerably due to the surface scattering.

Now let us consider the frequency dispersion of the conductivity of thin films $(d \ll l_F)$ in zero magnetic field and at $\omega \gg \nu_F$. For moderate frequencies, under the conditions $\nu_F \ll \omega \ll P_0/\tau_d$, we use Eqs. (14) and (20) and obtain

$$\sigma_d^\square = (1 + i)\frac{3\pi}{4}\frac{e^2 nd}{m}\sqrt{\frac{\tau_d}{\omega P_0}}. \tag{24}$$

In this limit, the frequency dispersion of the conductivity essentially differs from that of a bulk sample. We have $\sigma_d^\square \propto \omega^{-1/2}$ for both real and imaginary parts. For large frequencies, we use Eqs. (14) and (22) at $\omega_c = 0$ and find that, apart from the imaginary part $\text{Im } \sigma_d^\square = e^2 nd/m\omega$, the conductivity has a small real part containing the contributions of both bulk and surface scattering:

$$\text{Re } \sigma_d^\square = \frac{e^2 nd}{m\omega^2}\left(\nu_F + \frac{P_0}{10\tau_d}\right). \tag{25}$$

Since Eq. (22) is derived in the limit $v_F/d \gg \nu_F$, the surface scattering can give the main contribution even at $P_0 \ll 1$. For the intermediate frequencies, $\omega \sim \tau_d^{-1}$, we use Eqs. (14) and (21) leading to

$$\sigma_d^\square = \frac{e^2 nd}{m\omega}[i + F(\omega\tau_d)] + \frac{e^2 nd\nu_F}{m\omega^2},$$

$$F(y) = \frac{3}{2y} \int_0^1 d\zeta \frac{\zeta^2(1-\zeta^2)}{2/P_0 + i\zeta \cot(y/\zeta)}. \qquad (26)$$

In this case, the oscillations of $\cot(y/\zeta)$ cause the oscillations of the conductivity as a function of ω. The period of such oscillations is of the order of τ_d. At $P_0 \ll 1$, however, these oscillations are weak.

Consider now the components of the static conductivity tensor ($\omega = 0$) in the presence of a magnetic field. We restrict ourselves by the limit of thick films, when $d \gg l_F$. Using Eqs. (14) and (17), we have

$$\sigma_d^{\square} = \frac{\sigma_0 d}{1 + \omega_c^2/\nu_F^2} \left(1 - \frac{\beta l_F}{d} \frac{1 - \omega_c^2/\nu_F^2}{1 + \omega_c^2/\nu_F^2}\right),$$

$$\sigma_\perp^{\square} = \frac{\sigma_0 d\omega_c/\nu_F}{1 + \omega_c^2/\nu_F^2} \left(1 - \frac{\beta l_F}{d} \frac{2}{1 + \omega_c^2/\nu_F^2}\right). \qquad (27)$$

These expressions describe a weak negative magnetoresistance

$$\frac{\rho_d(H)}{\rho_d(0)} - 1 = -2 \left(\frac{\beta l_F}{d}\right)^2 \frac{\omega_c^2}{\nu_F^2 + \omega_c^2}, \qquad (28)$$

which decreases at $\omega_c \ll \nu_F$ as $-H^2$ and saturates at classically strong magnetic fields, when $\omega_c \gg \nu_F$. The appearance of a finite classical magnetoresistance can be qualitatively explained by the presence of two groups of carriers with different effective relaxation times: the first is the electrons moving perpendicular to the surface, and the other is the glancing electrons. The negative sign of the magnetoresistance is explained by a decrease in the number of electrons scattering on the surface as the magnetic field increases. For the same reasons, the Hall coefficient $R_H = \rho_{xy}(H)/H$ deviates from its classical value $1/|e|nc$ and depends on the magnetic field according to

$$\frac{R_H(H)}{R_H(0)} - 1 = 2 \left(\frac{\beta l_F}{d}\right)^3 \frac{\omega_c^2}{\nu_F^2 + \omega_c^2}. \qquad (29)$$

We remind that since $P_\zeta = P_0\zeta$, one should put $\beta = P_0/10$ in Eqs. (27)−(29). In the limit of thin films, $d \ll l_F$, the qualitative behavior of $\sigma_d^{\square}$ and $\sigma_\perp^{\square}$ is similar to those for the real and imaginary parts of the frequency-dependent conductivity obtained above. If $\omega_c \sim \tau_d^{-1}$, the components of the conductivity tensor oscillate as functions of the magnetic field.

If one assumes P_ζ other than $P_0\zeta$, the results given by Eqs. (23)−(26) are modified. For example, in the case of pure diffuse reflection, when $P_\zeta = 1$, the static conductivity in the limit of thin films, $d \ll l_F$, is proportional to $(d/l_F)\ln(l_F/d)$ (problem 9.4). Nevertheless, any choice of the function P_ζ does not affect the main qualitative features discussed above, such as the decrease of the conductivity with decreasing d, the modification of the frequency dispersion of the conductivity, the negative magnetoresistance, and the oscillations of the resistance as a function of both frequency and magnetic field. Finally, we remind that the validity of the algebraic boundary conditions (2) is limited, and one should generally describe the size effect by using the integral equation (44.10).

46. Thermal Conductivity of Bounded Insulators

According to Chapter 5, the low-temperature thermal conductivity of insulators increases exponentially as the temperature goes down. This is true under the assumption that the heat transfer is limited by the phonon-phonon interaction. At low enough temperatures, however, the thermal conductivity is observed to decrease with decreasing temperature in a power-law fashion. In clean samples, where the scattering of phonons by lattice imperfections can be neglected, this decrease is explained by diffuse scattering of phonons on rough surfaces. This mechanism appears to be rather efficient, because the mean free path lengths of phonons, even those associated with the normal collision processes, are fairly large and can exceed the transverse size of the sample.

To study the effect of boundaries, we introduce an algebraic boundary condition for the phonon distribution function $N_{\mathbf{rq}}^l$ at a non-ideal surface Γ in a similar way as it was done for the electrons in the previous section:

$$N_{\mathbf{rq}}^l|_{v_{ln}>0} = (1 - P_l)N_{\mathbf{rq}}^l|_{v_{ln}<0} , \quad \mathbf{r} \in \Gamma, \tag{1}$$

where P_l is the diffusivity of scattering for the l-th phonon mode and v_{ln} is the component of the phonon group velocity normal to the surface. The case of $v_{ln} > 0$ corresponds to reflected phonons. A microscopic justification of Eq. (1) is rather complicated. It requires, at least, a number of crude approximations such as conservation of the energy, phonon number, and mode index in the reflection processes. Below we treat Eq. (1) as a phenomenological boundary condition and are interested mostly in the case of pure diffuse reflection, when $P_l = 1$. Since the case of low temperatures is implied, we consider only three low-frequency acoustic phonon modes: $l = LA$, TA_1, and TA_2. The dispersion of the modes is assumed to be isotropic. It is described by the longitudinal and transverse phonon velocities, s_l and s_t. We consider the samples shaped as long rods, assuming that the length of the rod is large in comparison to its transverse size, and the modifications of heat transfer near the ends of the rod do not affect the energy flux through the cross-section of the sample. This flux is characterized by the integral of the energy flow density (19.32) over the surface of the cross-section

C:

$$\overline{G} = \sum_l \int \frac{d\mathbf{q}}{(2\pi)^3} \frac{1}{s_c} \int_{(C)} d\mathbf{s}_c \cdot \frac{\partial \omega_{ql}}{\partial \mathbf{q}} \hbar \omega_{ql} \Delta N_{\mathbf{rq}}^l \,, \qquad (2)$$

where $d\mathbf{s}_c$ is the differential of the square of this surface (the vector $\mathbf{s}_c$ is directed perpendicular to the surface) and $\Delta N_{\mathbf{rq}}^l$ is the non-equilibrium part of the phonon distribution function. Since we have divided the result by the square s_c of the cross-section, $\overline{G}$ is the averaged energy flow density along the rod. The kinetic equation (23.19), in the stationary regime and in the linear approximation with respect to the thermal gradient ∇T, is written as

$$\frac{\partial \omega_{ql}}{\partial \mathbf{q}} \cdot \frac{\partial \Delta N_{\mathbf{rq}}^l}{\partial \mathbf{r}} + \frac{\partial N_{l\mathbf{q}}^{(eq)}}{\partial T} \frac{\partial \omega_{ql}}{\partial \mathbf{q}} \cdot \nabla T = \Delta J_{ph,ph}(N|\mathbf{rq}l). \qquad (3)$$

In contrast to Eq. (24.2), it contains the term associated with coordinate dependence of $\Delta N_{\mathbf{rq}}^l$. The linearized collision integral depends on $\mathbf{r}$ through $\Delta N_{\mathbf{rq}}^l$. In the case of pure diffuse reflection, Eq. (1) gives us a simple boundary condition to this equation, $\Delta N_{\mathbf{rq}}^l|_{v_{ln}>0} = 0$ at the non-ideal surface Γ. If the characteristic transverse size of the sample is much smaller than the mean free path for normal collision processes, the collision integral on the right-hand side can be neglected, and Eq. (3) is solved exactly:

$$\Delta N_{\mathbf{rq}}^l = -\frac{\partial N_{l\mathbf{q}}^{(eq)}}{\partial T}(\mathbf{r} - \mathbf{r}_\Gamma) \cdot \nabla T \,. \qquad (4)$$

The constant of integration, $\mathbf{r}_\Gamma$, is dictated by the boundary conditions. It is the coordinate of the point on the surface Γ where the last reflection has occurred. This constant depends on the surface geometry, coordinate $\mathbf{r}$, and direction of the group velocity (the direction is given by the unit vector $\mathbf{q}/q$ in the isotropic approximation we use). To determine $\mathbf{r}_\Gamma$ for given $\mathbf{r}$ and $\mathbf{q}$, one should draw a straight line passing through the point $\mathbf{r}$ in the direction opposite to $\mathbf{q}$ until this line intersects with the surface; the point of the intersection is $\mathbf{r}_\Gamma$. The distribution function given by Eq. (4) is zero at the surface for the phonons with wave vectors $\mathbf{q}$ directed inside the sample. Substituting $\Delta N_{\mathbf{rq}}^l$ from Eq. (4) into Eq. (2), we find

$$\overline{G} = -\sum_l \int \frac{d\mathbf{q}}{(2\pi)^3} \frac{1}{s_c} \int_{(C)} d\mathbf{s}_c \cdot \mathbf{q} \frac{\hbar \omega_{ql}^2}{q^2} \frac{\partial N_{l\mathbf{q}}^{(eq)}}{\partial T}$$

$$\times (\mathbf{r} - \mathbf{r}_\Gamma) \cdot \nabla T = -\kappa \nabla T \,. \qquad (5)$$

The thermal conductivity κ is introduced here by taking into account that the vector $\mathbf{r} - \mathbf{r}_\Gamma$ is directed along $\mathbf{q}/q$. This thermal conductivity is given by the expression

$$\kappa = \frac{1}{s_c} \int_{(C)} ds_c \sum_l \int \frac{d\mathbf{q}}{(2\pi)^3} \frac{\partial N_{l\mathbf{q}}^{(eq)}}{\partial T} \frac{\hbar \omega_{ql}^2}{q} |\mathbf{r} - \mathbf{r}_\Gamma| \cos^2 \theta, \qquad (6)$$

where ds_c is the differential of the square of the cross-section perpendicular to the thermal gradient along the axis of the rod and θ is the angle between the vector $\mathbf{q}$ and the thermal gradient.

Transforming $\int d\mathbf{q} \ldots$ in Eq. (6) as $\int_0^\infty q^2 dq \int d\widetilde{\Omega} \ldots$, where $d\widetilde{\Omega}$ is the differential of the solid angle of $\mathbf{q}$, and taking into account that $\mathbf{r}_\Gamma$ is independent of both $q = |\mathbf{q}|$ and l, we rewrite this equation in the form

$$\kappa = \frac{1}{3} C \Lambda \bar{v}. \qquad (7)$$

In Eq. (7) we have introduced the specific heat $C = 3S$, where the entropy S is given by Eq. (24.17), the characteristic velocity

$$\bar{v} = \frac{s_l^{-2} + 2s_t^{-2}}{s_l^{-3} + 2s_t^{-3}}, \qquad (8)$$

and the characteristic length

$$\Lambda = \frac{3}{4\pi s_c} \int_{(C)} ds_c \int d\widetilde{\Omega} |\mathbf{r} - \mathbf{r}_\Gamma| \cos^2 \theta, \qquad (9)$$

which is entirely determined by the geometry of the sample and approximately equal to its transverse size. For a cylindrical sample of radius R, one has $\Lambda = 2R$ (problem 9.5). Equation (7) resembles the expression for the thermal conductivity of a gas of particles with mean free path length Λ. In the collisionless limit studied here, it is the transverse size of the sample that plays the role of the mean free path length. We stress that an estimate for the thermal conductivity of a bulk sample is also represented in the form of Eq. (7), where Λ should be treated as the mean free path length for umklapp scattering, $l_U = \bar{s}\tau_U$; see Eq. (24.19). Equation (7) can be improved to account for partly specular scattering, when P_l in Eq. (1) is not equal to 1. For the sake of simplicity, we assume that P_l does not depend on the angle of incidence and mode index, $P_l = P$. It is easy to find that Eq. (4) should be modified as

$$\Delta N_{\mathbf{rq}}^l = -\frac{\partial N_{l\mathbf{q}}^{(eq)}}{\partial T} P[(\mathbf{r} - \mathbf{r}_\Gamma) + (1 - P)(\mathbf{r} - \mathbf{r}_\Gamma')$$

$$+ (1 - P)^2 (\mathbf{r} - \mathbf{r}_\Gamma'') + \ldots] \cdot \nabla T, \qquad (10)$$

where $\mathbf{r}_\Gamma'$, $\mathbf{r}_\Gamma''$, ... are the coordinates of the points on the surface at which

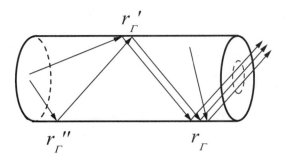

Figure 9.2. Multiple reflections of phonons in a cylinder.

we should arrive if we continued back beyond $\mathbf{r}_\Gamma$ to where the specularly reflected "particle" previously left the surface, and so on (Fig. 9.2). The extra terms represent the contributions of multiple reflections of various orders to the energy flow. To check that the function (10) satisfies Eq. (3) (without the right-hand side), we notice that $P[1 + (1 - P) + (1 - P)^2 + \ldots] = 1$. To check that it satisfies the boundary condition (1), we consider $\Delta N_{\mathbf{rq}}^l$ at $\mathbf{r} = \mathbf{r}_\Gamma$. For the phonons reflected at $\mathbf{r}_\Gamma$, the factor in the square brackets in Eq. (10) is equal to $(1 - P)[(\mathbf{r}_\Gamma - \mathbf{r}_\Gamma') + (1 - P)(\mathbf{r}_\Gamma - \mathbf{r}_\Gamma'') + \ldots]$. For the phonons incident at $\mathbf{r}_\Gamma$, one should make the substitutions $\mathbf{r}_\Gamma \to \mathbf{r}_\Gamma'$, $\mathbf{r}_\Gamma' \to \mathbf{r}_\Gamma''$, and so on in Eq. (10), according to the definition of the points $\mathbf{r}_\Gamma$, $\mathbf{r}_\Gamma'$, ... Therefore, the factor in the square brackets in Eq. (10) for these phonons becomes equal to $[(\mathbf{r}_\Gamma - \mathbf{r}_\Gamma') + (1 - P)(\mathbf{r}_\Gamma - \mathbf{r}_\Gamma'') + \ldots]$, and the boundary condition (1) is satisfied. Substituting the expression $P[\ldots]$ from Eq. (10) in place of $(\mathbf{r} - \mathbf{r}_\Gamma)$ into Eq. (9), we can calculate the integrals for a cylindrical sample like that in problem 9.5. This calculation demonstrates that the term $(\mathbf{r} - \mathbf{r}_\Gamma')$ contributes 3 times more than the term $(\mathbf{r} - \mathbf{r}_\Gamma)$, since the average length of this path is three times the average length of $(\mathbf{r} - \mathbf{r}_\Gamma)$. The term $(\mathbf{r} - \mathbf{r}_\Gamma'')$ contributes 5 times more, and so on. Calculating the sum of the series, $\sum_{k=0}^{\infty} z^k (2k + 1) = (1 + z)/(1 - z)^2$, where $z = 1 - P$, we find that the effective Λ we should substitute in Eq. (7) is

$$\Lambda = \frac{2 - P}{P} \Lambda|_{P=1}, \qquad (11)$$

where $\Lambda|_{P=1}$ is given by Eq. (9). Naturally, the thermal conductivity increases with decreasing P.

Now we consider another limiting case, when the transverse size of the sample is large in comparison to the mean free path length l_N for normal collision processes. There still can be an arbitrary relation between l_U and the transverse size, because $l_U \gg l_N$. The distribution function is determined by the normal collision processes. Therefore, the hydrodynamic transport regime is realized, and we can use the balance equations (25.19) and (25.20). Below we show that the last term in Eq. (25.19) should be neglected within the accuracy assumed. Thus, for the stationary case Eq. (25.19) is reduced to the continuity equation $\nabla \cdot \mathbf{u} = 0$, where $\mathbf{u}$ is the drift velocity of the phonon gas. Equation (25.20) is written as

$$S\frac{\partial T}{\partial r_\alpha} - \sum_{\beta\gamma\delta} \nu_{\alpha\beta\gamma\delta}\frac{\partial^2 u^\gamma}{\partial r_\beta \partial r_\delta} + TS^2 \sum_\beta (\kappa^{(0)\ -1})_{\alpha\beta} u^\beta = 0, \qquad (12)$$

where we have added the index (0) to the bulk thermal conductivity tensor in order to distinguish it from the effective thermal conductivity of a bounded medium. The approximation of pure diffuse reflection, in terms of the macroscopic variables, means that $\mathbf{u} = \mathbf{u}(\mathbf{r})$ satisfies the following boundary condition at the surface Γ:

$$\mathbf{u}|_{\mathbf{r} \in \Gamma} = 0. \qquad (13)$$

In the isotropic crystals, where the tensors $\nu_{\alpha\beta\gamma\delta}$ and $\kappa^{(0)}_{\alpha\beta}$ are reduced to the scalars $\delta_{\alpha\gamma}\delta_{\beta\delta}\nu$ and $\delta_{\alpha\beta}\kappa^{(0)}$, the velocity $\mathbf{u}$ is directed along the thermal gradient. Let us solve Eq. (12) for a sample of cylindrical geometry. Assuming that the axis of the cylinder is OZ, we obtain $\partial u^z/\partial z = 0$ from the continuity equation, which means that $|\mathbf{u}| = u = u^z$ does not depend on z. Introducing the cylindrical coordinates according to $\mathbf{r} = (\mathbf{r}_\perp, z)$, we rewrite Eq. (12) as

$$\nu\frac{1}{r_\perp}\frac{\partial}{\partial r_\perp}\left(r_\perp\frac{\partial u}{\partial r_\perp}\right) - \frac{TS^2}{\kappa^{(0)}}u = S\frac{\partial T}{\partial z}, \qquad (14)$$

where we have taken into account that u depends only on $r_\perp = |\mathbf{r}_\perp|$, due to cylindrical symmetry of the problem. As follows from Eq. (13), the boundary condition for $u = u(r_\perp)$ is $u(R) = 0$, where R is the radius of the cylinder. A finite solution of the inhomogeneous differential equation (14) is expressed through the modified Bessel function of the first kind, $\mathcal{I}_0(x)$:

$$u = -\frac{\kappa^{(0)}}{TS}\frac{\partial T}{\partial z} + c\mathcal{I}_0\left(\frac{r_\perp}{R_0}\right), \qquad (15)$$

where

$$R_0 = \sqrt{\frac{\kappa^{(0)}\nu}{TS^2}} \sim \sqrt{l_U l_N} \qquad (16)$$

is the characteristic scale of the phonon drift velocity distribution. The estimate for it is obtained with the aid of Eqs. (24.17), (24.19), and (25.23). The constant of integration, c, is to be found from the requirement $u(R) = 0$.

Finally, we remind that the energy flow density is given by $\mathbf{G} = TS\mathbf{u}$, see Eq. (24.15), and write the averaged energy flow density (2) through the cylindrical sample ($s_c = \pi R^2$) as

$$\overline{G} = \frac{1}{\pi R^2} ST \int_0^R 2\pi r_\perp dr_\perp u(r_\perp) = -\kappa^{(0)} \frac{2}{R^2} \int_0^R dr_\perp r_\perp$$

$$\times \left[1 - \frac{\mathcal{I}_0(r_\perp/R_0)}{\mathcal{I}_0(R/R_0)}\right] \nabla T. \tag{17}$$

Calculating the integral over $r_\perp$, we write the thermal conductivity through the Bessel functions $\mathcal{I}_0$ and $\mathcal{I}_1$:

$$\kappa = \kappa^{(0)} \left[1 - \frac{2R_0}{R} \frac{\mathcal{I}_1(R/R_0)}{\mathcal{I}_0(R/R_0)}\right]. \tag{18}$$

Similar expressions can be obtained for other geometries (problem 9.6). Using the asymptotic behavior of the Bessel functions at large arguments, we obtain

$$\kappa = \kappa^{(0)}(1 - 2R_0/R), \quad R \gg R_0, \tag{19}$$

which means that the size effect introduces a small correction to the thermal conductivity. The bulk regime of energy transfer, when this correction can be neglected, corresponds to $R \gg \sqrt{l_U l_N}$, when R still can be smaller than l_U. In the opposite limit,

$$\kappa = \kappa^{(0)} \frac{R^2}{8R_0^2} = \frac{TS^2 R^2}{8\nu}, \quad R \ll R_0. \tag{20}$$

Using the results (19) and (20), we can justify the neglect of the term $\mu \partial T/\partial \mathbf{r}$ in comparison to $TS\mathbf{u}$ in Eq. (25.19). In the limit $R \gg R_0$, when $\kappa \simeq \kappa^{(0)}$, the relative contribution from this term is of the order of $\mu/\kappa^{(0)}$, which is small as l_N/l_U. In the limit $R \ll R_0$, when κ is given by Eq. (20), this relative contribution is small as $(l_N/R)^2$.

Equation (20) demonstrates that the effective thermal conductivity of a sufficiently narrow cylinder does not depend at all on the bulk thermal conductivity coefficient. Instead, it is determined by the phonon-hydrodynamical viscosity coefficient ν and proportional to the rate of normal phonon-phonon scattering processes. Indeed, in the limit $R \ll R_0$ (or, equivalently, $R \ll \sqrt{l_U l_N}$, see Eq. (16)) the last term on the left-hand side of Eq. (14) can be neglected, and this equation is reduced to the one for a viscous liquid in a tube. By analogy to the theory of liquids, one may denote this limit, $l_N \ll R \ll \sqrt{l_u l_N}$, as the diffusive regime. On the other hand, the limit $R \ll l_N$ studied in the beginning of this section is analogous to the ballistic regime in the theory of gases. Comparing Eqs. (7) and (20), we find that in these regimes the temperature dependence of κ follows the same power law $\propto T^3$. However, the coefficient of proportionality is greater for the diffusive regime. The umklapp scattering does not influence the thermal conductivity in both regimes. To describe the transition region $R \sim l_N$ between these regimes, one needs to solve the kinetic equation (3) with the collision integral accounting for the normal processes only. Since this problem cannot be solved analytically, we do not consider it here. We just mention that assumptions about the phonon spectrum appear to be essential in such kind of calculations (problem 9.7).

47. Electron Relaxation by Near-Surface Phonons

The character of lattice vibrations is modified in the region near the surface. In particular, mixing of different vibrational modes takes place, and there can appear additional modes localized at the surface. Describing the electron-phonon interaction in the vicinity of the surface, one should take into account the modification of phonon dynamics. A detailed analysis of the incident, reflected, and localized phonon modes is a complex task. Below we apply a more compact approach using the Green's function for the phonons in the half-space $z > 0$. We consider the interaction of 2D electrons, whose confinement in a quantum well near the surface is described by the wave function ψ_z, with the deformation potential generated by the long-wavelength acoustic vibrations of the crystal lattice. The formalism of non-equilibrium Green's functions developed in Chapter 8 appears to be convenient for deriving the kinetic equation in the inhomogeneous medium defined by the half-space for phonons and by the quantum well for electrons.

We start from the generalized kinetic equation (41.27), where $\widehat{H}_0(1) = -(\hbar^2/2m)[\partial^2/\partial \mathbf{x}_1^2 + \partial^2/\partial z_1^2]$ is the Hamiltonian of free electrons, and $\Sigma^{s_1 s_2}(1,2)$ is the self-energy function for the case of electron-phonon interaction considered in Sec. 42; see Eq. (42.11). We note that the multi-indices $(1, 2, \ldots)$ include both time and coordinate variables. In the 2D approximation, we search for the Green's function of electrons in the form $G^{s_1 s_2}(1,2) = \psi_{z_1} \psi_{z_2}^* G_{t_1 t_2}^{s_1 s_2}(\mathbf{x}_1, \mathbf{x}_2)$, where $\mathbf{x} = (x, y)$ is the 2D coordinate. Let us substitute this function into Eq. (41.27), multiply this equation by $\psi_{z_1}^* \psi_{z_2}$, and integrate it over z_1 and z_2. As a result, we obtain

$$
-\left[i\hbar\left(\frac{\partial}{\partial t_1}+\frac{\partial}{\partial t_2}\right)+\frac{\hbar^2}{2m}(\nabla^2_{\mathbf{x}_1}-\nabla^2_{\mathbf{x}_2})\right]G^{-+}_{t_1 t_2}(\mathbf{x}_1,\mathbf{x}_2)
$$

$$
=-\sum_{s=\pm}\int dt_3\int d\mathbf{x}_3\left[\Sigma^{-s}_{t_1 t_3}(\mathbf{x}_1,\mathbf{x}_3)G^{s+}_{t_3 t_2}(\mathbf{x}_3,\mathbf{x}_2)\right. \tag{1}
$$

$$
\left.+G^{-s}_{t_1 t_3}(\mathbf{x}_1,\mathbf{x}_3)\Sigma^{s+}_{t_3 t_2}(\mathbf{x}_3,\mathbf{x}_2)\right],
$$

where the self-energy function of 2D electrons is introduced as

$$
\Sigma^{s_1 s_2}_{t_1 t_2}(\mathbf{x}_1,\mathbf{x}_2)=\int dz_1\int dz_2 \psi^*_{z_1}\psi_{z_2}\Sigma^{s_1 s_2}_{t_1 t_2}(\mathbf{r}_1,\mathbf{r}_2). \tag{2}
$$

The self-energy function standing under the integral in this equation depends on the 3D coordinates $\mathbf{r}=(\mathbf{x},z)$. According to Eq. (42.11), the main contribution to this function can be written within the second-order accuracy with respect to electron-phonon interaction:

$$
\Sigma^{s_1 s_2}_{t_1 t_2}(\mathbf{r}_1,\mathbf{r}_2)=s_1 s_2\hbar\mathcal{D}^2\psi^*_{z_1}\psi_{z_2}G^{s_1 s_2}_{t_1 t_2}(\mathbf{x}_1,\mathbf{x}_2)
$$

$$
\times\sum_{\alpha\beta}\nabla^\alpha_{\mathbf{r}_1}\nabla^\beta_{\mathbf{r}_2}iD^{\alpha\beta,s_1 s_2}_{t_1 t_2}(\mathbf{r}_1,\mathbf{r}_2), \tag{3}
$$

where $\mathcal{D}$ is the deformation potential and $D^{\alpha\beta,s_1 s_2}_{t_1 t_2}(\mathbf{r}_1,\mathbf{r}_2)$ is the Green's function of phonons defined in Sec. 42.

As we already know (see problem 8.8), the Wigner transformation of the left-hand side of Eq. (1) in the limit $t_1=t_2=t$ reduces it to the left-hand side of the quasi-classical kinetic equation for the electron distribution function $f_{\mathbf{xp}t}$, where $\mathbf{p}$ is the 2D momentum. Transforming the right-hand side (i.e., the generalized collision integral), we use the quasi-classical approximation, neglecting the dependence of both $G^{s_1 s_2}_{t_1 t_2}(\mathbf{x}_1,\mathbf{x}_2)$ and $\Sigma^{s_1 s_2}_{t_1 t_2}(\mathbf{x}_1,\mathbf{x}_2)$ on the average time $t=(t_1+t_2)/2$ and 2D coordinate $\mathbf{x}=(\mathbf{x}_1+\mathbf{x}_2)/2$ on the quantum scales $\hbar/\bar\varepsilon$ and $\hbar/\bar p$, respectively. Therefore, the right-hand side of Eq. (1) is transformed to

$$
-\sum_{s=\pm}\int\frac{d\varepsilon}{2\pi\hbar}\left[\Sigma^{-s}_{\varepsilon t}(\mathbf{x},\mathbf{p})G^{s+}_{\varepsilon t}(\mathbf{x},\mathbf{p})+G^{-s}_{\varepsilon t}(\mathbf{x},\mathbf{p})\Sigma^{s+}_{\varepsilon t}(\mathbf{x},\mathbf{p})\right]
$$

$$
=\int\frac{d\varepsilon}{2\pi\hbar}\left[\Sigma^{+-}_{\varepsilon t}(\mathbf{x},\mathbf{p})G^{-+}_{\varepsilon t}(\mathbf{x},\mathbf{p})-\Sigma^{-+}_{\varepsilon t}(\mathbf{x},\mathbf{p})G^{+-}_{\varepsilon t}(\mathbf{x},\mathbf{p})\right], \tag{4}
$$

as in problem 8.8. It is convenient to introduce the function

$$
K^{s_1 s_2}_{\omega t}(\mathbf{x},\mathbf{q})=is_1 s_2\mathcal{D}^2\int dz_1\int dz_2|\psi_{z_1}|^2|\psi_{z_2}|^2\int d\tau\int d\Delta\mathbf{x}
$$

$$\times e^{i\omega\tau - i\mathbf{q}\cdot\Delta\mathbf{x}} \sum_{\alpha\beta} \nabla_{\mathbf{r}_1}^{\alpha} \nabla_{\mathbf{r}_2}^{\beta} D_{t+\tau/2, t-\tau/2}^{\alpha\beta, s_1 s_2}(\mathbf{r}_1, \mathbf{r}_2), \tag{5}$$

where the coordinanes $\mathbf{r}_1$ and $\mathbf{r}_2$ are given by $\mathbf{r}_1 = (\mathbf{x} + \Delta\mathbf{x}/2, z_1)$ and $\mathbf{r}_2 = (\mathbf{x} - \Delta\mathbf{x}/2, z_2)$. This function is expressed through the Wigner transform of the factor $\sum_{\alpha\beta} \nabla_{\mathbf{r}_1}^{\alpha} \nabla_{\mathbf{r}_2}^{\beta} D_{t_1 t_2}^{\alpha\beta, s_1 s_2}(\mathbf{r}_1, \mathbf{r}_2)$. The collision integral (4) is rewritten as (the parametric arguments $\mathbf{x}$ and t are omitted below for brevity)

$$\int \frac{d\varepsilon}{2\pi} \int \frac{d\omega}{2\pi} \int \frac{d\mathbf{q}}{(2\pi)^2} \left[K_{\omega}^{+-}(\mathbf{q}) G_{\varepsilon-\hbar\omega}^{+-}(\mathbf{p}-\hbar\mathbf{q}) G_{\varepsilon}^{-+}(\mathbf{p}) \right.$$

$$\left. - K_{\omega}^{-+}(\mathbf{q}) G_{\varepsilon-\hbar\omega}^{-+}(\mathbf{p}-\hbar\mathbf{q}) G_{\varepsilon}^{+-}(\mathbf{p}) \right]. \tag{6}$$

To find the collision integral in the lowest (second) order in the interaction, it is sufficient to substitute the free-electron Green's functions $G_{\varepsilon}^{-+}(\mathbf{p}) \simeq g_{\varepsilon}^{-+}(\mathbf{p}) = 2\pi i f_{\mathbf{p}} \delta(\varepsilon - \varepsilon_{\mathbf{p}})$ and $G_{\varepsilon}^{+-}(\mathbf{p}) \simeq g_{\varepsilon}^{+-}(\mathbf{p}) = 2\pi i (f_{\mathbf{p}} - 1) \delta(\varepsilon - \varepsilon_{\mathbf{p}})$, see Eq. (40.33), into Eq. (6) and calculate the integral over ε. As a result, the kinetic equation is written as

$$\left(\frac{\partial}{\partial t} + \mathbf{v_p} \cdot \frac{\partial}{\partial \mathbf{x}} \right) f_{\mathbf{xp}t} = J_{e,ph}(f|\mathbf{xp}t), \tag{7}$$

where the electron-phonon collision integral is given in the form (34.26) with the following transition probabilities:

$$W_{\mathbf{pp}'} = -\frac{1}{L^2} \int d\omega K_{\omega}^{+-}[(\mathbf{p}-\mathbf{p}')/\hbar] \delta(\varepsilon_p - \varepsilon_{p'} - \hbar\omega),$$

$$W_{\mathbf{p}'\mathbf{p}} = -\frac{1}{L^2} \int d\omega K_{\omega}^{-+}[(\mathbf{p}-\mathbf{p}')/\hbar] \delta(\varepsilon_p - \varepsilon_{p'} - \hbar\omega). \tag{8}$$

The functions $K^{s_1 s_2}$ in this equation are determined by the phonon Green's functions $D^{\alpha\beta, s_1 s_2}$; see Eq. (5). Since K^{-+} and K^{+-} are already quadratic in the interaction, we use the free-phonon Green's functions $d^{\alpha\beta, s_1 s_2}$ instead of $D^{\alpha\beta, s_1 s_2}$ in Eq. (5). Thus, the probabilities of electron-phonon collisions are expressed through the Green's functions of elastic vibrations in the medium under consideration. Since this medium is inhomogeneous along OZ, there exists surface-induced mixing of different acoustic modes. For this reason, it is not convenient to express the Green's function according to Eq. (42.16).

To describe the equilibrium lattice vibrations, we write the definitions (42.3) by using the notations for equilibrium correlation functions introduced by Eq. (D.6). In these notations, the Green's functions of phonons in the energy representation are expressed as $d_{\omega}^{\alpha\beta, -+}(\mathbf{r}_1, \mathbf{r}_2) = -(i/\hbar) J_{u_{\mathbf{r}_2}^{\beta} u_{\mathbf{r}_1}^{\alpha}}(\omega)$ and $d_{\omega}^{\alpha\beta, +-}(\mathbf{r}_1, \mathbf{r}_2) = -(i/\hbar) J_{u_{\mathbf{r}_1}^{\alpha} u_{\mathbf{r}_2}^{\beta}}(\omega)$, where $\hat{\mathbf{u}}_{\mathbf{r}} \equiv$

$\hat{\mathbf{u}}_{ac}(\mathbf{r})$, see Eq.(6.29), is the displacement operator in the elastic medium. Employing the relations (D.7) and (D.11), we obtain

$$D_\omega^{\alpha\beta,\mp\pm}(\mathbf{r}_1,\mathbf{r}_2) = (N_\omega + 1/2 \mp 1/2)[d_\omega^{\alpha\beta,R}(\mathbf{r}_1,\mathbf{r}_2) - d_\omega^{\alpha\beta,A}(\mathbf{r}_1,\mathbf{r}_2)], \quad (9)$$

where $N_\omega = \left(e^{\hbar\omega/T} - 1\right)^{-1}$ is the Planck distribution function. The retarded and advanced Green's functions standing in Eq. (9) are the Green's functions of the theory of elasticity, in the usual mathematical sense. The retarded function is determined by

$$\sum_\beta \left[(\omega + i0)^2\delta_{\alpha\beta} + s_t^2\nabla_\mathbf{r}^2\delta_{\alpha\beta} + (s_l^2 - s_t^2)\nabla_\mathbf{r}^\alpha\nabla_\mathbf{r}^\beta\right] d_\omega^{\beta\gamma,R}(\mathbf{r},\mathbf{r}')$$

$$= \rho^{-1}\delta_{\alpha\gamma}\delta(\mathbf{r} - \mathbf{r}'), \qquad (10)$$

where ρ is the crystal density. The differential operator on the left-hand side is expressed through the longitudinal and transverse sound velocities (problems 9.8 and 9.9). The equation for the advanced Green's function differs from Eq. (10) by the factor $(\omega - i0)$ in place of $(\omega + i0)$. These functions are expressed through each other as $d_\omega^{\beta\gamma,A}(\mathbf{r},\mathbf{r}') = d_\omega^{\gamma\beta,R}{}^*(\mathbf{r}',\mathbf{r})$. Introducing the function

$$K_\omega(\mathbf{q}) = \int dz \int dz'|\psi_z|^2|\psi_{z'}|^2 K_\omega(\mathbf{q}|z,z') \qquad (11)$$

with

$$K_\omega(\mathbf{q}|z,z') = i\mathcal{D}^2 \int d\Delta\mathbf{x} e^{-i\mathbf{q}\cdot\Delta\mathbf{x}} \sum_{\alpha\beta} \nabla_\mathbf{r}^\alpha\nabla_{\mathbf{r}'}^\beta[d_\omega^{\alpha\beta,R}(\mathbf{r},\mathbf{r}') - d_\omega^{\alpha\beta,A}(\mathbf{r},\mathbf{r}')],$$

$$(12)$$

where $\mathbf{r} = (\mathbf{x} + \Delta\mathbf{x}/2, z)$ and $\mathbf{r}' = (\mathbf{x} - \Delta\mathbf{x}/2, z')$, we can rewrite the transition probabilities (8) as

$$W_{\mathbf{p}\mathbf{p}'} = \frac{1}{L^2} \int d\omega K_\omega[(\mathbf{p} - \mathbf{p}')/\hbar](N_\omega + 1)\delta(\varepsilon_p - \varepsilon_{p'} - \hbar\omega),$$

$$W_{\mathbf{p}'\mathbf{p}} = \frac{1}{L^2} \int d\omega K_\omega[(\mathbf{p} - \mathbf{p}')/\hbar]N_\omega\delta(\varepsilon_p - \varepsilon_{p'} - \hbar\omega). \qquad (13)$$

To find $d_\omega^{\alpha\beta,R}(\mathbf{r},\mathbf{r}')$ for a semi-infinite medium, it is necessary to solve Eq. (10) with proper boundary conditions at the surface of this medium. The Green's function $d_\omega^{\alpha\beta,R}(\mathbf{r},\mathbf{r}')$ depends only on $\Delta\mathbf{x} = \mathbf{x} - \mathbf{x}'$ because the medium is homogeneous in the plane XOY. This function can be written in the mixed $(\mathbf{q},z)$-representation as

$$d_\omega^{\alpha\beta}(\mathbf{q}|z,z') = \int d\Delta\mathbf{x} e^{-i\mathbf{q}\cdot\Delta\mathbf{x}} d_\omega^{\alpha\beta,R}(\mathbf{r},\mathbf{r}'), \qquad (14)$$

where $\omega = \omega + i0$ is implied and the index R is omitted for brevity. Using the isotropy of the medium in the plane XOY (which means that $d_\omega^{\alpha\beta}(\mathbf{q}|z,z') = d_\omega^{\alpha\beta}(q|z,z')$ does not depend on the direction of $\mathbf{q}$), one may direct the vector $\mathbf{q}$ along OX so that $\mathbf{q} = (q,0)$. Therefore, Eq. (10) is rewritten in the mixed representation as

$$\begin{bmatrix} \omega^2 - s_l^2 q^2 + s_t^2 d^2/dz^2 & iq(s_l^2 - s_t^2)d/dz \\ iq(s_l^2 - s_t^2)d/dz & \omega^2 - s_t^2 q^2 + s_l^2 d^2/dz^2 \end{bmatrix} \begin{vmatrix} d_\omega^{x\gamma}(q|z,z') \\ d_\omega^{z\gamma}(q|z,z') \end{vmatrix}$$

$$= \begin{vmatrix} \delta_{x\gamma} \\ \delta_{z\gamma} \end{vmatrix} \frac{1}{\rho}\delta(z-z'), \tag{15}$$

where γ is either x or z.

The boundary conditions for $d_\omega^{\alpha\beta}(q|z,z')$ are determined by the elastic properties of the surface $z = 0$. We consider two limiting cases: $i)$ a rigid surface, for which the displacement $\mathbf{u_{rt}}$ is zero at $z = 0$ so that

$$d_\omega^{\alpha\beta}(q|z,z')\Big|_{z=0} = 0, \tag{16}$$

and $ii)$ a free surface, for which the normal stress is zero at $z = 0$. The latter case implies the boundary conditions (problem 9.10)

$$\left[\frac{\partial}{\partial z}d_\omega^{xx}(q|z,z') + iqd_\omega^{zx}(q|z,z')\right]_{z=0} = 0,$$

$$\left[\frac{\partial}{\partial z}d_\omega^{zx}(q|z,z') + iq(1 - 2s_t^2/s_l^2)d_\omega^{xx}(q|z,z')\right]_{z=0} = 0,$$

$$\left[\frac{\partial}{\partial z}d_\omega^{zz}(q|z,z') + iq(1 - 2s_t^2/s_l^2)d_\omega^{xz}(q|z,z')\right]_{z=0} = 0, \tag{17}$$

$$\left[\frac{\partial}{\partial z}d_\omega^{xz}(q|z,z') + iqd_\omega^{zz}(q|z,z')\right]_{z=0} = 0.$$

A straightforward calculation of the components of the matrix Green's function from Eq. (15) leads to

$$\begin{pmatrix} d^{xx} \\ d^{zz} \\ d^{zx} \\ d^{xz} \end{pmatrix} = C_{ll}e^{-\kappa_l(z+z')}\begin{pmatrix} 1 \\ \kappa_l^2/q^2 \\ i\kappa_l/q \\ \kappa_l/iq \end{pmatrix} + C_{tt}e^{-\kappa_t(z+z')}\begin{pmatrix} 1 \\ q^2/\kappa_t^2 \\ iq/\kappa_t \\ q/i\kappa_t \end{pmatrix}$$

$$+ C_{lt}e^{-\kappa_l z - \kappa_t z'}\begin{pmatrix} 1 \\ \kappa_l/\kappa_t \\ i\kappa_l/q \\ q/i\kappa_t \end{pmatrix} + C_{tl}e^{-\kappa_t z - \kappa_l z'}\begin{pmatrix} 1 \\ \kappa_l/\kappa_t \\ iq/\kappa_t \\ \kappa_l/iq \end{pmatrix} \tag{18}$$

$$-\frac{e^{-\kappa_l|z-z'|}}{2\rho\omega^2}\begin{pmatrix} q^2/\kappa_l \\ -\kappa_l \\ iq\,\mathrm{sgn}(z-z') \\ iq\,\mathrm{sgn}(z-z') \end{pmatrix} + \frac{e^{-\kappa_t|z-z'|}}{2\rho\omega^2}\begin{pmatrix} \kappa_t \\ -q^2/\kappa_t \\ iq\,\mathrm{sgn}(z-z') \\ iq\,\mathrm{sgn}(z-z') \end{pmatrix},$$

where $\kappa_j = \sqrt{q^2 - \omega^2/s_j^2}$ can be either real or imaginary.

The coefficients $C_{jj'}$ $(j, j' = l, t)$ are to be found from the boundary conditions. To describe the electron-phonon scattering, we need only one of these coefficients, C_{ll}, because the function given by Eq. (12) is transformed to

$$K_\omega(q|z,z') = -2\mathcal{D}^2\mathrm{Im}\left[q^2 d_\omega^{xx}(q|z,z') + iq\frac{d}{dz'}d_\omega^{xz}(q|z,z')\right.$$

$$\left. -iq\frac{d}{dz}d_\omega^{zx}(q|z,z') + \frac{d^2}{dz\,dz'}d_\omega^{zz}(q|z,z')\right] \qquad (19)$$

$$= \mathrm{Im}\frac{\omega^2\mathcal{D}^2}{\rho s_l^4\kappa_l}\left[e^{-\kappa_l|z-z'|} + R_{\omega q}e^{-\kappa_l(z+z')}\right], \qquad R_{\omega q} = -C_{ll}\frac{2\rho\omega^2\kappa_l}{q^2}.$$

The term proportional to $e^{-\kappa_l|z-z'|}$ describes the bulk-phonon contribution (it depends only on the longitudinal sound velocity), while the term proportional to $R_{\omega q}e^{-\kappa_l(z+z')}$ describes a correction due to reflection of the sound waves from the surface. To determine the reflectance coefficient $R_{\omega q}$, we apply the boundary conditions (16) and (17). For the cases of rigid (i) and free (ii) surface, we obtain

$$R_{\omega q} = \begin{cases} \frac{\kappa_l\kappa_t+q^2}{\kappa_l\kappa_t-q^2} & (i) \\[2mm] \frac{4q^2\kappa_l\kappa_t+(\kappa_t^2+q^2)^2}{4q^2\kappa_l\kappa_t-(\kappa_t^2+q^2)^2} & (ii) \end{cases}. \qquad (20)$$

The rule of continuation of κ_j into the region $\omega > s_j q$ is determined by the fact that ω contains an infinitely small positive imaginary part $i0$ and $\mathrm{Re}\,\kappa_j > 0$. This implies $\kappa_j \to -i\,\mathrm{sgn}(\omega)k_j$, where $k_j = \sqrt{\omega^2/s_j^2 - q^2}$. The electron transitions may occur in the region $\omega > s_t q$, where $K_\omega(q|z,z')$ is non-zero.

As an application of the general formulas derived in this section, we investigate the energy and momentum relaxation of 2D electrons due to the scattering described above. The energy relaxation is characterized by the absorbed power, P, which is introduced by Eq. (36.20). One can also introduce another quantity, $\mathbf{U}$, to characterize the velocity relaxation (see the end of Sec. 36). The equations for these quantities in the 2D case are written in the following way:

$$P = -\frac{2}{L^2} \sum_{\mathbf{p}} \varepsilon_p J_{e,ph}(f|\mathbf{p}), \quad \mathbf{U} = -\frac{2}{L^2} \sum_{\mathbf{p}} \frac{\mathbf{p}}{m} J_{e,ph}(f|\mathbf{p}). \quad (21)$$

Assuming that the electrons are described by the non-equilibrium distribution function given by Eq. (31.25), we substitute the collision integral (34.26) with transition probabilities (13) into Eq. (21) and obtain

$$\left| \begin{array}{c} P \\ \mathbf{U} \end{array} \right| = \frac{2}{L^4} \sum_{\mathbf{pq}} \int_{-\infty}^{\infty} d\omega K_\omega(q) \left| \begin{array}{c} \varepsilon_p \\ \mathbf{p}/m \end{array} \right| \quad (22)$$

$$\times \frac{[1 - e^{\hbar\omega/T_e - \hbar\omega/T - \hbar\mathbf{q}\cdot\mathbf{u}/T}]\delta(\varepsilon_{\mathbf{p}} - \varepsilon_{\mathbf{p}-\hbar\mathbf{q}} - \hbar\omega)}{[1 - e^{-\hbar\omega/T}][e^{(\varepsilon_{\mathbf{p}} - \mathbf{p}\cdot\mathbf{u} - \mu)/T_e} + 1][e^{(-\varepsilon_{\mathbf{p}-\hbar\mathbf{q}} + (\mathbf{p}-\hbar\mathbf{q})\cdot\mathbf{u} + \mu)/T_e} + 1]},$$

where T_e, $\mathbf{u}$, and μ are the electron temperature, drift velocity, and chemical potential, respectively.

Below we assume that the deviation of the electron distribution function from the equilibrium one is small and search for the contributions linear in $T_e - T$ and $\mathbf{u}$. Taking into account the antisymmetry property $K_{-\omega}(q) = -K_\omega(q)$, we transform the integral over ω in Eq. (22) to the region of positive ω and rewrite Eq. (22) as

$$\left| \begin{array}{c} P \\ \mathbf{U} \end{array} \right| = \frac{1}{2L^4} \sum_{\mathbf{pq}} \int_0^\infty d\omega \frac{K_\omega(q)}{\sinh(\hbar\omega/2T)} \left| \begin{array}{c} (\hbar\omega/T)^2(T_e - T) \\ \hbar^2(\mathbf{q}\cdot\mathbf{u})\mathbf{q}/mT \end{array} \right|$$

$$\times \frac{\delta(\varepsilon_{\mathbf{p}} - \varepsilon_{\mathbf{p}-\hbar\mathbf{q}} - \hbar\omega)}{\cosh(\hbar\omega/2T) + \cosh[(\varepsilon_p - \mu)/T - \hbar\omega/2T]}. \quad (23)$$

The integral over the angle of $\mathbf{p}$ gives us

$$\int_0^{2\pi} \frac{d\varphi_p}{2\pi} \delta(\varepsilon_{\mathbf{p}} - \varepsilon_{\mathbf{p}-\hbar\mathbf{q}} - \hbar\omega) = \frac{1}{\pi\hbar} \left[\left(\frac{pq}{m}\right)^2 - \left(\omega + \frac{\hbar q^2}{2m}\right)^2 \right]^{-1/2}. \quad (24)$$

Using Eq. (24), we have

$$\frac{1}{L^2} \sum_{\mathbf{p}} \frac{\delta(\varepsilon_{\mathbf{p}} - \varepsilon_{\mathbf{p}-\hbar\mathbf{q}} - \hbar\omega)}{\cosh(\hbar\omega/2T) + \cosh[(\varepsilon_p - \mu)/T - \hbar\omega/2T]}$$

$$= \frac{mT}{2\pi^2\hbar^2} \int_{-(\mu+\hbar\omega/2)/T}^{\infty} \frac{dy}{\cosh(\omega/2T) + \cosh y}$$

$$\times \left[\frac{2q^2T}{m}y + \frac{q^2(2m\mu - \hbar^2q^2/4)}{m^2} - \omega^2 \right]^{-1/2} \quad (25)$$

$$\simeq \frac{m^2\omega}{\pi^2\hbar^2 q}\frac{1}{\sinh(\hbar\omega/2T)}\frac{1}{\sqrt{(2p_F)^2 - \hbar^2 q^2}}.$$

The final transformation of this expression has been carried out in the approximation of strongly degenerate electron gas, when $\mu = \varepsilon_F \gg T$. We also take into account that the characteristic energies $\hbar\omega$ contributing to the integral in Eq. (23) are either of the order of T or smaller than T. Let us introduce the energy and momentum relaxation rates, $\nu^{(e)}$ and $\nu^{(m)}$, according to $P = \nu^{(e)}n_{2D}(T_e - T)$ and $\mathbf{U} = \nu^{(m)}n_{2D}\mathbf{u}$, where n_{2D} is the 2D electron density. After substituting the result given by Eq. (25) into Eq. (23), the averaging over the angle of $\mathbf{q}$ transforms $(\mathbf{q}\cdot\mathbf{u})\mathbf{q}$ to $q^2\mathbf{u}/2$. We obtain $\nu^{(e)}$ and $\nu^{(m)}$ in the following form:

$$\left|\begin{array}{c}\nu^{(e)}\\ \nu^{(m)}\end{array}\right| = \frac{m^2}{4\pi^3 n_{2D}T}\int_0^\infty d\omega\frac{\omega}{\sinh^2(\hbar\omega/2T)}$$

$$\times\int_0^{2p_F/\hbar}dq\frac{K_\omega(q)}{\sqrt{(2p_F)^2 - \hbar^2 q^2}}\left|\begin{array}{c}\omega^2/T\\ q^2/2m\end{array}\right|. \tag{26}$$

Equation (26) is convenient for further analysis, where we apply the explicit form of $K_\omega(q|z,z')$ given by Eqs. (19) and (20).

Below we consider the 2D limit, $p_F \ll \pi\hbar/d$, where d is the quantum well width, and introduce two characteristic temperatures:

$$T_0 = 2s_l p_F, \quad T_1 = 2s_l\hbar/d \tag{27}$$

(one can see that $T_0 \ll T_1$). Let us study the region of relatively high temperatures, $T \gg T_0$. This means that the frequencies ω contributing to the integral in Eq. (26) are large in comparison to $s_j q$ (below we show that either $\hbar\omega \sim T$ or $\hbar\omega \sim T_1$) and $\kappa_j \simeq -i\omega/s_j$. The reflectance coefficient $R_{\omega q}$ in this case is presented in the most simple form: $R = 1$ for a rigid surface and $R = -1$ for a free surface. To find $K_\omega(q)$ from $K_\omega(q|z,z')$, we consider a rectangular potential well in the hard-wall model, with $\psi_z = \sqrt{2/d}\cos[\pi(z - z_0)/d]$, where z_0 is the distance of the center of the well from the surface. To take the integrals over z and z' in Eq. (11), one may transform the double integral to a squared single integral and use the result of problem 4.13 with q replaced by ω/s_l. We obtain

$$K_\omega(q) \simeq \frac{\omega\mathcal{D}^2}{\rho s_l^3}\frac{\sin^2(\omega d/2s_l)}{(\omega d/2s_l)^2[1 - (\omega d/2\pi s_l)^2]^2}[1 \pm \cos(2\omega z_0/s_l)], \tag{28}$$

where the upper and the lower signs correspond to the cases of rigid and free surface, respectively. The presence of the surface leads to the contribution proportional to $\cos(2\omega z_0/s_l)$. After substituting Eq. (28) into Eq. (26), we calculate the integral over q and obtain

$$\left|\begin{array}{c}\nu^{(e)}\\\nu^{(m)}\end{array}\right| = \bar{\nu}\frac{T^2}{T_0^2}\int_0^\infty dx\frac{x^2 F_T(x)}{\sinh^2(x/2)}\left|\begin{array}{c}Tx^2/2\varepsilon_F\\1\end{array}\right|[1\pm\cos(\Gamma x)], \qquad (29)$$

$$F_T(x) = \frac{\sin^2(Tx/T_1)}{(Tx/T_1)^2[1-(Tx/\pi T_1)^2]^2} \ ,$$

where we have introduced the relaxation rate

$$\bar{\nu} = \frac{\mathcal{D}^2 p_F^2 m}{\pi\hbar^4\rho s_l} \qquad (30)$$

and the dimensionless variables $x = \hbar\omega/T$ and $\Gamma = 4z_0 T/dT_1$.

If the case of high temperatures $T \gg T_1$ (corresponding to the equipartition condition for LA phonons), the main contribution to the integral in Eq. (29) comes from $x \sim T_1/T \ll 1$, and one may replace $x^2/\sinh^2(x/2)$ by 4. Calculating the integral over x in the limit of remote surface ($z_0 \gg d$), when the interference term $\cos(\Gamma x)$ can be neglected, we obtain $\nu^{(m)} = \nu_{ac}$ and $\nu^{(e)} = \nu_{ac}(\bar{\varepsilon}^2/\varepsilon_F T)$. The quantities ν_{ac} and $\bar{\varepsilon}$ are the quasielastic momentum relaxation rate for 2D electrons and characteristic energy introduced in Sec. 35. For the quantum well model considered here, these quantities are calculated in problem 7.5. The above expression for $\nu^{(e)}$ through ν_{ac} and $\bar{\varepsilon}$ can be obtained by integrating the product $\varepsilon J_{ac}(f|\varepsilon)$, where $J_{ac}(f|\varepsilon)$ is given by Eq. (35.6) with f_ε/T replaced by $f_\varepsilon(1 - f_\varepsilon)/T$, over ε. Therefore, Eqs. (29) and (30) at $z_0 \gg d$ describe the relaxation of electrons by bulk phonon modes, while at $z_0 \sim d$ there appears a correction caused by the presence of the surface.

At intermediate temperatures $T_1 \gg T \gg T_0$, when the main contribution comes from $x \sim 1$, one has $F_T(x) \simeq 1$. As a result, the relaxation rates do not depend on the well width d, and Eq. (29) with $F_T(x) = 1$ can be applied for a quantum well of an arbitrary shape. However, the dependence of the relaxation rates on z_0 exists and remains essential even if z_0 is considerably larger than d. If $z_0 \sim d$, one has $\Gamma \ll 1$ and the interference term $\cos(\Gamma x)$ is equal to unity at $x \sim 1$. For a rigid surface, it simply means that the rates $\nu^{(e)}$ and $\nu^{(m)}$ increase twice in comparison to the case of $z_0 \to \infty$, while for a free surface the rates go to zero. Only when $z_0 \gg (T_1/T)d$, the interference contribution becomes small because of the averaging of the fast-oscillating term $\cos(\Gamma x)$.

In the low-temperature regime, when $T \sim T_0$, one cannot neglect q in comparison to ω/s_j. Since in this case $K_\omega(q)$ essentially depends on q, the analysis of Eq. (26) becomes rather complicated and can be done only numerically. An interesting feature of this temperature regime is that the 2D electrons can effectively interact with the Rayleigh surface waves existing in the region $s_t q < \omega < s_l q$. These waves are characterized by an exponential decrease of their amplitudes away from the surface.

The formalism of Green's functions is also convenient for studying the interaction of electrons with optical phonons in the inhomogeneous media containing interfaces and surfaces. The guidelines for such applications are given in problems 9.11 and 9.12.

Problems

9.1. Write the boundary conditions describing the spin-flip processes on a surface.

Solution: Taking into account the spin-orbit interaction by analogy to Eq. (5.4), we write the Schroedinger equation for the two-component wave function (spinor) $\Psi_{\mathbf{r}}$ as

$$\left[\frac{\hat{\mathbf{p}}^2}{2m} + U(\mathbf{r}) + \alpha\hat{\boldsymbol{\sigma}}[\nabla U(\mathbf{r}) \times \hat{\mathbf{p}}] - E \right] \Psi_{\mathbf{r}} = 0,$$

where $U(\mathbf{r}) = \theta(-z)U_0$ is a step-like potential describing the surface and α is a coefficient describing the efficiency of the spin-orbit interaction. Inside of the crystal ($z > 0$), this equation in the $(\mathbf{p}, z)$-representation is written as

$$\left(\frac{\hat{p}_z^2}{2m} + \frac{\mathbf{p}^2}{2m} - E \right) \Psi_{\mathbf{p}z} = 0, \quad z > 0,$$

where $\mathbf{p}$ is the 2D momentum parallel to the surface. Let us integrate the equation for $\Psi_{\mathbf{r}}$ across the surface (from $-\delta$ to δ, where $\delta \to +0$ determines the width of the boundary region). Taking into account that at $z < 0$ one has $\Psi_{\mathbf{p}z} = \Psi_{\mathbf{p},z=0}e^{\kappa z}$ with $\hbar\kappa = \sqrt{2m(U_0 - E + p^2/2m)}$, we obtain the boundary conditions

$$\left[\hat{p}_z + i\chi\hat{\boldsymbol{\sigma}}[\mathbf{n}_z \times \mathbf{p}] + i\sqrt{2m(U_0 - E + p^2/2m)} \right] \Psi_{\mathbf{p}z}\big|_{z=0} = 0,$$

where $\chi = -2mU_0\alpha/\hbar$ and $\mathbf{n}_z$ is the unit vector along OZ. The components of $\Psi_{\mathbf{p}z}$ mix with each other at the surface, owing to the presence of the spin-dependent contribution $\propto \chi$.

9.2. Prove that $[\boldsymbol{\omega}_c \times \mathbf{p}] \cdot \nabla_{\mathbf{p}} = \omega_c \partial/\partial\varphi$.

Hint: Take into account that $\boldsymbol{\omega}_c$ is parallel to OZ and use the polar coordinates.

9.3. Estimate P_0 for the case of small correlation lengths by using Eq. (44.10).

Solution: From Eq. (44.10) it follows that $P_\zeta = P_0\zeta$, where

$$P_0 = \frac{\sqrt{p_z^2 + p^2}}{\pi^2\hbar^4} \int_0^{\sqrt{p_z^2+p^2}} p'dp' \int_0^{2\pi} d\varphi' W(|\mathbf{p} - \mathbf{p}'|)\sqrt{p_z^2 + p^2 - p'^2}$$

and φ' is the angle of the 2D vector $\mathbf{p}$. The total 3D momentum $\sqrt{p_z^2 + p^2}$ can be replaced by the Fermi momentum p_F. Let us approximate W by a Gaussian correlation function. Assuming that the correlation length is small, we have $W(|\mathbf{p} - \mathbf{p}'|) \simeq \pi l_c^2 \bar{\xi}^{-2}$. The integrals in this case are taken easily, with the result $P_0 \simeq (2/3)(p_F/\hbar)^4 l_c^2 \bar{\xi}^2$. The limit of small correlation length implies $(p_F l_c/\hbar)^2 \ll 1$. On the other hand, we have already assumed (see the beginning of Sec. 44) that $\bar{\xi} p_F/\hbar < 1$. Therefore, one always has $P_0 \ll 1$ in the limit of small correlation length.

9.4. Based on Eqs. (45.14) and (45.15), find the static conductivity in the limit of thin film $(d \ll l_F)$ for the case of diffuse scattering.

Hints: Substitute $P_\zeta = 1$ and $r_\pm = 1$ into Eq. (45.15) and calculate the integral over ζ under the assumption $d \ll l_F$, keeping the terms $\propto d/l_F$ and $\propto (d/l_F)^2 \ln(l_F/d)$.

9.5. Calculate the length Λ given by Eq. (46.9) for the samples with circular cross-section (cylinders).

Solution: Using the cylindrical coordinate system with $\mathbf{r} = (r_\perp \cos\alpha, r_\perp \sin\alpha, z)$, we obtain

$$\Lambda = \frac{3}{4\pi^2 R^2} \int_0^R dr_\perp r_\perp \int_0^{2\pi} d\alpha \int_0^{2\pi} d\varphi \int_0^\pi d\theta \sin\theta \cos^2\theta \, |\mathbf{r} - \mathbf{r}_\Gamma|,$$

where R is the radius of the cylinder and $d\widetilde{\Omega}$ is replaced by $d\theta \sin\theta d\varphi$. Next, we find that

$$|\mathbf{r} - \mathbf{r}_\Gamma| = \left[\sqrt{R^2 - r_\perp^2 \sin^2(\varphi - \alpha)} \pm r_\perp \cos(\varphi - \alpha) \right] / \sin\theta,$$

where the sign $+$ or $-$ depends on the direction of $\mathbf{q}$ (Λ does not depend on this sign). An elementary integration leads to the result $\Lambda = 2R$.

9.6. Find the thermal conductivity of a thin film of width d in the hydrodynamical regime and investigate the limiting cases $d \gg R_0$ and $d \ll R_0$.

Result: The analog of Eq. (46.18) in the film geometry is $\kappa = \kappa^{(0)}[1 - (2R_0/d) \times \tanh(d/2R_0)]$. In the limits $d \gg R_0$ and $d \ll R_0$, we have $\kappa = \kappa^{(0)}(1 - 2R_0/d)$ and $\kappa = Td^2 S^2/12\nu$, respectively.

9.7. Consider a phonon system containing only one branch with strictly linear dispersion law. Prove that the solution (46.4) obtained in the collisionless limit is always valid for such a system if we neglect the umklapp processes in the collision integral of Eq. (46.3).

Hint: One has to prove that a substitution of this solution into the linearized collision integral (24.4) makes the latter equal to zero. It is important to take into account that the collision integral for the system defined above describes only the processes for which the wave vectors of three participating phonons are in the same direction. As a result, the non-equilibrium parts of the distribution functions of these phonons contain the same $\mathbf{r}_\Gamma$.

9.8. Derive the elasticity equation

$$\frac{\partial^2}{\partial t^2}\mathbf{u_{r}}_t - s_t^2\nabla^2\mathbf{u_{r}}_t - (s_l^2 - s_t^2)\nabla(\nabla \cdot \mathbf{u_{r}}_t) = 0,$$

starting from the following expression for the elastic energy of a cubic crystal:

$$\mathcal{E}_{el} = \frac{\rho}{2}\int_{(V)} d\mathbf{r}\left[\left(\frac{\partial\mathbf{u_{r}}_t}{\partial t}\right)^2 + s_t^2\sum_{\alpha\beta}(\nabla^\alpha u_{\mathbf{r}t}^\beta)^2 + (s_l^2 - s_t^2)(\nabla \cdot \mathbf{u_{r}}_t)^2\right],$$

where $\mathbf{u_{r}}_t$ is the vector of lattice displacement, ρ is the crystal density, and s_l and s_t are the longitudinal and transverse sound velocities.

Hint: Using the expression for $\mathcal{E}_{el}$, find the Lagrangian density as a function of $\mathbf{u_{r}}_t$ and $\dot{\mathbf{u}}_{\mathbf{r}t} \equiv \partial\mathbf{u_{r}}_t/\partial t$. Compose the Lagrange equations according to problem 1.15.

9.9. Prove that the Green's function $D_{tt'}^{\alpha\beta,R}(\mathbf{r},\mathbf{r}') = \langle\langle\hat{u}_\mathbf{r}^\alpha|\hat{u}_{\mathbf{r}'}^\beta\rangle\rangle_{tt'}^R$ is determined by Eq. (47.10) if the Hamiltonian corresponds to the elastic energy defined in the previous problem.

Solution: The energy $\mathcal{E}_{el}$ is quantized by introducing the canonically conjugate variables $\mathbf{q_r} = \sqrt{\rho}\mathbf{u_r}$ and $\mathbf{p_r} = \sqrt{\rho}d\mathbf{u_r}/dt$. In terms of these variables, one has $\mathbf{p} = \dot{\mathbf{q}} = \partial\varepsilon_{el}/\partial\mathbf{p}$ and $\dot{\mathbf{p}} = \ddot{\mathbf{q}} = -\partial\varepsilon_{el}/\partial\mathbf{q}$, where $\varepsilon_{el} = \varepsilon_{el}(\mathbf{r})$ is the elastic energy density (the relation for $\dot{\mathbf{p}}$ becomes obvious when $\ddot{\mathbf{q}}$ is expressed according to the elasticity equation of the previous problem). Therefore, we replace $\mathbf{q_r}$ and $\mathbf{p_r}$ by the operators of generalized coordinate and momentum satisfying the commutation relation $[\hat{q}_\mathbf{r}^\alpha, \hat{p}_{\mathbf{r}'}^\beta] = i\hbar\delta_{\alpha\beta}\delta(\mathbf{r} - \mathbf{r}')$. The Hamiltonian is quadratic in these operators:

$$\widehat{H}_{el} = \frac{1}{2}\int_{(V)} d\mathbf{r}\left[\hat{\mathbf{p}}_\mathbf{r}^2 + s_t^2\sum_{\alpha\beta}(\nabla^\alpha\hat{q}_\mathbf{r}^\beta)^2 + (s_l^2 - s_t^2)(\nabla \cdot \hat{\mathbf{q}}_\mathbf{r})^2\right].$$

Using Eq. (D.13), one can find that $i\hbar\partial D_{tt'}^{\alpha\beta,R}(\mathbf{r},\mathbf{r}')/\partial t = \langle\langle[\hat{u}_\mathbf{r}^\alpha, \widehat{H}_{el}]|\hat{u}_{\mathbf{r}'}^\beta\rangle\rangle_{tt'}^R$. Taking the derivative of this equation over time, we obtain

$$\hbar^2\frac{\partial^2}{\partial t^2}D_{tt'}^{\alpha\beta,R}(\mathbf{r},\mathbf{r}') = -\delta(t - t')\langle\langle[[\hat{u}_\mathbf{r}^\alpha(t), \widehat{H}_{el}], \hat{u}_{\mathbf{r}'}^\beta(t)]\rangle\rangle + \langle\langle[\widehat{H}_{el}, [\hat{u}_\mathbf{r}^\alpha, \widehat{H}_{el}]]|\hat{u}_{\mathbf{r}'}^\beta\rangle\rangle_{tt'}^R.$$

Substituting the Hamiltonian given above into this equation, we transform the commutators with the use of the commutation relations for $\hat{\mathbf{q}}$ and $\hat{\mathbf{p}}$. Finally,

$$\rho\sum_\beta\left[\frac{\partial^2}{\partial t^2}\delta_{\alpha\beta} - s_t^2\nabla_\mathbf{r}^2\delta_{\alpha\beta} - (s_l^2 - s_t^2)\nabla_\mathbf{r}^\alpha\nabla_\mathbf{r}^\beta\right]D_{tt'}^{\beta\gamma,R}(\mathbf{r},\mathbf{r}') = -\delta_{\alpha\gamma}\delta(t - t')\delta(\mathbf{r} - \mathbf{r}').$$

In the energy representation, this equation is equivalent to Eq. (47.10).

9.10. Derive the boundary conditions (47.17) for a free surface.

Solution: The stress tensor $\sigma_{\alpha\beta}(\mathbf{r})$ is defined as

$$\sigma_{\alpha\beta}(\mathbf{r}) = \lambda(\nabla \cdot \mathbf{u_r})\delta_{\alpha\beta} + \mu\left(\frac{\partial u_\mathbf{r}^\alpha}{\partial x_\beta} + \frac{\partial u_\mathbf{r}^\beta}{\partial x_\alpha}\right),$$

where x_α are the Cartesian coordinates and λ and μ are the Lamé coefficients expressed through the sound velocities according to Eq. (6.28). By definition, the normal stress $\sigma_{z\beta}(\mathbf{r})$ should be zero at the surface $z = 0$. Since the in-plane Fourier transform of the derivative $\partial u_\mathbf{r}^\alpha / \partial x_\beta$ with $\beta = x, y$ is $iq_\beta u_\mathbf{qz}^\alpha$, we obtain the following boundary conditions for the displacement vectors in the $(\mathbf{p}, z)$-representation:

$$\left(\frac{du_\mathbf{qz}^x}{dz} + iq_x u_\mathbf{qz}^z \right)_{z=0} = 0, \quad \left(\frac{du_\mathbf{qz}^z}{dz} + iq_x \frac{\lambda}{\lambda + 2\mu} u_\mathbf{qz}^x \right)_{z=0} = 0.$$

The boundary conditions (47.17) for the Green's functions directly follow from these equations at $q_x = q$ and $q_y = 0$ if we notice that $\lambda/(\lambda + 2\mu) = 1 - 2(s_t/s_l)^2$.

9.11. Derive the kinetic equation for the electrons interacting with long-wavelength optical phonons in a spatially-inhomogeneous medium.

Solution: The interaction of electrons with electrostatic field generated by long-wavelength ionic vibrations is described by the potential energy $U(\mathbf{r})$ introduced in Sec. 6. In the homogeneous media, this energy is quantized according to Eq. (6.22). Considering spatially-inhomogeneous systems, it is convenient to introduce the operator $\widehat{U}_\mathbf{r}$ corresponding to the potential energy $U(\mathbf{r})$ and compose the Green's functions $d_{tt'}^{R,A}(\mathbf{r}, \mathbf{r}') = \langle\langle \widehat{U}_\mathbf{r} | \widehat{U}_{\mathbf{r}'} \rangle\rangle_{tt'}^{R,A}$. Further consideration is the same as in the case of deformation-potential interaction considered in Sec. 47, and $\widehat{U}_\mathbf{r}$ replaces the Hamiltonian $\mathcal{D}(\nabla \cdot \hat{\mathbf{u}}_\mathbf{r})$ in Eq. (47.12). Therefore, the function $K_\omega(\mathbf{q}|z, z)$ determining the transition probabilities (47.13) in the collision integral (34.26) is now defined as

$$K_\omega(\mathbf{q}|z, z') = i \int d\Delta\mathbf{x} e^{-i\mathbf{q}\cdot\Delta\mathbf{x}} [d_\omega^R(\mathbf{r}, \mathbf{r}') - d_\omega^A(\mathbf{r}, \mathbf{r}')]$$

$$= i[d_\omega^R(\mathbf{q}|z, z') - d_\omega^A(\mathbf{q}|z, z')].$$

To derive an equation for the Green's function, we write the mechanical energy of the long-wavelength longitudinal optical vibrations (see Sec. 6):

$$\mathcal{E}_{LO} = \frac{1}{2} \int_{(V)} d\mathbf{r} \left(\dot{\mathbf{w}}^2 + \omega_{LO}^2 \mathbf{w}^2 \right) = 2\pi \int_{(V)} d\mathbf{r} \, \epsilon^* \left(\dot{\mathbf{P}}^2/\omega_{LO}^2 + \mathbf{P}^2 \right)$$

$$= \frac{1}{8\pi e^2} \int_{(V)} d\mathbf{r} \, \epsilon^* \left[(\nabla\dot{U})^2/\omega_{LO}^2 + (\nabla U)^2 \right],$$

where $\mathbf{P}$ is the lattice polarization. To transform this expression, we have used Eq. (6.20) together with the Poisson equation $\nabla U/e = 4\pi\mathbf{P}$. The effective dielectric constant ϵ^* and LO phonon frequency ω_{LO} depend on coordinate $\mathbf{r}$ in spatially-inhomogeneous media. Using the last expression of $\mathcal{E}_{LO}$, one can compose the Lagrange equation of motion for U and the corresponding equation for the Green's function (compare $\mathcal{E}_{LO}$ to $\mathcal{E}_{el}$ of problem 9.8):

$$\left[(\omega + i0)^2 \nabla \frac{\epsilon^*}{\omega_{LO}^2} \nabla - \nabla\epsilon^*\nabla \right] d_\omega^R(\mathbf{r}, \mathbf{r}') = 4\pi e^2 \delta(\mathbf{r} - \mathbf{r}').$$

This equation replaces Eq. (47.10). Solving it with proper boundary conditions (see problem 9.12), one can find $K_\omega(\mathbf{q}|z, z')$ written above.

9.12. Calculate the Green's function $d_\omega^R(\mathbf{r}, \mathbf{r}')$ (see problem 9.11) for a system composed of two different materials occupying the regions $z > 0$ and $z < 0$.

Solution: The parameters ϵ^* and ω_{LO} are assumed to be constants in the regions $z > 0$ and $z < 0$, but they change abruptly at the interface. It is convenient to introduce the function $\gamma(z) = \epsilon^*[(\omega + i0)^2/\omega_{LO}^2 - 1]$ equal to γ_+ at $z > 0$ and to γ_- at $z < 0$. The Green's function can be written in the $(\mathbf{q}, z)$-representation as $d_\omega^R(\mathbf{q}|z, z')$. It is independent of the direction of $\mathbf{q}$ and satisfies the equation

$$\left[\frac{\partial^2}{\partial z^2} - q^2\right] d_\omega^R(q|z, z') = \frac{4\pi e^2}{\gamma_\pm} \delta(z - z')$$

in the regions $z > 0$ (+) and $z < 0$ (−). This equation is to be solved with the boundary conditions demanding that $d_\omega^R(q|z, z')$ and $\gamma(z)\partial d_\omega^R(q|z, z')/\partial z$ are continuous at $z = 0$ and that $d_\omega^R(q|z, z')$ is finite at $z = \pm\infty$. We obtain the following result:

$$d_\omega^R(q|z, z') = -\frac{2\pi e^2}{\gamma_\pm q} \left(e^{-q|z-z'|} + \frac{\gamma_\pm - \gamma_\mp}{\gamma_+ + \gamma_-} e^{-q|z| - q|z'|}\right),$$

where the upper and lower indices correspond to $z > 0$ and $z < 0$, respectively. Since this function is invariant with respect to the permutation of z and z', the function $K_\omega(q|z, z')$ determining the probability of electron-phonon interaction is equal to $-2\mathrm{Im} d_\omega^R(q|z, z')$. In addition to the poles $\gamma_\pm = 0$ at the frequencies of bulk LO phonon modes of the materials, the Green's function has one more pole determined by the equation $\gamma_+ + \gamma_- = 0$. This pole corresponds to a mode localized at the interface.

Chapter 10

QUANTUM MAGNETOTRANSPORT

A magnetic field causes electrons to rotate in the plane perpendicular to this field. As a result (see Sec. 11), the anisotropy of the response along the field and in the plane perpendicular to the field appears, and the kinetic coefficients are modified considerably if the cyclotron frequency exceeds the relaxation rate. Further qualitative modifications take place under the transition from the quasi-classical fields satisfying Eq. (9.35) to quantizing magnetic fields. Even when the cyclotron energy is still small in comparison to the Fermi energy, there appear oscillations of the conductivity as a function of the magnetic field. Another kind of oscillations takes place because of interaction of electrons with the dispersionless optical phonons. In the region where the cyclotron energy is comparable to the Fermi energy, the validity condition (7.21) of the kinetic equation (7.13) must be critically reconsidered because of accumulation of electrons at the bottoms of the Landau levels, where the density of states (5.16) increases according to the inverse square root dependence, leading to an increase in the relaxation rate. In the case of 2D electron gas in the magnetic field perpendicular to the 2D plane, the density of states of free electrons is represented by a set of δ-peaks associated with the Landau levels; see Sec. 16. For this reason, any description of electron transport based upon the quasi-classical kinetic equation (9.34) becomes invalid, and completely new kinetic phenomena, such as the quantum Hall effect, appear. The quantization of electron states in magnetic fields also has a considerable influence on the optical properties of dielectrics. There exists an anisotropy of the response due to virtual interband transitions, and the characteristics of exciton absorption are dramatically modified.

48. Method of Iterations

In strong fields, when the cyclotron frequency is much greater than the relaxation rate, it is possible to solve the kinetic equation (9.34) by iterations with respect to the collision integral, similar to what we did when considering the response to the high-frequency field in Secs. 10 and 37. For the stationary and spatially-homogeneous case, the distribution function is written as $f(\varepsilon_p) + \Delta f_{\mathbf{p}}$, where $f(\varepsilon)$ is the Fermi distribution function and $\Delta f_{\mathbf{p}}$ describes a linear response to the electric field $\mathbf{E}$. Under the conditions $\omega_c \tau_{tr} \gg 1$, one uses the expansion $\Delta f_{\mathbf{p}} = \Delta f_{\mathbf{p}}^{(0)} + \Delta f_{\mathbf{p}}^{(1)} + \ldots$, and the zero-order equation takes the form

$$\frac{e}{c}[\mathbf{v_p} \times \mathbf{H}] \cdot \frac{\partial \Delta f_{\mathbf{p}}^{(0)}}{\partial \mathbf{p}} = -e\mathbf{E} \cdot \mathbf{v_p} \frac{\partial f(\varepsilon_p)}{\partial \varepsilon_p}, \tag{1}$$

where we have neglected a small contribution of the collision integral. The exact solution of Eq. (1) is

$$\Delta f_{\mathbf{p}}^{(0)} = \frac{c}{H^2} \frac{\partial f(\varepsilon_p)}{\partial \varepsilon_p} [\mathbf{H} \times \mathbf{E}] \cdot \mathbf{p}. \tag{2}$$

If we substitute it into Eq. (9.11) for the electric current and consider the latter in the direction perpendicular to $\mathbf{H}$, we obtain the expression for the non-dissipative Hall current, $\mathbf{I} = |e|cn[\mathbf{H} \times \mathbf{E}]/H^2$, which can be also found from Eq. (11.19) in the limit $\omega_c \tau_{tr} \gg 1$. The correction $\Delta f_{\mathbf{p}}^{(1)}$ is governed by the kinetic equation of the first-order approximation:

$$\frac{e}{c}[\mathbf{v_p} \times \mathbf{H}] \cdot \frac{\partial \Delta f_{\mathbf{p}}^{(1)}}{\partial \mathbf{p}} = J_{im}(\Delta f^{(0)}|\mathbf{p}). \tag{3}$$

For degenerate electron gas, this correction leads to the contribution $\Delta \mathbf{I} = (e^2 n \tau_{tr}/m)(\omega_c \tau_{tr})^{-2}\mathbf{E}$ to the current density (problem 10.1). It is easy to identify this contribution with the density of dissipative current in the limit $\omega_c \tau_{tr} \gg 1$; see again Eq. (11.19). The corrections of higher order give higher-order terms to the expansion of the current in powers of $1/\omega_c \tau_{tr}$.

The iterations with respect to the factor $1/\omega_c$ can be also used for calculating the transverse conductivity in the strong-field regime, when the requirement (9.35) is violated and the quasi-classical kinetic equation is no longer applicable. Below we apply the Kubo formula to calculate the diagonal part of the transverse conductivity tensor in strong magnetic fields. Then we show how the conductivity can be found from a semi-classical consideration of the electrons hopping between the sites defined as cyclotron orbit centers. Finally, we generalize our results to describe the electrons interacting with phonons and consider the magnetophonon oscillations of the transverse conductivity. It is convenient to write the Kubo formula (13.18) as (problem 10.2)

$$\sigma_{\alpha\beta}(\omega) = i\frac{\hbar e^2}{V}\sum_{\delta\delta'}\langle\delta|\hat{v}_\beta|\delta'\rangle\langle\delta'|\hat{v}_\alpha|\delta\rangle\frac{f(\varepsilon_\delta) - f(\varepsilon_{\delta'})}{(\varepsilon_\delta - \varepsilon_{\delta'} - \hbar\omega - i\lambda)(\varepsilon_\delta - \varepsilon_{\delta'})}, \quad (4)$$

where the energies ε_δ and the states $|\delta\rangle$ are determined from the eigenstate problem $\widehat{H}|\delta\rangle = \varepsilon_\delta|\delta\rangle$. In the presence of the magnetic field $\mathbf{H} = [\nabla \times \mathbf{A}] = (0, 0, H)$, Eq. (4) contains the velocity operators $\hat{v}_\alpha = \hat{\pi}_\alpha/m = [\hat{p}_\alpha - (e/c)A_\alpha]/m$. In the basis of Landau states $|Np_yp_z\rangle$ described by the wave functions $\psi_{\mathbf{r}}^{(Np_yp_z)}$ given by Eqs. (5.13) and (5.15) [the vector potential is $\mathbf{A} = (0, Hx, 0)$], these operators have the following matrix elements:

$$\langle Np_yp_z| \begin{vmatrix} \hat{v}_x \\ \hat{v}_y \end{vmatrix} |N'p'_yp'_z\rangle = \sqrt{\frac{\hbar\omega_c}{2m}}\left(\sqrt{N}\delta_{N',N-1}\begin{vmatrix} i \\ 1 \end{vmatrix}\right.$$

$$\left. +\sqrt{N'}\delta_{N',N+1}\begin{vmatrix} -i \\ 1 \end{vmatrix}\right)\delta_{p_yp'_y}\delta_{p_zp'_z}, \quad (5)$$

and $\langle Np_yp_z|\hat{v}_z|N'p'_yp'_z\rangle = (p_z/m)\delta_{N'N}\delta_{p_yp'_y}\delta_{p_zp'_z}$. Using Eqs. (4) and (5), one can see that in the absence of scattering, when the Landau states are identified with the exact eigenstates, the real parts of the diagonal components σ_{xx} and σ_{yy} at $\omega = 0$ are equal to zero, since their expressions contain δ-functions of energies, $\delta(\varepsilon_\delta - \varepsilon_{\delta'})$. The energy conservation law $\varepsilon_\delta = \varepsilon_{\delta'}$ cannot be satisfied because the matrix elements (5) are diagonal in the longitudinal momentum p_z but non-diagonal in the Landau level numbers. In contrast, the non-diagonal components σ_{xy} and σ_{yx} remain non-zero in the absence of scattering, which corresponds to the non-dissipative transport.

We have emphasized that the transverse dissipative transport in a magnetic field is a scattering-assisted process. To find the diagonal conductivity from Eq. (13.18), one has to expand the density matrix $\hat{\eta}_{eq}$ and the exponential operators $e^{\pm i\widehat{H}\tau/\hbar}$, where $\widehat{H} = \hat{\pi}^2/2m + U_{sc}(\mathbf{r})$, in series with respect to the potential energy $U_{sc}(\mathbf{r})$ responsible for the scattering and search for the first nonvanishing contribution. The most convenient way of doing it is based upon the operator identities

$$\hat{v}_x = \frac{1}{m\omega_c}\frac{\partial U_{sc}(\mathbf{r})}{\partial y} - \frac{i}{\hbar\omega_c}[\hat{v}_y, \widehat{H}], \quad \hat{v}_y = -\frac{1}{m\omega_c}\frac{\partial U_{sc}(\mathbf{r})}{\partial x} + \frac{i}{\hbar\omega_c}[\hat{v}_x, \widehat{H}]. \quad (6)$$

To check these relations, one may calculate the commutators on the right-hand sides by taking into account that $[\hat{\pi}_x, \hat{\pi}_y^2]/2m = -i\hbar\omega_c\hat{\pi}_y$ and

$[\hat{\pi}_y, \hat{\pi}_x^2]/2m = i\hbar\omega_c\hat{\pi}_x$. Equations (6) have clear physical meaning. The velocity of the electron moving in the direction perpendicular to **H** is subdivided in two parts. The first part, containing the gradients of the potential energy, corresponds to the motion of the cyclotron orbit center, while the second part describes the rotational motion in the cyclotron orbit.

Let us substitute the velocity operators rewritten according to Eq. (6) into Eq. (13.18) for σ_{xx}. The commutator standing in Eq. (13.18) is written as a sum of three terms:

$$\left[e^{-i\widehat{H}\tau/\hbar}\hat{v}_x e^{i\widehat{H}\tau/\hbar}, \hat{v}_x\right] = -(\hbar\omega_c)^{-2}\left[e^{-i\widehat{H}\tau/\hbar}[\hat{v}_y, \widehat{H}]e^{i\widehat{H}\tau/\hbar}, [\hat{v}_y, \widehat{H}]\right]$$

$$-i(\hbar m\omega_c^2)^{-1}\left(\left[e^{-i\widehat{H}\tau/\hbar}[\hat{v}_y, \widehat{H}]e^{i\widehat{H}\tau/\hbar}, (\partial U_{sc}/\partial y)\right]\right.$$

$$\left.+\left[e^{-i\widehat{H}\tau/\hbar}(\partial U_{sc}/\partial y)e^{i\widehat{H}\tau/\hbar}, [\hat{v}_y, \widehat{H}]\right]\right) \tag{7}$$

$$+(m\omega_c)^{-2}\left[e^{-i\widehat{H}\tau/\hbar}(\partial U_{sc}/\partial y)e^{i\widehat{H}\tau/\hbar}, (\partial U_{sc}/\partial y)\right].$$

Only the third (last) term of this sum contributes to the conductivity at $\omega = 0$ (below we are interested only in $\text{Re}\sigma_{xx} = \sigma_{xx}$; the imaginary part vanishes at $\omega = 0$). This statement can be checked if one writes the Kubo formula in the basis of exact eigenstates. In this representation, the first and the second terms of the expansion (7) give the contributions proportional to $\omega[f(\varepsilon_\delta) - f(\varepsilon_\delta + \omega)]$ and $f(\varepsilon_\delta) - f(\varepsilon_\delta + \omega)$, respectively, which vanish when ω goes to zero. We have, therefore,

$$\sigma_{xx} = \frac{e^2}{\hbar V m^2\omega_c^2}\lim_{\omega\to 0}\frac{1}{\omega}\text{Re}\int_{-\infty}^0 d\tau e^{\lambda\tau - i\omega\tau}$$

$$\times\text{Sp}\hat{\eta}_{eq}\left[e^{-i\widehat{H}\tau/\hbar}(\partial U_{sc}/\partial y)e^{i\widehat{H}\tau/\hbar}, (\partial U_{sc}/\partial y)\right]. \tag{8}$$

This equation is exact, and its main advantage is that the velocity operators are expressed through the gradients of the potential energy. The conductivity given by Eq. (8) is proportional to the factor $(\overline{U}_{sc}/\hbar\omega_c)^2$, where $\overline{U}_{sc}$ gives an estimate of the random potential amplitude. An expansion of $e^{\pm i\widehat{H}\tau/\hbar}$ and $\hat{\eta}_{eq}$ in powers of U_{sc} demonstrates that all subsequent terms contain higher powers of this factor. Therefore, to find the conductivity in the limit of high fields, one may neglect the scattering-potential contribution to the Hamiltonian by substituting the Hamiltonian $\widehat{H} = \hat{\pi}^2/2m$ into Eq. (8) and calculating the trace over the Landau states which become the exact eigenstates in this approximation. The simplest application of this method is realized for the electron-impurity system. We obtain

$$\sigma_{xx} = \frac{2\pi e^2}{V m^2 \omega_c^2} \lim_{\omega \to 0} \frac{1}{\omega} \sum_{\delta\delta'} \left\langle\!\left\langle |\langle\delta|\partial U_{im}(\mathbf{r})/\partial y|\delta'\rangle|^2 \right\rangle\!\right\rangle$$

$$\times \delta(\varepsilon_\delta - \varepsilon_{\delta'} + \hbar\omega)[f(\varepsilon_\delta) - f(\varepsilon_{\delta'})], \tag{9}$$

where $|\delta\rangle$ and $|\delta'\rangle$ denote the Landau states $|Np_yp_z\rangle$ and $|N'p_y'p_z'\rangle$, respectively. Equation (9) can be also obtained from Eq. (4) by substituting there the matrix elements of $\hat{v}_x$ expressed according to Eq. (6). Since the spin splitting is neglected, the factor 2 in Eq. (9) accounts for the spin degeneracy (in the general case σ_{xx} can be represented as a sum of the contributions from spin-up and spin-down states, and the only difference between these contributions is a Zeeman shift of the energies; see the next section for more details). The double angular brackets denote the averaging over the random potential of impurities. This averaging gives us (see Sec. 14)

$$\left\langle\!\left\langle |\langle\delta|\partial U_{im}(\mathbf{r})/\partial y|\delta'\rangle|^2 \right\rangle\!\right\rangle = \frac{n_{im}}{V} \sum_{\mathbf{q}} q_y^2 |v(q)|^2 |\langle\delta|e^{i\mathbf{q}\cdot\mathbf{r}}|\delta'\rangle|^2. \tag{10}$$

Since the matrix elements of $e^{i\mathbf{q}\cdot\mathbf{r}}$ contain the δ-function $\delta_{p_y,p_y'+\hbar q_y}$, one can replace q_y^2 in Eq. (10) by $(X_{p_y} - X_{p_y'})^2/l_H^4$, where $X_{p_y} = -p_y c/|e|H$ is the "weight center" of the oscillatory wave function (the classical cyclotron orbit center). Finally, we obtain the following expression for the static conductivity:

$$\sigma_{xx} = \frac{2e^2}{TV} \sum_{\delta\delta'} f(\varepsilon_\delta)[1 - f(\varepsilon_{\delta'})]\nu_{\delta\delta'}(X_\delta - X_{\delta'})^2/2, \tag{11}$$

where $X_\delta = X_{p_y}$ does not depend on N and

$$\nu_{\delta\delta'} = \frac{2\pi}{\hbar} \frac{n_{im}}{V} \sum_{\mathbf{q}} |v(q)|^2 |\langle\delta|e^{i\mathbf{q}\cdot\mathbf{r}}|\delta'\rangle|^2 \delta(\varepsilon_\delta - \varepsilon_{\delta'}) \tag{12}$$

is the probability of the impurity-assisted transition between the states δ and δ'. Equation (11) is known as Adams-Holstein formula. It is written in the general form valid for describing any scattering-assisted hopping conductivity; see also Sec. 62. In our case, the positions of the sites between which the hopping occurs coincide with the centers of the classical cyclotron orbits, X_δ. To obtain Eq. (11), one can present a semiclassical consideration based upon the equation for the hopping current density (problem 10.3).

The matrix elements of the exponent $e^{i\mathbf{q}\cdot\mathbf{r}}$ in the basis of Landau states are given by (compare to Eq. (A.27))

$$\langle Np_y p_z | e^{i\mathbf{q}\cdot\mathbf{r}} | N' p'_y p'_z \rangle = \sqrt{\frac{N!2^{N'}}{N'!2^N}} [i(q_x + iq_y)/2l_H]^{N'-N} L_N^{N'-N}(q_\perp^2 l_H^2/2)$$

$$\times e^{-q_\perp^2 l_H^2/4} e^{-ip_y q_x l_H^2/\hbar + iq_y q_x l_H^2/2} \delta_{p_z, p'_z + \hbar q_z} \delta_{p_y, p'_y + \hbar q_y}, \qquad (13)$$

where $q_\perp^2 = q_x^2 + q_y^2$ and $L_N^M(x)$ are the Laguerre polynomials. The squared absolute value of this matrix element does not depend on the phase factor and can be presented as

$$|\langle Np_y p_z | e^{i\mathbf{q}\cdot\mathbf{r}} | N' p'_y p'_z \rangle|^2 = \Phi_{NN'}(q_\perp^2 l_H^2/2) \delta_{p_z, p'_z + \hbar q_z} \delta_{p_y, p'_y + \hbar q_y},$$

$$\Phi_{NN'}(u) = \frac{N!}{N'!} u^{N'-N} e^{-u} [L_N^{N'-N}(u)]^2. \qquad (14)$$

Equations (13) and (14) are written for $N' \geq N$ (if $N' < N$, one has to permute the indices, $N \leftrightarrow N'$). The impurity-scattering-assisted conductivity defined by Eqs. (11), (12), and (14) can be expressed in a more simple way in the case of degenerate electrons interacting with a short-range impurity potential. Since $\varepsilon_\delta = \varepsilon_{\delta'}$, we put $f(\varepsilon_\delta)[1 - f(\varepsilon_{\delta'})]/T \simeq \delta(\varepsilon_\delta - \varepsilon_F)$, where ε_F is the Fermi energy. Next, we replace $n_{im}|v(q)|^2$ by $n_{im}|v(0)|^2 \equiv w$ and substitute the expression (12) into Eq. (11). Taking into account that the energies of electron states do not depend on the variables p_y and p'_y, we integrate over these variables (note that $\int dp_y = L_x |e| H/c$, where L_x is the normalization length; see Eq. (5.16)) and over q_z. The integrals over p_z and p'_z are easily calculated by using the δ-functions of energies, since $\varepsilon_\delta = \varepsilon_{Np_z} = p_z^2/2m + \hbar\omega_c(N + 1/2)$. The remaining integrals over q_x and q_y are reduced to a single integral over $q_\perp$, which is transformed to $\int_0^\infty du\, u\Phi_{NN'}(u)$. This latter integral is equal to $N + N' + 1$ (problem 10.4), and the expression for σ_{xx} is finally written as

$$\sigma_{xx} = \frac{e^2 w m^2 \omega_c}{4\pi^3 \hbar^4} \sum_{NN'}^{N_{max}} \frac{N + N' + 1}{\sqrt{\varepsilon_F - \varepsilon_N}\sqrt{\varepsilon_F - \varepsilon_{N'}}}, \qquad (15)$$

where $\varepsilon_N = \hbar\omega_c(N + 1/2)$. Only the discrete sums over the Landau level numbers remain in Eq. (15). The upper limit of the sums is determined by the requirement that the expressions under the square roots must be positive, i.e., N_{max} is equal to the integer part of the expression $\varepsilon_F/\hbar\omega_c - 1/2$. Using Eq. (15), it is easy to carry out the quasi-classical limiting transition at $\varepsilon_F \gg \hbar\omega_c$. Introducing the continuous variables $\varepsilon = \hbar\omega_c(N + 1/2)$ and $\varepsilon' = \hbar\omega_c(N' + 1/2)$ instead of N and N', we take the integrals over these variables and obtain $\sigma_{xx} = 4e^2 w m^2 \varepsilon_F^2 / 3\pi^2 \hbar^7 \omega_c^2$. Introducing the relaxation time according to Eq. (8.21) as $1/\tau = \pi w \rho_{3D}(\varepsilon_F)/\hbar$, one can then express the Fermi energy through the electron density n and identify this result with the quasi-classical expression $\sigma_{xx} \simeq e^2 n / m \omega_c^2 \tau$ obtained earlier.

The scattering probability (12) can be written as a sum of partial contributions from all possible scattering mechanisms. So far we have considered the elastic scattering. Below we derive Eq. (11) for the case of electron-phonon interaction. We replace $U_{sc}(\mathbf{r})$ in Eq. (8) by the second-quantized electron-phonon interaction Hamiltonian $\widehat{H}_{e,ph}(\mathbf{r}) = \sum_{\mathbf{q}l} C_q^{(l)} e^{i\mathbf{q}\cdot\mathbf{r}} (\hat{b}_{\mathbf{q}l} + \hat{b}_{-\mathbf{q}l}^+)$ describing the energy of the interaction at the point $\mathbf{r}$; see Eqs. (6.22) and (6.30). The coupling energy $C_q^{(l)}$ for the interaction of electrons with the longitudinal optical ($l = LO$) and longitudinal acoustic ($l = LA$) phonons is given by Eq. (21.1). The gradient of the potential energy takes the following form:

$$\frac{\partial \widehat{H}_{e,ph}(\mathbf{r})}{\partial y} = i \sum_{\mathbf{q}l} q_y C_q^{(l)} e^{i\mathbf{q}\cdot\mathbf{r}} (\hat{b}_{\mathbf{q}l} + \hat{b}_{-\mathbf{q}l}^+). \tag{16}$$

Neglecting the contributions of higher order with respect to $(C_q^{(l)}/\hbar\omega_c)^2$, and using the second quantization representation, we obtain

$$\sigma_{xx} = \frac{e^2}{\hbar V m^2 \omega_c^2} \lim_{\omega \to 0} \frac{1}{\omega} \mathrm{Re} \sum_{\mathbf{q}\mathbf{q}'l} q_y q_y' C_q^{(l)} C_{q'}^{(l)}$$

$$\times \sum_{\delta\delta'\delta_1\delta_1'} \langle\delta|e^{i\mathbf{q}\cdot\mathbf{r}}|\delta'\rangle\langle\delta_1'|e^{-i\mathbf{q}'\cdot\mathbf{r}}|\delta_1\rangle \int_{-\infty}^0 d\tau e^{\lambda\tau - i\omega\tau} \tag{17}$$

$$\times \mathrm{Sp}\hat{\eta}_{eq} \left[e^{-i\widehat{H}\tau/\hbar} \hat{a}_\delta^+ \hat{a}_{\delta'} (\hat{b}_{\mathbf{q}l} + \hat{b}_{-\mathbf{q}l}^+) e^{i\widehat{H}\tau/\hbar}, \hat{a}_{\delta_1'}^+ \hat{a}_{\delta_1} (\hat{b}_{-\mathbf{q}'l} + \hat{b}_{\mathbf{q}'l}^+) \right],$$

where

$$\widehat{H} = \widehat{H}_e + \widehat{H}_{ph} = \sum_\delta \varepsilon_\delta \hat{a}_\delta^+ \hat{a}_\delta + \sum_{\mathbf{q}l} \hbar\omega_{ql} \left(\hat{b}_{\mathbf{q}l}^+ \hat{b}_{\mathbf{q}l} + \frac{1}{2} \right) \tag{18}$$

is the Hamiltonian describing non-interacting electrons and phonons, and the equilibrium statistical operator $\hat{\eta}_{eq}$ is determined by this Hamiltonian. Therefore, the traces over electron and phonon variables in Eq. (17) can be taken separately. Calculating the trace over phonon variables, see Sec. 34 for details, we obtain the following expression:

$$\text{Sp}\ldots = \delta_{\mathbf{qq}'}\left(\text{Sp}_e\hat{\eta}_{eq}\hat{a}_\delta^+(-\tau)\hat{a}_{\delta'}(-\tau)\hat{a}_{\delta_1'}^+\hat{a}_{\delta_1}\right)$$

$$\times[(N_q^l+1)e^{i\omega_{ql}\tau}+N_q^le^{-i\omega_{ql}\tau}]-\delta_{\mathbf{qq}'}\left(\text{Sp}_e\hat{\eta}_{eq}\hat{a}_{\delta_1'}^+\hat{a}_{\delta_1}\hat{a}_\delta^+(-\tau)\hat{a}_{\delta'}(-\tau)\right) \quad (19)$$

$$\times[(N_q^l+1)e^{-i\omega_{ql}\tau}+N_q^le^{i\omega_{ql}\tau}],$$

where $\hat{a}_\delta^+(t)=e^{i\widehat{H}_et/\hbar}\hat{a}_\delta^+e^{-i\widehat{H}_et/\hbar}$ and $\hat{a}_\delta(t)=e^{i\widehat{H}_et/\hbar}\hat{a}_\delta e^{-i\widehat{H}_et/\hbar}$ are the creation and annihilation operators in the Heisenberg representation. Next, N_q^l are the phonon occupation numbers (since we consider equilibrium phonons, the distribution over the wave vectors $\mathbf{q}$ is isotropic). The remaining trace $\text{Sp}_e\ldots$ is taken over the electron states described by the Hamiltonian $\widehat{H}_e$. Therefore, we have $\hat{a}_\delta^+(t)=\hat{a}_\delta^+e^{i\varepsilon_\delta t/\hbar}$, $\hat{a}_\delta(t)=\hat{a}_\delta e^{-i\varepsilon_\delta t/\hbar}$, and $\text{Sp}_e\hat{\eta}_{eq}\hat{a}_\delta^+\hat{a}_{\delta'}\hat{a}_{\delta_1'}^+\hat{a}_{\delta_1}=2\delta_{\delta\delta_1}\delta_{\delta'\delta_1'}f(\varepsilon_\delta)[1-f(\varepsilon_{\delta'})]$, where the factor of 2 appears because of spin degeneracy if the spin index is not included in δ. Expression (19) is rewritten as a product of $2\delta_{\mathbf{qq}'}\delta_{\delta\delta_1}\delta_{\delta'\delta_1'}$ by the factor

$$e^{i(\varepsilon_{\delta'}-\varepsilon_\delta+\hbar\omega_{ql})\tau/\hbar}\left\{f(\varepsilon_\delta)[1-f(\varepsilon_{\delta'})](N_q^l+1)-f(\varepsilon_{\delta'})[1-f(\varepsilon_\delta)]N_q^l\right\}$$

$$\tag{20}$$

$$+e^{i(\varepsilon_{\delta'}-\varepsilon_\delta-\hbar\omega_{ql})\tau/\hbar}\left\{f(\varepsilon_\delta)[1-f(\varepsilon_{\delta'})]N_q^l-f(\varepsilon_{\delta'})[1-f(\varepsilon_\delta)](N_q^l+1)\right\}.$$

We use Eq. (20) in order to calculate the integral over τ in Eq. (17). Carrying out the limiting transition $\omega\to0$ in Eq. (17), we obtain Eq. (11) with the electron-phonon scattering probability

$$\nu_{\delta\delta'}=\frac{2\pi}{\hbar}\sum_{\mathbf{q}l}|C_q^{(l)}|^2|\langle\delta|e^{i\mathbf{q}\cdot\mathbf{r}}|\delta'\rangle|^2 \quad (21)$$

$$\times\left[\delta(\varepsilon_\delta-\varepsilon_{\delta'}-\hbar\omega_{ql})(N_q^l+1)+\delta(\varepsilon_\delta-\varepsilon_{\delta'}+\hbar\omega_{ql})N_q^l\right].$$

Let us calculate the conductivity σ_{xx} of the electrons interacting with long-wavelength optical phonons whose dispersion can be neglected: $\omega_{ql}=\omega_{LO}$ and $N_q^l=N_{LO}=[e^{\hbar\omega_{LO}/T}-1]^{-1}$. Substituting Eq. (21) into Eq. (11), we calculate the integrals over p_y, p_y', p_z', and q_z by using the delta-functions $\delta_{p_z,p_z'+\hbar q_z}$ and $\delta(\varepsilon_{Np_z}-\varepsilon_{N'p_z'}\pm\hbar\omega_{LO})$. As a result,

$$\sigma_{xx}^{(LO)}=\frac{2e^2N_{LO}}{(2\pi)^4\hbar^2\omega_cT}\int d\varepsilon\sum_{NN'}A_{NN'}(\varepsilon)\frac{f(\varepsilon)[1-f(\varepsilon+\hbar\omega_{LO})]}{\sqrt{\varepsilon-\varepsilon_N}\sqrt{\varepsilon-\varepsilon_{N'}+\hbar\omega_{LO}}}, \quad (22)$$

where we employ the variable $\varepsilon=\varepsilon_{Np_z}$ instead of p_z. In Eq. (22) we have introduced a slowly varying function of energy and Landau level numbers according to

$$A_{NN'}(\varepsilon) = \frac{\pi}{2} \int_0^\infty dq_\perp^2 \left(V \sum_\pm |C_{q\pm}^{(LO)}|^2 \right) q_\perp^2 \Phi_{NN'}(q_\perp^2 l_H^2/2), \qquad (23)$$

where

$$q_\pm = \sqrt{q_{z\pm}^2(\varepsilon) + q_\perp^2},$$

$$q_{z\pm}^2(\varepsilon) = \frac{2m}{\hbar^2} \left[\sqrt{\varepsilon - \varepsilon_N} \pm \sqrt{\varepsilon - \varepsilon_{N'} + \hbar\omega_{LO}} \right]^2. \qquad (24)$$

The quantity $\hbar q_{z\pm}$ is the momentum transferred along the magnetic field for backscattering $(+)$ and forward-scattering $(-)$ electron transitions.

The energy dependence of the function under the integral in Eq. (22) has inverse-square-root divergences at $\varepsilon = \varepsilon_N$ and $\varepsilon = \varepsilon_{N'} - \hbar\omega_{LO}$. They appear due to the divergences in the density of electron states in the magnetic field; see Sec. 5. Under the conditions of magnetophonon resonance, $\omega_{LO} = M\omega_c$, when the optical phonon frequency is equal to an integer number of the cyclotron frequencies $(M = N' - N)$, the expression under the integral diverges as $(\varepsilon - \varepsilon_N)^{-1}$ and the conductivity diverges logarithmically. Therefore, each resonance gives a peak, and the conductivity oscillates as a function of the magnetic field. It is periodic in $(1/H)$ with the period $|e|/m\omega_{LO}c$ determined by the optical phonon frequency. This phenomenon is called the magnetophonon oscillations. The shape of the magnetophonon resonance at small deviations $\delta = \omega_{LO}/\omega_c - M \ll 1$ can be calculated analytically. Near the resonance, the main contribution to the integral in Eq. (22) comes from the energies $\varepsilon \simeq \varepsilon_N$ at $N' = N + M$. This means that the momentum transfer along the direction of the field is small, $q_{z\pm}^2(\varepsilon) \ll q_\perp^2$. In these conditions, the product $|C_{q\pm}^{(LO)}|^2 q_\perp^2$ does not depend on $q_\perp$, see Eq. (21.1), and $A_{NN'}$ is easily calculated analytically (see the normalization integral for the Laguerre polynomials in problem 10.4). The remaining integral over the energy ε is analytically calculated for the case of non-degenerate electrons, and the conductivity takes the form

$$\sigma_{xx}^{(LO)} = \frac{\alpha n e^2 N_{LO}}{2\sqrt{\pi}\omega_c m} \left(\frac{\hbar\omega_{LO}}{T} \right)^{3/2} \exp\left(-\frac{\delta\hbar\omega_c}{2T} \right) \mathcal{K}_0 \left(\frac{\delta\hbar\omega_c}{2T} \right), \qquad (25)$$

where α is the coupling constant defined by Eq. (21.28) and $\mathcal{K}_0(x)$ is the modified Bessel function of the second kind. If $\delta \ll 2T/\hbar\omega_c$, this function behaves as $-C + \ln(4T/\delta\hbar\omega_c)$, where $C \simeq 0.577$ is Euler's constant, so that the conductivity diverges at $\delta = 0$. This logarithmic divergence is suppressed because of the scattering-induced broadening of the Landau levels. To take this suppression into account, δ should be cut at $(\omega_c \tau)^{-1}$, where τ is the characteristic scattering time. If the temperature is considerably smaller than $\hbar\omega_{LO}$, this time is caused by elastic scattering mechanisms. However, the presence of the factor $N_{LO} \simeq \exp(-\hbar\omega_{LO}/T)$ in Eq. (26) suppresses the magnetophonon oscillations at low temperatures. The oscillations cannot be observed if the temperature is so low that the conductivity is dominated by elastic scattering. At high temperatures, when T is comparable to $\hbar\omega_{LO}$, the broadening of the Landau levels is dominated by LO phonon scattering and the magnetophonon oscillations are suppressed again. The most favorable region for observing these oscillations corresponds to intermediate temperatures, when the Landau level broadening depends on elastic scattering, but the resistivity is significantly influenced by LO phonon scattering. The magnetophonon oscillations also occur in the longitudinal part of the conductivity tensor, σ_{zz}.

The method of iterations leading to the Adams-Holstein formula essentially implies that the cyclotron frequency is much larger than the average scattering rate $1/\bar{\tau}$. For this reason, it does not describe the low-field conductivity at $\omega_c \bar{\tau} \leq 1$. Moreover, the Adams-Holstein formula is not applicable for describing the longitudinal conductivity σ_{zz}. To calculate all components of the conductivity tensor in a wide range of fields, it is convenient to apply a unified approach based on the Green's function formalism. This is done in the next sections.

49. Green's Function Approach

To calculate the conductivity tensor, we consider the model of elastic scattering and employ the Kubo formula (13.18) for linear response. Using the operator representation (16.3) of Green's functions, we express the exponents $e^{\pm i\hat{H}\tau/\hbar}$ in terms of the Green's function operators and rewrite Eq. (13.18) as

$$\sigma_{\alpha\beta}(\omega) = \frac{ie^2 n}{m\omega}\delta_{\alpha\beta} + \frac{\hbar e^2}{V\omega}\int_{-\infty}^{\infty} d\tau\, e^{-i\omega\tau} \tag{1}$$

$$\times \mathrm{Sp}\left[\hat{\eta}_{eq}(\widehat{G}_\tau^A - \widehat{G}_\tau^R)\hat{v}_\alpha \widehat{G}_{-\tau}^R \hat{v}_\beta + \hat{v}_\beta \widehat{G}_\tau^A \hat{v}_\alpha (\widehat{G}_{-\tau}^A - \widehat{G}_{-\tau}^R)\hat{\eta}_{eq}\right].$$

In contrast to Eq. (13.18), the integral over time in Eq. (1) is taken from $-\infty$ to ∞. The region from 0 to ∞ does not contribute to the integral due to the presence of $\widehat{G}_{-\tau}^R \propto \theta(-\tau)$ in the first term and $\widehat{G}_\tau^A \propto \theta(-\tau)$ in the second term under the integral. Transforming the Green's function operators to the energy representation according to $\widehat{G}_\tau^{R,A} =$

$(2\pi\hbar)^{-1} \int d\varepsilon e^{-i\varepsilon\tau/\hbar} \widehat{G}_\varepsilon^{R,A}$, we take the integral over time in Eq. (1) and obtain

$$\sigma_{\alpha\beta}(\omega) = \frac{ie^2 n}{m\omega} \delta_{\alpha\beta} + \frac{e^2}{2\pi\omega V} \int d\varepsilon \qquad (2)$$

$$\times \mathrm{Sp} \left[\hat{\eta}_{eq}(\widehat{G}_\varepsilon^A - \widehat{G}_\varepsilon^R)\hat{v}_\alpha \widehat{G}_{\varepsilon+\hbar\omega}^R \hat{v}_\beta + \hat{v}_\beta \widehat{G}_{\varepsilon-\hbar\omega}^A \hat{v}_\alpha (\widehat{G}_\varepsilon^A - \widehat{G}_\varepsilon^R)\hat{\eta}_{eq} \right].$$

Evaluating the trace in the exact eigenstate representation $|\delta\rangle$ defined by the eigenstate problem $\widehat{H}|\delta\rangle = \varepsilon_\delta|\delta\rangle$, one may notice that $\langle\delta|(\widehat{G}_\varepsilon^A - \widehat{G}_\varepsilon^R)|\delta\rangle \propto \delta(\varepsilon - \varepsilon_\delta)$. For this reason, the statistical operator $\hat{\eta}_{eq}$ in both terms of the integrand of Eq. (2) can be replaced by the equilibrium distribution function $f(\varepsilon)$, since this operator stands next to $\widehat{G}_\varepsilon^A - \widehat{G}_\varepsilon^R$. Doing this, we obtain the result of problem 8.5, which has been derived in a different way. To find the static conductivity, one needs to carry out the limiting transition $\omega \to 0$ in Eq. (2). Let us expand $\widehat{G}_{\varepsilon+\hbar\omega}^R$ and $\widehat{G}_{\varepsilon-\hbar\omega}^A$ in series of ω up to linear terms, as $\widehat{G}_{\varepsilon\pm\hbar\omega}^s = \widehat{G}_\varepsilon^s \pm \hbar\omega(\partial\widehat{G}_\varepsilon^s/\partial\varepsilon)$. The contribution coming from zero-order terms of such expansions appears to be imaginary and vanishes together with the first term on the right-hand side of Eq. (2). Thus, only the terms containing the derivatives of the Green's functions remain, and we finally obtain the general expression for the tensor of static conductivity:

$$\sigma_{\alpha\beta} = \frac{\hbar e^2}{2\pi V} \int d\varepsilon f(\varepsilon) \qquad (3)$$

$$\times \left\langle\!\left\langle \mathrm{Sp}\left\{ \hat{v}_\alpha \frac{\partial\widehat{G}_\varepsilon^R}{\partial\varepsilon} \hat{v}_\beta (\widehat{G}_\varepsilon^A - \widehat{G}_\varepsilon^R) - (\widehat{G}_\varepsilon^A - \widehat{G}_\varepsilon^R) \hat{v}_\beta \frac{\partial\widehat{G}_\varepsilon^A}{\partial\varepsilon} \hat{v}_\alpha \right\} \right\rangle\!\right\rangle.$$

The averaging $\langle\!\langle\ldots\rangle\!\rangle$ over the random potential is separated here from the quantum-mechanical averaging $\mathrm{Sp}\ldots$ including the trace over spin variables. Since the velocity operators are Hermitian and $\widehat{G}_\varepsilon^{A+} = \widehat{G}_\varepsilon^R$, the conductivity given by Eq. (3) is real. A more simple form of Eq. (3) is obtained for the diagonal components of the conductivity tensor. Assuming $\alpha = \beta$, we integrate over energy by parts and find

$$\sigma_{\alpha\alpha} = -\frac{\hbar e^2}{4\pi V} \int d\varepsilon \left(-\frac{df(\varepsilon)}{d\varepsilon}\right) \left\langle\!\left\langle \mathrm{Sp}\hat{v}_\alpha (\widehat{G}_\varepsilon^A - \widehat{G}_\varepsilon^R)\hat{v}_\alpha (\widehat{G}_\varepsilon^A - \widehat{G}_\varepsilon^R)\right\rangle\!\right\rangle. \quad (4)$$

Equation (4) is the Greenwood-Peierls formula (13.22) written in terms of the Green's functions in the operator form. The conductivity tensor is expressed through the correlation functions of certain combinations of the retarded and advanced Green's functions. In the case of degenerate electrons, the contribution to the diagonal conductivity comes only from the states near the Fermi level. In contrast, all the states below the Fermi level contribute to the non-diagonal part of $\sigma_{\alpha\beta}$. Applying Eqs. (3) and (4) to the case of electrons in a magnetic field, one should write the velocity operators as $\hat{v}_\alpha = [\hat{p}_\alpha - (e/c)A_\alpha]/m$.

The correlation functions in Eqs. (3) and (4) can be rewritten in the Landau level representation (see Appendix G). We have ($s, s' = R, A$)

$$\frac{1}{L_x L_y} \left\langle\!\!\left\langle \mathrm{Sp}\hat{v}_\alpha \widehat{G}_\varepsilon^s \hat{v}_\beta \widehat{G}_\varepsilon^{s'} \right\rangle\!\!\right\rangle = \frac{1}{\pi l_H^2} \sum_{N_1 - N_4} \sum_{p_z p_z'} v_{N_4 N_1}^\alpha v_{N_2 N_3}^\beta$$

$$\times K_{\varepsilon\varepsilon'}^{ss'}(N_1, N_4; p_z | N_2, N_3; p_z'), \tag{5}$$

where $v_{NN'}^\alpha$ is defined according to $\langle N p_y p_z | \hat{v}_\alpha | N' p_y' p_z' \rangle = v_{NN'}^\alpha \, \delta_{p_y p_y'} \delta_{p_z p_z'}$. The correlation function $K_{\varepsilon\varepsilon'}^{ss'}(N_1, N_4; p_z | N_2, N_3; p_z')$ describing the average of a product of Green's functions is defined by Eqs. (G.21) and (G.28). Below we consider the limit of short-range scattering potential, when this function is reduced to the product $G_\varepsilon^s(p_z, N_1) G_\varepsilon^{s'}(p_z, N_4) \delta_{N_1 N_2}$ $\times \delta_{N_4 N_3} \delta_{p_z p_z'}$, provided the higher-order contributions to the vertex part are neglected; see Eq. (G.31). Using the matrix elements (48.5), we obtain

$$\sigma_d \equiv \sigma_{xx} = \sigma_{yy} = \frac{e^2 \hbar \omega_c^2}{2\pi^2} \int d\varepsilon \left(-\frac{\partial f(\varepsilon)}{\partial \varepsilon} \right) \int \frac{dp_z}{2\pi\hbar} \sum_{N=0}^\infty (N+1)$$

$$\times \mathrm{Re}\left[G_\varepsilon^A(p_z, N) G_\varepsilon^R(p_z, N+1) - G_\varepsilon^A(p_z, N) G_\varepsilon^A(p_z, N+1) \right]. \tag{6}$$

Next, since $v_{NN'}^z = \delta_{NN'} p_z/m$, we find

$$\sigma_\| \equiv \sigma_{zz} = \frac{e^2 \omega_c}{2\pi^2 m} \int d\varepsilon \left(-\frac{\partial f(\varepsilon)}{\partial \varepsilon} \right) \int \frac{dp_z}{2\pi\hbar} p_z^2 \sum_{N=0}^\infty$$

$$\times \mathrm{Re}\left[G_\varepsilon^A(p_z, N) G_\varepsilon^R(p_z, N) - G_\varepsilon^A(p_z, N) G_\varepsilon^A(p_z, N) \right]. \tag{7}$$

Finally, the non-diagonal components of the conductivity tensor are expressed from Eq. (3) as

$$\sigma_\perp \equiv \sigma_{yx} = -\sigma_{xy} = \frac{e^2 \hbar \omega_c^2}{2\pi^2} \int d\varepsilon \int \frac{dp_z}{2\pi\hbar} \sum_{N=0}^\infty (N+1)$$

$$\times \mathrm{Im}\left[\frac{\partial f(\varepsilon)}{\partial \varepsilon} G_\varepsilon^A(p_z, N) G_\varepsilon^R(p_z, N+1) \right. \tag{8}$$

$$+ f(\varepsilon) \left(G_\varepsilon^R(p_z, N) \frac{\partial G_\varepsilon^R(p_z, N+1)}{\partial \varepsilon} - \frac{\partial G_\varepsilon^R(p_z, N)}{\partial \varepsilon} G_\varepsilon^R(p_z, N+1) \right) \right].$$

The term proportional to $\partial f(\varepsilon)/\partial\varepsilon$ in Eq. (8) is obtained after integrating by parts in Eq. (3). However, the term proportional to $f(\varepsilon)$, which describes the contribution of the states below the Fermi level, remains.

It is convenient to write Eqs. (6)–(8) through the self-energy function given by Eq. (G.19), with the use of Eq. (G.16). After some transformations, we have

$$\sigma_d = \frac{e^2\omega_c}{2\pi^2}\int d\varepsilon\left(-\frac{\partial f(\varepsilon)}{\partial\varepsilon}\right)\frac{(2\Sigma_\varepsilon'')^2}{(2\Sigma_\varepsilon'')^2+(\hbar\omega_c)^2}$$

$$\times\int\frac{dp_z}{2\pi\hbar}\sum_{N=0}^{\infty}\frac{\varepsilon-p_z^2/2m-\Sigma_\varepsilon'}{(\varepsilon-\varepsilon_N-p_z^2/2m-\Sigma_\varepsilon')^2+(\Sigma_\varepsilon'')^2}, \qquad (9)$$

$$\sigma_\parallel = \frac{e^2\omega_c}{2\pi^2}\int d\varepsilon\left(-\frac{\partial f(\varepsilon)}{\partial\varepsilon}\right)$$

$$\times\int\frac{dp_z}{2\pi\hbar}\sum_{N=0}^{\infty}\frac{\varepsilon-\varepsilon_N-\Sigma_\varepsilon'+p_z^2/2m}{(\varepsilon-\varepsilon_N-p_z^2/2m-\Sigma_\varepsilon')^2+(\Sigma_\varepsilon'')^2}, \qquad (10)$$

where $\Sigma_\varepsilon'' = \mathrm{Im}\Sigma_\varepsilon^A$ and $\Sigma_\varepsilon' = \mathrm{Re}\Sigma_\varepsilon^A$. To transform the contribution proportional to $f(\varepsilon)$ in Eq. (8), let us use the identity $\partial G_\varepsilon^s(p_z,N)/\partial\varepsilon = (\partial\Sigma_\varepsilon^s/\partial\varepsilon - 1)\,[G_\varepsilon^s(p_z,N)]^2$. As a result, this contribution is given by

$$\mathrm{Im}\sum_{N=0}^{\infty}\left[G_\varepsilon^R(p_z,N)\frac{\partial G_\varepsilon^R(p_z,N+1)}{\partial\varepsilon}-\frac{\partial G_\varepsilon^R(p_z,N)}{\partial\varepsilon}G_\varepsilon^R(p_z,N+1)\right]$$

$$=\frac{2}{(\hbar\omega_c)^2}\sum_{N=0}^{\infty}\mathrm{Im}\left[\left(1-\frac{\partial\Sigma_\varepsilon^A}{\partial\varepsilon}\right)G_\varepsilon^A(p_z,N)-\varepsilon_N\frac{\partial G_\varepsilon^A(p_z,N)}{\partial\varepsilon}\right], \qquad (11)$$

and we obtain the non-diagonal component of the conductivity as

$$\sigma_\perp = \frac{e^2}{\pi^2\hbar}\int d\varepsilon\int\frac{dp_z}{2\pi\hbar}\sum_{N=0}^{\infty}\left\{f(\varepsilon)\mathrm{Im}\frac{1-\partial\Sigma_\varepsilon^A/\partial\varepsilon}{\varepsilon-\varepsilon_N-p_z^2/2m-\Sigma_\varepsilon^A}\right. \qquad (12)$$

$$\left.-\frac{\partial f(\varepsilon)}{\partial\varepsilon}\frac{\Sigma_\varepsilon''}{(2\Sigma_\varepsilon'')^2+(\hbar\omega_c)^2}\frac{\hbar^2\omega_c^2(\varepsilon-\varepsilon_N-p_z^2/2m-\Sigma_\varepsilon')-(2\Sigma_\varepsilon'')^2\varepsilon_N}{(\varepsilon-\varepsilon_N-p_z^2/2m-\Sigma_\varepsilon')^2+(\Sigma_\varepsilon'')^2}\right\}.$$

The self-energy function, according to Eq. (G.19), is determined from the implicit equation

$$\Sigma_\varepsilon'+i\Sigma_\varepsilon''=\frac{w}{2\pi l_H^2}\sum_N\int\frac{dp_z}{2\pi\hbar}\frac{1}{\varepsilon-\varepsilon_N-p_z^2/2m-\Sigma_\varepsilon'-i\Sigma_\varepsilon''}. \qquad (13)$$

Using Eq. (13), we rewrite Eq. (12) as

$$\sigma_\perp = \frac{|e|cn}{H} - \frac{e^2}{2\pi^2\hbar} \int d\varepsilon \left(-\frac{\partial f(\varepsilon)}{\partial \varepsilon}\right) \frac{(2\Sigma''_\varepsilon)^3}{(2\Sigma''_\varepsilon)^2 + (\hbar\omega_c)^2}$$

$$\times \int \frac{dp_z}{2\pi\hbar} \sum_{N=0}^{\infty} \frac{\varepsilon - p_z^2/2m - \Sigma'_\varepsilon}{(\varepsilon - \varepsilon_N - p_z^2/2m - \Sigma'_\varepsilon)^2 + (\Sigma''_\varepsilon)^2}. \tag{14}$$

Apart from the classical Hall conductivity $|e|cn/H$, the non-diagonal component $\sigma_\perp$ contains a dissipative term which differs from σ_d of Eq. (9) only due to the presence of an extra factor $-2\Sigma''_\varepsilon/\hbar\omega_c$ under the integral over energy. In the transformations leading to Eq. (14), we have expressed the electron density as

$$n = \frac{2}{\pi w} \int d\varepsilon \Sigma''_\varepsilon f(\varepsilon). \tag{15}$$

This equation is obtained from $n = \int d\varepsilon \rho(\varepsilon) f(\varepsilon)$ with the use of the general expression for the density of states (see Eq. 14.30 and problem 3.10) written in the Landau level representation:

$$\rho(\varepsilon) = \frac{2}{\pi} \frac{1}{2\pi l_H^2} \int \frac{dp_z}{2\pi\hbar} \sum_{N=0}^{\infty} \text{Im} G_\varepsilon^A(p_z, N), \tag{16}$$

where $G_\varepsilon^A(p_z, N)$ should be expressed through the self-energy function. The spin splitting appearing due to the Pauli interaction, see Appendix B, is neglected in the equations presented above. Nevertheless, this splitting (Zeeman splitting) is directly taken into account by assuming the spin dependence of the Landau level energy, $\varepsilon_N \rightarrow \varepsilon_{N\sigma} = \hbar\omega_c(N+1/2) - \sigma\mu_B gH/2$, where g is the effective g-factor, μ_B is the Bohr magneton, and $\sigma = \pm 1$ is the spin number. Owing to this dependence, the Green's functions and self-energy functions themselves become spin-dependent, and the right-hand sides in Eqs. (9), (10), (14), (15), and (16) should include the spin sum $\sum_\sigma$ replacing the factor 2. There is no sum over spin in Eq. (13) because the spin number is conserved in the scattering processes.

Let us consider the case of degenerate electron gas by taking the spin splitting into account. Integrating over p_z in Eqs. (9), (10), and (14), we obtain

$$\sigma_d = \frac{e^2\omega_c m^{1/2}}{2^{1/2}\pi^2\hbar} \sum_\sigma \frac{\Sigma''_{\sigma\varepsilon_F}}{(2\Sigma''_{\sigma\varepsilon_F})^2 + (\hbar\omega_c)^2}$$

$$\times \sum_{N=0}^{\infty} \text{Re} \frac{\varepsilon_{N\sigma} + i\Sigma''_{\sigma\varepsilon_F}}{(\varepsilon_F - \varepsilon_{N\sigma} - \Sigma'_{\sigma\varepsilon_F} - i\Sigma''_{\sigma\varepsilon_F})^{1/2}}, \tag{17}$$

$$\sigma_{\parallel} = \frac{e^2 \omega_c m^{1/2}}{2^{3/2}\pi^2\hbar} \sum_{\sigma} \frac{1}{\Sigma''_{\sigma\varepsilon_F}} \sum_{N=0}^{\infty} \mathrm{Re} \frac{\varepsilon_F - \varepsilon_{N\sigma} - \Sigma'_{\sigma\varepsilon_F} + i\Sigma''_{\sigma\varepsilon_F}/2}{(\varepsilon_F - \varepsilon_{N\sigma} - \Sigma'_{\sigma\varepsilon_F} + i\Sigma''_{\sigma\varepsilon_F})^{1/2}}, \quad (18)$$

$$\sigma_{\perp} = \frac{|e|cn}{H} - \frac{2^{1/2}e^2 m^{1/2}}{\pi^2\hbar^2} \sum_{\sigma} \frac{(\Sigma''_{\sigma\varepsilon_F})^2}{(2\Sigma''_{\sigma\varepsilon_F})^2 + (\hbar\omega_c)^2}$$

$$\times \sum_{N=0}^{\infty} \mathrm{Re} \frac{\varepsilon_{N\sigma} + i\Sigma''_{\sigma\varepsilon_F}}{(\varepsilon_F - \varepsilon_{N\sigma} - \Sigma'_{\sigma\varepsilon_F} - i\Sigma''_{\sigma\varepsilon_F})^{1/2}}. \quad (19)$$

Using these results, let us find the conductivity of a 3D conductor in the ultraquantum limit, when only the ground-state spin-split Landau level is occupied. We put $N = 0$ in the expressions above and consider only one spin state ($\sigma = +1$ if the g-factor is positive). Let us denote $\Sigma_{\sigma\varepsilon_F}$ at $\sigma = +1$ as Σ_F and shift the reference point of energy according to $\widetilde{\varepsilon}_F = \varepsilon_F - \hbar\omega/2 + \mu_B g H/2 - \Sigma'_F$. Then we find

$$\sigma_d = \frac{e^2 m^{1/2}\Sigma''_F}{2^{3/2}\pi^2\hbar^2\sqrt{\widetilde{\varepsilon}_F}}, \quad \sigma_{\parallel} = \frac{e^2 m^{1/2}\omega_c\sqrt{\widetilde{\varepsilon}_F}}{2^{3/2}\pi^2\hbar\Sigma''_F}, \quad \sigma_{\perp} = \frac{|e|cn}{H} \quad (20)$$

in the leading order with respect to Σ''_F. To complete the description, one needs to calculate Σ''_F. From Eq. (13) we obtain

$$\Sigma''_F \simeq \frac{w}{2l_H^2} \int \frac{dp_z}{2\pi\hbar} \delta(\widetilde{\varepsilon}_F - p_z^2/2m) = \frac{wm^{3/2}\omega_c}{2^{3/2}\pi\hbar^2\sqrt{\widetilde{\varepsilon}_F}}, \quad (21)$$

since only the term with $N = 0$ and $\sigma = +1$ contributes to the imaginary part of the self-energy function. The expressions (20) and (21) are valid if $\widetilde{\varepsilon}_F \gg \Sigma''_F$, which means that the broadening of the ground Landau level is smaller than the kinetic energy of the electrons occupying this level. On the other hand, at a fixed electron density n the energy $\widetilde{\varepsilon}_F$ decreases with increasing H while Σ''_F increases. Therefore, the requirement of small broadening is violated in strong enough magnetic fields. Expressing the electron density in the ultraquantum limit as (problem 10.5)

$$n = \frac{(2m)^{3/2}\omega_c\sqrt{\widetilde{\varepsilon}_F}}{(2\pi)^2\hbar^2} = \frac{\hbar\omega_c}{2}\rho_{3D}(\widetilde{\varepsilon}_F), \quad (22)$$

one should restrict the magnetic field strength according to

$$\omega_c \ll \hbar^2 \left(\frac{(2n)^3\pi^7}{wm^6} \right)^{1/4}. \quad (23)$$

This is the condition of validity of the expressions (20).

Combining Eqs. (20), (21), and (22), we find

$$\sigma_d = \frac{e^2 w m^5 \omega_c^3}{16\pi^7 \hbar^8 n^2}. \tag{24}$$

This equation can be also obtained from the Adams-Holstein formula in the form (48.15) if we retain only one term in the sum, $N = N' = 0$. Next, we have

$$\sigma_\parallel = \frac{2\pi^3 \hbar^5 e^2 n^2}{w m^4 \omega_c^2}. \tag{25}$$

The difference between these equations, as concerns the dependence on the magnetic field and scattering potential correlation function $w = n_{im}|v(0)|^2$, is determined by the difference between the mechanisms of transport and can be understood as follows. The scattering helps the transport perpendicular to $\mathbf{H}$ and suppresses the transport along $\mathbf{H}$. The magnetic field produces the same effect, since it pushes the electrons closer to each other (note that the density of states is proportional to $1/\sqrt{\widetilde{\varepsilon}_F}$) thereby increasing the scattering probability.

The condition (23) is not the only one restriction on the magnetic field from the upper side. Another restriction comes from the limited applicability of the Born approximation for the scattering by a single impurity. Indeed, when the magnetic field increases, the kinetic energy of electrons becomes smaller, an electron spends most of its time near a single impurity, and the applicability of the Born approximation is violated. To introduce the diagram technique describing multiple scattering (beyond the Born approximation), we consider first the case of zero magnetic field. In the procedure of averaging described in Sec. 14, we neglected the terms $n_{im}\delta_{\mathbf{q}_1+\mathbf{q}_2+\mathbf{q}_3,0}$, $n_{im}\delta_{\mathbf{q}_1+\mathbf{q}_2+\mathbf{q}_3+\mathbf{q}_4,0}$, and so on; see Eq. (14.14) and below. If we take into account all such terms, the Born amplitudes, proportional to the squared Fourier transform $|v(q)|^2$ of the single-impurity scattering potential, will be replaced by the exact scattering amplitudes. Determining the exact scattering amplitudes for an arbitrary scattering potential is a complex problem which, however, is simplified for the case of short-range potential, when $v(q)$ does not depend on q.

Let us recall the expansion (14.10) and write it explicitly up to the fourth-order terms with respect to the impurity potential:

$$G_\varepsilon(\mathbf{p},\mathbf{p}') = \delta_{\mathbf{p}\mathbf{p}'}g_\varepsilon(\mathbf{p}) + g_\varepsilon(\mathbf{p})\Big\{ V^{-1}v[(\mathbf{p}-\mathbf{p}')/\hbar]\sum_\alpha e^{-i(\mathbf{p}-\mathbf{p}')\cdot\mathbf{R}_\alpha/\hbar}$$

$$+V^{-2}\sum_{\alpha_1\alpha_2}\sum_{\mathbf{q}_1\mathbf{q}_2}\delta_{\mathbf{q}_1+\mathbf{q}_2,(\mathbf{p}-\mathbf{p}')/\hbar}v(\mathbf{q}_1)g_\varepsilon(\mathbf{p}-\hbar\mathbf{q}_1)v(\mathbf{q}_2)e^{-i\mathbf{q}_1\cdot\mathbf{R}_{\alpha_1}-i\mathbf{q}_2\cdot\mathbf{R}_{\alpha_2}}$$

$$+V^{-3} \sum_{\alpha_1\alpha_2\alpha_3} \sum_{\mathbf{q}_1\mathbf{q}_2\mathbf{q}_3} \delta_{\mathbf{q}_1+\mathbf{q}_2+\mathbf{q}_3,(\mathbf{p}-\mathbf{p}')/\hbar} v(\mathbf{q}_1) g_\varepsilon(\mathbf{p}-\hbar\mathbf{q}_1) v(\mathbf{q}_2)$$

$$\times g_\varepsilon(\mathbf{p}-\hbar\mathbf{q}_1-\hbar\mathbf{q}_2) v(\mathbf{q}_3) e^{-i\mathbf{q}_1\cdot\mathbf{R}_{\alpha_1}-i\mathbf{q}_2\cdot\mathbf{R}_{\alpha_2}-i\mathbf{q}_3\cdot\mathbf{R}_{\alpha_3}} \qquad (26)$$

$$+V^{-4} \sum_{\alpha_1-\alpha_4} \sum_{\mathbf{q}_1-\mathbf{q}_4} \delta_{\mathbf{q}_1+\mathbf{q}_2+\mathbf{q}_3+\mathbf{q}_4,(\mathbf{p}-\mathbf{p}')/\hbar} v(\mathbf{q}_1) g_\varepsilon(\mathbf{p}-\hbar\mathbf{q}_1)$$

$$\times v(\mathbf{q}_2) g_\varepsilon(\mathbf{p}-\hbar\mathbf{q}_1-\hbar\mathbf{q}_2) v(\mathbf{q}_3) g_\varepsilon[\mathbf{p}-\hbar(\mathbf{q}_1+\mathbf{q}_2+\mathbf{q}_3)] v(\mathbf{q}_4)$$

$$\times e^{-i\mathbf{q}_1\cdot\mathbf{R}_{\alpha_1}-i\mathbf{q}_2\cdot\mathbf{R}_{\alpha_2}-i\mathbf{q}_3\cdot\mathbf{R}_{\alpha_3}-i\mathbf{q}_4\cdot\mathbf{R}_{\alpha_4}}+\dots\bigg\} g_\varepsilon(\mathbf{p}').$$

To average this expression, one should calculate the integrals over the impurity coordinates $\mathbf{R}_{\alpha_i}$; see Sec. 14. This procedure leads to the Dyson equation (14.24), where the self-energy function is given by

$$\Sigma_\varepsilon(\mathbf{p}) = \frac{n_{im}}{V} \sum_{\mathbf{q}_1} \bigg[|v(\mathbf{q}_1)|^2 g_\varepsilon(\mathbf{p}-\hbar\mathbf{q}_1) + \frac{1}{V} \sum_{\mathbf{q}_2} v(\mathbf{q}_1) v(\mathbf{q}_2) v(-\mathbf{q}_1-\mathbf{q}_2)$$

$$\times g_\varepsilon(\mathbf{p}-\hbar\mathbf{q}_1) g_\varepsilon(\mathbf{p}-\hbar\mathbf{q}_1-\hbar\mathbf{q}_2) + \frac{1}{V^2} \sum_{\mathbf{q}_2\mathbf{q}_3} v(\mathbf{q}_1) v(\mathbf{q}_2) v(\mathbf{q}_3) v(-\mathbf{q}_1-\mathbf{q}_2-\mathbf{q}_3)$$

$$\times g_\varepsilon(\mathbf{p}-\hbar\mathbf{q}_1) g_\varepsilon(\mathbf{p}-\hbar\mathbf{q}_1-\hbar\mathbf{q}_2) g_\varepsilon[\mathbf{p}-\hbar(\mathbf{q}_1+\mathbf{q}_2+\mathbf{q}_3)]+\dots\bigg]$$

$$+\frac{n_{im}^2}{V^2} \sum_{\mathbf{q}_1\mathbf{q}_2} \bigg[|v(\mathbf{q}_1)|^2 |v(\mathbf{q}_2)|^2 g_\varepsilon(\mathbf{p}-\hbar\mathbf{q}_1) g_\varepsilon(\mathbf{p}) g_\varepsilon(\mathbf{p}-\hbar\mathbf{q}_2) \qquad (27)$$

$$+|v(\mathbf{q}_1)|^2 |v(\mathbf{q}_2)|^2 g_\varepsilon(\mathbf{p}-\hbar\mathbf{q}_1) g_\varepsilon(\mathbf{p}-\hbar\mathbf{q}_1-\hbar\mathbf{q}_2) g_\varepsilon(\mathbf{p}-\hbar\mathbf{q}_1)$$

$$+|v(\mathbf{q}_1)|^2 |v(\mathbf{q}_2)|^2 g_\varepsilon(\mathbf{p}-\hbar\mathbf{q}_1) g_\varepsilon(\mathbf{p}-\hbar\mathbf{q}_1-\hbar\mathbf{q}_2) g_\varepsilon(\mathbf{p}-\hbar\mathbf{q}_2)+\dots\bigg]+\dots\,.$$

This expansion can be diagrammatically represented as

$$\Sigma_\varepsilon(\mathbf{p}) = \underset{\mathbf{p}-\hbar\mathbf{q}_1}{} \quad + \quad \underset{\mathbf{p}-\hbar\mathbf{q}_1 \;\; \mathbf{p}-\hbar\mathbf{q}_1-\hbar\mathbf{q}_2}{} \quad + \quad \quad + \dots$$

$$+ \quad \quad + \quad \quad + \quad \quad + \dots\,, \qquad (28)$$

where each vertex connecting two or more impurity-potential lines brings the factor n_{im}. Some of the terms of this expansion are reducible, and the other (the second, the third, and the last) are not. The second and the third terms contain more than two impurity lines attached to a single vertex and correspond to the contribution beyond the Born approximation. In the following, we neglect the contributions from the diagrams with crossed lines, like the last one. These diagrams describe the interference of the scattering by different impurities, and their contribution is small due to the parameter $\hbar/\tau\bar{\varepsilon}$, where $\bar{\varepsilon}$ is the mean kinetic energy of electrons. Applying the model of short-range potential $v(q) \simeq v(0) = v$ (note that v is real and $|v|^2 = v^2$), we can calculate the sum of all remaining diagrams. By taking into account all reducible diagrams, we replace the free-electron Green's functions $g_\varepsilon(\mathbf{p})$ by $G_\varepsilon(\mathbf{p})$. Then we write the expansion

$$\Sigma_\varepsilon(\mathbf{p}) = n_{im}\frac{v^2}{V}\sum_{\mathbf{q}_1} G_\varepsilon(\mathbf{p} - \hbar\mathbf{q}_1) \tag{29}$$

$$\times \left[1 + \frac{v}{V}\sum_{\mathbf{q}_2} G_\varepsilon(\mathbf{p} - \hbar\mathbf{q}_1 - \hbar\mathbf{q}_2)\left(1 + \frac{v}{V}\sum_{\mathbf{q}_3}\cdots\right)\right],$$

which is represented as an infinite embedded chain: the expression under each sum is a product of a Green's function by the factor $1+(v/V)\sum\cdots$, where an analogical expression stands under the sum, and so on. However, each sum appears to be momentum-independent, because one can shift the variables of summation. For example, $\sum_{\mathbf{q}_1} G_\varepsilon(\mathbf{p} - \hbar\mathbf{q}_1) = \sum_{\mathbf{p}} G_\varepsilon(\mathbf{p})$. Therefore, $\Sigma_\varepsilon(\mathbf{p})$ is momentum-independent as well. Finally, using the sum rule for geometric progressions, we find

$$\Sigma_\varepsilon = n_{im}\frac{v^2}{V}\left[\sum_{\mathbf{p}} G_\varepsilon(\mathbf{p})\right]\left\{1 + \frac{v}{V}\sum_{\mathbf{p}} G_\varepsilon(\mathbf{p}) + \left[\frac{v}{V}\sum_{\mathbf{p}} G_\varepsilon(\mathbf{p})\right]^2 + \ldots\right\}$$

$$= n_{im}v^2\left[\int \frac{d\mathbf{p}}{(2\pi\hbar)^3}G_\varepsilon(\mathbf{p})\right]\Big/\left[1 - v\int \frac{d\mathbf{p}}{(2\pi\hbar)^3}G_\varepsilon(\mathbf{p})\right]. \tag{30}$$

Although the above consideration is done for the Green's functions in the momentum representation in the absence of magnetic fields, it is rather straightforward to generalize the result for the Landau level representation, because this result depends only on the averaged Green's function integrated over the phase space. In the magnetic field, the integrals over p_x and p_y are transformed to the sum over the Landau level number N multiplied by the factor $(2\pi l_H^2)^{-1}$ appearing due to the orbital degeneracy of the Landau levels. Therefore, we obtain (spin indices of Σ, G, and $\mathcal{S}$ are omitted)

$$\Sigma'_\varepsilon + i\Sigma''_\varepsilon = \frac{n_{im}v\mathcal{S}_\varepsilon}{1 - \mathcal{S}_\varepsilon}, \quad \mathcal{S}_\varepsilon = \frac{v}{2\pi l_H^2}\sum_N \int \frac{dp_z}{2\pi\hbar}G_\varepsilon^A(p_z, N). \tag{31}$$

In the ultraquantum limit, $\mathcal{S}_{\varepsilon_F} \equiv \mathcal{S} \simeq ivm^3\omega_c^2/4\pi^3\hbar^4 n$. According to Eq. (31), the results (24) and (25) are valid at $|\mathcal{S}| \ll 1$. In the case of doped semiconductors, when $n \sim n_{im}$, the condition (23) is equivalent to the requirement $|\mathcal{S}| \ll 1$. In the case of metals, when $n \gg n_{im}$, there exists the region of fields where $|\mathcal{S}| \gg 1$ while the condition (23) is still valid. Taking into account that $\Sigma_F'' = n_{im}v|\mathcal{S}|/(1+|\mathcal{S}|^2)$, we substitute it into Eq. (20) and obtain a more general description of the ultraquantum limit. In particular, at $|\mathcal{S}| \gg 1$ one has

$$\sigma_d = \frac{|e|cn_{im}}{\pi H}, \quad \sigma_{\parallel} = \frac{1}{(2\pi)^3} \frac{e^4 H^2}{c^2 n_{im}}. \tag{32}$$

Both components of the conductivity in these conditions do not depend on the scattering potential v. Apart from the fundamental constants, only the magnitude of the field and the impurity concentration enter these equations.

50. Quasi-Classical Conductivity

The method of Green's functions can be applied at relatively weak magnetic fields, when ω_c is comparable to or less than the scattering rate. This corresponds to $\hbar\omega_c \ll \bar{\varepsilon}$, where $\bar{\varepsilon}$ is the mean electron energy estimated by the Fermi energy and temperature for the cases of degenerate and non-degenerate electrons, respectively. Therefore, the sums over N in Eqs. (49.17)−(49.19) and (49.9)−(49.14) can be roughly approximated by integrals, according to $\sum_N F(N) = (\hbar\omega_c)^{-1} \int d\varepsilon_N F(\varepsilon_N/\hbar\omega_c)$. Neglecting the terms of the order of $\Sigma_\varepsilon''/\varepsilon$, from Eq. (49.13) we find $\Sigma_\varepsilon'' \simeq \hbar/2\tau(\varepsilon)$, where $\tau^{-1}(\varepsilon)$ is the relaxation rate of electrons due to scattering by a short-range potential in the absence of the magnetic field. This rate is determined according to Eq. (8.21) as $(\pi w/\hbar)\rho_{\mathcal{D}}(\varepsilon)$. Therefore, $\tau^{-1}(\varepsilon) = wm^{3/2}\sqrt{2\varepsilon}/\pi\hbar^4$ for 3D electrons and $\tau^{-1}(\varepsilon) = wm/\hbar^3$ for 2D electrons. The time $\tau(\varepsilon_F)$ for the degenerate electron gas coincides with the transport time τ_{tr} introduced in Chapter 2. Evaluating the sums over N in Eqs. (49.17)−(49.19) as explained above and under the condition $\Sigma_\varepsilon'' \ll \varepsilon_F$, one can find (problem 10.6) the classical expression for the conductivity tensor in the magnetic field; see Eq. (11.19).

Apart from the classical contributions we just discussed, even at small H there exist small oscillating contributions both to the self-energy and to the conductivity. These contributions can be calculated if the sums over N are evaluated in a more careful way. We still assume $\hbar\omega_c \ll \bar{\varepsilon}$. Since the main contribution to the sum over N comes from the Landau levels with $N \gg 1$, we extend the sum to $-\infty$ and employ the identity

$$\sum_{N=-\infty}^{\infty} F(N) = \frac{1}{2i} \int_C dz \cot(\pi z) F(z), \tag{1}$$

where the contour C passes below the poles of $\cot(\pi z)$ (they are placed at real integer $z = N$) and above all peculiar points z_i of the function $F(z)$. To prove Eq. (1), one should shift the contour to the upper half-plane and note that the residue of $\cot(\pi z)$ in each pole, $[\mathrm{res}\,\cot(\pi z)]_{z=N}$, is equal to $1/\pi$. On the other hand, if the peculiar points z_i are poles, one can transform the integral in Eq. (1) into the sum over $[\mathrm{res}\,F(z)]_{z=z_i}$ by shifting the contour down to $-i\infty$, with the result $\int_C dz\,\cot(\pi z)F(z) = -2\pi i \sum_i [\mathrm{res}\,F(z)]_{z=z_i}\cot(\pi z_i)$. Considering such sums, it is convenient to use the Fourier expansion of the function $\cot(\pi z)$:

$$\cot(\pi z) = i + 2i \sum_{k=1}^{\infty} e^{-2ik\pi z}, \qquad \mathrm{Im}\, z < 0. \tag{2}$$

To consider the region $\mathrm{Im}\, z > 0$, one may use the expression which is complex conjugate to Eq. (2).

Let us apply this method of calculation in order to evaluate $\Sigma_\varepsilon \equiv \Sigma_\varepsilon^A$ described by Eq. (49.13). The sum over N is transformed to a contour integral according to Eq. (1). Since the expression under the integral has a single simple pole at $z = [\varepsilon - p_z^2/2m - \Sigma_\varepsilon' - i\Sigma_\varepsilon'']/\hbar\omega_c - 1/2$, we obtain

$$\Sigma_\varepsilon' + i\Sigma_\varepsilon'' = \frac{wm}{2\hbar^2} \int \frac{dp_z}{2\pi\hbar} \cot\left[\frac{\pi}{\hbar\omega_c}(\varepsilon - p_z^2/2m - \Sigma_\varepsilon' - i\Sigma_\varepsilon'') - \frac{\pi}{2} \right] \tag{3}$$

$$= i\frac{wm}{2\hbar^2} \int \frac{dp_z}{2\pi\hbar} \left\{ 1 + 2\sum_{k=1}^{\infty} (-1)^k \exp\left[-i\frac{2k\pi}{\hbar\omega_c}(\varepsilon - p_z^2/2m - \Sigma_\varepsilon' - i\Sigma_\varepsilon'') \right] \right\}.$$

The expression $\{\ldots\}$ contains both non-oscillating and oscillating parts. The integral over p_z on the right-hand side of Eq. (3) must be taken over the interval $[-p_\varepsilon, p_\varepsilon]$, where $p_\varepsilon = \sqrt{2m\varepsilon}$. This is because the sum in Eq. (49.13) rapidly decreases when $p_z^2/2m - \varepsilon$ exceeds either Σ_ε'' or $\hbar\omega_c$. This property is not reflected explicitly in Eq. (3), since the extension of the sum over N to $-\infty$ is not valid when $p_z^2/2m$ is close to ε. Thus, the non-oscillating part of $\Sigma_\varepsilon' + i\Sigma_\varepsilon''$ is estimated as $iwm^{3/2}\sqrt{2\varepsilon}/2\pi\hbar^3 = i\hbar/2\tau(\varepsilon)$. On the other hand, the oscillating part is not affected by the presence of the limits $\pm p_\varepsilon$ because it contains a factor which rapidly oscillates with p_z, and the integral over p_z converges before $|p_z|$ reaches p_ε. The integral over p_z in the oscillating part brings the factor $\sqrt{m\hbar\omega_c/2k}$ multiplied by an oscillating function of energy. Thus, the oscillating terms contain a small factor of the order of $(\hbar\omega_c/\varepsilon)^{1/2}$ with respect to the non-oscillating contribution. For this reason, we can substitute the main (non-oscillating) part of the self-energy, $\Sigma_\varepsilon^{(0)} = i\hbar/2\tau(\varepsilon)$, into these terms. In this way we obtain an approximate solution of the implicit equation (3). It contains a sum of oscillating terms (harmonics) characterized by their numbers k:

$$\Sigma'_\varepsilon = \frac{\hbar}{2\tau(\varepsilon)} \sum_{k=1}^{\infty} \sqrt{\frac{\hbar\omega_c}{2k\varepsilon}} (-1)^k \mathcal{D}_k(\varepsilon) \sin\left(\frac{2\pi k\varepsilon}{\hbar\omega_c} - \frac{\pi}{4}\right), \qquad (4)$$

$$\Sigma''_\varepsilon = \frac{\hbar}{2\tau(\varepsilon)} \left[1 + \sum_{k=1}^{\infty} \sqrt{\frac{\hbar\omega_c}{2k\varepsilon}} (-1)^k \mathcal{D}_k(\varepsilon) \cos\left(\frac{2\pi k\varepsilon}{\hbar\omega_c} - \frac{\pi}{4}\right)\right]. \qquad (5)$$

The coefficients

$$\mathcal{D}_k(\varepsilon) = \exp\left[-\frac{k\pi}{\omega_c\tau(\varepsilon)}\right], \qquad (6)$$

called the Dingle factors, describe the exponential decrease of the contributions of higher harmonics (with large k). In weak magnetic fields, when $\omega_c\tau(\varepsilon) < 1$, these factors are exponentially small and only the first harmonic ($k = 1$) is essential. Since the imaginary part of the self-energy is proportional to the inverse lifetime of an electron, Eq. (5) describes the oscillations of the scattering rate, as a function of energy, in the magnetic field.

Apart from the oscillations of the scattering rate, there exist weak oscillations of the chemical potential, since the electron density n is fixed (for example, by doping level) and stays independent of the magnetic field. As a consequence, different macroscopic equilibrium quantities characterizing the system also oscillate (we mention, for example, de Haas - van Alphen oscillations of the magnetization). Below we consider the case of strongly degenerate electron gas. One can show that the relative amplitudes of the Fermi energy oscillations contain an additional smallness with respect to the factor $(\hbar\omega_c/\varepsilon_F)^{1/2}\mathcal{D}_1(\varepsilon_F)$ characterizing the first harmonic of the oscillations of the scattering rate at the Fermi level. To show this, we express the electron density n in the presence of a weak magnetic field through the Fermi energy. Substituting Σ''_ε from Eq. (5) to Eq. (49.15), we calculate the integral over energy, neglecting, for simplicity, the energy dependence of the Dingle factors: $\mathcal{D}_k(\varepsilon) \simeq \mathcal{D}_k(\varepsilon_F) \equiv \mathcal{D}_k$. At zero temperature,

$$n = \frac{(2m\varepsilon_F)^{3/2}}{3\pi^2\hbar^3} \left\{1 + \frac{3}{2\pi} \sum_{k=1}^{\infty} \left(\frac{\hbar\omega_c}{2k\varepsilon_F}\right)^{3/2} \right.$$

$$\left. \times(-1)^k \mathcal{D}_k \left[\sin\left(\frac{2\pi k\varepsilon_F}{\hbar\omega_c} - \frac{\pi}{4}\right) + \frac{1}{\sqrt{2}}\right]\right\}. \qquad (7)$$

The magnetic-field-induced modification of the electron density as a function of the Fermi energy is described by the second term in this expression. The dependence of ε_F on the magnetic field can be found by iterations: we substitute $\varepsilon_F = \varepsilon_F^{(0)} + \delta\varepsilon_F$ to the first term and $\varepsilon_F = \varepsilon_F^{(0)}$ to the second, oscillating term, and determine $\delta\varepsilon_F$. It is easy to see that the relative correction, $\delta\varepsilon_F/\varepsilon_F^{(0)}$, is small as $(\hbar\omega_c/\varepsilon_F)^{3/2}\mathcal{D}_1$. Therefore, one may neglect the modification of the Fermi energy.

Now we calculate the components of the conductivity tensor given by the expressions (49.9), (49.10), and (49.14). Since the functions under the sums in all these expressions have two simple poles at $\varepsilon_N = \varepsilon - p_z^2/2m - \Sigma'_\varepsilon \mp \Sigma''_\varepsilon$, the sum over N and the integral over p_z are calculated in a similar way as above. Moreover, neglecting the oscillating contributions whose amplitude is of the order of $\hbar\omega_c/\varepsilon$ (they appear after calculating the integrals containing products of p_z^2 by the oscillating contribution of Σ_ε), one can express the conductivity through Σ''_ε:

$$\sigma_d = \sigma_0 \int d\varepsilon \left(-\frac{\partial f(\varepsilon)}{\partial\varepsilon} \right) \frac{2\Sigma''_\varepsilon[3\Sigma''_\varepsilon - \hbar/2\tau(\varepsilon_F)]}{(\hbar\omega_c)^2 + (2\Sigma''_\varepsilon)^2}, \qquad (8)$$

$$\sigma_\parallel = \sigma_0 \int d\varepsilon \left(-\frac{\partial f(\varepsilon)}{\partial\varepsilon} \right) \frac{\hbar}{2\tau\Sigma''_\varepsilon}, \qquad (9)$$

$$\sigma_\perp = \frac{|e|cn}{H} - \sigma_0 \int d\varepsilon \left(-\frac{\partial f(\varepsilon)}{\partial\varepsilon} \right) \frac{(2\Sigma''_\varepsilon)^2}{\hbar\omega_c} \frac{3\Sigma''_\varepsilon - \hbar/2\tau(\varepsilon_F)}{(\hbar\omega_c)^2 + (2\Sigma''_\varepsilon)^2}, \qquad (10)$$

where $\sigma_0 = e^2 n\tau(\varepsilon_F)/m$ is the static conductivity at zero magnetic field. To obtain these expressions, we have replaced ε by ε_F in the slowly varying functions of energy. Nevertheless, the integration over energy in Eqs. (8)−(10) is not trivial because Σ''_ε contains a rapidly oscillating function of energy, see Eq. (5), whose period can be smaller than the temperature T. Since we consider only the first-order contributions with respect to the small parameter $(\hbar\omega_c/k\varepsilon_F)^{1/2}\mathcal{D}_k(\varepsilon_F)$, the oscillating parts of the integrands in Eqs. (8)−(10) are proportional to the oscillating part of Σ''_ε. Thus, the averaging over energy in Eqs. (8)−(10) is reduced to the averaging of the function $\cos(2\pi k\varepsilon/\hbar\omega_c - \pi/4)$. With the aid of the identities $(-\partial f(\varepsilon)/\partial\varepsilon) = \{4T\cosh^2[(\varepsilon - \varepsilon_F)/2T]\}^{-1}$ and $\int_{-\infty}^{\infty} dx e^{\pm iax}/\cosh^2(x/b) = \pi ab^2/\sinh(\pi ab/2)$, we obtain

$$\int d\varepsilon \left(-\frac{\partial f(\varepsilon)}{\partial\varepsilon} \right) \cos\left(\frac{2\pi k\varepsilon}{\hbar\omega_c} - \frac{\pi}{4} \right) = \mathcal{B}_k \cos\left(\frac{2\pi k\varepsilon_F}{\hbar\omega_c} - \frac{\pi}{4} \right), \qquad (11)$$

where

$$\mathcal{B}_k = \frac{(2k\pi^2 T/\hbar\omega_c)}{\sinh(2k\pi^2 T/\hbar\omega_c)}. \qquad (12)$$

Equation (11) gives us the rule of integration over energy in Eqs. (8)−(10): the oscillating part of the expressions under the integrals should be taken at $\varepsilon = \varepsilon_F$ and multiplied by the temperature damping factor $\mathcal{B}_k$. This factor approaches unity at small temperatures, while at $T \gg \hbar\omega_c/2\pi^2$ it is exponentially small. Using Eqs. (5) and (11), we transform Eqs. (8)−(10) to

$$\sigma_d = \sigma_0 \frac{1}{1 + (\omega_c\tau)^2} \left[1 + \left(\frac{5}{2} - \frac{2}{1 + (\omega_c\tau)^2} \right) \mathcal{F}_{osc} \right], \tag{13}$$

$$\sigma_\| = \sigma_0 \left[1 - \mathcal{F}_{osc} \right], \tag{14}$$

and

$$\sigma_\perp = \sigma_0 \frac{\omega_c\tau}{1 + (\omega_c\tau)^2} \left[1 - \frac{1}{(\omega_c\tau)^2} \left(\frac{7}{2} - \frac{2}{1 + (\omega_c\tau)^2} \right) \mathcal{F}_{osc} \right], \tag{15}$$

where

$$\mathcal{F}_{osc} = \sum_{k=1}^{\infty} \sqrt{\frac{\hbar\omega_c}{2k\varepsilon_F}} (-1)^k \mathcal{D}_k \mathcal{B}_k \cos\left(\frac{2\pi k\varepsilon_F}{\hbar\omega_c} - \frac{\pi}{4} \right), \tag{16}$$

$\tau = \tau(\varepsilon_F)$, and $\mathcal{D}_k \equiv \mathcal{D}_k(\varepsilon_F)$. Therefore, at $\hbar\omega \ll \varepsilon_F$ all the components of the conductivity tensor show weak oscillations as functions of the magnetic field. They are known as Shubnikov - de Haas oscillations. The oscillations are periodic in H^{-1} (we remind that the deviations of the Fermi energy from its zero-field value $\varepsilon_F = (3\pi^2\hbar^3 n)^{2/3}/2m$ can be neglected). Since the period is determined entirely by $\varepsilon_F \propto n^{2/3}/m$, the measurements of these oscillations can serve as a source for determining the effective mass provided the electron density n is known from an independent experiment, for example, from the Hall measurements. The oscillations are suppressed due to the scattering-induced broadening of the Landau levels (described by the Dingle factors $\mathcal{D}_k$) and due to the thermal smearing of the Fermi distribution of electrons (described by the temperature damping factors $\mathcal{B}_k$). The presence of a large numerical factor $2\pi^2$ in the argument of hyperbolic sine in Eq. (12) indicates that one needs low enough temperatures in order to observe the oscillations of the conductivity: even at $T \sim \hbar\omega_c$ the oscillations are exponentially weak. In semimetals and doped semiconductors, the oscillations are typically observed in the magnetic fields of the order of one Tesla and at liquid helium temperatures. The oscillations of the diagonal component of the resistivity $\rho_d = \sigma_d/(\sigma_d^2 + \sigma_\perp^2)$ are shown in Fig. 10.1. The oscillations of $\sigma_\perp$ and of the Hall resistivity $\rho_\perp = \sigma_\perp/(\sigma_d^2 + \sigma_\perp^2)$ have much weaker amplitudes and should be neglected.

The spin splitting of electron states has been neglected in the calculations above. It it not difficult to take it into account if the amplitude of the oscillations is small. We have two groups of electrons whose

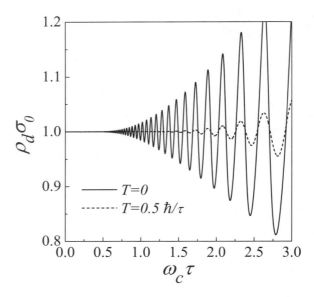

Figure 10.1. Magnetic-field dependence of the diagonal resistivity $\rho_d = \sigma_d/(\sigma_d^2 + \sigma_\perp^2)$ calculated according to Eqs. (50.13) and (50.15) at $\varepsilon_F \tau/\hbar = 20$ for $T = 0$ and $T = 0.5 \, \hbar/\tau$.

energies are shifted with respect to each other by the Zeeman energy $\hbar\Omega_H = g\mu_B H$; see the discussion after Eq. (49.16). Therefore, we obtain two sets of oscillations shifted by the phase $2\pi k\Omega_H/\omega_c$ with respect to each other. This means that the k-th harmonic of the function $\mathcal{F}_{osc}$ acquires an additional factor $\cos(2\pi k\Omega_H/\omega_c)$. Since both Ω_H and ω_c are linear in H, this factor does not modify the oscillation picture (it just reduces the oscillating term). However, in stronger magnetic fields, when the oscillating part of the conductivity is no longer small with respect to the non-oscillaring part, the oscillation picture is modified: the peaks of the conductivity become split due to spin splitting.

The Shubnikov - de Haas oscillations are also observed in the 2D systems, when **H** is perpendicular to the 2D plane. Equations (49.9), (49.13), and (49.14) can be formally applied to this case if we put $p_z = 0$ and remove the integrals $\int dp_z/(2\pi\hbar)$ from the expressions. Calculating the sums over N with the use of Eq. (1), we find that the self-energy function is determined by the implicit complex equation

$$\Sigma_\varepsilon' + i\Sigma_\varepsilon'' = \frac{\hbar}{2\tau} \cot\left[\frac{\pi}{\hbar\omega_c}(\varepsilon - \hbar\omega_c/2 - \Sigma_\varepsilon' - i\Sigma_\varepsilon'')\right], \qquad (17)$$

where $\tau^{-1} = wm/\hbar^3$ is the 2D scattering rate for the case of short-range scattering potentials. At weak enough magnetic fields, when $2\pi\Sigma''_\varepsilon \gg \hbar\omega_c$, Eq. (17) can be solved analytically, with the result

$$\Sigma''_\varepsilon \simeq \frac{\hbar}{2\tau} \left[1 - 2e^{-\pi/\omega_c\tau} \cos\left(\frac{2\pi\varepsilon}{\hbar\omega_c}\right) \right]. \qquad (18)$$

The oscillating contribution is small only when the Dingle factor $e^{-\pi/\omega_c\tau}$ is small (for this reason, only the principal harmonic is essential). If the Dingle factor is comparable to 1, Eq. (18) is not valid and one should use the more general Eq. (17). Similar to the 3D case, the oscillations of the Fermi energy at a fixed electron density appear to be small in comparison to the oscillations of Σ''_ε (problem 10.7). The conductivity is calculated in a similar way as in the 3D case. Below we present the result for σ_d:

$$\sigma_d = \frac{\sigma_0}{1 + (\omega_c\tau)^2} \left[1 - 4\frac{(\omega_c\tau)^2}{1 + (\omega_c\tau)^2} e^{-\pi/\omega_c\tau} \mathcal{B}_1 \cos\left(\frac{2\pi\varepsilon_F}{\hbar\omega_c}\right) \right]. \qquad (19)$$

With the increase of the field, the Dingle factor becomes comparable to 1, which means that the Landau levels become well-defined. Equations (18) and (19) are not valid in this region. They show only a tendency for the oscillating part of σ to increase and be comparable with the non-oscillating part provided that the temperature is low enough. Moreover, the theory given in this and previous sections is not valid for 2D electrons in this region of fields. For a correct description, it is necessary to take into account all higher-order corrections to the self-energy and to the vertex part in the Bethe-Salpeter equation. Though this cannot be done analytically, a description of the 2D electrons in strong magnetic fields is feasible for the case of smooth potentials; see the next section.

Let us consider the magnetophonon oscillations in the quasi-classical region of magnetic fields. We assume that the temperature is low enough and the broadening of the Landau levels is dominated by elastic scattering. The scattering rate of electrons by *LO* phonons is assumed to be smaller than the cyclotron frequency. These approximations allow us to consider the electron-phonon interaction by perturbations, i.e., to use Eq. (48.17) with the trace expressed according to Eq. (48.19). However, since we need to take into account the elastic-scattering-induced broadening, the free-electron Hamiltonian $\widehat{H}_e$ in Eq. (48.19) should be replaced by the Hamiltonian of electrons interacting with a random static potential. In summary, the *LO*-phonon-assisted part of the conductivity is given by the following expression:

$$\sigma_{xx}^{(LO)} = \frac{e^2}{\hbar V m^2 \omega_c^2} \lim_{\omega \to 0} \frac{1}{\omega} \text{Re} \sum_{\mathbf{q}} q_y^2 |C_q^{(LO)}|^2$$

$$\times \sum_{\delta\delta'\delta_1\delta_1'} \langle\delta|e^{i\mathbf{q}\cdot\mathbf{r}}|\delta'\rangle\langle\delta_1'|e^{-i\mathbf{q}\cdot\mathbf{r}}|\delta_1\rangle \int_{-\infty}^{0} d\tau e^{\lambda\tau - i\omega\tau} \qquad (20)$$

$$\times \left\{ \langle\langle \text{Sp}\hat{\eta}_{eq} e^{-i\widehat{H}\tau/\hbar}\hat{a}_\delta^+\hat{a}_{\delta'} e^{i\widehat{H}\tau/\hbar}\hat{a}_{\delta_1'}^+\hat{a}_{\delta_1}\rangle\rangle[(N_{LO}+1)e^{i\omega_{LO}\tau} + N_{LO}e^{-i\omega_{LO}\tau}] \right.$$

$$\left. - \langle\langle \text{Sp}\hat{\eta}_{eq}\hat{a}_{\delta_1'}^+\hat{a}_{\delta_1} e^{-i\widehat{H}\tau/\hbar}\hat{a}_\delta^+\hat{a}_{\delta'} e^{i\widehat{H}\tau/\hbar}\rangle\rangle[(N_{LO}+1)e^{-i\omega_{LO}\tau} + N_{LO}e^{i\omega_{LO}\tau}] \right\},$$

where $\widehat{H}$ is the Hamiltonian of electrons in the random static potential. The double angular brackets denote the averaging over this potential. If $|\delta\rangle$ are the exact eigenstates of the Hamiltonian $\widehat{H}$, we find

$$\text{Sp}\hat{\eta}_{eq} e^{-i\widehat{H}\tau/\hbar}\hat{a}_\delta^+\hat{a}_{\delta'} e^{i\widehat{H}\tau/\hbar}\hat{a}_{\delta_1'}^+\hat{a}_{\delta_1}$$

$$= \delta_{\delta\delta_1}\delta_{\delta'\delta_1'} e^{i(\varepsilon_{\delta'}-\varepsilon_\delta)\tau/\hbar} 2f(\varepsilon_\delta)[1 - f(\varepsilon_{\delta'})], \qquad (21)$$

and the second trace standing in Eq. (20) is expressed in a similar way. The product of the matrix elements of $e^{\pm i\mathbf{q}\cdot\mathbf{r}}$ in Eq. (20) is written as $|\langle\delta|e^{i\mathbf{q}\cdot\mathbf{r}}|\delta'\rangle|^2 = \int d\mathbf{r} \int d\mathbf{r}' e^{i\mathbf{q}\cdot(\mathbf{r}'-\mathbf{r})}\psi_{\mathbf{r}}^{(\delta)}\psi_{\mathbf{r}'}^{(\delta)*}\psi_{\mathbf{r}'}^{(\delta')}\psi_{\mathbf{r}}^{(\delta')*}$, where $\psi_{\mathbf{r}}^{(\delta)}$ is the wave function of the state $|\delta\rangle$. Integrating over τ in Eq. (20), we obtain, in the limit $\omega = 0$, the following expression:

$$\sigma_{xx}^{(LO)} = \frac{2\pi\hbar e^2 N_{LO}}{VT m^2 \omega_c^2} \sum_{\mathbf{q}} q_y^2 |C_q^{(LO)}|^2 \int d\mathbf{r} \int d\mathbf{r}' (e^{-i\mathbf{q}\cdot(\mathbf{r}-\mathbf{r}')} + c.c.)$$

$$\times \int d\varepsilon f(\varepsilon)[1 - f(\varepsilon + \hbar\omega_{LO})]\langle\langle A_\varepsilon(\mathbf{r},\mathbf{r}')A_{\varepsilon+\hbar\omega_{LO}}(\mathbf{r}',\mathbf{r})\rangle\rangle , \qquad (22)$$

where $A_\varepsilon(\mathbf{r},\mathbf{r}') = [G_\varepsilon^A(\mathbf{r},\mathbf{r}') - G_\varepsilon^R(\mathbf{r},\mathbf{r}')]/2\pi i$ is the spectral density function (13.26) in the coordinate representation. The expression of the Green's functions through $\psi_{\mathbf{r}}^{(\delta)}$ and ε_δ is given in problem 3.10. Below we consider the correlation functions of the Green's functions in the Landau level representation and approximate them, like in the beginning of the previous section, by the products of the averaged Green's functions, assuming the case of short-range scattering potential. We take into account Eqs. (G.8) and (G.9) and integrate over coordinates in Eq. (22) with the aid of Eq. (48.14). Then we integrate over p_y, p_y', and q_z. As a result, Eq. (22) is transformed to

$$\sigma_d^{(LO)} = \frac{e^2 N_{LO}}{16\pi^3 T m\omega_c} \sum_{NN'} \int \frac{dp_z}{2\pi\hbar} \int \frac{dp_z'}{2\pi\hbar} \int_0^\infty dq_\perp^2 \, q_\perp^2 V \qquad (23)$$

$$\times |C^{(LO)}_{\sqrt{q_\perp^2+(p_z-p_z')^2/\hbar^2}}|^2 \Phi_{NN'}(q_\perp^2 l_H^2/2) \int d\varepsilon f(\varepsilon)[1-f(\varepsilon+\hbar\omega_{LO})]$$

$$\times [G^A_\varepsilon(p_z,N)-G^R_\varepsilon(p_z,N)][G^R_{\varepsilon+\hbar\omega_{LO}}(p_z',N')-G^A_{\varepsilon+\hbar\omega_{LO}}(p_z',N')].$$

This equation generalizes the result described by Eqs. (48.22)−(48.24) since it takes into account the influence of elastic scattering on the electron spectrum. For free electrons, the product $[G^A_\varepsilon(p_z,N)-G^R_\varepsilon(p_z,N)]$ $\times[G^R_{\varepsilon+\hbar\omega_{LO}}(p_z',N')-G^A_{\varepsilon+\hbar\omega_{LO}}(p_z',N')]$ is equal to $(2\pi)^2\delta(\varepsilon-\varepsilon_{Np_z})\delta(\varepsilon+\hbar\omega_{LO}-\varepsilon_{N'p_z'})$ and, integrating over p_z and p_z', we recover Eqs. (48.22)−(48.24). Though the problem described by Eq. (23) is more complicated, a simple analytical consideration is possible if we neglect the dependence of the factor $\int_0^\infty dq_\perp^2\, q_\perp^2 |C^{(LO)}_{\sqrt{q_\perp^2+(p_z-p_z')^2/\hbar^2}}|^2 \Phi_{NN'}(q_\perp^2 l_H^2/2)$ on the Landau level numbers and on p_z-p_z'. This neglect is justified when $(p_z-p_z')^2/\hbar^2$ is much smaller than $q_\perp^2$ contributing to the integral, which corresponds to near-resonance conditions; see Sec. 48. Then we assume that $T \gg \hbar\omega_c$ and calculate the sums over N and N' in the quasi-classical approximation by using Eqs. (1) and (2). The sum over N gives us

$$\frac{1}{2\pi i}\sum_N [G^A_\varepsilon(p_z,N)-G^R_\varepsilon(p_z,N)] = \frac{1}{\pi}\sum_N \frac{\Sigma_\varepsilon''}{(\varepsilon-\varepsilon_{Np_z}-\Sigma_\varepsilon')^2+(\Sigma_\varepsilon'')^2}$$

$$\simeq \frac{1}{\hbar\omega_c}\left[1+2\sum_{k=1}^\infty (-1)^k \mathcal{D}_k(\varepsilon)\cos\left(\frac{2\pi k\varepsilon}{\hbar\omega_c}-\frac{\pi k p_z^2}{m\hbar\omega_c}\right)\right], \qquad (24)$$

where only the main (non-oscillating) part of the self-energy is substituted in the oscillating contribution; see Eqs. (4)−(6). The expression (24) should be integrated over p_z in a similar way as in the transition from Eq. (3) to Eqs. (4) and (5). Calculating also the sum over N' and the integral over p_z' of the factor $[G^A_{\varepsilon+\hbar\omega_{LO}}(p_z',N')-G^R_{\varepsilon+\hbar\omega_{LO}}(p_z',N')]$, we obtain

$$\frac{1}{(2\pi i)^2}\sum_N \int dp_z [G^A_\varepsilon(p_z,N)-G^R_\varepsilon(p_z,N)]$$

$$\times \sum_{N'} \int dp_z' [G^A_{\varepsilon+\hbar\omega_{LO}}(p_z',N')-G^R_{\varepsilon+\hbar\omega_{LO}}(p_z',N')]$$

$$\simeq \frac{4}{(\hbar\omega_c)^2}\left[\sqrt{2m\varepsilon}+\sum_{k=1}^\infty (-1)^k \sqrt{\hbar\omega_c m/k}\,\mathcal{D}_k(\varepsilon)\cos\left(\frac{2\pi k\varepsilon}{\hbar\omega_c}-\frac{\pi}{4}\right)\right]$$

$$\times\left[\sqrt{2m(\varepsilon+\hbar\omega_{LO})}+\sum_{k'=1}^\infty (-1)^{k'}\sqrt{\hbar\omega_c m/k'}\right.$$

$$\left.\times\mathcal{D}_{k'}(\varepsilon+\hbar\omega_{LO})\cos\left(\frac{2\pi k'\varepsilon}{\hbar\omega_c}+\frac{2\pi k'\omega_{LO}}{\omega_c}-\frac{\pi}{4}\right)\right]. \qquad (25)$$

The expression on the right-hand side of Eq. (25) contains the part which rapidly oscillates with energy due to the phase factors $2\pi k\varepsilon/\hbar\omega_c$ as well as the part which does not oscillate with energy. Under the assumed condition $T \gg \hbar\omega_c$, the oscillating part changes with energy much faster than the distribution function $f(\varepsilon)$ and, therefore, gives a small contribution to the integral over energy. For this reason, we should retain only the part which does not oscillate with energy. It is equal to

$$\frac{8m\sqrt{\varepsilon(\varepsilon + \hbar\omega_{LO})}}{(\hbar\omega_c)^2} + \frac{2m}{\hbar\omega_c}\sum_{k=1}^{\infty} k^{-1}\mathcal{D}_k(\varepsilon)\mathcal{D}_k(\varepsilon + \hbar\omega_{LO})\cos\left(\frac{2\pi k\omega_{LO}}{\omega_c}\right).$$
(26)

The principal contribution ($k = 1$) in the second term of this expression describes the magnetophonon oscillations. Keeping only this contribution, one may approximate the oscillating part $\delta\sigma_d^{(LO)}$ of the conductivity by the following expression:

$$\delta\sigma_d^{(LO)} \propto \exp\left(-\frac{2\pi\Gamma}{\hbar\omega_c}\right)\cos\left(2\pi\frac{\omega_{LO}}{\omega_c}\right),$$
(27)

where the energy dependence of the Dingle factors is neglected. The broadening energy Γ introduced in Eq. (27) is roughly estimated by the averaged electron-impurity scattering rate, $\Gamma \simeq \hbar/\bar{\tau}$. Equation (27) is consistent with experimental observations. The magnetophonon oscillations described by Eq. (27) are also observed in the 2D electron systems in the quasi-classical region of magnetic fields.

51. Quantum Hall Effect

With the increase of the magnetic field applied perpendicular to the plane of a 2D electron system, the amplitudes of Shubnikov - de Haas oscillations of σ_{xx} increase until the oscillation picture is transformed to a sequence of sharp peaks. In between the peaks, the diagonal part of the conductivity vanishes, $\sigma_{xx} = 0$. Moreover, when $\sigma_{xx} = 0$, the Hall resistivity $\rho_\perp = \sigma_{yx}/(\sigma_{xx}^2 + \sigma_{yx}^2)$ is constant and equal to $2\pi\hbar/N_f e^2$, where N_f is an integer. In other words, the Hall resistivity is quantized in units of $2\pi\hbar/e^2$. Its dependence on the magnetic field H (or on the 2D electron density n) looks like a staircase which replaces the well-known dependence $\rho_\perp = H/|e|cn$ following from Eq. (11.19). This experimental fact is called the quantum Hall effect (QHE). Since in the region of plateaus (where $\rho_\perp$ is constant) the diagonal part of the conductivity is zero, one may also write

$$\sigma_{yx} \equiv \sigma_\perp = \frac{e^2}{2\pi\hbar}N_f.$$
(1)

The integer N_f is identified with the number of filled Landau levels. Indeed, if N_f Landau levels are completely filled, the electron density n is equal to $N_f/2\pi l_H^2$. Substituting $N_f = 2\pi l_H^2 n$ into Eq. (1), we find $\sigma_\perp = |e|cn/H$, i.e., in these conditions $\sigma_\perp$ coincides with its classical value.

Can the quantization of the Hall conductivity be explained in the absence of any external (or random scattering) potential? To answer this question, let us consider a system of free 2D electrons and apply the Kubo formula in the form (48.4) at $\omega = 0$ in order to find the Hall conductivity. Substituting the Landau eigenstates $|Np_y\rangle$ of the 2D electrons with the wave functions $L_y^{-1/2} e^{ip_y y/\hbar} \varphi_x^{(Np_y)}$ in place of exact eigenstates $|\delta\rangle$, we use the matrix elements (48.5) of the velocity operator and the expression for electron spectrum $\varepsilon_{N\sigma} = \varepsilon_N - \sigma\hbar\Omega_H/2$, where $\hbar\Omega_H = g\mu_B H$. The expression for $\sigma_\perp$ takes the form

$$\sigma_\perp = -\frac{\hbar e^2 \omega_c^2}{2\pi} \sum_{\sigma=\pm 1} \sum_{NN'} f(\varepsilon_{N\sigma}) \frac{N\delta_{N',N-1} - N'\delta_{N',N+1}}{(\varepsilon_N - \varepsilon_{N'})^2} \tag{2}$$

$$= \frac{e^2}{2\pi\hbar} \sum_{\sigma=\pm 1} \sum_{N=0}^{\infty} (N+1)[f(\varepsilon_{N\sigma}) - f(\varepsilon_{N+1,\sigma})] = \frac{e^2}{2\pi\hbar} \sum_{\sigma=\pm 1} \sum_{N=0}^{\infty} f(\varepsilon_{N\sigma}).$$

If there are N_f completely filled levels (say, N_f^+ with spin $\sigma = 1$ and N_f^- with spin $\sigma = -1$), one has $f(\varepsilon_{N,\pm 1}) = 1$ for $N < N_f^\pm$, and $f(\varepsilon_{N,\pm 1}) = 0$ for $N \geq N_f^\pm$. Since $N_f^+ + N_f^- = N_f$, the Hall conductivity in this case is given by Eq. (1). This result seems to be consistent with the picture of the quantum Hall effect but does not really lead to plateaus as the magnetic field changes. To clarify this statement, let us analyze the dependence of the Fermi level on the magnetic field. The density of states for free 2D electrons in the magnetic field (see Sec. 16) is represented as a set of δ-shaped peaks corresponding to discrete Landau levels:

$$\rho_{2D}(E) = \frac{1}{2\pi l_H^2} \sum_{\sigma=\pm 1} \sum_{N=0}^{\infty} \delta(E - \varepsilon_N + \sigma\hbar\Omega_H/2). \tag{3}$$

Each Landau level is multiply degenerate and can accumulate the electron density up to $n_f = 1/2\pi l_H^2$ for each spin state. In a free-electron system, the Fermi level at $T = 0$ cannot stay in the gap between the Landau levels because there are no states in the gap. If the electron density is not equal to an integer number of n_f, the highest occupied Landau level is not completely filled. It can be characterized by the filling factor ν introduced according to

$$n = n_f(N_f + \nu), \quad \nu \in [0,1]. \tag{4}$$

If the magnetic field or electron density are changed, the Fermi level at $T = 0$ remains pinned to a Landau level as long as the filling factor lies in the region $[0, 1]$. Then, the Fermi level jumps to another Landau level but cannot stay in the gap. In conclusion, the conditions when one has completely filled levels ($\nu = 0$ or $\nu = 1$) are realized only for fixed values of the magnetic field, when, indeed, $|e|cn/H$ is equal to an integer number of $e^2/2\pi\hbar$. From the formal point of view, according to Eq. (3), the total electron density is given by

$$n = \int dE \rho_{2D}(E) f(E) = \frac{1}{2\pi l_H^2} \sum_{\sigma=\pm 1} \sum_{N=0}^{\infty} f(\varepsilon_{N\sigma}). \qquad (5)$$

Combining Eq. (5) with Eq. (2), we find $\sigma_\perp = |e|cn/H$. Therefore, in an infinite free-electron system, i.e., in the absence of any external potentials or boundaries, there is no quantization of the Hall conductivity.

One may also emphasize the role of the external potentials in the following way. Let us substitute the operator identities (48.6) into the Kubo formula (48.4) for σ_{yx} at $\omega = 0$. We remind that in Sec. 48 we have carried out a similar procedure for σ_{xx}. We obtain

$$\sigma_\perp = i \frac{e^2}{L^2 \hbar \omega_c^2} \sum_{\delta\delta'} \langle \delta | \hat{v}_x | \delta' \rangle \langle \delta' | \hat{v}_y | \delta \rangle \left[f(\varepsilon_\delta) - f(\varepsilon_{\delta'}) \right]$$

$$+ i \frac{e^2 l_H^4}{L^2 \hbar} \sum_{\delta\delta'} \left\langle \delta \left| \frac{\partial U}{\partial x} \right| \delta' \right\rangle \left\langle \delta' \left| \frac{\partial U}{\partial y} \right| \delta \right\rangle \frac{f(\varepsilon_\delta) - f(\varepsilon_{\delta'})}{(\varepsilon_\delta - \varepsilon_{\delta'})^2}$$

$$- \frac{e^2 l_H^4 m}{L^2 \hbar^2} \sum_{\delta\delta'} \left[\langle \delta | \hat{v}_x | \delta' \rangle \left\langle \delta' \left| \frac{\partial U}{\partial y} \right| \delta \right\rangle \right. \qquad (6)$$

$$\left. - \left\langle \delta \left| \frac{\partial U}{\partial x} \right| \delta' \right\rangle \langle \delta' | \hat{v}_y | \delta \rangle \right] \frac{f(\varepsilon_\delta) - f(\varepsilon_{\delta'})}{\varepsilon_\delta - \varepsilon_{\delta'}}.$$

The last (third) term in this expression is equal to zero (problem 10.8). The first term is proportional to the commutator $[\hat{v}_x, \hat{v}_y]$ averaged with the equilibrium distribution. Since this commutator is constant and equal to $-i\hbar\omega_c/m$, the first term of Eq. (6) is equal to $|e|cn/H$. Therefore, the Hall conductivity can be presented as $\sigma_\perp = |e|cn/H + \delta\sigma_\perp$, where $\delta\sigma_\perp$ is the second term on the right-hand side of Eq. (6). This term is proportional to the square of potential gradients. Thus, the very fact of the Hall quantization is caused by the existence of these gradients. From this consideration one may conclude that $\delta\sigma_\perp$ should depend on the scattering potential, at least on its averaged characteristics. Nevertheless, the conductivity (1) is determined only by the fundamental physical constants. In this connection, we notice that the quantum Hall effect is usually observed in clean samples, where the potentials are mostly macroscopic in the sense that they modify the electron dynamics but cannot be treated as scattering potentials. Below we prove the existence of the quantum Hall effect in the presence of a classically smooth potential, when the variations of the potential energy $U(\mathbf{r})$ with in-plane coordinate $\mathbf{r}$ are small on the scale of the magnetic length.

We first derive a convenient form for the kinetic equation describing motion of electrons in the smooth potential. The one-electron density matrix $\hat{n}_t$ satisfies Eq. (4.32). In the presence of a magnetic field, we write this equation by using the Landau level representation:

$$\frac{\partial n_{\sigma t}(Np, N'p')}{\partial t} + \frac{i}{\hbar}(\varepsilon_N - \varepsilon_{N'})n_{\sigma t}(Np, N'p') \tag{7}$$

$$+\frac{i}{\hbar}\sum_{N_1 p_1}\left[U_{NN_1}(p, p_1)n_{\sigma t}(N_1 p_1, N'p') - n_{\sigma t}(Np, N_1 p_1)U_{N_1 N'}(p_1, p')\right] = 0,$$

where p denotes the y-component of electron momentum (the index y is omitted for brevity here and below in this section), $n_{\sigma t}(Np, N'p') = \langle Np|\hat{n}_t|N'p'\rangle$, and $U_{NN_1}(p, p_1) = \langle Np|U(\mathbf{r})|N_1 p_1\rangle$. The spin index σ is not important here since the interaction we consider does not change the spin numbers provided the spin-orbit interaction is weak and can be neglected. Equation (7) is exact. However, for a smooth potential energy $U(\mathbf{r})$, one may neglect the mixing between different Landau levels in strong magnetic fields and consider only the diagonal contributions of $U_{NN'}(p, p')$ and $n_{\sigma t}(Np, N'p')$. In this approximation, $n_{\sigma t}(Np, N'p') \simeq \delta_{NN'}\rho_{N\sigma t}(p, p')$, where $\rho_{N\sigma t}(p, p')$ satisfies the equation

$$\frac{\partial \rho_{N\sigma t}(p, p')}{\partial t} + \frac{i}{\hbar}\sum_{p_1}\left[U_{NN}(p, p_1)\rho_{N\sigma t}(p_1, p')\right.$$

$$\left. - \rho_{N\sigma t}(p, p_1)U_{NN}(p_1, p')\right] = 0. \tag{8}$$

Since N and σ are conserved, the evolution of this density matrix is entirely described by the momentum variables. It is possible to introduce the Wigner distribution function as $\rho_{N\sigma t}(y, p) = \sum_k e^{iky}\rho_{N\sigma t}(p + \hbar k/2, p - \hbar k/2)$. The integral $\int dp\rho_{N\sigma t}(y, p)/2\pi\hbar$ is the local number of electrons in the subband $N\sigma$ per unit length along the OY axis. The same result should be obtained if we integrate the local electron density in this subband over x. Taking into account that the Wigner momentum p is related to the x-coordinate of the oscillator center of the electron wave function, we introduce the distribution function having the properties of the local electron density in the subband $N\sigma$ as

$$\rho_{N\sigma t}(x,y) = \int \frac{dp}{2\pi\hbar} \delta(x + l_H^2 p/\hbar) \sum_k e^{iky} \rho_{N\sigma t}\left(p + \frac{\hbar k}{2}, p - \frac{\hbar k}{2}\right). \quad (9)$$

If we consider x and y as classical coordinates of electrons in the plane, $\rho_{N\sigma t}(x,y) \equiv \rho_{N\sigma t}(\mathbf{r})$ can be treated as a quasi-classical distribution function of electrons in the coordinate representation. Below we will see that this function, indeed, satisfies a quasi-classical kinetic equation.

Transforming Eq. (8) according to Eq. (9), we obtain

$$\frac{\partial \rho_{N\sigma t}(x,y)}{\partial t} = \frac{2i}{\pi\hbar l_H^2} \int dx' \int dy' \rho_{N\sigma t}(x',y') \quad (10)$$

$$\times \sum_p \left[U_{NN}\left(-\frac{2\hbar x'}{l_H^2} - p, -\frac{2\hbar x}{l_H^2} - p\right) e^{2ip(y-y')/\hbar + 2i(xy - x'y')/l_H^2} - c.c. \right].$$

The matrix elements of the potential energy standing in this equation are expressed according to their definition given above:

$$U_{NN}\left(-\frac{2\hbar x'}{l_H^2} - p, -\frac{2\hbar x}{l_H^2} - p\right) = \frac{1}{L_y\sqrt{\pi}l_H 2^N N!} \int dx'' \int dy'' U(x'',y'')$$

$$\times e^{2iy''(x'-x)/l_H^2} e^{-[(x''-2x)/l_H - pl_H/\hbar]^2/2 - [(x''-2x')/l_H - pl_H/\hbar]^2/2} \quad (11)$$

$$\times H_N\left[\frac{x''-2x}{l_H} - \frac{pl_H}{\hbar}\right] H_N\left[\frac{x''-2x'}{l_H} - \frac{pl_H}{\hbar}\right].$$

Substituting this expression into Eq. (10), we calculate the sum over p by using the following identity: $\int_{-\infty}^{\infty} du\, e^{-u^2 - a^2 + 2ibu} H_N(u+a) H_N(u-a) = \sqrt{\pi} 2^N N! e^{-a^2 - b^2} L_N^0[2(a^2 + b^2)]$. As a result, Eq. (10) is written in the form

$$\frac{\partial \rho_{N\sigma t}(\mathbf{r})}{\partial t} = -\frac{2}{\pi^2 \hbar l_H^4} \int d\mathbf{r}_1 \int d\mathbf{r}_2 \rho_{N\sigma t}(\mathbf{r} + \mathbf{r}_1) U(\mathbf{r} + \mathbf{r}_2)$$

$$\times e^{-r_1^2/l_H^2} L_N^0\left(2r_1^2/l_H^2\right) \sin(2\mathbf{n}_z \cdot [\mathbf{r}_1 \times \mathbf{r}_2]/l_H^2), \quad (12)$$

where $\mathbf{n}_z$ is the unit vector in the direction of the magnetic field, $\mathbf{r} = (x,y)$, and the differential 2D coordinates are introduced as $\mathbf{r}_1 = (x' - x, y' - y)$ and $\mathbf{r}_2 = (x'' - x, y'' - y)$.

The function standing at $\rho_{N\sigma t}(\mathbf{r} + \mathbf{r}_1) U(\mathbf{r} + \mathbf{r}_2)$ under the integrals changes on the characteristic lengths $r_1, r_2 \sim l_H$. If the variations of the potential energy and distribution function on the scale of the magnetic length l_H are weak, one may expand $U(\mathbf{r} + \mathbf{r}_2)$ and $\rho_{N\sigma t}(\mathbf{r} + \mathbf{r}_1)$ in power series of $\mathbf{r}_2$ and $\mathbf{r}_1$. The contributions containing zero-order terms of such expansion (either $U(\mathbf{r})$ or $\rho_{N\sigma t}(\mathbf{r})$) vanish, and the first nonvanishing contribution on the right-hand side of Eq. (12) is

$$-\frac{2}{\pi^2 \hbar l_H^4} \int d\mathbf{r}_1 \int d\mathbf{r}_2 \left(\nabla_\mathbf{r} \rho_{N\sigma t}(\mathbf{r}) \cdot \mathbf{r}_1 \right) \left(\nabla_\mathbf{r} U(\mathbf{r}) \cdot \mathbf{r}_2 \right)$$

$$\times e^{-r_1^2/l_H^2} L_N^0 \left(2r_1^2/l_H^2 \right) \sin \left(\frac{2\mathbf{n}_z \cdot [\mathbf{r}_1 \times \mathbf{r}_2]}{l_H^2} \right). \tag{13}$$

Let us first calculate the integral over $\mathbf{r}_2$ in this expression:

$$\int d\mathbf{r}_2 \left(\nabla_\mathbf{r} U(\mathbf{r}) \cdot \mathbf{r}_2 \right) \sin \left(\frac{2\mathbf{n}_z \cdot [\mathbf{r}_1 \times \mathbf{r}_2]}{l_H^2} \right) = \frac{l_H^2}{2} \left(\nabla_\mathbf{r} U(\mathbf{r}) \cdot [\nabla_{\mathbf{r}_1} \times \mathbf{n}_z] \right)$$

$$\times \int d\mathbf{r}_2 \cos \left(\frac{2\mathbf{n}_z \cdot [\mathbf{r}_1 \times \mathbf{r}_2]}{l_H^2} \right) = \frac{\pi^2 l_H^6}{2} \left(\nabla_\mathbf{r} U(\mathbf{r}) \cdot [\nabla_{\mathbf{r}_1} \times \mathbf{n}_z] \right) \delta(\mathbf{r}_1). \tag{14}$$

The remaining integral over $\mathbf{r}_1$ is calculated by parts, using the properties of $\delta(\mathbf{r}_1)$. One should also take into account that $\nabla_{\mathbf{r}_1} e^{-r_1^2/l_H^2} L_N^0(2r_1^2/l_H^2)$ vanishes at $\mathbf{r}_1 = 0$. As a result, we obtain

$$\frac{\partial \rho_{N\sigma t}(\mathbf{r})}{\partial t} + \mathbf{v}_\perp \cdot \frac{\partial \rho_{N\sigma t}(\mathbf{r})}{\partial \mathbf{r}} = 0, \quad \mathbf{v}_\perp = \frac{1}{m\omega_c} [\nabla_\mathbf{r} U(\mathbf{r}) \times \mathbf{n}_z]. \tag{15}$$

This equation has a dynamical interpretation (compare to Eq. (9.36)) because it can be written as $\partial \rho_{N\sigma t}(\mathbf{R})/\partial t + (d\mathbf{R}/dt) \cdot \partial \rho_{N\sigma t}(\mathbf{R})/\partial \mathbf{R} = d\rho_{N\sigma t}(\mathbf{R})/dt = 0$, where $\mathbf{R}$ is the coordinate of the oscillator center satisfying the classical equation of motion $d\mathbf{R}/dt = (1/m\omega_c) [\nabla_\mathbf{R} U(\mathbf{R}) \times \mathbf{n}_z]$. This equation is consistent with the result of problem 1.13, where the classical drift velocity $d\mathbf{R}/dt$ has been calculated as a function of the external field $\mathbf{E} = |e|^{-1} \nabla_\mathbf{R} U(\mathbf{R})$. Equation (15) can be obtained in a simpler way by employing the property of smoothness of $U(\mathbf{r})$ directly in Eq. (8). As a result, the right-hand side of Eq. (10) is approximately written through the Poisson brackets; see Eqs. (C.15) and (C.16).

The quasi-classical kinetic equation (15) describes the particles whose group velocities $\mathbf{v}_\perp$ are determined entirely by the magnetic field and potential. It reflects the dynamical properties of 2D electrons in the magnetic field. If, apart from the smooth potential energy $U(\mathbf{r})$, there are additional scattering potentials (generated, for example, by phonons or impurities), their contribution to the quantum kinetic equation should give collision integrals on the right-hand side of Eq. (15). These collision integrals describe the jumps between the states with different coordinates $\mathbf{r}$ and $\mathbf{r}'$ and provide thermodynamic equilibrium in the system.

The probability of such jumps decreases exponentially if $|\mathbf{r} - \mathbf{r}'|$ exceeds the magnetic length. For this reason, the collision integrals have local forms, i.e., depend on $U(\mathbf{r})$ in a parametric way.

Below we consider a stationary problem. Equation (15) in this case is satisfied for any function depending on coordinate $\mathbf{r}$ through $U(\mathbf{r})$. However, if the thermalizing collisions are present, the equilibrium solution of Eq. (15) should be written as

$$\rho_{N\sigma}^{(eq)}(\mathbf{r}) = \frac{f_{N\sigma}^{(eq)}(\mathbf{r})}{2\pi l_H^2}, \quad f_{N\sigma}^{(eq)}(\mathbf{r}) = \left\{ e^{[\varepsilon_{N\sigma}+U(\mathbf{r})-\varepsilon_F]/T} + 1 \right\}^{-1}, \quad (16)$$

where $f_{N\sigma}^{(eq)}(\mathbf{r})$ is the Fermi distribution function which parametrically includes the variation of the Landau level energy due to the presence of the potential. The normalization factor $n_f = 1/2\pi l_H^2$ standing in Eq. (16) is determined from the requirement that the total electron density $(L_x L_y)^{-1} \int_0^{L_x} dx \int_0^{L_y} dy \rho_{N\sigma}(\mathbf{r})$ associated with a completely filled Landau level is equal to n_f (note that we consider a 2D system with dimensions L_x and L_y). In general, one can write $\rho_{N\sigma}(\mathbf{r})$ in the form similar to Eq. (16), $\rho_{N\sigma}(\mathbf{r}) = (2\pi l_H^2)^{-1} f_{N\sigma}(\mathbf{r})$, where $f_{N\sigma}(\mathbf{r})$ is a dimensionless local occupation number. Since the coordinate dependence of this function is reduced to a dependence on the energy $\varepsilon_{N\sigma} + U(\mathbf{r})$, the electrons can be characterized by their energy. The electrons belonging to each Landau level move along the equipotential lines determined by the relation $\varepsilon = \varepsilon_{N\sigma} + U(\mathbf{r})$, and their group velocities are $\mathbf{v}_\perp = (m\omega_c)^{-1}[\nabla_\mathbf{r} U(\mathbf{r}) \times \mathbf{n}_z]|_{U=\varepsilon-\varepsilon_{N\sigma}}$.

Let us assume that the amplitude of the potential energy $U(\mathbf{r})$ is smaller than both the cyclotron energy and Zeeman splitting energy. We also assume that the temperature is low enough (at least, T is much smaller than the amplitude of $U(\mathbf{r})$). The classical expression for the electron density

$$n = \frac{1}{L_x L_y} \sum_{N\sigma} \int_0^{L_x} \int_0^{L_y} \frac{dxdy}{2\pi l_H^2} f_{N\sigma}(\mathbf{r}) \qquad (17)$$

gives us Eq. (4) with the filling factor $\nu = S_f(U)/L_x L_y$, where S_f is the square of the area covered by the occupied states of the highest occupied Landau level denoted below by the indices N_m and σ_m. This area is determined from the relation $\varepsilon_{N_m\sigma_m} + U(\mathbf{r}) < \varepsilon_F$. The occupation of each Landau level by electrons begins with filling of the local minima of the function $\varepsilon_{N\sigma} + U(\mathbf{r})$. From the geometrical point of view, this looks like filling of "lakes", and each lake is separated from the other similar lakes. On the other hand, for an almost filled Landau level, the filled area covers almost all plane, except a number of "hills", also separated from each other. Electrons move in the lakes or around the hills along closed paths.

Now we proceed to calculating the Hall conductivity. If the electric field E_x is applied along OX, the total current I_y passing through the straight line between the points $(0, y)$ and (L_x, y) is given by the classical formula

$$I_y = e \sum_{N\sigma} \int_0^{L_x} \frac{dx}{2\pi l_H^2} v_{\perp y} f_{N\sigma}(\mathbf{r}), \qquad (18)$$

where we integrate the local currents, proportional to $v_{\perp y} \rho_{N\sigma}(\mathbf{r})$, over the length of the system. Due to current continuity, I_y must be independent of y. Note that the current (18) is zero in equilibrium (problem 10.9). The linear Hall conductivity $\sigma_\perp$ is equal to $L_x^{-1} \lim_{E_x \to 0} (I_y/E_x)$. In the presence of the homogeneous electric field, the potential energy becomes equal to $U_E(x, y) = U(x, y) - eE_x x$. The group velocity is also modified, $v_{\perp y} = -(m\omega_c)^{-1} \partial U(x, y)/\partial x + eE_x/m\omega_c$. Substituting this last expression into Eq. (18), we linearize the latter with respect to the field E_x and find the conductivity in the form

$$\sigma_\perp = \frac{e^2}{2\pi\hbar} \sum_{N\sigma} \frac{1}{L_x} \int_0^{L_x} dx$$

$$\times \left[f_{N\sigma}^{(eq)}(x, y) - e^{-1} \frac{\partial U(x, y)}{\partial x} \left. \frac{\partial f_{N\sigma}(x, y)}{\partial E_x} \right|_{E_x=0} \right], \qquad (19)$$

where the second term appears from the expansion of $f_{N\sigma}(x, y)$ in powers of E_x. The function $f_{N\sigma}(x, y)$ is equal to 1 for each completely filled Landau level. This means that the second term on the right-hand side of Eq. (19) vanishes, while the first term is reduced to $e^2/2\pi\hbar$. Therefore, summing the contributions of the completely filled Landau levels, we obtain Eq. (1).

Let us find the contribution of the partly filled Landau level ($N = N_m$ and $\sigma = \sigma_m$), assuming that this level is almost empty, which means that the electrons occupy local minima (lakes) of the potential energy $\varepsilon_{N_m \sigma_m} + U(\mathbf{r})$. Since each lake, numbered by the index λ, is a separate sub-system, the electrons there stay in the local equilibrium characterized by the local Fermi energy $\varepsilon_{F\lambda}$ and the distribution function

$$f_{N\sigma}^{(\lambda)}(\mathbf{r}) = \left\{ e^{[\varepsilon_{N\sigma} + U(\mathbf{r}) - eE_x x - \varepsilon_{F\lambda}]/T} + 1 \right\}^{-1}. \qquad (20)$$

The conductivity associated with this Landau level is represented as a sum over the contributions of the local minima:

$$\sigma_\perp|_{N=N_m,\sigma=\sigma_m} = \frac{e^2}{2\pi\hbar}\frac{1}{L_x}\sum_\lambda \int_{(\lambda)} dx$$

$$\times \left[f_{N\sigma}^{(eq,\lambda)}(x,y) + x\frac{\partial U(x,y)}{\partial x}\frac{\partial f_{N\sigma}^{(eq,\lambda)}(x,y)}{\partial U} \right]. \tag{21}$$

In this expression, $f_{N\sigma}^{(eq,\lambda)}(\mathbf{r})$ is given by Eq. (16) with ε_F replaced by $\varepsilon_{F\lambda}$. Taking into account that $(\partial U(x,y)/\partial x)(\partial f_{N\sigma}^{(eq,\lambda)}(x,y)/\partial U)$ is equal to $\partial f_{N\sigma}^{(eq,\lambda)}(x,y)/\partial x$, we transform the second term under the integral in Eq. (21) by using the integration by parts:

$$\int_{(\lambda)} dx\; x\frac{\partial f_{N\sigma}^{(eq,\lambda)}(x,y)}{\partial x} = -\int_{(\lambda)} dx\; f_{N\sigma}^{(eq,\lambda)}(x,y). \tag{22}$$

As a result, $\sigma_\perp|_{N=N_m,\sigma=\sigma_m} = 0$. We have demonstrated that when the filling of the $(N_f + 1)$-th level starts, the Hall conductivity remains constant and given by Eq. (1). This is a proof of the quantum Hall effect for the electrons moving in a smooth potential. In a similar way, one can consider an almost filled Landau level and, using the requirement of local equilibrium for the electron states moving around the hills, demonstrate that the Hall conductivity associated with this level is equal to $e^2/2\pi\hbar$, as though this level is completely filled (problem 10.10). It is important that the consideration given above does not employ any other assumptions about the potential energy $U(\mathbf{r})$ apart from its smoothness and finiteness. Besides, a logical consequence of this consideration shows us that in a certain range of the filling factors, when the map of the potential energy $\varepsilon_{N_m\sigma_m} + U(\mathbf{r})$ below (above) the Fermi energy cannot be represented as isolated lakes (hills), the Hall conductivity is not given by Eq. (1). As this range is passed, the Hall conductivity changes by a conductance quantum $e^2/2\pi\hbar$. The percolation theory considering classical motion of 2D particles proves that for an arbitrary potential there is only one energy at which the particles are delocalized. Therefore, at zero temperature the region of the filling factors corresponding to the change of the Hall conductivity by $e^2/2\pi\hbar$ is narrow, and the conductivity has sharp steps.

It is easy to explain the absence of σ_d in the conditions (1). For an almost empty (filled) Landau level, the electrons able to carry the current at low temperature, i.e., the electrons with energies near the Fermi level, move around the lakes (hills) and are localized. Since the jumps of electrons between the lakes (hills) are exponentially rare if the potential is smooth on the scale of l_H, the diagonal part of the conductivity is practically zero. It is possible to get a non-zero σ_d by applying a strong electric field which modifies the "lake-hill" map of the potential energy $\varepsilon_{Nm\sigma_m} + U(\mathbf{r})$. With the increase of the electric field, the conductivity (both σ_d and $\sigma_\perp$) changes abruptly and a breakdown of the quantum Hall effect takes place.

The broadening of the steps of the quantum Hall staircase is essential when the characteristic spatial scale of $U(\mathbf{r})$ becomes comparable to the magnetic length l_H. No analytical approach can describe this broadening in the general case. However, there exists a simple model which allows one to carry out such a description on the basis of a quasi-classical kinetic equation. Let us assume that the potential energy $U(\mathbf{r})$ contains a one-dimensional periodic component $V(x) = V(x + nd)$, where d is the period and n is an integer number. For convenience, we also assume the symmetry $V(x) = V(-x)$. The remaining part of $U(\mathbf{r})$ is the scattering potential $U_{sc}(\mathbf{r})$ caused by short-range potentials of randomly distributed impurities. The amplitude of the periodic component is assumed to be much smaller than the cyclotron energy $\hbar\omega_c$ so that one can neglect mixing between different Landau levels. Under this condition, one can also use the wave functions of zero-order approximation, $\psi_{\mathbf{r}}^{(Np)} \equiv \langle \mathbf{r}|Np\rangle = L_y^{-1/2} e^{ipy/\hbar} \varphi_x^{(Np)}$, where $\varphi_x^{(Np)}$ is given by Eq. (5.15), to calculate the matrix elements of the scattering potentials. Thus, the electron spectrum is found in the first-order approximation with respect to V:

$$E_{N\sigma p} = \varepsilon_{N\sigma} + \mathcal{E}_{Np}, \quad \mathcal{E}_{Np} = \langle Np|V|Np\rangle. \tag{23}$$

We remind that $p \equiv p_y$. The contribution $\langle Np|V|Np\rangle$ introduces a finite dispersion of the spectrum and produces a subband out of the N-th Landau level, the width of this subband is denoted below as $2A_N$. Therefore, the spectrum is no longer degenerate in p. Since $V(x)$ is periodic, the symmetry of $\varphi_x^{(Np)}$ makes the energy spectrum a periodic function of the cyclotron orbit center X_p. This function retains the symmetry properties of $V(x)$, in particular, it has the same period d. Taking into account the relation $X_p = -p l_H^2/\hbar$, one can state that the energy spectrum is a periodic and symmetric function of p, and its period is equal to $2p_H$, where $p_H = \hbar d/2l_H^2$. For example, if the modulation potential is harmonic, $V(x) = V_0 \cos(2\pi x/d)$, one has $\mathcal{E}_{Np} = A_N \cos(\pi p/p_H)$, where $A_N = V_0 L_N^0 \left(2\pi^2 l_H^2/d^2\right) \exp\left(-\pi^2 l_H^2/d^2\right)$. Below we also assume that the subband width $2A_N$ is smaller than the Zeeman splitting energy. This means that there is only one partly filled Landau-level subband with a given spin number, and the low-temperature transport properties are entirely determined by electron kinetics in this subband. For this reason, below we either omit the indices N and σ or imply that these indices correspond to the partly filled subband.

We apply a semiclassical kinetic theory for the electrons whose spectrum is given by Eq. (23). The electrons interact with an effective two-dimensional short-range potential $U_{sc}(\mathbf{r})$ formed by randomly distributed impurities. To justify the applicability of such approach, one must have the subband width much greater than the collision-broadening energy $\hbar/\tau$ determined by the scattering time τ introduced below. As seen from the estimates above, A_N rapidly decreases when l_H exceeds d. Therefore, only the region $d \geq l_H$ is important below. In the presence of a weak external electric field $\mathbf{E} = (E_x, E_y)$, one can write a linearized kinetic equation

$$eE_y\frac{\partial f_p^{(eq)}}{\partial p} = \frac{2\pi}{\hbar}\sum_{p'}w_N(p,p')\Big[\delta(\varepsilon_p - \varepsilon_{p'})(\Delta f_{p'} - \Delta f_p)$$

$$+\delta(\varepsilon_p - \varepsilon_{p'} + eE_x l_H^2(p-p')/\hbar)(f_{p'}^{(eq)} - f_p^{(eq)})\Big], \qquad (24)$$

where we have omitted the Landau-level index N of the energy spectra ($\mathcal{E}_{Np} \to \varepsilon_p$) and distribution functions. The contribution proportional to E_y enters this equation through the dynamical term, see Sec. 8, while the contribution proportional to E_x is associated with the shift of electron energy in the electric field, according to $\varepsilon_p \to \varepsilon_p - eE_x X_p$. Equation (24) allows us to determine the non-equilibrium part Δf_p of the distribution function $f_p = f_p^{(eq)} + \Delta f_p$, where $f_p^{(eq)}$ is the equilibrium Fermi distribution function which depends on the energy ε_p. The right-hand side of Eq. (24) contains the squared matrix element of the scattering potential averaged over the impurity distribution:

$$w_N(p,p') = \frac{w}{\sqrt{2\pi}l_H L_y}e^{-l_H^2(p-p')^2/2\hbar^2}Q_N[l_H^2(p-p')^2/\hbar^2]. \qquad (25)$$

Since we consider a short-range scattering potential, the 2D Fourier transform $w(q)$ of the correlation function $\langle\langle U_{sc}(\mathbf{r})U_{sc}(\mathbf{r}')\rangle\rangle$, see Eqs. (7.22) and (7.23), is independent of q and equal to $w = w(0)$. The function

$$Q_N(x) = \frac{1}{\sqrt{\pi}}\int_{-\infty}^{\infty}du e^{-u^2}[L_N^0(u^2 + x/2)]^2 \qquad (26)$$

is a polynomial equal to 1 for $N = 0$ and to $3/4 - x/2 + x^2/4$ for $N = 1$.

Since the kinetic equation (24) is one-dimensional, it is solved exactly; see Sec. 8. The non-equilibrium part of the distribution function is antisymmetric, $\Delta f_{-p} = -\Delta f_p = -\Delta f_{p+2p_Hm}$, where m is integer. Therefore, the contribution to the first term on the right-hand side of Eq. (24) comes from $p' = -p + 2mp_H$. This term can be written as $-\Delta f_p/\tau_p$, where the scattering time τ_p is given by $\tau_p^{-1} = (2\pi/\hbar)\sum_{p'}w_N(p,p')(1 - v_{p'}/v_p)\delta(\varepsilon_p - \varepsilon_{p'})$ and $v_p = \partial\varepsilon_p/\partial p$ is the group velocity. The second term on the right-hand side of Eq. (24) is written in the linear approximation as $eE_x\mathcal{V}_p(\partial f_p^{(eq)}/\partial\varepsilon_p)$, where $\mathcal{V}_p = (2\pi l_H^2/\hbar^2)\sum_{p'}w_N(p,p')(p - p')\delta(\varepsilon_p - \varepsilon_{p'})$ is a characteristic velocity. Taking into account the periodicity of ε_p, we find that the contribution to this sum also comes from $p' = -p + 2mp_H$. As a result,

$$\Delta f_p = e\frac{\partial f_p^{(eq)}}{\partial\varepsilon_p}\left[-E_y\tau_p v_p + E_x\frac{d}{2}\left(\frac{p}{p_H} - \frac{S_p^{(1)}}{S_p^{(0)}}\right)\right], \qquad (27)$$

and

$$\tau_p^{-1} = \sqrt{\frac{2}{\pi}}\frac{wS_p^{(0)}}{\hbar^2 l_H|v_p|}, \qquad (28)$$

where the quantities $S_p^{(k)}$ are the infinite sums over an integer variable:

$$S_p^{(k)} = \sum_{m=-\infty}^{\infty} m^k \exp\left[-\frac{d^2(p/p_H - m)^2}{2l_H^2}\right]Q_N[d^2(p/p_H - m)^2/l_H^2]. \quad (29)$$

The density of the dissipative current is expressed through the non-equilibrium part $\Delta\rho_N(p,p')$ of the one-electron density matrix $\rho_N(p,p') = \langle Np|\hat{\rho}|Np'\rangle$ and through the matrix elements of the velocity operator $\hat{\mathbf{v}}$ in the usual way: $\mathbf{I} = e(L_xL_y)^{-1}\sum_{pp'}\langle Np'|\hat{\mathbf{v}}|Np\rangle\,\Delta\rho_N(p,p')$. Only the diagonal in N matrix elements remain in this expression because the mixing between the Landau levels is neglected. The current can be expressed through the non-equilibrium part of the distribution function $\Delta f_p \equiv \Delta\rho_N(p,p)$. For example, the component of the current density perpendicular to the modulation axis is given as in the classical kinetic theory, $I_y = e(L_yL_x)^{-1}\sum_p v_p\Delta f_p$. This current is of the dynamical origin. On the other hand, the current along the modulation axis is a hopping current. It is given by an expression similar to that of problem 10.3 (problem 10.11). Apart from the dissipative currents, there are also non-dissipative Hall currents $I_x^{(H)} = -|e|cnE_y/H$ and $I_y^{(H)} = |e|cnE_x/H$. These currents are not related to Δf_p and do not affect the equilibrium in the system. The non-dissipative currents can be found from a consideration of the non-diagonal in N components of the electron density matrix in the absence of any potential, in a similar way as in the beginning of this section; see Eqs. (2) and (5). Let us substitute the solution of the kinetic equation into the expression for I_y given above and add the non-dissipative part $I_y^{(H)}$ there. The components of the conductivity tensor found from this consideration are

$$\sigma_{yx} = \frac{e^2 d}{2L_x L_y} \sum_p v_p \left(\frac{p}{p_H} - \frac{S_p^{(1)}}{S_p^{(0)}} \right) \frac{\partial f_p^{(eq)}}{\partial \varepsilon_p} + \frac{|e|cn}{H} \qquad (30)$$

and

$$\sigma_{yy} = \frac{e^2}{L_x L_y} \sum_p \tau_p v_p^2 \left(-\frac{\partial f_p^{(eq)}}{\partial \varepsilon_p} \right). \qquad (31)$$

Other components can be found by considering I_x (problem 10.12). The non-diagonal components $\sigma_\perp = \sigma_{yx}$ are independent of the scattering potential characteristics and of the amplitude of the periodic potential, but are sensitive to both d and l_H. In the limit $d \gg l_H$, the Hall conductivity has sharp steps at half-filling of the Landau-level subbands and quantized Hall plateaus between the steps. Below we assume that the temperature is zero. Expressing the filling factor introduced by Eq. (4) through the Fermi momentum p_F in the Landau-level subband as $\nu = 1 - p_F/p_H$ (we define p_F within the first Brillouin zone, $0 < p_F < p_H$), one gets a simple relation

$$\sigma_\perp = \frac{e^2}{2\pi\hbar} \left[N_f + \frac{g_{Nm}(1-\nu)}{g_{Nm}(\nu) + g_{Nm}(1-\nu)} \right], \qquad (32)$$

$$g_N(x) = Q_N(d^2 x^2 / l_H^2) e^{-d^2 x^2 / 2l_H^2}.$$

We remind that N_m is the Landau level number for the partly filled subband contributing to the transport. Equation (32) clearly demonstrates the quantization of $\sigma_\perp$, since the second term in the brackets is equal to $1/2$ at half-filling ($\nu = 1/2$), small in the region $\nu < 1/2$, and close to 1 at $\nu > 1/2$. The behavior of the Hall conductivity calculated numerically according to Eq. (30) is shown in Fig. 10.2 for the region of magnetic fields corresponding to the occupation of up to three lowest Landau-level subbands. For the sake of simplicity, only the subbands with the same spin number are assumed to be occupied. The figure shows us how the quantized Hall conductance picture is improved with the increase of d.

The model of the 2D electron gas modulated by a one-dimensional periodic potential can be directly applied to the cases when 2D layers are formed on crystal substrates with high Miller indices in order to provide the modulation period d much greater than the lattice constant (as a result, a large-scale periodic potential appears at the interface), or when the 2D gas is electrostatically depleted by a periodic surface Schottky gate. It is not surprising that this model shows quantization of the Hall conductivity, because it reflects the main features of the 2D electron dynamics in the presence of a smooth (large-scale) potential discussed above. We have considered the plane-state approach to the problem

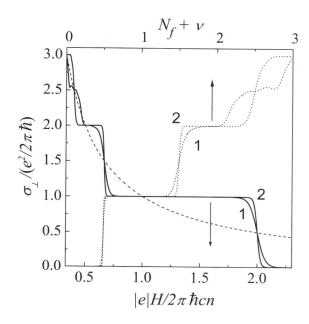

Figure 10.2. Dependence of the Hall conductivity $\sigma_\perp$ of a periodically modulated 2D electron gas on the magnetic field for two different periods of the modulation: $d = \sqrt{60/2\pi n}$ (1) and $d = \sqrt{150/2\pi n}$ (2). The magnetic field changes in the region where up to three Landau-level subbands are populated, and only one orientation of electron spin is considered. The dashed curve shows the classical Hall conductivity $|e|cn/H$. The dotted curves (corresponding to the upper scale) demonstrate the dependence of $\sigma_\perp$ on the filling factor ν.

of the quantum Hall effect. However, this effect, especially in small samples, can be also viewed as a manifestation of the edge transport, when the conductivity is determined by the electron states localized near the edges of the 2D layer in the presence of a strong magnetic field. This approach will be considered in Sec. 59.

Without regard to the subject of this section, we mention that the quasi-classical (low-field) transport of 2D electrons in the presence of a periodic one-dimensional potential is characterized by a special kind of oscillations whose properties are determined by the magnetic length, Fermi wavelength, and modulation period d (problem 10.13).

52. Magnetooptics

Below we consider the influence of magnetic fields on the optical properties of insulators due to interband transitions of electrons. The magnetic field considerably modifies the electron states near the edges of the conduction (c) and valence (v) bands but does not affect considerably the deep valence-band states. Therefore, one can calculate the magnetic-field-induced anisotropy of dielectric permittivity according to the general expression (17.2), by using the eigenstates of the two-band model described by Eqs. (B.16)−(B.19). According to Eqs. (17.2) and (17.3), the contribution to the conductivity tensor proportional to ω^{-1} is given by

$$\frac{ie^2}{\omega V} \sum_{\delta\delta'} \left[\frac{\delta_{\delta\delta'}\delta_{\alpha\beta}}{m_e} + \frac{\langle v\delta|\hat{v}_\alpha|c\delta'\rangle\langle c\delta'|\hat{v}_\beta|v\delta\rangle + (\alpha \leftrightarrow \beta)}{\varepsilon_{v\delta} - \varepsilon_{c\delta'}} \right] \, , \tag{1}$$

where δ is the quantum number of the eigenstates near the extrema of the bands, and we assume that the valence band is completely filled while the conduction band is empty, $f_{v\delta} = 1$ and $f_{c\delta} = 0$. The matrix elements of the velocity operator can be expressed through the matrix elements of the coordinate operator according to the relation $\hat{v}_\alpha = i[\hat{h}, \hat{x}_\alpha]/\hbar$, and the second term in Eq. (1) is rewritten as

$$\frac{ie^2}{m_e \omega V} \sum_{\delta} \frac{1}{i\hbar} \langle v\delta|[\hat{p}_\alpha, \hat{x}_\beta]|v\delta\rangle \, . \tag{2}$$

The sum of this term with the first term of Eq. (1) is zero, which means that the contribution $\propto \omega^{-1}$ vanishes for insulators.

The last term on the right-hand side of Eq. (17.2) is transformed in a similar way (problem 10.14), and the conductivity tensor is reduced to the following form:

$$\sigma_{\alpha\beta}(\omega) = -i\hbar \frac{e^2}{V} \sum_{\delta\delta'} \frac{\langle c\delta|\hat{v}_\beta|v\delta'\rangle\langle v\delta'|\hat{v}_\alpha|c\delta\rangle}{(\varepsilon_{c\delta} - \varepsilon_{v\delta'})(\varepsilon_{c\delta} - \varepsilon_{v\delta'} - \hbar\omega - i\lambda)}$$

$$+(c.c., \omega \to -\omega) \, , \tag{3}$$

which can be also obtained from Eq. (48.4). In the low-frequency region $\hbar\omega \ll \varepsilon_g$, the non-diagonal component of the conductivity tensor is written as

$$\sigma_{xy} \simeq i\hbar \frac{e^2}{V} \sum_{\delta\delta'} \frac{\langle c\delta|\hat{v}_x|v\delta'\rangle\langle v\delta'|\hat{v}_y|c\delta\rangle - c.c.}{(\varepsilon_{c\delta} - \varepsilon_{v\delta'})^2}. \tag{4}$$

Again, expressing the matrix elements of velocity through the matrix elements of coordinate, we transform the sum in Eq. (4) to $\sum_\delta \langle c\delta|[\hat{x}, \hat{y}]|c\delta\rangle$ and prove that the right-hand side of Eq. (4) is zero. This means that the Faraday effect is absent in the absence of free carriers (this effect appears, however, in the high-frequency region). Considering the diagonal components, we also use the condition $\hbar\omega_c \ll \varepsilon_g$ which is normally valid (apart from the case of narrow-gap semiconductors in strong enough magnetic fields). Under these conditions, the magnetic-field-induced modifications of the diagonal conductivity are small. Nevertheless, the electron responses to the electromagnetic radiation polarized along the magnetic field (along the OZ axis) and perpendicular to it are different. Thus, even in the case $\omega, \omega_c \ll \varepsilon_g/\hbar$, there exists an anisotropy of the dielectric permittivity described by the difference $\delta\epsilon_H = (4\pi i/\omega)(\sigma_\parallel - \sigma_d)$. Using Eq. (3), we find

$$\delta\epsilon_H = 8\pi \frac{(e\hbar)^2}{V} \sum_{\delta\delta'} \frac{|\langle c\delta|\hat{v}_z|v\delta'\rangle|^2 - |\langle c\delta|\hat{v}_x|v\delta'\rangle|^2}{(\varepsilon_{c\delta} - \varepsilon_{v\delta'})^3} \tag{5}$$

at $\hbar\omega \ll \varepsilon_g$.

To evaluate the anisotropy (5), we use the symmetric two-band model described by the Hamiltonian (B.18) with $M \to \infty$, where the kinematic momentum operator $\hat{\boldsymbol{\pi}} = \hat{\mathbf{p}} - (e/c)\mathbf{A}$ stands in place of $\mathbf{p}$. The solutions of the eigenstate problem for c- and v- bands (problem 10.15) depend on the Landau level number N, spin number $\sigma = \pm 1$, and two components of the momentum determining the position of the oscillator center and the kinetic energy along the magnetic field; see Eqs. (5.12)−(5.15). The interband matrix elements of the longitudinal velocity operator $\hat{v}_z$ are proportional to $\delta_{NN'}$ for the transitions without spin flip, while the spin-flip processes assume a change in the Landau level number:

$$|\langle cN\sigma p_y p_z|\hat{v}_z|vN'\sigma p_y p_z\rangle|^2 = \delta_{NN'} s^2 \left(\frac{ms^2 + E_{N,\sigma p_z}}{2E_{N,\sigma p_z}}\right)^2$$

$$\times \left[1 + \frac{(sp_0)^2(N+1/2+\sigma/2) - (sp_z)^2}{(ms^2 + E_{N,\sigma p_z})^2}\right]^2, \tag{6}$$

$$|\langle cN\sigma p_y p_z|\hat{v}_z|vN', -\sigma p_y p_z\rangle|^2 = \delta_{N',N+\sigma} \frac{(s^3 p_0 p_z)^2(N+1/2+\sigma/2)}{(ms^2 + E_{N,\sigma p_z})^2 E_{N,\sigma p_z}^2},$$

where $p_0 = \sqrt{2}\hbar/l_H$. On the other hand, the matrix elements of the transverse velocities are proportional to $\delta_{N,N'\pm 1}$ for spin-conserving transitions, while the spin-flip processes contain both $\propto \delta_{NN'}$ and $\propto \delta_{N,N'\pm 2}$ contributions:

$$|\langle cN\sigma p_y p_z|\hat{v}_x|vN'\sigma p_y p_z\rangle|^2$$

$$= \frac{[\delta_{N',N-1}N + \delta_{N',N+1}N'](s^3 p_0 p_z)^2}{4E_{N,\sigma p_z} E_{N',\sigma p_z}(ms^2 + E_{N,\sigma p_z})(ms^2 + E_{N',\sigma p_z})},$$

$$|\langle cN\sigma p_y p_z|\hat{v}_x|vN', -\sigma p_y p_z\rangle|^2 = \delta_{NN'} s^2 \frac{(ms^2 + E_{N,\sigma p_z})(ms^2 + E_{N,-\sigma p_z})}{4E_{N,\sigma p_z}E_{N,-\sigma p_z}}$$

$$\times \left[1 + \frac{(sp_z)^2}{(ms^2 + E_{N,\sigma p_z})(ms^2 + E_{N,-\sigma p_z})}\right]^2 \qquad (7)$$

$$+\delta_{N',N+2\sigma} s^2 \frac{(sp_0)^4(N+\sigma)(N+1+\sigma)}{4E_{N,\sigma p_z}E_{N+\sigma,\sigma p_z}(ms^2 + E_{N,\sigma p_z})(ms^2 + E_{N+\sigma,\sigma p_z})}.$$

The squared matrix elements of $\hat{v}_y$ are also given by Eq. (7), for obvious symmetry reasons. The energy $E_{N,\sigma p_z}$ entering Eqs. (6) and (7) is defined in problem 10.15. Introducing the dimensionless variables $x = sp_z/\varepsilon_g$ and $h = \hbar\omega_c/\varepsilon_g$ describing the momentum and magnetic field, we rewrite this energy as

$$E_{N,\sigma p_z} = \frac{\varepsilon_g}{2}\sqrt{1 + (2x)^2 + 4h(N + 1/2 + \sigma/2)} \equiv \frac{\varepsilon_g}{2}\eta_{N,\sigma x}, \qquad (8)$$

where $\eta_{N,\sigma x}$ is the dimensionless energy introduced by analogy to η_p of Appendix B. The energy (8) satisfies the property $E_{N,+1p_z} = E_{N+1,-1p_z}$.

We calculate $\delta\epsilon_H$ by substituting the matrix elements (6) and (7) into Eq. (5) with $|c\delta\rangle = |vN\sigma p_y p_z\rangle$ and $|v\delta'\rangle = |vN'\sigma' p_y p_z\rangle$. The energies $\varepsilon_{c\delta}$ and $\varepsilon_{v\delta'}$ in Eq. (5) should be replaced by $\varepsilon_{cN\sigma p_z} = E_{N,\sigma p_z}$ and $\varepsilon_{vN'\sigma' p_z} = -E_{N',\sigma' p_z}$. Since the energies and the matrix elements are independent of p_y, the sum $(L_x L_y)^{-1}\sum_{p_y}\ldots$ is reduced to the Landau-level degeneracy factor $(2\pi l_H^2)^{-1}$; see Eq. (5.16). The sums over N, N', σ, and σ' are transformed to a sum over N and σ by using the selection rules presented by Eqs. (6) and (7). After careful but straightforward transformations, we obtain

$$\delta\epsilon_H = \frac{e^2}{\pi\hbar s}h\sum_{N\sigma}\int_{-\infty}^{\infty}dx[\Phi_{N\sigma}^{(\parallel)}(x) - \Phi_{N\sigma}^{(\perp)}(x)], \qquad (9)$$

where the dimensionless functions

$$\Phi_{N\sigma}^{(\parallel)}(x) = [1 + 4h(N + 1/2 + \sigma/2)]/\eta_{N,\sigma x}^5 \qquad (10)$$

and

$$\Phi_{N\sigma}^{(\perp)}(x) = 4\frac{1 + \eta_{N,\sigma x}\eta_{N+1,\sigma x} + 4x^2 - 8\sigma hx^2/[(1 + \eta_{N,\sigma x})(1 + \eta_{N+1,\sigma x})]}{(\eta_{N,\sigma x} + \eta_{N+1,\sigma x})^3\eta_{N,\sigma x}\eta_{N+1,\sigma x}}$$

$$(11)$$

describe the contributions of parallel and perpendicular polarizations. Note that the last (proportional to σ) term in the nominator of Eq. (11)

can be safely neglected because the contributions from N, $\sigma = +1$ and $N + 1$, $\sigma = -1$ states to this term cancel each other in the sum so that only the contribution from the state with $N = 0$ and $\sigma = -1$ remains there.

To evaluate the integral over x and the sum over N in Eq. (9), a careful consideration is required. In the case of zero magnetic field studied in Sec. 17, the zero-frequency contribution to the dielectric permittivity described by the symmetric two-band model appears to be divergent. For the same reason, the contributions in Eq. (9) coming from $\Phi^{(\parallel)}$ and $\Phi^{(\perp)}$ are divergent, if considered separately. Their difference, which determines $\delta\epsilon_H$, is not divergent. However, the result for $\delta\epsilon_H$ depends on the way we calculate the integral and the sum in Eq. (9). The only correct, physically justified way is to introduce an upper limit for the energy, $\eta_{N,\sigma x} < \eta_m$, where $\eta_m \gg 1$, see also Eq. (17.9), and calculate the integral over x and the sum over N by using this restriction. Then, to obtain the result independent of η_m, one should carry out the limiting transition $\eta_m \to \infty$.

Applying the calculation method described above, one can prove that $\delta\epsilon_H$ goes to zero at $h \to 0$, though the functions $\Phi^{(\parallel)}$ and $\Phi^{(\perp)}$ are not equal to each other in these conditions. Indeed, at $h \to 0$ one can neglect the difference between $\eta_{N,\sigma x}$ and $\eta_{N+1,\sigma x}$, ignore the spin splitting, and introduce a continuous variable $\zeta^2 = hN$ so that $h \sum_N \ldots = 2 \int_0^\infty \zeta d\zeta \ldots$. We obtain

$$h \sum_{N=0}^{N_m} \int_{-x_m}^{x_m} dx \left| \begin{array}{c} \Phi_{N\sigma}^{(\parallel)}(x) \\ \Phi_{N\sigma}^{(\perp)}(x) \end{array} \right|$$

$$\simeq \int_0^{\zeta_m} \zeta d\zeta \int_0^{x_m} dx \frac{4}{[1 + 4(x^2 + \zeta^2)]^{5/2}} \left| \begin{array}{c} 1 + 4\zeta^2 \\ 1 + 2\zeta^2 + 4x^2 \end{array} \right| \quad (12)$$

with $x_m = \sqrt{(\eta_m^2 - 1)/4 - \zeta^2}$, $N_m = (\eta_m^2 - 1)/4h$, and $\zeta_m = \sqrt{\eta_m^2 - 1}/2$. An elementary calculation of the integrals shows us that the upper and the lower contributions are equal to each other. To find the first non-vanishing ($\propto h$) contributions into $\delta\epsilon_H$, we expand $\eta_{N+1,\sigma x}$ as $\eta_{N,\sigma x} + 2h/\eta_{N,\sigma x}$ and $\eta_{N,\sigma x}$ as $\sqrt{1 + 4x^2 + 4hN} + h(1+\sigma)/\sqrt{1 + 4x^2 + 4hN}$. Neglecting the terms proportional to σh in this expansion for the reason explained above, we find

$$\delta\epsilon_H = \frac{e^2 C_0}{\pi \hbar s} h . \quad (13)$$

The numerical constant C_0 is defined as

$$C_0 = \lim_{h \to 0} 4 \sum_{N=0}^{N_m} \int_0^{x_m} dx \left[\frac{2hN - 4x^2}{(1 + 4x^2 + 4hN)^{5/2}} \right.$$

$$\left. + 5h \frac{1 + 8x^2}{(1 + 4x^2 + 4hN)^{7/2}} \right]. \tag{14}$$

The contribution from the second term in the brackets of this expression can be calculated by introducing a continuous variable instead of N. The result is $4/3$. If we calculate the first term in this approximation, we obtain zero. One needs, therefore, to take into account the discrete nature of the sum over N. It is done in problem 10.16, and the contribution from the first term is $-1/3$. As a result, we obtain $C_0 = 1$. Numerical calculations according to Eqs. (9)−(11) show a sublinear deviation from the linear dependence (13), though the relative deviation does not exceed 10% up to $h = 0.05$. One can calculate the anisotropy of the frequency-dependent part of the dielectric permittivity in a similar way (problem 10.17).

Let us consider the interband absorption. To describe the influence of the magnetic field on the absorption spectrum in the one-particle approximation, one should take into account the corresponding modification of the joint density of states; see Eq. (17.6). Near the edge of fundamental absorption, where the non-parabolicity effects can be neglected, the joint density of states is given by Eq. (5.16) containing the reduced mass μ^* in place of m. Considering the Coulomb interaction between photoexcited electrons and holes, one obtains an additional contribution due to exciton absorption, which is also modified by the magnetic field. This contribution is studied below in the frames of the general approach developed in Sec. 30. To take into account the magnetic field in Eq. (30.14), one should substitute the kinematic momenta $\hat{\mathbf{p}} - (e/c)\mathbf{A_r}$ and $\hat{\mathbf{p}}' + (e/c)\mathbf{A_{r'}}$ to the Hamiltonians $\hat{h}_c$ and $\hat{h}'_v$, respectively. The spectral dependence of the absorption is expressed through the retarded Green's function $G_{\varepsilon\omega}(\mathbf{r}, \mathbf{r}')$ according to Eq. (30.15). Below we use the variables $\mathbf{R}$ and $\Delta\mathbf{r}$ introduced by Eq. (30.16). Applying the symmetric gauge $\mathbf{A_r} = [\mathbf{H} \times \mathbf{r}]/2$, we search for the Green's function in the form

$$G_{\varepsilon_\omega}(\mathbf{r}, \mathbf{r}') = \exp\left(-\frac{ie}{2\hbar c}[\mathbf{H} \times \Delta\mathbf{r}] \cdot \mathbf{R}\right) \widetilde{G}_{\varepsilon_\omega}(\mathbf{R}, \Delta\mathbf{r}), \tag{15}$$

where we have separated the phase factor introduced in Eq. (16.24) and Appendix G (note that $\mathbf{H} \cdot [\mathbf{r} \times \mathbf{r}'] = \mathbf{H} \cdot [\Delta\mathbf{r} \times \mathbf{R}] = [\mathbf{H} \times \Delta\mathbf{r}] \cdot \mathbf{R}$). The function $\widetilde{G}_{\varepsilon_\omega}$ should be translation-invariant. To write an equation for this function, one has to carry out a canonical transformation

$$\exp\left(\frac{ie}{2\hbar c}[\mathbf{H} \times \Delta\mathbf{r}] \cdot \mathbf{R}\right)(\hat{h}_c - \hat{h}_v')\exp\left(-\frac{ie}{2\hbar c}[\mathbf{H} \times \Delta\mathbf{r}] \cdot \mathbf{R}\right). \quad (16)$$

As a result of this transformation, $\mathbf{R}$ drops out of the operator of kinetic energy (problem 10.18). In the absence of external potentials, we obtain

$$\left(\frac{\hat{p}_{\mathbf{R}}^2}{2M} - \frac{e}{Mc}[\mathbf{H} \times \Delta\mathbf{r}] \cdot \hat{\mathbf{p}}_{\mathbf{R}} + \frac{\hat{p}_{\Delta\mathbf{r}}^2}{2\mu^*} - \frac{e\gamma}{2\mu^*c}[\mathbf{H} \times \Delta\mathbf{r}] \cdot \hat{\mathbf{p}}_{\Delta\mathbf{r}}\right.$$

$$\left.+ \frac{e^2}{8\mu^*c^2}[\mathbf{H} \times \Delta\mathbf{r}]^2 - v_{|\Delta\mathbf{r}|} - \varepsilon_\omega - i\lambda\right)\widetilde{G}_{\varepsilon_\omega}(\mathbf{R}, \Delta\mathbf{r}) = -\delta(\Delta\mathbf{r}), \quad (17)$$

where $\lambda \to +0$ and $\gamma = (m_v - m_c)/(m_v + m_c)$. Transforming the Green's function according to Eq. (30.18), we find

$$\left(\frac{\hat{p}_{\Delta\mathbf{r}}^2}{2\mu^*} - \frac{e\gamma}{2\mu^*c}[\mathbf{H} \times \Delta\mathbf{r}] \cdot \hat{\mathbf{p}}_{\Delta\mathbf{r}} + \frac{e^2}{8\mu^*c^2}[\mathbf{H} \times \Delta\mathbf{r}]^2\right.$$

$$\left.- v_{|\Delta\mathbf{r}|} - \varepsilon_\omega - i\lambda\right)\widetilde{G}_{\varepsilon_\omega}(\Delta\mathbf{r}) = -\delta(\Delta\mathbf{r}) \quad (18)$$

for $\widetilde{G}_{\varepsilon_\omega}(\Delta\mathbf{r}) \equiv \widetilde{G}_{\varepsilon_\omega}(\mathbf{P} = 0, \Delta\mathbf{r})$. The spectral dependence of the absorption is expressed in terms of this function as $\Psi(E) = -\mathrm{Im}\widetilde{G}_E(\Delta\mathbf{r} = 0)$. The motion of the exciton does not affect the absorption in a spatially homogeneous case.

The magnetic field acts differently on the particles with different effective masses. That is why the second term on the left-hand side of Eq. (18) appears. Using the cylindrical coordinates, one can show that this term is equal to $-i(\gamma\hbar\widetilde{\omega}_c/2)\partial/\partial\varphi$, where $\widetilde{\omega}_c = |e|H/\mu^*c$ and φ is the polar angle in the plane perpendicular to the magnetic field. This term is linear in H, but it does not contribute to the energy of the ground state of the exciton in the first order with respect to $\hbar\widetilde{\omega}_c/\varepsilon_B$, where ε_B is the Bohr energy of the exciton. The main correction to the ground-state energy comes from the third term, which is quadratic in H. This correction is equal to $(\hbar\widetilde{\omega}_c)^2/2\varepsilon_B$. The second term on the left-hand side of Eq. (18) does not contribute to this correction in the second order, because this term does not couple the ground state to the other eigenstates of the hydrogen atom problem. On the other hand, this term, in the first order, contributes to the $\propto H$ corrections to the energies of the excited states with non-zero magnetic quantum number m, because the wave functions of these states are proportional to $e^{im\varphi}$. Denoting the states of the hydrogen atom problem by their quantum indices n, l, m, where

n is the main quantum number, l is the orbital number ($l < n$), and m is the magnetic number ($|m| \leq l$), we have

$$\varepsilon_\nu \equiv \varepsilon_{nlm} \simeq -\varepsilon_B/n^2 + m\gamma\hbar\widetilde{\omega}_c/2 \qquad (19)$$

so that the degeneracy of the excited states is removed and, according to Eq. (30.21), this leads to a fine structure of the absorption spectrum.

Consider now the limit of high magnetic fields, when the ratio $\varepsilon_B/\hbar\widetilde{\omega}_c$ is small. Using the cylindrical coordinate system with $(\Delta \mathbf{r})_x = \rho\cos\varphi$, $(\Delta \mathbf{r})_y = \rho\sin\varphi$, and $(\Delta \mathbf{r})_z = z$, we search for the excitonic wave function $\psi_{\Delta \mathbf{r}}^{(\nu)}$ in the form $\psi_{\Delta \mathbf{r}}^{(\nu)} = (2\pi)^{-1/2}e^{im\varphi}F_m(\rho, z)$. We remind that the excitonic wave function and the Green's function are related as $\widetilde{G}_{\varepsilon_\omega}(\Delta \mathbf{r}) = \sum_\nu \psi_{\Delta \mathbf{r}}^{(\nu)}\psi_0^{(\nu)*}/[\varepsilon_\omega - \varepsilon_\nu + i\lambda]$; see Eqs. (30.19), (30.20), and problem 3.10. The function F_m satisfies the equation

$$\left[\widehat{H}_m(\rho) - \frac{\hbar^2}{2\mu^*}\frac{d^2}{dz^2} - \frac{e^2}{\epsilon\sqrt{\rho^2 + z^2}} - \varepsilon\right]F_m(\rho, z) = 0, \qquad (20)$$

$$\widehat{H}_m(\rho) = -\frac{\hbar^2}{2\mu^*}\left(\frac{d^2}{d\rho^2} + \frac{1}{\rho}\frac{d}{d\rho} - \frac{m^2}{\rho^2}\right) + \frac{e^2H^2\rho^2}{8\mu^*c^2} + m\frac{\gamma\hbar\widetilde{\omega}_c}{2},$$

where the Hamiltonian $\widehat{H}_m(\rho)$ depends only on the radial coordinate. One can search for $F_m(\rho, z)$ in the form $F_m(\rho, z) = \sum_n R_{nm}(\rho)W_{nm}(z)$, where $R_{nm}(\rho)$ are the eigenfunctions of $\widehat{H}_m(\rho)$ (problem 10.19). They are characterized by the quantum number n corresponding to the Landau level number. Let us multiply Eq. (20) by $\rho R_{n'm}^*(\rho)$ from the left and integrate the equation obtained over ρ, taking into account the orthogonality and normalization condition $\int_0^\infty d\rho\,\rho R_{n'm}^*(\rho)R_{nm}(\rho) = \delta_{nn'}$. We obtain a system of coupled differential equations for $W_{nm}(z)$:

$$\left(-\frac{\hbar^2}{2\mu^*}\frac{d^2}{dz^2} + \varepsilon_{nm} - \varepsilon\right)W_{nm}(z) + \sum_{n'}U_{nn'}^{(m)}(z)W_{n'm}(z) = 0, \qquad (21)$$

$$U_{nn'}^{(m)}(z) = -\frac{e^2}{\epsilon}\int_0^\infty d\rho\,\rho\frac{R_{nm}^*(\rho)R_{n'm}(\rho)}{\sqrt{\rho^2 + z^2}},$$

where ε_{nm} are the eigenvalues of $\widehat{H}_m(\rho)$. The mixing between the Landau-like states with different numbers n and n' is caused by the non-diagonal matrix elements $U_{nn'}^{(m)}(z)$. So far the treatment was exact. The approximation we use below corresponds to the case of high magnetic fields, when such mixing is neglected. Indeed, the radial coordinates ρ contributing into the integral in Eq. (21) are of the order of l_H. In the case of high magnetic fields, they are much smaller than the characteristic axial coordinates $|z|$, and the non-diagonal part of $U_{nn'}^{(m)}(z)$ becomes small due to the orthogonality condition for $R_{nm}(\rho)$.

Therefore, for each number n we obtain a differential equation similar to that appearing in the theory of 1D exciton; see problem 6.11. Such equations describe the fine structure of the magnetoexciton spectrum existing because the degeneracy of each Landau-like level is lifted by the Coulomb interaction. The equation for the lowest level ($n = m = 0$) is

$$\left[-\frac{\hbar^2}{2\mu^*}\frac{d^2}{dz^2} - \frac{e^2}{\epsilon l_H^2}\int_0^\infty d\rho\, \rho\frac{\exp(-\rho^2/2l_H^2)}{\sqrt{\rho^2 + z^2}} - \Delta\varepsilon \right] W(z) = 0, \qquad (22)$$

where we have put $\varepsilon - \varepsilon_{00} \equiv \Delta\varepsilon^{(0,0)}$ and omitted the indices $(n,m) = (0,0)$. Since at large z the effective potential energy behaves like $1/|z|$, we have an infinite number of quantized states. However, in contrast to the Schroedinger equation with the potential energy $-e^2/\epsilon|z|$ (see problem 6.11), the effective potential energy in Eq. (22) is finite at $z = 0$, and the energy of the ground state is also finite. Let us search for this energy in the form $\Delta\varepsilon_0 = -\varepsilon_B/\lambda^2$, where λ is a dimensionless parameter. The wave function for the case of an infinite magnetic field is written as $W(z) = (a_B\lambda)^{-1/2}e^{-|z|/a_B\lambda}$, where $a_B = \hbar^2\epsilon/\mu^*e^2$ is the Bohr radius of exciton; see problem 6.11. Using this function in order to calculate the matrix element of the interaction potential in Eq. (22) in the limit $\lambda a_B \gg \sqrt{2}l_H$, we find that λ satisfies an implicit algebraic equation

$$\lambda^{-1} = 2\ln(a_B\lambda/\sqrt{2}l_H) - C, \qquad (23)$$

where $C \simeq 0.577$ is Euler's constant. The Sommerfeld factor for the ground state is given by

$$|\psi_{\Delta\mathbf{r}=0}^{(0)}|^2 = \frac{1}{2\pi l_H^2 a_B\lambda}. \qquad (24)$$

The energies of excited levels, which are given by $-\varepsilon_B/l^2$ ($l = 1, 2, \ldots$) in the limit of infinite H, see problem 6.11, also acquire small corrections. These corrections can be found directly, with the aid of the perturbation theory with respect to the difference between $-e^2/\epsilon|z|$ and the effective potential energy entering Eq. (22). We obtain

$$\Delta\varepsilon_l = -\varepsilon_B\left(\frac{1}{l^2} - \frac{b_l}{l^3}\right), \quad b_l^{-1} = \ln\left(\frac{a_B l}{\sqrt{2}l_H}\right) - \psi(l) - \frac{1}{2l} - \frac{3C}{2}, \qquad (25)$$

where $\psi(x)$ is the logarithmic derivative of the Gamma function. Apart from these solutions, which correspond to the even (symmetric in z) wave functions, there exist odd (antisymmetric in z) solutions. As we know from problem 6.11, their energies coincide with $\Delta\varepsilon_l$ of Eq. (25) in the case of infinite H. The shift of the energies of odd states at finite H, however, is small in comparison to that given by Eq. (25), because the latter shift is small only in the logarithmic sense. Therefore, the degeneracy of even and odd states no longer exists in the case of finite magnetic fields. The odd states are not important for exciton absorption, since the Sommerfeld factors for them are zeros. The zero-order wave functions of the even (symmetric) excited states are also equal to zero at $z = 0$. The first-order corrections to these wave functions are finite at $z = 0$, and the corresponding Sommerfeld factors are

$$|\psi^{(l)}_{\Delta\mathbf{r}=0}|^2 = \frac{b_l^2}{2\pi l_H^2 a_B l^3}, \quad (l \neq 0). \tag{26}$$

A schematic picture of magnetoexciton absorption in strong magnetic fields is shown in Fig. 10.3. Each absorption line associated with the level with quantum numbers n and m, see problem 10.19, has a fine structure characterized by the quasi-one-dimensional exciton spectrum $\Delta\varepsilon_l^{(n,m)}$. The absorption peaks occur at $\varepsilon_\omega = \varepsilon_{nm} + \Delta\varepsilon_l^{(n,m)}$. To avoid confusions, we remind that the numbers n, m, and l of the problem of exciton in a strong magnetic field differ from the quantum numbers of the hydrogen atom problem (i.e., from the numbers standing in Eq. (19)).

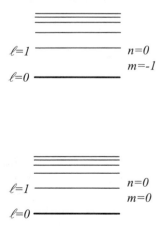

Figure 10.3. Schematic picture of exciton levels in the case of strong magnetic fields.

Let us consider the effect of a spatially-smooth external potential on the exciton absorption. If we neglect the potential gradients (see problem 6.9), only one additional term, $\Delta U(\mathbf{R})\widetilde{G}_{\varepsilon_\omega}(\mathbf{R}, \Delta\mathbf{r})$, enters the left-hand side of Eq. (17). We expand the Green's function in this equation by using the eigenstates $\psi_{\Delta\mathbf{r}}^{(\nu)}$ of the problem for relative motion of electron and hole considered above, $\widetilde{G}_E(\mathbf{R}, \Delta\mathbf{r}) = \sum_\nu G_E^{(\nu)}(\mathbf{R})\psi_{\Delta\mathbf{r}}^{(\nu)}$. Because of the presence of the second term on the left-hand side of Eq. (17), the motion of the exciton as a whole is coupled to the relative motion (in the absence of magnetic fields, see problem 6.9, these motions are separated). Substituting the expansion given above into Eq. (17), multiplying this equation by $\psi_{\Delta\mathbf{r}}^{(\nu)*}$ from the left, and integrating it over $\Delta\mathbf{r}$, we obtain the following system of equations:

$$\left[-\frac{\hbar^2}{2M}\frac{\partial^2}{\partial\mathbf{R}^2} + \Delta U(\mathbf{R}) + \varepsilon_\nu - E - i\lambda \right] G_E^{(\nu)}(\mathbf{R})$$

$$-i\frac{\hbar|e|}{Mc}\sum_{\nu'}[\mathbf{H}\times(\Delta\mathbf{r})_{\nu\nu'}]\cdot\frac{\partial}{\partial\mathbf{R}}G_E^{(\nu')}(\mathbf{R}) = -\psi_{\Delta\mathbf{r}=0}^{(\nu)*}. \tag{27}$$

The index ν defines the set (n, m, l). The matrix elements $(\Delta\mathbf{r})_{\nu\nu'} \equiv \langle\nu|\Delta\mathbf{r}|\nu'\rangle$ are non-zero only for $m' = m\pm 1$. Using a method of perturbations based upon the assumed smallness of the derivatives $\partial G_E^{(\nu)}(\mathbf{R})/\partial\mathbf{R}$ (this method is analogous to the **kp**-approach described in Appendix B), we find

$$\left\{ -\frac{\hbar^2}{2M}\frac{\partial^2}{\partial\mathbf{R}^2} + \sum_{\beta\beta'}\frac{\hbar^2\alpha_{\beta\beta'}^{(\nu)}}{2M^2c^2}\left[\mathbf{H}\times\frac{\partial}{\partial\mathbf{R}}\right]_\beta\left[\mathbf{H}\times\frac{\partial}{\partial\mathbf{R}}\right]_{\beta'} \right.$$

$$\left. +\Delta U(\mathbf{R}) + \varepsilon_\nu - E - i\lambda \right\} G_E^{(\nu)}(\mathbf{R}) = -\psi_{\Delta\mathbf{r}=0}^{(\nu)*}, \tag{28}$$

where

$$\alpha_{\beta\beta'}^{(\nu)} = 2e^2\sum_{\nu'}\frac{\langle\nu|\Delta r_\beta|\nu'\rangle\langle\nu'|\Delta r_{\beta'}|\nu\rangle}{\varepsilon_{\nu'} - \varepsilon_\nu} \tag{29}$$

is the polarizability of an electron-hole pair. Since $\alpha_{xx}^{(\nu)} = \alpha_{yy}^{(\nu)}$ and $\alpha_{xy}^{(\nu)} = -\alpha_{yx}^{(\nu)}$, the second term in Eq. (28) does not lead to an anisotropy of the effective mass of exciton in the plane XOY. However, the anisotropy of this mass with respect to the directions parallel and perpendicular to the magnetic field appears to be strong. Indeed, calculating the matrix elements in Eq. (29) for the ground state $|\nu\rangle = |n = 0, m = 0, l\rangle$, we find that only the states $|\nu'\rangle = |n' = 0, m' = \pm 1, l'\rangle$ contribute to the polarizability, and

$$\alpha_{xx}^{(00l)} = \sum_\pm\sum_{l'}\frac{e^2 l_H^2 I_{ll'}^\pm}{\hbar\widetilde{\omega}_c(1\pm\gamma)/2 + \Delta\varepsilon_{l'}^{(0,\pm 1)} - \Delta\varepsilon_l^{(0,0)}}, \tag{30}$$

where the squared overlap factor of quasi-one-dimensional exciton states is defined as $I_{ll'}^{\pm} = |\int dz W_{00l}(z) W_{0,\pm 1,l'}(z)|^2$. It is essential to take into account the fine structure of the levels. If we neglect it by using $\Delta\varepsilon_l^{(n,m)} = 0$ and $I_{ll'}^{\pm} = \delta_{ll'}$, we obtain $\alpha_{xx}^{(00l)} = Mc^2/H^2$ (note that $2/(1-\gamma) + 2/(1+\gamma) = M/\mu^*$). Substituting this result into Eq. (28), we find that the effective mass corresponding to the motion perpendicular to the field is infinite. A finite transverse mass occurs only when the Coulomb interaction is taken into account. In summary, the motion of the exciton as a whole in a strong magnetic field is described by a Schroedinger equation with anisotropic effective mass. The mass in the direction perpendicular to the field, $M_{\perp}$, is much larger than the free-exciton mass M. If $\Delta U(\mathbf{R})$ is a regular potential of bonding nature (as the potentials of quantum dots; see Sec. 38), the Green's function $G_E^{(\nu)}(\mathbf{R})$, according to Eq. (28), contains a number of poles corresponding to a set of discrete levels whose energy spacing is estimated as $(\hbar/l_c)^2/M_{\perp}$, where l_c is the characteristic spatial scale of the potential. These levels contribute to a more fine structure of the absorption spectrum.

Problems

10.1. Find the distribution function $\Delta f_{\mathbf{p}}^{(1)}$ from Eq. (48.3) and calculate the current density $\Delta\mathbf{I}$ associated with it.

Hints: Use the cylindrical coordinates to express the left-hand side of Eq. (48.3) as $\omega_c \partial \Delta f_{\mathbf{p}}^{(1)}/\partial\varphi$, where φ is the polar angle of $\mathbf{p}$. Then expand $\Delta f_{\mathbf{p}}^{(1)}$ in the basis $e^{im\varphi}$.

10.2. Derive Eq. (48.4) from Eq. (13.18).

Hints: Integrate over τ by parts, using the identities $\hbar\hat{v}_\alpha = i[\hat{H}, \hat{x}_\alpha]$ and $[\hat{x}_\alpha, \hat{v}_\beta] = \delta_{\alpha\beta} i\hbar/m$. Then use the exact eigenstate representation.

10.3. Obtain the conductivity (48.11) by using a semiclassical consideration of electron hopping.

Solution: A semiclassical expression for the current density I_x can be written in the form

$$I_x = -\frac{e}{V}\frac{2\pi}{\hbar} \sum_{\delta\delta'(X_{\delta'} > X_\delta)} (X_{\delta'} - X_\delta) \left\langle\left\langle |\langle\delta|U_{sc}|\delta'\rangle|^2 \right\rangle\right\rangle$$

$$\times \delta\left[\varepsilon_\delta - \varepsilon_{\delta'} + eE_x(X_{\delta'} - X_\delta)\right][f(\varepsilon_\delta) - f(\varepsilon_{\delta'})],$$

where E_x is the electric field along the current. The current is expressed through the probability of hopping between the centers δ' and δ with x-coordinates $X_{\delta'}$ and X_δ in unit time. The probability is calculated according to Fermi's golden rule. To obtain the current density, one should multiply this probability by the distance of hopping in the given direction and by the electron charge e, take the sum over all centers, and divide the result by the normalization volume V. This is the way we obtain the equation given above. Next, in the linear regime,

$$\delta\left[\varepsilon_\delta - \varepsilon_{\delta'} + eE_x(X_{\delta'} - X_\delta)\right]\left[f(\varepsilon_\delta) - f(\varepsilon_{\delta'})\right] = \frac{df(\varepsilon_\delta)}{d\varepsilon_\delta}eE_x(X_\delta - X_{\delta'}).$$

Replacing $\sum_{\delta\delta'(X_{\delta'} > X_\delta)}\cdots$ by $(1/2)\sum_{\delta\delta'}\cdots$, we find the conductivity $\sigma_{xx} = I_x/E_x$ in the form (48.11).

10.4. Calculate the integral $\int_0^\infty du \, u\Phi_{NN'}(u)$, where $\Phi_{NN'}(u)$ is given by Eq. (48.14).

Solution: The Laguerre polynomials satisfy the identity $L_n^{\alpha-1}(u) = L_n^\alpha(u) - L_{n-1}^\alpha(u)$ so that $[L_N^{N'-N}(u)]^2 = [L_N^{N'-N+1}(u)]^2 + [L_{N-1}^{N'-N+1}(u)]^2 - 2L_N^{N'-N+1}(u) \times L_{N-1}^{N'-N+1}(u)$. Applying the normalization condition

$$\int_0^\infty du \, u^\alpha e^{-u} L_n^\alpha(u) L_m^\alpha(u) = \delta_{nm}\frac{(\alpha+n)!}{n!}$$

for $\alpha = N' - N + 1$, we find $\int_0^\infty du \, u\Phi_{NN'}(u) = N + N' + 1$.

10.5. Derive the expression for the electron density in the ultraquantum limit from Eq. (5.16).

Hint: Consider only one term ($N = 0$) in the sum in Eq. (5.16).

10.6. Obtain the classical expressions for the components of the conductivity tensor from Eqs. (49.17)−(49.19).

Hint: Replace the sums over N by integrals and $\Sigma_{\sigma\varepsilon_F}''$ by $\hbar/2\tau_{tr}$.

10.7. Find the oscillations of the Fermi energy in the 2D case by assuming that the electron density n is constant and the oscillations of the scattering rate are small ($\omega_c\tau/\pi \ll 1$).

Result:

$$\varepsilon_F = \varepsilon_F^{(0)}\left[1 + e^{-\pi/\omega_c\tau}(\hbar\omega_c/\pi\varepsilon_F^{(0)})\sin(2\pi\varepsilon_F^{(0)}/\hbar\omega_c)\right],$$

where $\varepsilon_F^{(0)} = n/\rho_{2D}$.

10.8. Prove that the last term in Eq. (51.6) is equal to zero.

Solution: Using the operator identity

$$[e^{-\beta\widehat{H}}, \widehat{A}] = e^{-\beta\widehat{H}}\int_0^\beta d\lambda \, e^{\widehat{H}\lambda}[\widehat{A}, \widehat{H}]e^{-\widehat{H}\lambda},$$

where $\beta = 1/T$ (see problem 3.5), we find

$$\mathrm{Sp}[e^{-\beta\widehat{H}}, \widehat{A}]\widehat{B} = \mathrm{Sp}e^{-\beta\widehat{H}}[\widehat{A}, \widehat{B}] = \int_0^\beta d\lambda \, \mathrm{Sp}e^{-\beta\widehat{H}}e^{\widehat{H}\lambda}[\widehat{A}, \widehat{H}]e^{-\widehat{H}\lambda}\widehat{B},$$

for arbitrary operators $\widehat{A}$ and $\widehat{B}$. Writing this equation in the exact eigenstate representation, we obtain

$$\sum_\delta f(\varepsilon_\delta)\langle\delta|[\widehat{A}, \widehat{B}]|\delta\rangle = -\sum_{\delta\delta'}\langle\delta|[\widehat{A}, \widehat{H}]|\delta'\rangle\langle\delta'|\widehat{B}|\delta\rangle\frac{f(\varepsilon_\delta) - f(\varepsilon_{\delta'})}{\varepsilon_\delta - \varepsilon_{\delta'}}.$$

Consider the Hamiltonian $\widehat{H} = (\hat{\mathbf{p}} - e\mathbf{A}/c)^2/2m + U(\mathbf{r})$, where $\mathbf{A} = (0, Hx, 0)$. Let us substitute the operator of x-coordinate of the cyclotron orbit center in place of $\widehat{A}$ (so that $\widehat{A} = il_H^2 \partial/\partial y$) and the velocity operator $\hat{v}_x = -(i\hbar/m)\partial/\partial x$ in place of $\widehat{B}$. Since these operators commute, the left-hand side of the equation above is zero. Considering the right-hand side, we take into account that $\widehat{A}$ commutes with the kinetic-energy part of $\widehat{H}$ so that $[\widehat{A}, \widehat{H}] \propto \partial U/\partial y$. Therefore, the equation above is reduced to

$$\sum_{\delta\delta'} \langle\delta|\partial U/\partial y|\delta'\rangle \langle\delta'|\hat{v}_x|\delta\rangle \frac{f(\varepsilon_\delta) - f(\varepsilon_{\delta'})}{\varepsilon_\delta - \varepsilon_{\delta'}} = 0,$$

which means that the first part of the last term in Eq. (51.6) is zero. To prove that the second part of the last term in Eq. (51.6) is also zero, we substitute the operator of y-coordinate of the cyclotron orbit center in place of $\widehat{A}$ (so that $\widehat{A} = y - il_H^2 \partial/\partial x$) and the velocity operator $\hat{v}_y = (\hbar/m)[-i\partial/\partial y + x/l_H^2]$ in place of $\widehat{B}$. We have again $[\widehat{A}, \widehat{B}] = 0$ and $[\widehat{A}, \widehat{H}] \propto \partial U/\partial x$, which gives us the required proof.

10.9. Prove that I_y given by Eq. (51.18) is equal to zero in equilibrium, when $E_x = 0$.

Solution: In equilibrium, $v_{\perp y} = -(m\omega_c)^{-1}\partial U(x, y)/\partial x$ and $f_{N\sigma}(\mathbf{r})$ is given by Eq. (51.16). Therefore, the contribution to the current density from each completely filled Landau level (when $f_{N\sigma}(\mathbf{r}) = 1$ everywhere) is proportional to $[U(L_x, y) - U(0, y)]/L_x$. This value goes to zero after the limiting transition $L_x \to \infty$. For the partly filled Landau level $N_m\sigma_m$, the current density is proportional to

$$\sum_\lambda \int_{x_1^\lambda(y)}^{x_2^\lambda(y)} dx \; \partial U(x, y)/\partial x = \sum_\lambda [U(x_2^\lambda(y), y) - U(x_1^\lambda(y), y)],$$

where λ numbers the intervals occupied by electrons, and $x_1^\lambda(y)$ and $x_2^\lambda(y)$ are the coordinates of the left and right ends of these intervals. Since these coordinates are determined by the equation $E_{N_m\sigma_m} + U(x_{1,2}^\lambda(y), y) = \varepsilon_F$, the expression above is zero. As a result, the current is zero.

10.10. Prove that the Hall conductivity associated with an almost filled Landau level is equal to $e^2/2\pi\hbar$.

Hint: The problem can be reduced to the solved one for an almost empty Landau level, if one treats the empty electron states of an almost filled Landau level as "hole" states.

10.11. Find the expression of the current parallel to the modulation axis in the 2D gas placed in a strong magnetic field and one-dimensional periodic potential.

Solution: Let us express the current density through the one-electron density matrix in the momentum representation, as $I_x = e(L_x L_y)^{-1} \sum_{pp'} \langle Np'|\hat{v}_x|Np\rangle \rho_N(p, p')$. The non-diagonal in p part of the density matrix is found from Eq. (4.32) written in this representation as

$$i\hbar \frac{\partial \rho_N(p, p')}{\partial t} = (\mathcal{E}_p - \mathcal{E}_{p'})\rho_N(p, p') + \langle Np|U_{sc}|Np'\rangle(f_{p'} - f_p),$$

where the right-hand side is written with the accuracy up to the terms linear in the scattering potential U_{sc}. The diagonal part of the density matrix is the distribution function $f_p \equiv \rho_N(p, p)$, and $\mathcal{E}_p = E_{N\sigma p} + eE_x l_H^2 p/\hbar$ is the diagonal matrix element of the Hamiltonian in the presence of electric field. The energy spectrum $E_{N\sigma p}$ is given by Eq. (51.23). To calculate the matrix elements of velocity, we use the first operator identity (48.6), taking into account that the term containing $[\hat{v}_y, \hat{H}]$ does not contribute to the matrix element because of $\langle Np'|\hat{v}_y|Np \rangle \propto \delta_{pp'}$. Solving the equation above, we find $\rho_N(p, p')$ and substitute it into the expression for I_x. Finally, we average this expression over the random scattering potential and represent f_p as a sum of equilibrium and non-equilibrium parts, $f_p^{(eq)} + \Delta f_p$. As a result,

$$I_x = \frac{e}{L_x L_y} \frac{\pi}{\hbar} \sum_{pp'} \frac{p - p'}{m\omega_c} w_N(p, p')[\delta(\varepsilon_p - \varepsilon_{p'})(\Delta f_p - \Delta f_{p'})$$

$$+ \delta(\varepsilon_p - \varepsilon_{p'} + eE_x l_H^2 (p - p')/\hbar)(f_p^{(eq)} - f_{p'}^{(eq)})],$$

where $w_N(p, p')$ is given by Eqs. (51.25) and (51.26). The non-equilibrium part of the distribution function is expressed through E_x and E_y according to Eqs. (51.27)–(51.29). When describing the response to E_y, one should also add the non-dissipative current $-|e|cnE_y/H$ to I_x.

10.12. Using the result of problem 10.11, find σ_{xx} and prove, by a direct calculation, that $\sigma_{xy} = -\sigma_{yx}$.

Result:

$$\sigma_{xx} = \frac{e^2 d^2}{4L_x L_y} \sum_p \frac{1}{\tau_p} \left[\left(\frac{S_p^{(1)}}{S_p^{(0)}} \right)^2 - \frac{S_p^{(2)} + S_0^{(2)}}{S_p^{(0)}} \right] \frac{\partial f_p^{(eq)}}{\partial \varepsilon_p}.$$

10.13. Analyze the conductivity σ_{yy} of the 2D electrons modulated by a harmonic potential $V_0 \cos(2\pi x/d)$ in the presence of a weak magnetic field. Assume that the collision broadening of Landau levels is much smaller that the Landau level separation $\hbar\omega_c$.

Solution: The conductivity tensor is given by Eq. (48.4), where we put a finite broadening energy $\hbar/\tau$ in place λ. The energy spectrum in the basis of Landau states is $\varepsilon_\delta \equiv E_{N\sigma p}$, as given by Eq. (51.23). For the harmonic potential given above (see the discussion of Eq. (51.23) in Sec. 51),

$$E_{N\sigma p} = \hbar\omega_c(N + 1/2) - \sigma g\mu_B H/2 + A_N \cos(\pi p/p_H),$$

$$A_N = V_0 L_N^0 (2\pi^2 l_H^2/d^2) \exp(-\pi^2 l_H^2/d^2),$$

where $p_H = \hbar d/2l_H^2$. The matrix elements of the velocity operator $\hat{v}_y$ are given by Eq. (48.5) for $N' \neq N$ and are equal to the group velocity $v_{Np} = \partial E_{N\sigma p}/\partial p$ for $N = N'$. The existence of the diagonal in N matrix elements is caused by the presence of the periodic potential. Thus, the conductivity (48.4) has two terms: the first one (with non-diagonal matrix elements) is reduced to the Drude conductivity proportional to $1/\omega_c^2$, while the second one (with diagonal matrix elements) gives us the contribution

$$\Delta\sigma_{yy} = \frac{e^2}{L_x L_y} \sum_{N\sigma p} \tau v_{Np}^2 \left[-\frac{\partial f(E_{N\sigma p})}{\partial E_{N\sigma p}} \right],$$

which can be formally obtained from Eq. (51.31) by introducing the sums over the Landau-level index N and spin index σ. Using the equation for the spectrum, neglecting the spin splitting, and assuming zero temperature, we find

$$\Delta\sigma_{yy} \propto \sum_{Np} \tau A_N^2 \sin^2(\pi p/p_H)\delta\left[\varepsilon_F - \hbar\omega_c(N+1/2) - A_N\cos(\pi p/p_H)\right].$$

This expression shows us that N is fixed by the Fermi energy, according to $N = N_F$, where $N_F \simeq \varepsilon_F/\hbar\omega_c$, and the contribution $\Delta\sigma_{yy}$ is proportional to $A_{N_F}^2$. This contributions goes to zero when $A_{N_F}^2 = 0$. Using the asymptotic form of the Laguerre polynomials at large N, we find

$$A_N \simeq \frac{V_0}{\pi}\left(\frac{d^2}{2Nl_H^2}\right)^{1/4}\cos\left[\sqrt{\frac{8N\pi^2l_H^2}{d^2}} - \frac{\pi}{4}\right].$$

Substituting $N = N_F$, one can see that this expression oscillates as a function of the magnetic field and goes to zero when $p_F = p_H(k - 1/4)$. Therefore, the conductivity demonstrates the oscillations periodic in $1/H$. Their period, $|e|d/2cp_F$, is different from the period of Shubnikov - de Haas oscillations.

10.14. Obtain Eq. (52.3) by calculating the integral over time in Eq. (17.2).

Hint: Use the identity $[1/x - 1/(z+x)]/z = 1/[x(z+x)]$.

10.15. Find the energy spectrum and eigenstates for the symmetric two-band model in a magnetic field.

Solution: We search for the eigenstate of the matrix equation $(\hat{h} - E)\mathbf{\Psi} = 0$, where $\hat{h}$ is the Hamiltonian (B.18) in the magnetic field, in the form

$$\mathbf{\Psi} = \begin{pmatrix} F_+(x) \\ F_-(x) \\ \Phi_+(x) \\ \Phi_-(x) \end{pmatrix} \frac{1}{(L_zL_y)^{1/2}}e^{i(p_zz+p_yy)/\hbar},$$

where the components F_σ and Φ_σ describe $\sigma = \pm 1$ states of the conduction and valence bands, respectively. Substituting this solution into the matrix equation, one can express Φ_σ through F_σ and obtain two independent equations for F_+ and F_-. They are written below as a single equation:

$$\left[(ms^2)^2 - E^2 + (sp_z)^2 + (sp_y)^2 - (\hbar s)^2\frac{\partial^2}{\partial x^2}\right.$$

$$\left. +2\frac{\hbar s^2 p_y}{l_H^2}x + \left(\frac{\hbar s}{l_H^2}\right)^2 x^2 + \sigma\left(\frac{\hbar s}{l_H}\right)^2\right]F_\sigma(x) = 0.$$

Since this equation has the same form as Eq. (5.14), we easily find the eigenvalues $E_{cN\sigma p_z} = E_{N,\sigma p_z}$ and $E_{vN\sigma p_z} = -E_{N,\sigma p_z}$, where

$$E_{N,\sigma p_z} = \sqrt{(ms^2)^2 + (sp_z)^2 + 2ms^2\hbar\omega_c(N+1/2+\sigma/2)}.$$

According to this equation, the Zeeman splitting energy in the symmetric two-band model coincides with the cyclotron energy. This property leads to an additional degeneracy of the spectrum, $E_{N,+1p_z} = E_{N+1,-1p_z}$. The wave functions ($j = c, v$) are written as $(L_y L_z)^{-1/2} e^{i(p_z z + p_y y)/\hbar} \langle x|jN\sigma p_y p_z \rangle$, where

$$\langle x|cN, +1p_y p_z \rangle = C_{E_{N,+1p_z}} \begin{pmatrix} (E_{N,+1p_z} + ms^2)\varphi_x^{(Np_y)} \\ 0 \\ sp_z \varphi_x^{(Np_y)} \\ sp_0 \sqrt{N+1} \varphi_x^{(N+1p_y)} \end{pmatrix},$$

$$\langle x|cN, -1p_y p_z \rangle = C_{E_{N,-1p_z}} \begin{pmatrix} 0 \\ (E_{N,-1p_z} + ms^2)\varphi_x^{(Np_y)} \\ sp_0 \sqrt{N} \varphi_x^{(N-1p_y)} \\ -sp_z \varphi_x^{(Np_y)} \end{pmatrix},$$

$$\langle x|vN, +1p_y p_z \rangle = C_{E_{N,+1p_z}} \begin{pmatrix} -sp_z \varphi_x^{(Np_y)} \\ -sp_0 \sqrt{N+1} \varphi_x^{(N+1p_y)} \\ (E_{N,+1p_z} + ms^2)\varphi_x^{(Np_y)} \\ 0 \end{pmatrix},$$

$$\langle x|vN, -1p_y p_z \rangle = C_{E_{N,-1p_z}} \begin{pmatrix} -sp_0 \sqrt{N} \varphi_x^{(N-1p_y)} \\ sp_z \varphi_x^{(Np_y)} \\ 0 \\ (E_{N,-1p_z} + ms^2)\varphi_x^{(Np_y)} \end{pmatrix}.$$

Here $p_0 = \sqrt{2}\hbar/l_H$, $C_E = 1/\sqrt{2E(E + ms^2)}$, and the oscillator function $\varphi_x^{(Np_y)}$ is given by Eq. (5.15).

10.16. Calculate the contribution of the first term in the square brackets of Eq. (52.14).

Hint: To calculate the sum, use the identity

$$\lim_{a \to 0} \left[\sum_{N=N_1}^{N_2} F(aN) - \int_{N_1}^{N_2} dN F(aN) \right] = F(aN_2) - \frac{1}{2} \sum_{N=N_1}^{N_2-1} \frac{dF(aN)}{dN}$$

which leads to

$$\sum_{N=N_1}^{N_2} F(aN) = \frac{1}{a} \int_{aN_1}^{aN_2} dy F(y) + \frac{1}{2} [F(aN_1) + F(aN_2)] + O(a).$$

10.17. Find the magnetic-field-induced anisotropy of the frequency-dependent contribution to the real part of the dielectric permittivity at small ω.

Result: According to Eq. (52.3), the first frequency-dependent term in the expansion of $\delta \epsilon_H$ in powers of $\hbar\omega/\varepsilon_g$ is

$$\delta \epsilon_H(\omega) = 8\pi \frac{(e\hbar^2 \omega)^2}{V} \sum_{\delta\delta'} \frac{|\langle c\delta|\hat{v}_z|v\delta'\rangle|^2 - |\langle c\delta|\hat{v}_x|v\delta'\rangle|^2}{(\varepsilon_{c\delta} - \varepsilon_{v\delta'})^5}.$$

If $\hbar\omega_c \ll \varepsilon_g$, this term is reduced to

$$\delta\epsilon_H(\omega) = \frac{\hbar^2 e^2 \omega^2 \omega_c}{6\pi s \varepsilon_g^3}.$$

10.18. Transform the kinetic energy in the equation for the Green's function of magnetoexciton to the translation-invariant form.

Hint: Use the relations

$$\exp\left(\frac{ie}{2\hbar c}[\mathbf{H} \times \Delta\mathbf{r}] \cdot \mathbf{R}\right) \hat{\mathbf{p}}_{\mathbf{R}} \exp\left(-\frac{ie}{2\hbar c}[\mathbf{H} \times \Delta\mathbf{r}] \cdot \mathbf{R}\right) = \hat{\mathbf{p}}_{\mathbf{R}} - \frac{e}{2c}[\mathbf{H} \times \Delta\mathbf{r}],$$

$$\exp\left(\frac{ie}{2\hbar c}[\mathbf{H} \times \Delta\mathbf{r}] \cdot \mathbf{R}\right) \hat{\mathbf{p}}_{\Delta\mathbf{r}} \exp\left(-\frac{ie}{2\hbar c}[\mathbf{H} \times \Delta\mathbf{r}] \cdot \mathbf{R}\right) = \hat{\mathbf{p}}_{\Delta\mathbf{r}} + \frac{e}{2c}[\mathbf{H} \times \mathbf{R}].$$

10.19. Solve the Schroedinger equation $[\widehat{H}_m(\rho) - \varepsilon]R(\rho) = 0$, where $\widehat{H}_m(\rho)$ is defined by Eq. (52.20).

Solution: Let us search for $R(\rho)$ in the form $R(\rho) = \rho^{|m|}\exp(-\rho^2/4l_H^2)\chi(\rho)$. Introducing a new variable $x = \rho^2/2l_H^2$, we obtain

$$xd^2\chi/dx^2 + (1 + |m| - x)d\chi/dx - \alpha\chi = 0, \quad \alpha = -\varepsilon/\hbar\widetilde{\omega}_c + (|m| + 1 + \gamma m)/2.$$

The solutions of this equation are the confluent hypergeometric functions $\Phi(\alpha, 1 + |m|; x)$, where α is zero or negative integer. Therefore, the energy spectrum is

$$\varepsilon = \varepsilon_{nm} = \hbar\widetilde{\omega}_c[n + 1/2 + (|m| + \gamma m)/2],$$

and the normalized eigenfunctions $R(\rho) = R_{nm}(\rho)$ can be written through the Laguerre polynomials:

$$R_{nm}(\rho) = \sqrt{\frac{n!}{(n+|m|)!}}\frac{x^{|m|/2}}{l_H}e^{-x/2}L_n^{|m|}(x), \quad x = \rho^2/2l_H^2.$$

If $\gamma = 1$ (the case of infinite hole mass), this solution represents the spectrum and radial wave functions of electrons in the magnetic field described in the symmetric gauge. The Landau levels (with the numbers $N = n$ for $m < 0$ and $N = n + m$ for $m > 0$) are multiply degenerate over the quantum number m describing the angular momentum quantization. This degeneracy is analogous to the degeneracy over p_y in the Landau gauge. Since $0 < \gamma < 1$, the degeneracy over m is lifted for excitons.

Chapter 11

PHOTOEXCITATION

The evolution of the system excited by a high-frequency electromagnetic field can be investigated by averaging the statistical operator over the period of perturbation. Applying this approach to electrons, one can introduce the rate of photogeneration, which is proportional to the intensity of the radiation and describes electron redistribution between the ground and excited states. Below, as examples of such approach, we discuss the photon drag current under the intersubband transitions and the response to ultrafast photoexcitation. The distribution of photoexcited electrons is essentially non-equilibrium, and this property causes some peculiar features of the linear response to an additional probe field. For example, the negative transient conductivity appears. If the electromagnetic field is strong enough, the radiation cannot be treated as a perturbation. Beyond the second-order response, the non-Markovian photogeneration leads to the Rabi oscillations. A more careful approach is applied for studying the mixing of electron and hole states by the field of the radiation, when one should introduce new quasiparticles. We also consider the coherent response of the phonon system to the radiation and describe the relaxation of coherent phonon oscillations due to phonon-phonon interaction.

53. Photogeneration Rate

The Hamiltonian of the electrons placed in the electric field $\mathbf{E}_t e^{-i\omega t} + c.c.$ can be presented as

$$\hat{\overline{H}} + \left(\widehat{\Delta H}_t e^{-i\omega t} + H.c. \right),$$

$$\hat{\overline{H}} = \frac{\omega}{2\pi} \int_{-\pi/\omega}^{\pi/\omega} dt \, \widehat{H}_t \,, \qquad \widehat{\Delta H}_t = \frac{ie}{\omega}(\mathbf{E}_t \cdot \hat{\mathbf{v}}). \tag{1}$$

The interaction of the electrons with the field is written here in the dipole approximation. The term $\widehat{\bar{H}}$ determines the quasienergy spectrum, see Eqs. (5.30)–(5.33), while the perturbation $\widehat{\Delta H}_t$ describes interband or intersubband electron transitions and accounts for slow variations of the field with time. We have neglected the terms proportional to $e^{\pm 2i\omega t}$, which do not excite the transitions under consideration. Using Eq. (1), we rewrite Eq. (1.20) for the density matrix $\hat{\eta}_t$ as

$$\frac{\partial \hat{\eta}_t}{\partial t} + \frac{i}{\hbar}[\widehat{\bar{H}}, \hat{\eta}_t] = \frac{1}{i\hbar}\left[\left(\widehat{\Delta H}_t e^{-i\omega t} + H.c.\right), \hat{\eta}_t\right]. \tag{2}$$

With the initial condition $\hat{\eta}_{t \to -\infty} = \hat{\eta}_{-\infty}$, we transform this equation to the integral form $(\lambda \to +0)$

$$\hat{\eta}_t - \hat{\eta}_{-\infty} = \frac{1}{i\hbar}\int_{-\infty}^{t} dt'\, e^{\lambda t' - i\omega t'} e^{-i\widehat{\bar{H}}(t-t')/\hbar}[\widehat{\Delta H}_{t'}, \hat{\eta}_{t'}] e^{i\widehat{\bar{H}}(t-t')/\hbar} + H.c.$$

$$= \int_{-\infty}^{0} \frac{d\tau}{i\hbar} e^{\lambda\tau - i\omega(t+\tau)} e^{i\widehat{\bar{H}}\tau/\hbar}[\widehat{\Delta H}_{t+\tau}, \hat{\eta}_{t+\tau}] e^{-i\widehat{\bar{H}}\tau/\hbar} + H.c. , \tag{3}$$

which is more convenient for describing the interband (or intersubband) transitions.

Representing $\hat{\eta}_t$ in Eq. (3) as a sum of slowly varying contribution $\bar{\hat{\eta}}_t$ and high-frequency part $[\widehat{\Delta\eta}_t e^{-i\omega t} + H.c.]$, we write the first-order expression for the latter:

$$\widehat{\Delta\eta}_t = \int_{-\infty}^{0} \frac{d\tau}{i\hbar} e^{\lambda\tau - i\omega\tau} e^{i\widehat{\bar{H}}\tau/\hbar}[\widehat{\Delta H}_{t+\tau}, \bar{\hat{\eta}}_{t+\tau}] e^{-i\widehat{\bar{H}}\tau/\hbar}. \tag{4}$$

Substituting $\hat{\eta}_t = \bar{\hat{\eta}}_t + [\widehat{\Delta\eta}_t e^{-i\omega t} + H.c.]$ with $\widehat{\Delta\eta}_t$ from Eq. (4) into the right-hand side of Eq. (2), we average Eq. (2) over the period and obtain the following equation for $\bar{\hat{\eta}}_t$:

$$\frac{\partial \bar{\hat{\eta}}_t}{\partial t} + \frac{i}{\hbar}[\widehat{\bar{H}}, \bar{\hat{\eta}}_t] = \frac{1}{\hbar^2}\int_{-\infty}^{0} d\tau\, e^{\lambda\tau - i\omega\tau}$$

$$\times \left[e^{i\widehat{\bar{H}}\tau/\hbar}[\widehat{\Delta H}_{t+\tau}, \bar{\hat{\eta}}_{t+\tau}] e^{-i\widehat{\bar{H}}\tau/\hbar}, \widehat{\Delta H}_t^{+} \right] + H.c. \equiv \widehat{G}_t. \tag{5}$$

The high-frequency contributions $\propto e^{\pm 2i\omega t}$ disappear after the averaging procedure. The right-hand side of Eq. (5) forms the operator of interband photogeneration rate, $\widehat{G}_t$. Substituting $\widehat{\Delta H}_t$ from Eq. (1) into Eq. (5), we find

$$\widehat{G}_t = \left(\frac{e}{\hbar\omega}\right)^2 \int_{-\infty}^{0} d\tau\, e^{\lambda\tau - i\omega\tau} \tag{6}$$

$$\times \left[e^{i\hat{\bar{H}}\tau/\hbar} \left[(\mathbf{E}_{t+\tau} \cdot \hat{\mathbf{v}}), \hat{\bar{\eta}}_{t+\tau} \right] e^{-i\hat{\bar{H}}\tau/\hbar}, (\mathbf{E}_t \cdot \hat{\mathbf{v}})^+ \right] + H.c.$$

In this form, the generation rate still depends on the interaction of electrons with impurities, phonons, and other electrons, since $\hat{\bar{H}}$ contains the corresponding interaction terms. If these processes are neglected in Eq. (6), the slowly varying part $\hat{\bar{\rho}}_t$ of the one-electron density matrix satisfies the quantum kinetic equation

$$\frac{\partial \hat{\bar{\rho}}_t}{\partial t} + \frac{i}{\hbar}[\hat{\bar{h}}, \hat{\bar{\rho}}_t] = \hat{G}_t + \hat{J}(\hat{\bar{\rho}}|t), \tag{7}$$

where $\hat{J}(\hat{\bar{\rho}}|t)$ is the collision integral in the operator form and $\hat{G}_t$ is given by Eq. (6) with $\hat{\bar{H}}$ replaced by the one-electron Hamiltonian $\hat{\bar{h}}$. We point out the non-Markovian temporal dependence of the generation rate. This dependence appears because of exclusion of the high-frequency contribution describing the perturbation-induced polarization in the equation averaged over the period. This mechanism of memory is purely dynamical and, for this reason, is qualitatively different from the non-Markovian temporal dependence of the collision integrals describing the memory effects in the scattering processes; see Secs. 7, 19, 23, and 34.

Using the basis of quasienergy states, $\hat{\bar{h}}|\delta\rangle = \varepsilon_\delta|\delta\rangle$, we rewrite Eq. (7) as a system of kinetic equations for the matrix elements $\langle\delta'|\hat{\bar{\rho}}_t|\delta\rangle$. The non-diagonal (interband or intersubband) components of this matrix are small if the characteristic frequencies $|\varepsilon_\delta - \varepsilon_{\delta'}|/\hbar$ are large in comparison to both the generation rate and the collision relaxation rate $1/\bar{\tau}$ (problem 11.1). The diagonal components, $f_{\delta t} = \langle\delta|\hat{\bar{\rho}}_t|\delta\rangle$, are the distribution functions whose evolution is described by the kinetic equation

$$\frac{\partial f_{\delta t}}{\partial t} = G_{\delta t} + J(f|\delta t), \tag{8}$$

where $J(f|\delta t)$ and $G_{\delta t} = \langle\delta|\hat{G}_t|\delta\rangle$ are the collision integral and photogeneration rate for the state δ, respectively. If the external field is varying slower than both generation and relaxation occur, one can calculate $G_{\delta t}$ by using the approximations $\mathbf{E}_{t+\tau} \simeq \mathbf{E}_t$ and $f_{\delta,t+\tau} \simeq f_{\delta t}$. The integral over τ in this Markovian approximation gives us the energy conservation laws, and the photogeneration rate is written as

$$G_{\delta t} = \frac{2\pi}{\hbar}\left(\frac{e}{\omega}\right)^2 \sum_{\delta'} (f_{\delta't} - f_{\delta t})\left\{|\langle\delta|(\mathbf{E}_t \cdot \hat{\mathbf{v}})|\delta'\rangle|^2\delta(\varepsilon_\delta - \varepsilon_{\delta'} - \hbar\omega)\right.$$

$$\left. + |\langle\delta|(\mathbf{E}_t \cdot \hat{\mathbf{v}})^+|\delta'\rangle|^2\delta(\varepsilon_\delta - \varepsilon_{\delta'} + \hbar\omega)\right\}, \tag{9}$$

where we have separated the stimulated emission and the absorption (the first and the second contributions, respectively).

The equations given above are applicable for describing the photoexcitation processes in spatially-homogeneous systems, where the index δ includes a discrete quantum number l (it can be, for example, a combination of the band and spin indices, as in Appendix B) and the momentum $\mathbf{p}$. If there exist smooth (with respect to the quantum scale $\hbar/\bar{p}$) inhomogeneities, one should carry out the Wigner transformation of Eq. (7) as described in Sec. 9. As a result, we obtain a kinetic equation for the distribution function $f_{l\mathbf{rp}t}$:

$$\frac{\partial f_{l\mathbf{rp}t}}{\partial t} + \mathbf{v}_{l\mathbf{p}} \cdot \frac{\partial f_{l\mathbf{rp}t}}{\partial \mathbf{r}} = G_{l\mathbf{rp}t} + J(f|l\mathbf{rp}t), \tag{10}$$

where the group velocity $\mathbf{v}_{l\mathbf{p}}$ depends on the quantum number l. This kinetic equation is written under the approximation that the force term (which appears, for example, in Eq. (9.23)) can be neglected. The generation rate $G_{l\mathbf{rp}t}$ standing in Eq. (10) is obtained from Eq. (9) when a parametric dependence of the electron distribution on coordinate is taken into account. The same procedure is applied to obtain the collision integral $J(f|l\mathbf{rp}t)$, which is given by an equation of the kind of Eq. (9.33). If the inhomogeneities are smooth on the scale of characteristic relaxation lengths, one may use the balance equations (see Secs. 11 and 36) to describe the photoexcitation processes. Introducing the density and flow density of electrons in the state l as $n_{l\mathbf{r}t} = (1/V)\sum_\mathbf{p} f_{l\mathbf{rp}t}$ and $\mathbf{i}_{l\mathbf{r}t} = (1/V)\sum_\mathbf{p} \mathbf{v}_{l\mathbf{p}} f_{l\mathbf{rp}t}$, respectively, we obtain the density balance equation in the form

$$\frac{\partial n_{l\mathbf{r}t}}{\partial t} + \nabla \cdot \mathbf{i}_{l\mathbf{r}t} = G_{l\mathbf{r}t} - R_{l\mathbf{r}t}, \tag{11}$$

where the terms $G_{l\mathbf{r}t} = (1/V)\sum_\mathbf{p} G_{l\mathbf{rp}t}$ and $R_{l\mathbf{r}t}$ are the local generation and recombination rates per unit volume for the l-th state. The flow density is expressed through the drift and diffusion currents in the usual way; see Sec. 36. For a spatially-homogeneous system, Eq. (11) has the same form as Eq. (38.12) describing the density balance in quantum wells. Equations (9)−(11), together with the electrodynamical equations describing the distribution of electromagnetic fields, allow one to consider the photoelectric phenomena in semiconductors and insulators.

To describe the momentum transfer between photons and electrons in the processes of interband or intersubband transitions, one should go beyond the dipole approximation and use the operator of interaction determined by Eqs. (13.6) and (13.13) instead of $\widehat{\Delta H}_t$ in Eq. (1):

$$\widehat{\Delta h}_\omega = \frac{ie}{2\omega}\mathbf{E} \cdot \left(\hat{\mathbf{v}}e^{i\mathbf{q}\cdot\mathbf{r}} + e^{i\mathbf{q}\cdot\mathbf{r}}\hat{\mathbf{v}}\right) . \tag{12}$$

In other words, one should replace $\hat{\mathbf{v}}$ in Eqs. (1), (6), and (9) by $\left(\hat{\mathbf{v}}e^{i\mathbf{q}\cdot\mathbf{r}} + e^{i\mathbf{q}\cdot\mathbf{r}}\hat{\mathbf{v}}\right)/2$. As a result, the photogeneration rate (9) contains the $\mathbf{q}$-dependent matrix elements determining the transfer of photon momentum to electrons. This transfer leads to the effect called the photon drag. It is considered below for the case of intersubband transitions of electrons in a quantum well. In the basis $|n\mathbf{p}\rangle$, where n is the subband number, the matrix element of the perturbation (12) is

$$|\langle n\mathbf{p}|\widehat{\Delta h_\omega}|n'\mathbf{p}'\rangle|^2 = \delta_{\mathbf{p},\mathbf{p}'+\hbar\mathbf{q}}\left(\frac{e}{\omega}E_\perp\right)^2|\langle n|\hat{v}_z|n'\rangle|^2 , \qquad (13)$$

where $\langle n|\hat{v}_z|n'\rangle$ is the matrix element of the velocity operator along the confinement direction OZ and $E_\perp$ is the transverse electric field exciting the intersubband transitions ($E_\perp$ is assumed to be real). For the transitions between the ground ($n = 1$) and first excited ($n = 2$) states, the generation rate (6) is written as

$$G_{1\mathbf{p}} = \frac{2\pi}{\hbar}\left(\frac{e}{\omega}E_\perp\right)^2|v_{21}^z|^2(f_{2\mathbf{p}+\hbar\mathbf{q}} - f_{1\mathbf{p}})\delta_\gamma(\varepsilon_{1\mathbf{p}} - \varepsilon_{2\mathbf{p}+\hbar\mathbf{q}} + \hbar\omega),$$

$$G_{2\mathbf{p}} = \frac{2\pi}{\hbar}\left(\frac{e}{\omega}E_\perp\right)^2|v_{21}^z|^2(f_{1\mathbf{p}-\hbar\mathbf{q}} - f_{2\mathbf{p}})\delta_\gamma(\varepsilon_{2\mathbf{p}} - \varepsilon_{1\mathbf{p}-\hbar\mathbf{q}} - \hbar\omega). \qquad (14)$$

The notation $v_{21}^z \equiv \langle 2|\hat{v}_z|1\rangle$ is taken from Sec. 29. To make the expressions finite, we have introduced a broadening energy γ into the energy conservation law, as in Eq. (22.15), according to the definition of the δ-function used in problem 1.4. Below we neglect redistribution of electrons between the subbands and substitute the equilibrium distribution functions into Eq. (14).

Considering the photon drag of strongly degenerate 2D electrons occupying the ground-state subband, we take into account that the photon momentum $\hbar\mathbf{q}$ is small in comparison to the Fermi momentum. Therefore, we separate the small (linear in $\mathbf{q}$) anisotropic contributions in Eq. (14) and write them in the following way:

$$\delta G_{1\mathbf{p}} \simeq -2\pi\mathbf{q}\cdot\mathbf{v_p}\left(\frac{e}{\omega}E_\perp\right)^2|v_{21}^z|^2\theta(\varepsilon_F - \varepsilon_p)\delta_\gamma'(\varepsilon_{21} - \hbar\omega),$$

$$\delta G_{2\mathbf{p}} \simeq 2\pi\mathbf{q}\cdot\mathbf{v_p}\left(\frac{e}{\omega}E_\perp\right)^2|v_{21}^z|^2\left[\theta(\varepsilon_F - \varepsilon_p)\delta_\gamma'(\varepsilon_{21} - \hbar\omega)\right.$$

$$\left. +\delta(\varepsilon_p - \varepsilon_F)\delta_\gamma(\varepsilon_{21} - \hbar\omega)\right], \qquad (15)$$

where $\mathbf{v_p} = \mathbf{p}/m$ and $\delta_\gamma'(E) \equiv d\delta_\gamma(E)/dE$. The anisotropic parts of the distribution functions, $\delta f_{1\mathbf{p}}$ and $\delta f_{2\mathbf{p}}$, caused by the generation terms (15) are determined from the equations

$$\delta G_{1\mathbf{p}} - \delta f_{1\mathbf{p}}\nu_{1p} = 0, \qquad \delta G_{2\mathbf{p}} - \delta f_{2\mathbf{p}}\left[\nu_{2p} + \nu_p^{(21)}\right] = 0 \qquad (16)$$

obtained from Eq. (10) after representing the collision integrals through the isotropic intrasubband and intersubband relaxation rates ν_{1p}, ν_{2p}, and $\nu_p^{(21)}$; see Secs. 8 and 38. For the case of electrons interacting with a short-range static potential, these rates are momentum-independent (problem 11.2). Equations (15) and (16) explicitly define the anisotropic parts of stationary distribution of electrons under intersubband photoexcitation.

The photoinduced density of stationary current is calculated from the general formula $\mathbf{I}_{\mathbf{R}t} = \mathrm{Sp}\hat{\mathbf{I}}_{\mathbf{R}t}\hat{\eta}_t$, where the current density operator is given by Eq. (4.15). This operator contains field-dependent terms entering through the vector potential. Owing to these terms, not only the stationary density matrix $\hat{\bar{\rho}}$, but also the high-frequency corrections to $\hat{\rho}$ contribute to the current density averaged over the period $2\pi/\omega$. The averaging over the 2D layer width leads to the current density

$$\mathbf{I} = \frac{\omega}{2\pi} \int_{-\pi/\omega}^{\pi/\omega} dt \int dz \mathbf{I}_{\mathbf{R}t} = \frac{e}{L^2}\mathrm{sp}\hat{\mathbf{v}}\hat{\bar{\rho}} + \Delta\mathbf{I}. \tag{17}$$

The additional part $\Delta\mathbf{I}$ is found after averaging the product of the external field by $\widehat{\Delta\eta}_t$ of Eq. (4) over the period:

$$\Delta\mathbf{I} = -\frac{e^2/m}{\hbar\omega}\mathbf{E}^* e^{-i\mathbf{q}\cdot\mathbf{X}} \int_{-\infty}^{0} d\tau e^{-i(\omega+i\lambda)\tau}$$

$$\times \mathrm{sp}\delta(\mathbf{x} - \mathbf{X})e^{i\hat{h}\tau/\hbar}[\widehat{\Delta h}_\omega, \hat{\bar{\rho}}]e^{-i\hat{h}\tau/\hbar} + c.c. \tag{18}$$

This expression is independent of the macroscopic coordinate $\mathbf{X}$ along the 2D layer because the system is translation-invariant in the 2D plane. Further calculations are done by using the one-electron basis $|n\mathbf{p}\rangle$. Taking into account only the contributions linear in $\mathbf{q}$, we obtain

$$\Delta\mathbf{I} = \frac{2e^2\mathbf{E}^*}{m\hbar\omega L^2} \int_{-\infty}^{0} d\tau e^{-i(\omega+i\lambda)\tau} \sum_{n\mathbf{p}} \langle n\mathbf{p}|\frac{ie}{\omega}\mathbf{E}\cdot\hat{\mathbf{v}}|n\mathbf{p}\rangle(f_{n\mathbf{p}+\hbar\mathbf{q}} - f_{n\mathbf{p}})$$

$$+c.c. \simeq -\frac{4e^2\mathbf{E}^*}{m\omega^2} \int \frac{d\mathbf{p}}{(2\pi\hbar)^2} \left(\frac{e}{\omega}\mathbf{E}\cdot\mathbf{v}_\mathbf{p}\right) \delta(\varepsilon_p - \varepsilon_F)(\mathbf{q}\cdot\mathbf{v}_\mathbf{p}). \tag{19}$$

The current density $\Delta\mathbf{I}$ is associated with intrasubband transitions and excited by the field parallel to the 2D layer.

The first term in $\mathbf{I}$ is written through the anisotropic contributions $\delta f_{n\mathbf{p}}$ in the usual way, and the total current density is given by

$$\mathbf{I} = 2e \int \frac{d\mathbf{p}}{(2\pi\hbar)^2}\mathbf{v}_\mathbf{p}(\delta f_{1\mathbf{p}} + \delta f_{2\mathbf{p}}) + \Delta\mathbf{I}. \tag{20}$$

Note that the factor of 2 in Eqs. (19) and (20) comes from spin summation. Comparing the first term on the right-hand side of Eq. (20) to the second one, given by Eq. (19), we find that the latter contains a quantity of the order of $(\mathbf{q} \cdot \mathbf{v_p}/m\omega)(eE/\omega)^2\delta(\varepsilon_p - \varepsilon_F)$ in place of $\delta f_{1\mathbf{p}} + \delta f_{2\mathbf{p}}$ in the first term. Therefore, $\Delta \mathbf{I}$ is small as $(\omega\bar{\tau})^{-1}$ and can be neglected. Using Eqs. (15) and (16), we finally obtain

$$\mathbf{I} \simeq 2\pi\mathbf{q}\frac{en_{2D}}{m}\left(\frac{e}{\omega}E_\perp\right)^2 |\langle 2|\hat{v}_z|1\rangle|^2$$

$$\times \left[-\varepsilon_F\delta'_\gamma(\varepsilon_{21} - \hbar\omega)\tau_1 + \delta_\gamma(\varepsilon_{21} - \hbar\omega)\tau_2\right], \qquad (21)$$

where n_{2D} is the 2D electron density. The characteristic times are introduced as $\tau_1 = \varepsilon_F^{-2} \int_0^{\varepsilon_F} d\varepsilon_p \, \varepsilon_p[\nu_{1p}^{-1} - (\nu_{2p} + \nu_p^{(21)})^{-1}]$ and $\tau_2 = (\nu_{2p_F} + \nu_{p_F}^{(21)})^{-1}$. The photon drag current is essential at $\omega \simeq \varepsilon_{21}/\hbar$ and depends on the frequency of detuning, $\Delta\omega = \omega - \varepsilon_{21}/\hbar$, according to

$$\mathbf{I} \propto \Phi_{\Delta\omega} \equiv \delta_\gamma(\hbar\Delta\omega) + \varepsilon_F\frac{\tau_1}{\tau_2}\delta'_\gamma(\hbar\Delta\omega). \qquad (22)$$

The function $\Phi_{\Delta\omega}$ characterizes the spectral dependence of the photon drag. Since the broadening energy is much smaller than the Fermi energy, the second term in Eq. (22) dominates. Since this term changes its sign together with $\Delta\omega$, the drag current also changes its sign. At small $\Delta\omega$, the first term of $\Phi_{\Delta\omega}$ also becomes important so that the point of sign inversion is slightly shifted away from the exact resonance $\Delta\omega = 0$.

The photon drag effect is caused by the same elementary processes which are responsible for the drag of electromagnetic field by a current (see the last part of Sec. 17, where this drag has been considered for the case of interband transitions). The order-of-value estimate of $\mathbf{I}$ can be written as $\mu\xi_\omega S_\omega\mathbf{q}/\omega$, where $\mu = |e|\bar{\tau}/m$ is the electron mobility, ξ_ω is the relative absorption coefficient given by Eq. (29.25), and $S_\omega = |E|^2c\sqrt{\epsilon}/2\pi$ is the absolute value of the Poynting vector describing the energy flow of monochromatic photons. Since $\xi_\omega S_\omega\mathbf{q}/\omega$ is the photon momentum flow absorbed by the 2D electrons, this estimate has a clear physical meaning. Note that the contribution of the excited states (where the density of electrons is small) to the photon drag current is of the same order as the contribution of the ground states. The presence of the spectral inversion of the drag current is explained as follows. The photons with $\Delta\omega > 0$ induce the intersubband transitions only for the electrons whose momenta have positive components in the direction of $\mathbf{q}$. Selective excitation of such electrons gives rise to a current within the ground subband, corresponding to the motion of electrons in the direction opposite to $\mathbf{q}$. At the same time, electrons excited into the subband

2 for $\Delta\omega > 0$ have positive velocities in the direction of $\mathbf{q}$. When the frequency $\Delta\omega$ becomes negative, the currents within the subbands 1 and 2 (and the total current) change their signs.

54. Response to Ultrafast Excitation

Let us consider the response to an ultrashort pulse of electromagnetic radiation. The electric field of this pulse is written as $\mathbf{E}_t e^{-i\omega t} + c.c.$, where $\mathbf{E}_t = \mathbf{E}w_t$. The envelope form-factor w_t describes the pulse of duration τ_p, and we assume that $\tau_p \gg 2\pi/\omega$, though τ_p can be smaller than other characteristic times of the system. The linear response to this perturbation is described by the general expression (13.18) for the complex conductivity. Applying it, one should take into account that the Fourier transform of the electric field includes the form-factor w_t, and the Fourier components of the induced current are excited in a narrow interval around ω. To describe the second-order response, it is convenient to use the generation rate introduced in the previous section, where the time-dependent amplitude $\mathbf{E}_t$ is taken into account; see Eq. (53.6). Below we consider this response for the case of interband excitation of electrons in insulators or non-doped semiconductors. We also study the nonlinear ultrafast excitation of intersubband transitions in quantum wells, including the case of coherent control under two-pulse pumping.

To write the photogeneration rate due to electron transitions from the fully occupied valence band to the empty conduction band within the accuracy up to E^2, we use the one-particle approximation in the general equation (53.6) by substituting the density matrix of the valence-band electrons, $\hat{\rho}_v$, into this equation:

$$\widehat{G}_t = \left(\frac{e}{\hbar\omega}\right)^2 \int_{-\infty}^0 d\tau e^{\lambda\tau - i\omega\tau}$$

$$\times \left[e^{i\hat{h}\tau/\hbar} \left[(\mathbf{E}_{t+\tau} \cdot \hat{\mathbf{v}}), \hat{\rho}_v \right] e^{-i\hat{h}\tau/\hbar}, (\mathbf{E}_t \cdot \hat{\mathbf{v}})^+ \right] + H.c. \tag{1}$$

The constant λ in this equation can be treated phenomenologically as a small, though still finite, relaxation rate which causes a finite broadening (see the beginning of the next section for a more careful consideration). Assuming that the operators $\hat{h}$ and $\hat{\mathbf{v}}$ are the one-particle Hamiltonian and velocity operator of the two-band model, respectively, we write all operators in the basis $|n\sigma\mathbf{p}\rangle$ and introduce the generation rate in the c-band as $G_{\mathbf{p}t}^{(c)} = (1/2) \sum_\sigma \langle c\sigma\mathbf{p}|\widehat{G}_t|c\sigma\mathbf{p}\rangle$. By analogy to the consideration given in Sec. 17, we obtain

$$G_{\mathbf{p}t}^{(c)} = \left(\frac{e}{\hbar\omega}\right)^2 \int_{-\infty}^0 d\tau e^{\lambda\tau - i\omega\tau} \left(\mathbf{E}_{t+\tau} \cdot \widehat{M} \cdot \mathbf{E}_t^* \right)$$

$$\times \left[e^{-i(\varepsilon_{cp}-\varepsilon_{vp})\tau/\hbar} + c.c. \right] + c.c. \,, \tag{2}$$

where the tensor $M_{\alpha\beta} = (1/2)\sum_{\sigma\sigma'}\langle c\sigma|\hat{v}_\alpha|v\sigma'\rangle\langle v\sigma'|\hat{v}_\beta|c\sigma\rangle$ is written through the spin-dependent matrix elements of the velocity operator near the extrema of the bands at $\mathbf{p}=0$. Although we assume that the field $\mathbf{E}_t$ is linearly polarized, the spin averaging of $G_{\mathbf{p}t}^{(c)}$ makes the result independent of whether the field is linearly or elliptically polarized. Using also $\mathbf{E}_t = \mathbf{E}w_t$ and $M_{\alpha\beta} = \delta_{\alpha\beta}s^2$, we transform Eq. (2) to the form

$$G_{\mathbf{p}t}^{(c)} = \left(\frac{eEs}{\hbar\omega} \right)^2 w_t \int_{-\infty}^0 d\tau e^{\lambda\tau - i\omega\tau} w_{t+\tau} \left[e^{-i(\varepsilon_{cp}-\varepsilon_{vp})\tau/\hbar} + c.c. \right] + c.c. \tag{3}$$

In the resonance approximation, only the contributions containing the exponential factors with the energy $\Delta_p = \varepsilon_{cp} - \varepsilon_{vp} - \hbar\omega$ should be taken into account. Then, the kinetic equation for the c-band distribution function $f_{c\mathbf{p}t}$ is written as

$$\frac{\partial f_{c\mathbf{p}t}}{\partial t} = G_{\mathbf{p}t}^{(c)}, \qquad G_{\mathbf{p}t}^{(c)} = 2\left(\frac{eEs}{\hbar\omega} \right)^2 w_t \mathrm{Re} W_{\Delta_p}(t). \tag{4}$$

The generation rate is expressed through the function

$$W_\Delta(t) = \int_{-\infty}^0 d\tau w_{t+\tau} e^{-i\Delta\tau/\hbar}, \tag{5}$$

where we omit the term $e^{\lambda\tau}$ under the integral (see, however, problem 11.3). Using the initial condition $f_{c\mathbf{p}t\to-\infty} = 0$, we find the photoexcited electron distribution as $f_{c\mathbf{p}t} = \int_{-\infty}^t d\tau G_{\mathbf{p}\tau}^{(c)}$. Its momentum dependence is entirely determined by the momentum dependence of the generation rate through Δ_p.

At $|\Delta|\tau_p/\hbar \gg 1$ (and $\lambda\tau_p \ll 1$), the function (5), with the use of integration by parts, is transformed to

$$W_\Delta(t) \simeq \frac{i\hbar}{\Delta} w_t + \left(\frac{\hbar}{\Delta} \right)^2 \frac{dw_t}{dt}, \tag{6}$$

which means that away from the resonance the photogeneration rate is

$$G_{\mathbf{p}t}^{(c)} = 2\left(\frac{eEs}{\omega\Delta_p} \right)^2 \frac{1}{2}\frac{dw_t^2}{dt}. \tag{7}$$

In the exact resonance, when $\Delta_p = 0$, the function (5) is equal to $\int_{-\infty}^t d\tau w_\tau$ and we obtain

$$G_{\mathbf{p}t}^{(c)}\big|_{\Delta_p=0} = 2\left(\frac{eEs}{\hbar\omega} \right)^2 \frac{1}{2}\frac{d}{dt}\left(\int_{-\infty}^t d\tau w_\tau \right)^2. \tag{8}$$

As a result, the distribution function in these limiting cases is written as

$$f_{c\mathbf{p}t} = \left(\frac{eEs}{\hbar\omega}\right)^2 \begin{cases} (w_t\hbar/\Delta_p)^2, & \Delta_p\tau_p \gg \hbar \\ \left[\int_{-\infty}^t d\tau\, w_\tau\right]^2, & \Delta_p\tau_p \ll \hbar \end{cases}, \tag{9}$$

This equation demonstrates that the temporal dependences of the distribution in the resonance and away from the resonance are essentially different. The non-equilibrium electrons with momenta corresponding to the non-resonant regime are generated during the action of the short pump pulse. The density of these electrons goes to zero when the pulse is terminated. For the region of momenta close to the resonance, the total density of the photoexcited electrons at $t \gg \tau_p$ is proportional to the intensity of the radiation, which is written through the integral of the amplitude of the field over time, $f_{c\mathbf{p}t\to\infty} \propto \left(\int_{-\infty}^\infty d\tau\, w_\tau\right)^2$. If, however, the amplitude of the field changes with time slower than the relaxation occurs, i.e., $\tau_p \gg \hbar/\lambda$, one should take into account that λ is finite, which gives us $\mathrm{Re}\, W_\Delta(t) \simeq \pi\hbar\delta_{\hbar\lambda}(\Delta)w_t$ (problem 11.3). The generation rate (4) in this case depends on time through w_t^2, being proportional to the intensity of the radiation, and the electrons are excited only in the resonance region.

The general picture of photoexcited electron distribution becomes clear from the limiting cases analyzed above. However, to consider the temporal evolution of the electron density, we need to sum Eq. (4) over $\mathbf{p}$. This sum is analytically calculated in the case of 2D electrons, when the interband excitation in a quantum well is considered. The electron density is given by $n_t = \widetilde{\rho}_{2D}\int_0^\infty d\xi f_{\xi t}$, where $\widetilde{\rho}_{2D} = \mu^*/\pi\hbar^2$ is the 2D density of states with the reduced mass μ^* introduced in Sec. 17, and we write the energy distribution function $f_{\xi t}$, with the corresponding kinetic energy $\xi = p^2/2\mu^*$, in place of $f_{c\mathbf{p}t}$. Integrating Eq. (4) with $W_{\Delta_p}(t)$ from Eq. (5) over ξ, we use the identity

$$\int_0^\infty d\xi \exp[-i(\tau/\hbar - i\lambda)\xi] = \pi\hbar\delta(\tau) - i\hbar\frac{\mathcal{P}}{\tau}, \tag{10}$$

where $\lambda \to +0$ enters due to relaxation. In order to calculate the remaining integral $\int_{-\infty}^0 d\tau\ldots$, one should take into account that only a half of the delta-function $\delta(\tau)$ from Eq. (10) contributes to this integral. The expression for n_t is obtained from Eq. (4) in the form

$$\frac{dn_t}{dt} = \widetilde{\rho}_{2D}\int_0^\infty d\xi\, G_{\xi t}^{(c)} \equiv g_t,$$

$$g_t = \frac{N_{2D}}{2\tau_p}\left[w_t^2 + \frac{2w_t}{\pi}\int_{-\infty}^0 \frac{d\tau}{\tau} w_{t+\tau}\sin\frac{(\hbar\omega - \overline{\varepsilon}_g)\tau}{\hbar}\right], \tag{11}$$

where $N_{2D} = 2\pi\tilde{\rho}_{2D}(eE\tilde{s}/\omega)^2\tau_p/\hbar$ is a characteristic density determined by the parameters of the pulse. The velocity $\tilde{s}$ is equal to the interband velocity s multiplied by the overlap integral of the envelope functions describing confinement of c- and v-band states in the quantum well, and $\bar{\varepsilon}_g$ is the energy gap renormalized due to this confinement.

The density generation rate g_t entering Eq. (11) is considerably simplified in the case of very short pulses, when $|\hbar\omega - \bar{\varepsilon}_g|\tau_p \ll \hbar$ and the second term in the brackets in Eq. (11) can be neglected. As a result, the total density of photoexcited electrons, $n_{t\to\infty} \equiv n_\infty = \int_{-\infty}^{\infty} dt g_t$, is equal to $N_{2D}\int_{-\infty}^{\infty} dt w_t^2/2\tau_p$. In particular, for a properly normalized Gaussian pulse

$$w_t = \left(\frac{2}{\pi}\right)^{1/4} e^{-(t/\tau_p)^2}, \tag{12}$$

one finds $\int_{-\infty}^{\infty} dt w_t^2 = \tau_p$ and $n_\infty = N_{2D}/2$. For the pulses of finite duration, n_∞ depends on the dimensionless parameter $z = (\hbar\omega - \bar{\varepsilon}_g)\tau_p/\hbar$. For the Gaussian pulses (12), this dependence (see Fig. 11.1) is given by the analytical expression $n_\infty(z) = N_{2D}[1 + \text{erf}(z/\sqrt{2})]/2$, where erf is the error function. For the excitation above the threshold of interband absorption, $z > 0$, the density n_∞ increases and approaches the saturation value N_{2D} at $z \gg 1$. For the excitation below the threshold, $z < 0$, the density n_∞ decreases but remains finite. This dependence is a manifestation of the general principle of quantum mechanics, the energy-time uncertainty.

Consider now the resonant excitation of electrons from the ground-state subband of a quantum well to a higher subband. Using the dipole approximation and the basis $|n\mathbf{p}\rangle$, where $n = 1, 2$ are the numbers of ground ($n = 1$) and excited ($n = 2$) subbands, we obtain the following generation rates from Eq. (53.6):

$$G_{1\mathbf{p}t} = -G_{2\mathbf{p}t} = \left(\frac{e|v_{21}^z|}{\hbar\omega}\right)^2 E_t^* \int_{-\infty}^{0} d\tau e^{\lambda\tau} e^{i(\varepsilon_{21}-\hbar\omega)\tau/\hbar}$$

$$\times E_{t+\tau}(f_{2\mathbf{p}t+\tau} - f_{1\mathbf{p}t+\tau}) + c.c. , \tag{13}$$

where the distribution function for n-th state is introduced as $f_{n\mathbf{p}t} = \langle n\mathbf{p}|\hat{\eta}_t|n\mathbf{p}\rangle$, and $v_{21}^z = \langle 2|\hat{v}_z|1\rangle$ is the intersubband matrix element of the velocity operator. The distribution function satisfies the kinetic equation $\partial f_{n\mathbf{p}t}/\partial t = G_{n\mathbf{p}t} + J(f|n\mathbf{p}t)$, where $J(f|n\mathbf{p}t)$ is a collision integral. Summing both sides of this equation over the 2D momentum, we obtain the balance equations

$$\frac{dn_{1t}}{dt} = G_t + \left(\frac{\partial n_1}{\partial t}\right)_{sc},$$

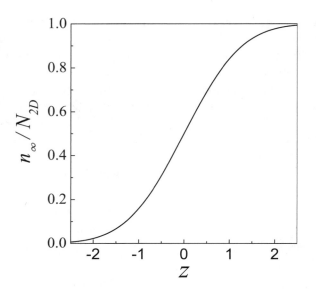

Figure 11.1. Dependence of the photoexcited electron density n_∞ on the parameter $z = (\hbar\omega - \bar{\varepsilon}_g)\tau_p/\hbar$ for the case of interband excitation of electrons in quantum wells by the Gaussian pulse (54.12).

$$\frac{dn_{2t}}{dt} = -G_t + \left(\frac{\partial n_2}{\partial t}\right)_{sc}, \qquad (14)$$

where $n_{1,2t} = (2/L^2)\sum_{\mathbf{p}} f_{1,2\mathbf{p}t}$ are the electron densities in the subbands 1 and 2. The density generation rate under intersubband transitions is introduced as $G_t = (2/L^2)\sum_{\mathbf{p}} G_{1\mathbf{p}t}$, and the terms $(\partial n_1/\partial t)_{sc} = -(\partial n_2/\partial t)_{sc}$ describe the collision-induced relaxation of population in the subbands (note that the total density $n_{1t} + n_{2t} = n_{2D}$ is conserved).

Let us consider first the case of weak pumping, when $n_{2t} \ll n_{1t} \simeq n_{2D}$. The equation describing the density evolution in the excited subband takes the form

$$\frac{dn_{2t}}{dt} = n_{2D}\left(\frac{e|v_{21}^z|}{\hbar\omega}\right)^2 E_t^* \int_{-\infty}^0 d\tau e^{\lambda\tau - i\Delta\omega\tau} E_{t+\tau} + c.c. , \qquad (15)$$

where $\Delta\omega = \omega - \varepsilon_{21}/\hbar$ and $n_{2,t\to-\infty} = 0$. Under the pumping by a single pulse $E_t = E w_t$, where w_t is real, the excited density $n_{2\infty} \equiv n_{2,t\to\infty}$ is given by

$$\frac{n_{2\infty}}{n_{2D}} = 2\left(\frac{eE|v_{21}^z|}{\hbar\omega}\right)^2 \int_{-\infty}^\infty dt w_t \int_{-\infty}^t dt' w_{t'} \cos[\Delta\omega(t - t')]. \qquad (16)$$

If $\Delta\omega = 0$, one has $n_{2\infty}/n_{2D} = \sqrt{2\pi}(eE|v_{21}^z|\tau_p/\hbar\omega)^2$ for the Gaussian pulse (12). With increasing $|\Delta\omega|$, the density $n_{2\infty}$ decreases but remains finite according to the energy-time uncertainty principle. The condition of small redistribution, $n_{2\infty} \ll n_{2D}$, is equivalent to $eE|v_{21}^z|\tau_p \ll \hbar\omega$.

Consider the response to the field $E_t = E(w_t + w_{t-t_d}e^{i\phi})$ describing a pair of identical coherent pulses with relative phase shift ϕ and delay time t_d (we assume that w_t is real). For such a case, Eq. (15) can be rewritten in the following way:

$$\frac{dn_{2t}}{dt} = 2n_{2D}\left(\frac{eE|v_{21}^z|}{\hbar\omega}\right)^2 \left\{ w_t \int_{-\infty}^{t} dt' w_{t'} \cos[\Delta\omega(t-t')] \right.$$

$$+ w_{t-t_d}\int_{-\infty}^{t} dt' w_{t'-t_d}\cos[\Delta\omega(t-t')] + w_t \int_{-\infty}^{t} dt' w_{t'-t_d} \quad (17)$$

$$\left. \times \cos[\Delta\omega(t-t')+\phi] + w_{t-t_d}\int_{-\infty}^{t} dt' w_{t'}\cos[\Delta\omega(t-t')-\phi] \right\}.$$

In the resonance, $\Delta\omega = 0$, the expression in the braces $\{\ldots\}$ is written as $(dT_t^2/dt)/2 + (dT_{t-t_d}^2/dt)/2 + [d(T_t T_{t-t_d})/dt]\cos\phi$, where $T_t = \int_{-\infty}^{t} dt' w_{t'}$. By aiming t to ∞, it is easy to find

$$\frac{n_{2\infty}}{n_{2D}} = 2\left(\frac{eE|v_{21}^z|T_\infty}{\hbar\omega}\right)^2 (1+\cos\phi), \quad (18)$$

where $T_\infty = \int_{-\infty}^{\infty} dt\, w_t$ is equal to $(2\pi)^{1/4}\tau_p$ for the Gaussian pulse (12). The excited density depends on the phase shift ϕ as $n_{2\infty} \propto (1+\cos\phi)$. Therefore, a coherent phase control of the ultrafast response is possible. At finite $\Delta\omega$, the phase of the oscillations of $n_{2\infty}$ is changed by $\Delta\omega t_d$ (problem 11.4), and their amplitude decreases when $|\Delta\omega|$ exceeds $1/\tau_p$.

To investigate the non-linear regime of coherent response, we use, instead of Eq. (14), the balance equation describing redistribution of electrons between the ground and excited states, $\Delta n_t = n_{1t} - n_{2t}$. The density generation rate in this case is written as

$$G_t = -2\left(\frac{eE|v_{21}^z|}{\hbar\omega}\right)^2 w_t \int_{-\infty}^{0} d\tau\, w_{t+\tau}\cos(\Delta\omega\tau)\Delta n_{t+\tau}, \quad (19)$$

and the generation at the instant t depends on the density distribution at the previous moments of time. The collision-induced relaxation at low temperatures takes place mostly due to transitions of electrons from the excited state to the ground state by spontaneous emission of optical phonons; see Sec. 38. Thus, we have $(\partial n_2/\partial t)_{sc} = -(\partial n_1/\partial t)_{sc} = -\nu_{LO}^{(21)}n_{2t}$, where the relaxation rate $\nu_{LO}^{(nn')}$ is given by Eqs. (38.8) and (38.9). As a result, we obtain the following equation for the density redistribution:

$$\frac{d\Delta n_t}{dt} + \omega_R^2 w_t \int_{-\infty}^{t} dt w_{t'} \cos[\Delta\omega(t-t')]\Delta n_{t'} + \nu_{LO}^{(21)}(\Delta n_t - n_{2D}) = 0, \quad (20)$$

where $\omega_R = \sqrt{2}|eEv_{21}^z|/\hbar\omega$ is a characteristic frequency proportional to the electric field strength. This equation should be solved with the initial condition $\Delta n_{t\to-\infty} = n_{2D}$. The non-linear regime of the response is realized at $\omega_R\tau_p \geq 1$ and can be observed in asymmetric quantum wells by the transient radiation caused by the oscillations of the total dipole moment after the excitation. This dipole moment is introduced according to

$$D_t = (n_{1t}d_1 + n_{2t}d_2)/n_{2D} \propto (1 - d_2/d_1)\Delta n_t/n_{2D}, \quad (21)$$

where d_1 and d_2 are the dipole moments of the subbands 1 and 2, respectively.

In the case of resonant excitation, $\Delta\omega = 0$, and in the absence of the relaxation (the latter approximation corresponds to the time scale $t \ll 1/\nu_{LO}^{(21)}$), Eq. (20) is reduced to a differential equation of the second order,

$$\frac{d^2\Delta n_t}{dt^2} - \frac{1}{w_t}\frac{dw_t}{dt}\frac{d\Delta n_t}{dt} + \omega_R^2 w_t^2 \Delta n_t = 0, \quad (22)$$

with an additional initial condition, $[w_t^{-1}(d\Delta n_t/dt)]_{t\to-\infty} = 0$ (problem 11.5). The general solution of this equation is written in the form

$$\Delta n_t = n_{2D} \cos\left(\omega_R \int_{-\infty}^{t} dt' w_{t'}\right). \quad (23)$$

The redistribution at $t \to \infty$ is determined by the integral of the envelope form-factor of electric field. According to Eq. (23), the final phase is written as $A_p = \omega_R \int_{-\infty}^{\infty} dt w_t = \omega_R T_\infty$. The effect of periodic modulation of $\Delta n_{t\to\infty}$ with A_p is caused by the Rabi oscillations, because during the excitation the electrons oscillate between the subbands with a characteristic frequency ω_R (problem 11.6). The complete transfer of electrons into the upper (excited) subband takes place at $A_p = (2k+1)\pi$, where k is an integer number. If $A_p = 2k\pi$, the excited subband remains empty. The case of equal occupation, $\Delta n_t = 0$, occurs at $A_p = (2k+1/2)\pi$. The temporal evolution of the density distribution is presented in Fig. 11.2 for $\Delta\omega = 0$. If the frequency ω is shifted away from the resonance ($\Delta\omega \neq 0$), the amplitude of the modulation becomes smaller because of the oscillating kernel in the integral term of Eq. (20). In such a case, the integro-differential equation (20) should be solved numerically.

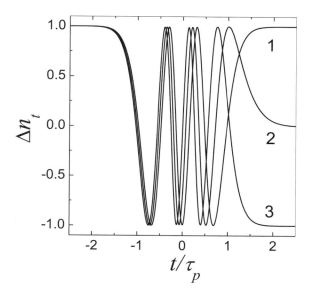

Figure 11.2. Temporal evolution of the density distribution in the two-level system under a resonant excitation by the Gaussian pulse (54.12) for $A_p = 6\pi$ (1), $A_p = 6.5\pi$ (2), and $A_p = 7\pi$ (3).

55. Partially Inverted Electron Distribution

In this section we describe the relaxation of the electrons excited by a short laser pulse. As follows from Eq. (54.9), in the collisionless approximation the energy distribution of such electrons is a narrow peak whose position is determined by the frequency of laser radiation. The width of this peak is determined by the pulse duration. As the pulse duration increases, the peak of the energy distribution becomes narrower, and one has to take into account the collision-induced broadening which determines the width of the peak in the limit of a long pulse. Because of the energy-time uncertainty principle, the elastic scattering also contributes to the broadening of the peak. Therefore, at the initial moments of time, when the energy relaxation is not yet important, there exists a partially inverted electron distribution f_ε with decreasing high-energy part and increasing low-energy part. According to Eqs. (8.26), (8.27), and (8.30), the linear response of the electrons with partially inverted distribution becomes essentially different in comparison to the quasi-equilibrium case: since the sign of $(-df_\varepsilon/d\varepsilon)$ can be negative, it is possible to obtain a negative transient conductivity and negative absorption of electromagnetic radiation. Describing temporal evolution of this response, it is necessary to consider a quasielastic relaxation of the initial excited distribution.

We consider first the photoexcitation of electrons in the presence of elastic scattering by including the interaction of electrons with static inhomogeneities into the Hamiltonian $\hat{h}$ in the expression (54.1) for the interband photogeneration rate. Using the basis of exact eigenstates, $\hat{h}|\delta\rangle = \varepsilon_\delta|\delta\rangle$, we write the matrix elements of the c-band photogeneration rate in the form

$$\langle\delta|\widehat{G}_t|\delta'\rangle = \left(\frac{e}{\hbar\omega}\right)^2 \int_{-\infty}^0 d\tau e^{\lambda\tau - i\omega\tau} \sum_{\delta_v} e^{i(\varepsilon_\delta - \varepsilon_{\delta_v})\tau/\hbar}$$

$$\times \langle\delta|\mathbf{E}_{t+\tau}\cdot\hat{\mathbf{v}}|\delta_v\rangle \langle\delta_v|\mathbf{E}_t^*\cdot\hat{\mathbf{v}}|\delta'\rangle + (c.c., \delta \leftrightarrow \delta'), \qquad (1)$$

where $|\delta_v\rangle$ is the set of eigenstates of the completely filled valence band. In this way, the scattering of electrons is taken into account exactly so that the density matrix of c-band electrons is governed by Eq. (53.7) without the collision integral. Such an equation should be solved with the initial condition $\hat{\rho}_{t\to-\infty} = 0$, and the result is written below:

$$\rho_{\delta\delta'}(t) = \int_{-\infty}^t dt' e^{i(\varepsilon_\delta - \varepsilon_{\delta'})(t'-t)/\hbar}\langle\delta|\widehat{G}_{t'}|\delta'\rangle. \qquad (2)$$

One can use the wave functions $\psi_{\mathbf{p}}^{(\delta)} \equiv \langle\mathbf{p}|\delta\rangle$ to write the density matrix in the momentum representation:

$$\rho_{\mathbf{p}\mathbf{p}'}(t) = \sum_{\delta\delta'} \psi_{\mathbf{p}}^{(\delta)} \rho_{\delta\delta'}(t) \psi_{\mathbf{p}'}^{(\delta')*}. \qquad (3)$$

Below we consider the photoexcitation of electrons in the quantum wells with non-ideal interfaces, when the 2D electrons of c-band scatter because of the fluctuations of quantization energy caused by the variations of the well width. These variations have much less effect on the valence-band 2D electrons in view of larger effective mass of these electrons. Therefore, the 2D electrons of v-band can be treated as almost ideal and characterized by their in-plane momentum. We assume that the electrons are photoexcited from the highest 2D subband of the valence band, where the eigenstates are the plane waves $|\mathbf{p}_1\rangle$ with the dispersion law $\varepsilon_{\delta_v} = -\overline{\varepsilon}_g - p_1^2/2m_v$, to the lowest 2D subband of the conduction band. After calculating the sums over spin numbers, by analogy to Eqs. (54.2) and (54.3), we rewrite the photogeneration matrix (1) in the form

$$\langle\delta|\widehat{G}_t|\delta'\rangle = \left(\frac{eE\tilde{s}}{\hbar\omega}\right)^2 w_t \int_{-\infty}^0 d\tau e^{\lambda\tau - i\omega\tau} w_{t+\tau}$$

$$\times \sum_{\mathbf{p}_1} e^{i(\varepsilon_\delta + \overline{\varepsilon}_g + p_1^2/2m_v)\tau/\hbar} \psi_{\mathbf{p}_1}^{(\delta)*} \psi_{\mathbf{p}_1}^{(\delta')} + (c.c., \delta \leftrightarrow \delta'). \qquad (4)$$

We remind that $\tilde{s}$ is the interband velocity s renormalized due to confinement of c- and v-band states. After substituting this photogeneration matrix into Eq. (2), we use Eq. (3) and obtain the density matrix in the momentum representation,

$$\rho_{\mathbf{pp}'}(t) = \left(\frac{eE\tilde{s}}{\hbar\omega}\right)^2 \int_{-\infty}^{t} dt' w_{t'} \int_{-\infty}^{0} d\tau w_{t'+\tau} \sum_{\delta\delta'\mathbf{p}_1} e^{i(\varepsilon_\delta - \varepsilon_{\delta'})(t'-t)/\hbar}$$

$$\times e^{i(\varepsilon_\delta - \hbar\Delta\omega + p_1^2/2m_v)\tau/\hbar} \psi_{\mathbf{p}}^{(\delta)} \psi_{\mathbf{p}_1}^{(\delta)*} \psi_{\mathbf{p}_1}^{(\delta')} \psi_{\mathbf{p}'}^{(\delta')*} + (c.c., \mathbf{p} \leftrightarrow \mathbf{p}') , \qquad (5)$$

where the elastic scattering of electrons is taken into account exactly. In Eq. (5) and below, $\Delta\omega = \omega - \bar{\varepsilon}_g/\hbar$.

The averaged, over the random potential of the inhomogeneities, density matrix is diagonal in $\mathbf{p}$ because of macroscopic homogeneity of the system, $\langle\langle\rho_{\mathbf{pp}'}(t)\rangle\rangle = \delta_{\mathbf{pp}'} f_{\mathbf{p}t}$. When t becomes great in comparison to both the pulse duration and the scattering time, a quasi-stationary distribution $f_{\mathbf{p}}^{(0)} = f_{\mathbf{p},t\to\infty}$ is formed. It is convenient to write the result through the retarded and advanced Green's functions defined in the operator representation by Eq. (16.3). Taking into account that both $t'-t$ and τ are negative, we obtain

$$\langle\langle\rho_{\mathbf{pp}'}(t)\rangle\rangle = \left(\frac{eE\tilde{s}}{\omega}\right)^2 \int_{-\infty}^{t} dt' w_{t'} \int_{-\infty}^{0} d\tau w_{t'+\tau} \sum_{\mathbf{p}_1} e^{i(p_1^2/2m_v - \hbar\Delta\omega)\tau/\hbar}$$

$$\times \langle\langle G_{t-t'-\tau}^{R}(\mathbf{p}, \mathbf{p}_1) G_{t'-t}^{A}(\mathbf{p}_1, \mathbf{p}')\rangle\rangle + (c.c., \mathbf{p} \leftrightarrow \mathbf{p}'). \qquad (6)$$

According to the results of Sec. 15, the averaging of such a pair of Green's functions gives us $\delta_{\mathbf{pp}'}$. It is convenient to use the energy representation of the Green's functions, when

$$\langle\langle G_{t-t'-\tau}^{R}(\mathbf{p}, \mathbf{p}_1) G_{t'-t}^{A}(\mathbf{p}_1, \mathbf{p}')\rangle\rangle = \delta_{\mathbf{pp}'} \int \frac{dE}{2\pi\hbar} \int \frac{d\varepsilon}{2\pi\hbar} \qquad (7)$$

$$\times e^{i(E+\varepsilon/2)\tau/\hbar} e^{i\varepsilon(t'-t)/\hbar} \langle\langle G_{E+\varepsilon/2}^{R}(\mathbf{p}, \mathbf{p}_1) G_{E-\varepsilon/2}^{A}(\mathbf{p}_1, \mathbf{p})\rangle\rangle .$$

Below we search for the correlation function $\langle\langle G_{E+\varepsilon/2}^{R}(\mathbf{p}, \mathbf{p}_1) G_{E-\varepsilon/2}^{A}(\mathbf{p}_1, \mathbf{p})\rangle\rangle$ by using the Bethe-Salpeter equation (15.8) in the ladder approximation, $\Gamma^{RA} \simeq w(q)$. This approximation is justified for the photoexcitation far above the fundamental threshold, where the kinetic energy of electrons, $\varepsilon_{\mathbf{p}}$, is much greater than the broadening energy $\hbar/\tau_s$ associated with the scattering under consideration. In the case of short-range correlated inhomogeneities, when $w(q) \simeq w$, the Bethe-Salpeter equation has the following solution:

$$\left\langle\left\langle G^R_{E+\varepsilon/2}(\mathbf{p},\mathbf{p}_1)G^A_{E-\varepsilon/2}(\mathbf{p}_1,\mathbf{p})\right\rangle\right\rangle = G^R_{E+\varepsilon/2}(\mathbf{p})G^A_{E-\varepsilon/2}(\mathbf{p})$$

$$\times\left[\delta_{\mathbf{p}\mathbf{p}_1} + \frac{w}{1-S_\varepsilon}\frac{1}{V}G^R_{E+\varepsilon/2}(\mathbf{p}_1)G^A_{E-\varepsilon/2}(\mathbf{p}_1)\right], \qquad (8)$$

where

$$S_\varepsilon = \frac{w}{V}\sum_{\mathbf{p}} G^R_{E+\varepsilon/2}(\mathbf{p})G^A_{E-\varepsilon/2}(\mathbf{p}) \simeq \frac{i\hbar/\tau_s}{\varepsilon + i\hbar/\tau_s}. \qquad (9)$$

The last equation is based upon the expression (14.29) for the averaged Green's function, where we have neglected the irrelevant real part of the self-energy function and put $\tau(E) = \tau_s$. We also remind the relation $\tau_s^{-1} = mw/\hbar^3$ for the 2D case. The presence of the second (correlation) term on the right-hand side of Eq. (8) is essential. It is the term that contains the pole at $\varepsilon = 0$ leading to a stationary distribution of electrons at $t \gg \tau_p, \tau_s$ (see the factor $e^{i\varepsilon(t'-t)/\hbar}$ in Eq. (7)). We have

$$\int_{-\infty}^{\infty}\frac{dE}{2\pi}e^{i(E+\varepsilon/2)\tau/\hbar}\sum_{\mathbf{p}_1}e^{ip_1^2\tau/2m_v\hbar}\left\langle\left\langle G^R_{E+\varepsilon/2}(\mathbf{p},\mathbf{p}_1)G^A_{E-\varepsilon/2}(\mathbf{p}_1,\mathbf{p})\right\rangle\right\rangle$$

$$= \frac{i}{\varepsilon}\exp\left[\frac{i}{\hbar}\varepsilon_p\left(1+\frac{m_c}{m_v}\right)\tau + \frac{\tau}{2\tau_s}\right] \qquad (10)$$

$$\times\left\{1 + \frac{i\hbar/\tau_s}{\varepsilon + i\hbar/\tau_s}\left(\exp\left[\left(-\frac{i}{\hbar}\varepsilon\tau + \frac{\tau}{\tau_s}\right)\frac{m_c}{m_v}\right] - 1\right)\right\},$$

where $\varepsilon_p = p^2/2m_c$. According to Eqs. (6) and (7), in order to find the distribution function $f_{\mathbf{p}t}$, one has to multiply the right-hand side of Eq. (10) by $e^{i\varepsilon(t'-t)/\hbar}$ and integrate the result over ε. Since $t' < t$, this integral is calculated by shifting the contour of integration into the lower half-plane of complex ε. Therefore, the contour should pass above the pole $\varepsilon = 0$ to provide a finite distribution function at $t \to \infty$. As a result,

$$f_{\mathbf{p}t} = 2\left(\frac{eE\tilde{s}}{\hbar\omega}\right)^2\int_{-\infty}^{t}dt'w_{t'}\int_{-\infty}^{0}d\tau w_{t'+\tau}$$

$$\times e^{\tau/2\tau_s}\cos\left[\frac{\varepsilon_p\tau}{\hbar}\left(1+\frac{m_c}{m_v}\right) - \Delta\omega\tau\right] \qquad (11)$$

$$\times\left[\exp\left(\frac{\tau m_c}{\tau_s m_v}\right)\theta\left(t-t'+\frac{m_c}{m_v}\tau\right) + \exp\left(\frac{t'-t}{\tau_s}\right)\theta\left(-t+t'-\frac{m_c}{m_v}\tau\right)\right].$$

Only the first term in this expression is important at $t \gg \tau_p, \tau_s$, and the quasi-stationary isotropic distribution $f_{\mathbf{p}}^{(0)} = f_{\varepsilon_p}^{(0)}$ is given by

$$f_\varepsilon^{(0)} = 2\left(\frac{eE\tilde{s}}{\hbar\omega}\right)^2\int_{-\infty}^{\infty}dt'w_{t'}\int_{-\infty}^{0}d\tau w_{t'+\tau}$$

$$\times e^{\tau/\tau^*} \cos\left[\frac{\varepsilon\tau}{\hbar}\left(1 + \frac{m_c}{m_v}\right) - \Delta\omega\tau\right] \tag{12}$$

$$= \frac{n_{ex}}{\pi\hbar\widetilde{\rho}_{2D}}\int_{-\infty}^{0} d\tau\, e^{\tau/\tau^*} e^{-\tau^2/2\tau_p^2}\cos\left[\frac{\varepsilon\tau}{\hbar}\left(1 + \frac{m_c}{m_v}\right) - \Delta\omega\tau\right],$$

where $\tau^* = \tau_s/(1/2 + m_c/m_v)$. The final expression in Eq. (12) is written for the case of excitation by the Gaussian pulse (54.12) and normalized by the total photogenerated density $n_{ex} = n_{t\to\infty}$.

We remind that we consider the photoexcitation far above the threshold, which means that the energy width of the distribution (12) is much smaller than the photoexcitation energy $\varepsilon_{ex} = \hbar\Delta\omega/(1 + m_c/m_v)$. In these conditions, the total photogenerated density n_{ex} is independent of $\Delta\omega$ and equal to N_{2D} introduced in Sec. 54. The shape of the distribution (12) is determined by a competition of the collision-induced broadening, described by the time τ^*, and dynamical broadening existing due to a finite pulse duration τ_p. If the dynamical broadening is more essential ($\tau^* \gg \tau_p$), one can neglect the term e^{τ/τ^*} in the integrand and obtain $f_\varepsilon^{(0)} = (n_{ex}/\sqrt{\pi}\rho_{2D}\Delta\varepsilon)\exp[-(\varepsilon - \varepsilon_{ex})^2/(\Delta\varepsilon)^2]$, where $\Delta\varepsilon = \sqrt{2}\hbar/\tau_p$. In the opposite case, $\tau^* \ll \tau_p$, the shape of the peak is determined by the collision-induced broadening: $f_\varepsilon^{(0)} = (n_{ex}/\pi\rho_{2D})\Delta\varepsilon/[(\varepsilon - \varepsilon_{ex})^2 + (\Delta\varepsilon)^2]$, where $\Delta\varepsilon = \hbar/\tau^*$. In the general case, the distribution (12) is more complicated, though it can be expressed analytically through the error function [erf(...)] of complex argument, and the broadening energy is roughly estimated as $\Delta\varepsilon \sim \hbar/\tau^* + \hbar/\tau_p$.

The relaxation of the peak-shaped distribution (12) occurs due to quasielastic scattering of electrons by acoustic phonons. This is the case when the peak is formed in the passive region (in other words, the photoexcitation energy in the c-band, ε_{ex}, is smaller than $\hbar\omega_{LO}$). If $\varepsilon_{ex} > \hbar\omega_{LO}$, the photoexcitation is followed by fast emission of a cascade of optical phonons until the peak is transferred into the passive region. The broadening associated with this process is weak because the dispersion of the optical phonons is negligible. Therefore, for both cases of photoexcitation, the evolution of electron distribution can be represented as follows. During a short time interval (about tens of picoseconds), a narrow peak is formed in the passive region. Then, this peak relaxes on a nanosecond scale of time due to quasielastic scattering of electrons by acoustic phonons. A special consideration, however, is necessary in the case when ε_{ex} is close to a multiple of $\hbar\omega_{LO}$ so that the high-energy electrons emit one more optical phonon during the initial step of the relaxation. As a result, the electron distribution in the passive region looks like two half-peaks localized near zero energy and near $\hbar\omega_{LO}$.

The temporal evolution in the passive region is governed by the equation describing drift and diffusion along the energy axis:

$$\frac{\partial f_{\varepsilon t}}{\partial t} = \frac{\partial}{\partial \varepsilon}\left(D_\varepsilon \frac{\partial f_{\varepsilon t}}{\partial \varepsilon} + V_\varepsilon f_{\varepsilon t}\right). \tag{13}$$

It is considered with the initial condition $f_{\varepsilon, t=0} = f_\varepsilon^{(0)}$. The diffusion coefficient D_ε and drift velocity V_ε can be assumed energy-independent in the case of 2D electrons. If the characteristic energies of acoustic phonons are small in comparison to the temperature T, the right-hand side of Eq. (13) is given by Eq. (35.6) and $D = VT$. In the general case, D and V are connected by a more complicated relation (problem 11.7). The boundary conditions to Eq. (13) are

$$f_{\varepsilon t}|_{\varepsilon \to \infty} = 0, \quad \left(D_\varepsilon \frac{\partial f_{\varepsilon t}}{\partial \varepsilon} + V_\varepsilon f_{\varepsilon t}\right)_{\varepsilon=0} = 0. \tag{14}$$

The condition of zero energy flow at $\varepsilon = 0$ implies that the recombination of photoexcited electrons (see Sec. 38) is neglected. The condition at $\varepsilon \to \infty$ is valid if we neglect penetration of electrons into the active region, assuming that the energy broadening of the distribution function during the whole period of evolution remains small in comparison to $\hbar\omega_{LO}$.

If the coefficients D and V are energy-independent, Eq. (13) with the boundary conditions (14) is solved as

$$f_{\varepsilon t} = \int_0^\infty d\varepsilon' \, G_t(\varepsilon, \varepsilon') f_{\varepsilon'}^{(0)}, \tag{15}$$

where $G_t(\varepsilon, \varepsilon')$ is the Green's function of the drift-diffusion equation. This function is defined in the quadrant $\varepsilon > 0$, $\varepsilon' > 0$ and satisfies the boundary conditions (14). We obtain the expression

$$G_t(\varepsilon, \varepsilon') = \frac{1}{\sqrt{4\pi Dt}}\left[e^{-(\varepsilon-\varepsilon'+Vt)^2/4Dt} + e^{-V\varepsilon/D}e^{-(\varepsilon+\varepsilon'-Vt)^2/4Dt}\right]$$

$$+\frac{V}{2D}e^{-V\varepsilon/D}\left[1 - \text{erf}\left(\frac{\varepsilon+\varepsilon'-Vt}{\sqrt{4Dt}}\right)\right], \tag{16}$$

which can be derived by using the Green's function of the ordinary diffusion equation (problem 11.8). During the initial steps of the relaxation, when the boundary condition at zero energy is still not important, $G_t(\varepsilon, \varepsilon')$ is reduced to the fundamental solution of the one-dimensional drift-diffusion equation: $G_t(\varepsilon, \varepsilon') \simeq (4\pi Dt)^{-1/2}e^{-(\varepsilon-\varepsilon'+Vt)^2/4Dt}$. The solution (15) in this case describes the photoexcited distribution peak which drifts towards low energies and, simultaneously, becomes broadened due to diffusion. If the initial distribution is a Gaussian function, $f_\varepsilon^{(0)} = (n_{ex}/\sqrt{\pi}\rho_{2D}\Delta\varepsilon)\exp[-(\varepsilon - \varepsilon_{ex})^2/(\Delta\varepsilon)^2]$, which is the case when $\tau^* \gg \tau_p$ in Eq. (12), the initial evolution of such distribution is given by the analytical expression

$$f_{\varepsilon t} = \frac{n_{ex}}{\sqrt{\pi}\rho_{2D}\Delta\varepsilon_t} \exp\left[-\frac{(\varepsilon - \varepsilon_{ex} + Vt)^2}{(\Delta\varepsilon_t)^2}\right]. \qquad (17)$$

The maximum of this distribution moves down with the velocity V, while the broadening is described by the function $\Delta\varepsilon_t = \sqrt{(\Delta\varepsilon)^2 + 4Dt}$.

At large t, only the second term on the right-hand side of Eq. (16) becomes important, and $1 - \text{erf}(\ldots) \simeq 2$. Therefore, the final steady-state distribution is given by

$$f_{\varepsilon, t\to\infty} = \frac{n_{ex}}{\rho_{2D}}\frac{V}{D}\exp\left(-\frac{V}{D}\varepsilon\right). \qquad (18)$$

This function is reduced to the equilibrium Boltzmann distribution at $D/V = T$. At low temperatures, when $D/V \sim \hbar s_l/d > T$ (here s_l is the longitudinal sound velocity and d is the well width; see problem 11.7), the distribution (18) does not coincide with the equilibrium one. We stress, however, that the condition of quasielasticity implies that the energy scale of the distribution function is large in comparison to the energy of acoustic phonons, while the distribution (18) does not satisfy this condition at $D/V \sim \hbar s_l/d$. Therefore, the drift-diffusion equation (13) is not valid for describing the final part of the relaxation if the temperature T is so low that the equipartition condition for the phonons is violated. A more careful consideration is necessary in this case. The general expressions (15) and (16) allow one to follow the transformation of the high-energy peak (17) to the monotonic low-energy distribution (18). Carrying out a numerical integration in Eq. (15), we find the distribution functions shown in Fig. 11.3.

Consider now the evolution of two peaks at the boundaries of the passive region, which is the case when the excitation energy ε_{ex} is close to a multiple of $\hbar\omega_{LO}$. The low-energy electrons remain localized near the bottom of the passive region, and their mean energy changes from the value determined by the half-width of the excited peak to the temperature T. The high-energy part is modified due to drift and diffusion towards lower energies as well as due to the diffusion-induced penetration of electrons into the active region and subsequent fast relaxation of these electrons by emission of optical phonons. The diffusion maintains a finite electron density at the boundary $\varepsilon = \hbar\omega_{LO}$ during the initial steps of the evolution. To study the region of energies in the vicinity of

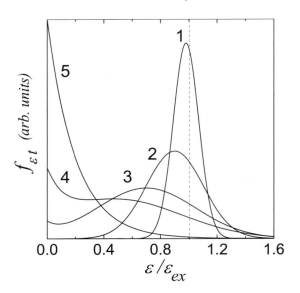

Figure 11.3. Temporal evolution of the energy distribution function $f_{\varepsilon t}$ in the case of infinitely narrow initial distribution $f_{\varepsilon}^{(0)} \propto \delta(\varepsilon - \varepsilon_{ex})$ at $D/V = \varepsilon_{ex}/5$. The curves 1-5 correspond to $t/t_V = 0.02, 0.1, 0.3, 0.5,$ and 5, where $t_V = \varepsilon_{ex}/V$ is the drift time.

$\hbar\omega_{LO}$, we introduce a new variable, $\xi = \varepsilon - \hbar\omega_{LO}$, and the distribution functions of electrons in the passive and active regions, $f_{\xi t}^<$ and $f_{\xi t}^>$, respectively. The spontaneous emission of optical phonons is described by the relaxation rate $\nu = \nu_{LO}$ introduced in Sec. 35. Thus, at $|\xi| \ll \hbar\omega_{LO}$ we have the following system of equations:

$$\frac{\partial f_{\xi t}^<}{\partial t} = D\frac{\partial^2 f_{\xi t}^<}{\partial \xi^2} + V\frac{\partial f_{\xi t}^<}{\partial \xi}, \qquad \xi < 0$$

$$\frac{\partial f_{\xi t}^>}{\partial t} = D\frac{\partial^2 f_{\xi t}^>}{\partial \xi^2} + V\frac{\partial f_{\xi t}^>}{\partial \xi} - \nu f_{\xi t}^>, \quad \xi > 0 . \qquad (19)$$

It should be solved with the boundary conditions expressing the continuity of the distribution functions and their derivatives at $\xi = 0$, $f_{\xi=0,t}^> = f_{\xi=0,t}^<$ and $(\partial f_{\xi t}^</\partial \xi)_{\xi=0} = (\partial f_{\xi t}^>/\partial \xi)_{\xi=0}$, as well as with the boundary conditions $f_{\xi \to \infty,t}^> = 0$ and $f_{\xi \to -\infty,t}^< = 0$. The initial condition is $f_{\xi,t=0} = f_{\xi+\hbar\omega_{LO}}^{(0)}$.

By fixing the value of the distribution function at the boundary, $f_{\xi=0,t}^> = f_{\xi=0,t}^< \equiv f_{\xi=0,t}$, one can solve Eq. (19) as

$$f_{\xi t}^< = \frac{e^{-V^2 t/4D}}{\sqrt{4\pi D t}} \int_{-\infty}^{0} d\xi' e^{-V(\xi-\xi')/2D} \left[e^{-\frac{(\xi-\xi')^2}{4Dt}} - e^{-\frac{(\xi+\xi')^2}{4Dt}} \right] f_{\xi',t=0}$$

$$-\frac{\xi e^{-V\xi/2D}}{\sqrt{4\pi D}}\int_0^t dt' e^{-V^2(t-t')/4D}\frac{e^{-\xi^2/4D(t-t')}}{(t-t')^{3/2}}f_{\xi=0,t'}\,,\qquad(20)$$

and

$$f_{\xi t}^> = \frac{e^{-V^2 t/4D-\nu t}}{\sqrt{4\pi Dt}}\int_0^\infty d\xi' e^{-V(\xi-\xi')/2D}\left[e^{-\frac{(\xi-\xi')^2}{4Dt}}-e^{-\frac{(\xi+\xi')^2}{4Dt}}\right]f_{\xi',t=0}$$

$$+\frac{\xi e^{-V\xi/2D}}{\sqrt{4\pi D}}\int_0^t dt' e^{-V^2(t-t')/4D-\nu(t-t')}\frac{e^{-\xi^2/4D(t-t')}}{(t-t')^{3/2}}f_{\xi=0,t'}\,.\qquad(21)$$

The function $f_{\xi=0,t}$ is yet to be found. To do this, we calculate the derivatives $\partial f_{\xi t}^</\partial\xi$ and $\partial f_{\xi t}^>/\partial\xi$ of the functions (20) and (21). By equating these derivatives to each other at $\xi=0$, we obtain

$$\int_0^t dt' K(t-t')\widetilde{f}_{\xi=0,t'}=R(t),\quad \widetilde{f}_{\xi=0,t'}=e^{V^2 t'/4D}f_{\xi=0,t'}$$

$$K(\tau)=(e^{-\nu\tau}+1)\left(\frac{\partial}{\partial\tau}\tau^{-1/2}e^{-\xi^2/4D\tau}\right)_{\xi\to 0},\qquad(22)$$

$$R(t)=\frac{1}{2Dt^{3/2}}\int_{-\infty}^\infty d\xi\, f_{\xi,t=0}\xi e^{V\xi/2D}e^{-\xi^2/4Dt}\left[e^{-\nu t}\theta(\xi)-\theta(-\xi)\right].$$

This integral equation is reduced to an algebraic one by means of the Laplace transformation of both its sides. The result is expressed through the inverse Laplace transformation of the ratio R_s/K_s, where R_s and K_s are the Laplace transforms (images) of the functions $R(t)$ and $K(t)$:

$$f_{\xi=0,t}=\frac{e^{-V^2 t/4D}}{2\pi i\sqrt{D}}\int_C ds\frac{e^{st}}{\sqrt{s}+\sqrt{s+\nu}}\int_{-\infty}^\infty d\xi\, f_{\xi,t=0}e^{V\xi/2D}$$

$$\times\left[\exp\left(-\sqrt{\frac{s+\nu}{D}}\xi\right)\theta(\xi)+\exp\left(\sqrt{\frac{s}{D}}\xi\right)\theta(-\xi)\right].\qquad(23)$$

To obtain this expression, we have taken into account that the Laplace image of $e^{-\nu t}(\xi/2\sqrt{\pi D}t^{3/2})e^{-\xi^2/4Dt}$ is equal to $\exp(-\sqrt{(s+\nu)/D}\xi)$ and that $K_s=\sqrt{\pi}(\sqrt{s+\nu}+\sqrt{s})$. The contour of integration, C, goes from $-i\infty$ to $+i\infty$ and passes the peculiar points of the integrand, $s=0$ and $s=-\nu$, from the right. The solution (23) can be obtained by means of direct Laplace transformation of Eq. (19) with the use of the above boundary conditions (problem 11.9).

The integral over s in Eq. (23) is calculated analytically in the limiting case $t\gg\nu^{-1}$, and we obtain

$$f_{\xi=0,t} = \frac{1}{\sqrt{\pi D t}} \int_{-\infty}^{0} d\xi \frac{f_{\xi,t=0}}{1 + \sqrt{1 + 4D\nu t^2/\xi^2}} \exp\left[-\frac{(\xi - Vt)^2}{4Dt}\right]$$

$$+ \frac{e^{-\nu_{qe}t}}{\sqrt{4\pi D}\nu t^{3/2}} \int_{0}^{\infty} d\xi \, f_{\xi,t=0} \exp\left(-\sqrt{\frac{\nu}{D}}\xi\right), \tag{24}$$

where $\nu_{qe} = V^2/4D$ is the quasielastic relaxation rate. It is related to the energy relaxation rate $\nu^{(e)}$ of quasi-equilibrium non-degenerate 2D electrons interacting with acoustic phonons under equipartition condition as $\nu_{qe} = \nu^{(e)}/4$ (note that $\nu^{(e)}$ can be found by equating the power loss term $-\rho_{2D} \int_0^\infty d\varepsilon \, \varepsilon J_{ac}(f|\varepsilon)$, where $J_{ac}(f|\varepsilon)$ is given by Eq. (35.6), to $\nu^{(e)} n_{2D}(T_e - T)$, where T_e is the effective temperature of the quasi-equilibrium electrons). This quasielastic relaxation rate is much smaller than ν. Below we restrict ourselves by the case when, during the time ν^{-1}, the diffusion-induced broadening of the peak, $2\sqrt{D/\nu}$, is much larger than the drift-induced shift of the peak, V/ν, and much smaller than the width $\Delta\varepsilon$ of the initial excited distribution. These conditions are reasonable because ν is large. In this case, there exists a time interval

$$\nu^{-1} \ll t \ll \tau_D, \nu_{qe}^{-1}, \tag{25}$$

where $\tau_D = (\Delta\varepsilon)^2/4D$ is the time of diffusive broadening of the initial distribution. The condition $t \ll \tau_D$ follows from $t \ll \nu_{qe}^{-1}$ if $\Delta\varepsilon$ is not small in comparison to $4D/V$. Under the conditions (25), the characteristic $|\xi|$ contributing to the first integral of Eq. (24) is estimated as $\sqrt{4Dt}$ so that $4D\nu t^2/\xi^2 \sim \nu t \gg 1$ and the integral can be taken analytically. Moreover, the conditions (25) allow one to expand the exponent in powers of V and replace $f_{\xi,t=0}$ by its value at $\xi = 0$. The contribution to the second integral in Eq. (24) comes from $\xi \sim \sqrt{D/\nu} \ll \Delta\varepsilon$, and one can replace $f_{\xi,t=0}$ by $f_{\xi=0,t=0}$ in this term as well. If $t \gg \nu^{-1}$, the first integral in Eq. (24) gives the main contribution to $f_{\xi=0,t}$, and we obtain a simple result

$$f_{\xi=0,t} \simeq f_{\xi=0,t=0}\left(\frac{1}{\sqrt{\pi\nu t}} - \frac{V}{2\sqrt{\nu D}}\right) \tag{26}$$

valid under the conditions (25). We remind that $f_{\xi=0,t=0} = f_{\hbar\omega_{LO}}^{(0)}$ is the initial (determined by the photoexcitation) distribution function of electrons at the boundary of the active region. Though the time scale for the decrease of $f_{\xi=0,t}$ is given by $\nu = \nu_{LO}$, this decrease is not exponential because of arrival of electrons from the passive region due to diffusion. Therefore, the high-energy electrons, for a long enough period of time, have a distribution which essentially differs from a simple peak moving in the passive region; see Eq. (17).

Below we discuss some features of the transient linear response of the electrons with the partially inverted distribution described by Eqs. (20), (21), and (26). This distribution depends on time parametrically. Using Eq. (8.27), we write the time-dependent conductivity of 2D electrons as

$$\sigma_t = \frac{e^2 \rho_{2D}}{m} \int_0^\infty d\varepsilon \varepsilon \tau_\varepsilon^{tot} \left(-\frac{\partial f_{\varepsilon t}}{\partial \varepsilon} \right), \quad \frac{1}{\tau_\varepsilon^{tot}} = \frac{1}{\tau_\varepsilon} + \frac{1}{\tau_\varepsilon^{opt}}. \tag{27}$$

We take into account both the elastic scattering and the scattering by optical phonons described by the momentum relaxation times τ_ε and τ_ε^{opt}, respectively. The total relaxation time is a sum of the partial relaxation rates, because the right-hand side of the kinetic equation contains a sum of the collision integrals for different scattering mechanisms; see Eq. (34.28). If the temperature is small in comparison to $\hbar \omega_{LO}$, a strong relaxation of momentum by electron-phonon scattering is essential only due to spontaneous LO phonon emission in the active region. Since in the passive region $\tau_\varepsilon^{tot} = \tau_\varepsilon \gg \tau_\varepsilon^{opt}$, the main contribution to the conductivity comes from this region, and the upper limit of integration in Eq. (27) can be set at $\hbar \omega_{LO}$. Integrating by parts, we obtain

$$\sigma_t = \frac{e^2 \rho_{2D}}{m} \left[\int_0^{\hbar \omega_{LO}} d\varepsilon f_{\varepsilon t} \frac{d}{d\varepsilon} (\varepsilon \tau_\varepsilon) - (f_{\varepsilon t} \varepsilon \tau_\varepsilon)_{\varepsilon = \hbar \omega_{LO}} \right]. \tag{28}$$

The transient conductivity σ_t contains a positive contribution from the passive region (usually $d(\varepsilon \tau_\varepsilon)/d\varepsilon > 0$) and a negative contribution from the transition region between the passive and active regions. If the electrons are photoexcited into the passive region, σ_t remains positive. However, if the excitation occurs near the boundary of the active region, the negative contribution becomes essential.

A simple expression for σ_t is obtained when the relaxation time is energy-independent, $\tau_\varepsilon = \tau$. This situation takes place for the 2D electrons interacting either with short-range correlated inhomogeneities or with acoustic phonons under the equipartition condition. The first term in Eq. (28) then gives us the electron density $n_{ex} = \rho_{2D} \int_0^{\hbar \omega_{LO}} d\varepsilon f_{\varepsilon t}$, while the second term can be transformed according to Eq. (26). Neglecting the drift-induced term in Eq. (26), we have

$$\sigma_t = \sigma \left(1 - \frac{\rho_{2D} \hbar \omega_{LO}}{n_{ex}} f_{\xi=0,t} \right) \simeq \sigma \left(1 - \sqrt{\frac{\tau_u}{t}} \right), \tag{29}$$

where $\sigma = e^2 n_{ex} \tau / m$. In the case of a Gaussian initial distribution formed at $\varepsilon_{ex} = \hbar \omega_{LO}$ (so that $f_{\xi=0,t=0} = n_{ex}/\sqrt{\pi} \rho_{2D} \Delta \varepsilon$), the characteristic time introduced in Eq. (29) is given by $\tau_u = \pi^{-2}(\hbar \omega_{LO}/\Delta \varepsilon)^2 \nu^{-1}$. Since this time is assumed to be much greater than ν^{-1}, the transient conductivity σ_t can be negative in the interval $\nu^{-1} \ll t < \tau_u$. Under these conditions, there should arise an instability of the transient response. A description of this instability requires a special investigation. One may expect a similar transient behavior for the other quantities characterizing the system. For example, the theory given above suggests a negative cyclotron absorption in the transient state (problem 11.10).

56. Photoinduced Interband Hybridization

The picture of linear interband absorption given in Sec. 17 is based upon the approximation that the electric field E of the electromagnetic wave is so small that the rate of interband transitions remains much smaller than the energy relaxation rates of conduction- and valence-band electrons. In these conditions, the interband transitions do not modify the electron distribution so that one can use the equilibrium Fermi distribution function to describe the absorption. As E increases, this approximation is no longer valid, and the distribution becomes distorted: the density of the valence-band electrons with momenta close to p_ω determined from the condition $p_\omega^2/2\mu^* = \hbar\omega - \varepsilon_g$ decreases, while the density of the conduction-band electrons with such momenta increases. The situation becomes more complicated if the frequency of interband Rabi oscillations of electron density, determined as $|e|E\overline{v}_{cv}/\hbar\omega$, where $\overline{v}_{cv}$ is a characteristic interband velocity, becomes comparable to or larger than the average scattering rate. In this case, the interband coherence takes place and, instead of the conduction- and valence-band electrons, one has to introduce new quasiparticles with a more sophisticated energy spectrum. These quasiparticles appear as a result of hybridization of c- and v-band states coupled by the electromagnetic field, and the mechanism of interband absorption becomes qualitatively different from the linear absorption. This mechanism is now determined by the scattering of the quasiparticles by impurities and phonons, because this scattering destroys the interband coherence.

Consider a semiconductor described by the two-band model, see Appendix B, in the high-frequency electric field $\mathbf{E}\cos\omega t$. The interaction of electrons with this field is written below in the dipole approximation, and the Hamiltonian is diagonal in the momentum representation:

$$\widehat{H}_t(\mathbf{p}) = \hat{h}_p - \frac{e}{c}\hat{\mathbf{v}}_{\mathbf{p}} \cdot \mathbf{A}_t . \tag{1}$$

This Hamiltonian is a 4×4 matrix with respect to band and spin indices. Here and below we use the symmetric two-band model with $M \to \infty$ in Eq. (B.18). It is convenient to employ the diagonal representation, when the matrices of the free-electron Hamiltonian $\hat{h}_p$ and velocity $\hat{\mathbf{v}}_{\mathbf{p}}$ are given by the right-hand sides of Eqs. (B.21) and (B.23). The vector potential corresponding to the electric field defined above is $\mathbf{A}_t = -i(c\mathbf{E}/2\omega)e^{-i\omega t} + c.c.$ Let us write the Hamiltonian (1) in the second quantization representation, introducing the operators of annihilation (creation) of electrons in the conduction and valence bands as $\hat{a}_{c\sigma\mathbf{p}}$ and $\hat{a}_{v\sigma\mathbf{p}}$ ($\hat{a}^+_{c\sigma\mathbf{p}}$ and $\hat{a}^+_{v\sigma\mathbf{p}}$), respectively:

$$\widehat{H}_t = \sum_{\sigma\mathbf{p}} \left[E_c(\mathbf{p},t)\hat{a}^+_{c\sigma\mathbf{p}}\hat{a}_{c\sigma\mathbf{p}} + E_v(\mathbf{p},t)\hat{a}^+_{v\sigma\mathbf{p}}\hat{a}_{v\sigma\mathbf{p}} \right]$$

$$- \sum_{\sigma\sigma'\mathbf{p}} W_{\sigma\sigma'}(\mathbf{p},t) \left(\hat{a}^+_{c\sigma\mathbf{p}}\hat{a}_{v\sigma'\mathbf{p}} + \hat{a}^+_{v\sigma\mathbf{p}}\hat{a}_{c\sigma'\mathbf{p}} \right). \tag{2}$$

The time-dependent energies are given by

$$E_{c,v}(\mathbf{p},t) = \pm ms^2\eta_p \mp \frac{e\mathbf{p}\cdot\mathbf{A}_t}{c\eta_p m}, \tag{3}$$

where the upper and lower signs stand for the c- and v-bands, respectively. The interband matrix elements of the perturbation operator are defined as

$$W_{\sigma\sigma'}(\mathbf{p},t) = \frac{es}{c}\mathbf{A}_t \cdot \left[\hat{\boldsymbol{\sigma}} - \frac{\eta_p - 1}{\eta_p}\frac{\mathbf{p}(\hat{\boldsymbol{\sigma}}\cdot\mathbf{p})}{p^2} \right]_{\sigma\sigma'}. \tag{4}$$

Since the interband coupling may occur between different spin states, the matrix $W_{\sigma\sigma'}$, in general, is not diagonal in the spin index $\sigma = \pm 1$.

Below we considerably simplify the problem by using the resonance approximation for the interaction of electrons with the field. We also assume that the excitation quantum $\hbar\omega$ is close to the energy gap ε_g so that the energies of the conduction- and valence-band electrons interacting with the field are not far from the band extrema, and the non-parabolicity effects can be neglected. The Hamiltonian (2) in these approximations is reduced to (problem 11.11)

$$\widehat{H}_t = \sum_{\sigma\mathbf{p}} \left[E_c(\mathbf{p})\hat{a}^+_{c\sigma\mathbf{p}}\hat{a}_{c\sigma\mathbf{p}} + E_v(\mathbf{p})\hat{a}^+_{v\sigma\mathbf{p}}\hat{a}_{v\sigma\mathbf{p}} \right]$$

$$+ \sum_{\sigma\sigma'\mathbf{p}} i\frac{es}{2\omega}\mathbf{E} \cdot \langle\sigma|\hat{\boldsymbol{\sigma}}|\sigma'\rangle \left(e^{-i\omega t}\hat{a}^+_{c\sigma\mathbf{p}}\hat{a}_{v\sigma'\mathbf{p}} - e^{i\omega t}\hat{a}^+_{v\sigma\mathbf{p}}\hat{a}_{c\sigma'\mathbf{p}} \right), \tag{5}$$

where $E_{c,v}(\mathbf{p}) \simeq \pm[\varepsilon_g/2 + p^2/2m]$. It is convenient to apply the electric field $\mathbf{E}$ along OZ axis to make the second term of the Hamiltonian (5) diagonal in the spin index, since $\langle\sigma|\hat{\sigma}_z|\sigma'\rangle = \sigma\delta_{\sigma\sigma'}$. The time-dependent

Schroedinger equation $i\hbar\partial\psi(t)/\partial t = \widehat{H}_t\psi(t)$ can be transformed by the unitary transformation $\psi(t) = \hat{\mathcal{U}}_t\phi(t)$ with

$$\hat{\mathcal{U}}_t = \exp\left[-\frac{i\omega t}{2}\sum_{\sigma\mathbf{p}}(\hat{a}_{c\sigma\mathbf{p}}^{+}\hat{a}_{c\sigma\mathbf{p}} - \hat{a}_{v\sigma\mathbf{p}}^{+}\hat{a}_{v\sigma\mathbf{p}})\right]. \tag{6}$$

As a result, we obtain the Schroedinger equation $i\hbar\partial\phi(t)/\partial t = \widehat{\overline{H}}\phi(t)$ for the wave function $\phi(t)$, where the Hamiltonian $\widehat{\overline{H}}$ is time-independent:

$$\widehat{\overline{H}} = \hat{\mathcal{U}}_t^{+}\widehat{H}_t\hat{\mathcal{U}}_t - i\hbar\hat{\mathcal{U}}_t^{+}\frac{\partial\hat{\mathcal{U}}_t}{\partial t}. \tag{7}$$

Calculating the right-hand side of Eq. (7) with the use of Eqs. (5) and (6), we find (problem 11.12)

$$\widehat{\overline{H}} = \sum_{\sigma\mathbf{p}}\left[\xi_p\left(\hat{a}_{c\sigma\mathbf{p}}^{+}\hat{a}_{c\sigma\mathbf{p}} - \hat{a}_{v\sigma\mathbf{p}}^{+}\hat{a}_{v\sigma\mathbf{p}}\right) - i\sigma\beta\left(\hat{a}_{c\sigma\mathbf{p}}^{+}\hat{a}_{v\sigma\mathbf{p}} - \hat{a}_{v\sigma\mathbf{p}}^{+}\hat{a}_{c\sigma\mathbf{p}}\right)\right], \tag{8}$$

where $\beta = |e|Es/2\omega$ and $\xi_p = p^2/2m - (\hbar\omega - \varepsilon_g)/2$.

The Hamiltonian (8) can be diagonalized by introducing new operators of creation and annihilation according to

$$\hat{c}_{+,\sigma\mathbf{p}} = \lambda_{+p}\hat{a}_{c\sigma\mathbf{p}} - i\sigma\lambda_{-p}\hat{a}_{v\sigma\mathbf{p}},$$

$$\hat{c}_{-,\sigma\mathbf{p}} = \lambda_{-p}\hat{a}_{c\sigma\mathbf{p}} + i\sigma\lambda_{+p}\hat{a}_{v\sigma\mathbf{p}}. \tag{9}$$

The coefficients describing the transformation (9) are assumed to be real, and they are normalized according to $\lambda_{+p}^2 + \lambda_{-p}^2 = 1$. The inverse transformation from quasiparticles to electrons is given by

$$\hat{a}_{c\sigma\mathbf{p}} = \lambda_{+p}\hat{c}_{+,\sigma\mathbf{p}} + \lambda_{-p}\hat{c}_{-,\sigma\mathbf{p}},$$

$$\hat{a}_{v\sigma\mathbf{p}} = i\sigma(\lambda_{-p}\hat{c}_{+,\sigma\mathbf{p}} - \lambda_{+p}\hat{c}_{-,\sigma\mathbf{p}}). \tag{10}$$

Choosing the coefficients as

$$\lambda_{+p} = \sqrt{\frac{1}{2} + \frac{\xi_p}{2\sqrt{\xi_p^2 + \beta^2}}}, \quad \lambda_{-p} = \sqrt{\frac{1}{2} - \frac{\xi_p}{2\sqrt{\xi_p^2 + \beta^2}}}, \tag{11}$$

we find

$$\widehat{\overline{H}} = \sum_{i=\pm}\sum_{\sigma\mathbf{p}}\varepsilon_{ip}\hat{c}_{i\sigma\mathbf{p}}^{+}\hat{c}_{i\sigma\mathbf{p}}, \quad \varepsilon_{\pm,p} = \pm\sqrt{\xi_p^2 + \beta^2}. \tag{12}$$

The new spin-degenerate quasiparticle states described by the operators $\hat{c}_{+,\sigma\mathbf{p}}$ and $\hat{c}_{-,\sigma\mathbf{p}}$ are obtained as a result of hybridization of c-

and v- states under interband photoexcitation. These quasiparticles are quasienergy states (we remind that the notion of quasienergy appears as a generalization of the notion of stationary state for the case when the Hamiltonian is a periodic function of time). The spectrum of the new quasiparticles, $\varepsilon_{\pm,p}$, is shown in Fig. 11.4 (a). Due to a finite strength of electric field of the electromagnetic wave, the degeneracy of the quasienergy states at $|\mathbf{p}| = p_\omega \equiv \sqrt{m(\hbar\omega - \varepsilon_g)}$ is lifted. As a result, a gap opens in the energy spectrum in the vicinity of p_ω. The magnitude of the gap, $2\beta = |e|Es/\omega$, is proportional to the field. The new quasiparticles are fermions, and they obey the anticommutation relations $\hat{c}_{i\sigma\mathbf{p}}\hat{c}^+_{i'\sigma'\mathbf{p}'} + \hat{c}^+_{i'\sigma'\mathbf{p}'}\hat{c}_{i\sigma\mathbf{p}} = \delta_{ii'}\delta_{\sigma\sigma'}\delta_{\mathbf{p}\mathbf{p}'}$.

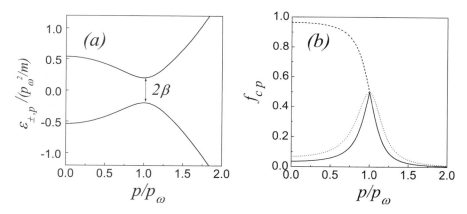

Figure 11.4. (a) Spectrum of quasiparticles at $\beta = 0.2\, p_\omega^2/m$. (b) Stationary distribution function f_{cp} for the cases when we take into account only excitation (solid), see Eq. (56.14), excitation and electron-phonon scattering (dashed), see Eq. (56.15), and excitation and recombination (dotted), see Eq. (56.20).

Since the Hamiltonian (12) is diagonal, the quasiparticle densities n_+ and n_- are conserved. Moreover, if the excitation is turned on adiabatically, the distribution functions of quasiparticles, $f_{i\mathbf{p}} = \langle\langle \hat{c}^+_{i\sigma\mathbf{p}}\hat{c}_{i\sigma\mathbf{p}}\rangle\rangle$, as well as the non-diagonal components ($i \neq i'$) of their density matrix $\langle\langle \hat{c}^+_{i\sigma\mathbf{p}}\hat{c}_{i'\sigma\mathbf{p}}\rangle\rangle$ should remain the same as in the absence of the excitation, i.e., at $\beta = 0$. Therefore, they are found according to

$$\langle\langle \hat{c}^+_{+,\sigma\mathbf{p}}\hat{c}_{+,\sigma\mathbf{p}}\rangle\rangle = \langle\langle \hat{c}^+_{+,\sigma\mathbf{p}}\hat{c}_{+,\sigma\mathbf{p}}\rangle\rangle_{\beta=0} = \lambda^2_{-p}|_{\beta=0} = \theta(p_\omega - p),$$

$$\langle\langle \hat{c}^+_{-,\sigma\mathbf{p}}\hat{c}_{-,\sigma\mathbf{p}}\rangle\rangle = \langle\langle \hat{c}^+_{-,\sigma\mathbf{p}}\hat{c}_{-,\sigma\mathbf{p}}\rangle\rangle_{\beta=0} = \lambda^2_{+p}|_{\beta=0} = \theta(p - p_\omega), \qquad (13)$$

$$\langle\langle \hat{c}^+_{+,\sigma\mathbf{p}}\hat{c}_{-,\sigma\mathbf{p}}\rangle\rangle = \langle\langle \hat{c}^+_{-,\sigma\mathbf{p}}\hat{c}_{+,\sigma\mathbf{p}}\rangle\rangle = -\lambda_{+p}\lambda_{-p}|_{\beta=0} = 0.$$

To obtain Eq. (13), we have taken into account that $\langle\langle \hat{a}^+_{v\sigma\mathbf{p}}\hat{a}_{v\sigma\mathbf{p}}\rangle\rangle_{\beta=0} = 1$ and $\langle\langle \hat{a}^+_{c\sigma\mathbf{p}}\hat{a}_{c\sigma\mathbf{p}}\rangle\rangle_{\beta=0} = \langle\langle \hat{a}^+_{c\sigma\mathbf{p}}\hat{a}_{v\sigma\mathbf{p}}\rangle\rangle_{\beta=0} = \langle\langle \hat{a}^+_{v\sigma\mathbf{p}}\hat{a}_{c\sigma\mathbf{p}}\rangle\rangle_{\beta=0} = 0$. Equations (10) and (13) allow us to find the distribution functions of electrons in the case of a finite excitation:

$$f_{c\mathbf{p}} = \langle\langle \hat{a}^+_{c\sigma\mathbf{p}} \hat{a}_{c\sigma\mathbf{p}} \rangle\rangle = \frac{1}{2} - \frac{|\xi_p|}{2\sqrt{\xi_p^2 + \beta^2}}, \qquad f_{v\mathbf{p}} = 1 - f_{c\mathbf{p}}. \qquad (14)$$

The distribution looks like a kink around p_ω, with a characteristic energy width β; see Fig. 11.4 (b). If $p = p_\omega$, we have $f_{c\mathbf{p}} = f_{v\mathbf{p}} = 1/2$, which means that an electron can be found in the c- and v-bands with equal probabilities, in accordance with the dynamical picture of an electron undergoing interband oscillations and spending half of its time in each band.

Of course, the behavior discussed above may take place only in the absence of any relaxation. If the electron-phonon scattering is taken into consideration, the electrons coming into the conduction band have a finite probability to scatter away from the resonance region and remain for a long time in this band instead of going back to the valence band. Eventually, a steady-state distribution corresponding to saturated optical absorption is reached, when (at $T \to 0$ and $\beta \to 0$) the c-band is completely filled at $p < p_\omega$, while the v-band is empty in this region. The distribution functions of quasiparticles, $f_{i\mathbf{p}} = \langle\langle \hat{c}^+_{i\sigma\mathbf{p}} \hat{c}_{i\sigma\mathbf{p}} \rangle\rangle$, in this case are the ordinary Fermi functions with common chemical potential. At low temperatures, $T \ll 2\beta$, the distribution functions are $f_{+\mathbf{p}} \simeq 0$ and $f_{-\mathbf{p}} \simeq 1$. Using Eq. (10), we find the distribution functions of electrons in these conditions:

$$f_{c\mathbf{p}} = 1 - f_{v\mathbf{p}} = \frac{1}{2} - \frac{\xi_p}{2\sqrt{\xi_p^2 + \beta^2}}. \qquad (15)$$

This function looks like a Fermi distribution function broadened around p_ω with a characteristic energy β, and $f_{c\mathbf{p}} = f_{v\mathbf{p}} = 1/2$ at $p = p_\omega$.

Let us forget for a while about electron-phonon scattering and other intraband scattering mechanisms, and consider the radiative interband recombination. This process is important because it changes the occupation numbers in the c- and v-bands. The Hamiltonian describing the interaction of electrons with photons in the dipole approximation, see Sec. 19 and Eq. (20.1), is written in the second quantization representation as

$$\hat{H}_{e,pht} = \sum_{\mu\sigma\sigma'} \sum_{\mathbf{pq}} \sqrt{\frac{2\pi\hbar e^2}{\omega_q \epsilon V}} \left[\langle c\sigma\mathbf{p} | \hat{\mathbf{v}} \cdot \mathbf{e}_{\mathbf{q}\mu} | v\sigma'\mathbf{p} \rangle \hat{a}^+_{c\sigma\mathbf{p}} \hat{a}_{v\sigma'\mathbf{p}} \right.$$

$$\left. + \langle v\sigma\mathbf{p} | \hat{\mathbf{v}} \cdot \mathbf{e}_{\mathbf{q}\mu} | c\sigma'\mathbf{p} \rangle \hat{a}^+_{v\sigma\mathbf{p}} \hat{a}_{c\sigma'\mathbf{p}} \right] \hat{b}_{\mathbf{q}\mu} + H.c. , \qquad (16)$$

where $\hat{b}_{\mathbf{q}\mu}$ is the operator of annihilation of the photon with wave vector $\mathbf{q}$ and polarization index μ ($\mu = 1, 2$). We consider only the interband matrix elements of the velocity operator in this Hamiltonian. The unitary transformation defined by the operator (6) brings the factors $e^{i\omega t}$ at $\hat{a}_{c\sigma\mathbf{p}}^+ \hat{a}_{v\sigma'\mathbf{p}}$ and $e^{-i\omega t}$ at $\hat{a}_{v\sigma\mathbf{p}}^+ \hat{a}_{c\sigma'\mathbf{p}}$ thereby making the Hamiltonian time-dependent: $\widehat{H}_{e,pht} \to \widehat{H}_{e,pht}(t)$. In the resonance approximation, only the terms proportional to $e^{i\omega t} b_{\mathbf{q}\mu} \hat{a}_{c\sigma\mathbf{p}}^+ \hat{a}_{v\sigma'\mathbf{p}}$ and $e^{-i\omega t} b_{\mathbf{q}\mu}^+ \hat{a}_{v\sigma\mathbf{p}}^+ \hat{a}_{c\sigma'\mathbf{p}}$ are essential in this Hamiltonian. Let us apply the canonical transformation (10) to $\widehat{\overline{H}} + \widehat{H}_{e,pht}(t)$. The Hamiltonian obtained in this way describes the interaction of quasiparticles with photons. Assuming that the quasiparticles are well-defined, which means that $\beta/\hbar$ is large in comparison to the radiative recombination rate ν_R, one can derive a kinetic equation for quasiparticles in the diagonal approximation (by neglecting the non-diagonal components of the density matrix) in a similar way as in Sec. 19; see also problems 7.3 and 11.1. Neglecting also the occupation numbers of equilibrium photons under the condition when only the spontaneous recombination is essential, we write the kinetic equation as (problem 11.13)

$$\frac{\partial f_{+\mathbf{p}}}{\partial t} = -\frac{\partial f_{-\mathbf{p}}}{\partial t} = 2\sum_{\mathbf{q}} \frac{(2\pi e s)^2}{V \omega_q \epsilon} \left\{ \lambda_{-p}^4 \delta\left(2\sqrt{\xi_p^2 + \beta^2} + \hbar\omega_q - \hbar\omega\right) \right.$$

$$\left. \times f_{-\mathbf{p}}(1 - f_{+\mathbf{p}}) - \lambda_{+p}^4 \delta\left(2\sqrt{\xi_p^2 + \beta^2} - \hbar\omega_q + \hbar\omega\right) f_{+\mathbf{p}}(1 - f_{-\mathbf{p}}) \right\}. \quad (17)$$

The sum over $\mathbf{q}$ is calculated in a similar way as in Eqs. (38.4) and (38.5). Neglecting the ratio $2\sqrt{\xi_p^2 + \beta^2}/\hbar\omega$, we reduce Eq. (17) to

$$\frac{\partial f_{+\mathbf{p}}}{\partial t} = -\frac{\partial f_{-\mathbf{p}}}{\partial t} = J_R(f| + \mathbf{p})$$

$$= \nu_R[\lambda_{-p}^4 f_{-\mathbf{p}}(1 - f_{+\mathbf{p}}) - \lambda_{+p}^4 f_{+\mathbf{p}}(1 - f_{-\mathbf{p}})], \quad (18)$$

where the recombination rate ν_R is given by Eq. (38.5). Equation (18) can be solved by using the relation $f_{-\mathbf{p}} = 1 - f_{+\mathbf{p}}$, which is just a requirement of conservation of the total number of quasiparticles with a given momentum $\mathbf{p}$ (the radiative recombination in the dipole approximation conserves the momentum). In the stationary case, there is a simple solution

$$f_{+\mathbf{p}} = \lambda_{-p}^2, \quad f_{-\mathbf{p}} = \lambda_{+p}^2, \quad (19)$$

which also gives us the distribution functions of electrons

$$f_{c\mathbf{p}} = 1 - f_{v\mathbf{p}} = 2\lambda_{+p}^2 \lambda_{-p}^2 = \frac{1}{2}\frac{\beta^2}{\xi_p^2 + \beta^2}. \quad (20)$$

This is a Lorentz function of the variable ξ_p; see Fig. 11.4 (b). The radiative recombination smooths the sharp kink of the electron distribution function (14).

The scattering due to different mechanisms smears the narrow stationary distribution (20) over a wide region of momenta. As we already noted, the electron-phonon scattering, if it occurs much faster than the recombination, establishes the distribution (15). The electron-impurity scattering alone cannot establish such a distribution, because it is elastic. Nevertheless, this kind of scattering tends to a similar smearing. Below we consider the kinetic equation which accounts for both electron-impurity scattering and recombination. The second-quantized Hamiltonian of electron-impurity interaction, $\widehat{H}_{e,im} = V^{-1} \sum_{\sigma \mathbf{pp'}} U_{im}(\mathbf{p} - \mathbf{p'})[\hat{a}^+_{c\sigma\mathbf{p}}\hat{a}_{c\sigma\mathbf{p'}} + \hat{a}^+_{v\sigma\mathbf{p}}\hat{a}_{v\sigma\mathbf{p'}}]$, where U_{im} is given by Eq. (14.3), is not modified by the unitary transformation defined by the operator (6). Applying the canonical transformation (10), we rewrite this Hamiltonian as

$$\widehat{H}_{e,im} = \frac{1}{V} \sum_{\sigma \mathbf{pp'}} U_{im}(\mathbf{p} - \mathbf{p'}) \left\{ (\lambda_{+p}\lambda_{+p'} + \lambda_{-p}\lambda_{-p'})[\hat{c}_{+,\sigma\mathbf{p}}\hat{c}_{+,\sigma\mathbf{p'}} \quad (21) \right.$$

$$\left. +\hat{c}_{-,\sigma\mathbf{p}}\hat{c}_{-,\sigma\mathbf{p'}}] + (\lambda_{+p}\lambda_{-p'} - \lambda_{-p}\lambda_{+p'})[\hat{c}_{+,\sigma\mathbf{p}}\hat{c}_{-,\sigma\mathbf{p'}} - \hat{c}_{-,\sigma\mathbf{p}}\hat{c}_{+,\sigma\mathbf{p'}}] \right\} .$$

To derive the electron-impurity collision integral based on the Hamiltonian (21), one should act in a similar way as in Chapter 2. The weight factors containing λ_+ and λ_- manifest themselves in the matrix elements determining the scattering probabilities for the quasiparticles. In the diagonal approximation, we neglect interband scattering of the quasiparticles, because it is forbidden by the energy conservation law. As a result (compare to Eq. (8.7)), we obtain ($i = \pm$)

$$J_{im}(f|i\mathbf{p}) = \frac{2\pi n_{im}}{\hbar V} \sum_{\mathbf{p'}} |v(|\mathbf{p} - \mathbf{p'}|/\hbar)|^2 \delta(\varepsilon_{ip} - \varepsilon_{ip'})$$

$$\times (\lambda_{+p}\lambda_{+p'} + \lambda_{-p}\lambda_{-p'})^2 (f_{i\mathbf{p'}} - f_{i\mathbf{p}}). \quad (22)$$

The scattering described by Eq. (22) is elastic: the quasiparticle transitions occur between the states with equal energies. If $p^2 > 2p^2_\omega$, there is only one branch of the isoenergetic surfaces in the spectrum of quasiparticles of each kind; see Fig. 11.4 (a). The collision integral (22) for this region is equal to zero because the quasiparticle distribution is isotropic. In the region $p^2 < 2p^2_\omega$, there are two branches corresponding to $\xi_p < 0$ (or, equivalently, $p < p_\omega$) and $\xi_p > 0$ ($p > p_\omega$). The collision integral is non-zero only if $\mathbf{p}$ and $\mathbf{p'}$ belong to the different branches. This scattering corresponds to the electron scattering between c- and

v-bands in the presence of a high-frequency field, and its probability rapidly decreases away from the resonance $p = p_\omega$. The isotropy of electron distribution allows one to average $|v(|\mathbf{p} - \mathbf{p}'|/\hbar)|^2$ over the angle of $\mathbf{p}'$. Below, however, we use the model of short-range impurity potential, when $v(|\mathbf{p} - \mathbf{p}'|/\hbar) \simeq v$ is independent of $\mathbf{p} - \mathbf{p}'$.

The stationary kinetic equations $J_R(f|i\mathbf{p}) + J_{im}(f|i\mathbf{p}) = 0$, where J_R is given by Eq. (18), can be written for $i = +$ and $i = -$, under the assumption that the momentum $\mathbf{p}$ belongs either to the branch $\xi_p < 0$ ("<" branch) or to the branch $\xi_p > 0$ (">" branch). Introducing four functions, $f_\pm^<$ and $f_\pm^>$, which are equal to $f_{\pm\mathbf{p}}$ for < and > branches, respectively, we obtain four algebraic equations connecting these functions:

$$a_-(f_+^> - f_+^<) + b_p^{>2} f_-^<(1 - f_+^<) - b_p^{<2} f_+^<(1 - f_-^<) = 0,$$

$$a_-(f_-^> - f_-^<) + b_p^{<2} f_+^<(1 - f_-^<) - b_p^{>2} f_-^<(1 - f_+^<) = 0,$$

$$a_+(f_+^< - f_+^>) + b_p^{<2} f_-^>(1 - f_+^>) - b_p^{>2} f_+^>(1 - f_-^>) = 0, \qquad (23)$$

$$a_+(f_-^< - f_-^>) + b_p^{>2} f_+^>(1 - f_-^>) - b_p^{<2} f_-^>(1 - f_+^>) = 0,$$

where $b_p^> = 1/2 + \eta_p$, $b_p^< = 1/2 - \eta_p$, $\eta_p = |\xi_p|/2\sqrt{\xi_p^2 + \beta^2}$, and $a_\pm(p) = r(\beta^2/\sqrt{\xi_p^2 + \beta^2}|\xi_p|)\sqrt{1 \mp 2m|\xi_p|/p_\omega^2}$. According to Eq. (11), $b_p^> = \lambda_{+p}^2$ and $b_p^< = \lambda_{-p}^2$ at $\xi_p > 0$ (in a similar way, $b_p^> = \lambda_{-p}^2$ and $b_p^< = \lambda_{+p}^2$ at $\xi_p < 0$). The coefficients $a_\pm(p)$ are proportional to the ratio $r = \nu_{im}/\nu_R$, where $\nu_{im} = n_{im}|v|^2 mp_\omega/\pi\hbar^4$ is the electron-impurity scattering rate at $p = p_\omega$. The square-root factor $\sqrt{1 \mp 2m|\xi_p|/p_\omega^2}$ describes the momentum dependence of the scattering rate corresponding to the energy dependence of the density of states. This factor can be neglected in the region $|\xi_p| \ll p_\omega^2/2m$.

Equations (23) should be accompanied by the normalization condition $f_+^< + f_+^> + f_-^< + f_-^> = 2$ expressing the particle conservation. As a result, we have $f_-^> = 1 - f_+^>$ and $f_-^< = 1 - f_+^<$, and the system is reduced to two equations

$$x + y = \frac{2\eta}{a_-}\left(x^2 - \frac{1 + 4\eta^2}{4\eta}x + \frac{1}{4}\right),$$

$$x + y = \frac{2\eta}{a_+}\left(y^2 - \frac{1 + 4\eta^2}{4\eta}y + \frac{1}{4}\right), \qquad (24)$$

where $\eta = \eta_p$, $x = f_+^< - 1/2$ and $y = -f_+^> + 1/2$. In the vicinity of the point $\xi_p = 0$, where $a_+ = a_-$, one has $x = y$, and an approximate solution at $r \gg 1$ can be written as $x = 0$. In other words, the electron-impurity scattering tends to make the distribution function of quasiparticles flat (equal to $1/2$) in some region around $p \simeq p_\omega$. To describe the whole region of p, one should solve Eq. (24) numerically. The results of such calculations for several different r are shown in Fig. 11.5, where we plot the distribution functions of the conduction-band electrons, $f_{c\mathbf{p}} = 1/2 - 2\eta x$ at $p < p_\omega$ and $f_{c\mathbf{p}} = 1/2 - 2\eta y$ at $p > p_\omega$, instead of the distribution functions of quasiparticles. As the impurity scattering becomes stronger, $f_{c\mathbf{p}}$ comes closer to the dependence given by Eq. (15) rather than to the one of Eq. (20).

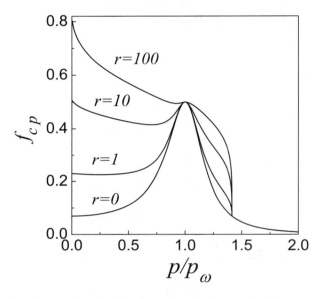

Figure 11.5. Stationary distribution function f_{cp} determined by mutual action of recombination and electron-impurity scattering. It is assumed that $\beta = 0.2\ p_\omega^2/m$, and the curves correspond to different ratios $r = \nu_{im}/\nu_R$.

So far we have considered the case when the quasiparticles are well-defined. However, if the high-frequency field is not very strong, the gap energy, 2β, can become comparable to the energy of collision-induced broadening. In this case the coherence is suppressed, and the gap in the spectrum may vanish. This effect is demonstrated below on the example of linear absorption of an additional weak electromagnetic wave $\mathbf{E}_1 e^{-i\omega_1 t} + c.c.$ whose frequency ω_1 is close to ω. We consider the interaction of electrons with the impurity potential described by the Hamiltonian $\widehat{H}_{e,im}$ given above. The absorption coefficient $\alpha_{\omega_1}^{(E)} = 4\pi \mathrm{Re}\sigma(\omega_1)/c\sqrt{\epsilon}$, where the superscript (E) indicates the absorption in the presence of interband pumping, is expressed through the real part of the conductivity $\sigma(\omega_1)$. The latter is defined according to $\mathbf{I}_t =$

$(e/V)\mathrm{Sp}\widehat{\Delta\eta}_t\hat{\mathbf{v}} = \sigma(\omega_1)\mathbf{E}_1 e^{-i\omega_1 t} + c.c.$ To apply the formalism of Sec. 13 for calculating the non-equilibrium part $\widehat{\Delta\eta}_t$ of the statistical operator, one should carry out the unitary transformation described by the operator (6) and making the equilibrium Hamiltonian time-independent. The transformed Hamiltonian is $\widehat{H} = \widehat{\tilde{H}} + \widehat{H}_{e,im}$, and we remind that $\widehat{H}_{e,im}$ is not changed by the unitary transformation. The operator of the interband perturbation, $i(es/\omega_1)\mathbf{E}_1 \cdot \hat{\boldsymbol{\sigma}}\hat{\rho}_1 e^{-i\omega_1 t} + H.c. \equiv \widehat{\Delta H}_{\omega_1} e^{-i\omega_1 t} + H.c.$, should be transformed accordingly (we remind that the matrices $\hat{\rho}_i$, $i = 1, 2, 3$ are defined by Eq. (B.19)). Finally, $\widehat{\Delta\eta}_t$ is determined by the inverse unitary transformation of the expression obtained in this way (compare the equation given below to Eq. (13.3)):

$$\widehat{\Delta\eta}_t = \hat{\mathcal{U}}_t \frac{1}{i\hbar} \int_{-\infty}^{t} dt' e^{\lambda t'} e^{i\widehat{H}(t'-t)/\hbar}$$

$$\times \left[\hat{\mathcal{U}}_{t'}^+ \left(\widehat{\Delta H}_{\omega_1} e^{-i\omega_1 t'} + H.c. \right) \hat{\mathcal{U}}_{t'}, \hat{\eta}_{eq} \right] e^{-i\widehat{H}(t'-t)/\hbar} \hat{\mathcal{U}}_t^+. \qquad (25)$$

Applying this expression in order to calculate the current density $\mathbf{I}_t$, we use the identity $\hat{\mathcal{U}}_t^+ \hat{\boldsymbol{\sigma}}\hat{\rho}_1 \hat{\mathcal{U}}_t = e^{i\omega t}\hat{\boldsymbol{\sigma}}\widehat{P}_{cv} + e^{-i\omega t}\hat{\boldsymbol{\sigma}}\widehat{P}_{vc}$, where the interband projection operators are defined as $\widehat{P}_{cv} = (\hat{\rho}_1 + i\hat{\rho}_2)/2$ and $\widehat{P}_{vc} = (\hat{\rho}_1 - i\hat{\rho}_2)/2$. With $\tau = t' - t$, we write the conductivity in the form

$$\sigma(\omega_1) = \frac{(es)^2}{\hbar\omega_1 V} \int_{-\infty}^{0} d\tau e^{\lambda\tau - i\delta\omega\tau} \mathrm{Sp}\hat{\eta}_{eq} \left[e^{-i\widehat{H}\tau/\hbar}\widehat{P}_{vc} e^{i\widehat{H}\tau/\hbar}, \widehat{P}_{cv} \right], \qquad (26)$$

where $\delta\omega = \omega_1 - \omega$.

Evaluating the trace in Eq. (26) in the exact eigenstate representation, one should use $\langle\delta|\widehat{P}_{jj'}|\delta'\rangle = \sum_{\mathbf{p}}\langle\delta|j\mathbf{p}\rangle\langle j'\mathbf{p}|\delta'\rangle$, where $j, j' = c, v$. Acting in this way, we express the real part of the conductivity through the two-particle averages of the retarded and advanced Green's functions in the momentum representation. These functions are defined as

$$G_\varepsilon^{R,A}(j\mathbf{p}, j'\mathbf{p}') = \sum_\delta \frac{\psi_{j\mathbf{p}}^{(\delta)}\psi_{j'\mathbf{p}'}^{(\delta)*}}{\varepsilon - \varepsilon_\delta \pm i\lambda}, \qquad (27)$$

where the upper and lower signs stand for R and A, respectively. This equation is an obvious generalization of Eq. (14.1) to the case of several bands. With the use of Eq. (27), the absorption coefficient is written as

$$\alpha_{\omega_1}^{(E)} = \frac{2(es)^2}{c\sqrt{\epsilon}\omega_1 V} \int d\varepsilon [f(\varepsilon) - f(\varepsilon + \hbar\delta\omega)] \sum_{s,s'=R,A} (-1)^l$$

$$\times \sum_{\mathbf{pp}'} \langle\langle G_{\varepsilon+\hbar\delta\omega}^s(c\mathbf{p}, c\mathbf{p}') G_\varepsilon^{s'}(v\mathbf{p}', v\mathbf{p})\rangle\rangle, \qquad (28)$$

where $l = 0$ for $s \neq s'$, $l = 1$ for $s = s'$, and $f(\varepsilon)$ is the equilibrium Fermi distribution function. The sign of $\alpha_{\omega_1}^{(E)}$ is determined by the sign of $[f(\varepsilon) - f(\varepsilon + \hbar\delta\omega)]$. We stress that the equilibrium state of the electron system under a strong interband excitation is a saturated state with inverted population of electrons at $p < p_\omega$. It is for this reason that the absorption coefficient is positive for $\delta\omega > 0$ and negative for $\delta\omega < 0$. For a perfect crystal, when the electron-impurity interaction is absent and the exact eigenstates are the quasiparticle states introduced above, we have $G_\varepsilon^{R,A}(v\mathbf{p}', v\mathbf{p}) = \delta_{\mathbf{pp}'}[\lambda_{-p}^2/(\varepsilon - \varepsilon_{+,p} \pm i\lambda) + \lambda_{+p}^2/(\varepsilon - \varepsilon_{-,p} \pm i\lambda)]$ and $G_\varepsilon^{R,A}(c\mathbf{p}, c\mathbf{p}') = \delta_{\mathbf{pp}'}[\lambda_{+p}^2/(\varepsilon - \varepsilon_{+,p} \pm i\lambda) + \lambda_{-p}^2/(\varepsilon - \varepsilon_{-,p} \pm i\lambda)]$ according to Eqs. (27) and (12). Since at $T \ll 2\beta$ $f(\varepsilon_{+,p}) = 0$ and $f(\varepsilon_{-,p}) = 1$, we write the absorption coefficient as

$$\alpha_{\omega_1}^{(E)} = \frac{(2es)^2}{c\sqrt{\epsilon}\hbar^3\omega_1} \int_0^\infty dp \, p^2 \left[\lambda_{+p}^4 \delta(2\sqrt{\xi_p^2 + \beta^2} - \hbar\delta\omega) \right.$$
$$\left. - \lambda_{-p}^4 \delta(2\sqrt{\xi_p^2 + \beta^2} + \hbar\delta\omega) \right]. \tag{29}$$

The absorption is an antisymmetric function of $\delta\omega$. It has a gap, which means that the absorption is absent at $|\delta\omega| < 2\beta/\hbar$. Near the edges $\delta\omega = \pm 2\beta/\hbar$, there is a simple relation $\alpha_{\omega_1}^{(E)} = (\alpha_\omega/2)\delta\omega/\sqrt{(\delta\omega)^2 - (2\beta/\hbar)^2}$, where α_ω is given by Eq. (17.6) (note that $\mu^* = m/2$ for the symmetric two-band model we use). The inverse-square-root divergence of the absorption at the edges reflects effective one-dimensional behavior of the density of states near the extremum for the quasiparticles whose energy spectrum is shown in Fig. 11.4 (a). The divergence of this kind should be washed out if we take into account scattering of the quasiparticles.

To find the absorption in the presence of impurities, one has to evaluate the two-particle average in Eq. (28). Without going into details of such calculation, we put some general remarks concerning properties of the Green's functions. The basis of c- and v-states appears to be convenient, because the matrix elements of electron-impurity interaction are diagonal with respect to the band indices. Therefore, the diagram technique developed in Secs. 14 and 15 can be applied in a straightforward way when calculating such averages. The only difference is that all relevant equations, in particular, the Dyson and Bethe-Salpeter equations, should be written for 2×2 matrices with $(jj') = (cc)$, (cv), (vc), and (vv) elements (problem 11.14). The one-particle Green's function, averaged over the impurity ensemble, is a matrix $\widehat{G}_\varepsilon^s(\mathbf{p})$ whose elements are $[G_\varepsilon^s(\mathbf{p})]_{jj'}$. Since the Hamiltonian (8) in this matrix representation can be written through the Pauli matrices as $\hat{\sigma}_z\xi_p + \hat{\sigma}_y\sigma\beta$, the averaged Green's function is governed by the following equation:

$$[\varepsilon - \hat{\sigma}_z\xi_p - \hat{\sigma}_y\beta - \widehat{\Sigma}_\varepsilon^s(\mathbf{p})]\widehat{G}_\varepsilon^s(\mathbf{p}) = \hat{1}, \tag{30}$$

which is equivalent to four scalar equations. Strictly speaking, β in Eq. (30) should be multiplied by $\sigma = \pm 1$ for different spin states. This substitution, however, does not change any observable value, while spin summation leads only to a factor of 2 in the expressions for observable values. For this reason, we omit the spin index in the Green's functions. The matrix of the self-energy functions is given by an expansion similar to Eq. (14.26). In the lowest order in the perturbation,

$$\widehat{\Sigma}_\varepsilon^s(\mathbf{p}) = \frac{n_{im}}{V} \sum_{\mathbf{p}'} |v(|\mathbf{p} - \mathbf{p}'|/\hbar)|^2 \widehat{G}_\varepsilon^s(\mathbf{p}'). \tag{31}$$

Using this relation together with Eq. (30), we express the components of $\widehat{\Sigma}_\varepsilon^s(\mathbf{p})$ as $[\Sigma_\varepsilon^s(\mathbf{p})]_{vc} = -[\Sigma_\varepsilon^s(\mathbf{p})]_{cv} \equiv \widetilde{\Sigma}_\varepsilon^s(\mathbf{p})$ and $[\Sigma_\varepsilon^s(\mathbf{p})]_{cc} = [\Sigma_\varepsilon^s(\mathbf{p})]_{vv} \equiv \Sigma_\varepsilon^s(\mathbf{p})$. The Green's functions are expressed as (we take into account that the averaged Green's functions are isotropic in $\mathbf{p}$)

$$[G_\varepsilon^s(p)]_{cc} = [\varepsilon + \xi_p - \Sigma_\varepsilon^s(p)]/D_\varepsilon^s(p),$$

$$[G_\varepsilon^s(p)]_{vv} = [\varepsilon - \xi_p - \Sigma_\varepsilon^s(p)]/D_\varepsilon^s(p),$$

$$[G_\varepsilon^s(p)]_{vc} = -[G_\varepsilon^s(p)]_{cv} = [i\beta + \widetilde{\Sigma}_\varepsilon^s(p)]/D_\varepsilon^s(p), \tag{32}$$

$$D_\varepsilon^s(p) = [\varepsilon - \Sigma_\varepsilon^s(p)]^2 - \xi_p^2 - [\beta - i\widetilde{\Sigma}_\varepsilon^s(p)]^2.$$

Since we are interested in the region $p \simeq p_\omega$ and $|\varepsilon| \simeq \beta$, the contribution to the sum over $\mathbf{p}'$ in Eq. (31) comes from the region $p' \simeq p_\omega$, where the denominator $D_\varepsilon^s(p')$ is small. We may, therefore, transform this sum to an integral over ξ_p, setting the limits of the integration at $\pm\infty$. In this approximation and under the assumption of short-range scattering potential, Σ^s and $\widetilde{\Sigma}^s$ appear to be independent of momentum. Substituting the expressions (32) into Eq. (31), we find a couple of equations for Σ^s and $\widetilde{\Sigma}^s$ in this approximation. It is convenient to write these equations through the function $u(\varepsilon) = (\varepsilon - \Sigma_\varepsilon^R)/(\beta - i\widetilde{\Sigma}_\varepsilon^R)$:

$$\Sigma_\varepsilon^R = -\frac{\hbar\nu_{im}}{2} \frac{u}{\sqrt{1 - u^2}}, \qquad \widetilde{\Sigma}_\varepsilon^R = -i\frac{\hbar\nu_{im}}{2\sqrt{1 - u^2}}. \tag{33}$$

The system of equations (33) is reduced to a single equation for $u(\varepsilon)$:

$$\frac{\varepsilon}{\beta} = u - \frac{u}{\chi\sqrt{1 - u^2}}, \qquad \chi = \frac{\beta}{\hbar\nu_{im}}. \tag{34}$$

This equation is to be solved numerically. The function $u(\varepsilon)$ can be used to express the density of states (problem 11.15) whose energy dependence is shown in Fig. 11.6. The spectral dependence of the absorption $\alpha_{\omega_1}^{(E)}$ is similar to the energy dependence of the density of states. In particular, the absorption has a gap only at $\chi > 1$.

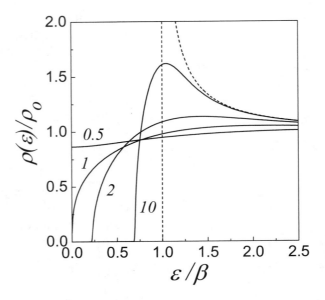

Figure 11.6. Density of electron states under interband pumping in the presence of impurities for $\chi = \beta/\hbar\nu_{im} = 0.5$, 1, 2, and 10 ($\rho_0$ is defined in problem 11.15). The dashed line shows the density of states in the absence of impurities, when $\chi \to \infty$.

57. Excitation of Coherent Phonons

The description of the lattice vibration modes carried out in Secs. 19, 21, 22, and 27 was based upon the formalism of phonon density matrix. This matrix is quadratic in atomic displacements or, equivalently, in the operators of creation and annihilation of phonons. Such quadratic forms were also used in the definition of the Green's functions of phonons in Sec. 42 and Appendix D. However, the coherent vibration of the lattice involves non-zero average displacements, and one has to consider equations of motion for $\langle\langle \hat{b}_{\mathbf{q}} \rangle\rangle_t$ and $\langle\langle \hat{b}_{\mathbf{q}}^+ \rangle\rangle_t$ by taking into account electron-phonon and phonon-phonon interactions. The coherent phonon oscillations can be excited in semiconductors or insulators by ultrashort laser pulses causing abrupt redistribution of electron density. The electric fields and elastic displacements appearing as a result of such redistribution lead to the lattice vibration characterized by the coherent amplitudes of atomic displacements. Experimental observation of the coherent phonon oscillations is possible with the use of the double-pulse technique. A powerful pump pulse creates the lattice oscillations which modify the optical characteristics of the medium. Then, a weak probe pulse is used to measure these modified characteristics. Below we consider the photoexcitation of coherent phonon oscillations by an ultrashort laser pulse and their damping owing to the relaxation caused by the phonon-phonon interaction, without going into details of the detection of coherent phonons.

The operator of atomic displacement is expressed in terms of the normal coordinate $\widehat{Q}_{\mathbf{q}l}$ describing the l-th phonon mode with wave vector $\mathbf{q}$, according to Eq. (6.5). The normal coordinate is expressed through the creation and annihilation operators according to Eq. (6.9). The long-wavelength acoustic phonons are described by the coordinate-dependent lattice displacement $\hat{\mathbf{u}}_{ac}(\mathbf{r})$, see Eq. (6.29), which is written as

$$\hat{\mathbf{u}}_{ac}(\mathbf{r}) = \sum_{\mathbf{q}l} \frac{\mathbf{e}_{\mathbf{q}l}}{\sqrt{\rho V}} e^{i\mathbf{q}\cdot\mathbf{r}} \widehat{Q}_{\mathbf{q}l}. \tag{1}$$

The relative ionic displacement $\hat{\mathbf{w}}(\mathbf{r})$ characterizing the long-wavelength optical phonons has the same form; see Eq. (6.19). The difference between these two cases is taken into account just by a formal substitution, $\rho \to \bar{\rho}$, where $\bar{\rho}$ for the case of optical phonons is expressed through the reduced mass of crystal cell, $\overline{M}$, which enters Eq. (6.14). The averaged displacement is given in terms of the averaged normal coordinate, $\langle\langle \widehat{Q}_{\mathbf{q}l} \rangle\rangle_t$, which satisfies the following equation:

$$\frac{\partial \langle\langle \widehat{Q}_q \rangle\rangle_t}{\partial t} = -\frac{i}{\hbar} \mathrm{Sp}[\widehat{H}, \hat{\eta}_t]\widehat{Q}_q = -\frac{i}{\hbar} \langle\langle [\widehat{Q}_q, \widehat{H}_{ph} + \widehat{H}_{e,ph} + \widehat{H}_{ph,ph}] \rangle\rangle_t . \tag{2}$$

Here and below we use the multi-indices $q \equiv \mathbf{q}l$ and $-q \equiv -\mathbf{q}l$. The free-phonon Hamiltonian, $\widehat{H}_{ph}$, the Hamiltonian of electron-phonon interaction, $\widehat{H}_{e,ph}$, and the Hamiltonian of phonon-phonon interaction, $\widehat{H}_{ph,ph}$, are given by Eqs. (3.12), (19.2), and (6.31), respectively. After calculating the commutators, one can transform Eq. (2) to the form

$$\frac{\partial \langle\langle \widehat{Q}_q \rangle\rangle_t}{\partial t} = \langle\langle \widehat{P}_q \rangle\rangle_t - \frac{i}{\sqrt{2\hbar\omega_q}} \left\langle\left\langle \sum_j (\hat{\chi}_q^{(j)+} - \hat{\chi}_{-q}^{(j)}) \right\rangle\right\rangle_t , \tag{3}$$

where $\widehat{P}_q \equiv \widehat{P}_{\mathbf{q}l}$ is the operator of normal momentum. This operator appears from the commutator $[\widehat{Q}_q, \widehat{H}_{ph}]$ and is expressed in terms of creation and annihilation operators according to Eq. (6.9). The commutation relation for $\widehat{P}_q$ and $\widehat{Q}_q$ is given by Eq. (6.7). Since $[\widehat{Q}_q, \widehat{H}_{ph,ph}] = 0$, the phonon-phonon collisions do not influence the evolution of the averaged coordinate in a direct way. However, the contribution of higher-order correlation functions leads to the relaxation of the averaged coordinate due to these collisions, see below. We also note that the second term on the right-hand side of Eq. (3) is equal to zero because the Hamiltonian of electron-phonon interaction is Hermitian and, as a consequence, $\hat{\chi}_q^{(j)+} = \hat{\chi}_{-q}^{(j)}$.

Calculating the second derivative of the normal coordinate over time, one obtains the equation of motion

$$\frac{\partial^2 \langle\langle \hat{Q}_q \rangle\rangle_t}{\partial t^2} = \frac{1}{i\hbar} \langle\langle [\hat{P}_q, \hat{H}_{ph} + \hat{H}_{e,ph} + \hat{H}_{ph,ph}] \rangle\rangle_t , \tag{4}$$

which is conveniently rewritten in the form

$$\left(\frac{\partial^2}{\partial t^2} + \omega_q^2 \right) \langle\langle \hat{Q}_q \rangle\rangle_t = F_{qt} + \frac{1}{i\hbar} \langle\langle [\hat{P}_q, \hat{H}_{ph,ph}] \rangle\rangle_t . \tag{5}$$

The term $\propto \omega_q^2$ follows from the commutator with $\hat{H}_{ph}$. In the right-hand side of Eq. (5), apart from the relaxation contribution due to phonon-phonon collisions, we have introduced the driving force caused by electron-phonon interaction:

$$F_{qt} = \frac{1}{i\hbar} \langle\langle [\hat{P}_q, \hat{H}_{e,ph}] \rangle\rangle_t = -\sqrt{\frac{\omega_q}{2\hbar}} \left\langle\left\langle \sum_j (\hat{\chi}_q^{(j)+} + \hat{\chi}_{-q}^{(j)}) \right\rangle\right\rangle_t . \tag{6}$$

This force is proportional to the constants of electron-phonon coupling. Consider, for example, a simple two-band model, where

$$\sum_j \hat{\chi}_{\mathbf{q}l}^{(j)} = 2 \sum_{\mathbf{pq}} C_{\mathbf{q}l}^{(c)} \hat{a}_{c\mathbf{p}+\hbar\mathbf{q}}^+ \hat{a}_{c\mathbf{p}} + 2 \sum_{\mathbf{pq}} C_{\mathbf{q}l}^{(v)} \hat{a}_{v\mathbf{p}+\hbar\mathbf{q}}^+ \hat{a}_{v\mathbf{p}}. \tag{7}$$

The factor of 2 comes from the sum over spin variables. Note that, in the case of interaction with long-wavelength phonons, the coupling constants $C_{\mathbf{q}l}^{(c)}$ and $C_{\mathbf{q}l}^{(v)}$ do not depend on $\mathbf{p}$. Equation (7) is a straightforward generalization of the single-band model, and the constants $C_{\mathbf{q}l}^{(c)} = C_q$ for the cases of deformation-potential interaction of conduction-band electrons with longitudinal acoustic phonons and Froelich interaction with longitudinal optical phonons are given by Eq. (21.1). Taking into account Eqs. (6) and (7), we obtain

$$F_{qt} \equiv F_{\mathbf{q}lt} = -\sqrt{\frac{2\omega_{\mathbf{q}l}}{\hbar}} \left(C_{\mathbf{q}l}^{(c)} - C_{\mathbf{q}l}^{(v)} \right) \frac{\rho_{\mathbf{q}}(t)}{e}, \tag{8}$$

where the Fourier component of the charge density in the conduction band, $\rho_{\mathbf{q}}(t) = \langle\langle \hat{\rho}_{\mathbf{q}} \rangle\rangle_t$, appears after averaging the charge density operator $\hat{\rho}_{\mathbf{q}} = 2e \sum_{\mathbf{p}} \hat{a}_{c\mathbf{p}}^+ \hat{a}_{c\mathbf{p}+\hbar\mathbf{q}}$.

Since the density of electrons generated in the conduction band by an optical pulse is equal to the density of electrons coming from the valence band, we obtain the difference of the coupling constants in Eq. (8). Considering the deformation mechanism of electron-phonon interaction, one can always assume that this difference is not equal to zero, because the microscopic fields acting on electrons in the conduction and valence bands are different. For example, expressing the constants of electron coupling to longitudinal acoustic phonons in terms of the deformation potentials, we obtain

$$C_{\mathbf{q}LA}^{(c)} - C_{\mathbf{q}LA}^{(v)} = \sqrt{\frac{\hbar q}{2s_l\rho V}}(\mathcal{D}^{(c)} - \mathcal{D}^{(v)}), \tag{9}$$

where s_l is the longitudinal sound velocity. Therefore, there exists a driving force directly proportional to the Fourier component of the photoexcited electron density. If we consider the excitation of phonons by macroscopic electric fields, the driving force is equal to zero, because the total generated charge is zero (the number of conduction-band electrons is equal to the number of holes in the valence band) and the electrostatic field is absent. However, this is true only for homogeneous media. In real samples the excitation takes place in the depletion region near the surface. There, owing to the presence of a built-in electric field, the excited electrons and holes start to drift in the opposite directions thereby creating a macroscopic electrostatic field which acts on the ionic sublattices. The excitation of the longitudinal optical phonons becomes possible in this way (problem 11.16).

It is important that a single optical pulse creates carriers in a macroscopically uniform state, and the photoexcited electron density can contain only the components with very small q, of the order of inverse size of the laser beam spot or inverse absorption length. Optical excitation with two pulses may result in a transient density grating with $|\mathbf{q}|$ of the order of inverse light wavelength. In any case, the driving force $F_{\mathbf{q}lt}$ excites only the phonons with $\mathbf{q}$ much smaller than the Brillouin zone. Microscopically, one can regard them as phonons with $\mathbf{q} = 0$, expressing the driving force according to Eq. (8) with $\rho_{\mathbf{q}}(t) = \delta_{\mathbf{q},0}eVn_{ct}$, where $n_{ct} = (2/V)\sum_{\mathbf{p}} f_{c\mathbf{p}t}$ is the electron density. Therefore, according to Eq. (54.4), the temporal dependence of the driving force is determined by the ultrafast photogeneration. A strong optical pulse creates a macroscopically large number of coherent phonons in the center of the Brillouin zone.

If we neglect the phonon-phonon relaxation term in Eq. (5), this equation, with the initial condition $\langle\langle\widehat{Q}_q\rangle\rangle_{t\to-\infty} = 0$, is solved as

$$\langle\langle\widehat{Q}_q\rangle\rangle_t = \int_{-\infty}^{t} dt'\, k(t-t')F_{qt}, \quad k(\tau) = \frac{\sin\omega_q\tau}{\omega_q}. \tag{10}$$

Because of the oscillating nature of the kernel $k(t - t')$ in this integral expression, the coherent phonon oscillations can be efficiently excited only under an ultrafast pump, when the characteristic time scale of F_{qt} is smaller than ω_q^{-1}.

Let us consider the damping of coherent phonons due to their interaction with equilibrium phonons. First, we simply replace the relaxation contribution in Eq. (5) by $-2\gamma_q\langle\langle\widehat{P}_q\rangle\rangle_t = -2\gamma_q(\partial\langle\langle\widehat{Q}_q\rangle\rangle_t/\partial t)$, where γ_q is a phenomenological relaxation rate. As a result, we obtain the equation of motion for the classical oscillator damped by a friction force:

$$\left(\frac{\partial^2}{\partial t^2} + 2\gamma_q\frac{\partial}{\partial t} + \omega_q^2\right)\langle\langle\widehat{Q}_q\rangle\rangle_t = F_{qt}. \tag{11}$$

Its solution is also given by Eq. (10), where, however, the kernel $k(\tau) = \widetilde{\omega}_q^{-1}e^{-\gamma_q\tau}\sin\widetilde{\omega}_q\tau$ contains an exponentially decreasing factor and the renormalized frequency $\widetilde{\omega}_q = \sqrt{\omega_q^2 - \gamma_q^2}$. If $\gamma_q \ll \omega_q$, this solution describes weakly damped oscillations, and the frequency decreases because the friction slows down the response of the system. A microscopic description of the damping of coherent oscillations cannot be directly reduced to the description given in Chapter 5 for the relaxation of the phonon density matrix. Below we consider the damping based on a microscopic theory.

Calculating the commutator in Eq. (5), we reduce the last term on the right-hand side of this equation to the form

$$-\frac{1}{2\sqrt{\rho V}}\sum_{q_1 q_2}\beta(-q, q_1, q_2)\langle\langle\widehat{Q}_{q_1}\widehat{Q}_{q_2}\rangle\rangle_t \equiv -\sum_{q_1 q_2}C_{QQ}, \tag{12}$$

where $C_{QQ} \equiv \beta(-q, q_1, q_2)\langle\langle\widehat{Q}_{q_1}\widehat{Q}_{q_2}\rangle\rangle_t/2\sqrt{\rho V}$. The correlation function $\langle\langle\widehat{Q}_{q_1}\widehat{Q}_{q_2}\rangle\rangle_t$ satisfies the following equation:

$$\frac{\partial\langle\langle\widehat{Q}_{q_1}\widehat{Q}_{q_2}\rangle\rangle_t}{\partial t} = \frac{1}{i\hbar}\langle\langle[\widehat{Q}_{q_1}\widehat{Q}_{q_2}, \widehat{H}_{ph} + \widehat{H}_{ph,ph}]\rangle\rangle_t . \tag{13}$$

Neglecting the mixing of electron-phonon and phonon-phonon contributions, we do not include the Hamiltonian $\widehat{H}_{e,ph}$ in the commutator. Equation (13) couples $\langle\langle\widehat{Q}_{q_1}\widehat{Q}_{q_2}\rangle\rangle_t$ to the correlation functions $\langle\langle\widehat{Q}_{q_1}\widehat{P}_{q_2}\rangle\rangle_t$ and $\langle\langle\widehat{P}_{q_1}\widehat{Q}_{q_2}\rangle\rangle_t$. They, in turn, satisfy equations of motion similar to Eq. (13) and are coupled to $\langle\langle\widehat{P}_{q_1}\widehat{P}_{q_2}\rangle\rangle_t$. By analogy to Eq. (12), we introduce three other coefficients,

$$\begin{vmatrix} C_{QP} \\ C_{PQ} \\ C_{PP} \end{vmatrix} = \frac{\beta(-q, q_1, q_2)}{2\sqrt{\rho V}}\begin{vmatrix} \langle\langle\widehat{Q}_{q_1}\widehat{P}_{q_2}\rangle\rangle_t \\ \langle\langle\widehat{P}_{q_1}\widehat{Q}_{q_2}\rangle\rangle_t \\ \langle\langle\widehat{P}_{q_1}\widehat{P}_{q_2}\rangle\rangle_t \end{vmatrix}, \tag{14}$$

and compose the system of equations

$$\partial C_{QQ}/\partial t = C_{QP} + C_{PQ}$$
$$\partial C_{QP}/\partial t = C_{PP} - \omega_{q_2}^2 C_{QQ} - M_{QP}$$
$$\partial C_{PQ}/\partial t = C_{PP} - \omega_{q_1}^2 C_{QQ} - M_{PQ}$$
$$\partial C_{PP}/\partial t = -\omega_{q_1}^2 C_{QP} - \omega_{q_2}^2 C_{PQ} - M_{PP}$$

(15)

with zero initial conditions at $t = 0$. This system has the same form as Eq. (27.23).

The terms M_{QP}, M_{PQ}, and M_{PP} come from the commutators of $\widehat{H}_{ph,ph}$ with $\widehat{Q}_{q_1}\widehat{P}_{q_2}$, $\widehat{P}_{q_1}\widehat{Q}_{q_2}$, and $\widehat{P}_{q_1}\widehat{P}_{q_2}$, respectively. The system (15) becomes closed if we calculate these terms within the second-order accuracy with respect to phonon-phonon coupling coefficients β. This implies that the three-operator averages appearing in M_{QP}, M_{PQ}, and M_{PP} after calculating the commutators are evaluated in the free-phonon approximation. For example,

$$\langle\langle \widehat{Q}_{q_1}\widehat{Q}_{q_2}\widehat{Q}_{q_3}\rangle\rangle_t \simeq \langle\langle\widehat{Q}_{q_1}\rangle\rangle_t\langle\langle\widehat{Q}_{q_2}\widehat{Q}_{q_3}\rangle\rangle_t + \langle\langle\widehat{Q}_{q_2}\rangle\rangle_t\langle\langle\widehat{Q}_{q_1}\widehat{Q}_{q_3}\rangle\rangle_t$$

$$+\langle\langle\widehat{Q}_{q_3}\rangle\rangle_t\langle\langle\widehat{Q}_{q_1}\widehat{Q}_{q_2}\rangle\rangle_t = \langle\langle\widehat{Q}_{q_1}\rangle\rangle_t\delta_{-q_2,q_3}(2N_{q_2}+1)\hbar/2\omega_{q_2}$$

(16)

$$+\langle\langle\widehat{Q}_{q_2}\rangle\rangle_t\delta_{-q_3,q_1}(2N_{q_3}+1)\hbar/2\omega_{q_3} + \langle\langle\widehat{Q}_{q_3}\rangle\rangle_t\delta_{-q_1,q_2}(2N_{q_1}+1)\hbar/2\omega_{q_1},$$

where N_q is the occupation number of phonons ($N_q = N_{-q}$ since the system is isotropic). Neglecting possible heating of the phonon system during the photoexcitation, one can treat N_q as the equilibrium Planck distribution given by Eq. (3.28). Applying the procedure of averaging given by Eq. (16) to the commutators of $\widehat{H}_{ph,ph}$ with $\widehat{Q}_{q_1}\widehat{P}_{q_2}$, $\widehat{P}_{q_1}\widehat{Q}_{q_2}$, and $\widehat{P}_{q_1}\widehat{P}_{q_2}$, one should omit the terms which contain the averages of two phonon operators standing in the Hamiltonian $\widehat{H}_{ph,ph}$. The calculation done in this way leads us to the following expressions:

$$M_{QP}(t) = \frac{\hbar(N_{\mathbf{q_1}}^{l_1} + 1/2)}{2\rho V \omega_{\mathbf{q_1}l_1}} \sum_{l'} \beta_{ll_1l_2}(-\mathbf{q}, \mathbf{q_1}, \mathbf{q_2})$$

$$\times \beta_{l'l_1l_2}(\mathbf{q}, -\mathbf{q_1}, -\mathbf{q_2})\langle\langle\widehat{Q}_{l'\mathbf{q}}\rangle\rangle_t \;,$$

(17)

$$M_{PQ}(t) = \frac{\hbar(N_{\mathbf{q_2}}^{l_2} + 1/2)}{2\rho V \omega_{\mathbf{q_2}l_2}} \sum_{l'} \beta_{ll_1l_2}(-\mathbf{q}, \mathbf{q_1}, \mathbf{q_2})$$

$$\times \beta_{l'l_1l_2}(\mathbf{q}, -\mathbf{q_1}, -\mathbf{q_2})\langle\langle\widehat{Q}_{l'\mathbf{q}}\rangle\rangle_t \;,$$

(18)

and $M_{PP}(t) = 0$. To obtain these expressions, we have used the property $\beta_{l_1 l_2 l_3}(\mathbf{q}_1, \mathbf{q}_2, \mathbf{q}_3) \propto \Delta_{\mathbf{q}_1 + \mathbf{q}_2 + \mathbf{q}_3, 0}$ (see Sec. 6 and Chapter 5), and the umklapp processes have been neglected because $\mathbf{q}$ is small.

The exact solution of the system (15) with respect to C_{QQ} is written as (problem 11.17)

$$
\begin{aligned}
C_{QQ}(t) = \int_{-\infty}^{t} dt' \bigg\{ & M_{QP}(t') \frac{\sin[\omega_{q_2}(t' - t)]}{\omega_{q_2}} \cos[\omega_{q_1}(t' - t)] \\
& + M_{PQ}(t') \frac{\sin[\omega_{q_1}(t' - t)]}{\omega_{q_1}} \cos[\omega_{q_2}(t' - t)] \\
& - M_{PP}(t') \frac{\sin[\omega_{q_1}(t' - t)]}{\omega_{q_1}} \frac{\sin[\omega_{q_2}(t' - t)]}{\omega_{q_2}} \bigg\}.
\end{aligned}
\tag{19}
$$

Since $M_{PP}(t) = 0$ under the approximations we use, the last term in this expression vanishes. The terms containing M_{QP} and M_{PQ} can be transformed by integrating over t' by parts. As a result, only the derivatives of the averaged coordinate remain in Eq. (19) after substituting the expressions (17) and (18) there. Therefore, the general form of the relaxation term (12) is

$$
\sum_{l'} \left[K_{\mathbf{q}}^{(ll')}(0) \langle\langle \widehat{Q}_{l'\mathbf{q}} \rangle\rangle_t - \int_{-\infty}^{t} dt' K_{\mathbf{q}}^{(ll')}(t - t') \frac{\partial \langle\langle \widehat{Q}_{l'\mathbf{q}} \rangle\rangle_{t'}}{\partial t'} \right],
\tag{20}
$$

where the kernel $K_{\mathbf{q}}^{(ll')}(\tau)$ depends on the absolute value of τ. The explicit form of this kernel is deduced from the expressions (17)−(19):

$$
\begin{aligned}
K_{\mathbf{q}}^{(ll')}(\tau) = \sum_{l_1 l_2} \sum_{\mathbf{q}_1 \mathbf{q}_2} & \frac{\hbar \beta_{ll_1 l_2}(-\mathbf{q}, \mathbf{q}_1, \mathbf{q}_2) \beta_{l' l_1 l_2}(\mathbf{q}, -\mathbf{q}_1, -\mathbf{q}_2)}{4\rho V \omega_{\mathbf{q}_1 l_1} \omega_{\mathbf{q}_2 l_2}} \\
& \times \left[(N_{\mathbf{q}_1}^{l_1} + N_{\mathbf{q}_2}^{l_2} + 1) \frac{\cos[(\omega_{\mathbf{q}_1 l_1} + \omega_{\mathbf{q}_2 l_2})\tau]}{\omega_{\mathbf{q}_1 l_1} + \omega_{\mathbf{q}_2 l_2}} \right. \\
& \left. + (N_{\mathbf{q}_2}^{l_2} - N_{\mathbf{q}_1}^{l_1}) \frac{\cos[(\omega_{\mathbf{q}_1 l_1} - \omega_{\mathbf{q}_2 l_2})\tau]}{\omega_{\mathbf{q}_1 l_1} - \omega_{\mathbf{q}_2 l_2}} \right].
\end{aligned}
\tag{21}
$$

The first term in this expression corresponds to a decay of the coherent phonon $\mathbf{q}l$ into two phonons, $\mathbf{q}_1 l_1$ and $\mathbf{q}_2 l_2$, while the second term corresponds to a fusion of this phonon with another one. Both these processes are responsible for phonon relaxation; see Chapter 5.

Substituting the expression (20) in place of the last term of the right-hand side of Eq. (5), one can see that the phonon-phonon collisions lead to non-Markovian contribution to phonon dynamics. Apart from this, different modes with the same wave vector mix because of these collisions. Neglecting this mixing, i.e., considering only the terms with $l' = l$ in $K_{\mathbf{q}}^{(ll')}(\tau)$, we obtain a single-mode integro-differential equation

$$\left(\frac{\partial^2}{\partial t^2} + \overline{\omega}_q^2\right) \langle\langle\widehat{Q}_q\rangle\rangle_t + \int_{-\infty}^{t} dt'\, K_q(t - t')\frac{\partial\langle\langle\widehat{Q}_q\rangle\rangle_{t'}}{\partial t'} = F_{qt}, \qquad (22)$$

where we again use the multi-indices $q \equiv \mathbf{q}l$ and the definition $K_q(\tau) \equiv K_{\mathbf{q}}^{(ll)}(\tau)$. According to Eq. (21), $K_q(\tau)$ is real since $\beta_{ll_1l_2}(-\mathbf{q}, \mathbf{q}_1, \mathbf{q}_2) = \beta_{ll_1l_2}^*(\mathbf{q}, -\mathbf{q}_1, -\mathbf{q}_2)$. The frequency $\overline{\omega}_q = \sqrt{\omega_q^2 - K_q(0)}$ is slightly smaller than ω_q due to the contribution of the first term from the expression (20). To prove that the integral term in Eq. (22) leads to dissipation, one can carry out a Fourier transformation of Eq. (22). As a result of this transformation, Eq. (22) becomes an algebraic equation, and the Fourier component $\langle\langle\widehat{Q}_q\rangle\rangle_\omega = \int_{-\infty}^{\infty} dt e^{i\omega t}\langle\langle\widehat{Q}_q\rangle\rangle_t$ is given by

$$\langle\langle\widehat{Q}_q\rangle\rangle_\omega = -\frac{F_{q\omega}}{\omega^2 - \overline{\omega}_q^2 + i\omega\Gamma_{q\omega}}, \qquad (23)$$

where $F_{q\omega}$ is the Fourier transform of the driving force. Next, $\Gamma_{q\omega} = \int_{-\infty}^{0} d\tau\, e^{-i\omega\tau + \lambda\tau} K_q(\tau)$, where $\lambda \to +0$. Since the real part of $\Gamma_{q\omega}$ is normally positive (problem 11.18), the expression (23) corresponds to a damped mode (check with the corresponding results of Sec. 27).

To prove that the microscopically derived equation (22) is related to the phenomenological equation (11), one needs to investigate the kernel $K_q(t - t')$. We have already noted that $K_q(t - t')$ depends only on the absolute value of its argument. One can expect that $K_q(t-t')$ decreases with increasing $|t - t'|$, because it is formed as a sum of the oscillating functions $\cos[(\omega_{q_1} \pm \omega_{q_2})(t - t')]$ over the modes q_1 and q_2 with different frequencies (problem 11.19). The characteristic scale of the decrease is estimated as $\Delta\omega^{-1}$, where $\Delta\omega$ is the width of the frequency band of the phonons participating in the collisions. In the case of phonon decay processes (the first term in Eq. (21)), $\Delta\omega$ is of the order of optical phonon frequency. For the dynamical processes whose characteristic times (the period of oscillations and decay time) are large in comparison to $\Delta\omega^{-1}$, the kernel $K_q(t - t')$ can be effectively approximated by a δ-function. In particular, when $K_q(t-t') = 4\gamma_q\delta(t-t')$, Eq. (22) is directly reduced to Eq. (11). Therefore, the phenomenological equation (11) is microscopically justified for the acoustic phonons whose frequencies ω_q are small. However, this equation cannot be justified for application to the coherent oscillations of optical phonon modes.

Problems

11.1. Prove that the non-diagonal components of the density matrix, $\langle\delta|\hat{\bar{\rho}}_t|\delta'\rangle \equiv \bar{\rho}_{\delta\delta'}$, are small if $|\varepsilon_\delta - \varepsilon_{\delta'}|/\hbar$ is large in comparison to the rates of generation and collision-induced relaxation.

<u>Solution</u>: In the basis $|\delta\rangle$, Eq. (53.7) is written as

$$\frac{\partial\bar{\rho}_{\delta\delta'}}{\partial t} + \frac{i}{\hbar}(\varepsilon_\delta - \varepsilon_{\delta'})\bar{\rho}_{\delta\delta'} = \langle\delta|\hat{G}_t + \hat{J}(\hat{\bar{\rho}}|t)|\delta'\rangle.$$

Estimating $\langle\delta|\hat{J}(\hat{\bar{\rho}}|t)|\delta'\rangle$ as $-\bar{\rho}_{\delta\delta'}/\bar{\tau}$ and $\langle\delta|\hat{G}_t|\delta'\rangle$ as $\nu_G\bar{\rho}_{\delta\delta'}$, where ν_G is a characteristic rate of generation, we obtain the required proof.

11.2. Calculate the rates of intra- and intersubband transitions of the 2D electrons interacting with impurities.

<u>Solution</u>. Consider a quantum well in the plane XOY. The relaxation rates (see Sec. 8) due to electron transitions from the 2D state $|n\mathbf{p}\rangle$, described by the wave function $\psi_z^{(n)}L^{-1}e^{i\mathbf{p}\cdot\mathbf{r}/\hbar}$, to the subband n' are given by the following expression:

$$\nu_p^{(nn')} = \frac{2\pi}{\hbar}\int\frac{d\mathbf{p}'}{(2\pi\hbar)^2}w_{nn'}(|\mathbf{p}-\mathbf{p}'|/\hbar)\delta(\varepsilon_{np} - \varepsilon_{n'p'}),$$

where $\varepsilon_{np} = \varepsilon_n + p^2/2m$ is the energy of the 2D state and $w_{nn'}(q)$ is the Fourier transform of the effective 2D correlation function of the random potential of impurities. Assuming that the impurities are homogeneously distributed in space, we obtain

$$w_{nn'}(q) = n_{im}\int_{-\infty}^{\infty}\frac{dq_z}{2\pi}|v(\mathbf{Q})|^2|\langle n|e^{iq_z z}|n'\rangle|^2,$$

where $\mathbf{Q} = (\mathbf{q}, q_z)$ is the 3D wave vector, $v(\mathbf{Q})$ is the 3D Fourier transform of the single-impurity potential, and n_{im} is the concentration of impurities. The characteristic q_z contributing to the integral are of the order of inverse width of the well, $1/d$. If one can neglect the Q-dependence of $v(\mathbf{Q})$ on this scale, $w_{nn'}(q)$ is equal to $n_{im}|v(0)|^2$ multiplied by the overlap factor $\int dz|\psi_z^{(n)}|^2|\psi_z^{(n')}|^2$ roughly estimated as $1/d$. In this case, corresponding to a short-range impurity potential, $w_{nn'}(q)$ is independent of q. Therefore, $\nu_p^{(nn')}$ becomes momentum-independent and equal to $mw_{nn'}/\hbar^3$.

11.3. Prove that $\text{Re}W_\Delta(t) \simeq \pi\hbar\delta_{\hbar\lambda}(\Delta)w_t$ in the case of slowly varying electric field Ew_t.

<u>Solution</u>: Instead of Eq. (54.5), we have

$$W_\Delta(t) = \int_{-\infty}^0 d\tau w_{t+\tau}e^{\lambda\tau - i\Delta\tau/\hbar} \simeq w_t\int_{-\infty}^0 d\tau e^{\lambda\tau - i\Delta\tau/\hbar} = w_t(\lambda - i\Delta/\hbar)^{-1},$$

and the real part of this expression is equal to $w_t\hbar^2\lambda/[(\hbar\lambda)^2 + \Delta^2] = \pi\hbar\delta_{\hbar\lambda}(\Delta)w_t$, where we use the first representation of the broadened δ-function from problem 1.4. Note that a substitution of a finite broadening energy in place of an infinitely small $\hbar\lambda$ gives us only an order-of-value estimate.

11.4. Consider the coherent control with two infinitely short pulses.

Solution: Substituting $w_t \propto \delta(t)$ into Eq. (54.17), we calculate the integrals over time and obtain $n_{2\infty} \propto 1 + \cos(\phi - \Delta\omega t_d)$. The amplitude of the oscillations is independent of $\Delta\omega$ in this approximation. The model of δ-pulse excitation is valid in the limit $\tau_p \ll (\Delta\omega)^{-1}, t_d$.

11.5. Derive Eq. (54.22) from Eq. (54.20) at $\Delta\omega = 0$ and $\nu_{LO}^{(21)} = 0$.

Hint: Differentiate Eq. (54.20) over time.

11.6. Consider the non-linear regime of intersubband response to a stationary excitation which is abruptly turned on at $t = 0$.

Solution: In the collisionless approximation and at $\Delta\omega = 0$, we use Eq. (54.22) with $w_t = 1$ at $t > 0$,

$$\frac{d^2 \Delta n_t}{dt^2} + \omega_R^2 \Delta n_t = 0,$$

and solve it with the initial condition $\Delta n_{t=0} = n_{2D}$. The solution $\Delta n_t = n_{2D} \cos \omega_R t$ describes the Rabi oscillations. In the general case, a substitution of $w_t = \theta(t)$ into Eq. (54.20) and subsequent Laplace transformation of this equation with the above initial condition leads us to an algebraic equation for $\Delta n_s = \int_0^\infty dt e^{-st} \Delta n_t$:

$$s\Delta n_s - n_{2D} + \omega_R^2 \frac{s\Delta n_s}{s^2 + (\Delta\omega)^2} + \nu_{LO}^{(21)}\left(\Delta n_s - \frac{n_{2D}}{s}\right) = 0.$$

The evolution is determined by three frequencies, ω_R, $\Delta\omega$, and $\nu_{LO}^{(21)}$. The final distribution is given by the limiting transition $\Delta n_{t\to\infty} = \lim_{s\to 0}(s\Delta n_s)$. In the exact resonance, $\Delta\omega = 0$, the electrons occupy both levels with equal densities, $\Delta n_{t\to\infty} = 0$, while at $\Delta\omega \neq 0$ only the lowest level remains occupied at $t > |\Delta\omega|^{-1}$.

11.7. Using the general equations (35.2) and (35.3), find the coefficients D_ε and V_ε entering Eq. (55.13).

Solution: Averaging Eq. (35.2) with the transition probabilities (35.3) over the angle of $\mathbf{p}$, we obtain the collision integral

$$J_{ac}(f|\varepsilon t) = \frac{\rho_{2D}}{2}\int_0^\infty d\varepsilon' \Delta\mathcal{K}_{\varepsilon\varepsilon'}\delta''\left(\varepsilon' - \varepsilon\right)(f_{\varepsilon't} - f_{\varepsilon t})$$

$$-\frac{\rho_{2D}}{4}\int_0^\infty d\varepsilon' \delta\mathcal{K}_{\varepsilon\varepsilon'}\delta'\left(\varepsilon' - \varepsilon\right)(f_{\varepsilon't} + f_{\varepsilon t})$$

describing the energy relaxation of isotropic electron distribution ($\varepsilon = p^2/2m$ and $\varepsilon' = p'^2/2m$). The coefficients in this expression are given by

$$\left|\begin{array}{c}\Delta\mathcal{K}_{\varepsilon\varepsilon'} \\ \delta\mathcal{K}_{\varepsilon\varepsilon'}\end{array}\right| = \frac{2\pi}{\hbar}L^2\int_0^{2\pi}\frac{d\varphi}{2\pi}\sum_{q_\perp}|C_Q|^2|\langle 1|e^{iq_\perp z}|1\rangle|^2\left|\begin{array}{c}(2N_Q + 1)(\hbar\omega_Q)^2/2 \\ 2\hbar\omega_Q\end{array}\right|,$$

where $Q = \sqrt{q_\perp^2 + (\mathbf{p} - \mathbf{p}')^2/\hbar^2}$. Integrating the expression for J_{ac} by parts, we write it in the form of the right-hand side of Eq. (55.13), where $D_\varepsilon = \rho_{2D}\Delta\mathcal{K}_{\varepsilon\varepsilon}/2$ and $V_\varepsilon = \rho_{2D}\delta\mathcal{K}_{\varepsilon\varepsilon}/4$. The characteristic $q_\perp$ are of the order of $1/d$, where d is the quantum well width. Therefore, at $\hbar/d \gg \sqrt{2m\varepsilon}$ the transverse components of the phonon wave numbers are much larger than the in-plane components, $Q \simeq q_\perp$, and the energy dependence of the coefficients can be neglected:

$$\left|\begin{array}{c} D \\ V \end{array}\right| = \frac{\pi\rho_{2D}}{\hbar} \int_{-\infty}^{\infty} \frac{dq_\perp}{2\pi} \frac{\mathcal{D}^2\hbar\omega_{q_\perp}}{2s_l^2\rho} |\langle 1|e^{iq_\perp z}|1\rangle|^2 \left| \begin{array}{c} \coth(\hbar s_l q_\perp/2T) \; (\hbar s_l q_\perp)^2/2 \\ \hbar s_l q_\perp \end{array} \right|,$$

where we assume the equilibrium distribution of phonons. If $\hbar s_l/d \ll T$, the scattering by equipartition phonons occurs, and $D/V = T$. In the low-temperature case, when $\hbar s_l/d > T$, one obtains $D/V \sim \hbar s_l/d$. For a more detailed calculation, one may use the expression of $\langle 1|e^{iq_\perp z}|1\rangle$ given in problem 4.13.

11.8. Find the Green's function of the drift-diffusion equation (55.13) with the boundary conditions (55.14).

Solution: This Green's function is determined by the equation

$$\left[\frac{\partial}{\partial t} - \frac{\partial}{\partial\varepsilon}\left(D\frac{\partial}{\partial\varepsilon} + V\right)\right] G_t(\varepsilon, \varepsilon') = \delta(t)\delta(\varepsilon - \varepsilon'),$$

which is reduced to

$$\left(\frac{\partial}{\partial t} - D\frac{\partial^2}{\partial\varepsilon^2}\right)\widetilde{G}_t(\varepsilon, \varepsilon') = \delta(t)\delta(\varepsilon - \varepsilon')$$

after the substitution $G_t(\varepsilon, \varepsilon') = \exp[-V^2 t/4D - V(\varepsilon - \varepsilon')/2D]\widetilde{G}_t(\varepsilon, \varepsilon')$. The boundary condition at zero energy is transformed by this substitution as

$$\left(D\frac{\partial}{\partial\varepsilon} + V\right)G_t(\varepsilon, \varepsilon')\Big|_{\varepsilon=0} = 0 \quad \Rightarrow \quad \left(D\frac{\partial}{\partial\varepsilon} + \frac{V}{2}\right)\widetilde{G}(\varepsilon, \varepsilon')\Big|_{\varepsilon=0} = 0.$$

Finally, the solution $\widetilde{G}_t(\varepsilon, \varepsilon')$ satisfying the above boundary condition is written in the form

$$\widetilde{G}(\varepsilon, \varepsilon') = \frac{1}{\sqrt{4\pi Dt}}\left[e^{-(\varepsilon-\varepsilon')^2/4Dt} + e^{-(\varepsilon+\varepsilon')^2/4Dt}\right.$$
$$\left. + \frac{V}{D}\int_0^\infty d\varepsilon'' e^{-(\varepsilon+\varepsilon'+\varepsilon'')^2/4Dt + V\varepsilon''/2D}\right],$$

where the integral can be expressed through the error function, erf(...).

11.9. Apply the Laplace transformation to the system (55.19) and solve the transformed system.

Solution: Searching for the distribution function in the form $f_{\xi t} = \exp(-V^2 t/4D - V\xi/2D)w_{\xi t}$, we carry out the Laplace transformation $\int_0^\infty dt e^{-st}\ldots$ of both sides of Eq. (55.19) and obtain

$$sw_{\xi s}^< - D\frac{\partial^2 w_{\xi s}^<}{\partial\xi^2} = w_{\xi,t=0}^<, \quad \xi < 0$$

$$(s+\nu)w_{\xi s}^> - D\frac{\partial^2 w_{\xi s}^>}{\partial\xi^2} = w_{\xi,t=0}^>, \quad \xi > 0$$

with the boundary conditions

$$w_{\xi=0,s}^> = w_{\xi=0,s}^<, \quad (\partial w_{\xi s}^</\partial\xi)_{\xi=0} = (\partial w_{\xi s}^>/\partial\xi)_{\xi=0},$$

$$w^>_{\xi \to \infty, s} = 0 , \qquad w^<_{\xi \to -\infty, s} = 0.$$

The initial condition is transformed as $w_{\xi, t=0} = e^{V\xi/2D} f_{\xi, t=0}$. The solutions of the non-homogeneous differential equations written above are expressed in terms of the fundamental solutions $\exp(\pm\sqrt{(s+\nu)/D}\xi)$ and $\exp(\pm\sqrt{s/D}\xi)$ as

$$w^<_{\xi s} = C_< e^{\sqrt{s/D}\xi} + \int_{-\infty}^0 \frac{d\xi'}{2\sqrt{Ds}} e^{-\sqrt{s/D}|\xi - \xi'|} w_{\xi', t=0} , \quad \xi < 0,$$

$$w^>_{\xi s} = C_> e^{-\sqrt{(s+\nu)/D}\xi} + \int_0^\infty \frac{d\xi'}{2\sqrt{D(s+\nu)}} e^{-\sqrt{(s+\nu)/D}|\xi - \xi'|} w_{\xi', t=0} , \quad \xi > 0.$$

Let us take the spatial derivatives of these solutions and exclude the constants $C_>$ and $C_<$. We obtain

$$\frac{\partial w^<_{\xi s}}{\partial \xi}\bigg|_{\xi=0} - \sqrt{\frac{s}{D}} w^<_{\xi=0, s} = -\int_{-\infty}^0 \frac{d\xi'}{D} e^{\sqrt{s/D}\xi'} w_{\xi', t=0} ,$$

$$\frac{\partial w^>_{\xi s}}{\partial \xi}\bigg|_{\xi=0} + \sqrt{\frac{s+\nu}{D}} w^>_{\xi=0, s} = \int_0^\infty \frac{d\xi'}{D} e^{-\sqrt{(s+\nu)/D}\xi'} w_{\xi', t=0} .$$

Finally, applying the boundary conditions, we find

$$w_{\xi=0, s} = \frac{1}{\sqrt{D(s+\nu)} + \sqrt{Ds}} \left[\int_0^\infty d\xi e^{-\sqrt{(s+\nu)/D}\xi} w_{\xi, t=0} + \int_{-\infty}^0 d\xi e^{\sqrt{s/D}\xi} w_{\xi, t=0} \right],$$

which leads to Eq. (55.23) for $f_{\xi=0, t}$.

11.10. Consider the transient cyclotron absorption of photoexcited 2D electrons.

Solution: The conductivity tensor is found from the quasi-classical kinetic equation (9.34), where the collision integral is written in the form $-f_{\mathbf{p}t}/\tau_\varepsilon$:

$$\sigma_{\alpha\beta} = 2e^2 \int \frac{d\mathbf{p}}{(2\pi\hbar)^2} v_\alpha \frac{v_\beta(\tau_\varepsilon^{-1} - i\omega) + [\boldsymbol{\omega}_c \times \mathbf{v}]_\beta}{(\tau_\varepsilon^{-1} - i\omega)^2 + \omega_c^2} \left(-\frac{\partial f_{\varepsilon t}}{\partial \varepsilon} \right)$$

with $\mathbf{v} = \mathbf{p}/m$. Since the angular averaging results in $v_\alpha v_\beta \to \delta_{\alpha\beta}\varepsilon/m$, and $v_\alpha[\boldsymbol{\omega}_c \times \mathbf{v}]_\alpha \to 0$, the real part of the diagonal component of this tensor is written as

$$\mathrm{Re}\,\sigma_d = \frac{e^2\rho_{2D}}{m} \int_0^\infty d\varepsilon \, \varepsilon \, \mathrm{Re}\frac{\tau_\varepsilon^{-1} - i\omega}{(\tau_\varepsilon^{-1} - i\omega)^2 + \omega_c^2} \left(-\frac{\partial f_{\varepsilon t}}{\partial \varepsilon} \right) \simeq \frac{\sigma_t}{2} \frac{1}{1 + (\Delta\omega\tau)^2},$$

where σ_t is the transient conductivity in the absence of magnetic fields and at $\omega = 0$. The last equation is written for the case of energy-independent relaxation time τ and under the conditions $|\Delta\omega| = |\omega - \omega_c| \ll \omega_c$ and $\omega_c\tau \ll 1$.

11.11. Justify the resonance approximation for interband excitation in the symmetric two-band model.

Solution: The columnar wave functions $\psi(\mathbf{p}, t)$ of the time-dependent Schroedinger equation $i\hbar\partial\psi(\mathbf{p}, t)/\partial t = \hat{H}_t(\mathbf{p})\psi(\mathbf{p}, t)$ with $\hat{H}_t(\mathbf{p})$ from Eq. (56.1) (the vector potential is $\mathbf{A}_t = -(c\mathbf{E}/\omega)\sin\omega t$) can be represented according to Eq. (5.30): $\psi(\mathbf{p}, t) = e^{-i\varepsilon t/\hbar} u_{\mathbf{p}\varepsilon}(t)$. Expanding the periodic columnar amplitude as $u_{\mathbf{p}\varepsilon}(t) =$

$\sum_k e^{-ik\omega t} u_{\mathbf{p}\varepsilon}(k)$, one can write the following equation for the Fourier components $u_{\mathbf{p}\varepsilon}(k)$ (see also problem 1.14):

$$\left(\varepsilon + k\hbar\omega - \hat{h}_p\right) u_{\mathbf{p}\varepsilon}(k) = \frac{ie}{2\omega}\mathbf{E}\cdot\hat{\mathbf{v}}_{\mathbf{p}}[u_{\mathbf{p}\varepsilon}(k-1) - u_{\mathbf{p}\varepsilon}(k+1)].$$

The four-component columns $u_{\mathbf{p}\varepsilon}(k)$ are represented below as combinations of two-component spinors $u_{c\mathbf{p}\varepsilon}(k)$ and $u_{v\mathbf{p}\varepsilon}(k)$ describing conduction and valence bands. In the following, the kinetic energy $p^2/2m$ is assumed to be small in comparison to $\varepsilon_g/2$ so that the parabolic approximation is valid. Assuming also that $\hbar\omega = \varepsilon_g + \hbar\Delta\omega$, where $|\hbar\Delta\omega|$ is small in comparison to $\varepsilon_g/2$, we rewrite the equation above as a system of equations coupling $u_{c\mathbf{p}\varepsilon}(k)$ to $u_{c\mathbf{p}\varepsilon}(k')$:

$$\left(\frac{p^2}{2m} - k\hbar\Delta\omega + \frac{\varepsilon_g}{2} - \varepsilon - k\varepsilon_g\right) u_{c\mathbf{p}\varepsilon}(k) + \frac{ie}{2m\omega}\mathbf{E}\cdot\mathbf{p}[u_{c\mathbf{p}\varepsilon}(k-1) - u_{c\mathbf{p}\varepsilon}(k+1)]$$

$$+\frac{ies}{2\omega}\mathbf{E}\cdot\hat{\boldsymbol{\sigma}}[u_{v\mathbf{p}\varepsilon}(k-1) - u_{v\mathbf{p}\varepsilon}(k+1)] = 0,$$

$$\left(-\frac{p^2}{2m} - k\hbar\Delta\omega - \frac{\varepsilon_g}{2} - \varepsilon - k\varepsilon_g\right) u_{v\mathbf{p}\varepsilon}(k) - \frac{ie}{2m\omega}\mathbf{E}\cdot\mathbf{p}[u_{v\mathbf{p}\varepsilon}(k-1) - u_{v\mathbf{p}\varepsilon}(k+1)]$$

$$+\frac{ies}{2\omega}\mathbf{E}\cdot\hat{\boldsymbol{\sigma}}[u_{c\mathbf{p}\varepsilon}(k-1) - u_{c\mathbf{p}\varepsilon}(k+1)] = 0.$$

Considering the quasienergies ε in the region $\varepsilon = \pm\varepsilon_g/2 + \delta\varepsilon$, where $|\delta\varepsilon| \ll \varepsilon_g/2$, we write this system for $k = 0$ and $k = \pm 1$. Such a procedure shows us that the system couples $u_{c\mathbf{p}\varepsilon}(0)$ to $u_{v\mathbf{p}\varepsilon}(-1)$ and $u_{c\mathbf{p}\varepsilon}(1)$ to $u_{v\mathbf{p}\varepsilon}(0)$ only. The other amplitudes are small as $eE\hbar s/2\varepsilon_g^2$ and can be neglected. Carrying out the inverse Fourier transformation from $u_{c\mathbf{p}\varepsilon}(k)$ and $u_{v\mathbf{p}\varepsilon}(k)$ to $u_{c\mathbf{p}\varepsilon}(t)$ and $u_{v\mathbf{p}\varepsilon}(t)$, it is easy to show that this resonance approximation is equivalent to a usage of the truncated Hamiltonian (56.5).

11.12. Carry out the unitary transformation (56.7).

Hint: Expand the exponent in $\hat{\mathcal{U}}_t$ of Eq. (56.6) in series.

11.13. Derive the collision integral standing in Eq. (56.17).

Hint: Take into account that

$$\hat{a}^+_{v\sigma\mathbf{p}}\hat{a}_{c\sigma'\mathbf{p}} = -i\sigma[\lambda_{+p}\lambda_{-p}(\hat{c}^+_{+,\sigma\mathbf{p}}\hat{c}_{+,\sigma'\mathbf{p}} - \hat{c}^+_{-,\sigma\mathbf{p}}\hat{c}_{-,\sigma'\mathbf{p}})$$

$$+\lambda^2_{-p}\hat{c}^+_{+,\sigma\mathbf{p}}\hat{c}_{-,\sigma'\mathbf{p}} - \lambda^2_{+p}\hat{c}^+_{-,\sigma\mathbf{p}}\hat{c}_{+,\sigma'\mathbf{p}}].$$

Only the last two terms, which lead to transitions between + and − states, are important in the collision integral.

11.14. Write the Bethe-Salpeter equation for the correlation function $K^{ss'}_{ii',jj'}(\mathbf{p}, \mathbf{p}') = \langle\langle G^s_\varepsilon(i\mathbf{p}, j\mathbf{p}')G^s_{\varepsilon'}(i' - \mathbf{p}, j' - \mathbf{p}')\rangle\rangle$ in the ladder approximation.

Result: The Bethe-Salpeter equation is

$$K^{ss'}_{ii',jj'}(\mathbf{p}, \mathbf{p}') = \delta_{\mathbf{p}\mathbf{p}'}[G^s_\varepsilon(\mathbf{p})]_{ij}[G^{s'}_{\varepsilon'}(\mathbf{p})]_{i'j'}$$

$$+ \sum_{kk'} [G_\varepsilon^s(\mathbf{p})]_{ik} [G_{\varepsilon'}^s(\mathbf{p})]_{i'k'} \frac{1}{V} \sum_{\mathbf{p_1}} n_{im} |v(|\mathbf{p}-\mathbf{p_1}|/\hbar)|^2 K_{kk',jj'}^{ss'}(\mathbf{p_1},\mathbf{p'}).$$

For short-range scattering potential, when $n_{im}|v(|\mathbf{p}-\mathbf{p_1}|/\hbar)|^2 \simeq n_{im}|v(0)|^2 = w$ is constant, this equation is reduced to a system of four algebraic equations,

$$M_{ii',jj'}^{ss'}(\varepsilon,\varepsilon') = L_{ii',jj'}^{ss'}(\varepsilon,\varepsilon') + w \sum_{kk'} L_{ii',kk'}^{ss'}(\varepsilon,\varepsilon') M_{kk',jj'}^{ss'}(\varepsilon,\varepsilon'),$$

where $M_{ii',jj'}^{ss'}(\varepsilon,\varepsilon') = V^{-1} \sum_{\mathbf{pp'}} K_{ii',jj'}^{ss'}(\mathbf{p},\mathbf{p'})$ and

$$L_{ii',jj'}^{ss'}(\varepsilon,\varepsilon') = \frac{1}{V} \sum_{\mathbf{p}} [G_\varepsilon^s(\mathbf{p})]_{ij} [G_{\varepsilon'}^{s'}(\mathbf{p})]_{i'j'}.$$

According to Eq. (56.28), the absorption coefficient is obtained by integrating the functions $M_{cv,cv}^{ss'}(\varepsilon+\hbar\delta\omega,\varepsilon)[f(\varepsilon)-f(\varepsilon+\hbar\delta\omega)]$ over ε.

11.15. Calculate and analyze the density of states of electrons in the presence of interband pumping and elastic scattering by impurities.

Solution: The density of states $\rho(\varepsilon)$ is given by $-(2/\pi V)\mathrm{Im}\sum_{\mathbf{p}} \mathrm{tr}\widehat{G}_\varepsilon^R(\mathbf{p})$, where $\mathrm{tr}\dots$ denotes the matrix trace. This expression also can be written as

$$\rho(\varepsilon) = -\frac{4mp_\omega}{\pi^2\hbar^3}\mathrm{Im}[\Sigma_\varepsilon^R/\hbar\nu_{im}] = \rho_0 \mathrm{Im}\frac{u}{\sqrt{1-u^2}}, \qquad \rho_0 = 2\frac{m^{3/2}\sqrt{\hbar\omega-\varepsilon_g}}{\pi^2\hbar^3}.$$

Analyzing Eq. (56.34), we find that $\rho(\varepsilon=0)$ is zero at $\chi>1$ and non-zero at $\chi<1$. Therefore, the gap exists at $\chi>1$, or, equivalently, at $\beta>\hbar\nu_{im}$. If $\chi>1$, the threshold energy, where the density of states starts to be non-zero, is given by Eq. (56.34) with $u = u_{min} = \sqrt{1-\chi^{-2/3}}$.

11.16. Derive and analyze the equations describing the longitudinal vibration of ionic sublattices in the presence of an external electric field and free charges.

Solution: In the presence of an external electric field and free charges, the longitudinal electric field $\mathbf{E}_L$ entering the system of equations (6.16) should be found from the equation $\epsilon_\infty \mathbf{E}_L + 4\pi(\mathbf{P}_{latt} + \mathbf{P}_{ext} + \mathbf{P}_e) = 0$. In this equation, $\mathbf{P}_{latt}$ is the lattice polarization given by (see the second equation of the system (6.16))

$$\mathbf{P}_{latt} = \mathbf{P} - \frac{\epsilon_\infty - 1}{4\pi}\mathbf{E}_L = \gamma_{12}\mathbf{w},$$

where $\gamma_{12} = \omega_{TO}\sqrt{(\epsilon_0-\epsilon_\infty)/4\pi}$ and $\mathbf{P}$ is the total polarization in the absence of external electric fields and free charges. Next, $\mathbf{P}_{ext}$ is the polarization due to external static charges (for example, surface charges and doping), which is related to the external electric field as $\epsilon_\infty E_{ext} + 4\pi\mathbf{P}_{ext} = 0$. Finally, $\mathbf{P}_e$ is the polarization associated with free carriers (electrons and holes) with total charge density $\rho_{\mathbf{rt}} = n_{\mathbf{rt}}^e - n_{\mathbf{rt}}^h$. It satisfies the equation $\nabla \cdot \mathbf{P}_e = -\rho_{\mathbf{rt}}$.

Substituting $\mathbf{E}_L$ found in this way into the first equation of the system (6.16), we obtain the equation of motion

$$\ddot{\mathbf{w}} + \omega_{LO}^2\mathbf{w} = -\frac{4\pi\gamma_{12}}{\epsilon_\infty}(\mathbf{P}_{ext} + \mathbf{P}_e), \qquad (I)$$

where the right-hand side contains the polarization of free charges. To complete the description, one should derive another equation for $\mathbf{P}_e$. This can be done by applying the first pair of balance equations (11.11) to electrons and holes. If we neglect the spatial gradients $\sum_\beta \nabla^\beta Q_{\mathbf{rt}}^{\alpha\beta}$ and estimate the collision-integral terms in the transport time approximation, with the same τ_{tr} for electrons and holes, we obtain

$$e\frac{\partial^2 n_{\mathbf{rt}}^e}{\partial t^2} + \frac{e}{\tau_{tr}}\frac{\partial n_{\mathbf{rt}}^e}{\partial t} + \frac{e^2 n_{\mathbf{rt}}^e}{m_e}\nabla\cdot\mathbf{E}_L = 0$$

and

$$e\frac{\partial^2 n_{\mathbf{rt}}^h}{\partial t^2} + \frac{e}{\tau_{tr}}\frac{\partial n_{\mathbf{rt}}^h}{\partial t} - \frac{e^2 n_{\mathbf{rt}}^h}{m_h}\nabla\cdot\mathbf{E}_L = 0.$$

Let us consider the case of strong excitation, when $n_{\mathbf{rt}}^e + n_{\mathbf{rt}}^h \equiv 2N_{\mathbf{rt}} \gg |n_{\mathbf{rt}}^e - n_{\mathbf{rt}}^h|$. After subtracting the first equation from the second one, we take into account the definition of $\mathbf{P}_e$ and arrive at

$$\ddot{\mathbf{P}}_e + \tau_{tr}^{-1}\dot{\mathbf{P}}_e = \frac{e^2 N_{\mathbf{rt}}}{\mu^*}\mathbf{E}_L,$$

where μ^* is the reduced mass. Finally, expressing the electric field in terms of polarizations, we find the equation of motion for electronic polarization:

$$\ddot{\mathbf{P}}_e + \tau_{tr}^{-1}\dot{\mathbf{P}}_e + \omega_p^{*2}\mathbf{P}_e = -\omega_p^{*2}(\mathbf{P}_{ext} + \gamma_{12}\mathbf{w}), \qquad (II)$$

where $\omega_p^* = \sqrt{4\pi e^2 N_{\mathbf{rt}}/\epsilon_\infty \mu^*}$ is the frequency of the electron-hole plasma. Equations (I) and (II) form a closed system describing coupled dynamics of relative ionic displacement $\mathbf{w}$ and electronic polarization $\mathbf{P}_e$. In the absence of external field and collision-induced damping, the solutions of the system are coupled plasmon-phonon modes; see problem 6.20. In the presence of an external electric field, Eq. (I) has a stationary solution before the excitation, when $\mathbf{P}_e = 0$: $\mathbf{w} = -(4\pi\gamma_{12}/\epsilon_\infty\omega_{LO}^2)\mathbf{P}_{ext}$. Once a carrier density $N_{\mathbf{rt}}$ is created by the excitation pulse, the current flows in the system until the electronic polarization compensates the external field, $\mathbf{P}_e = -\mathbf{P}_{ext}$, and the sublattices shift to another equilibrium position, $\mathbf{w} = 0$. Because both the lattice and the electrons possess a certain inertia represented by the second-derivative terms in Eqs. (I) and (II), the system will oscillate around the new equilibrium position. Both plasmon-phonon modes should be involved in the oscillatory transient. However, since the plasmon frequency ω_p^* depends on the coordinate-dependent carrier density $N_{\mathbf{rt}}$, the plasmon features are washed out by inhomogeneous broadening, while the density-independent LO-phonon features remain preserved.

11.17. Solve the system (57.15).

Hint: Using the Fourier transformation, reduce Eq. (57.15) to a system of four algebraic equations, then solve it and carry out the inverse Fourier transformation of the result.

11.18. Using Eq. (57.21), find the sign of the real part of the integral $\int_{-\infty}^0 d\tau e^{-i\omega\tau + \lambda\tau} K_{\mathbf{q}}^{(ll)}(\tau)$.

Solution: According to Eq. (57.21), the function $K_{\mathbf{q}}^{(ll)}(\tau)$ can be represented as a sum of the expression $A_{q_1 q_2}^{(+)} \cos[(\omega_{q_1} + \omega_{q_2})\tau] + A_{q_1 q_2}^{(-)} \cos[(\omega_{q_1} - \omega_{q_2})\tau]$ over phonon modes q_1 and q_2. The coefficients $A^{(+)}$ and $A^{(-)}$, whose explicit form can be deduced from Eq. (57.21), are real and positive. Substituting the form described above into the integral, one can check that the real part of the result is positive.

The coefficient $A^{(+)}$ is positive for an arbitrary phonon distribution. However, the coefficient $A^{(-)}$ is positive under the condition that the phonon occupation numbers decrease with increasing phonon frequencies. This implies a normal situation, with no inversion of phonon population. If the phonon population is inverted, the real part of the integral $\int_{-\infty}^{0} d\tau e^{-i\omega\tau + \lambda\tau} K_{\mathbf{q}}^{(ll)}(\tau)$, in principle, can be negative for some regions of ω. This means that the coherent phonon oscillations are amplified due to stimulated phonon emission, which leads to phonon instability.

11.19. Calculate the integral $\int_0^{\omega_m} d\omega\, \omega^k \cos(\omega\tau)$, where k is a positive integer number, in the limit $\tau \gg \omega_m^{-1}$.

Result: In this limiting case, the result is $\omega_m^k \sin(\omega_m \tau)/\tau$. The function $\sin(\omega_m \tau)/\tau$ rapidly decreases at $|\tau| > \pi/\omega_m$ and can be approximated by $\pi\delta(\tau)$ at $\omega_m \to \infty$.

Chapter 12

BALLISTIC AND HOPPING TRANSPORT

The picture of collision-limited (diffusive) transport fails in many cases, as we have seen on the example of electron transport in magnetic fields; see Chapter 10. The other examples considered in this chapter are *i)* the ballistic transport under the conditions when the size of the sample is comparable to the mean free path length so that electrons experience a few or any scattering events, and *ii)* the transport in mesoscopic samples at low temperatures, when the inelastic scattering is practically absent. In the absence of inelastic scattering, when the phase coherence takes place, the conductance of the sample is determined by the quantum-mechanical transmission probabilities and depends both on the geometry of the sample and on the scattering potential distribution. The phase memory leads to such phenomena as the localization of electrons in one-dimensional conductors and the Aharonov-Bohm oscillations. The low-temperature current between the regions separated by potential barriers is limited by the quantum-mechanical probability of tunneling transmission. In more complex cases, when the energy and momentum conservation requirements cannot be satisfied simultaneously without involvement of scattering, the tunneling becomes scattering-assisted, and the current is inversely proportional to the scattering time. Next, the tunneling current through small metallic islands appears to be sensitive to the electric charge quantization leading to Coulomb blockade phenomena. The electron transport in all these cases is conveniently treated by introducing the Hamiltonian of tunnel-coupled systems, which describes the low-probability hopping of electrons between the regions where the electrons are in local equilibrium. A similar Hamiltonian describes the localization of electrons in the crystal lattice in the presence of a strong electron-phonon interaction (the polaronic effect) and allows one to calculate the hopping current. This current demonstrates thermal-activation behavior, and its dependence on the strength and frequency of the applied electric field is essentially different from that considered in previous chapters.

58. Quantized Conductance

Considering transport of electrons in non-homogeneous media, one can meet the situation when low-resistivity regions (contact regions, or leads) are connected to each other through a short region whose resistance is much higher than that of the leads. This region may contain, for example, potential barriers through which the electrons have to be transmitted by tunneling or thermal activation. The regions of small size, such as the microcontacts studied in Sec. 12, also have high resistance in comparison to the leads. The leads in this situation should be considered as independent sub-systems, each having its own electrochemical potential, since the current densities inside these regions are small and, owing to scattering processes, the quasi-equilibrium distributions are established there. For this reason, the transport is determined mostly by the properties of the high-resistance regions. In these regions, if they are small enough, the electrons can move either without scattering, when it is said that the transport is ballistic, or experience a few scattering events. If the temperature is low enough and the inelastic scattering is suppressed, the scattering events are mostly elastic even in the regions whose size is large in comparison to the mean free path length. Thus, the conduction electrons maintain quantum phase coherence, and this property causes a variety of interesting interference phenomena. Such regions are called the mesoscopic systems, to emphasize that their size is intermediate between microscopic and macroscopic sizes. We point out that modern microfabrication techniques make it possible to create a great variety of nanostructures where the above-mentioned transport regimes are realized.

Based on the qualitative picture given above, one may write a rather general formula describing the current through a high-resistance region:

$$I = \sum_{\delta\delta'} i_{\delta\delta'}^{(+)} f^{(eq)}(\varepsilon_\delta - eV/2)[1 - f^{(eq)}(\varepsilon_{\delta'} + eV/2)]$$

$$- \sum_{\delta\delta'} i_{\delta'\delta}^{(-)} f^{(eq)}(\varepsilon_{\delta'} + eV/2)[1 - f^{(eq)}(\varepsilon_\delta - eV/2)]. \tag{1}$$

The coefficients $i_{\delta\delta'}^{(+)}$ are the microscopic currents describing electron transmission from the state δ on the left to the state δ' on the right. In a similar way, $i_{\delta'\delta}^{(-)}$ describe the backward transmission from the right to the left. In the case of ballistic transport through a microcontact modelled by a hole in an unpenetrable plane, the current is given by Eq. (12.10), which is a particular case of Eq. (1). The current (12.10) is proportional to the square of the hole, since we assume that the hole size is large in comparison to the electron wavelength λ. To find what happens if this size becomes comparable to λ, we introduce a finite thickness L of the unpenetrable region and assume $L \gg \lambda$ so that the microcontact is represented by a wire connected to the leads at the points $z = 0$ and $z = L$. Let us find $i_{\delta\delta'}^{(\pm)}$ for such a case, assuming that the electrons move through the wire ballistically and neglecting, for simplicity, the reflection of the electron waves at the ends $z = 0$ and $z = L$. This neglect corresponds to adiabatic transport in the vicinity of $z = 0$ and $z = L$, and we are going to discuss the conditions of the adiabaticity later. Now we can treat the states δ and δ' as eigenstates of a quasi-one-dimensional system (quantum wire) and describe them in terms of the one-dimensional momentum p and 1D subband number n, see Sec. 5, which are conserved in the ballistic transport. The microscopic current is described by the semi-classical expressions $i_{\delta\delta'}^{(+)} = i_{np,n'p'}^{(+)} = 2ev_{np}\theta(v_{np})\delta_{nn'}\delta_{pp'}$ and $i_{\delta\delta'}^{(-)} = -2ev_{np}\theta(-v_{np})\delta_{nn'}\delta_{pp'}$, where $v_{np} = \partial\varepsilon_{np}/\partial p$ is the group velocity of electrons in the subband n and the factor of 2 stands because of the assumed spin degeneracy. Taking into account that $\varepsilon_{np} = \varepsilon_{n,-p}$, we find from Eq. (1)

$$I = 2e \sum_n \int_{-\infty}^{\infty} \frac{dp}{2\pi\hbar} v_{np}\theta(v_{np})[f^{(eq)}(\varepsilon_{np} - eV/2) - f^{(eq)}(\varepsilon_{np} + eV/2)]. \quad (2)$$

In the linear case, when the energy $|eV|$ is much smaller than the temperature T, one can introduce the conductance G according to $I = GV$. It is calculated by expanding the distribution functions in series of $eV/2$ (here and below $f(\varepsilon) = f^{(eq)}(\varepsilon)$):

$$G = \frac{e^2}{\pi\hbar} \sum_n \int_{-\infty}^{\infty} dp\, v_{np}\theta(v_{np}) \left[-\frac{\partial f(\varepsilon_{np})}{\partial \varepsilon_{np}} \right]. \quad (3)$$

Since $v_{np}\left[-\partial f(\varepsilon_{np})/\partial\varepsilon_{np}\right] = -\partial f(\varepsilon_{np})/\partial p$, we obtain

$$G = \frac{e^2}{\pi\hbar} \sum_n f(\varepsilon_n) = \frac{e^2}{2\pi\hbar} N_f, \quad (4)$$

where $\varepsilon_n = \varepsilon_{n,p=0}$. Equation (4) is valid not only for the electrons with parabolic spectrum ($\varepsilon_{np} = \varepsilon_n + p^2/2m$), but also in any case when ε_{np} monotonically increases with increasing $|p|$. The right-hand side of Eq. (4) corresponds to zero temperature, and N_f is the total number of occupied 1D subbands (for which $f(\varepsilon_n) = 1$) multiplied by the spin degeneracy factor 2. The conductance quantization is determined by the fundamental conductance quantum, $G_0 = e^2/2\pi\hbar$. Note that the final result (4) has the same form as Eq. (51.1) for the quantized Hall

conductivity. As we will see in the next section, this is not a simple co-
incidence. In the quasi-classical case, when N_f is large, one has, approximately, $N_f/2 = p_F^2 S/4\pi\hbar^2$ for the 3D case and $N_f/2 = p_F S/\pi\hbar$ for the
2D case, where p_F is the Fermi momentum and S is the cross-section of
the wire. This estimate provides a link between the conductance (4) and
the conductance of classical microcontacts under collisionless regime; see
Eq. (12.11) and its discussion in Sec. 12.

A rigorous consideration of the linear conductance is based upon the
linear response theory discussed in Chapter 3. Below we apply this
formalism to the case when one has an arbitrary number of leads connected to a high-resistance region, the latter is called below the mesoscopic sample or, simply, the sample. It is assumed that the voltage
V_N is maintained at the N-th lead so that the boundary conditions are
$V_{\mathbf{r} \in S_N} = V_N$, where S_N is the contact area where the N-th lead is connected to the sample. Another boundary condition is the requirement
that the current is equal to zero at the boundary of the sample, where
there are no contacts. Using Eq. (13.9), where the electric fields are
expressed as $E_\beta(\mathbf{r}) = -\partial V_{\mathbf{r}}/\partial r_\beta$, we integrate over $\mathbf{r}'$ by parts. Applying the boundary conditions and the current continuity requirement, we
obtain the current density inside the sample in the form

$$I_\alpha(\mathbf{r}) = \sum_N V_N \sum_\beta \int_{\mathbf{r}' \in S_N} d\mathbf{r}' \sigma_{\alpha\beta}(\mathbf{r}, \mathbf{r}') n_{S_N}^\beta(\mathbf{r}'), \qquad (5)$$

where the integral is taken over the surface of the N-th contact and
$\mathbf{n}_{S_N}$ is the unit vector normal to this surface and directed inside the
sample. To find the total current I_M entering the sample through the
M-th contact, we use

$$I_M = \sum_\alpha \int_{\mathbf{r} \in S_M} d\mathbf{r}\, n_{S_M}^\alpha(\mathbf{r}) I_\alpha(\mathbf{r}) \qquad (6)$$

and finally obtain

$$I_M = -\sum_N G_{MN} V_N, \qquad (7)$$

where the multi-terminal conductance is given by

$$G_{MN} = -\sum_{\alpha\beta} \int_{\mathbf{r} \in S_M} d\mathbf{r} \int_{\mathbf{r}' \in S_N} d\mathbf{r}'\, n_{S_M}^\alpha(\mathbf{r}) \sigma_{\alpha\beta}(\mathbf{r}, \mathbf{r}') n_{S_N}^\beta(\mathbf{r}'). \qquad (8)$$

Using the current continuity requirement, one has $\sum_M I_M = 0$. Since this
relation should be valid for arbitrary voltages V_N, we obtain $\sum_M G_{MN} =$
0, which is also represented as

$$G_{NN} = -\sum_{M\ (M\neq N)} G_{MN} = -\sum_{M\ (M\neq N)} G_{NM}. \tag{9}$$

The last equation follows from Onsager's symmetry; see Eq. (23) below. Equation (9) shows that the diagonal components of the conductance are entirely determined by the off-diagonal components. This property also means that the currents I_M can depend only on the differences between the potentials V_N (there are no currents proportional to $\sum_N V_N$).

Equations (7) and (8) tell us that the linear conductance of a sample can be introduced rigorously through the non-local conductivity $\sigma_{\alpha\beta}(\mathbf{r}, \mathbf{r}')$. The latter is a well-defined quantity, without regard to the properties of the transport inside the sample (the transport may be ballistic or not). It can be calculated from the Kubo formula (13.10). Let us write Eq. (13.10) in the exact eigenstate representation at $\omega \to 0$ and zero magnetic field. Acting in a similar way as in the derivation of Eq. (13.22) from Eq. (13.18), and using the property $\sigma_{\alpha\beta}(\mathbf{r}, \mathbf{r}') = \sigma_{\beta\alpha}(\mathbf{r}', \mathbf{r})$ valid at $\mathbf{H} = 0$, we obtain

$$\sigma_{\alpha\beta}(\mathbf{r}, \mathbf{r}') = 2\pi\hbar \sum_{\delta\delta'} \langle\delta|\hat{I}_\alpha(\mathbf{r})|\delta'\rangle\langle\delta'|\hat{I}_\beta(\mathbf{r}')|\delta\rangle\delta(\varepsilon_\delta - \varepsilon_{\delta'})\left[-\frac{\partial f(\varepsilon_\delta)}{\partial \varepsilon_\delta}\right]. \tag{10}$$

In a similar way as in Eq. (13.22), the quantum numbers δ and δ' do not include the spin, and the spin degeneracy leads to the factor of 2 in the right-hand side. It is not difficult to apply Eq. (10) to the quasi-1D wire considered above. The eigenstate indices δ and δ' describe 1D subband numbers, n and n', as well as one-dimensional momenta, p and p'. Using Eqs. (4.15) for the current density operator and Eq. (5.25) for the 1D electron wave function and spectrum (note that in a homogeneous wire $\psi_x^{(p)} = L_x^{-1/2} e^{ipx/\hbar}$), we obtain

$$\langle np|\hat{I}_\alpha(\mathbf{r})|n'p'\rangle = \delta_{\alpha x}\psi_{y,z}^{(n)*}\psi_{y,z}^{(n')}\frac{e(p + p')}{2mL_x}e^{i(p'-p)x/\hbar}, \tag{11}$$

where $\psi_{y,z}^{(n)}$ is the wave function describing the confinement in the YOZ plane. Calculating the component σ_{xx} according to Eq. (10) in the case of low temperatures $T \to 0$, we find

$$\sigma_{xx}(\mathbf{r}, \mathbf{r}') = \frac{e^2}{\pi\hbar}\Bigg\{\sum_n |\psi_{y,z}^{(n)}|^2|\psi_{y',z'}^{(n)}|^2\theta(\varepsilon_F - \varepsilon_n)$$

$$+ \sum_{nn'(n\neq n')} \psi_{y,z}^{(n)*}\psi_{y,z}^{(n')}\psi_{y',z'}^{(n')*}\psi_{y',z'}^{(n)} \sum_\pm \frac{(v_n^{(F)} \pm v_{n'}^{(F)})^2}{4v_n^{(F)}v_{n'}^{(F)}} \tag{12}$$

$$\times \cos[(p_n^{(F)} \mp p_{n'}^{(F)})(x - x')/\hbar]\theta(\varepsilon_F - \varepsilon_n)\theta(\varepsilon_F - \varepsilon_{n'})\Bigg\},$$

where $p_n^{(F)}$ and $v_n^{(F)}$ are the Fermi momentum and Fermi velocity in the subband n. Finally, denoting the left and the right sides of the wire by the indices 1 and 2, respectively, we apply Eq. (8) for calculating the conductance $G_{21} = -G_{11}$. The integrals over the transverse coordinates y, z, y', and z' are calculated according to the property of orthogonality and normalization, $\int dy \int dz \psi_{y,z}^{(n)*} \psi_{y,z}^{(n')} = \delta_{nn'}$. Therefore, only the first term in Eq. (12) contributes to the conductance, and we find $G_{21} = e^2 N_f / 2\pi\hbar$, in accordance with Eq. (4). The above calculation demonstrates the usefulness of the non-local conductivity tensor. The latter can be applied as well for calculating the local response (problem 12.1).

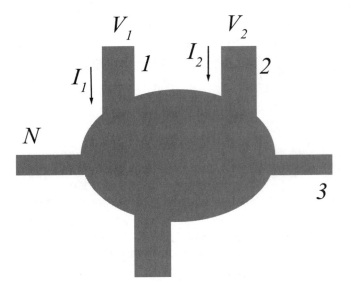

Figure 12.1. A mesoscopic sample connected to the leads $1, 2, 3, \ldots, N$. The voltages and currents for the contacts 1 and 2 are also shown.

The conductance G_{MN} is directly expressed through the quantum-mechanical transmission coefficients of electrons. To prove this, let us consider the region to which the currents are fed by ideal wires (Fig. 12.1). The asymptotic behavior of the wave functions in these wires is determined by

$$\varphi_{Nn\varepsilon}^{(\pm)}(\mathbf{r}) = \frac{1}{\sqrt{v_{n\varepsilon}}} \psi_{\mathbf{r}_{\perp N}}^{(Nn)} \exp(\pm i k_{n\varepsilon} r_{\parallel N}), \tag{13}$$

where N is the wire number and n is the subband number. Next, $v_{n\varepsilon}$ and $k_{n\varepsilon}$ are the group velocity and wave number of electrons, which depend on the subband number and energy. The representation based upon the quantum numbers n and ε instead of n and p is more convenient for our purposes. By $r_{\|N}$ we denote the coordinate along the wire N, while $\mathbf{r}_{\perp N}$ is the coordinate perpendicular to this wire. The signs $+$ and $-$ stand for incoming and outgoing electrons, respectively. Using the functions (13), one can construct a complete set of orthogonal eigenstates whose asymptotic behavior in the M-th wire is determined by

$$\Psi_{Nn\varepsilon}(\mathbf{r})|_{\mathbf{r}\in M} = \delta_{NM}\varphi_{Mn\varepsilon}^{(+)}(\mathbf{r}) + \sum_m S_{Mm,Nn}(\varepsilon)\varphi_{Mm\varepsilon}^{(-)}(\mathbf{r}). \quad (14)$$

The coefficient $S_{Mm,Nn}$ is the quantum-mechanical amplitude of scattering from the state n of wire N to the state m of wire M. The coefficient $S_{Nm,Nn}$ describes the backscattering in the wire N. The scattering amplitude is a unitary matrix which satisfies the symmetry relation

$$S_{Nn,Mm}(\varepsilon) = S_{Mm,Nn}(\varepsilon)|_{\mathbf{H}\to-\mathbf{H}}. \quad (15)$$

In the case of zero magnetic field we consider, this equation means that $S_{Nn,Mm}$ is symmetric. The unitarity condition, therefore, can be written as

$$\sum_{Mm} S_{Mm,N_1n_1}^+(\varepsilon)S_{Mm,N_2n_2}(\varepsilon) \quad (16)$$

$$= \sum_{Mm} S_{Mm,N_1n_1}^*(\varepsilon)S_{Mm,N_2n_2}(\varepsilon) = \delta_{N_1N_2}\delta_{n_1n_2}.$$

The choice of normalization of $\varphi_{Nn\varepsilon}^{(\pm)}(\mathbf{r})$ implies that the functions (14) are normalized according to (problem 12.2)

$$\int d\mathbf{r}\Psi_{Nn\varepsilon}^*(\mathbf{r})\Psi_{N'n'\varepsilon'}(\mathbf{r}) = 2\pi\hbar\delta_{NN'}\delta_{nn'}\delta(\varepsilon - \varepsilon'). \quad (17)$$

With this normalization rule, the Kubo formula (10) is transformed to

$$\sigma_{\alpha\beta}(\mathbf{r},\mathbf{r}') = \frac{1}{2\pi\hbar}\sum_{N_1N_2}\sum_{n_1n_2}\int d\varepsilon \int d\varepsilon' \langle N_1n_1\varepsilon|\hat{I}_\alpha(\mathbf{r})|N_2n_2\varepsilon'\rangle$$

$$\times\langle N_2n_2\varepsilon'|\hat{I}_\beta(\mathbf{r}')|N_1n_1\varepsilon\rangle\delta(\varepsilon - \varepsilon')\left[-\frac{\partial f(\varepsilon)}{\partial\varepsilon}\right], \quad (18)$$

where the matrix elements are calculated by using the eigenstates (14). Therefore, according to Eq. (8),

$$G_{MN} = -\frac{1}{2\pi\hbar}\sum_{N_1N_2}\sum_{n_1n_2}\int d\varepsilon I_{N_1n_1,N_2n_2}^{(M)}(\varepsilon,\varepsilon)I_{N_2n_2,N_1n_1}^{(N)}(\varepsilon,\varepsilon)\left[-\frac{\partial f(\varepsilon)}{\partial\varepsilon}\right],$$

$$I^{(M)}_{N_1 n_1, N_2 n_2}(\varepsilon, \varepsilon') = \int_{\mathbf{r} \in S_M} d\mathbf{r} \langle N_1 n_1 \varepsilon | \hat{\mathbf{I}}(\mathbf{r}) | N_2 n_2 \varepsilon' \rangle \cdot \mathbf{n}_{S_M}(\mathbf{r}). \quad (19)$$

Calculating the matrix elements of the current density operator (4.15), we obtain

$$I^{(M)}_{N_1 n_1, N_2 n_2}(\varepsilon, \varepsilon') = e \left[\delta_{N_1 M} \delta_{N_2 M} \delta_{n_1 n_2} \right.$$

$$\left. - \sum_m S^*_{Mm, N_1 n_1}(\varepsilon) S_{Mm, N_2 n_2}(\varepsilon') \right]. \quad (20)$$

Substituting this into the expression (19) for the conductance, we find that the terms which do not proportional to S contain δ_{NM}. The terms of the fourth order in S contain the sum $\sum_{N_1 n_1} S^*_{Mm, N_1 n_1} S_{Nn, N_1 n_1}$ multiplied by a similar sum over N_2 and n_2. As a result, these terms are also proportional to δ_{NM}, according to Eqs. (15) and (16). Therefore, the only contribution to the off-diagonal components of the conductance matrix comes from the terms quadratic in S. Finally, we obtain

$$G_{MN} = \frac{e^2}{\pi \hbar} T_{MN} \quad (M \neq N), \quad (21)$$

where T_{MN} is the probability of transmission from the lead N to the lead M averaged over the energy near the Fermi surface:

$$T_{MN} = \int d\varepsilon \left[-\frac{\partial f(\varepsilon)}{\partial \varepsilon} \right] T_{MN}(\varepsilon), \quad T_{MN}(\varepsilon) = \sum_{mn} T^{(mn)}_{MN}(\varepsilon),$$

$$T^{(mn)}_{MN}(\varepsilon) = S^+_{Mm, Nn}(\varepsilon) S_{Mm, Nn}(\varepsilon). \quad (22)$$

At zero temperature, $T_{MN} = T_{MN}(\varepsilon_F)$. The coefficient $T^{(mn)}_{MN}$ is the probability of transmission from the state n of lead N to the state m of lead M, it is also called the transmission coefficient. Since the electron states are assumed to be spin-degenerate, the sum over spin indices is accounted for by a factor of 2. In the general case, one should include the spin indices into the indices of the states and omit this factor. The diagonal components G_{NN} are found from Eq. (21) according to Eq. (9).

Introducing $\widehat{S}_{MN}$ as the operator representation of the matrix $S_{Mm, Nn}$ with respect to the indices of the states, one may rewrite the lead-to-lead transmission probability from Eq. (22) as $T_{MN}(\varepsilon) = \text{tr} \widehat{S}_{MN} \widehat{S}^+_{MN}$, where tr $\ldots$ indicates the matrix trace. The matrix $(\widehat{S}_{MN} \widehat{S}^+_{MN})_{nn'}$ standing under the trace can be diagonalized by a unitary transformation to the form $\delta_{nn'} T^{(n)}_{MN}(\varepsilon)$. In this way, a set of independent quantum channels characterized by the transmission coefficients $T^{(n)}_{MN}$ is defined. This formalism is called the channel representation. The transmission probability in this representation is a sum of the transmission coefficients of the channels, $T_{MN}(\varepsilon) = \sum_n T^{(n)}_{MN}(\varepsilon)$.

Although Eqs. (21) and (22) have been derived from the Kubo formula at $\mathbf{H} = 0$, they remain valid in the presence of magnetic fields, and the symmetry property (15) leads to similar properties of the transmission probability and conductance:

$$T_{MN}^{(mn)} = T_{NM}^{(nm)}|_{\mathbf{H} \to -\mathbf{H}}, \quad G_{MN} = G_{NM}|_{\mathbf{H} \to -\mathbf{H}} . \tag{23}$$

Equations (21) and (22) form the main result of this section. They relate the conductivity of a mesoscopic region to simple quantum-mechanical properties of electron waves. The derivation of Eqs. (21) and (22) from the Kubo formula implies the absence of inelastic scattering in this region and in the wires. The inelastic scattering, however, should be present in the reservoirs (leads) to which these wires are connected, to maintain local equilibrium in each of the reservoirs. The presence of the wires connected, from the one side, to the mesoscopic region and, from the other side, to reservoirs with uniform electrochemical potentials is a necessary element of the derivation, since it allows us to divide, formally, the entire system by the mesoscopic region (sample) and macroscopic regions (leads).

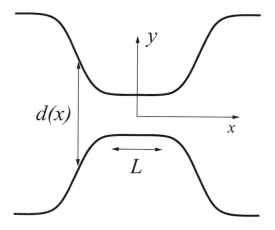

Figure 12.2. Adiabatic constriction for 2D electrons.

Having expressed the conductance through the transmission coefficients, we discuss the conductance of 2D microcontacts at zero temperature on simple examples. Consider first a constriction (narrow part) formed in a wide waveguide with hard-wall confinement. The width $d(x)$ of the waveguide is coordinate-dependent (Fig. 12.2), and the wave function can be searched for in the form

$$\Psi(x,y) = \sum_n \varphi_n(x)\psi_y^{(n)}, \qquad (24)$$

$$\psi_y^{(n)} = \sqrt{2/d(x)} \sin\left\{\pi n[2y + d(x)]/2d(x)\right\}, \qquad n = 1, 2, \ldots .$$

Substituting this function into the Schroedinger equation, we find a matrix equation for $\varphi_n(x)$ (problem 12.3). The non-diagonal terms in this matrix equation describe the mixing of different 1D-subband states. This mixing exists due to the parametric x-dependence of $\psi_y^{(n)}$ through $d(x)$. If the logarithmic derivative $d'(x)/d(x)$ is small in comparison to the wave numbers of electrons, the non-diagonal terms can be neglected. This is the case of adiabatic transport, when the electron motion in the subband n can be separated from the motion in the other subbands and, therefore, becomes effectively one-dimensional. This motion occurs in the presence of an effective potential energy $V_n(x) = [\pi\hbar n/d(x)]^2/2m$ representing a potential barrier. The electrons from higher subbands, whose energies are smaller than the barrier height $V_n(0) = [\pi\hbar n/d(0)]^2/2m$ (we assume that the narrowest part of the waveguide is an ideal wire placed at $-L/2 < x < L/2$), cannot pass through this barrier, since the probability of transmission exponentially decreases with increasing L. In contrast, the electrons from lower subbands travel above the potential barrier without backscattering. Formally, one can introduce the partial transmission coefficients $T^{(nm)}(\varepsilon_F) \propto \delta_{nm}$ as $T^{(nn)}(\varepsilon_F) = 1$ for $n \leq n_0$ and $T^{(nn)}(\varepsilon_F) = 0$ for $n > n_0$, where n_0 is defined by $V_{n_0}(0) < \varepsilon_F < V_{n_0+1}(0)$ and ε_F is the Fermi energy. The conductance in these conditions is equal to $G = e^2 n_0/\pi\hbar$. By varying $d(0)$ continuously, one gets a staircase-like modulation of the conductance, since n_0 is a discrete variable. However, if we decrease the wire length L, the electrons in the subband n_0 can be reflected back under the condition that ε_F is close enough to $V_{n_0}(0)$. In a similar way, the electrons in the subband $n_0 + 1$ have a finite probability to penetrate through the barrier. For this reason, one cannot have ideal (sharp) steps of the conductance: a finite smearing of the steps always occurs. A small-size constriction connecting two macroscopic reservoirs and demonstrating the quantization of the conductance is usually called the quantum point contact.

Another example is the exactly solvable quantum-mechanical problem of the 2D electron transmission through the region where the potential energy is approximated by a saddle point:

$$V(x,y) = V_0 + \frac{m}{2}(\omega_y^2 y^2 - \omega_x^2 x^2). \qquad (25)$$

Since $V(x, y)$ is a sum of the x-dependent component and y-dependent one, the Schroedinger equation is solved exactly. Its solution (problem 12.4) leads to the conductance

$$G = \frac{e^2}{\pi\hbar} \sum_{nm} T^{(nm)}(\varepsilon_F) = \frac{e^2}{\pi\hbar} \sum_{n=0}^{\infty} \frac{1}{1 + \exp(-2\pi\epsilon_n)}, \qquad (26)$$

$$\epsilon_n = [(\varepsilon_F - V_0)/\hbar - (n + 1/2)\omega_y]/\omega_x.$$

This conductance is exponentially small when $\varepsilon_F - V_0 < \hbar\omega_y/2$ and $|\varepsilon_F - V_0 - \hbar\omega_y/2| \gg \hbar\omega_x$, i.e., when the dimensionless variable ϵ_n is large and negative for arbitrary n. With increasing Fermi energy (or with decreasing V_0), the conductance shows steps each time when $\varepsilon_F - V_0$ increases by $\hbar\omega_y$. The steps are clearly seen at $\omega_y/\omega_x > 2$, and the quantized conductance staircase is improved when the ratio ω_y/ω_x increases, i.e., when the curvature of the equipotential lines $\varepsilon_F = V(x, y)$ decreases. The smearing of the steps in the energy scale is estimated as $\hbar\omega_x$. When ω_x goes to zero, the quantum point contact becomes an ideal wire and its conductance shows the ideal steps described by Eq. (4).

These two simple examples demonstrate that Eqs. (21) and (22) not only predict the conductance quantization, but also describe the effects of finite size of the quantum point contacts and finite curvature of their boundaries. These effects are responsible for the deviation of the conductance quantization from the ideal staircase. More complex examples of application of this formalism can be found in the next section. The methods discussed above can be used for calculating various kinetic coefficients in the ballistic transport regime. For example, the thermal conductance is calculated in problem 12.5.

The formalism introduced above can be developed to describe the intrinsic conductance of one-dimensional samples. This conductance should be defined in such a way that it is no longer determined by the presence of the leads. Let us consider a small piece of a disordered 1D conductor whose ends are assumed to be ideal 1D conductors. If the electron transmission probability through the given piece is small, Eq. (21) can be applied directly for the conductance of this piece, because there is local equilibrium at the ends of the piece. In the opposite case, when the transmission probability is not small, one still has to define the electrochemical potentials at the ends. Below we assume that there are no longitudinal electric fields so that the consideration will be done in terms of chemical potentials instead of electrochemical ones. One may introduce the exact distribution function $f_{1,2}(p)$ at the points 1 (left end) and 2 (right end). Since this function depends on the sign of p, it is more convenient to work with the energy distribution functions $f_{1,2}^{(\pm)}(\varepsilon)$ for the right-moving $(+)$ and left-moving $(-)$ electrons. In the linear transport regime and for a degenerate electron gas, these functions differ from the equilibrium distribution function $f(\varepsilon)$ only in a narrow energy interval near the Fermi energy. Therefore, one can write $f_{1,2}(p) = f(\varepsilon) + \Delta f_{1,2}(p)$ and $f_{1,2}^{(\pm)}(\varepsilon) = f(\varepsilon) + \Delta f_{1,2}^{(\pm)}(\varepsilon)$, where the non-equilibrium parts Δf are non-zero in the vicinity of the Fermi energy. The current through the point 1 is written as

$$I_1 = \frac{e}{\pi \hbar} \int d\varepsilon \left\{ [1 - R_1(\varepsilon)] f_1^{(+)}(\varepsilon) - T_{12}(\varepsilon) f_2^{(-)}(\varepsilon) \right\}$$

$$\simeq \frac{e}{\pi \hbar} \int d\varepsilon [(1 - R_1) \Delta f_1^{(+)}(\varepsilon) - T_{12} \Delta f_2^{(-)}(\varepsilon)]. \qquad (27)$$

where $T_{12}(\varepsilon)$ is the backward transmission coefficient and $R_1(\varepsilon)$ is the reflection coefficient for the point 1 (in the second equation these energy-dependent coefficients are approximated by the energy-independent coefficients $T_{12} = T_{12}(\varepsilon_F)$ and $R_1 = R_1(\varepsilon_F)$). A similar equation involving R_2 and T_{21} can be written for the current I_2 through the point 2. However, since $I_2 = I_1 = I$, $R_1 + T_{21} = 1$, $R_2 + T_{12} = 1$, $T_{12} = T_{21} = T$ and $R_1 = R_2 = R$, such an equation gives us nothing new in comparison to Eq. (27).

Let us formally introduce the chemical potentials for the right- and left-moving electrons according to

$$\mu_{1,2}^{(\pm)} = \mu_0 + \int dp \, |v_p| \theta(\pm v_p) \Delta f_{1,2}(p) = \mu_0 + \int d\varepsilon \Delta f_{1,2}^{(\pm)}(\varepsilon), \qquad (28)$$

where μ_0 is the equilibrium chemical potential. Using Eqs. (27) and (28) together with the identity $1 - R = T$, we have

$$I = \frac{e}{\pi \hbar} T(\mu_1^{(+)} - \mu_2^{(-)}). \qquad (29)$$

In other words, if we define the conductance of a piece of 1D conductor as $I/\Delta V_{inc}$, where $\Delta V_{inc} = (\mu_1^{(+)} - \mu_2^{(-)})/e$ is the difference in electro-chemical potentials for the electrons incoming to this piece from both sides, the conductance is given by Eq. (21). The definition (28) also means that

$$I = \frac{e}{\pi \hbar} (\mu_1^{(+)} - \mu_1^{(-)}) = \frac{e}{\pi \hbar} (\mu_2^{(+)} - \mu_2^{(-)}). \qquad (30)$$

This equation expresses the 1D current through the local differences in the chemical potentials for the right- and left-moving electrons, and the proportionality coefficient is the conductance of an ideal 1D wire divided by e. We also introduce the averaged chemical potentials $\mu_1 = (\mu_1^{(+)} + \mu_1^{(-)})/2$ and $\mu_2 = (\mu_2^{(+)} + \mu_2^{(-)})/2$. These potentials are related to the electron densities at the points 1 and 2, because these local densities are expressed as

$$n_{1,2} = \frac{1}{\pi\hbar} \int d\varepsilon_p |v_p|^{-1} [f_{1,2}^{(+)}(\varepsilon_p) + f_{1,2}^{(-)}(\varepsilon_p)]$$

$$\simeq n_0 + \frac{1}{\pi\hbar v_F} \int d\varepsilon [\Delta f_{1,2}^{(+)}(\varepsilon) + \Delta f_{1,2}^{(-)}(\varepsilon)], \qquad (31)$$

where n_0 is the equilibrium density and v_F is the Fermi velocity. Therefore,

$$n_1 - n_2 = 2\frac{\mu_1 - \mu_2}{\pi\hbar v_F} = \frac{\mu_1 - \mu_2}{\rho_{1D}(\varepsilon_F)}. \qquad (32)$$

Applying Eqs. (29) and (30) in order to express the current through the averaged chemical potentials, we obtain

$$I = \widetilde{G}(\mu_1 - \mu_2)/e, \quad \widetilde{G} = \frac{e^2}{\pi\hbar}\frac{T}{R}. \qquad (33)$$

The conductance $\widetilde{G}$, which describes the intrinsic response of a 1D conductor, coincides with the 1D conductance $e^2 T/\pi\hbar$ only in the case of low transmission, when $R = 1 - T$ is close to 1. In the case of high transmission, when $R \to 0$, $\widetilde{G}$ is much greater than $e^2 T/\pi\hbar$.

59. One-Dimensional Conductors

Considering the weak localization of 2D electrons in Sec. 15, we have found that the corrections to the 2D conductivity are proportional to the fundamental conductance quantum $e^2/2\pi\hbar$. These corrections have interference origin, they occur because the phases of electron waves are not destroyed in the elastic scattering processes. In the case of metallic 2D conductivity, the relative correction (15.28) is small. This is not true for 1D conductors. Indeed, if we formally rewrite Eq. (15.26) for the 1D case, the quantum correction to the conductivity is estimated as

$$\delta\sigma_{1D} \sim -\frac{2e^2}{\pi^2\hbar} l_D, \qquad (1)$$

where l_D is the diffusion length introduced in Sec. 15. On the other hand, the conductivity of a piece of 1D conductor of length L can be estimated according to Eq. (58.33) as $(e^2/\pi\hbar)(T/R)L$ (the dependence of (T/R) on L is discussed below in this section). This simple calculation shows us that, if the phase relaxation length due to inelastic scattering is large enough, $\delta\sigma_{1D}$ can be comparable to or larger than the conductivity itself. Moreover, with the increase of this length, the absolute value of the quantum correction increases rapidly (not in a slow logarithmic fashion as in the 2D case). Therefore, the weak localization becomes, in fact, a strong mechanism of electron localization in 1D conductors. The 1D conductivity is considerably influenced by the quantum-mechanical interference which eventually leads to complete localization of electrons if the sample is long enough and the inelastic scattering is absent. We discuss this phenomenon first, and then we consider the magnetotransport properties of 1D conductors.

Let us consider a piece of 1D conductor at $0 \leq x \leq l$ containing point-like obstacles (scatterers) at its ends $x = 0$ and $x = l$. Formulating the quantum-mechanical transmission-reflection problem for this sample, we write the wave function in the form $e^{ikx} + Ae^{-ikx}$ at $x < 0$, $Be^{ikx} + Ce^{-ikx}$ at $0 < x < l$, and $De^{ik(x-l)}$ at $x > l$, where k is the wave number of the electron. To connect the four unknown coefficients, we introduce the complex transmission and reflection amplitudes, t and r, characterizing separately each scatterer (1 or 2) and assume that the primed amplitudes correspond to the left-moving electrons. Introducing also the phase $\varphi = kl$ associated with the electron path between $x = 0$ and $x = l$, we obtain the linear relations

$$A = r_1 + Ct_1', \quad B = t_1 + Cr_1' \ , \quad Ce^{-i\varphi} = Be^{i\varphi}r_2, \quad D = Be^{i\varphi}t_2. \quad (2)$$

For example, r_1' describes reflection of the left-moving electrons from the point 1 while t_1 describes transmission of the right-moving electrons through this point. Because of the symmetry (58.15), $t_1' = t_1$ and $t_2 = t_2'$. Solving the system (2), we find the transmission coefficient

$$T \equiv T_{21} = |D|^2 = \frac{|t_1|^2 |t_2|^2}{1 + |r_1|^2 |r_2|^2 - 2|r_1||r_2|\cos\theta}, \quad (3)$$

where θ depends on the phase φ as well as on the arguments of the complex reflection amplitudes, $\theta = 2\varphi + \arg(r_1') + \arg(r_2)$. We have taken into account that $|r_1'|^2 = |r_1|^2$. Denoting the individual transmission and reflection coefficients of the scatterers 1 and 2 as $|t_1|^2 = T_1$, $|t_2|^2 = T_2$, $|r_1|^2 = R_1 = 1 - T_1$ and $|r_2|^2 = R_2 = 1 - T_2$, we use the result (3) to find the ratio of the reflection coefficient $R = 1 - T$ to the transmission coefficient T:

$$\frac{R}{T} = \frac{R_1 + R_2 - 2\sqrt{R_1 R_2}\cos\theta}{T_1 T_2}. \quad (4)$$

This is the intrinsic resistance expressed in units of the resistance quantum $\pi\hbar/e^2$, see Eq. (58.33), for a piece of the 1D conductor. Let us consider an ensemble of similar pieces. By averaging the intrinsic resistance over the random phases θ so that the cosine term disappears, one gets

$$\left\langle \frac{R}{T} \right\rangle = \frac{R_1 + R_2}{(1 - R_1)(1 - R_2)}. \tag{5}$$

In contrast, the result which one may expect by series addition of two resistances $R_1/(1 - R_1)$ and $R_2/(1 - R_2)$ is $(R_1 + R_2 - 2R_1 R_2)/[(1 - R_1)(1 - R_2)]$. Therefore, Ohm's law is not valid for 1D conductors: the dimensionless intrinsic resistance $\langle R/T \rangle$ is larger than expected. As long as we consider series addition of good-transmittance regions ($R_{1,2} \ll 1$), this fact makes no difference, because $2R_1 R_2 \ll R_1, R_2$. However, if we consider series addition of many resistances $R_i/(1 - R_i)$, the situation dramatically changes. In the limit $R_i \ll 1$, one can show (problem 12.6) that the averaged resistance of the 1D conductor containing N scatterers increases as

$$\left\langle \frac{R}{T} \right\rangle = \frac{R_1 + R_2 + \ldots + R_N}{(1 - R_1)(1 - R_2) \ldots (1 - R_N)}. \tag{6}$$

The denominator in this expression decreases with each addition in a non-linear fashion. Denoting the average on the left-hand side of Eq. (6) as $\mathcal{R}_{N+1}$, one may compose a finite-difference equation:

$$\mathcal{R}_{N+1} - \mathcal{R}_N = \frac{R_N}{1 - R_N} \left[\mathcal{R}_N + \frac{1}{(1 - R_1) \ldots (1 - R_{N-1})} \right] \simeq R_N(\mathcal{R}_N + 1). \tag{7}$$

While the approximation $1/(1 - R_N) \simeq 1$ is straightforward because of $R_N \ll 1$, the approximation of the second term in the square brackets by unity is not obvious because this term becomes larger than unity at large N. However, when this happens, $\mathcal{R}_N$ is much larger than this term so that the latter can be neglected. Therefore, the approximation is justified for arbitrary N. Equation (7) can be also written as a differential equation, $d\mathcal{R}_N/dN = \overline{R}(\mathcal{R}_N + 1)$, where $\overline{R} \ll 1$ is the averaged reflection coefficient for one piece in the series. This differential equation gives us a scaling law in the form

$$\mathcal{R} = e^{L/L_0} - 1, \tag{8}$$

where $L = Nl$ is the length of the 1D wire and L_0 is the localization length equal to $l/\overline{R}$. The normalization coefficient at the exponent is chosen to have the result $\mathcal{R} \simeq L/L_0$ at $L \ll L_0$. The intrinsic resistance linearly increases with increasing L at small L. However, when L becomes larger than L_0, the resistance increases exponentially. This is the phenomenon of localization in one-dimensional systems.

Now we consider two 1D conductors, labeled as 1 and 2, connected in parallel. Such a system can be also viewed as a ring connected to two leads, Fig. 12.3. Each branch ($i = 1, 2$) of the ring is characterized

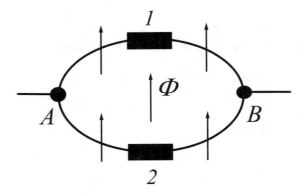

Figure 12.3.   One-dimensional ring with two contacts.

by the transmission and reflection amplitudes t_i, t_i', r_i, and r_i', which connect the amplitudes B_i and C_i of the forward- and backward-moving electron waves near the point A with similarly defined amplitudes B_i' and C_i' near the point B. If the branch i is an ideal ballistic wire, one has simply $r_i = r_i' = 0$ and $t_i = t_i' = e^{i\varphi_i}$, where $\varphi_i = k_i l_i$ is the phase acquired by the electron passing the length l_i of this branch with the wave number k_i. The relevant equations are

$$B_i' = B_i t_i + C_i' r_i', \quad C_i = B_i r_i + C_i' t_i' \quad (i = 1, 2). \qquad (9)$$

The electrons also experience scattering in the contacts A and B, where three 1D channels converge. Each such contact is characterized by a three-terminal scattering matrix. For the contact A, we denote the transmission amplitudes from the external channel to the branch 1(2) as $t_{1A}(t_{2A})$, the backward transmissions as $t_{A1}(t_{A2})$, and the transmission between the branches 1 and 2 as t_{21}^A and t_{12}^A. The reflection amplitudes near this contact are denoted as r_A, r_1^A, and r_2^A. Applying similar notations to the scattering amplitudes in the contact B, we compose a system of equations connecting the incoming wave $e^{ik(x-x_A)} + Ae^{-ik(x-x_A)}$ with the transmitted wave $De^{ik(x-x_B)}$ through the waves in the branches:

$$A = r_A + C_1 t_{A1} + C_2 t_{A2}, \quad B_1 = t_{1A} + C_1 r_1^A + C_2 t_{12}^A,$$

$$B_2 = t_{2A} + C_1 t_{21}^A + C_2 r_2^A, \quad C_1' = B_1' r_1^B + B_2' t_{12}^B, \qquad (10)$$

$$C_2' = B_1' t_{21}^B + B_2' r_2^B, \quad D = B_1' t_{B1} + B_2' t_{B2}.$$

Using these equations, one can find the conductance of the ring, which is proportional to $|D|^2$. For the case of ideal ballistic wires, this conductance oscillates as a function of the phases φ_1 and φ_2.

Suppose that there is an external magnetic field $\mathbf{H_r}$. The electron wave function in this field acquires an additional phase shift, whose magnitude, however, depends on the gauge of the vector potential (see Appendix G). In contrast, the phase acquired by the electron which has completed a closed path in this field is gauge-invariant. Indeed, expressing the phase associated with the vector potential as

$$\phi = \frac{e}{\hbar c} \oint_{\mathbf{r} \in l} d\mathbf{r} \mathbf{A_r} \cdot \mathbf{n}_l, \tag{11}$$

where the integral is calculated over a closed electron path l and $\mathbf{n}_l$ is the unit vector in the direction along this path, one may apply the Stokes theorem to transform the contour integral to the integral of $[\nabla \times \mathbf{A_r}] = \mathbf{H_r}$ over the area S encircled by the path (we assume that the motion occurs in a plane). This transformation gives us the following result:

$$\phi = 2\pi \frac{\Phi}{\Phi_0}, \quad \Phi = \int_{\mathbf{r} \in S} d\mathbf{r} \, \mathbf{H_r} \cdot \mathbf{n}_S, \tag{12}$$

where $\mathbf{n}_S$ is the unit vector perpendicular to the plane of S and

$$\Phi_0 = 2\pi \hbar c / |e|. \tag{13}$$

The direction of $\mathbf{n}_S$ is determined from the following rule: if the plane where the electron moves is XOY, then $\mathbf{n}_S$ is directed along OZ for clockwise motion and in the opposite direction for counterclockwise motion. The quantity Φ is the magnetic flux through the area S, and the quantity Φ_0, which depends only on the universal constants, is called the magnetic flux quantum. The phase changes periodically with the magnetic field, each period corresponds to penetration of one flux quantum into the area encircled by the electron path. Of course, ϕ is meaningful only modulo 2π so that the phases associated with Φ and $\Phi + n\Phi_0$, where n is integer, cannot be distinguished. One may show that the phase (12) enters the Schroedinger equation for the electron moving in an ideal ring (problem 12.7), and the electron energy is a continuous function of this phase.

Now we are ready to modify Eq. (9) in the presence of a magnetic flux. The electrons moving from the point A to the point B gain the phase ϕ_1 in the branch 1 (clockwise motion) and ϕ_2 in the branch 2 (counterclockwise motion). The difference $\phi = \phi_1 - \phi_2$ is the phase (12). To take these phases into account, one should make the substitutions

$$t_1 \to t_1 e^{i\phi_1}, \quad t_2 \to t_2 e^{i\phi_2}, \quad t_1' \to t_1 e^{-i\phi_1}, \quad t_2' \to t_2 e^{-i\phi_2}. \tag{14}$$

These substitutions satisfy the symmetry (58.15). The reflection amplitudes r_i and r_i' are not modified in the presence of the flux. The physically meaningful quantity $|D|^2$ depends only on $\phi = \phi_1 - \phi_2$. Below, for the sake of simplicity, we consider a model of symmetric, totally transparent contacts, where $r_1^A = r_2^A = r_1^B = r_2^B \equiv r$, $t_{12}^A = t_{21}^A = t_{12}^B = t_{21}^B \equiv t$, $t_{A2} = t_{B2} = t_{A1} = t_{B1} = t_{2B} = t_{2A} = t_{1B} = t_{1A} \equiv t'$ and $r_A = r_B = 0$. The unitarity conditions for the scattering matrices of the three-terminal contacts lead to $|t'|^2 = 1/2$, $|t|^2 = 1/4$, and $r = -t$. The phases of t and t' still remain indefinite. Choosing them in such a way that $t = -1/2$ and $t' = -1/\sqrt{2}$, we express the intrinsic conductance as

$$G = \frac{e^2}{\pi\hbar} \frac{T(\phi)}{1 - T(\phi)}, \qquad T(\phi) = 4\frac{|P_+ e^{i\phi/2} + P_- e^{-i\phi/2}|^2}{|P_0 + P\cos\phi|^2}, \qquad (15)$$

where

$$P = 2t_1 t_2, \quad P_0 = t_1^2 + t_2^2 - (2 - r_1 - r_2)(2 - r_1' - r_2'),$$

$$P_+ = t_1[t_2^2 - (1 - r_2)(1 - r_2')], \quad P_- = t_2[t_1^2 - (1 - r_1)(1 - r_1')], \qquad (16)$$

and one should also take into account the unitarity conditions for the scattering matrices $\widehat{S}_i = \begin{pmatrix} r_i & t_i \\ t_i & r_i' \end{pmatrix}$ leading to $|t_i|^2 + |r_i|^2 = 1$ and $r_i' = -r_i^* t_i / t_i^*$. The last equation is equivalent to

$$|r_i'| = |r_i|, \quad \arg(r_i') = 2\arg(t_i) - \arg(r_i) + \pi. \qquad (17)$$

Even in the absence of the magnetic flux, the conductance shows quantum effects. For example, making the transmission amplitude through one of the branches (say 1) zero, $t_1 = 0$, one still can change the transmission coefficient from 0 to 1 by modifying the phases of r_1 and r_1'. The system in this case is equivalent to a single wire 2 with two laterally attached dead ends (stubs), and the control of G is caused by the interference between the transmitted electron wave and the waves reflected in the stubs. In the presence of the magnetic flux, the conductance (15) oscillates as a function of ϕ. The period of these oscillations corresponds to a change of the flux Φ by Φ_0. This phenomenon is called the Aharonov-Bohm oscillations. In the simplest case of two equivalent ballistic wires, when $t_1 = t_2 = e^{i\varphi}$ and $r_1 = r_2 = r_1' = r_2' = 0$, the transmission coefficient given by Eq. (15) is reduced to

$$T(\phi) = \frac{4(1 - \cos 2\varphi)(1 + \cos\phi)}{5 - 4\cos 2\varphi + \cos^2\phi + 2\cos\phi(1 - 2\cos 2\varphi)}. \qquad (18)$$

Though the conductance is a periodic function of ϕ, it is not a harmonic function of this variable. The higher harmonics, whose periods are given by $\Delta\phi = 2\pi/n$ with integer n, are also present.

A configuration of scatterers in the ring is unique so that the coefficients t_i, r_i, and r_i' are well-defined. Suppose, however, that one has an ensemble of the rings with different distributions of the same scatterers. To obtain the average transmission coefficient $\langle T \rangle$ for such ensemble, one has to average the transmission coefficient T of Eq. (15) over four random phases $\arg(t_1)$, $\arg(t_2)$, $\arg(r_1)$, and $\arg(r_2)$, while the phases of r_1' and r_2' are expressed through these four according to Eq. (17). It seems that such an averaging is expected to destroy the Aharonov-Bohm oscillations. The averaging, indeed, destroys the $\Delta\phi = 2\pi$ periodicity of the transmission, but the $\Delta\phi = \pi$ periodicity remains. To understand why it happens, let us consider $T(\phi)$ and $T(\phi + \pi)$ given by Eq. (15). For a given configuration $\{\delta\}$ of the scatterers, with fixed $\arg(t_i)_{\{\delta\}}$ and $\arg(r_i)_{\{\delta\}}$, these transmission coefficients are different. However, for each $\{\delta\}$ one can find another (non-equivalent) configuration $\{\delta'\}$, for which $T_{\{\delta'\}}(\phi + \pi)$ is equal to $T_{\{\delta\}}(\phi)$. For example, this is the case when $\arg(t_1)_{\{\delta'\}} = \arg(t_1)_{\{\delta\}} + \pi$, $\arg(t_2)_{\{\delta'\}} = \arg(t_2)_{\{\delta\}}$, and $\arg(r_i)_{\{\delta'\}} = \arg(r_i)_{\{\delta\}}$. One may check this property directly from Eq. (15), using Eqs. (16) and (17). In other words, $T(\phi)$ and $T(\phi + \pi)$, as functions of $\arg(t_1)$, differ from each other just by a linear translation $\arg(t_1) \rightarrow \arg(t_1) + \pi$, and the procedure of averaging over this argument makes them equal to each other. The averaged quantity

$$\langle T(\phi) \rangle_{\arg(t_1)} = \int_0^{2\pi} \frac{d\arg(t_1)}{2\pi} T(\phi) = \int_0^{2\pi} \frac{d\arg(t_1)}{2\pi} T(\phi + \pi) \qquad (19)$$

is a periodic function of ϕ (and of $\arg(t_2)$), and its period is equal to π. The averaging over $\arg(t_2)$, $\arg(r_1)$, and $\arg(r_2)$ conserves this periodicity so that finally we have $\langle T(\phi) \rangle = \langle T(\phi + \pi) \rangle$. Thus, the ensemble-averaged conductance of a system of 1D rings remains periodic in the magnetic flux, and the period is equal to $\Phi_0/2$, in contrast to the conductance of a single 1D ring whose periodicity is Φ_0.

The periodicity $\Phi_0/2$ takes place in macroscopic rings, where the self-averaging takes place. The oscillating part of the conductivity in this case is associated with the quantum (weak-localization) correction (see Secs. 15 and 43), and the periodicity $\Phi_0/2$ instead of Φ_0 formally appears because the effective charge entering Eq. (43.19) for the Cooperon is $2e$ instead of e. One may consider, for example, a 2D layer folded into a cylinder along the magnetic field so that Eq. (43.19) should be solved with periodic boundary conditions (problem 12.8). This consideration shows us that the $\Phi_0/2$ oscillations are caused by the self-interference of electron waves associated with closed paths around the ring.

Consider now an isolated ring (without leads). In the presence of a finite magnetic flux, a diamagnetic current should flow around the ring in equilibrium. Indeed, according to Eq. (4.16), the energy $\Delta\mathcal{E}$ associated with the vector potential is $\Delta\mathcal{E} = -c^{-1}\int d\mathbf{r}\,\mathbf{I_r}\cdot\mathbf{A_r}$. In application to the 1D ring, this integral is a contour integral. The steady-state current $\mathbf{I_r}$ in the 1D case is constant, because of the continuity requirement. Since this current is directed along the circumference of the ring, we can write $\mathbf{I_r} = I\mathbf{n}_l$. Using the Stokes theorem, we obtain

$$\Delta\mathcal{E} = -\frac{I}{c}\oint_{\mathbf{r}\in l} d\mathbf{r}\,\mathbf{A_r}\cdot\mathbf{n}_l = -\frac{I\Phi}{c}. \tag{20}$$

Here we assume the clockwise direction of the current. For the counter-clockwise direction, the sign of $\Delta\mathcal{E}$ should be changed. The current is obtained as a derivative of the total energy of the electron system, $\mathcal{E}$, over the flux:

$$I = -c\frac{\partial\mathcal{E}}{\partial\Phi}. \tag{21}$$

In the case of ideal transmission of electrons around a circular ring of radius ρ_0, the energy is given as a sum over occupied discrete states, $\mathcal{E} = 2\sum_{n=-\infty}^{\infty}\varepsilon_n\theta(\varepsilon_F - \varepsilon_n)$, where $\varepsilon_n = (\hbar^2/2m\rho_0^2)[n + \Phi/\Phi_0]^2$; see problem 12.7. Owing to the alternating signs of the derivative $\partial\varepsilon_n/\partial\Phi$ for consecutive levels, there is a strong cancellation, and the sum is of the order of the last term (with ε_n nearest to the Fermi energy ε_F). This leads to an estimate

$$I \sim \frac{2|e|\varepsilon_F}{\pi\hbar N} \simeq \frac{|e|v_F}{L}, \tag{22}$$

where $L = 2\pi\rho_0$ is the length of the circumference, $N \simeq \sqrt{2mL^2\varepsilon_F}/\pi\hbar$ is the number of occupied discrete states, and $v_F = \sqrt{2\varepsilon_F/m}$ is the Fermi velocity. Thus, I is the current associated with the motion of a single electron around the ring. We stress that I is an equilibrium current. It is often called the persistent current.

In the presence of elastic scattering, the persistent current flows without dissipation. To prove this, let us calculate the energy spectrum of the electrons in a non-ideal ring. The relation between the amplitudes of forward- and backward-moving electron waves on both sides of a disordered one-dimensional conductor are given by Eq. (9), where the index i should be omitted since we consider a single conductor. Let us transform the conductor to a ring, which implies the boundary conditions $B' = B$ and $C' = C$. This immediately gives us the equation

$$(1 - t)(1 - t') = rr'. \tag{23}$$

We remind that the primed letters describe the transmission and reflection of backward-moving electrons. In the presence of a magnetic flux (assuming that the forward motion of electrons in the conductor becomes the clockwise motion in the ring) one should make the substitutions $t \rightarrow te^{i\phi}$ and $t' \rightarrow te^{-i\phi}$, where the new variable t is the transmission amplitude at zero flux. The reflection amplitudes are not modified by the flux, and they are described according to Eq. (17). Taking these properties into account, we find the relation

$$\cos[\arg(t)] = |t| \cos \phi, \qquad (24)$$

which directly follows from Eq. (23). The quantity $\arg(t) + 2\pi n$, where n is an integer, is the phase acquired by the electron in one revolution around the ring (the total phase is $\arg(t) + 2\pi n + \phi$). The electron energy associated with this state is proportional to $[n + \arg(t)/2\pi]^2$. Equation (24) is, therefore, a dispersion relation. Only the waves whose phases satisfy Eq. (24) can propagate in the ring. As a result of this restriction, there appears a miniband structure of the electron spectrum as a function of the flux. The situation is analogous to the case of electrons with a characteristic wave number $k = \phi/L$ in a one-dimensional crystal. For nearly ideal rings, where $|t| \simeq 1$, the bands are broad, and there are narrow forbidden gaps near $\phi = 0$ and $\phi = \pm\pi$. In the case of strongly localized electrons, when $|t| \ll 1$, the bands are narrow and their positions correspond to $\arg(t) = \pi/2$ and $\arg(t) = 3\pi/2$. The energy spectrum of a circular ring of radius ρ_0 in this case is quasi-discrete,

$$\varepsilon_n \simeq \frac{\hbar^2 \pi^2}{2mL^2} \left[\left(n + \frac{1}{2} \right)^2 - \frac{(-1)^n}{\pi} (2n+1)|t| \cos \phi \right], \qquad (25)$$

and the flux dependence is harmonic, since it corresponds to the tight-binding situation. The persistent current still exists, but its magnitude is much smaller than the one given by Eq. (21). The inelastic scattering cannot destroy the persistent current, but it causes its fluctuations with time. Such fluctuations become important in non-stationary processes, as described in the next paragraph.

If the magnetic field changes with time, there appears an electric field E directed along the circumference of the ring. Indeed, applying the Stokes theorem to the Maxwell equation $[\nabla \times \mathbf{E}] = -c^{-1} \partial \mathbf{H}/\partial t$, we obtain the electromotive force

$$V = \oint_{\mathbf{r} \in l} d\mathbf{r} \; \mathbf{E} \cdot \mathbf{n}_l = -\frac{1}{c} \frac{\partial \Phi}{\partial t}, \qquad (26)$$

which is constant for a linear variation of the flux with t. Since the properties of the ring are periodic in Φ, the constant electromotive force induces Bloch oscillations of the current I. The frequency of these oscillations, $\omega = |e|V/\hbar$, is determined from the periodicity Φ_0 given by

Eq. (13). This frequency is proportional to V. In the presence of inelastic scattering, which transfers an electron between the minibands at a constant flux Φ, the Bloch motion leads to a stationary dissipative current. To find it, we assume the case of low temperatures, when the fluctuations of the occupation numbers are important only for the two minibands, n and $n+1$, closest to the Fermi level. Denoting these occupation numbers as f_n and f_{n+1}, we write the time-averaged current (21) as

$$\langle I \rangle = -2c \left\langle f_n \frac{\partial \varepsilon_n}{\partial \Phi} + f_{n+1} \frac{\partial \varepsilon_{n+1}}{\partial \Phi} \right\rangle. \tag{27}$$

Since the occupation numbers satisfy the equation $f_n + f_{n+1} = 1$ following from the particle conservation requirement, one can eliminate f_{n+1} from Eq. (27). To determine f_n, let us write a balance equation

$$\frac{\partial f_n}{\partial t} = -\frac{1}{\tau}(f_n - f_n^{(eq)}), \tag{28}$$

where $f_n^{(eq)}$ is the equilibrium occupation number. The functions f_n, $f_n^{(eq)}$, and ε_n depend on the flux, while the relaxation time τ, for the sake of simplicity, is assumed to be independent of the flux. The problem is solved after substituting a solution of Eq. (28) into Eq. (27). To facilitate the averaging over time, which is defined as $(\omega/2\pi) \int_0^{2\pi/\omega} dt \dots$, one may introduce the Fourier expansions $f_n^{(eq)} = \sum_k f_k \cos(k\omega t)$ and $\varepsilon_n - \varepsilon_{n+1} = \sum_k \Delta_k \cos(k\omega t)$. The averaged current is expressed through the coefficients in these expansions as

$$\langle I \rangle = -\frac{|e|}{\hbar} \sum_{k=1}^{\infty} f_k \Delta_k \frac{k^2 \omega \tau}{1 + (k\omega \tau)^2}. \tag{29}$$

This current, as a function of ω, is maximal at $\omega \sim 1/\tau$. If $\omega \ll 1/\tau$, the current is a linear function of the voltage $V = \hbar\omega/|e|$. This corresponds to a dissipative conductance proportional to $e^2 \tau$.

A special case of one-dimensional transport is realized in the 2D samples of finite size in the presence of a magnetic field. If this field is strong enough, the Fermi electrons move along the well-defined equipotential lines, according to Eq. (51.15). In small (but still macroscopic) samples this motion occurs near the edges of the sample and has a classical interpretation in terms of skipping cyclotron orbits. In the quantum-mechanical interpretation, the wave functions of electrons in the presence of a magnetic field and smooth boundary potential become localized. An example of such situation is the exactly solvable problem of a 2D electron whose motion in the XOY plane is restricted by the parabolic potential energy $U(x) = m\omega_0^2 x^2/2$. Using the gauge $\mathbf{A} = (0, Hx, 0)$, we search for the electron wave function in the form $L_y^{-1/2} e^{ipy/\hbar} \psi(x)$ and write the Schroedinger equation for $\psi(x)$ as (spin splitting is neglected)

$$\left[-\frac{\hbar^2}{2m} \frac{\partial^2}{\partial x^2} + \frac{m\omega^2}{2}(x - X_p)^2 + \widetilde{\varepsilon}_p - \varepsilon \right] \psi(x) = 0, \qquad (30)$$

$$\omega = \sqrt{\omega_c^2 + \omega_0^2}, \quad \widetilde{\varepsilon}_p = \left(\frac{\omega_0}{\omega} \right)^2 \frac{p^2}{2m}, \quad X_p = - \left(\frac{\omega_c}{\omega} \right)^2 \frac{p l_H^2}{\hbar}.$$

This is an equation for the harmonic oscillator with shifted center and renormalized frequency. Therefore, the energy spectrum is $\varepsilon = \varepsilon_{Np} = \hbar\omega(N + 1/2) + \widetilde{\varepsilon}_p$. It contains a renormalized kinetic energy describing the drift motion of electrons along OY. The wave functions are given by Eq. (5.15) with $X_{p_y} = X_p$ and $l_H \to \ell = \sqrt{\hbar/m\omega}$. By equating ε_{Np} to the Fermi energy ε_F, one can define the position of the state originating from the Landau level N as X_{p_N}, where $|p_N| = (\omega/\omega_0)\sqrt{2m[\varepsilon_F - \hbar\omega(N + 1/2)]}$. The states with smaller Landau level numbers N are placed closer to the boundary. The states are localized if the size ℓ of the wave function is small in comparison to the distance $|X_{p_N} - X_{p_{N+1}}|$. In the model under consideration, this occurs when the ratio ω_c/ω_0 becomes much greater than the square root of the number of occupied Landau levels (estimated as $\sqrt{\varepsilon_F/\hbar\omega}$). This consideration proves us that the localization always occurs with increasing magnetic field. The positions of the localized states in this model are distributed over the parabolic potential well. However, if one considers an edge potential, for example, $U(x) = m\omega_0^2 x^2/2$ at $x > 0$ and $U(x) = 0$ at $x < 0$, the localized states occupied by electrons exist only near the edge, since their number is equal to the number of the filled Landau levels in the bulk of the sample. For this reason, these states are called the magnetic edge states.

The electrons in the magnetic edge states can move only in one direction determined by the potential gradient. Their group velocity, in the physically reasonable case of smooth edge potentials, is given by Eq. (51.15). In the example given above, the electrons at $x < 0$ are moving in the positive direction, while those at $x > 0$ are moving in the negative direction. The group velocity (51.15) for the parabolic potential $U(x) = m\omega_0^2 x^2/2$ always coincides with $d\varepsilon_{Np}/dp = (\omega_0^2/\omega^2)(p/m)$, without regard to the value of ω_0 (problem 12.9). The ability to move only in one direction, called the chirality, is an important property of the magnetic edge states. This property is essential for describing the magnetotransport in small 2D samples. First of all, the current carried by an edge state is a usual 1D current, and the consideration leading to

Eq. (58.4) is applicable to a system of N_f edge-state channels, which behave like effective 1D wires spatially separated from each other. In other words, if we have a sample connected to two leads, 1 and 2, the lead-to-lead transmission probability for the electrons at the Fermi surface, T_{21}, is equal to the number N_f of the edge-state channels connecting the leads; the electrons in these states move in the same direction, from 1 to 2. The zero-temperature conductance of such a sample is given by Eq. (58.4). The backscattering processes, which cause deviation of the transmission coefficients from unity, are exponentially suppressed in the case of edge-state transport, because of spatial separation of forward- and backward-moving states. The scattering between the states moving in the same direction is also exponentially suppressed. This leads to two important consequences. The first one is related to small-size constrictions, where the ballistic transport is considerably improved by application of a strong enough magnetic field. It is instructive to check this property on the example of the saddle-point potential (58.25), where there is an exact solution for the transmission in the magnetic field (problem 12.10). The other consequence is related to macroscopic 2D samples, where the inelastic scattering is present, but only within each edge channel. Since this scattering does not change the number of electrons and the direction of electron motion, the coefficient T_{MN} describing the total transmission probability from the lead N to the lead M is simply equal to the number of magnetic edge channels in which the electrons travel from N to M. The multi-terminal conductance in this case is described by Eq. (58.21). In other words, since we have isolated chiral 1D channels, the phase coherence becomes non-essential for the lead-to-lead transmission, and the consideration of electron transport in terms of quantized conductance appears to be valid for macroscopic samples.

The above consideration allows one to describe the quantum Hall effect as a manifestation of the edge-state transport. The relevant geometry for the Hall measurements includes at least four leads, see Fig. 12.4, two of them (1 and 3) are source and drain, and the other two (2 and 4) are voltage probes. Let us apply the formalism of Sec. 58 to such four-terminal system. The four equations for the current follow from Eq. (58.7) with $M, N =$ 1, 2, 3, and 4. Strictly speaking, we need only three equations, since $I_1 + I_2 + I_3 + I_4 = 0$, and the result should depend only on the differences of the voltages. Thus, we consider a system of three equations by setting $V_4 = 0$ and omitting the equation for I_1:

$$I_2 = G_{21}(V_2 - V_1) + G_{23}(V_2 - V_3) + G_{24}V_2,$$

$$I_3 = G_{31}(V_3 - V_1) + G_{32}(V_3 - V_2) + G_{34}V_3, \qquad (31)$$

$$I_4 = -G_{41}V_1 - G_{42}V_2 - G_{43}V_3.$$

We need to find V_2 under the condition that $I_2 = I_4 = 0$ and divide it by $I = I_1 = -I_3$. The ratio V_2/I is the Hall resistance $R_\perp$. Since we have the edge-state transport, the non-zero G_{MN} are $G_{21} = G_{32} = G_{43} = G_{14} = e^2 N_f/2\pi\hbar$, and we obtain $R_\perp = (2\pi\hbar/e^2)1/N_f$. The number of the edge channels, N_f, is equal to the number of filled Landau levels, which is determined by the magnetic field. This result correlates with Eq. (51.1) discussed in Sec. 51 (we point out that the Hall resistivity $\rho_\perp$ in the 2D case is formally equal to the Hall resistance $R_\perp$). Therefore, the quantization of the Hall resistance is obtained by considering the one-dimensional transport of electrons in the edge states. One can find also $V_1 = V_2$ and $V_3 = V_4$. A similar consideration, when the current is fed through the leads 1 and 2 and the voltages are measured at the leads 3 and 4, shows us that $V_3 - V_4 = 0$, which means the absence of the longitudinal resistance in the quantum Hall effect regime.

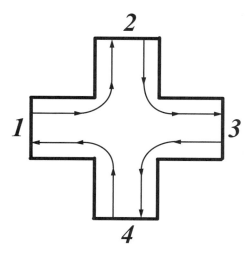

Figure 12.4. Edge-state transport in the four-terminal geometry.

If the leads are connected to the sample through narrow constrictions, the number of the edge channels penetrating into these leads is not, in general, equal to the number N_f of filled Landau levels, it can be smaller. However, if at least two of the leads can accommodate all (N_f) edge channels, the quantization of the Hall resistance is again determined only by the number of the Landau levels. If there is only one such lead, the upper edge channel(s) coming from it form loop(s) around the sample and return back to the lead, without contributing to the transport. A similar situation takes place when no one of the leads can accommodate all edge channels; in this case the upper edge channel(s) form closed loop(s) around the sample without penetrating into any lead. Since all such upper channel(s) should be excluded from the consideration, the Hall resistance shows anomalous behavior: it is not determined by the number N_f, but by the number of the edge states contributing to the transport. The number of these edge states can be varied not only by the magnetic field, but also by the gate voltages controlling the widths of the constrictions. Considering the four-terminal geometry of Fig. 12.4, one can find the Hall resistance as

$$R_\perp = \frac{2\pi\hbar}{e^2 N_{max}}, \tag{32}$$

where $N_{max} \leq N_f$ is the maximal number of the edge states capable to transfer electrons between leads (problem 12.11). To determine N_{max}, one should consider two leads whose contacts transmit the highest numbers of edge states, say M and M', and take the minimum transmission of these two, $N_{max} = \min(M, M')$.

60. Tunneling Current

The tunneling, as a quantum-mechanical property of particles to pass through the classically unpenetrable potential barriers, has numerous manifestations in physics. Since the tunneling processes, as usual, have exponentially small probabilities, one can describe the tunneling current according to the basic equation (58.1). The description of the coefficients $i_{\delta\delta'}^{(\pm)}$ in each particular case requires solution of a quantum-mechanical problem. This is sufficient if the electron transport through the barriers can be treated as ballistic. In more complicated cases, such as the tunneling between low-dimensional systems, it is also necessary to consider scattering of electrons. In this section we study different manifestations of the tunneling in the case of one-dimensional potential energy $U(z)$. The potential diagram in Fig. 12.5 shows us that even in this simple case there are several variants of tunneling transport of electrons. First, there exists tunneling between 3D states, which can be described by calculating the quantum-mechanical transmission coefficients. This tunneling, when associated with the presence of a quasibound 2D state in the barrier, requires a more complex description, especially when the scattering causes electron relaxation in this quasibound state. The tunneling between 2D states requires another description, because the direction of electron motion (in-plane motion) is different from the direction in which they tunnel. The tunneling between 2D and 3D states can be viewed as a tunneling decay of a quasibound state, with a characteristic decay rate. We will show that a unified description of all these cases can be done by using the method of tunneling Hamiltonian (Appendix H). The tunneling processes shown in Fig. 12.5 correspond to a single-band model. Apart from this, we will consider the Zener tunneling between the 3D conduction- and valence-band electron states, which occurs in the presence of a strong electric field; see Fig. 12.6 below.

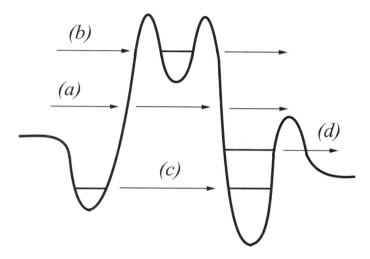

Figure 12.5. Different manifestations of tunneling in the presence of one-dimensional potential energy $U(x)$: tunneling between bulk (3D) states (a), resonant tunneling through a quasibound 2D state (b), tunneling between 2D states (c), and tunneling between 3D and 2D states (d).

To calculate the tunneling conductance through a potential barrier between 3D regions, one may use the formalism described in Sec. 58; see Eqs. (1), (21), and (22). Since the eigenstates for transverse motion (parallel to the barrier) are the plane waves described by the 2D momentum **p** which is conserved in the tunneling transitions, the conductance per unit square is given by

$$G = \frac{e^2}{2\pi\hbar} \int d\varepsilon \left[-\frac{\partial f(\varepsilon)}{\partial \varepsilon} \right] \int \frac{d\mathbf{p}}{(2\pi\hbar)^2} T_{\mathbf{p}\varepsilon}. \tag{1}$$

The tunneling current per unit square is written as

$$I_T = \frac{e}{2\pi\hbar} \int d\varepsilon \int \frac{d\mathbf{p}}{(2\pi\hbar)^2} \left[f_l(\varepsilon) - f_r(\varepsilon) \right] T_{\mathbf{p}\varepsilon}, \tag{2}$$

where $f_r(\varepsilon)$ and $f_r(\varepsilon)$ are the quasi-equilibrium electron distribution functions on the left (l) and right (r) sides. We use the coordinate system where the OZ axis is perpendicular to the barrier. The coefficient $T_{\mathbf{p}\varepsilon} \equiv 2T_{rl}^{(k_z'\mathbf{P}, k_z\mathbf{P})}$ is the probability of transmission between the states with the wave numbers k_z on the left and k_z' on the right (see the general definition of $T_{MN}^{(mn)}$ in Eq. (58.22)) summed over spin (we assume that the electron spectrum is spin-degenerate and the spin is conserved in the tunneling transitions). These wave numbers, as well as the group velocities v_z and v_z' on both sides, are functions of the electron energy ε and momentum $\mathbf{p}$. For the particles with isotropic parabolic spectrum moving in the potential with the asymptotic behavior $U(z \to -\infty) = 0$ and $U(z \to +\infty) = -\Delta U$, one has

$$\hbar k_z = \sqrt{2m(\varepsilon - p^2/2m)}, \quad \hbar k_z' = \sqrt{2m(\varepsilon + \Delta U - p^2/2m)}. \quad (3)$$

The transmission coefficient should be calculated from the quantum-mechanical problem assuming incident and reflected electron waves on the left and transmitted waves on the right,

$$\Psi_1 = (e^{ik_z z} + re^{-ik_z z})/\sqrt{v_z}, \quad \Psi_2 = te^{ik_z' z}/\sqrt{v_z'}, \quad (4)$$

where $v_z = v_z(\varepsilon, p) = \hbar k_z/m$ and $v_z' = v_z'(\varepsilon, p) = \hbar k_z'/m$. This transmission coefficient is expressed through the amplitude of the transmitted wave as $T_{p\varepsilon} = 2|t|^2$. Such a consideration implies the introduction of the wave functions proportional to $e^{\kappa z}$ and $e^{-\kappa z}$ in the barrier region, where κ^{-1} is the underbarrier penetration length (problem 12.12). Assuming that the potential barrier $U(z)$ is high and thick enough to provide exponentially small transmission probability, one can always find the transmission coefficient with the exponential accuracy:

$$T_{p\varepsilon} \sim \exp\left[-2\int_{z_0}^{z_0'} dz \kappa(z)\right], \quad \hbar\kappa(z) = \sqrt{2m[U(z) - \varepsilon] + p^2}, \quad (5)$$

where z_0 and z_0' are the classical return points determined by the equation $\kappa(z) = 0$. In this limit, the main contribution to the tunneling current comes from the electrons whose transverse momenta are small, $p \ll p_F$, so that the exponentially-accurate expression for the tunneling conductance is

$$G \propto \int d\varepsilon \left[-\frac{\partial f(\varepsilon)}{\partial \varepsilon}\right] \exp\left[-2\frac{\sqrt{2m}}{\hbar}\int_{z_0}^{z_0'} dz\sqrt{U(z) - \varepsilon}\right]. \quad (6)$$

At zero temperature, the tunneling conductance is proportional to $T_{p\varepsilon}$ calculated at $p = 0$ and $\varepsilon = \varepsilon_F$. The conductance stays independent of temperature in the low-temperature region. However, if the integral in the exponent of Eq. (6) decreases with increasing ε faster than linearly, the thermal-activated tunneling appears to be important even at relatively low temperatures. In particular, this is the case of triangular barriers, when $U(z) = 0$ at $z < 0$ and $U(z) = U_0(1 - z/d)$ at $z > 0$ so that $\int_{z_0}^{z_0'} dz \sqrt{U(z) - \varepsilon} = (2d/3U_0)(U_0 - \varepsilon)^{3/2}$. The characteristic energy of the tunneling electrons in the degenerate case is ε_F. As the temperature T increases and the energy $U_0 - E_d(U_0/T)^2$, where $E_d = \hbar^2/8md^2$, becomes greater than ε_F, the characteristic tunneling energy becomes equal to $U_0 - E_d(U_0/T)^2$. The conductance in these conditions shows a thermal-activation behavior according to $G \propto \exp[-(U_0 - \varepsilon_F)/T + E_d U_0^2/3T^3]$. The dependence on the barrier width d exists in a narrow temperature interval where the current is switched from the pure tunneling current to the pure thermal-activated one.

In the situation denoted as (b) in Fig 12.5, one should consider a barrier containing quasibound states. The simplest example of this kind is a double-barrier structure with $U(z) = U_0$ at $0 < z < z_1$ and $z_2 < z < z_3$, and $U(z) = 0$ elsewhere. Since the transverse momentum $\mathbf{p}$ is conserved in the transitions, the quantum-mechanical transmission coefficient for such a structure is analogous to the coefficient already calculated for the one-dimensional conductors with two obstacles; see Eq. (59.3). The coefficients $|t_i|^2$ and $|r_i|^2 = 1 - |t_i|^2$ are the transmission and reflection coefficients of individual barriers calculated for fixed values of the energy and transverse momentum. The phase θ, apart from the slowly varying arguments $\arg(r_1')$ and $\arg(r_2)$, contains the dynamical phase $2\varphi = 2k_z(z_2 - z_1)$ which is a fast function of both energy and momentum. If this phase changes, for example, as a function of the Fermi energy, the transmission coefficient varies from a small value $|t_1|^2|t_2|^2/4$ to $4|t_1|^2|t_2|^2/(|t_1|^2 + |t_2|^2)^2$; see Eq. (59.3). The peaks in the transmission correspond to $\cos\theta = 1$, they are narrow, and their line shape is close to a Lorentz function (this statement can be proved by expanding the cosine in Eq. (59.3) around its resonant value). In the case of identical barriers, $|t_1|^2 = |t_2|^2$, the peak transmission coefficient is equal to unity even if the transmission coefficients of individual barriers are small. The appearance of such peaks corresponds to resonant tunneling of electrons through the quasibound states. We have considered the case of coherent tunneling. If the quantum coherence is suppressed by scattering (see the description of the sequential tunneling below), the peak transmission coefficient becomes small, though the peak structure of the transmission remains.

The electrons in crystals can tunnel between the energy bands in the presence of a homogeneous electric field $\mathbf{E}$. To consider this phenomenon (Zener tunneling), it is convenient to apply the time-dependent Schroedinger equation

$$i\hbar \frac{\partial \Psi}{\partial t} = \left(\hat{h}_{cr} - e\mathbf{E} \cdot \mathbf{r} \right) \Psi, \tag{7}$$

where $\hat{h}_{cr}$ is the Hamiltonian (5.4) which contains a periodic potential. In the absence of the electric field, the solution of this equation would be $\psi_{l\mathbf{p}}(\mathbf{r})e^{-iE_l(\mathbf{p})t/\hbar}$, where the Bloch function $\psi_{l\mathbf{p}}(\mathbf{r})$ of the l-th band corresponding to the energy $E_l(\mathbf{p})$ is given by Eq. (5.5). If the electric field is applied, the electrons are accelerating according to $\mathbf{p} \to \mathbf{p}_t = \mathbf{p} + e\mathbf{E}t$. However, since $\mathbf{p}$ is a quasimomentum, they experience Bragg reflection at the edge of the Brillouin zone and start moving again across this zone. This process can be also viewed as Bloch oscillations, when the electron energy is a periodic function of time, $E_l(\mathbf{p}_t)$. The wave function which takes into account this motion in the adiabatic approximation is

$$\psi_l(\mathbf{r}\mathbf{p}t) = L^{-3/2} u_{l\mathbf{p}_t}(\mathbf{r}) e^{i\mathbf{p}_t \cdot \mathbf{r}/\hbar} \exp\left[-\frac{i}{\hbar} \int^t dt' E_l(\mathbf{p}_{t'}) \right]. \tag{8}$$

It satisfies Eq. (7) except for the term $\propto \mathbf{E} \cdot \partial u_{l\mathbf{p}_t}(\mathbf{r})/\partial \mathbf{p}_t$. The contribution of this term is responsible for the tunneling between the bands. It can be taken into account by expanding the exact wave function in the form $\Psi = \sum_l C_l(t)\psi_l(\mathbf{r}\mathbf{p}t)$. Substituting this expansion into Eq. (7), multiplying the latter by $\psi_{l'}^*(\mathbf{r}\mathbf{p}t)$ from the left, and integrating over $\mathbf{r}$ by using the orthogonality of the Bloch functions, we obtain an equation for $C_l(t)$ in the form $\partial C_l(t)/\partial t = \sum_{l'} A_{ll'}(t)C_{l'}(t)$. This gives us the probability $P_{ll'} = |\int_0^{T_B} dt A_{ll'}(t)|^2$ of the transition between the bands l' and l during the period of the Bloch oscillations, T_B. Taking into account that the index l (l') includes both the band number n (n') and the spin number σ (σ'), we introduce the total probability summed over all spins (the bands are assumed to be spin-degenerate) according to

$$P_{nn'} = \sum_{\sigma\sigma'} \left| \int_0^{T_B} dt X_{n\sigma,n'\sigma'}(\mathbf{p}_t) \exp\left\{ \frac{i}{\hbar} \int^t dt' [E_n(\mathbf{p}_{t'}) - E_{n'}(\mathbf{p}_{t'})] \right\} \right|^2,$$

$$X_{n\sigma,n'\sigma'}(\mathbf{p}) = \frac{|e|}{V_c} \int_{V_c} d\mathbf{r}\, u_{n\sigma\mathbf{p}}^*(\mathbf{r})\mathbf{E} \cdot \frac{\partial}{\partial \mathbf{p}} u_{n'\sigma'\mathbf{p}}(\mathbf{r}). \tag{9}$$

The integral over $\mathbf{r}$ in the last equation is taken over the volume of the crystal cell, V_c.

The integral over time t in Eq. (9) contains a fast-oscillating factor, and the main contribution to this integral comes from the region where the bands n and n' are closest to each other. This is the region of the band extrema, where the bands are separated by a direct gap ε_g. Therefore, one may extend the limits in the integral over t to $\pm\infty$ and use the **kp**-formalism to describe the interband transitions. The quantity $P_{nn'}$ in this case describes transmission of the n'-th band electron to the adjacent band n as this electron is decelerating and reflecting back from the potential wall created by the electric field. Therefore, we again obtain a transmission-reflection problem, as in the case of single-band electron tunneling through the potential barriers considered above. Below we use the symmetric two-band model of Appendix B to calculate the probability of transmission from the conduction band to the valence one. Assuming that the field is directed along OZ, we have

$$P_{vc} \equiv T_p = \sum_{\sigma\sigma'} \left| \int_{-\infty}^{\infty} \frac{dp_z}{|e|E} X_{v\sigma,c\sigma'}(p_z, \mathbf{p}) \exp\left[\frac{2i}{\hbar|e|E} \int_{0}^{p_z} dp'_z \mathcal{E}(p'_z, p) \right] \right|^2,$$

$$\mathcal{E}(p'_z, p) = \sqrt{(\varepsilon_g/2)^2 + s^2 p^2 + s^2 p'^2_z}. \tag{10}$$

This transmission coefficient is independent of energy because the electric field is homogeneous. Since T_p decreases exponentially with increasing $\mathbf{p} = (p_x, p_y)$, one can put $\mathbf{p} = 0$ in $X_{v\sigma,c\sigma'}(p_z, \mathbf{p})$ and obtain $X_{v\sigma,c\sigma}(p_z, \mathbf{p} = 0) = \delta_{\sigma\sigma'}\sigma s|e|E\varepsilon_g/[2\mathcal{E}(p_z,0)]^2$. This expression is found by a direct calculation taking into account that in the symmetric two-band model $X_{v\sigma,c\sigma'}(p_z, \mathbf{p}) = |e|E\mathbf{\Psi}^*_{v\sigma}(p_z, \mathbf{p}) \cdot \partial\mathbf{\Psi}_{c\sigma'}(p_z, \mathbf{p})/\partial p_z$, where $\mathbf{\Psi}_{n\sigma}(p_z, \mathbf{p})$ are the columnar eigenstates (eigenvectors) of the 4×4 matrix Hamiltonian (B.18) with $M \to \infty$, similar to the eigenstates introduced in problem 10.15. The integral over p_z in Eq. (10) is taken analytically by shifting the contour of integration into the upper half-plane of complex p_z up to the pole $ip_0 = i\sqrt{(\varepsilon_g/2s)^2 + p^2}$, where $\mathcal{E}(p_z, p) = 0$ (problem 12.13). The result is

$$T_p = \frac{2\pi^2}{9} \exp\left\{ -\frac{\pi m^{1/2}(\varepsilon_g/2)^{3/2}}{\hbar|e|E} \left(1 + \frac{2p^2}{m\varepsilon_g} \right) \right\}, \tag{11}$$

where we use the mass $m = \varepsilon_g/2s^2$ at the bottom of the band rather than the velocity s. The exponent in Eq. (11) at $p = 0$ is similar to that of Eq. (18.9) at $\omega = 0$, the difference is only in a numerical factor.

The probability of Zener tunneling calculated above can be used to write the tunneling current through the n-i-p diode, whose energy band diagram is represented in Fig. 12.6. With the aid of Eq. (2), we find the density of the tunneling current between the n- and p-doped regions:

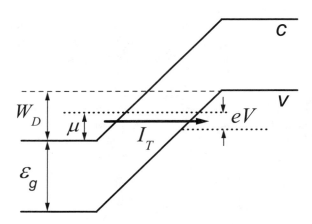

Figure 12.6.   Zener tunneling in the *n-i-p* diode structure.

$$I_T = \frac{em}{18\hbar^3} \exp\left[-\frac{\pi m^{1/2}(\varepsilon_g/2)^{3/2}}{\hbar|e|E} \right]$$

$$\times \int d\varepsilon \int_0^{\varepsilon_{max}} d\varepsilon_\perp e^{-\varepsilon_\perp/\varepsilon_0} [f_l(\varepsilon) - f_r(\varepsilon)], \tag{12}$$

where $\varepsilon_\perp = p^2/2m$ is the energy of transverse motion and ε_0 is a characteristic energy equal to $\hbar|e|E/\pi\sqrt{2m\varepsilon_g}$. Since the total energy and transverse momentum are conserved, the energy $\varepsilon_\perp$ is limited by $\varepsilon_{max} = \min(\varepsilon, W_D - \varepsilon)$, where $W_D = \Delta - \varepsilon_g > 0$ and Δ is the total drop of the potential energy across the structure. We assume $W_D \ll \varepsilon_g$ so that the parabolic approximation is justified. Integrating over $\varepsilon_\perp$, we obtain $\int_0^{\varepsilon_{max}} d\varepsilon_\perp e^{-\varepsilon_\perp/\varepsilon_0} = \varepsilon_0[1 - e^{-\varepsilon_{max}/\varepsilon_0}]$. If a bias eV is applied to the *n-i-p* diode, the energy W_D changes as $W_D = W_D^{(eq)} - eV$ because the electron density on the left and the hole density on the right are fixed by the doping level. The current at zero temperature is given by

$$I_T = e\frac{m|e|E}{18\pi\hbar^2\sqrt{2m\varepsilon_g}} \exp\left[-\frac{\pi m^{1/2}(\varepsilon_g/2)^{3/2}}{\hbar|e|E} \right] D, \tag{13}$$

$$D = \int_{\max(0,\mu-eV)}^{\min(W_D^{(eq)}-eV,\mu)} d\varepsilon \left\{ 1 - \exp\left[-\frac{\min(\varepsilon, W_D^{(eq)} - eV - \varepsilon)}{\varepsilon_0} \right] \right\},$$

where μ is the equilibrium chemical potential counted from the bottom of the conduction band in the left (*n*-doped) side of the diode. The modification of the exponential factor in I_T associated with the dependence of the electric field E on eV can be neglected if $|eV| \ll 4\varepsilon_0$, and the current-voltage characteristics in this case are entirely determined by the function D. At a small bias, when the upper and lower limits of integration in Eq. (13) are μ and $\mu - eV$, respectively, the current I_T is linear in V. If $|eV|$ becomes comparable to $W_D^{(eq)}$, the current-voltage characteristics essentially depend on the sign of V. In the case of negative bias, the tunneling current monotonically increases with $|V|$. In the region of positive bias, the current increases, goes through a maximum, and, finally, decreases to zero at $eV = W_D^{(eq)}$, when the tunneling becomes forbidden. The tunneling current is exponentially small at $\varepsilon_0 \ll \varepsilon_g$. If ε_0 is small in comparison to the Fermi energies of electrons and holes, μ and $W_D^{(eq)} - \mu$, respectively, only the electrons with $\varepsilon_\perp \simeq 0$ contribute to the current, and the exponential term in the expression for D can be neglected. The voltage dependence of the function D in this case has the most simple form (in particular, this function is always equal to eV in the case of negative bias).

Having considered the tunneling between 3D regions, we now study the tunneling decay of a metastable 2D state into a continuum of 3D states, the process (d) in Fig. 12.5. In the 3D region the wave function of this state should behave as an outgoing plane wave. This requirement leads to a special dispersion relation for the state quantized in the well so that the energy of this state acquires a small negative imaginary part, $-i\Gamma$, to ensure conservation of the electron flow across the structure, $\psi_z^*(d\psi_z/dz) - (d\psi_z^*/dz)\psi_z =$const. One may, therefore, introduce a characteristic rate $\nu = 2\Gamma/\hbar$ describing an exponential decay of the metastable state. If the electron density in such "leaky" quantum well is maintained by an external source, a steady-state current flows between the well and the continuum. According to Eq. (58.1), the tunneling current per unit square can be written as a sum of partial contributions from all discrete 2D states characterized by the level number k and 2D momentum $\mathbf{p}$. Taking into account the energy conservation law, we obtain

$$I_T = 2e \int \frac{d\mathbf{p}}{(2\pi\hbar)^2} \sum_k \nu_k [f_l(\varepsilon_{lk} + \varepsilon_p) - f_r(\varepsilon_{lk} + \varepsilon_p)], \qquad (14)$$

where $f_l(\varepsilon)$ is the quasi-equilibrium Fermi distribution function in the well (left region), which depends on the sum of the in-plane kinetic energy $\varepsilon_p = p^2/2m$ and discrete size-quantization energy ε_{lk}. The function $f_r(\varepsilon)$ is the quasi-equilibrium Fermi distribution function in the 3D (right) region. The usage of the quasi-equilibrium distribution functions in Eq. (14) is justified if ν_k is small in comparison to the energy relaxation rates in the well and in the continuum. This is easy to achieve, since ν_k is proportional to an exponentially small factor similar to the one standing in Eq. (6). The decay rate ν_k is calculated in problem 12.14 for the electrons escaping from a rectangular quantum well through a rectangular potential barrier. Since we assume a parabolic band with a coordinate-independent effective mass, ν_k is fixed by the potential $U(z)$ and does not depend on ε_p. Therefore, the tunneling current (14) can be expressed through the tunneling conductance given by

$$G = e^2 \rho_{2D} \nu, \quad \nu = \sum_k \nu_k \theta(\varepsilon_F - \varepsilon_{lk}), \tag{15}$$

where $\rho_{2D} = m/\pi\hbar^2$ is the 2D density of states and ν is the sum of the tunneling rates over all populated discrete states. Although we have considered the 3D-2D tunneling, one can introduce the tunneling rate ν_k in any case when there is a coupling between a metastable (quasibound) state and a continuum of states.

In the general case, the tunneling currents described by Eqs. (2) and (14) can be written through the tunneling matrix elements introduced in Appendix H. Using Eq. (H.15) for the tunneling probability in unit time, we write the tunneling current according to Eq. (58.1) as

$$I_T = \frac{4\pi e}{\hbar} \int \frac{d\mathbf{p}}{(2\pi\hbar)^2} \sum_{kk'} \int d\varepsilon |t_{lk',rk}|^2 \delta(\varepsilon_{lk'\mathbf{p}} - \varepsilon)\delta(\varepsilon - \varepsilon_{rk\mathbf{p}})[f_l(\varepsilon) - f_r(\varepsilon)]. \tag{16}$$

This equation can be also obtained with the use of $I_T = \mathrm{Sp}\hat{\rho}_t \hat{I}_T$, where the statistical operator $\hat{\rho}_t$ satisfies the equation $i\hbar\partial\hat{\rho}_t/\partial t = [\hat{H}_T, \hat{\rho}_t]$, the Hamiltonian $\hat{H}_T$ is given by Eq. (H.13), and the operator of the tunneling current, $\hat{I}_T$, is defined according to Eq. (H.18). If both k and k' are continuous variables, from a comparison of Eqs. (2) and (16) we obtain

$$T_{\mathbf{p}\varepsilon} = 8\pi^2 \sum_{kk'} |t_{lk',rk}|^2 \delta(\varepsilon_{lk'\mathbf{p}} - \varepsilon)\delta(\varepsilon - \varepsilon_{rk\mathbf{p}}). \tag{17}$$

To check this relation on a simple example, one may consider the tunneling of 3D electrons through a rectangular potential barrier (problem 12.15). For the case of tunneling between a single discrete (2D) state k' on the left and a continuum of states k on the right, we find simply $\nu_{k'} = \sum_k W_{lk'\mathbf{p},rk\mathbf{p}}$, where the probability of transition in unit time is determined according to Fermi's golden rule; see Eqs. (H.15) and (H.30).

The tunneling between 2D states, which occurs when two 2D electron layers are brought close together in real space, requires a special consideration. If we directly apply Eq. (H.15) to this case, using the 2D states with the energies $\varepsilon_{lp} = \Delta/2 + \varepsilon_p$ and $\varepsilon_{rp} = -\Delta/2 + \varepsilon_p$, where $\varepsilon_p = p^2/2m$, we find that the probability of tunneling is infinite in the resonance (at $\Delta = 0$) and zero out of the resonance, when $\Delta \neq 0$. The absence of out-resonance tunneling is associated with the conservation of the in-plane (2D) momentum **p**. To cancel this restriction, one should include the scattering potentials into the Hamiltonian. For the double-layer system, in the basis of l- and r- orbitals, this Hamiltonian is written as an effective 2×2 matrix 2D Hamiltonian derived in Appendix H; see Eqs. (H.25) and (H.26). The one-electron statistical operator is also written in the matrix form,

$$\hat{\rho}_t = \left| \begin{array}{cc} \hat{\rho}_{lt} & \tilde{\rho}_t \\ \tilde{\rho}_t^+ & \hat{\rho}_{rt} \end{array} \right|, \qquad (18)$$

which is used below to express the tunneling current through the operator of this current in the matrix representation, see Eq. (H.27):

$$I_T = -\frac{et_{lr}}{\hbar} \frac{2}{L^2} \text{Sp}(\hat{\sigma}_y \hat{\rho}_t). \qquad (19)$$

The trace in this equation includes both the 2×2 matrix trace and the trace over the 2D variables (the trace over the spin variables is already taken and gives the factor of 2 in Eq. (19)). The equation $i\hbar \partial \hat{\rho}_t / \partial t = [\widehat{H}_T, \hat{\rho}_t]$, when written in the matrix representation, takes the following form:

$$\frac{\partial \hat{\rho}_{jt}}{\partial t} + \frac{i}{\hbar} \left[\hat{h}_j, \hat{\rho}_{jt} \right] = \pm \frac{it_{lr}}{\hbar} \left(\tilde{\rho}_t - \tilde{\rho}_t^+ \right),$$

$$\frac{\partial \tilde{\rho}_t}{\partial t} + \frac{i}{\hbar} \left(\hat{h}_l \tilde{\rho}_t - \tilde{\rho}_t \hat{h}_r \right) = \frac{it_{lr}}{\hbar} \left(\hat{\rho}_{lt} - \hat{\rho}_{rt} \right), \qquad (20)$$

where $\hat{h}_{jt}$ is the Hamiltonian of the layer j, and the signs $+$ and $-$ on the right-hand side of the first equation stand for $j = l$ and $j = r$, respectively. The second equation in the set (20) allows one to express the non-diagonal components through the diagonal ones:

$$\tilde{\rho}_t = \frac{it_{lr}}{\hbar} \int_{-\infty}^{t} dt' e^{\lambda t'} e^{-i\hat{h}_l(t-t')/\hbar} (\hat{\rho}_{lt'} - \hat{\rho}_{rt'}) e^{i\hat{h}_r(t-t')/\hbar}, \qquad (21)$$

where $\lambda \to +0$ describes an adiabatic turn-on of the tunneling at $t \to -\infty$.

The trace in Eq. (19) should be taken over exact eigenstates of tunnel-coupled layers. However, since t_{lr} is small, we are searching for the $\propto t_{lr}^2$ contributions to the tunneling current and neglect all higher-order contributions. In this approximation, the matrix trace in Eq. (19) can be taken separately from the trace over the variables characterizing the layers. Using a set of exact single-layer eigenstates $|j\delta\rangle$, we rewrite Eq. (19) as

$$I_T = -i\frac{et_{lr}}{\hbar}\frac{2}{L^2}\left\langle\!\!\left\langle\sum_\delta\langle j\delta|(\tilde{\rho}_t - \tilde{\rho}_t^+)|j\delta\rangle\right\rangle\!\!\right\rangle. \tag{22}$$

The result does not depend on whether we put $j = l$ or $j = r$ in this expression. The double angular brackets in Eq. (22) denote the average over effective random 2D potentials of the l- and r-layers. After substituting Eq. (21) into Eq. (22), we calculate the integral over time and obtain

$$I_T = \frac{2\pi et_{lr}^2}{\hbar}\frac{2}{L^2}\left\langle\!\!\left\langle\sum_{\delta\delta'}|\langle r\delta|l\delta'\rangle|^2\delta(\varepsilon_{r\delta} - \varepsilon_{l\delta'})(f_{l\delta'} - f_{r\delta})\right\rangle\!\!\right\rangle, \tag{23}$$

where $f_{j\delta} = \langle j\delta|\hat{\rho}_{jt}|j\delta\rangle$ are the distribution functions in the layers. This expression can be rewritten through the correlation function of the spectral density functions of the layers in the momentum representation, see Sec. 13:

$$I_T = \frac{2\pi et_{lr}^2}{\hbar}\int d\varepsilon[f_l(\varepsilon) - f_r(\varepsilon)]\frac{2}{L^2}\sum_{\mathbf{pp}'}\langle\!\langle A_{l\varepsilon}(\mathbf{p},\mathbf{p}')A_{r\varepsilon}(\mathbf{p}',\mathbf{p})\rangle\!\rangle. \tag{24}$$

The spectral density function $A_{j\varepsilon}(\mathbf{p},\mathbf{p}')$ is expressed through the Green's functions $G_{j\varepsilon}^{R,A}(\mathbf{p},\mathbf{p}')$ according to Eq. (14.5). These Green's functions satisfy single-layer equations of the kind of Eq. (14.6), where ε_p is replaced by ε_{jp} ($\varepsilon_{lp} = \Delta/2 + \varepsilon_p$ and $\varepsilon_{rp} = -\Delta/2 + \varepsilon_p$) and $U_{im}(\mathbf{p} - \mathbf{p}_1)$ is replaced by the spatial Fourier transform of the effective potential energy $V^{(j)}(\mathbf{x})$ introduced in Eq. (H.26).

One can reasonably assume that there is no correlation between the scattering potentials of the different layers. This is always true when the scattering is caused by short-range potentials. In this approximation, the correlation function in Eq. (24) is expressed through the averaged Green's functions, $\langle\!\langle A_{l\varepsilon}(\mathbf{p},\mathbf{p}')A_{r\varepsilon}(\mathbf{p}',\mathbf{p})\rangle\!\rangle = -\delta_{\mathbf{pp}'}[G_{l\varepsilon}^A(\mathbf{p}) - G_{l\varepsilon}^R(\mathbf{p})][G_{r\varepsilon}^A(\mathbf{p}) - G_{r\varepsilon}^R(\mathbf{p})]/(2\pi)^2$. These functions, in the Born approximation, are given by Eq. (14.29) generalized to include the layer index:

$$G_{j\varepsilon}^R(\mathbf{p}) = G_{j\varepsilon}^{A*}(\mathbf{p}) = [\varepsilon - \varepsilon_{jp} + i\hbar/2\tau_j(\varepsilon)]^{-1}, \tag{25}$$

The real part of the self-energy, $\text{Re}\Sigma_{j\varepsilon}$, is omitted, since it always can be taken into account by proper shifts of ε and Δ. After using Eq. (25) in Eq. (24), we write the tunneling current as

$$I_T = \frac{2et_{lr}^2}{(2\pi)^2\hbar\tau_l\tau_r}\int d\varepsilon[f_l(\varepsilon) - f_r(\varepsilon)] \tag{26}$$

$$\times \int_0^\infty \frac{p\,dp}{[(\varepsilon - \varepsilon_{lp})^2 + (\hbar/2\tau_l)^2][(\varepsilon - \varepsilon_{rp})^2 + (\hbar/2\tau_r)^2]}.$$

We have taken into account that the 2D scattering times τ_j are energy-independent in the case of short-range scattering potentials. The integral over p in Eq. (26) can be calculated easily. Then, substituting the Fermi distribution functions with the quasi-Fermi energies ε_{Fl} and $\varepsilon_{Fr} = \varepsilon_{Fl} - eV$ in place of $f_l(\varepsilon)$ and $f_r(\varepsilon)$, we integrate over ε and obtain the following expression for the tunneling conductance $G = I_T/V$:

$$G = \frac{e^2 \rho_{2D}}{2\tau} \frac{(2t_{lr})^2}{\Delta^2 + (\hbar/\tau)^2}, \qquad \frac{1}{\tau} = \frac{1}{2}\left(\frac{1}{\tau_l} + \frac{1}{\tau_r}\right). \tag{27}$$

The conductance is a Lorentz function of the energy shift Δ. In the linear transport regime, when ε_{Fl} and ε_{Fr} are close to each other, this shift is expressed through the difference in the 2D electron densities in the layers, $\Delta = (n_r - n_l)/\rho_{2D}$. In the resonance, when the electron densities are matched and $\Delta = 0$, the tunneling current is proportional to τ, while far from the resonance it is proportional to $1/\tau$, indicating that the tunneling in this regime is a scattering-assisted process. Introducing the scattering-dependent tunneling rate according to $\nu = \tau^{-1} 2t_{lr}^2/[\Delta^2 + (\hbar/\tau)^2]$, one may express the conductance (27) in the form (15).

In practice, the contacts to 2D layers are made laterally, which means that the current in the double-layer system has both the tunneling (perpendicular to the layers) component I_T and the in-plane components $\mathbf{I}_j = (I_j^x, I_j^y)$, $j = l, r$. As a result, the resistance of the system is not entirely determined by the tunneling conductance: it depends also on the 2D conductivities of the layers and on the size of the system. To investigate this effect, we write the continuity equations for the layers l and r as $\nabla \mathbf{I}_l(\mathbf{r}) + I_T(\mathbf{r}) = 0$ and $\nabla \mathbf{I}_r(\mathbf{r}) - I_T(\mathbf{r}) = 0$, where the local current densities are expressed through the 2D conductivities, σ_j, and local electrochemical potentials, $w_{j\mathbf{r}}$, according to $\mathbf{I}_j(\mathbf{r}) = -\sigma_j \nabla w_{j\mathbf{r}}$ (see Sec. 36). The ratio $I_T(\mathbf{r})/G = w_{l\mathbf{r}} - w_{r\mathbf{r}} \equiv V_\mathbf{r}$ represents the local interlayer voltage. These expressions are written in the linear approximation, when the coordinate dependence of the 2D conductivities σ_j and of the tunneling conductance G is neglected. Substituting $\mathbf{I}_j$ and I_T into the continuity equations, we find a closed system of equations for the electrochemical potentials $w_{j\mathbf{r}}$, which is conveniently written as

$$\nabla[\sigma_l \nabla w_{l\mathbf{r}} + \sigma_r \nabla w_{r\mathbf{r}}] = 0,$$

$$\nabla^2 V_\mathbf{r} - l_T^{-2} V_\mathbf{r} = 0, \qquad l_T^{-1} = \sqrt{G\left(\frac{1}{\sigma_l} + \frac{1}{\sigma_r}\right)}. \tag{28}$$

The equation for $V_\mathbf{r}$ introduces a characteristic tunneling length l_T. Since the tunneling conductance is exponentially small, this length is much greater than the mean free path lengths. This property justifies the introduction of the local electrochemical potentials in the layers. However, l_T can be comparable to or smaller than the size of the 2D tunneling resistor.

Consider, for example, a double-layer system of length L and width L_y, where the voltage $V/2$ is applied to the edge $x = -L/2$ of the left layer and the voltage $-V/2$ is applied to the edge $x = L/2$ of the right layer. The continuity of the potentials assumes $w_{l,x=-L/2} = V/2$ and $w_{r,x=L/2} = -V/2$, while the termination of the layers at $x = \pm L/2$ implies the boundary conditions for the currents, $I_l^x(x = L/2) = 0$ and $I_r^x(x = -L/2) = 0$. This is the case when the currents and the potentials depend only on x, the transverse currents I_j^y are equal to zero, and the general solution of the system (28) is written as

$$w_{lx} = w_0 - \frac{Ix - \sigma_r V_x}{\sigma_l + \sigma_r}, \quad w_{rx} = w_0 - \frac{Ix + \sigma_l V_x}{\sigma_l + \sigma_r}, \qquad (29)$$

$$V_x = c_1 e^{x/l_T} + c_2 e^{-x/l_T},$$

where w_0 is the reference point for the potentials (non-essential in the following) and $I = I_l^x(x) + I_r^x(x)$ is the total current per unit width. Let us introduce the electrical resistance per square LL_y according to $R_T = VL/I$. Applying the boundary conditions specified above to the solution (29), we determine the coefficients $c_{1,2}$ and obtain

$$R_T = \frac{L^2}{\sigma_l + \sigma_r} \left[1 + \frac{(\sigma_l + \sigma_r)^2}{2\sigma_l \sigma_r} \frac{l_T}{L} \coth \frac{L}{2l_T} \right.$$

$$\left. + \frac{(\sigma_l - \sigma_r)^2}{2\sigma_l \sigma_r} \frac{l_T}{L} \tanh \frac{L}{2l_T} \right]. \qquad (30)$$

If $L \ll l_T$, we have $R_T = G^{-1}$, while at $L \gg l_T$ we find $R_T = L^2/(\sigma_l + \sigma_r)$ corresponding to the ohmic resistance of two layers connected in parallel. In the intermediate region, $L \sim l_T$, the resistance depends on the size in a more complex way.

61. Coulomb Blockade

If the probability of tunneling through the quasibound states in a potential well is small enough, the electrons experience multiple inelastic scattering in these states and lose their coherence before getting out of the potential well. In this case, the intermediate (potential-well) region can be considered as a local reservoir where the electrons are characterized by a quasi-equilibrium distribution function. The tunneling is viewed as a two-step process. First, an electron tunnels through the left barrier into the intermediate region containing quasibound states and is thermalized there because of the scattering. Then, an electron from the intermediate region tunnels through the right barrier. An equivalent process is realized when an electron first tunnels from the intermediate region to the right, and then another electron tunnels to this region from the left. Such transport is called the sequential tunneling. Assuming that a voltage V is applied between the left and right leads, and V_i is the electrochemical potential of electrons in the intermediate region, counted from the right-lead potential, we write the ohmic current as $I = G_l(V - V_i) = G_r V_i$, where G_l (G_r) are the tunneling conductances characterizing the transitions between the continuum states in the left (right) lead and the quasibound states in the intermediate region. This current is rewritten as $I = (G_l^{-1} + G_r^{-1})V$, and the effective tunneling conductance is given by

$$G = \frac{G_l G_r}{G_l + G_r} = e^2 \rho_i \frac{\nu_l \nu_r}{\nu_l + \nu_r}. \tag{1}$$

Note that we have expressed the tunneling conductances through the tunneling rates, $G_j = e^2 \rho_i \nu_j$, $j = l, r$, where ρ_i is the density of electron states in the intermediate region; see also Eq. (60.15). Equation (1) describes the ohmic conductance for two resistances connected in series.

The sequential tunneling through finite-size regions is particularly interesting because of discrete nature of the electric charge. If the energy associated with addition of a single electron to this region is comparable to or higher than the temperature T, the conductance appears to be considerably affected by the charging effects. The conductance in this case is no longer described by Eq. (1). To give an elementary consideration of the charging effects, we consider a small conducting region (island) containing N electrons and separated from the conducting environment (current-carrying leads and other electrodes). Assuming that the Coulomb energy of such a system does not depend on the distribution of electrons in the island and is determined only by the number N, we can treat the system in terms of effective capacitors as shown in Fig. 12.7. Since the electric charge Q_k at the capacitor k is given by $Q_k = C_k(V_i - V_k)$, the total electron charge of the island, $-|e|N = \sum_k Q_k$, is related to the voltage V_i (electrochemical potential of the island) as $-|e|N = CV_i - Q$, where $Q = \sum_k C_k V_k$ and $C = \sum_k C_k$ is the total capacitance. Therefore, the Coulomb energy of the system containing the island with N electrons is written as

$$E_N = \frac{(-|e|N + Q)^2}{2C}, \qquad (2)$$

The variable Q can be viewed as an excess charge of the environment, which is equal to a sum of excess charges over all regions coupled to the island by electrostatic fields and having the voltages V_k. This variable can be continuously varied by the applied voltages. On the other hand, the charge $-|e|N$ is discrete. According to Eq. (2), one can introduce a single-electron charging energy

$$E_{N+1} - E_N = \left(N + \frac{1}{2} - \frac{Q}{|e|}\right)\frac{e^2}{C}. \qquad (3)$$

Since the electron charge is small, the charging effects can manifest themselves only if the size of the intermediate region is so small that the capacitance C characterizing its electrostatic coupling to the leads is comparable to or smaller than e^2/T.

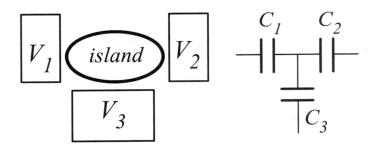

Figure 12.7. Isolated conductor surrounded by three electrodes and its schematic representation in terms of effective capacitors.

Let us assume that the island initially contains N electrons and the quasi-Fermi energy of the electrons in the left lead, ε_{Fl}, is higher than that in the right lead, ε_{Fr}. We also assume that the island is macroscopic and the electron distribution there is a Fermi distribution with a chemical potential μ determined by the number N (since $N \gg 1$, one can ignore the difference in μ for the cases of N and $N \pm 1$ electrons in the island). At low temperature, an electron can tunnel from the left lead to the island only if the difference $\varepsilon_{Fl} - \mu$ is greater than the charging energy (3). Next, the tunneling from the island (which now contains $N+1$ electrons) to the right lead is possible if $\mu - \varepsilon_{Fr}$ is greater than the charging energy (3) with the opposite sign. To have a current through the structure, both these conditions have to be satisfied simultaneously, $\varepsilon_{Fl} - \mu > E_{N+1} - E_N > \varepsilon_{Fr} - \mu$. A similar situation takes place when an electron first tunnels from the island to the right lead, and then another electron tunnels to the island from the left lead, $\varepsilon_{Fl} - \mu > E_N - E_{N-1} > \varepsilon_{Fr} - \mu$. Using these inequalities together with Eq. (3), we arrive at the conclusion that, to have a non-zero current, one should apply a finite bias $eV = \varepsilon_{Fl} - \varepsilon_{Fr}$ determined from the condition

$$eV > \frac{2e^2}{C} \left| N_{min} + \frac{1}{2} - \xi \right|, \tag{4}$$

where ξ is a continuous dimensionless variable defined as

$$\xi = Q/|e| + (\varepsilon_F - \mu)C/e^2, \tag{5}$$

$\varepsilon_F = (\varepsilon_{Fl} + \varepsilon_{Fr})/2$, and N_{min} is the integer number for which the expression $|N_{min} + 1/2 - \xi|$ is minimal. According to this definition, this expression can be varied from 0 (for half-integer ξ) to $1/2$ (for integer ξ). The threshold bias eV_{th}, which is defined by the right-hand side of Eq. (4), varies from 0 to e^2/C as ξ varies from half-integer to integer values. The existence of the threshold for the electron transport through the small islands is called the Coulomb blockade, and the sequential tunneling under the condition $eV_{th} + 2e^2/C > eV > eV_{th}$ is called the single-electron tunneling. Since ξ is a linear function of the voltages of all the regions electrostatically coupled to the island (we remind that $Q = \sum_k C_k V_k$), it can be linearly varied by each of these voltages. It is often convenient to have fixed voltages of the source and drain electrodes (so that eV is constant) and change the voltage of the gate electrode (V_3 in Fig. 12.7) to control the threshold bias through ξ. The above consideration suggests that the linear-response current ($eV \to 0$) at zero temperature exists only for certain discrete ξ equal to half-integers. The current, as a function of ξ, looks like a periodic sequence of sharp peaks, each peak corresponds to addition of an electron to the island. This phenomenon is called the Coulomb blockade oscillations.

Below we study the Coulomb blockade oscillations by using the methods of quantum kinetic theory. Let us write Eq. (1.20) in the form

$$i\hbar \frac{\partial \hat{\eta}_t}{\partial t} = [\widehat{H} + \widehat{\delta H}, \hat{\eta}_t], \quad \widehat{H} = \widehat{H}_0 + \widehat{H}_C, \tag{6}$$

where $\widehat{H}_0$ comprises the one-electron Hamiltonians describing l- and r-leads and the island,

$$\widehat{H}_0 = \sum_{jp} \varepsilon_{jp} \hat{a}_{jp}^+ \hat{a}_{jp} + \sum_q \varepsilon_q \hat{c}_q^+ \hat{c}_q \tag{7}$$

with $j = l, r$. We assume that the electrons in the leads are characterized by a continuous set of quantum numbers p, and those in the island by discrete quantum numbers q. Next, $\widehat{H}_C$ is the Hamiltonian of Coulomb interaction,

$$\widehat{H}_C = \frac{e^2}{2C}\left(\hat{n} - \frac{Q}{|e|}\right)^2, \quad \hat{n} = \sum_q \hat{c}_q^+ \hat{c}_q , \qquad (8)$$

written through the operator of the number of electrons in the island, $\hat{n}$. This Hamiltonian is formally obtained from the expression (2) for the Coulomb energy if the number of electrons, N, is replaced by its operator, $\hat{n}$. One can also derive the Hamiltonian (8) by using the general form of the interaction Hamiltonian, $(1/2)\sum_{qq'q_1q_1'} W_i(qq';q_1q_1')\hat{c}_q^+ \hat{c}_{q'}^+ \hat{c}_{q_1'}\hat{c}_{q_1} + \sum_{qq'} U_i(q,q')\hat{c}_q^+\hat{c}_{q'}$ (see Sec. 4 and Chapter 6), where the first and the second terms describe the interaction between the particles in the island and the interaction of these particles with external fields, respectively. Under the simplifying approximations $W_i(qq';q_1q_1') = W_i\delta_{qq_1}\delta_{q'q_1'}$ and $U_i(q,q') = U_i\delta_{qq'}$, which correspond to the mean-field interaction sensitive only to the number of the electrons, we obtain a model interaction Hamiltonian, $(W_i/2)\sum_{qq'}\hat{c}_q^+\hat{c}_{q'}^+\hat{c}_{q'}\hat{c}_q + U_i\sum_q \hat{c}_q^+\hat{c}_q$, which is easily reduced to the quadratic form with respect to $\hat{n}$, similar to $\widehat{H}_C$ of Eq. (8). Finally, $\widehat{\delta H}$ in Eq. (6) is the tunneling Hamiltonian

$$\widehat{\delta H} = \widehat{\delta H}_l + \widehat{\delta H}_r, \quad \widehat{\delta H}_j = \sum_{pq}\left[t_{jp,q}\hat{a}_{jp}^+\hat{c}_q + t_{q,jp}\hat{c}_q^+\hat{a}_{jp}\right]. \qquad (9)$$

It describes the electron transfer between the leads and the island.

The Hamiltonian $\widehat{H}_C$ is diagonal in the representation using a many-electron state $|N\rangle$ with N electrons in the island, $\langle N|\widehat{H}_C|N'\rangle = \delta_{NN'}E_N$. Therefore, it is convenient to employ this representation. We introduce the probability to have N electrons in the island as

$$\rho_{Nt} = \mathrm{Sp}_N\hat{\eta}_t = \mathrm{Sp}\langle N|\hat{\eta}_t|N\rangle = \sum_\delta \langle N\delta|\hat{\eta}_t|N\delta\rangle, \qquad (10)$$

where $\mathrm{Sp}_N\ldots$ means the trace over all states with N electrons, while $\mathrm{Sp}\ldots$ denotes the full trace. After solving Eq. (6) by perturbations up to the terms quadratic in the tunneling Hamiltonian, we calculate the trace Sp_N of the equation obtained and find

$$\frac{d\rho_{Nt}}{dt} = -\frac{1}{\hbar^2}\int_{-\infty}^t dt'e^{\lambda t'}\mathrm{Sp}_N[\widehat{\delta H}, [\widehat{\delta H}(t'-t), \hat{\eta}_{t'}]], \quad \lambda \to +0, \qquad (11)$$

where $\widehat{\delta H}(\tau) = e^{i\widehat{H}\tau/\hbar}\widehat{\delta H}e^{-i\widehat{H}\tau/\hbar}$ is the tunneling Hamiltonian $\widehat{\delta H}$ in the Heisenberg representation. The operator $\hat{\eta}_t$ in Eq. (11) satisfies the equation $i\hbar\partial\hat{\eta}_t/\partial t = [\widehat{H}, \hat{\eta}_t]$.

The tunneling Hamiltonian changes the number of electrons in the island by one, while the Hamiltonian $\widehat{H}$ does not change this number. Therefore, the trace of the commutator in Eq. (11) is calculated as

$$
\mathrm{Sp}_N[\widehat{\delta H}, [\widehat{\delta H}(t'-t), \hat{\eta}_{t'}]] = \mathrm{Sp} \sum_{N'=N\pm1} \left\{ \langle N|\hat{\eta}_{t'}|N\rangle \left[\langle N|\widehat{\delta H}|N'\rangle \right. \right.
$$

$$
\times \langle N'|\widehat{\delta H}(t'-t)|N\rangle + \langle N|\widehat{\delta H}(t'-t)|N'\rangle\langle N'|\widehat{\delta H}|N\rangle \Big] - \langle N'|\hat{\eta}_{t'}|N'\rangle
$$

$$
\times \left[\langle N'|\widehat{\delta H}|N\rangle\langle N|\widehat{\delta H}(t'-t)|N'\rangle + \langle N'|\widehat{\delta H}(t'-t)|N\rangle\langle N|\widehat{\delta H}|N'\rangle \right] \Big\} .
$$

(12)

Since the trace on the right-hand side is complete, one may carry out cyclic permutations of the operators $\langle N|\hat{\eta}_{t'}|N\rangle$ and $\langle N|\widehat{\delta H}|N'\rangle$ under the sign of Sp. Moreover, according to the properties of the Hamiltonian $\widehat{H}_C$, one has $\langle N|\widehat{\delta H}(\tau)|N'\rangle = e^{i(E_N-E_{N'})\tau/\hbar}\widehat{\delta H}'(\tau)$, where $\widehat{\delta H}'(\tau) = e^{i\widehat{H}_0\tau/\hbar}\widehat{\delta H}e^{-i\widehat{H}_0\tau/\hbar}$ and the energy E_N given by Eq. (2) is the eigenvalue of $\widehat{H}_C$. Then, according to Eq. (D.16),

$$
\widehat{\delta H}'(\tau) = \sum_{jpq} \left[t_{jp,q}\hat{a}_{jp}^+\hat{c}_q e^{i(\varepsilon_{jp}-\varepsilon_q)\tau/\hbar} + H.c. \right].
$$

(13)

With these substitutions, we rewrite the right-hand side of Eq. (12) as

$$
\sum_{jpq} |t_{jp,q}|^2 \mathrm{Sp} \left\{ \langle N|\hat{\eta}_{t'}|N\rangle \left[\langle N|\hat{a}_{jp}^+\hat{c}_q\hat{c}_q^+\hat{a}_{jp}|N\rangle \left(e^{\frac{i}{\hbar}(\varepsilon_q-\varepsilon_{jp}+E_{N+1}-E_N)(t'-t)} \right. \right. \right.
$$

$$
\left. +c.c. \right) + \langle N|\hat{c}_q^+\hat{a}_{jp}\hat{a}_{jp}^+\hat{c}_q|N\rangle \left(e^{\frac{i}{\hbar}(\varepsilon_q-\varepsilon_{jp}+E_N-E_{N-1})(t'-t)} + c.c. \right) \Big]
$$

$$
- \langle N-1|\hat{\eta}_{t'}|N-1\rangle\langle N-1|\hat{a}_{jp}^+\hat{c}_q\hat{c}_q^+\hat{a}_{jp}|N-1\rangle \left(e^{\frac{i}{\hbar}(\varepsilon_q-\varepsilon_{jp}+E_N-E_{N-1})(t'-t)} \right.
$$

$$
\left. +c.c. \right) - \langle N+1|\hat{\eta}_{t'}|N+1\rangle\langle N+1|\hat{c}_q^+\hat{a}_{jp}\hat{a}_{jp}^+\hat{c}_q|N+1\rangle
$$

(14)

$$
\times \left(e^{\frac{i}{\hbar}(\varepsilon_q-\varepsilon_{jp}+E_{N+1}-E_N)(t'-t)} + c.c. \right) \Big\} .
$$

The variables describing the leads and the island are separated under the trace, since $\hat{\eta}_t$ in Eqs. (11), (12), and (14) is the statistical operator for uncoupled sub-systems (we remind that the accuracy of our consideration is restricted by the second order in the coupling).

The trace over the variables of the leads is independent of the number of electrons in the island. Therefore, if we substitute the expression (14) into Eq. (11) and integrate over time in the Markovian approximation, $\hat{\eta}_{t'} \simeq \hat{\eta}_t$, we obtain the following kinetic equation:

$$\frac{d\rho_{Nt}}{dt} = -\frac{2\pi}{\hbar} \sum_{jpq} |t_{jp,q}|^2 \left[\delta(\varepsilon_q - \varepsilon_{jp} + E_N - E_{N-1})\rho_{Nt} f_{qt}^{(N)}(1 - f_{jp}) \right.$$

$$+ \delta(\varepsilon_q - \varepsilon_{jp} + E_{N+1} - E_N)\rho_{Nt}(1 - f_{qt}^{(N)})f_{jp}$$

$$- \delta(\varepsilon_q - \varepsilon_{jp} + E_N - E_{N-1})\rho_{N-1t}(1 - f_{qt}^{(N-1)})f_{jp} \qquad (15)$$

$$\left. - \delta(\varepsilon_q - \varepsilon_{jp} + E_{N+1} - E_N)\rho_{N+1t} f_{qt}^{(N+1)}(1 - f_{jp}) \right].$$

Here $f_{jp} = \mathrm{Sp}\{\hat{\eta}_t \hat{a}_{jp}^+ \hat{a}_{jp}\}$ is the distribution function of electrons in the lead j. This is a quasi-equilibrium Fermi distribution function: $f_{jp} = f_j(\varepsilon_{jp}) = \left[e^{(\varepsilon_{jp} - \varepsilon_{Fj})/T} + 1 \right]^{-1}$. Next,

$$f_{qt}^{(N)} = \mathrm{Sp}_N\{\hat{\eta}_t \hat{c}_q^+ \hat{c}_q\}/\rho_{Nt} \qquad (16)$$

is the average occupation number of the state q under the condition that the island contains N electrons. The first two terms of the collision integral (i.e., of the right-hand side of Eq. (15)) describe, respectively, the departure of an electron from the island with N electrons due to tunneling to the leads and the arrival of an electron to this island from the leads. The second two terms describe the opposite processes, the arrival of an electron to the island with $N-1$ electrons and the departure of an electron from the island with $N+1$ electrons.

The kinetic equation (15) in the stationary case, $d\rho_{Nt}/dt = 0$, has a solution corresponding to the detailed balance of electron occupation. To prove this statement, we rewrite the stationary kinetic equation (15) as $\sum_q [J_N(q) - J_{N+1}(q)] = 0$, where $J_N(q)$ is the part of the collision integral originating from the terms containing $\delta(\varepsilon_q - \varepsilon_{jp} + E_N - E_{N-1})$ in Eq. (15) (these are the first and the third terms). Since the relation $\sum_q [J_N(q) - J_{N+1}(q)] = 0$ is valid for arbitrary N, there should be a detailed balance, $J_N(q) = 0$. Let us introduce the densities of electron states in the leads, $\mathcal{N}_j(\varepsilon_{jp})$, and the tunneling rates near the quasi-Fermi levels, ν_{jq}, as

$$\sum_{jp} \ldots = \sum_j \int d\varepsilon_{jp} \mathcal{N}_j(\varepsilon_{jp}) \ldots ,$$

$$\nu_{jq} = \frac{2\pi}{\hbar} \left[\mathcal{N}_j(\varepsilon_{jp}) |t_{jp,q}|^2 \right]_{\varepsilon_{jp} = \varepsilon_{Fj}} . \qquad (17)$$

With the aid of these definitions, the detailed balance equation, $J_N(q) = 0$, is rewritten in the form

$$\rho_N f_q^{(N)} \sum_{j=l,r} \nu_{jq} [1 - f_j(\varepsilon_q + E_N - E_{N-1})]$$

$$= \rho_{N-1}(1 - f_q^{(N-1)}) \sum_{j=l,r} \nu_{jq} f_j(\varepsilon_q + E_N - E_{N-1}). \tag{18}$$

In equilibrium, when $f_l(\varepsilon) = f_r(\varepsilon) = f(\varepsilon) = [e^{(\varepsilon - \varepsilon_F)/T} + 1]^{-1}$, Eq. (18) is reduced to $\rho_N f_q^{(N)} = \rho_{N-1}(1 - f_q^{(N-1)})e^{-(\varepsilon_q + E_N - E_{N-1} - \varepsilon_F)/T}$. Its solution is (problem 12.16)

$$\rho_N = \sum_{\{n_\gamma\}} P_N(\{n_\gamma\}), \quad f_q^{(N)} = \sum_{\{n_\gamma\}} P_N(\{n_\gamma\})n_q, \tag{19}$$

where $n_\gamma = 0, 1$ is the occupation number of the state γ in the island and

$$P_N(\{n_\gamma\}) = Z^{-1} \exp\left[-\frac{1}{T}\left(\sum_\gamma \varepsilon_\gamma n_\gamma + E_N - \varepsilon_F N\right)\right]_{\sum_\gamma n_\gamma = N} \tag{20}$$

is the equilibrium probability of realization of the set of occupation numbers $\{n_\gamma\}$, see Sec. 4, under the condition that the island contains N electrons. Equations (19) and (20) are obtained when Eqs. (10) and (16) with $\hat{\eta}_t = \hat{\eta}_{eq}$ are rewritten in the occupation number representation. Since only the variables describing the island are relevant, one may use the equilibrium statistical operator of the island, $\hat{\eta}_{eq}^{(i)} = Z^{-1} \exp[-(\sum_\gamma \varepsilon_\gamma \hat{c}_\gamma^+ \hat{c}_\gamma + \widehat{H}_C - \varepsilon_F \hat{n})/T]$, instead of $\hat{\eta}_{eq}$. The partition function Z is defined in the usual way, $Z = \mathrm{Sp}\hat{\eta}_{eq}^{(i)}$, which means that $Z = \sum_N \sum_{\{n_\gamma\}} P_N(\{n_\gamma\})$ in the occupation number representation.

Considering Eq. (18) for the case of small deviation from equilibrium, when an infinitely small bias eV is applied between the leads, we put $\varepsilon_{Fl} = \varepsilon_F + eV/2$ and $\varepsilon_{Fr} = \varepsilon_F - eV/2$. Expanding the distribution functions f_j in Eq. (18) up to the terms linear in V, we obtain

$$\rho_N f_q^{(N)} - \rho_{N-1}(1 - f_q^{(N-1)})e^{-(\varepsilon_q + E_N - E_{N-1} - \varepsilon_F)/T}$$

$$= \frac{eV}{2T} \frac{\nu_{lq} - \nu_{rq}}{\nu_{lq} + \nu_{rq}} (\rho_N f_q^{(N)})_{eq}. \tag{21}$$

By convention, the functions inside $(\ldots)_{eq}$ are the equilibrium ones. In equilibrium, when $V = 0$, the expression on the left-hand side of Eq. (21) is equal to zero, as already mentioned. Equation (21) is used below for calculating the linear-response current through the island. This current is equal to $e \sum_N (d\rho_N/dt)_{i \to r}$, where $(d\rho_N/dt)_{i \to r}$ is the net number of electrons going from the island with N electrons to the right lead in unit time. Extracting the corresponding terms from the collision integral in Eq. (15), we find

$$I = \frac{2\pi e}{\hbar} \sum_{qNp} |t_{rp,q}|^2 \delta(\varepsilon_q + E_N - E_{N-1} - \varepsilon_{rp})$$

$$\times \left[\rho_N f_q^{(N)} (1 - f_{rp}) - \rho_{N-1}(1 - f_q^{(N-1)}) f_{rp} \right]. \tag{22}$$

To consider the linear regime, we expand $f_{rp} = [e^{(\varepsilon_{rp} - \varepsilon_F + eV/2)/T} + 1]^{-1}$ up to the terms linear in eV. Applying the definition (17), we obtain

$$I = e \sum_{qN} \nu_{rq} \left\{ \rho_N f_q^{(N)} [1 - f(\varepsilon_q + E_N - E_{N-1})] - \rho_{N-1}(1 - f_q^{(N-1)}) \right.$$

$$\times f(\varepsilon_q + E_N - E_{N-1}) + \frac{eV}{2T} [\rho_N f_q^{(N)} + \rho_{N-1}(1 - f_q^{(N-1)})]_{eq} \tag{23}$$

$$\left. \times f(\varepsilon_q + E_N - E_{N-1})[1 - f(\varepsilon_q + E_N - E_{N-1})] \right\}.$$

Using the relation (21) in this equation, we rewrite the latter as $I = GV$, where the linear conductance is

$$G = \frac{e^2}{T} \sum_{qN} \frac{\nu_{lq} \nu_{rq}}{\nu_{lq} + \nu_{rq}} (\rho_N f_q^{(N)})_{eq} [1 - f(\varepsilon_q + E_N - E_{N-1})]. \tag{24}$$

Let us consider some limiting cases described by the general expression (24). If $N \gg 1$ and the level separation in the island, $\Delta_q = \varepsilon_q - \varepsilon_{q-1}$, is much smaller than the temperature T, the function $f_q^{(N)}$, at arbitrary N, is represented in the form

$$f_q^{(N)} \simeq f_i(\varepsilon_q) = [e^{(\varepsilon_q - \mu)/T} + 1]^{-1}, \quad \Delta_q \ll T, \quad N \gg 1. \tag{25}$$

The equilibrium function ρ_N in these conditions is equal to $\exp\{-[E_N - (\varepsilon_F - \mu)N]/T\} / \sum_N \exp\{-[E_N - (\varepsilon_F - \mu)N]/T\}$. This relation can be obtained by substituting the distribution (25) into the left-hand side of Eq. (21) and taking into account that this side is equal to zero in equilibrium. Equivalently, one can find ρ_N directly from Eqs. (19) and (20). We can put $\mu = \varepsilon_F$ in the equilibrium expressions for ρ_N and $f_q^{(N)}$. The discrete sum over q in Eq. (24) is replaced by the integral over ε_q multiplied by the density of states in the island, $\rho_i(\varepsilon_q)$. The factor $f_i(\varepsilon_q)[1 - f(\varepsilon_q + E_N - E_{N-1})]$ is non-zero in a narrow interval near the Fermi level, where the q-dependence of $\rho_i(\varepsilon_q)$ and ν_{jq} can be neglected. Therefore, the integral over ε_q is calculated according to

$$\frac{1}{T} \int d\varepsilon f(\varepsilon)[1 - f(\varepsilon + x)] = \frac{x/T}{1 - e^{-x/T}}. \tag{26}$$

Finally, using Eq. (2) for E_N, we find

$$G = e^2 \rho_i \frac{\nu_l \nu_r}{\nu_l + \nu_r} \sum_N \frac{e^{-\kappa(N - \xi - 1/2)^2} \kappa(N - \xi - 1/2)}{\sinh[\kappa(N - \xi - 1/2)]}$$

$$\times \left[\sum_N e^{-\kappa(N-\xi)^2+\kappa/4} \right]^{-1}, \tag{27}$$

where ξ is defined by Eq. (5) with $\mu = \varepsilon_F$. The dimensionless variable $\kappa = e^2/2CT$ characterizes the charging effects. The regime described by Eqs. (25) and (27) is called the classical Coulomb blockade. At high temperatures, when $\kappa \ll 1$, one can replace the sum over N in Eq. (27) by the integral over $N - \xi - 1/2$, and the main contribution to this integral comes from the region $|N - \xi - 1/2| \ll \kappa^{-1}$, where $\kappa(N - \xi - 1/2)/\sinh[\kappa(N - \xi - 1/2)] \simeq 1$. Therefore, we obtain the conductance G in the form of Eq. (1). In the opposite case of strong Coulomb blockade, $\kappa \gg 1$, we retain only one or two terms in the sum over N, those for which $|N - \xi - 1/2|$ is minimal. The conductance, as a function of the variable ξ, shows a periodic sequence of sharp peaks at $\xi = \xi_N = N - 1/2$. The periodicity $\Delta\xi = 1$ corresponds to the addition energy e^2/C (we remind that each peak signifies the addition of one electron to the island). In the vicinity of the peaks,

$$\frac{G}{G_{\kappa=0}} = \frac{\kappa(\xi - \xi_N)}{\sinh[2\kappa(\xi - \xi_N)]}. \tag{28}$$

Therefore, the maximum of the ratio $G/G_{\kappa=0}$ at $\kappa \gg 1$ is fixed at $1/2$, and the peaks are narrowing with decreasing temperature. In the minima (at integer ξ), the conductance is proportional to $e^{-\kappa}$. Figure 12.8 shows three periods of the Coulomb blockade oscillations calculated according to Eq. (27) for different κ.

Consider now the case when one of the conditions $\Delta_q \ll T$ and $N \gg 1$ is violated. The function $f_q^{(N)}$ in this regime is not reduced to the form (25) (problem 12.17) and cannot be written in any simple way if N is large. The shape of the Coulomb blockade oscillations can be represented in terms of elementary functions in the limit $\Delta_q \gg T$, when, for each N, there is only one realization of the occupation numbers (for simplicity, we assume that the energy levels are not degenerate). This implies that there are N completely occupied levels, $f_q^{(N)} = 1$ at $q \leq N$ and $f_q^{(N)} = 0$ at $q > N$, and the conductance is proportional to $\sum_N \sum_{q \leq N} \rho_N [1 - f(\varepsilon_q + E_N - E_{N-1})]$. Next, $\rho_N = Z^{-1} \exp[-(\sum_{\gamma=1}^N \varepsilon_\gamma + E_N - \varepsilon_F N)/T]$; see Eqs. (19) and (20). The partition function is calculated as

$$Z = \sum_{N'=1}^{\infty} \exp \left[-\frac{1}{T} \left(\sum_{\gamma=1}^{N'} \varepsilon_\gamma + E_{N'} - \varepsilon_F N' \right) \right]$$

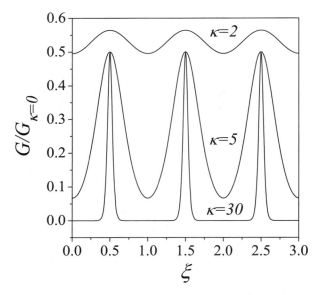

Figure 12.8. Classical Coulomb blockade oscillations at $e^2/2CT =$ 2, 5, and 30.

$$= \exp\left[-\frac{1}{T}\left(\sum_{\gamma=1}^{N}\varepsilon_{\gamma} + E_N - \varepsilon_F N\right)\right] \tag{29}$$

$$\times \sum_{N'=1}^{\infty}\exp\left[\frac{1}{T}\left(\sum_{\gamma=1}^{N}\varepsilon_{\gamma} - \sum_{\gamma=1}^{N'}\varepsilon_{\gamma} + E_N - E_{N'} - \varepsilon_F(N - N')\right)\right].$$

The exponential factor which enters ρ_N is taken out of the infinite sum over N'. Using Eq. (29) together with the expression for ρ_N given above, one can identify $(\rho_N)^{-1}$ with the sum from the right-hand side of Eq. (29). The main contribution to this sum comes from the terms with N' close to N. Indeed, if we rewrite this sum in more detail, we obtain

$$\rho_N = \Big[\ldots + e^{2\mathcal{E}_N/T}e^{-\Delta_N/T - e^2/CT}$$

$$+ e^{\mathcal{E}_N/T} + 1 + e^{-\mathcal{E}_N/T}e^{-\Delta_{N+1}/T - e^2/CT} + \ldots\Big]^{-1}, \tag{30}$$

where $\mathcal{E}_N = \varepsilon_N + E_N - E_{N-1} - \varepsilon_F$ and $\Delta_N = \varepsilon_N - \varepsilon_{N-1}$. The terms with $N' = N - 2, N - 1, N,$ and $N + 1$ under the sum in Eq. (29) are written in Eq. (30) explicitly, while the other terms (with $N' = 1, \ldots, N - 3$ and $N' = N + 2, \ldots$) are denoted by the dots. These terms contain products of the exponentially small expressions $e^{-\Delta_M/T - e^2/CT}$ multiplied by the powers of $e^{\pm \mathcal{E}_N/T}$. Since only the terms with $|\mathcal{E}_N| \sim T$ contribute to the conductance, all the terms containing $e^{-\Delta_M/T - e^2/CT}$ should be neglected in the sum in Eq. (30). Therefore, $\rho_N = [e^{\mathcal{E}_N/T} + 1]^{-1}$ and $\rho_N[1 - f(\varepsilon_N + E_N - E_{N-1})] = [e^{\mathcal{E}_N/2T} + e^{-\mathcal{E}_N/2T}]^{-2}$.

In summary, if the level separation Δ_q is much larger than the temperature T, only the terms with $q = N$ contribute into the sum in Eq. (24), and the product $(\rho_N f_q^{(N)})_{eq}[1 - f(\varepsilon_q + E_N - E_{N-1})]$ should be replaced by $[e^{\mathcal{E}_N/2T} + e^{-\mathcal{E}_N/2T}]^{-2}$, where $\mathcal{E}_N = \varepsilon_N + E_N - E_{N-1} - \varepsilon_F$. Expressing $E_N - E_{N-1}$ according to Eq. (3), we finally obtain

$$G = \frac{e^2}{4T} \sum_N \frac{\nu_{lN}\nu_{rN}}{\nu_{lN} + \nu_{rN}} \cosh^{-2}\left[\frac{\varepsilon_N - \varepsilon_F}{2T} + \frac{e^2}{2CT}\left(N - \frac{1}{2} - \frac{Q}{|e|}\right)\right]. \quad (31)$$

This equation describes the regime of quantum Coulomb blockade. The conductance (31), similar to that of Eq. (27), shows series of thermally broadened peaks as the variable Q is varied continuously. The separation of the peaks increases in comparison to the case of classical Coulomb blockade, since the energy level separation, Δ_N, is added to the charging energy e^2/C. The energy Δ_N depends on N, which means that the Coulomb blockade oscillations are not exactly periodic. Next, the N-dependence of ν_{lN} and ν_{rN} results in different heights of the peaks. The peak heights are proportional to $1/T$, in contrast to the classical Coulomb blockade, when they are temperature-independent. This property allows one to separate the two regimes experimentally. Usually, the classical Coulomb blockade takes place in metallic islands, while the quantum Coulomb blockade is realized in small semiconductor quantum dots, where the energy level separation is large enough. Equation (31) can be viewed as an expression for the resonant tunneling conductance, which takes into account the effects of thermal broadening and the influence of the charging energies on the resonance condition.

62. Polaronic Transport

Considering the electron-phonon interaction in previous chapters, we assumed that this interaction was sufficiently weak to be treated as a perturbation causing electron transitions with emission or absorption of phonons. The situation becomes more complicated in a number of ionic and molecular crystals, where the interaction is strong and must be treated in a non-perturbative way. Equivalently, one may say that an electron interacts simultaneously with many phonons. As a result of this interaction, the lattice atoms surrounding the electron are displaced to new positions thereby decreasing the energy of the electron-lattice system. The induced displacements provide a potential well for the electron, where the latter occupies a bound state. The electron is unable to move unless accompanied by the well, that is, by the induced lattice deformation. The quasiparticle consisting of the electron and its self-induced lattice deformation is called the polaron.

To introduce the polaron, we expand the wave function of the electron in the crystal by using the basis of single-site orbitals $\varphi(\mathbf{r} - \mathbf{R_n})$, where $\mathbf{R_n}$ is the radius-vector of the crystal cell numbered by the integer vector $\mathbf{n}$,

$$\Psi(\mathbf{r}) = \sum_{\mathbf{n}} a_{\mathbf{n}} \varphi(\mathbf{r} - \mathbf{R_n}), \tag{1}$$

and the function φ is normalized according to $\int d\mathbf{r} |\varphi(\mathbf{r})|^2 = 1$. Equation (1) defines the wave function in the site representation. Considering the electron-phonon interaction described by the Hamiltonian $\widehat{H}_{e,ph}$, we calculate the matrix elements of the Hamiltonian $\widehat{H} = \hat{h}_{cr} + \widehat{H}_{ph} + \widehat{H}_{e,ph}$ in this basis. Here $\hat{h}_{cr}$ is the Hamiltonian of the crystal, see Eq. (5.4), and $\widehat{H}_{ph}$ is the Hamiltonian of free phonons. In the second quantization representation, the Hamiltonian $\widehat{H}$ takes the following form:

$$\widehat{H} = \sum_{\mathbf{n}} \varepsilon_0 \hat{a}_{\mathbf{n}}^+ \hat{a}_{\mathbf{n}} + \sum_{\mathbf{nm} \, (\mathbf{m} \neq 0)} J_{\mathbf{m}} \hat{a}_{\mathbf{n+m}}^+ \hat{a}_{\mathbf{n}} + \sum_{\mathbf{q}} \hbar\omega_{\mathbf{q}} (\hat{b}_{\mathbf{q}}^+ \hat{b}_{\mathbf{q}} + 1/2)$$

$$- \sum_{\mathbf{nmq}} \hbar\omega_{\mathbf{q}} \left[u_{\mathbf{nmq}} \hat{a}_{\mathbf{n+m}}^+ \hat{a}_{\mathbf{n}} \hat{b}_{\mathbf{q}}^+ + H.c. \right]. \tag{2}$$

In this equation,

$$\varepsilon_0 = \int d\mathbf{r} \varphi^*(\mathbf{r}) \hat{h}_{cr}(\mathbf{r}) \varphi(\mathbf{r}), \quad J_{\mathbf{m}} = \int d\mathbf{r} \varphi^*(\mathbf{r} - \mathbf{R_m}) \hat{h}_{cr}(\mathbf{r}) \varphi(\mathbf{r}), \tag{3}$$

and

$$u_{\mathbf{nmq}} = -\frac{\gamma_{\mathbf{q}}^*}{\sqrt{2}} e^{-i\mathbf{q} \cdot \mathbf{R_n}} \int d\mathbf{r} e^{-i\mathbf{q} \cdot \mathbf{r}} \varphi^*(\mathbf{r} - \mathbf{R_m}) \varphi(\mathbf{r}). \tag{4}$$

For the sake of brevity, we do not write spin indices at the electron operators, though the spin sums are implicitly assumed, leading to a factor of 2 in the expressions for the current below. The phonon mode index is omitted since we consider the interaction with only one mode. The dimensionless factor $\gamma_{\mathbf{q}}$ describes the strength of such interaction. For the interaction of electrons with long-wavelength phonons in cubic crystals, this factor is isotropic, $\gamma_{\mathbf{q}} = \gamma_q = \sqrt{2} C_q / \hbar\omega_q$; see the expressions for C_q in Eq. (21.1). For example, $\gamma_q = q^{-1} \sqrt{4\pi e^2 / \epsilon^* \hbar\omega_{LO} V}$ in the case of Froelich interaction with long-wavelength LO phonons. In the transformations leading to Eqs. (2)–(4), we have used the periodicity of the crystal Hamiltonian. The index $\mathbf{n} + \mathbf{m}$ at the operator $\hat{a}^+$ corresponds to creation of an electron in the crystal cell with coordinate $\mathbf{R}_{\mathbf{n+m}} = \mathbf{R}_{\mathbf{n}} + \mathbf{R}_{\mathbf{m}}$. Usually, a crystal cell contains more than one atom. Nevertheless, in ionic crystals with high degree of ionicity the electron is localized near the ions for which the affinity to the electron charge is largest. Therefore, the vector $\mathbf{R}_{\mathbf{n}}$ characterizes uniquely the required ion in the $\mathbf{n}$-th cell. This means that the diagonal matrix element of coordinate in the basis under consideration is reduced to this vector: $\langle \mathbf{n}|\hat{\mathbf{x}}|\mathbf{n}\rangle = \int d\mathbf{r}|\varphi(\mathbf{r} - \mathbf{R}_{\mathbf{n}})|^2 \mathbf{r} \simeq \mathbf{R}_{\mathbf{n}}$.

The coefficients $J_{\mathbf{m}}$ describe the probabilities of electron jumps between the sites. These probabilities can be sufficiently high and extend over several neighboring sites. This case is described by the model of nearly free electrons and corresponds to wide bands and small band gaps. In the opposite case, the probabilities of jumps to the nearest-neighbor sites are small ($J_{\mathbf{m}} \ll \varepsilon_0$), and the jumps to the further-neighbor sites can be neglected at all. This is the tight-binding model, which results in narrow bands (problem 12.18). The basis functions $\varphi(\mathbf{r} - \mathbf{R}_{\mathbf{n}})$ in this case are strongly localized and the integral over $\mathbf{r}$ in Eq. (4) is close to unity for $\mathbf{m} = 0$, because q^{-1} is much larger than the size of the wave function. If $\mathbf{m} \neq 0$, this integral is reduced to a small overlap integral of the neighboring orbitals. Neglecting these small contributions appearing in the last term of the Hamiltonian (2) in comparison to the second term, we obtain a Hamiltonian which is very similar to the effective Hamiltonian of tunnel-coupled systems; see Eq. (H.14). Each two nearest-neighbor sites now play the role of two tunnel-coupled regions, and there is a local interaction with the scattering potential (created, in our case, by the lattice vibrations) on each site. A similar Hamiltonian describes the superlattice, i.e., a periodic system of quantum wells coupled by tunneling (problem 12.19).

Therefore, we write the last term in Eq. (2) in the local approximation, as $-\sum_{\mathbf{nq}} \hbar\omega_{\mathbf{q}} \hat{a}_{\mathbf{n}}^+ \hat{a}_{\mathbf{n}} [u_{\mathbf{nq}}^* \hat{b}_{\mathbf{q}} + u_{\mathbf{nq}} \hat{b}_{\mathbf{q}}^+]$ with $u_{\mathbf{nq}} = -\gamma_{\mathbf{q}}^* e^{-i\mathbf{q}\cdot\mathbf{R}_{\mathbf{n}}}/\sqrt{2}$. Let us carry out the canonical transformation

$$\widetilde{H} = e^{-\hat{S}}\widehat{H}e^{\hat{S}}, \quad \widehat{S} = \sum_{\mathbf{nq}} \hat{a}_{\mathbf{n}}^+ \hat{a}_{\mathbf{n}}[u_{\mathbf{nq}}\hat{b}_{\mathbf{q}}^+ - u_{\mathbf{nq}}^*\hat{b}_{\mathbf{q}}] \tag{5}$$

implying the following exact relations for the transformed operators:

$$\tilde{a}_{\mathbf{n}} = e^{-\hat{S}}\hat{a}_{\mathbf{n}}e^{\hat{S}} = \hat{a}_{\mathbf{n}} \exp\left[\sum_{\mathbf{q}} \left(u_{\mathbf{nq}}\hat{b}_{\mathbf{q}}^+ - u_{\mathbf{nq}}^*\hat{b}_{\mathbf{q}}\right)\right] \tag{6}$$

and

$$\tilde{b}_{\mathbf{q}} = e^{-\hat{S}}\hat{b}_{\mathbf{q}}e^{\hat{S}} = \hat{b}_{\mathbf{q}} + \sum_{\mathbf{n}} u_{\mathbf{nq}}\hat{a}_{\mathbf{n}}^+ \hat{a}_{\mathbf{n}}, \tag{7}$$

while the operators $\tilde{a}_\mathbf{n}^+$ and $\tilde{b}_\mathbf{q}^+$ are described by Hermitian conjugate expressions. Note that the transformed operators $\tilde{a}_\mathbf{n}$ and $\tilde{b}_\mathbf{q}$ obey the usual fermionic and bosonic commutation relations, respectively. The product $\hat{a}_\mathbf{n}^+\hat{a}_\mathbf{n}$ is invariant with respect to this transformation so that all single-site operators related to the electron system (such as the coordinate operator $\hat{\mathbf{x}} = \sum_\mathbf{n} \mathbf{R}_\mathbf{n}\hat{a}_\mathbf{n}^+\hat{a}_\mathbf{n}$) remain unchanged. The Hamiltonian is transformed to

$$\widetilde{H} = \sum_\mathbf{n}(\varepsilon_0 - \Delta\varepsilon)\hat{a}_\mathbf{n}^+\hat{a}_\mathbf{n} + \sum_{\mathbf{n}'\mathbf{n}} J_{\mathbf{n}'-\mathbf{n}}\widetilde{\Phi}_{\mathbf{n}'\mathbf{n}}\hat{a}_{\mathbf{n}'}^+\hat{a}_\mathbf{n} + \sum_\mathbf{q} \hbar\omega_\mathbf{q}(\hat{b}_\mathbf{q}^+\hat{b}_\mathbf{q} + 1/2)$$

$$- {\sum_{\mathbf{n}'\mathbf{n}}}' \hat{a}_\mathbf{n}^+\hat{a}_\mathbf{n}\hat{a}_{\mathbf{n}'}^+\hat{a}_{\mathbf{n}'} \sum_\mathbf{q} \hbar\omega_\mathbf{q}\mathrm{Re}[u_\mathbf{nq}^*u_{\mathbf{n}'\mathbf{q}}], \qquad (8)$$

where $\Delta\varepsilon = \sum_\mathbf{q} \hbar\omega_\mathbf{q}|\gamma_\mathbf{q}|^2/2$ is the polaronic shift (renormalization) of the electron energy (problem 12.20). The prime sign at the sum in the last term of the Hamiltonian means that $\mathbf{n}' \neq \mathbf{n}$. Next,

$$\widetilde{\Phi}_{\mathbf{n}'\mathbf{n}} = \exp\left[\sum_\mathbf{q}\left(v_{\mathbf{n}'\mathbf{nq}}^*\hat{b}_\mathbf{q}^+ - v_{\mathbf{n}'\mathbf{nq}}\hat{b}_\mathbf{q}\right)\right], \qquad (9)$$

$$v_{\mathbf{n}'\mathbf{nq}} = \frac{1}{\sqrt{2}}(e^{i\mathbf{q}\cdot\mathbf{R}_{\mathbf{n}'}} - e^{i\mathbf{q}\cdot\mathbf{R}_\mathbf{n}})\gamma_\mathbf{q}$$

is the multi-phonon operator. The last term in Eq. (8) describes the interaction of the electrons occupying different sites by exchange of virtual phonons. Since this intersite term is quadratic in the interaction, we do not keep it in the following. We have already neglected a more significant (linear) term, the intersite contribution to the last term in the Hamiltonian (2).

The Hamiltonian (8) can be generalized in the presence of electric and magnetic fields of arbitrary strengths (problem 12.21) and takes the final form

$$\widetilde{H} = \widetilde{H}_0 + \widetilde{\delta H}, \quad \widetilde{H}_0 = \sum_\mathbf{n} \varepsilon_\mathbf{n}\hat{a}_\mathbf{n}^+\hat{a}_\mathbf{n} + \sum_\mathbf{q} \hbar\omega_\mathbf{q}(\hat{b}_\mathbf{q}^+\hat{b}_\mathbf{q} + 1/2),$$

$$\widetilde{\delta H} = \sum_{\mathbf{n}'\mathbf{n}} J_{\mathbf{n}'\mathbf{n}}\widetilde{\Phi}_{\mathbf{n}'\mathbf{n}}\hat{a}_{\mathbf{n}'}^+\hat{a}_\mathbf{n}, \qquad (10)$$

where $\varepsilon_\mathbf{n} = \varepsilon_0 - \Delta\varepsilon - e\mathbf{E}\cdot\mathbf{R}_\mathbf{n}$ and $J_{\mathbf{n}'\mathbf{n}} = J_{\mathbf{n}'-\mathbf{n}} \exp\{ie\mathbf{H}\cdot[\mathbf{R}_\mathbf{n}\times\mathbf{R}_{\mathbf{n}'}]/2\hbar c\}$. The Hamiltonian $\widetilde{H}_0$ describes two independent sub-systems, the electrons localized at the sites and the phonons. The Hamiltonian $\widetilde{\delta H}$ describes the jumps of the electron surrounded by the phonon "cloud" between the sites. This Hamiltonian can be represented as a sum of two parts:

$$\widetilde{\delta H} = \widetilde{\delta H}_0 + \widetilde{\delta H}_1, \qquad \widetilde{\delta H}_0 = \sum_{\mathbf{n'n}} J_{\mathbf{n'n}} \overline{\Phi}_{\mathbf{n'n}} \hat{a}_{\mathbf{n'}}^+ \hat{a}_{\mathbf{n}},$$

$$\widetilde{\delta H}_1 = \sum_{\mathbf{n'n}} J_{\mathbf{n'n}} \widetilde{\delta\Phi}_{\mathbf{n'n}} \hat{a}_{\mathbf{n'}}^+ \hat{a}_{\mathbf{n}}, \qquad (11)$$

where $\overline{\Phi}_{\mathbf{n'n}} = \langle\langle \widetilde{\Phi}_{\mathbf{n'n}} \rangle\rangle_{ph}$ is the operator (9) averaged over the phonon variables and $\widetilde{\delta\Phi}_{\mathbf{n'n}} = \widetilde{\Phi}_{\mathbf{n'n}} - \overline{\Phi}_{\mathbf{n'n}}$. The Hamiltonian $\widetilde{\delta H}_0$ describes the jumps without emission or absorption of real phonons. However, the probabilities of such jumps are renormalized by the factor $\overline{\Phi}_{\mathbf{n'n}}$ because of emission and absorption of virtual phonons. In other words, when hopping between the sites, the electron carries with it the entire load of atomic displacements. Below we will see that $\overline{\Phi}_{\mathbf{n'n}}$ is exponentially small if the electron-phonon interaction is strong. As a result, the hopping caused by the contribution $\widetilde{\delta H}_0$ is considerably suppressed, and the effective band width, estimated as $2\overline{\Phi}J$ (see problem 12.18), becomes narrow. This is another manifestation of the polaronic effect, apart from the renormalization $-\Delta\varepsilon$ of the electron energy. One can say that $\widetilde{\delta H}_0$ describes the tunneling of the polaron.

The second term in Eq. (11), $\widetilde{\delta H}_1$, describes the phonon-assisted hopping, when electrons emit or absorb phonons to move between the sites. With the increase of temperature, this mechanism becomes considerably more important than the jumps described by the contribution $\widetilde{\delta H}_0$. The physical reason for this is the following. The electron localized at a site is separated from the other sites by potential barriers. By absorbing the phonons, the electron increases its energy and overcomes the potential barriers. At high temperatures, this activation process appears to be more efficient than the tunneling process.

The localization of electrons at the sites is characterized by the parameter $J/\Delta\varepsilon$, which is roughly the ratio of the kinetic energy of free electron to the depth of the potential well created by the polaronic effect. When this parameter is much less than unity, the polaron is called small, since its radius does not extend the lattice period. In the theory of small polaron the term $\widetilde{\delta H}_1$ is treated as a small perturbation with respect to other terms. Depending on the value of the term $\widetilde{\delta H}_0$, there are two important limiting cases characterizing the kinetics of small polaron: the hopping transport and the band transport. The hopping transport takes place when the localization is so strong that $\widetilde{\delta H}_0$ can be neglected. The case of band transport assumes weak enough electron-phonon interaction and non-perturbative treatment of $\widetilde{\delta H}_0$.

Below we consider the hopping transport regime. Let us obtain a kinetic equation for this case. We introduce the distribution function $\rho_{\mathbf{n}t} = \mathrm{Sp}\{\widetilde{\eta}_t \widetilde{a}_{\mathbf{n}}^+ \widetilde{a}_{\mathbf{n}}\} = \mathrm{Sp}\{\widetilde{\eta}_t \hat{a}_{\mathbf{n}}^+ \hat{a}_{\mathbf{n}}\}$, which describes the probability of occupation of the site $\mathbf{n}$ (note that $\widetilde{\eta}_t = e^{-\hat{S}} \hat{\eta}_t e^{\hat{S}}$). The evolution of the distribution function is governed by δH_1, while δH_0 is neglected. Within the accuracy up to the terms quadratic in $\widetilde{\delta H}_1$, we obtain ($\lambda \to +0$)

$$\frac{d\rho_{\mathbf{n}t}}{dt} = -\frac{1}{\hbar^2} \int_{-\infty}^{t} dt'\, e^{\lambda t'} \mathrm{Sp}\left\{ \widetilde{\eta}_{t'}^{(0)} [\widetilde{\delta H}_1(t'-t), [\widetilde{\delta H}_1, \hat{a}_{\mathbf{n}}^+ \hat{a}_{\mathbf{n}}]] \right\}, \qquad (12)$$

where $\widetilde{\delta H}_1(\tau) = e^{i\widetilde{H}_0\tau/\hbar} \widetilde{\delta H}_1 e^{-i\widetilde{H}_0\tau/\hbar}$ denotes the Heisenberg representation. The statistical operator $\widetilde{\eta}_{t'}^{(0)}$ in Eq. (12) satisfies Eq. (1.20) with the Hamiltonian $\widetilde{H}_0$, i.e., describes independent systems of electrons and phonons. Therefore, the averages over electron and phonon variables are separated from each other. Calculating the trace over the electron variables by using Eq. (D.16), we have

$$\mathrm{Sp}\left\{ \widetilde{\eta}_{t'}^{(0)} [\widetilde{\delta H}_1(t'-t), [\widetilde{\delta H}_1, \hat{a}_{\mathbf{n}}^+ \hat{a}_{\mathbf{n}}]] \right\} = \mathrm{Sp}_{ph}\, \widetilde{\eta}_{ph} \sum_{\mathbf{m}} J_{\mathbf{nm}} J_{\mathbf{mn}}$$

$$\times \left\{ \rho_{\mathbf{n}t'}(1 - \rho_{\mathbf{m}t'})[\widetilde{\delta\Phi}_{\mathbf{nm}} \widetilde{\delta\Phi}_{\mathbf{mn}}(t'-t) e^{i(\varepsilon_{\mathbf{m}} - \varepsilon_{\mathbf{n}})(t'-t)/\hbar} + H.c.] \qquad (13) \right.$$

$$\left. - \rho_{\mathbf{m}t'}(1 - \rho_{\mathbf{n}t'})[\widetilde{\delta\Phi}_{\mathbf{mn}} \widetilde{\delta\Phi}_{\mathbf{nm}}(t'-t) e^{i(\varepsilon_{\mathbf{n}} - \varepsilon_{\mathbf{m}})(t'-t)/\hbar} + H.c.] \right\},$$

where $\widetilde{\eta}_{ph}$ is the phonon part of the statistical operator.

The remaining trace over the phonon variables in Eq. (13) is more complicated because $\widetilde{\delta\Phi}_{\mathbf{nm}}$ contains the phonon operators in the exponent. To proceed with this calculation, we first note that the Heisenberg representation of $\widetilde{\Phi}$ is reduced to a substitution of the Heisenberg operators $\hat{b}_{\mathbf{q}}(t)$ and $\hat{b}_{\mathbf{q}}^+(t)$ in place of the time-independent ones in Eq. (9). This property is checked directly by expanding the operator exponent in Eq. (9) in series. Therefore, $\widetilde{\Phi}_{\mathbf{mn}}(t)$ can be represented as

$$\widetilde{\Phi}_{\mathbf{mn}}(t) = \exp\left\{ \sum_{\mathbf{q}} \left[v_{\mathbf{mnq}}^*(t) \hat{b}_{\mathbf{q}}^+ - v_{\mathbf{mnq}}(t) \hat{b}_{\mathbf{q}} \right] \right\}, \qquad (14)$$

where we have transferred the time dependence from the operators to the coefficients by using $\hat{b}_{\mathbf{q}}(t) = \hat{b}_{\mathbf{q}} e^{-i\omega_{\mathbf{q}}t}$ and introducing a new variable according to $v_{\mathbf{mnq}}(t) = v_{\mathbf{mnq}} e^{-i\omega_{\mathbf{q}}t}$. Next, the commutators of the two operators entering the exponents in Eqs. (9) and (14) are reduced to scalars. We may, therefore, use the operator identity (A.22) to simplify the calculation of phonon averages. According to Eqs. (9), (14), and (A.22),

$$\widetilde{\Phi}_{\mathbf{nm}}\widetilde{\Phi}_{\mathbf{mn}}(\tau) = \exp\left\{-\sum_{\mathbf{q}}\left[\frac{|v_{\mathbf{nmq}}|^2}{2} + \frac{|v_{\mathbf{mnq}}|^2}{2} + v_{\mathbf{nmq}}v^*_{\mathbf{mnq}}(\tau)\right]\right\}$$

$$\times \exp\left\{\sum_{\mathbf{q}}[v^*_{\mathbf{nmq}} + v^*_{\mathbf{mnq}}(\tau)]\hat{b}^+_{\mathbf{q}}\right\}\exp\left\{-\sum_{\mathbf{q}}[v_{\mathbf{nmq}} + v_{\mathbf{mnq}}(\tau)]\hat{b}_{\mathbf{q}}\right\},$$

(15)

and the other operator products standing in Eq. (13) are transformed in a similar way.

Let us calculate the average of the product of two operator exponents of the form (15), assuming the equilibrium phonon distribution:

$$\mathrm{Sp}_{\mathrm{ph}}\left[\widetilde{\eta}_{ph}\exp\left(\sum_{\mathbf{q}}\mu^*_{\mathbf{q}}\hat{b}^+_{\mathbf{q}}\right)\exp\left(-\sum_{\mathbf{q}}\mu_{\mathbf{q}}\hat{b}_{\mathbf{q}}\right)\right]$$

$$= \overline{Z}^{-1}\prod_{\mathbf{q}}\sum_{n_{\mathbf{q}}=0}^{\infty}e^{-\hbar\omega_{\mathbf{q}}n_{\mathbf{q}}/T}\langle n_{\mathbf{q}}|\exp(\mu^*_{\mathbf{q}}\hat{b}^+_{\mathbf{q}})\exp(-\mu_{\mathbf{q}}\hat{b}_{\mathbf{q}})|n_{\mathbf{q}}\rangle,$$

(16)

where $n_{\mathbf{q}}$ is the occupation number of the phonon state $\mathbf{q}$, $\mu_{\mathbf{q}} \equiv v_{\mathbf{nmq}} + v_{\mathbf{mnq}}(\tau)$, and $\overline{Z} = \prod_{\mathbf{q}}(1 - e^{-\hbar\omega_{\mathbf{q}}/T})^{-1}$ is the bosonic partition function (3.26). To calculate the matrix elements in Eq. (16), we expand the operator exponents in series and use Eq. (3.13) for the matrix elements of the bosonic operators. The result is a sum over integer number n:

$$\langle n_{\mathbf{q}}|\exp(\mu^*_{\mathbf{q}}\hat{b}^+_{\mathbf{q}})\exp(-\mu_{\mathbf{q}}\hat{b}_{\mathbf{q}})|n_{\mathbf{q}}\rangle = \sum_{n=0}^{n_{\mathbf{q}}}(-1)^n\frac{|\mu_{\mathbf{q}}|^{2n}n_{\mathbf{q}}!}{(n!)^2(n_{\mathbf{q}} - n)!}.$$

(17)

Calculating the sums over n and $n_{\mathbf{q}}$ (problem 12.22), we finally obtain

$$\mathrm{Sp}_{\mathrm{ph}}\left[\widetilde{\eta}_{ph}\exp\left(\sum_{\mathbf{q}}\mu^*_{\mathbf{q}}\hat{b}^+_{\mathbf{q}}\right)\exp\left(-\sum_{\mathbf{q}}\mu_{\mathbf{q}}\hat{b}_{\mathbf{q}}\right)\right]$$

$$= \exp\left(-\sum_{\mathbf{q}}|\mu_{\mathbf{q}}|^2 N_{\mathbf{q}}\right),$$

(18)

where $N_{\mathbf{q}}$ is the equilibrium (Planck) distribution function of phonons.

With the aid of Eqs. (15) and (18), one can calculate the bosonic averages $\langle\langle\dots\rangle\rangle_{ph} \equiv \mathrm{Sp}_{\mathrm{ph}}\{\widetilde{\eta}_{ph}\dots\}$ of the operator products standing in Eq. (13) and determine the function $\overline{\Phi}_{\mathbf{nm}}$ standing in $\widetilde{\delta H}_0$. Using the expression of $v_{\mathbf{nmq}}$ from Eq. (9), we find

$$\overline{\Phi}_{\mathbf{nm}} \equiv \langle\langle\widetilde{\Phi}_{\mathbf{nm}}\rangle\rangle_{ph} = \langle\langle\widetilde{\Phi}_{\mathbf{mn}}\rangle\rangle_{ph}$$

$$= \exp\left[-\sum_{\mathbf{q}}|v_{\mathbf{nmq}}|^2(N_{\mathbf{q}}+1/2)\right] = e^{-\overline{S}_{\mathbf{m-n}}/2},$$

$$\langle\langle\widetilde{\delta\Phi}_{\mathbf{nm}}\widetilde{\delta\Phi}_{\mathbf{mn}}(\tau)\rangle\rangle_{ph} = \langle\langle\widetilde{\delta\Phi}_{\mathbf{mn}}\widetilde{\delta\Phi}_{\mathbf{nm}}(\tau)\rangle\rangle_{ph} = e^{-\overline{S}_{\mathbf{m-n}}}$$

$$\times\left[\exp\left\{\sum_{\mathbf{q}}S_{\mathbf{m-n}}(\mathbf{q})[N_{\mathbf{q}}e^{-i\omega_{\mathbf{q}}\tau}+(N_{\mathbf{q}}+1)e^{i\omega_{\mathbf{q}}\tau}]\right\}-1\right], \qquad (19)$$

$$\langle\langle\widetilde{\delta\Phi}_{\mathbf{nm}}(\tau)\widetilde{\delta\Phi}_{\mathbf{mn}}\rangle\rangle_{ph} = \langle\langle\widetilde{\Phi}_{\mathbf{mn}}(\tau)\widetilde{\Phi}_{\mathbf{nm}}\rangle\rangle_{ph} = e^{-\overline{S}_{\mathbf{m-n}}}$$

$$\times\left[\exp\left\{\sum_{\mathbf{q}}S_{\mathbf{m-n}}(\mathbf{q})[N_{\mathbf{q}}e^{i\omega_{\mathbf{q}}\tau}+(N_{\mathbf{q}}+1)e^{-i\omega_{\mathbf{q}}\tau}]\right\}-1\right].$$

The factors appearing in these expressions are

$$S_{\mathbf{m-n}}(\mathbf{q}) = |v_{\mathbf{nmq}}|^2 = |\gamma_{\mathbf{q}}|^2\{1-\cos[\mathbf{q}\cdot(\mathbf{R}_{\mathbf{m}}-\mathbf{R}_{\mathbf{n}})]\},$$

$$\overline{S}_{\mathbf{m-n}} = \sum_{\mathbf{q}}S_{\mathbf{m-n}}(\mathbf{q})\coth(\hbar\omega_{\mathbf{q}}/2T). \qquad (20)$$

We point out that the term $\delta\widetilde{H}_0$ in the Hamiltonian (11) is exponentially suppressed at $\overline{S}_{\mathbf{m-n}} \gg 1$, as already mentioned. The small factor $e^{-\overline{S}_{\mathbf{m-n}}}$ is also present in the averages $\langle\langle\widetilde{\delta\Phi}_{\mathbf{nm}}\widetilde{\delta\Phi}_{\mathbf{mn}}(\tau)\rangle\rangle_{ph}$, see Eq. (19), but it is partly compensated by the large exponential factor standing in the square brackets.

Collecting the results (13), (19), and (20), we finally write the kinetic equation (12) in the Markovian approximation as

$$\frac{d\rho_{\mathbf{n}t}}{dt} = \sum_{\mathbf{m}}\left[W_{\mathbf{mn}}\rho_{\mathbf{m}t}(1-\rho_{\mathbf{n}t}) - W_{\mathbf{nm}}\rho_{\mathbf{n}t}(1-\rho_{\mathbf{m}t})\right], \qquad (21)$$

where the probability of transition (hopping) in unit time is given by the integral

$$W_{\mathbf{mn}} = e^{(\varepsilon_{\mathbf{m}}-\varepsilon_{\mathbf{n}})/T}W_{\mathbf{nm}} = \frac{1}{\hbar^2}J_{\mathbf{nm}}J_{\mathbf{mn}}e^{-\overline{S}_{\mathbf{m-n}}}$$

$$\times \int_{-\infty}^{\infty}d\tau e^{i(\varepsilon_{\mathbf{m}}-\varepsilon_{\mathbf{n}})\tau/\hbar}P_{\mathbf{m-n}}(\tau+i\hbar/2T), \qquad (22)$$

$$P_{\mathbf{n-m}}(t) = \exp\left[\sum_{\mathbf{q}}S_{\mathbf{m-n}}(\mathbf{q})\frac{\cos(\omega_{\mathbf{q}}t)}{\sinh(\hbar\omega_{\mathbf{q}}/2T)}\right] - 1.$$

The variable of integration is $\tau = t' - t$, and we have noticed that the expression under the integral $\int_{-\infty}^{0} d\tau \ldots$ is symmetric in τ so that this integral is transformed to $\int_{-\infty}^{\infty} d\tau \ldots$. The quantity $P_{\mathbf{n-m}}(t)$ depends only on $|\mathbf{R_m} - \mathbf{R_n}|$. The product $J_{\mathbf{nm}}J_{\mathbf{mn}}$ also depends only on $|\mathbf{R_m} - \mathbf{R_n}|$, even in the presence of a magnetic field. Therefore, the hopping probability is isotropic unless an electric field is present.

If an electric field is applied, a current flows in the system because the hopping probability becomes anisotropic. To calculate this current, one may use the general expression for the current density, $\mathbf{I}_t = (2e/V)\mathrm{Sp}(\hat{\eta}_t \hat{\mathbf{v}})$. It is convenient to employ the identity $\hat{\mathbf{v}} = (i/\hbar)[\widehat{H}, \hat{\mathbf{x}}]$ expressing the velocity operator through the coordinate operator. Using the transformed Hamiltonian (10), we obtain $\mathbf{I}_t = (2ei/\hbar V)\mathrm{Sp}\{\widetilde{\eta}_t [\widetilde{H}, \hat{\mathbf{x}}]\}$ (we note that $\hat{\mathbf{x}}$ is invariant with respect to the transformation (5)). Since $\widetilde{H}_0$ commutes with $\hat{\mathbf{x}}$, and the contribution of $\delta \widetilde{H}_0$ can be neglected in comparison to the contribution of $\delta \widetilde{H}_1$, we replace $\widetilde{H}$ in the commutator by $\delta \widetilde{H}_1$. To find $\widetilde{\eta}_t$, we solve the equation $i\hbar\partial\widetilde{\eta}_t/\partial t = [\widetilde{H}_0 + \delta\widetilde{H}_1, \widetilde{\eta}_t]$ with the initial condition $\widetilde{\eta}_{t\to-\infty} = 0$ by perturbations up to the terms linear in $\delta\widetilde{H}_1$. Substituting the result into the expression for $\mathbf{I}_t$, we obtain

$$\mathbf{I}_t = -\frac{2e}{\hbar^2 V} \int_{-\infty}^{0} dt' e^{\lambda t'} \mathrm{Sp}\left\{\widetilde{\eta}_{t'}^{(0)}[\delta\widetilde{H}_1(t' - t), [\delta\widetilde{H}_1, \hat{\mathbf{x}}]]\right\}. \qquad (23)$$

Since $\hat{\mathbf{x}} = \sum_{\mathbf{n}} \mathbf{R_n}\hat{a}_{\mathbf{n}}^{+}\hat{a}_{\mathbf{n}}$, the trace in this expression is the same as in Eq. (12), and $\mathbf{I}_t = (2e/V)\sum_{\mathbf{n}} \mathbf{R_n}(d\rho_{\mathbf{n}t}/dt)$. With the aid of Eq. (21), we obtain the stationary current density in the form

$$\mathbf{I} = \frac{2e}{V} \sum_{\mathbf{nm}} W_{\mathbf{nm}}\rho_{\mathbf{n}}(1 - \rho_{\mathbf{m}})(\mathbf{R_m} - \mathbf{R_n})$$

$$= \frac{e}{V} \sum_{\mathbf{nm}} [W_{\mathbf{nm}}\rho_{\mathbf{n}}(1 - \rho_{\mathbf{m}}) - W_{\mathbf{mn}}\rho_{\mathbf{m}}(1 - \rho_{\mathbf{n}})] (\mathbf{R_m} - \mathbf{R_n}). \qquad (24)$$

This equation has clear physical meaning: the current is proportional to the number of electrons hopping between the sites in unit time multiplied by the radius-vector of the hop (see also Sec. 48).

In a perfectly ordered crystal placed in a homogeneous electric field, $\rho_{\mathbf{n}}$ is independent of $\mathbf{n}$ and equal to its equilibrium value ρ_{eq}. It describes the average occupation of the site (per spin) and is related to the total electron density n as $n/2 = \rho_{eq}N/V$, where N is the number of elementary cells (or, equivalently, the number of sites). The hopping probability $W_{\mathbf{nm}}$, however, becomes anisotropic and field-dependent. Taking into account that $\varepsilon_{\mathbf{n}} - \varepsilon_{\mathbf{m}} = e\mathbf{E} \cdot (\mathbf{R_m} - \mathbf{R_n})$, we find

$$\mathbf{I} = \frac{en(1 - \rho_{eq})}{N\hbar^2} \sum_{\mathbf{nm}} (\mathbf{R_n} - \mathbf{R_m}) J_{\mathbf{mn}} J_{\mathbf{nm}} e^{-\overline{S}_{\mathbf{m-n}}} \qquad (25)$$

$$\times \sinh[e\mathbf{E} \cdot (\mathbf{R_n} - \mathbf{R_m})/2T] \int_{-\infty}^{\infty} dt \, e^{ie\mathbf{E} \cdot (\mathbf{R_n} - \mathbf{R_m})t/\hbar} P_{\mathbf{n-m}}(t).$$

To obtain this expression, we have put $t = \tau + i\hbar/2T$ in Eq. (22) and shifted the contour of integration accordingly, taking into account that $P_{\mathbf{n-m}}(t) = P_{\mathbf{n-m}}(-t)$. The linear current is obtained by expanding the hyperbolic sine in Eq. (25) in series of $\mathbf{E}$ up to the first (linear) term. The hopping conductivity, introduced as $I_\alpha = \sum_\beta \sigma_{\alpha\beta} E_\beta$, is given by (compare to Eq. (48.11))

$$\sigma_{\alpha\beta} = \frac{e^2 n(1 - \rho_{eq})}{2TN} \sum_{\mathbf{nm}} W_{\mathbf{nm}}^{(0)} (R_{\mathbf{n}} - R_{\mathbf{m}})_\alpha (R_{\mathbf{n}} - R_{\mathbf{m}})_\beta, \qquad (26)$$

where $W_{\mathbf{nm}}^{(0)} = W_{\mathbf{mn}}^{(0)} = \hbar^{-2} J_{\mathbf{mn}} J_{\mathbf{nm}} e^{-\overline{S}_{\mathbf{m-n}}} \int_{-\infty}^{\infty} dt P_{\mathbf{n-m}}(t)$ is the equilibrium hopping probability.

To evaluate the current density (25), we use the approximation of nearest-neighbor hopping in the simple cubic lattice, when $J_{\mathbf{mn}} J_{\mathbf{nm}} = J^2$ and $\overline{S}_{\mathbf{m-n}} = \overline{S}$ are constants for arbitrary $\mathbf{n} \neq \mathbf{m}$. Assuming that the field is directed along one of the axes of the lattice (say OX), we find the absolute value of the current density in the form

$$I = \frac{2ea}{\hbar^2} n(1 - \rho_{eq}) J^2 e^{-\overline{S}} \sinh \frac{eEa}{2T} \int_{-\infty}^{\infty} dt \, e^{i(eEa/\hbar)t}$$

$$\times \left\{ \exp \left[\sum_{\mathbf{q}} S_{\mathbf{q}} \frac{\cos(\omega_{\mathbf{q}} t)}{\sinh(\hbar\omega_{\mathbf{q}}/2T)} \right] - 1 \right\}, \qquad (27)$$

where $S_{\mathbf{q}} = |\gamma_{\mathbf{q}}|^2 [1 - \cos(q_x a)]$ and a is the lattice constant. Only the hops along the field contribute to the current. The case of strong electron-phonon coupling is realized when the factor $S_{\mathbf{q}}/\sinh(\hbar\omega_{\mathbf{q}}/2T)$ is large. The main contribution to the integral over time comes from a narrow region near the maximum of the function

$$F(t) = \mathrm{Re} \left\{ \sum_{\mathbf{q}} S_{\mathbf{q}} \frac{\cos(\omega_{\mathbf{q}} t)}{\sinh(\hbar\omega_{\mathbf{q}}/2T)} + i\frac{eEa}{\hbar}t \right\} \qquad (28)$$

in the plane of the complex variable t. Strictly speaking, there are many local maxima of this function associated with $\mathrm{Re}\, t = 2\pi k/\omega_{\mathbf{q}}$, where k is integer. These maxima are well separated from each other in the case of optical phonons. Assuming a finite dispersion of these phonons, one can show that the maximum at Re $t = 0$ is the highest. For this reason, it is the only one that should be considered. The position of the maximum is given by $t = i\tau_0$, where τ_0 is determined from the equation $dF(t)/dt = 0$. In the limit $\tau_0\omega_{\mathbf{q}} \ll 1$, we find $\tau_0 = eEa/[\sum_{\mathbf{q}}(\hbar S_{\mathbf{q}}\omega_{\mathbf{q}}^2/\sinh(\hbar\omega_{\mathbf{q}}/2T))]$ and obtain

$$I = \sigma_h E \frac{\sinh(eEa/2T)}{(eEa/2T)} \exp\left[-\frac{(eEa)^2}{16\mathcal{E}^2}\right], \qquad (29)$$

where

$$\mathcal{E}^2 = \frac{\hbar^2}{8} \sum_{\mathbf{q}} \frac{S_{\mathbf{q}}\omega_{\mathbf{q}}^2}{\sinh(\hbar\omega_{\mathbf{q}}/2T)} \qquad (30)$$

and

$$\sigma_h = e^2 n(1 - \rho_{eq}) \frac{\sqrt{\pi}a^2 J^2}{2\hbar T \mathcal{E}} e^{-E_A/T}, \quad E_A = T \sum_{\mathbf{q}} S_{\mathbf{q}} \tanh\left(\frac{\hbar\omega_{\mathbf{q}}}{4T}\right). \qquad (31)$$

The quantity σ_h is the linear static conductivity in the hopping regime, which can also be obtained from the Kubo formula (problem 12.23). This conductivity is proportional to the exponential factor $e^{-E_A/T}$, which becomes an activation exponent in the high-temperature limit, $T \gg \hbar\omega_{\mathbf{q}}/2$, when $E_A \simeq \sum_{\mathbf{q}} S_{\mathbf{q}}\hbar\omega_{\mathbf{q}}/4$ is temperature-independent. In this limit $\mathcal{E} = \sqrt{E_A T}$. Beyond the ohmic region, when E reaches $2T/ea$, the current starts to increase exponentially, as $\exp(eEa/2T)$. When the energy eEa reaches $4\sqrt{E_A T}$, the increase slows down. The current density I has a maximum at $eEa = 4E_A$ and decreases at $eEa > 4E_A$. Near the maximum, I follows a Gaussian dependence on the field, $I \propto \exp[-(eEa - 4E_A)^2/16E_A T]$.

To complete this section, we consider the case of weak electron-phonon coupling, when $S_{\mathbf{m-n}}(\mathbf{q}) \ll \sinh(\hbar\omega_{\mathbf{q}}/2T)$. Even when $\overline{S}_{\mathbf{m-n}} \gg 1$, this case can be realized at low enough temperatures. After expanding the exponent in $P_{\mathbf{n-m}}(t)$ from Eq. (25) in series, the integral over time gives us the δ-functions expressing the energy conservation in the phonon-assisted hopping:

$$\mathbf{I} = \frac{2\pi e n(1 - \rho_{eq})}{N\hbar} \sum_{\mathbf{nm}} (\mathbf{R_n} - \mathbf{R_m}) J_{\mathbf{mn}} J_{\mathbf{nm}}$$

$$\times e^{-\overline{S}_{\mathbf{m-n}}} \sum_{\mathbf{q}} \frac{|\gamma_{\mathbf{q}}|^2 \{1 - \cos[\mathbf{q} \cdot (\mathbf{R_n} - \mathbf{R_m})]\}}{1 - e^{-\hbar\omega_{\mathbf{q}}/T}} \qquad (32)$$

$$\times \left\{ \delta[e\mathbf{E} \cdot (\mathbf{R_n} - \mathbf{R_m}) - \hbar\omega_{\mathbf{q}}] + e^{-\hbar\omega_{\mathbf{q}}/T} \delta[e\mathbf{E} \cdot (\mathbf{R_n} - \mathbf{R_m}) + \hbar\omega_{\mathbf{q}}] \right\}.$$

In the case of nearest-neighbor hopping along the main axis of the cubic lattice,

$$I = \frac{2\pi en(1 - \rho_{eq})}{\hbar} aJ^2 e^{-\overline{S}} \sum_{\mathbf{q}} S_{\mathbf{q}}[\delta(eEa - \hbar\omega_{\mathbf{q}}) - \delta(eEa + \hbar\omega_{\mathbf{q}})]. \quad (33)$$

This current shows the electrophonon resonance when the energy drop between the nearest-neighbor sites, eEa, coincides with the LO phonon energy. Higher-order (multi-phonon) resonances are described by the next terms in the expansion of $P_{\mathbf{n}-\mathbf{m}}(t)$.

Finally, we note that the magnetic-field dependence of the current (in particular, the Hall effect) in the hopping regime cannot be described within the approximation of the second order in the phonon-assisted hopping, $\widetilde{\delta H_1}$. This is because the magnetic field disappears from the expression $J_{\mathbf{nm}}J_{\mathbf{mn}}$ if the modification of the basis functions by the magnetic field is neglected. To study the hopping magnetotransport, it is necessary to include the terms of the third order in $\widetilde{\delta H_1}$ in the consideration. The corresponding process is called the three-site hopping.

Problems

12.1. Using the real part of the non-local conductivity tensor for the 1D case, find the real and imaginary parts of the local 1D conductivity tensor $\sigma_{xx}(\omega)$ at small ω.

<u>Solution</u>: From Eq. (13.10) at small ω, we obtain

$$\sigma_{\alpha\beta}(\mathbf{r}, \mathbf{r}'|\omega \to 0) = \pi\hbar \sum_{\delta\delta'} \langle\delta|\hat{I}_\alpha(\mathbf{r})|\delta'\rangle\langle\delta'|\hat{I}_\beta(\mathbf{r}')|\delta\rangle[\delta(\varepsilon_\delta - \varepsilon_{\delta'} + \hbar\omega)$$

$$+ \delta(\varepsilon_\delta - \varepsilon_{\delta'} - \hbar\omega)]\left(-\frac{\partial f(\varepsilon_\delta)}{\partial\varepsilon_\delta}\right) + O(\omega),$$

where $O(\omega)$ denotes the terms which are proportional to ω and do not contain singularities. To find the real part of the local conductivity along OX for the electrons occupying a single 1D subband, one should express the matrix elements of the currents according to Eq. (58.11), with the result $\mathrm{Re}\sigma_{xx}(\mathbf{r}, \mathbf{r}'|\omega \to 0) = (e^2/\pi\hbar)|\psi_{y,z}|^2|\psi_{y',z'}|^2 \cos[\omega(x - x')/v_F]$, and integrate $\mathrm{Re}\sigma_{xx}(\mathbf{r}, \mathbf{r}'|\omega \to 0)$ over y, z, y', z', and $(x - x')$. The result is $\mathrm{Re}\sigma_{xx}(\omega) = (2e^2v_F/\hbar)\delta(\omega) = e^2n\pi\delta(\omega)/m$, where n is the electron density. The imaginary part is found according to the Kramers-Kronig dispersion relation (13.15) leading to the free-electron conductivity $\mathrm{Im}\sigma_{xx}(\omega) = e^2n/m\omega$.

12.2. Check the normalization condition (58.17).

<u>Hints</u>: Using the definitions (58.13) and (58.14), calculate the integral over coordinate in the half-space $r_{\parallel N} < 0$, taking into account the orthogonality and normalization conditions for $\psi_{\mathbf{r}_{\perp N}}^{(Nn)}$. Then use Eq. (58.16).

12.3. Write a matrix equation for $\varphi_n(x)$ in Eq. (58.24) corresponding to the problem of an electron waveguide with hard-wall confinement and coordinate-dependent width $d(x)$.

Result:

$$-\frac{\hbar^2}{2m}\frac{\partial^2\varphi_n(x)}{\partial x^2} - \sum_{n'}\left[A_{nn'}(x)\frac{\partial\varphi_{n'}(x)}{\partial x} + E_{nn'}(x)\varphi_{n'}(x)\right] = 0,$$

where

$$A_{nn'}(x) = \frac{\hbar^2}{m}\int dy\,\psi_y^{(n)}\frac{\partial\psi_y^{(n')}}{\partial x},$$

$$E_{nn'}(x) = \delta_{nn'}\left\{\varepsilon - \frac{[\hbar\pi n/d(x)]^2}{2m}\right\} + \frac{\hbar^2}{2m}\int dy\,\psi_y^{(n)}\frac{\partial^2\psi_y^{(n')}}{\partial x^2}.$$

The non-diagonal elements of the matrices $A_{nn'}(x)$ and $E_{nn'}(x)$ contain the derivatives of $d(x)$ over x.

12.4. Find the transmission probability of a 2D electron through the saddle-point constriction defined by the potential energy (58.25).

Solution: Substituting the wave function $\Psi(x,y) = \sum_n \varphi_n(x)\psi_n(y\sqrt{m\omega_y/\hbar})$, where $\psi_n(q)$ is the harmonic oscillator function defined in Appendix A, into the Schroedinger equation

$$\left[-\frac{\hbar^2}{2m}\frac{\partial^2}{\partial\mathbf{r}^2} + V(x,y) - E\right]\Psi = 0,$$

where $\mathbf{r} = (x,y)$ and E is the electron energy, we obtain an ordinary differential equation for $\varphi_n(x)$:

$$\left[-\frac{\hbar^2}{2m}\frac{d^2}{dx^2} + \hbar\omega_y\left(n + \frac{1}{2}\right) + V_0 - \frac{m\omega_x^2 x^2}{2} - E\right]\varphi_n(x) = 0.$$

Therefore, the problem is reduced to a calculation of the transmission probability through the parabolic potential barrier in one dimension. Introducing the dimensionless coordinate $\xi = x\sqrt{m\omega_x/\hbar}$ and dimensionless energy ϵ_n defined in Eq. (58.26), we find the asymptotes of the wave function in the form

$$\varphi_n(\xi) = e^{-i\xi^2/2}|\xi|^{-i\epsilon_n-1/2} + r_n e^{i\xi^2/2}|\xi|^{i\epsilon_n-1/2}, \quad x \to -\infty,$$

$$\varphi_n(\xi) = t_n e^{i\xi^2/2}|\xi|^{i\epsilon_n-1/2}, \quad x \to \infty,$$

where t_n and r_n are the transmission and reflection amplitudes. The wave function at an arbitrary x is expressed as a linear combination of confluent hypergeometric functions. By matching this wave function to the asymptotic forms given above, we find $r_n = -it_n e^{-\pi\epsilon_n}$. Finally, using the requirement $|r_n|^2 + |t_n|^2 = 1$, we obtain

$$|t_n|^2 = 1/(1 + e^{-2\pi\epsilon_n}).$$

This is the transmission coefficient for an electron in the state n described by the function ψ_n. Since this state is conserved in the transmission, $T^{(nm)} = \delta_{nm}|t_n|^2$.

12.5. Find the thermal conductance due to electron transmission through an ideal 1D quantum wire.

Solution: Consider a quantum wire connecting electron reservoirs 1 and 2 with different temperatures T_1 and T_2. If only one spin-degenerate 1D subband in the wire is occupied, the energy flow between the reservoirs is given by the expression

$$2 \int \frac{dp}{2\pi\hbar} v_p \theta(v_p) \varepsilon_p [f_1(\varepsilon_p) - f_2(\varepsilon_p)],$$

where v_p and ε_p are the group velocity and energy of the 1D electrons, and f_1 and f_2 are the quasi-equilibrium Fermi distribution functions with effective temperatures T_1 and T_2 (the electrochemical potentials on both sides are assumed to be equal to each other). In the linear transport regime, when the difference $|T_1 - T_2|$ is small in comparison to $(T_1 + T_2)/2 \equiv T$, the energy flow is equal to $\kappa(T_1 - T_2)$, where κ is the thermal conductance. For degenerate electrons, $\kappa = \pi T/3\hbar = (\pi^2/3e^2)GT$, where $G = e^2/\pi\hbar$ is the electric conductance of the 1D quantum wire. We point out that the relation (36.23) between the thermal and electric conductivities remains valid in the ballistic transport regime.

12.6. Calculate the dimensionless resistance R/T of the piece of 1D conductor with three obstacles (point scatterers) and generalize the result to the case of N obstacles.

Solution: The electrons in the piece containing three point scatterers at $x = 0, l_1, l_2$ are described by the wave functions $e^{ikx} + Ae^{-ikx}$ at $x < 0$, $B_1 e^{ikx} + C_1 e^{-ikx}$ at $0 < x < l_1$, $B_2 e^{ik(x-l_1)} + C_2 e^{-ik(x-l_1)}$ at $l_1 < x < l_2$, and $De^{ik(x-l_1-l_2)}$ at $x > l_1 + l_2$. To find the six coefficients entering these expressions, we compose equations similar to Eq. (59.2). In the matrix form,

$$\begin{pmatrix} 1 & 0 & -t_1' & 0 & 0 & 0 \\ 0 & 1 & -r_1' & 0 & 0 & 0 \\ 0 & -e^{i\phi_1}r_2 & e^{-i\phi_1} & 0 & -t_2' & 0 \\ 0 & -e^{i\phi_1}t_2 & 0 & 1 & -r_2' & 0 \\ 0 & 0 & 0 & -e^{i\phi_2}r_3 & e^{-i\phi_2} & 0 \\ 0 & 0 & 0 & -e^{i\phi_2}t_3 & 0 & 1 \end{pmatrix} \begin{pmatrix} A \\ B_1 \\ C_1 \\ B_2 \\ C_2 \\ D \end{pmatrix} = \begin{pmatrix} r_1 \\ t_1 \\ 0 \\ 0 \\ 0 \\ 0 \end{pmatrix}$$

where ϕ_1 and ϕ_2 are the phases acquired by the electron waves propagating from 0 to l_1 and from l_1 to l_2, respectively. An extension of this equation to the case of an arbitrary number of obstacles is obvious. Using the symmetry $t_i' = t_i$ and the unitarity property which gives us $t_i/t_i'^* = -r_i/r_i'^*$, we find

$$D = \frac{t_1 t_2 t_3 e^{i(\phi_1+\phi_2)}}{(1 - |r_1||r_2|e^{i\theta_1})(1 - |r_2||r_3|e^{i\theta_1}) + |r_1||r_3|(1 - |r_2|^2)e^{i(\theta_1+\theta_2)}},$$

where $\theta_1 = 2\phi_1 + \arg(r_1') + \arg(r_2)$ and $\theta_2 = 2\phi_2 + \arg(r_2') + \arg(r_3)$. Calculating the absolute value of D, we express the intrinsic resistance of the piece as

$$\frac{R}{T} = \frac{R_1 + R_2 + R_3 + R_1 R_2 R_3 + \dots}{T_1 T_2 T_3},$$

where the dots indicate the oscillating terms proportional to $\cos\theta_1$, $\cos\theta_2$, and $\cos(\theta_1 \pm \theta_2)$, which disappear after averaging over random phases. If R_i are small, one can write, with the accuracy up to $\overline{R}^2$, $\langle R/T \rangle = (R_1 + R_2 + R_3)/T_1 T_2 T_3$. Acting by induction, we find Eq. (59.6) for the case of N scatterers.

12.7. Find the energy spectrum of electrons in an ideal ring placed in the magnetic field **H** perpendicular to the plane of the ring.

<u>Solution:</u> Applying **H** along OZ, we use the gauge $\mathbf{A} = (-Hy/2, Hx/2, 0)$ and write the Schroedinger equation in the cylindrical coordinate system:

$$\left\{ -\frac{\hbar^2}{2m} \left[\frac{\partial^2}{\partial z^2} + \frac{\partial^2}{\partial \rho^2} + \frac{1}{\rho}\frac{\partial}{\partial \rho} + \frac{1}{\rho^2}\frac{\partial^2}{\partial \varphi^2} \right] - i\frac{\hbar\omega_c}{2}\frac{\partial}{\partial \varphi} + \frac{e^2 H^2 \rho^2}{8mc^2} + U(\rho, z) - E \right\} \Psi = 0,$$

where ρ is the radial coordinate, φ is the polar angle, and $U(\rho, z)$ is a confinement potential defining the ring at $\rho \simeq \rho_0$ and $z \simeq 0$. We represent the wave function in the form of the expansion $\Psi = \sum_n \chi_n(\rho, z) \psi_n(\varphi)$, where χ_n is the eigenfunction of the Hamiltonian

$$-\frac{\hbar^2}{2m} \left[\frac{\partial^2}{\partial z^2} + \frac{\partial^2}{\partial \rho^2} + \frac{1}{\rho}\frac{\partial}{\partial \rho} \right] + U(\rho, z),$$

which is independent of the magnetic field. The eigenvalues of this Hamiltonian are ε_n. After substituting the expansion introduced above into the Schroedinger equation, we multiply the latter by $\chi_n^*(\rho, z)$ from the left and integrate it over ρ and z. Neglecting the non-diagonal matrix elements of ρ^2 and $1/\rho^2$ in the adiabatic approximation, when the ring is narrow, we obtain

$$\left[-\frac{\hbar^2}{2m}\left(\frac{1}{\rho^2}\right)_{nn}\frac{\partial^2}{\partial \varphi^2} - i\frac{\hbar\omega_c}{2}\frac{\partial}{\partial \varphi} + \frac{e^2 H^2 (\rho^2)_{nn}}{8mc^2} + \varepsilon_n - E \right] \psi_n(\varphi) = 0.$$

Finally, approximating the ground-state ($n = 1$) wave function as $|\chi_1(\rho, z)|^2 \sim \delta(\rho - \rho_0)\delta(z)$, we find the following equation:

$$\left[\hbar\omega\left(-i\frac{\partial}{\partial \varphi} + \frac{\Phi}{\Phi_0} \right)^2 - \varepsilon \right] \psi(\varphi) = 0,$$

where $\Phi = \pi\rho_0^2 H$ is the magnetic flux, $\varepsilon = E - \varepsilon_1$, and $\omega = \hbar/2m\rho_0^2$ is a characteristic frequency describing the energy quantization. This equation has a simple solution $\psi_l(\varphi) = e^{il\varphi}$ and $\varepsilon_l = \hbar\omega[l + \Phi/\Phi_0]^2$, where $l = 0, \pm 1, \pm 2, \ldots$. This solution describes the quantization of the angular momentum.

12.8. Find the quantum correction to the conductivity of a two-dimensional cylinder of radius ρ_0 and length L_z placed along the magnetic field **H**.

<u>Solution:</u> The stationary ($\omega = 0$) correction is given by $\delta\sigma = -(2\sigma/\pi\hbar\rho_{2D})C(\mathbf{r}, \mathbf{r})$, see Eq. (43.23), where $\mathbf{r}$ is the 2D coordinate on the cylindrical surface. Assuming that the axis of the cylinder is along OZ, we use the coordinates z and $\rho_0\varphi$, where ρ_0 is the radius of the cylinder and $\varphi \in [0, 2\pi]$ is the polar angle. Since ρ_0 is constant, we write the Cooperon $C(\mathbf{r}, \mathbf{r}')$ as a function of the variables z and φ, i.e., $C(\mathbf{r}, \mathbf{r}') = C(z\varphi, z'\varphi')$. This function satisfies the diffusion equation following from the general equation (43.19) in the stationary case:

$$\left[\frac{1}{\rho_0^2}\left(-i\frac{\partial}{\partial \varphi} + 2\frac{\Phi}{\Phi_0} \right)^2 - \frac{\partial^2}{\partial z^2} + \frac{1}{l_D^2} \right] C(z\varphi, z'\varphi') = \frac{1}{\rho_0 D}\delta(z - z')\delta(\varphi - \varphi'),$$

where $l_D = \sqrt{D\tau_\varphi}$ is the diffusion length. Assuming that $-L_z/2 < z < L_z/2$, we use the boundary conditions $[\partial C(z\varphi, z'\varphi')/\partial z]_{z=\pm L_z/2} = 0$. There is also a periodic boundary condition, $C(z\ \varphi + 2\pi, z'\varphi') = C(z\varphi, z'\varphi')$. The equation for $C(z\varphi, z'\varphi')$ with these boundary conditions has the following solution:

$$C(z\varphi, z'\varphi') = \frac{\rho_0}{2\pi D L_z} \sum_{m=-\infty}^{\infty} \sum_{n=-\infty}^{\infty} \exp[in(\varphi - \varphi')]$$

$$\times \frac{\cos[\pi m(z/L_z - 1/2)]\cos[\pi m(z'/L_z - 1/2)]}{(n + 2\Phi/\Phi_0)^2 + (\pi m \rho_0/L_z)^2 + (\rho_0/l_D)^2}.$$

To find $\delta\sigma$, one should calculate the discrete sums in this expression at $z = z'$ and $\varphi = \varphi'$. Since

$$\sum_{n=-\infty}^{\infty} \frac{1}{(n + 2\Phi/\Phi_0)^2 + b^2} = \frac{\pi}{b} \frac{\sinh(2\pi b)}{\cosh(2\pi b) - \cos(4\pi\Phi/\Phi_0)},$$

the result is always periodic in Φ with the period $\Phi_0/2$. In particular, for a short cylinder with $L_z \ll l_D$ only the term with $m = 0$ is essential in the sum over m, and

$$\delta\sigma = -\frac{e^2}{\pi\hbar} \frac{l_D}{L_z} \frac{\sinh(2\pi\rho_0/l_D)}{\cosh(2\pi\rho_0/l_D) - \cos(4\pi\Phi/\Phi_0)}.$$

If the circumference of the cylinder, $2\pi\rho_0$, becomes greater than l_D, the oscillations are suppressed exponentially.

12.9. Calculate the group velocity of the edge-state electrons in the parabolic potential $U(x) = m\omega_0^2 x^2/2$.

Solution: Using Eq. (51.15), one has $|\mathbf{v}_\perp| = (\omega_0^2/\omega_c)|x|$. In the quasi-classical approximation, x is identified with the center of the oscillator, X_p, entering Eq. (59.30). Thus, the absolute value of the group velocity is equal to $(\omega_0/\omega)^2(p/m)$.

12.10. Calculate the transmission probability of a 2D electron with energy E through the saddle-point potential constriction (58.25) in the presence of a magnetic field directed perpendicular to the 2D plane.

Result: The transmission coefficient is given by the same equation as in problem 12.4, but the dimensionless energy ϵ_n is defined in a different way: $\epsilon_n = [E - E_2(n + 1/2) - V_0]/E_1$, where $E_1 = \hbar[(\omega^4 + 4\omega_x^2\omega_y^2)^{1/2} - \omega^2]^{1/2}/\sqrt{2}$, $E_2 = \hbar[(\omega^4 + 4\omega_x^2\omega_y^2)^{1/2} + \omega^2]^{1/2}/\sqrt{2}$, and $\omega^2 = \omega_c^2 + \omega_y^2 - \omega_x^2$.

12.11. Consider the anomalous quantum Hall effect and prove Eq. (59.32) for the geometry of Fig. 12.4.

Hints: Assuming that the leads 1, 2, 3, and 4 can accommodate N_1, N_2, N_3, and N_4 edge channels, respectively, use Eq. (59.31) together with the relations $\sum_{M'} G_{M'M} = \sum_{M'} G_{MM'} = G_0 N_M$, where $M' \neq M$ and $G_0 = e^2/2\pi\hbar$ is the conductance quantum. Apply the equations $G_{21} = G_0 \min(N_1, N_2)$, $G_{32} = G_0 \min(N_2, N_3)$, $G_{43} = G_0 \min(N_3, N_4)$, and $G_{14} = G_0 \min(N_4, N_1)$.

12.12. Calculate the probability of electron tunneling through a rectangular potential defined by $U(z) = U$ at $0 < z < d$, $U(z) = 0$ at $z < 0$, and $U(z) = -\Delta U$ at $z > d$.

Solution: The wave functions at $z < 0$ and $z > d$ are given by Eq. (60.3). The wave function in the barrier is $C_1 e^{\kappa z} + C_2 e^{-\kappa z}$, where $\hbar\kappa = \sqrt{2m(U - \varepsilon) + p^2}$. The boundary conditions assuming the continuity of the wave function and its derivative at $z = 0$ and $z = d$ give us four equations, from which one can find the amplitude t of the transmitted wave. Calculating the square of the absolute value of t, we obtain

$$T_{p\varepsilon} = 2\frac{16 k_z k_z' \kappa^2}{(k_z^2 + \kappa^2)(k_z'^2 + \kappa^2)} e^{-2\kappa d},$$

where the factor of 2 appears due to spin degeneracy.

12.13. Calculate the probability of Zener tunneling in the two-band model.

Solution: The probability is given by Eq. (60.10). Shifting the contour of integration in the upper half-plane of complex p_z up to the point $ip_0 = i\sqrt{(\varepsilon_g/2s)^2 + p^2}$, we get a sum of two contributions. The first one is associated with a semi-circle around ip_0, denoted as C_{sc},

$$\int_{C_{sc}} dp_z \frac{1}{p_z^2 + p_0^2} \exp\left\{\frac{2i}{\hbar|e|E}\int_0^{p_z} dp_z' \mathcal{E}(p_z', p)\right\}$$

$$= \frac{\pi}{2p_0} \exp\left\{\frac{2i}{\hbar|e|E}\int_0^{ip_0} dp_z' \mathcal{E}(p_z', p)\right\},$$

while the second contribution is due to a straight-path integral, $\int_{C_{sp}} dp_z \ldots$, over $\delta p_z = p_z - ip_0$. If the field E is not very strong, this last integral converges rapidly in the vicinity of $\delta p_z = 0$, and one can expand $\int_0^{p_z} dp_z' \mathcal{E}(p_z', p)$ as $\int_0^{ip_0} dp_z' \mathcal{E}(p_z', p) + (2s/3)\sqrt{2ip_0}(\delta p_z)^{3/2}$ at $\delta p_z > 0$ and as $\int_0^{ip_0} dp_z' \mathcal{E}(p_z', p) + (2s/3)\sqrt{2ip_0}i|\delta p_z|^{3/2}$ at $\delta p_z < 0$. Combining the contributions from the regions $\delta p_z > 0$ and $\delta p_z < 0$, substituting $\xi = |\delta p_z|^{3/2}$, and using the integral $\int_0^\infty d\xi\, \xi^{-1} e^{-b\xi}\sin(b\xi) = \pi/4$, we find

$$\int_{C_{sp}} dp_z \frac{1}{p_z^2 + p_0^2} \exp\left\{\frac{2i}{\hbar|e|E}\int_0^{p_z} dp_z' \mathcal{E}(p_z', p)\right\}$$

$$= \frac{1}{3}\int_{C_{sc}} dp_z \frac{1}{p_z^2 + p_0^2} \exp\left\{\frac{2i}{\hbar|e|E}\int_0^{p_z} dp_z' \mathcal{E}(p_z', p)\right\}.$$

Combining together these two contributions, we finally notice that $\int_0^{ip_0} dp_z' \mathcal{E}(p_z', p) = i\pi s p_0^2/4$ and obtain Eq. (60.11).

12.14. Find the decay time of electron states in the leaky quantum well created by the potential energy $U(z) = \infty$ at $z < 0$, $U(z) = V$ at $a < z < b$, and $U(z) = 0$ at $0 < z < a$ and $z > a + b$.

Solution: Let us consider the Schroedinger equation describing the motion of the particles with effective mass m and in-plane kinetic energy ε_p in the potential specified above. Representing the wave function in the form $A\sin(kz)$ at $z < a$, $c_+ e^{\kappa(z-a)} +$

$c_- e^{-\kappa(z-a)}$ at $a < z < a+b$, and $t e^{ik(z-a-b)}$ at $z > a+b$, where $k = \sqrt{2m(\varepsilon - \varepsilon_p)/\hbar^2}$ and $\kappa = \sqrt{2mV/\hbar^2 - k^2}$, we derive the dispersion relation

$$\kappa \sin(ka) + k \cos(ka) = e^{-2\kappa b} \frac{\kappa + ik}{\kappa - ik} [\kappa \sin(ka) - k \cos(ka)],$$

which gives us the imaginary part of the energy. In the approximation $e^{-\kappa b} \ll 1$ we find

$$\nu = -\frac{2\text{Im}\varepsilon}{\hbar} = \frac{8\hbar k^3 \kappa^3}{m(\kappa^2 + k^2)^2(1 + \kappa a)} \exp(-2\kappa b),$$

while the wave number k, which is related to the quantization energy $\hbar^2 k^2 / 2m$ of the state in the well, is found from the algebraic equation $\kappa \sin(ka) + k \cos(ka) = 0$.

12.15. Using Eqs. (60.17) and (H.19), find the transmission coefficient $T_{p\varepsilon}$ for the case of tunneling through a rectangular potential barrier. Compare it to the result of problem 12.12.

Solution: We use Eqs. (60.17) and (H.19) together with the relations $\varepsilon_{lk'\mathbf{p}} = U_{0l} + p^2/2m + \hbar^2 k'^2/2m$ and $\varepsilon_{rk\mathbf{p}} = U_{0r} + p^2/2m + \hbar^2 k^2/2m$. Taking into account that the sums over continuous variables are transformed to integrals as $\sum_{k'} \ldots = (L_l/2\pi) \int dk' \ldots$ and $\sum_k \ldots = (L_r/2\pi) \int dk' \ldots$, we find

$$T_{p\varepsilon} = 8\pi^2 \frac{16\hbar^4}{m^2} \int \frac{dk}{2\pi} \int \frac{dk'}{2\pi} \frac{(kk')^2 e^{-2\kappa d}}{(k^2 + \kappa^2)(k'^2 + \kappa^2)}$$

$$\times \delta(\varepsilon - U_{0l} - p^2/2m - \hbar^2 k'^2/2m) \delta(\varepsilon - U_{0r} - p^2/2m - \hbar^2 k^2/2m).$$

The integrals over k and k' are calculated with the use of the δ-functions, and we obtain the result of problem 12.12, where $k_z \equiv k$, $k_z' \equiv k'$, and κ are expressed through ε and $p^2/2m$.

12.16. Show that the distribution functions given by Eqs. (61.19) and (61.20) satisfy the principle of detailed balance in equilibrium.

Hint: Prove that

$$\sum_{\{n_\gamma\}} \exp\left(-\frac{1}{T}\sum_\gamma \varepsilon_\gamma n_\gamma + \frac{\varepsilon_q}{T}\right) n_q \Bigg|_{\sum_\gamma n_\gamma = N}$$

$$= \sum_{\{n_\gamma\}} \exp\left(-\frac{1}{T}\sum_\gamma \varepsilon_\gamma n_\gamma\right) (1 - n_q) \Bigg|_{\sum_\gamma n_\gamma = N-1} .$$

12.17. Find the equilibrium occupation numbers $f_q^{(N)}$ for a two-level system $(q = 1, 2)$ at $N = 1$.

Result: $f_1^{(1)} = [1 + e^{(\varepsilon_1 - \varepsilon_2)/T}]^{-1}$ and $f_2^{(1)} = [1 + e^{(\varepsilon_2 - \varepsilon_1)/T}]^{-1}$.

12.18. Find the spectrum of electrons in the simple cubic lattice by using the tight-binding approximation.

Solution: In the absence of electron-phonon interaction, the Hamiltonian of electrons, $\widehat{H}_e$, is given by the first two terms of the Hamiltonian (62.2). Let us make the transformation

$$\hat{a}_{\mathbf{n}} = N^{-1/2} \sum_{\mathbf{k}} \hat{a}_{\mathbf{k}} e^{i\mathbf{k}\cdot\mathbf{R_n}},$$

which relates the site representation (62.1) to the representation of Bloch waves with wave vectors $\mathbf{k}$. The corresponding basis functions of the new representation are $\psi_{\mathbf{k}}(\mathbf{r}) = V^{-1/2} u_{\mathbf{k}}(\mathbf{r}) e^{i\mathbf{k}\cdot\mathbf{r}}$, where $u_{\mathbf{k}}(\mathbf{r}) = V_c^{1/2} \sum_n \varphi(\mathbf{r}-\mathbf{R_n}) e^{-i\mathbf{k}\cdot(\mathbf{r}-\mathbf{R_n})}$ are the Bloch amplitudes and $V_c = V/N$ is the volume of elementary cell. Using the identity given in problem 1.16, we write the Hamiltonian in the form

$$\widehat{H}_e = \sum_{\mathbf{k}} \varepsilon_{\mathbf{k}} \hat{a}_{\mathbf{k}}^+ \hat{a}_{\mathbf{k}}, \quad \varepsilon_{\mathbf{k}} = \varepsilon_0 + \sum_{\mathbf{m}} J_{\mathbf{m}} e^{-i\mathbf{k}\cdot\mathbf{R_m}}.$$

In the tight-binding approximation the sum over $\mathbf{m}$ is finite and accounts only for the nearest-neighbor sites. Since $J_{\mathbf{m}} = J$ for all nearest-neighbor sites in the simple cubic lattice, we obtain $\varepsilon_{\mathbf{k}} = \varepsilon_0 + 2J[\cos(ak_x) + \cos(ak_y) + \cos(ak_z)]$, where a is the lattice period and the wave vector $\mathbf{k}$ is defined in the Cartesian coordinate system oriented along the main axes.

12.19. Consider electron transport along the biased one-dimensional superlattice described by the Hamiltonian

$$\widehat{H} = \sum_{n\mathbf{p}} (\varepsilon_n + \varepsilon_p) \hat{a}_{n\mathbf{p}}^+ \hat{a}_{n\mathbf{p}} + \sum_{nn'\mathbf{p}}{}' t_{nn'} \hat{a}_{n\mathbf{p}}^+ \hat{a}_{n'\mathbf{p}}$$

$$+ \frac{1}{L^2} \sum_{n\mathbf{pp}'} U_n[(\mathbf{p}-\mathbf{p}')/\hbar] \hat{a}_{n\mathbf{p}}^+ \hat{a}_{n\mathbf{p}'},$$

where n is the layer index, $\varepsilon_n = \varepsilon_0 - eEdn$ is the potential energy of the layer n determined by the electric field E and superlattice period d, $\varepsilon_p = p^2/2m$ is the kinetic energy of in-plane motion, $t_{nn'}$ is the tunneling matrix element describing hopping between the layers, and $U_n(\mathbf{q})$ is the Fourier transform of the effective random scattering potential in the layer n.

Solution: The density of electric current is given by $I_t = (2e/V)\mathrm{Sp}(\hat{v}_z \hat{\rho}_t)$, where $\hat{v}_z$ is the operator of velocity in the direction z perpendicular to the layer planes and $\hat{\rho}_t$ is the one-electron statistical operator averaged over the scattering potential; see Sec. 7. In the basis $|n\mathbf{p}\rangle$, the current density is expressed as

$$I_t = \frac{2e}{Nd} \sum_{nn'} \int \frac{d\mathbf{p}}{(2\pi\hbar)^2} \langle n'|\hat{v}_z|n\rangle \rho_{nn'}(\mathbf{p}t) = \frac{2ie}{\hbar N} \sum_{nn'} t_{n'n}(n-n') \int \frac{d\mathbf{p}}{(2\pi\hbar)^2} \rho_{nn'}(\mathbf{p}t),$$

where $N = L_z/d$ is the total number of layers. We have used the identity $\langle n'|\hat{v}_z|n\rangle = (i/\hbar)\langle n'|[\widehat{H}, \hat{z}]|n\rangle = (i/\hbar) t_{n'n}(nd - n'd)$. The matrix $\rho_{nn'}(\mathbf{p}t) = \langle n\mathbf{p}|\hat{\rho}_t|n'\mathbf{p}\rangle$ satisfies the quantum kinetic equation

$$\frac{\partial \rho_{nn'}(\mathbf{p}t)}{\partial t} + \frac{i}{\hbar}(\varepsilon_n - \varepsilon_{n'})\rho_{nn'}(\mathbf{p}t) + \frac{i}{\hbar} t_{nn'}(f_{n'\mathbf{p}} - f_{n\mathbf{p}}) = \langle n\mathbf{p}|\widehat{J}(\hat{\rho}|t)|n'\mathbf{p}\rangle,$$

where $f_{n\mathbf{p}} = \langle n\mathbf{p}|\hat{\rho}_t|n\mathbf{p}\rangle$ is the distribution function in the layer n. The collision integral in the operator representation is defined by (compare to Eq. (8.3))

$$\hat{J}(\hat{\rho}|t) = \frac{1}{\hbar^2 L^2} \sum_{mm'\mathbf{q}} w_{mm'}(q) \int_{-\infty}^{0} d\tau e^{\lambda\tau} \left[e^{i\hat{h}\tau/\hbar} [\hat{P}_m e^{i\mathbf{q}\cdot\mathbf{x}}, \hat{\rho}_{t+\tau}] e^{-i\hat{h}\tau/\hbar}, \hat{P}_{m'} e^{-i\mathbf{q}\cdot\mathbf{x}} \right],$$

where $\hat{h}$ is the Hamiltonian of superlattice in the absence of the scattering potential, $w_{mm'}(q) = L^{-2}\langle\langle U_m(\mathbf{q})U_{m'}(-\mathbf{q})\rangle\rangle$ is the correlation function of the potentials in the layers m and m', and $\hat{P}_m$ is the operator of projection on the layer m. This operator is defined by $\langle n|\hat{P}_m|n'\rangle = \delta_{nm}\delta_{n'm}$.

Below we neglect the interlayer correlation of the scattering potentials and take into account that the layers are macroscopically identical. As a result, $w_{mm'}(q) = \delta_{mm'}w(q)$. Next, we assume weak interlayer coupling and calculate the matrix elements of $\hat{J}(\hat{\rho}|t)$ by neglecting the non-diagonal in the layer index components of $\hat{h}$. As a result, the kinetic equation in the stationary case can be written as

$$\frac{i}{\hbar}(\varepsilon_n - \varepsilon_{n'})\rho_{nn'}(\mathbf{p}) + \nu_{nn'}(p)\rho_{nn'}(\mathbf{p}) = \frac{i}{\hbar}t_{nn'}(f_{n\mathbf{p}} - f_{n'\mathbf{p}}),$$

where

$$\hbar\nu_{nn'}(p) = \int \frac{d\mathbf{p'}}{(2\pi\hbar)^2} w(|\mathbf{p}-\mathbf{p'}|/\hbar) \left[\pi\delta(\varepsilon_p - \varepsilon_{p'} + \varepsilon_n - \varepsilon_{n'}) + \pi\delta(\varepsilon_{p'} - \varepsilon_p + \varepsilon_n - \varepsilon_{n'}) \right.$$

$$\left. -i\mathcal{P}\frac{1}{\varepsilon_p - \varepsilon_{p'} + \varepsilon_n - \varepsilon_{n'}} - i\mathcal{P}\frac{1}{\varepsilon_{p'} - \varepsilon_p + \varepsilon_n - \varepsilon_{n'}} \right].$$

The imaginary part of $\nu_{nn'}(p)$ describes a renormalization of the potential energies in the layers and can be neglected, while the real part describes the relaxation of the non-diagonal component of the density matrix. In the simplest case of short-range scattering potential, when $w(q) \simeq w$, and under the condition that $|\varepsilon_n - \varepsilon_{n'}|$ does not exceed typical kinetic energies of electrons, the real part of $\nu_{nn'}(p)$ is equal to $1/\tau = mw/\hbar^3$. Substituting this result into the stationary kinetic equation, one can find $\rho_{nn'}(\mathbf{p}) = t_{nn'}(f_{n\mathbf{p}} - f_{n'\mathbf{p}})/(\varepsilon_n - \varepsilon_{n'} - i\hbar/\tau)$. Therefore, the stationary current density is given by

$$I = \frac{2ie}{\hbar N} \sum_{nn'} \frac{|t_{nn'}|^2(n-n')}{\varepsilon_n - \varepsilon_{n'} - i\hbar/\tau} \int \frac{d\mathbf{p}}{(2\pi\hbar)^2}(f_{n\mathbf{p}} - f_{n'\mathbf{p}})$$

$$= -\frac{2ie}{\hbar} \sum_{k=\pm 1} \frac{t^2}{keEd + i\hbar/\tau} \int \frac{d\mathbf{p}}{(2\pi\hbar)^2}k(f_{n\mathbf{p}} - f_{n-k,\mathbf{p}}).$$

The last equation corresponds to the tight-binding approximation with $|t_{n,n\pm 1}|^2 \equiv t^2$. Assuming that the electron distribution in each layer is quasi-equilibrium, with quasi-Fermi energies $\varepsilon_{Fn} = \varepsilon_F - eEnd$, we estimate the integral over $\mathbf{p}$ as $-\rho_{2D}eEd/2$ and finally obtain

$$I = \frac{e^2\rho_{2D}Ed}{2\tau} \frac{(2t)^2}{(eEd)^2 + (\hbar/\tau)^2}.$$

This current density can be represented as $I = GV$, where $V = Ed$ is the voltage per period and G is the tunneling conductance given by Eq. (60.27) with $\Delta = eEd$ and $t_{lr} = t$. At small bias, the current linearly increases with the applied field, while at $(eEd)^2 > (\hbar/\tau)^2$ the current decreases because the field drives the 2D states out of resonance.

12.20. Estimate $\Delta\varepsilon$ by assuming the Froelich interaction of electrons with *LO* phonons.

<u>Result</u>: $\Delta\varepsilon \sim e^2/\epsilon^* r_0$ with $r_0 = \pi/q_0$, where q_0 is the characteristic wave number at which the Froelich interaction ($\propto 1/q^2$ at $q \ll \pi/a$) is cut off. Since r_0 is of the order of the lattice constant a, the energy $\Delta\varepsilon$ can be of 1 eV scale, especially in the crystals with high ionicity, where ϵ_0 and ϵ_∞ considerably differ from each other.

12.21. Prove that the Hamiltonian (62.8) (without the last term) is written in the form (62.10) in the presence of uniform electric and magnetic fields.

<u>Hints</u>: To describe the effect of electric field, add the potential energy $-e\mathbf{E} \cdot \mathbf{r}$ to $\hat{h}_{cr}(\mathbf{r})$ in Eq. (62.3). To describe the effect of magnetic field, take into account that the wave function of the electron which jumps from the site $\mathbf{n}$ to the site $\mathbf{n}'$ acquires the phase $(e/2\hbar c)\mathbf{H} \cdot [\mathbf{R_n} \times \mathbf{R_{n'}}]$, according to the general consideration given in Appendix G.

12.22. Prove Eq. (62.18).

<u>Solution</u>: To calculate the sums over $n_\mathbf{q}$ and n in Eqs. (62.16) and (62.17), it is convenient to use a new variable $l = n_\mathbf{q} - n$ rather than $n_\mathbf{q}$. With the aid of the integral representation of the factorial, $(l+n)! = \Gamma(l+n+1) = \int_0^\infty dx e^{-x} x^{l+n}$, the expression (62.16) is transformed as

$$\prod_\mathbf{q} \left(1 - e^{-\hbar\omega_\mathbf{q}/T}\right) \sum_{n=0}^\infty \sum_{l=0}^\infty e^{-\hbar\omega_\mathbf{q}(n+l)/T}(-1)^n \frac{|\mu_\mathbf{q}|^{2n}(n+l)!}{(n!)^2 l!}$$

$$= \prod_\mathbf{q} \left(1 - e^{-\hbar\omega_\mathbf{q}/T}\right) \sum_{n=0}^\infty e^{-\hbar\omega_\mathbf{q}n/T}(-1)^n \frac{|\mu_\mathbf{q}|^{2n}}{(n!)^2} \int dx\, x^n e^{-x} \sum_{l=0}^\infty \frac{x^l e^{-\hbar\omega_\mathbf{q}l/T}}{l!}.$$

The sum over l is, in fact, a serial expansion of the exponent $\exp(xe^{-\hbar\omega_\mathbf{q}/T})$. Then, the integral over x is calculated elementary, and the remaining sum over n is again a serial expansion of an exponent. The final transformation of the expression (62.16) is

$$\prod_\mathbf{q} \sum_{n=0}^\infty e^{-\hbar\omega_\mathbf{q}n/T}(-1)^n \frac{|\mu_\mathbf{q}|^{2n}}{n!\left(1 - e^{-\hbar\omega_\mathbf{q}/T}\right)^n} = \exp\left(-\sum_\mathbf{q} |\mu_\mathbf{q}|^2 N_\mathbf{q}\right).$$

12.23. Find the frequency-dependent polaronic conductivity in the hopping regime by using the Kubo formula (13.18).

<u>Solution</u>: Let us replace the velocity operators in Eq. (13.18) by $(i/\hbar)[\widehat{H}, \hat{\mathbf{x}}]$ and apply the transformation (62.5) to the operators. As a result, the Kubo formula is rewritten as

$$\sigma_{\alpha\beta}(\omega) = \frac{ie^2 n}{m\omega}\delta_{\alpha\beta} - \frac{e^2}{V\hbar^3\omega}\int_{-\infty}^0 d\tau\, e^{\lambda\tau - i\omega\tau}$$

$$\times \mathrm{Sp}\,\widetilde{\eta}_{eq}\left[e^{-i\widetilde{H}\tau/\hbar}[\widetilde{H}, \hat{x}_\alpha]e^{i\widetilde{H}\tau/\hbar}, [\widetilde{H}, \hat{x}_\beta]\right],$$

where Sp... includes the spin trace and $\widetilde{H}$ is defined by Eqs. (62.10) and (62.11). The matrix elements of the commutators in the site representation are written as $\langle\mathbf{m}|[\widetilde{H},\hat{\mathbf{x}}]|\mathbf{n}\rangle = \langle\mathbf{m}|\widetilde{H}|\mathbf{n}\rangle(\mathbf{R_n}-\mathbf{R_m}) = \langle\mathbf{m}|\widetilde{\delta H_0}+\widetilde{\delta H_1}|\mathbf{n}\rangle(\mathbf{R_n}-\mathbf{R_m})$. Therefore, in the hopping regime, when $\widetilde{\delta H_0}$ is neglected, the expression under the integral is already quadratic in $\widetilde{\delta H_1}$. This means that, with the required accuracy, one may replace the Hamiltonian $\widetilde{H}$ standing in $e^{\pm i\widetilde{H}\tau/\hbar}$ by $\widetilde{H}_0$ and use the equilibrium statistical operator $\widetilde{\eta}_{eq}$ determined by $\widetilde{H}_0$. The trace over electron and phonon variables is calculated as described in Sec. 62, with the result

$$\sigma_{\alpha\beta}(\omega) = \frac{ie^2n}{m\omega}\delta_{\alpha\beta} + \frac{2e^2}{V}\sum_{\mathbf{nm}}\rho_{\mathbf{n}}(1-\rho_{\mathbf{m}})\frac{J_{\mathbf{nm}}J_{\mathbf{mn}}}{\hbar^3\omega}(\mathbf{R_n}-\mathbf{R_m})_{\alpha}(\mathbf{R_n}-\mathbf{R_m})_{\beta}$$

$$\times e^{-\overline{S}_{\mathbf{m-n}}}\int_{-\infty}^{0}d\tau e^{\lambda\tau-i\omega\tau}\left[e^{i(\varepsilon_{\mathbf{m}}-\varepsilon_{\mathbf{n}})\tau/\hbar}P_{\mathbf{m-n}}(\tau-i\hbar/2T)-c.c.\right],$$

where the factor of 2 appears due to spin degeneracy. Taking into account that all quantities in this equation are the equilibrium ones, we put $\varepsilon_{\mathbf{n}}=\varepsilon_{\mathbf{m}}$ and $\rho_{\mathbf{n}}=\rho_{\mathbf{m}}=\rho_{eq}$. Transforming the integral over τ by introducing $t = \tau - i\hbar/2T$, we find the real part of the conductivity:

$$\text{Re}\,\sigma_{\alpha\beta}(\omega) = \frac{e^2n(1-\rho_{eq})}{2\hbar^2TN}\sum_{\mathbf{nm}}(\mathbf{R_n}-\mathbf{R_m})_{\alpha}(\mathbf{R_n}-\mathbf{R_m})_{\beta}$$

$$\times J_{\mathbf{mn}}J_{\mathbf{nm}}e^{-\overline{S}_{\mathbf{m-n}}}\frac{\sinh(\hbar\omega/2T)}{(\hbar\omega/2T)}\int_{-\infty}^{\infty}dt e^{i\omega t}P_{\mathbf{n-m}}(t).$$

Note that in the limit $\omega = 0$ we obtain Eq. (62.26). The integral over time is very similar to that in Eq. (62.25). Therefore, considering the simple cubic lattice in the nearest-neighbor approximation, we obtain the diagonal frequency-dependent conductivity in the form

$$\text{Re}\,\sigma(\omega) = \sigma_h\frac{\sinh(\hbar\omega/2T)}{(\hbar\omega/2T)}\exp\left[-\frac{(\hbar\omega)^2}{16\mathcal{E}^2}\right],$$

where the static hopping conductivity σ_h and the characteristic energy $\mathcal{E}$ are given by Eqs. (62.31) and (62.30), respectively. At high temperatures, $T \gg \hbar\omega_{\mathbf{q}}/2$, there exists a resonant absorption at $\omega = 4E_A/\hbar$, with a Gaussian line shape.

Chapter 13

MULTI-CHANNEL KINETICS

The transitions of electrons, or other quasiparticles, between the states belonging to different branches of energy spectrum are often met in transport theory and can be described with the use of coupled kinetic equations for the distribution functions of these quasiparticles. There are, however, the cases when this approach is not valid, because the branches of energy spectrum are close to each other, and the consideration of non-diagonal (with respect to the branch index) components of the density matrix becomes essential. The formalism which adequately describes such cases is based upon matrix kinetic equations. In this chapter we study two examples of this kind: the electrons in different spin states coupled by spin-orbit interaction, and a pair of two-dimensional electron layers weakly coupled by tunneling. We also consider other important examples of multi-channel transport: the relaxation of electrons due to the interband transitions caused by electron-electron interaction (Auger processes) and the anomalies of electrical conductivity due to spin-dependent scattering of electrons by localized states of magnetic impurities (Kondo effect). The description of these processes is based upon a detailed consideration of the multi-channel scattering.

63. Spin-Flip Transitions

In most cases, the spin degree of freedom of electrons manifests itself only in Zeeman splitting of the electron energy spectrum in magnetic fields. However, one should also take into account that the spin-orbit contribution to the Hamiltonian leads to a spin-dependent interaction of electrons with the potentials caused by imperfections of the crystal lattice or with external electric fields. This interaction provides an additional channel of electron scattering due to spin-flip transitions and is responsible for the spin relaxation in the crystals which do not contain magnetic impurities. Near the extremum of the conduction band, when the mean electron energy $\bar{\varepsilon}$ is much smaller than the gap energy ε_g, electrons experience an additional, spin-dependent interaction described by the term

$$\frac{\chi\hbar}{m\varepsilon_g}\left(\hat{\boldsymbol{\sigma}}\cdot[\nabla U_{\mathbf{r}}\times\hat{\mathbf{p}}]\right), \tag{1}$$

where $\hat{\boldsymbol{\sigma}}$ is the vector of Pauli matrices, $U_{\mathbf{r}}$ is the potential energy, and m is the effective mass near the extremum. The coefficient χ is determined by the details of the band structure and equal to $1/2$ for the two-band model discussed in Appendix B (problem 13.1). This coefficient is small in the materials with weak spin-orbit interaction.

Consider the relaxation of spin caused by spin-flip transitions of electrons due to their interaction with randomly distributed impurities. The total Hamiltonian of such interaction is obtained by adding the term (1) to the potential energy $U_{\mathbf{r}} = U_{im}(\mathbf{r})$ from Eq. (7.1). Acting in the same way as described in Sec. 7, we obtain the kinetic equation (7.13) with the collision integral (7.17), where $e^{i\mathbf{q}\cdot\mathbf{r}}$ is replaced by $e^{i\mathbf{q}\cdot\mathbf{r}}\{1 + i(\chi\hbar/m\varepsilon_g)$ $\times(\hat{\boldsymbol{\sigma}}\cdot[\mathbf{q}\times\hat{\mathbf{p}}])\}$. In the momentum representation, one may introduce the distribution function $\hat{f}_{\mathbf{p}t} = \langle\mathbf{p}|\hat{\rho}_t|\mathbf{p}\rangle$, which is a 2×2 matrix with respect to the spin indices. This function satisfies the matrix kinetic equation

$$\frac{\partial\hat{f}_{\mathbf{p}t}}{\partial t} = \frac{2n_{im}}{\hbar^2 V}\sum_{\mathbf{q}}|v(\mathbf{q})|^2\int_{-\infty}^{0}d\tau e^{\lambda\tau}\cos\left[(\varepsilon_p - \varepsilon_{\mathbf{p}-\hbar\mathbf{q}})\,\tau/\hbar\right]$$

$$\times\left(\hat{\xi}_{\mathbf{q}}\hat{f}_{\mathbf{p}-\hbar\mathbf{q},t+\tau}\hat{\xi}_{\mathbf{q}}^+ - \hat{f}_{\mathbf{p}t+\tau}\hat{\xi}_{\mathbf{q}}\hat{\xi}_{\mathbf{q}}^+\right), \tag{2}$$

where the 2×2 matrix $\hat{\xi}_{\mathbf{q}}$ is defined according to

$$\hat{\xi}_{\mathbf{q}} = 1 + i\frac{\chi\hbar}{m\varepsilon_g}\left(\hat{\boldsymbol{\sigma}}\cdot[\mathbf{q}\times\mathbf{p}]\right). \tag{3}$$

In the transformations leading to Eq. (2), we have taken into account that $\hat{\xi}_{-\mathbf{q}} = \hat{\xi}_{\mathbf{q}}^+$ and that $\hat{\xi}_{\mathbf{q}}\hat{\xi}_{\mathbf{q}}^+ = 1 + (\chi\hbar/m\varepsilon_g)^2[\mathbf{q}\times\mathbf{p}]^2$ is a scalar. Here and below in this chapter, the scalar part of 2×2 matrix expressions of the kind of Eq. (3) means the contribution standing at the unit matrix $\hat{1}$. The collision integral in Eq. (2) is a generalization of the collision integral (8.4) to the case of interaction with the effective potential containing the matrix contribution (1). Because of the presence of the term $\hat{\xi}_{\mathbf{q}}\hat{f}_{\mathbf{p}-\hbar\mathbf{q},t+\tau}\hat{\xi}_{\mathbf{q}}^+$, this collision integral mixes different components of the matrix $\hat{f}_{\mathbf{p}t+\tau}$.

Let us represent the matrix distribution function as a combination of scalar and vector parts:

$$\hat{f}_{\mathbf{p}t} = f_{\mathbf{p}t} + \hat{\boldsymbol{\sigma}}\cdot\mathbf{f}_{\mathbf{p}t}, \quad f_{\mathbf{p}t} = \frac{1}{2}\mathrm{tr}_\sigma\hat{f}_{\mathbf{p}t}, \quad \mathbf{f}_{\mathbf{p}t} = \frac{1}{2}\mathrm{tr}_\sigma\hat{\boldsymbol{\sigma}}\hat{f}_{\mathbf{p}t}, \tag{4}$$

where $\text{tr}_\sigma \ldots$ defines the trace over the spin variable, $f_{\mathbf{p}t}$ is the distribution function averaged over spin, while the vector function $\mathbf{f}_{\mathbf{p}t}$ describes the distribution of the electrons whose spins are polarized along the direction of $\mathbf{f}$. For example, z-component of this function, $f_{\mathbf{p}t}^z = [\langle \sigma = +1|\hat{f}_{\mathbf{p}t}|\sigma = +1\rangle - \langle \sigma = -1|\hat{f}_{\mathbf{p}t}|\sigma = -1\rangle]/2$, is expressed through the diagonal components and equal to the half-difference of the distributions of electrons with up and down spins with respect to the spin quantization axis OZ. In the Markovian approximation, the spin-averaged distribution function $f_{\mathbf{p}t}$ is determined by Eq. (8.7), where the transition probability is multiplied by the factor $1 + (\chi/m\varepsilon_g)^2 [\mathbf{p} \times \mathbf{p}']^2$. This factor is close to unity since $\bar{\varepsilon} \ll \varepsilon_g$.

The spin distribution is governed by the following vector equation obtained from Eq. (2) with the use of Eqs. (3) and (4):

$$\frac{\partial \mathbf{f}_{\mathbf{p}t}}{\partial t} = \frac{2\pi n_{im}}{\hbar V} \sum_{\mathbf{q}} |v(\mathbf{q})|^2 \, \delta\left(\varepsilon_p - \varepsilon_{\mathbf{p}-\hbar\mathbf{q}}\right)$$

$$\times \left\{ \left[1 - \left(\frac{\chi\hbar}{m\varepsilon_g}\right)^2 [\mathbf{q} \times \mathbf{p}]^2 \right] \mathbf{f}_{\mathbf{p}-\hbar\mathbf{q}t} \right.$$

$$+ 2\left(\frac{\chi\hbar}{m\varepsilon_g}\right)^2 (\mathbf{f}_{\mathbf{p}-\hbar\mathbf{q}t} \cdot [\mathbf{q} \times \mathbf{p}]) [\mathbf{q} \times \mathbf{p}] \tag{5}$$

$$+ 2\frac{\chi\hbar}{m\varepsilon_g}[\mathbf{f}_{\mathbf{p}-\hbar\mathbf{q}t} \times [\mathbf{q} \times \mathbf{p}]] - \left[1 + \left(\frac{\chi\hbar}{m\varepsilon_g}\right)^2 [\mathbf{q} \times \mathbf{p}]^2 \right] \mathbf{f}_{\mathbf{p}t} \right\}.$$

If we consider the times which are much greater than the momentum relaxation time, the distribution $\mathbf{f}_{\mathbf{p}t}$ becomes isotropic with respect to the momentum $\mathbf{p}$. Replacing $\mathbf{f}_{\mathbf{p}t}$ by $\mathbf{f}_{\varepsilon t}$, where $\varepsilon = \varepsilon_p$, we average the kinetic equation (5) over the angles of $\mathbf{p}$ and express the collision integral through the spin relaxation time:

$$\frac{\partial \mathbf{f}_{\varepsilon t}}{\partial t} = -\frac{\mathbf{f}_{\varepsilon t}}{\tau_s(\varepsilon)}, \tag{6}$$

$$\tau_s^{-1}(\varepsilon) = \frac{4\pi n_{im}}{\hbar V} \sum_{\mathbf{q}} |v(\mathbf{q})|^2 \left(\frac{\chi\hbar}{m\varepsilon_g}\right)^2 [\mathbf{q} \times \mathbf{p}]_\perp^2 \, \delta\left(\varepsilon_p - \varepsilon_{\mathbf{p}-\hbar\mathbf{q}}\right),$$

where $[\mathbf{q} \times \mathbf{p}]_\perp$ is the component of the vector $[\mathbf{q} \times \mathbf{p}]$ perpendicular to $\mathbf{f}_{\varepsilon t}$. In the case of scattering by point defects, when $v(\mathbf{q}) \simeq v(0)$, the calculation of the sum over $\mathbf{q}$ in Eq. (6) is reduced to the angular averaging of $[\mathbf{q} \times \mathbf{p}]_\perp^2$. As a result,

$$\tau_s^{-1}(\varepsilon) = \frac{32}{9} \left(\frac{\chi\varepsilon}{\varepsilon_g}\right)^2 \nu_\varepsilon, \tag{7}$$

where ν_ε is given by Eq. (8.21). Therefore, the spin relaxation rate is equal to the momentum relaxation rate multiplied by a small factor characterizing the efficiency of spin flip.

Another mechanism of spin relaxation takes place if the potential energy $U_{\mathbf{r}}$ is quasi-classically smooth. The contribution (1) containing the gradient of this potential energy can be viewed as a Pauli interaction of electrons with an effective magnetic field whose direction is determined by the electron momentum $\mathbf{p}$. The axis of precession of electron spin in this field changes frequently because of scattering of electrons by impurities, and this leads to spin relaxation. To describe the spin precession in the presence of random scattering, let us introduce the Wigner distribution function $\hat{f}_{\mathbf{rp}t} = \int d\Delta \mathbf{r} e^{-i\mathbf{p}\cdot\Delta\mathbf{r}/\hbar} \langle \mathbf{r}+\Delta\mathbf{r}/2|\hat{\rho}_t|\mathbf{r}-\Delta\mathbf{r}/2\rangle$ and generalize the formalism of Sec. 9 to the case of matrix distribution functions. One can derive the following quasi-classical matrix kinetic equation for this function (problem 13.2):

$$\frac{\partial \hat{f}_{\mathbf{rp}t}}{\partial t} + \frac{1}{2}\left[\mathbf{v_p} + \hbar\frac{\partial\hat{\boldsymbol{\sigma}}\cdot\boldsymbol{\Omega}_{\mathbf{rp}}}{\partial\mathbf{p}}, \frac{\partial\hat{f}_{\mathbf{rp}t}}{\partial\mathbf{r}}\right]_+ + i\left[(\hat{\boldsymbol{\sigma}}\cdot\boldsymbol{\Omega}_{\mathbf{rp}}), \hat{f}_{\mathbf{rp}t}\right]$$

$$-\frac{1}{2}\left[\frac{\partial U_{\mathbf{r}}}{\partial\mathbf{r}} + \hbar\frac{\partial\hat{\boldsymbol{\sigma}}\cdot\boldsymbol{\Omega}_{\mathbf{rp}}}{\partial\mathbf{r}}, \frac{\partial\hat{f}_{\mathbf{rp}t}}{\partial\mathbf{p}}\right]_+ = \hat{J}(\hat{f}|\mathbf{rp}t), \qquad (8)$$

where $\mathbf{v_p} = \mathbf{p}/m$ and the brackets $[\ldots,\ldots]_+$ define the anticommutators. The frequency characterizing the spin precession, $\boldsymbol{\Omega}_{\mathbf{rp}} = (\chi/m\varepsilon_g) \times [\nabla U_{\mathbf{r}} \times \mathbf{p}]$, is obtained from Eq. (1). It is assumed that the vectors standing in the anticommutators in Eq. (8) form the scalar products. The spin-dependent contributions into the collision integral can be neglected so that this integral is written in the form (9.33), where the scalar $f_{\mathbf{rp}t}$ is replaced by the matrix $\hat{f}_{\mathbf{rp}t}$.

Let us represent $\hat{f}_{\mathbf{rp}t}$ as a sum of symmetric and antisymmetric contributions, according to $\hat{f}_{\mathbf{rp}t} = \hat{f}_{\mathbf{r}\varepsilon t} + \widehat{\Delta f}_{\mathbf{rp}t}$, where $\varepsilon = \varepsilon_p$. We search for the symmetric part $\hat{f}_{\mathbf{r}\varepsilon t}$ in the form $\hat{f}_t(\varepsilon + U_{\mathbf{r}})$ so that its coordinate dependence is purely parametric. This parametric representation is valid in the fields smooth on the scale of the mean free path length and in the absence of gradients of electrochemical potential and temperature, when there are no drift and diffusion currents; see Sec. 36. Then we neglect weak temporal and spatial variations of $\widehat{\Delta f}_{\mathbf{rp}t}$ on the scale of the momentum relaxation time and mean free path length determined by the collision integral. Considering the antisymmetric part of the kinetic equation (8) under the assumed condition $\hat{f}_{\mathbf{r}\varepsilon t} \simeq \hat{f}_t(\varepsilon + U_{\mathbf{r}})$, we obtain $\widehat{\Delta f}_{\mathbf{rp}t} \simeq [(\hat{\boldsymbol{\sigma}}\cdot\boldsymbol{\Omega}_{\mathbf{rp}}), \hat{f}_{\mathbf{r}\varepsilon t}]/i\nu_\varepsilon$. Using this relation in the symmetric (averaged over the angle of $\mathbf{p}$) part of Eq. (8), we write the latter as

$$\frac{\partial \hat{f}_{\mathbf{r}\varepsilon t}}{\partial t} + \int \frac{d\widetilde{\Omega}_{\mathbf{p}}}{4\pi\nu_\varepsilon} \left[(\hat{\boldsymbol{\sigma}} \cdot \boldsymbol{\Omega}_{\mathbf{rp}}), \left[(\hat{\boldsymbol{\sigma}} \cdot \boldsymbol{\Omega}_{\mathbf{rp}}), \hat{f}_{\mathbf{r}\varepsilon t} \right] \right] = 0, \qquad (9)$$

where $d\widetilde{\Omega}_{\mathbf{p}}$ is the differential of the solid angle of the vector $\mathbf{p}$. All the terms associated with drift and diffusion have disappeared from Eq. (9) after the angular averaging.

The second term on the left-hand side of Eq. (9) describes the spin relaxation of electrons. To prove this, we write the matrix distribution function as a combination of scalar and vector parts, $\hat{f}_{\mathbf{r}\varepsilon t} = f_{\mathbf{r}\varepsilon t} + \hat{\boldsymbol{\sigma}} \cdot \mathbf{f}_{\mathbf{r}\varepsilon t}$, as in Eq. (4). Then we make use of the identity $[(\hat{\boldsymbol{\sigma}} \cdot \mathbf{A}), (\hat{\boldsymbol{\sigma}} \cdot \mathbf{B})] = 2i\hat{\boldsymbol{\sigma}} \cdot [\mathbf{A} \times \mathbf{B}]$ and transform the double commutator in Eq. (9) to the form $-4[(\hat{\boldsymbol{\sigma}} \cdot \boldsymbol{\Omega}_{\mathbf{rp}})(\mathbf{f}_{\mathbf{r}\varepsilon t} \cdot \boldsymbol{\Omega}_{\mathbf{rp}}) - (\hat{\boldsymbol{\sigma}} \cdot \mathbf{f}_{\mathbf{r}\varepsilon t})\boldsymbol{\Omega}_{\mathbf{rp}}^2]$. The angular averaging of $\Omega_{\mathbf{rp}}^\alpha \Omega_{\mathbf{rp}}^\beta$ gives us $(\chi p/m\varepsilon_g)^2[(\nabla U_{\mathbf{r}})^2\delta_{\alpha\beta} - \nabla_\alpha U_{\mathbf{r}}\nabla_\beta U_{\mathbf{r}}]/3$, and we rewrite the second term on the left-hand side of Eq. (9) as

$$\frac{4}{3}\nu_\varepsilon^{-1} \left(\frac{\chi p}{m\varepsilon_g} \right)^2 \sum_{\alpha\beta} \hat{\sigma}_\alpha \mathrm{f}_{\mathbf{r}\varepsilon t}^\beta [(\nabla U_{\mathbf{r}})^2\delta_{\alpha\beta} + \nabla_\alpha U_{\mathbf{r}}\nabla_\beta U_{\mathbf{r}}]. \qquad (10)$$

Finally, let us average Eq. (9) over realizations of the random potential $U_{\mathbf{r}}$ under the condition $\langle\langle U_{\mathbf{r}}\rangle\rangle = 0$. As a result, the averaged distribution function $\mathbf{f}_{\varepsilon t} = \langle\langle \mathbf{f}_{\mathbf{r}\varepsilon t}\rangle\rangle$ satisfies the following equation:

$$\frac{\partial \mathbf{f}_{\varepsilon t}}{\partial t} = -\frac{\mathbf{f}_{\varepsilon t}}{\tau_{sp}(\varepsilon)}, \qquad \tau_{sp}^{-1}(\varepsilon) = \frac{32}{9}\frac{\chi^2\varepsilon\psi}{m\nu_\varepsilon\varepsilon_g^2}, \qquad (11)$$

where $\psi = \langle\langle(\nabla U_{\mathbf{r}})^2\rangle\rangle$ is the mean square of the potential gradient; see Eq. (18.18). By comparing the times τ_s and τ_{sp} given by Eqs. (7) and (11), respectively, we obtain $\tau_s/\tau_{sp} \sim l_\varepsilon^2\psi/\varepsilon^2$, where $l_\varepsilon = \sqrt{2\varepsilon/m}/\nu_\varepsilon$ is the mean free path length of the electrons with energy ε. Estimating ψ as $\langle\langle U_{\mathbf{r}}^2\rangle\rangle/l_c^2$, where l_c is the correlation length (characteristic spatial scale) of the potential, we find that in the hydrodynamic approximation, when $l_\varepsilon \ll l_c$, the mechanism of relaxation caused by the spin precession is much less important than the spin-flip scattering. However, a similar mechanism of relaxation becomes significant in the crystals without center of inversion (non-centrosymmetric crystals), where, instead of $\boldsymbol{\Omega}_{\mathbf{rp}}$ introduced in Eq. (8), one should use a coordinate-independent frequency $\boldsymbol{\Omega}_{\mathbf{p}}$ proportional to the third power of momentum (problem 13.3).

The spin-polarized electron distribution can appear in various nonequilibrium processes which are sensitive to spin orientation. Let us discuss the interband photoexcitation of electrons by a circular-polarized electromagnetic wave. This process is described by the general expression for the photogeneration rate $\widehat{G}_t$, see Eq. (54.1), where the electric field of the wave is given by $\mathbf{E}_t^{(\pm)} = \mathbf{E}_\pm w_t e^{-i\omega t} + c.c.$ and the field strength for the wave propagating along OZ is chosen as $\mathbf{E}_\pm = (E/\sqrt{2}, \pm iE/\sqrt{2}, 0)$. The envelope form-factor w_t should be set to unity in the case of a stationary photoexcitation. Calculating the matrix elements of $\widehat{G}_t$ in the basis of conduction-band states, one gets a 2×2 matrix of generation rate whose elements $\langle c\sigma\mathbf{p}|\widehat{G}_t|c\sigma'\mathbf{p}\rangle$ are proportional to the spin-dependent factors

$$\Psi_{\sigma\sigma'}^{(\pm)} = 2 \frac{\sum_{\sigma_v} (\mathbf{E}_\pm \cdot \mathbf{v}_{c\sigma,v\sigma_v})(\mathbf{v}_{v\sigma_v,c\sigma'} \cdot \mathbf{E}_\pm^*)}{\sum_{\sigma\sigma_v} (\mathbf{E}_\pm \cdot \mathbf{v}_{c\sigma,v\sigma_v})(\mathbf{v}_{v\sigma_v,c\sigma} \cdot \mathbf{E}_\pm^*)}. \tag{12}$$

For the excitation near the band edge, the matrix elements of the velocity operator are obtained from Eq. (B.23) in the limit $\eta_p \simeq 1$, with the result $\mathbf{v}_{c\sigma,v\sigma_v} \simeq s\langle\sigma|\hat{\boldsymbol{\sigma}}|\sigma_v\rangle$. Using the identity $[\mathbf{E}_\pm \times \mathbf{E}_\pm^*]/|\mathbf{E}_\pm|^2 = (0,0,\mp i)$, we obtain the polarization factor (12) in the form of a 2×2 matrix:

$$\widehat{\Psi}^{(\pm)} = 1 \pm \hat{\sigma}_z . \tag{13}$$

The circular-polarized waves $\mathbf{E}_t^{(+)}$ and $\mathbf{E}_t^{(-)}$ excite only spin-up and spin-down electrons, respectively. If the complex field strength $\mathbf{E}_\pm$ is chosen in a different way, for example, $\mathbf{E}_\pm = (E_1/\sqrt{2}, \pm iE_2/\sqrt{2}, 0)$ (elliptic polarization if $E_1 \neq E_2$), the spin polarization of the photoexcited electrons is not complete. The matrix of generation rate, $\langle c\mathbf{p}|\widehat{G}_t|c\mathbf{p}\rangle$, is represented as $G_{\mathbf{p}t}^{(c)}\widehat{\Psi}^{(\pm)}$, where $G_{\mathbf{p}t}^{(c)}$ is the spin-averaged generation rate given by Eq. (54.3). In the problems of relaxation of photoexcited spin-polarized electrons, this matrix should be added to the right-hand side of the corresponding 2×2 matrix kinetic equation, for example, to the right-hand side of Eq. (2). Calculating the matrix trace of the equation obtained in this way (note that $(1/2)\mathrm{tr}_\sigma \widehat{\Psi}^{(\pm)} = 1$), one can check that the distribution function $f_{\mathbf{p}t}$ satisfies Eq. (54.4) in the absence of scattering.

The intraband spin-flip optical transitions also become allowed if the terms of the order of $\bar{\varepsilon}/\varepsilon_g$ are taken into account. Using the general expression for the coefficient of absorption (see Eqs. (10.23) and (48.4)),

$$\alpha_\omega = \frac{(2\pi e)^2}{c\sqrt{\epsilon}\omega V} \sum_{\delta\delta'} |\langle\delta|\hat{v}_\alpha|\delta'\rangle|^2 \delta(\varepsilon_\delta - \varepsilon_{\delta'} + \hbar\omega)(f_\delta - f_{\delta'}), \tag{14}$$

where $f_\delta = f(\varepsilon_\delta)$ is the distribution function of electrons in the state δ, we consider the transitions between the Landau levels in the ultraquantum limit, when both the cyclotron energy $\hbar\omega_c$ and the Zeeman splitting energy $|g\mu_B H|$ exceed the mean energy of electrons, $\bar{\varepsilon}$. Below we assume that the magnetic field is directed along OZ and use the gauge $\mathbf{A} = (0, Hx, 0)$. Let us calculate the absorption (14) by employing the basis of the conduction-band states of the two-band model. Since only the lowest state, $|N = 0, \sigma = -1, p_y p_z\rangle$, is occupied by electrons, Eq. (14) is reduced to

$$\alpha_\omega = \frac{(2\pi e)^2}{c\sqrt{\epsilon}\omega V} \sum_{N'\sigma' p_y p_z} f_{0,-1p_z} \left| \langle c0, -1p_y p_z | \hat{v}_\alpha | cN', \sigma' p_y p_z \rangle \right|^2$$

$$\times \delta(E_{0,-1p_z} - E_{N',\sigma' p_z} + \hbar\omega), \tag{15}$$

where the eigenstates are presented in problem 10.15 and the energy spectrum $E_{N,\sigma p_z}$ is given by Eq. (52.8). The electron transitions between these levels are caused by the electromagnetic waves polarized perpendicular to the magnetic field, and we should consider either $\alpha = x$ or $\alpha = y$ in Eq. (15). The corresponding matrix elements presented below are obtained in the approximation $\bar{\varepsilon}/\varepsilon_g \ll \hbar\omega_c/\varepsilon_g \ll 1$:

$$\left| \langle c0, -1p_y p_z | \hat{v}_x | cN, -1p_y p_z \rangle \right|^2 = \delta_{N1} \frac{\hbar\omega_c}{2m},$$

$$\left| \langle c0, -1p_y p_z | \hat{v}_x | cN, +1p_y p_z \rangle \right|^2 = \delta_{N0} \left(\frac{p_z \hbar\omega_c}{2m\varepsilon_g} \right)^2. \tag{16}$$

We also present the expressions for the energy spectra obtained from Eq. (52.8) in this approximation:

$$E_{0,+1p_z} = E_{1,-1p_z} = \frac{\varepsilon_g}{2} \sqrt{1 + \left(\frac{2sp_z}{\varepsilon_g} \right)^2 + 4\frac{\hbar\omega_c}{\varepsilon_g}} \simeq \frac{\varepsilon_g}{2} + \hbar\omega_c + \frac{p_z^2}{2m_H},$$

$$E_{0,-1p_z} = \frac{\varepsilon_g}{2} \sqrt{1 + \left(\frac{2sp_z}{\varepsilon_g} \right)^2} \simeq \frac{\varepsilon_g}{2} + \frac{p_z^2}{2m}, \tag{17}$$

where $m_H \simeq m(1 + 2\hbar\omega_c/\varepsilon_g)$.

According to Eq. (16), the high-frequency electric field perpendicular to $\mathbf{H}$ causes the transitions between the states with different N and the same σ (cyclotron resonance, CR), as well as the transitions between the states with the same N and different σ (combined resonance, cr). The absorption coefficients corresponding to these two types of optical transitions are

$$\left| \begin{array}{c} \alpha_\omega^{(CR)} \\ \alpha_\omega^{(cr)} \end{array} \right| = \frac{(2\pi e)^2}{c\sqrt{\epsilon}\omega V} \sum_{p_y p_z} f_{0,-1p_z}$$

$$\times \delta_\gamma \left(\hbar\omega - \hbar\omega_c + \frac{p_z^2}{2m} \frac{2\hbar\omega_c}{\varepsilon_g} \right) \left| \begin{array}{c} \hbar\omega_c/2m \\ (p_z/2m)^2 (\hbar\omega_c/\varepsilon_g)^2 \end{array} \right|, \tag{18}$$

where $\delta_\gamma(E) = \pi^{-1}\gamma/(E^2 + \gamma^2)$. By applying this function, we take into account the collision-induced broadening of the energy levels, with the broadening energy γ, in addition to the broadening caused by the difference in the energy dispersion of the states due to non-parabolicity (note that $m^{-1} - m_H^{-1} \simeq 2\hbar\omega_c/m\varepsilon_g$ is already used in Eq. (18)). For the two-band model, the energies of Zeeman splitting and Landau-level splitting coincide, though it is not the case for the majority of materials, where the CR and cr peaks in the absorption spectrum appear at different energies. The calculation of the shape of the peaks based upon Eq. (18) can be done with the use of $V^{-1}\sum_{p_y p_z} \ldots = [|e|H/(2\pi\hbar)^2 c] \int_{-\infty}^{\infty} dp_z \ldots$. In the case of non-degenerate electrons described by the Maxwell distribution $f_{0,-1p_z} = (n/\mathcal{N}) \exp(-p_z^2/2mT)$, where n is the electron density and $\mathcal{N}$ is the normalization constant, the result is written in the following way:

$$\left| \begin{array}{c} \alpha_\omega^{(CR)} \\ \alpha_\omega^{(cr)} \end{array} \right| = \frac{\pi^{3/2} e^2 n \varepsilon_g}{c\sqrt{\epsilon}\omega mT} \left| \begin{array}{c} \Phi_\lambda^{(CR)}(\Delta) \\ (\hbar\omega_c T/\varepsilon_g^2)\ \Phi_\lambda^{(cr)}(\Delta) \end{array} \right|, \tag{19}$$

where

$$\left| \begin{array}{c} \Phi_\lambda^{(CR)}(\Delta) \\ \Phi_\lambda^{(cr)}(\Delta) \end{array} \right| = \int_0^\infty dx\, e^{-x} \delta_\lambda(x + \Delta) \left| \begin{array}{c} x^{-1/2} \\ x^{1/2} \end{array} \right|. \tag{20}$$

In these equations, $\Delta = (\varepsilon_g/2T)(\omega/\omega_c - 1)$ and $\lambda = (\varepsilon_g/2T)(\gamma/\hbar\omega_c)$ are the dimensionless parameters describing the shift from the resonance and the broadening, respectively. If $\lambda \ll 1$ (weak broadening), the function $\delta_\lambda(x + \Delta)$ can be replaced by the true δ-function. Both $\Phi_\lambda^{(CR)}(\Delta)$ and $\Phi_\lambda^{(cr)}(\Delta)$ in this case show strongly asymmetric peaks. The absorption is absent at $\Delta > 0$, while at $\Delta < 0$ the absorption is proportional to $|\Delta|^{\mp 1/2} e^{-|\Delta|}$ (the upper and lower signs stand for the CR and cr absorption, respectively). In the opposite case of $\lambda \gg 1$, the collision-induced broadening is more important than the non-parabolicity effect. As a result, both $\Phi_\lambda^{(CR)}(\Delta)$ and $\Phi_\lambda^{(cr)}(\Delta)$ are proportional to $\delta_\lambda(\Delta) = \pi^{-1}\lambda/(\lambda^2 + \Delta^2)$. The absorption in this case is described by symmetric Lorentzian peaks. In any case, since we assume $\varepsilon_g \gg T$ and $\gamma \ll \hbar\omega_c$, the absorption takes place in the vicinity of the resonance, where $|\omega - \omega_c| \ll \omega_c$.

Apart from the combined resonance caused by the presence of the spin-orbit interaction, there exists another mechanism of spin-flip optical transitions. This mechanism, called the electron spin resonance, is caused by the Pauli contribution to the Hamiltonian due to the magnetic field of the electromagnetic wave. The absorption coefficient associated with this mechanism does not contain the factor ε_g^{-2}. On the other hand, this absorption coefficient is proportional to c^{-2} and often can be neglected in comparison to the coefficient of cr absorption (problem 13.4).

In low-dimensional electron systems the electron spectrum can become split even in the absence of magnetic fields, owing to the spin-orbit interaction. The existence of these spin-split states and the interplay of the spin-orbit splitting with Zeeman splitting considerably influence the transport phenomena. Consider a two-dimensional electron gas confined in the XOY plane by an asymmetric potential $U(z)$. The asymmetry of the potential leads to the presence of a spin-orbit term

$$\hat{\mathcal{H}}_{so} = \mathbf{v}_s \cdot [\hat{\boldsymbol{\sigma}} \times \hat{\mathbf{p}}] \qquad (21)$$

in the Hamiltonian of the 2D electrons. The velocity $\mathbf{v}_s$ is directed perpendicular to the plane and determined by the shape of the confinement potential and by the wave functions of electrons (problem 13.5). To consider the 2D electron gas in the presence of external fields, one should use the Hamiltonian

$$\widehat{H} = \frac{\hat{\boldsymbol{\pi}}_{\mathbf{x}t}^2}{2m} + \mathbf{v}_s \cdot [\hat{\boldsymbol{\sigma}} \times \hat{\boldsymbol{\pi}}_{\mathbf{x}t}] - \frac{\hbar}{2}\boldsymbol{\Omega}_H \cdot \hat{\boldsymbol{\sigma}}, \qquad (22)$$

where $\hat{\boldsymbol{\pi}}_{\mathbf{x}t} = \hat{\mathbf{p}} - (e/c)\mathbf{A}_{\mathbf{x}t}$ is the operator of kinematic momentum and $\mathbf{A}_{\mathbf{x}t}$ is the vector potential describing external electric and magnetic fields (the gauge is chosen in such a way that $\mathbf{A}$ in Eq. (22) is a 2D vector depending only on the in-plane coordinate $\mathbf{x}$; see problem 13.6). The last term in Eq. (22) describes Zeeman splitting, and the corresponding frequency vector $\boldsymbol{\Omega}_H$ is equal to $g\mu_B\mathbf{H}/\hbar$, where μ_B is the Bohr magneton and g is the effective g-factor (see Appendix B and Chapter 10). In the absence of external fields, one can use the basis of plane waves, $|\mathbf{p}\rangle$, to represent the Hamiltonian (22) in the form of an algebraic 2×2 matrix $\hat{h}_{\mathbf{p}} = \langle\mathbf{p}|\widehat{H}|\mathbf{p}\rangle$. By diagonalizing this matrix with the aid of the unitary transformation $\hat{\mathcal{U}}_{\mathbf{p}}\hat{h}_{\mathbf{p}}\hat{\mathcal{U}}_{\mathbf{p}}^+$, where $\hat{\mathcal{U}}_{\mathbf{p}} = [1 + i(\hat{\boldsymbol{\sigma}} \cdot \mathbf{p})/p]/\sqrt{2}$, we obtain the spin-split electron spectrum $p^2/2m \pm v_s p$ consisting of two isotropic branches.

Let us consider the spin-flip transitions between the branches described above in the presence of a high-frequency electric field $\mathbf{E}e^{-i\omega t} + c.c.$ The Hamiltonian of the perturbation caused by this field is written as $(ie/\omega)(\mathbf{E}\cdot\hat{\mathbf{v}}_{\mathbf{p}})e^{-i\omega t}+H.c.$, where $\hat{\mathbf{v}}_{\mathbf{p}} = \partial\hat{h}_{\mathbf{p}}/\partial\mathbf{p} = \mathbf{p}/m+[\mathbf{v}_s \times \hat{\boldsymbol{\sigma}}]$ is the matrix of the group velocity of 2D electrons. The high-frequency component of the density matrix of electrons, $\widehat{\delta f}_{\mathbf{p}}e^{-i\omega t} + H.c.$, is governed by the linearized equation

$$(\hbar\omega + i\gamma)\widehat{\delta f}_{\mathbf{p}} - \left[\hat{h}_{\mathbf{p}}, \widehat{\delta f}_{\mathbf{p}}\right] - i\frac{e}{\omega}\left[(\mathbf{E} \cdot \hat{\mathbf{v}}_{\mathbf{p}}), \hat{f}_{\mathbf{p}}\right] = 0, \qquad (23)$$

where a weak broadening energy γ comes from the collision-integral contribution. The equilibrium density matrix $\hat{f}_{\mathbf{p}}$ is given by

$$\hat{f}_{\mathbf{p}} = f_p^{(+)} + \frac{\hat{\boldsymbol{\sigma}} \cdot [\mathbf{p} \times \mathbf{v}_s]}{v_s p} f_p^{(-)}, \tag{24}$$

where $f_p^{(\pm)} = [\theta(\varepsilon_F - \varepsilon_p - v_s p) \pm \theta(\varepsilon_F - \varepsilon_p + v_s p)]/2$ for the case of zero temperature (here $\varepsilon_p = p^2/2m$). Since $\mathrm{tr}_\sigma \widehat{\delta f}_{\mathbf{p}} = 0$ (as it follows from Eq. (23)), one can search for $\widehat{\delta f}_{\mathbf{p}}$ in the form $\widehat{\delta f}_{\mathbf{p}} = \hat{\boldsymbol{\sigma}} \cdot \delta \mathbf{f}_{\mathbf{p}}$. The real part of the high-frequency Fourier component of the current density is expressed as

$$\mathrm{Re}\, \mathbf{I}_\omega = \mathrm{Re} \frac{e}{L^2} \sum_{\mathbf{p}} \mathrm{tr}_\sigma \hat{\mathbf{v}}_{\mathbf{p}} \cdot \widehat{\delta f}_{\mathbf{p}} = \mathrm{Re} \frac{2e}{L^2} \sum_{\mathbf{p}} [\mathbf{v}_s \times \delta \mathbf{f}_{\mathbf{p}}]. \tag{25}$$

After multiplying Eq. (23) by $\hat{\boldsymbol{\sigma}}$, we take the trace of the equation obtained and find an equation for $\delta \mathbf{f}_{\mathbf{p}}$. Since $\mathbf{v}_s$ is perpendicular to the 2D plane, we write this equation as a system of two equations connecting the in-plane vector component $\delta \mathbf{f}_{\mathbf{p}}^{(\|)} = (\delta \mathrm{f}_{\mathbf{p}}^x, \delta \mathrm{f}_{\mathbf{p}}^y)$ with the scalar component $\delta \mathrm{f}_{\mathbf{p}}^{(\perp)} \equiv \delta \mathrm{f}_{\mathbf{p}}^z$ in the direction of $\mathbf{v}_s$:

$$(\hbar\omega + i\gamma)\, \delta \mathbf{f}_{\mathbf{p}}^{(\|)} = -2i\mathbf{p} v_s \delta \mathrm{f}_{\mathbf{p}}^{(\perp)},$$

$$(\hbar\omega + i\gamma)\, \delta \mathrm{f}_{\mathbf{p}}^{(\perp)} = 2i v_s (\mathbf{p} \cdot \delta \mathbf{f}_{\mathbf{p}}^{(\|)}) - \frac{2e}{\omega p} f_p^{(-)}(\mathbf{E} \cdot [\mathbf{p} \times \mathbf{v}_s]). \tag{26}$$

The current (25) is expressed through $\delta \mathbf{f}_{\mathbf{p}}^{(\|)}$ and appears to be proportional to $\mathbf{E}$ after averaging the vector product $[\mathbf{v}_s \times \delta \mathbf{f}_{\mathbf{p}}]$ over the angle of $\mathbf{p}$. Therefore, one may find the real part of the conductivity introduced as $\mathrm{Re}\mathbf{I}_\omega = \mathrm{Re}\sigma_\omega \mathbf{E}$:

$$\mathrm{Re}\sigma_\omega = -\mathrm{Re} \frac{4e^2}{\omega L^2} \sum_{\mathbf{p}} \frac{v_s}{p} \frac{i(v_s p)^2 f_p^{(-)}}{(\hbar\omega + i\gamma)^2 - (2v_s p)^2}. \tag{27}$$

Substituting the above-given expression for $f_p^{(-)}$ into Eq. (27), we find the relative absorption coefficient defined by Eq. (29.24) in the form

$$\xi_\omega = \frac{e^2}{\hbar c \sqrt{\epsilon}} \frac{4v_s^3}{\hbar\omega} \mathrm{Re} \int_{p_0 - mv_s}^{p_0 + mv_s} dp \frac{ip^2}{(\hbar\omega + i\gamma)^2 - (2v_s p)^2}, \tag{28}$$

where $p_0 = \sqrt{2m\varepsilon_F + (mv_s)^2} = \hbar\sqrt{2\pi n_{2D}}$ is the effective Fermi momentum in the presence of the spin splitting of electron spectrum.

To calculate the integral in Eq. (28), we assume that $mv_s \ll p_0$ and approximately replace $dp\, p^2$ by $(p_0/2)dp^2$, with the result

$$\xi_\omega = \frac{e^2}{\hbar c\sqrt{\epsilon}} \frac{v_s p_0}{2\hbar\omega} \arctan \frac{(2v_s p)^2 - (\hbar\omega)^2 + \gamma^2}{2\hbar\omega\gamma} \bigg|_{p=p_0-mv_s}^{p=p_0+mv_s}. \qquad (29)$$

This equation describes the frequency dependence of the absorption which takes place in the vicinity of $\hbar\omega = 2v_s p_0$. We remind that the broadening energy γ is related to the relaxation rate. If $\gamma \ll 2mv_s^2$ (weak relaxation), the function $\arctan(\mathcal{E}/\gamma)$ in Eq. (29) can be approximated by a step function of the energy $\mathcal{E}$, as $-\pi/2 + \pi\theta(\mathcal{E})$. As a result, the absorption ξ_ω is equal to $(\pi e^2/2\hbar c\sqrt{\epsilon})(v_s p_0/\hbar\omega)$ in the interval $|\hbar\omega - 2v_s p_0| < 2mv_s^2$ (in these conditions $\mathrm{Re}\,\sigma_\omega \simeq e^2/16\hbar$ is of the order of the fundamental conductance quantum) and goes to zero elsewhere. Therefore, the absorption line shape is given by a narrow rectangle. In the opposite case, when $\gamma \gg 2mv_s^2$, one can expand the expression (29) in series of mv_s/p_0 so that the derivative of the arctangent appears there. In these conditions, assuming also that $\gamma \ll \hbar\omega$, we write the result in the form of a Lorentzian peak

$$\xi_\omega = \frac{e^2}{\hbar c\sqrt{\epsilon}} \frac{mv_s^2}{\hbar\omega} \frac{2v_s p_0 \gamma}{(\hbar\omega - 2v_s p_0) + \gamma^2}. \qquad (30)$$

In the exact resonance, the relative absorption described by this expression is equal to $(e^2/\hbar c\sqrt{\epsilon})(mv_s^2/\gamma)$. It is considerably smaller than the fundamental constant $e^2/\hbar c \simeq 1/137$. For this reason, the absorption caused by the transitions between spin-split branches is difficult to observe. Nevertheless, such transitions can be studied experimentally owing to their influence on the plasmon spectrum of 2D electron gas.

64. Spin Hydrodynamics

A simple description of the spin degree of freedom determining the magnetic moment (and other spin-dependent quantities) of electron systems can be carried out in the case of smooth spatial and slow temporal variations of the parameters of the systems. To do this, it is necessary to generalize the hydrodynamic approach developed in Secs. 11 and 36 by introducing additional equations for the spin density $\mathbf{s}_{\mathbf{r}t}$ and related spin-dependent currents. Owing to the weakness of the coupling between the spin and kinematic degrees of freedom (for this reason, the spin relaxation time often appears to be much greater than the momentum relaxation time), this approach proves to be satisfactory for describing various spin-dependent transport phenomena. In this section we consider the hydrodynamical equations for 3D and 2D electrons describing the evolution of spin density and spin flow density.

The quasi-classical kinetic equation derived in Sec. 9 can be generalized by including the Pauli term $-\hbar\mathbf{\Omega}_H \cdot \hat{\boldsymbol{\sigma}}/2$, where $\mathbf{\Omega}_H = g\mu_B \mathbf{H}/\hbar$, see Eq. (63.22), into the Hamiltonian. This term describes the interaction of electron spins with the external magnetic field $\mathbf{H}$. Next, the density matrix in the coordinate representation, $\rho_t(\mathbf{r}_1, \mathbf{r}_2)$, should be replaced by a 2×2 matrix. As a result of the Wigner transformation (9.6) applied to this matrix, we obtain a quasi-classical matrix kinetic equation

$$\left(\frac{\partial}{\partial t} + \mathbf{v_p} \cdot \frac{\partial}{\partial \mathbf{r}} + \mathbf{F_{rp}}_t \cdot \frac{\partial}{\partial \mathbf{p}}\right)\hat{f}_{\mathbf{rp}t} - \frac{i}{2}[\mathbf{\Omega}_H \cdot \hat{\boldsymbol{\sigma}}, \hat{f}_{\mathbf{rp}t}] = \widehat{J}(\hat{f}|\mathbf{rp}t), \quad (1)$$

where $\mathbf{v_p} = \mathbf{p}/m$ and $\mathbf{F_{rp}}_t$ is the Lorentz force (9.23). The spin precession in the magnetic field enters Eq. (1) through the commutator, as in Eq. (63.8). The collision integral in Eq. (1) is given by Eq. (8.7) or Eq. (34.27) without spin-flip scattering. If there is a need to include such scattering, the collision integral can be taken from Eq. (63.2). Equation (1) can be used in the case of coordinate-dependent magnetic field if this dependence is smooth enough to neglect the terms proportional to $\partial(\hat{\boldsymbol{\sigma}} \cdot \mathbf{\Omega}_H)/\partial \mathbf{r}$. A transition to the hydrodynamic description is done by introducing the 2×2 matrices of electron density and flow. Summing both sides of Eq. (1) over the momentum $\mathbf{p}$, we obtain

$$\frac{\partial \hat{n}_{\mathbf{r}t}}{\partial t} + \mathrm{div}\,\widehat{\mathbf{i}}_{\mathbf{r}t} - \frac{i}{2}[(\mathbf{\Omega}_H \cdot \hat{\boldsymbol{\sigma}}), \hat{n}_{\mathbf{r}t}] = \frac{1}{V}\sum_{\mathbf{p}} \widehat{J}(\hat{f}|\mathbf{rp}t), \quad (2)$$

$$\hat{n}_{\mathbf{r}t} = \frac{1}{V}\sum_{\mathbf{p}} \hat{f}_{\mathbf{rp}t}, \quad \widehat{\mathbf{i}}_{\mathbf{r}t} = \frac{1}{V}\sum_{\mathbf{p}} \mathbf{v_p}\hat{f}_{\mathbf{rp}t}.$$

The right-hand side of this balance equation is not equal to zero if the spin-flip scattering is taken into account. However, $\mathrm{tr}_\sigma \sum_{\mathbf{p}} \widehat{J}(\hat{f}|\mathbf{rp}t) = 0$ because of particle conservation. Calculating the trace of both sides of Eq. (2), one should notice that $\mathrm{tr}_\sigma \hat{n}_{\mathbf{r}t} = n_{\mathbf{r}t}$ and $\mathrm{tr}_\sigma \widehat{\mathbf{i}}_{\mathbf{r}t} = \mathbf{i}_{\mathbf{r}t}$ are the electron density and electron flow density introduced by Eq. (11.2). As a result, we obtain the continuity equation (11.5).

Let us represent the matrices $\hat{n}$ and $\widehat{\mathbf{i}}$ in the form

$$\hat{n}_{\mathbf{r}t} = \frac{1}{2}[n_{\mathbf{r}t} + \mathbf{s}_{\mathbf{r}t} \cdot \hat{\boldsymbol{\sigma}}], \quad \hat{i}^\beta_{\mathbf{r}t} = \frac{1}{2}[i^\beta_{\mathbf{r}t} + \sum_\alpha \hat{\sigma}_\alpha q^{\alpha\beta}_{\mathbf{r}t}], \quad (3)$$

where the spin density vector and the spin flow density tensor are given by $\mathbf{s}_{\mathbf{r}t} = \mathrm{tr}_\sigma \hat{\boldsymbol{\sigma}} \hat{n}_{\mathbf{r}t}$ and $q^{\alpha\beta}_{\mathbf{r}t} = \mathrm{tr}_\sigma \hat{\sigma}_\alpha \hat{i}^\beta_{\mathbf{r}t}$, respectively. Substituting Eq. (3)

into Eq. (2), we find the following equation for $\mathbf{s_{rt}}$:

$$\frac{\partial s_{\mathbf{rt}}^{\alpha}}{\partial t} + \sum_{\beta} \nabla_{\mathbf{r}}^{\beta} q_{\mathbf{rt}}^{\alpha\beta} + [\boldsymbol{\Omega}_H \times \mathbf{s_{rt}}]_{\alpha} = \frac{1}{V}\mathrm{tr}_{\sigma}\hat{\sigma}_{\alpha}\sum_{\mathbf{p}} \hat{J}(\hat{f}|\mathbf{rp}t). \qquad (4)$$

To find $q_{\mathbf{rt}}^{\alpha\beta}$, we represent the matrix distribution function $\hat{f}_{\mathbf{rp}t}$ as a sum of symmetric and antisymmetric parts, $\hat{f}_{\mathbf{r}\varepsilon t} + \widehat{\Delta f}_{\mathbf{rp}t}$. The collision integral in the equation for $\widehat{\Delta f}_{\mathbf{rp}t}$ can be written in the approximation when the spin-flip scattering is neglected, as $\hat{J}(\widehat{\Delta f}|\mathbf{rp}t) \simeq -\widehat{\Delta f}_{\mathbf{rp}t}/\tau_{tr}(\varepsilon_p)$, where $\tau_{tr}(\varepsilon)$ is the transport time; see Secs. 8, 11, and 36. In the case of weak magnetic field, when $\omega_c\tau_{tr} \ll 1$ and $\Omega_H\tau_{tr} \ll 1$, we have $\widehat{\Delta f}_{\mathbf{rp}t} = -\tau_{tr}(\varepsilon)\mathbf{v_p} \cdot \left(\partial\hat{f}_{\mathbf{r}\varepsilon t}/\partial\mathbf{r} + e\mathbf{E_{rt}}\partial\hat{f}_{\mathbf{r}\varepsilon t}/\partial\varepsilon\right)$, where $\varepsilon = \varepsilon_p$. In the hydrodynamic approximation, the symmetric part of the distribution function can be searched for in the form

$$\hat{f}_{\mathbf{r}\varepsilon t} = \left[\exp\left(\frac{\varepsilon - \hbar\boldsymbol{\Omega}_H \cdot \hat{\boldsymbol{\sigma}}/2 - \hat{\mu}_{\mathbf{rt}}}{T}\right) + 1\right]^{-1} \qquad (5)$$

corresponding to the local equilibrium (here and below we assume that the temperature is constant). The term $\hbar\boldsymbol{\Omega}_H\cdot\hat{\boldsymbol{\sigma}}/2$ in the exponent can be neglected in comparison to the mean kinetic energy of electrons (in fact, we have already assumed a stronger inequality, $\Omega_H \ll \tau_{tr}^{-1}$). The matrix of the local chemical potential introduced in Eq. (5) is reduced to a scalar $\mu_{\mathbf{rt}}$ if the distribution of the spin density is relaxed to the equilibrium one, that is, to $\mathbf{s_{rt}} = 0$ (at $\Omega_H \ll \bar{\varepsilon}$). To have an idea of how the distribution (5) looks like, we stress that, according to Eq. (H.28), $f(\varepsilon + \mathbf{A} \cdot \hat{\boldsymbol{\sigma}}) = f^{(+)} + (\mathbf{A} \cdot \hat{\boldsymbol{\sigma}}/|\mathbf{A}|)f^{(-)}$, where $f^{(\pm)} = [f(\varepsilon + |\mathbf{A}|) \pm f(\varepsilon - |\mathbf{A}|)]/2$, for an arbitrary vector $\mathbf{A}$ (see also Eq. (63.24)).

Using Eq. (5) and the relation between $\widehat{\Delta f}_{\mathbf{rp}t}$ and $\hat{f}_{\mathbf{r}\varepsilon t}$ given above, one can find linearized expressions for the matrix of the flow density:

$$\widehat{\mathbf{i}}_{\mathbf{rt}} = -\frac{1}{2e}\sigma_{\mathbf{rt}}\nabla_{\mathbf{r}}\widehat{w}_{\mathbf{rt}}, \quad \widehat{w}_{\mathbf{rt}} = e^{-1}\hat{\mu}_{\mathbf{rt}} + \boldsymbol{\Phi}_{\mathbf{rt}}, \qquad (6)$$

and

$$\widehat{\mathbf{i}}_{\mathbf{rt}} = -D_{\mathbf{rt}}\nabla_{\mathbf{r}}\hat{n}_{\mathbf{rt}} + \mathbf{u_{rt}}\hat{n}_{\mathbf{rt}}, \qquad (7)$$

which generalize Eqs. (36.8) and (36.11) written in the absence of temperature gradients. The matrix of the local electrochemical potential, $\widehat{w}_{\mathbf{rt}}$, is defined in a similar way as in Eq. (36.7), and the local conductivity, diffusion coefficient, and drift velocity ($\sigma_{\mathbf{rt}}$, $D_{\mathbf{rt}}$, and $\mathbf{u_{rt}}$) are described in Sec. 36. Equation (5) also leads to

$$\hat{n}_{\mathbf{rt}} = \frac{e}{2}\frac{\partial n_{\mathbf{rt}}}{\partial\mu_{\mathbf{rt}}}\widehat{w}_{\mathbf{rt}}, \qquad (8)$$

where the derivative of $n_{\mathbf{r}t}$ over $\mu_{\mathbf{r}t}$ is the same as in Eq. (36.10). Substituting the expression (7) into $q_{\mathbf{r}t}^{\alpha\beta} = \mathrm{tr}_\sigma \hat\sigma_\alpha \hat{i}_{\mathbf{r}t}^\beta$, we find a relation between $q_{\mathbf{r}t}^{\alpha\beta}$ and $\mathbf{s}_{\mathbf{r}t}$. Therefore, Eq. (4) can be rewritten as

$$\frac{\partial \mathbf{s}_{\mathbf{r}t}}{\partial t} - (\nabla_{\mathbf{r}} \cdot D_{\mathbf{r}t}\nabla_{\mathbf{r}})\mathbf{s}_{\mathbf{r}t} + (\nabla_{\mathbf{r}} \cdot \mathbf{u}_{\mathbf{r}t})\mathbf{s}_{\mathbf{r}t} + [\mathbf{\Omega}_H \times \mathbf{s}_{\mathbf{r}t}] = -\frac{\mathbf{s}_{\mathbf{r}t}}{\tau_s}. \qquad (9)$$

The right-hand side of this equation is written through the spin relaxation time τ_s considered in Sec. 63. Though $1/\tau_s$ is much smaller than the momentum relaxation rate, it can be comparable to Ω_H. Therefore, both relaxation and precession of the spin density vector are important. In strong magnetic fields ($\Omega_H \sim \bar\varepsilon$) or in ferromagnetic materials, when the equilibrium spin density is not zero and the relaxation is anisotropic, one should use a more complicated form of the right-hand side of Eq. (9), namely $-\nu_{s\perp}(\mathbf{s}_{\mathbf{r}t} - \mathbf{s}_{eq}) - (\nu_{s\parallel} - \nu_{s\perp})\mathbf{h}\left[(\mathbf{s}_{\mathbf{r}t} - \mathbf{s}_{eq}) \cdot \mathbf{h}\right]$, where $\mathbf{s}_{eq}$ is the equilibrium spin density and $\mathbf{h}$ is the unit vector in the direction of the magnetic field (or in the direction of the magnetization in ferromagnetic materials). The quantities $\nu_{s\perp}$ and $\nu_{s\parallel}$ are called the transverse and longitudinal relaxation rates.

In the presence of photoexcitation by a circularly polarized electromagnetic wave, one should add the photogeneration term $\mathbf{g}_t = V^{-1}\sum_{\mathbf{p}} \times G_{\mathbf{p}t}^{(c)}\mathrm{tr}_\sigma \hat{\boldsymbol{\sigma}}\widehat{\Psi}^{(\pm)}$, see Sec. 63, to the right-hand side of Eq. (9). If the diffusion and drift of the spin density are absent, such an equation describes both precession and relaxation of the excited spin density. Because of this precession, the excitation of the component of the spin density vector normal to the magnetic field is suppressed. For example, the absolute value of this component, created by a stationary excitation, decreases with Ω_H as $|\mathbf{s}_\perp| = |\mathbf{g}|\tau_s/\sqrt{1 + (\Omega_H\tau_s)^2}$ (problem 13.7). The spin precession in a weak magnetic field also suppresses polarization of the photoluminescence caused by the interband recombination of spin-polarized electrons and holes.

Apart from the relaxation and precession, Eq. (9) describes the diffusion and drift of the spin density. These processes are caused by the spatial gradients. We remind that the corresponding contributions in Eq. (9) are evaluated for the case of weak magnetic field, when $\omega_c\tau_{tr} \ll 1$ and $\Omega_H\tau_{tr} \ll 1$. The relaxation of a non-equilibrium electron density perturbation (created, for example, by photoexcitation) is fast because of the screening effects discussed in Sec. 36 (see Eq. (36.17) and problem 7.13), and the gradients of the electric field become screened out. For this reason, one can neglect the gradients of the drift velocity and write the term $(\nabla_{\mathbf{r}} \cdot \mathbf{u}_{\mathbf{r}t})\mathbf{s}_{\mathbf{r}t}$ in Eq. (9) as $(\mathbf{u} \cdot \nabla_{\mathbf{r}})\mathbf{s}_{\mathbf{r}t}$, where $\mathbf{u}$ is a constant drift velocity. For the same reasons, the diffusion coefficient in Eq. (9) is also constant, and the second term on the left-hand side is written as

$-D\nabla_{\mathbf{r}}^2\mathbf{s}_{\mathbf{r}t}$. In the absence of magnetic fields and in the stationary regime, it is convenient to write Eq. (9) as an equation describing spatial distribution of electrochemical potentials. Indeed, using the representation $\widehat{w}_{\mathbf{r}} = w_{\mathbf{r}} + \mathbf{w}_{\mathbf{r}} \cdot \hat{\boldsymbol{\sigma}}$ and Eqs. (6) and (8), we obtain

$$\nabla_{\mathbf{r}}^2\mathbf{w}_{\mathbf{r}} - l_s^{-2}\mathbf{w}_{\mathbf{r}} = 0, \quad l_s = \sqrt{D\tau_s}. \tag{10}$$

The scalar component satisfies the Laplace equation $\nabla_{\mathbf{r}}^2 w_{\mathbf{r}} = 0$ because of current continuity. The characteristic spatial scale of relaxation of the non-equilibrium spin distribution is determined by the spin diffusion length l_s (problem 13.8). Equation (10) has the same form as Eq. (60.28) for the distribution of electrochemical potentials in the tunnel-coupled layers. This is understandable because both these equations describe relaxation processes in weakly coupled sub-systems within the hydrodynamic approach.

Now we consider a more complicated case, the two-dimensional electron gas confined in the XOY plane by an asymmetric potential in a heterostructure. Apart from the Zeeman term considered in Eq. (1), one should include the spin-orbit term, and the Hamiltonian in the presence of external fields is given by Eq. (63.22). If the external fields are quasi-classically smooth and slowly varying with time, one can derive the following matrix kinetic equation for the Wigner distribution function $\hat{f}_{\mathbf{r}\mathbf{p}t}$:

$$\frac{\partial\hat{f}_{\mathbf{r}\mathbf{p}t}}{\partial t} - \frac{i}{2}\left[(2[\mathbf{v}_s\times\mathbf{p}]/\hbar + \boldsymbol{\Omega}_H)\cdot\hat{\boldsymbol{\sigma}}, \hat{f}_{\mathbf{r}\mathbf{p}t}\right] + \frac{1}{2}\left[\hat{\mathbf{v}}_{\mathbf{p}}, \frac{\partial\hat{f}_{\mathbf{r}\mathbf{p}t}}{\partial\mathbf{r}}\right]_+$$

$$+\frac{1}{2}\left[e\mathbf{E}_{\mathbf{r}t} + \frac{e}{c}[\hat{\mathbf{v}}_{\mathbf{p}}\times\mathbf{H}_{\mathbf{r}t}] + \frac{\hbar\,\partial(\boldsymbol{\Omega}_H\cdot\hat{\boldsymbol{\sigma}})}{2\ \ \partial\mathbf{r}}, \frac{\partial\hat{f}_{\mathbf{r}\mathbf{p}t}}{\partial\mathbf{p}}\right]_+ = \widehat{J}(\hat{f}|\mathbf{r}\mathbf{p}t). \tag{11}$$

The procedure of derivation of this equation is similar to that of Eqs. (63.8) and (1) (see, however, problem 13.9). Similar as in Eq. (63.8), it is assumed that the vectors standing in the anticommutators form the scalar products. In contrast to Eq. (1), the commutator describing the spin precession contains a momentum-dependent precession frequency. The matrix of the group velocity, $\hat{\mathbf{v}}_{\mathbf{p}} = \mathbf{p}/m + [\mathbf{v}_s\times\hat{\boldsymbol{\sigma}}]$, is already introduced in Sec. 63. With the use of $\hat{f}_{\mathbf{r}\mathbf{p}t} = f_{\mathbf{r}\mathbf{p}t} + \mathbf{f}_{\mathbf{r}\mathbf{p}t}\cdot\hat{\boldsymbol{\sigma}}$, we transform Eq. (11) to a system of scalar and vector equations:

$$\left(\frac{\partial}{\partial t} + \mathbf{v}_{\mathbf{p}}\cdot\frac{\partial}{\partial\mathbf{r}} + \mathbf{F}_{\mathbf{r}\mathbf{p}t}\cdot\frac{\partial}{\partial\mathbf{p}}\right)f_{\mathbf{r}\mathbf{p}t} - \left[\mathbf{v}_s\times\frac{\partial}{\partial\mathbf{r}}\right]\cdot\mathbf{f}_{\mathbf{r}\mathbf{p}t}$$

$$-\frac{e}{c}\left[\mathbf{v}_s\times\left[\mathbf{H}\times\frac{\partial}{\partial\mathbf{p}}\right]\right]\cdot\mathbf{f}_{\mathbf{r}\mathbf{p}t} + \frac{\hbar}{2}\sum_\gamma\frac{\partial\boldsymbol{\Omega}_H}{\partial r_\gamma}\cdot\frac{\partial\mathbf{f}_{\mathbf{r}\mathbf{p}t}}{\partial p_\gamma} = \frac{1}{2}\mathrm{tr}_\sigma\widehat{J}(f,\mathbf{f}|\mathbf{r}\mathbf{p}t), \tag{12}$$

and

$$\left(\frac{\partial}{\partial t} + \mathbf{v_p} \cdot \frac{\partial}{\partial \mathbf{r}} + \mathbf{F_{rpt}} \cdot \frac{\partial}{\partial \mathbf{p}}\right) \mathbf{f_{rpt}} + [(2[\mathbf{v}_s \times \mathbf{p}]/\hbar + \mathbf{\Omega}_H) \times \mathbf{f_{rpt}}]$$

$$- \left[\mathbf{v}_s \times \frac{\partial}{\partial \mathbf{r}}\right] f_{\mathbf{rpt}} - \frac{e}{c} \left[\mathbf{v}_s \times \left[\mathbf{H} \times \frac{\partial}{\partial \mathbf{p}}\right]\right] f_{\mathbf{rpt}} + \frac{\hbar}{2} \sum_\gamma \frac{\partial \mathbf{\Omega}_H}{\partial r_\gamma} \frac{\partial f_{\mathbf{rpt}}}{\partial p_\gamma}$$

$$= \frac{1}{2} \mathrm{tr}_\sigma \hat{\boldsymbol{\sigma}} \hat{J}(f, \mathbf{f}|\mathbf{rpt}). \tag{13}$$

Apart from the contributions containing the sum of the temporal derivative, spatial gradient, and Lorentz force term, these equations contain the terms responsible for the mixing of spatial motion and spin dynamics due to both spin-orbit interaction and inhomogeneous magnetic field. These are three last terms on the left-hand sides of Eqs. (12) and (13). For spatially-homogeneous magnetic fields, the last terms on the left-hand sides of these equations disappear, and the mixing of spatial motion and spin dynamics exists only due to the spin-orbit interaction. If the magnetic field $\mathbf{H}$ is parallel to the 2D plane, the terms with $[\mathbf{v}_s \times [\mathbf{H} \times (\partial/\partial \mathbf{p})]]$ disappear as well. The collision integrals on the right-hand sides of Eqs. (12) and (13), in principle, can be evaluated for any scattering mechanism. To facilitate the evaluation of such integrals (see below), one should use the condition (7.21) and make additional assumptions concerning the characteristic energies of spin splitting:

$$\hbar\Omega_H \ll \bar{\varepsilon}, \quad v_s \bar{p} \ll \bar{\varepsilon}, \quad \hbar\nu \ll \bar{\varepsilon}, \tag{14}$$

where $\bar{\varepsilon}$ and $\bar{p}$ are the mean energy and momentum of electrons, and ν is the momentum relaxation rate. The conditions (9.35) are also assumed. In other words, the mean energy of electrons is large in comparison to the other characteristic energies appearing in the problem.

In the following, we employ the method of moments, similar to the one discussed in Sec. 11. The local density $n_{\mathbf{r}t}$ and spin density $\mathbf{s}_{\mathbf{r}t}$ are introduced in the same way as in Eqs. (2) and (3). The local flow density and spin flow density, however, require a more elaborate definition, because of the matrix nature of the group velocity in the presence of the spin-orbit term:

$$\mathbf{i_{r}}_t = \frac{1}{L^2} \sum_{\mathbf{p}} \mathrm{tr}_\sigma \hat{\mathbf{v}}_{\mathbf{p}} \hat{f}_{\mathbf{rpt}} = \frac{2}{L^2} \sum_{\mathbf{p}} \{\mathbf{v_p} f_{\mathbf{rpt}} + [\mathbf{v}_s \times \mathbf{f_{rpt}}]\}, \tag{15}$$

$$q_{\mathbf{r}t}^{\alpha\beta} = \frac{1}{L^2} \sum_{\mathbf{p}} \mathrm{tr}_\sigma \hat{u}_{\mathbf{p}}^{\alpha\beta} \hat{f}_{\mathbf{rpt}} = \frac{2}{L^2} \sum_{\mathbf{p}} \left[v_{\mathbf{p}}^\beta f_{\mathbf{rpt}}^\alpha + v_s e_{\alpha\beta z} f_{\mathbf{rpt}}\right]. \tag{16}$$

We note that the antisymmetric unit tensor $e_{\alpha\beta\gamma}$ has been introduced in Sec. 11 and problem 2.14. The matrix of the spin-velocity tensor used in Eq. (16) is defined as an anticommutator: $\hat{u}_{\mathbf{p}}^{\alpha\beta} = [\hat{\sigma}_\alpha, \hat{v}_{\mathbf{p}}^\beta]_+/2$. To introduce the spin flow density $q^{\alpha\beta}$, one may equivalently employ the matrix of flow density, $\hat{i}_{\mathbf{r}t}^\beta = L^{-2}\sum_{\mathbf{p}}[\hat{v}_{\mathbf{p}}^\beta, \hat{f}_{\mathbf{r}\mathbf{p}t}]_+/2$, multiply it by $\hat{\sigma}_\alpha$, and take the matrix trace. The local magnetic moment $\mathbf{M}_{\mathbf{r}t}$ and spin current density tensor $I_{\mathbf{r}t}^{\alpha\beta}$ of the electron system are expressed according to $M_{\mathbf{r}t}^\alpha = g\mu_B s_{\mathbf{r}t}^\alpha$ and $I_{\mathbf{r}t}^{\alpha\beta} = e q_{\mathbf{r}t}^{\alpha\beta}$. In the definitions of $q^{\alpha\beta}$, $\hat{u}^{\alpha\beta}$, and $I^{\alpha\beta}$, we place the Cartesian coordinate indices in such a way that the first one indicates the direction of spin polarization, which can be x, y, or z, while the second one indicates the direction of motion, which can be x or y.

Summing Eq. (12) over the momentum $\mathbf{p}$, we obtain the continuity equation (11.5). Then, summing Eq. (13) over the momentum, we find

$$\frac{\partial s_{\mathbf{r}t}^\alpha}{\partial t} + \sum_\beta \nabla_{\mathbf{r}}^\beta q_{\mathbf{r}t}^{\alpha\beta} + [\mathbf{\Omega}_H \times \mathbf{s}_{\mathbf{r}t}]_\alpha$$

$$+ \frac{2m v_s}{\hbar}\{q_{\mathbf{r}t}^{z\alpha} - \delta_{\alpha z}(q_{\mathbf{r}t}^{xx} + q_{\mathbf{r}t}^{yy})\} = -\frac{s_{\mathbf{r}t}^\alpha - s_{eq}^\alpha}{\tau_s}. \tag{17}$$

This equation differs from Eq. (4) by the presence of the last term on the left-hand side. This term, proportional to the velocity v_s, originates from the momentum-dependent precession due to spin-orbit interaction. The relaxation term (the right-hand side) is written in the isotropic approximation through the spin relaxation time, as in Eq. (9). The equilibrium spin density $\mathbf{s}_{eq}$, which exists because of Zeeman splitting of electron spectrum, is proportional to $\mathbf{\Omega}_H$ (note that the equilibrium spin density has been neglected in Eq. (9)). To describe the second-order moments (i.e., the flow densities (15) and (16)), it is convenient to multiply the matrix equation (11) by $\hat{\mathbf{v}}_{\mathbf{p}}$ and $\hat{u}_{\mathbf{p}}^{\alpha\beta}$, respectively, take the matrix trace, and sum the equations obtained over $\mathbf{p}$. Using also Eqs. (17) and (11.5), we find the next pair of equations:

$$\frac{\partial i_{\mathbf{r}t}^\beta}{\partial t} + v_s \sum_\gamma e_{\beta\gamma z}\frac{\partial s_{\mathbf{r}t}^\gamma}{\partial t} + \sum_{\alpha\gamma} e_{\beta\alpha\gamma}\omega_c^\gamma i_{\mathbf{r}t}^\alpha + \sum_\gamma \frac{\partial Q_{\mathbf{r}t}^{\beta\gamma}}{\partial r_\gamma}$$

$$- \frac{eE_\beta}{m}n_{\mathbf{r}t} - \frac{\hbar}{2m}\sum_\gamma \frac{\partial \Omega_H^\gamma}{\partial r_\beta}s_{\mathbf{r}t}^\gamma = -\nu i_{\mathbf{r}t}^\beta \tag{18}$$

and

$$\frac{\partial q_{\mathbf{rt}}^{\alpha\beta}}{\partial t} + v_s e_{\beta\alpha z}\frac{\partial n_{\mathbf{rt}}}{\partial t} + \sum_{\delta\gamma}(e_{\beta\delta\gamma}\omega_c^{\gamma}q_{\mathbf{rt}}^{\alpha\delta} + e_{\alpha\gamma\delta}\Omega_H^{\gamma}q_{\mathbf{rt}}^{\delta\beta}) + \sum_{\gamma}\frac{\partial P_{\mathbf{rt}}^{\beta\alpha\gamma}}{\partial r_{\gamma}}$$

$$-\frac{eE_{\beta}}{m}s_{\mathbf{rt}}^{\alpha} - \frac{\hbar}{2m}\frac{\partial\Omega_H^{\alpha}}{\partial r_{\beta}}n_{\mathbf{rt}} + \frac{2mv_s}{\hbar}\left[P_{\mathbf{rt}}^{\beta z\alpha} - \delta_{\alpha z}(P_{\mathbf{rt}}^{\beta xx} + P_{\mathbf{rt}}^{\beta yy})\right] \qquad (19)$$

$$-v_s(\delta_{\alpha\beta}\Omega_s^z - \delta_{\alpha z}\Omega_s^{\beta})n_{\mathbf{rt}} = -\nu q_{\mathbf{rt}}^{\alpha\beta}.$$

We remind that $\boldsymbol{\omega}_c = |e|\mathbf{H}/mc$ is the vector whose absolute value is equal to the cyclotron frequency ω_c. The collision-integral contributions standing on the right-hand sides of Eqs. (18) and (19) have been evaluated in the approximation of elastic scattering by impurities with short-range potential, and the spin-flip scattering has been neglected. The terms of the order of $mv_s^2/\overline{\varepsilon}$ and $(\hbar\Omega_H/\overline{\varepsilon})^2$ have been neglected as well. The scattering rate ν obtained in this way is given by Eq. (8.21) for the 2D case (problem 13.10).

In Eqs. (18) and (19) we have introduced two new moments:

$$Q_{\mathbf{rt}}^{\beta\gamma} = \frac{1}{L^2}\sum_{\mathbf{p}}v_{\mathbf{p}}^{\beta}\mathrm{tr}_{\sigma}\hat{v}_{\mathbf{p}}^{\gamma}\hat{f}_{\mathbf{rpt}} = \frac{2}{L^2}\sum_{\mathbf{p}}v_{\mathbf{p}}^{\beta}\left(v_{\mathbf{p}}^{\gamma}f_{\mathbf{rpt}} + v_s\sum_{\alpha}e_{\alpha\gamma z}f_{\mathbf{rpt}}^{\alpha}\right),$$

$$P_{\mathbf{rt}}^{\beta\alpha\gamma} = \frac{1}{L^2}\sum_{\mathbf{p}}v_{\mathbf{p}}^{\beta}\mathrm{tr}_{\sigma}\hat{u}_{\mathbf{p}}^{\alpha\gamma}\hat{f}_{\mathbf{rpt}} = \frac{2}{L^2}\sum_{\mathbf{p}}v_{\mathbf{p}}^{\beta}\left(v_{\mathbf{p}}^{\gamma}f_{\mathbf{rpt}}^{\alpha} + v_s e_{\alpha\gamma z}f_{\mathbf{rpt}}\right). \qquad (20)$$

The first equation generalizes the definition (11.7). To find these moments, one should compose the equations which express them through higher-order moments. This leads to an infinite chain of coupled equations, as in Sec. 11. In our case, however, the number of equations increases because the spin degree of freedom is taken into account. At this point, to avoid consideration of higher-order moments, one may cut the chain by using the approximate relations

$$Q_{\mathbf{rt}}^{\beta\gamma} \simeq \delta_{\beta\gamma}\frac{\overline{\varepsilon}}{m}n_{\mathbf{rt}}, \qquad P_{\mathbf{rt}}^{\beta\alpha\gamma} \simeq \delta_{\beta\gamma}\frac{n_{2D}}{m\rho_{2D}}s_{\mathbf{rt}}^{\alpha}, \qquad (21)$$

where n_{2D} is the equilibrium electron density and $\overline{\varepsilon}$ is the mean energy of the electron system, which is related to the energy density $\mathcal{E}_0$ (introduced by Eq. (11.21) for the 3D case) as $\overline{\varepsilon} = \mathcal{E}_0/n_{2D}$. The relations (21) are justified, for example, if we represent the distribution function in the form similar to Eq. (5), $\hat{f}_{\mathbf{rpt}} = \{\exp([\varepsilon_p + ([\mathbf{p}\times\mathbf{v}_s] - \hbar\boldsymbol{\Omega}_H/2)\cdot\hat{\boldsymbol{\sigma}} - \hat{\mu}_{\mathbf{rt}}]/T) + 1\}^{-1}$, and take into account the strong inequalities (14).

Substituting $P_{\mathbf{r}t}^{\beta\alpha\gamma}$ from Eq. (21) into Eq. (19), we find the spin polarization in equilibrium, $\mathbf{s}_{eq} = \rho_{2D}\hbar\mathbf{\Omega}_H/2$, which can also be obtained from a simple physical consideration. The equilibrium spin polarization exists due to Zeeman splitting and disappears at $\mathbf{H} = 0$.

Equations (17)$-$(19), accompanied by the approximate relations (21) and the continuity equation (11.5), form a closed set, which can be written as a system of twelve differential equations with partial derivatives. A general analysis of this system would be complicated enough. Below we consider the spatially homogeneous case, when the terms with spatial derivatives are dropped out. First, let us find characteristic frequencies of the system in the absence of the external fields. The Fourier components s_ω^α and $q_\omega^{z\alpha}$, where $\alpha = x, y$, are coupled by the following equations:

$$
\begin{aligned}
(-i\omega + \nu_s)s_\omega^\alpha + (2mv_s/\hbar)q_\omega^{z\alpha} = 0, \\
-(2v_s n_{2D}/\hbar\rho_{2D})s_\omega^\alpha + (-i\omega + \nu)q_\omega^{z\alpha} = 0,
\end{aligned}
\tag{22}
$$

where we put $\nu_s \equiv 1/\tau_s$. A similar system of equations which couples s_ω^z and $(q_\omega^{xx} + q_\omega^{yy})$ can be obtained from Eq. (22) after formal substitutions $v_s \to -v_s$, $n_{2D} \to 2n_{2D}$, $s_\omega^\alpha \to s_\omega^z$, and $q_\omega^{z\alpha} \to (q_\omega^{xx} + q_\omega^{yy})$. The characteristic frequencies are determined from the dispersion relation $\omega^2 + i\omega(\nu + \nu_s) - \nu\nu_s - 4\pi n_{2D}v_s^2 = 0$ which has two solutions:

$$
\omega_\pm^{(\|)} = -i\frac{\nu + \nu_s}{2} \mp i\sqrt{\frac{(\nu - \nu_s)^2}{4} - \omega_s^2}, \quad \omega_s^2 = 4\pi n_{2D}v_s^2.
\tag{23}
$$

The superscript $(\|)$ indicates that this expression describes the frequencies of the in-plane components of spin density, s_ω^x and s_ω^y. The characteristic frequencies of the perpendicular component s_ω^z are described by a similar expression with $\omega_s^2 \to 2\omega_s^2$. If the spin-orbit interaction is weak, $\omega_s \ll \nu$, the solutions (23) are imaginary: $\omega_+^{(\|)} \simeq -i\nu$ and $\omega_-^{(\|)} \simeq -i(\nu_s + \omega_s^2/\nu)$. They describe fast relaxation with the rate ν and slow relaxation with the rate $\nu_s + \omega_s^2/\nu$. It is not hard to identify ω_s^2/ν with the relaxation rate due to a spin-precession mechanism similar to that considered in Sec. 63 and problem 13.3. Indeed, if we substitute the frequency $[\mathbf{p} \times \mathbf{v}_s]/\hbar$ in place of $\mathbf{\Omega}_{\mathbf{rp}}$ in Eq. (63.9) (compare the commutators in Eqs. (63.8) and (11)) and take into account that the non-equilibrium spin distribution occurs near the Fermi surface ($p \simeq \hbar\sqrt{2\pi n_{2D}} \simeq p_F$), we obtain the relaxation rate ω_s^2/ν for the components s^x and s^y (and $2\omega_s^2/\nu$ for the component s^z). If ω_s is comparable to ν, both characteristic frequencies $\omega_\pm^{(\|)}$ describe fast relaxation. Finally, if $\omega_s > (\nu - \nu_s)/2$, one has $\omega_\pm^{(\|)} \simeq -i\nu/2 \pm \Omega_s$, where

$\Omega_s = \sqrt{\omega_s^2 - (\nu - \nu_s)^2/4}$. The relaxation of spin density in this case is accompanied by the oscillations due to spin precession. The absence of slow spin relaxation at $\omega_s \sim \nu$ is explained by the fact that the branches of spin-split spectrum of 2D electrons are well defined under these conditions, and (since the electron spin orientation in these branches depends on momenta) the momentum relaxation cannot be separated from the spin relaxation.

Consider now a linear response of the electron system to a harmonic field perturbation $\mathbf{E}_t = \mathbf{E}e^{-i\omega t} + c.c.$ in the presence of a magnetic field. This perturbation excites non-equilibrium densities and flows with the Fourier components $\delta n_\omega = n_\omega - n_{2D}$, $\delta s_\omega^\alpha = s_\omega^\alpha - s_{eq}^\alpha$, i_ω^β, and $q_\omega^{\alpha\beta}$. Assuming that the magnetic field is directed along OZ (perpendicular to the 2D layer), we obtain a system of 12 linear equations for these quantities. The first one, following from the continuity equation, gives simply $\delta n_\omega = 0$. The remaining 11 equations are split in two subsystems. The first sub-system couples 5 quantities, $q^{x\beta}$, $q^{y\beta}$, and δs^z, while the second one couples 6 quantities, i^β, $q^{z\beta}$, δs^x, and δs^y ($\beta = x, y$). The equations of the first sub-system do not contain the electric-field terms in the linear approximation. Therefore, $q^{x\beta} = q^{y\beta} = \delta s^z = 0$. The equations of the second sub-system are (the index ω is omitted for brevity)

$$(-i\omega + \nu_s)\delta s^x - \Omega_H \delta s^y + (2mv_s/\hbar)q^{zx} = 0,$$
$$(-i\omega + \nu_s)\delta s^y + \Omega_H \delta s^x + (2mv_s/\hbar)q^{zy} = 0, \tag{24}$$

$$(-i\omega + \nu)i^x + \omega_c i^y - iv_s\omega\delta s^y = en_{2D}E_x/m,$$
$$(-i\omega + \nu)i^y - \omega_c i^x + iv_s\omega\delta s^x = en_{2D}E_y/m, \tag{25}$$

and

$$(-i\omega + \nu)q^{zx} + \omega_c q^{zy} - (2v_s n_{2D}/\hbar\rho_{2D})\delta s^x = es_{eq}^z E_x/m,$$
$$(-i\omega + \nu)q^{zy} - \omega_c q^{zx} - (2v_s n_{2D}/\hbar\rho_{2D})\delta s^y = es_{eq}^z E_y/m, \tag{26}$$

each pair follows, respectively, from Eqs. (17), (18), and (19). We remind that $s_{eq}^z = \rho_{2D}\hbar\Omega_H/2$ is the equilibrium spin density existing due to Zeeman splitting. Solving the system of four equations given by Eqs. (24) and (26), we obtain

$$\delta s^x = -\frac{2ev_s}{\hbar}s_{eq}^z\frac{E_xR_1 + E_yR_2}{R_1^2 + R_2^2},$$

$$\delta s^y = -\frac{2ev_s}{\hbar}s_{eq}^z\frac{E_yR_1 - E_xR_2}{R_1^2 + R_2^2}. \tag{27}$$

In these expressions,

$$R_1 = -(\omega + i\nu)(\omega + i\nu_s) + \omega_c\Omega_H + \omega_s^2,$$

$$R_2 = i\omega_c(\omega + i\nu_s) - i\Omega_H(\omega + i\nu), \qquad (28)$$

and ω_s^2 is introduced in Eq. (23). The spin flow densities q^{zx} and q^{zy} are expressed through δs^x and δs^y of Eq. (27) with the use of Eq. (24).

Substituting δs^x and δs^y into Eq. (25), we find the current density $I_\omega^\beta = e i_\omega^\beta$ in the form $I_\omega^\beta = \sum_\gamma \sigma_{\beta\gamma} E_\gamma$, where $\sigma_{xx} = \sigma_{yy} \equiv \sigma_d$ and $\sigma_{yx} = -\sigma_{xy} \equiv \sigma_\perp$ are the diagonal and non-diagonal components of the conductivity tensor; see Sec. 11. The absorbed power of the electromagnetic radiation, which is equal to Re $\overline{\mathbf{I}_t \cdot \mathbf{E}_t}$ = Re $\mathbf{I}_\omega \cdot \mathbf{E}/2$ = Re$\sigma_d(\omega)E^2/2$, is determined by the diagonal component. The latter is represented as

$$\sigma_d(\omega) = \sigma_d^{(CR)}(\omega) + \sigma_d^{(cr)}(\omega). \qquad (29)$$

The first term is the conductivity in the absence of the spin-orbit interaction. This term is given by the first expression of Eq. (11.18) with $\tau_{tr} = \nu^{-1}$ and describes the cyclotron absorption. The second term represents a correction to the conductivity due to the spin-orbit interaction:

$$\sigma_d^{(cr)}(\omega) = -i\sigma_0^{(cr)}[\Delta_+(\omega) - \Delta_-(\omega)],$$

$$\Delta_\pm(\omega) = \frac{\nu\Omega_H\omega}{(R_1 \pm iR_2)(\omega + i\nu \pm \omega_c)}, \qquad (30)$$

where $\sigma_0^{(cr)} = e^2 v_s^2 \rho_{2D}/2\nu$. In a similar way, the non-diagonal part is written as $\sigma_\perp(\omega) = \sigma_\perp^{(CR)}(\omega) + \sigma_\perp^{(cr)}(\omega)$, where $\sigma_\perp^{(CR)}$ is given by the second expression of Eq. (11.18) and $\sigma_\perp^{(cr)}(\omega) = -\sigma_0^{(cr)}[\Delta_+(\omega) + \Delta_-(\omega)]$. The terms $\sigma_d^{(cr)}$ and $\sigma_\perp^{(cr)}$ are caused by the mixing of the spin dynamics and spatial motion of electrons. These terms contain new resonances. In particular, when the relaxation is negligible, the denominators $R_1 + isR_2$, where $s = \pm 1$, go to zero at the frequencies

$$\omega_\pm^{(s)} = s\frac{\Omega_H - \omega_c}{2} \pm \sqrt{\frac{(\Omega_H + \omega_c)^2}{2} + \omega_s^2}. \qquad (31)$$

If the spin-orbit interaction is weak, $\omega_s \ll \Omega_H, \omega_c$, the resonances appear at $\omega \simeq \Omega_H$ and $\omega \simeq \omega_c$. The first one is the combined resonance corresponding to the spin-flip transitions caused by the presence of the spin-orbit term (63.21) in the Hamiltonian of 2D electrons. The frequency dependence of $\sigma_d^{(cr)}(\omega)$ is shown in Fig 13.1 for some chosen parameters. The broadening of the combined resonance is determined by the spin relaxation rate ν_s. We have considered the simplest case when the magnetic field is directed perpendicular to the 2D plane. If the magnetic field is tilted, the combined-resonance absorption depends on the direction of the electric field.

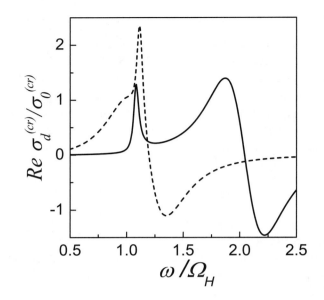

Figure 13.1. Line shape of the combined resonance at $\omega_s = 0.5\ \Omega_H$, $\nu = 0.3\ \Omega_H$, and $\nu_s = 0.02\ \Omega_H$ for two cyclotron frequencies: $\omega_c = 2\ \Omega_H$ (solid) and $\omega_c = 1.1\ \Omega_H$ (dashed).

We point out that, because of crudeness of the approximation (21), the balance equations considered above do not describe the effect of absorption due to electron transitions between the spin-split subbands considered in the end of Sec. 63 in the absence of magnetic fields. Another important effect, the spin polarization of electrons by a steady-state current, is also left beyond. This effect, however, can be described with the aid of Eq. (11) (problem 13.11). From a qualitative point of view, the spin polarization appears as a result of the field-induced redistribution of electrons in the momentum space. If the electron spectrum is split due to spin-orbit interaction, the anisotropic distribution of electrons in the momentum space causes the spin polarization. For the same reasons, a non-equilibrium spin density (created, for example, by photoexcitation) induces the electric current along the 2D layer.

65. Coupled Quantum Wells

The consideration of the tunneling between 2D electron systems in Sec. 60 has been based upon the approximation of weak tunnel coupling, when the 2D electrons in the left (l) and right (r) quantum wells can be considered as separate sub-systems and the tunneling current is calculated according to the perturbation theory. However, if the tunnel barrier is narrow enough, one needs to take into account the coupling (hybridization) of the states belonging to different wells. The 2D Hamiltonian describing this hybridization is obtained in the basis of *l*- and *r*-well eigenstates, $|l\rangle$ and $|r\rangle$, as (see Appendix H)

$$\widehat{H} = \frac{\hat{\pi}_{\mathbf{x}t}^2}{2m} + \frac{\Delta}{2}\hat{\sigma}_z + t_{lr}\hat{\sigma}_x + \widehat{V}_{\mathbf{x}}, \qquad (1)$$

where $\hat{\pi}_{\mathbf{x}t} = \hat{\mathbf{p}} - (e/c)\mathbf{A}_{\mathbf{x}t}$ is the operator of kinematic momentum, Δ is the level splitting energy in the absence of the tunnel coupling, and t_{lr} is the tunneling matrix element describing the strength of this coupling. It is assumed that the magnetic field, which enters Eq. (1) through the vector potential, is directed perpendicular to the 2D planes (along OZ) and is weak enough to neglect the Zeeman splitting. In a similar way as in Secs. 63 and 64, we use the Pauli matrices to express the Hamiltonian of two coupled sub-systems (so-called isospin representation). To introduce the matrix of potential energy, $\widehat{V}_{\mathbf{x}t}$, which describes the scattering of electrons by impurities (or other static inhomogeneities), one should take into account that the impurity distribution across the coupled quantum wells can be substantially inhomogeneous, for example, the densities of impurities in the wells are different. The potential energy is written as a diagonal matrix

$$\widehat{V}_{\mathbf{x}} = \sum_{\alpha} \left[\widehat{P}_l v_l(\mathbf{x} - \mathbf{X}_\alpha | Z_\alpha) + \widehat{P}_r v_r(\mathbf{x} - \mathbf{X}_\alpha | Z_\alpha) \right], \qquad (2)$$

where $\widehat{P}_l = (1 + \hat{\sigma}_z)/2$ and $\widehat{P}_r = (1 - \hat{\sigma}_z)/2$ are the projection matrices, the index α numbers the impurities with coordinates $(\mathbf{X}_\alpha, Z_\alpha)$, and $v_j(\mathbf{x} - \mathbf{X}_\alpha | Z_\alpha) = \langle j | v(\mathbf{x} - \mathbf{X}_\alpha, z - Z_\alpha) | j \rangle$ is the effective two-dimensional potential created in the layer j by a single impurity. In the absence of the magnetic field and potential energy $\widehat{V}_{\mathbf{x}}$, the Hamiltonian (1) in the momentum representation is written as $p^2/2m + (\Delta/2)\hat{\sigma}_z + t_{lr}\hat{\sigma}_x$ and can be diagonalized by a unitary transformation to the form $p^2/2m + (\Delta_T/2)\hat{\sigma}_z$, where $\Delta_T = \sqrt{\Delta^2 + 4t_{lr}^2}$ is the level splitting energy renormalized due to tunnel coupling. This Hamiltonian describes two parabolic branches of electron energy spectrum shifted by Δ_T with respect to each other, and the density of states is formed by two step functions: $\rho(\varepsilon) = \rho_{2D}[\theta(\varepsilon + \Delta_T/2) + \theta(\varepsilon - \Delta_T/2)]$.

The presence of a non-symmetric scattering potential considerably modifies the density of states of 2D electrons near the edge. To consider this effect, let us introduce a matrix Green's function satisfying the following equation (at zero magnetic field):

$$\left[\varepsilon + \frac{\hbar^2}{2m}\frac{\partial^2}{\partial \mathbf{x}^2} - \frac{\Delta}{2}\hat{\sigma}_z - t_{lr}\hat{\sigma}_x - \widehat{V}_{\mathbf{x}} \right] \widehat{G}_\varepsilon(\mathbf{x}, \mathbf{x}') = \delta(\mathbf{x} - \mathbf{x}'). \qquad (3)$$

The averaged (retarded, $s = R$, or advanced, $s = A$) Green's function in the momentum representation satisfies the matrix Dyson equation (compare to Eq. (56.30)):

$$\left[\varepsilon - \varepsilon_p - (\Delta/2)\hat{\sigma}_z - t_{lr}\hat{\sigma}_x - \widehat{\Sigma}^s_\varepsilon(\mathbf{p})\right]\widehat{G}^s_\varepsilon(\mathbf{p}) = 1. \tag{4}$$

The self-energy matrix in the Born approximation is given by

$$\widehat{\Sigma}^s_\varepsilon(\mathbf{p}) = \frac{1}{L^2}\sum_{jj'}\sum_{\mathbf{p}'} w_{jj'}(|\mathbf{p} - \mathbf{p}'|/\hbar)\widehat{P}_j\widehat{G}^s_\varepsilon(\mathbf{p}')\widehat{P}_{j'}, \tag{5}$$

where the correlation function $w_{jj'}(q)$ is defined according to ($\Delta\mathbf{x} = \mathbf{x} - \mathbf{x}'$)

$$w_{jj'}(q) = \int d\Delta\mathbf{x}\, e^{-i\mathbf{q}\cdot\Delta\mathbf{x}}\langle\langle\sum_{\alpha\beta} v_j(\mathbf{x} - \mathbf{X}_\alpha|Z_\alpha)v_{j'}(\mathbf{x}' - \mathbf{X}_\beta|Z_\beta)\rangle\rangle. \tag{6}$$

To obtain Eqs. (4)–(6), it is sufficient to write the diagram equations of the kind of Eqs. (14.21)–(14.23) for the matrix Green's function and notice that each vertex associated with the scattering potential brings the matrix contribution (2). Since $w_{jj'} = w_{j'j}$, one may permute the indices of the projection matrices in Eq. (5). In the approximation of short-range scattering potential, the correlation function becomes diagonal in the layer index and independent of q. The self-energy matrix in these conditions is diagonal and independent of momentum. Representing it as $\widehat{\Sigma}^s_\varepsilon = \Sigma^{s(+)}_\varepsilon + \hat{\sigma}_z\Sigma^{s(-)}_\varepsilon$, we can compose a pair of coupled equations following from Eqs. (4) and (5):

$$\Sigma^{s(+)}_\varepsilon - \mu\Sigma^{s(-)}_\varepsilon = \frac{mw_+}{4\pi\hbar^2}(1 - \mu^2)(L^{(+)}_\varepsilon + L^{(-)}_\varepsilon),$$

$$\Sigma^{s(-)}_\varepsilon - \mu\Sigma^{s(+)}_\varepsilon = \frac{mw_+}{4\pi\hbar^2}(1 - \mu^2)\frac{(\Delta/2 + \Sigma^{s(-)}_\varepsilon)(L^{(+)}_\varepsilon - L^{(-)}_\varepsilon)}{\sqrt{(\Delta_T/2)^2 + \Sigma^{s(-)}_\varepsilon\Delta + (\Sigma^{s(-)}_\varepsilon)^2}}, \tag{7}$$

where

$$L^{(\pm)}_\varepsilon = \ln\left[-\varepsilon + \Sigma^{s(+)}_\varepsilon \pm \sqrt{(\Delta_T/2)^2 + \Sigma^{s(-)}_\varepsilon\Delta + (\Sigma^{s(-)}_\varepsilon)^2}\right] - \ln E_0. \tag{8}$$

These equations contain the averaged correlation function $w_+ = (w_{ll} + w_{rr})/2$ and the degree of scattering asymmetry, $\mu = (w_{ll} - w_{rr})/(w_{ll} + w_{rr})$. The cut-off energy E_0 is introduced in order to make the real part of $\Sigma^{s(\pm)}_\varepsilon$ finite; see the discussion at the end of Sec. 14.

Although the system of equations (7) is rather complicated, one can get a general idea about behavior of the density of states without a detailed consideration. Indeed, the density of states is given by

$$\rho(\varepsilon) = -\frac{2}{\pi L^2} \sum_{\mathbf{p}} \text{Im tr } \widehat{G}_\varepsilon^R(\mathbf{p}) = -\frac{4}{\pi w_+(1-\mu^2)} \text{Im}(\Sigma_\varepsilon^{R(+)} - \mu \Sigma_\varepsilon^{R(-)}). \quad (9)$$

In the limit of weak non-symmetric scattering, when $|\Sigma_\varepsilon^{s(-)}| \ll \Delta_T/2$, we expand the square root in Eqs. (7) and (8) as $\Delta_T/2 + (\Delta/\Delta_T)\Sigma_\varepsilon^{s(-)}$. The density of states (9) in these conditions is written as

$$\rho(\varepsilon) = \rho_{2D}[F_-(\varepsilon) + F_+(\varepsilon)], \quad (10)$$

$$F_\pm(\varepsilon) = \frac{1}{2} + \frac{1}{\pi} \arctan \frac{\varepsilon \mp \Delta_T/2 - \text{Re}[\Sigma_\varepsilon^{(+)} \pm (\Delta/\Delta_T)\Sigma_\varepsilon^{(-)}]}{\text{Im}[\Sigma_\varepsilon^{(+)} \pm (\Delta/\Delta_T)\Sigma_\varepsilon^{(-)}]},$$

where $\Sigma_\varepsilon^{(\pm)} \equiv \Sigma_\varepsilon^{A(\pm)}$ (we point out that $\text{Im}[\Sigma^{A(+)} \pm (\Delta/\Delta_T)\Sigma^{A(-)}]$ is positive). In comparison to Eq. (14.32), this density of states is composed of two broadened steps, and the broadening energies are different. Besides, the broadening near the edge (at $\varepsilon \simeq -\Delta_T/2$) is sensitive to the sign of Δ. If, for example, the left well is more doped than the right one, the broadening at the edge is more considerable at negative Δ, when the wave function of the ground state is localized mostly in the left well. If, however, the levels are aligned ($\Delta = 0$), the broadening of both steps is described by a single quantity, $\text{Im}\Sigma_\varepsilon^{(+)}$ (since this quantity depends on ε, the actual broadening energies of the steps are different from each other). This behavior is qualitatively different from that at weak tunnel coupling ($t_{lr} \ll \text{Im}\Sigma_\varepsilon^{(-)}$), when the density of states (9) is formed merely as a superposition of independent densities of electron states in the l and r wells.

The interband optical absorption in double quantum well structures is considerably influenced by the non-symmetric scattering. The coefficient of this absorption is expressed according to Eq. (18.10) through a correlation function of the Green's functions of conduction- and valence-band electrons. In a similar way as in Sec. 55, one may consider the approximation of almost ideal valence-band 2D electrons, because the valence-band effective mass m_v is large. The frequency dependence of the relative absorption ξ_ω in this case is given by

$$\xi_\omega \propto -\frac{1}{L^4} \text{Im} \int d\mathbf{x} \int d\mathbf{x}' \sum_{\mathbf{p}} e^{-i\mathbf{p}\cdot(\mathbf{x}-\mathbf{x}')/\hbar} \text{tr}\widehat{P}_v \langle\langle \widehat{G}_{\varepsilon_\omega - p^2/2m_v}^R(\mathbf{x},\mathbf{x}')\rangle\rangle$$

$$= -\frac{1}{L^2} \sum_{\mathbf{p}} \text{Im tr}\widehat{P}_v \widehat{G}_{\varepsilon_\omega - p^2/2m_v}^R(\mathbf{p}), \quad (11)$$

where $\varepsilon_\omega = \hbar\omega - \bar{\varepsilon}_g$ is the excess energy of optical quanta and $\bar{\varepsilon}_g$ is the effective gap renormalized by the quantization energies of c- and v-band states. The projection matrix $\widehat{P}_v$ introduced in Eq. (11) is expressed through the overlap factors of v-band states with l- and r-well conduction-band states: $\langle j|\widehat{P}_v|j'\rangle = \langle j|v\rangle\langle v|j'\rangle$. If, for example, the highest valence-band electron states are localized in the left well so that $\langle l|v\rangle \simeq 1$ and $\langle r|v\rangle \simeq 0$, the matrix $\widehat{P}_v$ is reduced to the projection matrix $\widehat{P}_l$ introduced above. In this case and under the approximation $|\Sigma_\varepsilon^{s(-)}| \ll \Delta_T/2$, we obtain

$$\xi_\omega \propto \left(1 - \frac{\Delta}{\Delta_T}\right) F_-(\varepsilon_\omega) + \left(1 + \frac{\Delta}{\Delta_T}\right) F_+(\varepsilon_\omega). \tag{12}$$

The frequency dependence of the interband absorption coefficient, therefore, is similar to the energy dependence of the density of states (10). Since both the broadening energy and the absolute value of the absorption coefficient near the edge $\varepsilon_\omega = -\Delta_T/2$ are sensitive to Δ, there exists a possibility to control the spectral characteristics of interband absorption and photoluminescence by an external bias applied across the double quantum well structure.

Consider now the in-plane transport properties of coupled quantum wells. Introducing the matrix Wigner distribution function $\hat{f}_{\mathbf{r}\mathbf{p}t}$, one can derive a matrix kinetic equation similar to Eqs. (64.1) and (64.11). Below we consider the spatially-homogeneous case, when the matrix kinetic equation is written as

$$\frac{\partial \hat{f}_{\mathbf{p}t}}{\partial t} + \frac{i}{\hbar}\left[\frac{\Delta}{2}\hat{\sigma}_z + t_{lr}\hat{\sigma}_x, \hat{f}_{\mathbf{p}t}\right] + \mathbf{F}_{\mathbf{p}t} \cdot \frac{\partial \hat{f}_{\mathbf{p}t}}{\partial \mathbf{p}} = \widehat{J}(\hat{f}|\mathbf{p}t). \tag{13}$$

The matrix structure of the Hamiltonian (1) leads to the commutator on the left-hand side of this equation and dictates the matrix form of the collision integral. The latter is written below in the Markovian approximation:

$$\widehat{J}(\hat{f}|\mathbf{p}t) = \frac{1}{\hbar^2 L^2}\sum_{\mathbf{p}'}\sum_{jj'} w_{jj'}(|\mathbf{p} - \mathbf{p}'|/\hbar) \int_{-\infty}^{0} d\tau e^{\lambda\tau}\left\{e^{i\hat{h}_p\tau/\hbar}\right. \tag{14}$$

$$\left. \times(\widehat{P}_j\hat{f}_{\mathbf{p}'t} - \hat{f}_{\mathbf{p}t}\widehat{P}_j)e^{-i\hat{h}_{p'}\tau/\hbar}\widehat{P}_{j'} - \widehat{P}_{j'}e^{i\hat{h}_{p'}\tau/\hbar}(\widehat{P}_j\hat{f}_{\mathbf{p}t} - \hat{f}_{\mathbf{p}'t}\widehat{P}_j)e^{-i\hat{h}_p\tau/\hbar}\right\},$$

where $\hat{h}_p = p^2/2m + (\Delta/2)\hat{\sigma}_z + t_{lr}\hat{\sigma}_x$. Equation (14) can be obtained from Eq. (7.14) in the way described in Secs. 7 and 8. It is valid under the usual quasi-classical assumptions that both the momentum relaxation rate and cyclotron frequency are small in comparison to $\bar{\varepsilon}/\hbar$.

Below we omit the time index and write the matrices $\hat{f}_{\mathbf{p}}$ and $\widehat{J}(\hat{f}|\mathbf{p})$ as (problem 13.12)

$$\hat{f}_{\mathbf{p}} = \widehat{P}_l f_{\mathbf{p}}^l + \widehat{P}_r f_{\mathbf{p}}^r + \hat{\sigma}_x f_{\mathbf{p}}^x + \hat{\sigma}_y f_{\mathbf{p}}^y$$

$$\widehat{J}(\hat{f}|\mathbf{p}) = \widehat{P}_l J_l(\hat{f}|\mathbf{p}) + \widehat{P}_r J_r(\hat{f}|\mathbf{p}) + \hat{\sigma}_x J_x(\hat{f}|\mathbf{p}) + \hat{\sigma}_y J_y(\hat{f}|\mathbf{p}). \quad (15)$$

To find a simple expression for $\widehat{J}$, let us assume that $\Delta_T \ll \bar{\varepsilon}$. Expressing the operator exponents according to the relation (H.28), we obtain

$$J_j(\hat{f}|\mathbf{p}) = \frac{2\pi}{\hbar L^2} \sum_{\mathbf{p}'} w_{jj}(|\mathbf{p} - \mathbf{p}'|/\hbar)(f_{\mathbf{p}'}^j - f_{\mathbf{p}}^j)\delta(\varepsilon_p - \varepsilon_{p'}), \quad (j = l, r) \quad (16)$$

and

$$J_k(\hat{f}|\mathbf{p}) = \frac{2\pi}{\hbar L^2} \sum_{\mathbf{p}'} \left[w_{lr}(|\mathbf{p} - \mathbf{p}'|/\hbar)f_{\mathbf{p}'}^k - w_+(|\mathbf{p} - \mathbf{p}'|/\hbar)f_{\mathbf{p}}^k \right]$$

$$\times \delta(\varepsilon_p - \varepsilon_{p'}), \quad (k = x, y), \quad (17)$$

where $\varepsilon_p = p^2/2m$ and $w_+(q) = [w_{ll}(q) + w_{rr}(q)]/2$. The matrix kinetic equation (13) is written as four scalar equations

$$\left[\frac{\partial}{\partial t} + \left(e\mathbf{E} + \frac{e}{c}[\mathbf{v}_{\mathbf{p}} \times \mathbf{H}] \right) \cdot \frac{\partial}{\partial \mathbf{p}} \right] \left| \begin{array}{c} f_{\mathbf{p}}^l \\ f_{\mathbf{p}}^r \end{array} \right| - \frac{2t_{lr}}{\hbar} \left| \begin{array}{c} 1 \\ -1 \end{array} \right| f_{\mathbf{p}}^y = \left| \begin{array}{c} J_l \\ J_r \end{array} \right| \quad (18)$$

and

$$\left[\frac{\partial}{\partial t} + \left(e\mathbf{E} + \frac{e}{c}[\mathbf{v}_{\mathbf{p}} \times \mathbf{H}] \right) \cdot \frac{\partial}{\partial \mathbf{p}} \right] \left| \begin{array}{c} f_{\mathbf{p}}^x \\ f_{\mathbf{p}}^y \end{array} \right|$$

$$+ \frac{\Delta}{\hbar} \left| \begin{array}{c} f_{\mathbf{p}}^y \\ -f_{\mathbf{p}}^x \end{array} \right| + \frac{t_{lr}}{\hbar} \left| \begin{array}{c} 0 \\ 1 \end{array} \right| (f_{\mathbf{p}}^l - f_{\mathbf{p}}^r) = \left| \begin{array}{c} J_x \\ J_y \end{array} \right|. \quad (19)$$

Let us apply Eqs. (18) and (19) for calculating the linear response to a homogeneous electric field $\mathbf{E}e^{-i\omega t} + c.c.$ The density of electric current along the layers is determined by

$$\mathbf{I} = \frac{2e}{L^2} \sum_{\mathbf{p}} \mathbf{v}_{\mathbf{p}} \mathrm{tr} \widehat{\delta f}_{\mathbf{p}} , \quad (20)$$

where $\widehat{\delta f}_{\mathbf{p}}$ is the non-equilibrium part of the matrix distribution function. We search for this part in the form

$$\widehat{\delta f}_{\mathbf{p}} = -e\mathbf{E} \cdot \mathbf{v}_{\mathbf{p}} \frac{\partial f_{\varepsilon_p}^{(eq)}}{\partial \varepsilon_p} [\hat{\xi}_+ e^{i\varphi} + \hat{\xi}_- e^{-i\varphi}], \quad (21)$$

where $f_{\varepsilon_p}^{(eq)}$ is the equilibrium distribution function and φ is the polar angle of the 2D momentum $\mathbf{p}$. The matrices $\hat{\xi}_{\pm}$ are independent of this angle. In the case of degenerate electron gas, when $-(\partial f_{\varepsilon_p}^{(eq)}/\partial \varepsilon_p) = \delta(\varepsilon_p - \varepsilon_F)$, we express the components of the linearized collision integral through the transport times for diagonal and non-diagonal parts:

$$\frac{1}{\tau_{tr}^j} = \frac{m}{\hbar^3} \int_0^{2\pi} \frac{d\theta}{2\pi} w_{jj}(2p_F|\sin(\theta/2)|/\hbar)(1 - \cos\theta), \tag{22}$$

$$\frac{1}{\tau_1} = \frac{m}{\hbar^3} \int_0^{2\pi} \frac{d\theta}{2\pi} \left[w_+(2p_F|\sin(\theta/2)|/\hbar) - w_{lr}(2p_F|\sin(\theta/2)|/\hbar)\cos\theta \right].$$

The matrix kinetic equation is reduced to a system of linear algebraic equations for the components of the matrices $\hat{\xi}_{\pm}$:

$$(-i\omega \pm \omega_c + 1/\tau_{tr}^l)\xi_{\pm}^l - (2t_{lr}/\hbar)\xi_{\pm}^y = 1,$$

$$(-i\omega \pm \omega_c + 1/\tau_{tr}^r)\xi_{\pm}^r + (2t_{lr}/\hbar)\xi_{\pm}^y = 1, \tag{23}$$

and

$$(-i\omega \pm \omega_c + 1/\tau_1) \left| \begin{array}{c} \xi_{\pm}^x \\ \xi_{\pm}^y \end{array} \right| + \frac{\Delta}{\hbar} \left| \begin{array}{c} \xi_{\pm}^y \\ -\xi_{\pm}^x \end{array} \right| = \frac{t_{lr}}{\hbar} \left| \begin{array}{c} 0 \\ \xi_{\pm}^r - \xi_{\pm}^l \end{array} \right|. \tag{24}$$

Solving the system of equations (23) and (24), we describe the linear response in terms of the complex conductivity tensor whose components, $\sigma_{xx} = \sigma_{yy} = \sigma_d$ and $\sigma_{yx} = -\sigma_{xy} = \sigma_{\perp}$, are given by

$$\sigma_d = \sigma_0[\Psi(\Omega - \Omega_c) + \Psi(\Omega + \Omega_c)]/2,$$

$$\sigma_{\perp} = \sigma_0[\Psi(\Omega - \Omega_c) - \Psi(\Omega + \Omega_c)]/2. \tag{25}$$

In these expressions, $\sigma_0 = e^2 n \tau_{tr}/m$ is the averaged static conductivity of the 2D layers, $n = 2\rho_{2D}\varepsilon_F$ is the doubled (because of the presence of two layers) 2D electron density, and $\tau_{tr} = 2\tau_{tr}^l \tau_{tr}^r/(\tau_{tr}^l + \tau_{tr}^r)$ is the averaged transport time. Next, $\Omega_c = \tau_{tr}\omega_c$ and $\Omega = \tau_{tr}\omega$ are the dimensionless cyclotron frequency and perturbation field frequency. The function Ψ is given by the expression

$$\Psi(x) = \frac{1}{1 - ix} \left[1 + \mu^2 \frac{\lambda_x^2 + \delta^2}{[(1 - ix)^2 - \mu^2](\lambda_x^2 + \delta^2) + \Omega_T^2(1 - ix)\lambda_x} \right] \tag{26}$$

with $\lambda_x = 1 - ix\tau_1/\tau_{tr}$, $\mu = (\tau_{tr}^r - \tau_{tr}^l)/(\tau_{tr}^r + \tau_{tr}^l)$, $\delta = \tau_1\Delta/\hbar$, and $\Omega_T = 2t_{lr}\sqrt{\tau_{tr}\tau_1}/\hbar$. In the presence of the scattering asymmetry described by the parameter μ, the behavior of the conductivity tensor is rather sophisticated. The conductivity tensor in these conditions depends on the level splitting Δ and on the strength of the tunnel coupling characterized by the tunneling matrix element t_{lr}.

Let us analyze the behavior of the static conductivity at zero magnetic field. Equations (25) and (26) at $\Omega = \Omega_c = 0$ give us

$$\sigma_d = \sigma_0 \left(1 + \frac{\mu^2}{(1 - \mu^2) + \Omega_T^2/(1 + \delta^2)} \right). \qquad (27)$$

The conductivity is minimal at $\delta = 0$ and increases up to $\sigma_0/(1 - \mu^2)$ at $|\delta| \gg \Omega_T$. Therefore, the resistivity $\rho = 1/\sigma_d$ has a peak at $\Delta = 0$, when the levels are aligned. This phenomenon is called the resistance resonance. Its physical explanation is the following. When we connect two uncoupled ($|\Delta| \gg t_{lr}$) layers with different resistances in parallel, the current flows mostly through the layer with lower resistivity, the one which is less doped by impurities. When the tunnel-coupled states are considered, the wave functions of electrons in the resonance condition $\Delta \simeq 0$ are spread over both layers, and all electrons "feel" the scattering potential in average. The resistivity is larger in this case. It is important that the effect described above is suppressed when $\Omega_T = 2t_{lr}\sqrt{\tau_{tr}\tau_1}/\hbar$ decreases. This is because the scattering suppresses the tunnel coherence. The characteristic time for this suppression is $\sqrt{\tau_{tr}\tau_1}$.

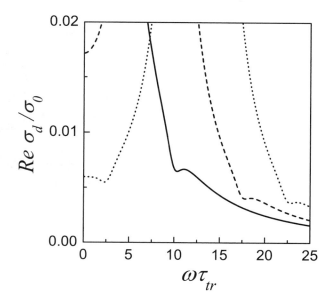

Figure 13.2. Weak resonant features in the frequency dependence of absorption coefficient in the coupled quantum wells with non-symmetric scattering at $2t_{lr}\tau_{tr}/\hbar = 10$, $\Delta = 0$, $\mu = 0.8$, and $\tau_1 = \tau_{tr}$. The curves are plotted for $\omega_c = 0$ (solid), $\omega_c\tau_{tr} = 7.5$ (dashed), and $\omega_c\tau_{tr} = 12.5$ (dotted).

In the presence of a magnetic field, both σ_d and $\sigma_\perp$ are non-zero. Considering the stationary case, it is convenient to present the dissipative resistivity ρ_d and Hall coefficient $R_H = \rho_{xy}/H$, since they are independent of the magnetic field for a single layer (or for the double-layer system at $\mu = 0$):

$$\rho_d \sigma_0 = 1 - \frac{\mu^2}{1 + \Omega_c^2} \left[1 - \Omega_T^2 \frac{1 + \delta^2 + \Omega_T^2 - 3\Omega_c^2}{(1 + \delta^2 + \Omega_T^2 - \Omega_c^2)^2 + 4\Omega_c^2} \right], \qquad (28)$$

$$\frac{R_H}{R_H^{(0)}} = 1 + \frac{\mu^2}{1 + \Omega_c^2} \left[1 - \Omega_T^2 \frac{3 + \delta^2 + \Omega_T^2 - \Omega_c^2}{(1 + \delta^2 + \Omega_T^2 - \Omega_c^2)^2 + 4\Omega_c^2} \right], \qquad (29)$$

where $R_H^{(0)} = 1/|e|nc$. The non-symmetric scattering causes a magnetic-field dependence of ρ_d and R_H. If $\Omega_T \to 0$, this dependence is given by a Lorentz function and can be explained by the presence of two groups of carriers with different mobilities. However, if the tunnel coherence is present, this dependence is more complicated, and one may even expect a negative magnetoresistance (problem 13.13). In this case, both ρ_d and R_H depend also on Δ and t_{lr}.

Finally, we consider the coefficient of absorption of electromagnetic waves, which is proportional to the real part of σ_d given by the general expressions (25) and (26). These expressions describe the cyclotron resonance absorption peak. In the resonance approximation, when $|\Omega - \Omega_c| \ll \Omega_c$ and $\Omega_c \gg 1$, this peak is symmetric and its maximum is placed at $\Omega = \Omega_c$. The real part of the conductivity in the maximum is equal to $\sigma_0 \Psi(0)/2 = \sigma_d/2$, where σ_d is given by Eq. (27). Therefore, the height of the cyclotron absorption peak is minimal at $\Delta = 0$. The relative width of the peak, which is estimated as $1/\Psi(0)$, is maximal in the resonance conditions $\Delta = 0$. The effect of resonant broadening of the cyclotron absorption peak has the same origin as the resistance resonance. Apart from this peak, one should also mention the resonant features of absorption appearing at combined frequencies, when

$$\omega = |\omega_c \pm \Delta_T/\hbar|, \qquad (30)$$

see Fig. 13.2. These features exist only in the case of non-symmetric scattering.

66. Auger Processes

The electron-electron collisions conserve the total number of electrons as well as their total energy and momentum. However, if the electrons of several bands (or subbands) are involved in such collisions, the number of electrons in each band may change due to interband transitions. Let us consider the kinetic equation describing electron-electron scattering in a two-band model, when the indices of quantum states in the electron-

electron collision integral given by Eqs. (31.19) and (31.20) include the band index j ($j = c, v$). Summing this kinetic equation over momentum and spin, we find

$$\frac{\partial n_j}{\partial t} = \frac{1}{V} \sum_{\sigma \mathbf{p}} J_{ee}(f|j\sigma \mathbf{p}), \tag{1}$$

where n_j is the electron density in the band j. The balance equations of this kind describe the Auger recombination process. Owing to conservation of the total number of electrons, one has $(\partial n_c/\partial t) = -(\partial n_v/\partial t)$. Therefore, it is sufficient to consider Eq. (1) only for $j = c$. Below we assume that the electron bands are spin-degenerate and isotropic: $\varepsilon_{j\sigma \mathbf{p}} = \varepsilon_{jp}$. In these conditions,

$$\sum_{\sigma} J_{ee}(f|j\sigma \mathbf{p}) = \frac{2\pi}{\hbar} \sum_{j'j_1j_1'} \sum_{\mathbf{p}'\mathbf{p}_1\mathbf{p}_1'} \delta_{\mathbf{p}+\mathbf{p}',\mathbf{p}_1+\mathbf{p}_1'}$$

$$\times V(j\mathbf{p}\ j'\mathbf{p}'|j_1\mathbf{p}_1\ j_1'\mathbf{p}_1')\delta(\varepsilon_{jp} + \varepsilon_{j'p'} - \varepsilon_{j_1p_1} - \varepsilon_{j_1'p_1'}) \tag{2}$$

$$\times [f_{j_1\mathbf{p}_1}f_{j_1'\mathbf{p}_1'}(1 - f_{j\mathbf{p}})(1 - f_{j'\mathbf{p}'}) - f_{j\mathbf{p}}f_{j'\mathbf{p}'}(1 - f_{j_1\mathbf{p}_1})(1 - f_{j_1'\mathbf{p}_1'})].$$

The matrix element describing the scattering is written as

$$V(j\mathbf{p}\ j'\mathbf{p}'|j_1\mathbf{p}_1\ j_1'\mathbf{p}_1') = \left(\frac{4\pi e^2 \hbar^2}{\epsilon V}\right)^2 \sum_{\sigma\sigma'\sigma_1\sigma_1'}$$

$$\times \left\{ \frac{|\varphi_{j\sigma,j_1\sigma_1}(\mathbf{p},\mathbf{p}_1)|^2 |\varphi_{j'\sigma',j_1'\sigma_1'}(\mathbf{p}',\mathbf{p}_1')|^2}{|\mathbf{p}-\mathbf{p}_1|^4} \right. \tag{3}$$

$$\left. -\mathrm{Re}\frac{\varphi_{j\sigma,j_1\sigma_1}(\mathbf{p},\mathbf{p}_1)\varphi_{j_1\sigma_1,j'\sigma'}(\mathbf{p}_1,\mathbf{p}')\varphi_{j'\sigma',j_1'\sigma_1'}(\mathbf{p}',\mathbf{p}_1')\varphi_{j_1'\sigma_1',j\sigma}(\mathbf{p}_1',\mathbf{p})}{|\mathbf{p}-\mathbf{p}_1|^2|\mathbf{p}'-\mathbf{p}_1|^2} \right\},$$

where $\varphi_{j\sigma,j_1\sigma_1}(\mathbf{p},\mathbf{p}_1)$ is the overlap factor defined according to

$$\langle j\sigma\mathbf{p}|e^{-i\mathbf{q}\cdot\mathbf{x}}|j_1\sigma_1\mathbf{p}_1\rangle = \delta_{\mathbf{p}+\hbar\mathbf{q},\mathbf{p}_1}\varphi_{j\sigma,j_1\sigma_1}(\mathbf{p},\mathbf{p}_1). \tag{4}$$

To express $\varphi_{j\sigma,j_1\sigma_1}(\mathbf{p},\mathbf{p}_1)$ through the parameters of the two-band model of Appendix B, one should use Eq. (B.24).

Equation (1) with the collision integral defined by Eqs. (2)–(4) is written as ($j = c$)

$$\frac{\partial n_c}{\partial t} = \frac{2\pi}{\hbar V} \sum_{j'j_1j_1'} \sum_{\mathbf{p}\mathbf{p}'\mathbf{p}_1\mathbf{p}_1'} \delta_{\mathbf{p}+\mathbf{p}',\mathbf{p}_1+\mathbf{p}_1'}$$

$$\times V(c\mathbf{p}\ j'\mathbf{p}'|j_1\mathbf{p}_1\ j_1'\mathbf{p}_1')\delta(\varepsilon_{cp} + \varepsilon_{j'p'} - \varepsilon_{j_1p_1} - \varepsilon_{j_1'p_1'}) \tag{5}$$

$$\times [f_{j_1\mathbf{p}_1}f_{j_1'\mathbf{p}_1'}(1 - f_{c\mathbf{p}})(1 - f_{j'\mathbf{p}'}) - f_{c\mathbf{p}}f_{j'\mathbf{p}'}(1 - f_{j_1\mathbf{p}_1})(1 - f_{j_1'\mathbf{p}_1'})].$$

Since each of the band indices j', j_1, and j'_1 can be either c or v, there are eight possible terms in the sum in Eq. (5), corresponding to the processes $cj' \leftrightarrow j_1 j'_1$. Some of these processes, however, are either forbidden by the energy conservation law, such as $cc \leftrightarrow vv$, or do not contribute to the sum, such as $cc \leftrightarrow cc$, $cv \leftrightarrow cv$, and $cv \leftrightarrow vc$. Each of the processes $cc \leftrightarrow vc$ and $cv \leftrightarrow cc$ contributes to the sum separately, but their total contribution is equal to zero (problem 13.14). Therefore, there are only two essential contributions, $cc \leftrightarrow cv$ and $cv \leftrightarrow vv$. In the first process, the electron drops from the conduction band to the valence one, and its energy is spent to excite a transition of the other electron inside the conduction band. The second process is similar: when an electron drops from the conduction band to the valence one, the other electron jumps up inside the valence band. Denoting these processes by the indices Ac and Av, we write Eq. (5) in the form

$$\frac{\partial n_c}{\partial t} = -J_{Ac} - J_{Av}, \qquad (6)$$

$$J_{Aj} = \frac{2\pi}{\hbar V} \sum_{\mathbf{p}\mathbf{p}'\mathbf{p}_1\mathbf{p}'_1} \delta_{\mathbf{p}+\mathbf{p}',\mathbf{p}_1+\mathbf{p}'_1} V(c\mathbf{p}j\mathbf{p}'|j\mathbf{p}_1 v\mathbf{p}'_1) \delta(\varepsilon_{cp} + \varepsilon_{jp'} - \varepsilon_{jp_1} - \varepsilon_{vp'_1})$$

$$\times [f_{c\mathbf{p}} f_{j\mathbf{p}'}(1 - f_{j\mathbf{p}_1})(1 - f_{v\mathbf{p}'_1}) - f_{j\mathbf{p}_1} f_{v\mathbf{p}'_1}(1 - f_{c\mathbf{p}})(1 - f_{j\mathbf{p}'})].$$

Using the quasi-equilibrium Fermi distribution functions with chemical potentials μ_c and μ_v, $f_{j\mathbf{p}} = [e^{(\varepsilon_{j\mathbf{p}} - \mu_j)/T} + 1]^{-1}$, we rewrite the factor in the square brackets as

$$[\dots] = f_{c\mathbf{p}} f_{j\mathbf{p}'}(1 - f_{j\mathbf{p}_1})(1 - f_{v\mathbf{p}'_1}) \left[1 - e^{(\mu_v - \mu_c)/T} \right]. \qquad (7)$$

In the case of small deviations of the chemical potentials from the equilibrium ones, when $\mu_c = \mu + \Delta\mu_c$ and $\mu_v = \mu + \Delta\mu_v$, one approximately has $1 - e^{(\mu_v - \mu_c)/T} \simeq (\Delta\mu_c - \Delta\mu_v)/T$, where $|\Delta\mu_c - \Delta\mu_v| \ll T$ is assumed. Using $\Delta\mu_c = \Delta n_c/(dn_c/d\mu_c)$ and $\Delta\mu_v = \Delta n_h/(dn_h/d\mu_v)$, where Δn_c and Δn_h are the excess densities of electrons and holes, we employ the electric neutrality equation $\Delta n_c = \Delta n_h \equiv \Delta n$ and obtain

$$1 - e^{(\mu_v - \mu_c)/T} \simeq \frac{\Delta n}{T} \left[\left(\frac{dn_c}{d\mu_c} \right)^{-1} - \left(\frac{dn_h}{d\mu_v} \right)^{-1} \right]. \qquad (8)$$

This equation is simplified for the case of non-degenerate electrons, when $dn_c/d\mu_c = n_c/T$ and $dn_h/d\mu_v = -n_h/T$. In the intrinsic (non-doped) semiconductor, when $n_c = n_h \equiv n$, Eq. (8) takes the most simple form $1 - e^{(\mu_v - \mu_c)/T} \simeq 2\Delta n/n$, and the right-hand side of the balance equation can be represented as $-J_{Ac} - J_{Av} = -\nu_A \Delta n$ while the left-hand side is written as $\partial \Delta n / \partial t$. Such equation describes an exponential decrease of the excess density of electrons with a characteristic rate ν_A.

Calculating the Auger recombination rate ν_A for realistic (many-band) models is a complicated problem, which requires a correct account of non-parabolicity of electron spectrum and momentum dependence of the overlap factors standing in Eq. (3) (problem 13.15). The sums over momenta in Eq. (6) are usually calculated numerically. Below we consider the simplest approximation, when the momentum dependence of the overlap factors is neglected so that b_{jj_1} and $B_{jj_1j'j'_1}$ introduced by the relations $\sum_{\sigma \sigma_1} |\varphi_{j\sigma,j_1\sigma_1}(\mathbf{p}, \mathbf{p}_1)|^2 = 2b_{jj_1}$ and $\sum_{\sigma \sigma' \sigma_1 \sigma'_1}$ Re $\varphi_{j\sigma,j_1\sigma_1}(\mathbf{p}, \mathbf{p}_1)\varphi_{j_1\sigma_1,j'\sigma'}(\mathbf{p}_1, \mathbf{p}')\varphi_{j'\sigma',j'_1\sigma'_1}(\mathbf{p}', \mathbf{p}'_1)\varphi_{j'_1\sigma'_1,j\sigma}(\mathbf{p}'_1, \mathbf{p}) = 2B_{jj_1j'j'_1}$ are constants. We also consider conduction and valence bands with $m_c = m$ and $m_v \to \infty$. In the model with infinitely large m_v, only the contribution J_{Ac} is non-zero. It is determined by

$$J_{Ac} = \frac{4\pi}{\hbar V^3} \left(\frac{4\pi e^2 \hbar^2}{\epsilon} \right)^2 \sum_{\mathbf{p}\mathbf{p}'\mathbf{p}_1} \left[\frac{2b_{cc}b_{cv}}{|\mathbf{p} - \mathbf{p}_1|^4} - \frac{B_{cccv}}{|\mathbf{p} - \mathbf{p}_1|^2 |\mathbf{p}' - \mathbf{p}_1|^2} \right]$$

$$\times \delta(\varepsilon_{cp} + \varepsilon_{cp'} - \varepsilon_{cp_1} + \varepsilon_g) f_{c\mathbf{p}} f_{c\mathbf{p}'} (1 - f_{c\mathbf{p}_1}) e^{-(\mu_v + \varepsilon_g)/T} (1 - e^{(\mu_v - \mu_c)/T}), \quad (9)$$

To obtain this expression, we have taken into account that the holes at $m_v \to \infty$ are always non-degenerate (note that n_h must be finite) so that $1 - f_{v\mathbf{p}'_1}$ has been replaced by $e^{(\varepsilon_{vp'_1} - \mu_v)/T}$. Next, $\varepsilon_{vp'_1}$ has been replaced by $-\varepsilon_g$ (the energy is counted from the bottom of the conduction band), and the integral over $\mathbf{p}'_1$ has been taken by using the momentum conservation law. The expression (9) does not contain any parameters related to the energy spectrum and distribution of the valence-band electrons apart from the factor $e^{-\mu_v/T}(1 - e^{(\mu_v - \mu_c)/T})$ which is finite. Integrating over the angles of $\mathbf{p}$ and $\mathbf{p}'$ in Eq. (9), we obtain

$$J_{Ac} = \frac{8e^4}{\pi^3 \epsilon^2 \hbar^6} \int_0^\infty p^2 dp \int_0^\infty p'^2 dp' \int_0^\infty p_1^2 dp_1 \left\{ \frac{2b_{cc}b_{cv}}{(p^2 - p_1^2)^2} \right.$$

$$\left. - \frac{B_{cccv}}{4pp'p_1^2} \ln \frac{p_1 + p}{p_1 - p} \ln \frac{p_1 + p'}{p_1 - p'} \right\} \delta \left(\frac{p^2}{2m} + \frac{p'^2}{2m} - \frac{p_1^2}{2m} + \varepsilon_g \right) \quad (10)$$

$$\times f_{c\mathbf{p}} f_{c\mathbf{p}'} (1 - f_{c\mathbf{p}_1}) e^{-(\mu_v + \varepsilon_g)/T} (1 - e^{(\mu_v - \mu_c)/T}).$$

Below we take into account that the temperature (or the Fermi energy of electrons) is small in comparison to the gap energy ε_g. In these conditions, p^2 and p'^2 are small in comparison to p_1^2, the factor $1 - f_{c\mathbf{p}_1}$ can be replaced by unity, and the integrals in Eq. (10) are taken elementary. Finally, using $1 - e^{(\mu_v - \mu_c)/T} \simeq 2\Delta n/n$, we express the Auger recombination rate as

$$\nu_A = \frac{4\sqrt{2}(2b_{cc}b_{cv} - B_{cccv})\pi e^4 n}{\epsilon^2 m^{1/2}\varepsilon_g^{3/2}} e^{-(\mu_v + \varepsilon_g)/T}$$

$$= \frac{4(2b_{cc}b_{cv} - B_{cccv})e^4 m T^{3/2}}{\pi^{1/2}\hbar^3 \epsilon^2 \varepsilon_g^{3/2}} e^{-\varepsilon_g/T}. \tag{11}$$

To obtain the last equation, we have taken into account that $\mu_v \simeq \mu_c$ near equilibrium and expressed μ_c through the equilibrium electron density n for the case of non-degenerate electrons, $ne^{-\mu_c/T} \simeq (mT)^{3/2}/\sqrt{2}\pi^{3/2}\hbar^3$. The rate (11) is estimated as $(\varepsilon_B/\hbar)(\bar{\varepsilon}/\varepsilon_g)^{3/2}e^{-\varepsilon_g/T}$, where ε_B is the Bohr energy and $\bar{\varepsilon}$ is the mean kinetic energy of electrons. The presence of the exponentially small factor, $e^{-\varepsilon_g/T}$, strongly suppresses the Auger recombination in wide-gap materials. In narrow-gap semiconductors, the Auger processes give a considerable contribution to the interband recombination at room temperature.

The exponential factor $e^{-\varepsilon_g/T}$ is modified if the hole mass m_v is finite. Without giving the details of calculations for this case, we notice that as the ratio $r = m_c/m_v$ increases, the momenta p and p' effectively contributing to the sum in Eq. (6) become comparable to p_1 so that the effective gap for electron transitions increases. Assuming that the electron gas is non-degenerate, we write the exponential factor standing in the expression for J_{Ac} (see Eqs. (6)−(8)) as $\exp(-p_1^2/2m_c T)$. The momentum p_1 is determined from the conservation law

$$p^2 + p'^2 - p_1^2 + 2m_c\varepsilon_g + r(\mathbf{p}_1 - \mathbf{p} - \mathbf{p}')^2 = p^2 + p'^2 - p_1^2 + 2m_c\varepsilon_g$$

$$+r(p_1^2 + p^2 + p'^2 + 2pp'\cos\theta_{pp'} - 2pp_1\cos\theta_{pp_1} - 2p'p_1\cos\theta_{p'p_1}) = 0, \tag{12}$$

where $\theta_{pp'}$ is the angle between the vectors $\mathbf{p}$ and $\mathbf{p}'$, and the other angles are defined by the same notation. Let us minimize the function p_1^2 implicitly defined by the conservation law (12). This function depends on five variables: $p, p', \theta_{pp'}, \theta_{pp_1}$, and $\theta_{p'p_1}$. By noting that $\mathbf{p}$ and $\mathbf{p}'$ enter the conservation law symmetrically, we assume them equal to each other. This reduces the number of variables of the function p_1^2 to two, p and θ_{pp_1}. Solving the quadratic equation obtained in this way, we find p_1:

$$p_1 = -\frac{2rp\cos\theta_{pp_1}}{1-r} + \sqrt{2\frac{1 + r - 2r^2\sin^2\theta_{pp_1}}{(1-r)^2}p^2 + \frac{2m_c\varepsilon_g}{1-r}}. \tag{13}$$

This function decreases with increasing $\cos\theta_{pp_1}$ for arbitrary p. Thus, we put $\cos\theta_{pp_1} = 1$. After minimizing p_1 as a function of p, we obtain

$p_{1min}^2 = 2m_c \varepsilon_g (1 + 2r)/(1 + r)$. Therefore, the exponential temperature dependence of the Auger recombination rate is given by

$$\nu_A \propto \exp\left(-\frac{\varepsilon_g}{T}\frac{1 + 2r}{1 + r}\right). \tag{14}$$

One should remember that there is also a contribution of the Av process, which brings a similar exponent with m_c/m_v replaced by m_v/m_c, as follows from the mutual symmetry of the Ac and Av processes. As long as $m_c < m_v$, there is no need to account for the Av process. In the case of equal c- and v-band masses, the effective gap is maximal and equal to $3\varepsilon_g/2$.

A more detailed calculation based upon many-band models and accounting for the momentum dependence of the overlap integrals can give another temperature dependence of the Auger recombination rate in the prefactor of the exponent. The effects of non-parabolicity also modify, to some extent, the exponential factor. We do not consider such calculations in this book. It is important to mention, however, that for the symmetric variant ($M \to \infty$) of the two-band model described in Appendix B the Auger transitions appear to be forbidden because the conservation of energy and momentum cannot be satisfied simultaneously (problem 13.16). Since the Hamiltonian of the symmetric two-band model is formally equivalent to the Dirac Hamiltonian, the absence of the Auger transitions in this model correlates with a known result of the relativistic quantum theory: the annihilation of an electron-positron pair with energy transferred to another electron or positron is forbidden.

Let us study the intersubband Auger transitions in a quantum well. Employing the general form of electron-electron collision integral given by Eqs. (31.19) and (31.20), we write the kinetic equation as follows:

$$\frac{\partial f_{j\mathbf{p}}}{\partial t} = \frac{(2\pi\hbar)^3}{m^2 L^4} \sum_{j_1 j' j_1'} \sum_{\mathbf{p}_1 \mathbf{p}' \mathbf{p}_1'} \delta_{\mathbf{p}+\mathbf{p}',\mathbf{p}_1+\mathbf{p}_1'} \delta(\varepsilon_{jp} + \varepsilon_{j'p'} - \varepsilon_{j_1 p_1} - \varepsilon_{j_1' p_1'})$$

$$\times \left[2M_{jj_1 j' j_1'}^2 (|\mathbf{p} - \mathbf{p}_1|/\hbar) - M_{jj_1 j' j_1'}(|\mathbf{p} - \mathbf{p}_1|/\hbar) M_{j_1' j j_1 j'}(|\mathbf{p}' - \mathbf{p}_1|/\hbar) \right]$$

$$\times (f_{j_1 \mathbf{p}_1} f_{j_1' \mathbf{p}_1'} - f_{j\mathbf{p}} f_{j'\mathbf{p}'}), \tag{15}$$

where the sum over spin variables is already calculated. We have employed the coefficients $M_{abcd}(q)$ defined by Eq. (29.11) (these coefficients are real since the wave functions describing the confinement are chosen to be real) and assumed that electrons are non-degenerate. The index j is now a subband number, and the electron spectrum is given by $\varepsilon_{jp} = \varepsilon_j + p^2/2m$. Considering only the ground and first excited subbands, we apply Eq. (15) for describing the relaxation of electrons from the excited subband to the ground one. Let us put $j = 2$ and sum Eq. (15) over $\mathbf{p}$, with the result $\partial n_2/\partial t = -J_A$. The Auger processes contributing to such relaxation are $21 \leftrightarrow 11$, $22 \leftrightarrow 21$, $22 \leftrightarrow 11$, $22 \leftrightarrow 12$, and $21 \leftrightarrow 22$. The contributions of the last two processes cancel one another, as we already know from the consideration of interband Auger transitions. Note that the conservation rules do not forbid the process $22 \leftrightarrow 11$, since the electron spectra for each subband are described by the same effective mass. Nevertheless, if the subband 2 is much less populated than the ground one, the processes $22 \leftrightarrow 21$ and $22 \leftrightarrow 11$ give a small contribution, and only one process, $21 \leftrightarrow 11$, remains essential. Taking it into account, we obtain

$$
J_A = 2\frac{(2\pi\hbar)^3}{m^2 L^6} \sum_{\mathbf{p}\mathbf{p}_1\mathbf{p}'\mathbf{p}'_1} \delta_{\mathbf{p}+\mathbf{p}',\mathbf{p}_1+\mathbf{p}'_1} \left[2M_{2111}^2(q) - M_{2111}(q)M_{2111}(q')\right]
$$

$$
\times\delta(\varepsilon_{2p} + \varepsilon_{1p'} - \varepsilon_{1p_1} - \varepsilon_{1p'_1})(f_{2\mathbf{p}}f_{1\mathbf{p}'} - f_{1\mathbf{p}_1}f_{1\mathbf{p}'_1}), \tag{16}
$$

where $\mathbf{q} = (\mathbf{p} - \mathbf{p}_1)/\hbar$ and $\mathbf{q}' = (\mathbf{p}' - \mathbf{p}_1)/\hbar$. It is convenient to calculate the sums by using the variables $\mathbf{q}$, $\mathbf{q}'$, and $\mathbf{P} = (\mathbf{p} + \mathbf{p}')/2$. Below we neglect the arrival term proportional to $f_{1\mathbf{p}_1}f_{1\mathbf{p}'_1}$ because of the assumed strong inequality $T \ll \varepsilon_{21}$, where $\varepsilon_{21} = \varepsilon_2 - \varepsilon_1$ is the energy level separation. Approximating $f_{2\mathbf{p}}$ and $f_{1\mathbf{p}'}$ by the Maxwell distribution functions with effective chemical potentials μ_2 and μ_1, we find

$$
J_A \simeq \frac{2\pi^2\hbar^3}{m^3 T^2}n_1 n_2 \int_0^\infty P dP \int_0^\infty q dq \int_0^\infty q' dq'
$$

$$
\times \int_0^\pi \frac{d\varphi}{\pi} \left[2M_{2111}^2(q) - M_{2111}(q)M_{2111}(q')\right] \tag{17}
$$

$$
\times\delta\left(\frac{m\varepsilon_{21}}{\hbar^2} - qq'\cos\varphi\right)\exp\left[-\frac{4P^2 + \hbar^2(q^2 + q'^2 - 2qq'\cos\varphi)}{4mT}\right],
$$

where φ is the angle between $\mathbf{q}$ and $\mathbf{q}'$. Note that we have expressed the factors $e^{\mu_1/T}$ and $e^{\mu_2/T}$ through the densities of the 2D electrons in the subbands 1 and 2 according to $n_j = e^{\mu_j/T}\rho_{2D}T$, where $j = 1,2$ and $\rho_{2D} = m/\pi\hbar^2$. Integrating over φ with the use of the δ-function and, independently, over P, we reduce Eq. (17) to

$$
J_A \simeq \frac{\pi\hbar^3}{m^2 T}n_1 n_2 e^{\varepsilon_{21}/2T} \int_0^\infty q dq \int_0^\infty q' dq'
$$

$$
\times\frac{\exp\left[-\hbar^2(q^2 + q'^2)/4mT\right]}{\sqrt{(qq')^2 - (m\varepsilon_{21}/\hbar^2)^2}}[2M_{2111}^2(q) - M_{2111}(q)M_{2111}(q')], \tag{18}
$$

where the expression under the square root must be positive. Analyzing Eq. (18) at low temperatures, when $2T \ll \varepsilon_{21}$, we find that the main contribution to the integrals comes from $q \simeq q' \simeq \sqrt{m\varepsilon_{21}}/\hbar$. Calculating the remaining integrals, we finally obtain

$$J_A \simeq \frac{\pi^2 \hbar}{m} M_{2111}^2 (\sqrt{m\varepsilon_{21}}/\hbar) n_1 n_2. \qquad (19)$$

Since $q \simeq q'$, the exchange contribution reduces the result by half.

The details of the calculation make it clear that the main contribution to the Auger relaxation comes from the process when an electron from the bottom of the excited subband descends to the state with the energy $\varepsilon_{21}/2$ and momentum $\mathbf{p}$ in the ground-state subband, while an electron from the bottom of the ground-state subband jumps to the state with the same energy and the opposite momentum, $-\mathbf{p}$. Therefore, Eq. (19) remains valid for any distribution of electrons in the subbands, provided that the mean kinetic energies of electrons in both subbands are small in comparison to $\varepsilon_{21}/2$. Since there is no energy gap, the intersubband Auger transitions considered above lead to a fast depopulation of the second subband, with the rate $\nu_A^{(21)} \simeq \pi M_{2111}^2 (\sqrt{m\varepsilon_{21}}/\hbar)(\bar{\varepsilon}/\hbar)$, where $\bar{\varepsilon} = T$ for the non-degenerate case and $\bar{\varepsilon} = n_1/\rho_{2D} = \varepsilon_F$ for the degenerate case. However, if the quantum well is symmetric, the coefficient $M_{2111}(q)$ is zero, which means that the Auger transitions involving the states with different parity of their wave functions appear to be forbidden. In asymmetric wells, the Auger processes can give a substantial contribution to intersubband relaxation. For typical values of n_1, this contribution is comparable to the contribution (38.9) due to spontaneous *LO* phonon emission.

We now return to consideration of the interband Auger transitions. If there is an excess of high-energy electrons or holes, as in the case when the electron gas is heated by an applied electric field, the Auger processes can increase the electron and hole densities. This phenomenon is called the impact ionization. To describe it, on can use the balance equation (6). Let us consider this equation under the assumption that the distribution function of electrons in the conduction band is controlled by electron-electron scattering and represented in the form (31.25). This function is characterized by the effective temperature T_e and drift velocity $\mathbf{u}$. In what follows, we again use the limit of infinite valence-band mass, retaining only the term with $j = c$ in Eq. (6). Since the acceleration of the valence-band electrons by the electric field is not effective, their effective temperature is equal to the equilibrium temperature T, and their drift velocity is zero. Both electrons and holes are assumed to be non-degenerate. Substituting the distribution functions described above into the expression for J_{Ac} given by Eq. (9), we rewrite the balance equation as

$$\frac{\partial n_c}{\partial t} = \frac{4\pi}{\hbar V^3} \left(\frac{4\pi e^2 \hbar^2}{\epsilon} \right)^2 \sum_{\mathbf{p}\mathbf{p}'\mathbf{p}_1} \left[\frac{2b_{cc}b_{cv}}{|\mathbf{p} - \mathbf{p}_1|^4} - \frac{B_{cccv}}{|\mathbf{p} - \mathbf{p}_1|^2 |\mathbf{p}' - \mathbf{p}_1|^2} \right]$$

$$\times \delta(p^2/2m + p'^2/2m - p_1^2/2m + \varepsilon_g) \exp[-(p_1^2/2mT_e - \mu_c)/T_e] \qquad (20)$$

$$\times \left\{ e^{\mathbf{u}\cdot\mathbf{p}_1/T_e} - e^{\mathbf{u}\cdot(\mathbf{p}+\mathbf{p}')/T_e} e^{(\mu_c+\varepsilon_g)/T_e} e^{-(\mu_v+\varepsilon_g)/T} \right\}.$$

Evaluating the term in the braces $\{\dots\}$ in the case of small deviations from equilibrium ($\mu_c = \mu + \Delta\mu_c$, $\mu_v = \mu + \Delta\mu_v$ and $T_e = T + \Delta T$), we use $\Delta n_h = -\Delta\mu_v n_h/T$ and the following relation for the deviation of electron density obtained from Eq. (31.25):

$$\Delta n_c = \Delta\mu_c \frac{n_c}{T} + \Delta T \frac{n_c}{T} \left(\frac{3}{2} - \frac{\mu + mu^2/2}{T} \right). \qquad (21)$$

This expression contains the contribution quadratic in u. After the angular averaging in Eq. (20), the terms containing $\mathbf{u}$ in the factor $\{\dots\}$ also lead to a contribution quadratic in u, since the electron spectrum is isotropic. In the linear approximation with respect to small deviations of electron densities and temperature, all such quadratic terms are neglected. Therefore,

$$\{\dots\} \simeq (\mu + \Delta\mu_v + \varepsilon_g)/T - (\mu + \Delta\mu_c + \varepsilon_g)/(T + \Delta T)$$

$$\simeq -2\frac{\Delta n}{n} + \frac{\Delta T}{T} \left(\frac{\varepsilon_g}{T} + \frac{3}{2} \right), \qquad (22)$$

where the last equation is written for an intrinsic material (with $n_c = n_h$). Employing our results for the Auger recombination rate in the linear regime under the conditions $T \ll \varepsilon_g$, we write the linearized balance equation as

$$\frac{\partial \Delta n}{\partial t} = -\nu_A [\Delta n - (n\varepsilon_g/2T^2)\Delta T], \qquad (23)$$

where ν_A is given by Eq. (11). If the second, proportional to ΔT, term on the right-hand side becomes greater than the first term, the electron density increases with time. This process, however, saturates if ΔT is constant. Expressing the electron temperature through the electric field E according to Eq. (36.27), one can find the steady-state excess density which depends on the energy relaxation rate $\nu^{(e)}$:

$$\Delta n = \frac{e^2 E^2 \tau_{tr} \varepsilon_g}{2m\nu^{(e)} T^2}. \qquad (24)$$

Since the result is proportional to E^2, and the square of the drift velocity is also proportional to E^2 (note that $u = eE\tau_{tr}/m$ according to Eq. (36.12)), one may improve the consideration by including the terms $\propto u^2$ into the right-hand side of Eq. (22) (problem 13.17). However, under the conditions $\nu^{(e)} \ll \tau_{tr}^{-1}$ which are often met in semiconductors, the square of the drift velocity increases with increasing field much slower than the effective temperature. Therefore, Eq. (24) is justified. If, apart from the Auger processes, there are other mechanisms of interband recombination, the steady-state excess density depends on the ratio of the corresponding recombination rates to ν_A. Equations (23) and (24) describe the case of quasi-isotropic electron distribution.

To investigate the behavior of electron system beyond the linear approximation, we calculate the sums over momenta in Eq. (20), assuming that $T_e \ll \varepsilon_g$, when $p^2 \ll p_1^2$ and $p'^2 \ll p_1^2$. The balance equation becomes

$$\frac{\partial n_c}{\partial t} = n_c \frac{2(2b_{cc}b_{cv} - B_{cccv})e^4 m T_e^{3/2}}{\pi^{1/2}\hbar^3 \epsilon^2 \varepsilon_g^{3/2}} e^{-\varepsilon_g/T_e} \tag{25}$$

$$\times \left[\frac{\sinh \sqrt{2m\varepsilon_g}u/T_e}{\sqrt{2m\varepsilon_g}u/T_e} e^{-mu^2/2T_e} - e^{(\mu_c+\varepsilon_g)/T_e} e^{-(\mu_v+\varepsilon_g)/T} e^{mu^2/2T_e} \right].$$

The first term on the right-hand side of this equation is calculated in the approximation $mu^2 \ll \varepsilon_g$. This term can be presented as $\nu_{ii} n_c$, where ν_{ii} is the rate of impact ionization under the condition that the second term, which describes Auger recombination of hot electrons, can be neglected in comparison to the first one. Owing to the presence of the factor $\sinh \sqrt{2m\varepsilon_g}u/T_e$, this rate exponentially increases with increasing drift velocity at $u > T_e/\sqrt{2m\varepsilon_g}$. Still, if $2mu^2 \ll \varepsilon_g$, the dependence of the impact ionization rate on the electric field is determined mostly by the factor $e^{-\varepsilon_g/T_e}$. Assuming that the electron temperature is proportional to E^2, one gets the following law:

$$\nu_{ii} \propto \exp\left(-E_0^2/E^2\right), \tag{26}$$

where E_0 is a threshold field. To determine this field, one should consider a system of energy balance and momentum balance equations for conduction-band electrons, taking into account the contribution of the Auger processes in these equations.

Application of the distribution (31.25) for calculating the impact ionization rate in strong electric fields is not enough justified. In fact, a conduction-band electron must have a very high energy ($\varepsilon \sim \varepsilon_g$) to excite the interband transition of another electron, and it is not evident that Eq. (31.25) describes such high-energy electrons as well. There exists a different approach to the problem. It is assumed that the impact ionization is produced by the electrons which have avoided collisions and gained a high energy. The number of these electrons is small. If l is the mean free path between the collisions, considered as a constant, and ε_g is the threshold energy for ionization in the intrinsic semiconductor, the probability for an electron to avoid a collision is proportional to $\exp(-\varepsilon_g/|eE|l)$. Therefore, the ionization rate, in the simplest approximation, takes the form

$$\nu_{ii} \propto \exp\left(-\varepsilon_g/|eE|l\right). \tag{27}$$

The field dependence in this expression is distinctly different from the one given by Eq. (26). In practice, the interval of the fields over which ν_{ii} is measured is often too small to decide between the two predicted field dependences, and both types have been reported in the literature.

67. Kondo Effect

The atoms of the transition metals, which have in the isolated state unfilled d- or f- shells and a non-zero electron spin, can retain this property when embedded in a non-magnetic metal. The interaction of electrons with such magnetic impurities involves manifestation of the spin degree of freedom which leads to certain peculiarities of the conductivity. The most important feature of the electron-impurity scattering in this system is that this scattering becomes spin-dependent. Indeed, when the incoming electron with momentum $\mathbf{p}$ has the opposite projection of its spin with respect to the electron spin of the impurity atom, the scattering can proceed in the following ways: *i)* the electron localized at the magnetic impurity is transformed into a free electron with momentum $\mathbf{p}'$, and then the incoming electron is transformed into a localized one, and *ii)* the incoming electron is transformed into a localized one so that the intermediate state is a filled shell containing two electrons with opposite spins, then one of the localized electrons is transformed into a free electron with momentum $\mathbf{p}'$. The process *(ii)* does not exist in the case when the localized and incoming electrons have the same projection of spin, because of the Pauli principle. Therefore, the scattering probabilities for these two cases are different, and the Hamiltonian of electron-impurity interaction should have an additional term depending on the spin polarization of the free electron with respect to that of the localized electron.

The term which we add to the potential energy $U_{im}(\mathbf{r})$ of Eq. (7.1) can be written as a matrix

$$-\sum_n J(\mathbf{r} - \mathbf{R}_n)\hat{\boldsymbol{\sigma}} \cdot \widehat{\mathbf{S}}_n \tag{1}$$

where $\mathbf{R}_n$ is the coordinate of the n-th impurity, $\widehat{\mathbf{S}}_n$ is the impurity spin operator (if the absolute value of spin is $S = 1/2$, this operator is equal to the vector of Pauli matrices $\hat{\boldsymbol{\sigma}}$ divided by 2), and the energy $J(\mathbf{r} - \mathbf{R}_n)$ describes the strength of the spin-dependent contribution to the interaction. The spin-dependent factor $\hat{\boldsymbol{\sigma}} \cdot \widehat{\mathbf{S}}_n$ makes the expression (1) essentially different from $U_{im}(\mathbf{r})$ of Eq. (7.1). In the second quantization representation, the term (1) is written as a spin-dependent Hamiltonian of the interaction,

$$\widehat{\delta H}_{e,im} = - \sum_{n\sigma\sigma'} \int d\mathbf{r} \, J(\mathbf{r} - \mathbf{R}_n)\hat{\Psi}^+_{\sigma'\mathbf{r}}(\hat{\boldsymbol{\sigma}})_{\sigma'\sigma} \cdot \widehat{\mathbf{S}}_n \hat{\Psi}_{\sigma\mathbf{r}}. \qquad (2)$$

The index σ numbers the spin states of electrons ($+1$ and -1, corresponding to the spinors $\left|\begin{array}{c}1\\0\end{array}\right|$ and $\left|\begin{array}{c}0\\1\end{array}\right|$, respectively), $(\hat{\boldsymbol{\sigma}})_{\sigma'\sigma} \equiv \langle\sigma'|\hat{\boldsymbol{\sigma}}|\sigma\rangle$ are the matrix elements of the Pauli matrices, and $\hat{\Psi}_{\sigma\mathbf{r}}$ is the field operator of electrons. Since the spin correlation of the free and localized electrons manifests itself at rather small distances, we make an approximation of short-range interaction, $J(\mathbf{r} - \mathbf{R}_n) \simeq (J/\mathcal{N})\delta(\mathbf{r} - \mathbf{R}_n)$, where $\mathcal{N}$ is the density of the host-metal atoms introduced for normalization purpose (the constant J has the dimensionality of energy). Although we have introduced the Hamiltonian $\widehat{\delta H}_{e,im}$ phenomenologically, it can be rigorously derived from a Hamiltonian which allows hopping of electrons between free and localized states (problem 13.18).

To consider the scattering by magnetic impurities, we add the Hamiltonian (2) to the electron-impurity interaction Hamiltonian $\widehat{H}_{e,im} = \sum_\sigma \int d\mathbf{r} U_{im}(\mathbf{r})\hat{\Psi}^+_{\sigma\mathbf{r}}\hat{\Psi}_{\sigma\mathbf{r}}$ describing the spin-independent scattering. Let us write Eq. (D.13) for the retarded Green's function of electrons in the energy-momentum representation:

$$(\varepsilon - \varepsilon_{\mathbf{p}})G^R_\varepsilon(\sigma\mathbf{p}, \sigma'\mathbf{p}') = \delta_{\sigma\sigma'}\delta_{\mathbf{p}\mathbf{p}'} + \langle\langle[\hat{a}_{\sigma\mathbf{p}}, \widehat{H}_{e,im} + \widehat{\delta H}_{e,im}]|\hat{a}^+_{\sigma'\mathbf{p}'}\rangle\rangle^R_\varepsilon, \quad (3)$$

where we have introduced the creation and annihilation operators according to $\hat{\Psi}_{\sigma\mathbf{r}} = V^{-1/2} \sum_{\mathbf{p}} e^{i\mathbf{p}\cdot\mathbf{r}/\hbar}\hat{a}_{\sigma\mathbf{p}}$. The Green's function is expressed through such operators according to Eq. (D.15): $G^R_\varepsilon(\sigma\mathbf{p}, \sigma'\mathbf{p}') = \langle\langle\hat{a}_{\sigma\mathbf{p}}|\hat{a}^+_{\sigma'\mathbf{p}'}\rangle\rangle^R_\varepsilon$. The last term on the right-hand side of Eq. (3) is transformed as

$$\frac{1}{V}\sum_{\mathbf{p}_1} U_{im}(\mathbf{p} - \mathbf{p}_1)G^R_\varepsilon(\sigma\mathbf{p}_1, \sigma'\mathbf{p}')$$

$$-\frac{J}{\mathcal{N}}\sum_{\mathbf{p}_1}\sum_n e^{-i(\mathbf{p}-\mathbf{p}_1)\cdot\mathbf{R}_n/\hbar}Q^R_{\varepsilon n}(\sigma\mathbf{p}_1, \sigma'\mathbf{p}'), \qquad (4)$$

where the first and the second terms come, respectively, from the commutators with $\widehat{H}_{e,im}$ and $\widehat{\delta H}_{e,im}$, and $N = \mathcal{N}V$ is the number of the host atoms in the crystal. The Green's function $Q^R_{\varepsilon n}$ in the second term is defined as $Q^R_{\varepsilon n}(\sigma \mathbf{p}_1, \sigma'\mathbf{p}') = \sum_{\sigma_1}\langle\langle(\hat{\boldsymbol{\sigma}} \cdot \widehat{\mathbf{S}}_n)_{\sigma\sigma_1}\hat{a}_{\sigma_1\mathbf{p}_1}|\hat{a}^+_{\sigma'\mathbf{p}'}\rangle\rangle^R_\varepsilon$. Since $\widehat{\mathbf{S}}_n$ is the operator of atomic spin, it does not act on the electron spin variables. The matrix element $(\hat{\boldsymbol{\sigma}} \cdot \widehat{\mathbf{S}}_n)_{\sigma\sigma_1} = (\hat{\boldsymbol{\sigma}})_{\sigma\sigma_1} \cdot \widehat{\mathbf{S}}_n$ is an operator acting on the spin variables of n-th atom. For this reason, the second term in the expression (4) cannot be written in the same form as the first term, where the Fourier component $U_{im}(\mathbf{p} - \mathbf{p}_1) = \sum_n v(|\mathbf{p} - \mathbf{p}_1|/\hbar)e^{-i(\mathbf{p}-\mathbf{p}_1)\cdot\mathbf{R}_n/\hbar}$ is recognized as the scattering amplitude (14.3). Instead, one should try to get an approximate expression for the total scattering amplitude by using Eq. (D.13) for $Q^R_{\varepsilon n}(\sigma\mathbf{p}_1, \sigma'\mathbf{p}')$:

$$(\varepsilon - \varepsilon_{\mathbf{p}_1})Q^R_{\varepsilon n}(\sigma\mathbf{p}_1, \sigma'\mathbf{p}') = \delta_{\mathbf{p}_1\mathbf{p}'}(\hat{\boldsymbol{\sigma}} \cdot \widehat{\mathbf{S}}_n)_{\sigma\sigma'} + \frac{1}{V}\sum_{\mathbf{p}_2}U_{im}(\mathbf{p}_1 - \mathbf{p}_2)$$

$$\times Q^R_{\varepsilon n}(\sigma\mathbf{p}_2, \sigma'\mathbf{p}') - \frac{J}{N}\sum_{\sigma_1\sigma_2\sigma'_2}\sum_{\mathbf{p}_2\mathbf{p}'_2}\sum_{n'}e^{-i(\mathbf{p}'_2-\mathbf{p}_2)\cdot\mathbf{R}_{n'}/\hbar}$$

$$\times\langle\langle(\hat{\boldsymbol{\sigma}} \cdot \widehat{\mathbf{S}}_n)_{\sigma\sigma_1}(\hat{\boldsymbol{\sigma}} \cdot \widehat{\mathbf{S}}_{n'})_{\sigma'_2\sigma_2}\hat{a}_{\sigma_1\mathbf{p}_1}\hat{a}^+_{\sigma'_2\mathbf{p}'_2}\hat{a}_{\sigma_2\mathbf{p}_2}\tag{5}$$

$$-(\hat{\boldsymbol{\sigma}} \cdot \widehat{\mathbf{S}}_{n'})_{\sigma'_2\sigma_2}(\hat{\boldsymbol{\sigma}} \cdot \widehat{\mathbf{S}}_n)_{\sigma\sigma_1}\hat{a}^+_{\sigma'_2\mathbf{p}'_2}\hat{a}_{\sigma_2\mathbf{p}_2}\hat{a}_{\sigma_1\mathbf{p}_1}|\hat{a}^+_{\sigma'\mathbf{p}'}\rangle\rangle^R_\varepsilon.$$

This equation is exact.

The correlation function standing in the last term on the right-hand side of Eq. (5) cannot be reduced to a simpler one because the operators $\widehat{\mathbf{S}}_n$ and $\widehat{\mathbf{S}}_{n'}$ do not commute at $n' = n$. To find this function, one should substitute it in Eq. (D.13), which couples this function to higher-order correlation functions through the terms proportional to J, because of the spin-dependent interaction. Below we consider an approximation of weak spin-dependent interaction, when the electron creation and annihilation operators in Eq. (5) can be averaged separately from the operators of atomic spin. The products of four electron operators are averaged according to the following approximate transformations:

$$\langle\langle\hat{a}_{\sigma_1\mathbf{p}_1}\hat{a}^+_{\sigma'_2\mathbf{p}'_2}\hat{a}_{\sigma_2\mathbf{p}_2}|\hat{a}^+_{\sigma'\mathbf{p}'}\rangle\rangle^R_\varepsilon \simeq \langle\langle\hat{a}_{\sigma_1\mathbf{p}_1}\hat{a}^+_{\sigma'_2\mathbf{p}'_2}\rangle\rangle\langle\langle\hat{a}_{\sigma_2\mathbf{p}_2}|\hat{a}^+_{\sigma'\mathbf{p}'}\rangle\rangle^R_\varepsilon$$

$$\simeq \delta_{\sigma_1\sigma'_2}\delta_{\mathbf{p}_1\mathbf{p}'_2}(1 - f_{\mathbf{p}_1})G^R_\varepsilon(\sigma_2\mathbf{p}_2, \sigma'\mathbf{p}'),\tag{6}$$

$$\langle\langle\hat{a}^+_{\sigma'_2\mathbf{p}'_2}\hat{a}_{\sigma_2\mathbf{p}_2}\hat{a}_{\sigma_1\mathbf{p}_1}|\hat{a}^+_{\sigma'\mathbf{p}'}\rangle\rangle^R_\varepsilon \simeq -\langle\langle\hat{a}^+_{\sigma'_2\mathbf{p}'_2}\hat{a}_{\sigma_1\mathbf{p}_1}\rangle\rangle\langle\langle\hat{a}_{\sigma_2\mathbf{p}_2}|\hat{a}^+_{\sigma'\mathbf{p}'}\rangle\rangle^R_\varepsilon$$

$$\simeq -\delta_{\sigma_1\sigma'_2}\delta_{\mathbf{p}_1\mathbf{p}'_2}f_{\mathbf{p}_1}G^R_\varepsilon(\sigma_2\mathbf{p}_2, \sigma'\mathbf{p}'),$$

where $f_{\mathbf{p}_1}$ is the distribution function of electrons. Since we assume that the electrons are not spin-polarized in average, this function is spin-independent. The terms proportional to $\langle\langle \hat{a}^+_{\sigma'_2\mathbf{p}'_2}\hat{a}_{\sigma_2\mathbf{p}_2}\rangle\rangle \simeq \delta_{\sigma_2\sigma'_2}\delta_{\mathbf{p}_2\mathbf{p}'_2}f_{\mathbf{p}_2}$ are intentionally omitted on the right-hand sides of the equations (6), because these terms do not contribute into Eq. (5). Applying Eq. (6), we write the last term on the right-hand side of Eq. (5) as

$$-\frac{J}{N}\sum_{\sigma_1\sigma_2}\sum_{\mathbf{p}_2}\sum_{n'}e^{-i(\mathbf{p}_1-\mathbf{p}_2)\cdot\mathbf{R}_{n'}/\hbar}\Big[(\hat{\boldsymbol{\sigma}}\cdot\widehat{\mathbf{S}}_n)_{\sigma\sigma_1}(\hat{\boldsymbol{\sigma}}\cdot\widehat{\mathbf{S}}_{n'})_{\sigma_1\sigma_2}(1-f_{\mathbf{p}_1})$$

$$+(\hat{\boldsymbol{\sigma}}\cdot\widehat{\mathbf{S}}_{n'})_{\sigma_1\sigma_2}(\hat{\boldsymbol{\sigma}}\cdot\widehat{\mathbf{S}}_n)_{\sigma\sigma_1}f_{\mathbf{p}_1}\Big]G^R_\varepsilon(\sigma_2\mathbf{p}_2,\sigma'\mathbf{p}'). \qquad (7)$$

Below we use Eqs. (3), (4), and (5) to find $Q^R_{\varepsilon n}(\sigma\mathbf{p}_1,\sigma'\mathbf{p}')$ by iterations, as $Q^R_{\varepsilon n}(\sigma\mathbf{p}_1,\sigma'\mathbf{p}') = Q^{R(0)}_{\varepsilon n}(\sigma\mathbf{p}_1,\sigma'\mathbf{p}') + Q^{R(1)}_{\varepsilon n}(\sigma\mathbf{p}_1,\sigma'\mathbf{p}') + \ldots$. In the zero order in J, we neglect the last term on the right-hand side of Eq. (5) and search for $Q^R_{\varepsilon n}(\sigma\mathbf{p}_1,\sigma'\mathbf{p}')$ in the form

$$Q^{R(0)}_{\varepsilon n}(\sigma\mathbf{p}_1,\sigma'\mathbf{p}') = \sum_{\sigma_1}(\hat{\boldsymbol{\sigma}}\cdot\widehat{\mathbf{S}}_n)_{\sigma\sigma_1}G^{R(0)}_\varepsilon(\sigma_1\mathbf{p}_1,\sigma'\mathbf{p}'), \qquad (8)$$

where $G^{R(0)}_\varepsilon$ satisfies Eq. (3) written without the spin-dependent part of the interaction Hamiltonian. Without a loss of accuracy, $G^{R(0)}_\varepsilon$ in Eq. (8) can be replaced by the exact Green's function $G^R_\varepsilon(\sigma_1\mathbf{p}_1,\sigma'\mathbf{p}')$. The next iteration leads us to the following equation:

$$(\varepsilon-\varepsilon_{\mathbf{p}_1})Q^{R(1)}_{\varepsilon n}(\sigma\mathbf{p}_1,\sigma'\mathbf{p}') = \frac{1}{V}\sum_{\mathbf{p}_2}U_{im}(\mathbf{p}_1-\mathbf{p}_2)Q^{R(1)}_{\varepsilon n}(\sigma\mathbf{p}_2,\sigma'\mathbf{p}')$$

$$-\frac{J}{N}\sum_{\sigma_1\sigma_2\mathbf{p}_2}\sum_{n'}e^{-i(\mathbf{p}_1-\mathbf{p}_2)\cdot\mathbf{R}_{n'}/\hbar}\Big[(\hat{\boldsymbol{\sigma}}\cdot\widehat{\mathbf{S}}_n)_{\sigma\sigma_1}(\hat{\boldsymbol{\sigma}}\cdot\widehat{\mathbf{S}}_{n'})_{\sigma_1\sigma_2}(1-f_{\mathbf{p}_1}) \qquad (9)$$

$$+(\hat{\boldsymbol{\sigma}}\cdot\widehat{\mathbf{S}}_{n'})_{\sigma_1\sigma_2}(\hat{\boldsymbol{\sigma}}\cdot\widehat{\mathbf{S}}_n)_{\sigma\sigma_1}f_{\mathbf{p}_1}\Big]G^{R(0)}_\varepsilon(\sigma_2\mathbf{p}_2,\sigma'\mathbf{p}'),$$

where the second term on the right-hand side contains the zero-order Green's function of electrons. Again, we replace this function by the exact Green's function. Neglecting the first term on the right-hand side of Eq. (9), we can find $Q^{R(1)}_{\varepsilon n}(\sigma\mathbf{p}_1,\sigma'\mathbf{p}')$ explicitly. In summary, we obtain an approximate solution in the form

$$Q^R_{\varepsilon n}(\sigma\mathbf{p}_1,\sigma'\mathbf{p}') \simeq \sum_{\sigma_2\mathbf{p}_2}\Big\{(\hat{\boldsymbol{\sigma}}\cdot\widehat{\mathbf{S}}_n)_{\sigma\sigma_2}\delta_{\mathbf{p}_1\mathbf{p}_2} - \frac{J}{N}\sum_{n'}e^{-i(\mathbf{p}_1-\mathbf{p}_2)\cdot\mathbf{R}_{n'}/\hbar}$$

$$\times\frac{1}{\varepsilon-\varepsilon_{\mathbf{p}_1}}\sum_{\sigma_1}\Big[(\hat{\boldsymbol{\sigma}}\cdot\widehat{\mathbf{S}}_n)_{\sigma\sigma_1}(\hat{\boldsymbol{\sigma}}\cdot\widehat{\mathbf{S}}_{n'})_{\sigma_1\sigma_2}(1-f_{\mathbf{p}_1}) \qquad (10)$$

$$+(\hat{\boldsymbol{\sigma}} \cdot \widehat{\mathbf{S}}_{n'})_{\sigma_1 \sigma_2} (\hat{\boldsymbol{\sigma}} \cdot \widehat{\mathbf{S}}_n)_{\sigma \sigma_1} f_{\mathbf{p}_1} \Big] \Big\} G_{\varepsilon}^R(\sigma_2 \mathbf{p}_2, \sigma' \mathbf{p}').$$

Substituting this result to the second term of the expression (4), we notice the appearance of a double sum over the impurity coordinates, $\sum_{nn'} e^{-i(\mathbf{p}-\mathbf{p}_1)\cdot \mathbf{R}_n/\hbar} e^{-i(\mathbf{p}_1-\mathbf{p}_2)\cdot \mathbf{R}_{n'}/\hbar} \dots$. Assuming that the positions of different impurities are not correlated, we put $n' = n$ in this sum. This means that we consider scattering of an electron by a single impurity atom independently of the scattering by other atoms. Moreover, if the spins of different impurity atoms are not correlated, only the terms with $n' = n$ are essential for the effects studied in this section. To calculate the sum over σ_1 in Eq. (10) at $n' = n$, we use the operator identities (problem 13.19)

$$(\hat{\boldsymbol{\sigma}} \cdot \widehat{\mathbf{S}}_n)(\hat{\boldsymbol{\sigma}} \cdot \widehat{\mathbf{S}}_n) = S(S+1) - (\hat{\boldsymbol{\sigma}} \cdot \widehat{\mathbf{S}}_n),$$

$$\sum_{\alpha\beta} \hat{\sigma}_\alpha \hat{\sigma}_\beta \widehat{S}_{n\beta} \widehat{S}_{n\alpha} = S(S+1) + (\hat{\boldsymbol{\sigma}} \cdot \widehat{\mathbf{S}}_n) \qquad (11)$$

and find $\sum_{\sigma_1} [(\hat{\boldsymbol{\sigma}} \cdot \widehat{\mathbf{S}}_n)_{\sigma\sigma_1} (\hat{\boldsymbol{\sigma}} \cdot \widehat{\mathbf{S}}_n)_{\sigma_1 \sigma_2} (1-f_{\mathbf{p}_1}) + (\hat{\boldsymbol{\sigma}} \cdot \widehat{\mathbf{S}}_n)_{\sigma_1 \sigma_2} (\hat{\boldsymbol{\sigma}} \cdot \widehat{\mathbf{S}}_n)_{\sigma\sigma_1} f_{\mathbf{p}_1}] = S(S+1)\delta_{\sigma\sigma_2} - (1 - 2f_{\mathbf{p}_1})(\hat{\boldsymbol{\sigma}} \cdot \widehat{\mathbf{S}}_n)_{\sigma\sigma_2}$. After some transformations, we obtain

$$-\frac{J}{N} \sum_{\mathbf{p}_1} \sum_n e^{-i(\mathbf{p}-\mathbf{p}_1)\cdot \mathbf{R}_n/\hbar} Q_{\varepsilon n}^R(\sigma \mathbf{p}_1, \sigma' \mathbf{p}')$$

$$\simeq \frac{1}{V} \sum_{\sigma_1 \mathbf{p}_1} \sum_n e^{-i(\mathbf{p}-\mathbf{p}_1)\cdot \mathbf{R}_n/\hbar} A_{\varepsilon n}(\sigma, \sigma_1) G_{\varepsilon}^R(\sigma_1 \mathbf{p}_1, \sigma' \mathbf{p}'), \qquad (12)$$

where

$$A_{\varepsilon n}(\sigma, \sigma_1) = -\frac{J}{\mathcal{N}} \left\{ (\hat{\boldsymbol{\sigma}} \cdot \widehat{\mathbf{S}}_n)_{\sigma\sigma_1} [1 - J(2\gamma_\varepsilon - \gamma_\varepsilon^{(0)})] - JS(S+1)\delta_{\sigma\sigma_1} \gamma_\varepsilon^{(0)} \right\}, \qquad (13)$$

and the factors $\gamma^{(0)}$ and γ are defined as

$$\gamma_\varepsilon^{(0)} = \frac{1}{N} \sum_{\mathbf{p}} \frac{1}{\varepsilon - \varepsilon_{\mathbf{p}}}, \quad \gamma_\varepsilon = \frac{1}{N} \sum_{\mathbf{p}} \frac{f_{\mathbf{p}}}{\varepsilon - \varepsilon_{\mathbf{p}}}. \qquad (14)$$

Expressions (12)−(14) describe the spin-dependent interaction of electrons with magnetic impurities in terms of its contribution into the scattering amplitude, $\sum_n e^{-i(\mathbf{p}-\mathbf{p}_1)\cdot \mathbf{R}_n/\hbar} A_{\varepsilon n}(\sigma, \sigma_1)$. This scattering amplitude depends on electron energy because of the terms proportional to J^2 in $A_{\varepsilon n}(\sigma, \sigma_1)$. Let us analyze these J^2-terms in more detail. They are determined by the sums (14) and contain an extra factor $Jn/\varepsilon_F \mathcal{N}$ with respect to the main term proportional to J, where n is the electron density and ε_F is the Fermi energy. Should we keep the J^2-terms under the condition $Jn/\varepsilon_F\mathcal{N} \ll 1$? At first glance, it seems worth to retain the terms proportional to $\gamma_\varepsilon^{(0)}$, because they have a logarithmic divergence. However, this divergence is caused by the approximation of short-range spin-dependent interaction and can be formally removed by introducing a cutoff momentum. In any case, it appears that the energy dependence of $\gamma_\varepsilon^{(0)}$ in the region close to the Fermi energy ε_F is not essential and can be neglected. On the other hand, the function γ_ε diverges at $\varepsilon = \varepsilon_F$ if the temperature is zero (problem 13.20). Therefore, we should keep the term containing γ_ε rather than the terms containing $\gamma_\varepsilon^{(0)}$, because of the strong energy dependence of the former. Retaining only this particular term, we write the following expression for the total scattering amplitude:

$$U_{\sigma\sigma'}(\mathbf{p} - \mathbf{p}') = \sum_n e^{-i(\mathbf{p}-\mathbf{p}')\cdot\mathbf{R}_n/\hbar}$$

$$\times \left[\delta_{\sigma\sigma'} v(|\mathbf{p} - \mathbf{p}'|/\hbar) - (\hat{\boldsymbol{\sigma}})_{\sigma\sigma'} \cdot \widehat{\mathbf{S}}_n \frac{J}{\mathcal{N}}(1 - 2J\gamma_\varepsilon) \right]. \qquad (15)$$

This scattering amplitude depends on spins of electrons. Moreover, it is an operator with respect to spin states of impurity atoms. Below we assume that the spins of different impurity atoms are not correlated so that the average over the atomic spin can be carried out separately for each atom. For this reason, we omit the index n at the atomic spin operator $\widehat{\mathbf{S}}$.

To derive the collision integral based upon the scattering amplitude (15) in this case, one should merely repeat the derivation given in Secs. 7 and 8 by using the basis $|\sigma\mathbf{p}\rangle$ instead of the plane-wave basis $|\mathbf{p}\rangle$. The kinetic equation is written for the spin-dependent distribution functions $f_{\sigma\mathbf{p}t} = \langle \sigma\mathbf{p}|\hat{\rho}_t|\sigma\mathbf{p}\rangle$. Instead of Eq. (8.7), we obtain

$$\frac{\partial f_{\sigma\mathbf{p}t}}{\partial t} = \frac{2\pi}{\hbar} \frac{n_{im}}{V} \sum_{\sigma'\mathbf{p}'} \overline{v_{\sigma\sigma'}^{MM'}(\varepsilon_\mathbf{p}, q) v_{\sigma'\sigma}^{M'M}(\varepsilon_\mathbf{p}, q)} \delta(\varepsilon_\mathbf{p} - \varepsilon_{\mathbf{p}'})(f_{\sigma'\mathbf{p}'t} - f_{\sigma\mathbf{p}t}),$$

$$(16)$$

where $q = |\mathbf{p} - \mathbf{p}'|/\hbar$. This equation describes evolution of the diagonal components of the averaged density matrix. The non-diagonal components, $\langle \sigma_1\mathbf{p}|\hat{\rho}_t|\sigma_2\mathbf{p}\rangle$, are zero if the initial distribution of electrons is diagonal in the spin index (problem 13.21). The matrix element $v_{\sigma\sigma'}^{MM'}(\varepsilon_\mathbf{p}, q)$ describes scattering of an electron by a single atom when the projection of the atomic spin changes between M and M':

$$v_{\sigma\sigma'}^{MM'}(\varepsilon, q) = \delta_{\sigma\sigma'}\delta_{MM'}v(q) - (\hat{\boldsymbol{\sigma}})_{\sigma\sigma'} \cdot (\widehat{\mathbf{S}})_{MM'} \frac{J}{\mathcal{N}}[1 - 2J\gamma_\varepsilon]. \qquad (17)$$

The line in Eq. (16) denotes the average of the product of the matrix elements over the projection of the localized spin. Before calculating this average, let us write this product by using Eq. (17):

$$v_{\sigma\sigma'}^{MM'}(\varepsilon_{\mathbf{p}}, q)v_{\sigma'\sigma}^{M'M}(\varepsilon_{\mathbf{p}}, q) = \delta_{\sigma\sigma'}\delta_{MM'}|v(q)|^2 + (\hat{\boldsymbol{\sigma}})_{\sigma\sigma'} \cdot (\widehat{\mathbf{S}})_{MM'}(\hat{\boldsymbol{\sigma}})_{\sigma'\sigma} \cdot (\widehat{\mathbf{S}})_{M'M}$$

$$\times \frac{J^2}{\mathcal{N}^2}[1 - 2J\gamma_{\varepsilon_{\mathbf{p}}}]^2 - 2v(q)\frac{J}{\mathcal{N}}[1 - 2J\gamma_{\varepsilon_{\mathbf{p}}}]\delta_{\sigma\sigma'}\delta_{MM'}(\hat{\boldsymbol{\sigma}})_{\sigma\sigma} \cdot (\widehat{\mathbf{S}})_{MM} . \quad (18)$$

Denoting, for brevity, the electron spin indices $\sigma = \pm 1$ as $+$ and $-$, we make use of the identities

$$(\hat{\boldsymbol{\sigma}})_{++} \cdot (\widehat{\mathbf{S}})_{MM'} = (\widehat{S}_z)_{MM'} = \delta_{MM'}M,$$

$$(\hat{\boldsymbol{\sigma}})_{--} \cdot (\widehat{\mathbf{S}})_{MM'} = -(\widehat{S}_z)_{MM'} = -\delta_{MM'}M, \quad (19)$$

and

$$(\hat{\boldsymbol{\sigma}})_{+-} \cdot (\widehat{\mathbf{S}})_{MM'} = (\widehat{S}_-)_{MM'} = \delta_{M',M+1}\sqrt{(S + M')(S - M' + 1)},$$

$$(\hat{\boldsymbol{\sigma}})_{-+} \cdot (\widehat{\mathbf{S}})_{MM'} = (\widehat{S}_+)_{MM'} = \delta_{M',M-1}\sqrt{(S + M)(S - M + 1)}, \quad (20)$$

where $\widehat{S}_\pm = \widehat{S}_x \pm i\widehat{S}_y$.

Taking into account that the localized spins are randomly oriented (the magnetization of the crystal is zero), we find

$$\overline{(\widehat{S}_z)_{MM}(\widehat{S}_z)_{MM}} = \overline{M^2} = S(S + 1)/3,$$

$$\overline{(\widehat{S}_+)_{M,M-1}(\widehat{S}_-)_{M-1,M}} = \overline{S^2 + S - M^2 + M} = 2S(S + 1)/3, \quad (21)$$

$$\overline{(\widehat{S}_-)_{M,M+1}(\widehat{S}_+)_{M+1,M}} = \overline{S^2 + S - M^2 - M} = 2S(S + 1)/3.$$

The average of the expression (18) is

$$\overline{v_{\sigma\sigma'}^{MM'}(\varepsilon_{\mathbf{p}}, q)v_{\sigma'\sigma}^{M'M}(\varepsilon_{\mathbf{p}}, q)} = \delta_{\sigma\sigma'}\left[|v(q)|^2 + (J/\mathcal{N})^2(1 - 4J\gamma_{\varepsilon_{\mathbf{p}}})S(S + 1)/3\right]$$

$$+ \delta_{\sigma,-\sigma'}(J/\mathcal{N})^2(1 - 4J\gamma_{\varepsilon_{\mathbf{p}}})2S(S + 1)/3, \quad (22)$$

where we have neglected the terms proportional to J^4. Note that the interference term proportional to $v(q)J$ in Eq. (18) disappears after the averaging over the impurity spins, because $\overline{(\widehat{S}_z)_{MM}} = \overline{M} = 0$. As a result, the contributions caused by the interaction Hamiltonians $\widehat{H}_{e,im}$ and $\widehat{\delta H}_{e,im}$ enter the kinetic equation as independent scattering mechanisms. Unless the electrons are spin-polarized by some external perturbation, there are no reasons to have different distribution functions for different σ, and $f_{\sigma\mathbf{p}} = f_{\mathbf{p}}$. We have already used this property in Eq. (6). Therefore, the terms proportional to $\delta_{\sigma\sigma'}$ and $\delta_{\sigma,-\sigma'}$ in Eq. (22) can be combined together, according to $\sum_{\sigma'}(\delta_{\sigma\sigma'}B_1 + \delta_{\sigma,-\sigma'}B_2) = (\delta_{\sigma,+1} + \delta_{\sigma,-1})(B_1 + B_2) = B_1 + B_2$.

Let us consider the response of the electron system to a stationary and homogeneous electric field $\mathbf{E}$. The stationary kinetic equation is

$$e\mathbf{E} \cdot \frac{\partial f_{\mathbf{p}}}{\partial \mathbf{p}} = J_{im}(f|\mathbf{p}) + \delta J_{im}(f|\mathbf{p}), \qquad (23)$$

where $J_{im}(f|\mathbf{p})$ is the electron-impurity collision integral considered in Sec. 8; see Eqs. (8.7) and (8.8). The other collision integral is caused by the spin-dependent scattering:

$$\delta J_{im}(f|\mathbf{p}) = \frac{2\pi}{\hbar} \frac{J^2 S(S+1) n_{im}}{\mathcal{N}^2 V} \sum_{\mathbf{p}'} (1 - 4J\gamma_{\varepsilon_{\mathbf{p}}}) \delta(\varepsilon_{\mathbf{p}} - \varepsilon_{\mathbf{p}'})(f_{\mathbf{p}'} - f_{\mathbf{p}}). \quad (24)$$

Below we assume the isotropic dispersion law, $\varepsilon_{\mathbf{p}} = \varepsilon_p = p^2/2m$. The conductivity characterizing the linear response of the electron system is found in a straightforward way, according to Eq. (8.27) with

$$\tau_{tr}(\varepsilon) = [\nu_\varepsilon + \delta\nu_\varepsilon]^{-1},$$

$$\delta\nu_\varepsilon = \tilde{\nu}(1 - 4J\gamma_\varepsilon), \quad \tilde{\nu} = \frac{3\pi J^2 S(S+1)}{2\varepsilon_F \hbar} \frac{n n_{im}}{\mathcal{N}^2}. \qquad (25)$$

To find the transport time $\tau_{tr}(\varepsilon)$, we have used Eq. (8.21), taking into account that the density of electron states is expressed through the electron density n and Fermi energy ε_F as $\rho_{3D}(\varepsilon_F) = (3/2)n/\varepsilon_F$. In the case of strongly degenerate electrons, the static conductivity is equal to $\sigma = e^2 n/m(\nu_{\varepsilon_F} + \tilde{\nu}) + \delta\sigma$. The correction $\delta\sigma$ is temperature-dependent:

$$\delta\sigma = \frac{e^2 n}{m} \frac{4J\tilde{\nu}}{(\nu_{\varepsilon_F} + \tilde{\nu})^2} \frac{m^{3/2}}{\sqrt{2}\pi^2 \hbar^3 \mathcal{N}} \int_0^\infty d\varepsilon \frac{\partial f_\varepsilon}{\partial \varepsilon} \int_0^\infty d\varepsilon' \sqrt{\varepsilon'} \frac{f_{\varepsilon'}}{\varepsilon' - \varepsilon}, \qquad (26)$$

where f_ε is the equilibrium Fermi distribution function. To estimate the double integral in Eq. (26), we integrate over ε' by parts:

$$\int_0^\infty d\varepsilon' \sqrt{\varepsilon'} \frac{f_{\varepsilon'}}{\varepsilon' - \varepsilon} = -\int_0^\infty d\varepsilon' \frac{\partial f_{\varepsilon'}}{\partial \varepsilon'} \left(2\sqrt{\varepsilon'} + \sqrt{\varepsilon} \ln \left| \frac{\sqrt{\varepsilon} - \sqrt{\varepsilon'}}{\sqrt{\varepsilon} + \sqrt{\varepsilon'}} \right| \right). \quad (27)$$

Substituting this transformation into Eq. (26), we obtain

$$\delta\sigma = -\frac{e^2 n^2}{m\mathcal{N}} \frac{3J\tilde{\nu}}{\varepsilon_F (\nu_{\varepsilon_F} + \tilde{\nu})^2}$$

$$\times \left(2 + \int_{-\infty}^\infty dx \int_{-\infty}^\infty dx' \frac{\ln(|x - x'|T/4\varepsilon_F)}{16 \cosh^2(x/2) \cosh^2(x'/2)} \right), \qquad (28)$$

where $x = (\varepsilon - \varepsilon_F)/T$, and we have extended the lower limits of integration to $-\infty$ because of $\varepsilon_F \gg T$. We finally may neglect all temperature-independent terms in the expression (28) and write the resistivity $\rho = \sigma^{-1}$ in the form

$$\rho = \rho_0 + \rho_1 \left(1 + \frac{3Jn}{\varepsilon_F \mathcal{N}} \ln \frac{T}{\varepsilon_F} \right), \qquad (29)$$

where $\rho_0 = m\nu_{\varepsilon_F}/e^2 n$ is the static resistivity obtained in Sec. 8 and $\rho_1 = m\tilde{\nu}/e^2 n$ is the temperature-independent part of the correction to this resistivity caused by the scattering by magnetic impurities. Equation (29) describes the temperature dependence of the resistivity for the metals containing magnetic impurities. This unusual temperature behavior is called the Kondo effect. The resistivity decreases with increasing temperature if the sign of J is negative. If this sign is positive, the resistivity increases with increasing temperature. Moreover, both kinds of behavior are experimentally observed, though the first case, corresponding to negative J, is met more often.

If the temperature is so small that the temperature-dependent correction becomes comparable to the main contribution to the resistivity, the theory presented above is not valid. This characteristic temperature, estimated as

$$T_K \sim \varepsilon_F \exp \left(-\frac{\varepsilon_F \mathcal{N}}{3Jn} \right) \qquad (30)$$

is called the Kondo temperature. To describe the conductivity in this case, one should take into account not only the lowest-order term in the expansion of the scattering amplitude, but a sum of all relevant terms. Such a calculation requires a special diagram technique involving the Green's functions composed from atomic spin operators.

Problems

13.1. Using the two-band model, find the effective Hamiltonian of the spin-orbit interaction for the conduction-band electrons in the presence of a potential energy $U_{\mathbf{r}}$.

<u>Solution</u>: The effective interaction term for the c-band electrons is determined by the matrix elements of the potential energy $U_{\mathbf{r}}$ in the two-band model:

$$\langle c\mathbf{p} | U_{\mathbf{r}} | c\mathbf{p}' \rangle = U_{(\mathbf{p} - \mathbf{p}')/\hbar} \langle c | \hat{\mathcal{U}}_{\mathbf{p}} \hat{\mathcal{U}}_{\mathbf{p}'}^+ | c \rangle \simeq U_{(\mathbf{p} - \mathbf{p}')/\hbar} \left[1 + \frac{(\hat{\boldsymbol{\sigma}} \cdot \mathbf{p})(\hat{\boldsymbol{\sigma}} \cdot \mathbf{p}')}{(2ms)^2} \right].$$

To obtain this equation, we have employed Eq. (B.24) and the expansion $\eta_p \simeq 1 + (p/ms)^2/2$ valid near the extremum of the band. Using the identity $(\hat{\boldsymbol{\sigma}} \cdot \mathbf{p})(\hat{\boldsymbol{\sigma}} \cdot \mathbf{p}') = \mathbf{p} \cdot \mathbf{p}' + i\hat{\boldsymbol{\sigma}} \cdot [\mathbf{p} \times \mathbf{p}']$, we can neglect the spin-independent contribution proportional to $\mathbf{p} \cdot \mathbf{p}'$ because of a small parameter $(p/2ms)^2$. Therefore, introducing $\mathbf{q} = (\mathbf{p} - \mathbf{p}')/\hbar$, we write the Fourier transform of the effective potential as

$$U_{\mathbf{q}} \left[1 - i\hbar \frac{\hat{\boldsymbol{\sigma}} \cdot [\mathbf{p} \times \mathbf{q}]}{2m\varepsilon_g} \right],$$

where $\varepsilon_g = 2ms^2$. The second term in this expression is a 2×2 matrix with respect to spin variables. Applying the inverse Fourier transformation according to Eq. (7.15), we obtain this term in the coordinate representation:

$$\frac{\hbar}{2m\varepsilon_g} \left(\hat{\boldsymbol{\sigma}} \cdot [\nabla U_{\mathbf{r}} \times \mathbf{p}] \right).$$

Therefore, $\chi = 1/2$ in Eq. (63.1) for the model under consideration.

13.2. Derive the quasi-classical matrix kinetic equation (63.8).

Hint: Take into account that the commutator of the Hamiltonian (63.1) with the density matrix $\hat{\rho}_t$ in the coordinate representation is equal to

$$-\frac{\chi i \hbar^2}{m \varepsilon_g} \left\{ \left[\hat{\boldsymbol{\sigma}} \times \frac{\partial U_{\mathbf{r}_1}}{\partial \mathbf{r}_1} \right] \cdot \frac{\partial \hat{\rho}_t(\mathbf{r}_1, \mathbf{r}_2)}{\partial \mathbf{r}_1} + \frac{\partial \hat{\rho}_t(\mathbf{r}_1, \mathbf{r}_2)}{\partial \mathbf{r}_2} \cdot \left[\hat{\boldsymbol{\sigma}} \times \frac{\partial U_{\mathbf{r}_2}}{\partial \mathbf{r}_2} \right] \right\}$$

and carry out the Wigner transformation $\hat{f}_{\mathbf{rpt}} = \int d\Delta\mathbf{r} e^{-i\mathbf{p}\cdot\Delta\mathbf{r}/\hbar} \hat{\rho}_t(\mathbf{r} + \Delta\mathbf{r}/2, \mathbf{r} - \Delta\mathbf{r}/2)$.

13.3. Consider the spin-precession mechanism of spin relaxation in non-centrosymmetric crystals.

Solution: The symmetry property of the equations of motion with respect to time reversal allows the corrections of the third power in quasimomenta, $\propto \hat{\sigma}_\alpha p_\beta p_\gamma p_\delta$, to the effective-mass Hamiltonian of electrons in such crystals. In the isotropic approximation, these corrections give zero contribution, because the spectrum should depend only on $|\mathbf{p}|$. In the cubic crystals, there are three invariants,

$$\hat{\sigma}_x p_x (p_y^2 - p_z^2), \quad \hat{\sigma}_y p_y (p_z^2 - p_x^2), \quad \hat{\sigma}_z p_z (p_x^2 - p_y^2),$$

and the spin-dependent term of the Hamiltonian in the momentum representation can be written as a scalar product $\hbar(\hat{\boldsymbol{\sigma}} \cdot \boldsymbol{\Omega}_{\mathbf{p}})$, where $\hbar\boldsymbol{\Omega}_{\mathbf{p}} = \alpha(2m\sqrt{2m\varepsilon_g})^{-1}\boldsymbol{\kappa}_{\mathbf{p}}$, $\boldsymbol{\kappa}_{\mathbf{p}} = \left[p_x(p_y^2 - p_z^2), p_y(p_z^2 - p_x^2), p_z(p_x^2 - p_y^2) \right]$, and α is a dimensionless constant determining the magnitude of the spin-orbit splitting. The distribution function $\hat{f}_{\varepsilon t}$ satisfies the kinetic equation

$$\frac{\partial \hat{f}_{\varepsilon t}}{\partial t} + \int \frac{d\widetilde{\Omega}_{\mathbf{p}}}{4\pi\nu_\varepsilon} \left[(\hat{\boldsymbol{\sigma}} \cdot \boldsymbol{\Omega}_{\mathbf{p}}), \left[(\hat{\boldsymbol{\sigma}} \cdot \boldsymbol{\Omega}_{\mathbf{p}}), \hat{f}_{\varepsilon t} \right] \right] = 0$$

similar to Eq. (63.9). The angular average of $\kappa_{\mathbf{p}}^\alpha \kappa_{\mathbf{p}}^\beta$ is equal to $\delta_{\alpha\beta}(4p^6/105)$, and the kinetic equation is reduced to

$$\frac{\partial \mathbf{f}_{\varepsilon t}}{\partial t} = -\frac{\mathbf{f}_{\varepsilon t}}{\tau_{sp}(\varepsilon)}, \quad \tau_{sp}^{-1}(\varepsilon) = \frac{32}{105} \frac{\alpha^2 \varepsilon^3}{\hbar^2 \nu_\varepsilon \varepsilon_g^2}.$$

Because of ε^3 dependence of the spin relaxation rate $\tau_{sp}^{-1}(\varepsilon)$, the efficiency of spin relaxation rapidly increases with increasing electron energy.

13.4. Calculate the absorption coefficient $\alpha_\omega^{(esr)}$ due to electron spin resonance and compare it to the absorption coefficient due to combined resonance.

Solution: The Hamiltonian of the perturbation causing the electron spin resonance is written as $-(g\mu_B/2)\hat{\boldsymbol{\sigma}} \cdot \mathbf{H}_1 e^{-i\omega t} + H.c.$, where $\mathbf{H}_1$ is the amplitude of the magnetic field of the wave. According to the Maxwell equations, $\mathbf{H}_1 = \sqrt{\epsilon}[\mathbf{n} \times \mathbf{E}]$, where $\mathbf{E}$ is the electric field of the wave and $\mathbf{n}$ is the unit vector in the direction of propagation of the wave. The Hamiltonian describing the perturbation due to electric field is $(ie/\omega)\hat{\mathbf{v}} \cdot \mathbf{E} e^{-i\omega t} + H.c.$. Therefore, to calculate $\alpha_\omega^{(esr)}$, one can use Eq. (10.23) and, for example, Eq. (48.4), with a formal substitution $\hat{\mathbf{v}} \Rightarrow (i\omega g\mu_B \sqrt{\epsilon}/2e)[\hat{\boldsymbol{\sigma}} \times \mathbf{n}]$. In the resonance approximation, we find (the non-parabolicity of the energy spectrum is neglected and the spin splitting energy is introduced as $\hbar\Omega_H = |g\mu_B H|$)

$$\alpha_\omega^{(esr)} = \frac{(2\pi e)^2}{c\sqrt{\epsilon}\omega V} \sum_{p_y p_z} f_{0,-1 p_z} \delta_\gamma \left(\hbar\omega - \hbar\Omega_H\right) \frac{\epsilon(\omega g\mu_B)^2}{4e^2}.$$

Comparing this to the result for $\alpha_\omega^{(cr)}$ given by Eq. (63.18), we find the ratio $\alpha_\omega^{(esr)}/\alpha_\omega^{(cr)} \sim (\epsilon g^2 m/m_e)(\varepsilon_g/T)(\varepsilon_g/m_e c^2)$, where m_e is the free-electron mass, and the kinetic energy $p_z^2/2m$ in Eq. (63.18) is estimated by the temperature T. This ratio is small because the energy $2m_e c^2$ is very large. In the two-band model, where $|g| = 2m_e/m$, one has $\alpha_\omega^{(esr)}/\alpha_\omega^{(cr)} \sim \epsilon(\varepsilon_g/T)(s/c)^2$.

13.5. Consider the origin of the spin-orbit term in the Hamiltonian of 2D electrons.

Solution: Let us add an asymmetric confinement potential $U(z)$ to $U_{cr}(\mathbf{r})$ in the Hamiltonian (5.4) and search for the wave function in the form $\psi(\mathbf{x})\phi(z)u_{n\sigma}(\mathbf{r})$, where $\mathbf{x}$ is the 2D coordinate, $u_{n\sigma}(\mathbf{r})$ is the Bloch amplitude, and $\phi(z)$ is the envelope function of electrons in the ground-state subband. Multiplying the Schroedinger equation by $u_{n\sigma}^*(\mathbf{r})\phi^*(z)$ from the left, and integrating it over the lattice cell volume and over z, one can obtain a 2D Schroedinger equation for $\psi(\mathbf{x})$, which contains a spin-orbit term in the form (63.21). We stress that, in the effective mass approximation, $\phi(z)$ is the eigenfunction of the Hamiltonian $\hat{p}_z^2/2m + U(z)$. As a consequence, $\int dz |\phi(z)|^2 \nabla_z U(z) = 0$ for arbitrary $U(z)$. This property can be checked by using the integration by parts. For this reason, to obtain a non-zero spin-orbit term (63.21), one should either assume that $U(z)$ changes considerably on the scale of the lattice constant or go beyond the effective mass approximation.

13.6. Find a suitable gauge to introduce the vector potential $\mathbf{A}$ describing the external magnetic field $\mathbf{H}$ in the two-dimensional Hamiltonian (63.22).

Solution: One can use the vector potential $(H_y z, H_z x - H_x z, 0)$ in the three-dimensional Hamiltonian. After transforming the 3D Schroedinger equation to the 2D one, the coordinate z is transformed to $\langle 1|z|1\rangle$, where $|1\rangle$ is the ground eigenstate, and should be considered as a coordinate of the weight center of the wave function of this eigenstate. Since this coordinate can be set to zero, we finally obtain the effective vector potential in the form of a 2D vector, $\mathbf{A} = (0, H_z x)$.

13.7. Write the general solution of the vector equation $\partial \mathbf{s}_t / \partial t + [\boldsymbol{\Omega}_H \times \mathbf{s}_t] = -\mathbf{s}_t / \tau_s + \mathbf{g}_t$. In the stationary case, find the absolute value of the component $\mathbf{s}_\perp$ perpendicular to $\boldsymbol{\Omega}_H$.

<u>Results</u>: If $\boldsymbol{\Omega}_H$ is directed along OZ,

$$s_t^x = \int_{-\infty}^t dt' e^{-(t-t')/\tau_s} \left[g_{t'}^x \cos \Omega_H(t - t') + g_{t'}^y \sin \Omega_H(t - t') \right],$$

$$s_t^y = \int_{-\infty}^t dt' e^{-(t-t')/\tau_s} \left[g_{t'}^y \cos \Omega_H(t - t') - g_{t'}^x \sin \Omega_H(t - t') \right],$$

and $s_t^z = \int_{-\infty}^t dt' e^{-(t-t')/\tau_s} g_{t'}^z$. In the case of a stationary excitation, $s^x = \tau_s(g^x + g^y \Omega_H \tau_s)/[1 + (\Omega_H \tau_s)^2]$ and $s^y = \tau_s(g^y - g^x \Omega_H \tau_s)/[1 + (\Omega_H \tau_s)^2]$, and we obtain $|\mathbf{s}_\perp| = |\mathbf{g}| \tau_s / \sqrt{1 + (\Omega_H \tau_s)^2}$ so that $\mathbf{s}_\perp$ is suppressed when $\Omega_H \tau_s \gg 1$.

13.8. Find the matrix of electrochemical potential in a half-space ($x > 0$) if a homogeneous spin-polarized current $\hat{I} = [I + \Delta I \hat{\sigma}_z]/2$ is injected through the interface at $x = 0$.

<u>Solution</u>: The electrochemical potential depend only on x. Therefore, the solution of Eq. (64.10) satisfying the requirement of finiteness of the gradients of $\mathbf{w_r} = \mathbf{w}_x$ at $x \to \infty$ is $\mathbf{w}_x = \mathbf{c} e^{-x/l_s}$. The Laplace equation for the scalar component is solved as $w_x = w_0 - (I/\sigma)x$, where σ is the conductivity of the medium. The total current through the interface, $I = \mathrm{tr}_\sigma \hat{I}$, is conserved. The constant $\mathbf{c}$ is determined from the boundary condition expressing current continuity under the assumption that no spin flip occurs at the interface: $\mathrm{tr}_\sigma \hat{\boldsymbol{\sigma}} \hat{I} = -\sigma (d\mathbf{w}_x/dx)_{x=0}$. As a result, we obtain $\hat{w}_x = w_0 - (I/\sigma)x + \hat{\sigma}_z (\Delta I l_s/\sigma) e^{-x/l_s}$. The spin polarization of the current decreases inside the sample and vanishes at $x \gg l_s$.

13.9. Using the Hamiltonian (63.22), derive the matrix kinetic equation (64.11) for the Wigner distribution function

$$\hat{f}_{\mathbf{rp}t} = \int d\Delta \mathbf{r} \exp\left(-i\mathbf{P}_{\mathbf{r}t} \cdot \Delta \mathbf{r}/\hbar\right) \hat{\rho}_t(\mathbf{r} + \Delta \mathbf{r}/2, \mathbf{r} - \Delta \mathbf{r}/2),$$

where $\mathbf{P}_{\mathbf{r}t} = \mathbf{p} + (e/c)\mathbf{A}_{\mathbf{r}t}$.

<u>Solution</u>: The consideration of the kinetic-energy contribution is done in the same way as in Sec. 9. Let us consider the contribution of the spin-orbit term. Substituting the spin-orbit part of the Hamiltonian (63.22) into the integral term of Eq. (9.2), we transform this term to

$$\frac{i}{\hbar} \left\{ [\mathbf{v}_s \times \hat{\boldsymbol{\sigma}}](-i\hbar\nabla_{\mathbf{x}_1} - \frac{e}{c}\mathbf{A}_{\mathbf{x}_1 t})\hat{\rho}(\mathbf{x}_1, \mathbf{x}_2) - (i\hbar\nabla_{\mathbf{x}_2} - \frac{e}{c}\mathbf{A}_{\mathbf{x}_2 t})\hat{\rho}(\mathbf{x}_1, \mathbf{x}_2)[\mathbf{v}_s \times \hat{\boldsymbol{\sigma}}] \right\}.$$

Using the new coordinates $\mathbf{r} = (\mathbf{x}_1 + \mathbf{x}_2)/2$ and $\Delta \mathbf{r} = \mathbf{x}_1 - \mathbf{x}_2$, we expand the components of the vector potential as $A_{\mathbf{x}_{1,2}t}^\alpha \simeq A_{\mathbf{r}t}^\alpha \pm (\partial A_{\mathbf{r}t}^\alpha/\partial \mathbf{r}) \cdot \Delta \mathbf{r}/2$. Writing the spatial derivatives according to Eq. (9.9), one can see that the terms proportional to $\partial \rho/\partial \Delta \mathbf{r}$ and $\partial \rho/\partial \mathbf{r}$ contribute to commutators and anticommutators, respectively. After multiplying the term transformed in this way by $\exp(-i\mathbf{P}_{\mathbf{r}t} \cdot \Delta \mathbf{r}/\hbar)$, we integrate it over $\Delta \mathbf{r}$ and obtain

$$\frac{i}{\hbar}[\mathbf{p} \cdot [\mathbf{v}_s \times \hat{\boldsymbol{\sigma}}], \hat{f}_{\mathbf{rp}t}] + \frac{1}{2}\left[[\mathbf{v}_s \times \hat{\boldsymbol{\sigma}}], \frac{\partial \hat{f}_{\mathbf{rp}t}}{\partial \mathbf{r}}\right]_+ + \frac{e}{2c}\left[[[\mathbf{v}_s \times \hat{\boldsymbol{\sigma}}] \times [\nabla_{\mathbf{r}} \times \mathbf{A}_{\mathbf{r}t}]], \frac{\partial \hat{f}_{\mathbf{rp}t}}{\partial \mathbf{p}}\right]_+,$$

which is the spin-orbit contribution standing in Eq. (64.11).

13.10. Calculate the collision-integral contributions in Eqs. (64.18) and (64.19).

Solution: To find the collision integral $\widehat{J}(\hat{f}|\mathbf{rp}t)$, let us use Eq. (8.3) in the momentum representation and take into account the matrix nature of the problem by replacing $\langle\mathbf{p}|\hat{h}|\mathbf{p}\rangle$ and $\langle\mathbf{p}|\hat{\rho}_t|\mathbf{p}\rangle$ by $\hat{h}_\mathbf{p} = p^2/2m + \mathbf{v}_s\cdot[\hat{\boldsymbol{\sigma}}\times\mathbf{p}] - \hbar\boldsymbol{\Omega}_H\cdot\hat{\boldsymbol{\sigma}}/2$ and $\hat{f}_{\mathbf{rp}t}$, respectively. In the Markovian approximation,

$$\widehat{J}(\hat{f}|\mathbf{rp}t) = \frac{n_{im}}{\hbar^2 L^2}\sum_{\mathbf{p}'}|v(|\mathbf{p}-\mathbf{p}'|/\hbar)|^2\int_{-\infty}^0 d\tau e^{\lambda\tau}$$

$$\times\left[e^{i\hat{h}_\mathbf{p}\tau/\hbar}(\hat{f}_{\mathbf{rp}'t}-\hat{f}_{\mathbf{rp}t})e^{-i\hat{h}_{\mathbf{p}'}\tau/\hbar} + e^{i\hat{h}_{\mathbf{p}'}\tau/\hbar}(\hat{f}_{\mathbf{rp}'t}-\hat{f}_{\mathbf{rp}t})e^{-i\hat{h}_\mathbf{p}\tau/\hbar}\right].$$

To transform the matrix exponents containing the Pauli matrices, one should use the exact relation (H.28). To simplify calculation of the integrals over momenta, one may assume the case of short-range scattering potential, $v(|\mathbf{p}-\mathbf{p}'|/\hbar)\simeq v(0) = v$. After lengthy but straightforward calculations, one finds that the collision-integral contributions in Eqs. (64.18) and (64.19) are reduced to $-\nu i_{\mathbf{rt}}^\beta$ and $-\nu q_{\mathbf{rt}}^{\alpha\beta}$ if the terms proportional to $mv_s^2/\overline{\varepsilon}$ and $(\hbar\Omega_H/2\overline{\varepsilon})^2$ are neglected. The momentum relaxation rate ν is given by Eq. (8.21).

13.11. Calculate the spin density appearing in the 2D system with a stationary electric current.

Solution: The stationary kinetic equation for the matrix distribution function $\hat{f}_\mathbf{p} + \widehat{\delta f}_\mathbf{p}$ is obtained from Eq. (64.11) written in the absence of magnetic fields and spatial gradients. The linearized kinetic equation is

$$\frac{i}{\hbar}\left[\mathbf{v}_s\cdot[\hat{\boldsymbol{\sigma}}\times\mathbf{p}], \widehat{\delta f}_\mathbf{p}\right] + e\mathbf{E}\cdot\frac{\partial\hat{f}_\mathbf{p}}{\partial\mathbf{p}} = \widehat{J}(\widehat{\delta f}|\mathbf{p}),$$

where $\mathbf{E}$ is the in-plane electric field, the equilibrium distribution function $\hat{f}_\mathbf{p}$ is given by Eq. (63.24), and the collision integral is given in problem 13.10 (one should replace $\hat{f}_{\mathbf{rp}t}$ by $\widehat{\delta f}_\mathbf{p}$ and put $\hat{h}_\mathbf{p} = p^2/2m + \mathbf{v}_s\cdot[\hat{\boldsymbol{\sigma}}\times\mathbf{p}]$ since there is no magnetic field). Keeping the terms of the order $mv_s/\overline{p}$ (and neglecting the terms of higher order) in the collision integral, we can write the linearized matrix kinetic equation as a set of equations for the scalar and vector parts of the matrix distribution function:

$$e\mathbf{E}\cdot\frac{\partial f_\mathbf{p}}{\partial\mathbf{p}} = \frac{2\pi n_{im}|v|^2}{\hbar V}\sum_{\mathbf{p}'}\left[(\delta f_{\mathbf{p}'}-\delta f_\mathbf{p})\delta(\varepsilon_p - \varepsilon_{p'})\right.$$

$$\left. + \sum_{\alpha\beta}e_{\alpha\beta z}v_s(p_\beta - p'_\beta)(\delta f_{\mathbf{p}'}^\alpha - \delta f_\mathbf{p}^\alpha)\delta'(\varepsilon_p - \varepsilon_{p'})\right],$$

$$\frac{2v_s}{\hbar}[p_\alpha\delta f_\mathbf{p}^z - \delta_{\alpha z}(p_x\delta f_\mathbf{p}^x + p_y\delta f_\mathbf{p}^y)] + e\mathbf{E}\cdot\frac{\partial f_\mathbf{p}^\alpha}{\partial\mathbf{p}} = \frac{2\pi n_{im}|v|^2}{\hbar V}$$

$$\times\sum_{\mathbf{p}'}\left[(\delta f_{\mathbf{p}'}^\alpha - \delta f_\mathbf{p}^\alpha)\delta(\varepsilon_p - \varepsilon_{p'}) + \sum_\beta e_{\alpha\beta z}v_s(p_\beta - p'_\beta)(\delta f_{\mathbf{p}'} - \delta f_\mathbf{p})\delta'(\varepsilon_p - \varepsilon_{p'})\right],$$

where $\varepsilon_p = p^2/2m$. We have formally introduced the derivative $\delta'(x) = d\delta(x)/dx$. In the first equation, the correction $\propto v_s$ to the collision integral can be neglected, while in the second one such a correction should be kept, because the field term containing $f_{\mathbf{p}}^\alpha$ is also proportional to v_s. Excluding the scalar part of the distribution function, we get the following equation for the vector part:

$$\frac{2v_s}{\hbar}\left[p_\alpha \delta f_{\mathbf{p}}^z - \delta_{\alpha z}(p_x \delta f_{\mathbf{p}}^x + p_y \delta f_{\mathbf{p}}^y)\right] + e\sum_{\beta\gamma} E_\beta v_s e_{\alpha\gamma z}\left(\frac{p_\beta p_\gamma}{m} - \delta_{\beta\gamma}\varepsilon_p\right)\frac{\partial^2}{\partial \varepsilon_p^2}f(\varepsilon_p)$$

$$= \frac{2\pi n_{im}|v|^2}{\hbar V}\sum_{\mathbf{p}'}(\delta f_{\mathbf{p}'}^\alpha - \delta f_{\mathbf{p}}^\alpha)\delta(\varepsilon_p - \varepsilon_{p'}),$$

where $f(\varepsilon_p)$ is the equilibrium distribution function. Solving this equation, we find the vector part of the distribution function:

$$\delta \mathbf{f_p} = -\frac{e}{m\nu}(\mathbf{E}\cdot\mathbf{p})[\mathbf{p}\times\mathbf{v}_s]\frac{\partial^2 f(\varepsilon_p)}{\partial \varepsilon_p^2},$$

where the momentum relaxation rate ν is given by Eq. (8.21). Note that, since this solution is symmetric in $\mathbf{p}$, the first term on the left-hand side of the kinetic equation can be left out. Finally, the spin density is $\mathbf{s} = L^{-2}\sum_{\mathbf{p}}\text{tr}_\sigma\hat{\boldsymbol{\sigma}}\widehat{\delta f}_{\mathbf{p}} = (2/L^2)\sum_{\mathbf{p}}\delta \mathbf{f_p} = e[\mathbf{v}_s\times\mathbf{E}]\rho_{2D}/\nu$. The spin polarization is perpendicular to the direction of the current determined by the applied field.

13.12. Find the components f^l, f^r, f^x, and f^y of the electron distribution function in the coupled quantum wells in thermodynamic equilibrium.

<u>Solution</u>: The equilibrium matrix distribution function is $\hat{f}_p = \{\exp[(\hat{h}_p - \mu)/T] + 1\}^{-1}$. It is convenient to express its elements by using the representation of exact (hybridized) states $|+\rangle$ and $|-\rangle$. The distribution function is diagonal in this representation: $f_p^{(\pm)} = \{\exp[(\varepsilon_p \pm \Delta_T/2 - \mu)/T] + 1\}^{-1}$ and $f_p^{(+-)} = f_p^{(-+)} = 0$. Expressing the states $|l\rangle$ and $|r\rangle$ in the basis $|+\rangle$ and $|-\rangle$, we find

$$f_p^l = [f_p^{(+)} + f_p^{(-)}]/2 + [f_p^{(+)} - f_p^{(-)}](\Delta/2\Delta_T),$$

$$f_p^r = [f_p^{(+)} + f_p^{(-)}]/2 - [f_p^{(+)} - f_p^{(-)}](\Delta/2\Delta_T),$$

$f_p^x = (t_{lr}/\Delta)(f_p^l - f_p^r)$, and $f_p^y = 0$.

13.13. Find a criterion for negative magnetoresistance by using Eq. (65.28).

<u>Result</u>: The derivative $d\rho_d/d\omega_c$ at $\omega_c \to 0$ is negative at $1 + \delta^2 < \Omega_T\sqrt{\Omega_T^2 + 4}$.

13.14. Prove that the sum of the contributions of the processes $j = c, j' = c \leftrightarrow j_1 = v, j'_1 = c$ and $j = c, j' = v \leftrightarrow j_1 = c, j'_1 = c$ is zero.

<u>Hint</u>: Carry out the permutations $\mathbf{p} \leftrightarrow \mathbf{p}'_1$ and $\mathbf{p}' \leftrightarrow \mathbf{p}_1$, and use the symmetry property $V(j\mathbf{p}\ j'\mathbf{p}'|j_1\mathbf{p}_1\ j'_1\mathbf{p}'_1) = V(j'_1\mathbf{p}_1\ j_1\mathbf{p}_1|j'\mathbf{p}'\ j\mathbf{p})$ following from the symmetry of the matrix elements (31.20).

13.15. Consider the overlap factors $\varphi_{j\sigma, j'\sigma'}(\mathbf{p}, \mathbf{p}')$ for the two-band model described in Appendix B. Calculate the expressions

$$\sum_{\sigma\sigma'} |\varphi_{c\sigma, c\sigma'}(\mathbf{p}, \mathbf{p}')|^2 \quad \text{and} \quad \sum_{\sigma\sigma'} |\varphi_{c\sigma, v\sigma'}(\mathbf{p}, \mathbf{p}')|^2.$$

<u>Solution:</u> The overlap factor $\varphi_{j\sigma, j'\sigma'}(\mathbf{p}, \mathbf{p}')$ for the two-band model is equal to the matrix element $\langle j\sigma| \ldots |j'\sigma'\rangle$ of the product $\hat{U}_{\mathbf{p}}\hat{U}_{\mathbf{p}'}^+$ given by Eq. (B.24). Thus, the first two terms in the braces of Eq. (B.24) describe intraband transitions, while the third term, proportional to $\hat{\rho}_2$, describes interband transitions. Taking the sums over spin variables, we find that the first required expression is

$$\frac{(\eta_p + 1)(\eta_{p'} + 1)}{2\eta_p \eta_{p'}} \left\{ \left[1 + \left(\frac{\eta_p - 1}{\eta_p + 1} \frac{\eta_{p'} - 1}{\eta_{p'} + 1} \right)^{1/2} \frac{\mathbf{p} \cdot \mathbf{p}'}{pp'} \right]^2 + \frac{\eta_p - 1}{\eta_p + 1} \frac{\eta_{p'} - 1}{\eta_{p'} + 1} \frac{[\mathbf{p} \times \mathbf{p}']^2}{(pp')^2} \right\}$$

and the second one is

$$\frac{(\eta_p + 1)(\eta_{p'} + 1)}{2\eta_p \eta_{p'}} \left\{ \frac{\eta_p - 1}{\eta_p + 1} + \frac{\eta_{p'} - 1}{\eta_{p'} + 1} - 2 \left(\frac{\eta_p - 1}{\eta_p + 1} \frac{\eta_{p'} - 1}{\eta_{p'} + 1} \right)^{1/2} \frac{\mathbf{p} \cdot \mathbf{p}'}{pp'} \right\}.$$

13.16. Prove that for the two-band model described in Appendix B the Auger transitions are forbidden at $M \to \infty$.

<u>Solution:</u> To write the energy conservation law (for example, for Ac processes), we introduce new variables: $\mathbf{P} = (\mathbf{p} + \mathbf{p}')/2$, $\Delta\mathbf{P} = (\mathbf{p} - \mathbf{p}')/2$, and $\mathbf{P}_1 = \mathbf{p}_1 - (\mathbf{p} + \mathbf{p}')/2$ so that $\mathbf{p}_1' = \mathbf{P}_1 - (\mathbf{p} + \mathbf{p}')/2$. The conservation law becomes:

$$\sqrt{(ms)^2 + (\mathbf{P} + \Delta\mathbf{P})^2} + \sqrt{(ms)^2 + (\mathbf{P} - \Delta\mathbf{P})^2}$$

$$= \sqrt{(ms)^2 + (\mathbf{P} + \mathbf{P}_1)^2} - \sqrt{(ms)^2 + (\mathbf{P} - \mathbf{P}_1)^2}.$$

If one calculates the squares of the left- and right-hand sides of this equation and repeats the procedure for the equation obtained to eliminate the square roots, one can find that the equation expressing the energy conservation law has no solutions.

13.17. Write the density balance equation (66.23) with the accuracy including the terms $\propto u^2$.

<u>Hint:</u> Expand the terms containing u in Eq. (66.25) in powers of u.

13.18. Consider the Anderson Hamiltonian which describes the interaction of free electrons with the electrons localized in a single atomic orbital. Derive the interaction term (67.2) by considering the interaction as a perturbation.

<u>Solution:</u> The Anderson Hamiltonian has the following form:

$$\hat{H} = \sum_{\sigma\mathbf{p}} \varepsilon_p \hat{a}_{\sigma\mathbf{p}}^+ \hat{a}_{\sigma\mathbf{p}} + \sum_{\sigma} \varepsilon_d \hat{d}_\sigma^+ \hat{d}_\sigma + U \hat{d}_{+1}^+ \hat{d}_{+1} \hat{d}_{-1}^+ \hat{d}_{-1}$$

$$+ \sum_{\sigma\mathbf{p}} \left(V_{\mathbf{p}} \hat{a}_{\sigma\mathbf{p}}^+ \hat{d}_\sigma + V_{\mathbf{p}}^* \hat{d}_\sigma^+ \hat{a}_{\sigma\mathbf{p}} \right),$$

where $\hat{d}_\sigma^+$ and $\hat{d}_\sigma$ are the creation and annihilation operators of localized electrons, ε_d is the energy of these electrons, U is the Coulomb repulsion energy between the electrons with opposite spins localized in the same orbital of the impurity atom, and $V_{\mathbf{p}}$ is the energy describing the interaction of free and localized electrons. Let us carry out the unitary transformation $\widehat{\overline{H}} = e^{\hat{S}} \hat{H} e^{-\hat{S}}$, where

$$\hat{S} = \sum_{\sigma \mathbf{p}} \sum_{\delta = \pm} \frac{V_{\mathbf{p}}}{\varepsilon_p - \varepsilon_\alpha} n_{d,-\sigma}^{(\alpha)} \hat{a}_{\sigma \mathbf{p}}^+ \hat{d}_\sigma - H.c. \ ,$$

$\varepsilon_+ = \varepsilon_d + U$, $\varepsilon_- = \varepsilon_d$, $n_{d,-\sigma}^{(+)} = \hat{d}_{-\sigma}^+ \hat{d}_{-\sigma}$, and $n_{d,-\sigma}^{(-)} = 1 - \hat{d}_{-\sigma}^+ \hat{d}_{-\sigma}$. In the general case, the transformed Hamiltonian is complicated. The problem is simplified if we consider the last term in the Anderson Hamiltonian as a perturbation (we denote this term as $\hat{H}_1$). Using the Born approximation, we retain only the terms of the second order in $V_{\mathbf{p}}$. Then the transformation gives us the Hamiltonian $\widehat{\overline{H}} = \hat{H}_0 + \hat{H}_2$, where $\hat{H}_0$ is the sum of the first three terms of the Anderson Hamiltonian, while

$$\hat{H}_2 = \frac{1}{2} [\hat{S}, \hat{H}_1] = \hat{H}_{ex} + \hat{H}_{dir} + \hat{H}_0' + \hat{H}_{ch}.$$

In this expression,

$$\hat{H}_{ex} = -\frac{1}{2} \sum_{\mathbf{p}\mathbf{p}'} J_{\mathbf{p}'\mathbf{p}} \sum_{\sigma\sigma'} \hat{a}_{\sigma'\mathbf{p}'}^+ (\hat{\boldsymbol{\sigma}})_{\sigma'\sigma} \hat{a}_{\sigma\mathbf{p}} \cdot \sum_{\sigma_d \sigma_d'} \hat{d}_{\sigma_d'}^+ (\hat{\boldsymbol{\sigma}})_{\sigma_d'\sigma_d} \hat{d}_{\sigma_d}$$

is the spin-dependent (exchange) contribution to electron-impurity interaction,

$$\hat{H}_{dir} = \sum_{\mathbf{p}\mathbf{p}'} \left(W_{\mathbf{p}'\mathbf{p}} + \frac{1}{2} J_{\mathbf{p}'\mathbf{p}} \sum_{\sigma_d} \hat{d}_{\sigma_d}^+ \hat{d}_{\sigma_d} \right) \sum_\sigma \hat{a}_{\sigma\mathbf{p}'}^+ \hat{a}_{\sigma\mathbf{p}}$$

is the spin-independent (direct) contribution to electron-impurity interaction, where

$$J_{\mathbf{p}'\mathbf{p}} = \frac{1}{2} V_{\mathbf{p}'} V_{\mathbf{p}}^* [(\varepsilon_p - \varepsilon_+)^{-1} + (\varepsilon_{p'} - \varepsilon_+)^{-1} - (\varepsilon_p - \varepsilon_-)^{-1} - (\varepsilon_{p'} - \varepsilon_-)^{-1}]$$

and

$$W_{\mathbf{p}'\mathbf{p}} = \frac{1}{2} V_{\mathbf{p}'} V_{\mathbf{p}}^* [(\varepsilon_p - \varepsilon_-)^{-1} + (\varepsilon_{p'} - \varepsilon_-)^{-1}].$$

Next,

$$\hat{H}_0' = -\sum_{\sigma\mathbf{p}} \left(W_{\mathbf{p}\mathbf{p}} + J_{\mathbf{p}\mathbf{p}} \hat{d}_{-\sigma}^+ \hat{d}_{-\sigma} \right) \hat{d}_\sigma^+ \hat{d}_\sigma$$

and

$$\hat{H}_{ch} = \frac{1}{2} \sum_{\sigma\mathbf{p}\mathbf{p}'} J_{\mathbf{p}'\mathbf{p}} \hat{a}_{-\sigma\mathbf{p}'}^+ \hat{a}_{\sigma\mathbf{p}}^+ \hat{d}_\sigma \hat{d}_{-\sigma} + H.c.$$

The part $\hat{H}_0'$ can be appended to $\hat{H}_0$ by proper renormalization of the energies ε_+ and ε_-. The contribution $\hat{H}_{ch}$ changes the number of localized electrons by two and, therefore, should be neglected. Therefore, only $\hat{H}_{ex}$ and $\hat{H}_{dir}$ are essential. The latter is analogous to the spin-independent electron-impurity interaction considered in Chapter 2, while the former is analogous to the Hamiltonian (67.2) with $S = 1/2$. Indeed, let us take into account that $\hat{\Psi}_{\sigma\mathbf{r}} = V^{-1/2} \sum_{\mathbf{p}} e^{i\mathbf{p}\cdot\mathbf{r}/\hbar} \hat{a}_{\sigma\mathbf{p}}$ and $(1/2) \sum_{\sigma_d \sigma_d'} \hat{d}_{\sigma_d'}^+ (\hat{\boldsymbol{\sigma}})_{\sigma_d'\sigma_d} \hat{d}_{\sigma_d} = \hat{\mathbf{S}}_n$. If we define the matrix elements $J_{\mathbf{p}'\mathbf{p}}$ as

$J_{\mathbf{p'p}} = V^{-1} \int d\mathbf{r} J(\mathbf{r} - \mathbf{R}_n) \exp\left[-i(\mathbf{p'} - \mathbf{p}) \cdot \mathbf{r}/\hbar\right]$ and sum $\widehat{H}_{ex}$ over the impurity number n, we obtain $\widehat{\delta H}_{e,im}$ of Eq. (67.2).

13.19. Prove the operator identities (67.11).

Hint: Consider the matrix elements of the operators on the left- and right-hand sides of these identities.

13.20. Calculate the factor γ_ε given by Eq. (67.14) in the isotropic approximation ($\varepsilon_{\mathbf{p}} = p^2/2m$) at $T = 0$.

Result:

$$\gamma_\varepsilon = -\frac{3z}{2\varepsilon_F}\left(1 + \sqrt{\frac{\varepsilon}{4\varepsilon_F}} \ln\left|\frac{\sqrt{\varepsilon} - \sqrt{\varepsilon_F}}{\sqrt{\varepsilon} + \sqrt{\varepsilon_F}}\right|\right),$$

where $z = n/\mathcal{N}$ is the ratio of the electron density to the host atom density.

13.21. Prove that the scattering of electrons by magnetic impurities does not mix the non-diagonal in the spin index components of the averaged density matrix with the diagonal ones.

Solution: The kinetic equation for the non-diagonal components $\rho_{\sigma_1\sigma_2}(\mathbf{p}, t) = \langle\sigma_1\mathbf{p}|\hat{\rho}_t|\sigma_2\mathbf{p}\rangle$ is derived in a similar way as Eq. (67.16) (consider a generalization of the collision integral (7.14) when the scattering amplitude (67.15) is taken into account) and, in the Markovian approximation, has the following form:

$$\frac{\partial\rho_{\sigma_1\sigma_2}(\mathbf{p}, t)}{\partial t} = \frac{\pi}{\hbar}\frac{n_{im}}{V}\sum_{\mathbf{p'}}\sum_{\sigma\sigma'}\left\{\overline{2v_{\sigma_1\sigma}^{MM'}(\varepsilon_{\mathbf{p}}, q)v_{\sigma'\sigma_2}^{M'M}(\varepsilon_{\mathbf{p}}, q)}\rho_{\sigma\sigma'}(\mathbf{p'}, t)\right.$$

$$\left. - \left[\overline{v_{\sigma\sigma'}^{MM'}(\varepsilon_{\mathbf{p}}, q)v_{\sigma'\sigma_2}^{M'M}(\varepsilon_{\mathbf{p}}, q)}\rho_{\sigma_1\sigma}(\mathbf{p}, t) + \overline{v_{\sigma_1\sigma}^{MM'}(\varepsilon_{\mathbf{p}}, q)v_{\sigma\sigma'}^{M'M}(\varepsilon_{\mathbf{p}}, q)}\rho_{\sigma'\sigma_2}(\mathbf{p}, t)\right]\right\}$$

$$\times\delta(\varepsilon_{\mathbf{p}} - \varepsilon_{\mathbf{p'}}),$$

where the matrix elements are given by Eq. (67.17). Calculating the averages as described in Eqs. (67.18)−(67.21), one can prove that $\overline{v_{\sigma\sigma'}^{MM'}(\varepsilon_{\mathbf{p}}, q)v_{\sigma'\sigma_2}^{M'M}(\varepsilon_{\mathbf{p}}, q)} \propto \delta_{\sigma\sigma_2}$ and that $\overline{v_{\sigma_1\sigma}^{MM'}(\varepsilon_{\mathbf{p}}, q)v_{\sigma'\sigma_2}^{M'M}(\varepsilon_{\mathbf{p}}, q)}$ is proportional to $\delta_{\sigma_1\sigma_2}$ only if $\sigma = \sigma'$ (if $\sigma \neq \sigma'$, this average is non-zero only for $\sigma_1 \neq \sigma_2$). In conclusion, the kinetic equation written above does not couple diagonal and non-diagonal components of the density matrix. If the initial distribution of electrons is diagonal in the spin index, it remains diagonal in the presence of the scattering by magnetic impurities.

Chapter 14

FLUCTUATIONS

Each system is characterized by a number of physical quantities, which are often described by their mean values obtained as a result of the averaging of the operators of these quantities according to Eq. (1.18). Such a description is of the main interest in physics, because these mean values are the observable values. Nevertheless, one can obtain a more detailed information about the system, taking into account that the physical quantities defined at a given moment of time or/and in a given spatial point deviate from their mean values. The behavior of these deviations with time or from one small volume to another allows one to consider them as fluctuations of the physical quantities around their mean values. A detailed theoretical description of the evolution of the physical quantities fluctuating with time is not possible because, even if we take into account all interactions, we still do not have initial conditions. Fortunately, such a description is not necessary, since the physical measurements cannot follow this evolution. Instead of the detailed description of the fluctuating quantities, one can characterize them by correlation functions, which have direct physical meaning (for example, the noise in physical systems) and can be investigated experimentally. In this chapter we define the correlation functions of the fluctuations of physical quantities, derive a relation between these functions and generalized susceptibilities in thermodynamic equilibrium, and consider the kinetic equations for fluctuations, which can be applied for calculating the correlation functions of non-equilibrium systems. Then we describe the light scattering due to fluctuations of electron and phonon systems in solids and discuss the current and conductance fluctuations in mesoscopic systems. Finally, we demonstrate how to calculate the correlation functions by using the methods of non-equilibrium diagram technique. Since we do not consider the systems undergoing phase transitions, only the fluctuations whose amplitudes are small in comparison to the mean values are studied below.

663

68. Non-Equilibrium Fluctuations

Consider a quantity Q_t which behaves classically and undergo small variations with time, fluctuating around its mean value $\overline{Q}$. To have an example of such fluctuations, one can imagine a classical system of particles and count the number of particles N_V in a given volume V inside of the system. This number changes with time, though the relative change decreases with increasing V, and the average over time (or over many identical volumes inside of the system) gives us the averaged value $\overline{N}_V$ which is related to the particle density, $n = \overline{N}_V/V$. Of course, if we consider a closed system, the total number of particles and the total energy of the system do not change with time. In practice, however, completely isolated systems do not exist. For example, the electrons in solids interact with phonons, and the physical quantities characterizing the system of electrons (the energy of electrons, the electric current, etc.) inevitably fluctuate.

There exists a correlation between the values of Q_t at different instants. This means that the value of Q at a given instant t affects the probabilities of its various values at a later instant t'. One can characterize this correlation by composing the product $\Delta Q_{t'} \Delta Q_t$, where $\Delta Q_t = Q_t - \overline{Q}$ is the deviation from the mean value, and average this product over all possible values (realizations) of the quantity Q at the instants t and t'. The result of this averaging depends only on the difference $t' - t$ provided that the system is in a stationary state. Therefore, we introduce the correlation function as

$$\Phi(t' - t) = \langle\langle \Delta Q_{t'} \Delta Q_t \rangle\rangle. \tag{1}$$

It is worth mentioning that the statistical averaging in Eq. (1) is equivalent to the averaging over time t (or t') when the difference $t' - t$ is kept constant: $\Phi(\Delta t) = \lim_{\tau \to \infty} \tau^{-1} \int_0^\tau dt Q_{t+\Delta t} Q_t$.

To generalize the definition (1) to the case of quantum variables, we replace the function Q_t by its operator $\widehat{Q}(t)$ in the Heisenberg representation; see Eq. (D.1). For convenience, we consider the symmetrized product of these operators. The correlation function is defined as

$$\Phi(t' - t) = \frac{1}{2}\langle\langle \widehat{\Delta Q}(t')\widehat{\Delta Q}(t) + \widehat{\Delta Q}(t)\widehat{\Delta Q}(t') \rangle\rangle$$

$$\equiv \frac{1}{2}\mathrm{Sp}\{\hat{\eta}[\widehat{\Delta Q}(t'), \widehat{\Delta Q}(t)]_+\}. \tag{2}$$

We remind that the statistical operator $\hat{\eta}$ is independent of time in the Heisenberg representation (see Appendix D), and the dependence of $\Phi(t' - t)$ on time enters through the operators $\widehat{\Delta Q}$. The symmetry of the correlation function with respect to time reversal, $\Phi(\Delta t) = \Phi(-\Delta t)$, is obvious from the definition (2). The correlation function $\Phi(0)$ expresses the mean square of the fluctuations (variance), which is denoted also as $\overline{(\Delta Q)^2}$. Since $\overline{\Delta Q} = 0$, one can write $\Phi(0) \equiv \overline{(\Delta Q)^2} = \overline{Q^2} - (\overline{Q})^2$ (problem 14.1). To introduce correlation functions of different quantities (for example, electric current and electric field, or different components of electric current), we denote the operators of these quantities as $\widehat{Q}_i$ and define the correlation function as

$$\Phi_{ij}(t' - t) = \frac{1}{2}\langle\langle\widehat{\Delta Q_i}(t')\widehat{\Delta Q_j}(t) + \widehat{\Delta Q_j}(t)\widehat{\Delta Q_i}(t')\rangle\rangle. \tag{3}$$

This equation is a generalization of Eq. (2). The subscripts i and j should be considered as multi-indices including the index of the physical quantity itself (current, density, field, etc.) as well as the index of both discrete and continuous (for example, macroscopic coordinate) variables on which this quantity depends. It is seen directly from Eq. (3) that $\Phi_{ij}(\Delta t) = \Phi_{ji}(-\Delta t)$. To study temporal dependence of the correlation function (3), it is often convenient to consider its spectral (Fourier) representation

$$\Phi_{ij}(\omega) = \int_{-\infty}^{\infty} d\Delta t\, e^{i\omega\Delta t}\Phi_{ij}(\Delta t), \tag{4}$$

which has the symmetry property $\Phi_{ij}(\omega) = \Phi_{ji}(-\omega)$.

In the case of thermodynamic equilibrium, the statistical operator is determined by the temperature T and by the Hamiltonian of the system. Let us show that in this case there exists a simple expression relating $\Phi_{ij}(\omega)$ to the generalized susceptibility of the system with respect to the variable Q. Assuming that the mean values of Q_i and Q_j are zeros so that $\widehat{\Delta Q_i}(t) = \widehat{Q}_i(t)$, we use Eqs. (D.2), (D.3), (D.7), and (D.11) with $\widehat{A} = \widehat{Q}_i$ and $\widehat{B} = \widehat{Q}_j$. The correlation function is represented as

$$\Phi_{ij}(\omega) = \frac{1}{2}\coth\frac{\hbar\omega}{2T}\int dt\, e^{i\omega t}\langle\langle\widehat{Q}_i(t)\widehat{Q}_j(0) - \widehat{Q}_j(0)\widehat{Q}_i(t)\rangle\rangle. \tag{5}$$

Using the second quantization representation of the operators, $\widehat{Q}_i = \sum_{\delta\delta'}\langle\delta|\hat{q}_i|\delta'\rangle\hat{a}_\delta^+\hat{a}_{\delta'}$, we assume that the ket-vectors $|\delta\rangle$ are the exact eigenstates and employ Eq. (D.16) in order to calculate the correlation functions. The commutators of the pairs of creation and annihilation operators are reduced to products of one creation and one annihilation operators, and we obtain

$$\langle\langle\widehat{Q}_i(t)\widehat{Q}_j(0) - \widehat{Q}_j(0)\widehat{Q}_i(t)\rangle\rangle$$
$$= 2\sum_{\delta\delta'} e^{i(\varepsilon_\delta - \varepsilon_{\delta'})t/\hbar}\langle\delta|\hat{q}_i|\delta'\rangle\langle\delta'|\hat{q}_j|\delta\rangle[f(\varepsilon_\delta) - f(\varepsilon_{\delta'})], \tag{6}$$

where $f(\varepsilon_\delta) = \mathrm{Sp}\{\hat{\eta}\hat{a}_\delta^+\hat{a}_\delta\}$ is the one-particle distribution function in equilibrium. The sum over spin variables is taken out of the sum over δ and δ' and gives the factor of 2 in Eq. (6). Calculating the integral over time in Eq. (5), we finally obtain

$$\Phi_{ij}(\omega) = 2\pi\hbar \coth\frac{\hbar\omega}{2T} \sum_{\delta\delta'} \langle\delta|\hat{q}_i|\delta'\rangle\langle\delta'|\hat{q}_j|\delta\rangle$$

$$\times \delta(\varepsilon_\delta - \varepsilon_{\delta'} + \hbar\omega)[f(\varepsilon_\delta) - f(\varepsilon_{\delta'})]. \qquad (7)$$

The symmetry properties $\Phi_{ij}(\omega) = \Phi_{ji}(-\omega) = \Phi_{ji}^*(\omega)$ are seen directly from this expression. The symmetry of the equations of motion with respect to time reversal, if the sign of the magnetic field $\mathbf{H}$ is changed simultaneously, imposes the symmetry property

$$\Phi_{ij}(\omega) = \Phi_{ji}(\omega)|_{\mathbf{H}\to-\mathbf{H}} , \qquad (8)$$

which is related to Onsager's symmetry of linear kinetic coefficients discussed in Sec. 13.

Substituting the operators of electric current density, $\hat{I}_\alpha(\mathbf{r})$ and $\hat{I}_\beta(\mathbf{r}')$, in place of $\hat{q}_i$ and $\hat{q}_j$, one may compare the result (7) to the expression (13.10), writing the trace in the latter in the exact eigenstate representation. We find

$$\frac{1}{2}\left[\mathcal{S}_{\alpha\beta}(\mathbf{r},\mathbf{r}'|\omega) + \mathcal{S}_{\beta\alpha}(\mathbf{r}',\mathbf{r}|\omega)\right] = \hbar\omega \coth\frac{\hbar\omega}{2T}\mathrm{Re}\,\sigma_{\alpha\beta}^{(s)}(\mathbf{r},\mathbf{r}'|\omega), \qquad (9)$$

$$\mathcal{S}_{\alpha\beta}(\mathbf{r},\mathbf{r}'|\omega) \equiv \frac{1}{2}\langle\langle\hat{I}_\alpha(\mathbf{r},t)\hat{I}_\beta(\mathbf{r}',0) + \hat{I}_\beta(\mathbf{r}',0)\hat{I}_\alpha(\mathbf{r},t)\rangle\rangle_\omega ,$$

where $\sigma_{\alpha\beta}^{(s)}(\mathbf{r},\mathbf{r}'|\omega) \equiv [\sigma_{\alpha\beta}(\mathbf{r},\mathbf{r}'|\omega) + \sigma_{\beta\alpha}(\mathbf{r}',\mathbf{r}|\omega)]/2$ is the symmetric part of the conductivity tensor and $\langle\langle\ldots\rangle\rangle_\omega \equiv \int dt e^{i\omega t}\langle\langle\ldots\rangle\rangle$ defines the temporal Fourier transform of the correlation function $\langle\langle\ldots\rangle\rangle$. The current-current correlation function $\mathcal{S}_{\alpha\beta}(\mathbf{r},\mathbf{r}'|\omega)$ is called the noise power (problem 14.2). In a spatially homogeneous system, where $\sigma_{\alpha\beta}(\mathbf{r},\mathbf{r}'|\omega) = \sigma_{\alpha\beta}(\mathbf{r}-\mathbf{r}'|\omega)$, this correlation function depends only on the differential coordinate $\mathbf{r} - \mathbf{r}'$. The equilibrium fluctuations of low frequency, $\omega \ll T/\hbar$, are classical. Indeed, the factor $\hbar\omega \coth(\hbar\omega/2T)$ in this case is reduced to $2T$, and the Planck constant no longer enters the equation. In addition to the above example of the current-current correlation functions, one can consider a more general case. Comparing Eq. (7) to the expression (13.31) describing the symmetric part of the generalized susceptibility, we find that

$$\Phi_{ij}^{(s)}(\omega) = -\hbar\coth\frac{\hbar\omega}{2T}\mathrm{Im}\chi_{ij}^{(s)}(\omega), \qquad (10)$$

where the left-hand side contains the symmetric part of the correlation function (7). This simple equation is known as the fluctuation-dissipation theorem. A particular case of it is given by Eq. (9) (because of the factor of i standing in the relation (13.6) between the perturbation Hamiltonian and the current, the correlation function of the fluctuations is proportional to the real part of the conductivity). Using Eq. (10), one can study, for example, the equilibrium fluctuations of the density of electron gas (problem 14.3).

To study fluctuations in non-equilibrium conditions, one cannot use the approach described above. Instead, one should take into account the characteristic features of the system under consideration: energy spectrum, interactions, etc. Similar as in Eqs. (6) and (7), we assume that the operators $\widehat{Q}_i$ of the fluctuating quantities Q_i are additive, i.e., representable through the products of one creation and one annihilation operators, as in Eq. (4.23). Therefore, to find the correlation function (3), one has to average a product of four (two creation and two annihilation) operators. This product can be found from the equation of motion for Heisenberg operators (see Appendix D):

$$\frac{d}{dt}\hat{A}(t) = \frac{i}{\hbar}[\widehat{H}, \hat{A}(t)], \qquad (11)$$

where $\widehat{H}$ is the Hamiltonian of the system and $\hat{A}$ is an arbitrary operator in the Heisenberg representation. Let us choose this operator as a product of the Heisenberg creation and annihilation operators taken at the same instant t. Consider, for example, the electron-boson system described by the Hamiltonian (19.1). We assume that the electrons occupy a single band with an isotropic parabolic energy spectrum in a macroscopically homogeneous crystal and are characterized by the momentum $\mathbf{p}$. These electrons interact with a single phonon mode characterized by the wave vector $\mathbf{q}$. Therefore, the Hamiltonian under consideration is written as

$$\widehat{H} = \sum_{\mathbf{p}} \varepsilon_{\mathbf{p}} \hat{a}_{\mathbf{p}}^{+}\hat{a}_{\mathbf{p}} + \sum_{\mathbf{q}} \hbar\omega_{\mathbf{q}}\left(\hat{b}_{\mathbf{q}}^{+}\hat{b}_{\mathbf{q}} + \frac{1}{2}\right) + \sum_{\mathbf{pq}} C_q \hat{a}_{\mathbf{p}+\hbar\mathbf{q}}^{+}\hat{a}_{\mathbf{p}}(\hat{b}_{\mathbf{q}} + \hat{b}_{-\mathbf{q}}^{+}), \quad (12)$$

where the matrix element of electron-phonon interaction, C_q, is assumed to be real. For the deformation-potential interaction of electrons with *LA* phonons or Froelich interaction with *LO* phonons, this coefficient is given by Eq. (21.1). Strictly speaking, the creation and annihilation operators of electrons should bear spin indices σ, and the sums over these indices in the first and third terms of the Hamiltonian (12) are implied. However, since below we study the equations for the operator products

with coinciding spin indices and do not consider spin-dependent inter-
actions, these indices are omitted. Substituting $\hat{a}_{\mathbf{p}'}^{+}(t)\hat{a}_{\mathbf{p}}(t) \equiv \left\{\hat{a}_{\mathbf{p}'}^{+}\hat{a}_{\mathbf{p}}\right\}_t$
in place of $\hat{A}(t)$ in Eq. (11) with the Hamiltonian (12), we obtain the
equation

$$\left[\frac{d}{dt} + \frac{i}{\hbar}(\varepsilon_{\mathbf{p}} - \varepsilon_{\mathbf{p}'})\right] \hat{a}_{\mathbf{p}'}^{+}\hat{a}_{\mathbf{p}} = -\frac{i}{\hbar}\sum_{\mathbf{q}} C_q(\hat{b}_{\mathbf{q}} + \hat{b}_{-\mathbf{q}}^{+})(\hat{a}_{\mathbf{p}'}^{+}\hat{a}_{\mathbf{p}-\hbar\mathbf{q}} - \hat{a}_{\mathbf{p}'+\hbar\mathbf{q}}^{+}\hat{a}_{\mathbf{p}}),$$

(13)

where the coinciding time arguments of the operators are omitted for
brevity. Its right-hand side contains the products of three operators: one
bosonic and two fermionic. The equations of motion for these operator
products are

$$\left[\frac{d}{dt} - \frac{i}{\hbar}\Delta_{\pm}(\mathbf{p}_1, \mathbf{p}_2)\right] \hat{A}_{\pm}(\mathbf{p}_1, \mathbf{p}_2) = \widehat{F}_{\pm}(\mathbf{p}_1, \mathbf{p}_2),$$
(14)

where $\hat{A}_{-}(\mathbf{p}_1, \mathbf{p}_2) = \hat{b}_{\mathbf{q}}\hat{a}_{\mathbf{p}_1}^{+}\hat{a}_{\mathbf{p}_2}$, $\hat{A}_{+}(\mathbf{p}_1, \mathbf{p}_2) = \hat{b}_{-\mathbf{q}}^{+}\hat{a}_{\mathbf{p}_1}^{+}\hat{a}_{\mathbf{p}_2}$, and $\Delta_{\pm}(\mathbf{p}_1, \mathbf{p}_2)$
$= \varepsilon_{\mathbf{p}_1} - \varepsilon_{\mathbf{p}_2} \pm \hbar\omega_{\mathbf{q}}$. The expression for $\widehat{F}_{+}$ and $\widehat{F}_{-}$ standing on the
right-hand side of Eq. (14) is written as

$$\left| \begin{array}{c} \widehat{F}_{+}(\mathbf{p}_1, \mathbf{p}_2) \\ F_{-}(\mathbf{p}_1, \mathbf{p}_2) \end{array} \right| = \pm\frac{i}{\hbar}C_q\hat{a}_{\mathbf{p}_1}^{+}\hat{a}_{\mathbf{p}_2}\sum_{\mathbf{p}'}\hat{a}_{\mathbf{p}'-\hbar\mathbf{q}}^{+}\hat{a}_{\mathbf{p}'} \qquad (15)$$

$$-\frac{i}{\hbar}\sum_{\mathbf{q}'}C_{q'}(\hat{b}_{\mathbf{q}'} + \hat{b}_{-\mathbf{q}'}^{+}) \left| \begin{array}{c} \hat{b}_{-\mathbf{q}}^{+} \\ \hat{b}_{\mathbf{q}} \end{array} \right| \left(\hat{a}_{\mathbf{p}_1}^{+}\hat{a}_{\mathbf{p}_2-\hbar\mathbf{q}'} - \hat{a}_{\mathbf{p}_1+\hbar\mathbf{q}'}^{+}\hat{a}_{\mathbf{p}_2}\right),$$

Omitting, for brevity, the arguments of Δ, $\hat{A}$, and $\widehat{F}$, we write the general
solution of Eq. (14) as

$$\hat{A}_{\pm}(t) = e^{i\Delta_{\pm}(t-t_0)/\hbar}\hat{A}_{\pm}(t_0) + \int_{t_0}^{t} dt' e^{i\Delta_{\pm}(t-t')/\hbar}\widehat{F}_{\pm}(t'), \qquad (16)$$

where t_0 is an arbitrary time. The first term on the right-hand side of
this expression describes the influence of the initial conditions (at $t = t_0$)
on the evolution of the three-operator products denoted here by $\hat{A}_{\pm}$. The
second term, which is linear in the electron-phonon interaction, describes
the influence of the scattering on this evolution. After substituting the
solution (16) into Eq. (13), one can see that this term leads to the
contributions quadratic in the interaction.

According to the above consideration, we rewrite the operator equation (13) as

$$\left(\frac{d}{dt} + i\mathbf{v}_{\mathbf{p}}\cdot\mathbf{k}\right)\left\{\hat{a}_{\mathbf{p}-\frac{\hbar\mathbf{k}}{2}}^{+}\hat{a}_{\mathbf{p}+\frac{\hbar\mathbf{k}}{2}}\right\}_t = \widehat{K}_t(\mathbf{p}, \mathbf{k}) + \widehat{M}_t(\mathbf{p}, \mathbf{k}), \qquad (17)$$

where $\mathbf{v_p} = \mathbf{p}/m$ is the electron velocity. We have done the substitutions $\mathbf{p'} \to \mathbf{p} - \hbar\mathbf{k}/2$ and $\mathbf{p} \to \mathbf{p} + \hbar\mathbf{k}/2$. The operators $\widehat{K}_t(\mathbf{p},\mathbf{k})$ and $\widehat{M}_t(\mathbf{p},\mathbf{k})$ denote all the contributions coming, respectively, from the first and second terms on the right-hand side of Eq. (16). Let us neglect the term $\widehat{M}_t(\mathbf{p},\mathbf{k})$, since it is quadratic in the electron-phonon interaction, while the term $\widehat{K}_t(\mathbf{p},\mathbf{k})$ is linear in this interaction. This neglect is justified if $|\omega - \mathbf{v_p} \cdot \mathbf{k}|$, where ω is the frequency of the Fourier transform of $\{\hat{a}^+_{\mathbf{p}-\hbar\mathbf{k}/2}\hat{a}_{\mathbf{p}+\hbar\mathbf{k}/2}\}_t$, is large in comparison to the electron-phonon scattering rate. Equation (17) can be solved in the form (16). Introducing the temporal Fourier transform $\widehat{K}_\omega(\mathbf{p},\mathbf{k})$ of the operator $\widehat{K}_t(\mathbf{p},\mathbf{k})$, we present the solution as

$$\left\{\hat{a}^+_{\mathbf{p}-\frac{\hbar\mathbf{k}}{2}}\hat{a}_{\mathbf{p}+\frac{\hbar\mathbf{k}}{2}}\right\}_t = \widehat{K}^0_t(\mathbf{p},\mathbf{k}) + \int \frac{d\omega}{2\pi}e^{-i\omega t}\frac{i\widehat{K}_\omega(\mathbf{p},\mathbf{k})}{\omega - \mathbf{v_p}\cdot\mathbf{k}}[1-e^{i(\omega-\mathbf{v_p}\cdot\mathbf{k})(t-t_1)}], \tag{18}$$

where

$$\widehat{K}^0_t(\mathbf{p},\mathbf{k}) = \left\{\hat{a}^+_{\mathbf{p}-\frac{\hbar\mathbf{k}}{2}}\hat{a}_{\mathbf{p}+\frac{\hbar\mathbf{k}}{2}}\right\}_{t_1} e^{-i\mathbf{v_p}\cdot\mathbf{k}(t-t_1)}. \tag{19}$$

As follows from Eqs. (13) and (16),

$$\widehat{K}_\omega(\mathbf{p},\mathbf{k}) = -2\pi i e^{i\omega t_0} \sum_{\mathbf{q}} C_q \tag{20}$$

$$\times \left[\left\{\hat{b}_{\mathbf{q}}\hat{a}^+_{\mathbf{p}-\frac{\hbar\mathbf{k}}{2}}\hat{a}_{\mathbf{p}+\frac{\hbar\mathbf{k}}{2}-\hbar\mathbf{q}}\right\}_{t_0} \delta\left(\varepsilon_{\mathbf{p}-\frac{\hbar\mathbf{k}}{2}} - \varepsilon_{\mathbf{p}+\frac{\hbar\mathbf{k}}{2}-\hbar\mathbf{q}} - \hbar\omega_{\mathbf{q}} + \hbar\omega\right)\right.$$

$$+ \left\{\hat{b}^+_{-\mathbf{q}}\hat{a}^+_{\mathbf{p}-\frac{\hbar\mathbf{k}}{2}}\hat{a}_{\mathbf{p}+\frac{\hbar\mathbf{k}}{2}-\hbar\mathbf{q}}\right\}_{t_0} \delta\left(\varepsilon_{\mathbf{p}-\frac{\hbar\mathbf{k}}{2}} - \varepsilon_{\mathbf{p}+\frac{\hbar\mathbf{k}}{2}-\hbar\mathbf{q}} + \hbar\omega_{\mathbf{q}} + \hbar\omega\right)$$

$$- \left\{\hat{b}_{\mathbf{q}}\hat{a}^+_{\mathbf{p}-\frac{\hbar\mathbf{k}}{2}+\hbar\mathbf{q}}\hat{a}_{\mathbf{p}+\frac{\hbar\mathbf{k}}{2}}\right\}_{t_0} \delta\left(\varepsilon_{\mathbf{p}-\frac{\hbar\mathbf{k}}{2}+\hbar\mathbf{q}} - \varepsilon_{\mathbf{p}+\frac{\hbar\mathbf{k}}{2}-\hbar\omega_{\mathbf{q}}} + \hbar\omega\right)$$

$$\left. - \left\{\hat{b}^+_{-\mathbf{q}}\hat{a}^+_{\mathbf{p}-\frac{\hbar\mathbf{k}}{2}+\hbar\mathbf{q}}\hat{a}_{\mathbf{p}+\frac{\hbar\mathbf{k}}{2}}\right\}_{t_0} \delta\left(\varepsilon_{\mathbf{p}-\frac{\hbar\mathbf{k}}{2}+\hbar\mathbf{q}} - \varepsilon_{\mathbf{p}+\frac{\hbar\mathbf{k}}{2}} + \hbar\omega_{\mathbf{q}} + \hbar\omega\right)\right].$$

We point out that the expression under the integral in Eq. (18) does not diverge at $\omega = \mathbf{v_p} \cdot \mathbf{k}$ because of the presence of the expression in the square brackets which goes to zero in these conditions. Nevertheless, since we assume that the main contribution to the integral comes from the region where $|\omega - \mathbf{v_p} \cdot \mathbf{k}|$ is large in comparison to the electron-phonon scattering rate, one should neglect the contribution of the second term in these square brackets. This contribution has the same temporal dependence ($\propto e^{-i\mathbf{v_p}\cdot\mathbf{k}t}$) as the collisionless term $\widehat{K}^0$, but also contains the smallness associated with weak electron-phonon coupling.

The operators (19) and (20) are called the Langevin sources. The first of them describes a collisionless (dynamical) evolution of the pair product of fermionic operators. The second one describes the influence of the electron-phonon scattering on this evolution. Below we are interested in the functions of the kind $\langle\langle\{\hat{a}^+_{\mathbf{p}-\hbar\mathbf{k}/2}\hat{a}_{\mathbf{p}+\hbar\mathbf{k}/2}\}_t\{\hat{a}^+_{\mathbf{p'}-\hbar\mathbf{k'}/2}\hat{a}_{\mathbf{p'}+\hbar\mathbf{k'}/2}\}_{t'}\rangle\rangle$, which are determined by the correlation functions of the operators (19) and (20). As already mentioned, the operators $\hat{a}$ and $\hat{a}^+$ bear spin indices, which are not written explicitly since we consider the products $\hat{a}^+_{\mathbf{p}-\hbar\mathbf{k}/2}\hat{a}_{\mathbf{p}+\hbar\mathbf{k}/2}$ of the operators with coinciding spin indices. In a similar way, the four-operator correlation function written above and the correlation functions of the Langevin sources (19) and (20) (see Eqs. (22) and (23) below) are considered for the case of coinciding spin indices (for this reason we also omit the spin indices at the Langevin sources). The temporal dependence of $\widehat{K}^0_t(\mathbf{p},\mathbf{k})$ and the frequency dependence of $\widehat{K}_\omega(\mathbf{p},\mathbf{k})$ are determined by the initial moments of time, t_1 and t_0, respectively, which can be chosen in an arbitrary way. However, under the assumption of weak interaction, the correlation functions $\langle\langle\widehat{K}^0_t(\mathbf{p},\mathbf{k})\widehat{K}^0_{t'}(\mathbf{p'},\mathbf{k'})\rangle\rangle$ and $\langle\langle\widehat{K}_\omega(\mathbf{p},\mathbf{k})\widehat{K}_{\omega'}(\mathbf{p'},\mathbf{k'})\rangle\rangle$ become independent of t_1 and t_0, because they are proportional to $\delta_{\mathbf{k},-\mathbf{k'}}\delta_{\mathbf{p},\mathbf{p'}}$ and $\delta(\omega+\omega')$, respectively. Indeed, since we have already taken into account the terms of the lowest order in the electron-phonon interaction (note that the product of two Langevin sources $\widehat{K}_\omega(\mathbf{p},\mathbf{k})$ is quadratic in the interaction), we can calculate these correlation functions by averaging over the states of non-interacting system. Using the temporal dependence of the creation and annihilation operators of free quasiparticles, see Eq. (D.16), we have, for example,

$$\left\langle\left\langle\left\{\hat{b}_{\mathbf{q}}\hat{a}^+_{\mathbf{p}-\frac{\hbar\mathbf{k}}{2}}\hat{a}_{\mathbf{p}+\frac{\hbar\mathbf{k}}{2}-\hbar\mathbf{q}}\right\}_{t_0}\left\{\hat{b}^+_{-\mathbf{q'}}\hat{a}^+_{\mathbf{p'}-\frac{\hbar\mathbf{k'}}{2}}\hat{a}_{\mathbf{p'}+\frac{\hbar\mathbf{k'}}{2}-\hbar\mathbf{q'}}\right\}_{t_0}\right\rangle\right\rangle$$

$$\simeq \delta_{\mathbf{q},-\mathbf{q'}}\delta_{\mathbf{k},-\mathbf{k'}}\delta_{\mathbf{p'},\mathbf{p}-\hbar\mathbf{q'}}(N_{\mathbf{q}}+1)f_{\mathbf{p}-\frac{\hbar\mathbf{k}}{2}}\left(1-f_{\mathbf{p}+\frac{\hbar\mathbf{k}}{2}-\hbar\mathbf{q}}\right), \qquad (21)$$

where $N_{\mathbf{q}}$ and $f_{\mathbf{p}}$ are the distribution functions of phonons and electrons, respectively.

Considering the other relevant terms in a similar way, we obtain the expressions

$$\langle\langle\widehat{K}^0_t(\mathbf{p},\mathbf{k})\widehat{K}^0_{t'}(\mathbf{p'},\mathbf{k'})\rangle\rangle = \delta_{\mathbf{k},-\mathbf{k'}}\delta_{\mathbf{p},\mathbf{p'}}f_{\mathbf{p}-\frac{\hbar\mathbf{k}}{2}}\left(1-f_{\mathbf{p}+\frac{\hbar\mathbf{k}}{2}}\right)e^{-i\mathbf{v_p}\cdot\mathbf{k}(t-t')}$$

$$(22)$$

and

$$\langle\langle \widehat{K}_\omega(\mathbf{p},\mathbf{k})\widehat{K}_{\omega'}(\mathbf{p}',\mathbf{k}')\rangle\rangle = \frac{(2\pi)^2}{\hbar}\delta(\omega+\omega')\delta_{\mathbf{k},-\mathbf{k}'}\sum_{\mathbf{q}}|C_q|^2$$

$$\times(\delta_{\mathbf{p}',\mathbf{p}}-\delta_{\mathbf{p}',\mathbf{p}-\hbar\mathbf{q}})\left[\delta\left(\varepsilon_{\mathbf{p}-\frac{\hbar\mathbf{k}}{2}}-\varepsilon_{\mathbf{p}+\frac{\hbar\mathbf{k}}{2}-\hbar\mathbf{q}}-\hbar\omega_{\mathbf{q}}+\hbar\omega\right)\right.$$

$$\times(N_{\mathbf{q}}+1)f_{\mathbf{p}-\frac{\hbar\mathbf{k}}{2}}\left(1-f_{\mathbf{p}+\frac{\hbar\mathbf{k}}{2}-\hbar\mathbf{q}}\right)+\delta\left(\varepsilon_{\mathbf{p}-\frac{\hbar\mathbf{k}}{2}-\hbar\mathbf{q}}\right.$$

$$\left.-\varepsilon_{\mathbf{p}+\frac{\hbar\mathbf{k}}{2}}-\hbar\omega_{\mathbf{q}}+\hbar\omega\right)(N_{-\mathbf{q}}+1)f_{\mathbf{p}-\frac{\hbar\mathbf{k}}{2}-\hbar\mathbf{q}}\left(1-f_{\mathbf{p}+\frac{\hbar\mathbf{k}}{2}}\right)\qquad(23)$$

$$+\delta\left(\varepsilon_{\mathbf{p}-\frac{\hbar\mathbf{k}}{2}}-\varepsilon_{\mathbf{p}+\frac{\hbar\mathbf{k}}{2}-\hbar\mathbf{q}}+\hbar\omega_{\mathbf{q}}+\hbar\omega\right)N_{-\mathbf{q}}f_{\mathbf{p}-\frac{\hbar\mathbf{k}}{2}}\left(1-f_{\mathbf{p}+\frac{\hbar\mathbf{k}}{2}-\hbar\mathbf{q}}\right)$$

$$+\delta\left(\varepsilon_{\mathbf{p}-\frac{\hbar\mathbf{k}}{2}-\hbar\mathbf{q}}-\varepsilon_{\mathbf{p}+\frac{\hbar\mathbf{k}}{2}}+\hbar\omega_{\mathbf{q}}+\hbar\omega\right)N_{\mathbf{q}}f_{\mathbf{p}-\frac{\hbar\mathbf{k}}{2}-\hbar\mathbf{q}}\left(1-f_{\mathbf{p}+\frac{\hbar\mathbf{k}}{2}}\right)\right],$$

which can be used for calculating various correlation functions. The times t_0 and t_1 do not enter the correlation functions. A consideration of the other scattering mechanisms brings additional terms into the Langevin sources (problem 14.4). It is important to notice that the mixed correlation functions $\langle\langle \widehat{K}^0\widehat{K}\rangle\rangle$ are equal to zero, since $\widehat{K}$ is linear in the bosonic operators (we do not consider the case of coherent phonons studied in Sec. 57). In the case of electron-impurity interaction considered in problem 14.4, such mixed correlation functions are also equal to zero because they are linear in the random potential whose average is zero.

To find the symmetrized correlation function of the electric current densities (noise power) defined in Eq. (9), it is convenient to write the current density operator $\hat{\mathbf{I}}(\mathbf{k},t)=\int d\mathbf{r}e^{-i\mathbf{k}\cdot\mathbf{r}}\hat{\mathbf{I}}(\mathbf{r},t)$ in the second quantization representation (problem 14.5). Then we obtain

$$\mathcal{S}_{\alpha\beta}(\mathbf{r},\mathbf{r}'|\omega)=\frac{e^2}{2m^2}\frac{1}{V^2}\sum_{\mathbf{p}\mathbf{p}'}p_\alpha p_\beta'\sum_{\mathbf{k}\mathbf{k}'}e^{i\mathbf{k}\cdot\mathbf{r}+i\mathbf{k}'\cdot\mathbf{r}'}$$

$$\times\sum_{\sigma\sigma'}\left\langle\left\langle\left[\left\{\hat{a}^+_{\sigma\mathbf{p}-\frac{\hbar\mathbf{k}}{2}}\hat{a}_{\sigma\mathbf{p}+\frac{\hbar\mathbf{k}}{2}}\right\}_t,\left\{\hat{a}^+_{\sigma'\mathbf{p}'-\frac{\hbar\mathbf{k}'}{2}}\hat{a}_{\sigma'\mathbf{p}'+\frac{\hbar\mathbf{k}'}{2}}\right\}_{t'}\right]_+\right\rangle\right\rangle_\omega.\qquad(24)$$

The spin indices at the operators $\hat{a}$ and $\hat{a}^+$ are written explicitly since the correlation function in Eq. (24) contains the operators of electrons with different spins. It is easy to see that this correlation function is proportional to $\delta_{\sigma\sigma'}$, and the sum over spin gives an extra factor of 2 in the equations below. Let us substitute the solution (18) (without the second term in the square brackets under the integral, as discussed above) into Eq. (24) and employ Eqs. (22) and (23). We obtain, after some transformations, the following result:

$$\mathcal{S}_{\alpha\beta}(\mathbf{r}, \mathbf{r}'|\omega) = \frac{1}{V} \sum_{\mathbf{k}} e^{i\mathbf{k}\cdot(\mathbf{r}-\mathbf{r}')} \left[\mathcal{S}_{\alpha\beta}^{(0)}(\mathbf{k}, \omega) + \mathcal{S}_{\alpha\beta}^{(1)}(\mathbf{k}, \omega) \right], \qquad (25)$$

where

$$\mathcal{S}_{\alpha\beta}^{(0)}(\mathbf{k}, \omega) = \frac{2\pi e^2}{m^2 V} \sum_{\mathbf{p}} p_\alpha p_\beta \delta(\omega - \mathbf{v_p} \cdot \mathbf{k})$$

$$\times \left[f_{\mathbf{p} - \frac{\hbar\mathbf{k}}{2}} \left(1 - f_{\mathbf{p} + \frac{\hbar\mathbf{k}}{2}} \right) + f_{\mathbf{p} + \frac{\hbar\mathbf{k}}{2}} \left(1 - f_{\mathbf{p} - \frac{\hbar\mathbf{k}}{2}} \right) \right] \qquad (26)$$

comes from the correlation function (22) and describes the free-electron fluctuations, while the contribution $\mathcal{S}_{\alpha\beta}^{(1)}(\mathbf{k}, \omega)$ comes from the correlation function (23) and describes the influence of electron-phonon scattering on the fluctuations of current density. Below we write $\mathcal{S}_{\alpha\beta}^{(1)}(\mathbf{k}, \omega)$ in the local approximation, when ω is large enough and the spatial dispersion can be neglected, $\mathcal{S}_{\alpha\beta}^{(1)}(\mathbf{k}, \omega) \simeq \mathcal{S}_{\alpha\beta}^{(1)}(\mathbf{k} = 0, \omega) \equiv \mathcal{S}_{\alpha\beta}^{(1)}(\omega)$:

$$\mathcal{S}_{\alpha\beta}^{(1)}(\omega) = \frac{2\pi\hbar e^2}{m^2 \omega^2 V} \sum_{\mathbf{pq}} |C_q|^2 q_\alpha q_\beta$$

$$\times [\delta(\varepsilon_\mathbf{p} - \varepsilon_{\mathbf{p}-\hbar\mathbf{q}} - \hbar\omega_\mathbf{q} + \hbar\omega) + \delta(\varepsilon_\mathbf{p} - \varepsilon_{\mathbf{p}-\hbar\mathbf{q}} - \hbar\omega_\mathbf{q} - \hbar\omega)] \qquad (27)$$

$$\times [(N_\mathbf{q} + 1)f_\mathbf{p}(1 - f_{\mathbf{p}-\hbar\mathbf{q}}) + N_\mathbf{q} f_{\mathbf{p}-\hbar\mathbf{q}}(1 - f_\mathbf{p})].$$

We point out that $\mathcal{S}_{\alpha\beta}(\mathbf{r}, \mathbf{r}'|\omega) \simeq \delta(\mathbf{r} - \mathbf{r}')\mathcal{S}_{\alpha\beta}^{(1)}(\omega)$ in this approximation, since the collisionless part (26) is equal to zero at $\mathbf{k} \to 0$ and $\omega \neq 0$.

In the case of isotropic electron and phonon systems, $\mathcal{S}_{\alpha\beta}^{(1)}(\omega) = \delta_{\alpha\beta} \times \mathcal{S}^{(1)}(\omega)$ is diagonal in the coordinate index (the factor $q_\alpha q_\beta$ is replaced by $\delta_{\alpha\beta}q^2/3$ as a result of angular averaging in Eq. (27)). In equilibrium, one can prove the identity (9) by substituting the Fermi distribution function for electrons and Planck distribution function for phonons into Eq. (27). Indeed, Eq. (9) is written in the local approximation as $\mathcal{S}_{\alpha\beta}^{(1)}(\omega) = \hbar\omega \coth(\hbar\omega/2T)\mathrm{Re}\sigma_{\alpha\beta}(\omega)$ (note also that $\sigma_{\alpha\beta}(\omega) = \delta_{\alpha\beta}\sigma(\omega)$ since there are no magnetic fields). In order to compare $\mathcal{S}_{\alpha\beta}^{(1)}(\omega)$ and $\mathrm{Re}\sigma_{\alpha\beta}(\omega)$ (problem 14.6), the real part of the high-frequency conductivity under electron-phonon scattering should be extracted from Eq. (37.9) written in the limit $E \to 0$, when $[\mathcal{J}_{\pm 1}(\mathbf{q} \cdot \mathbf{v}_\omega/\omega)]^2 = (e\mathbf{E} \cdot \mathbf{q}/2m\omega^2)^2$ (we remind that the absorbed power $\mathcal{P}_{ph}$ given by the contribution proportional to $k\omega$ in the second term on the left-hand side of Eq. (37.9) is expressed through the conductivity as $\mathcal{P}_{ph} = \mathrm{Re}\sigma(\omega)E^2/2$):

$$\mathrm{Re}\sigma(\omega) = \frac{2\pi e^2}{3m^2\omega^3 V} \sum_{\mathbf{pq}} |C_q|^2 q^2 [(N_\mathbf{q}+1)f_\mathbf{p}(1 - f_{\mathbf{p}-\hbar\mathbf{q}}) - N_\mathbf{q} f_{\mathbf{p}-\hbar\mathbf{q}}(1 - f_\mathbf{p})]$$

$$\times[\delta(\varepsilon_{\mathbf{p}} - \varepsilon_{\mathbf{p}-\hbar\mathbf{q}} - \hbar\omega_{\mathbf{q}} + \hbar\omega) - \delta(\varepsilon_{\mathbf{p}} - \varepsilon_{\mathbf{p}-\hbar\mathbf{q}} - \hbar\omega_{\mathbf{q}} - \hbar\omega)]. \quad (28)$$

In a similar way, one can prove (problem 14.6) that the collisionless contribution to the correlation function (25) in thermodynamic equilibrium satisfies the identity (9), where the real part of the conductivity in the collisionless limit (which is responsible for the Landau damping) is given in problem 6.21.

If the electron-phonon system is out of equilibrium, the deviations from the fluctuation-dissipation theorem can be substantial. In particular, when the distribution function of electrons is anisotropic, the non-diagonal ($\alpha \neq \beta$) components of the current-current correlation function are not equal to zero. If the distribution is isotropic, it is convenient to characterize the non-equilibrium systems by the noise temperature T_n introduced as

$$T_n = \frac{\langle\langle[\hat{I}_\alpha(\mathbf{r}, t), \hat{I}_\alpha(\mathbf{r}', 0)]_+\rangle\rangle_\omega}{4\mathrm{Re}\sigma_{\alpha\alpha}(\mathbf{r}, \mathbf{r}'|\omega)} = \frac{\mathcal{S}_{\alpha\alpha}^{(1)}(\omega)}{2\mathrm{Re}\sigma_{\alpha\alpha}(\omega)}. \quad (29)$$

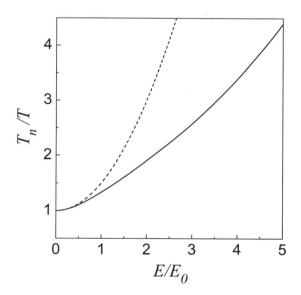

Figure 14.1. Field dependence of the high-frequency ($\tau_{LA}^{-1} \ll \omega \ll 2T/\hbar$) noise temperature in the model of quasielastic scattering of electrons by acoustic phonons. The dashed curve shows the function $1 + (E/E_0)^2/2$ approximating the ratio T_n/T at low fields.

This temperature is equal to T in equilibrium and at $\hbar\omega \ll 2T$. If the electrons are heated, for example, by an external homogeneous electric field **E**, the noise temperature increases. Below we demonstrate this effect on the example of non-degenerate electrons interacting with acoustic phonons via deformation potential in the quasielastic conditions, when the temperature is high in comparison to the characteristic energies of the phonons. The electron distribution function is calculated for this case in problem 7.8, where we have introduced a characteristic energy $\zeta_E \propto E$. Neglecting the phonon energy $\hbar\omega_{\mathbf{q}}$ under the δ-functions in Eqs. (27) and (28), and taking into account that the factor $|C_q|^2(N_q + 1/2)$ is independent of the phonon wave number q, we obtain

$$T_n = \frac{\hbar\omega}{2} \frac{\sum_{\mathbf{pq}} q^2 (f_{\mathbf{p}} + f_{\mathbf{p}-\hbar\mathbf{q}}) \delta(\varepsilon_{\mathbf{p}} - \varepsilon_{\mathbf{p}-\hbar\mathbf{q}} + \hbar\omega)}{\sum_{\mathbf{pq}} q^2 (f_{\mathbf{p}} - f_{\mathbf{p}-\hbar\mathbf{q}}) \delta(\varepsilon_{\mathbf{p}} - \varepsilon_{\mathbf{p}-\hbar\mathbf{q}} + \hbar\omega)}. \tag{30}$$

The sums in Eq. (30) are calculated easily in the limit $\hbar\omega \ll 2T$. Introducing a characteristic field E_0 according to $\zeta_E/T = E/E_0$, we have

$$\frac{T_n}{T} = \kappa \frac{\int_0^\infty dx \; x^2 (1 + x/\kappa)^\kappa e^{-x}}{\int_0^\infty dx \; x^3 (1 + x/\kappa)^{\kappa-1} e^{-x}}, \tag{31}$$

where $\kappa = (E/E_0)^2$. The dependence of the noise temperature on the electric field E calculated according to Eq. (31) is shown in Fig. 14.1.

69. Quasi-Classical Approach

In this section we study the long-wavelength fluctuations with $|\mathbf{k}| \ll \bar{p}/\hbar$, where $\bar{p}$ is the characteristic momentum of electrons. Our aim is to linearize Eq. (68.17) with respect to small fluctuations in this quasi-classical region. We need, however, to give some preliminary remarks. Let us introduce the operator of the number of electrons with a given spin σ at the point $(\mathbf{r}, \mathbf{p})$ of the phase space at the instant t:

$$\hat{f}_{\sigma\mathbf{rp}t} = \int d\Delta\mathbf{r} e^{-i\mathbf{p}\cdot\Delta\mathbf{r}/\hbar} \hat{\Psi}^+_{\sigma\mathbf{r}-\frac{\Delta\mathbf{r}}{2}}(t) \hat{\Psi}_{\sigma\mathbf{r}+\frac{\Delta\mathbf{r}}{2}}(t)$$

$$= \sum_{\mathbf{k}} e^{i\mathbf{k}\cdot\mathbf{r}} \left\{ \hat{a}^+_{\sigma\mathbf{p}-\frac{\hbar\mathbf{k}}{2}} \hat{a}_{\sigma\mathbf{p}+\frac{\hbar\mathbf{k}}{2}} \right\}_t. \tag{1}$$

Using this definition together with Eqs. (4.29) and (9.6) (at zero vector potential), one can find that the average $\langle\langle \hat{f}_{\sigma\mathbf{rp}t} \rangle\rangle \equiv \mathrm{Sp}\hat{\eta}\hat{f}_{\sigma\mathbf{rp}t}$ is the Wigner distribution function $f_t(\mathbf{r}, \mathbf{p}) \equiv f_{\mathbf{rp}t}$. Since below we consider spin-degenerate electron states only, and the Hamiltonian of the system does not mix the states with different spins, this distribution function does not depend on the spin number σ. For the sake of brevity, we will also omit the spin index at the operators $\hat{a}_{\sigma\mathbf{p}}$ and $\hat{f}_{\sigma\mathbf{rp}t}$ whenever possible and retain this index only when it is necessary, for example, when there appear correlation functions of two operators $\hat{f}_{\sigma\mathbf{r}\mathbf{p}t}$ with different spins (see Eqs. (21)−(31)).

We first consider a stationary and spatially homogeneous electron system, where the Wigner distribution function does not depend on time and coordinate, $f_{\mathbf{r}\mathbf{p}t} = f_{\mathbf{p}}$. The fluctuating part of the operator (1) is given by $\widehat{\Delta f}_{\mathbf{r}\mathbf{p}t} = \hat{f}_{\mathbf{r}\mathbf{p}t} - f_{\mathbf{p}}$. These fluctuations are not small. Indeed, the mean square $\langle\langle(\widehat{\Delta f}_{\mathbf{r}\mathbf{p}t})^2\rangle\rangle$ appears to be large in comparison to $f_{\mathbf{p}}^2$ (problem 14.7). This property reflects the fact that $\widehat{\Delta f}_{\mathbf{r}\mathbf{p}t}$ describes the fluctuation of the number of electrons in the infinitely small (point-like) region of the phase space. To obtain the operator whose fluctuations are small, one should average $\widehat{\Delta f}_{\mathbf{r}\mathbf{p}t}$ over a considerably large volume of the phase space, according to

$$\widehat{\Delta f}_{\mathbf{r}\mathbf{p}t}^{(av)} = \frac{1}{(\Delta L \Delta K)^3} \int_{|p'_\alpha - p_\alpha| < \pi\hbar\Delta K} \frac{d\mathbf{p}'}{(2\pi\hbar)^3} \int_{|r'_\alpha - r_\alpha| < \Delta L/2} d\mathbf{r}' \widehat{\Delta f}_{\mathbf{r}'\mathbf{p}'t} , \tag{2}$$

where $\alpha = x, y, z$ is the Cartesian coordinate index, ΔL is a length in the coordinate space, and ΔK is a length in the space of the variable $p_\alpha/(2\pi\hbar)$. Accordingly, $(\Delta L \Delta K)^3$ is a dimensionless parameter called below the normalized phase volume. We choose ΔL and ΔK in such a way that variations of the distribution function $f_{\mathbf{p}}$ and energy $\varepsilon_{\mathbf{p}}$ are small if p increases by $\pi\hbar/\Delta L$ or $\pi\hbar\Delta K$. Calculating the mean square of the fluctuation, we obtain

$$\langle\langle(\widehat{\Delta f}_{\mathbf{r}\mathbf{p}t}^{(av)})^2\rangle\rangle = \frac{1}{(\Delta L \Delta K)^6} \int \frac{d\mathbf{p}_1}{(2\pi\hbar)^3} \int \frac{d\mathbf{p}_2}{(2\pi\hbar)^3} \int d\mathbf{r}_1 \int d\mathbf{r}_2$$

$$\times \sum_{\mathbf{k}_1\mathbf{k}_2} e^{i\mathbf{k}_1\cdot\mathbf{r}_1 + i\mathbf{k}_2\cdot\mathbf{r}_2} \left\langle\left\langle \hat{a}^+_{\mathbf{p}_1 - \frac{\hbar\mathbf{k}_1}{2}} \hat{a}_{\mathbf{p}_1 + \frac{\hbar\mathbf{k}_1}{2}} \hat{a}^+_{\mathbf{p}_2 - \frac{\hbar\mathbf{k}_2}{2}} \hat{a}_{\mathbf{p}_2 + \frac{\hbar\mathbf{k}_2}{2}} \right\rangle\right\rangle \tag{3}$$

$$- \frac{2f_{\mathbf{p}}}{(\Delta L \Delta K)^3} \int \frac{d\mathbf{p}_1}{(2\pi\hbar)^3} \int d\mathbf{r}_1 \sum_{\mathbf{k}_1} e^{i\mathbf{k}_1\cdot\mathbf{r}_1} \left\langle\left\langle \hat{a}^+_{\mathbf{p}_1 - \frac{\hbar\mathbf{k}_1}{2}} \hat{a}_{\mathbf{p}_1 + \frac{\hbar\mathbf{k}_1}{2}} \right\rangle\right\rangle + f_{\mathbf{p}}^2 ,$$

where the integrals over $\mathbf{p}_1$, $\mathbf{p}_2$, $\mathbf{r}_1$, and $\mathbf{r}_2$ have the same limits as in Eq. (2). Under the approximation of non-interacting electrons, the correlation function of four electron operators in Eq. (3) is equal to $\delta_{\mathbf{k}_1, -\mathbf{k}_2} \delta_{\mathbf{p}_1, \mathbf{p}_2} f_{\mathbf{p}_1 - \hbar\mathbf{k}_1/2} (1 - f_{\mathbf{p}_1 + \hbar\mathbf{k}_1/2}) + \delta_{\mathbf{k}_1, 0} \delta_{\mathbf{k}_2, 0} f_{\mathbf{p}_1} f_{\mathbf{p}_2}$. Further calculations are based upon the assumed smallness of $1/\Delta L$ and ΔK. As a result,

$$\langle\langle(\widehat{\Delta f}_{\mathbf{r}\mathbf{p}t}^{(av)})^2\rangle\rangle \simeq \frac{f_{\mathbf{p}}(1 - f_{\mathbf{p}})}{(\Delta L \Delta K)^3}. \tag{4}$$

Therefore, the mean square of the fluctuations of $\widehat{\Delta f}_{\mathbf{rp}t}^{(av)}$ is small in comparison to $f_{\mathbf{p}}^2$ when $(\Delta L\Delta K)^3$ is large in comparison to $1/f_{\mathbf{p}}$. In other words, the normalized phase volume $(\Delta L\Delta K)^3$ must contain many particles. In any case $(\Delta L\Delta K)^3 \gg 1$, which means that the phase volume over which the averaging occurs is large in comparison to the quantum $(2\pi\hbar)^3$ given by the momentum-coordinate uncertainty principle. One can always choose ΔL large enough to fulfil this condition together with the assumed smallness of the interval ΔK. However, by averaging the operator $\widehat{\Delta f}_{\mathbf{rp}t}$ over the large volume $(\Delta L)^3$, we can describe only the fluctuations whose characteristic lengths are larger than ΔL. Applying the averaging procedure (2) to Eq. (1), we calculate the integral over coordinate and find

$$\hat{f}_{\mathbf{rp}t}^{(av)} = \sum_{\mathbf{k}} e^{i\mathbf{k}\cdot\mathbf{r}} \left(\prod_{\alpha=x,y,z} \frac{\sin k_\alpha \Delta L/2}{k_\alpha \Delta L/2} \right) \tag{5}$$

$$\times \int_{|p'_\alpha - p_\alpha| < \pi\hbar\Delta K} \frac{d\mathbf{p}'}{(2\pi\hbar\Delta K)^3} \left\{ \hat{a}_{\mathbf{p}'-\frac{\hbar\mathbf{k}}{2}}^{+} \hat{a}_{\mathbf{p}'+\frac{\hbar\mathbf{k}}{2}} \right\}_t .$$

The function $(\prod \ldots)$ is (roughly) close to unity at $|k_\alpha| < \pi/\Delta L$, while beyond this region $(\prod \ldots)$ is small and oscillating. We also notice that the integral $(\Delta L/2\pi)^3 \int d\mathbf{k} \ldots$ of this function is equal to unity. In other words, Eq. (2) defines a truncation procedure which makes the wave vectors $\mathbf{k}$ entering the terms in Eq. (68.17) shrink into the region of about $|\mathbf{k}| < \pi/\Delta L = k_0$. If $|\mathbf{k}|$ lies within this region, the averaging does not change these terms. For larger $|\mathbf{k}|$, the averaged operator product $\hat{a}_{\mathbf{p}-\hbar\mathbf{k}/2}^{+}\hat{a}_{\mathbf{p}+\hbar\mathbf{k}/2}$ becomes small. Approximately, the presence of the function $(\prod \ldots)$ in Eq. (5) is equivalent to the truncation $\sum_{\mathbf{k}} \to \sum_{|\mathbf{k}|<k_0}$.

Applying the averaging procedure (2) to Eq. (68.17), let us write this equation in the Wigner representation. The left-hand side of Eq. (68.17) is reduced to

$$\left(\frac{d}{dt} + \mathbf{v}_{\mathbf{p}} \cdot \frac{\partial}{\partial \mathbf{r}} \right) \widehat{\Delta f}_{\mathbf{rp}t}^{(av)} , \tag{6}$$

and the averaged Langevin source in this representation is given by

$$\widehat{K}_t^{(av)}(\mathbf{p},\mathbf{r}) = \int \frac{d\omega}{2\pi} \sum_{|\mathbf{k}|<k_0} e^{i\mathbf{k}\cdot\mathbf{r}-i\omega t} \int_{|p'_\alpha - p_\alpha| < \pi\hbar\Delta K} \frac{d\mathbf{p}'}{(2\pi\hbar\Delta K)^3} \widehat{K}_\omega(\mathbf{p}',\mathbf{k}).$$

$$\tag{7}$$

Taking into account the assumed smallness of $|\mathbf{k}|$ and ΔK, one can see that the correlation functions of the averaged sources are the same as for the non-averaged ones; see Eq. (68.20). Therefore, we obtain the same expression (68.23) for the correlation function of the averaged Langevin sources. Moreover, in this expression one can neglect $\mathbf{k}$ in the energies and distribution functions. To consider the remaining term $\widehat{M}_t(\mathbf{p}, \mathbf{k})$ in Eq. (68.17), we need to introduce the operator of the number of phonons:

$$\widehat{N}_{\mathbf{rq}t} = \sum_{\mathbf{k}} e^{i\mathbf{k}\cdot\mathbf{r}} \left\{ \hat{b}^+_{\mathbf{q}-\frac{\mathbf{k}}{2}} \hat{b}_{\mathbf{q}+\frac{\mathbf{k}}{2}} \right\}_t . \qquad (8)$$

The average $\mathrm{Sp}\hat{\eta}\widehat{N}_{\mathbf{rq}t}$ is the Wigner distribution function $N_{\mathbf{rq}t}$ given by Eq. (19.26) (since we consider a single phonon mode, the mode indices of this distribution function are omitted), and $N_{\mathbf{rq}t} = N_{\mathbf{q}}$ in the stationary and spatially homogeneous case. The fluctuations of the number of phonons are described by the operator $\widehat{\Delta N}_{\mathbf{rq}t} = \widehat{N}_{\mathbf{rq}t} - N_{\mathbf{q}}$, and the average of this operator over the phase space is defined by

$$\widehat{\Delta N}^{(av)}_{\mathbf{rq}t} = \frac{1}{(\Delta L \Delta Q)^3} \int_{|q'_\alpha - q_\alpha| < \pi\Delta Q} \frac{d\mathbf{q}'}{(2\pi)^3} \int_{|r'_\alpha - r_\alpha| < \Delta L/2} d\mathbf{r}' \widehat{\Delta N}_{\mathbf{r}'\mathbf{q}'t}. \quad (9)$$

Equations (8) and (9) are written by analogy to Eqs. (1) and (2). Moreover, ΔQ for phonons has the same meaning as ΔK for electrons.

According to the consideration presented in Sec. 68, the function $\widehat{M}_t(\mathbf{p}, \mathbf{r})$ can be written as

$$\widehat{M}_t(\mathbf{p}, \mathbf{r}) \simeq -\frac{1}{\hbar^2} \sum_{\mathbf{k}} e^{i\mathbf{k}\cdot\mathbf{r}} \int_{t_0}^t dt' \left\{ \sum_{\mathbf{q}\mathbf{p}_1} |C_q|^2 \right.$$

$$\times \left[\hat{a}^+_{\mathbf{p}-\frac{\hbar\mathbf{k}}{2}} \hat{a}_{\mathbf{p}+\frac{\hbar\mathbf{k}}{2}-\hbar\mathbf{q}} \hat{a}^+_{\mathbf{p}_1-\hbar\mathbf{q}} \hat{a}_{\mathbf{p}_1} \left(e^{i\Omega_-(t-t')} - e^{i\Omega_+(t-t')} \right) \right.$$

$$\left. + \hat{a}^+_{\mathbf{p}-\frac{\hbar\mathbf{k}}{2}-\hbar\mathbf{q}} \hat{a}_{\mathbf{p}+\frac{\hbar\mathbf{k}}{2}} \hat{a}^+_{\mathbf{p}_1+\hbar\mathbf{q}} \hat{a}_{\mathbf{p}_1} \left(e^{-i\Omega_-(t-t')} - e^{-i\Omega_+(t-t')} \right) \right]$$

$$+ \sum_{\mathbf{q}\mathbf{q}'} C_q C_{q'} \left[\left(\hat{a}^+_{\mathbf{p}-\frac{\hbar\mathbf{k}}{2}} \hat{a}_{\mathbf{p}+\frac{\hbar\mathbf{k}}{2}-\hbar(\mathbf{q}+\mathbf{q}')} - \hat{a}^+_{\mathbf{p}-\frac{\hbar\mathbf{k}}{2}+\hbar\mathbf{q}'} \hat{a}_{\mathbf{p}+\frac{\hbar\mathbf{k}}{2}-\hbar\mathbf{q}} \right) \right. \qquad (10)$$

$$\times \left(\hat{b}^+_{-\mathbf{q}'} \hat{b}_{\mathbf{q}} e^{i\Omega_-(t-t')} + \hat{b}_{\mathbf{q}'} \hat{b}^+_{-\mathbf{q}} e^{i\Omega_+(t-t')} \right)$$

$$+ \left(\hat{a}^+_{\mathbf{p}-\frac{\hbar\mathbf{k}}{2}-\hbar(\mathbf{q}'+\mathbf{q})} \hat{a}_{\mathbf{p}+\frac{\hbar\mathbf{k}}{2}} - \hat{a}^+_{\mathbf{p}-\frac{\hbar\mathbf{k}}{2}-\hbar\mathbf{q}} \hat{a}_{\mathbf{p}+\frac{\hbar\mathbf{k}}{2}+\hbar\mathbf{q}'} \right)$$

$$\left. \left. \times \left(\hat{b}_{-\mathbf{q}'} \hat{b}^+_{\mathbf{q}} e^{-i\Omega_-(t-t')} + \hat{b}^+_{\mathbf{q}'} \hat{b}_{-\mathbf{q}} e^{-i\Omega_+(t-t')} \right) \right] \right\},$$

where $\Omega_\pm = (\varepsilon_{\mathbf{p}} - \varepsilon_{\mathbf{p}-\hbar\mathbf{q}})/\hbar \pm \omega_{\mathbf{q}}$, and we have used the approximation $\varepsilon_{\mathbf{p}\pm\hbar\mathbf{k}/2} \simeq \varepsilon_{\mathbf{p}}$, since $|\mathbf{k}|$ is small. The products of Heisenberg creation and annihilation operators in Eq. (10) depend on t'. Considering slow temporal variations of these operators in comparison to the fast oscillations of the exponential factors, one can replace the argument t' by t, which corresponds to the Markovian approximation applied to derive the collision integrals in previous chapters. Next, the time t_0 can be aimed to $-\infty$, and the factor $e^{\lambda(t'-t)}$ (where $\lambda \to +0$), which describes the weakening of correlations at $t' \to -\infty$, should be placed under the integral to make it convergent. Having applied these approximations, we express the operator products through the fluctuations of the operators (1) and (8). For example,

$$\hat{a}^+_{\mathbf{p}-\frac{\hbar\mathbf{k}}{2}}\hat{a}_{\mathbf{p}+\frac{\hbar\mathbf{k}}{2}-\hbar(\mathbf{q}+\mathbf{q}')}\hat{b}^+_{-\mathbf{q}'}\hat{b}_{\mathbf{q}} = \delta_{\mathbf{q},-\mathbf{q}'}\delta_{\mathbf{k},0}f_{\mathbf{p}}N_{\mathbf{q}} \qquad (11)$$

$$+\delta_{\mathbf{q},-\mathbf{q}'}N_{\mathbf{q}}\frac{1}{V}\int d\mathbf{r}e^{-i\mathbf{k}\cdot\mathbf{r}}\widehat{\Delta f}_{\mathbf{r}\mathbf{p}t} + \delta_{\mathbf{q},\mathbf{k}-\mathbf{q}'}f_{\mathbf{p}-\frac{\hbar\mathbf{k}}{2}}\frac{1}{V}\int d\mathbf{r}e^{-i\mathbf{k}\cdot\mathbf{r}}\widehat{\Delta N}_{\mathbf{r}\mathbf{q}-\frac{\mathbf{k}}{2}t}$$

$$+\frac{1}{V^2}\int d\mathbf{r}_1 e^{i(\mathbf{q}+\mathbf{q}'-\mathbf{k})\cdot\mathbf{r}_1}\widehat{\Delta f}_{\mathbf{r}_1\mathbf{p}-\frac{\hbar(\mathbf{q}+\mathbf{q}')}{2}t}\int d\mathbf{r}_2 e^{-i(\mathbf{q}+\mathbf{q}')\cdot\mathbf{r}_2}\widehat{\Delta N}_{\mathbf{r}_2\frac{\mathbf{q}-\mathbf{q}'}{2}t}.$$

The other operator products are written in a similar way. Each one of them contain a term proportional to the distribution functions (the first one), two terms linear in the fluctuations of $\widehat{\Delta f}$ or $\widehat{\Delta N}$ (the second and the third terms in Eq. (11)), and a term quadratic in the fluctuations (the last one). Substituting all such terms into Eq. (10), we average $\widehat{M}_t(\mathbf{p},\mathbf{r})$ according to Eq. (2) and find that the terms proportional just to the distribution functions form together the electron-phonon collision integral $J_{e,ph}(f|\mathbf{p})$ given by Eq. (34.25). This collision integral is equal to zero, since the stationary kinetic equation in the absence of external fields is written as $J_{e,ph}(f|\mathbf{p}) = 0$. Next, the averaging of the terms linear in the fluctuations $\widehat{\Delta f}$ and $\widehat{\Delta N}$ results merely in the substitutions $\widehat{\Delta f}_{\mathbf{r}\mathbf{p}t} \to \widehat{\Delta f}^{(av)}_{\mathbf{r}\mathbf{p}t}$ and $\widehat{\Delta N}_{\mathbf{r}\mathbf{q}t} \to \widehat{\Delta N}^{(av)}_{\mathbf{r}\mathbf{q}t}$. Collecting all such terms, we find that their sum forms the operator $\widehat{\Delta J}_{e,ph}(\widehat{\Delta f}^{(av)},\widehat{\Delta N}^{(av)}|\mathbf{r}\mathbf{p}t)$, which can be formally obtained by a linearization of the collision integral $J_{e,ph}(f|\mathbf{p})$, where one substitutes $f_{\mathbf{p}} \to f_{\mathbf{p}} + \widehat{\Delta f}^{(av)}_{\mathbf{r}\mathbf{p}t}$ and $N_{\mathbf{q}} \to N_{\mathbf{q}} + \widehat{\Delta N}^{(av)}_{\mathbf{r}\mathbf{q}t}$ everywhere, with respect to the fluctuations $\widehat{\Delta f}^{(av)}_{\mathbf{r}\mathbf{p}t}$ and $\widehat{\Delta N}^{(av)}_{\mathbf{r}\mathbf{q}t}$. When carrying out such a transformation, one should take into account that $C_{|\mathbf{q}\pm\mathbf{k}|} \simeq C_{\mathbf{q}}$ because $|\mathbf{k}|$ is small. Finally, the last terms, quadratic in the fluctuations $\widehat{\Delta f}$ and $\widehat{\Delta N}$, are reduced to the terms quadratic in $\widehat{\Delta f}^{(av)}$ and $\widehat{\Delta N}^{(av)}$ (problem 14.8), and should be neglected.

Summarizing the results obtained above, we can write the kinetic equation for the fluctuations of electron distribution in the form

$$\left(\frac{\partial}{\partial t} + \mathbf{v_p} \cdot \frac{\partial}{\partial \mathbf{r}}\right) \widehat{\Delta f}^{(av)}_{\mathbf{rp}t} - \widehat{\Delta J}_{e,ph}(\widehat{\Delta f}^{(av)}, \widehat{\Delta N}^{(av)}|\mathbf{rp}t) = \widehat{K}^{(av)}_t(\mathbf{p}, \mathbf{r}).$$

(12)

The usage of the quasi-classical linearized collision integrals $\widehat{\Delta J}_{e,ph}$ in Eq. (12) essentially implies that $\widehat{\Delta f}^{(av)}_{\mathbf{rp}t}$ slowly varies with time on the quantum scale $\hbar/\bar{\varepsilon}$. In other words, the fluctuations $\widehat{\Delta f}^{(av)}_{\mathbf{rp}t}$ are quasi-classical not only by their coordinate dependence (as it was initially assumed), but also by their dependence on time. To describe the quantum region of frequencies, one should consider the non-Markovian term $\widehat{M}_t(\mathbf{p}, \mathbf{r})$ given by Eq. (10). Fortunately, if $\widehat{\Delta f}^{(av)}_{\mathbf{rp}t}$ varies with time much faster than the relaxation occurs, the contribution $\widehat{M}_t(\mathbf{p}, \mathbf{r})$ can be neglected in comparison to the term given by the temporal derivative in Eq. (12). Moreover, in this high-frequency region one may also neglect the term proportional to the spatial gradient in Eq. (12). Therefore, a high-frequency solution for the Fourier component $\widehat{\Delta f}^{(av)}_{\mathbf{kp}\omega} = \int dt e^{i\omega t} \int d\mathbf{r} e^{-i\mathbf{q}\cdot\mathbf{r}} \widehat{\Delta f}^{(av)}_{\mathbf{rp}t}$ is written merely as $\widehat{\Delta f}^{(av)}_{\mathbf{kp}\omega} = iV\omega^{-1}\widehat{K}^{(av)}_\omega(\mathbf{p}, \mathbf{k})$. Using this solution, one can obtain, for example, the noise power in the form (68.27).

The collision-integral contribution in Eq. (12) depends on the operator $\widehat{\Delta N}^{(av)}_{\mathbf{rq}t}$. Therefore, the description of the fluctuations is still incomplete. To give a complete description, one should write an additional equation for $\widehat{\Delta N}^{(av)}_{\mathbf{rq}t}$. Let us compose an operator equation similar to Eq. (68.13):

$$\left[\frac{d}{dt} + i(\omega_{\mathbf{q}} - \omega_{\mathbf{q}'})\right]\hat{b}^+_{\mathbf{q}'}\hat{b}_{\mathbf{q}} = \frac{i}{\hbar}\sum_{\sigma\mathbf{p}}(C_{q'}\hat{b}_{\mathbf{q}}\hat{a}^+_{\sigma\mathbf{p}+\hbar\mathbf{q}'}\hat{a}_{\sigma\mathbf{p}} - C_q\hat{b}^+_{\mathbf{q}'}\hat{a}^+_{\sigma\mathbf{p}-\hbar\mathbf{q}}\hat{a}_{\sigma\mathbf{p}}).$$

(13)

The spin indices are written explicitly in this equation to emphasize that both spin states of electrons contribute to the evolution of the phonon operators through the electron-phonon interaction (we remind that this interaction is spin-independent). Considering equations of motion for the operators $\hat{b}^+_{\mathbf{q}'}\hat{a}^+_{\sigma\mathbf{p}-\hbar\mathbf{q}}\hat{a}_{\sigma\mathbf{p}}$ and $\hat{b}_{\mathbf{q}}\hat{a}^+_{\sigma\mathbf{p}+\hbar\mathbf{q}'}\hat{a}_{\sigma\mathbf{p}}$, we obtain the kinetic equation describing the fluctuations of phonon distribution in the quasi-classical region,

$$\left(\frac{\partial}{\partial t} + \frac{\partial\omega_{\mathbf{q}}}{\partial\mathbf{q}} \cdot \frac{\partial}{\partial\mathbf{r}}\right)\widehat{\Delta N}^{(av)}_{\mathbf{rq}t} - \widehat{\Delta J}_{ph,e}\left(\widehat{\Delta N}^{(av)}, \widehat{\Delta f}^{(av)}|\mathbf{rq}t\right) = \hat{\mathcal{K}}^{(av)}_t(\mathbf{q}, \mathbf{r}),$$

(14)

where $\widehat{\Delta J}_{ph,e}\left(\widehat{\Delta N}^{(av)}, \widehat{\Delta f}^{(av)}|\mathbf{rq}t\right)$ is the linearized phonon-electron collision integral; see Sec. 19. The Fourier transform of the exact Langevin source for phonons is

$$\hat{\mathcal{K}}_\omega(\mathbf{q}, \mathbf{k}) = -2\pi i e^{i\omega t_0} \sum_{\sigma\mathbf{p}} \left[C_{|\mathbf{q}+\frac{\mathbf{k}}{2}|} \left\{ \hat{b}^+_{\mathbf{q}-\frac{\mathbf{k}}{2}} \hat{a}^+_{\sigma\mathbf{p}-\hbar(\mathbf{q}+\frac{\mathbf{k}}{2})} \hat{a}_{\sigma\mathbf{p}} \right\}_{t_0} \right.$$

$$\times \delta\left(\varepsilon_{\mathbf{p}-\hbar(\mathbf{q}+\frac{\mathbf{k}}{2})} - \varepsilon_{\mathbf{p}} + \hbar\omega_{\mathbf{q}-\frac{\mathbf{k}}{2}} + \hbar\omega\right) - C_{|\mathbf{q}-\frac{\mathbf{k}}{2}|} \qquad (15)$$

$$\left. \times \left\{ \hat{b}_{\mathbf{q}+\frac{\mathbf{k}}{2}} \hat{a}^+_{\sigma\mathbf{p}+\hbar(\mathbf{q}-\frac{\mathbf{k}}{2})} \hat{a}_{\sigma\mathbf{p}} \right\}_{t_0} \delta\left(\varepsilon_{\mathbf{p}+\hbar(\mathbf{q}-\frac{\mathbf{k}}{2})} - \varepsilon_{\mathbf{p}} - \hbar\omega_{\mathbf{q}+\frac{\mathbf{k}}{2}} + \hbar\omega\right) \right],$$

where the initial time t_0 has the same meaning as in Eq. (68.20). The phase factor $e^{i\omega t_0}$ disappears in the correlation function written below:

$$\langle\langle \hat{\mathcal{K}}_\omega(\mathbf{q}, \mathbf{k}) \hat{\mathcal{K}}_{\omega'}(\mathbf{q}', \mathbf{k}') \rangle\rangle = 2\frac{(2\pi)^2}{\hbar} \delta(\omega + \omega') \delta_{\mathbf{k},-\mathbf{k}'} \delta_{\mathbf{q},\mathbf{q}'}$$

$$\times \sum_{\mathbf{p}} \left[|C_{|\mathbf{q}+\frac{\mathbf{k}}{2}|}|^2 \delta\left(\varepsilon_{\mathbf{p}-\hbar(\mathbf{q}+\frac{\mathbf{k}}{2})} - \varepsilon_{\mathbf{p}} + \hbar\omega_{\mathbf{q}-\frac{\mathbf{k}}{2}} + \hbar\omega\right) N_{\mathbf{q}-\frac{\mathbf{k}}{2}} \right.$$

$$\times f_{\mathbf{p}-\hbar(\mathbf{q}+\frac{\mathbf{k}}{2})}(1 - f_{\mathbf{p}}) + |C_{|\mathbf{q}-\frac{\mathbf{k}}{2}|}|^2 \delta\left(\varepsilon_{\mathbf{p}-\hbar(\mathbf{q}-\frac{\mathbf{k}}{2})} - \varepsilon_{\mathbf{p}} + \hbar\omega_{\mathbf{q}+\frac{\mathbf{k}}{2}} - \hbar\omega\right)$$

$$\left. \times \left(N_{\mathbf{q}+\frac{\mathbf{k}}{2}} + 1\right) f_{\mathbf{p}} \left(1 - f_{\mathbf{p}-\hbar(\mathbf{q}-\frac{\mathbf{k}}{2})}\right) \right]. \qquad (16)$$

The factor of 2 comes from the sum over electron spin. The correlation function of the averaged Langevin sources standing in Eq. (14) is described by Eq. (16) where one can neglect the small wave number $\mathbf{k}$ in the energies, distribution functions, and matrix elements C. Equations (12) and (14) with the right-hand sides described by Eqs. (68.20) and (15), respectively, form a complete system.

So far we neglected external electric and magnetic fields, $\mathbf{E}_{\mathbf{r}t}$ and $\mathbf{H}_{\mathbf{r}t}$. These fields modify the dynamics of the fluctuations for charged particles. The external fields are considered below as quasi-classical in the sense explained in Sec. 9. They are described by the scalar and vector potentials, $\Phi_{\mathbf{r}t}$ and $\mathbf{A}_{\mathbf{r}t}$. In the presence of the vector potential, the definition of $\hat{f}_{\mathbf{rp}t}$ should be modified as $\hat{f}_{\mathbf{rp}t} = \int d\Delta\mathbf{r} e^{-i[\mathbf{p}+(e/c)\mathbf{A}_{\mathbf{r}t}]\cdot\Delta\mathbf{r}/\hbar}$ $\times \hat{\Psi}^+_{\mathbf{r}-\Delta\mathbf{r}/2}(t)\hat{\Psi}_{\mathbf{r}+\Delta\mathbf{r}/2}(t)$. Therefore, according to Eqs. (4.29) and (9.6), the statistical averaging $\mathrm{Sp}\hat{\eta}\hat{f}_{\mathbf{rp}t}$ of the operator $\hat{f}_{\mathbf{rp}t}$ defined in this way again gives us the Wigner distribution function $f_{\mathbf{rp}t}$; see the discussion of Eq. (1) in the beginning of this section. We point out that the definition of $\hat{f}_{\mathbf{rp}t}$ in the presence of the vector potential formally coincides with Eq. (1) if the momentum $\mathbf{p}$ in the argument of $\hat{f}_{\mathbf{rp}t}$ is treated as the kinematic momentum $\mathbf{p} - (e/c)\mathbf{A}_{\mathbf{r}t}$ which depends on both time and coordinate. For this reason, instead of the partial derivatives standing in Eq. (12), one should put the full derivatives. Next, one should take into account that the Hamiltonian (68.12) acquires an additional contribution

$$\frac{1}{V}\sum_{\mathbf{pq}}\left[-\frac{e}{mc}\left(\mathbf{p}+\frac{\hbar\mathbf{q}}{2}\right)\cdot\mathbf{A}_{\mathbf{q}t}+\frac{e^2}{2mc^2}\left(\frac{1}{V}\sum_{\mathbf{q}'}\mathbf{A}_{\mathbf{q}/2+\mathbf{q}'\,t}\cdot\mathbf{A}_{\mathbf{q}/2-\mathbf{q}'\,t}\right)\right.$$

$$\left.+e\Phi_{\mathbf{q}t}\right]\hat{a}^+_{\mathbf{p}+\hbar\mathbf{q}}\hat{a}_{\mathbf{p}} \tag{17}$$

expressed here through the spatial Fourier transforms of the scalar and vector potentials. After calculating the commutators of $\hat{a}^+_{\mathbf{p}-\hbar\mathbf{k}/2}\hat{a}_{\mathbf{p}+\hbar\mathbf{k}/2}$ with the Hamiltonian (17) (problem 14.9), we neglect the contributions quadratic in $(e/c)\mathbf{A}$ by using the quasi-classical conditions (9.35). As a result, we obtain the following operator equation:

$$\left(\frac{\partial}{\partial t}+\mathbf{v}_{\mathbf{p}}\cdot\frac{\partial}{\partial\mathbf{r}}-\frac{e}{c}\frac{\partial\mathbf{A}_{\mathbf{r}t}}{\partial t}\cdot\frac{\partial}{\partial\mathbf{p}}-\frac{e}{c}\sum_{\alpha\beta}v^\beta_{\mathbf{p}}\frac{\partial A^\alpha_{\mathbf{r}t}}{\partial r_\beta}\frac{\partial}{\partial p_\alpha}\right)\hat{f}_{\mathbf{r}\mathbf{p}t}$$

$$-\frac{ie}{\hbar mcV^2}\sum_{\mathbf{kq}}\int d\mathbf{r}'e^{i\mathbf{k}\cdot\mathbf{r}-i(\mathbf{k}-\mathbf{q})\cdot\mathbf{r}'}\mathbf{A}_{\mathbf{q}t}\cdot\left[\left(\mathbf{p}-\frac{\hbar\mathbf{q}}{2}+\frac{\hbar\mathbf{k}}{2}\right)\hat{f}_{\mathbf{r}'\mathbf{p}-\frac{\hbar\mathbf{q}}{2}t}\right.$$

$$\left.-\left(\mathbf{p}+\frac{\hbar\mathbf{q}}{2}-\frac{\hbar\mathbf{k}}{2}\right)\hat{f}_{\mathbf{r}'\mathbf{p}+\frac{\hbar\mathbf{q}}{2}t}\right]+\frac{ie}{\hbar V}\sum_{\mathbf{q}}\Phi_{\mathbf{q}t}e^{i\mathbf{q}\cdot\mathbf{r}}\left(\hat{f}_{\mathbf{r}\mathbf{p}-\frac{\hbar\mathbf{q}}{2}t}-\hat{f}_{\mathbf{r}\mathbf{p}+\frac{\hbar\mathbf{q}}{2}t}\right)$$

$$=\widehat{K}_t(\mathbf{p},\mathbf{r})+\widehat{M}_t(\mathbf{p},\mathbf{r}). \tag{18}$$

Applying the statistical averaging $\mathrm{Sp}\hat{\eta}\ldots$ to this equation under the quasi-classical conditions (9.29) and (9.35), we obtain the kinetic equation (9.34) for the distribution function $f_{\mathbf{r}\mathbf{p}t} = \mathrm{Sp}\hat{\eta}\hat{f}_{\mathbf{r}\mathbf{p}t}$. This function slowly varies in space and time because of the presence of external fields. Introducing the fluctuating part of the operator $\hat{f}_{\mathbf{r}\mathbf{p}t} = f_{\mathbf{r}\mathbf{p}t} + \widehat{\Delta f}_{\mathbf{r}\mathbf{p}t}$, we can apply the averaging procedure (2) to Eq. (18). The contributions coming from the external fields are easily averaged in this way, because they are linear in $\hat{f}_{\mathbf{r}\mathbf{p}t}$. After lengthy but straightforward transformations under the assumption that $\widehat{\Delta f}^{(av)}_{\mathbf{r}\mathbf{p}t}$ weakly depends on $\mathbf{r}$ on the quantum scale, we use Eq. (4.3) to express the potentials through $\mathbf{E}_{\mathbf{r}t}$ and $\mathbf{H}_{\mathbf{r}t}$. Finally, we obtain a linearized kinetic equation called the Boltzmann-Langevin equation:

$$\sum_{\mathbf{p}_1}\widehat{L}_{\mathbf{r}t}(\mathbf{p},\mathbf{p}_1)\widehat{\Delta f}^{(av)}_{\mathbf{r}\mathbf{p}_1t}=\widehat{K}^{(av)}_t(\mathbf{p},\mathbf{r}), \tag{19}$$

$$\sum_{\mathbf{p}_1} \widehat{L}_{\mathbf{r}t}(\mathbf{p}, \mathbf{p}_1) \widehat{\Delta f}^{(av)}_{\mathbf{r}\mathbf{p}_1 t} \equiv \left(\frac{\partial}{\partial t} + \mathbf{v}_{\mathbf{p}} \cdot \frac{\partial}{\partial \mathbf{r}} + \mathbf{F}_{\mathbf{r}\mathbf{p}t} \cdot \frac{\partial}{\partial \mathbf{p}} \right) \widehat{\Delta f}^{(av)}_{\mathbf{r}\mathbf{p}t}$$

$$- \widehat{\Delta J}_{e,ph} \left(\widehat{\Delta f}^{(av)}, \widehat{\Delta N}^{(av)} | \mathbf{r}\mathbf{p}t \right),$$

where $\mathbf{F}_{\mathbf{r}\mathbf{p}t}$ is the Lorentz force (9.23). Equation (19) is a generalization of Eq. (12). Its left-hand side is represented as an action of the integro-differential operator $\widehat{L}$ on the fluctuating part of the electron distribution function. If the fluctuations of phonon distribution are neglected so that $\widehat{\Delta J}_{e,ph}$ is a linear functional of $\widehat{\Delta f}^{(av)}$, one can apply Eq. (19) in order to obtain the equation

$$\sum_{\mathbf{p}_1 \mathbf{p}_1'} \widehat{L}_{\mathbf{r}t}(\mathbf{p}, \mathbf{p}_1) \widehat{L}_{\mathbf{r}'t'}(\mathbf{p}', \mathbf{p}_1') \langle \langle \widehat{\Delta f}^{(av)}_{\mathbf{r}\mathbf{p}_1 t} \widehat{\Delta f}^{(av)}_{\mathbf{r}'\mathbf{p}_1' t'} \rangle \rangle$$

$$= \langle \langle \widehat{K}^{(av)}_t (\mathbf{p}, \mathbf{r}) \widehat{K}^{(av)}_{t'} (\mathbf{p}', \mathbf{r}') \rangle \rangle \qquad (20)$$

directly describing the correlation function of the fluctuations of electron distribution. The right-hand side of this equation contains the correlation function of Langevin sources. If only the electron-phonon interaction is considered, this right-hand side is determined by the double Fourier transformation of the expression (68.23) over the wave vectors and frequencies.

Apart from the external fields contributing into the Lorentz force in Eq. (19), we consider the random fields appearing as a result of the fluctuations of the charge density and electric current. The fluctuations of the magnetic field may be neglected as small relativistic corrections. The fluctuations of the electric field can be described, in the regular way, if we include the Coulomb interaction term (28.2) into the Hamiltonian (68.12) and compose the operator equations in a similar way as above. However, taking into account that these fluctuations are of large scale, it is convenient to apply another approach, when the fluctuations of the electrostatic potential energy, $\Delta U_{\mathbf{r}t}$, are expressed through the fluctuations of the electron density $n_{\mathbf{r}t}$ with the use of the Poisson equation. To simplify the consideration, we neglect in the following both the electron-phonon scattering and the external fields. The linearized quasi-classical kinetic equation is written as

$$\left(\frac{\partial}{\partial t} + \mathbf{v}_{\mathbf{p}} \cdot \frac{\partial}{\partial \mathbf{r}} \right) \widehat{\Delta f}^{(av)}_{\sigma \mathbf{r}\mathbf{p}t} - \frac{\partial \widehat{\Delta U}_{\mathbf{r}t}}{\partial \mathbf{r}} \cdot \frac{\partial f_{\mathbf{p}}}{\partial \mathbf{p}} = 0, \qquad (21)$$

where

$$\epsilon_\infty \nabla^2_{\mathbf{r}} \widehat{\Delta U}_{\mathbf{r}t} = -4\pi e^2 \widehat{\Delta n}_{\mathbf{r}t}, \quad \widehat{\Delta n}_{\mathbf{r}t} = \frac{1}{V} \sum_{\sigma \mathbf{p}} \widehat{\Delta f}^{(av)}_{\sigma \mathbf{r}\mathbf{p}t}. \qquad (22)$$

The spin index at $\widehat{\Delta f}_{\sigma \mathbf{rpt}}^{(av)}$ is written explicitly to emphasize that both spin-degenerate states contribute to the fluctuations of the electron density and electrostatic potential energy.

The solution of Eq. (21), obtained in a similar way as Eq. (68.18), is written for the Fourier components $\widehat{\Delta f}_{\sigma \mathbf{kp}\omega}^{(av)} = \int dt e^{i\omega t} \int d\mathbf{r} e^{-i\mathbf{q}\cdot\mathbf{r}} \widehat{\Delta f}_{\sigma \mathbf{rpt}}^{(av)}$:

$$\widehat{\Delta f}_{\sigma \mathbf{kp}\omega}^{(av)} = \widehat{\Delta f}_{\sigma \mathbf{kp}t_0}^{(av)} 2\pi e^{i\omega t_0} \delta(\omega - \mathbf{v_p}\cdot\mathbf{k}) - \frac{\mathbf{k}\cdot(\partial f_\mathbf{p}/\partial \mathbf{p})\widehat{\Delta U}_{\mathbf{k}\omega}}{\omega - \mathbf{v_p}\cdot\mathbf{k} + i\lambda}, \quad (23)$$

where the factor $i\lambda$ with $\lambda \to +0$ can be viewed as a collision-integral contribution in the limit when the scattering is negligible. Let us sum both sides of Eq. (23) over σ and $\mathbf{p}$. With $\widehat{\Delta U}_{\mathbf{k}\omega} = (4\pi e^2/\epsilon_\infty k^2)\widehat{\Delta n}_{\mathbf{k}\omega}$ following from Eq. (22), we express the fluctuation of the Fourier component of electron density as

$$\widehat{\Delta n}_{\mathbf{k}\omega} = \frac{\epsilon_\infty}{\epsilon(\mathbf{k},\omega)} \frac{1}{V} \sum_{\sigma \mathbf{p}} \widehat{\Delta f}_{\sigma \mathbf{kp}t_0}^{(av)} 2\pi e^{i\omega t_0} \delta(\omega - \mathbf{v_p}\cdot\mathbf{k}), \quad (24)$$

where

$$\epsilon(\mathbf{k},\omega) = \epsilon_\infty + \frac{4\pi e^2}{k^2} \frac{2}{V} \sum_{\mathbf{p}} \frac{\mathbf{k}\cdot(\partial f_\mathbf{p}/\partial \mathbf{p})}{\omega - \mathbf{v_p}\cdot\mathbf{k} + i\lambda} \quad (25)$$

is the RPA dielectric permittivity at $|\mathbf{k}| \ll \bar{p}/\hbar$; see Sec. 33. Taking into account that $\langle\langle\widehat{\Delta f}_{\sigma \mathbf{kp}t_0}^{(av)}\widehat{\Delta f}_{\sigma'\mathbf{k}'\mathbf{p}'t_0}^{(av)}\rangle\rangle = V^2\delta_{\sigma\sigma'}\delta_{\mathbf{k},-\mathbf{k}'}\delta_{\mathbf{p},\mathbf{p}'}f_{\mathbf{p}-\hbar\mathbf{k}/2}(1 - f_{\mathbf{p}+\hbar\mathbf{k}/2}) \simeq \delta_{\sigma\sigma'}V^2\delta_{\mathbf{k},-\mathbf{k}'}\delta_{\mathbf{p},\mathbf{p}'}f_\mathbf{p}(1 - f_\mathbf{p})$ and $\epsilon(\mathbf{k},\omega) = \epsilon^*(-\mathbf{k},-\omega)$, we find the correlation function of the fluctuating potential energies:

$$\langle\langle\widehat{\Delta U}_{\mathbf{k}\omega}\widehat{\Delta U}_{\mathbf{k}'\omega'}\rangle\rangle = 2\pi\delta(\omega + \omega')\delta_{\mathbf{k},-\mathbf{k}'}\frac{64\pi^3 e^4}{k^4|\epsilon(\mathbf{k},\omega)|^2}$$

$$\times \sum_{\mathbf{p}} f_\mathbf{p}(1 - f_\mathbf{p})\delta(\omega - \mathbf{v_p}\cdot\mathbf{k}). \quad (26)$$

One can find also the function describing the correlations of the potential energy and the fluctuating electron distribution:

$$\langle\langle\widehat{\Delta U}_{\mathbf{k}\omega}\widehat{\Delta f}_{\sigma\mathbf{k}'\mathbf{p}\omega'}^{(av)}\rangle\rangle = \frac{\mathbf{k}\cdot(\partial f_\mathbf{p}/\partial \mathbf{p})}{-\omega + \mathbf{v_p}\cdot\mathbf{k} + i\lambda}\langle\langle\widehat{\Delta U}_{\mathbf{k}\omega}\widehat{\Delta U}_{\mathbf{k}'\omega'}\rangle\rangle$$

$$+2\pi V\delta(\omega + \omega')\delta_{\mathbf{k},-\mathbf{k}'}\frac{8\pi^2 e^2}{k^2\epsilon(\mathbf{k},\omega)}f_\mathbf{p}(1 - f_\mathbf{p})\delta(\omega - \mathbf{v_p}\cdot\mathbf{k}). \quad (27)$$

This correlation function is spin-independent. Finally, we present the correlation function of the fluctuations of electron distribution:

$$\langle\langle\widehat{\Delta f}_{\sigma\mathbf{kp}\omega}^{(av)}\widehat{\Delta f}_{\sigma'\mathbf{k}'\mathbf{p}'\omega'}^{(av)}\rangle\rangle = 2\pi V\delta(\omega + \omega')\delta_{\mathbf{k},-\mathbf{k}'}\left\{2\pi V\delta_{\sigma\sigma'}\delta_{\mathbf{p},\mathbf{p}'}\delta(\omega - \mathbf{v_p}\cdot\mathbf{k})\right.$$

$$\times f_{\mathbf{p}}(1 - f_{\mathbf{p}}) - \frac{8\pi^2 e^2}{k^2} \left[\frac{\mathbf{k} \cdot (\partial f_{\mathbf{p}}/\partial \mathbf{p}) f_{\mathbf{p}'}(1 - f_{\mathbf{p}'}) \delta(\omega - \mathbf{v}_{\mathbf{p}'} \cdot \mathbf{k})}{\epsilon(\mathbf{k}, \omega)(\omega - \mathbf{v}_{\mathbf{p}} \cdot \mathbf{k} + i\lambda)} \right. \tag{28}$$

$$\left. + (\mathbf{p} \leftrightarrow \mathbf{p}', \ c.c.) \right] \bigg\} + \frac{\mathbf{k} \cdot (\partial f_{\mathbf{p}}/\partial \mathbf{p}) \, \mathbf{k} \cdot (\partial f_{\mathbf{p}'}/\partial \mathbf{p}') \, \langle\langle \widehat{\Delta U}_{\mathbf{k}\omega} \widehat{\Delta U}_{\mathbf{k}'\omega'} \rangle\rangle}{(\omega - \mathbf{v}_{\mathbf{p}} \cdot \mathbf{k} + i\lambda)(\omega - \mathbf{v}_{\mathbf{p}'} \cdot \mathbf{k} - i\lambda)} \, .$$

The possibility to express the fluctuations of the electrostatic field through the dielectric permittivity describing the dynamical screening is related to quasi-classical (long-wavelength) nature of the fluctuations, when the characteristic length π/k is large in comparison to the screening length and the collective effects are important; see Sec. 33. The formalism of kinetic equations for fluctuations allows one to obtain the electron-electron collision integral, which describes the screened Coulomb interaction, in the most elegant way. Indeed, the equation for the operator $\hat{f}_{\sigma\mathbf{rpt}}^{(av)} = f_{\mathbf{p}} + \widehat{\Delta f}_{\sigma\mathbf{rpt}}^{(av)}$ in the presence of the fluctuating electrostatic potential has the following form (compare to Eq. (21)):

$$\left(\frac{\partial}{\partial t} + \mathbf{v}_{\mathbf{p}} \cdot \frac{\partial}{\partial \mathbf{r}} \right) \hat{f}_{\sigma\mathbf{rpt}}^{(av)} = \frac{\partial \widehat{\Delta U}_{\mathbf{rt}}}{\partial \mathbf{r}} \cdot \frac{\partial \hat{f}_{\sigma\mathbf{rpt}}^{(av)}}{\partial \mathbf{p}}. \tag{29}$$

The averaging $\langle\langle \ldots \rangle\rangle$ transforms the operator equation (29) into a kinetic equation for the distribution function $f_{\mathbf{p}}$. The collision integral in this kinetic equation is expressed through the correlation function (27):

$$J_{e,e}(f|\mathbf{p}) = \left\langle\left\langle \frac{\partial \widehat{\Delta U}_{\mathbf{rt}}}{\partial \mathbf{r}} \cdot \frac{\partial \widehat{\Delta f}_{\sigma\mathbf{rpt}}^{(av)}}{\partial \mathbf{p}} \right\rangle\right\rangle \tag{30}$$

$$= \frac{1}{V^2} \sum_{\mathbf{kk'}} i\mathbf{k} \cdot \frac{\partial}{\partial \mathbf{p}} \int \int \frac{d\omega \, d\omega'}{2\pi \, 2\pi} \left\langle\left\langle \widehat{\Delta U}_{\mathbf{k}\omega} \widehat{\Delta f}_{\sigma\mathbf{k'p}\omega'}^{(av)} \right\rangle\right\rangle \, .$$

We have already taken into account that the correlation function on the right-hand side is proportional to $\delta(\omega + \omega')$. Using Eqs. (27) and (26) for non-degenerate electrons, we obtain

$$J_{e,e}(f|\mathbf{p}) = \frac{2\pi}{V^2} \sum_{\alpha\beta} \frac{\partial}{\partial p_\alpha} \sum_{\mathbf{p'k}} \frac{(4\pi e^2)^2}{|\epsilon(\mathbf{k}, \mathbf{v}_{\mathbf{p}} \cdot \mathbf{k})|^2 k^4} \delta(\mathbf{v}_{\mathbf{p}} \cdot \mathbf{k} - \mathbf{v}_{\mathbf{p}'} \cdot \mathbf{k})$$

$$\times k_\alpha k_\beta \left(f_{\mathbf{p}'} \frac{\partial f_{\mathbf{p}}}{\partial p_\beta} - f_{\mathbf{p}} \frac{\partial f_{\mathbf{p}'}}{\partial p'_\beta} \right) . \tag{31}$$

This quasi-classical collision integral describes the case of small momentum transfer in the collisions of non-degenerate electrons. It is called the Balescu-Lenard collision integral. The general form of the electron-electron collision integral derived in Chapter 6 can be reduced to the form (31) in the limit of small momentum transfer (problem 14.10).

70. Light Scattering

Below we consider the scattering of light by fluctuations of electron system. The physical reason for the fact that the light scattering is determined by the fluctuations is related to random spatial and temporal dependence of the physical properties of the system. For example, in the medium with smooth spatial variations of the dielectric permittivity (opaque medium), there exists elastic scattering of electromagnetic waves. This scattering is described by the equations of classical electrodynamics (problem 14.11). If the temporal variations of the dielectric permittivity are taken into account, the inelastic scattering takes place. If these spatial and temporal variations are comparable to characteristic quantum lengths and quantum times of the system under consideration, the quantum effects become essential. In particular, the spectrum of inelastic light scattering contains a number of peaks corresponding to characteristic discrete frequencies of the system, for example, the frequency of the optical phonons which interact with electrons or the frequency of intersubband transitions of electrons. Such effects are the subject of the quantum theory of light scattering.

The scattering is usually described by a cross-section. This quantity is introduced as a ratio of the energy loss of incident radiation in unit time to the intensity (the absolute value of the energy density flux) of this radiation. The intensity is equal to $\tilde{c}\hbar\omega_I N_I/V$, where ω_I and N_I are the frequency and the number of quanta in the incident beam, $\tilde{c} = c/\sqrt{\epsilon}$ is the velocity of light in the medium with dielectric permittivity ϵ, and V is the normalization volume. Taking into account that the departure rate from the initial state i to all other states f is written through the transition probabilities W_{if}, we express the energy loss in unit time as $\hbar\omega_I \sum_{if} W_{if} f_i(1 - f_f)$, where f_i and f_f are the occupation numbers of electrons in the initial and final states, respectively. As a result, the cross section is given by

$$\sigma = \frac{V}{N_I \tilde{c}} \sum_{if} W_{if} f_i(1 - f_f). \tag{1}$$

The differential cross-section, which depends on the wave vectors $\mathbf{q}_S$ and polarizations μ_S of the scattered photon modes, is introduced as

$$\sigma = \frac{1}{V} \sum_{\mathbf{q}_s \mu_s} \left(\frac{d\sigma}{d\mathbf{q}_S}\right)_{\mu_s} = \sum_{\mu_s} \int_0^\infty d\omega_S \int d\Omega \left(\frac{\partial^2 \sigma}{\partial \omega_S \partial \Omega}\right)_{\mu_s}, \tag{2}$$

where Ω is the solid angle in the space of wave vectors of the scattered photons. The equations written above are general in the sense that they are valid for arbitrary systems and arbitrary mechanisms of interaction between the scattering quasiparticles and the system.

To find the cross-section from a microscopic theory, we take into account both $\propto e$ and $\propto e^2$ terms in the operator of interaction of electrons with the second-quantized electromagnetic field described by the vector potential $\hat{\mathbf{A}}_{\mathbf{r}}$ given by Eq. (3.18). The Hamiltonian of the interaction is written in the form $\widehat{V} = \widehat{H}^{(1)} + \widehat{H}^{(2)}$, where

$$\widehat{H}^{(1)} = -\frac{e}{2c}\sum_j \left(\hat{\mathbf{v}}_j \cdot \hat{\mathbf{A}}_{\mathbf{r}_j} + \hat{\mathbf{A}}_{\mathbf{r}_j} \cdot \hat{\mathbf{v}}_j \right) \ , \quad \widehat{H}^{(2)} = \frac{(e/c)^2}{2m_e}\sum_j \hat{\mathbf{A}}_{\mathbf{r}_j}^2 \ , \quad (3)$$

$\mathbf{r}_j$ and $\hat{\mathbf{v}}_j$ are the operators of coordinate and velocity of j-th electron, and m_e is the mass of free electron. Below we are interested in the probability of transitions between the states with the same total number of photons. Since the operator $\hat{\mathbf{A}}$ changes the number of photons by ± 1, the matrix elements of such transitions should be at least quadratic in the vector potential. Therefore, these matrix elements are to be calculated within the second-order accuracy in the electron-photon interaction. Taking into account that the second-order correction to the matrix element $\langle f|\widehat{V}|i\rangle$ is given by Eq. (34.29), we write the total transition probability (2.13) as

$$W_{if} = \frac{2\pi}{\hbar}\left| \langle f|\widehat{H}^{(2)}|i\rangle + \sum_k{}' \frac{\langle f|\widehat{H}^{(1)}|k\rangle\langle k|\widehat{H}^{(1)}|i\rangle}{\varepsilon_i - \varepsilon_k} \right|^2 \delta(\varepsilon_f - \varepsilon_i), \quad (4)$$

where k denotes an intermediate state which does not coincide with the states i and f. Since the operator $\widehat{H}^{(1)}$ does not couple the states with a given number of photons, its contribution enters only the second-order term in the matrix element in Eq. (4). The states i, k, and f are the states of the electron-photon system in the absence of electron-photon interaction. Therefore, their wave functions are represented in the factorized form, as products of electron and photon components. For example, the initial state is described by the wave function $|\{N_i\}\rangle|\delta_i\rangle$. The wave function of photons, $|\{N_i\}\rangle$, is characterized by the set of occupation numbers N_i. It is given by Eq. (3.16). The wave function of electrons, $|\delta_i\rangle$, is characterized by the set of quantum numbers δ_i of many-electron states. The initial state contains N_I incident photons and zero scattered photons, while the final state contains $N_I - 1$ incident photons and one scattered photon.

The matrix element of $\widehat{H}^{(2)}$ is written with the use of Eq. (3.18) as

$$\langle f|\widehat{H}^{(2)}|i\rangle = \frac{\pi\hbar e^2}{V\sqrt{\omega_{\mathbf{q}\mu}\omega_{\mathbf{q}'\mu'}}\epsilon m_e} \sum_{\mathbf{q}\mu\mathbf{q}'\mu'} \left\langle \delta_f \left| \sum_j e^{i(\mathbf{q}-\mathbf{q}')\cdot\mathbf{r}_j} \right| \delta_i \right\rangle$$

$$\times \left\langle N_I - 1, 1_S \left| \left(\mathbf{e}_{\mathbf{q}\mu} \hat{b}_{\mathbf{q}\mu} + \mathbf{e}^*_{-\mathbf{q}\mu} \hat{b}^+_{-\mathbf{q}\mu} \right) \cdot \left(\mathbf{e}_{-\mathbf{q}'\mu'} \hat{b}_{-\mathbf{q}'\mu'} \right. \right.\right. \tag{5}$$

$$\left.\left.\left. + \mathbf{e}^*_{\mathbf{q}'\mu'} \hat{b}^+_{\mathbf{q}'\mu'} \right) \right| N_I, 0_S \right\rangle = \frac{2\pi\hbar e^2 \sqrt{N_I} (\mathbf{e}_I \cdot \mathbf{e}^*_S)}{V \sqrt{\omega_I \omega_S} \epsilon m_e} \left\langle \delta_f \left| \sum_j e^{i(\mathbf{q}_I - \mathbf{q}_S) \cdot \mathbf{r}_j} \right| \delta_i \right\rangle .$$

Here and below, $\omega_I \equiv \omega_{\mathbf{q}_I \mu_I}$, $\omega_S \equiv \omega_{\mathbf{q}_S \mu_S}$, $\mathbf{e}_I \equiv \mathbf{e}_{\mathbf{q}_I \mu_I}$, and $\mathbf{e}^*_S \equiv \mathbf{e}^*_{\mathbf{q}_S \mu_S}$. Calculating the matrix elements of the bosonic operators entering $\widehat{H}^{(1)}$, one should take into account that the intermediate state contains either N_I incident and one scattered photons or $N_I - 1$ incident and zero scattered photons. We obtain

$$\sum_k \frac{\langle f | \widehat{H}^{(1)} | k \rangle \langle k | \widehat{H}^{(1)} | i \rangle}{\varepsilon_i - \varepsilon_k} = \frac{2\pi\hbar \sqrt{N_I}}{V \sqrt{\omega_I \omega_S} \epsilon} \sum_{\delta_k}$$

$$\times \left\{ \frac{\langle \delta_f | \hat{\mathbf{I}}(\mathbf{q}_S) \cdot \mathbf{e}^*_S | \delta_k \rangle \langle \delta_k | \hat{\mathbf{I}}(-\mathbf{q}_I) \cdot \mathbf{e}_I | \delta_i \rangle}{\varepsilon_{\delta_i} - \varepsilon_{\delta_k} + \hbar\omega_I} \right. \tag{6}$$

$$\left. + \frac{\langle \delta_f | \hat{\mathbf{I}}(-\mathbf{q}_I) \cdot \mathbf{e}_I | \delta_k \rangle \langle \delta_k | \hat{\mathbf{I}}(\mathbf{q}_S) \cdot \mathbf{e}^*_S | \delta_i \rangle}{\varepsilon_{\delta_i} - \varepsilon_{\delta_k} - \hbar\omega_S} \right\},$$

where $\hat{\mathbf{I}}(\mathbf{q})$ is the operator of the current density given by Eq. (13.13). After substituting the contributions (5) and (6) into the transition probability (4), we find the differential cross-section according to Eqs. (1) and (2):

$$\left(\frac{d\sigma}{d\mathbf{q}_S} \right)_{\mu_s} = \frac{(2\pi)^3 \hbar}{\omega_I \omega_S \epsilon^2 \tilde{c}} \sum_{\delta_i \delta_f} f_{\delta_i} (1 - f_{\delta_f}) \delta \left(\varepsilon_{\delta_i} - \varepsilon_{\delta_f} + \hbar\omega_I - \hbar\omega_S \right)$$

$$\times \left| \frac{e^2}{m_e} (\mathbf{e}_I \cdot \mathbf{e}^*_S) \left\langle \delta_f \left| \sum_j e^{i(\mathbf{q}_I - \mathbf{q}_S) \cdot \mathbf{r}_j} \right| \delta_i \right\rangle \right.$$

$$+ \sum_{\delta_k} \left\{ \frac{\langle \delta_f | \hat{\mathbf{I}}(\mathbf{q}_S) \cdot \mathbf{e}^*_S | \delta_k \rangle \langle \delta_k | \hat{\mathbf{I}}(-\mathbf{q}_I) \cdot \mathbf{e}_I | \delta_i \rangle}{\varepsilon_{\delta_i} - \varepsilon_{\delta_k} + \hbar\omega_I} \right. \tag{7}$$

$$\left. \left. + \frac{\langle \delta_f | \hat{\mathbf{I}}(-\mathbf{q}_I) \cdot \mathbf{e}_I | \delta_k \rangle \langle \delta_k | \hat{\mathbf{I}}(\mathbf{q}_S) \cdot \mathbf{e}^*_S | \delta_i \rangle}{\varepsilon_{\delta_i} - \varepsilon_{\delta_k} - \hbar\omega_S} \right\} \right|^2 .$$

To obtain this expression, we have taken into account that $\sum_{if} f_i (1 - f_f) \ldots = \sum_{\delta_i \delta_f} \sum_{\mathbf{q}_s \mu_s} f_{\delta_i} (1 - f_{\delta_f}) \ldots$.

Let us apply the general equation (7) to describe the light scattering by free electrons. The velocity operator is $-(i\hbar/m_e)\partial/\partial\mathbf{r}$, and the wave functions of electrons are the plane waves characterized by momenta.

Denoting the momenta of the initial and final states by $\mathbf{p}$ and $\mathbf{p}'$, respectively, we calculate the expression standing inside $|\dots|^2$ in Eq. (7). The result is

$$\delta_{\mathbf{p}-\mathbf{p}',\hbar(\mathbf{q}_S-\mathbf{q}_I)}\frac{e^2}{m_e}\sum_{\alpha\beta}e_I^\alpha e_S^{\beta*}\left\{\delta_{\alpha\beta}+\frac{1}{m_e}\frac{(p+\hbar q_I/2)_\alpha(p'+\hbar q_S/2)_\beta}{\varepsilon_p-\varepsilon_{\mathbf{p}+\hbar\mathbf{q}_I}+\hbar\omega_I}\right.$$

$$\left.+\frac{1}{m_e}\frac{(p'-\hbar q_I/2)_\alpha(p-\hbar q_S/2)_\beta}{\varepsilon_p-\varepsilon_{\mathbf{p}-\hbar\mathbf{q}_S}-\hbar\omega_S}\right\}. \tag{8}$$

The matrix element of the scattering contains the Kronecker symbol expressing the momentum conservation law. The energy conservation law (the δ-function in Eq. (7)) gives us $p^2-p'^2=2m_e\hbar(\omega_S-\omega_I)$. Since the dispersion laws for photons contain the velocity of light (which is much greater than the characteristic group velocity $\bar{v}$ of non-relativistic electrons), the conservation rules can be simultaneously satisfied only for nearly elastic scattering, when $\omega_S\simeq\omega_I$. In these conditions, the second and the third terms in Eq. (8), which originate from the second-order contribution (6), nearly cancel each other. Their sum contains a small factor of the order $\bar{v}/c$ (problem 14.12) and should be neglected in comparison to the first term originating from the first-order contribution (5). Therefore, only the contribution (5) should be taken into account in the case of free electrons.

Now we turn to a more complex example, the light scattering by electrons in crystals. The electron states in the vicinity of the band extrema are described by the many-band $\mathbf{kp}$ approach; see Appendix B and Sec. 5. The wave functions of non-interacting electrons are products of the plane waves $V^{-1/2}\exp(i\mathbf{p}\cdot\mathbf{r}/\hbar)$ by the Bloch amplitudes $u_l(\mathbf{r})$ which retain the periodicity of the crystal potential (we remind that the quantum number l includes both the band number n and the spin number σ). Near the band extrema, the plane waves are smooth on the scale of the lattice constant. Using these functions for calculating the matrix elements, we obtain the expression standing inside $|\dots|^2$ in Eq. (7) in the form

$$\delta_{\mathbf{p}-\mathbf{p}',\hbar(\mathbf{q}_S-\mathbf{q}_I)}e^2\sum_{\alpha\beta}e_I^\alpha e_S^{\beta*}\left\{\frac{\delta_{ll'}\delta_{\alpha\beta}}{m_e}\right. \tag{9}$$

$$\left.+\sum_{l_1}\left(\frac{v_{l'l_1}^\alpha v_{l_1l}^\beta}{\varepsilon_{lp}-\varepsilon_{l_1\mathbf{p}-\hbar\mathbf{q}_S}-\hbar\omega_S}+\frac{v_{l'l_1}^\beta v_{l_1l}^\alpha}{\varepsilon_{lp}-\varepsilon_{l_1\mathbf{p}+\hbar\mathbf{q}_I}+\hbar\omega_I}\right)\right\},$$

where the initial and final states are denoted by the sets of quantum numbers $l\mathbf{p}$ and $l'\mathbf{p}'$, respectively, and the index l_1 includes both spin and band numbers of the intermediate state. We have already neglected the second-order contributions originating from the action of the momentum operator $-i\hbar\partial/\partial\mathbf{r}$ on the plane-wave factors $\exp(i\mathbf{p}\cdot\mathbf{r}/\hbar)$. These contributions couple the states in the same band with the same spin $(l' = l_1 = l)$ and are analogous to those written in Eq. (8). All such contributions appear to be small as relativistic corrections; see the discussion following Eq. (8). The remaining second-order contributions correspond to interband coupling $(l_1 \neq l, l')$ and are expressed through the matrix elements (5.7). If the energies of the incident and scattered light quanta are smaller than the interband gaps, the energy conservation law forbids the interband transitions of electrons so that the initial and final electron states belong to the same band (the possibility of spin flip, however, remains and $l \neq l'$ in general). In this case, the electron transitions occur near the extremum of a partially filled band (conduction band for n-type and valence band for p-type materials). Moreover, if the energies of the quanta are much smaller than the energy gap, the spin-flip processes can be neglected, which means $l = l'$. One can also neglect the momentum dependence of the electron energies and remove small energies $\hbar\omega_I$ and $\hbar\omega_S$ from the denominators in Eq. (9). As a result of these approximations, the total contribution inside the braces $\{\dots\}$ in Eq. (9) is reduced to $\delta_{ll'}m_{\alpha\beta}^{-1}$, where the effective mass tensor is given by Eq. (5.11).

Although only the non-interacting electrons have been considered above, the calculations can be repeated for any electrons whose wave functions are represented as products of envelope (smooth on the scale of the lattice constant) functions $\psi_{\mathbf{r}}^{(n\delta)}$ of the electron states belonging to the band n by the Bloch amplitudes $u_{n\sigma}(\mathbf{r})$. Denoting the initial, final, and intermediate envelope states as $|n\delta\rangle$, $|n'\delta'\rangle$, and $|n_1\delta_1\rangle$, we rewrite Eq. (7) in the form

$$\left(\frac{\partial^2\sigma}{\partial\omega_S\partial\Omega}\right)_{\mu_s} = \frac{\omega_S^2}{(2\pi\tilde{c})^3}\left(\frac{d\sigma}{d\mathbf{q}_S}\right)_{\mu_s}$$

$$= \frac{\hbar e^4}{c^4}\frac{\omega_S}{\omega_I}\sum_{\sigma\sigma'}\sum_{nn'}\sum_{\delta\delta'}\delta\left(\varepsilon_{n\delta} - \varepsilon_{n'\delta'} + \hbar\omega_I - \hbar\omega_S\right)f_{n\delta}(1 - f_{n'\delta'})$$

$$\times\left|\sum_{\alpha\beta}e_I^{\alpha}e_S^{\beta*}\left\{\frac{\delta_{\alpha\beta}\delta_{nn'}\delta_{\sigma\sigma'}}{m_e}\left\langle n\delta'\left|\sum_j e^{i(\mathbf{q}_I-\mathbf{q}_S)\cdot\mathbf{r}_j}\right|n\delta\right\rangle\right.\right. \tag{10}$$

$$+ \sum_{n_1\sigma_1\delta_1}\left(\frac{v_{n'\sigma',n_1\sigma_1}^{\alpha}v_{n_1\sigma_1,n\sigma}^{\beta}}{\varepsilon_{n\delta} - \varepsilon_{n_1\delta_1} - \hbar\omega_S} + \frac{v_{n'\sigma',n_1\sigma_1}^{\beta}v_{n_1\sigma_1,n\sigma}^{\alpha}}{\varepsilon_{n\delta} - \varepsilon_{n_1\delta_1} + \hbar\omega_I}\right)$$

$$\left.\left.\times\left\langle n'\delta'\left|\sum_j e^{i\mathbf{q}_I\cdot\mathbf{r}_j}\right|n_1\delta_1\right\rangle\left\langle n_1\delta_1\left|\sum_j e^{-\mathbf{q}_S\cdot\mathbf{r}_j}\right|n\delta\right\rangle\right\}\right|^2,$$

where $d\mathbf{q}/(2\pi)^3$ is written as $\omega^2 d\omega d\Omega/(2\pi\tilde{c})^3$. If the energies of the quanta are much smaller than the energy gaps, one may remove $\hbar\omega_I$ and $\hbar\omega_S$ from the denominators and put $n = n'$, $\sigma = \sigma'$, $\varepsilon_{n\delta} \simeq \varepsilon_n$, and $\varepsilon_{n_1\delta_1} \simeq \varepsilon_{n_1}$ (however, the energy spectra standing in the δ-function should remain as they are). Since the envelope functions form complete orthogonal sets for each band, the sum $\sum_{\delta_1}\langle n\delta'|\sum_j e^{i\mathbf{q}_I\cdot\mathbf{r}_j}|n_1\delta_1\rangle\langle n_1\delta_1|$ $\sum_j e^{-i\mathbf{q}_S\cdot\mathbf{r}_j}|n\delta\rangle$ over the intermediate envelope states is reduced to $\langle n\delta'|$ $\sum_j e^{i(\mathbf{q}_I-\mathbf{q}_S)\cdot\mathbf{r}_j}|n\delta\rangle$. This matrix element is the same one that stands in the first-order contribution, and the combined first-order and second-order contributions again form the effective mass tensor for the band n. In the case of a scalar effective mass, $m_{\alpha\beta}^{-1} = m^{-1}\delta_{\alpha\beta}$, one may write (the band index n is omitted)

$$\left(\frac{\partial^2\sigma}{\partial\omega_S\partial\Omega}\right)_{\mu_s} = r_o^2\frac{\omega_S}{\omega_I}|\mathbf{e}_I\cdot\mathbf{e}_S^*|^2\,2\hbar\sum_{\delta\delta'}\delta\left(\varepsilon_\delta - \varepsilon_{\delta'} + \hbar\Delta\omega\right)$$

$$\times f_\delta(1 - f_{\delta'})\left|\left\langle\delta'\left|\sum_j e^{i\Delta\mathbf{q}\cdot\mathbf{r}_j}\right|\delta\right\rangle\right|^2, \qquad (11)$$

where $\Delta\mathbf{q} = \mathbf{q}_I - \mathbf{q}_S$, $\Delta\omega = \omega_I - \omega_S$, and the factor 2 comes from the sum over the spin index σ. We have introduced the classical radius of electron with mass m, $r_o = e^2/mc^2$, so that the dimensionality of the differential cross-section (11) (square over frequency) is seen explicitly. The most efficient light scattering occurs when the polarizations of the incident and scattered electromagnetic waves are parallel or antiparallel.

To study the light scattering under conditions when $\hbar\omega_I$ and $\hbar\omega_S$ are comparable to the interband gap energy ε_g, we use the two-band model of electron states described in Appendix B. If the electron transitions take place near the band extrema (the kinetic energies of electrons in the bands are considerably smaller than $\varepsilon_g = 2ms^2$), the components of the velocity matrix (B.23) are large for interband transitions only. Therefore, one should consider the case when the initial and final states of electrons belong to the same (partially filled) band, while the intermediate state belongs to the other band. Let us assume that the material is of n-type so that the initial and final states (whose spins are σ and σ', respectively) belong to the conduction (c) band. Consequently, the intermediate state belongs to the valence (v) band. The differential cross-section (10) in these conditions is written as

$$\left(\frac{\partial^2\sigma}{\partial\omega_S\partial\Omega}\right)_{\mu_s} = \frac{\hbar e^4}{c^4}\frac{\omega_S}{\omega_I}\sum_{\delta\delta'}f_{c\delta}(1 - f_{c\delta'})\left|\left\langle c\delta'\left|\sum_j e^{i\Delta\mathbf{q}\cdot\mathbf{r}_j}\right|c\delta\right\rangle\right|^2 \quad (12)$$

$$\times \delta \left(\varepsilon_{c\delta} - \varepsilon_{c\delta'} + \hbar \Delta \omega \right) \sum_{\sigma\sigma'} \left| \frac{(\mathbf{e}_I \cdot \mathbf{e}_S^*)}{m_e} \delta_{\sigma\sigma'} + \frac{\varepsilon_g A_+(\sigma,\sigma') + \hbar \omega A_-(\sigma,\sigma')}{\varepsilon_g^2 - (\hbar \omega)^2} \right|^2 .$$

The spin-dependent factors

$$A_\pm(\sigma,\sigma') = \sum_{\sigma_v} \left[\langle c\sigma' | \hat{\mathbf{v}} \cdot \mathbf{e}_I | v\sigma_v \rangle \langle v\sigma_v | \hat{\mathbf{v}} \cdot \mathbf{e}_S^* | c\sigma \rangle \right.$$

$$\left. \pm \langle c\sigma' | \hat{\mathbf{v}} \cdot \mathbf{e}_S^* | v\sigma_v \rangle \langle v\sigma_v | \hat{\mathbf{v}} \cdot \mathbf{e}_I | c\sigma \rangle \right] \qquad (13)$$

contain the sums over the spin index of the intermediate state. Taking into account that $\hbar \Delta \omega \ll \varepsilon_g$, we have put $\hbar \omega_S \simeq \hbar \omega_I = \hbar \omega$ and $\varepsilon_{c\delta} - \varepsilon_{v\gamma} \simeq \varepsilon_g$ in the denominators of Eq. (10). Let us consider the velocity matrix in the parabolic approximation, when $\hat{\mathbf{v}} \simeq s \hat{\boldsymbol{\sigma}} \hat{\rho}_1$. We obtain

$$A_+(\sigma,\sigma') = 2s^2 (\mathbf{e}_I \cdot \mathbf{e}_S^*) \delta_{\sigma'\sigma}, \quad A_-(\sigma,\sigma') = i2s^2 \boldsymbol{\sigma}_{\sigma'\sigma} \cdot [\mathbf{e}_I \times \mathbf{e}_S^*]. \qquad (14)$$

The coefficients A_- and A_+ characterize the contributions to the cross-section due to virtual interband transitions with and without spin flip, respectively. The term with these coefficients in Eq. (12) describes a resonant enhancement of the cross-section in the vicinity of $\hbar \omega = \varepsilon_g$. Since $2m_e s^2 \gg \varepsilon_g$, the non-resonant term proportional to m_e^{-1} in Eq. (12) can be neglected in comparison to the resonant term containing A_+. Let us substitute A_- and A_+ from Eq. (14) into Eq. (12) and calculate the sums over σ and σ'. We obtain the cross-section in the form

$$\left(\frac{\partial^2 \sigma}{\partial \omega_S \partial \Omega} \right)_{\mu_s} = r_o^2 \frac{\omega_S}{\omega_I} P(\mathbf{e}_I, \mathbf{e}_S) 2\hbar \sum_{\delta\delta'} \delta \left(\varepsilon_\delta - \varepsilon_{\delta'} + \hbar \Delta \omega \right)$$

$$\times f_\delta (1 - f_{\delta'}) \left| \left\langle \delta' \left| \sum_j e^{i\Delta \mathbf{q} \cdot \mathbf{r}_j} \right| \delta \right\rangle \right|^2, \qquad (15)$$

where the band index c is omitted. The polarization factor

$$P(\mathbf{e}_I, \mathbf{e}_S) = \frac{|\mathbf{e}_I \cdot \mathbf{e}_S^*|^2}{(1 - \Omega^2)^2} + \left(\frac{\Omega}{1 - \Omega^2} \right)^2 |[\mathbf{e}_I \times \mathbf{e}_S^*]|^2 \qquad (16)$$

depends on the dimensionless frequency $\Omega = \hbar \omega / \varepsilon_g$. Equation (15) generalizes Eq. (11) to the case of large energies of light quanta. If $\Omega \to 0$, the factor $P(\mathbf{e}_I, \mathbf{e}_S)$ is reduced to $|\mathbf{e}_I \cdot \mathbf{e}_S^*|^2$, and the results (15) and (11) are equivalent to each other. When Ω becomes comparable to 1, the light scattering can occur for crossed polarizations, $\mathbf{e}_I \perp \mathbf{e}_S^*$, because of spin-flip processes. The formal divergence of the cross-section (15) at $\Omega = 1$ should be cut off at $1 - \Omega \simeq \bar{\varepsilon}/\varepsilon_g$, where $\bar{\varepsilon}$ is the mean kinetic energy of electrons in the conduction band; see the analogous calculations leading to Eq. (17.16).

The δ-function in Eqs. (11) and (15) can be represented in the form of an integral over time, as in Eq. (8.6). Then, it is convenient to write the sum $\sum_{\delta\delta'} f_\delta(1 - f_{\delta'})e^{i(\varepsilon_\delta - \varepsilon_{\delta'})t/\hbar}\langle\delta'|\sum_j e^{i\Delta\mathbf{q}\cdot\mathbf{r}_j}|\delta\rangle\langle\delta|\sum_j e^{-i\Delta\mathbf{q}\cdot\mathbf{r}_j}|\delta'\rangle$ as a correlation function of the spatial Fourier transforms of the operators of electron density, $\hat{n}_\mathbf{q} = \int d\mathbf{r} e^{-i\mathbf{q}\cdot\mathbf{r}}\hat{n}(\mathbf{r}) = \sum_j e^{-i\mathbf{q}\cdot\mathbf{r}_j}$, in the Heisenberg representation. Indeed, in the second quantization representation $\hat{n}_{\Delta\mathbf{q}}(t) = \sum_{\delta\delta'}\langle\delta|\sum_j e^{-i\Delta\mathbf{q}\cdot\mathbf{r}_j}|\delta'\rangle\hat{a}_\delta^+(t)\hat{a}_{\delta'}(t)$, where $\hat{a}_\delta^+(t) = e^{i\varepsilon_\delta t/\hbar}\hat{a}_\delta^+$. Introducing the operator of the fluctuations of electron density as $\widehat{\Delta n}_{\Delta\mathbf{q}}(t) = \hat{n}_{\Delta\mathbf{q}}(t) - \langle\langle\hat{n}_{\Delta\mathbf{q}}(t)\rangle\rangle$, one can see that the correlation function $\langle\langle\widehat{\Delta n}_{\Delta\mathbf{q}}(t)\widehat{\Delta n}_{-\Delta\mathbf{q}}(0)\rangle\rangle$ is equal to $\sum_{\delta\delta'}\sum_{\delta_1\delta_1'}\langle\delta|\sum_j e^{-i\Delta\mathbf{q}\cdot\mathbf{r}_j}|\delta'\rangle$
$\times\langle\delta_1'|\sum_j e^{i\Delta\mathbf{q}\cdot\mathbf{r}_j}|\delta_1\rangle e^{i(\varepsilon_\delta - \varepsilon_{\delta'})t/\hbar}\left[\langle\langle\hat{a}_\delta^+\hat{a}_{\delta'}\hat{a}_{\delta_1'}^+\hat{a}_{\delta_1}\rangle\rangle - \langle\langle\hat{a}_\delta^+\hat{a}_{\delta'}\rangle\rangle\langle\langle\hat{a}_{\delta_1'}^+\hat{a}_{\delta_1}\rangle\rangle\right]$,
and the factor in the square brackets is reduced to $f_\delta(1 - f_{\delta'})\delta_{\delta\delta_1}\delta_{\delta'\delta_1'}$. Accordingly, Eq. (15) is rewritten as

$$\left(\frac{\partial^2\sigma}{\partial\omega_S\partial\Omega}\right)_{\mu_s} = \frac{r_o^2}{2\pi}\frac{\omega_S}{\omega_I}P(\mathbf{e}_I, \mathbf{e}_S)\langle\langle\widehat{\Delta n}_{\Delta\mathbf{q}}(t)\widehat{\Delta n}_{-\Delta\mathbf{q}}(0)\rangle\rangle_{\Delta\omega} . \qquad (17)$$

Therefore, the problem of light scattering is solved by calculating the Fourier transform of the density-density correlation function, $\langle\langle\widehat{\Delta n}_{\Delta\mathbf{q}}(t)\widehat{\Delta n}_{-\Delta\mathbf{q}}(0)\rangle\rangle_{\Delta\omega} = \int_{-\infty}^\infty dt e^{i\Delta\omega t}\langle\langle\widehat{\Delta n}_{\Delta\mathbf{q}}(t)\widehat{\Delta n}_{-\Delta\mathbf{q}}(0)\rangle\rangle$. In thermodynamic equilibrium, this correlation function is expressed through the polarizability $\alpha(\Delta\mathbf{q}, \Delta\omega)$, see Sec. 33 and problems 3.3 and 14.3, and can be calculated with the aid of various diagram methods described in this book. Expressing the polarizability through the dielectric permittivity according to Eq. (33.5), we obtain

$$\left(\frac{\partial^2\sigma}{\partial\omega_S\partial\Omega}\right)_{\mu_s} = \frac{r_o^2 V \epsilon_\infty^2 |\Delta\mathbf{q}|^2 \hbar}{4\pi^2 e^2}\frac{\omega_S}{\omega_I}\frac{P(\mathbf{e}_I, \mathbf{e}_S)}{1 - e^{-\hbar\Delta\omega/T}}\text{Im}\left(-\frac{1}{\epsilon(\Delta\mathbf{q}, \Delta\omega)}\right).$$
$$(18)$$

This equation relates, under the condition of thermodynamic equilibrium, the differential cross-section to the dielectric properties of the medium. The cross-section (18) is always positive since Im $\epsilon^{-1}(\mathbf{q}, \omega)$ is negative at $\omega > 0$ and positive at $\omega < 0$. The inelastic light scattering with absorption (emission) of the energy of electromagnetic waves corresponds to $\Delta\omega > 0$ ($\Delta\omega < 0$) and is called the Stokes (anti-Stokes) scattering. Owing to the property Im $\epsilon(\Delta\mathbf{q}, -\Delta\omega) = -\text{Im }\epsilon(\Delta\mathbf{q}, \Delta\omega)$, the probability of the Stokes scattering in equilibrium is $e^{\hbar|\Delta\omega|/T}$ times greater than that of anti-Stokes scattering.

The relation of the differential cross-section to the linear response to the perturbation with frequency $\Delta\omega$ is already seen from Eqs. (11) and

(15). Indeed, in equilibrium one can write $f_\delta(1 - f_{\delta'}) = (f_\delta - f_{\delta'})\,[1 - \exp(-\hbar\Delta\omega/T)]^{-1}$, and the differential cross-section contains the sum $\sum_{\delta\delta'} W_{\delta\delta'}(\Delta\omega)\,(f_\delta - f_{\delta'})$, where the transition probability $W_{\delta\delta'}(\Delta\omega)$ is proportional to $\delta\,(\varepsilon_\delta - \varepsilon_{\delta'} + \hbar\Delta\omega)\,\left|\langle\delta'|\sum_j e^{i\Delta\mathbf{q}\cdot\mathbf{r}_j}|\delta\rangle\right|^2$. Thus, the differential cross-section can be treated in terms of the absorbed power $U_{\Delta\omega}$ defined by Eq. (2.23).

Let us calculate the differential cross-section (18) by using different approximations for $\epsilon(\Delta\mathbf{q}, \Delta\omega)$. The free-electron approximation takes place if we substitute the Hartree-Fock (HF) polarizability (33.8) into Eq. (18):

$$\frac{\epsilon_\infty^2 |\Delta\mathbf{q}|^2}{4\pi^2 e^2} \mathrm{Im}\left(-\frac{1}{\epsilon(\Delta\mathbf{q}, \Delta\omega)}\right) \simeq -\frac{1}{\pi e^2} \mathrm{Im}\ \alpha_{HF}(\Delta\mathbf{q}, \Delta\omega) \qquad (19)$$

$$= 2 \int \frac{d\mathbf{p}}{(2\pi\hbar)^3} (f_{\mathbf{p}-\hbar\Delta\mathbf{q}/2} - f_{\mathbf{p}+\hbar\Delta\mathbf{q}/2}) \delta(\varepsilon_{\mathbf{p}+\hbar\Delta\mathbf{q}/2} - \varepsilon_{\mathbf{p}-\hbar\Delta\mathbf{q}/2} - \hbar\Delta\omega).$$

Since the dispersion laws for photons contain the velocity of light, which is large in comparison to the group velocities of electrons, the scattering is nearly elastic, and the δ-function in Eq. (19) can be approximated by $\delta(\hbar\Delta\omega)$. This δ-function is not sensitive to electron momenta, and the differential cross-section (18) in this approximation is expressed as

$$\left(\frac{\partial^2\sigma}{\partial\omega_S\partial\Omega}\right)_{\mu_s} \simeq r_o^2 N P(\mathbf{e}_I, \mathbf{e}_S)\delta(\omega_I - \omega_S). \qquad (20)$$

This equation is written for non-degenerate electron gas and contains the total number of electrons, $N = Vn$, where n is the electron density. Using Eq. (20), one can calculate the integral cross-section of light scattering per one electron (problem 14.13).

A more sophisticated approach, which takes into account the collective behavior of electron system, is the random phase approximation (RPA) considered in Sec. 33. As follows from the results of Sec. 33,

$$\left(\frac{\partial^2\sigma}{\partial\omega_S\partial\Omega}\right)_{\mu_s}^{RPA} = \left(\frac{\partial^2\sigma}{\partial\omega_S\partial\Omega}\right)_{\mu_s}^{HF} \left|\frac{\epsilon_\infty}{\epsilon(\Delta\mathbf{q}, \Delta\omega)}\right|^2. \qquad (21)$$

Therefore, to obtain the RPA result, one should divide the result of the free-electron approximation by the squared absolute value of the RPA dielectric permittivity normalized by ϵ_∞. The relation (21) has two important consequences. First, if $|\Delta\mathbf{q}|$ is small in comparison to the Thomas-Fermi screening length, see Eq. (33.25), the quasielastic scattering is screened out. Second, since $\epsilon(\Delta\mathbf{q}, \Delta\omega)$ has plasmon poles at $\Delta\omega = \pm\omega_p$, see Eq. (33.26), there exists inelastic (Raman) scattering with Stokes $(\omega_S = \omega_I - \omega_p)$ and anti-Stokes $(\omega_S = \omega_I + \omega_p)$ components.

In the general case, the correlation function $\langle\langle\widehat{\Delta n}_{\Delta\mathbf{q}}(t)\,\widehat{\Delta n}_{-\Delta\mathbf{q}}(0)\rangle\rangle_{\Delta\omega}$ has to be found from the equations of motion for pair products of creation and annihilation operators $(\hat{n}_{\mathbf{q}}(t) = \sum_{\delta\delta'}\langle\delta|\sum_j e^{-i\mathbf{q}\cdot\mathbf{r}_j}|\delta'\rangle\{\hat{a}_\delta^+\hat{a}_{\delta'}\}_t$ in the second quantization representation). The methods of such calculations are developed in Secs. 68 and 69. Considering the correlation function which enters Eq. (17), we notice that the difference in the wave vectors, $\Delta\mathbf{q}$, is small in comparison to the typical inverse lengths of electron waves, since the group velocities of electrons are small in comparison to the velocity of light (the maximal $|\Delta\mathbf{q}|$ corresponds to backscattering of light and is equal to $(\omega_I + \omega_S)/\tilde{c}$). In other words, the fluctuations leading to the light scattering belong to the quasi-classical length scale. To describe them, one may use the equations of Sec. 69 for the averaged operator $\widehat{\Delta f}_{\sigma\mathbf{r}\mathbf{p}t}^{(av)}$. Let us assume that the electrons interact only with the self-consistent electric field appearing due to the fluctuations of electron density. According to Eqs. (69.21) and (69.22), the Fourier transform of the electron density operator is given by Eq. (69.24). We obtain

$$\langle\langle\widehat{\Delta n}_{\Delta\mathbf{q}}(t)\widehat{\Delta n}_{-\Delta\mathbf{q}}(0)\rangle\rangle_{\Delta\omega} = 2\pi\left|\frac{\epsilon_\infty}{\epsilon(\Delta\mathbf{q},\Delta\omega)}\right|^2 \tag{22}$$

$$\times 2\sum_{\mathbf{p}} f_{\mathbf{p}-\hbar\Delta\mathbf{q}/2}(1 - f_{\mathbf{p}+\hbar\Delta\mathbf{q}/2})\delta(\Delta\omega - \mathbf{v_p}\cdot\Delta\mathbf{q}),$$

where $\epsilon(\Delta\mathbf{q},\Delta\omega)$ is given by Eq. (69.25). Using Eq. (22) in equilibrium, one may notice that $f_{\mathbf{p}-\hbar\Delta\mathbf{q}/2}(1-f_{\mathbf{p}+\hbar\Delta\mathbf{q}/2}) = (f_{\mathbf{p}-\hbar\Delta\mathbf{q}/2} - f_{\mathbf{p}+\hbar\Delta\mathbf{q}/2})[1 - e^{-\hbar\Delta\omega/T}]^{-1} \simeq -\hbar\Delta\mathbf{q}\cdot(\partial f_{\mathbf{p}}/\partial\mathbf{p})[1-e^{-\hbar\Delta\omega/T}]^{-1}$, which leads to the result (18) with the RPA dielectric permittivity calculated in the limit of small $\Delta\mathbf{q}$. Equation (22) is valid for electrons with an arbitrary distribution function, and this distribution function should be as well substituted into Eq. (69.25) for $\epsilon(\Delta\mathbf{q},\Delta\omega)$ entering Eq. (22). Since the differential cross-section is sensitive to $f_{\mathbf{p}}$, the inelastic light scattering can be used to probe the electron distribution. In the case of inverted electron distribution (for example, when non-equilibrium electrons are created in the conduction band by a strong laser pulse), the anti-Stokes components of inelastic light scattering become stronger than the Stokes components. The anisotropy of electron distribution (for example, in the presence of a strong electric field) can be studied by both frequency and angular dependence of the scattered radiation. The inelastic light scattering is one of the most important tools for experimental probing of collective excitations not only in bulk media, but also in low-dimensional electron systems. In particular, it allows one to study the two-dimensional plasmon modes whose dispersion is different from that of the usual three-dimensional plasma waves (problem 14.14).

So far we have neglected the electron-phonon interaction. One of the effects of this interaction is the contribution of the lattice vibrations in ionic crystals into the frequency-dependent dielectric permittivity; see Sec. 27 and problem 6.20. As a result, the operator of the fluctuating potential energy entering Eq. (69.21) includes effects of both electron and lattice polarization. Its Fourier transform is written as $\widehat{\Delta U}_{\mathbf{k}\omega} = (4\pi e/\epsilon_\infty k^2)[e\widehat{\Delta n}_{\mathbf{k}\omega} - \gamma_{12} i\mathbf{k} \cdot \widehat{\mathbf{w}}_{\mathbf{k}\omega}]$, where $\gamma_{12} = \omega_{TO}\sqrt{(\epsilon_0 - \epsilon_\infty)/4\pi}$ and $\widehat{\mathbf{w}}_{\mathbf{k}\omega}$ is the Fourier transform of the operator of relative displacement found from the equation $[\omega^2 - \omega_{LO}^2](i\mathbf{k} \cdot \widehat{\mathbf{w}}_{\mathbf{k}\omega}) = -(4\pi e\gamma_{12}/\epsilon_\infty)\widehat{\Delta n}_{\mathbf{k}\omega}$ which follows from the consideration given in problem 11.16. As a consequence of these modifications, the constant ϵ_∞ entering Eqs. (69.24) and (69.25) for $\widehat{\Delta n}_{\mathbf{k}\omega}$ is replaced by the dielectric permittivity of the ionic crystal lattice, $\kappa(\omega)$, given by Eq. (27.17). The correlation function of the electron densities is again given by Eq. (22), where ϵ_∞ is replaced by $\kappa(\Delta\omega)$ and the total dielectric permittivity $\epsilon(\Delta\mathbf{q}, \Delta\omega)$ also contains $\kappa(\Delta\omega)$ in place of ϵ_∞. The equilibrium cross-section (18) undergoes the same modifications. If $\Delta\mathbf{q} \to 0$, the RPA expression for the real part of the total dielectric permittivity has zeros at coupled plasmon-phonon frequencies $\omega_\pm$; see problem 6.20. Thus, in the absence of the Landau damping, when the imaginary part of the dielectric permittivity goes to zero (this is the case of strongly degenerate electron gas at $\Delta\omega > v_F|\Delta\mathbf{q}|$), the differential cross-section has sharp peaks at $|\Delta\omega| = \omega_\pm$ corresponding to excitation of coupled plasmon-phonon modes (problem 14.15).

The plasmon-phonon peaks observed in experiments are, however, considerably broadened due to collision-induced relaxation of the electron density fluctuations. If the Landau damping is weak, the relaxation gives the main contribution to the imaginary part of $\epsilon(\Delta\mathbf{q}, \Delta\omega)$. Consider, for example, electron-phonon scattering. If the spatial dispersion is neglected, one finds Im $\epsilon(\Delta\mathbf{q}, \Delta\omega) \simeq$ Im $\epsilon(\Delta\omega) = (4\pi/\omega)$Re $\sigma(\Delta\omega)$, where the function Re $\sigma(\omega)$ is given by Eq. (68.28). One may also use a simplified approach based upon the Drude formula $\epsilon(\omega) \simeq \kappa(\omega) - \epsilon_\infty\omega_p^2/[\omega(\omega + i\nu)]$ with a frequency-independent relaxation rate ν; see Eq. (8.31). This approach is rigorously justified at $\varepsilon_F \gg \hbar\omega$. Then, the differential cross-section is given by

$$\left(\frac{\partial^2\sigma}{\partial\omega_S\partial\Omega}\right)_{\mu_s} = \frac{r_o^2 V\epsilon_\infty|\Delta\mathbf{q}|^2\hbar}{4\pi^2 e^2}\frac{\omega_S}{\omega_I}\frac{P(\mathbf{e}_I, \mathbf{e}_S)F(\Delta\omega)}{1 - \exp(-\hbar\Delta\omega/T)}, \qquad (23)$$

$$F(\omega) = \frac{(\omega^2 - \omega_{LO}^2)^2\omega_p^2\omega\nu}{[\omega^4 - \omega^2(\omega_{LO}^2 + \omega_p^2) + \omega_{TO}^2\omega_p^2]^2 + (\omega^2 - \omega_{LO}^2)^2\omega^2\nu^2}.$$

The frequency-dependent form-factor of light scattering, $F(\Delta\omega)$, contains broadened plasmon-phonon resonances shown in Fig. 14.2. The cross-section (23) goes to zero at $\Delta\omega = \pm\omega_{LO}$ because the fluctuations of electron density are completely screened out at LO phonon frequency by ionic motion. Equation (23) is valid for the equilibrium case only. To take into account the collision-induced relaxation of density fluctuations in the general (non-equilibrium) case, one should use Eq. (69.21) with linearized collision integrals and corresponding Langevin sources on the right-hand side.

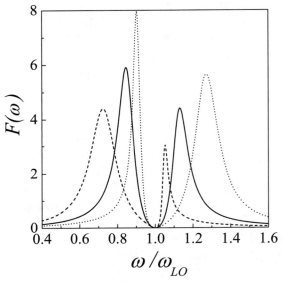

Figure 14.2. Frequency dependence of the differential cross-section of light scattering by coupled plasmon-phonon excitations at $\omega_p = \omega_{LO}$, $\omega_p = 0.8\,\omega_{LO}$, and $\omega_p = 1.2\,\omega_{LO}$ (solid, dashed, and dotted curves, respectively). The calculation is carried out according to Eq. (70.23) with $\epsilon_0 = 15.69$, $\epsilon_\infty = 14.44$ ($\omega_{TO}/\omega_{LO} \simeq 0.96$), and $\nu = 0.2\,\omega_{LO}$.

Apart from the light scattering by the electron density fluctuations, there exists another inelastic scattering mechanism associated with the lattice vibrations. It is called the phonon-assisted Raman scattering. The light interacts with the lattice not directly but rather through the electron sub-system. The lattice vibrations produce in the medium a transient optical superlattice and it is from the latter that the scattering occurs. The efficiency of the scattering is connected with the intensity of the corresponding fluctuations of the dielectric permittivity, $\delta\epsilon_{\alpha\beta}(\mathbf{r}, t)$. The differential cross-section can be represented in the form (problem

14.16)

$$\left(\frac{\partial^2 \sigma}{\partial \omega_S \partial \Omega}\right)_{\mu_s} = \frac{(\omega_S/c)^4}{32\pi^3} \sum_{\alpha\beta\gamma\delta} e_S^\delta e_I^{\gamma*} e_S^{\alpha*} e_I^\beta \langle\langle \delta\epsilon_{\delta\gamma}(\Delta\mathbf{q}, t)\delta\epsilon_{\alpha\beta}(-\Delta\mathbf{q}, 0)\rangle\rangle_{\Delta\omega} .$$

$$(24)$$

This expression is derived from the classical electrodynamics. In order to apply it to the quantum case, one should consider the fluctuations of the dielectric permittivity standing in Eq. (24) as operators in the Heisenberg representation. The quantity $\delta\epsilon$ contains the terms linear in the deformation tensor (B.12): $\delta\epsilon_{\alpha\beta}(\mathbf{r}) = \sum_{\alpha'\beta'}(\partial\epsilon_{\alpha\beta}/\partial\varepsilon_{\alpha'\beta'})\varepsilon_{\alpha'\beta'}(\mathbf{r})$. The partial derivative in this expression is treated merely as a proportionality coefficient (a constant dimensionless tensor of the fourth rank). Expressing the deformation tensor through the quantized displacement vector (6.29), one can write the correlation function in Eq. (24) through the correlation functions of two bosonic creation and annihilation operators. After simple transformations, we obtain the differential cross-section for the light scattering by acoustic phonons:

$$\left(\frac{\partial^2 \sigma}{\partial \omega_S \partial \Omega}\right)_{\mu_s} = \frac{(\omega_S/c)^4}{(4\pi)^2} \frac{\hbar V |\Delta\mathbf{q}|^2}{2\rho\omega_{\Delta\mathbf{q}l}} \left|\sum_{\alpha\beta\alpha'\beta'}\left(\frac{\partial\epsilon_{\alpha\beta}}{\partial\varepsilon_{\alpha'\beta'}}\right) e_S^{\alpha*} e_I^\beta e_{\Delta\mathbf{q}l}^{\alpha'} \frac{\Delta q_{\beta'}}{\Delta q}\right|^2$$

$$\times \left[(N_{\Delta\mathbf{q}}^l + 1)\delta(\Delta\omega - \omega_{\Delta\mathbf{q}l}) + N_{\Delta\mathbf{q}}^l \delta(\Delta\omega + \omega_{\Delta\mathbf{q}l})\right]. \qquad (25)$$

The Stokes and anti-Stokes contributions correspond to the emission and absorption of phonons, respectively.

In the ionic crystals, one uses the expansion

$$\delta\epsilon_{\alpha\beta}(\mathbf{r}) = \sum_\gamma \left[\left(\frac{\partial\epsilon_{\alpha\beta}}{\partial w_\gamma}\right) w_\gamma(\mathbf{r}) + \left(\frac{\partial\epsilon_{\alpha\beta}}{\partial E_{L\gamma}}\right) E_{L\gamma}(\mathbf{r})\right] \qquad (26)$$

in terms of the relative ionic displacement $\mathbf{w}$ and longitudinal electric field $\mathbf{E}_L = -4\pi\mathbf{P}$ associated with this displacement; see Eqs. (6.14)–(6.21). Employing the expression (6.19) for the quantized relative displacement together with the relation (6.20), we calculate the correlation functions of two bosonic operators and find

$$\left(\frac{\partial^2 \sigma}{\partial \omega_S \partial \Omega}\right)_{\mu_s} = \frac{(\omega_S/c)^4}{(4\pi)^2} \frac{2\pi\hbar V \omega_{LO}}{\epsilon^*} |\mathcal{A}_{\Delta\mathbf{q}}|^2$$

$$\times \left[(N_{LO} + 1)\delta(\Delta\omega - \omega_{LO}) + N_{LO}\delta(\Delta\omega + \omega_{LO})\right], \qquad (27)$$

where the *LO* phonon dispersion is neglected. The factor $\mathcal{A}_{\Delta\mathbf{q}}$ is given by the following expression:

$$\mathcal{A}_{\Delta \mathbf{q}} = \sum_{\alpha\beta\gamma} \left(\frac{\partial \epsilon_{\alpha\beta}}{\partial E_{L\gamma}} \right) e_S^{\alpha *} e_I^\beta \frac{\Delta q_\gamma}{\Delta q}$$

$$- \frac{\epsilon_\infty}{\omega_{TO} \sqrt{4\pi(\epsilon_0 - \epsilon_\infty)}} \sum_{\alpha\beta\gamma} \left(\frac{\partial \epsilon_{\alpha\beta}}{\partial w_\gamma} \right) e_S^{\alpha *} e_I^\beta \frac{\Delta q_\gamma}{\Delta q}. \tag{28}$$

The inelastic light scattering described by Eq. (27) brings the resonant contribution at $\Delta\omega = \pm\omega_{LO}$. Such a contribution is missing in the spectrum of collective plasmon-phonon excitations; see Eq. (23). To describe the broadening of the LO phonon peak, one should consider the damping due to phonon-phonon interaction; see the final part of Sec. 27. Since this damping is weak in comparison to the damping of the electron density fluctuations, the LO phonon peaks observed in experiments are sharper than the plasmon-phonon peaks.

71. Fluctuations in Mesoscopic Conductors

We remind that mesoscopic conductors are the systems whose size is small in comparison to the inelastic scattering length so that only the elastic scattering of electrons has to be taken into account. The averaged current through such systems is described in Sec. 58 within the scattering-matrix formalism. Let us use this formalism in order to calculate the correlation function of the currents in the leads contacted to a mesoscopic sample. The correlation function of the currents in the leads N and M (noise power) is defined by analogy to Eq. (68.9):

$$\mathcal{S}_{NM}(\omega) = \frac{1}{2} \langle\langle \hat{I}_N(t)\hat{I}_M(0) + \hat{I}_M(0)\hat{I}_N(t) \rangle\rangle_\omega. \tag{1}$$

The Heisenberg operator of the current is given by

$$\hat{I}_M(t) = \int_{\mathbf{r} \in S_M} d\mathbf{r} \ \hat{\mathbf{I}}(\mathbf{r}, t) \cdot \mathbf{n}_{S_M}(\mathbf{r}), \tag{2}$$

where $\hat{\mathbf{I}}(\mathbf{r}, t) = e^{i\widehat{H}t/\hbar}\hat{\mathbf{I}}(\mathbf{r})e^{-i\widehat{H}t/\hbar}$ with $\hat{\mathbf{I}}(\mathbf{r})$ defined by Eq. (4.15), and $\mathbf{n}_{S_M}(\mathbf{r})$ is the unit vector perpendicular to the surface S_M of the contact M and directed inside the sample. Employing the basis $|Nn\varepsilon\rangle$ described by the wave functions (58.14), we introduce the operators of creation and annihilation of the electrons which have the energy ε in the state n and come to the sample through the lead N: $\hat{a}_{Nn\varepsilon}^+$ and $\hat{a}_{Nn\varepsilon}$. These operators satisfy the fermionic anticommutation rule similar to Eq. (4.21):

$$\hat{a}_{Nn\varepsilon}\hat{a}_{Mm\varepsilon'}^+ + \hat{a}_{Mm\varepsilon'}^+\hat{a}_{Nn\varepsilon} = \delta_{NM}\delta_{nm}\delta(\varepsilon - \varepsilon'). \tag{3}$$

Expressing the operator $\hat{\mathbf{I}}(\mathbf{r}, t)$ in the second quantization representation, we find

$$\hat{I}_M(t) = \frac{1}{2\pi\hbar} \sum_{N_1 N_2} \sum_{n_1 n_2} \int d\varepsilon \int d\varepsilon' e^{i(\varepsilon-\varepsilon')t/\hbar} I^{(M)}_{N_1 n_1, N_2 n_2}(\varepsilon, \varepsilon') \hat{a}^+_{N_1 n_1 \varepsilon} \hat{a}_{N_2 n_2 \varepsilon'},$$
(4)

where $I^{(M)}_{N_1 n_1, N_2 n_2}(\varepsilon, \varepsilon')$ is defined in Eq. (58.19). Let us substitute Eq. (4) into Eq. (1) and average the products of four second-quantization operators (a similar averaging is already considered in Eq. (68.24)). We obtain

$$\mathcal{S}_{NM}(\omega) = \frac{1}{2\pi\hbar} \sum_{N_1 N_2} \sum_{n_1 n_2} \int d\varepsilon I^{(N)}_{N_1 n_1, N_2 n_2}(\varepsilon, \varepsilon + \hbar\omega) I^{(M)}_{N_2 n_2, N_1 n_1}(\varepsilon + \hbar\omega, \varepsilon)$$

$$\times \left\{ f_{N_1}(\varepsilon)[1 - f_{N_2}(\varepsilon + \hbar\omega)] + f_{N_2}(\varepsilon + \hbar\omega)[1 - f_{N_1}(\varepsilon)] \right\}, \quad (5)$$

where $f_N(\varepsilon)$ is the electron distribution function in the lead N. It is introduced according to $\langle\langle \hat{a}^+_{Nn\varepsilon} \hat{a}_{N'n'\varepsilon'} \rangle\rangle = \delta_{NN'}\delta_{nn'}\delta(\varepsilon - \varepsilon') f_N(\varepsilon)$. This function depends only on the energy ε, because each lead is assumed to be in local equilibrium (since the chemical potentials of electrons in different leads are different, we retain the index N). Note that we have multiplied $\mathcal{S}_{NM}(\omega)$ by the spin degeneracy factor 2 because the spin index is not written explicitly in $\hat{a}^+_{Nn\varepsilon}$ and $\hat{a}_{Nn\varepsilon}$. The correlation function (5) has the symmetry property $\mathcal{S}_{NM}(\omega) = \mathcal{S}_{MN}(-\omega)$.

Let us consider the static noise, $\omega = 0$. Substituting the expressions (58.20) of the matrix elements $I^{(M)}_{N_1 n_1, N_2 n_2}(\varepsilon, \varepsilon')$ into Eq. (5), we find

$$\mathcal{S}_{NM} = \frac{e^2}{\pi\hbar} \int d\varepsilon \left[\mathcal{N}_N f_N(1 - f_N)\delta_{NM} - T_{MN} f_N(1 - f_N) - T_{NM} f_M(1 - f_M) \right.$$

$$\left. + \frac{1}{2} \sum_{N_1 N_2} \text{tr}\left(\widehat{S}^+_{NN_1} \widehat{S}_{NN_2} \widehat{S}^+_{MN_2} \widehat{S}_{MN_1} \right) (f_{N_1} + f_{N_2} - 2 f_{N_1} f_{N_2}) \right]. \quad (6)$$

We note (see Sec. 58) that $\widehat{S}_{NN'}$ is the matrix whose elements are $S_{Nn,N'n'}(\varepsilon)$, and tr ... defines the matrix trace over the states $n, n', \ldots$. Next, $\mathcal{N}_N$ is the total number of channels in the lead N. The scattering matrices, transmission coefficients, and electron distribution functions in Eq. (6) have the same argument, ε, which is omitted for brevity. In the equilibrium case, when the distribution function $f(\varepsilon)$ is independent of the lead index, the last term on the right-hand side of Eq. (6) is transformed according to Eq. (58.16) to a term proportional to $\mathcal{N}_N \delta_{NM}$. The non-diagonal ($N \neq M$) correlation function in the case of equilibrium electron distribution with temperature T_e is expressed with the use of $f(\varepsilon)[1 - f(\varepsilon)] = -T_e[\partial f(\varepsilon)/\partial \varepsilon]$. Employing also Eqs. (58.21) and (58.22) together with the symmetry property $G_{NM} = G_{MN}$ in the absence of magnetic fields, one can see that

$$\mathcal{S}_{NM} = -\frac{e^2 T_e}{\pi \hbar} \int d\varepsilon \left(-\frac{\partial f(\varepsilon)}{\partial \varepsilon} \right) [T_{NM}(\varepsilon) + T_{MN}(\varepsilon)] = -2T_e G_{NM}. \quad (7)$$

Therefore, the current-current correlation function is proportional to the conductance between the leads N and M. Taking into account the linear relation (58.7), we find that Eq. (7) is a manifestation of the fluctuation-dissipation theorem (68.9) at $\hbar\omega \ll T_e$. Note that the sign of $\mathcal{S}_{NM}$ is negative. One can prove that the correlation functions of the currents in the different leads are always negative (problem 14.17). Consequently, the correlation functions of the currents in the same lead are always positive, because of current conservation; see Eq. (58.9).

Consider a two-terminal conductor, where there are only left (L) and right (R) leads, and $\mathcal{S} \equiv \mathcal{S}_{LL} = \mathcal{S}_{RR} = -\mathcal{S}_{LR} = -\mathcal{S}_{RL}$ since the current is conserved. Applying Eq. (6) to this case, it is convenient to consider non-diagonal components, for example, $N = L$ and $M = R$. By using Eq. (58.16) (see also problem 14.17) and the channel representation (see the discussion after Eq. (58.22)), we obtain

$$\mathcal{S} = \frac{e^2}{\pi \hbar} \int d\varepsilon \left(T(\varepsilon) \left\{ f_L(\varepsilon)[1 - f_L(\varepsilon)] + f_R(\varepsilon)[1 - f_R(\varepsilon)] \right\} \right.$$

$$\left. + \sum_n T_n(\varepsilon)[1 - T_n(\varepsilon)][f_L(\varepsilon) - f_R(\varepsilon)]^2 \right), \quad (8)$$

where $T_n(\varepsilon) \equiv T_{RL}^{(n)}(\varepsilon)$ is the channel transmission probability and $T(\varepsilon) = T_{RL}(\varepsilon) = \sum_n T_n(\varepsilon)$ is the total transmission probability at the energy ε. The first term in Eq. (8) describes the quasi-equilibrium noise due to fluctuations in the leads, the latter are considered as local-equilibrium sub-systems. The second term is associated with the current and called the shot noise. This term is of the second order in the distribution functions. It increases the total noise power and cannot be neglected if the applied bias eV is comparable to the temperature T_e. On the other hand, one may often neglect the energy dependence of $T(\varepsilon)$ and $T_n(\varepsilon)$ on the scale of T_e and eV and replace the transmission coefficients by those taken at the Fermi energy. Then, the integral over energy gives us

$$\mathcal{S} = \frac{e^2}{\pi \hbar} \left[2T_e \sum_n T_n^2 + eV \coth \left(\frac{eV}{2T_e} \right) \sum_n T_n(1 - T_n) \right]. \quad (9)$$

The noise power (9) is a complicated function of temperature and voltage rather than a superposition of equilibrium and shot noises. For low voltages, $eV \ll T_e$, we obtain the equilibrium formula $\mathcal{S} = 2T_e G$, in accordance with the general result (7). One can derive a similar formula for frequency-dependent noise (problem 14.18).

For zero temperature, when $\coth(eV/2T_e) = 1$, the shot noise contribution in Eq. (9) is much greater than the equilibrium one and proportional to $eV \sum_n T_n(1 - T_n)$. The factor $1 - T_n$ describes the reduction of the shot noise associated with a finite transparency of the mesoscopic sample and originates from the correlations imposed by the Pauli principle. As a result, the ratio of the shot noise power to the Poissonian noise power $\mathcal{S}_P = eI$ (which corresponds to low-transparent conductors such as tunneling junctions) is less than 1. Depending on the transparency of the sample, this ratio, denoted below as $\mathcal{F}$, varies between 0, for high-transmitting channels, and 1, for low-transmitting channels. Under the approximation of energy-independent transmission, as in Eq. (9), the factor $\mathcal{F}$ is given by

$$\mathcal{F} = \frac{\sum_n T_n(1 - T_n)}{\sum_n T_n}. \tag{10}$$

In the classical limit of disordered conductors containing many channels, one has a universal value $\mathcal{F} = 1/3$, without regard to the degree of disorder, size, and any other properties of the sample. This important result can be obtained by considering quasi-classical fluctuations of the current through the microcontacts described in Sec. 12 (problem 14.19). To find a link between the classical result and Eq. (10), one may notice that in the limit of many channels the sums in Eq. (10) are equivalent to averages over the transmission coefficient distribution. Thus, $\mathcal{F} = \int dT\, P(T)T(1-T)/\int dT\, P(T)T$, where $P(T)$ is the distribution function of transmission coefficients. In the case of a disordered conductor, when the characteristic length L of the sample is large in comparison to the mean free path length l_F of Fermi electrons but small in comparison to the localization length L_0 (see Sec. 59), one has

$$P(T) = \frac{l_F}{2L_0} \frac{1}{T\sqrt{1-T}}, \quad T_{min} < T < 1, \quad T_{min} = 4e^{-2L_0/l_F}, \tag{11}$$

and the result $\mathcal{F} = 1/3$ follows immediately. Equation (11) is presented here without a derivation. Such a derivation is based upon the random matrix theory which we do not consider in this book. The most remarkable feature of $P(T)$ is its bimodal form: almost open ($T \to 1$) and almost closed ($T \to 0$) channels are preferred. In spite of $L \gg l_F$, the sample is assumed to be mesoscopic, the inelastic scattering is absent. If the inelastic scattering becomes significant, the shot noise is suppressed.

The conductance of a mesoscopic sample is unique and determined by the sample geometry and by a given distribution (configuration) of the scattering centers in the sample. If one takes several samples of the same shape and with the same number of identical scatterers, the conductances of these samples will be different, because the configurations of the scatterers are different. Therefore, apart from the fluctuations of electric current described above, there exist fluctuations of the conductance. Let us express the conductance G in fundamental units and introduce a dimensionless function g according to $G = (e^2/2\pi\hbar)g$. One may introduce the variance $\overline{(\Delta g)^2} = \overline{g^2} - (\overline{g})^2$, where $\overline{g}$ is the dimensionless conductance averaged over configurations of the scattering potential. The impurities and defects in the sample can move between the sites in the process of diffusion. This process produces fluctuations of the conductance of a single sample with time. Next, by applying a magnetic field H or by changing the Fermi energy ε_F of electrons in the sample (for example, by biasing a gate electrode), one considerably rearranges electron paths in the sample. Therefore, a fluctuating dependence of the conductance on the external fields should be expected.

One may naively assume that the relative variance of the conductance fluctuations, $\overline{(\Delta g)^2}/(\overline{g})^2$, decreases with increasing sample size or with the number of the scattering centers. This statement is not correct, because the electrons keep phase coherence over large parts of the sample, and the concept of self-averaging, which works well for classical variables (see problem 14.1), is not valid for the mesoscopic samples. The remarkable result discussed below is that the variance of g is a universal constant of the order of unity, independent of the size of the system and its dimensionality. Thus, taking into account that the conductance of a d-dimensional sample is proportional to L^{d-2}, where L is the linear size of the sample, we obtain

$$\frac{\overline{(\Delta g)^2}}{(\overline{g})^2} \propto \frac{1}{L^{2(d-2)}}. \tag{12}$$

If $d \leq 2$, this quantity does not decrease with increasing L, which means that no self-averaging occurs. By rearranging the scattering potential pattern, we close some channels and open other channels, but the universality of $\overline{(\Delta g)^2}$ means that the fluctuations of the effective number of open channels remain of the order of unity. A rigorous proof of this property is presented below with the use of the Green's function formalism developed in Secs. 14 and 15.

To describe the conductance fluctuations, let us introduce the correlation function

$$F(\Delta E, \Delta H) = \langle\langle \Delta g(\varepsilon_F, H) \Delta g(\varepsilon_F + \Delta E, H + \Delta H) \rangle\rangle \qquad (13)$$

which depends on the variations of the Fermi energy, ΔE, and magnetic field, ΔH. Here $\langle\langle \ldots \rangle\rangle$ denotes the averaging over the random potentials. The conductances entering Eq. (13) are expressed through the Green's functions as described in Sec. 13. We consider metallic samples with dimensions L_x, L_y, and L_z, assuming that the current flows along OZ. In the absence of magnetic fields, the dimensionless conductance is expressed through the static conductivity σ given by Eq. (13.27), where, however, the averaging over the random potentials should be omitted:

$$g(\varepsilon_F) = \frac{2\pi\hbar L_x L_y}{e^2 L_z}\sigma = \left(\frac{\hbar}{mL_z}\right)^2 \sum_{ss'}\sum_{\mathbf{pp'}} p_z p'_z (-1)^l G^s_{\varepsilon_F}(\mathbf{p}, \mathbf{p'}) G^{s'}_{\varepsilon_F}(\mathbf{p'}, \mathbf{p}).$$

$$(14)$$

It is assumed that the temperature is zero. The indices s and s' denote retarded (R) or advanced (A) Green's function, and $l = 1$ (or 0) for $s = s'$ (or $s \neq s'$). Therefore, Eqs. (13) and (14) lead to

$$F(\Delta E) = \left(\frac{\hbar}{mL_z}\right)^4 \sum_{s_1 s'_1 s_2 s'_2}\sum_{\mathbf{p_1 p_2 p'_1 p'_2}} p_{1z} p'_{1z} p_{2z} p'_{2z} (-1)^{l_1+l_2} \langle\langle G^{s_1}_{\varepsilon_F}(\mathbf{p_1}, \mathbf{p'_1})$$

$$(15)$$

$$\times G^{s'_1}_{\varepsilon_F}(\mathbf{p'_1}, \mathbf{p_1}) G^{s_2}_{\varepsilon_F + \Delta E}(\mathbf{p_2}, \mathbf{p'_2}) G^{s'_2}_{\varepsilon_F + \Delta E}(\mathbf{p'_2}, \mathbf{p_2}) \rangle\rangle - g(\varepsilon_F)g(\varepsilon_F + \Delta E).$$

The first term of this expression contains the contribution $g(\varepsilon_F)g(\varepsilon_F + \Delta E)$ which corresponds to independent averaging of the first and second pairs of the Green's functions, $\langle\langle G^{s_1}_{\varepsilon_F} G^{s'_1}_{\varepsilon_F}\rangle\rangle\langle\langle G^{s_2}_{\varepsilon_F+\Delta E} G^{s'_2}_{\varepsilon_F+\Delta E}\rangle\rangle$. This contribution cancels with the second term in Eq. (15). Thus, only the averages including the Green's functions both of the first and of the second pair contribute to $F(\Delta E)$. The simplest contribution of this kind, denoted below as $F_1(\Delta E)$, is written through the correlation functions defined by Eq. (15.1):

$$F_1(\Delta E) = \left(\frac{\hbar}{mL_z}\right)^4 \sum_{s_1 s'_1 s_2 s'_2}\sum_{\mathbf{p_1 p_2 p'_1 p'_2}} p_{1z} p'_{1z} p_{2z} p'_{2z} (-1)^{l_1+l_2}$$

$$\times \left[K^{s_1 s'_2}_{EE'}(\mathbf{p_1}, \mathbf{p'_2}|\mathbf{p'_1}, \mathbf{p_2}) K^{s'_1 s_2}_{EE'}(\mathbf{p'_1}, \mathbf{p_2}|\mathbf{p_1}, \mathbf{p'_2}) \qquad (16)\right.$$

$$\left. + K^{s_1 s_2}_{EE'}(\mathbf{p_1}, \mathbf{p_2}|\mathbf{p'_1}, \mathbf{p'_2}) K^{s'_1 s'_2}_{EE'}(\mathbf{p'_1}, \mathbf{p'_2}|\mathbf{p_1}, \mathbf{p_2}) \right],$$

where $E = \varepsilon_F$ and $E' = \varepsilon_F + \Delta E$.

The methods of calculation of the correlation functions entering Eq. (16) are described in detail in Sec. 15. These functions satisfy Eq. (15.5) containing the irreducible vertex part Γ. Consider, for example, the correlation function $K_{EE'}^{s_1 s_2'}(\mathbf{p}_1, \mathbf{p}_2' | \mathbf{p}_1', \mathbf{p}_2)$. Since we are interested in quantum interference effects, we evaluate the vertex part by considering the contribution of maximally crossed diagrams; see Eqs. (15.16)–(15.19). An obvious generalization of these equations by the substitutions $\mathbf{p} \to \mathbf{p}_1$ and $-\mathbf{p} \to \mathbf{p}_2'$ leads to (see Eq. (15.20))

$$\Gamma_{EE'}^{s_1 s_2'}(\mathbf{p}_1, \mathbf{p}_2' | \mathbf{p}_1 - \hbar\mathbf{q}, \mathbf{p}_2' + \hbar\mathbf{q}) \tag{17}$$

$$= w \left[1 - \frac{w}{V} \sum_{\mathbf{p}} G_E^{s_1}\left(\mathbf{p} + \frac{\Delta\mathbf{p}}{2}\right) G_{E'}^{s_2'}\left(\mathbf{p} - \frac{\Delta\mathbf{p}}{2}\right) \right]^{-1} \equiv \widetilde{\Gamma}_{EE'}^{s_1 s_2'}(\Delta\mathbf{p}),$$

where $\Delta\mathbf{p} = \mathbf{p}_1 - \mathbf{p}_2' - \hbar\mathbf{q}$ and $G_E^s(\mathbf{p})$ is the averaged Green's function. Here and below in this section, we assume the case of short-range scattering potentials, when the Fourier transform of the random potential correlation function is a constant, w. We consider the metallic case $\varepsilon_F \tau / \hbar \gg 1$, where τ is the scattering time defined, according to Eq. (8.21), by $\hbar/\tau = \pi w \rho_D(\varepsilon_F)$. In these conditions, the divergent contribution to Γ from Eq. (17) appears at $s_1 \neq s_2'$; see Sec. 15. Calculating the sum over $\mathbf{p}$ in Eq. (17) for $q \ll \hbar/l_F$ and $\Delta E \ll \hbar/\tau$, we obtain

$$\widetilde{\Gamma}_{EE'}^{AR}(\hbar\mathbf{q}) = \frac{w}{\tau} C_{\Delta E/\hbar}(\mathbf{q}), \quad C_{\Delta E/\hbar}(\mathbf{q}) = \frac{1}{Dq^2 - i\Delta E/\hbar}, \tag{18}$$

where $D = v_F^2 \tau / d$ is the diffusion coefficient. This vertex part has been calculated in Sec. 15 for the two-dimensional case ($d = 2$) at $\Delta E = 0$; see Eqs. (15.23) and (15.24). The function $C_{\Delta E/\hbar}(\mathbf{q})$ can be viewed as a spatial Fourier transform of the Cooperon $C_{\Delta E/\hbar}(\mathbf{r}, \mathbf{r}')$ satisfying the diffusion equation (43.23) for a spatially homogeneous system.

Therefore, to find $K_{EE'}^{s_1 s_2'}(\mathbf{p}_1, \mathbf{p}_2' | \mathbf{p}_1', \mathbf{p}_2)$, one should solve Eq. (15.5) by iterations, with the use of Eq. (17). However, in contrast to the results of Sec. 15, such a solution is more complicated because the singular contribution may appear at $\mathbf{p}_1 - \mathbf{p}_2 \to 0$ as well as at $\mathbf{p}_1 + \mathbf{p}_2' \to 0$ (problem 14.20):

$$K_{EE'}^{s_1 s_2'}(\mathbf{p}_1, \mathbf{p}_2' | \mathbf{p}_1', \mathbf{p}_2) = \delta_{\mathbf{p}_1 \mathbf{p}_1'} \delta_{\mathbf{p}_2 \mathbf{p}_2'} G_E^{s_1}(\mathbf{p}_1) G_{E'}^{s_2'}(\mathbf{p}_2')$$

$$+ \frac{\delta_{\mathbf{p}_1 - \mathbf{p}_2, \mathbf{p}_1' - \mathbf{p}_2'}}{V} G_E^{s_1}(\mathbf{p}_1) G_{E'}^{s_2'}(\mathbf{p}_2') \left[\widetilde{\Gamma}_{EE'}^{s_1 s_2'}(\mathbf{p}_1 - \mathbf{p}_2) \right. \tag{19}$$

$$+\widetilde{\Gamma}^{s_1 s_2'}_{EE'}(\mathbf{p}_1 + \mathbf{p}_2') - w\Big] G^{s_1}_E(\mathbf{p'}_1)G^{s_2'}_{E'}(\mathbf{p}_2).$$

The factor w in the square brackets should be neglected because the corresponding term is not associated with singularities and gives, in the metallic case we consider, a small contribution in comparison to the first term on the right-hand side of Eq. (19).

Considering the other correlation functions on the right-hand side of Eq. (16) in a similar way, one may conclude that the main divergent contribution to F_1 comes from the terms where s-indices of each correlation function are different. This necessarily implies $(-1)^{l_1+l_2} = 1$ so that all contributing terms are positive. The second term in the square brackets of Eq. (16) gives the same contribution as the first one (this can be checked by the permutation $\mathbf{p}_2' \leftrightarrow \mathbf{p}_2$). Next, since only the terms containing products of two vertex parts with the same arguments (either $\mathbf{p}_1 - \mathbf{p}_2$ or $\mathbf{p}_1 + \mathbf{p}_2'$) are essential, the presence of the two different vertex parts in Eq. (19) leads to doubling of the results. Therefore, using Eqs. (18) and (19), we transform Eq. (16) to

$$F_1(\Delta E) = 8 \left(\frac{\hbar}{mL_z}\right)^4 \frac{w^2}{\tau^2} \frac{1}{V^2} \sum_{\mathbf{p}\mathbf{p'}\mathbf{q}} p_z p_z'(p_z - \hbar q_z)(p_z' - \hbar q_z)$$

$$\times \left\{ \left| C_{\Delta E/\hbar}(\mathbf{q}) G^R_{\varepsilon_F}(\mathbf{p}) G^R_{\varepsilon_F}(\mathbf{p'}) G^R_{\varepsilon_F+\Delta E}(\mathbf{p} - \hbar\mathbf{q}) G^R_{\varepsilon_F+\Delta E}(\mathbf{p'} - \hbar\mathbf{q}) \right|^2 \quad (20)$$

$$+ \mathrm{Re}\left[C_{\Delta E/\hbar}(\mathbf{q}) G^A_{\varepsilon_F}(\mathbf{p}) G^A_{\varepsilon_F}(\mathbf{p'}) G^R_{\varepsilon_F+\Delta E}(\mathbf{p} - \hbar\mathbf{q}) G^R_{\varepsilon_F+\Delta E}(\mathbf{p'} - \hbar\mathbf{q}) \right]^2 \right\}.$$

The region of interest is associated with small q and ΔE. Since the singular behavior for $q \to 0$ and $\Delta E \to 0$ is already present in $C_{\Delta E/\hbar}(\mathbf{q})$, one may neglect q and ΔE in the Green's functions. As a result, Eq. (20) is reduced to

$$F_1(\Delta E) = 4 \left(\frac{\hbar}{mL_z}\right)^4 \frac{w^2}{\tau^2} \sum_{\mathbf{q}} [\mathrm{Re} C_{\Delta E/\hbar}(\mathbf{q})]^2 \quad (21)$$

$$\times \left[\frac{2}{V} \sum_{\mathbf{p}} \frac{p_z^2}{[(\varepsilon_F - p^2/2m)^2 + (\hbar/2\tau)^2]^2}\right]^2.$$

The contribution $(2/V)\sum_{\mathbf{p}}\ldots$ standing in the square brackets is equal to $\pi p_F^2 \rho_{\mathcal{D}}(\varepsilon_F)(2\tau/\hbar)^3/2d$, and we finally obtain

$$F_1(\Delta E) = 4 \left(\frac{2}{\pi}\right)^4 \sum_{\mathbf{q}} \left(\mathrm{Re}\left[\left(\frac{qL_z}{\pi}\right)^2 - i\frac{\Delta E}{E_c}\right]^{-1}\right)^2, \quad (22)$$

where the energy $E_c = \pi^2 \hbar D / L_z^2$ is associated with the time of diffusion through the sample, L_z^2/D. This energy defines the correlation range so that the conductances $g(\varepsilon_F)$ and $g(\varepsilon_F + \Delta E)$ are statistically independent at $\Delta E \gg E_c$.

Considering the variance $\overline{(\Delta g)^2} = F(0)$, we have to introduce a cutoff for $\mathbf{q}$ to make the result convergent. In the absence of inelastic scattering (the case we consider), this cutoff is determined by the characteristic size L of the sample, according to $q > \pi/L$. This means that $F_1(0)$ given by Eq. (22) at $\Delta E = 0$ is a constant independent of the size of the sample. To take into account the shape of the sample, one needs a more careful consideration. We have already noticed that $C_0(\mathbf{q})$ is a spatial Fourier transform of the Cooperon $C_0(\mathbf{r}, \mathbf{r}')$. Consequently, $\sum_{\mathbf{q}} [\mathrm{Re} C_0(\mathbf{q})]^2$ in Eq. (21) should be replaced by $\int d\mathbf{r} \int d\mathbf{r}' [\mathrm{Re} C_0(\mathbf{r}, \mathbf{r}')]^2$. Let us find the Cooperon from the diffusion equation (43.23) with physically reasonable boundary conditions. Namely, at the boundaries $x = \pm L_x/2$ and $y = \pm L_y/2$, where the diffusion stops, the normal derivatives $dC_0(\mathbf{r}, \mathbf{r}')/dx$ and $dC_0(\mathbf{r}, \mathbf{r}')/dy$, respectively, are equal to zero. At the current-carrying boundaries $z = \pm L_z/2$, the function $C_0(\mathbf{r}, \mathbf{r}')$ should be equal to zero, because the leads are assumed to be good conductors, where the diffusion coefficients are large in comparison to the diffusion coefficient in the mesoscopic sample. Imposing these boundary conditions, we find

$$C_0(\mathbf{r}, \mathbf{r}') = \sum_{\mathbf{m}} \frac{\chi_{\mathbf{mr}} \chi_{\mathbf{mr}'}^*}{\nu_{\mathbf{m}}}, \quad \mathbf{m} = (m_x, m_y, m_z), \qquad (23)$$

where m_x, m_y, and m_z are integer numbers. The eigenfunctions $\chi_{\mathbf{mr}}$ and eigenvalues $\nu_{\mathbf{m}}$ are given by

$$\chi_{\mathbf{mr}} = \sqrt{\frac{8}{L_x L_y L_z}} \cos\left[\pi m_x \left(\frac{x}{L_x} - \frac{1}{2}\right)\right]$$

$$\times \cos\left[\pi m_y \left(\frac{y}{L_y} - \frac{1}{2}\right)\right] \sin\left[\pi m_z \left(\frac{z}{L_z} - \frac{1}{2}\right)\right],$$

$$\nu_{\mathbf{m}} = D\frac{\pi^2}{L_z^2}\left[m_z^2 + \frac{L_z^2}{L_x^2}m_x^2 + \frac{L_z^2}{L_y^2}m_y^2\right] \equiv D\frac{\pi^2}{L_z^2}\lambda_{\mathbf{m}} . \qquad (24)$$

The integers m_x and m_y run from 0 to infinity, while m_z runs from 1 to infinity. With the aid of the dimensionless quantity $\lambda_{\mathbf{m}}$ introduced by Eq. (24), we rewrite Eq. (22) at $\Delta E = 0$ as

$$F_1(0) = 4\left(\frac{2}{\pi}\right)^4 \sum_{m_x, m_y=0}^{\infty} \sum_{m_z=1}^{\infty} \left(\mathrm{Re}\lambda_{\mathbf{m}}^{-1}\right)^2 . \qquad (25)$$

For the samples long in z direction (mesoscopic wires), the main contribution to the sum comes from $m_x = m_y = 0$. Therefore, one has $\lambda_{\mathbf{m}} = m_z^2$ and $F_1(0) = 32/45$.

We have calculated the contribution F_1 which comes from the correlations of two Green's functions in Eq. (15). The correlation function of four Green's functions, however, contains additional contributions which are not reduced to double correlation functions of the kind (15.1). One can find only two types of such contributions, presented in the diagram form as

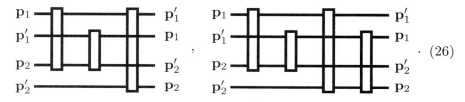

$$, \qquad \cdot \text{(26)} $$

As in Sec. 15, the rectangle stands for the irreducible vertex part. The two diagrams of Eq. (26) are irreducible. The other diagrams of such kind can be obtained by permutations of the vertex parts. Denoting the contribution coming from the irreducible diagrams containing three and four vertex parts by F_2 and F_3, respectively, we present the result without details of derivation:

$$ F_2(0) = -16 \left(\frac{2}{\pi}\right)^4 \mathrm{Re} \sum_{m_x,m_y=0}^{\infty} \sum_{n_x,n_y=0}^{\infty} \sum_{m_z=1,3,\dots}^{\infty} \sum_{n_z=2,4,\dots}^{\infty} $$

$$ \times \frac{(f_{m_z n_z})^2}{\lambda_{\mathbf{m}} \lambda_{\mathbf{n}}} \left(\frac{1}{\lambda_{\mathbf{m}}} + \frac{1}{\lambda_{\mathbf{n}}}\right) , \tag{27} $$

$$ F_3(0) = 48 \left(\frac{2}{\pi}\right)^4 \mathrm{Re} \sum_{m_x,m_y=0}^{\infty} \sum_{n_x,n_y=0}^{\infty} \sum_{l_x,l_y=0}^{\infty} \sum_{k_x,k_y=0}^{\infty} $$

$$ \times \sum_{m_z,l_z=1,3,\dots}^{\infty} \sum_{n_z,k_z=2,4,\dots}^{\infty} \frac{f_{m_z n_z} f_{n_z l_z} f_{l_z k_z} f_{k_z m_z}}{\lambda_{\mathbf{m}} \lambda_{\mathbf{n}} \lambda_{\mathbf{l}} \lambda_{\mathbf{k}}} , \tag{28} $$

where $f_{mn} = 4mn/[\pi(m^2 - n^2)]$. The total contribution is

$$ F(0) = F_1(0) + F_2(0) + F_3(0). \tag{29} $$

A numerical calculation of the sums gives us $[\overline{(\Delta g)^2}]^{1/2} = \sqrt{F(0)} = 0.729$ for a wire, 0.862 for a square, and 1.088 for a cube. The shape dependence becomes strong when either L_x or L_y are longer than L_z, and it is possible to obtain $[\overline{(\Delta g)^2}]^{1/2}$ considerably greater than 1.

The introduction of the phase relaxation due to inelastic scattering can be done by adding the phase relaxation rate $1/\tau_\varphi$ to $\nu_{\mathbf{m}}$ in Eq. (23), like in Eq. (43.25). The conductance fluctuations are suppressed if the diffusion length $\sqrt{D\tau_\varphi}$ associated with the phase relaxation time τ_φ becomes less than the size of the sample. The magnetic field suppresses the conductance fluctuations when the magnetic length becomes less than the sample size $L_\perp$ in the direction perpendicular to the field. This suppression, however, is not complete. Indeed, by adding the term proportional to the vector potential to each of the momenta in Eq. (15), we suppress only one of the two singular vertex part present in Eq. (19), namely $\widetilde{\Gamma}^{s_1 s_2'}_{EE'}(\mathbf{p}_1 + \mathbf{p}_2')$. Therefore, in strong enough magnetic field, when the magnetic length is much smaller than $L_\perp$, one may expect $F(\Delta E)$ twice smaller than in the absence of this field. The correlation field H_c, which characterizes the suppression of $F(\Delta E, \Delta H)$ with ΔH, is estimated as $H_c \sim \Phi_0/L_\perp^2$, where Φ_0 is the flux quantum (59.13).

72. NDT Formalism for Fluctuations

The central problem of the quantum theory of fluctuations is to find the fourth-order correlation function $\langle\langle \{\hat{a}^+_{p_1'}\hat{a}_{p_1}\}_{t_1}\{\hat{a}^+_{p_2'}\hat{a}_{p_2}\}_{t_2}\rangle\rangle$, see, for example, Eq. (68.24), or the function $\langle\langle \hat{\Psi}^+_{\mathbf{x}_1'}(t_1)\hat{\Psi}_{\mathbf{x}_1}(t_1)\hat{\Psi}^+_{\mathbf{x}_2'}(t_2)\hat{\Psi}_{\mathbf{x}_2}(t_2)\rangle\rangle$ expressed in terms of the field operators. The diagram technique considered in Appendix E allows one to derive closed equations for the correlation functions of arbitrary order at zero temperature provided that all time arguments are different so that one can use a formalism of Green's functions. Such equations cannot be, in general, written for the above-defined correlation functions depending only on two time variables. A similar situation has been encountered in Chapter 8. We remind that since the usual double-time Green's functions are not defined for coinciding time arguments, one cannot use them to obtain equations for the averages of the operators of physical quantities. It is necessary to introduce a two-branch time contour and consider four (instead of one) Green's functions, which leads to the non-equilibrium diagram technique.

One can act in a similar way to obtain the desired equations for the correlation functions of fluctuations. Below we generalize the non-equilibrium diagram technique developed in Chapter 8 to the case of two-particle correlation functions. We present a brief outline of this method without considering its possible applications. Let us introduce a four-branch time contour C_4, see Fig. 14.3, by adding one more forward and one more backward branches to the contour shown in Fig. 8.1.

By definition,

$$C^{s_1 s_1' s_2 s_2'}(11'; 22') = \langle\langle \hat{\mathcal{T}}_{C_4} \hat{\Psi}_{s_1}(1) \hat{\Psi}^+_{s_1'}(1') \hat{\Psi}_{s_2}(2) \hat{\Psi}^+_{s_2'}(2') \rangle\rangle$$

$$- \langle\langle \hat{\mathcal{T}}_{C_4} \hat{\Psi}_{s_1}(1) \hat{\Psi}^+_{s_1'}(1') \rangle\rangle \langle\langle \hat{\mathcal{T}}_{C_4} \hat{\Psi}_{s_2}(2) \hat{\Psi}^+_{s_2'}(2') \rangle\rangle, \qquad (1)$$

where $\hat{\mathcal{T}}_{C_4}$ is the operator of chronological ordering along the contour C_4. The contour branch indices (s_1, s_1', s_2, and s_2' in Eq. (1)) can take values from 1 to 4, in contrast to Sec. 41, where there are only two branch indices denoted as $-$ and $+$ (equivalent to 1 and 2 of the present technique, respectively).

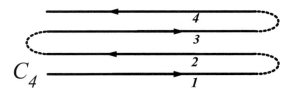

Figure 14.3. Four-branch time contour used in the calculations of the correlation functions of fluctuations.

The one-particle Green's functions defined through the double correlation functions of the field operators become 4×4 matrices in the four-branch representation. There are, however, only three linearly-independent components of such matrices, because of the obvious properties $G^{33} = G^{11} \equiv G^{--}$, $G^{44} = G^{22} \equiv G^{++}$, $G^{ss'}|_{s<s'} = G^{12} \equiv G^{-+}$, and $G^{ss'}|_{s>s'} = G^{21} \equiv G^{+-}$. Therefore, one may write the 4×4 matrix Green's function $\widetilde{G}$ in terms of 2×2 matrix blocks:

$$\widetilde{G} = \begin{pmatrix} \widehat{G} & G^{-+}(\hat{\sigma}_x + \hat{1}) \\ G^{+-}(\hat{\sigma}_x + \hat{1}) & \widehat{G} \end{pmatrix}, \qquad (2)$$

where $\widehat{G}$ is the 2×2 matrix (39.12). In a similar fashion, the 4×4 self-energy matrix is defined as $\Sigma^{33} = \Sigma^{11} \equiv \Sigma^{--}$, $\Sigma^{44} = \Sigma^{22} \equiv \Sigma^{++}$, $\Sigma^{ss'}|_{s<s'} = (-1)^{s+s'+1}\Sigma^{12} \equiv (-1)^{s+s'+1}\Sigma^{-+}$, and $\Sigma^{ss'}|_{s>s'} = (-1)^{s+s'+1}\Sigma^{21} \equiv (-1)^{s+s'+1}\Sigma^{+-}$, or, equivalently,

$$\widetilde{\Sigma} = \begin{pmatrix} \widehat{\Sigma} & \Sigma^{-+}(\hat{\sigma}_x - \hat{1}) \\ \Sigma^{+-}(\hat{\sigma}_x - \hat{1}) & \widehat{\Sigma} \end{pmatrix} \qquad (3)$$

with $\widehat{\Sigma}$ determined by Eqs. (39.17)–(39.20). Using these definitions, one can generalize the Dyson equations (39.21) for $\widetilde{G}$ and $\widetilde{\Sigma}$. These 4×4

matrix equations have exactly the same form as (39.21), though instead of $\hat{\sigma}_z$ one should substitute $\tilde{\sigma}_z = \begin{pmatrix} \hat{\sigma}_z & \hat{0} \\ \hat{0} & \hat{\sigma}_z \end{pmatrix}$ (problem 14.21):

$$\left(i\hbar \frac{\partial}{\partial t_1} - \hat{h}_1 \right) \widetilde{G}(1, 1') = \tilde{\sigma}_z \delta(1 - 1') + \int d3 \tilde{\sigma}_z \widetilde{\Sigma}(1, 3) \widetilde{G}(3, 1'),$$

$$\left(i\hbar \frac{\partial}{\partial t_1'} - \hat{h}_{1'} \right)^* \widetilde{G}(1, 1') = \tilde{\sigma}_z \delta(1 - 1') + \int d3 \widetilde{G}(1, 3) \widetilde{\Sigma}(3, 1') \tilde{\sigma}_z. \quad (4)$$

By subtracting the second equation in this set from the first one, we write for $\widetilde{G}(1, 1')$ an equation similar to the generalized kinetic equations (40.19) and (41.27):

$$\left[i\hbar \left(\frac{\partial}{\partial t_1} + \frac{\partial}{\partial t_1'} \right) - \hat{h}_1 + \hat{h}_{1'}^* \right] \widetilde{G}(1, 1')$$

$$= \int d3 \left[\tilde{\sigma}_z \widetilde{\Sigma}(1, 3) \widetilde{G}(3, 1') - \widetilde{G}(1, 3) \widetilde{\Sigma}(3, 1') \tilde{\sigma}_z \right]. \quad (5)$$

The correlation function (1) is a tensor of the fourth rank. Let us consider a particular form of this function, when the arguments 1 and $1'$ correspond to the second pair of the time contour axes $(s_1, s_1' = 3, 4)$ while the second pair of arguments (2 and $2'$) corresponds to the first pair of the axes $(s_2, s_2' = 1, 2)$. Such functions can be also written in the form

$$C^{s_1 s_1' s_2 s_2'}(11'; 22') = -\hbar^2 \langle\langle \widehat{\delta G}^{s_1 s_1'}(1, 1') \widehat{\delta G}^{s_2 s_2'}(2, 2') \rangle\rangle, \quad (6)$$

where $\widehat{\delta G}$ are the operators of fluctuations:

$$i\hbar \widehat{\delta G}^{s_1 s_1'}(1, 1') = \hat{\mathcal{T}}_C \hat{\Psi}_{s_1}(1) \hat{\Psi}_{s_1'}^+(1') - \langle\langle \hat{\mathcal{T}}_C \hat{\Psi}_{s_1}(1) \hat{\Psi}_{s_1'}^+(1') \rangle\rangle. \quad (7)$$

The ordering in Eq. (7) is carried out along the usual two-branch contour C, but one should have in mind that this contour is a part of the contour C_4, and the two operators of fluctuations entering Eq. (6) are ordered within different pairs of branches of the contour C_4. As follows from the definition (7), $\langle\langle \widehat{\delta G}^{s_1 s_1'}(1, 1') \rangle\rangle = 0$. There are obvious relations between the elements of the matrix $C^{s_1 s_1' s_2 s_2'}$. Such relations are quite similar to the relations (39.22) between the one-particle Green's functions:

$$C^{34 s_2 s_2'} + C^{43 s_2 s_2'} = C^{33 s_2 s_2'} + C^{44 s_2 s_2'},$$

$$C^{s_1 s_1' 12} + C^{s_1 s_1' 21} = C^{s_1 s_1' 11} + C^{s_1 s_1' 22}. \quad (8)$$

Using the first equation, we define three linearly-independent correlation functions

$$C^{R,s_2 s_2'} = C^{33 s_2 s_2'} - C^{34 s_2 s_2'},$$

$$C^{A,s_2 s_2'} = C^{33 s_2 s_2'} - C^{43 s_2 s_2'}, \qquad (9)$$

$$C^{F,s_2 s_2'} = C^{34 s_2 s_2'} + C^{43 s_2 s_2'}$$

by analogy with Eq. (40.3). Considering combinations of the functions with different second pairs of the branch indices, one can use the second identity of Eq. (8) and construct nine independent correlation functions of the type (6): C^{RR}, C^{RA}, C^{RF}, C^{AR}, C^{AA}, C^{AF}, C^{FR}, C^{FA}, and C^{FF}. For example, $C^{RR} = C^{R,11} - C^{R,12}$.

To develop the diagram technique, one should express the two-particle Green's functions in terms of the correlation functions of the field operators in the interaction representation, similar to Eq. (41.15):

$$\langle\langle \hat{\mathcal{T}}_{C_4} \hat{\Psi}_{s_1}(1) \hat{\Psi}^+_{s_1'}(1') \hat{\Psi}_{s_2}(2) \hat{\Psi}^+_{s_2'}(2') \rangle\rangle \qquad (10)$$

$$= \langle\langle \hat{\mathcal{T}}_{C_4} \hat{\mathcal{S}}_{C_4} \hat{\Psi}_{0 s_1}(1) \hat{\Psi}^+_{0 s_1'}(1') \hat{\Psi}_{0 s_2}(2) \hat{\Psi}^+_{0 s_2'}(2') \rangle\rangle .$$

The operator $\hat{\mathcal{S}}_{C_4}$ is defined by Eq. (41.14), where the integral is now taken along the contour C_4. Expanding this operator in series of the interaction Hamiltonian, one can obtain expressions for the two-particle Green's functions in a given order with respect to the interaction. Since we work with causal Green's functions defined on a special time contour, such expressions can be constructed in a similar way as in Appendix E, where many-particle Green's functions have been considered. In particular, one can write the Bethe-Salpeter equation similar to Eq. (E.29) by adding the contour branch indices ($s = 1 - 4$) to each variable and implying summation over internal branch indices, which is equivalent to multiplication of 4×4 matrices. In the following, to simplify the notations in the operations with 4×4 matrices, we include the contour branch indices into the multi-indices, i.e., write $C(11'; 22')$ instead of $C^{s_1 s_1' s_2 s_2'}(11'; 22')$ and so on. The expressions whose multi-indices do not include the branch indices will contain these indices in the superscripts, as in Eq. (6), or will be written in the form of 4×4 matrices denoted by the sign of tilde, as in Eqs. (4) and (5).

The correlation function (1) is related to the two-electron Green's function (E.6) according to $C(11'; 22') = -\hbar^2[G(1, 2; 1', 2') - G(1, 1') G(2, 2')]$. To find an equation for this correlation function, one can use, in principle, the Bethe-Salpeter equation (E.29) for $G(1, 2; 1', 2')$. For our purposes, however, it is convenient to use an alternative form of the Bethe-Salpeter equation, which is written as an integral relation between $G(1, 2; 1', 2')$ and $G(\tilde{2}', 2; \tilde{2}, 2')$, in contrast to the relation between $G(1, 2; 1', 2')$ and $G(\tilde{1}', \tilde{2}'; 1', 2')$ given by Eq. (E.29). The general integral relation of this kind can be written as an equation for the correlation function:

$$C(11'; 22') = \hbar^2 G(1, 2') G(2, 1') \tag{11}$$

$$+ \int d3 \int d3' \int d4 \int d4' G(1, 3) G(3', 1') I(33'44') C(4'4; 22').$$

To find the vertex part $I(33'44')$ standing in Eq. (11) for the systems with pair interaction, we first remind that, in the diagram representation, this interaction couples only the vertices with the same branch indices. In the diagram technique of Sec. 41, the broken line connecting the vertices $1s_1$ and $2s_2$ corresponds to the factor $is_1\delta_{s_1s_2}U_{1-2}$, where U_{1-2} is given by Eq. (41.24). In the present technique, this line corresponds to the factor $-i\overline{U}_{1-2}$, where $\overline{U}_{1-2}$ is given by $(-1)^{s_1+1}\delta_{s_1s_2}\hbar\delta(t_1-t_2)v(\mathbf{r}_1-\mathbf{r}_2)$. With this definition, one can carry out a diagrammatic expansion of $G(1, 2; 1', 2')$ according to the general rules described in Appendix E. As a result,

$$I(33'44') = -i\overline{U}_{3-4}[\delta_{33'}\delta_{44'} - \delta_{34'}\delta_{3'4}] + \overline{U}_{3-4}\overline{U}_{3'-4'}[G(3, 3')G(4, 4')$$

$$-G(3, 4')G(4, 3')] + \overline{U}_{3-4'}\overline{U}_{4-3'}G(3, 3')G(4, 4')$$

$$-\delta_{43'}\overline{U}_{3-4'} \int d5 \overline{U}_{4-5}G(3, 5)G(5, 4') \tag{12}$$

$$-\delta_{34'}\overline{U}_{4-3'} \int d5 \overline{U}_{3-5}G(4, 5)G(5, 3')$$

$$+\delta_{34'}\delta_{3'4} \int d5 \int d5' \overline{U}_{3-5}\overline{U}_{5'-3'}G(5, 5')G(5', 5) + \dots ,$$

where the dots . . . indicate higher-order terms with respect to the interaction. Other interactions can also be considered (problem 14.22). The vertex part satisfies the symmetry property $I(33'44') = I(44'33')$ and is presented as a functional derivative of the self-energy over the Green's function:

$$I(33'44') = \delta\Sigma(3, 3')/\delta G(4', 4). \tag{13}$$

This equation can be checked by comparing the expansions of $I(33'44')$ and $\Sigma(3, 3')$ in series with respect to the interaction.

The integrals over the internal variables in Eqs. (11) and (12) also imply the sums over the branch indices s_i. For a further simplification of notations, we do not write these integrals in the following, implicitly assuming integration over repeated variables, and skip the dividing commas in the expressions for v, G and Σ. Let us consider two equations:

$$C(11'; 22') = \hbar^2 G(12') G(21') + G(13) G(3'1') I(33'44') C(4'4; 22'),$$

$$C(11'; 22') = \hbar^2 G(12') G(21') + C(11'; 3'3) I(33'44') G(24) G(4'2'). \quad (14)$$

The first of them is Eq. (11) written in the simplified notations (the integrals over the variables $3, 3', 4$, and $4'$ are implied on the right-hand side), while the second one is obtained from Eq. (11) by using the symmetry properties $C(11'; 22') = C(22'; 11')$ and $I(33'44') = I(44'33')$.

Let us introduce the operator $\widehat{G}_{13}^{-1}$ by the relation $\widehat{G}_{13}^{-1} G(31') = \delta_{11'}$ (or $G(3'1) \widehat{G}_{13}^{-1} = \delta_{33'}$). As explained above, the integrals over the repeated variables 3 (or 1) are implied in these relations. According to Eq. (4), the operator $\widehat{G}_{13}^{-1}$ can be expressed as

$$\widehat{G}_{13}^{-1} = \sigma_z(1) \left(i\hbar \frac{\partial}{\partial t_1} - \hat{h}_1 \right) \delta_{13} - \Sigma(13), \quad (15)$$

where $\sigma_z(1) = (-1)^{s_1+1}$ (problem 14.23). Acting by $\widehat{G}_{13}^{-1}$ from the left on the first equation of Eq. (14) written for $C(31'; 22')$, then acting by $\widehat{G}_{3'1'}^{-1}$ from the right on the same equation for $C(13'; 22')$, we obtain a couple of equations

$$\widehat{G}_{13}^{-1} C(31'; 22') = G(5'1') I(15'3'3) C(33'; 22') + \delta_{12'} \hbar^2 G(21'),$$

$$C(13'; 22') \widehat{G}_{3'1'}^{-1} = G(15) I(51'3'3) C(33'; 22') + \delta_{1'2} \hbar^2 G(12'). \quad (16)$$

Let us act, in a similar way, on the second equation of Eq. (14) by $\widehat{G}_{24}^{-1}$ from the left and by $\widehat{G}_{4'2'}^{-1}$ from the right. We obtain another couple of equations:

$$\widehat{G}_{24}^{-1} C(11'; 42') = C(11'; 44') I(4'425') G(5'2') + \delta_{1'2} \hbar^2 G(12'),$$

$$C(11'; 24') \widehat{G}_{4'2'}^{-1} = C(11'; 44') I(4'452') G(25) + \delta_{12'} \hbar^2 G(21'). \quad (17)$$

The further step is to derive equations for the correlation function (6) as a function of $\bar{t}_1 = (t_1 + t_1')/2$ and $\bar{t}_2 = (t_2 + t_2')/2$ (we note that $\partial/\partial t_1 + \partial/\partial t_1' = \partial/\partial \bar{t}_1$). Let us multiply all terms in the first and second equations of Eq. (16) by $\sigma_z(1)$ and $\sigma_z(1')$, respectively, and subtract the second equation obtained in this way from the first one. The result is

$$\left(i\hbar \frac{\partial}{\partial \bar{t}_1} - \hat{h}_1 + \hat{h}_{1'}^* \right) C(11; 22') - \sigma_z(1) \Sigma(13) C(31'; 22') + C(13'; 22')$$

$$\times \Sigma(3'1') \sigma_z(1') - \left[\sigma_z(1) G(5'1') I(15'3'3) - \sigma_z(1') G(15) I(51'3'3) \right] \quad (18)$$

$$\times C(33'; 22') = \sigma_z(1)\delta_{12'}\hbar^2 G(21') - \sigma_z(1')\delta_{1'2}\hbar^2 G(12').$$

Transforming the equations of Eq. (17) in a similar way, we find

$$\left(i\hbar\frac{\partial}{\partial\bar{t}_2} - \hat{h}_2 + \hat{h}_{2'}^*\right) C(11; 22') - \sigma_z(2)\Sigma(24)C(11'; 42') + C(11'; 24')$$

$$\times\Sigma(4'2')\sigma_z(2') - \left[\sigma_z(2)I(4'425')G(5'2') - \sigma_z(2')I(4'452')G(25)\right] \quad (19)$$

$$\times C(11'; 44') = \sigma_z(2)\delta_{1'2}\hbar^2 G(12') - \sigma_z(2')\delta_{12'}\hbar^2 G(21').$$

The left-hand sides of Eqs. (18) and (19) can be viewed as a result of the action of integro-differential operators on the correlation function. In Eq. (18) these operators act on the first pair of variables, while in Eq. (19) they act on the second pair. Let us retain on the left-hand sides of these equations only the operators which realize the integral transformation within the same pair of the branch indices (as mentioned above, we consider $C(11'; 22')$ with s_1 and s_1' equal to either 3 or 4 and s_2 and s_2' equal to either 1 or 2). We denote this part as $\hat{L}$. The rest of the terms are transferred to the right-hand sides and denoted by M. In more detail,

$$\hat{L}(11'33')C(33'; 22') = M(11'33')C(33'; 22')$$

$$+\sigma_z(1)\delta_{12'}\hbar^2 G(21') - \sigma_z(1')\delta_{1'2}\hbar^2 G(12'),$$

$$\hat{L}(22'44')C(11'; 44') = M(22'44')C(11'; 44') \quad (20)$$

$$+\sigma_z(2)\delta_{1'2}\hbar^2 G(12') - \sigma_z(2')\delta_{12'}\hbar^2 G(21'),$$

where

$$\hat{L}(11'33') \equiv \left(i\hbar\frac{\partial}{\partial\bar{t}_1} - \hat{h}_1 + \hat{h}_{1'}^*\right)\delta_{13}\delta_{1'3'} - \sigma_z(1)\Sigma(13)\delta_{1'3'}(\delta_{s_33} + \delta_{s_34})$$

$$+\Sigma(3'1')\sigma_z(1')\delta_{13}(\delta_{s_3'3} + \delta_{s_3'4}) - \left[\sigma_z(1)G(5'1')I(15'3'3)\right.$$

$$\left. -\sigma_z(1')G(15)I(51'3'3)\right](\delta_{s_33} + \delta_{s_34})(\delta_{s_3'3} + \delta_{s_3'4}), \quad (21)$$

$$\hat{L}(22'44') \equiv \left(i\hbar\frac{\partial}{\partial\bar{t}_2} - \hat{h}_2 + \hat{h}_{2'}^*\right)\delta_{24}\delta_{2'4'} - \sigma_z(2)\Sigma(24)\delta_{2'4'}(\delta_{s_41} + \delta_{s_42})$$

$$+\Sigma(4'2')\sigma_z(2')\delta_{24}(\delta_{s_4'1} + \delta_{s_4'2}) - \left[\sigma_z(2)I(4'425')G(5'2')\right.$$

$$\left. -\sigma_z(2')I(4'452')G(25)\right](\delta_{s_41} + \delta_{s_42})(\delta_{s_4'1} + \delta_{s_4'2})$$

and

$$M(11'33') \equiv \sigma_z(1)\Sigma(13)\delta_{1'3'}(\delta_{s_31} + \delta_{s_32}) - \Sigma(3'1')\sigma_z(1')\delta_{13}(\delta_{s_3'1} + \delta_{s_3'2})$$

$$+ \left[\sigma_z(1)G(5'1')I(15'3'3) - \sigma_z(1')G(15)I(51'3'3) \right]$$

$$\times [1 - (\delta_{s33} + \delta_{s34})(\delta_{s'_33} + \delta_{s'_34})], \tag{22}$$

$$M(22'44') \equiv \sigma_z(2)\Sigma(24)\delta_{2'4'}(\delta_{s43} + \delta_{s44}) - \Sigma(4'2')\sigma_z(2')\delta_{24}(\delta_{s'_43} + \delta_{s'_44})$$

$$+ \left[\sigma_z(2)I(4'425')G(5'2') - \sigma_z(2')I(4'452')G(25) \right]$$

$$\times [1 - (\delta_{s41} + \delta_{s42})(\delta_{s'_41} + \delta_{s'_42})].$$

One has to emphasize that the internal branch indices s_5 and s'_5 in the expression for $\widehat{L}(11'33')$ take, in fact, only the values within the pair to which the indices s_1, s'_1, s_3, and s'_3 belong. If $s_5, s'_5 < s_1, s'_1$ (i.e., $s_5, s'_5 = 1, 2$), the Green's functions $G(15) = G^{+-}(1,5)$ and $G(5'1') = G^{-+}(5', 1')$ remain the same when the indices s_5 and s'_5, respectively, are changed (from 1 to 2 or from 2 to 1). Under these changes, the signs of $I(51'3'3)$ and $I(15'3'3)$, respectively, are reversed according to Eqs. (13) and (3). Therefore, the partial internal sums $\sum_{s_5=1,2} G(15)I(51'3'3)$ and $\sum_{s'_5=1,2} G(5'1')I(15'3'3)$ are equal to zero. A similar property can be checked for the operator $\widehat{L}(22'44')$. In conclusion, each of the operators $\widehat{L}(11'33')$ and $\widehat{L}(22'44')$ is defined within its own pair of the contour branch indices.

Let us find the result of the action of both $\widehat{L}(11'33')$ and $\widehat{L}(22'44')$ on the correlation function:

$$\widehat{L}(11'33')\widehat{L}(22'44')C(33'; 44') = \widehat{L}(11'33')[\delta_{23'}\hbar^2 G(32')\sigma_z(2)$$

$$-\delta_{2'3}\hbar^2 G(23')\sigma_z(2') + M(22'44')C(33'; 44')] = \widehat{L}(22'44') \tag{23}$$

$$\times [\delta_{14'}\hbar^2 G(41')\sigma_z(1) - \delta_{41'}\hbar^2 G(14')\sigma_z(1') + M(11'33')C(33'; 44')].$$

Since the numbers s_1, s'_1, s_3, and s'_3 are equal to 3 or 4, while s_2, s'_2, s_4, and s'_4 are equal to 1 or 2, the terms containing the δ-symbols vanish, and only the contributions containing M remain on the right-hand sides. Taking into account Eq. (20), we obtain the following equation for the correlation functions:

$$\widehat{L}(11'33')\widehat{L}(22'44')C(33'; 44') = \hbar^2 R(11'22')$$

$$+M(11'33')M(22'44')C(33'; 44'). \tag{24}$$

The first term on its right-hand side is introduced as

$$R(11'22') = M(22'41)G(41')\sigma_z(1) - M(22'1'4)G(14')\sigma_z(1')$$

$$= M(11'32)G(32')\sigma_z(2) - M(11'2'3')G(23')\sigma_z(2')$$

$$= -\sigma_z(2)\Sigma(21')\sigma_z(1')G(12') - \sigma_z(1)\Sigma(12')\sigma_z(2')G(21') \tag{25}$$

$$+[\sigma_z(1)G(5'1')I(15'23) - \sigma_z(1')G(15)I(51'23)]G(32')\sigma_z(2)$$

$$-[\sigma_z(1)G(5'1')I(15'3'2') - \sigma_z(1')G(15)I(51'3'2')]G(23')\sigma_z(2'),$$

where the last equation is obtained after substituting M from Eq. (22). The second term on the right-hand side of Eq. (24) is a linear functional of the correlation functions which do not belong to the kind (6). However, this term, according to the definition of M, contains higher powers of the interaction Hamiltonian as compared to $R(11'22')$ (problem 14.24). Therefore, this term can be neglected in the case of weak interaction. In this approximation, Eq. (24) becomes a closed equation for the correlation functions (6), because each of the operators $\widehat{L}(11'33')$ and $\widehat{L}(22'44')$ acts within its own pair of the contour branch indices. The approximate equation

$$\widehat{L}(11'33')\widehat{L}(22'44')C(33';44') \simeq \hbar^2 R(11'22') \tag{26}$$

will be considered below instead of Eq. (24).

The operators $\widehat{L}(11'33')$ and $\widehat{L}(22'44')$ on the left-hand side of Eq. (26) can be identified with the operators of linearized generalized kinetic equation in the ordinary non-equilibrium diagram technique. Indeed, subtracting the second equation of the set (39.21) from the first one, we can linearize the equation obtained and write it in the form

$$\left(i\hbar\frac{\partial}{\partial \bar{t}_1} - \hat{h}_1 + \hat{h}_{1'}^*\right)\delta G(11') - \sigma_z(1)\Sigma(13)\delta G(31') + \delta G(13')\Sigma(3'1')\sigma_z(1')$$

$$- \left[\sigma_z(1)\frac{\delta\Sigma(15')}{\delta G(33')}G(5'1') - \sigma_z(1')G(15)\frac{\delta\Sigma(51')}{\delta G(33')}\right]\delta G(33') = 0, \tag{27}$$

where the multi-indices include the branch indices of the two-branch contour. Applying Eq. (13), one can see that Eq. (27) is formally equivalent to $\widehat{L}(11'33')\delta G(33') = 0$. The equivalence of the relations obtained within four-branch and two-branch techniques takes place because the internal branch indices s_5 and s_5' in the expression for $\widehat{L}(11'33')$ and the indices s_1, s_1', s_3, and s_3' belong to the same pair of branches. In spite of the formal analogy, $G(11')$ in the generalized kinetic equation is a function, while $\widehat{\delta G}(11')$ entering the definition (6) is an operator, and the fluctuations of $\widehat{\delta G}(11')$ are not small. Nevertheless, by observing that $\widehat{\delta\Sigma}(11') = [\delta\Sigma(11')/\delta G(33')]\widehat{\delta G}(33')$, one can write

$$\sum_{s_3 s_3'}\widehat{L}^{34 s_3 s_3'}(11'33')\widehat{\delta G}^{s_3 s_3'}(33') = \left(i\hbar\frac{\partial}{\partial\bar{t}_1} - \hat{h}_1 + \hat{h}_{1'}^*\right)\widehat{\delta G}^{-+}(11')$$

$$-\sum_{s=\pm}\left[\Sigma^{-s}(13)\widehat{\delta G}^{s+}(31') + \widehat{\delta\Sigma}^{-s}(13)G^{s+}(31')\right. \tag{28}$$

$$\left.+\widehat{\delta G}^{-s}(13)\Sigma^{s+}(31') + G^{-s}(13)\widehat{\delta\Sigma}^{s+}(31')\right].$$

The multi-indices in this equation do not include the contour branch indices, the latter are written explicitly. The integration over the internal variable 3 is implied in the second term on the right-hand side of Eq. (28). Let us consider this integral in the quasi-classical approximation (see Sec. 40 and problem 8.8) and assume that $\hat{h}_1 = \hat{p}_1/2m$ (no external fields) and $t_1 = t_1' = t$. Then we transform Eq. (28) to

$$-\int d\Delta\mathbf{r}\,e^{-i\mathbf{p}\cdot\Delta\mathbf{r}/\hbar}\sum_{s_3 s_3'}\widehat{L}^{34s_3 s_3'}(11'33')\widehat{\delta G}^{s_3 s_3'}(33')$$

$$=\left(\frac{\partial}{\partial t} + \mathbf{v_p}\cdot\frac{\partial}{\partial\mathbf{r}}\right)\widehat{\Delta f}_{\mathbf{r}\mathbf{p}t} + \sum_{s=\pm}\int\frac{d\varepsilon}{2\pi\hbar}\left[\Sigma_{\varepsilon t}^{-s}(\mathbf{r},\mathbf{p})\widehat{\delta G}_{\varepsilon t}^{s+}(\mathbf{r},\mathbf{p})\right. \tag{29}$$

$$\left.+\widehat{\delta\Sigma}_{\varepsilon t}^{-s}(\mathbf{r},\mathbf{p})G_{\varepsilon t}^{s+}(\mathbf{r},\mathbf{p}) + \widehat{\delta G}_{\varepsilon t}^{-s}(\mathbf{r},\mathbf{p})\Sigma_{\varepsilon t}^{s+}(\mathbf{r},\mathbf{p}) + G_{\varepsilon t}^{-s}(\mathbf{r},\mathbf{p})\widehat{\delta\Sigma}_{\varepsilon t}^{s+}(\mathbf{r},\mathbf{p})\right],$$

where $\Delta\mathbf{r} = \mathbf{r}_1 - \mathbf{r}_1'$ and $\mathbf{r} = (\mathbf{r}_1 + \mathbf{r}_1')/2$. The transformations have been based upon the definitions (69.1) and (7) so that $-i\hbar\int d\Delta\mathbf{r}\,e^{-\frac{i}{\hbar}\mathbf{p}\cdot\Delta\mathbf{r}}$ $\times\widehat{\delta G}^{-+}(\mathbf{r}+\Delta\mathbf{r}/2\,t,\mathbf{r}-\Delta\mathbf{r}/2\,t) = \widehat{\Delta f}_{\mathbf{r}\mathbf{p}t}$. If we consider the lowest-order contributions to Σ for the case of electron-phonon interaction and take into account linear relations between the components of electron and phonon Green's functions, we can prove, like in problem 8.8, that the collision-integral part of Eq. (29) depends only on $\widehat{\delta G}^{-+}$ and $\widehat{\delta G}^{+-}$. Therefore, application of the averaging procedure (69.2) to Eq. (29) reduces its right-hand side to $\sum_{\mathbf{p}_1}\widehat{L}_{\mathbf{r}t}(\mathbf{p},\mathbf{p}_1)\widehat{\Delta f}_{\mathbf{r}\mathbf{p}_1 t}^{(av)}$, where $\widehat{L}_{\mathbf{r}t}(\mathbf{p},\mathbf{p}_1)$ is the operator introduced in Eq. (69.19). The expression $\sum_{s_4 s_4'}\widehat{L}^{12s_4 s_4'}(22'44')\widehat{\delta G}^{s_4 s_4'}(44')$ can be analyzed in a similar way.

The spatial Fourier transform of the left-hand side of Eq. (26) is written as

$$\int d\Delta\mathbf{r}\int d\Delta\mathbf{r}'\,e^{-i\mathbf{p}\cdot\Delta\mathbf{r}/\hbar - i\mathbf{p}'\cdot\Delta\mathbf{r}'/\hbar} \tag{30}$$

$$\times\sum_{s_3 s_3' s_4 s_4'}\widehat{L}^{34s_3 s_3'}(11'33')\widehat{L}^{12s_4 s_4'}(22'44')C^{s_3 s_3' s_4 s_4'}(33';44'),$$

where $\Delta\mathbf{r} = \mathbf{r}_1 - \mathbf{r}_1'$ and $\Delta\mathbf{r}' = \mathbf{r}_2 - \mathbf{r}_2'$, while $C^{s_3 s_3' s_4 s_4'}(33';44')$ is defined according to Eq. (6). Let us set $t_1 = t_1' = t$, $t_2 = t_2' = t'$, $(\mathbf{r}_1+\mathbf{r}_1')/2 = \mathbf{r}$, and $(\mathbf{r}_2 + \mathbf{r}_2')/2 = \mathbf{r}'$ in Eq. (30) and apply the averaging procedure

(69.2) to this equation in the quasi-classical limit. The expression (30) is reduced to $-\hbar^2 \sum_{\mathbf{p}_1} \sum_{\mathbf{p}_1'} \widehat{L}_{\mathbf{rt}}(\mathbf{p},\mathbf{p}_1)\widehat{L}_{\mathbf{r}'t'}(\mathbf{p}',\mathbf{p}_1')\langle\langle\widehat{\Delta f}_{\mathbf{rp}_1 t}^{(av)}\widehat{\Delta f}_{\mathbf{r}'\mathbf{p}_1't'}^{(av)}\rangle\rangle$, i.e., to the left-hand side of Eq. (69.20) multiplied by $-\hbar^2$.

Let us consider the expression on the right-hand side of Eq. (26) and prove that this expression is directly related to the correlation function of the Langevin sources at $t_1 = t_1' = t$ and $t_2 = t_2' = t'$. In other words, one has to check the identity

$$\langle\langle K_t(\mathbf{p},\mathbf{r})K_{t'}(\mathbf{p}',\mathbf{r}')\rangle\rangle = -\int d\Delta\mathbf{r}\int d\Delta\mathbf{r}' e^{-i\mathbf{p}\cdot\Delta\mathbf{r}/\hbar - i\mathbf{p}'\cdot\Delta\mathbf{r}'/\hbar} \qquad (31)$$

$$\times R^{3412}\left(\mathbf{r} + \frac{\Delta\mathbf{r}}{2}\,t, \mathbf{r} - \frac{\Delta\mathbf{r}}{2}\,t, \mathbf{r}' + \frac{\Delta\mathbf{r}'}{2}\,t', \mathbf{r}' - \frac{\Delta\mathbf{r}'}{2}\,t'\right).$$

Below we prove it directly, by considering the electron-phonon interaction in the lowest-order approximation. In this case, the Fourier transform of the correlation function on the right-hand side of Eq. (31),

$$\langle\langle K_\omega(\mathbf{p},\mathbf{k})K_{\omega'}(\mathbf{p}',\mathbf{k}')\rangle\rangle = \frac{1}{V^2}\int d\mathbf{r}\int d\mathbf{r}'\int dt\int dt'\, e^{i(\omega t + \omega't' - \mathbf{k}\cdot\mathbf{r} - \mathbf{k}'\cdot\mathbf{r}')}$$

$$\times\langle\langle K_t(\mathbf{p},\mathbf{r})K_{t'}(\mathbf{p}',\mathbf{r}')\rangle\rangle\,, \qquad (32)$$

is given by Eq. (68.23). On the other hand, using the result of problem 14.22, one can find R from Eq. (25):

$$R^{3412}(\mathbf{r}_1 t, \mathbf{r}_1' t, \mathbf{r}_2 t', \mathbf{r}_2' t') \simeq -iG_{t't}^{-+}(\mathbf{r}_2,\mathbf{r}_1')G_{tt'}^{+-}(\mathbf{r}_1,\mathbf{r}_2')\left[\bar{D}_{t't}^{-+}(\mathbf{r}_2,\mathbf{r}_1)\right.$$

$$\left. +\bar{D}_{t't}^{-+}(\mathbf{r}_2',\mathbf{r}_1) - \bar{D}_{t't}^{-+}(\mathbf{r}_2,\mathbf{r}_1) - \bar{D}_{t't}^{-+}(\mathbf{r}_2',\mathbf{r}_1')\right], \qquad (33)$$

where $\bar{D}$ is defined in problem 14.24 (this function is proportional to the Green's function of phonons introduced in Sec. 42). It is important to notice that $R^{s_1 s_1' s_2 s_2'}$ does not depend on its four contour branch indices in the lowest-order approximation. This invariance takes place in a more general case, when $I(33'44')$, as a matrix in the space of the indices s_3, s_3', s_4, and s_4', can be represented in the form $\overline{I}(33'44') + \widetilde{I}(33'44')$, where $\overline{I}$ contains only the terms with $s_3 = s_4'$ and $s_3' = s_4$, while $\widetilde{I}$ contains only the terms with $s_3 = s_4$ and $s_3' = s_4'$ (problem 14.25). This representation is always valid in the second-order approximation with respect to the interaction potential; see Eq. (12) and problem 14.22. We note that the contribution proportional to $\delta_{33'}\delta_{44'}$ in Eq. (12) can be safely omitted since it does not affect $\widehat{L}$, R, and M. For the same reason, one can omit the contributions to $\Sigma(12)$ proportional to δ_{12}. Because of the invariance discussed above, it is not necessary to show the contour branch indices of R, and such indices will be omitted in the following.

In the homogeneous case, the Green's functions depend only on the differences of their coordinate and time arguments. Expressing these functions in the energy-momentum representation, we apply the double spatial and temporal Fourier transformations to Eq. (33) and obtain

$$-\frac{1}{V^2}\int d\mathbf{r}\int d\mathbf{r}'\int dt\int dt'\, e^{i(\omega t+\omega' t'-\mathbf{k}\cdot\mathbf{r}-\mathbf{k}'\cdot\mathbf{r}')}$$

$$\times R\left(\mathbf{r}+\frac{\Delta\mathbf{r}}{2}\,t,\mathbf{r}-\frac{\Delta\mathbf{r}}{2}\,t,\mathbf{r}'+\frac{\Delta\mathbf{r}'}{2}\,t',\mathbf{r}'-\frac{\Delta\mathbf{r}'}{2}\,t'\right)$$

$$=i\delta(\omega+\omega')\delta_{\mathbf{k},-\mathbf{k}'}\int\frac{d\varepsilon_1}{2\pi}\int d\omega_1\frac{1}{V^2}\sum_{\mathbf{p}_1\mathbf{q}}|C_q|^2 e^{i(\mathbf{p}_1/\hbar+\mathbf{q}/2)\cdot(\Delta\mathbf{r}+\Delta\mathbf{r}')}$$

$$\times G^{-+}_{\varepsilon_1}\left(\mathbf{p}_1-\frac{\hbar\mathbf{k}}{2}\right)G^{+-}_{\varepsilon_1+\hbar(\omega_1+\omega)}\left(\mathbf{p}_1+\hbar\mathbf{q}+\frac{\hbar\mathbf{k}}{2}\right)\mathrm{D}^{-+}_{\omega_1}(\mathbf{q})\qquad(34)$$

$$\times\left[e^{i\mathbf{q}\cdot(\Delta\mathbf{r}+\Delta\mathbf{r}')/2}+e^{-i\mathbf{q}\cdot(\Delta\mathbf{r}+\Delta\mathbf{r}')/2}-e^{i\mathbf{q}\cdot(\Delta\mathbf{r}-\Delta\mathbf{r}')/2}-e^{-i\mathbf{q}\cdot(\Delta\mathbf{r}-\Delta\mathbf{r}')/2}\right],$$

where $|C_q|^2$ for the acoustic-phonon scattering is given by Eq. (21.1). The Green's function $\mathrm{D}^{-+}_\omega(\mathbf{q})$ is introduced in Sec. 42. Within the lowest-order accuracy, one can replace the Green's functions of electrons and phonons in Eq. (34) by those in the absence of interaction, $G^{ss'}_\varepsilon(\mathbf{p})\simeq g^{ss'}_\varepsilon(\mathbf{p})$ and $\mathrm{D}^{ss'}_\omega(\mathbf{q})\simeq \mathrm{d}^{ss'}_\omega(\mathbf{q})$. Finally, employing the expressions (40.33) and (42.23) for these functions, we find that the Fourier transform $\int d\Delta\mathbf{r}\int d\Delta\mathbf{r}' e^{-i\mathbf{p}\cdot\Delta\mathbf{r}/\hbar-i\mathbf{p}'\cdot\Delta\mathbf{r}'/\hbar}\ldots$ of the right-hand side of Eq. (34) coincides with the expression on the right-hand side of Eq. (68.23), i.e., with $\langle\langle K_\omega(\mathbf{p},\mathbf{k})K_{\omega'}(\mathbf{p}',\mathbf{k}')\rangle\rangle$. Therefore, Eq. (31) is proved. In summary, we have demonstrated that Eq. (26) with $s_1=3$, $s_1'=4$, $s_2=1$, and $s_2'=2$ is a quantum analog of Eq. (69.20).

Equation (26) is the central result of this section. It allows one to calculate the correlation functions (6) by taking into account external fields and interactions with a required accuracy. Let us use the invariance of $R(11'22')$ against all its four contour branch indices (valid in the case of weak interaction) and write Eq. (26) as a system of nine equations for nine linearly-independent correlation functions $C^{\bar s\bar s'}$, where the indices $\bar s$ take the values R, A, and F (see Eq. (9) and its discussion). Introducing the operators $\widehat{L}^{\bar s_1\bar s_1'}$ in a similar way as $C^{\bar s\bar s'}$ (for example, $\widehat{L}^{RR}=\widehat{L}^{3311}-\widehat{L}^{3411}-\widehat{L}^{3312}+\widehat{L}^{3412}$), we obtain

$$\widehat{L}^{\bar s_1\bar s_1'}(11'33')\widehat{L}^{\bar s_2\bar s_2'}(22'44')C^{\bar s_1'\bar s_2'}(33';44')\simeq\delta_{\bar s_1 F}\delta_{\bar s_2 F}4\hbar^2 R(11'22').\qquad(35)$$

These equations are, in general, coupled. The non-zero right-hand side exists only in one of these equations, where $\bar s_1=\bar s_2=F$.

The spectral densities of the fluctuations of physical quantities, for example, the noise power (68.9), are expressed through the Fourier transforms of the symmetrized correlation functions. Owing to the symmetry of the left-hand side of Eq. (26) with respect to permutation of the variables, the symmetrized correlation function $C_s(33'; 44') = [C(33'; 44') + C(44'; 33')]/2$ satisfies the equation $\widehat{L}(11'33')\widehat{L}(22'44')C_s(33'; 44') \simeq \hbar^2 R_s(11', 22')$ with the symmetrized right-hand side $R_s(11'22') = [R(11'22') + R(22'11')]/2$. This equation can be used for calculating any correlation functions of the kind (68.3). In order to consider correlation functions of N operators of physical quantities (these functions are representable through $2N$ field operators), it is possible to introduce $2N$-branch time contours. Such a technique would be a generalization of the approach used in Chapter 8 and in this section.

To conclude this section and the monograph itself, we stress that any physical interpretation of the double-branch or, in general, many-branch time contours (see Figs. 8.1 and 14.3) is currently missing. Using other words, one may ask: *does the multi-time dynamics introduced in order to formulate the non-equilibrium diagram technique lead to new physical consequences?* At the present state of development of the quantum kinetic theory, *the approach of many-branch time contours remains just a convenient formal trick.* It is not clear whether it can have more profound influence on such issues of modern physics as irreversibility, memory effects, and quantum theory of measurements.

Problems

14.1. Find the mean square of the fluctuations of the number of gas particles in a half-volume of a box if the total number of the particles in the box is N.

Solution: The probability that n particles out of N occupy a given half of the volume is $P_N(n) = C_N^n/2^N$, where $C_N^n = N!/n!(N-n)!$ is the binomial coefficient and $2^N = \sum_{n=0}^N C_N^n$ is the normalization constant. The probability satisfies the normalization condition $\sum_{n=0}^N P_N(n) = 1$. The mean value of the number of particles is $\overline{n} = \sum_{n=0}^N n P_N(n) = N/2$, where the line denotes the averaging with the probability $P_N(n)$. The mean square of n is $\overline{n^2} = \sum_{n=0}^N n^2 P_N(n) = N(N+1)/4$, and the mean square of the fluctuation $\Delta n = n - \overline{n}$ is $\overline{(\Delta n)^2} = \overline{n^2} - (\overline{n})^2 = N/4$. As N increases, the ratio $\overline{(\Delta n)^2}/(\overline{n})^2 = 1/N$ goes to zero.

14.2. Express the noise power $\mathcal{S}_{\alpha\beta}(\mathbf{r}, \mathbf{r}'|\omega)$ in terms of Fourier transforms of the current density operators.

Result:

$$2\pi\delta(\omega + \omega')\mathcal{S}_{\alpha\beta}(\mathbf{r}, \mathbf{r}'|\omega) = \frac{1}{2}\langle\langle \hat{I}_\alpha(\mathbf{r}, \omega)\hat{I}_\beta(\mathbf{r}', \omega') + \hat{I}_\beta(\mathbf{r}', \omega')\hat{I}_\alpha(\mathbf{r}, \omega)\rangle\rangle.$$

14.3. Express the correlation function $\langle\langle \widehat{\Delta n}_\mathbf{r}(t)\widehat{\Delta n}_{\mathbf{r}'}(t')\rangle\rangle$, which describes the density fluctuations in a homogeneous ideal electron gas, through the polarizability $\alpha(\mathbf{q},\omega)$. Calculate this correlation function at $t = t'$ in the Hartree-Fock approximation at zero temperature.

Solution: The polarizability (see problem 3.3) is defined as a generalized susceptibility describing the charge density response to the scalar potential (generalized force). The operator $\widehat{\Delta n}_\mathbf{r}(t)$ is equal to the Heisenberg operator of the charge density deviation divided by the electron charge e. Therefore, the fluctuation-dissipation theorem (68.10) can be applied directly. We obtain

$$\langle\langle \widehat{\Delta n}_\mathbf{r}(t)\widehat{\Delta n}_{\mathbf{r}'}(t')\rangle\rangle = -\frac{\hbar}{V}\sum_\mathbf{q} e^{i\mathbf{q}\cdot(\mathbf{r}-\mathbf{r}')}\int \frac{d\omega}{2\pi}e^{-i\omega(t-t')}\frac{2\mathrm{Im}\,\alpha(\mathbf{q},\omega)/e^2}{1-\exp(-\hbar\omega/T)}.$$

The Hartree-Fock polarizability calculated in Sec. 33 is $\alpha(\mathbf{q},\omega) = e^2\Pi_\omega^{(0)R}(\mathbf{q})$, and the polarization function $\Pi_\omega^{(0)R}(\mathbf{q})$ at zero temperature is given by Eq. (33.24). To calculate the integral over ω at $t = t'$, we use the expression for the imaginary part of this function and take into account that $\omega[1-\exp(-\hbar\omega/T)]^{-1} = \omega\theta(\omega)$ at $T\to 0$. The result is

$$\langle\langle \widehat{\Delta n}_\mathbf{r}(t)\widehat{\Delta n}_{\mathbf{r}'}(t)\rangle\rangle = \frac{1}{V}\sum_\mathbf{q} e^{i\mathbf{q}\cdot(\mathbf{r}-\mathbf{r}')}\frac{k_F^2 q}{4\pi^2}\left(1 - \frac{q^2}{12k_F^2}\right).$$

The main contribution to this expression comes from $q^2 \ll k_F^2$, since we are interested in the case when $|\mathbf{r} - \mathbf{r}'|$ is large in comparison to the wavelength of Fermi electrons. Finally,

$$\langle\langle \widehat{\Delta n}_\mathbf{r}(t)\widehat{\Delta n}_{\mathbf{r}'}(t)\rangle\rangle \simeq -\frac{k_F^2}{4\pi^4|\mathbf{r}-\mathbf{r}'|^4}.$$

The correlation function rapidly decreases as the distance $|\mathbf{r} - \mathbf{r}'|$ increases. The negative sign of the correlation function reflects an effective repulsion between the electrons due to Pauli principle.

14.4. Find the correlation function of the Langevin sources $\widehat{K}_\omega(\mathbf{p},\mathbf{k})$ for the electrons interacting with impurities.

Solution: The consideration is similar to the case of electron-boson interaction, since the Hamiltonian of electron-impurity interaction in the second quantization representation is given by $V^{-1}\sum_{\mathbf{pp}'} U_{im}(\mathbf{p} - \mathbf{p}')\hat{a}_\mathbf{p}^+\hat{a}_{\mathbf{p}'}$, where $U_{im}(\mathbf{p} - \mathbf{p}')$ given by Eq. (14.3) is the Fourier transform of the random potential created by all impurities. The Langevin source is

$$\widehat{K}_\omega(\mathbf{p},\mathbf{k}) = -\frac{2\pi i}{V}\sum_\mathbf{q} U_{im}(\hbar\mathbf{q})\Big[\hat{a}_{\mathbf{p}-\hbar\mathbf{k}/2}^+\hat{a}_{\mathbf{p}+\hbar\mathbf{k}/2-\hbar\mathbf{q}}\delta(\varepsilon_{\mathbf{p}-\hbar\mathbf{k}/2} - \varepsilon_{\mathbf{p}+\hbar\mathbf{k}/2-\hbar\mathbf{q}} + \hbar\omega)$$

$$-\hat{a}_{\mathbf{p}-\hbar\mathbf{k}/2+\hbar\mathbf{q}}^+\hat{a}_{\mathbf{p}+\hbar\mathbf{k}/2}\delta(\varepsilon_{\mathbf{p}-\hbar\mathbf{k}/2+\hbar\mathbf{q}} - \varepsilon_{\mathbf{p}+\hbar\mathbf{k}/2} + \hbar\omega)\Big],$$

and the correlation function is found in the form

$$\langle\langle \widehat{K}_\omega(\mathbf{p},\mathbf{k})\widehat{K}_{\omega'}(\mathbf{p}',\mathbf{k}')\rangle\rangle = \frac{(2\pi)^2}{\hbar V}\delta(\omega+\omega')\delta_{\mathbf{k},-\mathbf{k}'}\sum_\mathbf{q} n_{im}|v(q)|^2(\delta_{\mathbf{p}',\mathbf{p}} - \delta_{\mathbf{p}',\mathbf{p}-\hbar\mathbf{q}})$$

$$\times \left[\delta(\varepsilon_{\mathbf{p}-\hbar\mathbf{k}/2} - \varepsilon_{\mathbf{p}+\hbar\mathbf{k}/2-\hbar\mathbf{q}} + \hbar\omega) f_{\mathbf{p}-\hbar\mathbf{k}/2}(1 - f_{\mathbf{p}+\hbar\mathbf{k}/2-\hbar\mathbf{q}}) \right.$$

$$\left. + \delta(\varepsilon_{\mathbf{p}-\hbar\mathbf{k}/2-\hbar\mathbf{q}} - \varepsilon_{\mathbf{p}+\hbar\mathbf{k}/2} + \hbar\omega) f_{\mathbf{p}-\hbar\mathbf{k}/2-\hbar\mathbf{q}}(1 - f_{\mathbf{p}+\hbar\mathbf{k}/2}) \right].$$

The averaging in this expression includes both the trace with the statistical operator and the averaging over the random potential distribution.

14.5. Prove the expression $\hat{\mathbf{I}}(\mathbf{k}, t) = V^{-1} \sum_{\mathbf{p}} \frac{e\mathbf{p}}{m} \{\hat{a}^+_{\mathbf{p}-\hbar\mathbf{k}/2} \hat{a}_{\mathbf{p}+\hbar\mathbf{k}/2}\}t$ for the operator of current density.

Hint: Expand the field operators in Eq. (42.25) over the plane-wave basis and carry out a spatial Fourier transformation of this equation.

14.6. *i)* Check that the local current-current correlation function (68.27) in thermodynamic equilibrium is expressed according to the fluctuation-dissipation theorem (68.9). *ii)* Check the same statement for the collisionless part (68.26) of the correlation function.

Hints: *i)* Use the relation

$$[(N_{\mathbf{q}} + 1)f_{\mathbf{p}}(1 - f_{\mathbf{p}-\hbar\mathbf{q}}) + N_{\mathbf{q}} f_{\mathbf{p}-\hbar\mathbf{q}}(1 - f_{\mathbf{p}})]$$

$$= \coth\frac{\hbar\omega}{2T}[(N_{\mathbf{q}} + 1)f_{\mathbf{p}}(1 - f_{\mathbf{p}-\hbar\mathbf{q}}) - N_{\mathbf{q}} f_{\mathbf{p}-\hbar\mathbf{q}}(1 - f_{\mathbf{p}})]$$

valid at $\varepsilon_{\mathbf{p}} - \varepsilon_{\mathbf{p}-\hbar\mathbf{q}} - \hbar\omega_q + \hbar\omega = 0$ in equilibrium.
ii) Use the relation

$$f_{\mathbf{p}-\hbar\mathbf{k}/2}(1 - f_{\mathbf{p}+\hbar\mathbf{k}/2}) + f_{\mathbf{p}+\hbar\mathbf{k}/2}(1 - f_{\mathbf{p}-\hbar\mathbf{k}/2})$$

$$= \coth\frac{\hbar\omega}{2T}[f_{\mathbf{p}-\hbar\mathbf{k}/2} - f_{\mathbf{p}+\hbar\mathbf{k}/2}] \simeq \coth\frac{\hbar\omega}{2T} \hbar\mathbf{k} \cdot \mathbf{v_p} \left(-\frac{\partial f_{\mathbf{p}}}{\partial \varepsilon_{\mathbf{p}}}\right)$$

valid at $\varepsilon_{\mathbf{p}+\hbar\mathbf{k}/2} - \varepsilon_{\mathbf{p}-\hbar\mathbf{k}/2} = \hbar\mathbf{v_p} \cdot \mathbf{k} = \hbar\omega$ in equilibrium. Compare the result for $\mathcal{S}^{(0)}_{\alpha\beta}(\mathbf{k}, \omega)$ obtained in this way from Eq. (68.26) to the equation for $\sigma_{\alpha\beta}(\mathbf{q}, \omega)$ in problem 6.21.

14.7. Estimate $\langle\langle(\widehat{\Delta f}_{\mathbf{rp}t})^2\rangle\rangle$ for ideal electron gas.

Solution: According to the definition of $\hat{f}_{\mathbf{rp}t}$, we obtain

$$\langle\langle(\hat{f}_{\mathbf{rp}t})^2\rangle\rangle = \sum_{\mathbf{kk}'} e^{i(\mathbf{k}+\mathbf{k}')\cdot\mathbf{r}} \langle\langle\hat{a}^+_{\mathbf{p}-\hbar\mathbf{k}/2}\hat{a}_{\mathbf{p}+\hbar\mathbf{k}/2}\hat{a}^+_{\mathbf{p}-\hbar\mathbf{k}'/2}\hat{a}_{\mathbf{p}+\hbar\mathbf{k}'/2}\rangle\rangle$$

$$= \sum_{\mathbf{k}} f_{\mathbf{p}-\hbar\mathbf{k}/2}(1 - f_{\mathbf{p}+\hbar\mathbf{k}/2}) + f_{\mathbf{p}}^2,$$

where the correlation function of four fermionic operators is calculated under approximation of ideal (non-interacting) electron gas. The first term on the right-hand side is of the order of the total number of electrons in the system. Therefore, $\langle\langle(\widehat{\Delta f}_{\mathbf{rp}t})^2\rangle\rangle = \langle\langle(\hat{f}_{\mathbf{rp}t})^2\rangle\rangle - f_{\mathbf{p}}^2$ is large in comparison to $f_{\mathbf{p}}^2$, because $f_{\mathbf{p}}^2 < 1$.

14.8. Prove that the contribution of the last term on the right-hand side of Eq. (69.11) into $\widehat{M}_t(\mathbf{p}, \mathbf{r})$ can be neglected after averaging $\widehat{M}_t(\mathbf{p}, \mathbf{r})$ according to Eq. (69.2).

<u>Solution</u>: This contribution is written as

$$-\frac{1}{V} \sum_{\mathbf{q}\, \Delta \mathbf{q}} C_{|\mathbf{q}+\Delta \mathbf{q}/2|} C_{|\mathbf{q}-\Delta \mathbf{q}/2|} \int d\mathbf{r}' e^{i\Delta \mathbf{q} \cdot (\mathbf{r}-\mathbf{r}')} \widehat{\Delta f}_{\mathbf{r}\ \mathbf{p}-\hbar\Delta \mathbf{q}/2\ t} \widehat{\Delta N}_{\mathbf{r}'\mathbf{q}t}$$

$$\times \frac{1}{\hbar^2} \int_{t_0}^{t} dt' \exp\left\{ \frac{i}{\hbar} [\varepsilon_{\mathbf{p}} - \varepsilon_{\mathbf{p}-\hbar(\mathbf{q}-\Delta \mathbf{q}/2)} - \hbar\omega_{\mathbf{q}+\Delta \mathbf{q}/2}](t - t') \right\},$$

where we have done the substitutions $\mathbf{q} + \mathbf{q}' = \Delta \mathbf{q}$ and $(\mathbf{q} - \mathbf{q}')/2 \to \mathbf{q}$. Averaging over the phase volume, we assume that $|\Delta \mathbf{q}| \ll \pi/\Delta L$. This allows us to neglect $\Delta \mathbf{q}$ everywhere except the exponential factor $e^{i\Delta \mathbf{q} \cdot (\mathbf{r}-\mathbf{r}')}$, and the averaged contribution becomes

$$-\frac{1}{V} \sum_{\mathbf{q}\, \Delta \mathbf{q}} |C_q|^2 \int d\mathbf{r}' e^{i\Delta \mathbf{q} \cdot (\mathbf{r}-\mathbf{r}')} \widehat{\Delta f}_{\mathbf{r}\mathbf{p}t}^{(av)} \widehat{\Delta N}_{\mathbf{r}'\mathbf{q}t}^{(av)}$$

$$\times \frac{1}{\hbar^2} \int_{t_0}^{t} dt' \exp\left[\frac{i}{\hbar} (\varepsilon_{\mathbf{p}} - \varepsilon_{\mathbf{p}-\hbar\mathbf{q}} - \hbar\omega_{\mathbf{q}})(t - t') \right].$$

The integral over $\Delta \mathbf{q}$ in this expression converges at $|\Delta \mathbf{q}| < \pi/|\mathbf{r} - \mathbf{r}'|$, and $\widehat{\Delta N}_{\mathbf{r}'\mathbf{q}t}^{(av)}$ varies with $\mathbf{r}'$ on a scale much larger than ΔL. Therefore, $\Delta \mathbf{q}$ contributing to the integral is small so that the assumption $|\Delta \mathbf{q}| \ll \pi/\Delta L$ is justified. The averaged contribution given by the equation above is quadratic in small fluctuations and can be neglected.

14.9. Calculate the commutator of $\hat{a}^+_{\mathbf{p}-\hbar\mathbf{k}/2} \hat{a}_{\mathbf{p}+\hbar\mathbf{k}/2}$ with the first term of the perturbation Hamiltonian (69.17). Write the result through the operator $\hat{f}_{\mathbf{r}\mathbf{p}t}$ defined by Eq. (69.1).

<u>Result</u>:

$$\frac{e}{mcV^2} \sum_{\mathbf{q}} \int d\mathbf{r}' e^{-i(\mathbf{k}-\mathbf{q})\cdot \mathbf{r}'} \mathbf{A}_{\mathbf{q}t} \cdot \left[\left(\mathbf{p} - \frac{\hbar\mathbf{q}}{2} + \frac{\hbar\mathbf{k}}{2} \right) \hat{f}_{\mathbf{r}'\mathbf{p}-\hbar\mathbf{q}/2\ t} \right.$$

$$\left. - \left(\mathbf{p} + \frac{\hbar\mathbf{q}}{2} - \frac{\hbar\mathbf{k}}{2} \right) \hat{f}_{\mathbf{r}'\mathbf{p}+\hbar\mathbf{q}/2\ t} \right].$$

14.10. Transform the collision integral (31.21) with $V(\mathbf{p}\mathbf{p}'|\mathbf{p}_1\mathbf{p}_1')$ from Eq. (33.32) to the form (69.31).

<u>Hints</u>: In the limit of small momentum transfer $\hbar\mathbf{k} = \mathbf{p} - \mathbf{p}_1 = \mathbf{p}_1' - \mathbf{p}'$, expand the distribution functions in series of $\mathbf{k}$ up to the terms quadratic in $\mathbf{k}$. Also use the expansions

$$|\epsilon(\mathbf{k}, \mathbf{v}_{\mathbf{p}} \cdot \mathbf{k} - \hbar k^2/2m)|^{-2} \simeq |\epsilon(\mathbf{k}, \mathbf{v}_{\mathbf{p}} \cdot \mathbf{k})|^{-2} - \frac{\hbar\mathbf{k}}{2} \cdot \frac{\partial}{\partial \mathbf{p}} |\epsilon(\mathbf{k}, \mathbf{v}_{\mathbf{p}} \cdot \mathbf{k})|^{-2}$$

and

$$\delta(\mathbf{v}_{\mathbf{p}'} \cdot \mathbf{k} - \mathbf{v}_{\mathbf{p}} \cdot \mathbf{k} + \hbar k^2/m) \simeq \delta(\mathbf{v}_{\mathbf{p}'} \cdot \mathbf{k} - \mathbf{v}_{\mathbf{p}} \cdot \mathbf{k}) + \hbar\mathbf{k} \cdot \frac{\partial}{\partial \mathbf{p}'} \delta(\mathbf{v}_{\mathbf{p}'} \cdot \mathbf{k} - \mathbf{v}_{\mathbf{p}} \cdot \mathbf{k}).$$

Integrate over $\mathbf{p}'$ by parts to transform the second term in the latter expansion.

14.11. Determine the energy loss of a plane electromagnetic wave in the medium with small variations of the dielectric permittivity, $\epsilon_{\mathbf{r}} = \epsilon + \delta\epsilon_{\mathbf{r}}$.

Solution: The electric field of the electromagnetic wave is represented as $\mathbf{E}_{\mathbf{r}} = V^{-1} \sum_{\mathbf{k}} \mathbf{E}_{\mathbf{k}} e^{i\mathbf{k}\cdot\mathbf{r} - i\omega t} + c.c.$ Assuming a local isotropic relation between the field and electric induction, $\mathbf{D}_{\mathbf{r}} = \epsilon_{\mathbf{r}} \mathbf{E}_{\mathbf{r}}$, we find that the amplitude $\mathbf{E}_{\mathbf{k}}$ satisfies the wave equation

$$\mathbf{k}\left(\mathbf{k}\cdot\mathbf{E}_{\mathbf{k}}\right) - k^2\mathbf{E}_{\mathbf{k}} + \left(\frac{\omega}{c}\right)^2 \epsilon\mathbf{E}_{\mathbf{k}} + \frac{(\omega/c)^2}{V}\sum_{\mathbf{k}'}\delta\epsilon_{\mathbf{k}-\mathbf{k}'}\mathbf{E}_{\mathbf{k}'} = 0,$$

where $\delta\epsilon_{\mathbf{q}}$ is the spatial Fourier transform of the inhomogeneous part of the dielectric permittivity. Expressing the field as $\mathbf{E}_{\mathbf{k}} = \overline{\mathbf{E}}_{\mathbf{k}} + \delta\mathbf{E}_{\mathbf{k}}$, we average the wave equation over possible realizations of the spatial distribution of random inhomogeneities and obtain

$$\mathbf{k}\left(\mathbf{k}\cdot\overline{\mathbf{E}}_{\mathbf{k}}\right) - k^2\overline{\mathbf{E}}_{\mathbf{k}} + \left(\frac{\omega}{c}\right)^2 \epsilon\overline{\mathbf{E}}_{\mathbf{k}} = -\frac{(\omega/c)^2}{V}\sum_{\mathbf{k}'}\langle\delta\epsilon_{\mathbf{k}-\mathbf{k}'}\delta\mathbf{E}_{\mathbf{k}'}\rangle.$$

The right-hand side of this equation contains the correlation function of the variations of the field and dielectric permittivity. The random part of the field is determined from the linearized wave equation

$$\mathbf{k}\left(\mathbf{k}\cdot\delta\mathbf{E}_{\mathbf{k}}\right) - k^2\delta\mathbf{E}_{\mathbf{k}} + \left(\frac{\omega}{c}\right)^2 \epsilon\delta\mathbf{E}_{\mathbf{k}} = -\frac{(\omega/c)^2}{V}\sum_{\mathbf{k}'}\delta\epsilon_{\mathbf{k}-\mathbf{k}'}\overline{\mathbf{E}}_{\mathbf{k}'}.$$

This equation has the following solution:

$$\delta\mathbf{E}_{\mathbf{k}} = -\frac{1}{V}\sum_{\mathbf{k}'}\delta\epsilon_{\mathbf{k}-\mathbf{k}'}\left\{\frac{\mathbf{k}\left(\mathbf{k}\cdot\overline{\mathbf{E}}_{\mathbf{k}'}\right)}{k^2\epsilon} - \frac{[\mathbf{k}\times[\mathbf{k}\times\overline{\mathbf{E}}_{\mathbf{k}'}]]}{k^2\left[\epsilon - (ck/\omega)^2\right]}\right\}.$$

Substituting this expression into the right-hand side of the averaged wave equation, we find

$$k_\alpha\left(\mathbf{k}\cdot\overline{\mathbf{E}}_{\mathbf{k}}\right) - k^2\overline{E}_{\mathbf{k}}^\alpha + \left(\frac{\omega}{c}\right)^2\sum_{\beta}\left[\epsilon\delta_{\alpha\beta} - Q_{\alpha\beta}(\mathbf{k},\omega)\right]\overline{E}_{\mathbf{k}}^\beta = 0,$$

$$Q_{\alpha\beta}(\mathbf{k},\omega) = \frac{1}{V}\sum_{\mathbf{k}'}W_{|\mathbf{k}-\mathbf{k}'|}\left\{\frac{k_\alpha'k_\beta'}{k'^2\epsilon} - \frac{k_\alpha'k_\beta' - \delta_{\alpha\beta}k'^2}{k'^2[\epsilon - (ck'/\omega)^2]}\right\}.$$

The variations of the dielectric permittivity are described here by the Fourier transform of their correlation function, $W_k = \int d\Delta\mathbf{r}\exp(-i\mathbf{k}\cdot\Delta\mathbf{r})\langle\delta\epsilon_{\mathbf{r}}\delta\epsilon_{\mathbf{r}'}\rangle$, where $\Delta\mathbf{r} = \mathbf{r} - \mathbf{r}'$. Due to macroscopic homogeneity of the medium, the average $\langle\delta\epsilon_{\mathbf{r}}\delta\epsilon_{\mathbf{r}'}\rangle$ is independent of $\mathbf{r} + \mathbf{r}'$ so that W_k is equally introduced as $\langle\delta\epsilon_{\mathbf{k}}\delta\epsilon_{\mathbf{k}'}\rangle = \delta_{\mathbf{k},-\mathbf{k}'}VW_k$. In the case of short-scale (in comparison to the length of the electromagnetic wave) inhomogeneities, $W_k \simeq W$ is a constant and $Q_{\alpha\beta}(\mathbf{k},\omega) \simeq \delta_{\alpha\beta}Q_\omega$, where

$$Q_\omega = \frac{1}{V}\sum_{\mathbf{k}'}W\left[\frac{1}{3\epsilon} + \frac{2/3}{\epsilon - (ck'/\omega)^2}\right].$$

The dispersion of transverse waves is determined from the relation $k^2 - (\omega/c)^2(\overline{\epsilon} - i\text{Im}Q_\omega) = 0$, where $\overline{\epsilon} = \epsilon - \text{Re}Q_\omega$. The imaginary part of Q_ω describes the decrease in the amplitude of the field due to elastic scattering, with the decrement $\text{Im}k \simeq -(\omega/c)\text{Im}Q_\omega/2\sqrt{\overline{\epsilon}}$. The decrement of the energy losses is twice greater. The expression for $\mathrm{Im}Q_\omega$ is obtained from the general expression by using a formal substitution $\omega \to \omega + i0$. As a result,

$$\mathrm{Im}Q_\omega = -\frac{2\pi}{3V}\sum_{\mathbf{k}} W\delta\left[\epsilon - \left(\frac{ck}{\omega}\right)^2\right] = -\frac{\sqrt{\epsilon}}{6\pi}\left(\frac{\omega}{c}\right)^3 W,$$

and the decrement of the energy losses is equal to $(\omega/c)^4 W/6\pi$ (the difference between $\bar{\epsilon}$ and ϵ is neglected as a second-order effect).

14.12. Transform the expression (70.8) by taking into account that the group velocities of electrons are small in comparison to the velocity of light.

<u>Solution:</u> In this approximation, the expression standing inside the braces $\{\dots\}$ in Eq. (70.8) is equal to

$$\delta_{\alpha\beta} + \frac{p_{0\alpha}q_{0\beta} + p_{0\beta}q_{0\alpha}}{m_e\omega_0},$$

where $\mathbf{p}_0 = (\mathbf{p} + \mathbf{p}')/2$, $\mathbf{q}_0 = (\mathbf{q}_I + \mathbf{q}_S)/2$, and $\omega_0 = (\omega_I + \omega_S)/2$. Since $\omega_0 = \tilde{c}(q_I + q_S)/2$, the second term is of the order of $p_0/m_e\tilde{c} \ll 1$ and can be neglected in comparison to the first term.

14.13. Find the cross-section of elastic light scattering by free electrons at $\hbar\omega_I \ll \varepsilon_g$, when $P(\mathbf{e}_I, \mathbf{e}_S) = |(\mathbf{e}_I \cdot \mathbf{e}_S^*)|^2$.

<u>Solution:</u> One should integrate the result (70.20) over ω_S and over the solid angle of the vector $\mathbf{q}_S$. The first integration is trivial because of the presence of the δ-function, while the integral $\int d\Omega |(\mathbf{e}_I \cdot \mathbf{e}_S^*)|^2$ is equal to $8\pi/3$ (note that $\mathbf{e}_S$ is perpendicular to $\mathbf{q}_S$). The cross-section per one electron is equal to $(8\pi/3)r_o^2$.

14.14. Consider the light scattering by two-dimensional electrons.

<u>Solution:</u> Let us consider a quantum well in the plane XOY containing electrons in the single 2D state described by the envelope function $\psi_z\phi_{\mathbf{x}}^{(\delta)}$, where $\mathbf{x} = (x,y)$ is the in-plane coordinate. The fluctuations of electron distribution in the plane, $\widehat{\Delta f}_{\sigma\mathbf{xpt}}^{(av)}$, where $\mathbf{p}$ is the 2D momentum, satisfy an equation analogous to Eq. (69.21):

$$\left(\frac{\partial}{\partial t} + \mathbf{v}_\mathbf{p} \cdot \frac{\partial}{\partial\mathbf{x}}\right)\widehat{\Delta f}_{\sigma\mathbf{xpt}}^{(av)} - \frac{\partial\widehat{\Delta U}_{\mathbf{xt}}}{\partial\mathbf{x}} \cdot \frac{\partial f_\mathbf{p}}{\partial\mathbf{p}} = 0,$$

The fluctuating potential energy in the 2D plane is given by $\widehat{\Delta U}_{\mathbf{xt}} = \int dz |\psi_z|^2 \widehat{\Delta U}_{\mathbf{x}zt}$, where $\widehat{\Delta U}_{\mathbf{x}zt}$ is found from the Poisson equation $\epsilon_\infty\nabla^2\widehat{U}_{\mathbf{x}zt} = -4\pi e^2 \widehat{\Delta n}_{\mathbf{x}zt}$. Since the fluctuating 3D electron density $\widehat{\Delta n}_{\mathbf{x}zt}$ is equal to $|\psi_z|^2\widehat{\Delta n}_{\mathbf{xt}}$, where $\widehat{\Delta n}_{\mathbf{xt}} = L^{-2}\sum_{\sigma\mathbf{p}}\widehat{\Delta f}_{\sigma\mathbf{xpt}}^{(av)}$, we obtain

$$\widehat{\Delta U}_{\mathbf{k}\omega} = \int dt \int d\mathbf{x} e^{i\omega t - i\mathbf{k}\cdot\mathbf{x}}\widehat{U}_{\mathbf{xt}} = \frac{2\pi e^2}{\epsilon_\infty k}I(k)\widehat{\Delta n}_{\mathbf{k}\omega},$$

$$I(k) = \int dz \int dz' |\psi_z|^2 |\psi_{z'}|^2 e^{-k|z-z'|},$$

where $\mathbf{k}$ is the in-plane wave vector. Assuming that $\Delta\mathbf{q} = (\Delta\mathbf{k}, \Delta q_z)$ is small in comparison to the inverse width of the quantum well (in these conditions $I(k) \simeq 1$ and $\int dz |\psi_z|^2 e^{-i\Delta q_z z} \simeq 1$), we obtain (compare to Eqs. (69.24) and (69.25))

$$\widehat{\Delta n}_{\Delta\mathbf{q},\Delta\omega} = \frac{\epsilon_\infty}{\epsilon_{2D}(\Delta\mathbf{k},\Delta\omega)} \frac{1}{L^2} \sum_{\sigma\mathbf{p}} \widehat{\Delta f}^{(av)}_{\sigma\Delta\mathbf{k}\mathbf{p}t_0} 2\pi e^{i\Delta\omega t_0} \delta(\Delta\omega - \mathbf{v_p}\cdot\Delta\mathbf{k}),$$

$$\epsilon_{2D}(\mathbf{k},\omega) = \epsilon_\infty + \frac{2\pi e^2}{k} \frac{2}{L^2} \sum_{\mathbf{p}} \frac{\mathbf{k}\cdot(\partial f_\mathbf{p}/\partial\mathbf{p})}{\omega - \mathbf{v_p}\cdot\mathbf{k} + i\lambda},$$

where $\epsilon_{2D}(\mathbf{k},\omega)$ is the 2D dielectric permittivity. The density correlation function is independent of Δq_z. It is given by the expression

$$\langle\langle\widehat{\Delta n}_{\Delta\mathbf{q}}(t)\widehat{\Delta n}_{-\Delta\mathbf{q}}(0)\rangle\rangle_{\Delta\omega} = 2\pi\left|\frac{\epsilon_\infty}{\epsilon_{2D}(\Delta\mathbf{k},\Delta\omega)}\right|^2$$

$$\times 2\sum_{\mathbf{p}} f_{\mathbf{p}-\hbar\Delta\mathbf{k}/2}(1 - f_{\mathbf{p}+\hbar\Delta\mathbf{k}/2})\delta(\Delta\omega - \mathbf{v_p}\cdot\Delta\mathbf{k}),$$

which is similar to Eq. (70.22). Assuming $\omega \gg \mathbf{v_p}\cdot\mathbf{k}$, one can obtain $\epsilon_{2D}(\mathbf{k},\omega) \simeq \epsilon_\infty[1 - (\omega_k^{(2D)}/\omega)^2]$, where $\omega_k^{(2D)} = \sqrt{2\pi e^2 n_{2D} k/\epsilon_\infty m}$ is the 2D plasmon frequency expressed through the 2D electron density n_{2D}. Therefore, the light scattered by the 2D electron layer has peaks at $\Delta\omega = \pm\sqrt{2\pi e^2 n_{2D}|\Delta\mathbf{k}|/\epsilon_\infty m}$ corresponding to the 2D plasmon excitation.

14.15. Find the differential cross-section of light scattering by coupled plasmon-phonon modes at $T = 0$ and in the absence of relaxation.
    Result:

$$\left(\frac{\partial^2\sigma}{\partial\omega_S\partial\Omega}\right)_{\mu_S} = \frac{r_o^2 V|\Delta\mathbf{q}|^2\hbar}{8\pi e^2}\frac{\omega_S}{\omega_I}P(\mathbf{e}_I,\mathbf{e}_S)$$

$$\times\frac{\epsilon_\infty\omega_p^4}{(\omega_+^2 - \omega_-^2)}\left[\frac{\omega_+^2 - \omega_{TO}^2}{\omega_+^3}\delta(\Delta\omega - \omega_+) + \frac{\omega_{TO}^2 - \omega_-^2}{\omega_-^3}\delta(\Delta\omega - \omega_-)\right].$$

Only the Stokes components are present at $T = 0$.

14.16. Derive Eq. (70.24) for the differential cross-section of light scattering in the medium with spatial and temporal variations of the dielectric permittivity.
    Solution: We start from the wave equation

$$[\nabla\times[\nabla\times\mathbf{E}_{\mathbf{r}t}]] + \frac{1}{c^2}\frac{\partial^2}{\partial t^2}\mathbf{D}_{\mathbf{r}t} = 0, \quad \mathbf{D}_{\mathbf{r}t} = [\epsilon + \widehat{\delta\epsilon}(\mathbf{r},t)]\mathbf{E}_{\mathbf{r}t},$$

where $\mathbf{D}$ is the electric induction locally related to the electric field. The fluctuation of the dielectric permittivity, $\widehat{\delta\epsilon}(\mathbf{r},t)$, is assumed to be a tensor in the Cartesian coordinate space, and $\widehat{\delta\epsilon}(\mathbf{r},t)\mathbf{E}_{\mathbf{r}t}$ defines a vector whose components are $\sum_\beta \delta\epsilon_{\alpha\beta}(\mathbf{r},t)E_{\mathbf{r}t}^\beta$. The wave equation can be used to find the amplitude of the scattered waves, $\mathbf{E}'_{\mathbf{r}t}$, as a weak response to the perturbation described by the monochromatic plane wave

$\mathbf{E_{rt}} = E_0 \mathbf{e_{k\mu}} e^{i\mathbf{k}\cdot\mathbf{r} - i\omega t} + c.c.$ Substituting the total field (scattered wave + perturbation) to the wave equation, we take the first iteration and obtain

$$\nabla^2 \mathbf{D}'_{\mathbf{r}\omega'} + \left(\frac{\omega'}{c}\right)^2 \epsilon \mathbf{D}'_{\mathbf{r}\omega'} = -E_0 \int dt e^{i\omega' t} [\nabla \times [\nabla \times \widehat{\delta\epsilon}(\mathbf{r}, t) \mathbf{e_{k\mu}}]] e^{i\mathbf{k}\cdot\mathbf{r} - i\omega t}$$

$$= -E_0 [\nabla \times [\nabla \times \widehat{\delta\epsilon}(\mathbf{r}, \omega' - \omega) \mathbf{e_{k\mu}}]] e^{i\mathbf{k}\cdot\mathbf{r}}.$$

This equation describes the temporal Fourier transform of the electric induction of scattered wave as a response to the plane wave $E_0 \mathbf{e_{k\mu}} e^{i\mathbf{k}\cdot\mathbf{r} - i\omega t}$. We have taken into account that $(\nabla \cdot \mathbf{D}') = 0$, which leads to $[\nabla \times [\nabla \times \mathbf{D}']] = -\nabla^2 \mathbf{D}'$. The solution of the inhomogeneous equation for $\mathbf{D}'_{\mathbf{r}\omega'}$ is expressed with the aid of the Green's function

$$G(\mathbf{r}, \mathbf{r}') = -\frac{e^{ik'|\mathbf{r} - \mathbf{r}'|}}{4\pi|\mathbf{r} - \mathbf{r}'|}$$

satisfying the equation $[\nabla^2 + k'^2] G(\mathbf{r}, \mathbf{r}') = \delta(\mathbf{r} - \mathbf{r}')$. We obtain

$$\mathbf{D}'_{\mathbf{r}\omega'} = \frac{E_0}{4\pi} \int d\mathbf{r}' \frac{e^{ik'|\mathbf{r} - \mathbf{r}'|}}{|\mathbf{r} - \mathbf{r}'|} [\nabla' \times [\nabla' \times \widehat{\delta\epsilon}(\mathbf{r}', \omega' - \omega) \mathbf{e_{k\mu}}]] e^{i\mathbf{k}\cdot\mathbf{r}'},$$

where $\nabla' \equiv \partial/\partial\mathbf{r}'$ and $k' = \omega'\sqrt{\epsilon}/c$ is the wave number of the scattered wave. In the region far away from the sample, when $|\mathbf{r}| \gg |\mathbf{r}'|$, one has $k'|\mathbf{r} - \mathbf{r}'| \simeq k'|\mathbf{r}| - \mathbf{k}' \cdot \mathbf{r}'$, where $\mathbf{k}' = k'\mathbf{r}/|\mathbf{r}|$. Besides, in this region $\mathbf{D}' = \epsilon \mathbf{E}'$, and we have

$$\mathbf{E}'_{\mathbf{r}\omega'} = -\frac{E_0 e^{ik'|\mathbf{r}|}}{4\pi\epsilon|\mathbf{r}|} [\mathbf{k}' \times [\mathbf{k}' \times \widehat{\delta\epsilon}(\mathbf{k}' - \mathbf{k}, \omega' - \omega) \mathbf{e_{k\mu}}]].$$

This solution is represented in the form of a wave radiating from the sample in all directions.

The cross-section of light scattering is equal to the ratio of the averaged energy scattered in unit time, $-\int d\mathbf{r} \langle\langle \partial \mathcal{E}_{\mathbf{rt}}/\partial t \rangle\rangle$, where $\mathcal{E}_{\mathbf{rt}}$ is the energy density of the electromagnetic field, to the absolute value of the energy density flux (Poynting vector) of the incident radiation, $|\mathbf{S}^{(I)}|$. The latter is independent of coordinate since the incident radiation is a plane wave:

$$\mathbf{S}^{(I)} = \frac{c}{4\pi} \overline{[\mathbf{E_{rt}} \times \mathbf{H_{rt}}]} = \frac{c^2 |E_0|^2}{2\pi\omega} [\mathbf{e_{k\mu}} \times [\mathbf{k} \times \mathbf{e^*_{k\mu}}]] = \frac{c\sqrt{\epsilon}|E_0|^2}{2\pi} \frac{\mathbf{k}}{|\mathbf{k}|},$$

where the line denotes the averaging over time. Expressing the energy density $\mathcal{E}_{\mathbf{rt}}$ according to the continuity equation $\partial \mathcal{E}_{\mathbf{rt}}/\partial t + \nabla \cdot \mathbf{S}^{(S)}_{\mathbf{rt}} = 0$ through the Poynting vector of the scattered radiation, we obtain

$$\sigma = \frac{\int d\mathbf{r} \langle\langle \nabla \cdot \mathbf{S}^{(S)}_{\mathbf{rt}} \rangle\rangle}{|\mathbf{S}^{(I)}|} = \frac{\int_{\mathbf{r} \in \Gamma} d\mathbf{r} \langle\langle \mathbf{S}^{(S)}_{\mathbf{rt}} \cdot \mathbf{n}_\Gamma \rangle\rangle}{|\mathbf{S}^{(I)}|}.$$

The averaging $\langle\langle \ldots \rangle\rangle$ is carried out over the motion of the particles in the medium which scatters the light. We have transformed the integral of $\nabla \cdot \mathbf{S}^{(S)}_{\mathbf{rt}}$ over the volume to the integral over an arbitrary surface Γ enclosing this volume. It is convenient to consider a spherical surface, when the unit vector normal to the surface, $\mathbf{n}_\Gamma$, is equal to $\mathbf{r}/|\mathbf{r}| = \mathbf{k}'/k'$ and the integral $\int_{\mathbf{r} \in \Gamma} d\mathbf{r} \ldots$ is written as $\int d\Omega |\mathbf{r}|^2 \ldots$, where $d\Omega$ is the differential of the solid angle. Since $\mathbf{E}'_{\mathbf{r}\omega'} \propto |\mathbf{r}|^{-1}$, the integral $\int_{\mathbf{r} \in \Gamma} d\mathbf{r}(\mathbf{S}^{(S)}_{\mathbf{r}} \cdot \mathbf{n}_\Gamma)$

is independent of $\mathbf{r}$. Taking into account that $\nabla e^{ik'|\mathbf{r}|}/|\mathbf{r}| \simeq i\mathbf{k}' e^{ik'|\mathbf{r}|}/|\mathbf{r}|$ for large $|\mathbf{r}|$, we have

$$\mathbf{S}_{\mathbf{r}t}^{(S)} \cdot \mathbf{n}_\Gamma = \frac{c\sqrt{\epsilon}}{2\pi} \frac{\mathbf{k}'}{k'} \cdot \left[\mathbf{E}_{\mathbf{r}t}' \times \left[\frac{\mathbf{k}'}{k'} \times \mathbf{E}_{\mathbf{r}t}'^* \right] \right]$$

$$= \frac{c\sqrt{\epsilon}}{2\pi} \int \frac{d\omega''}{2\pi} \int \frac{d\omega'}{2\pi} e^{i(\omega'-\omega'')t} \left(\mathbf{E}_{\mathbf{r}\omega''}' \cdot \mathbf{E}_{\mathbf{r}\omega'}'^* \right),$$

where $\mathbf{E}_{\mathbf{r}\omega'}'$ is calculated above. Using also the identity $[\mathbf{k}' \times [\mathbf{k}' \times \mathbf{a}]] = -k'^2 \mathbf{a} + \mathbf{k}'(\mathbf{k}' \cdot \mathbf{a})$, where $\mathbf{a} = \widehat{\delta\epsilon}(\mathbf{k}' - \mathbf{k}, \omega' - \omega)\mathbf{e}_{\mathbf{k}\mu}$, we obtain

$$\int_{\mathbf{r}\in\Gamma} d\mathbf{r}(\mathbf{S}_{\mathbf{r}t}^{(S)} \cdot \mathbf{n}_\Gamma) = \frac{|\mathbf{S}^{(I)}|}{(4\pi\epsilon)^2} \int d\Omega \int d\omega' \int \frac{d\omega''}{(2\pi)^2} k''^2 e^{i(\omega'-\omega'')(t-|\mathbf{r}|\sqrt{\epsilon}/c)}$$

$$\times \left\{ k'^2 \left(\widehat{\delta\epsilon}(\mathbf{k}'' - \mathbf{k}, \omega'' - \omega)\mathbf{e}_{\mathbf{k}\mu} \cdot \widehat{\delta\epsilon}^*(\mathbf{k}' - \mathbf{k}, \omega' - \omega)\mathbf{e}_{\mathbf{k}\mu}^* \right) \right.$$

$$\left. - \left(\mathbf{k}' \cdot \widehat{\delta\epsilon}(\mathbf{k}'' - \mathbf{k}, \omega'' - \omega)\mathbf{e}_{\mathbf{k}\mu} \right) \left(\mathbf{k}' \cdot \widehat{\delta\epsilon}^*(\mathbf{k}' - \mathbf{k}, \omega' - \omega)\mathbf{e}_{\mathbf{k}\mu}^* \right) \right\}$$

with $\mathbf{k}'' = \mathbf{k}'(\omega''/\omega')$. The average $\langle\langle \int_{\mathbf{r}\in\Gamma} d\mathbf{r}(\mathbf{S}_{\mathbf{r}t}^{(S)} \cdot \mathbf{n}_\Gamma) \rangle\rangle$ is non-zero only at $\omega'' = \omega'$ so that the dependence on t and $|\mathbf{r}|$ vanishes. With the aid of the symmetry property $\epsilon_{\alpha\beta}(\mathbf{k}, \omega) = \epsilon_{\alpha\beta}^*(-\mathbf{k}, -\omega)$, the differential cross-section (see also Eq. (70.2)) is written as

$$\frac{\partial^2 \sigma}{\partial\omega'\partial\Omega} = \frac{(\omega'/c)^4}{32\pi^3} \sum_{\alpha\beta\gamma\delta} e_{\mathbf{k}\mu}^{\gamma*} e_{\mathbf{k}\mu}^\beta \left(\delta_{\delta\alpha} - \frac{k_\delta' k_\alpha'}{k'^2} \right)$$

$$\times \left\langle\left\langle \delta\epsilon_{\delta\gamma}(\mathbf{k} - \mathbf{k}', t)\delta\epsilon_{\alpha\beta}(\mathbf{k}' - \mathbf{k}, 0) \right\rangle\right\rangle_{\omega-\omega'},$$

where the spectral density is introduced as $\langle\langle \ldots \rangle\rangle_\omega \equiv \int dt e^{i\omega t} \langle\langle \ldots \rangle\rangle$. The equation above describes the differential cross-section summed over all polarizations of the scattered light. In the case of a scalar, time-independent dielectric permittivity, $\delta\epsilon_{\alpha\beta}(\mathbf{q}, t) = \delta_{\alpha\beta}\epsilon_{\mathbf{q}}$, one has $\langle\langle \delta\epsilon_{\delta\gamma}(\mathbf{k} - \mathbf{k}', t)\delta\epsilon_{\alpha\beta}(\mathbf{k}' - \mathbf{k}, 0) \rangle\rangle_{\omega-\omega'} = 2\pi\delta_{\gamma\delta}\delta_{\alpha\beta}\delta(\omega - \omega')VW_{|\mathbf{k}-\mathbf{k}'|}$, where the correlation function W_k is introduced in problem 14.11. The scattering in this case is elastic and

$$\frac{\partial\sigma}{\partial\Omega} = \frac{(\omega/c)^4 V}{(4\pi)^2} \sum_{\alpha\beta} W_{|\mathbf{k}-\mathbf{k}'|} e_{\mathbf{k}\mu}^{\alpha*} e_{\mathbf{k}\mu}^\beta \left(\delta_{\alpha\beta} - \frac{k_\alpha' k_\beta'}{k'^2} \right).$$

For the short-range inhomogeneities, when $W_{|\mathbf{k}-\mathbf{k}'|} \simeq W$, the integration over the solid angle gives us the integral cross-section $\sigma = (\omega/c)^4 VW/6\pi$ which does not depend on the polarization of incident light. The decrement of energy losses calculated in problem 14.11 is equal to σ/V, in agreement with the definition of σ and energy conservation requirement.

To find the cross-section for the scattering with a given polarization, $\mathbf{e}_{\mathbf{k}'\mu'}$, of the scattered light, one should calculate the Poynting vector $\mathbf{S}_{\mathbf{r}t}^{(S)}$ by using the field $\mathbf{E}_{\mathbf{r}\omega'\mu'}' = \mathbf{e}_{\mathbf{k}'\mu'}(\mathbf{e}_{\mathbf{k}'\mu'}^* \cdot \mathbf{E}_{\mathbf{r}\omega'}')$ instead of $\mathbf{E}_{\mathbf{r}\omega'}'$. This means that $(\mathbf{E}_{\mathbf{r}\omega''}' \cdot \mathbf{E}_{\mathbf{r}\omega'}'^*)$ in the expression for $\mathbf{S}_{\mathbf{r}t}^{(S)} \cdot \mathbf{n}_\Gamma$ should be replaced by $(\mathbf{e}_{\mathbf{k}''\mu'}^* \cdot \mathbf{E}_{\mathbf{r}\omega''}')(\mathbf{e}_{\mathbf{k}'\mu'} \cdot \mathbf{E}_{\mathbf{r}\omega'}'^*)$. The result is

$$\left(\frac{\partial^2 \sigma}{\partial\omega'\partial\Omega} \right)_{\mu'} = \frac{(\omega'/c)^4}{32\pi^3} \sum_{\alpha\beta\gamma\delta} e_{\mathbf{k}'\mu'}^\delta e_{\mathbf{k}\mu}^{\gamma*} e_{\mathbf{k}'\mu'}^{\alpha*} e_{\mathbf{k}\mu}^\beta$$

$$\times \left\langle\left\langle \delta\epsilon_{\delta\gamma}(\mathbf{k} - \mathbf{k}', t)\delta\epsilon_{\alpha\beta}(\mathbf{k}' - \mathbf{k}, 0) \right\rangle\right\rangle_{\omega-\omega'}.$$

We remind that ω, $\mathbf{k}$, and μ are the frequency, wave vector and polarization index of the incident light, while ω', $\mathbf{k}'$, and μ' correspond to the scattered light.

14.17. Prove that the correlations described by Eq. (71.6) are always negative at $N \neq M$.

Hint: Assuming $N \neq M$ and using the unitarity property of the scattering matrix, reduce the last term under the integral in Eq. (71.6) to

$$-\mathrm{tr}\left[\left(\sum_{N_1} \widehat{S}_{MN_1} \widehat{S}^+_{NN_1} f_{N_1}(\varepsilon) \right) \left(\sum_{N_2} \widehat{S}_{NN_2} \widehat{S}^+_{MN_2} f_{N_2}(\varepsilon) \right) \right].$$

14.18. Find the noise power in two-terminal mesoscopic conductors at finite frequencies $\omega \ll \varepsilon_F/\hbar$. Check the fluctuation-dissipation theorem for this case.

Results:

$$\mathcal{S}(\omega) = \frac{e^2}{2\pi\hbar} \left\{ 2\hbar\omega \coth\left(\frac{\hbar\omega}{2T_e} \right) \sum_n T_n^2 + \left[(\hbar\omega + eV) \coth\left(\frac{\hbar\omega + eV}{2T_e} \right) \right. \right.$$

$$\left. \left. + (\hbar\omega - eV) \coth\left(\frac{\hbar\omega - eV}{2T_e} \right) \right] \sum_n T_n(1 - T_n) \right\}.$$

In equilibrium, when $V = 0$, one has $\mathcal{S}(\omega) = \hbar\omega G \coth(\hbar\omega/2T_e)$.

14.19. Calculate the static noise power of a microcontact in the diffusive regime, when the size of the microcontact is much greater than the mean free path of electrons.

Solution: The fluctuating part of the current operator is given by

$$\widehat{\Delta I}_N(t) = e \int_{\mathbf{r} \in S_N} d\mathbf{r} \sum_\sigma \int \frac{d\mathbf{p}}{(2\pi\hbar)^d} \mathbf{v_p} \cdot \mathbf{n}_{S_N} \widehat{\Delta f}^{(av)}_{\sigma\mathbf{rp}t},$$

where $d = 3$ or $d = 2$ is the dimensionality of the contact. The operator $\widehat{\Delta f}^{(av)}_{\sigma\mathbf{rp}t}$ satisfies the Boltzmann-Langevin equation (69.19), where, however, the collision integral and Langevin source should be taken for the case of electron-impurity interaction. Because of the current conservation (continuity), the surface S_N can be chosen as an arbitrary cross-section of the microcontact, not necessarily at the leads $N = L$ or $N = R$. For this reason, we omit the index N and chose the surface S_N as a plane perpendicular to the contact axis OZ. Besides, if we integrate the current over an arbitrary interval of length L along OZ and divide the result by L, the current will remain the same. Thus, we represent the operator of fluctuating current in the following form:

$$\widehat{\Delta I}(t) = \frac{e}{L} \int_{V_c} d\mathbf{r} \sum_\sigma \int \frac{d\mathbf{p}}{(2\pi\hbar)^d} v_\mathbf{p}^z \widehat{\Delta f}^{(av)}_{\sigma\mathbf{rp}t}.$$

The integral is taken over the microcontact volume defined by the boundary of the contact and by $-L/2 < z < L/2$. The steady-state Boltzmann-Langevin equation in the diffusive regime, when the spatial gradient contribution can be neglected in comparison to the collision-integral contribution, is solved as $\widehat{\Delta f}^{(av)}_{\sigma\mathbf{rpt}} \simeq \tau_{tr}\widehat{K}^{(av)}_t(\mathbf{p},\mathbf{r})$, where τ_{tr} is the transport time at the Fermi level. We obtain the correlation function

$$\langle\langle\widehat{\Delta I}(t)\widehat{\Delta I}(0)\rangle\rangle_\omega = 2\frac{e^2\tau_{tr}^2}{L^2m^2}\int_{V_c}d\mathbf{r}\int_{V_c}d\mathbf{r}'\int\frac{d\mathbf{p}}{(2\pi\hbar)^d}\int\frac{d\mathbf{p}'}{(2\pi\hbar)^d}$$

$$\times p_z p_z' \langle\langle\widehat{K}^{(av)}_t(\mathbf{p},\mathbf{r})\widehat{K}^{(av)}_0(\mathbf{p}',\mathbf{r}')\rangle\rangle_\omega ,$$

where the sum over spin is already calculated. Using the expression for the correlation function given in problem 14.4, we take into account only the symmetric, with respect to $\mathbf{p}$, part of the electron distribution functions. After some transformations, we find the following formula for the noise power at $\omega = 0$:

$$S = \frac{4e^2\tau_{tr}}{L^2m^2d}\int_{V_c}d\mathbf{r}\int\frac{d\mathbf{p}}{(2\pi\hbar)^d}p^2 f_{\mathbf{rp}}(1-f_{\mathbf{rp}}).$$

The symmetric distribution function $f_{\mathbf{rp}}$ is given by Eqs. (12.13) and (12.14) with $\alpha_{\mathbf{rp}} = \alpha_0(\mathbf{r})$, and $\alpha_0(\mathbf{r})$ is defined according to Eq. (12.20). Therefore,

$$f_{\mathbf{rp}} = \alpha_0(\mathbf{r})f_R + [1 - \alpha_0(\mathbf{r})]f_L,$$

where the quasi-equilibrium distribution functions in the leads, f_L and f_R, exactly correspond to f^- and f^+ of Sec. 12. We search for $\alpha_0(\mathbf{r})$ in the simplest geometry, when the microcontact is a linear tube along OZ (or a linear stripe in the case of $d = 2$), so that $\alpha_0(\mathbf{r})$ depends only on z. The condition at the boundary Γ is automatically satisfied, while the condition at $|\mathbf{r}| \to \infty$ in Eq. (12.20) is reduced to $\alpha_0(z = L/2) = 1$ and $\alpha_0(z = -L/2) = 0$. The only solution satisfying both the Laplace equation and these boundary conditions is $\alpha_0 = 1/2 + z/L$. Substituting it into the expression for $f_{\mathbf{rp}}$, we find

$$\frac{1}{L}\int_{-L/2}^{L/2}dz f_{\mathbf{rp}}(1-f_{\mathbf{rp}}) = \frac{1}{2}f_L(1-f_L) + \frac{1}{2}f_R(1-f_R) + \frac{1}{6}(f_L - f_R)^2.$$

Let us calculate the sum over $\mathbf{p}$ in the expression for S, assuming that the electron temperature T_e and bias eV are much smaller than the Fermi energy. The result is

$$S = 2G[2T_e/3 + (eV/6)\coth(eV/2T_e)],$$

where $G = \sigma_0 S/L$ is the conductance expressed through the bulk conductivity $\sigma_0 = e^2n\tau_{tr}/m$ and cross-section S of the microcontact. Note that the electron density in the d-dimensional case is related to the Fermi energy as $n = 2\varepsilon_F\rho_\mathcal{D}(\varepsilon_F)/d$. In the limit $eV \gg T_e$, the noise power $S = eVG/3 = eI/3$ is three times smaller than the Poissonian noise power eI. This result is reproducible for other geometries, for example, for the hyperbolic boundary considered in Sec. 12, though the corresponding calculations are more complicated.

14.20. Assuming the case of short-range scattering, solve the Bethe-Salpeter equation (15.5) for $K^{s_1 s_2'}_{EE'}(\mathbf{p}_1, \mathbf{p}_2'|\mathbf{p}_1', \mathbf{p}_2)$.

   Solution: Let us write this equation in the form

$$K^{s_1 s_2'}_{EE'}(\mathbf{p}_1, \mathbf{p}_2'|\mathbf{p}_1', \mathbf{p}_2) = G^{s_1}_E(\mathbf{p}_1)G^{s_2'}_{E'}(\mathbf{p}_2')$$

$$\times \left[\delta_{\mathbf{p}_1 \mathbf{p}_1'} \delta_{\mathbf{p}_2 \mathbf{p}_2'} + \frac{w}{V} \sum_{\mathbf{q}} K_{EE'}^{s_1 s_2'}(\mathbf{p}_1 - \hbar\mathbf{q}, \mathbf{p}_2' + \hbar\mathbf{q}|\mathbf{p}_1', \mathbf{p}_2) \right.$$

$$\left. + \frac{1}{V} \sum_{\mathbf{q}} [\widetilde{\Gamma}_{EE'}^{s_1 s_2'}(\mathbf{p}_1 - \mathbf{p}_2' - \hbar\mathbf{q}) - w] K_{EE'}^{s_1 s_2'}(\mathbf{p}_1 - \hbar\mathbf{q}, \mathbf{p}_2' + \hbar\mathbf{q}|\mathbf{p}_1', \mathbf{p}_2) \right]$$

and note that the factor $\widetilde{\Gamma} - w$ in the last term is close to 0 everywhere, except for a narrow region $|\mathbf{p}_1 - \mathbf{p}_2' - \hbar\mathbf{q}| < \hbar/l_F$, where the vertex part $\widetilde{\Gamma}$ at $s_1 \neq s_2'$ has a singularity. Without this last term, the solution is written exactly, as

$$K_{EE'}^{s_1 s_2'}(\mathbf{p}_1, \mathbf{p}_2'|\mathbf{p}_1', \mathbf{p}_2) = G_E^{s_1}(\mathbf{p}_1) G_{E'}^{s_2'}(\mathbf{p}_2') \left[\delta_{\mathbf{p}_1 \mathbf{p}_1'} \delta_{\mathbf{p}_2 \mathbf{p}_2'} \right.$$

$$\left. + \frac{w}{V} \frac{G_E^{s_1}(\mathbf{p}_1') G_{E'}^{s_2'}(\mathbf{p}_2) \delta_{\mathbf{p}_1 - \mathbf{p}_2, \mathbf{p}_1' - \mathbf{p}_2'}}{1 - (w/V) \sum_{\mathbf{p}} G_E^{s_1}(\mathbf{p} + \Delta\mathbf{p}/2) G_{E'}^{s_2'}(\mathbf{p} - \Delta\mathbf{p}/2)} \right]$$

$$= G_E^{s_1}(\mathbf{p}_1) G_{E'}^{s_2'}(\mathbf{p}_2') \left[\delta_{\mathbf{p}_1 \mathbf{p}_1'} \delta_{\mathbf{p}_2 \mathbf{p}_2'} + \frac{1}{V} \delta_{\mathbf{p}_1 - \mathbf{p}_2, \mathbf{p}_1' - \mathbf{p}_2'} G_E^{s_1}(\mathbf{p}_1') G_{E'}^{s_2'}(\mathbf{p}_2) \widetilde{\Gamma}_{EE'}^{s_1 s_2'}(\Delta\mathbf{p}) \right],$$

where $\Delta\mathbf{p} = \mathbf{p}_1 + \mathbf{p}_2'$. Substituting this solution into the last term of the Bethe-Salpeter equation written above, we solve this equation by iterations and obtain Eq. (71.19).

14.21. Prove that the Dyson equations (39.21) and (72.4) have the same form.

<u>Hint</u>: Use Eqs. (39.22) and (39.24).

14.22. Find, in the lowest-order approximation, the vertex part $I(33'44')$ and the self-energy function $\Sigma(33')$ for the electrons interacting (a) with the random static potential $U_{\mathbf{r}}$ and (b) with the deformation potential of acoustic phonons.

<u>Result</u>:

$$(a) \quad I(33'44') = \delta_{34'} \delta_{3'4} (-1)^{s_3 + s_3'} \langle\langle U_{\mathbf{r}_3} U_{\mathbf{r}_3'} \rangle\rangle,$$

$$\Sigma(33') = G(33')(-1)^{s_3 + s_3'} \langle\langle U_{\mathbf{r}_3} U_{\mathbf{r}_3'} \rangle\rangle,$$

$$(b) \quad I(33'44') = \hbar\mathcal{D}^2 \delta_{34'} \delta_{3'4} (-1)^{s_3 + s_3'} \sum_{\alpha\beta} \nabla_{\mathbf{r}_3}^{\alpha} \nabla_{\mathbf{r}_3'}^{\beta} i D^{\alpha\beta}(33'),$$

$$\Sigma(33') = \hbar\mathcal{D}^2 G(33')(-1)^{s_3 + s_3'} \sum_{\alpha\beta} \nabla_{\mathbf{r}_3}^{\alpha} \nabla_{\mathbf{r}_3'}^{\beta} i D^{\alpha\beta}(33'),$$

where $D^{\alpha\beta}(33')$ is the Green's function of phonons introduced in Sec. 42 (the contour branch indices are included in the multi-indices). The expressions for Σ are obvious generalizations of Eqs. (39.20) and (42.11) to the case of four-branch contour. It is easy to check that Eq. (72.13) is valid for each kind of interaction.

14.23. Derive the expression (72.15) for $\widehat{G}_{13}^{-1}$.

Solution: In the matrix form, one can check that the 4×4 matrix

$$\widetilde{G}_{13}^{-1} = \widetilde{\sigma}_z \left(i\hbar \frac{\partial}{\partial t_1} - \hat{h}_1 \right) \delta_{13} - \widetilde{\Sigma}(1,3)$$

satisfies $\int d3 \widetilde{G}_{13}^{-1} \widetilde{G}(3,1') = \hat{1} \delta_{11'}$, according to Eq. (72.4). In the simplified notations, when the contour branch indices are included in the multi-indices, this expression is transformed into Eq. (72.15). Note that, instead of the matrix $\widetilde{\sigma}_z$ acting from the left on an arbitrary 4×4 matrix $\widetilde{A}(1,1')$, we have introduced a scalar quantity $\sigma_z(1) = (-1)^{s_1+1}$ so that this action is expressed as $\sigma_z(1)A(11')$. In a similar way, the action $\widetilde{A}(1,1')\widetilde{\sigma}_z$ is expressed as $\sigma_z(1')A(11')$.

14.24. Find $M(11'33')$ and $M(22'44')$ in the lowest order with respect to the interaction of electrons with deformation potential of acoustic phonons.

Solution: Substituting the result of problem 14.22 into Eq. (72.22), we obtain (the multi-indices and contour branch indices are written separately):

$$M^{s_1 s_1' s_3 s_3'}(11'33') = \delta_{3'1'}\delta_{s_3' s_1'}(\delta_{s_3 1} - \delta_{s_3 2})iG^{+-}(13)[\bar{D}^{-+}(31) - \bar{D}^{-+}(31')]$$

$$+\delta_{31}\delta_{s_3 s_1}(\delta_{s_3' 1} - \delta_{s_3' 2})iG^{-+}(3'1)[\bar{D}^{-+}(3'1) - \bar{D}^{-+}(3'1')],$$

$$M^{s_2 s_2' s_4 s_4'}(22'44') = \delta_{4'2'}\delta_{s_4' s_2'}(\delta_{s_4 3} - \delta_{s_4 4})iG^{-+}(24)[\bar{D}^{-+}(24) - \bar{D}^{-+}(2'4)]$$

$$+\delta_{42}\delta_{s_4 s_2}(\delta_{s_4' 3} - \delta_{s_4' 4})iG^{+-}(4'2)[\bar{D}^{-+}(24') - \bar{D}^{-+}(2'4')],$$

where $\bar{D}^{s_1 s_2}(12) = \hbar \mathcal{D}^2 \sum_{\alpha\beta} \nabla_{\mathbf{r}_1}^\alpha \nabla_{\mathbf{r}_2}^\beta D^{\alpha\beta, s_1 s_2}(12)$. It is taken into account that s_1 and s_1' can be either 3 or 4, while s_2 and s_2' can be either 1 or 2. We have also used the symmetry property (42.5).

14.25. Prove the invariance of $R^{s_1 s_1' s_2 s_2'}(11'22')$ against all its four contour branch indices under the condition $I(33'44') = \bar{I}(33'44') + \widetilde{I}(33'44')$, where $\bar{I}$ contains only the terms with $s_3 = s_4'$ and $s_3' = s_4$, while $\widetilde{I}$ contains only the terms with $s_3 = s_4$ and $s_3' = s_4'$.

Solution: Under the conditions given above and with the aid of Eqs. (72.2) and (72.3), we rewrite Eq. (72.25) in the following way:

$$R^{s_1 s_1' s_2 s_2'}(11'22') = \Sigma^{-+}(21')G^{+-}(12') + \Sigma^{+-}(12')G^{-+}(21')$$

$$+[G^{-+}(5'1')\sigma_z(1)\bar{I}(15'23)\sigma_z(2) - G^{+-}(15)\sigma_z(1')\bar{I}(51'23)\sigma_z(2)]G^{+-}(32')$$

$$-[G^{-+}(5'1')\sigma_z(1)\widetilde{I}(15'3'2')\sigma_z(2') - G^{+-}(15)\sigma_z(1')\bar{I}(51'3'2')\sigma_z(2')]G^{-+}(23').$$

We have taken into account that s_1 and s_1' belong to the pair $3,4$ and s_2 and s_2' belong to the pair $1,2$. The invariance becomes obvious when we write, for example,

$$\sigma_z(1)\bar{I}(15'23)\sigma_z(2) \equiv (-1)^{s_1+1}I^{s_1 s_2 s_2 s_1}(15'23)(-1)^{s_2+1}$$

$$= (-1)^{s_1+s_2}\delta\Sigma^{s_1 s_2}(15')/\delta G^{s_1 s_2}(32) = -\delta\Sigma^{+-}(15')/\delta G^{+-}(32)$$

according to Eqs. (72.13) and (72.3). Similar transformations can be carried out for all terms containing $\bar{I}$ and $\widetilde{I}$.

Appendix A
Harmonic Oscillator

Quantum-mechanical description of small vibrations of a particle in the one-dimensional potential $U(x)$ near the equilibrium position $x = x_0$ (determined by the requirement $dU(x)/dx = 0$) is carried out by expanding $U(x)$ in powers of $(x - x_0)$ with the accuracy up to the second-order terms. By choosing $x_0 = 0$, we have

$$U(x) = U(0) + \frac{1}{2}\left(\frac{d^2U}{dx^2}\right)_{x=0} x^2, \qquad (A.1)$$

and the eigenstate problem $\hat{h}_{osc}\psi(x) = E\psi(x)$ is determined by the Hamiltonian of the harmonic oscillator

$$\hat{h}_{osc} = \frac{\hat{p}^2}{2m} + \frac{m\omega^2}{2}x^2, \quad \omega = \sqrt{\frac{1}{m}\left(\frac{d^2U}{dx^2}\right)_{x=0}}, \qquad (A.2)$$

which is quadratic in both momentum and coordinate. The energy E is counted from $U(0)$, and ω is the vibration frequency of the classical oscillator near the minimum of the potential energy at $x = 0$. Introducing the dimensionless coordinate and energy according to $q = x\sqrt{m\omega/\hbar}$ and $\varepsilon = E/\hbar\omega$, we rewrite the eigenstate problem in the form

$$\left[\frac{1}{2}\left(-\frac{d^2}{dq^2} + q^2\right) - \varepsilon\right]\psi(q) = 0. \qquad (A.3)$$

If $|q|$ is large in comparison to unity, the wave function decreases as $\exp(-q^2/2)$, and the main contribution from the kinetic-energy term is equal to $-(q^2/2)\exp(-q^2/2)$ and cancels the potential-energy term. Therefore, such a solution satisfies Eq. (A.3) for $\varepsilon \sim 1$. In the general case, the wave function is written as a product of $\exp(-q^2/2)$ by a finite-order polynomial. The eigenstate problem (A.3) with the boundary condition $\psi|_{|q|\to\infty} = 0$ has the following solution:

$$\psi_n(q) = N_n e^{-q^2/2} H_n(q), \quad \varepsilon_n = n + \frac{1}{2}, \quad n = 0, 1, 2, \ldots, \qquad (A.4)$$

where the quantum number n numbers the levels, the normalization constant $N_n = [n!2^n\sqrt{\pi}]^{-1/2}$ is determined by using the identity $\int_{-\infty}^{\infty} dq\exp(-q^2)[H_n(q)]^2 = 2^n n!\sqrt{\pi}$, and the Hermite polynomial of n-th order is introduced according to

$$H_n(q) = (-1)^n e^{q^2} \frac{d^n}{dq^n} e^{-q^2}. \tag{A.5}$$

Since the Hamiltonian (A.2) is even (symmetric) with respect to x, the wave functions can be either even or odd. According to Eq. (A.4), their parity coincides with the parity of n, as it is seen from Eq. (A.5) or from the expressions

$$H_0(q) = 1, \quad H_1(q) = 2q, \quad H_2(q) = 4q^2 - 2,$$

$$H_3(q) = 8q^3 - 12q, \quad H_4(q) = 16q^4 - 48q^2 + 12, \dots . \tag{A.6}$$

Using the recurrence relations for the Hermite polynomials,

$$qH_n(q) = nH_{n-1}(q) + \frac{1}{2}H_{n+1}(q), \qquad \frac{dH_n}{dq} = 2nH_{n-1}(q), \tag{A.7}$$

we obtain the following connection between the wave functions of the neighboring states:

$$q\psi_n(q) = \sqrt{\frac{n}{2}}\psi_{n-1}(q) + \sqrt{\frac{n+1}{2}}\psi_{n+1}(q),$$

$$\frac{d\psi_n}{dq} = 2\sqrt{\frac{n}{2}}\psi_{n-1}(q) - q\psi_n(q) = \sqrt{\frac{n}{2}}\psi_{n-1}(q) - \sqrt{\frac{n+1}{2}}\psi_{n+1}(q). \tag{A.8}$$

Using them, one can show that the matrix elements of coordinate are non-zero for neighboring states only:

$$\langle n'|\hat{x}|n\rangle = \sqrt{\frac{\hbar}{2m\omega}} \left\{ \begin{array}{l} \sqrt{n}\delta_{n',n-1} \\ \sqrt{n+1}\delta_{n',n+1} \end{array} \right. . \tag{A.9}$$

The same property takes place for the matrix elements of the momentum operator:

$$\langle n'|\hat{p}|n\rangle = i\sqrt{\frac{m\hbar\omega}{2}} \left\{ \begin{array}{l} -\sqrt{n}\delta_{n',n-1} \\ \sqrt{n+1}\delta_{n',n+1} \end{array} \right. . \tag{A.10}$$

Using the operator identity $q^2 - d^2/dq^2 = (q-d/dq)(q+d/dq)+1$, one can reformulate the eigenstate problem (A.3) by introducing new operators

$$\hat{b} = \frac{1}{\sqrt{2}}\left(q + \frac{d}{dq}\right), \quad \hat{b}^+ = \frac{1}{\sqrt{2}}\left(q - \frac{d}{dq}\right), \tag{A.11}$$

which also connect the neighboring states only. The Hamiltonian $\hat{h}_{osc}$ is expressed in terms of these operators as

$$\hat{h}_{osc} = \frac{\hbar\omega}{2}\left(-\frac{d^2}{dq^2} + q^2\right) = \hbar\omega\left(\hat{b}^+\hat{b} + \frac{1}{2}\right). \tag{A.12}$$

The operators (A.11) are Hermitian conjugate and satisfy the commutation relation

$$[\hat{b}, \hat{b}^+] = 1. \tag{A.13}$$

The operator $\hat{b}^+$ increases the number n of the oscillator state, while the operator $\hat{b}$ decreases this number:

$$\hat{b}^+\psi_n(q) = \sqrt{n+1}\psi_{n+1}(q), \quad \hat{b}\psi_n(q) = \sqrt{n}\psi_{n-1}(q). \tag{A.14}$$

The coordinate and momentum operators, which mix the n-th and $(n \pm 1)$-th states, are expressed through $\hat{b}^+$ and $\hat{b}$ as

$$\hat{x} = \sqrt{\frac{\hbar}{2m\omega}}(\hat{b}^+ + \hat{b}), \quad \hat{p} = -i\hbar\sqrt{\frac{m\omega}{\hbar}}\frac{d}{dq} = i\sqrt{\frac{m\hbar\omega}{2}}(\hat{b}^+ - \hat{b}). \qquad (A.15)$$

Let us act by the operator $\hat{b}^+$ on the function $\psi_0(q)$ describing the ground ("vacuum") state. After n sequential actions, we obtain the wave function of the n-th state:

$$\psi_n(q) = \frac{(\hat{b}^+)^n}{\sqrt{n!}}\psi_0(q), \qquad (A.16)$$

while $\psi_0(q)$ is determined from the equation $\hat{b}\psi_0(q) = 0$. Using the explicit form of $\hat{b}$ from Eq. (A.11), we have $\psi_0(q) = N_0 \exp(-q^2/2)$.

Let us reformulate the harmonic oscillator problem by turning from the coordinate representation to the occupation number representation. We introduce a set of ket-vectors

$$|0\rangle, |1\rangle, \dots |n\rangle, \dots \equiv \{|n\rangle\} \qquad (A.17)$$

corresponding to the levels $0, 1, \dots n \dots$. They are connected to each other by the relations analogous to the ones introduced in Eq. (A.14):

$$\hat{b}^+|n\rangle = \sqrt{n+1}|n+1\rangle, \quad \hat{b}|n\rangle = \sqrt{n}|n-1\rangle. \qquad (A.18)$$

Each element of the set $\{|n\rangle\}$ is obtained from the ground-state ket-vector $|0\rangle$ after a number of actions of the operator $\hat{b}^+$, as in Eq. (A.16):

$$|n\rangle = \frac{(\hat{b}^+)^n}{\sqrt{n!}}|0\rangle. \qquad (A.19)$$

Therefore, the operators $\hat{b}^+$ ($\hat{b}$) describe creation (annihilation) of a quantum with energy $\hbar\omega$. If the Hamiltonian of the oscillatory type describes vibrational modes of electromagnetic field or small vibrations of crystal lattice, such quanta correspond to quasiparticles, photons or phonons. The Hermitian operator $\hat{N} = \hat{b}^+\hat{b}$ is called the quantum (particle) number operator, since $|n\rangle$ satisfies the eigenstate problem $\hat{N}|n\rangle = n|n\rangle$. To check it, one may use the operator equation $\hat{N}\hat{b}^+ = \hat{b}^+(\hat{N} + 1)$ and the explicit form of the ket-vector $|n\rangle$. This leads to a chain of n equations

$$\hat{N}|n\rangle = \frac{\hat{b}^+}{\sqrt{n!}}(\hat{N} + 1)(\hat{b}^+)^{n-1}|0\rangle = \dots = \frac{(\hat{b}^+)^n}{\sqrt{n!}}(\hat{N} + n)|0\rangle = n|n\rangle. \qquad (A.20)$$

One may rewrite the Hamiltonian (A.12) in terms of $\hat{N}$ as $\hat{h}_{osc} = \hbar\omega(\hat{N} + 1/2)$.

As an example of the operator algebra based on the properties of $\hat{b}^+$ and $\hat{b}$, we calculate the matrix element $\langle n|e^{ikx}|n'\rangle$. Such elements appear, for example, in calculation of the matrix elements of a potential $V(x)$ expressed through its Fourier transform according to $V(x) = (2\pi)^{-1}\int dk e^{ikx}V(k)$. Expressing the coordinate operator according to Eq. (A.15), we rewrite the matrix element as

$$\langle n|e^{ikx}|n'\rangle = \langle n|e^{i\kappa(\hat{b}^+ + \hat{b})}|n'\rangle, \quad \kappa = k\sqrt{\frac{\hbar}{2m\omega}}, \qquad (A.21)$$

where κ is a dimensionless wave number. Further, we use the operator identity (known as Weyl identity)

$$e^{\hat{A} + \hat{B}} = e^{\hat{A}}e^{\hat{B}}e^{-[\hat{A},\hat{B}]/2} \qquad (A.22)$$

which is true under the requirement that the commutator $[\hat{A}, \hat{B}]$ commutes with both $\hat{A}$ and $\hat{B}$. We substitute $\hat{A} = i\kappa\hat{b}^+$ and $\hat{B} = i\kappa\hat{b}$ and obtain $[\hat{A}, \hat{B}] = \kappa^2$ so that this requirement is fulfilled. Now we have

$$\langle n|e^{i\kappa(\hat{b}^+ + \hat{b})}|n'\rangle = e^{-\kappa^2/2}\sum_l \langle n|e^{i\kappa\hat{b}^+}|l\rangle\langle l|e^{i\kappa\hat{b}}|n'\rangle. \qquad (A.23)$$

The matrix elements of the exponential operators are calculated by expanding the exponents in series, $\exp(\hat{A}) = \sum_{p=0}^{\infty}(\hat{A})^p/p!$. Then we use Eq. (A.18) and obtain

$$\langle l|e^{i\kappa\hat{b}}|n'\rangle = \frac{(i\kappa)^{n'-l}}{(n'-l)!}\sqrt{\frac{n'!}{l!}}, \quad \langle n|e^{i\kappa\hat{b}^+}|l\rangle = \frac{(i\kappa)^{n-l}}{(n-l)!}\sqrt{\frac{n!}{l!}}, \qquad (A.24)$$

because only the terms with $p = n' - l$ and $p = n - l$ contribute to the matrix elements $\langle l|e^{i\kappa\hat{b}}|n'\rangle$ and $\langle n|e^{i\kappa\hat{b}^+}|l\rangle$, respectively. The sum over l in Eq. (A.23) runs up to $l = \min\{n, n'\}$. Below we put $n' > n$ and choose the variable of summation as $m = n - l$, where m runs from 0 to n. Using Eq. (A.24), we obtain

$$\langle n|e^{i\kappa(\hat{b}^+ + \hat{b})}|n'\rangle = e^{-\kappa^2/2}\sum_{m=0}^{n}\frac{\sqrt{n!n'!}(i\kappa)^{n'-n+2m}}{m!(n'-n+m)!(n-m)!}. \qquad (A.25)$$

The sum over m can be written through the Laguerre polynomials defined as

$$L_n^\alpha(x) = \sum_{m=0}^{n}(-1)^m\frac{(n+\alpha)!}{(n-m)!(\alpha+m)!m!}x^m. \qquad (A.26)$$

Substituting $\alpha = n' - n$ and $x = \kappa^2$, we finally find

$$\langle n|e^{ikx}|n'\rangle = \sqrt{\frac{n!}{n'!}}(i\kappa)^{n'-n}e^{-\kappa^2/2}L_n^{n'-n}(\kappa^2), \quad \kappa = k\sqrt{\frac{\hbar}{2m\omega}}. \qquad (A.27)$$

This result, of course, can be obtained directly after calculating the integrals with the wave functions (A.4). Equation (A.27) have numerous applications, in particular, in magnetotransport theory.

Appendix B
Many-Band KP-Approach

Let us consider the electron states in crystals under external fields. The dynamics of the electrons can be described by the **kp**-formalism if these fields are smooth on the scale of the lattice constant. The electric and magnetic field strengths, $\mathbf{E}_{\mathbf{r}t}$ and $\mathbf{H}_{\mathbf{r}t}$, are expressed through the vector potential $\mathbf{A}_{\mathbf{r}t}$ and scalar potential $\Phi_{\mathbf{r}t}$ according to Eq. (4.3). By including the field-induced contributions into the one-electron Hamiltonian (5.4), we obtain

$$\hat{h}_{cr}(t) = \hat{h}_{cr} - \frac{e}{2m_e c}(\hat{\mathbf{p}} \cdot \mathbf{A}_{\mathbf{r}t} + \mathbf{A}_{\mathbf{r}t} \cdot \hat{\mathbf{p}}) + \frac{(e/c)^2}{2m_e} A_{\mathbf{r}t}^2$$

$$+ U_{\mathbf{r}t} + \mu_B(\hat{\boldsymbol{\sigma}} \cdot [\nabla \times \mathbf{A}_{\mathbf{r}t}]), \qquad (B.1)$$

where m_e is the free-electron mass, $U_{\mathbf{r}t} = e\Phi_{\mathbf{r}t}$ is the potential energy proportional to the scalar potential, and $\mu_B = |e|\hbar/2m_e c$ is the Bohr magneton. The last term in Eq. (B.1) describes the Pauli interaction of electrons with the magnetic field, and we have neglected the contribution $(\mu_B/2m_e c^2)\hat{\boldsymbol{\sigma}} \cdot [\nabla U_{cr}(\mathbf{r}) \times \mathbf{A}_{\mathbf{r}t}]$ coming from the spin-orbit interaction term for the reason of its smallness. Near the extremum $\mathbf{p} = 0$ (a generalization to the case of an arbitrary extremum $\mathbf{p} = \mathbf{p}_0$ is straightforward), we write a complete set of eigenfunctions as $\psi_{l\mathbf{p}}(\mathbf{r}) = V^{-1/2} \exp(i\mathbf{p} \cdot \mathbf{r}/\hbar) u_l(\mathbf{r})$; see Eq. (5.5). We remind that the index l contains both the band number n and the spin number σ. The matrix elements of $\mathbf{A}_{\mathbf{r}t}$ and $U_{\mathbf{r}t}$ can be written through their spatial Fourier transforms, where $\mathbf{q} = (\mathbf{p} - \mathbf{p}')/\hbar$:

$$\langle l\mathbf{p}|\mathbf{A}_{\mathbf{r}t}|l'\mathbf{p}'\rangle \simeq \delta_{ll'}\mathbf{A}_{\mathbf{q}t}, \qquad \langle l\mathbf{p}|U_{\mathbf{r}t}|l'\mathbf{p}'\rangle \simeq \delta_{ll'}U_{\mathbf{q}t},$$

$$\langle l\mathbf{p}|A_{\mathbf{r}t}^2|l'\mathbf{p}'\rangle \simeq \delta_{ll'} A_{\mathbf{q}t}^2 . \qquad (B.2)$$

Therefore, within the **kp**-approach the matrix elements of the Hamiltonian (B.1) take the form

$$\langle l\mathbf{p}|\hat{h}_{cr}(t)|l'\mathbf{p}'\rangle = \delta_{\mathbf{p}\mathbf{p}'} H_{ll'}(\mathbf{p}) + \delta_{ll'}\left[-\frac{e}{2m_e c}(\mathbf{p} + \mathbf{p}') \cdot \mathbf{A}_{\mathbf{q}t}\right.$$

$$\left. + \frac{e^2}{2m_e c^2} A_{\mathbf{q}t}^2 + U_{\mathbf{q}t}\right] + \mu_B(\boldsymbol{\sigma}_{ll'} \cdot \mathbf{H}_{\mathbf{q}t}) - \frac{e}{c}\mathbf{A}_{\mathbf{q}t} \cdot \mathbf{v}_{ll'}, \qquad (B.3)$$

where $H_{ll'}(\mathbf{p})$ is introduced by Eq. (5.6), $\boldsymbol{\sigma}_{ll'}$ is the interband matrix element of the spin operator, and the matrix elements of the velocity operator, $\mathbf{v}_{ll'}$, are given by Eq. (5.7). The time-dependent envelope functions $\varphi_{l\mathbf{p}t}$ in the expansion (5.8) is determined from the system of equations

$$i\hbar\frac{\partial}{\partial t}\varphi_{l\mathbf{p}t} = \sum_{l'\mathbf{p}'}\langle l\mathbf{p}|\hat{h}_{cr}(t)|l'\mathbf{p}'\rangle\varphi_{l'\mathbf{p}'t} \qquad (B.4)$$

describing the dynamics of many-band electron states.

Since the problem is spatially inhomogeneous, it is convenient to use the coordinate representation of the envelope function instead of the momentum representation. Expanding the exact wave function as in Eq. (5.8), $\Psi(\mathbf{r}t) = \sum_{l\mathbf{p}}\varphi_{l\mathbf{p}t}\psi_{l\mathbf{p}}(\mathbf{r})$, we may write it as $\Psi(\mathbf{r}t) = \sum_l \varphi_{l\mathbf{r}t}u_l(\mathbf{r})$, where $\varphi_{l\mathbf{r}t}$ is connected to $\varphi_{l\mathbf{p}t}$ by the Fourier transformations

$$\varphi_{l\mathbf{r}t} = \frac{1}{\sqrt{V}}\sum_{\mathbf{p}}e^{i\mathbf{p}\cdot\mathbf{r}/\hbar}\varphi_{l\mathbf{p}t}, \qquad \varphi_{l\mathbf{p}t} = \frac{1}{\sqrt{V}}\int d\mathbf{r}\, e^{-i\mathbf{p}\cdot\mathbf{r}/\hbar}\varphi_{l\mathbf{r}t} . \qquad (B.5)$$

The sum over $\mathbf{p}$ must be taken inside the first Brillouin zone. The coordinate-dependent envelope function $\varphi_{l\mathbf{r}t}$ is governed by the following equation:

$$i\hbar\frac{\partial}{\partial t}\varphi_{l\mathbf{r}t} = \sum_{l'}\hat{H}_{ll'}\varphi_{l'\mathbf{r}t},$$

$$\hat{H}_{ll'} = \delta_{ll'}(\varepsilon_l + U_{\mathbf{r}t}) + \mathbf{v}_{ll'}\cdot\hat{\boldsymbol{\pi}}_{\mathbf{r}t} \qquad (B.6)$$

$$+\frac{1}{4}\sum_{\alpha\beta}D_{ll'}^{\alpha\beta}(\hat{\pi}_{\mathbf{r}t}^{\alpha}\hat{\pi}_{\mathbf{r}t}^{\beta} + \hat{\pi}_{\mathbf{r}t}^{\beta}\hat{\pi}_{\mathbf{r}t}^{\alpha}) + \mu_B\mathbf{G}_{ll'}\cdot\mathbf{H}_{\mathbf{r}t},$$

where $\hat{\boldsymbol{\pi}}_{\mathbf{r}t} = \hat{\mathbf{p}} - e\mathbf{A}_{\mathbf{r}t}/c$ is the kinematic momentum operator. Equation (B.6) describes the electron dynamics in external fields when there are several bands, numbered by the indices l and l', close to each other in energy. The contribution of the other, remote bands is taken into account through the symmetric inverse effective mass tensor

$$D_{ll'}^{\alpha\beta} = \frac{\delta_{ll'}\delta_{\alpha\beta}}{m_e} + \frac{1}{2}\sum_{s\neq l,l'}(v_{ls}^{\alpha}v_{sl'}^{\beta} + v_{ls}^{\beta}v_{sl'}^{\alpha})[(\varepsilon_l - \varepsilon_s)^{-1} + (\varepsilon_{l'} - \varepsilon_s)^{-1}] , \qquad (B.7)$$

which generalizes the tensor (5.11) to the many-band case. The effective spin vector,

$$G_{ll'}^{\alpha} = \sigma_{ll'}^{\alpha} - i\frac{m_e}{2}\sum_{s\neq l,l'}[\mathbf{v}_{ls}\times\mathbf{v}_{sl'}]_{\alpha}[(\varepsilon_{l'} - \varepsilon_s)^{-1} + (\varepsilon_l - \varepsilon_s)^{-1}] , \qquad (B.8)$$

is also determined by the remote band contributions and describes modification of the g-factor of electrons in the crystal. In order to describe the remote bands in the way given by Eqs. (B.6)–(B.8), one needs, apart from the condition $|\varepsilon_l - \varepsilon_s| \gg |\varepsilon_{l'} - \varepsilon_l|$, the requirements of smoothness of external fields on the scale of the lattice constant and of low frequency of these fields, to ensure $\omega \ll |\varepsilon_l - \varepsilon_s|/\hbar$.

The many-band current density operator $\hat{\mathbf{I}}_{ll'}(\mathbf{r}, t)$ is introduced as a proportionality coefficient determining the correction to the Hamiltonian due to a small variation of the vector potential, $\delta\mathbf{A}_{\mathbf{r}t}$, according to the expression

$$\delta\hat{H}_{ll'} = -\frac{1}{c}\int d\mathbf{r}\,\hat{\mathbf{I}}_{ll'}(\mathbf{r}, t)\cdot\delta\mathbf{A}_{\mathbf{r}t} . \qquad (B.9)$$

In other words, the current density operator is a functional derivative of the matrix Hamiltonian (B.3) over the vector potential. Comparing Eqs. (B.9) and (B.6), we obtain the explicit expression for $\hat{\mathbf{I}}_{ll'}(\mathbf{r}, t)$:

$$\hat{I}^\alpha_{ll'}(\mathbf{r}, t) = ev^\alpha_{ll'}\delta(\mathbf{x} - \mathbf{r}) + \frac{e}{2}\sum_\beta D^{\alpha\beta}_{ll'}[\hat{p}_\beta\delta(\mathbf{x} - \mathbf{r}) + \delta(\mathbf{x} - \mathbf{r})\hat{p}_\beta] \qquad (B.10)$$

$$-\frac{e^2}{c}\sum_\beta D^{\alpha\beta}_{ll'}A^\beta_{\mathbf{x}t}\delta(\mathbf{x} - \mathbf{r}) - \frac{ie}{2m_e}\{[\hat{\mathbf{p}} \times \mathbf{G}_{ll'}]_\alpha\delta(\mathbf{x} - \mathbf{r}) - \delta(\mathbf{x} - \mathbf{r})[\hat{\mathbf{p}} \times \mathbf{G}_{ll'}]_\alpha\}.$$

The first term is the contribution of the bands l and l' which are included in the matrix Hamiltonian (B.6). The next two terms describe the remote-band contributions. The last term is a spin-dependent contribution to the current. In the simplest case of a non-degenerate band, when there is only one state in the set l, the matrix $D^{\alpha\beta}_{ll'}$ is reduced to the inverse effective mass tensor (5.11), while $-2G^\alpha_{ll'}$ becomes a scalar effective g-factor multiplied by the vector of Pauli matrices so that the Pauli contribution to the electron Hamiltonian is written as $-g\mu_B\hat{\boldsymbol{\sigma}} \cdot \mathbf{H}_{\mathbf{r}t}/2$. The first term in the expression (B.10) is omitted in this case. In the dipole approximation, one has to consider only the Fourier component $\hat{\mathbf{I}}_{ll'}(\mathbf{q}, t) = \int d\mathbf{r}\hat{\mathbf{I}}_{ll'}(\mathbf{r}, t)\exp(-i\mathbf{q} \cdot \mathbf{r})$ at $\mathbf{q} = 0$. Since the spin contribution in Eq. (B.10) is equal to zero in this case, the current density operator $\hat{\mathbf{I}}_{ll'}(\mathbf{q} = 0, t) \equiv \hat{\mathbf{I}}_{ll'}(t)$ is written as

$$\hat{I}^\alpha_{ll'}(t) = ev^\alpha_{ll'} + e\sum_\beta D^{\alpha\beta}_{ll'}\left(\hat{p}_\beta - \frac{e}{c}A^\beta_t\right) , \qquad (B.11)$$

where only the vector potential of the homogeneous field remains. If the energies of electrons in the bands l and l' are small in comparison to the interband energy $|\varepsilon_l - \varepsilon_{l'}|$, only the first term in Eq. (B.11) is essential.

Consider the electron states in deformed crystals. Small deformations of elastic materials are described by a symmetric tensor of deformation, $\varepsilon_{\alpha\beta} = \varepsilon_{\alpha\beta}(\mathbf{r})$:

$$\varepsilon_{\alpha\beta} = \frac{1}{2}\left(\frac{\partial u_\alpha}{\partial r_\beta} + \frac{\partial u_\beta}{\partial r_\alpha}\right) , \qquad (B.12)$$

where $\mathbf{u} = \mathbf{u}(\mathbf{r})$ is the displacement vector at the point $\mathbf{r}$. The deformation changes the symmetry of the crystal lattice and the potential energy $W(\mathbf{r})$ in the elementary cell. Since the point $\mathbf{r}$ under the deformation is shifted to the point $(1 + \hat{\varepsilon})\mathbf{r}$, the momentum is transformed to $(1 - \hat{\varepsilon})\mathbf{p}$, within the linear accuracy. The linear in $\mathbf{u}(\mathbf{r})$ contribution to the crystal Hamiltonian (5.4) is written as

$$\widehat{\delta H}(\varepsilon) = -\sum_{\alpha\beta}\frac{\hat{p}_\alpha\varepsilon_{\alpha\beta}\hat{p}_\beta}{m_e} + \sum_{\alpha\beta}\mathcal{V}_{\alpha\beta}(\mathbf{r})\varepsilon_{\alpha\beta} . \qquad (B.13)$$

The spin-orbit contribution is neglected in this expression because the relativistic corrections to $\widehat{\delta H}(\varepsilon)$ are small. The matrix $\mathcal{V}_{\alpha\beta}$ describes the deformation-induced modification of the potential energy $W(\mathbf{r})$ according to

$$W_\varepsilon[(1 + \hat{\varepsilon})\mathbf{r}] - W(\mathbf{r}) = \sum_{\alpha\beta}\mathcal{V}_{\alpha\beta}(\mathbf{r})\varepsilon_{\alpha\beta} . \qquad (B.14)$$

Using the set of eigenfunctions (5.5), we find the deformation-induced contributions to the **kp**-Hamiltonian (5.6):

$$\delta H_{ll'}(\varepsilon) = \sum_{\alpha\beta}\Xi^{\alpha\beta}_{ll'}\varepsilon_{\alpha\beta}, \qquad \Xi^{\alpha\beta}_{ll'} = -\frac{(p_\alpha p_\beta)_{ll'}}{m_e} + \mathcal{V}^{\alpha\beta}_{ll'} . \qquad (B.15)$$

To estimate the deformation-potential tensor $\Xi_{ll'}^{\alpha\beta}$, one may use the models of crystal lattice with either rigid or deformed ions. In the deformed ion approximation, the change of the potential energy is small so that $\Xi_{ll'}^{\alpha\beta}$ is determined only by the first term of Eq. (B.15). In the rigid ion model, the crystal potential $W(\mathbf{r})$ is approximated by a sum of atomic potentials $V_a(\mathbf{r} - \mathbf{R}_i)$ placed at the lattice sites $\mathbf{R}_i$, and the deformation merely shifts the site positions to $(1+\hat{\varepsilon})\mathbf{R}_k$ without any change of $V_a(\mathbf{r})$. For both approximations, the tensors $\Xi_{ll'}^{\alpha\beta}$ appear to be of the order of atomic energies, though the deviations of their values from experimental data are considerable. One should use a more detailed description of the band structure in order to calculate $\Xi_{ll'}^{\alpha\beta}$. In the vicinity of a non-degenerate extremum of the conduction band, it is convenient to consider the deformation-potential tensor in the main axes whose orientation is determined by the crystal symmetry. According to Eq. (B.15), the symmetry of this tensor is the same as for the effective mass tensor (5.11). If the surfaces of equal energy are uniaxial ellipsoids, the deformation-induced contribution (B.15) to the Hamiltonian is expressed through two constants, the longitudinal, $d_\parallel$, and transverse, $d_\perp$, deformation potentials, as $\delta H_{cc}(\varepsilon) = d_\perp(\varepsilon_{xx} + \varepsilon_{yy}) + d_\parallel \varepsilon_{zz}$. In the spherically-symmetric case, one has $d_\parallel = d_\perp = \mathcal{D}$, and the induced energy $\delta H_{cc}(\varepsilon) = \mathcal{D}\sum_\alpha \varepsilon_{\alpha\alpha} = \mathcal{D}\mathrm{div}\mathbf{u}(\mathbf{r})$ is proportional to the change of the crystal volume due to the deformations. In other words, the isotropic conduction band is simply shifted in energy due to the hydrostatic component of the deformation and does not feel the displacements induced by uniaxial stresses.

Consider the simplest case described by the many-band **kp**-approach, when there are two spin-degenerate bands (conduction and valence bands) close in energy. Their contributions to the Hamiltonian (5.6) should be considered in the frames of a two-band **kp**-model. The envelope wave function has four components numbered by the band index $n = c, v$ and spin index $\sigma = \pm 1$. There are only two non-zero components of the velocity matrix (5.7), the interband velocities $\mathbf{v}_{cv}$ and $\mathbf{v}_{vc}$. Since the velocity operator is Hermitian, $\mathbf{v}_{cv}^* = \mathbf{v}_{vc}$. Below we consider the case of cubic crystals, when these velocities are isotropic (the isotropy, however, exists only if we consider an extremum in the center of the Brillouin zone). Defining $v_{cv} = v_{vc} \equiv s$, one may write the velocity (5.7) in the form of a 4×4 matrix

$$\hat{v}^\alpha = s \left| \begin{array}{cc} \hat{0} & \hat{\sigma}_\alpha \\ \hat{\sigma}_\alpha & \hat{0} \end{array} \right| , \qquad (B.16)$$

where $\alpha = x, y, z$ is the Cartesian coordinate index, $\hat{0}$ defines a 2×2 matrix whose elements are zeros, and $\hat{\sigma}_\alpha$ is a 2×2 matrix which acts on spin variables only. One should choose this matrix in such a way that the right-hand side of Eq. (5.11) becomes spherically-symmetric and independent of the spin quantum number σ. These conditions are fulfilled if the set of 2×2 matrices satisfies the anticommutation relations

$$\hat{\sigma}_\alpha \hat{\sigma}_\beta + \hat{\sigma}_\beta \hat{\sigma}_\alpha = 0, \quad \alpha \neq \beta, \quad \hat{\sigma}_\alpha \hat{\sigma}_\alpha = \hat{1}, \qquad (B.17)$$

where $\hat{1}$ is the unit 2×2 matrix. Therefore, $\hat{\sigma}_\alpha$ may be chosen as the Pauli matrices: $\hat{\sigma}_x = \left| \begin{array}{cc} 0 & 1 \\ 1 & 0 \end{array} \right|, \hat{\sigma}_y = i \left| \begin{array}{cc} 0 & -1 \\ 1 & 0 \end{array} \right|$, and $\hat{\sigma}_z = \left| \begin{array}{cc} 1 & 0 \\ 0 & -1 \end{array} \right|$. Let us set the reference point of energy in the middle of the gap between the bands and introduce the effective mass m according to $2ms^2 = \varepsilon_g$. Then we write 4×4 matrices of the Hamiltonian $\hat{h}$ and velocity operator $\hat{\mathbf{v}}$ as

$$\hat{h} = ms^2 \hat{\rho}_3 + (\mathbf{p} \cdot \hat{\mathbf{v}}) + \frac{p^2}{2M}, \qquad \hat{\mathbf{v}} = s\hat{\rho}_1 \hat{\boldsymbol{\sigma}} . \qquad (B.18)$$

The Hamiltonian in Eq. (B.18) differs from the Dirac Hamiltonian for a relativistic electron only due to the presence of the remote-band contribution $p^2/2M$, which appears since the tensor (B.7) in the case under consideration is reduced to a scalar denoted as M^{-1}. The 4×4 matrices $\hat{\rho}_i$ ($i = 1, 2, 3$) satisfy the commutation relations (B.17) and can be chosen in the form

$$\hat{\rho}_1 = \begin{vmatrix} \hat{0} & \hat{1} \\ \hat{1} & \hat{0} \end{vmatrix}, \quad \hat{\rho}_2 = i \begin{vmatrix} \hat{0} & -\hat{1} \\ \hat{1} & \hat{0} \end{vmatrix}, \quad \hat{\rho}_3 = \begin{vmatrix} \hat{1} & \hat{0} \\ \hat{0} & -\hat{1} \end{vmatrix}. \qquad (B.19)$$

The symbolic product of $\hat{\rho}_i$ by a Pauli matrix, used in Eq. (B.18) and below, simply means that each of the unit matrices in $\hat{\rho}_i$ should be replaced by the Pauli matrix. We also note that the scalar contributions to the matrix expressions, like $p^2/2M$ in $\hat{h}$ of Eq. (B.18), should be formally considered as the contributions standing at the unit matrices. If the contribution of the remote bands is not essential so that $M = \infty$, the expressions (B.18) and (B.19) describe a "relativistic" electron of mass m, while the interband velocity s plays the role of the velocity of light.

In a similar way as in the relativistic quantum theory, the Hamiltonian $\hat{h}$ is diagonalized by a **p**-dependent unitary transformation

$$\hat{\mathcal{U}}_{\mathbf{p}} = \left(\frac{\eta_p + 1}{2\eta_p} \right)^{1/2} + i\hat{\rho}_2 \frac{\hat{\boldsymbol{\sigma}} \cdot \mathbf{p}}{p} \left(\frac{\eta_p - 1}{2\eta_p} \right)^{1/2}, \quad \eta_p = \sqrt{1 + (p/ms)^2} \qquad (B.20)$$

according to

$$\hat{\mathcal{U}}_{\mathbf{p}} \hat{h} \hat{\mathcal{U}}_{\mathbf{p}}^+ = ms^2 \eta_p \hat{\rho}_3 + \frac{p^2}{2M}. \qquad (B.21)$$

Since $\hat{\boldsymbol{\sigma}}$ has dropped out of this expression, the electron states appear to be spin-degenerate. The energy spectra are determined from the eigenstate problem for the Hamiltonian (B.21):

$$\varepsilon_{cp} = \frac{p^2}{2M} + ms^2 \eta_p, \quad \varepsilon_{vp} = \frac{p^2}{2M} - ms^2 \eta_p. \qquad (B.22)$$

The energy spectrum is parabolic, $\pm[\varepsilon_g/2 + p^2/2m] + p^2/2M$, at $p \ll ms$ and becomes linear, $\pm sp$, at $p \gg ms$. This behavior, corresponding to a transition from non-relativistic to relativistic regimes for a Dirac electron, is called the non-parabolicity of energy spectrum. In the limiting case of zero energy gap, $\varepsilon_g = 0$, one has $m \to 0$ and the energy spectra at $M \to \infty$ are always linear, $\varepsilon_{c,vp} = \pm sp$. This situation takes place in some semiconductor alloys, where a change in the alloy composition leads to inversion of the sign of ε_g. Such materials are called the gapless semiconductors of type I. Another kind of gapless materials (type II) form $Cd_{1-x}Hg_xTe$ alloys, where the zero-gap situation is realized in a certain range of alloy composition, and the energy spectra are parabolic. To describe this case, one needs to take into account more sophisticated band models involving several energy bands, which is beyond the scope of this book. Nevertheless, the two-band Hamiltonian described above reflects essential features of narrow-gap and zero-gap materials.

When the diagonalization of the Hamiltonian by the unitary transformation (B.20) is carried out, the interband velocity matrix $\hat{\mathbf{v}}$ and the potential energy $U_{\mathbf{r}} = V^{-1} \times \sum_{\mathbf{p}} e^{i\mathbf{q} \cdot \mathbf{r}} U_{\mathbf{q}}$ are transformed accordingly. The velocity matrix becomes

$$\hat{\mathcal{U}}_{\mathbf{p}} \hat{\mathbf{v}} \hat{\mathcal{U}}_{\mathbf{p}}^+ = \frac{\mathbf{p}}{m\eta_p} \hat{\rho}_3 + s\hat{\rho}_1 \left[\hat{\boldsymbol{\sigma}} - \frac{\eta_p - 1}{\eta_p} \frac{\mathbf{p}(\hat{\boldsymbol{\sigma}} \cdot \mathbf{p})}{p^2} \right]. \qquad (B.23)$$

The interband (proportional to $\hat{\rho}_1$) part of this operator is of the order of s, while the diagonal, with respect to the band index, part changes from $\mathbf{p}/m$ for small p to $s\mathbf{p}/p$ for large p. To consider the transformation of the potential energy, we use the momentum representation. Namely, the matrix element $\langle \mathbf{p}|U_{\mathbf{r}}|\mathbf{p}'\rangle$ is transformed into $U_{(\mathbf{p}-\mathbf{p}')/\hbar}\hat{\mathcal{U}}_{\mathbf{p}}\hat{\mathcal{U}}_{\mathbf{p}'}^+$, where the matrix factor is

$$\hat{\mathcal{U}}_{\mathbf{p}}\hat{\mathcal{U}}_{\mathbf{p}'}^+ = \frac{1}{2}\sqrt{\frac{(\eta_p+1)(\eta_{p'}+1)}{\eta_p\eta_{p'}}}\left\{1 + \frac{(\hat{\boldsymbol{\sigma}}\cdot\mathbf{p})(\hat{\boldsymbol{\sigma}}\cdot\mathbf{p}')}{pp'}\sqrt{\frac{\eta_p-1}{\eta_p+1}\frac{\eta_{p'}-1}{\eta_{p'}+1}}\right.$$

$$\left. + i\hat{\rho}_2\left[\frac{(\hat{\boldsymbol{\sigma}}\cdot\mathbf{p})}{p}\sqrt{\frac{\eta_p-1}{\eta_p+1}} - \frac{(\hat{\boldsymbol{\sigma}}\cdot\mathbf{p}')}{p'}\sqrt{\frac{\eta_{p'}-1}{\eta_{p'}+1}}\right]\right\}. \qquad (B.24)$$

This factor is close to unity for small momenta. The second term on the right-hand side of Eq. (B.24) contains the contribution proportional to $\hat{\boldsymbol{\sigma}}[\mathbf{p}\times\mathbf{p}']$. This contribution is responsible for the intraband spin-flip scattering, which becomes important with increasing p/ms. The interband (proportional to $\hat{\rho}_2$ in Eq. (B.24)) contributions become essential when the momentum transfer $|\mathbf{p}-\mathbf{p}'|$ is comparable to ms, i.e., when the characteristic spatial scale of the potential energy $U_{\mathbf{r}}$ is comparable to the interband length $\hbar/ms$.

Appendix C
Wigner Transformation of Product

The product of operators, $\hat{c}_t = \hat{a}_t \hat{b}_t$, is written in the coordinate representation as

$$c_t(\mathbf{x}_1, \mathbf{x}_2) = \int d\mathbf{x}' a_t(\mathbf{x}_1, \mathbf{x}') b_t(\mathbf{x}', \mathbf{x}_2). \qquad (C.1)$$

Below we consider the transformation of Eq. (C.1) to the Wigner representation. According to the general definition of the Wigner transformation in Sec. 9, we have

$$c_t(\mathbf{r}, \mathbf{p}) = \int d\Delta\mathbf{r} \exp\left(-\frac{i}{\hbar}\mathbf{P}_{\mathbf{r}t} \cdot \Delta\mathbf{r}\right)$$
$$\times \int d\mathbf{x}' a_t\left(\mathbf{r} + \frac{\Delta\mathbf{r}}{2}, \mathbf{x}'\right) b_t\left(\mathbf{x}', \mathbf{r} - \frac{\Delta\mathbf{r}}{2}\right), \qquad (C.2)$$

where $a_t(\ldots)$ and $b_t(\ldots)$ can be written by using the inverse Wigner transformation (9.7):

$$a_t\left(\mathbf{r} + \frac{\Delta\mathbf{r}}{2}, \mathbf{x}'\right) = \int \frac{d\mathbf{p}_1}{(2\pi\hbar)^3} a_t\left(\frac{\mathbf{r} + \mathbf{x}'}{2} + \frac{\Delta\mathbf{r}}{4}, \mathbf{p}_1\right)$$
$$\times \exp\left[\frac{i}{\hbar}\left(\mathbf{p}_1 + \frac{e}{c}\mathbf{A}_{(\mathbf{r}+\mathbf{x}')/2+\Delta\mathbf{r}/4,t}\right) \cdot \left(\mathbf{r} - \mathbf{x}' + \frac{\Delta\mathbf{r}}{2}\right)\right], \qquad (C.3)$$

$$b_t\left(\mathbf{x}', \mathbf{r} - \frac{\Delta\mathbf{r}}{2}\right) = \int \frac{d\mathbf{p}_2}{(2\pi\hbar)^3} b_t\left(\frac{\mathbf{x}' + \mathbf{r}}{2} - \frac{\Delta\mathbf{r}}{4}, \mathbf{p}_2\right)$$
$$\times \exp\left[\frac{i}{\hbar}\left(\mathbf{p}_2 + \frac{e}{c}\mathbf{A}_{(\mathbf{x}'+\mathbf{r})/2-\Delta\mathbf{r}/4,t}\right) \cdot \left(\mathbf{x}' - \mathbf{r} + \frac{\Delta\mathbf{r}}{2}\right)\right]. \qquad (C.4)$$

Instead of the variables $\mathbf{x}'$ and $\Delta\mathbf{r}$ in Eq. (C.2), we introduce new coordinates $\mathbf{r}_1$ and $\mathbf{r}_2$ according to

$$\mathbf{r}_1 = \frac{\mathbf{r} + \mathbf{x}'}{2} + \frac{\Delta\mathbf{r}}{4}, \quad \mathbf{r}_2 = \frac{\mathbf{r} + \mathbf{x}'}{2} - \frac{\Delta\mathbf{r}}{4}, \qquad (C.5)$$

so that $\Delta\mathbf{r} = 2(\mathbf{r}_1 - \mathbf{r}_2)$ and $\mathbf{x}' = \mathbf{r}_1 + \mathbf{r}_2 - \mathbf{r}$. Using these variables and Eqs. (C.3) and (C.4), we rewrite Eq. (C.2) in the form

$$c_t(\mathbf{r}, \mathbf{p}) = \int \frac{d\mathbf{p}_1}{(2\pi\hbar)^3} \int \frac{d\mathbf{p}_2}{(2\pi\hbar)^3} \int d\mathbf{r}_1 \int d\mathbf{r}_2 |\mathcal{J}_3| a_t(\mathbf{r}_1, \mathbf{p}_1) b_t(\mathbf{r}_2, \mathbf{p}_2)$$

$$\times e^{-(2i/\hbar)\mathbf{P}_{\mathbf{r}t} \cdot (\mathbf{r}_1 - \mathbf{r}_2)} e^{(2i/\hbar)\mathbf{P}_{\mathbf{r}_1 t} \cdot (\mathbf{r} - \mathbf{r}_2)} e^{(2i/\hbar)\mathbf{P}_{\mathbf{r}_2 t} \cdot (\mathbf{r}_1 - \mathbf{r})}, \qquad (C.6)$$

where $\mathbf{P}_{\mathbf{r}_k t} \equiv \mathbf{p}_k + (e/c)\mathbf{A}_{\mathbf{r}_k t}$ is introduced in Eq. (9.6), and the Jacobian of the coordinate transformation is introduced in the usual way, as

$$|\mathcal{J}_3| = \frac{\partial(\Delta\mathbf{r}, \mathbf{x}')}{\partial(\mathbf{r}_1, \mathbf{r}_2)}. \qquad (C.7)$$

These expressions are written for the 3D case. To consider one- or two-dimensional problems ($d = 1$ or $d = 2$) one should write the phase volume $(2\pi\hbar)^d$ in the integrals over momenta. To calculate the Jacobian (C.7), we use

$$\frac{\partial\Delta r_\alpha}{\partial r_{1\beta}} = 2\delta_{\alpha\beta}, \quad \frac{\partial\Delta r_\alpha}{\partial r_{2\beta}} = -2\delta_{\alpha\beta}, \quad \frac{\partial x'_\alpha}{\partial r_{1\beta}} = \frac{\partial x'_\alpha}{\partial r_{2\beta}} = \delta_{\alpha\beta} \qquad (C.8)$$

and obtain $J_d = 2^{2d}$, $d = 1, 2, 3$. The exact formula for the operator product transformation takes the form

$$c_t(\mathbf{r}, \mathbf{p}) = \int \frac{d\mathbf{p}_1}{(2\pi\hbar)^d} \int \frac{d\mathbf{p}_2}{(2\pi\hbar)^d} \int d\mathbf{r}_1 \int d\mathbf{r}_2 |\mathcal{J}_d|$$

$$\times a_t(\mathbf{r}_1, \mathbf{p}_1) b_t(\mathbf{r}_2, \mathbf{p}_2) \exp\left[\frac{2i}{\hbar} S(\mathbf{rp}, \mathbf{r}_1\mathbf{p}_1, \mathbf{r}_2\mathbf{p}_2)\right], \qquad (C.9)$$

where the factor $S(\mathbf{rp}, \mathbf{r}_1\mathbf{p}_1, \mathbf{r}_2\mathbf{p}_2)$ in the exponent is determined by

$$S(\mathbf{rp}, \mathbf{r}_1\mathbf{p}_1, \mathbf{r}_2\mathbf{p}_2) = (\mathbf{P}_{\mathbf{r}t} - \mathbf{P}_{\mathbf{r}_2t}) \cdot (\mathbf{r} - \mathbf{r}_1) - (\mathbf{P}_{\mathbf{r}t} - \mathbf{P}_{\mathbf{r}_1t}) \cdot (\mathbf{r} - \mathbf{r}_2). \qquad (C.10)$$

In order to simplify Eq. (C.9) for smooth functions $a_t(\mathbf{r}, \mathbf{p})$ and $b_t(\mathbf{r}, \mathbf{p})$, we introduce new variables $\Delta\mathbf{r}_{1,2} = \mathbf{r}_{1,2} - \mathbf{r}$ and $\Delta\mathbf{p}_{1,2} = \mathbf{p}_{1,2} - \mathbf{p}$, and expand the function in the expression under the integrals of Eq. (C.9) by using $\mathbf{A}_\mathbf{r} - \mathbf{A}_{\mathbf{r}+\Delta\mathbf{r}_{1,2}} \simeq 0$. As a result, we have

$$c_t(\mathbf{r}, \mathbf{p}) = |\mathcal{J}_d| \int \frac{d\Delta\mathbf{p}_1}{(2\pi\hbar)^d} \int \frac{d\Delta\mathbf{p}_2}{(2\pi\hbar)^d} \int d\Delta\mathbf{r}_1 \int d\Delta\mathbf{r}_2 \left[a_t(\mathbf{r}, \mathbf{p}) \right.$$

$$\left. + \frac{\partial a_t}{\partial\mathbf{r}} \cdot \Delta\mathbf{r}_1 + \frac{\partial a_t}{\partial\mathbf{p}} \cdot \Delta\mathbf{p}_1 + \dots \right] \left[b_t(\mathbf{r}, \mathbf{p}) + \frac{\partial b_t}{\partial\mathbf{r}} \cdot \Delta\mathbf{r}_2 + \frac{\partial b_t}{\partial\mathbf{p}} \cdot \Delta\mathbf{p}_2 + \dots \right]$$

$$\times \exp\left[\frac{2i}{\hbar}(\Delta\mathbf{r}_1 \cdot \Delta\mathbf{p}_2 - \Delta\mathbf{r}_2 \cdot \Delta\mathbf{p}_1)\right]. \qquad (C.11)$$

The contribution proportional to $a_t b_t$ in Eq. (C.11) is determined by the integral

$$|\mathcal{J}_d| \int \frac{d\Delta\mathbf{p}_1}{(2\pi\hbar)^d} \int \frac{d\Delta\mathbf{p}_2}{(2\pi\hbar)^d} \int d\Delta\mathbf{r}_1 \int d\Delta\mathbf{r}_2 \exp\left[\frac{2i}{\hbar}(\Delta\mathbf{r}_1 \cdot \Delta\mathbf{p}_2 - \Delta\mathbf{r}_2 \cdot \Delta\mathbf{p}_1)\right]$$

$$= 2^{2d} \int d\Delta\mathbf{p}_1 \int d\Delta\mathbf{p}_2 \delta(2\Delta\mathbf{p}_1)\delta(2\Delta\mathbf{p}_2) = 1. \qquad (C.12)$$

The products of a_t and b_t by the derivatives of these functions vanish from Eq. (C.11), since they are multiplied by the integrals containing the contributions linear in $\Delta\mathbf{r}_{1,2}$ or $\Delta\mathbf{p}_{1,2}$. These integrals are equal to zero (this can be checked by the substitutions $\Delta\mathbf{r}_{1,2} \to -\Delta\mathbf{r}_{1,2}$ or $\Delta\mathbf{p}_{1,2} \to -\Delta\mathbf{p}_{1,2}$). If the products $\Delta r_{1,2}^\alpha \Delta r_{1,2}^\beta$ or $\Delta p_{1,2}^\alpha \Delta p_{1,2}^\beta$ stay under an integral of the type (C.12), they also vanish. This means that the products $(\partial a_t/\partial r_\alpha)(\partial b_t/\partial r_\beta)$ or $(\partial a_t/\partial p_\alpha)(\partial b_t/\partial p_\beta)$ drop out of Eq. (C.11). Therefore, only the integrals

$$|\mathcal{J}_d| \int \frac{d\Delta\mathbf{p}_1}{(2\pi\hbar)^d} \int \frac{d\Delta\mathbf{p}_2}{(2\pi\hbar)^d} \int d\Delta\mathbf{r}_1 \int d\Delta\mathbf{r}_2$$

$$\times \exp\left[\frac{2i}{\hbar}(\Delta\mathbf{r}_1 \cdot \Delta\mathbf{p}_2 - \Delta\mathbf{r}_2 \cdot \Delta\mathbf{p}_1)\right]\Bigg|\begin{array}{c}\Delta r_1^\alpha \Delta p_2^\beta \\ \Delta p_1^\alpha \Delta r_2^\beta\end{array}\Bigg| \tag{C.13}$$

remain. These integrals stay at the products of the first derivatives over coordinate and over momentum. If $\alpha \neq \beta$, we again have zero in Eq. (C.13). One may prove this statement by changing the signs of the variables. The contribution $\propto \delta_{\alpha\beta}$ is calculated analogous to Eq. (C.12), using the integration by parts. The expression (C.13) is equal to

$$\delta_{\alpha\beta}2^d \left(\int\frac{d\Delta p}{2\pi\hbar}\int d\Delta r e^{\pm(2i/\hbar)\Delta r \Delta p}\right)^{d-1}\int\frac{d\Delta p}{2\pi\hbar}\int d\Delta r e^{\pm(2i/\hbar)\Delta r \Delta p}\Delta r\Delta p$$

$$= \delta_{\alpha\beta}2^d\left[\int d\Delta p\,\delta(2\Delta p)\right]^{d-1}\int\frac{d\Delta p}{2\pi\hbar}\frac{\Delta p}{2}\int d\Delta r\left(\mp i\hbar\frac{\partial}{\partial\Delta p}e^{\pm(2i/\hbar)\Delta r\Delta p}\right)$$

$$= \pm\delta_{\alpha\beta}\frac{i\hbar}{2}. \tag{C.14}$$

Therefore, the quasi-classical expression for the operator product is given as follows:

$$c_t(\mathbf{r}, \mathbf{p}) = a_t(\mathbf{r}, \mathbf{p})b_t(\mathbf{r}, \mathbf{p}) + \frac{i\hbar}{2}(a_t, b_t)_{\mathbf{rp}} + \dots. \tag{C.15}$$

The contribution linear in the Planck constant $\hbar$ is written through the classical Poisson brackets

$$(a_t, b_t)_{\mathbf{rp}} = \left(\frac{\partial a_t}{\partial\mathbf{r}}\cdot\frac{\partial b_t}{\partial\mathbf{p}} - \frac{\partial a_t}{\partial\mathbf{p}}\cdot\frac{\partial b_t}{\partial\mathbf{r}}\right). \tag{C.16}$$

These expressions are consistent with Eq. (9.24), and the quantum correction determined by Eq. (C.16) can be neglected under the condition $\hbar/\bar{\lambda} \ll \bar{p}$, where $\bar{\lambda}$ and $\bar{p}$ are the characteristic spatial scale and momentum for the functions a_t and b_t. The $\hbar^2$-corrections to Eq. (C.15) can be written if the next terms of the expansion in Eq. (C.11) are taken into account.

Let us use the quasi-classical expression (C.15) in order to check the commutation relation for coordinate and momentum, $[\hat{r}_\alpha, \hat{p}_\beta] = i\hbar\delta_{\alpha\beta}$. According to Eq. (C.15), the products of the operators standing in the commutator take the form

$$\hat{r}_\alpha\hat{p}_\beta \to r_\alpha p_\beta + \frac{i\hbar}{2}\delta_{\alpha\beta}, \qquad \hat{p}_\beta\hat{r}_\alpha \to p_\beta r_\alpha - \frac{i\hbar}{2}\delta_{\alpha\beta}. \tag{C.17}$$

Therefore, the classical contributions to the commutator annihilate and the commutator is equal to $i\hbar\delta_{\alpha\beta}$. In a similar fashion, one may prove that the relation $\hat{v}_\alpha = (i/\hbar)[\hat{h}, \hat{r}_\alpha]$ connecting the coordinate and velocity operators is consistent with the classical expression for the group velocity. According to Eq. (C.15),

$$\hat{h}\hat{r}_\alpha \to \varepsilon_{\mathbf{rp}}r_\alpha - \frac{i\hbar}{2}\frac{\partial\varepsilon_{\mathbf{rp}}}{\partial p_\alpha}, \qquad \hat{r}_\alpha\hat{h} \to r_\alpha\varepsilon_{\mathbf{rp}} + \frac{i\hbar}{2}\frac{\partial\varepsilon_{\mathbf{rp}}}{\partial p_\alpha}. \tag{C.18}$$

Composing the commutator, we obtain the classical expression $\mathbf{v}_{\mathbf{rp}} = \partial\varepsilon_{\mathbf{rp}}/\partial\mathbf{p}$.

Appendix D
Double-Time Green's Functions

The introduction of the Green's functions of electrons in Chapter 3 is based upon the averaging, over the impurity distribution, of the Green's function of the Schroedinger equation. Below we present a more general definition of Green's functions, which is widely used in statistical physics. Let us first introduce the Heisenberg representation of an arbitrary operator $\hat{A}$:

$$\hat{A}(t) = e^{i\hat{H}t/\hbar} \hat{A} e^{-i\hat{H}t/\hbar}, \qquad (D.1)$$

where $\hat{H}$ is the time-independent Hamiltonian. The operator $\hat{A}$ can be expressed in terms of creation and annihilation operators obeying either bosonic or fermionic commutation rules. The retarded, advanced, and causal Green's functions, which are labeled by the indices R, A, and c, respectively, are defined through the correlation functions of a pair of Heisenberg operators $\hat{A}(t)$ and $\hat{B}(t')$ according to

$$G_{tt'}^{R} = -\frac{i}{\hbar}\theta(t - t')\langle\langle\hat{A}(t)\hat{B}(t') \pm \hat{B}(t')\hat{A}(t)\rangle\rangle \equiv \langle\langle\hat{A}|\hat{B}\rangle\rangle_{tt'}^{R}, \qquad (D.2)$$

$$G_{tt'}^{A} = \frac{i}{\hbar}\theta(t' - t)\langle\langle\hat{A}(t)\hat{B}(t') \pm \hat{B}(t')\hat{A}(t)\rangle\rangle \equiv \langle\langle\hat{A}|\hat{B}\rangle\rangle_{tt'}^{A}, \qquad (D.3)$$

$$G_{tt'}^{c} = -\frac{i}{\hbar}\theta(t - t')\langle\langle\hat{A}(t)\hat{B}(t')\rangle\rangle \pm \frac{i}{\hbar}\theta(t' - t)\langle\langle\hat{B}(t')\hat{A}(t)\rangle\rangle \equiv \langle\langle\hat{A}|\hat{B}\rangle\rangle_{tt'}^{c}. \qquad (D.4)$$

Here and below, the upper and lower signs in equations stand for the fermion and boson operators, respectively. This definition is made for the sake of convenience, to employ the commutation rules for fermions and bosons in the equations of motion for the Green's functions, see below. The double angular brackets denote the averaging in the sense of Eq. (1.18), $\langle\langle\ldots\rangle\rangle = \mathrm{Sp}(\hat{\eta}\ldots)$, and the statistical operator $\hat{\eta}$ is time-independent in the Heisenberg representation. In the case of thermodynamic equilibrium, when the statistical operator $\hat{\eta} = \hat{\eta}_{eq}$ commutes with the Hamiltonian $\hat{H}$, one can easily show that $G_{tt'}^{s}$ ($s = R, A, c$) depend only of $t - t'$, i.e., $G_{tt'}^{s} = G_{t-t'}^{s}$. Therefore, it is convenient to use the energy representation of the Green's functions according to

$$G_{t-t'}^{s} = \int_{-\infty}^{\infty} \frac{d\varepsilon}{2\pi\hbar} e^{-i\varepsilon(t-t')/\hbar} G_{\varepsilon}^{s}, \qquad G_{\varepsilon}^{s} = \int_{-\infty}^{\infty} dt\, e^{i\varepsilon t/\hbar} G_{t}^{s}. \qquad (D.5)$$

The energy representation of the Green's function is also denoted as $\langle\langle \hat{A}|\hat{B}\rangle\rangle_{\varepsilon}^{s}$.

In the case of thermodynamic equilibrium, the correlation function $\langle\langle \hat{A}(t)\hat{B}(t')\rangle\rangle$ can be expressed through the retarded and advanced Green's functions in the energy representation. Let us introduce

$$J_{AB}(\omega) = \int d(t - t') e^{i\omega(t-t')} \langle\langle \hat{A}(t)\hat{B}(t')\rangle\rangle,$$

$$J_{BA}(\omega) = \int d(t - t') e^{i\omega(t-t')} \langle\langle \hat{B}(t')\hat{A}(t)\rangle\rangle. \tag{D.6}$$

First we note that

$$J_{AB}(\omega) = e^{\hbar\omega/T} J_{BA}(\omega). \tag{D.7}$$

This identity can be checked easily if we calculate the traces in $\langle\langle \ldots \rangle\rangle$ in the exact eigenstate representation and use the equilibrium statistical operator which is expressed through the temperature T. Using Eq. (D.7), we rewrite G_{ε}^{R} and G_{ε}^{A}, where $\varepsilon = \hbar\omega$, as

$$\left| \begin{array}{c} G_{\varepsilon}^{R} \\ G_{\varepsilon}^{A} \end{array} \right| = \frac{i}{\hbar} \int_{-\infty}^{\infty} dt \left| \begin{array}{c} -\theta(t) \\ \theta(-t) \end{array} \right| e^{i\omega t} \int_{-\infty}^{\infty} \frac{d\omega'}{2\pi} (e^{\hbar\omega'/T} \pm 1) J_{BA}(\omega'). \tag{D.8}$$

The step function $\theta(t)$ can be represented as

$$\theta(t) = \frac{i}{2\pi} \int_{-\infty}^{\infty} dx \frac{e^{-ixt}}{x + i\lambda}, \quad \lambda \to +0. \tag{D.9}$$

We substitute this expression into Eq. (D.8), integrate this equation over t and x, and obtain

$$G_{\varepsilon}^{R,A} = \frac{1}{2\pi\hbar} \int_{-\infty}^{\infty} \frac{d\omega'}{\omega - \omega' + i\lambda} (e^{\hbar\omega'/T} \pm 1) J_{BA}(\omega'), \tag{D.10}$$

where $\lambda \to +0$ for the retarded (R) and $\lambda \to -0$ for the advanced (A) Green's function. Equation (D.10) leads to the exact relation

$$G_{\varepsilon}^{A} - G_{\varepsilon}^{R} = \frac{i}{\hbar}(e^{\varepsilon/T} \pm 1) J_{BA}(\omega). \tag{D.11}$$

Assuming that $J_{BA}(\omega)$ is real, we also find $G_{\varepsilon}^{A} = G_{\varepsilon}^{R*}$ so that the left-hand side of Eq. (D.11) can be rewritten as $2i\mathrm{Im}G_{\varepsilon}^{A}$ or $-2i\mathrm{Im}G_{\varepsilon}^{R}$. Let us express $[\omega - \omega' + i\lambda]^{-1}$ through the principal value $\mathcal{P}(\omega - \omega')^{-1}$ and delta-function $\delta(\omega - \omega')$ as in problem 1.4. We find that Eq. (D.10) leads to the following dispersion relations:

$$\mathrm{Re}G_{\varepsilon}^{R} = \frac{1}{\pi} \mathcal{P} \int_{-\infty}^{\infty} \frac{d\varepsilon'}{\varepsilon' - \varepsilon} \mathrm{Im}G_{\varepsilon'}^{R},$$

$$\mathrm{Re}G_{\varepsilon}^{A} = -\frac{1}{\pi} \mathcal{P} \int_{-\infty}^{\infty} \frac{d\varepsilon'}{\varepsilon' - \varepsilon} \mathrm{Im}G_{\varepsilon'}^{A}. \tag{D.12}$$

The spectral representation (D.10) and dispersion relations (D.12) are directly related to the spectral representation and dispersion relations for the kinetic coefficients describing the linear response of the system under consideration, since these kinetic coefficients can be expressed through the retarded Green's functions. We note that equations similar to Eqs. (D.10) and (D.12) can be written for the causal Green's function as well.

Since the Heisenberg operators $\hat{A}(t)$ satisfy the equation of motion $i\hbar d\hat{A}(t)/dt = \hat{A}(t)\hat{H} - \hat{H}\hat{A}(t)$, one can write the following equation of motion for the Green's functions:

$$i\hbar \frac{dG_{tt'}^s}{dt} = \delta(t - t')\langle\langle \hat{A}(t)\hat{B}(t) \pm \hat{B}(t)\hat{A}(t)\rangle\rangle + \langle\langle [\hat{A}, \hat{H}]|\hat{B}\rangle\rangle_{tt'}^s . \qquad (D.13)$$

The double-time Green's function $\langle\langle [\hat{A}, \hat{H}]|\hat{B}\rangle\rangle_{tt'}^s$ standing on the right-hand side of Eq. (D.13) is determined from a similar equation of motion and expressed through another double-time Green's function containing the commutator $[[\hat{A}, \hat{H}], \hat{H}]$. In this way one gets a chain of coupled equations which can be cut under appropriate approximations. In particular, for the systems with weak interaction one can retain only the terms of a given order (linear, quadratic, etc.) in the interaction part of the Hamiltonian.

The Green's functions of quasiparticles are defined by substituting the field operators $\hat{\Psi}_{\mathbf{x}}$ and $\hat{\Psi}_{\mathbf{x}}^+$ (or the annihilation and creation operators of these quasiparticles) in place of $\hat{A}$ and $\hat{B}$. The one-particle Green's function in the coordinate representation is introduced as

$$G_{tt'}^s(\mathbf{x}, \mathbf{x}') = \langle\langle \hat{\Psi}_{\mathbf{x}}|\hat{\Psi}_{\mathbf{x}'}^+\rangle\rangle_{tt'}^s. \qquad (D.14)$$

Expanding $\hat{\Psi}_{\mathbf{x}}$ over a set of quantum states as $\hat{\Psi}_{\mathbf{x}}(t) = \sum_k \psi_{\mathbf{x}}^{(k)} \hat{a}_k(t)$, where the index k numbers these states, we have $G_{tt'}^s(\mathbf{x}, \mathbf{x}') = \sum_{kk'} G_{tt'}^s(k, k')\psi_{\mathbf{x}}^{(k)}\psi_{\mathbf{x}'}^{(k')*}$. The one-particle Green's function in the k-state representation is given by

$$G_{tt'}^s(k, k') = \langle\langle \hat{a}_k|\hat{a}_{k'}^+\rangle\rangle_{tt'}^s. \qquad (D.15)$$

If $\psi_{\mathbf{x}}^{(k)}$ are exact eigenstates of the Hamiltonian $\hat{H}$, i.e., when $\hat{H} = \sum_k \varepsilon_k \hat{a}_k^+ \hat{a}_k$, the correlation functions are calculated with the use of the following identities:

$$\hat{a}_k(t) = e^{-i\varepsilon_k t/\hbar}\hat{a}_k, \quad \hat{a}_k^+(t) = e^{i\varepsilon_k t/\hbar}\hat{a}_k^+, \quad \langle\langle \hat{a}_k^+ \hat{a}_{k'}\rangle\rangle = n_k \delta_{kk'}, \qquad (D.16)$$

where n_k are the occupation numbers for the quasiparticles (fermions or bosons). The Green's function in the exact eigenstate representation is diagonal, $G_{tt'}^s(k, k') = \delta_{kk'} G_{t-t'}^s(k)$, where

$$G_{t-t'}^R(k) = -\frac{i}{\hbar}\theta(t - t')e^{-i\varepsilon_k(t-t')/\hbar}, \qquad (D.17)$$

$$G_{t-t'}^A(k) = \frac{i}{\hbar}\theta(t' - t)e^{-i\varepsilon_k(t-t')/\hbar}, \qquad (D.18)$$

and

$$G_{t-t'}^c(k) = \frac{i}{\hbar}[-\theta(t - t')(1 \mp n_k) \pm \theta(t' - t)n_k]e^{-i\varepsilon_k(t-t')/\hbar}. \qquad (D.19)$$

We note that the retarded and advanced Green's functions in the exact eigenstate representation are temperature-independent. The energy representation of these functions coincides with the one given for electrons in Chapter 3, see Eq. (14.9), where the momentum $\mathbf{p}$ stands in place of the quantum number k. In contrast, the causal Green's function depends on the occupation number n_k. One has

$$G_\varepsilon^R(k) = G_\varepsilon^{A*}(k) = \frac{1}{\varepsilon - \varepsilon_k + i\lambda},$$

$$G_\varepsilon^c(k) = \frac{\pm n_k}{\varepsilon - \varepsilon_k - i\lambda} + \frac{1 \mp n_k}{\varepsilon - \varepsilon_k + i\lambda} \qquad (D.20)$$

with $\lambda \to +0$.

Considering photons and phonons, it is convenient to introduce Green's functions in another way. We note that the observable physical values related to the photons and phonons are expressed through the vector potential of electromagnetic field and through the atomic displacement vectors, respectively. The Hamiltonians describing the interaction of photons and phonons with electrons, as well as the interaction of photons with phonons and phonon-phonon interaction, are also expressed in terms of these vectors. Since the spatial Fourier transforms of both the vector potential and the vectors of atomic displacement contain bosonic creation and annihilation operators in the combination $\hat{b}_{\mathbf{q}\mu} + \hat{b}^+_{-\mathbf{q}\mu}$, where μ is the mode index (polarization), it is natural to define the Green's function of photons (or phonons) as

$$D_{tt'}^{\mu\mu',s}(\mathbf{q}, \mathbf{q}') = \langle\langle \hat{b}_{\mathbf{q}\mu} + \hat{b}^+_{-\mathbf{q}\mu} | \hat{b}_{-\mathbf{q}'\mu'} + \hat{b}^+_{\mathbf{q}'\mu'} \rangle\rangle^s_{tt'}. \qquad (D.21)$$

We use the letter D instead of G to emphasize the difference of the definition (D.21) with respect to (D.15). If the photons (or phonons) are described by the free-boson Hamiltonian $\hat{H}_b = \sum_{\mathbf{q}\mu} \hbar\omega_{\mathbf{q}\mu}(\hat{b}^+_{\mathbf{q}\mu}\hat{b}_{\mathbf{q}\mu} + 1/2)$, we can use Eq. (D.16) rewritten for the boson operators $\hat{b}$ and $\hat{b}^+$. As a result, $D_{tt'}^{\mu\mu',s}(\mathbf{q}, \mathbf{q}') = \delta_{\mu\mu'}\delta_{\mathbf{qq}'}D_{t-t'}^{\mu,s}(\mathbf{q})$, where

$$D_{t-t'}^{\mu,R}(\mathbf{q}) = -\frac{i}{\hbar}\theta(t - t')[e^{-i\omega_{\mathbf{q}\mu}(t-t')} - e^{i\omega_{\mathbf{q}\mu}(t-t')}], \qquad (D.22)$$

$$D_{t-t'}^{\mu,A}(\mathbf{q}) = \frac{i}{\hbar}\theta(t' - t)[e^{-i\omega_{\mathbf{q}\mu}(t-t')} - e^{i\omega_{\mathbf{q}\mu}(t-t')}], \qquad (D.23)$$

and

$$D_{t-t'}^{\mu,c}(\mathbf{q}) = -\frac{i}{\hbar}\left\{\theta(t - t')[(N_{\mathbf{q}}^{\mu} + 1)e^{-i\omega_{\mathbf{q}\mu}(t-t')} + N_{-\mathbf{q}}^{\mu}e^{i\omega_{\mathbf{q}\mu}(t-t')}]\right.$$
$$\left. + \theta(t' - t)[N_{\mathbf{q}}^{\mu}e^{-i\omega_{\mathbf{q}\mu}(t-t')} + (N_{-\mathbf{q}}^{\mu} + 1)e^{i\omega_{\mathbf{q}\mu}(t-t')}]\right\}. \qquad (D.24)$$

Here $N_{\mathbf{q}}^{\mu}$ is the distribution function of photons or phonons, which becomes the Planck distribution function in equilibrium. In the energy representation,

$$D_{\omega}^{\mu,R}(\mathbf{q}) = D_{\omega}^{\mu,A}{}^*(\mathbf{q}) = \frac{1}{\hbar\omega - \hbar\omega_{\mathbf{q}\mu} + i\lambda} - \frac{1}{\hbar\omega + \hbar\omega_{\mathbf{q}\mu} + i\lambda}, \qquad (D.25)$$

where $\omega = \varepsilon/\hbar$ and $\lambda \to +0$. In Eqs. (D.22)–(D.25) we have used the symmetry property $\omega_{-\mathbf{q}\mu} = \omega_{\mathbf{q}\mu}$ following from the symmetry with respect to time reversal. For the same reason, $N_{-\mathbf{q}}^{\mu}$ in Eq. (D.24) can be replaced by $N_{\mathbf{q}}^{\mu}$.

Appendix E
Many-Electron Green's Functions

To describe a system of many electrons, one can use a set of n-particle Green's functions describing evolution of the system when n electrons are added to it at the instant t_1' and taken out at the instant t_1. The one-particle Green's function $G(\mathbf{r}_1 t_1, \mathbf{r}_1' t_1')$ is introduced as

$$G(\mathbf{r}_1 t_1, \mathbf{r}_1' t_1') = -i\langle \hat{\mathcal{T}} \hat{\Psi}_{\mathbf{r}_1}(t_1) \hat{\Psi}_{\mathbf{r}_1'}^+(t_1') \rangle_o$$

$$= \begin{cases} -i\langle \hat{\Psi}_{\mathbf{r}_1}(t_1) \hat{\Psi}_{\mathbf{r}_1'}^+(t_1') \rangle_o \,, & t_1 > t_1' \\ i\langle \hat{\Psi}_{\mathbf{r}_1'}^+(t_1') \hat{\Psi}_{\mathbf{r}_1}(t_1) \rangle_o \,, & t_1 < t_1' \end{cases} \,, \qquad (E.1)$$

where $\langle \ldots \rangle_o = \langle 0| \ldots |0 \rangle$ is the quantum-mechanical averaging over the ground state $|0\rangle$ of the system, $\hat{\Psi}_{\mathbf{r}}(t)$ is the electron field operator (see Sec. 4) in the Heisenberg representation (D.1), and $\hat{\mathcal{T}}$ is the operator of chronological ordering introduced in Sec. 2. The field operators and the Green's functions depend on times and coordinates as well as on spin variables. For the sake of brevity, we omit the corresponding spin indices. We also put $\hbar = 1$ in Eq. (E.1) and below. Since the averaging $\langle 0| \ldots |0 \rangle$ is equivalent to the averaging $\mathrm{Sp}\hat{\eta}_{eq} \ldots$ at zero temperature, the function $G(\mathbf{r}_1 t_1, \mathbf{r}_1' t_1')$ is the causal zero-temperature double-time Green's function (D.4), where the electron field operators $\hat{\Psi}_{\mathbf{r}_1}$ and $\hat{\Psi}_{\mathbf{r}_1'}^+$ stand in place of $\hat{A}$ and $\hat{B}$. The Hamiltonian $\hat{H} = \hat{H}_0 + \hat{H}_{ee}$ contains the free-electron and electron-electron interaction terms

$$\hat{H}_0 = \int d\mathbf{r}_1 \hat{\Psi}_{\mathbf{r}_1}^+ \hat{h}_1 \hat{\Psi}_{\mathbf{r}_1}, \qquad \hat{H}_{ee} = \frac{1}{2} \int d\mathbf{r}_1 \int d\mathbf{r}_2 \hat{\Psi}_{\mathbf{r}_1}^+ \hat{\Psi}_{\mathbf{r}_2}^+ v_{12} \hat{\Psi}_{\mathbf{r}_2} \hat{\Psi}_{\mathbf{r}_1}, \qquad (E.2)$$

where $\hat{h}_1 \equiv \hat{h}_{\mathbf{r}_1}$ is the one-electron Hamiltonian and $v_{12} = v_{21} = v(\mathbf{r}_1, \mathbf{r}_2)$ is the potential energy of electron-electron interaction; see Eqs. (4.27), (28.1), and (28.2). The field operators satisfy the relations

$$[\hat{\Psi}_{\mathbf{r}_k}, \hat{H}_0] = \hat{h}_k \hat{\Psi}_{\mathbf{r}_k}, \qquad [\hat{\Psi}_{\mathbf{r}_k}, \hat{H}_{ee}] = \int d\mathbf{r}_1 v_{k1} \hat{\Psi}_{\mathbf{r}_1}^+ \hat{\Psi}_{\mathbf{r}_1} \hat{\Psi}_{\mathbf{r}_k}. \qquad (E.3)$$

The equation of motion for $G(\mathbf{r}_1 t_1, \mathbf{r}_1' t_1')$ can be obtained from Eq. (D.13):

$$i\frac{\partial G(\mathbf{r}_1 t_1, \mathbf{r}_1' t_1')}{\partial t_1} = \delta_{11'} - i\langle \hat{\mathcal{T}}[\hat{\Psi}_{\mathbf{r}_1}(t_1), \hat{H}] \hat{\Psi}_{\mathbf{r}_1'}^+(t_1') \rangle_o \,, \qquad (E.4)$$

where $\delta_{11'} \equiv \delta(t_1 - t_1')\delta(\mathbf{r}_1 - \mathbf{r}_1')$. Substituting the relations (E.3) into Eq. (E.4), we find

$$\left(i\frac{\partial}{\partial t_1} - \hat{h}_1\right) G(\mathbf{r}_1 t_1, \mathbf{r}_1' t_1') = \delta_{11'} + i\int d\mathbf{r}_2 v_{12}\langle\hat{\mathcal{T}}\hat{\Psi}_{\mathbf{r}_1}(t_1)\hat{\Psi}_{\mathbf{r}_2}(t_1)$$

$$\times\hat{\Psi}_{\mathbf{r}_2}^+(t_1 + 0)\hat{\Psi}_{\mathbf{r}_1'}^+(t_1')\rangle_o = \delta_{11'} - i\int d\mathbf{r}_2 v_{12} G(\mathbf{r}_1 t_1, \mathbf{r}_2 t_1; \mathbf{r}_1' t_1', \mathbf{r}_2 t_1 + 0). \qquad (E.5)$$

The interaction contribution is expressed in this equation through the two-particle Green's function introduced as

$$G(\mathbf{r}_1 t_1, \mathbf{r}_2 t_2; \mathbf{r}_1' t_1', \mathbf{r}_2' t_2') = (-i)^2\langle\hat{\mathcal{T}}\hat{\Psi}_{\mathbf{r}_1}(t_1)\hat{\Psi}_{\mathbf{r}_2}(t_2)\hat{\Psi}_{\mathbf{r}_2'}^+(t_2')\hat{\Psi}_{\mathbf{r}_1'}^+(t_1')\rangle_o . \qquad (E.6)$$

The infinitely small factor $+0$ is introduced in Eq. (E.5) to provide correct ordering of the field operators with the same time argument t_1 under the operator $\hat{\mathcal{T}}$. The function (E.6) satisfies an equation similar to Eq. (E.5), where the interaction term is expressed through the three-particle Green's function which, in turn, satisfies an equation containing four-particle Green's function. Therefore, we obtain an infinite chain of equations.

The n-particle Green's function is defined as

$$G(\mathbf{r}_1 t_1, \ldots, \mathbf{r}_n t_n; \mathbf{r}_1' t_1', \ldots, \mathbf{r}_n' t_n') = (-i)^n$$

$$\times\langle\hat{\mathcal{T}}\hat{\Psi}_{\mathbf{r}_1}(t_1)\ldots\hat{\Psi}_{\mathbf{r}_n}(t_n)\hat{\Psi}_{\mathbf{r}_n'}^+(t_n')\ldots\hat{\Psi}_{\mathbf{r}_1'}^+(t_1')\rangle_o , \qquad (E.7)$$

which generalizes the definitions given by Eqs. (E.1) and (E.6). Let us prove that the Green's function (E.7) satisfies an equation similar to Eq. (E.5). Differentiating this function over t_1, we obtain the term $\langle\hat{\mathcal{T}}(\partial\hat{\Psi}_{\mathbf{r}_1}(t_1)/\partial t_1)\ldots\rangle_o$ as well as the contributions appearing due to permutations of $\hat{\Psi}$-operators when the times t_1 and t_k $(2 \leq k \leq n)$, or t_1 and t_k' $(1 \leq k \leq n)$, change their order. Such contributions are proportional to derivatives of the θ-functions and contain sets of paired terms whose signs are determined by the number of permutations leading to a given ordering of the $\hat{\Psi}$-operators:

$$\frac{\partial\theta(t_1 - t_k)}{\partial t_1}\langle\ldots\hat{\Psi}_{\mathbf{r}_1}(t_1)\hat{\Psi}_{\mathbf{r}_k}(t_k)\ldots\rangle_o - \frac{\partial\theta(t_k - t_1)}{\partial t_1}\langle\ldots\hat{\Psi}_{\mathbf{r}_k}(t_k)\hat{\Psi}_{\mathbf{r}_1}(t_1)\ldots\rangle_o$$

$$= \delta(t_1 - t_k)\langle\ldots(\hat{\Psi}_{\mathbf{r}_1}\hat{\Psi}_{\mathbf{r}_k} + \hat{\Psi}_{\mathbf{r}_k}\hat{\Psi}_{\mathbf{r}_1})\ldots\rangle_o = 0, \qquad (E.8)$$

$$\frac{\partial\theta(t_1 - t_k')}{\partial t_1}\langle\ldots\hat{\Psi}_{\mathbf{r}_1}(t_1)\hat{\Psi}_{\mathbf{r}_k'}^+(t_k')\ldots\rangle_o - \frac{\partial\theta(t_k' - t_1)}{\partial t_1}\langle\ldots\hat{\Psi}_{\mathbf{r}_k'}^+(t_k')$$

$$\times\hat{\Psi}_{\mathbf{r}_1}(t_1)\ldots\rangle_o = \delta(t_1 - t_k')\langle\ldots(\hat{\Psi}_{\mathbf{r}_1}\hat{\Psi}_{\mathbf{r}_k'}^+ + \hat{\Psi}_{\mathbf{r}_k'}^+\hat{\Psi}_{\mathbf{r}_1})\ldots\rangle_o = \langle\ldots\delta_{1k'}\ldots\rangle_o. \qquad (E.9)$$

Collecting such expressions with all possible permutations of the operators (the total number $n - 1 + n - k = 2(n - k) + k - 1$ of the permutations brings us the factor $(-1)^{k-1}$), we finally obtain the sum of all contributions coming from the derivatives of the θ-functions:

$$\sum_{k=1}^{n}(-1)^{k-1}\langle\hat{\mathcal{T}}\hat{\Psi}_2(t_2)\ldots\hat{\Psi}_{\mathbf{r}_n}(t_n)\hat{\Psi}_{\mathbf{r}_n'}^+(t_n')\ldots$$

$$\times\hat{\Psi}_{\mathbf{r}_{k+1}'}^+(t_{k+1}')\hat{\Psi}_{\mathbf{r}_{k-1}'}^+(t_{k-1}')\ldots\hat{\Psi}_{\mathbf{r}_1'}^+(t_1')\rangle_o\delta_{1k'}. \qquad (E.10)$$

Using Eq. (E.10), one can write, for example, an equation for the two-particle Green's function (E.6) in the form

$$\left(i\frac{\partial}{\partial t_1} - \hat{h}_1\right) G(\mathbf{r}_1 t_1, \mathbf{r}_2 t_2; \mathbf{r}_1' t_1', \mathbf{r}_2' t_2') = [\delta_{11'} G(\mathbf{r}_2 t_2; \mathbf{r}_2' t_2')]_{As}$$

$$-i\int d\mathbf{r}_3 v_{13} G(\mathbf{r}_1 t_1, \mathbf{r}_2 t_2, \mathbf{r}_3 t_1; \mathbf{r}_1' t_1', \mathbf{r}_2' t_2', \mathbf{r}_3 t_1 + 0), \qquad (E.11)$$

where the antisymmetrization of the inhomogeneous term with respect to the primed indices is defined as

$$[\delta_{11'} G(\mathbf{r}_2 t_2; \mathbf{r}_2' t_2')]_{As} \equiv \delta_{11'} G(\mathbf{r}_2 t_2; \mathbf{r}_2' t_2') - \delta_{12'} G(\mathbf{r}_2 t_2; \mathbf{r}_1' t_1'). \qquad (E.12)$$

For the n-particle Green's function we obtain

$$\left(i\frac{\partial}{\partial t_1} - \hat{h}_1\right) G(\mathbf{r}_1 t_1, \dots, \mathbf{r}_n t_n; \mathbf{r}_1' t_1', \dots, \mathbf{r}_n' t_n') \qquad (E.13)$$

$$= [\delta_{11'} G(\mathbf{r}_2 t_2, \dots, \mathbf{r}_n t_n; \mathbf{r}_2' t_2', \dots, \mathbf{r}_n' t_n')]_{As}$$

$$-i\int d\mathbf{r}_{n+1} v_{1,n+1} G(\mathbf{r}_1 t_1, \dots, \mathbf{r}_n t_n, \mathbf{r}_{n+1} t_1; \mathbf{r}_1' t_1', \dots, \mathbf{r}_n' t_n', \mathbf{r}_{n+1} t_1 + 0),$$

where the inhomogeneous term, according to Eq. (E.10), is antisymmetrized as

$$[\delta_{11'} G(\mathbf{r}_2 t_2, \dots, \mathbf{r}_n t_n; \mathbf{r}_2' t_2', \dots, \mathbf{r}_n' t_n')]_{As} \equiv \sum_{k=1}^{n} \delta_{1k'}(-1)^{k-1}$$

$$\times G(\mathbf{r}_2 t_2, \dots, \mathbf{r}_n t_n; \ \mathbf{r}_1' t_1', \dots, \mathbf{r}_{k-1}' t_{k-1}', \mathbf{r}_{k+1}' t_{k+1}', \dots, \mathbf{r}_n' t_n'). \qquad (E.14)$$

This equation should be also considered as a definition of the antisymmetrization operation $[\dots]_{As}$ if, instead of $\delta_{11'}$ and $\delta_{1k'}$, one substitutes arbitrary functions of $\mathbf{r}_1 t_1$ and $\mathbf{r}_1' t_1'$ and of $\mathbf{r}_1 t_1$ and $\mathbf{r}_k' t_k'$, respectively.

In the following, it is helpful to introduce the function $v(\mathbf{r}_1 t_1, \mathbf{r}_1' t_1') = v_{11'}\delta(t_1 - t_1')$. This allows one to write the integral on the right-hand side of Eq. (E.5) as $\int d\mathbf{r}_2 \int dt_2 v(\mathbf{r}_1 t_1, \mathbf{r}_2 t_2) G(\mathbf{r}_1 t_1, \mathbf{r}_2 t_2; \mathbf{r}_1' t_1', \mathbf{r}_2 t_2 + 0)$. The integrals on the right-hand side of Eqs. (E.11) and (E.13) are transformed in a similar way. At this point, it becomes convenient to unify the coordinate and time variables into the multi-indices $k = \mathbf{r}_k t_k$ so that $G(\mathbf{r}_1 t_1, \mathbf{r}_1' t_1') \equiv G(1, 1')$, $v(\mathbf{r}_1 t_1, \mathbf{r}_1' t_1') \equiv v(1, 1')$, and so on. To define $\mathbf{r}_k t_k + 0$, we use the multi-index k^+ so that $G(\mathbf{r}_1 t_1, \mathbf{r}_2 t_2; \ \mathbf{r}_1' t_1', \mathbf{r}_2 t_2 + 0)$ is written as $G(1, 2; 1', 2^+)$. The spin variables, which so far have been implicitly assumed, can be included in the multi-indices as well. Using these new notations, we rewrite Eq. (E.5) in the integral form

$$G(1, 1') = g(1, 1') - i\int d\tilde{1} g(1, \tilde{1}) \int d2 v(\tilde{1}, 2) G(\tilde{1}, 2; 1', 2^+), \qquad (E.15)$$

where we have introduced the one-particle Green's function $g(1, 1') = g(\mathbf{r}_1 t_1, \mathbf{r}_1' t_1')$ for non-interacting electrons, which satisfies Eq. (E.5) without the integral term:

$$\left(i\frac{\partial}{\partial t_1} - \hat{h}_1\right) g(1, 1') = \delta_{11'}. \qquad (E.16)$$

Since the Hamiltonian $\hat{h}_1$ is time-independent, $g(1, 1')$ depends only on $t_1 - t_1'$. Equation (E.11) is rewritten in the integral form in a similar fashion:

$$G(1,2;1',2') = [g(1,1')G(2,2')]_{As} \qquad (E.17)$$

$$-i \int d\tilde{1} g(1,\tilde{1}) \int d3 \; v(\tilde{1},3)G(\tilde{1},2,3;1',2',3^{+}).$$

With the use of Eqs. (E.13) and (E.16), the integral equations of such kind can be written for any n-particle Green's function. They form an infinite chain of integral equations. This chain can be cut in order to find the Green's functions with a given accuracy. For example, searching for the contributions quadratic in the interaction, one may neglect the interaction term in the equation for the three-particle Green's function. This leads to

$$G(1,2,3;1',2',3') \simeq [g(1,1')G(2,3;2',3')]_{As} \qquad (E.18)$$

$$= g(1,1')G(2,3;2',3') - g(1,2')G(2,3;1',3') + g(1,3')G(2,3;1',2').$$

Substituting this expression into Eq. (E.17), we obtain a closed system of two non-linear integral equations (E.15) and (E.17). The one-particle Green's function found from this system has the second-order accuracy with respect to the interaction. If we need to have just the first-order accuracy, we write an approximate solution of Eq. (E.17) as $G(1,2;1',2') \simeq [g(1,1')G(2,2')]_{As}$ and substitute it into Eq. (E.15), reducing the latter to

$$G(1,1') \simeq g(1,1') - i \int d\tilde{1} g(1,\tilde{1}) \int d2 v(\tilde{1},2) \qquad (E.19)$$

$$\times \left[g(\tilde{1},1')G(2,2^{+}) - g(\tilde{1},2)G(2,1') \right].$$

This is a single integral equation. The first-order correction to the one-electron Green's function is obtained by the substitutions $G(2,2^{+}) \to g(2,2^{+})$ and $G(2,1') \to g(2,1')$ in the integral term.

The infinite chain of integral equations discussed above can be solved by iterations with a given accuracy. Though this procedure is complicated, it can be simplified by introducing graphic images (diagrams). The one-electron Green's function $g(1,1')$ multiplied by i corresponds to a single thin solid line with the ends 1 and $1'$. The exact one-electron Green's function $G(1,1')$ multiplied by i is denoted by a bold solid line. The interaction-potential term $-iv(1,1')$ corresponds to a broken line with the ends 1 and $1'$. The two-particle (n-particle) Green's functions multiplied by i^{2} (i^{n}) are represented by two (n) lines entering and exiting a rectangle:

$$i^{n}G(1,2,\ldots n;1',2',\ldots n') = \qquad (E.20)$$

All terms in the iterative expansion of the one-particle Green's functions can be represented by combinations of solid and broken lines. Each vertex of such diagrams contains one entering and one exiting solid line (electron lines) and one broken line (potential line), similar to the case of the diagram technique for electron-impurity system considered in Secs. 14 and 15 (we note, however, that the broken lines in Secs. 14 and 15 have a different meaning). A summation over the multi-indices of the vertices is implied. There exist the diagrams containing loops of electron lines, each such loop gives a factor of -1 to the corresponding analytical expression of the diagram. For example, the perturbation term in Eq. (E.19) can be represented as a sum of two diagrams:

$$(E.21)$$

The loop in the first diagram of Eq. (E.21) is self-closed. Such self-closed loops, connected to the other part of the diagram by a single potential line, can appear in the diagrams of arbitrary order. Each of them gives a constant factor proportional to the electron density n, because $G(k, k^+) = G(\mathbf{r}_k t_k, \; \mathbf{r}_k t_k + 0) = i\langle \hat{\Psi}^+_{\mathbf{r}_k}(t_k)\hat{\Psi}_{\mathbf{r}_k}(t_k)\rangle_o = in$.

The diagrammatic expansion of $iG(1, 1')$ contains both reducible and irreducible diagrams. Separating these two kinds of diagrams in a similar way as in Sec. 14, we obtain the Dyson equation

$$G(1, 1') = g(1, 1') + \int d\tilde{1}g(1, \tilde{1}) \int d\tilde{1}'\Sigma(\tilde{1}, \tilde{1}')G(\tilde{1}', 1') , \qquad (E.22)$$

where $\Sigma(\tilde{1}, \tilde{1}')$ is the self-energy function formed as a sum of all irreducible diagrams. The structure of the leading terms of the diagrammatic expansion of $-i\Sigma(1, 1')$ can be understood from a comparison of Eq. (E.22) to Eq. (E.19) and diagrams (E.21): $-i\Sigma(1, 2) = -\delta_{12} \int d3[-iv(1, 3)]iG(3, 3^+) + [-iv(1, 2)]iG(1, 2^+) + \dots$. The terms up to the second order in the interaction are given by

$$(E.23)$$

Apart from the diagrams similar to those on the right-hand side of Eq. (14.23) (the second and the fourth), the self-energy function contains various diagrams with loops of electron lines. Some of these diagrams allow a further reduction if one can separate some blocks of lines out of them by cutting just two potential lines. The third diagram on the right-hand side of Eq. (E.23) is an example of such diagrams. The sums of all diagrams contributing into each of these blocks are denoted by bold broken lines and describe the effective (screened) interaction potential $V(1, 1')$. The bold broken line with the indices 1 and $1'$ corresponds to $-iV(1, 1')$ [to $-i\hbar V(1, 1')$ in the usual notations]. To describe the screened potential, one can also write the equation

$$V(1, 1') = v(1, 1') + \int d\tilde{1}v(1, \tilde{1}) \int d\tilde{1}'\Pi(\tilde{1}, \tilde{1}')V(\tilde{1}', 1'), \qquad (E.24)$$

or, in the diagram form,

$$\cdots\!\!\cdots \;=\; \cdots\cdots \;+\; \cdots\cdots\bullet\bigcirc\bullet\cdots\cdots \qquad (E.25)$$

The circle with the indices $\tilde{1}, \tilde{1}'$ corresponds to $i\Pi(\tilde{1}, \tilde{1}')$ [to $(i/\hbar)\Pi(\tilde{1}, \tilde{1}')$ in the usual notations]. The function $\Pi(\tilde{1}, \tilde{1}')$, which plays the role of the self-energy function in such equations, is called the polarization function. It is given by the infinite series containing all irreducible diagrams. Two leading-order terms of such expansion are shown below:

$$i\Pi(1,2) \equiv \bigcirc \;=\; \langle\rangle \;+\; \langle\vdots\rangle \;+\; \dots \; . \qquad (E.26)$$

The terms denoted by the dots ... may contain several loops connected to each other by two or more broken lines. All such diagrams are irreducible. On the other hands, the diagrams containing the loops connected to each other by a single broken line are reducible and, by definition, does not enter the polarization function. The polarization function and the screened potential are symmetric in their multi-indices: $\Pi(1, 1') = \Pi(1', 1)$ and $V(1, 1') = V(1', 1)$. The introduction of the screened interaction makes the third diagram on the right-hand side of Eq. (E.23) reducible. Replacing the thin broken lines in Eq. (E.23) by the bold broken lines, one should omit this diagram because it is united with the second one.

Consider now the diagrammatic expansion of two-particle Green's functions. According to Eq. (E.17),

$$\square \;=\; - \;-\; \times \;+\; \dots \; , \qquad (E.27)$$

where the dots ... define all possible terms containing the potential lines. Evaluating these terms by iterations as described above, we find that some of them describe all possible corrections to the one-particle Green's functions $g(1, 1')$ and $g(1, 2')$. Accounting for all such terms, we replace $g(1, 1')$ and $g(1, 2')$ by $G(1, 1')$ and $G(1, 2')$, respectively. In other words, the thin lines on the right-hand side of Eq. (E.27) should be replaced by the bold lines. The other terms in the sum denoted by the dots ... contain the potential lines connecting the upper and lower electron lines. All these terms can be described by the scattering amplitude $\mathcal{V}(\tilde{1}, \tilde{2}; \tilde{1}', \tilde{2}')$ defined in such a way that Eq. (E.17) is rewritten as

$$G(1, 2; 1', 2') = G(1, 1')G(2, 2') - G(1, 2')G(2, 1') \qquad (E.28)$$

$$+ \int d\tilde{1} \int d\tilde{2} G(1, \tilde{1})G(2, \tilde{2}) \int d\tilde{1}' \int d\tilde{2}' \mathcal{V}(\tilde{1}\tilde{2}; \tilde{1}'\tilde{2}')G(\tilde{1}', 1')G(\tilde{2}', 2').$$

We point out the symmetry properties of the scattering amplitude, $\mathcal{V}(1, 2; 1', 2') = \mathcal{V}(2, 1; 2', 1') = -\mathcal{V}(2, 1; 1', 2') = -\mathcal{V}(1, 2; 2', 1')$, following from the definition of the two-particle Green's function. The expansion of the scattering amplitude is written as $\mathcal{V}(1,2;1',2') = \delta_{11'}\delta_{22'}iV(1,2) - \delta_{12'}\delta_{21'}iV(1,2) + \ldots$, where the dots include the second-order and all higher-order terms with respect to the screened interaction. Equation (E.28) can be rewritten in the form of the Bethe-Salpeter equation (see Eqs. (15.3) and (15.4)) if we describe all irreducible diagrams contributing to the integral term of this equation by the irreducible vertex part $\Gamma(\tilde{1}, \tilde{2}; \tilde{1}', \tilde{2}')$:

$$G(1,2;1',2') = G(1,1')G(2,2') - G(1,2')G(2,1') \qquad (E.29)$$

$$+ \int d\tilde{1} \int d\tilde{2} G(1,\tilde{1})G(2,\tilde{2}) \int d\tilde{1}' \int d\tilde{2}' \Gamma(\tilde{1},\tilde{2};\tilde{1}',\tilde{2}')G(\tilde{1}',\tilde{2}';1',2').$$

The first- and second-order terms in the expansion of the vertex part with respect to the screened interaction look similar to those in Eq. (15.4):

$$-\Gamma(12;1'2') = \delta_{11'}\delta_{22'} \quad \vdots \quad + \quad \boxtimes \quad + \delta_{11'} \quad \vdots \quad + \delta_{22'} \quad \overline{} \quad + \ldots . \quad (E.30)$$

We have already mentioned that the Green's function of the electrons described by a stationary one-particle Hamiltonian depends only on the difference of its time arguments, $g(1,1') \equiv g(\mathbf{r}_1 t_1; \mathbf{r}_1' t_1') = g_{t_1 - t_1'}(\mathbf{r}_1, \mathbf{r}_1')$. As follows from the diagram equations discussed above, the same property is true for the functions G, Σ, V, and Π. Therefore, in many cases it is convenient to carry out the temporal Fourier transformation according to Eq. (D.5) and use the energy representation of these functions. Next, in a homogeneous system, when the one-particle Hamiltonian is translation-invariant, the functions G, Σ, V, and Π depend only of the difference of their coordinate arguments so that one may work with spatial Fourier transforms of these functions. Carrying out both temporal and spatial Fourier transformations, we obtain G, Σ, V, and Π in the energy-momentum representation. Equations (E.22) and (E.24) connecting these functions are transformed to the algebraic equations

$$G_\varepsilon(\mathbf{p}) = g_\varepsilon(\mathbf{p}) + g_\varepsilon(\mathbf{p})\Sigma_\varepsilon(\mathbf{p})G_\varepsilon(\mathbf{p}) \qquad (E.31)$$

and

$$V_\omega(\mathbf{q}) = v_q + v_q\Pi_\omega(\mathbf{q})V_\omega(\mathbf{q}). \qquad (E.32)$$

We point out the symmetry properties $\Pi_\omega(\mathbf{q}) = \Pi_{-\omega}(-\mathbf{q})$ and $V_\omega(\mathbf{q}) = V_{-\omega}(-\mathbf{q})$. According to Eq. (D.20), the one-electron Green's function of non-interacting electrons in the energy-momentum representation is

$$g_\varepsilon(\mathbf{p}) = \lim_{T \to 0} \left[\frac{f_p}{\varepsilon - \varepsilon_p - i\lambda} + \frac{1 - f_p}{\varepsilon - \varepsilon_p + i\lambda} \right] = \frac{1}{\varepsilon - \varepsilon_p + i\lambda \, \mathrm{sgn}(p - p_F)}, \qquad (E.33)$$

where $\lambda \to +0$ and one should take into account that $f_p = \theta(p_F - p)$ at zero temperature. Though the analytical properties of this function are simple, they are essentially determined by the absolute value of electron momentum. With the aid of Eq. (E.33), we rewrite Eq. (E.31) as $G_\varepsilon(\mathbf{p}) = [\varepsilon - \varepsilon_p - \Sigma_\varepsilon(\mathbf{p}) + i\lambda \mathrm{sgn}(p - p_F)]^{-1}$. It describes a quasiparticle with a renormalized spectrum. The Hartree-Fock approximation, when $\Sigma_\varepsilon(\mathbf{p})$ is independent of ε and equal to $-\Delta\varepsilon_p$ given by Eq. (28.22), is realized when one takes into account only the first two diagrams on the right-hand side of Eq. (E.23), replacing the bold solid lines by the thin ones. The first (Hartree) diagram describes a constant shift of the electron energy compensated by the positive background charge, and only the second (Fock) diagram contributes to the electron spectrum.

Appendix F
Equation for Cooperon

The Bethe-Salpeter equation, given in the diagrammatic form by Eq. (43.17), can be rewritten as an equation for the function $C_{\varepsilon\varepsilon'}\left(\mathbf{r}t,\mathbf{r}'t'\right)$ introduced by Eq. (43.16):

$$C_{\varepsilon\varepsilon'}\left(\mathbf{r}t,\mathbf{r}'t'\right) - \int d\tau \int d\tau' e^{i(\varepsilon\tau+\varepsilon'\tau')/\hbar} G^R\left(\mathbf{r}\,t+\frac{\tau}{2},\mathbf{r}'\,t'-\frac{\tau'}{2}\right) \qquad (F.1)$$

$$\times G^A\left(\mathbf{r}\,t'+\frac{\tau'}{2},\mathbf{r}'\,t-\frac{\tau}{2}\right) = w\int d\tau \int d\tau' e^{i(\varepsilon\tau+\varepsilon'\tau')/\hbar}\int d\Delta t \int d\Delta t' \int d\Delta \mathbf{x}$$

$$\times G^R\left(\mathbf{r}\,t+\frac{\tau}{2},\mathbf{r}+\Delta\mathbf{x}\,t+\frac{\tau}{2}+\Delta t\right) G^A\left(\mathbf{r}\,t'+\frac{\tau'}{2},\mathbf{r}+\Delta\mathbf{x}\,t'+\frac{\tau'}{2}+\Delta t'\right)$$

$$\times\left\langle\left\langle \mathcal{G}^R\left(\mathbf{r}+\Delta\mathbf{x}\,t+\frac{\tau}{2}+\Delta t,\mathbf{r}'\,t'-\frac{\tau'}{2}\right) \mathcal{G}^A\left(\mathbf{r}+\Delta\mathbf{x}\,t'+\frac{\tau'}{2}+\Delta t',\mathbf{r}'\,t-\frac{\tau}{2}\right)\right\rangle\right\rangle.$$

We use $\mathbf{r}+\Delta\mathbf{x}$ instead of $\mathbf{r}_3$, and the time variables are shifted according to $t_3 = t+\tau/2+\Delta t$ and $t_3' = t'+\tau'/2+\Delta t'$. Let us carry out the inverse Wigner transformation of the retarded and advanced Green's functions according to

$$G^{R,A}\left(\mathbf{r}t,\mathbf{r}+\Delta\mathbf{x}\,t+\Delta t\right) = \int \frac{d\varepsilon_1}{2\pi\hbar} e^{i\varepsilon_1\Delta t/\hbar}\int \frac{d\mathbf{p}_1}{(2\pi\hbar)^3} \qquad (F.2)$$

$$\times \exp\left[-\frac{i}{\hbar}\left(\mathbf{p}_1+\frac{e}{c}\mathbf{A}_{\mathbf{r}+\Delta\mathbf{x}/2,t+\Delta t/2}\right)\cdot\Delta\mathbf{x}\right] G^{R,A}_{\varepsilon_1}\left(\mathbf{r}+\frac{\Delta\mathbf{x}}{2},\mathbf{p}_1\right)$$

and rewrite the right-hand side of Eq. (F.1) as

$$w\int d\tau \int d\tau' e^{i(\varepsilon\tau+\varepsilon'\tau')/\hbar}\int d\Delta t \int d\Delta t'\int \frac{d\varepsilon_1}{2\pi\hbar}\int \frac{d\varepsilon_2}{2\pi\hbar} e^{i(\varepsilon_1\Delta t+\varepsilon_2\Delta t')/\hbar}$$

$$\times \int d\Delta\mathbf{x}\int \frac{d\mathbf{p}_1}{(2\pi\hbar)^3}\int \frac{d\mathbf{p}_2}{(2\pi\hbar)^3} e^{-i\left[(\mathbf{p}_1+\mathbf{p}_2)/\hbar+\boldsymbol{\mathcal{K}}_{\mathbf{r}+\Delta\mathbf{x}/2}\left(t+\frac{\Delta t+\tau}{2},t'+\frac{\Delta t'+\tau'}{2}\right)\right]\cdot\Delta\mathbf{x}}$$

$$\times G^R_{\varepsilon_1}\left(\mathbf{r}+\frac{\Delta\mathbf{x}}{2},\mathbf{p}_1\right) G^A_{\varepsilon_2}\left(\mathbf{r}+\frac{\Delta\mathbf{x}}{2},\mathbf{p}_2\right) \qquad (F.3)$$

$$\times\left\langle\left\langle \mathcal{G}^R\left(\mathbf{r}+\Delta\mathbf{x}\,t+\frac{\tau}{2}+\Delta t,\mathbf{r}'\,t'-\frac{\tau'}{2}\right) \mathcal{G}^A\left(\mathbf{r}+\Delta\mathbf{x}\,t'+\frac{\tau'}{2}+\Delta t',\mathbf{r}'\,t-\frac{\tau}{2}\right)\right\rangle\right\rangle,$$

where $\boldsymbol{\kappa}_\mathbf{r}(t,t')$ is defined in Sec. 43. Using the variables $\Delta\mathbf{p} = \mathbf{p}_1+\mathbf{p}_2$, $\bar{\varepsilon} = (\varepsilon_1+\varepsilon_2)/2$, and $\Delta\varepsilon = \varepsilon_1 - \varepsilon_2$, we transform the expression (F.3) to the following form:

$$
w \int d\Delta t \int d\Delta t' \int d\Delta\mathbf{x} \int \frac{d\bar{\varepsilon}}{2\pi\hbar} \int \frac{d\Delta\varepsilon}{2\pi\hbar} \exp\left\{ \frac{i}{\hbar}\left[\bar{\varepsilon}(\Delta t + \Delta t') + \frac{\Delta\varepsilon}{2}(\Delta t - \Delta t') \right] \right\}
$$

$$
\times \int \frac{d\Delta\mathbf{p}}{(2\pi\hbar)^3} \exp\left\{ -\frac{i}{\hbar}\left[\Delta\mathbf{p} + \hbar\boldsymbol{\kappa}_{\mathbf{r}+\Delta\mathbf{x}/2}\left(t + \frac{\Delta t}{2}, t' + \frac{\Delta t'}{2} \right) \right] \cdot \Delta\mathbf{x} \right\}
$$

$$
\times \int \frac{d\mathbf{p}_1}{(2\pi\hbar)^3} G^R_{\bar{\varepsilon}+\Delta\varepsilon/2}\left(\mathbf{r} + \frac{\Delta\mathbf{x}}{2}, \mathbf{p}_1 \right) G^A_{\bar{\varepsilon}-\Delta\varepsilon/2}\left(\mathbf{r} + \frac{\Delta\mathbf{x}}{2}, \Delta\mathbf{p} - \mathbf{p}_1 \right) \qquad (F.4)
$$

$$
\times \int d\tau \int d\tau' e^{i(\varepsilon\tau+\varepsilon'\tau')/\hbar} \left\langle\!\left\langle \mathcal{G}^R\left(\mathbf{r} + \Delta\mathbf{x}\ t + \frac{\tau}{2} + \Delta t, \mathbf{r}'\ t' - \frac{\tau'}{2} \right) \right.\right.
$$

$$
\left.\left. \times \mathcal{G}^A\left(\mathbf{r} + \Delta\mathbf{x}\ t' + \frac{\tau'}{2} + \Delta t', \mathbf{r}'\ t - \frac{\tau}{2} \right) \right\rangle\!\right\rangle .
$$

Since ε and ε' are of the order of the Fermi energy, only small τ and τ' are essential, and we have omitted corresponding temporal shifts in the vector potentials. After the substitutions $\tau \to \tau - \Delta t$ and $\tau' \to \tau' - \Delta t'$, Eq. (F.4) is rewritten as

$$
w \int d\Delta t \int d\Delta t' e^{-i(\varepsilon\Delta t+\varepsilon'\Delta t')/\hbar} \int \frac{d\bar{\varepsilon}}{2\pi\hbar} \int \frac{d\Delta\varepsilon}{2\pi\hbar} e^{i\varepsilon(\Delta t+\Delta t')/\hbar} e^{\Delta\varepsilon(\Delta t-\Delta t'))/2\hbar}
$$

$$
\times \int d\Delta\mathbf{x} \int \frac{d\Delta\mathbf{p}}{(2\pi\hbar)^3} \exp\left\{ -\frac{i}{\hbar}\left[\Delta\mathbf{p} + \hbar\boldsymbol{\kappa}_{\mathbf{r}+\Delta\mathbf{x}/2}\left(t + \frac{\Delta t}{2}, t' + \frac{\Delta t'}{2} \right) \right] \cdot \Delta\mathbf{x} \right\}
$$

$$
\times \int \frac{d\mathbf{p}_1}{(2\pi\hbar)^3} G^R_{\bar{\varepsilon}+\Delta\varepsilon/2}\left(\mathbf{r} + \frac{\Delta\mathbf{x}}{2}, \mathbf{p}_1 \right) G^A_{\bar{\varepsilon}-\Delta\varepsilon/2}\left(\mathbf{r} + \frac{\Delta\mathbf{x}}{2}, \Delta\mathbf{p} - \mathbf{p}_1 \right) \qquad (F.5)
$$

$$
\times C_{\varepsilon\varepsilon'}\left(\mathbf{r} + \Delta\mathbf{x}\ t + \frac{\Delta t}{2}, \mathbf{r}'\ t' + \frac{\Delta t'}{2} \right)
$$

so that Eq. (F.1) with the right-hand side given by Eq. (F.5) becomes a closed integral equation for the function $C_{\varepsilon\varepsilon'}(\mathbf{r}t, \mathbf{r}'t')$.

Substituting the Green's functions of Eq. (F.2) into the inhomogeneous term of Eq. (F.1), we transform this term by using the variables $\Delta\mathbf{p}$, $\bar{\varepsilon}$, and $\Delta\varepsilon$ given above. As a result, the inhomogeneous term becomes

$$
\int d\tau \int d\tau' e^{i(\varepsilon\tau+\varepsilon'\tau')/\hbar} \int \frac{d\bar{\varepsilon}}{2\pi\hbar} \int \frac{d\Delta\varepsilon}{2\pi\hbar} e^{-i[\bar{\varepsilon}(\tau+\tau')+\Delta\varepsilon(t-t')]/\hbar}
$$

$$
\times \int \frac{d\Delta\mathbf{p}}{(2\pi\hbar)^3} \exp\left[\frac{i}{\hbar}\left(\Delta\mathbf{p} + \frac{2e}{c}\mathbf{A}_{(\mathbf{r}+\mathbf{r}')/2,(t+t')/2} \right) \cdot (\mathbf{r} - \mathbf{r}') \right] \qquad (F.6)
$$

$$
\times \int \frac{d\mathbf{p}_1}{(2\pi\hbar)^3} G^R_{\bar{\varepsilon}+\Delta\varepsilon/2}\left(\frac{\mathbf{r}+\mathbf{r}'}{2}, \mathbf{p}_1 \right) G^A_{\bar{\varepsilon}-\Delta\varepsilon/2}\left(\frac{\mathbf{r}+\mathbf{r}'}{2}, \Delta\mathbf{p} - \mathbf{p}_1 \right) .
$$

In a similar way as in Eq. (F.4), a weak dependence of the vector potentials on τ and τ' is omitted here. The integrals over τ and τ' give us the factor $(2\pi\hbar)^2\delta(\varepsilon-\bar{\varepsilon})\delta(\bar{\varepsilon}-\varepsilon')$. After calculating the integral over $\bar{\varepsilon}$, we transform the expression (F.6) to

$$
\delta(\varepsilon - \varepsilon') \int d\Delta\varepsilon\, e^{-i\Delta\varepsilon(t-t')/\hbar} \int \frac{d\Delta\mathbf{p}}{(2\pi\hbar)^3} \qquad (F.7)
$$

$$
\times \exp\left[\frac{i}{\hbar}\left(\Delta\mathbf{p} + \frac{2e}{c}\mathbf{A}_{(\mathbf{r}+\mathbf{r}')/2,(t+t')/2} \right) \cdot (\mathbf{r} - \mathbf{r}') \right] w^{-1} I_{\varepsilon,\Delta\varepsilon}\left(\Delta\mathbf{p}, \frac{\mathbf{r}+\mathbf{r}'}{2} \right),
$$

where

$$I_{\varepsilon,\Delta\varepsilon}(\Delta\mathbf{p},\mathbf{r}) = w\int\frac{d\mathbf{p}_1}{(2\pi\hbar)^3}\left(\varepsilon+\frac{\Delta\varepsilon}{2}-\varepsilon_{p_1}-U_\mathbf{r}+\frac{i\hbar}{2\tau_\mathbf{r}}\right)^{-1}$$

$$\times\left(\varepsilon-\frac{\Delta\varepsilon}{2}-\varepsilon_{\Delta\mathbf{p}-\mathbf{p}_1}-U_\mathbf{r}-\frac{i\hbar}{2\tau_\mathbf{r}}\right)^{-1}. \qquad (F.8)$$

For spatially homogeneous 2D systems and at $\Delta\varepsilon = 0$, this integral is reduced to the function defined in Eq. (15.23). We remind that $U_\mathbf{r}$ is the potential energy which smoothly varies in space. In the hydrodynamical region, when $|\Delta\mathbf{p}|l_F/\hbar \ll 1$ and $|\Delta\varepsilon|\tau_F/\hbar \ll 1$, one can expand the integrand in Eq. (F.8) up to the contributions proportional to $\Delta\varepsilon$ and $(\Delta p)^2$ so that

$$I_{\varepsilon,\Delta\varepsilon}(\Delta\mathbf{p},\mathbf{r}) \simeq w\int\frac{d\mathbf{p}}{(2\pi\hbar)^3}\left[(E-\varepsilon_p)^2+\left(\frac{\hbar}{2\tau_\mathbf{r}}\right)^2\right]^{-1}\left\{1+\frac{\Delta\varepsilon}{2}\left[\left(E-\varepsilon_p-\frac{i\hbar}{2\tau_\mathbf{r}}\right)^{-1}\right.\right.$$

$$-\left.\left(E-\varepsilon_p+\frac{i\hbar}{2\tau_\mathbf{r}}\right)^{-1}\right]-\frac{(\Delta\mathbf{p}\cdot\mathbf{v_p})^2}{4}\left(E-\varepsilon_p-\frac{i\hbar}{2\tau_\mathbf{r}}\right)^{-1}\left(E-\varepsilon_p+\frac{i\hbar}{2\tau_\mathbf{r}}\right)^{-1}$$

$$+\frac{(\Delta\mathbf{p}\cdot\mathbf{v_p})^2}{4}\left[\left(E-\varepsilon_p-\frac{i\hbar}{2\tau_\mathbf{r}}\right)^{-2}+\left(E-\varepsilon_p+\frac{i\hbar}{2\tau_\mathbf{r}}\right)^{-2}\right]\right\}, \qquad (F.9)$$

where $E = \varepsilon-(\Delta p)^2/8m-U_\mathbf{r}$, $\mathbf{p} = \mathbf{p}_1-\Delta\mathbf{p}/2$, and $\mathbf{v_p} = \mathbf{p}/m$. The main contribution into the integral over momentum in Eq. (F.9) comes from $\varepsilon_p = E$, and we obtain

$$I_{\varepsilon,\Delta\varepsilon}(\Delta\mathbf{p},\mathbf{r}) \simeq 1+i\frac{\Delta\varepsilon\tau_\mathbf{r}}{\hbar}-\frac{1}{3}\left(\frac{\Delta p l_\mathbf{r}}{\hbar}\right)^2, \quad \frac{\Delta\varepsilon\tau_\mathbf{r}}{\hbar}<1, \quad \frac{\Delta p l_\mathbf{r}}{\hbar}<1, \qquad (F.10)$$

where $l_\mathbf{r} = v_\mathbf{r}\tau_\mathbf{r}$ is the mean free path length for the electrons with coordinate-dependent velocities $v_\mathbf{r}$ defined as $v_\mathbf{r} = \sqrt{2(\varepsilon-U_\mathbf{r})/m}$. A similar calculation for the 2D case gives the factor $1/2$ instead of $1/3$ in the last term of the right-hand side of Eq. (F.10); see Eq. (15.24). We substitute $I_{\varepsilon,\Delta\varepsilon} \simeq 1$ in the expression (F.7) for the inhomogeneous term of Eq. (F.1) and restrict the consideration by the regions $|\Delta\mathbf{p}| < \hbar/l_F$ and $|\Delta\varepsilon| < \hbar/\tau_F$. Below we also assume that the external fields are weak, $|2e\mathbf{A}/c| \ll \hbar/l_F$. In these conditions, the remaining integrals give us δ-functions, and the inhomogeneous term is finally written as

$$\frac{2\pi\hbar}{w}\delta(\varepsilon-\varepsilon')\delta(t-t')\delta(\mathbf{r}-\mathbf{r}'). \qquad (F.11)$$

Since the variables ε and ε' enter the integral term of the expression (F.5) as parameters, $C_{\varepsilon\varepsilon'}(\mathbf{r}t,\mathbf{r}'t')$ becomes proportional to $\delta(\varepsilon-\varepsilon')$, and one can search for this function in the form given by Eq. (43.18). Substituting this equation into Eq. (F.5), we use the expression (F.11) and write the integral equation for $C(\mathbf{r}t,\mathbf{r}'t')$:

$$C(\mathbf{r}t,\mathbf{r}'t') - \frac{\hbar\delta(t-t')\delta(\mathbf{r}-\mathbf{r}')}{\pi w\rho_{3D}(\varepsilon-U_\mathbf{r})} = \int d\Delta\mathbf{x}\int d\Delta t\int d\Delta t'\int\frac{d\bar{\varepsilon}}{2\pi\hbar}$$

$$\times e^{i(\bar{\varepsilon}-\varepsilon)(\Delta t+\Delta t')/\hbar}\int\frac{d\Delta\varepsilon}{2\pi\hbar}e^{i\Delta\varepsilon(\Delta t-\Delta t')/2\hbar}\int\frac{d\Delta\mathbf{p}}{(2\pi\hbar)^3}$$

$$\times\exp\left\{-\frac{i}{\hbar}\left[\Delta\mathbf{p}+\hbar\boldsymbol{\kappa}_{\mathbf{r}+\Delta\mathbf{x}/2}\left(t+\frac{\Delta t}{2},t'+\frac{\Delta t'}{2}\right)\right]\cdot\Delta\mathbf{x}\right\}I_{\bar{\varepsilon},\Delta\varepsilon}\left(\Delta\mathbf{p},\mathbf{r}+\frac{\Delta\mathbf{x}}{2}\right)$$

$$\times \frac{\rho_{3D}(\varepsilon - U_{\mathbf{r}+\Delta\mathbf{x}/2})}{\rho_{3D}(\varepsilon - U_{\mathbf{r}})} C\left(\mathbf{r} + \Delta\mathbf{x}\; t + \frac{\Delta t}{2}, \mathbf{r}'\; t' + \frac{\Delta t'}{2}\right). \tag{F.12}$$

Substituting the approximate expression (F.10) for I to the right-hand side of this equation, one may neglect its dependence on $\bar{\varepsilon}$ which comes only from the energy dependence of $\tau_{\mathbf{r}}$ and $l_{\mathbf{r}}$. Thus, the integral over $\bar{\varepsilon}$ gives $\delta(\Delta t + \Delta t')$ on the right-hand side of Eq. (F.12). After substituting $\Delta\mathbf{p} + \hbar\boldsymbol{\kappa}_{\mathbf{r}+\Delta\mathbf{x}/2}(t + \Delta t/2, t' - \Delta t/2) \to \Delta\mathbf{p}$, this equation is rewritten in the form

$$C\left(\mathbf{r}t, \mathbf{r}'t'\right) - \tau_{\mathbf{r}}\delta(t - t')\delta(\mathbf{r} - \mathbf{r}') = \int d\Delta\mathbf{x} \int d\Delta t \int \frac{d\Delta\varepsilon}{2\pi\hbar} e^{i\Delta\varepsilon\Delta t/\hbar}$$

$$\times \int \frac{d\Delta\mathbf{p}}{(2\pi\hbar)^3} e^{-i\Delta\mathbf{p}\cdot\Delta\mathbf{x}/\hbar} \frac{\rho_{3D}(\varepsilon - U_{\mathbf{r}+\Delta\mathbf{x}/2})}{\rho_{3D}(\varepsilon - U_{\mathbf{r}})} C\left(\mathbf{r} + \Delta\mathbf{x}\; t + \frac{\Delta t}{2}, \mathbf{r}'\; t' - \frac{\Delta t}{2}\right) \tag{F.13}$$

$$\times \left\{ 1 + i\frac{\Delta\varepsilon\tau_{\mathbf{r}+\Delta\mathbf{x}/2}}{\hbar} - \frac{1}{3}\left(\frac{l_{\mathbf{r}+\Delta\mathbf{x}/2}}{\hbar}\right)^2 \left[\Delta\mathbf{p} - \hbar\boldsymbol{\kappa}_{\mathbf{r}+\Delta\mathbf{x}/2}\left(t + \frac{\Delta t}{2}, t' - \frac{\Delta t}{2}\right)\right]^2 \right\}$$

with coordinate-dependent scattering time τ and mean free path length l. It is assumed that the integrals are taken in the hydrodynamical region of the variables, $|\Delta\varepsilon\tau_{\mathbf{r}+\Delta\mathbf{x}/2}| < \hbar$ and $|\Delta\mathbf{p} - \hbar\kappa_{\mathbf{r}+\Delta\mathbf{x}/2}(t + \Delta t/2, t' - \Delta t/2)|l_{\mathbf{r}+\Delta\mathbf{x}/2} < \hbar$.

One can replace $\int d\Delta\varepsilon\; e^{i\Delta\varepsilon\Delta t/\hbar}\ldots$ by $2\pi\hbar\delta(\Delta t)\ldots$ and $\int d\Delta\mathbf{p}\; e^{-i\Delta\mathbf{p}\cdot\Delta\mathbf{x}/\hbar}\ldots$ by $(2\pi\hbar)^3\delta(\Delta\mathbf{x})\ldots$ if the functions denoted here by the dots do not depend on $\Delta\varepsilon$ and $\Delta\mathbf{p}$, respectively. Therefore, the contribution from the unity in the braces $\{\ldots\}$ of the right-hand side of Eq. (F.13) gives $C\left(\mathbf{r}t, \mathbf{r}'t'\right)$, which annihilates with $C\left(\mathbf{r}t, \mathbf{r}'t'\right)$ of the left-hand side of this equation. Transforming the contribution proportional to $\Delta\varepsilon$, one may replace $\int d\Delta\varepsilon\; e^{i\Delta\varepsilon\Delta t/\hbar}(i\Delta\varepsilon/\hbar)\ldots$ by $2\pi\hbar[d\delta(\Delta t)/d\Delta t]\ldots$ and carry out the further transformations by using the integration by parts:

$$\int d\Delta\mathbf{x} \int d\Delta t\; \tau_{\mathbf{r}+\Delta\mathbf{x}/2} \frac{d\delta(\Delta t)}{d\Delta t} \delta(\Delta\mathbf{x}) C\left(\mathbf{r}\; t + \frac{\Delta t}{2}, \mathbf{r}'\; t' - \frac{\Delta t}{2}\right) \tag{F.14}$$

$$= -\tau_{\mathbf{r}} \frac{d}{d\Delta t} C\left(\mathbf{r}\; t + \frac{\Delta t}{2}, \mathbf{r}'\; t' - \frac{\Delta t}{2}\right)\bigg|_{\Delta t=0} = -\frac{\tau_{\mathbf{r}}}{2}\left(\frac{\partial}{\partial t} - \frac{\partial}{\partial t'}\right) C\left(\mathbf{r}t, \mathbf{r}'t'\right).$$

Considering the last term on the right-hand side of Eq. (F.13), we notice that $\tau_{\mathbf{r}+\Delta\mathbf{x}/2}\rho_{3D}(\varepsilon - U_{\mathbf{r}+\Delta\mathbf{x}/2})/\rho_{3D}(\varepsilon - U_{\mathbf{r}}) = \tau_{\mathbf{r}}$. Therefore, it is convenient to express $l_{\mathbf{r}}^2 = (v_{\mathbf{r}}\tau_{\mathbf{r}})^2$ through the diffusion coefficient $D_{\mathbf{r}} = v_{\mathbf{r}}^2\tau_{\mathbf{r}}/3$. Expanding $D_{\mathbf{r}+\Delta\mathbf{x}/2}$ as $D_{\mathbf{r}} + \Delta\mathbf{x} \cdot \nabla_{\mathbf{r}}D_{\mathbf{r}}/2$, we transform this term to

$$-\frac{\tau_{\mathbf{r}}}{\hbar} \int d\Delta\mathbf{x} \int \frac{d\Delta\mathbf{p}}{(2\pi\hbar)^3} e^{-i\Delta\mathbf{p}\cdot\Delta\mathbf{x}/\hbar} D_{\mathbf{r}+\Delta\mathbf{x}/2}[\Delta\mathbf{p} - \hbar\boldsymbol{\kappa}_{\mathbf{r}+\Delta\mathbf{x}/2}(t,t')]^2 \tag{F.15}$$

$$\times C\left(\mathbf{r} + \Delta\mathbf{x}\; t, \mathbf{r}'t'\right) \simeq -\tau_{\mathbf{r}}[-i\nabla_{\mathbf{r}} - \boldsymbol{\kappa}_{\mathbf{r}}(t,t')] \cdot D_{\mathbf{r}}[-i\nabla_{\mathbf{r}} - \boldsymbol{\kappa}_{\mathbf{r}}(t,t')]C\left(\mathbf{r}t, \mathbf{r}'t'\right),$$

where the coordinate-dependent diffusion coefficient stands between the gradients. Collecting together Eqs. (F.13)–(F.15), we obtain the equation for the Cooperon:

$$\left\{ \frac{1}{2}\left(\frac{\partial}{\partial t} - \frac{\partial}{\partial t'}\right) + [-i\nabla_{\mathbf{r}} - \boldsymbol{\kappa}_{\mathbf{r}}(t,t')] \cdot D_{\mathbf{r}}[-i\nabla_{\mathbf{r}} - \boldsymbol{\kappa}_{\mathbf{r}}(t,t')] \right\}$$

$$\times C\left(\mathbf{r}t, \mathbf{r}'t'\right) = \delta(t - t')\delta\left(\mathbf{r} - \mathbf{r}'\right). \tag{F.16}$$

In the absence of time-dependent fields, when $C\left(\mathbf{r}t, \mathbf{r}'t'\right)$ depends only on $t - t'$, this equation defines the Green's function of the diffusion equation with coordinate-dependent diffusion coefficient. Equation (F.16) is applicable to 2D electrons if $\mathbf{r}$ is treated as a 2D coordinate and the diffusion coefficient is given by its 2D expression.

Appendix G
Green's Function in Magnetic Field

To generalize the diagram technique described in Chapter 3 to the case when a magnetic field is present, one should take into account that the Green's function, even after the averaging over a macroscopically homogeneous and isotropic impurity distribution, is not translation-invariant in the plane perpendicular to the magnetic field. However, it is possible to separate the translation-invariant part of the Green's function in the way shown below. Let us consider first the Green's function of free electrons in the magnetic field $\mathbf{H}$ directed along OZ axis and described by the vector potential $\mathbf{A} = (0, Hx, 0)$. Using the coordinate representation (see problem 3.10), we have

$$g_E^s(\mathbf{r}, \mathbf{r}') = \sum_{p_z p_y N} g_E^s(p_z, N) \psi_{\mathbf{r}}^{(Np_y p_z)} \psi_{\mathbf{r}'}^{(Np_y p_z)*},$$

$$g_E^s(p_z, N) = [E - p_z^2/2m - \varepsilon_N \pm i\lambda]^{-1}, \qquad (G.1)$$

where $\lambda \to +0$, $\mathbf{r} = (x, y, z)$ is the 3D coordinate, $g_E^s(p_z, N)$ is the free-electron Green's function in the Landau level representation, and ε_N is the energy of the Landau level N. The upper and lower signs at $i\lambda$ correspond to the retarded ($s = R$) and advanced ($s = A$) Green's functions, respectively. The wave functions $\psi_{\mathbf{r}}^{(Np_y p_z)}$ describing the Landau eigenstates in the gauge chosen above are given by Eqs. (5.13) and (5.15). Using them, we calculate the sum over p_y in Eq. (G.1) and obtain

$$g_E^s(\mathbf{r}, \mathbf{r}') = e^{-(i/2l_H^2)(x+x')(y-y')} g_E^s(|\mathbf{x} - \mathbf{x}'|, |z - z'|),$$

$$g_E^s(|\mathbf{x} - \mathbf{x}'|, |z - z'|) = \int \frac{dp_z}{2\pi\hbar} e^{ip_z(z-z')/\hbar} g_E^s(|\mathbf{x} - \mathbf{x}'|, p_z), \qquad (G.2)$$

where $\mathbf{x} = (x, y)$ is the in-plane coordinate vector, l_H is the magnetic length,

$$g_E^s(|\mathbf{x} - \mathbf{x}'|, p_z) = \frac{1}{2\pi l_H^2} e^{-|\mathbf{x}-\mathbf{x}'|^2/4l_H^2} \sum_N L_N^0 \left(\frac{|\mathbf{x} - \mathbf{x}'|^2}{2l_H^2} \right) g_E^s(p_z, N), \qquad (G.3)$$

and $L_N^0(x)$ is the Laguerre polynomial introduced by Eq. (A.26). The integral over p_z in Eq. (G.2) can be calculated as

$$\int \frac{dp_z}{2\pi\hbar} e^{ip_z(z-z')/\hbar} g_E^s(p_z, N) = \frac{\sqrt{m} \exp\left[\pm i\sqrt{2m(E - \varepsilon_N \pm i\lambda)}|z - z'|/\hbar\right]}{\pm i\sqrt{2(E - \varepsilon_N \pm i\lambda)}\hbar}. \qquad (G.4)$$

The Green's function is presented as a product of the phase factor $e^{i\theta(\mathbf{x},\mathbf{x}')}$, where $\theta(\mathbf{x},\mathbf{x}') = -(1/2l_H^2)(x+x')(y-y')$, by a translation-invariant function $g_E^s(|\mathbf{x}-\mathbf{x}'|,|z-z'|)$, which is also gauge-invariant. On the other hand, any gauge transformation changes the phase factor of the Green's function, because it changes the phase of the wave functions (according to the transformation $\hat{\mathbf{p}} \to \hat{\mathbf{p}} - (e/c)\mathbf{A_r}$, the phase factor of the wave function $\psi_\mathbf{r}$ of the electron moving along the path l in the presence of the vector potential can be represented as $\exp\left[(ie/\hbar c)\int_{\mathbf{r}'\in l}^{\mathbf{r}} d\mathbf{r}' \mathbf{A}_{\mathbf{r}'} \cdot \mathbf{n}_l\right]$, where the integral is calculated over the path l and $\mathbf{n}_l$ is the unit vector along this path). Of course, this change does not modify any observable physical quantity. In the symmetric gauge $\mathbf{A} = [\mathbf{H} \times \mathbf{r}]/2 = (-Hy/2, Hx/2, 0)$, the phase is written in the form $\theta(\mathbf{x},\mathbf{x}') = -(e/2\hbar c)\mathbf{H} \cdot [\mathbf{r} \times \mathbf{r}'] = \mathbf{n}_z \cdot [\mathbf{x} \times \mathbf{x}']/2l_H^2$, where $\mathbf{n}_z$ is the unit vector in the direction of $\mathbf{H}$.

The averaged Green's function $G_E^s(\mathbf{r},\mathbf{r}')$ of the electrons interacting with a random scattering potential has the same property:

$$\langle\langle G_E^s(\mathbf{r},\mathbf{r}')\rangle\rangle = e^{i\theta(\mathbf{x},\mathbf{x}')}G_E^s(|\mathbf{x}-\mathbf{x}'|,|z-z'|). \tag{G.5}$$

To demonstrate this, let us consider the elastic scattering model and write the third term of the perturbation expansion of the Green's function in the coordinate representation (see, for example, problem 3.8) as

$$\int d\mathbf{r}_1 \int d\mathbf{r}_2 e^{i[\theta(\mathbf{x},\mathbf{x}_1)+\theta(\mathbf{x}_1,\mathbf{x}_2)+\theta(\mathbf{x}_2,\mathbf{x}')]}g_E^s(|\mathbf{x}-\mathbf{x}_1|,|z-z_1|)$$

$$\times U_{sc}(\mathbf{r}_1)g_E^s(|\mathbf{x}_1-\mathbf{x}_2|,|z_1-z_2|)U_{sc}(\mathbf{r}_2)g_E^s(|\mathbf{x}_2-\mathbf{x}'|,|z_2-z'|). \tag{G.6}$$

Averaging this expression over the random potential, we obtain the correlation function $\langle\langle U_{sc}(\mathbf{r}_1)U_{sc}(\mathbf{r}_2)\rangle\rangle = w(|\mathbf{r}_1-\mathbf{r}_2|)$ which depends on $|\mathbf{x}_1 - \mathbf{x}_2|$. Introducing new coordinates according to $\mathbf{r}_1 = \mathbf{r}_1' + \mathbf{r}$ and $\mathbf{r}_2 = \mathbf{r}_2' + \mathbf{r}'$, we rewrite (G.6) as

$$e^{i\theta(\mathbf{x},\mathbf{x}')} \int d\mathbf{r}_1' \int d\mathbf{r}_2' e^{i\mathbf{n}_z \cdot ([\mathbf{x}_1' \times \mathbf{x}_2'] + [(\mathbf{x}-\mathbf{x}') \times (\mathbf{x}_1'+\mathbf{x}_2')])/2l_H^2}g_E^s(|\mathbf{x}_1'|,|z_1'|) \tag{G.7}$$

$$\times g_E^s(|\mathbf{x}_1'-\mathbf{x}_2'+\mathbf{x}-\mathbf{x}'|,|z_1'-z_2'+z-z'|)w(|\mathbf{r}_1'-\mathbf{r}_2'+\mathbf{r}-\mathbf{r}'|)g_E^s(|\mathbf{x}_2'|,|z_2'|),$$

where the phase factor $e^{i\theta(\mathbf{x},\mathbf{x}')}$ is separated. Since the expression under the integral depends only on the differences $\mathbf{x} - \mathbf{x}'$ and $z - z'$, the first-order (with respect to $w(|\mathbf{r}_1 - \mathbf{r}_2|)$) contribution to the averaged Green's function satisfies the property (G.5). In a similar fashion, this property can be checked for the contributions of arbitrary order (and for different mechanisms of interaction) provided the system possesses macroscopic translational invariance in the plane perpendicular to the magnetic field. In conclusion, the symmetry properties of the free-electron Green's function are conserved for the averaged Green's function of the electron interacting with random scattering potentials.

The most natural and often convenient basis for representing the Green's functions in magnetic fields is the basis of the Landau states. Applying it, we obtain

$$G_E^s(\mathbf{r},\mathbf{r}') = \sum_{p_z p_y N} \psi_\mathbf{r}^{(Np_yp_z)}\psi_{\mathbf{r}'}^{(N'p_y'p_z')*} G_E^s(p_y,p_z,N|p_y',p_z',N'), \tag{G.8}$$

where $G_E^s(p_y,p_z,N|p_y',p_z',N')$ is the Green's function in the Landau level representation. For free electrons, when the Landau states are exact eigenstates, we have

$G_E^s(p_y, p_z, N | p_y', p_z', N') = \delta_{p_y p_y'} \delta_{p_z p_z'} \delta_{NN'} g_E^s(p_z, N)$ and Eq. (G.8) is reduced to Eq. (G.1). A similar property is valid for the averaged Green's function:

$$\langle\langle G_E^s(p_y, p_z, N | p_y', p_z', N') \rangle\rangle = \delta_{p_y p_y'} \delta_{p_z p_z'} \delta_{NN'} G_E^s(p_z, N). \qquad (G.9)$$

Below we prove Eq. (G.9) by using the general symmetry property (G.5). We apply a transformation from the coordinate representation to the Landau level representation:

$$G_E^s(p_y, p_z, N | p_y', p_z', N') = \int d\mathbf{r} \int d\mathbf{r}' \psi_{\mathbf{r}}^{(N p_y p_z)*} G_E^s(\mathbf{r}, \mathbf{r}') \psi_{\mathbf{r}'}^{(N' p_y' p_z')}. \qquad (G.10)$$

Then we average Eq. (G.10) and use the explicit form of the functions $\psi_{\mathbf{r}}^{(N p_y p_z)}$ in order to calculate the integrals over $\mathbf{r}$ and $\mathbf{r}'$. Let us introduce new coordinates according to $\mathbf{r}_+ = (\mathbf{r}+\mathbf{r}')/2$ and $\mathbf{r}_- = \mathbf{r}-\mathbf{r}'$. The integrals over z_+ and y_+ immediately give the factors $\delta_{p_z p_z'}$ and $\delta_{p_y p_y'}$. Next, the integral over z_- transforms $G_E^s(|\mathbf{x}-\mathbf{x}'|, |z-z'|)$ into $G_E^s(|\mathbf{x}_-|, p_z)$. Therefore, after the averaging, the expression on the right-hand side of Eq. (G.10) becomes proportional to

$$\delta_{p_z p_z'} \delta_{p_y p_y'} \int dx_- \int dy_- e^{-(x_-^2 + y_-^2)/4l_H^2} G_E^s(|\mathbf{x}_-|, p_z)$$

$$\times \int du_+ e^{-(u_+ + iy_-/2l_H)^2} H_N(u_+ + x_-/2l_H) H_N(u_+ - x_-/2l_H), \qquad (G.11)$$

where $u_+ = x_+/l_H + l_H p_y/\hbar$ and $H_N(x)$ is the Hermite polynomial used in Eq. (5.15). Note that the expression (G.11) does not depend on p_y. After calculating the integral over u_+, we introduce the polar coordinates $r = |\mathbf{x}_-|$ and $\varphi = \arctan(y_-/x_-)$ and obtain

$$\langle\langle G_E^s(p_y, p_z, N | p_y', p_z', N') \rangle\rangle = \delta_{p_z p_z'} \delta_{p_y p_y'} \int_0^\infty r\, dr \int_0^{2\pi} d\varphi\, e^{-r^2/4l_H^2}$$

$$\times G_E^s(r, p_z) \sqrt{\frac{2^{N'} N'!}{2^N N'!}} (-r/2l_H)^{N'-N} e^{i\varphi(N'-N)} L_N^{N'-N}(r^2/2l_H^2). \qquad (G.12)$$

Since $\int_0^{2\pi} d\varphi\, e^{i\varphi(N'-N)} = 2\pi \delta_{NN'}$, Eq. (G.12) is reduced to

$$\langle\langle G_E^s(p_y, p_z, N | p_y', p_z', N') \rangle\rangle = \delta_{p_z p_z'} \delta_{p_y p_y'} \delta_{NN'} G_E^s(p_z, N),$$

$$G_E^s(p_z, N) = 2\pi \int_0^\infty r\, dr\, e^{-r^2/4l_H^2} L_N^0(r^2/2l_H^2) G_E^s(r, p_z). \qquad (G.13)$$

Using the orthogonality property $\int_0^\infty du\, e^{-u} L_N^0(u) L_{N'}^0(u) = \delta_{NN'}$, we can express $G_E^s(r, p_z)$ through $G_E^s(p_z, N)$:

$$G_E^s(|\mathbf{x} - \mathbf{x}'|, p_z) = \frac{1}{2\pi l_H^2} e^{-|\mathbf{x}-\mathbf{x}'|^2/4l_H^2} \sum_N L_N^0 \left(\frac{|\mathbf{x}-\mathbf{x}'|^2}{2l_H^2} \right) G_E^s(p_z, N). \qquad (G.14)$$

This expression has the same form as Eq. (G.3) for the free-electron Green's function. Therefore, the averaged Green's function is diagonal in the Landau level representation.

To find the averaged Green's function, we need to sum diagram series. This leads us to the Dyson equation, which is conveniently written in the mixed $(\mathbf{x}, p_z)$-representation:

$$G_E^s(|\mathbf{x} - \mathbf{x}'|, p_z) = g_E^s(|\mathbf{x} - \mathbf{x}'|, p_z)$$

$$+ \int d\mathbf{x}_1 \int d\mathbf{x}_2 e^{i[\theta(\mathbf{x},\mathbf{x}_1)+\theta(\mathbf{x}_1,\mathbf{x}_2)+\theta(\mathbf{x}_2,\mathbf{x}')+\theta(\mathbf{x}',\mathbf{x})]} \qquad (G.15)$$

$$\times g_E^s(|\mathbf{x} - \mathbf{x}_1|, p_z)\Sigma_E^s(|\mathbf{x}_1 - \mathbf{x}_2|, p_z)G_E^s(|\mathbf{x}_2 - \mathbf{x}'|, p_z),$$

where the self-energy function $\Sigma_E^s(\mathbf{x}_1, \mathbf{x}_2, p_z)$ is written as $e^{i\theta(\mathbf{x}_1,\mathbf{x}_2)}\Sigma_E^s(|\mathbf{x}_1 - \mathbf{x}_2|, p_z)$. Using Eq. (G.15), one can derive the Dyson equation in the Landau level representation. First of all, we note that the self-energy function $e^{i\theta(\mathbf{x}_1,\mathbf{x}_2)}\Sigma_E^s(|\mathbf{x}_1 - \mathbf{x}_2|, p_z)$ in the Landau level representation is diagonal with respect to the Landau state indices and does not depend on p_y. This property is derived in the same way as the analogical property of the averaged Green's function; see Eq. (G.13). Transforming Eq. (G.15) with the use of the Landau basis, we obtain an algebraic Dyson equation

$$G_E^s(p_z, N) = g_E^s(p_z, N) + g_E^s(p_z, N)\Sigma_E^s(p_z, N)G_E^s(p_z, N). \qquad (G.16)$$

Let us consider the self-energy function. In the leading order with respect to the interaction, $\Sigma_E^s(p_z, N)$ is given by the expression

$$\int d\mathbf{r} \int d\mathbf{r}' \psi_\mathbf{r}^{(Np_yp_z)*} \psi_{\mathbf{r}'}^{(Np_yp_z)} \int \frac{dp_z'}{2\pi\hbar} \int \frac{d\mathbf{q}}{(2\pi)^3} w\left(\sqrt{q_\perp^2 + q_z^2}\right)$$

$$\times e^{i\mathbf{q}_\perp \cdot (\mathbf{x}-\mathbf{x}')} e^{i(q_z+p_z'/\hbar)(z-z')} e^{i\theta(\mathbf{x},\mathbf{x}')} G_E^s(|\mathbf{x} - \mathbf{x}'|, p_z'). \qquad (G.17)$$

We use the Fourier transformation $w(|\mathbf{r}|) = (2\pi)^{-3} \int d\mathbf{q} e^{i\mathbf{q}\cdot\mathbf{r}} w(|\mathbf{q}|)$ with $\mathbf{q} = (\mathbf{q}_\perp, q_z)$. Substituting $G_E^s(|\mathbf{x} - \mathbf{x}'|, p_z')$ from Eq. (G.14), we calculate the integrals and obtain

$$\Sigma_E^s(p_z, N) = \sum_{N'} \int \frac{d\mathbf{q}}{(2\pi)^3} \Phi_{NN'}(q_\perp^2 l_H^2/2)w(|\mathbf{q}|)G_E^s(p_z - \hbar q_z, N'), \qquad (G.18)$$

where $\Phi_{NN'}(u)$ is defined by Eq. (48.14). The self-energy defined by Eq. (G.18) loses its dependence on p_z and N for the scattering by a short-range potential, when $w(|\mathbf{q}|) \simeq w$ is independent of q:

$$\Sigma_E^s = \frac{w}{2\pi l_H^2} \sum_N \int \frac{dp_z}{2\pi\hbar} G_E^s(p_z, N). \qquad (G.19)$$

The next correction to $\Sigma_E^s(p_z, N)$ is a contribution of the second order with respect to $w(q)$. It is given by the diagram with two crossed impurity lines, similar to that in Eq. (14.23). Calculating this correction for the model of short-range potential, we obtain

$$\delta\Sigma_E^{s(2)}(p_z, N) = \left(\frac{w}{2\pi l_H^2}\right)^2 \sum_{N_1 N_2 N_3} \int_0^\infty du e^{-2u} L_N^0(u) L_{N_1}^0(u) L_{N_2}^0(u) L_{N_3}^0(u)$$

$$\times \int \frac{dq_z}{2\pi\hbar} \int \frac{dq_z'}{2\pi\hbar} G_E^s(p_z - \hbar q_z, N_1) G_E^s(p_z - \hbar q_z - \hbar q_z', N_2) G_E^s(p_z - \hbar q_z', N_3). \qquad (G.20)$$

The results given above can be easily modified for application to the 2D electron gas occupying the plane XOY. To do this, one should simply omit the variables p_z and q_z and remove the integrals $\int dp_z/(2\pi\hbar)$ and $\int dq_z/(2\pi)$ from the equations. The function $w(|\mathbf{q}|)$ in this case should be considered as an effective 2D correlation function of the scattering potentials. The equations obtained below can be modified in a similar way.

To average the correlation functions of two Green's functions, one should use the Bethe-Salpeter equation for the function

$$K_{EE'}^{ss'}(\mathbf{r}_1, \mathbf{r}_4 | \mathbf{r}_2, \mathbf{r}_3) \equiv \left\langle\!\left\langle G_E^s(\mathbf{r}_1, \mathbf{r}_2) G_{E'}^{s'}(\mathbf{r}_3, \mathbf{r}_4)\right\rangle\!\right\rangle$$

$$= \sum_{\gamma_1 - \gamma_4} K_{EE'}^{ss'}(\gamma_1, \gamma_4 | \gamma_2, \gamma_3) \psi_{\mathbf{r}_1}^{(\gamma_1)} \psi_{\mathbf{r}_3}^{(\gamma_3)} \psi_{\mathbf{r}_2}^{(\gamma_2)*} \psi_{\mathbf{r}_4}^{(\gamma_4)*}, \qquad (G.21)$$

where γ_i is a multi-index for the quantum numbers N_i, p_{yi}, and p_{zi}. Equation (G.21) defines the correlation function in the Landau level representation. The Bethe-Salpeter equation in this representation becomes

$$K_{EE'}^{ss'}(\gamma_1, \gamma_4 | \gamma_2, \gamma_3) = G_E^s(\gamma_1) G_{E'}^{s'}(\gamma_4)$$

$$\times \left[\delta_{\gamma_1\gamma_3} \delta_{\gamma_4\gamma_2} + \frac{1}{V} \sum_{\gamma_5\gamma_6} \Gamma_{EE'}^{ss'}(\gamma_1\gamma_4 | \gamma_5\gamma_6) K_{EE'}^{ss'}(\gamma_5, \gamma_6 | \gamma_2, \gamma_3) \right], \qquad (G.22)$$

where the vertex part is given by the expansion

$$\Gamma_{EE'}^{ss'}(\gamma_1\gamma_4 | \gamma_5\gamma_6) = \sum_{\mathbf{q}} w(q) \langle\gamma_1|e^{i\mathbf{q}\cdot\mathbf{r}}|\gamma_5\rangle \langle\gamma_6|e^{-i\mathbf{q}\cdot\mathbf{r}}|\gamma_4\rangle$$

$$+ \frac{1}{V} \sum_{\mathbf{q}\mathbf{q}'} w(q)w(q') \sum_{\gamma_7\gamma_8} \Big\{ \langle\gamma_1|e^{i\mathbf{q}\cdot\mathbf{r}}|\gamma_7\rangle \langle\gamma_7|e^{i\mathbf{q}'\cdot\mathbf{r}}|\gamma_5\rangle \langle\gamma_6|e^{-i\mathbf{q}\cdot\mathbf{r}}|\gamma_8\rangle \qquad (G.23)$$

$$\times \langle\gamma_8|e^{-i\mathbf{q}'\cdot\mathbf{r}}|\gamma_4\rangle G_E^s(\gamma_7) G_{E'}^{s'}(\gamma_8) + \langle\gamma_1|e^{i\mathbf{q}\cdot\mathbf{r}}|\gamma_5\rangle \langle\gamma_6|e^{iq\cdot\mathbf{r}}|\gamma_8\rangle \langle\gamma_8|e^{-i\mathbf{q}\cdot\mathbf{r}}|\gamma_7\rangle$$

$$\times \langle\gamma_7|e^{-i\mathbf{q}'\cdot\mathbf{r}}|\gamma_4\rangle G_{E'}^{s'}(\gamma_7) G_{E'}^{s'}(\gamma_8) + \langle\gamma_1|e^{i\mathbf{q}\cdot\mathbf{r}}|\gamma_7\rangle \langle\gamma_7|e^{i\mathbf{q}'\cdot\mathbf{r}}|\gamma_8\rangle$$

$$\times \langle\gamma_8|e^{-i\mathbf{q}\cdot\mathbf{r}}|\gamma_5\rangle \langle\gamma_6|e^{-i\mathbf{q}'\cdot\mathbf{r}}|\gamma_4\rangle G_E^s(\gamma_7) G_E^s(\gamma_8) \Big\} + \cdots .$$

All the contributions linear and quadratic in $w(q)$ are presented explicitly in Eq. (G.23). Equations (G.22) and (G.23) are analogous to Eqs. (15.5) and (15.6) written in the momentum representation in the absence of the magnetic field.

Calculating the conductivity in the magnetic field, one encounters the expressions of the kind

$$\sum_{\gamma_1 - \gamma_4} \langle\gamma_4|\hat{v}_\alpha|\gamma_1\rangle \langle\gamma_2|\hat{v}_\beta|\gamma_3\rangle K_{EE'}^{ss'}(\gamma_1, \gamma_4 | \gamma_2, \gamma_3) \qquad (G.24)$$

$$\equiv \sum_{N_1 - N_4} \sum_{p_y p_y'} \sum_{p_z p_z'} v_{N_4 N_1}^\alpha v_{N_2 N_3}^\beta K_{EE'}^{ss'}(N_1, N_4; p_y; p_z | N_2, N_3; p_y'; p_z'),$$

where we use the definitions $\langle Np_y p_z | \hat{v}_\alpha | N'p_y'p_z' \rangle = v_{NN'}^\alpha \delta_{p_y p_y'} \delta_{p_z p_z'}$ and $K_{EE'}^{ss'}(N_1, N_4; p_y; p_z | N_2, N_3; p_y'; p_z') = K_{EE'}^{ss'}(N_1 p_y p_z, N_4 p_y p_z | N_2 p_y' p_z', N_3 p_y' p_z')$. Since $v_{NN'}^x$ and $v_{NN'}^y$ are non-zero only for $N = N' \pm 1$ and $v_{NN'}^z$ is non-zero only for $N = N'$, the sum in Eq. (G.24) is really taken over two (instead of four) Landau level indices. Next, due to the momentum conservation law, the vertex part $\Gamma_{EE'}^{ss'}(\gamma_1\gamma_4 | \gamma_5\gamma_6)$ is non-zero only for $p_{y1} - p_{y4} = p_{y5} - p_{y6}$ and $p_{z1} - p_{z4} = p_{z5} - p_{z6}$. Therefore, one can rewrite the vertex part standing in the equation for the correlation functions $K_{EE'}^{ss'}(N_1, N_4; p_y; p_z | N_2, N_3; p_y'; p_z')$ according to

$$\Gamma_{EE'}^{ss'}(N_1 p_y p_z, N_4 p_y p_z | N_5 p_y' p_z', N_6 p_y'' p_z'') \equiv \delta_{p_y'' p_y'} \delta_{p_z'' p_z'} \qquad (G.25)$$

$$\times \sum_{Q_y Q_z} \delta_{p_y', p_y - \hbar Q_y} \delta_{p_z', p_z - \hbar Q_z} \Gamma_{EE'}^{ss'}(N_1, N_4; p_y; p_z | N_5, N_6; p_y - \hbar Q_y; p_z - \hbar Q_z),$$

where $\hbar Q_y$ and $\hbar Q_z$ are the components of the total momentum $\hbar \mathbf{Q}$ transmitted through the vertex from the upper to the lower line of the diagrams. The Bethe-Salpeter equation for the two-particle correlation functions introduced by Eq. (G.24) takes the form

$$K_{EE'}^{ss'}(N_1, N_4; p_y; p_z | N_2, N_3; p_y'; p_z')$$

$$= G_E^s(p_z, N_1) G_{E'}^{s'}(p_z, N_4) \left[\delta_{N_1 N_2} \delta_{N_4 N_3} \delta_{p_y p_y'} \delta_{p_z p_z'} \right. \qquad (G.26)$$

$$+ \frac{1}{V} \sum_{Q_y Q_z} \sum_{N_5 N_6} \Gamma_{EE'}^{ss'}(N_1, N_4; p_y; p_z | N_5, N_6; p_y - \hbar Q_y; p_z - \hbar Q_z)$$

$$\left. \times K_{EE'}^{ss'}(N_5, N_6; p_y - \hbar Q_y; p_z - \hbar Q_z | N_2, N_3; p_y'; p_z') \right].$$

We point out that $\Gamma_{EE'}^{ss'}(N_1, N_4; p_y; p_z | N_5, N_6; p_y - \hbar Q_y; p_z - \hbar Q_z)$ does not depend on p_y. This follows from the fact that the vertex part of an arbitrary order is built from a product of averaged Green's functions (which are independent on the y-momenta) multiplied by a product of the matrix elements $\langle N_j \, p_y - \hbar q_{yk} | \exp(i q_{xi} x) | N_{j'} \, p_y - \hbar q_{yk} - \hbar q_{yi} \rangle$. Since the oscillator functions $\langle \mathbf{x} | N \, p_y - \hbar q_{yk} \rangle \equiv \varphi_x^{(N \, p_y - \hbar q_{yk})}$ depend on x in the combination $x + l_H^2(p_y/\hbar - q_{yk})$, see Eq. (5.15), each matrix element of this kind depends on p_y only through the phase factor $\exp(-i l_H^2 q_{xi} p_y / \hbar)$. On the other hand, due to the conservation rule $\sum_i q_{xi} = 0$, the product of all such phase factors is equal to 1. Since $\Gamma_{EE'}^{ss'}(N_1, N_4; p_y; p_z | N_5, N_6; p_y - \hbar Q_y; p_z - \hbar Q_z)$ is independent of p_y, we can integrate Eq. (G.26) over p_y and p_y' and obtain

$$K_{EE'}^{ss'}(N_1, N_4; p_z | N_2, N_3; p_z') = G_E^s(p_z, N_1) G_{E'}^{s'}(p_z, N_4) \left[\delta_{N_1 N_2} \delta_{N_4 N_3} \delta_{p_z p_z'} \right.$$

$$+ \sum_{N_5 N_6} \int \frac{dQ_z}{2\pi} \Gamma_{EE'}^{ss'}(N_1, N_4; p_z | N_5, N_6; p_z - \hbar Q_z) \qquad (G.27)$$

$$\left. \times K_{EE'}^{ss'}(N_5, N_6; p_z - \hbar Q_z | N_2, N_3; p_z') \right],$$

where

$$K_{EE'}^{ss'}(N_1, N_4; p_z | N_2, N_3; p_z') = \frac{2\pi l_H^2}{L_x L_y} \sum_{p_y p_y'} K_{EE'}^{ss'}(N_1, N_4; p_y; p_z | N_2, N_3; p_y'; p_z'),$$

$$\Gamma_{EE'}^{ss'}(N_1, N_4; p_z | N_5, N_6; p_z - \hbar Q_z) \qquad (G.28)$$

$$= \frac{1}{L_x L_y} \sum_{Q_y} \Gamma_{EE'}^{ss'}(N_1, N_4; p_y; p_z | N_5, N_6; p_y - \hbar Q_y; p_z - \hbar Q_z).$$

Finally, we note that $\Gamma_{EE'}^{ss'}(N_1, N_4; p_z | N_5, N_6; p_z - \hbar Q_z)$ is non-zero only for $N_1 - N_4 = N_5 - N_6$. To prove this, one may use the matrix elements (48.13) and show that $\Gamma_{EE'}^{ss'}(N_1, N_4; p_y; p_z | N_5, N_6; p_y - \hbar Q_y; p_z - \hbar Q_z) = \sum_{Q_x} e^{i\Phi(N_1 - N_4 - N_5 + N_6)} \ldots$, where Φ is the polar angle of the 2D vector (Q_x, Q_y), and the contribution denoted here by the dots depends only on the absolute value of this vector. Then Eq. (G.27) takes its final form

$$K_{EE'}^{ss'}(N_1, N_4; p_z | N_2, N_3; p_z') = G_E^s(p_z, N_1) G_{E'}^{s'}(p_z, N_4) \left[\delta_{N_1 N_2} \delta_{N_4 N_3} \delta_{p_z p_z'} \right.$$

$$+ \sum_N \int \frac{dQ_z}{2\pi} \Gamma_{EE'}^{ss'}(N_1, N_4; p_z | N_1 - N, N_4 - N; p_z - \hbar Q_z) \qquad (G.29)$$

$$\times\, K_{EE'}^{ss'}(N_1 - N, N_4 - N; p_z - \hbar Q_z | N_2, N_3; p_z') \Big].$$

Since, according to Eqs. (G.24) and (G.28), the components of the conductivity tensor in the magnetic field are expressed through $K_{EE'}^{ss'}(N_1, N_4; p_z | N_2, N_3; p_z')$, Eq. (G.29) is the basic equation for evaluating the conductivity.

The Bethe-Salpeter equation (G.29) has a simple solution in the case when the higher-order terms of the vertex part can be neglected and the scattering is described by a short-range potential. Indeed, let us retain in the expansion (G.23) only the first term with $w(q) = w$ and use Eqs. (G.25) and (G.28). We obtain

$$\Gamma_{EE'}^{ss'}(N_1, N_4; p_z | N_5, N_6; p_z - \hbar Q_z) = \frac{w}{2\pi l_H^2} \delta_{N_1 N_4} \delta_{N_5 N_6}. \qquad (G.30)$$

The integral equation (G.29) in this case has the following solution:

$$K_{EE'}^{ss'}(N_1, N_4; p_z | N_2, N_3; p_z') \qquad (G.31)$$

$$= G_E^s(p_z, N_1) G_{E'}^{s'}(p_z, N_4) \delta_{N_1 N_2} \delta_{N_4 N_3} \delta_{p_z p_z'} + \delta K_{EE'}^{ss'}.$$

The term $\delta K_{EE'}^{ss'}$ is proportional to $\delta_{N_1 N_4}$. For this reason, it does not contribute to the transverse conductivity, because $v_{N_4 N_1}^x$ and $v_{N_4 N_1}^y$ are zero for $N_1 = N_4$. Next, since $\delta K_{EE'}^{ss'}$ does not depend on p_z, it does not contribute to the longitudinal conductivity, because the integral $\int dp_z v_{N_4 N_1}^z \delta K_{EE'}^{ss'} \propto \int_{-\infty}^{\infty} dp_z p_z$ is equal to zero. Therefore, the conductivity in this approximation is expressed through the product of averaged Green's functions.

The situation is not that simple when the higher-order terms of the vertex part must be taken into account. Let us present the expressions for the second-order corrections to the vertex part, under the approximation of short-range potential:

$$\delta\Gamma_{EE'}^{ss'(2)}(N_1, N_4; p_z | N_1 - N, N_4 - N; p_z - \hbar Q_z) = \left(\frac{w}{2\pi l_H^2} \right)^2 \sum_{N_7 N_8}$$

$$\times \int \frac{dq_z}{2\pi} \Big\{ \Lambda_{N_7 N_8}^{(1)}(N_1, N_4, N) G_E^s(p_z - \hbar q_z, N_7) G_{E'}^{s'}(p_z - \hbar(Q_z - q_z), N_8)$$

$$+ \Lambda_{N_7 N_8}^{(2)}(N_1, N_4, N) \Big[G_E^s(p_z - \hbar q_z, N_7) G_E^s(p_z - \hbar(Q_z + q_z), N_8) \qquad (G.32)$$

$$+ G_{E'}^{s'}(p_z - \hbar q_z, N_7) G_{E'}^{s'}(p_z - \hbar(Q_z + q_z), N_8) \Big] \Big\}.$$

The numerical coefficients $\Lambda^{(1)}$ and $\Lambda^{(2)}$ are given by

$$\left| \begin{array}{c} \Lambda_{N_7 N_8}^{(1)}(N_1, N_4, N) \\ \Lambda_{N_7 N_8}^{(2)}(N_1, N_4, N) \end{array} \right| = \frac{2}{\pi^2} \frac{1}{2^{N_1 + N_4 + N_7 + N_8 - N}}$$

$$\times \int \frac{du_1 du_2 dv_1 dv_2 \, e^{-2(u_1^2 + u_2^2 + v_1^2 + v_2^2)}}{N_7! N_8! \sqrt{N_1! N_4! (N_1 - N)! (N_4 - N)!}} \qquad (G.33)$$

$$\times \left| \begin{array}{c} H_{N_1}(u_1 + v_1) H_{N_4}(u_2 + v_1) H_{N_4 - N}(u_1 - v_1) H_{N_1 - N}(u_2 - v_1) \\ \times H_{N_8}(u_1 + v_2) H_{N_8}(u_2 + v_2) H_{N_7}(u_1 - v_2) H_{N_7}(u_2 - v_2) \\ \\ H_{N_1}(u_1 + v_1) H_{N_4}(u_2 + v_1) H_{N_4 - N}(u_2 - v_2) H_{N_1 - N}(u_1 - v_2) \\ \times H_{N_8}(u_1 - v_1) H_{N_8}(u_2 - v_1) H_{N_7}(u_1 + v_2) H_{N_7}(u_2 + v_2) \end{array} \right|.$$

The momentum space for p_z and $\hbar Q_z$ in Eq. (G.32) is reduced because of the integral over q_z. As a result, $\delta\Gamma^{ss'(2)}_{EE'}(N_1, N_4; p_z | N_1 - N, N_4 - N; p_z - \hbar Q_z)$ is small in comparison to the first-order vertex part given by Eq. (G.30) provided that the mean kinetic energy of electrons, $\overline{p_z^2/2m}$, is greater than the characteristic energy of collision-induced broadening of the Landau levels. The broadening energy is estimated as the imaginary part of the self-energy function given by Eq. (G.19). This situation can be violated only in very high magnetic fields, when the Fermi energy is close to the bottom of the lowest Landau level. On the other hand, for 2D electrons, when the motion along OZ is absent, the correction (G.32), and, therefore, all higher-order corrections, become important when the cyclotron energy exceeds the broadening of the Landau levels. Nevertheless, if the Fermi energy is much greater than the cyclotron energy (so that many Landau levels are populated), the mean (non-oscillating) part of the conductivity is not affected by the higher-order corrections.

Appendix H
Hamiltonian of Tunnel-Coupled Systems

Consider the electrons with a simple parabolic energy spectrum in the presence of a one-dimensional potential energy $U(z)$ which creates a barrier separating left (l) and right (r) regions of the coordinate space, as shown in Fig. H.1. It is convenient to choose $z = 0$ at the point where the potential is maximal and to count the energies from this maximum. The Schroedinger equation describing the electron states in this system is written as

$$(\hat{H} - \varepsilon)\Psi(\mathbf{x}, z) = 0, \quad \hat{H} = \frac{\hat{p}_z^2 + \hat{\mathbf{p}}_{\mathbf{x}}^2}{2m} + U(z) + V(\mathbf{x}, z), \qquad (H.1)$$

where $\mathbf{x} = (x, y)$ is the 2D coordinate. The electron energy is assumed to be negative, $\varepsilon < 0$, so that the barrier is classically unpenetrable. The Hamiltonian (H.1) also contains a random potential $V(\mathbf{x}, z)$ describing scattering of the electrons.

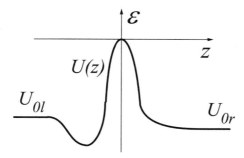

Figure H.1. Potential barrier $U(z)$ separating the left ($z < 0$) and right ($z > 0$) regions.

Let us decompose $U(z)$ into two parts, namely

$$U(z) = U_l(z) + U_r(z), \qquad (H.2)$$

$$U_l(z) = \left\{ \begin{array}{ll} U(z), & z < 0 \\ 0, & z > 0 \end{array} \right. , \quad U_r(z) = \left\{ \begin{array}{ll} 0, & z < 0 \\ U(z), & z > 0 \end{array} \right. ,$$

which means that U_l and U_r represent unpenetrable potential walls for the electrons in the left and right regions, respectively. For each region $(j = l, r)$, we introduce the basis functions $F_{jk}(z) \equiv \langle z | jk \rangle$ satisfying the following one-dimensional Schroedinger equations:

$$\left[\frac{\hat{p}_z^2}{2m} + U_j(z) - \varepsilon_{jk} \right] F_{jk}(z) = 0. \qquad (H.3)$$

These functions form complete sets for each region j. They belong either to continuous spectrum, when k is a continuous variable, or to discrete spectrum. The discrete electron states exist if the region j contains a potential well where the size quantization occurs. In any case, the functions $F_{jk}(z)$ are orthogonal, and they are normalized according to

$$\int_{-\infty}^{\infty} dz \, F_{jk'}^*(z) F_{jk}(z) = \delta_{kk'}. \qquad (H.4)$$

The functions from different regions, however, are not orthogonal. Their overlap integrals

$$S_{j'k',jk} \equiv \langle j'k' | jk \rangle = \int_{-\infty}^{\infty} dz \, F_{j'k'}^*(z) F_{jk}(z), \quad (j' \neq j) \qquad (H.5)$$

are not equal to zero.

Let us calculate the matrix elements of the Hamiltonian (H.1) in the basis $|jk\rangle$ described above. The elements diagonal in j are

$$\langle jk' | \hat{H} | jk \rangle = \delta_{kk'} \left[\frac{\hat{\mathbf{p}}_x^2}{2m} + \varepsilon_{jk} \right] + \langle jk' | V(\mathbf{x}, z) | jk \rangle + \langle jk' | U_{j'}(z) | jk \rangle_{j' \neq j}. \qquad (H.6)$$

The last term in this expression contains an exponential smallness of the second order with respect to the overlap factors (H.5). Indeed, this term is formed as an integral of a product of two wave functions of the region j over the region $j' \neq j$, where these functions decrease exponentially. We neglect such terms in the following. The remaining part of Eq. (H.6) represents the effective Hamiltonian of the region j. The non-diagonal elements of $\hat{H}$ are represented as $(j' \neq j)$

$$\langle j'k' | \hat{H} | jk \rangle = S_{j'k',jk} \left[\frac{\hat{\mathbf{p}}_x^2}{2m} + \frac{\varepsilon_{j'k'} + \varepsilon_{jk}}{2} \right] + t_{j'k',jk} + \langle j'k' | V(\mathbf{x}, z) | jk \rangle$$

$$+ \frac{1}{2} \langle j'k' | U_{j'}(z) - U_j(z) - \varepsilon_{j'k'} + \varepsilon_{jk} | jk \rangle, \qquad (H.7)$$

where we have separated the contribution

$$t_{j'k',jk} = \frac{1}{2} \langle j'k' | U_l(z) + U_r(z) | jk \rangle = \frac{1}{2} \langle j'k' | U(z) | jk \rangle \qquad (H.8)$$

called the tunneling matrix element. The last term in Eq. (H.7) is equal to zero. One can prove this statement by writing this term, with the use of Eq. (H.3), as

$$-\frac{1}{2} \langle j'k' | U_j(z) - \varepsilon_{jk} | jk \rangle + \frac{1}{2} \langle j'k' | U_{j'}(z) - \varepsilon_{j'k'} | jk \rangle$$

$$= \frac{1}{4m} \int_{-\infty}^{\infty} dz \left[F_{j'k'}^*(z) \hat{p}_z^2 F_{jk}(z) - F_{jk}(z) \hat{p}_z^2 F_{j'k'}^*(z) \right]. \qquad (H.9)$$

The integral over z in this expression should be transformed by parts. Taking into account that either $F_{jk}(z)$ or $F_{j'k'}^*(z)$ is equal to zero at $z = \pm\infty$, we complete the required proof.

Let us search for the wave function from Eq. (H.1) in the form

$$\Psi(\mathbf{x}, z) = \frac{1}{L} \sum_{j=l,r} \sum_{k\mathbf{p}} A_{jk\mathbf{p}} e^{i\mathbf{p}\cdot\mathbf{x}/\hbar} F_{jk}(z), \qquad (H.10)$$

where L is the normalization length in the plane XOY. Using the matrix elements (H.6) and (H.7), we write the following equation for the coefficients $A_{jk\mathbf{p}}$:

$$(\varepsilon_{jkp} - \varepsilon)A_{jk\mathbf{p}} + \frac{1}{L^2} \sum_{j'k'\mathbf{p}'} \langle jk|V[(\mathbf{p} - \mathbf{p}')/\hbar, z]|j'k'\rangle A_{j'k'\mathbf{p}'}$$

$$+ \sum_{k'(j'\neq j)} \left[t_{jk,j'k'} + S_{jk,j'k'} \left(\frac{\varepsilon_{j'k'} + \varepsilon_{jk}}{2} - \varepsilon \right) \right] A_{j'k'\mathbf{p}} = 0, \qquad (H.11)$$

where $\varepsilon_{jkp} = \varepsilon_{jk} + p^2/2m$ and $V(\mathbf{q}, z)$ is the spatial 2D Fourier transform of $V(\mathbf{x}, z)$. Equation (H.11) is exact, except for the fact that we have already neglected the terms containing higher exponential smallness (originating from the last term in Eq. (H.6)). It is often reasonable to neglect also the contribution non-diagonal in j in the second term on the left-hand side of Eq. (H.11), since the random potential $V(\mathbf{x}, z)$ is much smaller than the regular potential $U(z)$ determining the tunneling matrix element. The last term on the left-hand side of Eq. (H.11) describes the tunneling with conservation of the in-plane momentum $\mathbf{p}$. By noticing that $|t_{jk,j'k'}| \sim U_0|S_{jk,j'k'}|$, where U_0 is a characteristic energy scale of $U(z)$, we neglect the contribution proportional to $S_{jk,j'k'}$ in this term in the region

$$\left| \varepsilon - \frac{\varepsilon_{j'k'} + \varepsilon_{jk}}{2} \right| \ll U_0. \qquad (H.12)$$

It is the region of energies that is important for considering the tunneling. In these approximations, Eq. (H.11) is reduced to the effective eigenstate problem $(\hat{H}_T - \varepsilon)A = 0$, where the Hamiltonian $\hat{H}_T$ is defined by its matrix elements in the basis $|jk\mathbf{p}\rangle$ described by the wave functions $L^{-1}e^{i\mathbf{p}\cdot\mathbf{x}/\hbar}F_{jk}(z)$:

$$\langle j'k'\mathbf{p}'|\hat{H}_T|jk\mathbf{p}\rangle = \delta_{jj'} \left\{ \delta_{kk'}\delta_{\mathbf{p}\mathbf{p}'}\varepsilon_{jkp} + L^{-2}V_{k'k}^{(j)}[(\mathbf{p}' - \mathbf{p})/\hbar] \right\} + \delta_{\mathbf{p}\mathbf{p}'}t_{j'k',jk}, \quad (H.13)$$

where $V_{k'k}^{(j)}(\mathbf{q}) = \langle jk'|V(\mathbf{q}, z)|jk\rangle$. The Hamiltonian $\hat{H}_T$ is the effective Hamiltonian of tunnel-coupled systems. It is written with the use of the overfilled basis consisting of two complete sets of l- and r- states described by Eqs. (H.3)−(H.5). The non-diagonal part of $\hat{H}_T$, which describes the coupling of the left and right regions, is called the tunneling Hamiltonian.

In the second quantization representation, we rewrite Eq. (H.13) as follows:

$$\hat{H}_T = \sum_{jk\mathbf{p}} \varepsilon_{jkp}\hat{a}_{jk\mathbf{p}}^+ \hat{a}_{jk\mathbf{p}} + \frac{1}{L^2} \sum_{jkk'\mathbf{p}\mathbf{p}'} V_{k'k}^{(j)}[(\mathbf{p}' - \mathbf{p})/\hbar]\hat{a}_{jk'\mathbf{p}'}^+ \hat{a}_{jk\mathbf{p}}$$

$$+ \sum_{kk'\mathbf{p}} [t_{lk',rk}\hat{a}_{lk'\mathbf{p}}^+ \hat{a}_{rk\mathbf{p}} + t_{rk,lk'}\hat{a}_{rk\mathbf{p}}^+ \hat{a}_{lk'\mathbf{p}}], \qquad (H.14)$$

where $\hat{a}_{jk\mathbf{p}}^+$ and $\hat{a}_{jk\mathbf{p}}$ are the creation and annihilation operators of the electron in the state $|jk\mathbf{p}\rangle$. Note that the Hamiltonian (H.14) is Hermitian since $t_{lk',rk} = t_{rk,lk'}^*$.

Considering the tunneling Hamiltonian as a perturbation, one may write the probability of tunneling transition between the states $|jk\mathbf{p}\rangle$ and $|j'k'\mathbf{p}'\rangle$ in unit time according to Fermi's golden rule, see Eq. (2.16):

$$W_{j'k'\mathbf{p}',jk\mathbf{p}} = \delta_{\mathbf{p}\mathbf{p}'}\frac{2\pi}{\hbar}|t_{j'k',jk}|^2\delta(\varepsilon_{jkp} - \varepsilon_{j'k'p'}), \quad j' \neq j. \tag{H.15}$$

These transitions conserve the in-plane momentum $\mathbf{p}$. One may also consider higher-order contributions into the transition probability, which include both the tunneling and the scattering. Such contributions describe the scattering-assisted tunneling in which the in-plane momentum is not conserved.

Let us write the matrix elements of the operator of electric current through the barrier (the tunneling current). Using the general expression (4.15), we obtain the current operator at the point Z:

$$\hat{I}_T(Z) = \frac{e}{2m}[\hat{p}_z\delta(z - Z) + \delta(z - Z)\hat{p}_z]. \tag{H.16}$$

The matrix element of the operator of the tunneling current per unit square is calculated as

$$\langle j'k'\mathbf{p}|\hat{I}_T(Z)|jk\mathbf{p}\rangle = \frac{e}{2mL^2}\left[F_{j'k'}^*(z)\hat{p}_z F_{jk}(z) - (\hat{p}_z F_{j'k'}^*(z))F_{jk}(z)\right]_{z=Z}$$

$$= -\frac{ie\hbar}{2mL^2}\int_{-\infty}^Z dz[F_{j'k'}^*(z)\nabla_z^2 F_{jk}(z) - (\nabla_z^2 F_{j'k'}^*(z))F_{jk}(z)]$$

$$= -\frac{ie}{2\hbar L^2}\left\{\int_{-\infty}^Z dz[U_j(z) - U_{j'}(z) - \varepsilon_{jk} + \varepsilon_{j'k'}]F_{j'k'}^*(z)F_{jk}(z)\right. \tag{H.17}$$

$$\left. - \int_Z^\infty dz[U_j(z) - U_{j'}(z) - \varepsilon_{jk} + \varepsilon_{j'k'}]F_{j'k'}^*(z)F_{jk}(z)\right\}.$$

If we choose $Z = 0$ and take into account Eq. (H.3) and the energy conservation law from Eq. (H.15), we obtain

$$\langle lk'\mathbf{p}|\hat{I}_T|rk\mathbf{p}\rangle = \frac{ie}{\hbar L^2}t_{lk',rk}, \quad \langle rk\mathbf{p}|\hat{I}_T|lk'\mathbf{p}\rangle = -\frac{ie}{\hbar L^2}t_{rk,lk'}. \tag{H.18}$$

Therefore, the matrix elements of the tunneling current are expressed through the tunneling matrix elements (H.8). The current is independent of the choice of Z, according to the continuity equation (4.14). The latter can be written as $d\hat{I}_T(Z)/dZ = 0$ in the stationary case and in the absence of in-plane currents, when $\hat{\mathbf{I}}_{\mathbf{r}} = [0, 0, \hat{I}_T(Z)]$. The presence of infinite normalization lengths in Eq. (H.18), as well as in the tunneling matrix elements (H.19) and (H.21) below, should not create a confusion, because these lengths vanish in the observable quantities (such as the density of tunneling current).

Below we present the expressions for $t_{lk',rk}$ calculated directly for three simple potentials shown in Fig. H.2. This corresponds to the cases of 3D-3D *(a)*, 2D-3D *(b)*, and 2D-2D *(c)* tunneling. In the case *(a)*,

$$t_{lk',rk} = \frac{4\hbar^2}{m\sqrt{L_l L_r}}\frac{k'k\kappa e^{-\kappa d}}{(\kappa + ik)(\kappa - ik)}, \tag{H.19}$$

where L_j is the normalization length along z in the region j, which appears because of normalization of the wave function of continuous spectrum. The underbarrier penetration length, κ^{-1}, is related to the wave numbers k and k' according to

$$\hbar\kappa = \sqrt{2m|\varepsilon - p^2/2m|}, \quad \hbar k = \sqrt{2m(\varepsilon - U_{0r} - p^2/2m)},$$

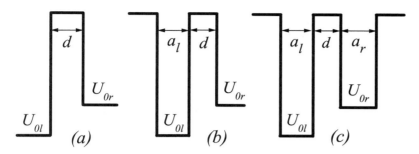

Figure H.2. Rectangular potentials defining tunnel-coupled systems of different dimensionalities.

$$\hbar k' = \sqrt{2m(\varepsilon - U_{0l} - p^2/2m)}, \tag{H.20}$$

where ε is the energy of the tunneling electron and U_{0j} is the potential energy in the region j. In the case *(b)*,

$$t_{lk',rk} = \frac{2\hbar^2}{m\sqrt{L_r}} \frac{k'k\kappa^{3/2}e^{-\kappa d}}{(\kappa - ik)\sqrt{(\kappa^2 + k'^2)(1 + a_l\kappa/2)}}. \tag{H.21}$$

The relations (H.20) remain valid, though k' is now discrete and determined by the dispersion relation

$$\cot(k'a_l) = \frac{k'^2 - \kappa^2}{2k'\kappa}. \tag{H.22}$$

The values of k and κ are fixed by k'. Finally, in the case *(c)*

$$t_{lk',rk} = \frac{\hbar^2}{m} \frac{k'k\kappa^2 e^{-\kappa d}}{\sqrt{(\kappa^2 + k'^2)(\kappa^2 + k^2)(1 + a_l\kappa/2)(1 + a_r\kappa/2)}}. \tag{H.23}$$

The wave number k in this case is also discrete. Therefore, apart from Eqs. (H.20) and (H.22), one has

$$\cot(ka_r) = \frac{k^2 - \kappa^2}{2k\kappa}. \tag{H.24}$$

In this case, the tunneling (without scattering) may occur only if the parameters of the system are adjusted to satisfy Eqs. (H.20), (H.22), and (H.24) simultaneously. This tunneling occurs between the states k and k' with matched energies. The values of $t_{lk',rk}$ given by Eqs. (H.19), (H.21), and (H.23) are defined with the accuracy up to a phase factor $e^{i\phi}$, since the phases of the wave functions $F_{lk'}$ and F_{rk} can be chosen in an arbitrary way. This phase factor is not essential, because the transition probabilities and tunneling currents are expressed through the squared absolute values of the tunneling matrix elements. If both $|lk'\rangle$ and $|rk\rangle$ are discrete states, the wave functions $F_{lk'}$ and F_{rk} can be chosen real. The tunneling matrix elements in this case are real and symmetric, $t_{lk',rk} = t_{lk',rk}^* = t_{rk,lk'}$.

In the case of tunneling between discrete states, one may consider only a pair of the states k and k' with closely matched energies. The indices k and k', therefore, can be omitted, and the Hamiltonian of such tunnel-coupled electron systems is written in the form of a 2×2 matrix in the basis $|l\rangle$ and $|r\rangle$:

$$\hat{H}_T = \hat{P}_l\hat{h}_l + \hat{P}_r\hat{h}_r + \hat{\sigma}_x t_{lr} , \tag{H.25}$$

where $\hat{P}_l = (1 + \hat{\sigma}_z)/2$ and $\hat{P}_r = (1 - \hat{\sigma}_z)/2$ are the operators of projection, $\hat{\sigma}_i$ are the Pauli matrices, and t_{lr} is the tunneling matrix element for the pair of states under consideration. The Hamiltonians of the left and right regions are

$$\hat{h}_j = \varepsilon_j + \frac{\hat{\mathbf{p}}_\mathbf{x}^2}{2m} + V^{(j)}(\mathbf{x}), \qquad (H.26)$$

where $V^{(j)}(\mathbf{x}) = \int dz\, |F_j(z)|^2 V(\mathbf{x}, z)$. According to Eq. (H.18), the operator of the tunneling current in this basis is expressed through the Pauli matrix $\hat{\sigma}_y$:

$$\hat{I}_T = -\frac{e t_{lr}}{\hbar} \hat{\sigma}_y. \qquad (H.27)$$

It is convenient to use the Pauli matrices for applications, since the matrix algebra (the commutation relations, etc.) for these matrices is well known. For example, we present a useful formula for the operator exponent:

$$\exp(i\hat{\boldsymbol{\sigma}} \cdot \mathbf{A}) = \cos|\mathbf{A}| + i\frac{\hat{\boldsymbol{\sigma}} \cdot \mathbf{A}}{|\mathbf{A}|} \sin|\mathbf{A}|\,, \qquad (H.28)$$

where $\mathbf{A}$ is an arbitrary vector.

Although we have considered the case of one-dimensional potential barriers separating two regions, the method used above suggests that introducing the tunneling matrix elements is feasible for the case of a system consisting of several weakly-coupled regions numbered by the index j. Assuming that each region is characterized by a set of eigenstates $|j\delta\rangle$, which are the exact eigenstates in the absence of tunneling, one can write the Hamiltonian of the system in the form

$$\hat{H}_T = \sum_{j\delta} \varepsilon_{j\delta} \hat{a}_{j\delta}^+ \hat{a}_{j\delta} + \sum_{j\delta j'\delta'} \left[t_{j'\delta',j\delta}\, \hat{a}_{j'\delta'}^+ \hat{a}_{j\delta} + H.c. \right]_{j\neq j'}. \qquad (H.29)$$

The probability of transitions between the regions in unit time is given by

$$W_{j\delta,j'\delta'} = \frac{2\pi}{\hbar} |t_{j'\delta',j\delta}|^2 \delta(\varepsilon_{j\delta} - \varepsilon_{j'\delta'}), \quad j' \neq j. \qquad (H.30)$$

Calculating the tunneling matrix elements for each particular case is a complicated procedure. It is convenient to choose them as parameters.

About the Authors

Fedir T. Vasko has received his PhD and Doctor of Science degrees in 1976 and 1986, respectively. He is now a Senior Research Scientist at the Institute of Semiconductor Physics, National Academy of Sciences of Ukraine. During a decade, he taught the Quantum Kinetic Theory at the Kiev State University. He has published over 170 papers on quantum kinetics and related subjects.

Oleg E. Raichev has received his PhD and Doctor of Science degrees in 1992 and 1998, respectively. He is now a Senior Research Scientist at the Institute of Semiconductor Physics, National Academy of Sciences of Ukraine. He has published about 60 papers on quantum kinetics and its applications.

Index